Springer Series in Translational Stroke Research

Series editor
John Zhang

For further volumes:
http://www.springer.com/series/10064

Paul A. Lapchak · John H. Zhang
Editors

Translational Stroke Research

From Target Selection to Clinical Trials

 Springer

Editors
Paul A. Lapchak, PhD, FAHA
Department of Neurology
Cedars-Sinai Medical Center
Los Angeles, CA, USA

John H. Zhang
Loma Linda University
Loma Linda, CA, USA

ISBN 978-1-4419-9529-2 e-ISBN 978-1-4419-9530-8
DOI 10.1007/978-1-4419-9530-8
Springer New York Dordrecht Heidelberg London

Library of Congress Control Number: 2012934161

Printed on acid-free paper

Springer is part of Springer Science+Business Media (www.springer.com)

Stroke Progress and Promises: Going Forward (Preface)

To ancient Greeks, the hero met his nemesis or tragic ending as a direct result of his hubris or pride. Gazing over the smoking battlefield of failed clinical trials [1] one cannot help but think of the armies lost to the hubris of those who launched campaigns with terrible certainty: this plan must work because it fits everything we know. Given the spectacular stroke-trial failures of the past 2 decades, one might reasonably give the stage over to the Greek chorus, lamenting the tragic end of neuroprotection, and moving on to other therapeutic areas. Reviewing the new and novel ideas contained in this volume, as well as new twists on old ideas, one can be ensured a renewed energy and optimism.

The industry-sponsored dextrophan trial in the United States [2] brought to the forefront the "one mechanism–one drug" principle that guided drug development for decades. The trial arose from an elegant theory that a single neurotransmitter, glutamate, acting on a single channel, the NMDA receptor, injuring a single cell-type, the neuron, unlocked the secret to reversing stroke. The notion that a single drug, acting on a single receptor to activate a single, defined mechanism of action seduced trialists, basic scientists, regulators, and funding agencies. The "single mechanism" theory came to permeate translational neurology, so much so that studies of pleiotropic therapies came to a virtual standstill. Twelve years later, the spectacular demise of the free-radical scavenging agent NXY-059 [3, 4] finally and (hopefully) permanently put an end to the naïve idea that a single magic bullet therapy could overcome the plethora of pathologic processes running in parallel during ischemia.

Even worse than naiveté, however, during the 2-decade search for the "single mechanism" magic bullet, a hubritic sense permeated the field: experts knew all that anyone needed to know. Study groups designed trials—and powered them—as if there were no unknowns among human stroke victims participating in trials. For example, during the development of the steroid tirilazad, an unrecognized gender effect on drug metabolism partially influenced the outcome of the pivotal clinical trial [5]. When designing the NINDS Trial of rt-PA for Acute Ischemic Stroke [6], we were advised by very experienced and senior neurologists to exclude lacunes: it was "well known" that the mechanisms of small vessel occlusion included cystic medial necrosis and lipohyalinosis but *not* thrombosis. Had we followed dogma and excluded lacunar stroke patients, we would have excluded the subgroup that benefited most from thrombolytic therapy [7].

Hubris of another kind affects the few companies interested in funding clinical trials in stroke. Large pharma will have little or nothing to do with stroke trials until positive results can be guaranteed; the field is left to brave start-up ventures willing to gamble on finding the next big winner in stroke. But limited funds drives these companies to design trials targeted at an idealized subgroup of patients. After retrospective review of failed trials, these businessmen conclude that we should study only the portion of the fraction of the sub-subgroup that would presumably benefit. This "threading-the-needle" principle assumes that tomorrow's treatment behaves exactly like yesterday's after we exclude patients who presumably could not have responded to therapy. The next few clinical trials will reveal whether this needle can be threaded successfully.

The next phase—and hopefully a more mature phase—of stroke clinical trials seems to be emerging and this volume seeks to illustrate two critical points relevant to this renaissance. First, the need for pleiotropic and combinatorial therapies is obvious. The search for the "magic bullet" is over and therapies like hypothermia, with multiple mechanisms, should receive priority [8]. Second, while a "thread-the-needle" approach to finding magic subgroups may succeed, such is doubtful. A viable alternative strategy includes a confession of humility: we do not know everything and we should power our trials to include the unforeseen responsive or nonresponsive populations. Larger trials may cost more, but new approaches can reduce risk by including stronger futility analyses, adaptive and sequential designs, and frequent interim analyses. Larger trials that can be pooled with other large trials allow for *rational* subgroup analysis and the rigorous confirmation of trends. Obviously, larger trials demand simplified designs and the collection of only the most salient data. Regulatory reform must arise on multiple continents.

The time is ripe for many of the ideas presented in this volume and the Editors have done a good job attracting new talent bringing novel, perhaps radical, ideas. In Part I, the neurovascular unit comes to the forefront, as it should. What were we thinking when we designed therapies targeted only at neurons, as if there were no other cells in the nervous system? Conceptually, the neurovascular unit seems obvious and simple—as do all major advances in science—but the implications may be complex and protean. The second part of this book offers the reader a potpourri of new therapeutic possibilities. Few, or perhaps none, of these proposals will pan out, but in investigating these and other new targets, we should stumble our way into a significant advance like thrombolysis. Some—like laser therapy—are far-fetched and hard to accept given our current understanding of ischemia. Others—like hypothermia—are very old ideas newly reformulated thanks to technical delivery advances. While we cannot predict which of these ideas will prove successful, we can assert that at least something here will move our field forward significantly. Parts III and IV tackle the very real but arcane details of modeling. No small part of our collective failure lies with the preclinical models, and a critical review will inform the reader on limitations of past and present animal stroke models.

Part V offers updates on therapies currently on the "hot list" and illustrates some of the pitfalls inherent in clinical trials. Part VI offers criticism and proposals for improving clinical trial design; these revisions are sorely needed.

Stroke research is not for the faint of heart. We fight a common, devastating disease and lose more often than we win. But every year around the world, more patients receive intravenous rt-PA, more stroke centers open up, and more Fellows are trained as Vascular Neurologists. Undoubtedly, a multi-mechanistic neuroprotective treatment—and likely something first glimpsed in this book—will emerge to compliment recanalization, but only if we humbly learn the lessons of the last 2 decades and thoughtfully plan for the unknown.

Patrick D. Lyden

References

1. O'Collins VE, Macleod MR, Donnan GA, Horky LL, van der Worp BH, Howells DW. 1,026 experimental treatments in acute stroke. Ann Neurol. 2006;59(3):467–77.
2. Albers GW, Atkinson RP, Kelley RE, Rosenbaum DM. Safety, tolerability, and pharmacokinetics of the N-menthyl-D-aspartate antagonist dextrorphan in patients with acute stroke. Stroke. 1995;26:254.
3. Shuaib A, Lees KR, Lyden P, et al. NXY-059 for the treatment of acute ischemic stroke. N Engl J Med. 2007;357(6):562–71.
4. Lyden PD, Shuaib A, Lees KR, et al. Safety and tolerability of NXY-059 for acute intracerebral hemorrhage: the CHANT Trial. Stroke. 2007;38(8):2262–9.
5. Haley EC, Jr. High-dose tirilazad for acute stroke (RANTTAS II). RANTTAS II Investigators. Stroke. 1998;29(6):1256–7.
6. NINDS rt-PA Stroke Study Group. Tissue plasminogen activator for acute ischemic stroke. N Engl J Med. 1995;333(24):1581–7.
7. NINDS rt-PA Stroke Study Group. Generalized efficacy of t-PA for acute stroke. Stroke. 1997;28:2119.
8. Hemmen TM, Lyden PD. Multimodal neuroprotective therapy with induced hypothermia after ischemic stroke. Stroke. 2008;40(3 Suppl):S126–8.
9. Lyden P. The future of basic science research and stroke: hubris and translational stroke research. Int J Stroke. 2011;6(5):412–3. DOI:10.1111/j.1747-4949.2011.00657.x.

Contents

Contributors

Golnaz Ahadi Department of Radiology, University of California, San Diego, CA, USA
gahadi@ucsd.edu

Nabil J. Alkayed Department of Anesthesiology and Perioperative Medicine, Oregon Health & Science University, Portland, OR, USA
alkayedn@ohsu.edu

Ken Arai Neuroprotection Research Laboratory, Massachusetts General Hospital/Harvard Medical School, Charlestown, MA, USA
karai@partners.org

Stephen Ashwal, MD Department of Pediatrics, Loma Linda University, Loma Linda, CA, USA
SAshwal@llu.edu

Frances Rena Bahjat, PhD Department of Molecular Microbiology and Immunology, Oregon Health and Science University, Portland, OR, USA
Bahjat@ohsu.edu

Ludmila Belayev, MD Neuroscience Center of Excellence, Louisiana State University Health Sciences Center, New Orleans, LA, USA
lbelay@lsuhsc.edu

Bir Bhanu, PhD Center for Research in Intelligent Systems, University of California, Riverside, CA, USA
bhanu@ee.ucr.edu

Alain César Biraboneye Laboratoire de Chimie Biomoleculaire, CNRS, IBDML-UMR-6216, Marseille Cedex, France
biraboneye@luminy.univ-mrs.fr

Fernanda Borges CIQUP/Departamento de Química e Bioquímica, Universidade do Porto, Porto, Portugal
fborges@fc.up.pt

Bo Chen, PhD Department of Neurology, Cedars-Sinai Medical Center, Los Angeles, CA, USA
Bo.Chen@cshs.org

Chunhua Chen Department of Anatomy and Embryo, Peking University Health Science Center, Haidian Qu, Beijing, China
cchhh1000@gmail.com

Heyu Chen School of Biomedical Engineering and Med-X Research Institute, Shanghai Jiao Tong University, Shanghai, P.R. China
chenheyucn@gmail.com

Jieli Chen, MD Department of Neurology, Henry Ford Hospital, Detroit, MI, USA
jieli@neuro.hfh.edu

Michael Chopp Department of Neurology, Henry Ford Hospital, Detroit, MI, USA

Department of Physics, Oakland University, Rochester, MI, USA
michael.chopp@gmail.com

Marilyn J. Cipolla, PhD Department of Neurology, Obstetrics, Gynecology and Reproductive Sciences and Pharmacology, University of Vermont, Burlington, VT, USA
Marilyn.Cipolla@uvm.edu

Steven C. Cramer, MD Departments of Neurology and Anatomy & Neurobiology, University of California, Irvine, CA, USA

UC Irvine Medical Center, Orange, CA, USA
scramer@uci.edu

Krishna A. Dani, MRCP Institute of Neurosciences & Psychology, University of Glasgow, Glasgow, Scotland
krishna.dani@glasgow.ac.uk

Todd Deveau Department of Anesthesiology, Emory University, Atlanta, GA, USA
tdeveau@emory.edu

Ashutosh Dharap Department of Neurological Surgery and Neuroscience Training Program, University of Wisconsin, Madison, WI, USA
dharap@wisc.edu

Murat Digicaylioglu, MD, PhD Departments of Neurosurgery and Physiology, University of Texas, Health Science Center, Interdisciplinary Graduate Training Program in Neuroscience, San Antonio, TX, USA
muratd@uthscsa.edu

Jens P. Dreier Center for Stroke Research Berlin, Translation in Stroke Research, Charité University Medicine Berlin, Berlin, Germany
jens.dreier@charite.de

David Fisher Department of Radiology, University of California, San Diego, CA, USA
djfisher@ucsd.edu

Gary M. Fiskum, PhD Department of Anesthesiology, University of Maryland School of Medicine, Baltimore, MD, USA
gfiskum@anes.umm.edu

Karen L. Furie, MD, MPH J. Philip Kistler MGH Stroke Research Center, Boston, MA, USA
kfurie@partners.org

Jorge Garrido CIQUP/Departamento de Engenharia Química, Instituto Superior de Engenharia, IPP, Porto, Portugal
jjg@isep.ipp.pt

Nirmalya Ghosh Department of Pediatrics, Loma Linda University, Loma Linda, CA, USA
NGhosh@llu.edu

Myron D. Ginsberg, MD Department of Neurology, University of Miami Miller School of Medicine, Miami, FL, USA
mginsberg@med.miami.edu

Yuxiang Gu, MD, PhD Department of Neurosurgery, University of Michigan, Ann Arbor, MI, USA
guyuxiang@umich.edu

Kama Guluma, MD Department of Emergency Medicine, University of California, San Diego, CA, USA
kguluma@ucsd.edu

Tomoki Hashimoto, MD Department of Anesthesia and Perioperative Care, Center for Cerebrovascular Research, University of California, San Francisco, CA, USA
hashimot@anesthesia.ucsf.edu

Thilo Hölscher, MD Department of Radiology, University of California, San Diego, CA, USA

Department of Neurosciences, University of California, San Diego, CA, USA
thoelscher@ucsd.edu

Masaaki Hokari, MD, PhD Department of Neurology, University of California, San Francisco and the San Francisco Veterans Affairs Medical Center, San Francisco, CA, USA
karimasa@med.hokudai.ac.jp

Bingren Hu, PhD Department of Anesthesiology, University of Maryland School of Medicine, Baltimore, MD, USA
bhu@anes.umm.edu

Ya Hua, MD Department of Neurosurgery, University of Michigan,
Ann Arbor, MI, USA
yahua@umich.edu

Muhammad Shazam Hussain Cerebrovascular Center, Cleveland Clinic,
Cleveland, OH, USA
msh593@hotmail.com

Bevyn Jarrott Brain Injury and Repair Program, Howard Florey Institute,
University of Melbourne, Parkville, VIC, Australia
bevyn.jarrott@florey.edu.au

Steven A. Kates Ischemix LLC, Maynard, MA, USA
skates@ischemix.com

Richard F. Keep, PhD Department of Neurosurgery, University of Michigan,
Ann Arbor, MI, USA
rKeep@umich.edu

Thomas A. Kent, MD Department of Neurology and The Stroke Outcomes
Laboratory (SOuL), Baylor College of Medicine and The Michael E. DeBakey VA
Medical Center Comprehensive Stroke Progra, Houston, TX, USA
tkent@bcm.edu

Jiming Kong Department of Human Anatomy and Cell Science,
University of Manitoba, Winnipeg, MB, Canada
kongj@cc.umanitoba.ca

Jean-Louis Kraus Laboratoire de Chimie Biomoleculaire,
CNRS, IBDML-UMR-6216, Marseille Cedex, France
kraus@luminy.univ-mrs.fr

Tibor Krisrian, PhD Department of Anesthesiology, University of Maryland
School of Medicine, Baltimore, MD, USA
tkristian@anes.umm.edu

Chase S. Krumpelman, PhD, MSEE Department of Neurology
and The Stroke Outcomes Laboratory (SOuL), Baylor College of Medicine
and The Michael E. DeBakey VA Medical Center Comprehensive Stroke Program,
Houston, TX, USA
krumpelm@bcm.edu

Yair Lampl, MD Neurological Department, Edith Wolfson Medical Center,
Holon Israel and Sackler Medical School, University of Tel-Aviv, Tel-Aviv, Israel
y_lampl@hotmail.com

Paul A. Lapchak, PhD, FAHA Department of Neurology, Cedars-Sinai Medical
Center, Los Angeles, CA, USA
Paul.Lapchak@cshs.org

Christopher C. Leonardo University of South Florida, College of Medicine,
Tampa, FL, USA
cleonardo@anest.ufl.edu

Elena I. Liang, BS Department of Anesthesia and Perioperative Care,
Center for Cerebrovascular Research, University of California,
San Francisco, CA, USA
eiliang710@gmail.com

Chunli H. Liu, MD Department of Anesthesiology, University of Maryland
School of Medicine, Baltimore, MD, USA
cliu@anes.umm.edu

Ning Liu Neuroprotection Research Laboratory, Departments of Neurology and
Radiology, Massachusetts General Hospital, Harvard Medical School,
Charlestown, MA, USA
liuning0731@yahoo.cn

Eng H. Lo Neuroprotection Research Laboratory, Massachusetts General
Hospital/Harvard Medical School, Charlestown, MA, USA
lo@helix.mgh.harvard.edu

Daniel Lotz Department of Radiology, University of California,
San Diego, CA, USA
dtlotz@ucsd.edu

Patrick D. Lyden, MD, FAAN, FAHA Department of Neurology,
Cedars-Sinai Medical Center, Los Angeles, CA, USA
lydenp@cshs.org

R. Loch Macdonald, MD, PhD Division of Neurosurgery, Labatt Family
Centre of Excellence in Brain Injury and Trauma Research, Keenan Research
Centre of the Li Ka Shing Knowledge Institute of St. Michael's Hospital,
and Department of Surgery, University of Toronto, Toronto, ON, Canada
macdonaldlo@smh.ca

Pamela Maher, PhD The Salk Institute for Biological Studies,
La Jolla, CA, USA
pmaher@salk.edu

Hiroshi Makino, MD Department of Anesthesia and Perioperative Care,
Center for Cerebrovascular Research, University of California,
San Francisco, CA, USA
Kizaigakari@hotmail.com

Pitchaiah Mandava, MD, PhD, MSEE Department of Neurology
and The Stroke Outcomes Laboratory (SOuL), Baylor College of Medicine
and The Michael E. DeBakey VA Medical Center Comprehensive Stroke Program,
Houston, TX, USA
pmandava@bcm.edu

Sarah McCann Cytoprotection Pharmacology Program, O'Brien Institute, University of Melbourne, Parkville, VIC, Australia
sarah.mccann3@gmail.com

Carlos A. Molina, MD, PhD Stroke Unit, Department of Neurosciences, Hospital Universitari Vall d'Hebron, Barcelona, Spain
cmolina@vhebron.net

Keith W. Muir Institute of Neurosciences & Psychology, University of Glasgow, Glasgow, Scotland
k.muir@clinmed.gla.ac.uk

Santosh B. Murthy, MD, MPH Department of Neurology
and The Stroke Outcomes Laboratory (SOuL), Baylor College of Medicine
and The Michael E. DeBakey VA Medical Center Comprehensive Stroke Program, Houston, TX, USA
bsantosh@bcm.edu

Venkata P. Nakka Department of Neurological Surgery and Neuroscience Training Program, University of Wisconsin, Madison, WI, USA
v.nakka@wisc.edu

Jonathan W. Nelson Department of Anesthesiology and Perioperative Medicine, Oregon Health & Science University, Portland, OR, USA

Department of Molecular and Medical Genetics, Oregon Health
& Science University, Portland, OR, USA
nelsonjo@ohsu.edu

Andre Obenaus, PhD Department of Pediatrics, Loma Linda University, Loma Linda, CA, USA

Department of Radiology, Loma Linda University, Loma Linda, CA, USA

Department of Radiation Medicine, Loma Linda University, Loma Linda, CA, USA

Department of Biophysics and Bioengineering, Loma Linda University, Loma Linda, CA, USA
AObenaus@llu.edu

Sara Morales Palomares, PhD Departments of Neurology, Obstetrics, Gynecology and Reproductive Sciences and Pharmacology, University of Vermont, Burlington, VT, USA

Michael K. Parides, PhD Director, Mount Sinai Center for Biostatistics, Director of Biostatistics of the International Center for Health Outcomes and Innovation Research (InCHOIR), Professor, Health Evidence and Policy, Mount Sinai School of Medicine, New York
Michael.parides@mountsinai.org

Keith R. Pennypacker University of South Florida, College of Medicine, Tampa, FL, USA
kpennypa@health.usf.edu

Loc-Duyen D. Pham Neuroprotection Research Laboratory, Massachusetts
General Hospital/Harvard Medical School, Charlestown, MA, USA
lpham1@partners.org

Brian Polster, PhD Department of Anesthesiology, University of Maryland
School of Medicine, Baltimore, MD, USA
bpolster@anes.umm.edu

Liren Qian Department of Hematology, Navy General Hospital,
Beijng, China
qlr2007@126.com

Clemens Reiffurth Center for Stroke Research Berlin, Translation in Stroke
Research, Charité University Medicine Berlin, Berlin, Germany
clemens.reiffurth@charite.de

Fernanda M.F. Roleira CEF/Grupo de Química Farmacêutica,
Universidade de Coimbra, Coimbra, Portugal
froleira@ff.uc.pt

Carli L. Roulston Cytoprotection Pharmacology Program, O'Brien Institute,
University of Melbourne, Parkville, VIC, Australia
carlir@unimelb.edu.au

Marta Rubiera, MD, PhD Stroke Unit, Department of Neurosciences,
Hospital Universitari Vall d'Hebron, Barcelona, Spain
mrubifu@hotmail.com

Aarti Sarwal Wake Forest School of Medicine, Reynolds M,
Winston Salem, NC, USA
aartisarwal@yahoo.com

Michael Scheel Department of Neuroradiology, Charité University
Medicine Berlin, Berlin, Germany
michael.scheel@charite.de

Cheryl Schendel Department of Radiology, University of California,
San Diego, CA, USA
cschendel@ucsd.edu

Hilary Seifert University of South Florida, College of Medicine,
Tampa, FL, USA
hseifert@health.usf.edu

Caibin Sheng School of Biomedical Engineering and Med-X Research Institute,
Shanghai Jiao Tong University, Shanghai, P.R. China
shengcaibin@gmail.com

Prativa Sherchan Division of Physiology, School of Medicine,
Loma Linda University, Loma Linda, CA , USA
psherchan@llu.edu

Ruoyang Shi Department of Human Anatomy and Cell Science,
University of Manitoba, Winnipeg, MB, Canada
umshir@cc.umanitoba.ca

Yejie Shi Neuroscience Training Program, University of Wisconsin,
Madison, WI, USA

Department of Neurology, University of Pittsburgh, Pittsburgh, PA, USA
yshi7@wisc.edu

Ashfaq Shuaib Division of Neurology, University of Alberta,
Edmonton, AB, Canada
ashfaq.shuaib@ualberta.ca

Ihsan Solaroglu, MD Department of Neurosurgery, Neuroscience Research
Laboratories, Koç University, School of Medicine, Rumelifeneri Yolu,
Sariyer, Istanbul, Turkey
isolaroglu@hotmail.com

Mary P. Stenzel-Poore, PhD Department of Molecular Microbiology
and Immunology, Oregon Health and Science University, Portland, OR, USA

Division of Neuroscience, Oregon National Primate Research Center,
Beaverton, OR, USA
poorem@ohsu.edu

Dandan Sun, MD, PhD Neuroscience Training Program,
University of Wisconsin, Madison, WI, USA

Department of Neurological Surgery, University of Wisconsin, Madison, WI, USA

Department of Neurology, University of Pittsburgh, Pittsburgh, PA, USA
sund@upmc.edu

Xuejun Sun Department of Hematology, Navy General Hospital, Beijng, China

Department of Diving Medicine, Second Military Medical University,
Shanghai, China
sunxjk@hotmail.com

Yu Sun, PhD Center for Research in Intelligent Systems,
University of California, Riverside, CA, USA
ysun005@student.ucr.edu

Paul Szelemej Department of Human Anatomy and Cell Science,
University of Manitoba, Winnipeg, MB, Canada
umszelem@cc.umanitoba.ca

Yoshiteru Tada, MD, PhD Department of Anesthesia and Perioperative Care,
Center for Cerebrovascular Research, University of California,
San Francisco, CA, USA
taday@anesthesia.ucsf.edu

Elisiário J. Tavares-da-Silva CEF/Grupo de Química Farmacêutica, Universidade de Coimbra, Coimbra, Portugal
etavares@ff.uc.pt

Katya Tsaioun Apredica, a Cyprotex Company, Watertown, MA, USA
k.tsaioun@Cyprotex.com

Christine Turenius, PhD Department of Pediatrics, Loma Linda University, Loma Linda, CA, USA
CTurenius@llu.edu

Keri B. Vartanian, PhD Department of Molecular Microbiology and Immunology, Oregon Health and Science University, Portland, OR, USA
Vartanik@ohsu.edu

Raghu Vemuganti, PhD Department of Neurological Surgery and Neuroscience Training Program, University of Wisconsin, Madison, WI, USA
vemuganti@neurosurgery.wisc.edu

Arne Voie Department of Radiology, University of California, San Diego, CA, USA
avoie@ucsd.edu

Kosuke Wada, MD Department of Anesthesia and Perioperative Care, Center for Cerebrovascular Research, University of California, San Francisco, CA, USA
wadak@anesthesia.ucsf.edu

Xiaoying Wang, MD, PhD Neuroprotection Research Laboratory, Departments of Neurology and Radiology, Massachusetts General Hospital, Harvard Medical School, Charlestown, MA, USA
wangxi@helix.mgh.harvard.edu

Ling Wei Department of Anesthesiology, Emory University, Atlanta, GA, USA
lwei7@emory.edu

Jiequn Weng Department of Human Anatomy and Cell Science, University of Manitoba, Winnipeg, MB, Canada
umwengj@cc.umanitoba.ca

G. Alexander West, MD, PhD Colorado Brain and Spine Institute, Neurotrauma Research Laboratory, Swedish Medical Center, Englewood, CO, USA
Awest@cbsi.md

Robert M. Weston Cytoprotection Pharmacology Program, O'Brien Institute, University of Melbourne, Parkville, VIC, Australia

Brain Injury and Repair Program, Howard Florey Institute, University of Melbourne, Parkville, VIC, Australia
robmweston@gmail.com

Dirk Wiesenthal Center for Stroke Research Berlin, Translation in Stroke Research, Charité University Medicine Berlin, Berlin, Germany
dirk.wiesenthal.dreier@charite.de

Maren Winkler Center for Stroke Research Berlin, Translation in Stroke Research, Charité University Medicine Berlin, Berlin, Germany
maren.winkler.dreier@charite.de

Guohua Xi, MD Department of Neurosurgery, University of Michigan, Ann Arbor, MI, USA
guohuaxi@umich.edu

Weiliang Xia School of Biomedical Engineering and Med-X Research Institute, Shanghai Jiao Tong University, Shanghai, P.R. China
weihaiy@sjtu.edu.cn

Yan Xu, PhD Departments of Anesthesiology, Pharmacology and Chemical Biology, Computational Biology, and Structural Biology, University of Pittsburgh School of Medicine, Pittsburgh, PA, USA
xuy@anes.upmc.edu

Midori A. Yenari, MD Department of Neurology, University of California, San Francisco and the San Francisco Veterans Affairs Medical Center, San Francisco, CA, USA
Yenari@alum.mit.edu

Weihai Ying School of Biomedical Engineering and Med-X Research Institute, Shanghai Jiao Tong University, Shanghai, P.R. China
weihaiy@sjtu.edu.cn

Shan Ping Yu Department of Anesthesiology, Emory University, Atlanta, GA, USA
spyu@emory.edu

Zhanyang Yu Neuroprotection Research Laboratory, Departments of Neurology and Radiology, Massachusetts General Hospital, Harvard Medical School, Charlestown, MA, USA
zyu@partners.org

Fan Zhang, PhD Department of Anesthesiology, University of Maryland School of Medicine, Baltimore, MD, USA
fzhang@anes.umm.edu

Changman Zhou, MD, PhD Department of Anatomy and Embryo, Peking University Health Science Center, Haidian Qu, Beijing, China
cmzhou@bjmu.edu.cn

Tommaso Zoerle, MD Department of Anesthesia and Critical Care Medicine, University of Milano, Neurosurgical Intensive Care Unit, Fondazione IRCCS Ca' Granda—Ospedale Maggiore, Policlinico, Milan, Italy
tommaso.zoerle@studenti.unimi.it

Part I
Theoretical Overview

Chapter 1
Vascular Targets for Ischemic Stroke Treatment

Sara Morales Palomares and Marilyn J. Cipolla

Abstract The vast majority of studies on stroke therapy have focused on protecting the neuron from hypoxic/ischemic injury. While undoubtedly important, stroke is a vascular disorder affecting not only neurons, but numerous other cell types in the brain including astrocytes, microglia, and vascular cells (endothelium and smooth muscle). In fact, the only effective treatment for ischemic stroke is a vascular one—dissolution of the clot with tissue plasminogen activator and rapid recanalization of an occluded vessel. The failure of every neuroprotective agent to make it into clinical trials highlights the complexity of ischemic stroke pathophysiology that involves inflammation, oxidative stress, and both macro- and micro-vascular dysregulation that causes brain injury itself and exacerbates the primary insult (i.e., secondary brain injury). The vasculature in the brain has a central role in defining stroke injury since the core infarction is dependent on the depth and duration of ischemia. In addition, any neuroprotective therapy for stroke depends on a patent and functional vasculature, further highlighting the importance of vascular protection as an important therapeutic approach to limiting stroke damage.

1 Introduction to Vascular Protection

The vasculature is a major target of ischemia/reperfusion (I/R) injury. Cerebrovascular damage occurs early during focal ischemia and is progressive. Prolonged ischemia causes hemorrhagic transformation and edema especially during reperfusion. Reperfusion after short durations of ischemia is important to salvage damaged brain

S. Morales Palomares, PhD • M.J. Cipolla, PhD (✉)
Departments of Neurology, Obstetrics, Gynecology and Reproductive
Sciences and Pharmacology, University of Vermont,
89 Beaumont Ave, Given C454, Burlington, VT 05405, USA
e-mail: Marilyn.Cipolla@uvm.edu

P.A. Lapchak and J.H. Zhang (eds.), *Translational Stroke Research*, Springer Series
in Translational Stroke Research, DOI 10.1007/978-1-4419-9530-8_1,
© Springer Science+Business Media, LLC 2012

tissue; however, postischemic reperfusion is not entirely beneficial. It has been shown to impair autoregulatory mechanisms and promote loss of control of cerebral blood flow (CBF) that exacerbates ischemic injury [1–7]. In addition to brain tissue injury, reperfusion clearly affects the cerebral arteries and arterioles that control both the extent of ischemia and the degree of reperfusion [2, 8]. In this chapter, we will review current knowledge, both clinical and experimental, on the effects of I/R on the structure and function of different vascular segments in the brain. In addition, we will highlight some of the more promising targets for vascular protection with a focus on hemodynamics and blood flow regulation.

2   Hemodynamics of Focal Ischemia

2.1   Collateral Vessels During Focal Ischemia

The collateral circulation in the brain is the subsidiary network of arteries and arterioles that maintain and stabilize CBF when the principal routes of flow are compromised or fail. Primary and secondary collateral pathways exist in the brain defined by their cerebrovascular anatomy and differences in contribution to vascular resistance. The arterial segments of the circle of Willis are primary collaterals and are responsible for redistribution of blood flow when extracranial or large intracranial vessels are occluded [9, 10]. This anastomotic loop is a low-resistance connection that allows for reversal of flow to provide collateral support to the anterior and posterior circulations. The anatomy of the circle of Willis varies considerably with species and with individuals. It is often asymmetric and not complete in ~1% of cases [10]. The leptomeningeal anastomoses are secondary collateral vessels that are responsible for redistribution of CBF when occlusion occurs distal to the circle of Willis [10]. This network of pial vessels is high-resistance distal anastomoses between vascular territories and is important for determining the severity of ischemia after distal occlusion to the circle of Willis.

   The collaterals, both primary and secondary, are critical to stroke outcome. The primary collaterals are low-resistance and thus provide immediate redistribution of CBF to ischemic regions through existing anastomoses. However, secondary collaterals are high-resistance and thus recruitment of flow is slower and involves invoking hemodynamic, metabolic, and neuronal mechanisms for dilation [9, 10]. Overall, the collateral circulation critically establishes perfusion pressure during ischemia and maintains perfusion to the penumbra (see Sect. 3). Both the caliber and number of these vessels determines the functional capacity of the collateral supply and therefore the severity of ischemia. These collateral vessels may also facilitate removal and clearance of thrombus fragments from more proximal locations [11, 12]. Although active dilation of secondary collaterals influences flow under acute conditions, inward and outward remodeling of these vessels, that occurs under normal (e.g., pregnancy) and pathological (e.g., hypertension) states, is also an important determinant of collateral flow. Outward remodeling occurs when there is chronic hypoperfusion of upstream vessels (e.g., carotid stenosis, atherosclerosis) that alters shear stress distally. Inward remodeling occurs during pathological states such as hypertension. Structural remodeling will ultimately determine the size of the collaterals and thus hemodynamic reserve capacity when maximally dilated during hypoxia/ischemia. Therefore, mechanisms that actively dilate collaterals as well as promote outward remodeling may be important therapeutic targets for salvaging the penumbra and limiting stroke damage.

2.2   Hemodynamic Reserve Capacity and Regulation of CBF During Ischemia and Reperfusion

The collateral supply to a brain region will maintain normal CBF during an occlusion initially, after which hypoxia/ischemia in the region will invoke hemodynamic mechanisms that increase flow. When collateral vessels cannot maintain perfusion pressure, myogenic vasodilation increases the caliber of resistance vessels, increasing flow to the ischemic region. Other dilatory mechanisms become invoked during prolonged ischemia [9]. Stimulation of anaerobic metabolism causes lactic acidosis and pH-induced vasodilation [13]. Carbon dioxide builds up due to decreased clearance, also causing vasodilation [9]. Once the brain vessels are fully dilated, autoregulation is lost. If recanalization does not occur under these conditions, oxygen demands by the brain tissue cannot be met and ischemic brain damage occurs. The passive diameters of the primary and secondary collaterals ultimately determine the depth and duration of ischemia during occlusion. Thus, the collateral vasculature becomes a major target for stroke therapy yet it is one of the least studied.

3   The Penumbra as a Target in Stroke Protection

Although reperfusion after focal brain ischemia is generally thought to be beneficial, the window of opportunity for improvement of stroke outcome is brief (<6 h), after which there is no longer benefit to reperfusion [9, 14]. One target of reperfusion is the ischemic penumbra, a region of constrained blood supply in which energy metabolism is preserved [15]. The penumbra in the ischemic brain is of considerable interest to stroke therapy because it is a region in which neurons are electrically silent (cannot fire action potentials), but retain a −70 mV membrane potential [16]. Unfortunately, this region is also the basis for progressive evolution of ischemic injury, i.e., the infarct expands to include the penumbra [17]. While many neuroprotective strategies have been employed to limit expansion of infarcted tissue with little success, the role of the collateral vessels in evolving infarction is critical and may provide an important therapeutic target to salvage the penumbra. Recently, the concept that repair mechanisms begin concurrent with stroke damage has led many to rethink stroke treatment, especially with regard to the penumbra [18]. Because many of the same processes appear to be involved in damage as well as repair, inhibiting only damaging processes may only be targeting half the problem or in some cases be harmful. This concept is probably true for the vasculature as well even though repair is also largely understudied in vascular protection.

Both leptomeningeal and lenticulostriate vessels are high-resistance arterioles that determine penumbral flow. Thus, selectively targeting these vessels structurally and functionally to increase flow during an occlusion is an important and largely understudied area for stroke therapy. For example, we recently found that the hormone relaxin causes selective outward remodeling of brain penetrating arterioles within the middle cerebral artery (MCA) territory without affecting upstream MCA [19]. Importantly, selectively increasing the caliber of these arterioles with relaxin treatment did not affect their ability to respond to changes in pressure (i.e., myogenic reactivity was intact), although they were larger under active conditions. This may provide a therapeutic approach to increasing collateral blood flow without affecting overall cerebrovascular resistance (CVR) or autoregulation of CBF. In fact, relaxin is currently in clinical trials for treatment of heart failure and preeclampsia.

4   Parenchymal Arterioles as Targets for Vascular Protection During Stroke

Penetrating arteries branch at right angles off the pial arteries and lie within the Virchow–Robin space. In the MCA territory, they are the lenticulostriate arteries and arterioles and provide an important secondary collateral supply to this brain region. The penetrating arteries turn into parenchymal arterioles once they penetrate into the brain tissue. There they are surrounded by astrocytic end feet and become innervated from within the brain tissue, i.e., intrinsically [20–22]. Parenchymal arterioles have only one layer of circumferentially oriented smooth muscle, but they have greater basal tone at lower pressures than upstream pial arteries (Fig. 1.1) [22]. While pial vessels are anatomically networked such that occlusion of one surface vessel does not appreciably decrease CBF [23], penetrating and parenchymal arterioles are long and largely unbranched. These vessels are less interconnected than pial arterioles and as such occlusion of an individual arteriole results in significant reductions in flow and damage to surrounding brain tissue (infarction) [23].

Parenchymal arterioles are part of the neurovascular unit and are anatomically and phenotypically distinct from the larger MCA. Because of their considerable basal tone, they contribute approximately 40% to CVR and provide an important protective function in the brain to limit transmission of arterial pressure to the capillary bed within the cortex. While the importance of the MCA to CVR and controlling blood flow is well established, the role of the parenchymal arterioles in ischemic brain injury may be of greater importance due to the fact that these small arterioles are high-resistance vessels in the cortex that connect the MCA to the microcirculation [23]. They have been shown to be a bottleneck for blood flow to the microcirculation, the occlusion of which allows little if any collateral flow due

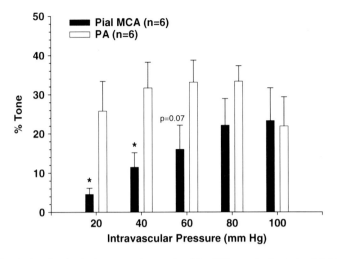

Fig. 1.1 Graph showing intrinsic myogenic tone in pial middle cerebral arteries (pial MCA) and parenchymal arterioles (PA) from the same animals, at all pressures studied. Notice that PA developed significantly greater tone at lower pressures than middle cerebral arteries (MCA). *$P<0.05$ vs. PA. From Cipolla et al. [22]. Used with permission

to their unbranched architecture [23]. Parenchymal arterioles are also functionally distinct from upstream cerebral arteries. For example, parenchymal arterioles have greater basal tone than upstream arteries and a larger influence of endothelium-derived hyperpolarizing factor (EDHF) on tone [22, 24, 25]. In addition, they are largely unresponsive to neurotransmitters such as norepinephrine and serotonin (Fig. 1.2) [22].

Our own study comparing MCA and parenchymal arterioles taken from the same animals showed that basal tone and EDHF responses are well preserved in parenchymal arterioles after I/R, but not in MCA (Fig. 1.3), suggesting there are significant differences in how vascular segments respond to stroke [25]. Similar to peripheral organ vasculature, pial arteries dilate to hypoxic conditions; however, parenchymal arterioles appear to lack this response [8, 25, 26]. The importance of parenchymal arterioles maintaining vascular tone during I/R is that they remain constricted under conditions in which blood flow is decreased, and therefore likely contribute to perfusion deficit and expansion of infarct into the penumbra.

5   Microcirculation as a Target for Vascular Protection During Stroke

The cerebral capillary bed is unique structurally and functionally compared to endothelium outside the central nervous system (CNS) and comprises the blood–brain barrier (BBB). It has a number of unique features that are highly protective of

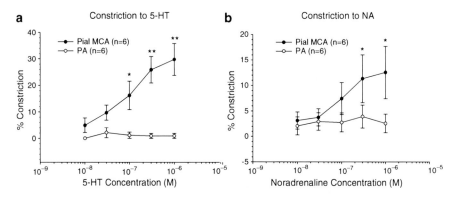

Fig. 1.2 (**a**) Graph showing the percent constriction in response to increasing concentrations of 5-hydroxytryptamine (5-HT) of pial MCA and parenchymal arterioles (PA) from the middle cerebral artery territory. Notice that PA had little response to the neurotransmitter; however, pial MCA were highly reactive. (**b**) Graph showing the percent constriction in response to increasing concentrations of noradrenaline (NA) of MCA and PA from the middle cerebral artery territory. Notice that PA had little response to NA whereas pial MCA were reactive. *$P < 0.05$, **$P < 0.01$ vs. PA and 10^{-8} M. From Cipolla et al. [22]. Used with permission

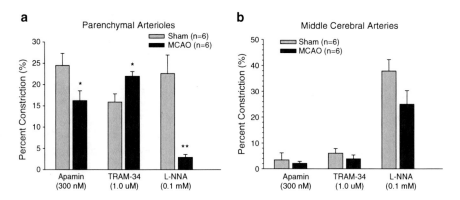

Fig. 1.3 Percent constriction to cumulative addition of apamin, TRAM-34, and L-NNA in (**a**) parenchymal arterioles (PA) and (**b**) MCA at 40 and 75 mmHg, respectively. Comparisons between sham controls (*gray bars*) and after ischemia and reperfusion (*red bars*) are shown. Notice that Sham PA constrict to apamin and TRAM-34, suggesting basal EDHF that inhibits tone in these vessels, but not MCA. In addition, constriction to apamin and TRAM-34 was preserved in PA after MCAO for 2 h of ischemia and 30 min of reperfusion, but constriction to NOS inhibition was decreased in both PA and MCA. *$P < 0.05$, **$P < 0.01$ vs. sham. From Cipolla et al. [25]

the brain milieu, including high electrical resistance tight junctions that restrict ion flux and paracellular movement of solutes, a low rate of transcellular transport, very low hydraulic conductivity, and high metabolic activity due to a large number of mitochondrion [27]. Another important feature of the cerebral microcirculation is that it is in close association with other cell types within the brain, including astrocytes, pericytes, and neurons [28]. Separating endothelial cells from other cell types in the brain is the basal lamina, a specialized extracellular matrix generated by both endothelial cells and astrocytes [29]. Taken together, the cellular and acellular components of the cerebral microcirculation comprise the neurovascular unit and are important therapeutic targets in stroke [30]. Microvessel integrity is compromised during ischemic stroke that has far-reaching implications for outcome and recovery. For example, endothelial and astrocyte cell swelling can inhibit flow and prevent therapeutic agents, such as neuroprotectants, from reaching their target [31]. In addition, vasogenic edema and hemorrhagic transformation also occur due to microvascular disturbances [28–31]. Thus, preserving microvascular structure and function during I/R is critical to stroke therapy.

During focal ischemia, significant alterations in microvessel integrity occur, including swelling of endothelium and surrounding astrocytic endfeet that can further obstruct flow, cause loss of permeability barriers, alter endothelium–astrocyte interaction, and change basal lamina structure [30, 31]. MCA occlusion (MCAO) causes microvascular disturbances that initiates inflammation, cellular edema, vascular permeability and vasogenic edema, and if prolonged, hemorrhagic transformation. The hypoxic tissue also activates microglia, astrocytes, and endothelial cells that react by secreting cytokines, chemokines, and matrix metalloproteinases (MMPs) [32, 33]. These inflammatory mediators lead to an upregulation of cell adhesion molecules by vascular endothelial cells that promote infiltration into the brain of blood-borne leukocytes, mainly neutrophils [31–34]. The infiltrated leukocytes themselves also secrete cytokines and release reactive oxygen and nitrogen species (RONS) promoting further brain injury [32, 34]. RONS in turn increase chemokine and cytokine expression [35]. In addition, cytokines can induce the expression of cytokine receptors and other cytokines as well as expression of several enzymes such as inducible NOS (iNOS) and cyclooxygenase (COX) that amplify the inflammatory response [36]. Thus, all these postischemic inflammatory processes lead to dysfunction of the neurovascular unit, causing cerebral edema and neural cell death, and eventually exacerbating the spread of damage to the ischemic penumbra [32, 33].

Inflammatory cytokines produced by the ischemic brain can also impact BBB integrity and exacerbate edema formation [37]. Experimental studies have shown that BBB disruption and hemorrhagic transformation result from the expression and activation of MMPs [38]. Tumor necrosis factor-α (TNF-α) activates MMPs that contributes to increased BBB permeability and promotes infiltration into the CNS of soluble and cellular mediators of inflammation causing neurodegeneration [39]. Moreover, higher levels of BBB disruption of cellular fibronectin (c-Fn), a component of basal lamina, and MMP-9 have been found in stroke patients [40]. In addition to proinflammatory cytokines, other circulating factors are released during I/R that increase BBB permeability and promote edema formation, including vascular endothelial growth factor (VEGF), histamine and thrombin [27].

In addition to inflammation, cytotoxic edema is also an important therapeutic target during focal ischemia [41]. Cytotoxic edema results from an influx of cations, most notably sodium, into the cell through cation channels. Most important to cell swelling during focal ischemia are nonselective cation (NC) channels because they allow flux of any monovalent cation and even a mixture of monovalent and divalent cations to pass [41]. One of the most important nonselective cation (NC) channels in ischemic stroke is the NC_{Ca-ATP} channel. This channel is not constitutively expressed, but is induced during focal ischemia in astrocytes, neurons, and endothelium [42]. Under ischemic conditions, depletion of ATP opens the channel, inducing a strong inward current that depolarizes the cells and causes cell swelling. Importantly, inhibitors of the regulatory subunit of NC_{Ca-ATP} channel, SUR1, prevent cytotoxic edema and improve stroke outcome [42].

6 Middle Cerebral Arteries as Targets for Vascular Protection During Stroke

The cerebral circulation is a unique vascular bed in that the large extracranial vessels (internal carotid and vertebral) and intracranial pial vessels such as MCA contribute approximately half of total CVR [43, 44]. This unusually prominent role of large arteries in vascular resistance helps to maintain CBF under conditions that change blood flow locally (e.g., functional hyperemia) but also protects downstream vessels during changes in systemic arterial pressure [45–47]. The significant contribution of large cerebral arteries (e.g., MCA) to vascular resistance in the brain makes understanding functional changes in these vessels during I/R important. The MCA has been studied most often after I/R likely because the MCAO model of focal ischemia provides a convenient methodology to do so. Because the MCA territory is the most commonly affected brain region during focal ischemia, this focus on the MCA is appropriate. One target for vascular protection is the regulation of arterial diameter because small changes in lumen diameter can have significant effects on CBF and CVR. In fact, diameter changes in response to pathological stimuli can occur rapidly and dramatically to alter regional and global CBF [47, 48].

6.1 Myogenic Mechanisms

A critical component of resistance artery function is the myogenic response which involves two phenomena: the alteration in tone in response to a change in pressure (myogenic reactivity) and the state of partial constriction at a constant pressure (myogenic tone) [49]. The vasoactive response of cerebral arteries to changes in pressure contributes to autoregulation of CBF and is facilitated by vascular smooth muscle that contracts to increased pressure and relaxes in response to decreased pressure [43, 46, 47, 50]. In normotensive adults, CBF is maintained at ~50 mL/100 g of brain tissue per minute, provided cerebral perfusion pressure is in the range of ~60–160 mmHg [51]. Above and below this limit, autoregulation is lost and CBF becomes dependent on mean arterial pressure in a linear fashion [48]. The importance of autoregulation in normal brain function is highlighted because significant tissue damage, including neuronal injury, BBB disruption, and edema formation occur when autoregulatory mechanisms are ineffective [52–55]. Autoregulation becomes lost during prolonged cerebral ischemia when several vasodilatory mechanisms are invoked in order to increase flow to the ischemic region. Initially, occlusion of an artery decreases perfusion pressure that invokes vasodilation of downstream arteries and arterioles due to myogenic mechanisms, followed by vasodilation in response to lactic acidosis and carbon dioxide (see Sect. 2).

Experimental models of acute focal cerebral ischemia demonstrate that reperfusion following transient cerebral ischemia can cause CBF autoregulatory failure and an increase blood flow (hyperemia) in some cases [56–59]. Studies of patients with stroke also show that focal cerebral hyperemia is a common and transient phenomenon most frequently found in the acute phase of stroke [52]. For example, the incidence of hyperemia is high (~40–50%) within 3 days after stroke and decreases to only 10% of the patients within 2–6 months after onset of stroke [52, 59]. Cerebral hyperemia can exacerbate neuronal injury during stroke and also promote BBB disruption and brain edema due to a diminished CVR that exposes the microcirculation to excessive perfusion pressure [52, 60]. Hyperemia during postischemic reperfusion occurs from vasodilatation of the cerebral vasculature and loss of myogenic responses. In fact, after 30 min of reperfusion, the caliber of MCA increases 172% and CBF by 588% of baseline [7]. Thus, studies have focused on the underlying mechanisms by which I/R affects MCA diameter regulation and loss of myogenic tone as this event can cause considerable secondary brain injury.

6.2   Threshold Duration of Ischemia and Reperfusion for Myogenic Responses

Postischemic reperfusion impairs myogenic mechanisms in MCA that is dependent on the duration of I/R. When the MCAO model of focal ischemia was used to induce 2 h of ischemia, myogenic tone in MCA was preserved after a brief reperfusion period of 1 min but was significantly lost after 24 h (Fig. 1.4a) [2]. In addition, the decrease in tone after 24 h of reperfusion was associated with a significant loss of myogenic reactivity (Fig. 1.4b). This initial study demonstrated that there was a threshold duration of ischemia and/or reperfusion for myogenic responses, and subsequent studies were focused on defining those thresholds. For example, when the reperfusion duration was held constant at 24 h and the ischemic duration was increased, there was a greater loss of tone in MCA with increased ischemic duration (Fig. 1.5a) [61]. In this manner, it was found that the threshold duration of ischemia to tone was between 15 and 30 min. It is important to note that this threshold is only for reperfusion durations of 24 h. This study also found that there was a significant positive correlation between the diminished myogenic tone and F-actin content of vascular smooth muscle (Fig. 1.5b), suggesting that loss of F-actin is an underlying mechanism by which tone is diminished with ischemia.

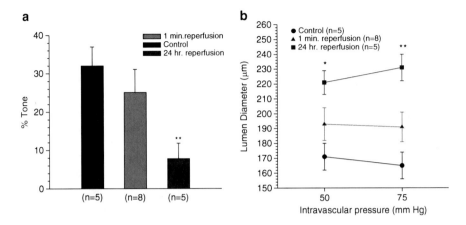

Fig. 1.4 The effect of different periods of reperfusion on MCA myogenic responses. (**a**) Percent basal tone present in MCA from sham control (*black bar*), after 1 min of reperfusion (*gray bar*), and 24 h of reperfusion (*red bar*) at a transmural pressure (TMP) of 75 mmHg. MCA exposed to (**a**) short reperfusion of 1 min had a similar level of basal tone as sham controls. Longer reperfusion caused arteries to have significantly diminished basal tone. (**b**) The effect of reperfusion duration on active diameter changes to a stepwise increase in TMP from 50 to 75 mmHg. MCA from sham control (*black circles*), after 1 min of reperfusion (*black triangles*), and after 24 h of reperfusion (*red squares*). Arteries from sham control and after 1 min of reperfusion responded to increased TMP with active vasoconstriction (i.e., myogenic reactivity); however, after 24 h of reperfusion arteries behaved passively and increased diameter with increased TMP. *$P < 0.05$, **$P < 0.01$ vs. control. From Cipolla et al. [2]

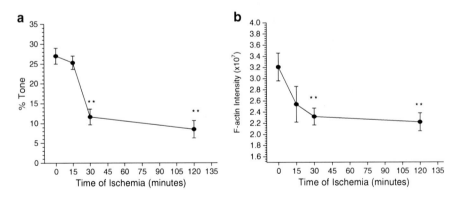

Fig. 1.5 The effect of ischemic duration of myogenic responses and filamentous (F-) actin content of MCA. (**a**) Myogenic tone development in MCA exposed to different periods of ischemia, all with 24 h of reperfusion: sham-operated control ($n=6$) and 15 ($n=6$), 30 ($n=7$), and 120 min ($n=8$) of ischemia. **$P < 0.01$ vs. control. (**b**) F-actin content in the same arteries shown in (**a**). Average F-actin content determined by fluorescence intensity of each group of ischemic arteries. **$P < 0.01$ vs. control. From Cipolla et al. [61]

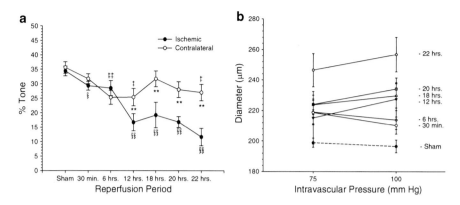

Fig. 1.6 The effect of reperfusion duration of myogenic responses of MCA. (**a**) Graph showing percent tone of MCA exposed to different periods of reperfusion all with 30 min of ischemia compared with sham-operated control animals. Shown are responses of MCA from both ipsilateral (*closed circles*) and contralateral (*open circles*) to occlusion. The amount of tone significantly diminished compared with sham-operated control in both contralateral and ischemic arteries as reperfusion duration increased. **$P < 0.01$ contralateral vs. ischemic; ‡$P < 0.05$ contralateral vs. sham control; ‡‡$P < 0.01$ contralateral vs. sham control; §$P < 0.05$ ischemic vs. sham control; §§$P < 0.01$ ischemic vs. sham control. (**b**) Graph showing diameter of MCA after step increases in transmural pressure from 75 to 125 mmHg after exposure to different periods of reperfusion all with 30 min of ischemia. Myogenic reactivity of MCA significantly diminished as reperfusion duration increased. From Cipolla et al. [62]

   The reperfusion threshold for both myogenic tone and reactivity was also determined in MCA when the ischemic duration was held constant at 30 min and reperfusion varied from 30 min to 22 h. A significant loss in myogenic tone was noted between 6 and 12 h of reperfusion (Fig. 1.6a) [62]. Interestingly, the contralateral MCA was also affected and had diminished tone as well, but not to the same extent as the ipsilateral MCA, suggesting there may be circulating factors that affect vascular function globally. In this study, myogenic reactivity was also assessed at each duration of I/R by measuring the change in lumen diameter in response to a step increase in pressure (Fig. 1.6b). Similar to the effect on tone, the reactivity to pressure was preserved at 6 h of reperfusion, but diminished after 12 h. In addition, the slope of the pressure diameter curve became more positive as the reperfusion duration increased, suggesting greater vascular damage with time that could impair autoregulation (Fig. 1.7).

6.3   Effect of tPA on Myogenic Responses and Vascular Function

Understanding how I/R affects MCA function during treatment may also be important for vascular protection. Currently, the only FDA-approved treatment for ischemic stroke is to restore CBF using recombinant tPA (rtPA) to recanalize an occluded vessel. A major limitation of this pharmacologic treatment is that it is restricted to

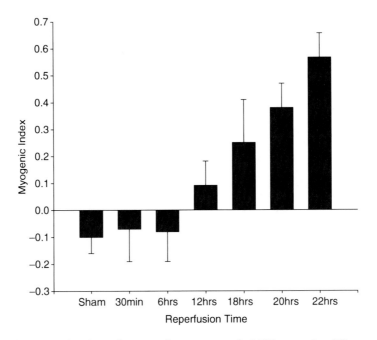

Fig. 1.7 Graph showing slope of pressure-diameter curves for MCA exposed to different periods of reperfusion. Arteries that produced a negative slope are considered myogenic (sham-operated control, 30 min and 6 h of reperfusion), whereas arteries that had a positive slope (≥12 h of reperfusion) had diminished myogenic behavior. From Cipolla et al. [62]

use within 4.5 h of symptom onset because of the potential for adverse vascular effects (intracerebral hemorrhage) outside this treatment window [63, 64]. Our own studies found that treatment with rtPA significantly diminished myogenic reactivity in isolated cerebral arteries, a result that was additive if arteries were exposed to ischemia (Fig. 1.8) [65]. In addition, exposure to I/R or rtPA intraluminally significantly impaired vasodilation to acetylcholine that was also additive when ischemic arteries were perfused with tPA (Fig. 1.9) [65]. These results suggest that rtPA administration adversely affects the cerebral vasculature, especially under ischemic conditions and that treatment to protect the vasculature with rtPA should be considered.

6.4 Mechanism of I/R-Induced Loss of Myogenic Tone

The mechanisms by which I/R causes loss of myogenic tone are not completely clear and may be a combination of several processes. Accumulation of acid metabolites in the ischemic region results in vasodilatation and a varying degree of vasomotor paresis depending on the degree of pH decrease [66]. Aggregation of blood elements and vasoactive metabolites that induce vasodilatation through relaxation of vascular smooth muscle may also have role in affecting myogenic tone during I/R [60, 67].

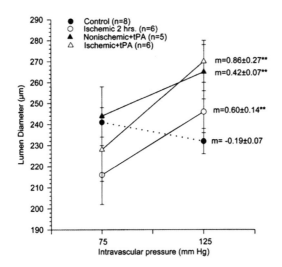

Fig. 1.8 The effect of recombinant tissue plasminogen activator (rtPA) and ischemia and reperfusion on myogenic responses of MCA. Graph shows the change in lumen diameter after an increase in intravascular pressure from 75 to 125 mmHg in MCA from the control (*filled circle*), ischemic 2 h with 24 h of reperfusion (*open circle*), nonischemic + rtPA (*filled triangle*), and ischemic + rtPA (*open triangle*) groups. Mean slope (m) of the pressure-diameter curve for each group is shown on the right next to each symbol. Notice that both rtPA and ischemia and reperfusion caused a decrease in myogenic reactivity that were potentiated in ischemic MCA exposed to rtPA. *$P < 0.05$ vs. control. From Cipolla et al. [65]

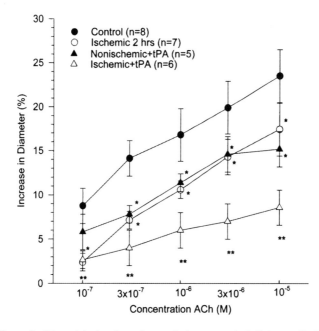

Fig. 1.9 Effect of rtPA and ischemia and reperfusion on endothelial vasodilation of MCA. Concentration-response to acetylcholine (ACh) in MCA from the control (*filled circle*), ischemic 2 h (*open circle*), nonischemic + rtPA (*filled triangle*), and ischemic + rtPA (*open triangle*) groups. Notice that both rtPA and ischemia and reperfusion caused a decrease in dilation to ACh that was worsened when ischemic MCA were exposed to rtPA. *$P < 0.05$, **$P < 0.01$ vs. control. From Cipolla et al. [65]

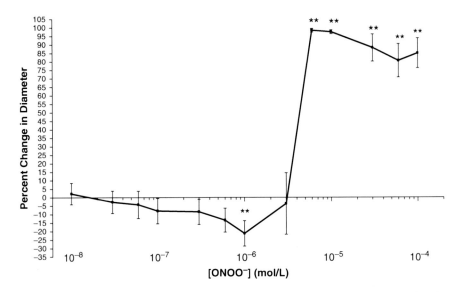

Fig. 1.10 The vasoactive effect of peroxynitrite (ONOO⁻) on cerebral arteries. Graph showing the concentration response of posterior cerebral arteries (PCA) to cumulative addition of ONOO⁻ (10^{-8} to 10^{-4} mol/L) at 75 mmHg. Addition of ONOO⁻ caused constriction at low concentrations (10^{-6} mol/L); however, arteries significantly dilated to ONOO⁻ at concentrations $>6 \times 10^{-6}$ mol/L ($n=7$; $*P<0.05$; $**P<0.01$ vs. no ONOO⁻). From Maneen et al. [77]

Although these compounds may have a role in the vasodilatation induced by I/R, they cannot completely explain the prolonged vasodilatation that occurs, particularly since it is demonstrated in vitro when these metabolites are not present and persist in humans for 2 or 3 weeks after stroke [60, 68]. Since the loss of myogenic tone induced by I/R is associated with actin cytoskeletal disruption of the vascular smooth muscle cells [61] and myogenic behavior has been shown to depend on an intact and dynamic actin cytoskeleton [69, 70], it is likely that loss of tone due to increased ischemic duration is caused by loss of F-actin content in vascular smooth muscle. In agreement with this, ischemia disrupts actin filaments in several cell types, which is considered a major contributor to ischemic damage [71, 72]. An underlying cause of F-actin disruption during ischemia is the decrease in brain ATP levels that occurs within the first 15 min of cerebral ischemia [73]. To compensate for decreased ATP during ischemia, ATP bound to actin is exchanged for ADP thereby releasing ATP for cellular use and causing actin polymers to dissociate [74].

Another mechanism by which tone is affected during I/R is through increased oxidative stress. Production of nitrogen and oxygen free radicals during I/R can affect tone, including superoxide and nitric oxide, which can react rapidly to produce peroxynitrite (ONOO⁻) [75]. In fact, most of the cytotoxic effects attributed to nitric oxide are due to ONOO⁻, including in the vasculature [76]. Our own studies have shown that high concentrations of ONOO⁻, similar to those produced during I/R, cause vasodilatation of cerebral arteries and loss of myogenic tone and reactivity whereas lower concentrations cause vasoconstriction (Fig. 1.10) [76, 77]. Similar to

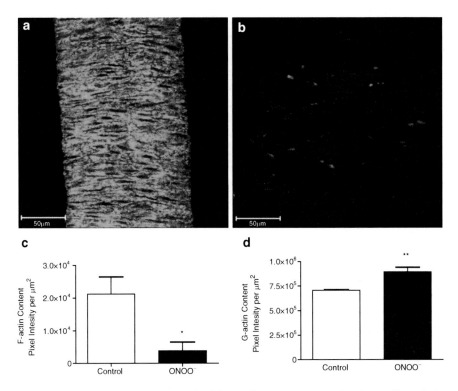

Fig. 1.11 The effect of peroxynitrite (ONOO⁻) on filamentous (F-) and globular (G-) actin in cerebral artery smooth muscle. Photomicrographs showing F-actin content in pressurized posterior cerebral arteries (PCA) with (**a**) spontaneous tone without ONOO⁻ exposure and (**b**) in the presence of 10^{-4} mol/L ONOO⁻ using phalloidin Oregon-Green. (**c**) F-actin pixel intensity per μm^2 in pressurized PCA with tone (control) and in 10^{-4} mol/L ONOO⁻ ($n=8$ per group). (**d**) G-actin pixel intensity per μm^2 in pressurized PCA with tone (control) and in 10^{-4} mol/L ONOO⁻ using DNase I, Alexa 488 ($n=5$ per group; *$P<0.05$, **$P<0.01$ vs. control). From Maneen et al. [77]

I/R, the loss of tone due to ONOO⁻ exposure is associated with a decrease in vascular smooth muscle F-actin and an increase in G-actin content, providing a mechanistic link between ONOO⁻ and vascular dysfunction during I/R (Fig. 1.11) [77]. ONOO⁻ interacts with tissues, causing nitrosylation of tyrosine residues. Nitrosylation is often used for ONOO⁻ detection, but it can also interfere with normal cellular function. We found that ONOO⁻ causes nitrosylation of F-actin in vascular smooth muscle which leads to depolymerization and the subsequent loss of myogenic tone [78]. Because ONOO⁻ affects myogenic tone and reactivity, the effects of ONOO⁻ on the cytoskeleton could be a mechanism by which autoregulation is disrupted after I/R and may be an important therapeutic target for vascular protection. In support of this, the loss VSM filamentous F-actin evoked by ONOO⁻ is similar to that seen after cerebral ischemia [61, 77]. However, in contrast with the effect of ischemic duration on vascular function where loss of myogenic tone is correlated with F-actin content, when the ischemic duration was kept constant and the reperfusion period increased, there was less correlation between the polymerization state of actin and the loss of myogenic activity. Loss of myogenic behavior was higher after 22 h of reperfusion, while the F-actin was significantly less only at 6 and 12 h of reperfusion [62].

It is likely that the mechanism by which ischemia affects the cerebral circulation is different from that of reperfusion. For example, during ischemia there is a loss of ATP [73] and upregulation of endothelial NOS (eNOS) and neuronal NOS (nNOS) activity that increases NO production within 10 min after focal ischemia and returns to normal at 60 min [79]. These events may explain why F-actin appears to be more susceptible to ischemic injury. However, during reperfusion other mechanisms are prominent, including production of superoxide and hydrogen peroxide, that also dilate cerebral vasculature by activating ATP-sensitive K^+ (K_{ATP}) and large-conductance Ca^{2+}-activated K^+ (BK) channels, respectively [80].

$ONOO^-$ can cause vasodilatation not only by nitrosylation of actin, but also via other mechanisms that are not as paralyzing to the vasculature and may be more related to transient responses seen during shorter durations of I/R. $ONOO^-$ has been shown to activate BK channels in vascular smooth muscle and cause hyperpolarization, increase production of cyclic guanosine 3′, 5′-monophosphate (cGMP), decrease smooth muscle cell calcium through sarcoplasmic reticulum Ca^{2+}-ATPase channel activity, and cause increased activation of myosin phosphatase activity [81, 82]. In addition, the hypoxic brain tissue synthesizes proinflammatory cytokines within a few minutes of stroke that can promote adhesion of circulating neutrophils to endothelial cells with subsequent release of additional proinflammatory mediators. Generation of RONS by the infiltrated neutrophils amplifies the inflammatory signal, causing damage to the brain and vasculature [34, 83, 84]. Thus, inflammatory responses may directly or indirectly affect the vasculature to cause loss of myogenic behavior and vasodilatation (see Sect. 9).

7   The Cerebral Endothelium as a Therapeutic Target During Focal Ischemia

The cerebral endothelium has a prominent role in many vascular functions including angiogenesis, inflammation, thrombosis, and regulation of vascular tone [85, 86]. In the brain, the endothelium has an unusually potent influence on resting vascular tone and CBF and is of considerable interest as a therapeutic target during disease states [86]. Although the cerebral endothelium can produce several vasoactive mediators, NO and EDHF are some of the most important and well studied. Under resting conditions, basal NO production by the cerebral endothelium inhibits resting tone and thus affects CBF [87]. This influence is demonstrated by effects of compounds that inhibit NOS, such as nitro-L-arginine (L-NNA), that cause significant constriction of cerebral arteries and arterioles [8, 87]. While the exact nature of EDHF in brain is not clear, its hyperpolarizing influence appears to depend on the activity of small- and intermediate-conductance calcium-activated potassium channels (SK and IK channels, also known as KCa2.3 and KCa3.1, respectively) due to the findings that specific blockade of these channels abrogates EDHF responses [88–92]. While EDHF does not appear to contribute to basal tone in cerebral (pial) arteries, inhibition of SK and IK channels causes constriction of parenchymal arterioles under control conditions, suggesting that unlike MCA, basal EDHF is present in these small vessels and inhibits basal tone (Fig. 1.3) [25, 93].

7.1   eNOS

Numerous studies have demonstrated that NO plays an important role in the pathogenesis of brain injury during cerebral ischemia. While both nNOS and eNOS are activated during ischemia, they appear to have opposing effects on stroke outcome [94–98]. Increased nNOS activation in neurons can cause neuronal injury whereas eNOS activation has been shown to be protective during stroke [95–97]. Mice lacking eNOS expression showed a greater degree of hemodynamic compromise with smaller penumbra areas compared to wildtype, suggesting that eNOS activity is protective of the brain during focal ischemia by improving blood flow in the penumbra [95].

NO is an important target for stroke therapy because it mediates vascular responses by causing vasodilatation, and inhibits platelet aggregation and leukocyte adhesion. Thus, it can improve postischemic blood flow by increasing collateral flow to the ischemic area and prevent microvascular plugging by platelets and leukocytes [99, 100]. However, as described above, NO can react with superoxide to produce ONOO$^-$ which interacts with proteins, lipids, and DNA, and promotes cytotoxic and proinflammatory responses [101]. In addition, during cerebral ischemia where the NOS substrate L-arginine and cofactors such as tetrahydrobiopterin (BH$_4$) are likely to be rate-limiting, eNOS generates not only NO but also superoxide and hydrogen peroxide promoting oxidative damage [102, 103].

Rapid upregulation of eNOS occurs during focal ischemia and remains upregulated on the ischemic side of the brain as long as 7 days of reperfusion [95, 98, 104]. While the effect of ischemia on eNOS expression has been well studied, postischemic reperfusion affects NO-mediated dilation. Our own studies [2, 8, 65] and numerous others [105–108] have consistently found that I/R diminishes the influence of basal NO and responsiveness of cerebral arteries to agonists that activate eNOS (e.g., ACh, bradykinin). These results importantly suggest that although eNOS may be activated by ischemia, reperfusion profoundly affects and diminishes NO-mediated vasodilation.

Diminished NO-mediated vasodilation after I/R might indicate that eNOS is uncoupled and producing a product other than NO like superoxide anion. A critical determinant of eNOS activity is the availability of the cofactor BH$_4$ [109]. Under conditions of limited BH$_4$ availability (due to oxidation or reduced formation), eNOS functions in an uncoupled state and nicotinamide adenine dinucleotide phosphate hydrogen (NAD(P)H)-derived electrons are added to molecular oxygen instead of L-arginine, leading to the production of superoxide [110]. Uncoupling of eNOS has been implicated in a number of diseases associated with decreased BH_4 levels, including atherosclerosis, diabetes, and hypertension [111–114]. Importantly, BH_4 supplementation in animals with these diseases reduces endothelial dysfunction [114]. Although data consistent with uncoupling of eNOS has been presented for cerebral vessels [111], the importance of this process as a mechanism of vascular dysfunction during I/R is not clear. Thus, it is possible that eNOS becomes uncoupled during I/R by cofactor depletion because of its enhanced activity in the ischemic brain that uses all available BH_4.

In addition to substrate and cofactor depletion during ischemia, eNOS can become uncoupled through oxidation and inactivation of BH_4 during reperfusion. The restoration of blood flow following ischemia causes a significant increase in production of $ONOO^-$ [115]. $ONOO^-$ is highly cell permeable where it promotes lipid peroxidation, nitrosylation of proteins and nucleic acids, and significant cell damage [115]. $ONOO^-$ readily produces oxidation of BH_4 and may also be an underlying mechanism by which eNOS is uncoupled during I/R [111, 116].

The activity of eNOS is also regulated by phosphorylation of the enzyme on specific amino acid residues that may be affected by I/R [117, 118]. Following Ser1177 phosphorylation, NO production is increased [117]. By contrast, eNOS activity is inhibited by Thr495 phosphorylation [118]. Little is known about how focal ischemia affects the phosphorylation state of eNOS; however, global forebrain ischemia for 10 min causes de-phosphorylation of Thr495 as early as 30 min of reperfusion [119]. In this study, Ser1177 phosphorylation was unchanged, but the effect on Thr495 suggests greater eNOS activation. eNOS phosphorylation has been shown to regulate superoxide production during eNOS uncoupling [120]. Under conditions of substrate depletion, phosphorylation of Ser1177 increased superoxide generation by eNOS >50% and altered the calcium sensitivity of the enzyme such that superoxide production was largely calcium-independent [120]. It is possible that I/R alters the phosphorylation state of eNOS that contributes to its diminished activity and/or production of superoxide if the enzyme is uncoupled.

7.2   EDHF

Some studies have shown that I/R affects receptor systems involved with endothelial-mediated dilations. Endothelial $P2Y_2$ purinoceptor-mediated dilations in MCA are potentiated following I/R via an upregulation of EDHF [105]. The basic mechanism of EDHF-mediated response may be that increased intracellular calcium in endothelium that activates SK and IK channels causing K^+ efflux and membrane hyperpolarization [90, 105]. Hyperpolarization of endothelium is then transferred to VSM via synthesis or generation of signals capable of diffusing through membranes or myoendothelial gap junctions [121]. In VSM, EDHF activates K^+-channels, causing hyperpolarization [93, 122]. The movement of K^+ from the cytoplasm to the extra-cellular space hyperpolarizes vascular smooth muscle membrane potential, closing voltage-gated Ca^{2+} channels that decreases cytoplasmic Ca^{2+} and relaxes vascular smooth muscle [123]. Thus, EDHF-induced opening or activation of potassium channels in vascular smooth muscle can decrease myogenic tone and may be a mechanism of cerebral vasodilatation after I/R. Since EDHF does not contribute to vascular tone of cerebral arteries under basal conditions, it is not likely a prominent target for vascular protection of MCA during I/R.

7.3 Calcium

Central to many signaling events during I/R is an increase in intracellular calcium ($[Ca^{2+}]_i$) in endothelial cells [124]. High levels of $[Ca^{2+}]_i$ not only impair endothelial barrier function and integrity of tight junctions [125] but also alter activity of enzymes and processes involved in regulation of vascular tone, including eNOS and EDHF. Thus, an increase in intracellular calcium in cerebral endothelium during I/R is pleiotropic and likely an important therapeutic target. Production of NO by eNOS is principally activated by calcium-dependent binding of calmodulin, making eNOS activation calcium-dependent [126]. During cerebral ischemia when $[Ca^{2+}]_i$ remains persistently elevated, eNOS becomes continuously active and produces potentially toxic amounts of NO [105, 127]. Moreover, after 6–12 h of MCAO, cerebral ischemia induces expression of iNOS in the vascular cells which produces large amounts of NO continuously for long periods [128]. The mechanism by which I/R increases endothelial cell calcium is not known, but may involve ROS-induced activation of SK and IK channels or transient receptor potential (TRP) channels.

7.4 Endothelin

Endothelin-1 is a potent vasoconstrictor peptide synthesized and released by endothelial cells in both the peripheral and cerebral vasculature [129]. ET-1 plays a significant role in vascular tone and CBF regulation [129]. Under physiological conditions, there is a balance between the vasoconstrictor effect induced by ET-1 and the vasodilator mechanism mediated by NO [130]. This balance is important for regulating vascular function and may be disrupted in cerebrovascular disease such as cerebral ischemia where ET-1 levels are increased [131]. Marked elevation of ET-1 in plasma, cerebrospinal fluid, and ischemic brain has been demonstrated in patients and animal models after ischemic stroke [131–133]. ET-1 regulates vascular tone through the activation of two specific receptor subtypes, ET_A and ET_B receptors. It is via the activation of ET_A receptors, expressed in smooth muscle cells, that ET-1 evokes contractions [134]. ET_A receptor antagonists have been shown to improve outcome after experimental cerebral ischemia [135–137]. For example, ET_A receptor antagonist S-0139 attenuated the increase in brain water and infarct size observed after transient MCAO [137], and Ro 61–1790 was able to improve microvascular perfusion and reduced ischemic lesion volume [136]. In addition, a study using transgenic mice with endothelial ET-1 overexpression showed that ET_A receptor activation contributes to the increased oxidative stress, water accumulation, and BBB breakdown after transient MCAO [138]. The mechanisms of action of ET-1 in stroke outcome are not well understood. Several mechanisms have been proposed for the detrimental actions of ET-1 during stroke, including activating superoxide production through NAD(P)H oxidase and affecting the BBB to increase permeability, and promoting edema [138–140].

In contrast to the role of ET_A receptor, ET_B receptors, primarily expressed in endothelial cells, may have a protective role in ET-1 provoked cerebral ischemic injury. Thus, ET_B blockade increased infarct volume after transient MCAO [138]. These results support the concept of ET-1 receptors as potential therapeutic targets for the vascular protection during ischemic stroke.

8   Hyperglycemia as a Therapeutic Target During Stroke

Hyperglycemia is common in acute stroke. High glucose levels are present in up to 40% of ischemic stroke patients, often without a preexisting diagnosis of diabetes [141]. Different etiologies have been proposed for explaining this phenomenon. Possibilities include hyperglycemia caused by a generalized stress response to acute brain injury, impaired glucose tolerance, and first presentation of previously unrecognized diabetes [142–144]. Stress reaction is characterized by hypersecretion of glucocorticoids which inhibit glucose uptake into peripheral tissues, causing an increase in serum glucose levels [145]. Patients who have suffered transient ischemic attacks or minor ischemic stroke have impaired glucose tolerance, suggesting that the stroke itself impairs glucose metabolism [143]. A recent study suggests a potential role of adipose proinflammatory cytokines/chemokines such as TNF-α and monocyte chemoattractant protein-1 (MCP-1) in the pathogenesis of a glucose metabolic disorder during stroke [146]. Lastly, a study by Gray et al. found that acute stroke may unmask deficient glucose regulation in patients with insulin resistance, developing diabetes mellitus after 12 weeks post stroke [144].

Hyperglycemia during acute stroke is associated with significantly worsened outcome, including larger infarction, edema formation, and a higher risk of mortality [147, 148]. The exacerbated damage with hyperglycemia is especially prevalent in reperfusion ischemic stroke models and occurs less with permanent occlusion [149], suggesting reperfusion may be an important factor in hyperglycemic brain injury. One explanation may be that during reperfusion ischemic tissue is exposed to oxygen-rich blood which provides an abundant substrate to NAD(P)H oxidase which can produce superoxide [150]. In agreement with this, oxidative stress found in hyperglycemic stroke is higher than in normoglycemic conditions [76]. Because both diabetic and nondiabetic patients are adversely affected by hyperglycemia, it is elevated glucose and not diabetic complications that increase stroke damage [147, 148]. In contrast with the association between hyperglycemia and poor outcome, it has been proposed that hyperglycemia represent a stress response which may not have any direct effect in the prognosis of hyperglycemic stroke patients [151, 152]. However, the fact that glucose administration before onset of ischemia enhances brain injury and edema, and therapy approaches that lower glucose levels such as insulin and overexpression of the glucose transporter gene have a protective effect reducing ischemic brain injury in animal models of stroke [147, 153, 154] support the concept that hyperglycemia plays an important role in the pathophysiology of stroke.

Hyperglycemia can exacerbate ischemic damage by multiple physiologic mechanisms such as brain acidosis, vascular inflammation, increased BBB permeability, hemorrhagic transformation, accumulation of free radicals, and impaired vascular reactivity [155]. Much attention has been devoted to the neuronal and astrocytic effect of brain lactate accumulation, followed by a further decrease in intracellular and extracellular pH, as a mechanism of enhanced brain injury during hyperglycemic stroke. However, lactate was not able to increase ischemic brain damage in either animal models or pilot clinical stroke trial, and thus the role of lactate has been questioned [156, 157].

Compared with normoglycemia, hyperglycemia fundamentally alters vascular responses [158]. Thus, studies have focused on the effect of hyperglycemia on the vasculature. Extracellular hyperglycemia activates Glut-1 transporters in cerebral endothelial cells which transport glucose into the cell causing endothelial intracellular hyperglycemia [159]. It has been reported that intracellular hyperglycemia can induce a variety of detrimental changes in vascular cerebral cells within hours [150, 158]. For example, it can decrease NAD(P)H-production, an important intracellular antioxidant. Intracellular high glucose can also induce posttranslational modifications of intracellular proteins via hexosamine pathway and activate protein kinase C (PKC) which increases RONS via NAD(P)H oxidase, stimulate proinflammatory gene expression through nuclear factor-kappa B (NFkB) leading to upregulation of intracellular adhesion molecules, impair NO-mediated vasodilation via eNOS dysfunction, and enhance plasminogen activator inhibitor-1 (PAI-1) expression causing thrombus stabilization and microvascular plugging (for review see [150]). All these proinflammatory, pro-thrombotic, and pro-vasoconstrictive mechanisms activate in response to intracellular hyperglycemia and can have significant effects on vascular function including impaired autoregulation and in ineffective reperfusion. Some studies also found that hyperglycemia increases perfusion deficit and impairs reperfusion compared to normoglycemic animals after longer durations of I/R [149, 160, 161]. By contrast, other studies did not find an effect of hyperglycemia on reperfusion blood flow after 30 min of ischemia with 1 or 2 h of reperfusion, suggesting that the differences may be related to the duration of I/R or severity of hyperglycemia [162, 163].

One of the most detrimental consequences of the effect of hyperglycemia and stroke on CBF is the development of brain edema. In a retrospective review, Bergen and Hakim reported that hyperglycemic stroke patients develop more pronounced cerebral edema and have a worse clinical outcome compared to normoglycemic stroke patients [164]. Experimental studies have also shown that hyperglycemia causes edema formation after cerebral ischemia that is more pronounced than in normoglycemic animals [147, 165, 166]. Increased BBB permeability occurs during hyperglycemic stroke and causes cerebral edema [165, 167, 168]. Ennis and

Keep reported that both mild and severe hyperglycemia (blood glucose ~150 and 400 mg/dL, respectively) induced BBB disruption and permeability during permanent and transient MCAO [167]. Different mechanisms such as oxidative stress, MMP activation, lactic acidosis, and PKC activation have been implicated in BBB dysfunction after hyperglycemic I/R injury [165, 166, 168]. For example, PKC activation by hyperglycemia can directly affect BBB permeability through its ability to phosphorylate zona occluden-1 (ZO-1) affecting the integrity of the tight junction as well as can promote calcium/calmodulin-dependent endothelial cell contraction, increasing paracellular permeability [169, 170].

Although hyperglycemia occurs with all stroke subtypes, the detrimental effects of high admission glucose appear to occur only with cortical stroke and not lacunar stroke. Lacunar strokes are small subcortical infarcts that result from the occlusion of one of the penetrating lenticulostriate arterioles that supply blood to the brain's deep structures [171]. Several clinical and experimental studies have shown that hyperglycemia does not worsen outcome from lacunar stroke and in some studies, patients with moderate hyperglycemia had better outcome than normoglycemic patients [172–175]. For example, in animals with end-artery infarcts hyperglycemia decreased infarct size and improved outcome [172, 173]. Furthermore, clinical studies showed that hyperglycemia worsened the outcome in non-lacunar stroke, but not in lacunar stroke where the outcome was better than in normoglycemic patients [174, 175]. In addition, in cortical ischemic strokes, but not in lacunar strokes, hyperglycemia increased serum levels of neuron-specific enolase (NSE), a peripheral indicator for neural damage [176].

The underlying mechanism by which hyperglycemia improves lacunar stroke outcome is not clear. It has been proposed that lactate production in astrocytes and oligodendrocytes may protect axons in white matter [175]. Another theory is that hyperglycemia and/or I/R has a different effect on the vasculature involved in lacunar vs. cortical strokes. We studied the effect of hyperglycemia on the function of lenticulostriate arterioles after MCAO and found there was no effect of hyperglycemia on their function compared to normoglycemic animals that underwent similar MCAO [162]. Thus, presence of hyperglycemia during focal ischemia does not seem to affect the lenticulostriate arterioles. This preservation of function of the lenticulostriate arterioles during hyperglycemic I/R may maintain CBF and improve outcome during lacunar stroke [162].

9 Inflammation and Circulating Immune Factors as a Target for Vascular Protection During Stroke

Inflammation represents an important mechanism in the pathogenesis of stroke [32–34, 177]. Clinical stroke and experimental cerebral ischemia elicit an inflammatory response that limits the efficacy and time window of stroke treatments such as restoration of perfusion with t-PA [37]. Production of proinflammatory cytokines and chemokines increases after stroke. The most studied are those related to inflammation in acute ischemic stroke and include the cytokines TNF-α, the interleukins (IL) IL-1β, IL-6, IL-10, IL-20, and transforming growth factor (TGF)-β and the chemokines IL-8, interferon inducible protein-10 (IP-10), and monocyte chemoattraction protein-1 (MCP-1) [32, 83, 178]. IL-1 β, TNF-α, IL-8, and MCP-1 appear to exacerbate ischemic injury, whereas the anti-inflammatory cytokines IL-10, TGF-β, and IL-1ra appear to be neuroprotective [178]. High levels of inflammatory cytokines and lower levels of anti-inflammatory cytokines are associated with greater infarction and poorer clinical outcome. Inflammation therefore becomes an important target to limit ischemic tissue damage. In fact, some studies have shown that the use of anti-inflammatory agents such as IL-1r antagonist and IL-10 is associated with better stroke outcome [179, 180]. Furthermore, inflammatory biomarkers such as high-sensitivity C-reactive protein (hsCRP) have been identified as predictors of stroke and prognosis after stroke in certain populations [181].

The inflammatory processes induced by stroke are not confined to the damaged brain, but manifest as systemic responses that can affect other organs. Offner and coworkers were the first to demonstrate that the inflammatory changes occurring in the damaged brain are dynamically reflected in the peripheral immune organs [83]. In particular, they showed that stroke activated splenocytes and lymphoid tissue which release inflammatory cytokines [83]. In agreement with this, circulating levels of inflammatory mediators such as IL-6, TNF-α, intercellular adhesion molecule-1 (ICAM-1), and interferon gamma (IFN-γ) have been shown to be increased in plasma from stroke patients and cerebral ischemic animals [83, 182–184].

The elevations in plasma levels of proinflammatory cytokines after stroke may affect vascular function, including modulating vascular tone and the signaling pathways of vasoconstriction and vasodilation, by acting on the endothelial and vascular smooth muscle [35]. Endothelium and vascular smooth muscle express receptors for TNF-α, IL-1β, IL-6, and IL-10 [185, 186]. Moreover, there is evidence that cytokines can affect organ vascular resistance and thus blood flow regulation. For example, TNF-α, IL-1β, and IL-6 cause vasoconstriction, increased vasoconstrictor responses induced by different agonists, and impaired endothelium-dependent vasodilatation [187]. It has been also shown that injection of TNF-α into the cisterna magna reduces CBF in the rabbit [188]. The underlying mechanism by which inflammatory cytokines affects vascular tone is not well understood, but may involve cytokine-induced RONS production. Indeed, TNF-α can stimulate RONS production in endothelium by activating oxidative enzymes including COX, xanthine oxidase, and NAD(P)H oxidase [39]. In addition, cytokines may also increase iNOS expression in vascular smooth muscle and endothelial cells, which produces an excess of NO that can impair vasomotor responses and react with superoxide to produce ONOO$^-$ [87, 187].

We recently studied the effect of plasma from animals that underwent hyperglycemic MCAO on myogenic activity and endothelial function in nonischemic MCA. We found that circulating factors present in plasma during postischemic reperfusion increased myogenic tone and impaired endothelial function in nonischemic MCA (Fig. 1.12) [189]. These findings confirm that circulating factors produced in response to stroke can affect cerebrovascular function. A systemic effect of stroke

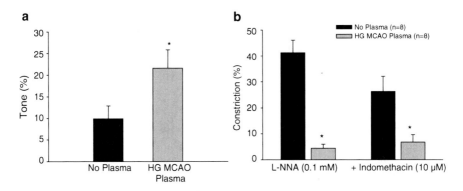

Fig. 1.12 Effect of circulating factors on myogenic tone and constriction to nitric oxide synthase (NOS) and cyclooxygenase (COX) inhibition in MCA. MCA were from nonischemic animals and perfused with either physiological buffer or plasma (20% in buffer) from hyperglycemic (HG) animal that underwent middle cerebral artery occlusion (MCAO) for 2 h with 2 h of reperfusion. (**a**) Myogenic tone was significantly increased in arteries that had plasma in their lumen compared to no plasma controls, suggesting circulating factors after stroke can influence vascular function of cerebral arteries. (**b**) Constriction to NOS inhibition with L-NNA and COX inhibition with indomethacin were impaired in arteries that had plasma in their lumen, demonstrating impaired endothelial function that could be responsible for increasing tone. *$P < 0.05$ vs. no plasma

has been shown in some studies that found impaired endothelium-dependent vasodilation in cerebral and brachial arteries in stroke patients, suggesting that both cerebral and peripheral endothelial function are affected by a systemic reaction in response to cerebral ischemia [190–192]. The role of circulating inflammatory mediators in systemic and brain vascular dysfunction is largely understudied but may be an important target for vascular protection during stroke.

Increasing evidence from experimental and clinical studies indicates a link between inflammatory mechanisms and high blood pressure [193, 194]. It has been reported that hypertension is present in up to 80% of patients with acute ischemic stroke and is associated with poor outcome, mainly due to promoting stroke recurrence and increasing cerebral edema [195]. Hypertension may stimulate inflammation through an increase in cytokines and inflammatory mediator expression. It is well established that angiotensin II and endothelin-1, which levels are increased in hypertension, can upregulate cytokines expression and adhesion molecules [194]. Indeed, elevated levels of circulating IL-6, macrophage inflammatory protein-1alpha (MIP-1alpha), soluble vascular cell adhesion molecule-1 (sVCAM-1), and sICAM-1 are associated with high blood pressure levels [193, 196]. Furthermore, increasing blood pressure after stroke is correlated with elevated circulating levels of the C-reactive protein (CRP), an inflammatory protein which expression can be induced by IL-6, suggesting that hypertension during acute stroke may worsen outcome via proinflammatory effects [197].

On the other hand, as it has been discussed above, hyperglycemia can also enhance inflammatory responses to I/R, exacerbating the ischemic injury. For example, hyperglycemia triggers neutrophil infiltration in brain after transient focal ischemia

[198]. Furthermore, it has been reported that insulin infusion therapy causes a potent anti-inflammatory effect suppressing several proinflammatory transcription factors and reducing plasma concentration of inflammation mediators including ICAM-1, MCP-1, VEGF, and MMP-9 [199]. Thus, inflammatory processes after stroke, that are more intense under conditions of hypertension and hyperglycemia, may provide another target for vascular protection and treatment of secondary brain injury.

10 Conclusions

Ischemic stroke has a significant effect on the cerebral vasculature that is progressive with increasing durations of I/R. All segments of the vasculature are affected by I/R. In MCA, prominent effects include loss of myogenic responses, vascular paralysis due to cytoskeletal disruption in vascular smooth muscle, and effects on endothelial function (eNOS, endothelin, and calcium). By contrast, downstream parenchymal arterioles have considerable basal tone despite exposure to I/R and appear to be the bottleneck to flow to the microcirculation. These arterioles are secondary collaterals that are high-resistance and thus an important target to stroke therapy. While EDHF responses are well preserved in parenchymal arterioles during I/R, NO is diminished similar to MCA. Decreased NO responsiveness may be due to decreased bioavailability, eNOS uncoupling, or changes in eNOS phosphorylation state during I/R. Microvascular dysregulation during stroke also contributes to poor outcome through several mechanisms. Cytotoxic edema can cause secondary ischemia due to compression of swollen endothelial cells and astrocytes, whereas BBB disruption can cause vasogenic edema and hemorrhagic transformation. Hyperglycemia, which is common in acute stroke often without preexisting diabetes, impairs vascular function and worsens stroke outcome. Hyperglycemia therefore becomes an important target for vascular protection during I/R. Inflammation and RONS production during postischemic reperfusion, that is enhanced by hyperglycemia, also have significant effects on vascular structure and function in all segments and are also important therapeutic targets to protect the vasculature during I/R.

Acknowledgments We gratefully acknowledge the support of the NIH, NINDS grants RO1 NS043316, RO1 NS 045940, The Neural Environment Cluster Supplement 3RO1 NS045940-06S1, ARRA Supplement 3RO1 NS045940-05S1, and NHLBI grant PO1 HL095488.

References

1. Nishigaya K, Yoshida Y, Sasuga M, Nukui H, Ooneda G. Effect of recirculation on exacerbation of ischemic vascular lesions in rat brain. Stroke. 1991;22(5):635–42.
2. Cipolla MJ, McCall AL, Lessov N, Porter JM. Reperfusion decreases myogenic reactivity and alters middle cerebral artery function after focal cerebral ischemia in rats. Stroke. 1997;28(1):176–80.

3. Kagstrom E, Smith ML, Siesjo BK. Local cerebral blood flow in the recovery period following complete cerebral ischemia in the rat. J Cereb Blood Flow Metab. 1983;3(2):170–82.
4. Takahashi A, Park HK, Melgar MA, Alcocer L, Pinto J, Lenzi T, et al. Cerebral cortex blood flow and vascular smooth muscle contractility in a rat model of ischemia: a correlative laser Doppler flowmetric and scanning electron microscopic study. Acta Neuropathol. 1997;93(4):354–68.
5. Siesjo BK. Pathophysiology and treatment of focal cerebral ischemia. Part II: mechanisms of damage and treatment. J Neurosurg. 1992;77(3):337–54.
6. Siesjo BK. Pathophysiology and treatment of focal cerebral ischemia. Part I: pathophysiology. J Neurosurg. 1992;77(2):169–84.
7. Tasdemiroglu E, Macfarlane R, Wei EP, Kontos HA, Moskowitz MA. Pial vessel caliber and cerebral blood flow become dissociated during ischemia-reperfusion in cats. Am J Physiol. 1992;263(2 Pt 2):H533–6.
8. Cipolla MJ, Bullinger LV. Reactivity of brain parenchymal arterioles after ischemia and reperfusion. Microcirculation. 2008;15(6):495–501.
9. Hossmann KA. Pathophysiology and therapy of experimental stroke. Cell Mol Neurobiol. 2006;26(7–8):1057–83.
10. Liebeskind DS. Collateral circulation. Stroke. 2003;34(9):2279–84.
11. Caplan LR, Hennerici M. Impaired clearance of emboli (washout) is an important link between hypoperfusion, embolism, and ischemic stroke. Arch Neurol. 1998;55(11):1475–82.
12. Wang CX, Todd KG, Yang Y, Gordon T, Shuaib A. Patency of cerebral microvessels after focal embolic stroke in the rat. J Cereb Blood Flow Metab. 2001;21(4):413–21.
13. Santa N, Kitazono T, Ago T, Ooboshi H, Kamouchi M, Wakisaka M, et al. ATP-sensitive potassium channels mediate dilatation of basilar artery in response to intracellular acidification in vivo. Stroke. 2003;34(5):1276–80.
14. Hoehn-Berlage M, Norris DG, Kohno K, Mies G, Leibfritz D, Hossmann KA. Evolution of regional changes in apparent diffusion coefficient during focal ischemia of rat brain: the relationship of quantitative diffusion NMR imaging to reduction in cerebral blood flow and metabolic disturbances. J Cereb Blood Flow Metab. 1995;15(6):1002–11.
15. Hossmann KA. Viability thresholds and the penumbra of focal ischemia. Ann Neurol. 1994;36(4):557–65.
16. Symon L, Branston NM, Strong AJ, Hope TD. The concepts of thresholds of ischaemia in relation to brain structure and function. J Clin Pathol Suppl (R Coll Pathol). 1977;11:149–54.
17. Hakim AM. The cerebral ischemic penumbra. Can J Neurol Sci. 1987;14(4):557–9.
18. Lo EH. A new penumbra: transitioning from injury into repair after stroke. Nat Med. 2008;14(5):497–500.
19. Chan SL, Cipolla MJ. Relaxin causes selective outward remodeling of brain parenchymal arterioles via activation of peroxisome proliferator-activated receptor-{gamma}. FASEB J. 2011;25(9):3229–39. doi:10.1096.
20. Rennels ML, Nelson E. Capillary innervation in the mammalian central nervous system: an electron microscopic demonstration. Am J Anat. 1975;144(2):233–41.
21. Cohen Z, Bonvento G, Lacombe P, Hamel E. Serotonin in the regulation of brain microcirculation. Prog Neurobiol. 1996;50(4):335–62.
22. Cipolla MJ, Li R, Vitullo L. Perivascular innervation of penetrating brain parenchymal arterioles. J Cardiovasc Pharmacol. 2004;44(1):1–8.
23. Nishimura N, Schaffer CB, Friedman B, Lyden PD, Kleinfeld D. Penetrating arterioles are a bottleneck in the perfusion of neocortex. Proc Natl Acad Sci USA. 2007;104(1):365–70.
24. You J, Johnson TD, Marrelli SP, Bryan Jr RM. Functional heterogeneity of endothelial P2 purinoceptors in the cerebrovascular tree of the rat. Am J Physiol. 1999;277(3 Pt 2):H893–900.
25. Cipolla MJ, Smith J, Kohlmeyer MM, Godfrey JA. SKCa and IKCa channels, myogenic tone, and vasodilator responses in middle cerebral arteries and parenchymal arterioles: effect of ischemia and reperfusion. Stroke. 2009;40(4):1451–7.
26. Shih AY, Friedman B, Drew PJ, Tsai PS, Lyden PD, Kleinfeld D. Active dilation of penetrating arterioles restores red blood cell flux to penumbral neocortex after focal stroke. J Cereb Blood Flow Metab. 2009;29(4):738–51.

27. Cipolla MJ. Stroke and the blood–brain interface. In: Dermietzel R, Spray D, Nedergaard M, editors. Blood–brain barrier interfaces. Weinheim: Wiley; 2006.

28. del Zoppo GJ. Stroke and neurovascular protection. N Engl J Med. 2006;354(6):553–5.

29. Wang CX, Shuaib A. Critical role of microvasculature basal lamina in ischemic brain injury. Prog Neurobiol. 2007;83(3):140–8.

30. del Zoppo GJ. Virchow's triad: the vascular basis of cerebral injury. Rev Neurol Dis. 2008;5 Suppl 1:S12–21.

31. del Zoppo GJ, Mabuchi T. Cerebral microvessel responses to focal ischemia. J Cereb Blood Flow Metab. 2003;23(8):879–94.

32. Lakhan SE, Kirchgessner A, Hofer M. Inflammatory mechanisms in ischemic stroke: therapeutic approaches. J Transl Med. 2009;7:97.

33. Denes A, Thornton P, Rothwell NJ, Allan SM. Inflammation and brain injury: acute cerebral ischaemia, peripheral and central inflammation. Brain Behav Immun. 2010;24(5):708–23.

34. Wang CX, Shuaib A. Involvement of inflammatory cytokines in central nervous system injury. Prog Neurobiol. 2002;67(2):161–72.

35. Sprague AH, Khalil RA. Inflammatory cytokines in vascular dysfunction and vascular disease. Biochem Pharmacol. 2009;78(6):539–52.

36. Peters K, Unger RE, Brunner J, Kirkpatrick CJ. Molecular basis of endothelial dysfunction in sepsis. Cardiovasc Res. 2003;60(1):49–57.

37. Ahmad M, Graham SH. Inflammation after stroke: mechanisms and therapeutic approaches. Transl Stroke Res. 2010;1(2):74–84.

38. Rosell A, Lo EH. Multiphasic roles for matrix metalloproteinases after stroke. Curr Opin Pharmacol. 2008;8(1):82–9.

39. Hallenbeck JM. The many faces of tumor necrosis factor in stroke. Nat Med. 2002;8(12): 1363–8.

40. Serena J, Blanco M, Castellanos M, Silva Y, Vivancos J, Moro MA, et al. The prediction of malignant cerebral infarction by molecular brain barrier disruption markers. Stroke. 2005;36(9):1921–6.

41. Liang D, Bhatta S, Gerzanich V, Simard JM. Cytotoxic edema: mechanisms of pathological cell swelling. Neurosurg Focus. 2007;22(5):E2.

42. Simard JM, Chen M, Tarasov KV, Bhatta S, Ivanova S, Melnitchenko L, et al. Newly expressed SUR1-regulated NC(Ca-ATP) channel mediates cerebral edema after ischemic stroke. Nat Med. 2006;12(4):433–40.

43. Paulson OB, Strandgaard S, Edvinsson L. Cerebral autoregulation. Cerebrovasc Brain Metab Rev. 1990;2(2):161–92.

44. Shapiro HM, Stromberg DD, Lee DR, Wiederhielm CA. Dynamic pressures in the pial arterial microcirculation. Am J Physiol. 1971;221(1):279–83.

45. Faraci FM, Heistad DD. Regulation of large cerebral arteries and cerebral microvascular pressure. Circ Res. 1990;66(1):8–17.

46. Mellander S. Functional aspects of myogenic vascular control. J Hypertens Suppl. 1989;7(4):S21–30; discussion S31.

47. Johansson B. Myogenic tone and reactivity: definitions based on muscle physiology. J Hypertens Suppl. 1989;7(4):S5–8; discussion S9.

48. Faraci FM, Baumbach GL, Heistad DD. Myogenic mechanisms in the cerebral circulation. J Hypertens Suppl. 1989;7(4):S61–4; discussion S65.

49. Osol G, Brekke JF, McElroy-Yaggy K, Gokina NI. Myogenic tone, reactivity, and forced dilatation: a three-phase model of in vitro arterial myogenic behavior. Am J Physiol Heart Circ Physiol. 2002;283(6):H2260–7.

50. Kontos HA, Wei EP, Navari RM, Levasseur JE, Rosenblum WI, Patterson Jr JL. Responses of cerebral arteries and arterioles to acute hypotension and hypertension. Am J Physiol. 1978;234(4):H371–83.

51. Phillips SJ, Whisnant JP. Hypertension and the brain. The national high blood pressure education program. Arch Intern Med. 1992;152(5):938–45.

52. Olsen TS, Larsen B, Skriver EB, Herning M, Enevoldsen E, Lassen NA. Focal cerebral hyperemia in acute stroke. Incidence, pathophysiology and clinical significance. Stroke. 1981;12(5):598–607.
53. Lassen NA, Agnoli A. The upper limit of autoregulation of cerebral blood flow—on the pathogenesis of hypertensive encepholopathy. Scand J Clin Lab Invest. 1972;30(2):113–6.
54. Euser AG, Cipolla MJ. Cerebral blood flow autoregulation and edema formation during pregnancy in anesthetized rats. Hypertension. 2007;49(2):334–40.
55. Johansson B, Li CL, Olsson Y, Klatzo I. The effect of acute arterial hypertension on the blood–brain barrier to protein tracers. Acta Neuropathol. 1970;16(2):117–24.
56. Gourley JK, Heistad DD. Characteristics of reactive hyperemia in the cerebral circulation. Am J Physiol. 1984;246(1 Pt 2):H52–8.
57. Sundt Jr TM, Waltz AG. Cerebral ischemia and reactive hyperemia. Studies of cortical blood flow and microcirculation before, during, and after temporary occlusion of middle cerebral artery of squirrel monkeys. Circ Res. 1971;28(4):426–33.
58. Hayakawa T, Waltz AG, Hansen T. Relationships among intracranial pressure, blood pressure, and superficial cerebral vasculature after experimental occlusion of one middle cerebral artery. Stroke. 1977;8(4):426–32.
59. Skinhoj E, Hoedt-Rasmussen K, Paulson OB, Lassen NA. Regional cerebral blood flow and its autoregulation in patients with transient focal cerebral ischemic attacks. Neurology. 1970;20(5):485–93.
60. Macfarlane R, Moskowitz MA, Sakas DE, Tasdemiroglu E, Wei EP, Kontos HA. The role of neuroeffector mechanisms in cerebral hyperperfusion syndromes. J Neurosurg. 1991;75(6):845–55.
61. Cipolla MJ, Lessov N, Hammer ES, Curry AB. Threshold duration of ischemia for myogenic tone in middle cerebral arteries: effect on vascular smooth muscle actin. Stroke. 2001;32(7):1658–64.
62. Cipolla MJ, Curry AB. Middle cerebral artery function after stroke: the threshold duration of reperfusion for myogenic activity. Stroke. 2002;33(8):2094–9.
63. Kleindorfer D, Xu Y, Moomaw CJ, Khatri P, Adeoye O, Hornung R. US geographic distribution of rt-PA utilization by hospital for acute ischemic stroke. Stroke. 2009;40(11):3580–4.
64. Koudstaal PJ, Stibbe J, Vermeulen M. Fatal ischaemic brain oedema after early thrombolysis with tissue plasminogen activator in acute stroke. BMJ. 1988;297(6663):1571–4.
65. Cipolla MJ, Lessov N, Clark WM, Haley Jr EC. Postischemic attenuation of cerebral artery reactivity is increased in the presence of tissue plasminogen activator. Stroke. 2000;31(4):940–5.
66. Lassen NA. The luxury-perfusion syndrome and its possible relation to acute metabolic acidosis localised within the brain. Lancet. 1966;2(7473):1113–5.
67. Yamaguchi T, Waltz AG, Okazaki H. Hyperemia and ischemia in experimental cerebral infarction: correlation of histopathology and regional blood flow. Neurology. 1971;21(6):565–78.
68. Olsen TS. Regional cerebral blood flow after occlusion of the middle cerebral artery. Acta Neurol Scand. 1986;73(4):321–37.
69. Cipolla MJ, Gokina NI, Osol G. Pressure-induced actin polymerization in vascular smooth muscle as a mechanism underlying myogenic behavior. FASEB J. 2002;16(1):72–6.
70. Cipolla MJ, Osol G. Vascular smooth muscle actin cytoskeleton in cerebral artery forced dilatation. Stroke. 1998;29(6):1223–8.
71. Banan A, Fields JZ, Zhang Y, Keshavarzian A. iNOS upregulation mediates oxidant-induced disruption of F-actin and barrier of intestinal monolayers. Am J Physiol Gastrointest Liver Physiol. 2001;280(6):G1234–46.
72. Schwartz N, Hosford M, Sandoval RM, Wagner MC, Atkinson SJ, Bamburg J, et al. Ischemia activates actin depolymerizing factor: role in proximal tubule microvillar actin alterations. Am J Physiol. 1999;276(4 Pt 2):F544–51.
73. Hsu SS, Meno JR, Gronka R, Kushmerick M, Winn HR. Moderate hyperglycemia affects ischemic brain ATP levels but not intracellular pH. Am J Physiol. 1994;266(1 Pt 2):H258–62.
74. Gisselsson LL, Matus A, Wieloch T. Actin redistribution underlies the sparing effect of mild hypothermia on dendritic spine morphology after in vitro ischemia. J Cereb Blood Flow Metab. 2005;25(10):1346–55.

75. Pacher P, Beckman JS, Liaudet L. Nitric oxide and peroxynitrite in health and disease. Physiol Rev. 2007;87(1):315–424.
76. Bemeur C, Ste-Marie L, Montgomery J. Increased oxidative stress during hyperglycemic cerebral ischemia. Neurochem Int. 2007;50(7–8):890–904.
77. Maneen MJ, Hannah R, Vitullo L, DeLance N, Cipolla MJ. Peroxynitrite diminishes myogenic activity and is associated with decreased vascular smooth muscle F-actin in rat posterior cerebral arteries. Stroke. 2006;37(3):894–9.
78. Maneen MJ, Cipolla MJ. Peroxynitrite diminishes myogenic tone in cerebral arteries: role of nitrotyrosine and F-actin. Am J Physiol Heart Circ Physiol. 2007;292(2):H1042–50.
79. Iadecola C. Bright and dark sides of nitric oxide in ischemic brain injury. Trends Neurosci. 1997;20(3):132–9.
80. Wei EP, Kontos HA, Beckman JS. Mechanisms of cerebral vasodilation by superoxide, hydrogen peroxide, and peroxynitrite. Am J Physiol. 1996;271(3 Pt 2):H1262–6.
81. Li J, Li W, Altura BT, Altura BM. Peroxynitrite-induced relaxation in isolated rat aortic rings and mechanisms of action. Toxicol Appl Pharmacol. 2005;209(3):269–76.
82. Cohen RA, Adachi T. Nitric-oxide-induced vasodilatation: regulation by physiologic s-gluta-thiolation and pathologic oxidation of the sarcoplasmic endoplasmic reticulum calcium ATPase. Trends Cardiovasc Med. 2006;16(4):109–14.
83. Offner H, Subramanian S, Parker SM, Afentoulis ME, Vandenbark AA, Hurn PD. Experimental stroke induces massive, rapid activation of the peripheral immune system. J Cereb Blood Flow Metab. 2006;26(5):654–65.
84. Shreeniwas R, Koga S, Karakurum M, Pinsky D, Kaiser E, Brett J, et al. Hypoxia-mediated induction of endothelial cell interleukin-1 alpha. An autocrine mechanism promoting expression of leukocyte adhesion molecules on the vessel surface. J Clin Invest. 1992;90(6):2333–9.
85. Andresen J, Shafi NI, Bryan Jr RM. Endothelial influences on cerebrovascular tone. J Appl Physiol. 2006;100(1):318–27.
86. Faraci FM. Protecting against vascular disease in brain. Am J Physiol Heart Circ Physiol. 2011;300(5):H1566–82.
87. Faraci FM, Brian Jr JE. Nitric oxide and the cerebral circulation. Stroke. 1994;25(3):692–703.
88. Marrelli SP, Eckmann MS, Hunte MS. Role of endothelial intermediate conductance KCa channels in cerebral EDHF-mediated dilations. Am J Physiol Heart Circ Physiol. 2003;285(4):H1590–9.
89. McNeish AJ, Sandow SL, Neylon CB, Chen MX, Dora KA, Garland CJ. Evidence for involvement of both IKCa and SKCa channels in hyperpolarizing responses of the rat middle cerebral artery. Stroke. 2006;37(5):1277–82.
90. Dora KA, Gallagher NT, McNeish A, Garland CJ. Modulation of endothelial cell KCa3.1 channels during endothelium-derived hyperpolarizing factor signaling in mesenteric resistance arteries. Circ Res. 2008;102(10):1247–55.
91. Zygmunt PM, Hogestatt ED. Role of potassium channels in endothelium-dependent relaxation resistant to nitroarginine in the rat hepatic artery. Br J Pharmacol. 1996;117(7):1600–6.
92. Si H, Heyken WT, Wolfle SE, Tysiac M, Schubert R, Grgic I, et al. Impaired endothelium-derived hyperpolarizing factor-mediated dilations and increased blood pressure in mice deficient of the intermediate-conductance Ca2+-activated K+ channel. Circ Res. 2006;99(5):537–44.
93. Ledoux J, Werner ME, Brayden JE, Nelson MT. Calcium-activated potassium channels and the regulation of vascular tone. Physiology (Bethesda). 2006;21:69–78.
94. Dalkara T, Moskowitz MA. The complex role of nitric oxide in the pathophysiology of focal cerebral ischemia. Brain Pathol. 1994;4(1):49–57.
95. Lo EH, Hara H, Rogowska J, Trocha M, Pierce AR, Huang PL, et al. Temporal correlation mapping analysis of the hemodynamic penumbra in mutant mice deficient in endothelial nitric oxide synthase gene expression. Stroke. 1996;27(8):1381–5.
96. Huang Z, Huang PL, Ma J, Meng W, Ayata C, Fishman MC, et al. Enlarged infarcts in endothelial nitric oxide synthase knockout mice are attenuated by nitro-L-arginine. J Cereb Blood Flow Metab. 1996;16(5):981–7.

97. Panahian N, Yoshida T, Huang PL, Hedley-Whyte ET, Dalkara T, Fishman MC, et al. Attenuated hippocampal damage after global cerebral ischemia in mice mutant in neuronal nitric oxide synthase. Neuroscience. 1996;72(2):343–54.

98. Zhang ZG, Chopp M, Bailey F, Malinski T. Nitric oxide changes in the rat brain after transient middle cerebral artery occlusion. J Neurol Sci. 1995;128(1):22–7.

99. Iadecola C. Regulation of the cerebral microcirculation during neural activity: is nitric oxide the missing link? Trends Neurosci. 1993;16(6):206–14.

100. Kanwar S, Kubes P. Nitric oxide is an antiadhesive molecule for leukocytes. New Horiz. 1995;3(1):93–104.

101. Szabo C. The pathophysiological role of peroxynitrite in shock, inflammation, and ischemia-reperfusion injury. Shock. 1996;6(2):79–88.

102. Heinzel B, John M, Klatt P, Bohme E, Mayer B. Ca2+/calmodulin-dependent formation of hydrogen peroxide by brain nitric oxide synthase. Biochem J. 1992;281(Pt 3):627–30.

103. Pou S, Pou WS, Bredt DS, Snyder SH, Rosen GM. Generation of superoxide by purified brain nitric oxide synthase. J Biol Chem. 1992;267(34):24173–6.

104. Veltkamp R, Rajapakse N, Robins G, Puskar M, Shimizu K, Busija D. Transient focal ischemia increases endothelial nitric oxide synthase in cerebral blood vessels. Stroke. 2002;33(11): 2704–10.

105. Marrelli SP, Khorovets A, Johnson TD, Childres WF, Bryan Jr RM. P2 purinoceptor-mediated dilations in the rat middle cerebral artery after ischemia-reperfusion. Am J Physiol. 1999;276(1 Pt 2):H33–41.

106. Mayhan WG, Amundsen SM, Faraci FM, Heistad DD. Responses of cerebral arteries after ischemia and reperfusion in cats. Am J Physiol. 1988;255(4 Pt 2):H879–84.

107. Rosenblum WI. Selective impairment of response to acetylcholine after ischemia/reperfusion in mice. Stroke. 1997;28(2):448–51; discussion 451–2.

108. Nelson CW, Wei EP, Povlishock JT, Kontos HA, Moskowitz MA. Oxygen radicals in cerebral ischemia. Am J Physiol. 1992;263(5 Pt 2):H1356–62.

109. Tayeh MA, Marletta MA. Macrophage oxidation of L-arginine to nitric oxide, nitrite, and nitrate. Tetrahydrobiopterin is required as a cofactor. J Biol Chem. 1989;264(33):19654–8.

110. Xia Y, Tsai AL, Berka V, Zweier JL. Superoxide generation from endothelial nitric-oxide synthase. A Ca2+/calmodulin-dependent and tetrahydrobiopterin regulatory process. J Biol Chem. 1998;273(40):25804–8.

111. Katusic ZS. Vascular endothelial dysfunction: does tetrahydrobiopterin play a role? Am J Physiol Heart Circ Physiol. 2001;281(3):H981–6.

112. Fukai T. Endothelial GTPCH in eNOS uncoupling and atherosclerosis. Arterioscler Thromb Vasc Biol. 2007;27(7):1493–5.

113. Landmesser U, Dikalov S, Price SR, McCann L, Fukai T, Holland SM, et al. Oxidation of tetrahydrobiopterin leads to uncoupling of endothelial cell nitric oxide synthase in hypertension. J Clin Invest. 2003;111(8):1201–9.

114. Pannirselvam M, Simon V, Verma S, Anderson T, Triggle CR. Chronic oral supplementation with sepiapterin prevents endothelial dysfunction and oxidative stress in small mesenteric arteries from diabetic (db/db) mice. Br J Pharmacol. 2003;140(4):701–6.

115. Beckman JS, Chen J, Crow JP, Ye YZ. Reactions of nitric oxide, superoxide and peroxynitrite with superoxide dismutase in neurodegeneration. Prog Brain Res. 1994;103:371–80.

116. Milstien S, Katusic Z. Oxidation of tetrahydrobiopterin by peroxynitrite: implications for vascular endothelial function. Biochem Biophys Res Commun. 1999;263(3):681–4.

117. Bauer PM, Fulton D, Boo YC, Sorescu GP, Kemp BE, Jo H, et al. Compensatory phosphorylation and protein-protein interactions revealed by loss of function and gain of function mutants of multiple serine phosphorylation sites in endothelial nitric-oxide synthase. J Biol Chem. 2003;278(17):14841–9.

118. Dudzinski DM, Michel T. Life history of eNOS: partners and pathways. Cardiovasc Res. 2007;75(2):247–60.

119. Hashiguchi A, Yano S, Morioka M, Hamada J, Kochi M, Fukunaga K. Dephosphorylation of eNOS on Thr495 after transient forebrain ischemia in gerbil hippocampus. Brain Res Mol Brain Res. 2005;133(2):317–9.

120. Chen CA, Druhan LJ, Varadharaj S, Chen YR, Zweier JL. Phosphorylation of endothelial nitric-oxide synthase regulates superoxide generation from the enzyme. J Biol Chem. 2008;283(40):27038–47.
121. Luksha L, Nisell H, Luksha N, Kublickas M, Hultenby K, Kublickiene K. Endothelium-derived hyperpolarizing factor in preeclampsia: heterogeneous contribution, mechanisms, and morphological prerequisites. Am J Physiol Regul Integr Comp Physiol. 2008;294(2):R510–9.
122. McNeish AJ, Dora KA, Garland CJ. Possible role for K+ in endothelium-derived hyperpolarizing factor-linked dilatation in rat middle cerebral artery. Stroke. 2005;36(7):1526–32.
123. Kitazono T, Faraci FM, Taguchi H, Heistad DD. Role of potassium channels in cerebral blood vessels. Stroke. 1995;26(9):1713–23.
124. Park SL, Lee DH, Yoo SE, Jung YS. The effect of Na(+)/H(+) exchanger-1 inhibition by sabiporide on blood–brain barrier dysfunction after ischemia/hypoxia in vivo and in vitro. Brain Res. 2010;1366:189–96.
125. Brown RC, Davis TP. Hypoxia/aglycemia alters expression of occludin and actin in brain endothelial cells. Biochem Biophys Res Commun. 2005;327(4):1114–23.
126. Yun HY, Dawson VL, Dawson TM. Neurobiology of nitric oxide. Crit Rev Neurobiol. 1996;10(3–4):291–316.
127. Garthwaite J, Boulton CL. Nitric oxide signaling in the central nervous system. Annu Rev Physiol. 1995;57:683–706.
128. Iadecola C, Zhang F, Casey R, Clark HB, Ross ME. Inducible nitric oxide synthase gene expression in vascular cells after transient focal cerebral ischemia. Stroke. 1996;27(8):1373–80.
129. Rubanyi GM, Polokoff MA. Endothelins: molecular biology, biochemistry, pharmacology, physiology, and pathophysiology. Pharmacol Rev. 1994;46(3):325–415.
130. Ehrenreich H, Schilling L. New developments in the understanding of cerebral vasoregulation and vasospasm: the endothelin-nitric oxide network. Cleve Clin J Med. 1995;62(2):105–16.
131. Ziv I, Fleminger G, Djaldetti R, Achiron A, Melamed E, Sokolovsky M. Increased plasma endothelin-1 in acute ischemic stroke. Stroke. 1992;23(7):1014–6.
132. Barone FC, Globus MY, Price WJ, White RF, Storer BL, Feuerstein GZ, et al. Endothelin levels increase in rat focal and global ischemia. J Cereb Blood Flow Metab. 1994;14(2):337–42.
133. Bian LG, Zhang TX, Zhao WG, Shen JK, Yang GY. Increased endothelin-1 in the rabbit model of middle cerebral artery occlusion. Neurosci Lett. 1994;174(1):47–50.
134. D'Orleans-Juste P, Claing A, Warner TD, Yano M, Telemaque S. Characterization of receptors for endothelins in the perfused arterial and venous mesenteric vasculatures of the rat. Br J Pharmacol. 1993;110(2):687–92.
135. Barone FC, Willette RN, Yue TL, Feurestein G. Therapeutic effects of endothelin receptor antagonists in stroke. Neurol Res. 1995;17(4):259–64.
136. Dawson DA, Sugano H, McCarron RM, Hallenbeck JM, Spatz M. Endothelin receptor antagonist preserves microvascular perfusion and reduces ischemic brain damage following permanent focal ischemia. Neurochem Res. 1999;24(12):1499–505.
137. Matsuo Y, Mihara S, Ninomiya M, Fujimoto M. Protective effect of endothelin type A receptor antagonist on brain edema and injury after transient middle cerebral artery occlusion in rats. Stroke. 2001;32(9):2143–8.
138. Leung JW, Chung SS, Chung SK. Endothelial endothelin-1 over-expression using receptor tyrosine kinase tie-1 promoter leads to more severe vascular permeability and blood brain barrier breakdown after transient middle cerebral artery occlusion. Brain Res. 2009;1266:121–9.
139. Kawai N, McCarron RM, Spatz M. Endothelins stimulate sodium uptake into rat brain capillary endothelial cells through endothelin A-like receptors. Neurosci Lett. 1995;190(2):85–8.
140. Stanimirovic DB, Bertrand N, McCarron R, Uematsu S, Spatz M. Arachidonic acid release and permeability changes induced by endothelins in human cerebromicrovascular endothelium. Acta Neurochir Suppl (Wien). 1994;60:71–5.
141. Williams LS, Rotich J, Qi R, Fineberg N, Espay A, Bruno A, et al. Effects of admission hyperglycemia on mortality and costs in acute ischemic stroke. Neurology. 2002;59(1):67–71.
142. Matz K, Keresztes K, Tatschl C, Nowotny M, Dachenhausenm A, Brainin M, et al. Disorders of glucose metabolism in acute stroke patients: an underrecognized problem. Diabetes Care. 2006;29(4):792–7.

143. Kernan WN, Viscoli CM, Inzucchi SE, Brass LM, Bravata DM, Shulman GI, et al. Prevalence of abnormal glucose tolerance following a transient ischemic attack or ischemic stroke. Arch Intern Med. 2005;165(2):227–33.
144. Gray CS, Scott JF, French JM, Alberti KG, O'Connell JE. Prevalence and prediction of unrecognised diabetes mellitus and impaired glucose tolerance following acute stroke. Age Ageing. 2004;33(1):71–7.
145. Horner HC, Packan DR, Sapolsky RM. Glucocorticoids inhibit glucose transport in cultured hippocampal neurons and glia. Neuroendocrinology. 1990;52(1):57–64.
146. Wang YY, Lin SY, Chuang YH, Chen CJ, Tung KC, Sheu WH. Adipose proinflammatory cytokine expression through sympathetic system is associated with hyperglycemia and insulin resistance in a rat ischemic stroke model. Am J Physiol Endocrinol Metab. 2011;300(1):E155–63.
147. Pulsinelli WA, Waldman S, Rawlinson D, Plum F. Moderate hyperglycemia augments ischemic brain damage: a neuropathologic study in the rat. Neurology. 1982;32(11):1239–46.
148. Capes SE, Hunt D, Malmberg K, Pathak P, Gerstein HC. Stress hyperglycemia and prognosis of stroke in nondiabetic and diabetic patients: a systematic overview. Stroke. 2001;32(10):2426–32.
149. Quast MJ, Wei J, Huang NC, Brunder DG, Sell SL, Gonzalez JM, et al. Perfusion deficit parallels exacerbation of cerebral ischemia/reperfusion injury in hyperglycemic rats. J Cereb Blood Flow Metab. 1997;17(5):553–9.
150. Martini SR, Kent TA. Hyperglycemia in acute ischemic stroke: a vascular perspective. J Cereb Blood Flow Metab. 2007;27(3):435–51.
151. Woo E, Ma JT, Robinson JD, Yu YL. Hyperglycemia is a stress response in acute stroke. Stroke. 1988;19(11):1359–64.
152. Murros K, Fogelholm R, Kettunen S, Vuorela AL, Valve J. Blood glucose, glycosylated haemoglobin, and outcome of ischemic brain infarction. J Neurol Sci. 1992;111(1):59–64.
153. Hamilton MG, Tranmer BI, Auer RN. Insulin reduction of cerebral infarction due to transient focal ischemia. J Neurosurg. 1995;82(2):262–8.
154. Lawrence MS, Sun GH, Kunis DM, Saydam TC, Dash R, Ho DY, et al. Overexpression of the glucose transporter gene with a herpes simplex viral vector protects striatal neurons against stroke. J Cereb Blood Flow Metab. 1996;16(2):181–5.
155. Bruno A, Liebeskind D, Hao Q, Raychev R. Diabetes mellitus, acute hyperglycemia, and ischemic Stroke. Curr Treat Options Neurol. 2010;12(6):492–503.
156. McCormick MT, Muir KW, Gray CS, Walters MR. Management of hyperglycemia in acute stroke: how, when, and for whom? Stroke. 2008;39(7):2177–85.
157. McCormick M, Hadley D, McLean JR, Macfarlane JA, Condon B, Muir KW. Randomized, controlled trial of insulin for acute poststroke hyperglycemia. Ann Neurol. 2010;67(5):570–8.
158. Brownlee M. Biochemistry and molecular cell biology of diabetic complications. Nature. 2001;414(6865):813–20.
159. Mandarino LJ, Finlayson J, Hassell JR. High glucose downregulates glucose transport activity in retinal capillary pericytes but not endothelial cells. Invest Ophthalmol Vis Sci. 1994;35(3):964–72.
160. Kawai N, Keep RF, Betz AL. Hyperglycemia and the vascular effects of cerebral ischemia. Stroke. 1997;28(1):149–54.
161. Venables GS, Miller SA, Gibson G, Hardy JA, Strong AJ. The effects of hyperglycaemia on changes during reperfusion following focal cerebral ischaemia in the cat. J Neurol Neurosurg Psychiatry. 1985;48(7):663–9.
162. Cipolla MJ, Godfrey JA. Effect of hyperglycemia on brain penetrating arterioles and cerebral blood flow before and after ischemia/reperfusion. Transl Stroke Res. 2010;1(2):127–34.
163. Gisselsson L, Smith ML, Siesjo BK. Hyperglycemia and focal brain ischemia. J Cereb Blood Flow Metab. 1999;19(3):288–97.
164. Berger L, Hakim AM. The association of hyperglycemia with cerebral edema in stroke. Stroke. 1986;17(5):865–71.
165. Dietrich WD, Alonso O, Busto R. Moderate hyperglycemia worsens acute blood–brain barrier injury after forebrain ischemia in rats. Stroke. 1993;24(1):111–6.

166. Cipolla MJ, Huang Q, Sweet JG. Inhibition of PKCβ prevents increased blood–brain barrier permeability and edema formation during hyperglycemic stroke. Stroke. 2011;42(11):3252–7.
167. Ennis SR, Keep RF. Effect of sustained-mild and transient-severe hyperglycemia on ischemia-induced blood–brain barrier opening. J Cereb Blood Flow Metab. 2007;27(9):1573–82.
168. Kamada H, Yu F, Nito C, Chan PH. Influence of hyperglycemia on oxidative stress and matrix metalloproteinase-9 activation after focal cerebral ischemia/reperfusion in rats: relation to blood–brain barrier dysfunction. Stroke. 2007;38(3):1044–9.
169. Clarke H, Marano CW, Peralta Soler A, Mullin JM. Modification of tight junction function by protein kinase C isoforms. Adv Drug Deliv Rev. 2000;41(3):283–301.
170. Yuan Y, Huang Q, Wu HM. Myosin light chain phosphorylation: modulation of basal and agonist-stimulated venular permeability. Am J Physiol. 1997;272(3 Pt 2):H1437–43.
171. Feekes JA, Cassell MD. The vascular supply of the functional compartments of the human striatum. Brain. 2006;129(Pt 8):2189–201.
172. Ginsberg MD, Prado R, Dietrich WD, Busto R, Watson BD. Hyperglycemia reduces the extent of cerebral infarction in rats. Stroke. 1987;18(3):570–4.
173. Prado R, Ginsberg MD, Dietrich WD, Watson BD, Busto R. Hyperglycemia increases infarct size in collaterally perfused but not end-arterial vascular territories. J Cereb Blood Flow Metab. 1988;8(2):186–92.
174. Bruno A, Biller J, Adams Jr HP, Clarke WR, Woolson RF, Williams LS, et al. Acute blood glucose level and outcome from ischemic stroke. Trial of ORG 10172 in Acute Stroke Treatment (TOAST) Investigators. Neurology. 1999;52(2):280–4.
175. Uyttenboogaart M, Koch MW, Stewart RE, Vroomen PC, Luijckx GJ, De Keyser J. Moderate hyperglycaemia is associated with favourable outcome in acute lacunar stroke. Brain. 2007;130(Pt 6):1626–30.
176. Sulter G, Elting JW, De Keyser J. Increased serum neuron specific enolase concentrations in patients with hyperglycemic cortical ischemic stroke. Neurosci Lett. 1998;253(1):71–3.
177. Muir KW, Tyrrell P, Sattar N, Warburton E. Inflammation and ischaemic stroke. Curr Opin Neurol. 2007;20(3):334–42.
178. Emsley HC, Smith CJ, Tyrrell PJ, Hopkins SJ. Inflammation in acute ischemic stroke and its relevance to stroke critical care. Neurocrit Care. 2008;9(1):125–38.
179. Banwell V, Sena ES, Macleod MR. Systematic review and stratified meta-analysis of the efficacy of interleukin-1 receptor antagonist in animal models of stroke. J Stroke Cerebrovasc Dis. 2009;18(4):269–76.
180. Spera PA, Ellison JA, Feuerstein GZ, Barone FC. IL-10 reduces rat brain injury following focal stroke. Neurosci Lett. 1998;251(3):189–92.
181. Elkind MS. Impact of innate inflammation in population studies. Ann N Y Acad Sci. 2010;1207:97–106.
182. Fassbender K, Rossol S, Kammer T, Daffertshofer M, Wirth S, Dollman M, et al. Proinflammatory cytokines in serum of patients with acute cerebral ischemia: kinetics of secretion and relation to the extent of brain damage and outcome of disease. J Neurol Sci. 1994;122(2):135–9.
183. Zaremba J, Skrobanski P, Losy J. Tumour necrosis factor-alpha is increased in the cerebrospinal fluid and serum of ischaemic stroke patients and correlates with the volume of evolving brain infarct. Biomed Pharmacother. 2001;55(5):258–63.
184. Rodriguez-Gonzalez R, Sobrino T, Rodriguez-Yanez M, Millan M, Brea D, Miranda E, et al. Association between neuroserpin and molecular markers of brain damage in patients with acute ischemic stroke. J Transl Med. 2011;9(1):58.
185. Iversen PO, Nicolaysen A, Kvernebo K, Benestad HB, Nicolaysen G. Human cytokines modulate arterial vascular tone via endothelial receptors. Pflugers Arch. 1999;439(1–2):93–100.
186. van der Poll T, Lowry SF. Tumor necrosis factor in sepsis: mediator of multiple organ failure or essential part of host defense? Shock. 1995;3(1):1–12.
187. Vila E, Salaices M. Cytokines and vascular reactivity in resistance arteries. Am J Physiol Heart Circ Physiol. 2005;288(3):H1016–21.

188. Tureen J. Effect of recombinant human tumor necrosis factor-alpha on cerebral oxygen uptake, cerebrospinal fluid lactate, and cerebral blood flow in the rabbit: role of nitric oxide. J Clin Invest. 1995;95(3):1086–91.

189. Cipolla MJ, Sweet JG, Gardner-Morse I. Effect of circulating factors on cerebral artery function during hyperglycemic stroke. FASEB J. 2011;25:1024–6.

190. Pretnar-Oblak J, Sabovic M, Pogacnik T, Sebestjen M, Zaletel M. Flow-mediated dilatation and intima-media thickness in patients with lacunar infarctions. Acta Neurol Scand. 2006;113(4):273–7.

191. Pretnar-Oblak J, Sabovic M, Sebestjen M, Pogacnik T, Zaletel M. Influence of atorvastatin treatment on L-arginine cerebrovascular reactivity and flow-mediated dilatation in patients with lacunar infarctions. Stroke. 2006;37(10):2540–5.

192. Pretnar-Oblak J, Zaletel M, Zvan B, Sabovic M, Pogacnik T. Cerebrovascular reactivity to L-arginine in patients with lacunar infarctions. Cerebrovasc Dis. 2006;21(3):180–6.

193. Chae CU, Lee RT, Rifai N, Ridker PM. Blood pressure and inflammation in apparently healthy men. Hypertension. 2001;38(3):399–403.

194. Virdis A, Schiffrin EL. Vascular inflammation: a role in vascular disease in hypertension? Curr Opin Nephrol Hypertens. 2003;12(2):181–7.

195. Bath P. High blood pressure as risk factor and prognostic predictor in acute ischaemic stroke: when and how to treat it? Cerebrovasc Dis. 2004;17 Suppl 1:51–7.

196. Parissis JT, Korovesis S, Giazitzoglou E, Kalivas P, Katritsis D. Plasma profiles of peripheral monocyte-related inflammatory markers in patients with arterial hypertension. Correlations with plasma endothelin-1. Int J Cardiol. 2002;83(1):13–21.

197. Di Napoli M, Papa F. Association between blood pressure and C-reactive protein levels in acute ischemic stroke. Hypertension. 2003;42(6):1117–23.

198. Lin B, Ginsberg MD, Busto R, Li L. Hyperglycemia triggers massive neutrophil deposition in brain following transient ischemia in rats. Neurosci Lett. 2000;278(1–2):1–4.

199. Garg R, Chaudhuri A, Munschauer F, Dandona P. Hyperglycemia, insulin, and acute ischemic stroke: a mechanistic justification for a trial of insulin infusion therapy. Stroke. 2006;37(1):267–73.

Chapter 2
Identifying Vascular Targets to Treat Hemorrhagic Stroke

Paul A. Lapchak

Abstract Hemorrhagic stroke is a devastating type of stroke that affects 20% of all stroke patients. In spite of the severity of morbidity and the high rate of mortality associated with hemorrhagic stroke, current treatment methods are quite insufficient to reduce long-term morbidity and high mortality rate, up to 50%, associated with bleeding into critical brain structures, the ventricles, and the subarachnoid space. Preclinical and translational research programs have led to significant advances in the understanding of important mechanisms that contribute to cell death and clinical deficits. The most important findings revolve around a key set of basic mechanisms intimately involved in brain bleeding, including activation of matrix membrane metalloproteinases (MMP), specifically MMP-2 and MMP-9, enhanced free radical production and oxidative stress, and both altered inflammatory and coagulation pathways. It is now becoming apparent and accepted that brain bleeding may result in the activation of a "hemorrhagic stroke cascade" in many cell types including neurons, glia, and vasculature. Moreover, in hemorrhage there is also activation of many components of the "ischemic stroke cascade." The activation of multiple hemorrhage and ischemia pathways is a two-edged sword: it results in the activation of series of detrimental pathways, some of which are duplicated in each cascade, but they may allow for useful pharmacological intervention.

Because there is parallel or simultaneous activation of pathways, future therapies may have to rely on combining drugs or pleiotropic compounds with multi-target activities. This chapter will comprehensively review some of the possible targets for pharmacological intervention as well as address some new approaches to induce metabolic downregulation or inhibition of multiple pathways, with the goal of clinical improvement.

P.A. Lapchak, PhD, FAHA (✉)
Department of Neurology, Cedars-Sinai Medical Center, Davis Research Building,
D-2091, 110 N. George Burns Road, Los Angeles, CA 90048, USA
e-mail: Paul.Lapchak@cshs.org

P.A. Lapchak and J.H. Zhang (eds.), *Translational Stroke Research*, Springer Series
in Translational Stroke Research, DOI 10.1007/978-1-4419-9530-8_2,
© Springer Science+Business Media, LLC 2012

1 Hemorrhage: Incidence and Pathophysiology

Hemorrhagic stroke represents approximately 20% of all stroke patients [1–3]. Hemorrhagic strokes are usually classified as intracranial hemorrhage including epidural hematomas (epidural, subdural) and subarachnoid hemorrhages (SAH). Patients also present with intracerebral hemorrhage (ICH) including intraparenchymal hemorrhage (IPH) and intraventricular hemorrhage (IVH). It is important to note that hemorrhagic stroke subtypes can be distinguished from hemorrhagic transformations associated with ischemic stroke. ICH has a higher morbidity and mortality rate than ischemic stroke. For example, the 30-day mortality rate for hemorrhagic stroke is estimated to be around 50–60%, whereas mortality in ischemic stroke patients is 8–12%, but there is a high morbidity, often requiring institutionalization for care [4–7]. Patients with ICH may require extensive evaluation to detect and repair the source of bleeding, with possible surgical intervention to remove the clot to prevent hemorrhage expansion and secondary repercussions [7, 8]. There are many causes for ICH, which may occur in small arteries (arterioles). ICH is common in patients with hypertension, intracranial vascular malformations, or cerebral amyloid angiopathy [9]. Vascular damage due to cerebral amyloidosis primarily affects the elderly and represents up to 10% of hemorrhagic strokes. In addition, commonly used therapeutics such as anticoagulants, platelet inhibitors [10, 11], and thrombolytics (tissue plasminogen activator; tPA, Alteplase™) are typically used to treat acute myocardial infarction (AMI) or acute ischemic stroke (AIS) [12].

Numerous important preliminary translational steps toward the development of effective therapeutics to treat hemorrhage have been taken. As described below, research from many groups have consistently identified a few key processes may targeted with small molecules of antibodies to reduce the deleterious effects of blood accumulation in brain matter. Drugs aimed at reducing the hemorrhage and ischemic cascade surrounding a hemorrhage may attenuate edema, apoptosis, and necrotic cell death. Since there are key processes involved in hemorrhage and the clinical deficits that occur subsequent to hemorrhage, this section will provide an overview of a few targets that should be further considered.

2 Inflammatory Pathways and Vascular Damage

2.1 Coagulation Factors and Pathway

Brain hemorrhage results in the production of enzymes involved in blood clotting and clot lysis [13–16]. Many of the mediators that are produced are potentially toxic to cells. Primary hemostasis is initiated when vascular injury triggers adherence of platelets to proteins within the vascular endothelium using glycoprotein (GP) Ia/IIa receptor docking. The sequence of platelet aggregation is mediated by von Willebrand factor (vWF), which links platelet GP Ib/IX/V and collagen [17–19].

Binding and activation causes structural and conformational changes in platelets. Secondary hemostasis is comprised of two specific pathways: an "intrinsic" or "contact activation" pathway and the "extrinsic" or "tissue factor" pathway. This results in the local activation of plasma coagulation factors and the generation of a fibrin clot that reinforces the platelet aggregate. The classical blood coagulation pathway involves a "cascade" of zymogen-activated reactions involving six coagulation factors in the intrinsic pathway (factors VIII, IX, XI, XII, prekallikrein, and high-molecular weight kininogen), one coagulation factor in the extrinsic pathway (factor VII), a part of a series of hemostatic defense mechanisms [18, 19], and four factors in a common pathway (factors II, V, X, and fibrinogen). The extrinsic pathway generates thrombin to convert soluble fibrinogen into insoluble fibrin strands, the clot backbone. Following damage to the blood vessel, endothelium tissue factor (ETF) is released, forming a complex with Factor VIIa, which is present in excess of any other activated coagulation factor could be a potential target for ICH therapy. Factor VIIa then activates Factor IX and X. Thrombin can activate Factor VII resulting in a cyclic complex including the activation of prothrombin to thrombin. However, clinical results in hemorrhage trials with Novoseven targeting Factor VII have failed [20], because there was an increased incidence of thromboembolism.

2.2 Thrombin and Coagulation

Thrombin is an important mediator produced in the brain immediately after ICH. Thrombin is a pleiotropic molecule that causes vascular damage, an inflammatory response, oxidative stress, and also has direct cellular toxicity, which are mediated in part by thrombin protease-activated receptors (PARs) [21–23]. Thrombin when injected at high concentrations can activate numerous pathways that may exacerbate brain damage [13, 24, 25]. For instance, high doses of thrombin injected into brain cause inflammatory cell infiltration and edema associated with a breach of the blood–brain barrier (BBB). Thrombin mediates endothelial cell permeability, an effect that can be blocked by administration of the small molecule direct thrombin antagonist argatroban [26]. Moreover, thrombin stimulates PARs expressed on microglia/macrophages to activate these cells via recruitment of mitogen-activated protein kinases (MAPKs), and produces several inflammatory mediators, which contribute to edema formation through disruption of the BBB [24]. It has been suggested that thrombin-induced edema is mediated by stimulating PARs to activate src family kinases, which are a family of proto-oncogene tyrosine kinases [27]. Src family kinase members mediate BBB permeability changes and brain edema by phosphorylating matrix membrane metalloproteinases (MMPs), tight junction proteins, and other BBB-related proteins [28]. Thus, these kinases may be appropriate targets for further development to attenuate the downstream detrimental effects of thrombin.

2.3 Inflammation and Cytokines

Inflammatory cells present in the CNS after ICH are primarily blood-derived leukocytes, macrophages, and resident microglia "brain macrophage" [29, 30], which may be the first non-neuronal cells to react following CNS injury. Microglia are activated to undergo morphologic including upregulation of pro-inflammatory cytokines, migration, proliferation, and phagocytic behavior [31]. Activated microglia release a variety of cytokines [32–34], reactive oxygen species (ROS) [35–37], and other potentially toxic factors, suggesting that activated microglia/macrophages might contribute to hemorrhage-induced early brain injury [38, 39]. Infiltrating leukocytes are also believed to play a role in ICH-induced brain injury. Neutrophils are the earliest leukocyte cell subtype to infiltrate into the hemorrhagic brain, and these may damage brain tissue directly also by producing ROS, releasing pro-inflammatory proteases [40], and modulating BBB permeability [41]. Leukocytes, macrophage, and activated microglia are CNS sources of cytokines, chemokines, prostaglandins, proteases, and other immunoactive molecules resulting from a brain bleed [29, 42–45].

Cytokine changes after ICH have been studies in detail. Two primary molecules, TNF-α and IL-1β, have been shown to be increased in various experimental models of brain injury and hemorrhage. TNF-α is a pleiotropic cytokine that received a lot of attention because it has multiple biologic activities that are temporally organized including the stimulation of acute phase protein secretion, vascular permeability, and post-ICH brain edema formation [46, 47]. TNF-α expression is increased in response to ICH when presented with either autologous blood or thrombin [47–49]. TNF also mediates hemorrhage in brain following embolic strokes [50]. Because of the importance of TNF and its family of signaling receptors, this is a prime target for intervention to reduce hemorrhage damage or hemorrhage-induced ischemic damage [51, 52]. In addition, IL-1β has been measured in brain following autologous blood injection [44, 53]. Similarly, expression of the TNF receptor and IL-1β was upregulated following intrastriatal blood infusion [54].

2.4 Other Anti-inflammatory Drug Approaches

When an ICH occurs, blood components including erythrocytes, leukocytes, macrophages, and plasma proteins such as thrombin (see above) and plasmin have access to the brain. Notably, there are also detrimental effects of microglial activation in ICH-induced brain injury [55]. This may be an opportunity for therapeutic intervention. In rodent studies, macrophage/microglial inhibitory factor (MIF) was shown to inhibit microglial activation and macrophage infiltration following collagenase-induced ICH [37, 39]. Importantly, the treatment also reduced stroke injury volume and improved behavior. These findings further support that microglial activation promotes inflammatory reactions after ICH, and MIF could be a valuable neuroprotective agent for the treatment of ICH. Furthermore, another study using a rodent ICH model showed that antileukocyte intervention reduced neutrophil infiltration, behavioral deficits, and neuronal damage [56]. Therefore, the strategies targeting leukocytes and microglial activation may merit further evaluation as either alternative or adjunctive therapeutic approaches to treat hemorrhage.

Interleukin-1 (IL-1) has also gained some attention as a therapeutic target for stroke because extensive evidence supports the direct involvement of IL-1 in the neuronal injury that occurs in acute neurodegeneration [57]. Studies have shown that inhibiting IL-1 release or activity markedly reduces ischemic and hemorrhagic damage [58]. Moreover, previous studies have reported that overexpression of interleukin-1 receptor antagonist (IL-1ra) attenuated brain edema formation and thrombin-induced intracerebral inflammation in a rat autologous blood injection model of ICH [59, 60]. Therefore, these studies suggested that IL-1ra could be considered as a potential therapeutic agent for patients with ICH and could be the focus of additional preclinical and clinical research.

3 Free Radicals and Vascular Damage

Free radicals appear to be important in the progression of cellular toxicity and have been a primary focus of stroke research. Free radicals are highly reactive molecules that have one or more unpaired electrons. The molecular species are often divided into two groups, oxygen and nitrogen species. ROS usually refers to superoxide, hydrogen peroxide (H_2O_2), hydroxyl radical, and singlet oxygen, whereas a reactive nitrogen species can include nitric oxide (NO) and peroxynitrite. Free radicals and their related non-free radical reactive species have been implicated in stroke pathophysiology as an important contributor to cell and tissue injury [61–63]. Increased levels of free radicals can cause damage to virtually all cellular components, including DNA, lipids, and proteins, which then leads to injury of neurons, glial cells, blood vessels, and the vasculature. Free radicals can exert effects directly on cellular components and regulate cellular molecular pathways which contribute to the development of brain edema and cell death [29].

Brain hemorrhage has also been shown to produce a robust induction of heme oxygenase (HO) in microglia/macrophages. HO catalyzes degradation of heme into iron, carbon monoxide (CO), and biliverdin, which is then converted to bilirubin by biliverdin reductase [64]. Rodent studies have shown that hemorrhage increases iron accumulation in brain, and non-heme iron has been shown to increase threefold after ICH in rats [65]. Interestingly, on the same topic, intracerebral infusion of iron causes brain injury and deferoxamine, a chelator, reduces ICH-induced brain damage, suggesting that iron plays an important part in brain injury after ICH [66–68]. Iron and other products contribute to pathological changes such as the formation of edema, infiltration of neutrophils, and induction of neuron death [24]. Moreover, Iron and iron-related products increase oxidative stress and catalyze hydroxyl radical production and lipid peroxidation [69, 70], which cause cellular damage. The production of ROS is a consequence of normal oxidative metabolism, but high ROS levels can be lethal [71–73] because they are involved in a series of processes such as contributing to brain edema by triggering the induction and activation of MMP family members both directly and indirectly [62]. There are direct and indirect processes affected by free radicals. The direct process involves the oxidation or nitrosylation of MMP, resulting in MMP activation [74]. The indirect process may involve redox-sensitive elements of transcription factors such as nuclear factor kappa-light-chain-enhancer of activated B cells (NF-κB) and activator protein 1 (AP-1), which is known to be an integral part of the binding sites for MMP transcription [75, 76]. Excessive H_2O_2 production has been suggested to induce MMP-1 mRNA expression in fibroblasts, and sublethal exposure to H_2O_2 has been found to increase the expression and activation of MMP-2 in human endothelial cells [77]. The treatment of fibroblasts with xanthine/xanthine oxide (XO) results in the induction of MMP-2 and MMP-9; furthermore, superoxide-stimulated extracellular signal-regulated kinase activation mediates MMP-9 [78]. Animal studies have suggested that superoxide and/or H_2O_2 are involved in the induction and activation of MMPs [79] and have implied that both molecules mediate BBB disruption through the activation of MMPs. Excessive production of superoxide radicals can also result in increased water and sodium content in the brain (i.e., edema) and the extravasation of Evans blue, suggesting the development of vasogenic edema [80]. Moreover, superoxide radicals have been identified as mediators of increased vascular permeability and edema development in various disease models [81]. Based on an ever increasing amount of data, it is clear that free radicals and oxidative stress are involved in BBB disruption and cellular injury after stroke. Thus, oxidative stress and free radical mediator remain appropriate targets for intervention.

3.1 Free Radical Scavengers

Free radicals have been proposed to mediate an array of injuries following a stroke [72, 82–87]. ROS cause brain injury via many different pathways. Compounds that can counteract or reverse the effects of free radicals have received a great deal of attention in recent years. Edaravone (3-methyl-1-phenyl-2-pyrazolin-5-one) is a potent lipid soluble hydroxyl and peroxyl radical scavenger used clinically for treatment of ischemic stroke in Japan. Previous study has shown that edaravone attenuated ICH-induced brain edema, neurologic deficits, and oxidative injury and also reduced iron- and thrombin-induced brain injury, suggesting that edaravone is a potential therapeutic agent for ICH [88]. Moreover, the edaravone clinical trial [89] reported that there was a significant improvement in functional outcome as evaluated using the modified Rankin Scale, when treatment was started within 72 h of onset. Furthermore, another study reported preliminary findings of a clinical trial showing that patients treated with edaravone prior to administration of intravenous tPA had a reduced incidence of ICH compared with placebo-treated tPA-treated patients [90]. Thus, edaravone or other lipophilic free radical scavengers may be useful to treat either AIS or hemorrhagic stroke.

Also, other studies have attempted to target pro-oxidant heme or iron to reduce a potential source of ROS production during hemorrhage. It is postulated that the regulation of HO might decrease ICH-induced toxicity because the enzyme metabolizes heme to release iron [91]. Several studies have also shown that nonselective inhibitors of HO (tin-mesoporphyrin IX, tin-protoporphyrin, and zinc protoporphyrin) decreased ICH-induced brain edema and neurologic deficits [67, 92–94]. Moreover, a ferric iron chelator (Deferoxamine) was shown to have a similar neuroprotective effect after ICH [66, 95], suggesting that ROS could be a potential target for ICH therapy. Furthermore, the effect of melatonin, a potent antioxidant and free-radical scavenger, on outcomes was investigated in rat collagenase-induced ICH model. The results showed that brain edema and neurological function at 24 h were unchanged in spite of oxidative stress reductions. However, repeated treatment with the lower dose of melatonin (5 mg/kg) given at 1 h and every 24 h thereafter for 3 days after ICH led to normalization of striatal function, normalized memory tasks, and reduced brain atrophy, suggesting that melatonin is safe for use after ICH and is protective [96].

4   Regulation of MMPs in Hemorrhage

MMPs are a group of proteolytic enzymes; zinc- and calcium-dependent endopeptidases which degrade many components of the extracellular matrix (ECM) including fibronectin, laminin, proteoglycans, and type IV collagen [97, 98]. The enzymes are categorized into four groups based upon protein structure: collagenases, stromelysins, gelatinases, and membrane-type MMPs [99]. These enzymes are secreted as a pro-protein in a latent inactive form, but can be activated due to a variety of physiological signals. Transcription of MMPs is also regulated by a variety of signals such as TNFα and IL-1, which induce the transcription of MMP-3 and MMP-9 and are involved in acute and chronic neuroinflammation. Several MMP activation mechanisms have been suggested including other proteases and free radicals [74, 100] and activation may either be through activation of the latent forms or by induction of mRNA through signaling via the nuclear factor-kappaB site [101].

Many investigators have emphasized the role MMP-2, 3, 9, and 12 in hemorrhage-mediated processes [38, 54, 102–105]. One of the first studies demonstrated the activation of MMP-2 and MMP-9 following injection of collagenase in rats to promote hemorrhage and ECM damage [103]. Other studies have also found that brain MMP-2, 3, 7, 9, and 12 mRNA levels were increased in the collagenase-induced hemorrhage rat model [38]. These results have been confirmed in other animal ICH models including mouse and pig [102, 104]. Increased MMP-9 activity has also been observed by gel and in situ zymography in the same model [105]. It is important to note that several clinical studies have also reported an elevation of MMP-9 levels in blood of patients with ICH [106–108]. Taken together, animal and human studies support the view that some members of the MMP family, especially MMP-3, 9, and 12, play an important role in the pathophysiology of ICH, BBB, and vascular damage.

Studies have shown that astrocytes, neurons, oligodendroglia, endothelial cells, pericytes, and microglia produce MMPs [109, 110]. In primary culture, mixed microglia and astrocytes appear to produce an active form of MMP-9, whereas astrocytes alone in culture induce pro-MMP-9 but do not produce the active form of MMP-9. Immunostaining of mixed glial cultures with an antibody directed against MMP-2 showed that MMP-3 was expressed by microglia, but not by astrocytes. This finding suggested that the microglia-derived MMP-3 may be important for activation of MMP-9 during the inflammatory response. In cell cultures, microglia interact with the pericytes, endothelial cells, and astrocytes to activate MMPs. Therefore, microglia may be necessary for the activation of the proMMP-9 which could be mediated by MMP-3.

As previously mentioned, TNF-α and IL-1β can induce the production of the MMPs. In addition, activation processes of MMPs involve proteases and free radicals. Therefore, microglia and tissue macrophages play a critical role in the inflammatory response both by releasing MMPs and by forming inflammatory mediators that activate them. Animal studies have shown that TNF-α stimulates cells to produce active MMPs, which facilitate leukocyte extravasation and brain edema by degradation of ECM components and the opening of the BBB that could be blocked by the use of the MMP inhibitor [111]. MMPs affect the function of the neurovascular structures by degrading the components of the basal lamina around the cerebral vessels to increase the permeability of the BBB, thereby contributing to brain edema and hemorrhagic brain injury [62, 103]. Overall, MMPs cause an increase in BBB by targeting membrane matrix proteins, resulting in activation of a final common pathway downstream of acute neuroinflammation and can induce vasogenic edema. Preclinical evidence suggests that the development and use of specific MMP inhibitors may reduce hemorrhage expansion, vascular damage, and also ischemic damage.

4.1 TNF and MMPs Are Useful Drug Targets

Cytokine production and release is an important component of the stroke cascade that causes damage following a stroke. One cytokine, TNFα, has received a substantial amount of preclinical attention, but has still not been evaluated in a clinical trial. The TNFα precursor, pro-TNFα, can be cleaved to biologically active TNFα by MMPs, primarily TNF-alpha-converting enzyme (TACE) [112–116]. While MMPs are involved in the processing of pro-TNFα, mature TNFα can also induce or activate MMP-9 [103, 117–120] which perpetuates the cycle of TNFα production that results in BBB and membrane damage [110]. Synthesis of TNFα from pro-TNFα can be blocked by nonselective MMP inhibitors such as BB-2284 [121] and BB-94 [122–124] as well as specific TACE inhibitors such as Ro32-7315 [125] and DPH-067517 [126].

Previous animal studies have shown that ICH can be reduced by administration of a nonspecific MMP inhibitor BB-94 in a rat collagenase model [103]. In the RLCEM, the MMP inhibitor BB-94 also lowered the rate of tPA-induced hemorrhage, while not affecting significantly hemorrhage rate in the absence of tPA administration [127]. BB-94 could also reduce TNFα levels in rabbit brain even though BB-94 is a nonselective MMP inhibitor [128]. This has also been reported by other groups [122–124]. The reduction in hemorrhage following BB-94 administration is consistent with the idea that MMPs regulate BBB vascular function and ECM remodeling following a stroke [109] and also implicates MMPs in the processing of proTNFα. Of particular promise are MMP-9 inhibitors that have recently been shown to reduce brain injury and apoptosis following SAH [129, 130].

There is some new evidence from ischemic and hemorrhagic stroke patients that MMP-2 and MMP-9 may be involved in BBB breakdown or remodeling following the injury. In ischemic stroke patients, serum MMP-2 and MMP-9 levels increased during the course of ischemia [131, 132]. However, in SAH patients, serum MMP-2 levels are significantly decreased while MMP-9 levels are increased. Temporally, MMP-2 levels remain decreased up to 12 days post-SAH, but MMP-9 levels appeared to recover [131, 132]. There is also evidence linking MMP-2 to SAH from intracranial aneurysm rupture [133], which substantiates the hypothesis that MMP-2 and/or MMP-9 may be directly involved in the progression of stroke and hemorrhage. In a postmortem study, there was increased endothelial expression of both MMP-2 and MMP-9 suggesting that endothelial expression of MMPs may affect vascular matrix stability and contribute to hemorrhage [134]. The results from preclinical studies suggest that MMPs and possibly TNFα are directly involved in BBB breakdown and hemorrhage in brain following a variety of insults. MMP-2, MMP-9, and TACE are all valid targets for the development of small molecules to reduce hemorrhage.

5 New Treatment Options

Sections 2–4 highlighted research areas where consistent findings indicate that key cascade processes may be targets for drug development. The following section will review new therapies that are still in the early stages of development. All of the novel treatments have the potential to reduce hemorrhage and may be the hemorrhage treatment of the future.

5.1 Hydrogen Gas (HG)

A novel and quite interesting observation was recently made by Chen et al. [135] and Zhang et al. [136]. The authors provided evidence that hydrogen gas is an effective antioxidant that may reduce cell death resulting from lipid peroxidation and protein modification. In a rodent model, hydrogen gas improved behavior and decreased both hemorrhage and infarct volume. The beneficial effects of hydrogen were related to a reduction of 8-Hydroxyguanosine (8-OHG), 4-Hydroxy-2-Nonenal

(HNE), and nitrotyrosine and reduced MMP-9 activity. This novel discovery brings together many of the key mechanisms described above including free radicals, oxidative stress, and MMPs. The authors showed that hydrogen gas by inhalation can exert neuroprotective effects and reduce hemorrhagic transformation following MCA occlusion.

5.2   Hyperbaric Oxygen (HBO)

Another gaseous method of neuroprotection has been studied in preclinical rodent models [137–142]. Hyperbaric oxygen (HBO), 100% O_2 at 3 bar or ATA [143], has been shown to decrease BBB damage, reduce hemorrhagic transformation, and reduce hemoglobin extravasation in ischemic zones following embolic stroke [143]. HBO has been found to reduce thrombolytic-induced hemorrhage [143]. HBO is a pleiotropic treatment [140, 141] that reduces edema (nonthrombin-mediated and ribosomal protein S6 kinases (p70 S6 K)-activated), implicating new protein synthesis in the process [140]. Ostrowski et al. [138, 139] have studied mechanisms of HBO in SAH models and have shown that HBO improves behavior and reduces SAH mortality. Both effects were correlated with reduced lipid peroxidation measured by detecting malondialdehyde, the degradation product of polyunsaturated lipids, and reduced expression of the superoxide (free radical) producing enzyme, NADPH oxidase (NOX) [139]. The study suggested that HBO decreases oxidative stress via an early inhibition at the level of NOX. Matchett et al. [137] reviewed the use of HBO and suggested that HBO treatment has other positive effects in animal models, since HBO reduces BBB breakdown, decreases inflammation and oxidative stress, reduces edema, and suppresses apoptosis [144].

5.3   Normobaric Oxygen (NBO)

Normobaric oxygen (NBO) therapy a third noninvasive gaseous technique has been studied by Lo and colleagues to treat both ischemic stroke and hemorrhagic stroke [145–147]. Like HBO described in Sect. 5.2, NBO (100% O_2) has some neuroprotective effects. Sun et al. [143] used a thrombin-induced hemorrhage model to show that NBO decreased infarct size and hemorrhage in response to tPA administration. In the study, HBO therapy also reduced hemoglobin extravasation in the ischemic brain suggesting deceased hemorrhage rate and/or volume. Moreover, both NBO and HBO treatment decreased BBB damage and the incidence of hemorrhagic transformation. In a collagenase-induced hemorrhage model, NBO did not affect collagenase-induced blood volume or edema and was ineffective at reducing neurological deficits [145]. Taken together, early preclinical data suggests that NBO may reduce stroke-induced deficits, but the effects may be specific to certain types of stroke (i.e., not collagenase-induced ECM disruption).

5.4   Brain Hypothermia (BH)

Preclinical brain hypothermia (BH) has been used as an effective neuroprotective treatment in experimental brain ischemia and injury [148–151]. BH is classified by the amount of core cooling below normal body temperature (i.e., 37–38°C). Mild hypothermia reduces body temperature by 3–6°C, whereas deep hypothermia reduces body temperature by 10°C [152]; the largest neuroprotective benefit is obtained with 34°C [153]. Neuroprotection by mild hypothermia is associated with metabolic downregulation, mitochondrial preservation, and the suppression of apoptosis [154]. Significant metabolic downregulation including attenuation of the inflammatory response and reduced ROS may be the basis for most neuroprotection [154, 155]. Suppression of both mechanisms can reduce the activity of MMPs and prevent BBB damage and edema [156–158]. Clinical trials are warranted to define the specific operating conditions for effective hypothermia including the therapeutic window and temperature regimen [152] that should be used for optimal neuroprotection. Unfortunately, clinical trials of hypothermia for ischemic stroke (ICTuS-L trial) have had mixed results. While a recent study provided feasibility, there were significant side effects related to the treatment. Pneumonia occurred in 50% of hypothermic patients compared to 10% of normothermic patients ($p=0.001$); however, there was an almost equal number of deaths in both groups ($p>0.05$) [159]. Additional clinical studies will be required to determine the usefulness of BH.

6   Conclusion

Hemorrhagic stroke is a condition that represents an important problem that must be solved, especially as the population ages worldwide. There are many suitable primary targets for drug development including specific MMPs, TNFα, thrombin, inflammation, oxidative stress, and the coagulation pathway. With hemorrhagic stroke, as is the case with ischemic stroke [160], there is simultaneous activation of a "hemorrhage cascade" and a "stroke cascade" with some overlapping key mediators [13, 29, 63, 161, 162]. There is a growing consensus that the treatment of ischemic stroke will require pleiotropic drugs or combination therapy [63, 160, 163–169]. It is my view that this may be even more crucial when one considers the effect of brain blood on tissue, in addition to the ischemic and edema effects that are observed in hemorrhage. Currently, the following drugs and drug classes hold promise for an eventual treatment: anti-inflammatory drugs, antioxidants, MMP inhibitors, thrombin inhibitors, and HO inhibitors. Some of the most effective preclinical treatments are noninvasive therapies such as hyperthermia, HBO, NBO, and HG. The novel treatments should continue to be developed since they are all "pleiotropic" therapies that have multiple beneficial physiological consequences. Translational and mechanism-based research will no doubt provide a clearer perspective of the important mechanisms involved in hemorrhagic stroke so that they can be targeted and translated into clinically useful therapeutics.

Disclosure There are no conflicts of interest to disclose.

Acknowledgments This article was supported by a U01 Translational research grant NS060685 to PAL.

References

1. Lloyd-Jones D, Adams R, Carnethon M, De Simone G, Ferguson TB, Flegal K, et al. Heart disease and stroke statistics—2009 update: a report from the American Heart Association Statistics Committee and Stroke Statistics Subcommittee. Circulation. 2009;119(3):480–6.
2. Lloyd-Jones D, Adams RJ, Brown TM, Carnethon M, Dai S, De Simone G, et al. Executive summary: heart disease and stroke statistics—2010 update: a report from the American Heart Association. Circulation. 2010;121(7):948–54.
3. Lloyd-Jones D, Adams RJ, Brown TM, Carnethon M, Dai S, De Simone G, et al. Heart disease and stroke statistics—2010 update: a report from the American Heart Association. Circulation. 2010;121(7):e46–215.
4. Lapchak PA. Hemorrhagic transformation following ischemic stroke: significance, causes, and relationship to therapy and treatment. Curr Neurol Neurosci Rep. 2002;2(1):38–43.
5. Lyden PD, Zivin JA. Hemorrhagic transformation after cerebral ischemia: mechanisms and incidence. Cerebrovasc Brain Metab Rev. 1993;5(1):1–16.
6. Bernstein RA, Del-Signore M. Recent advances in the management of acute intracerebral hemorrhage. Curr Neurol Neurosci Rep. 2005;5(6):483–7.
7. van Gijn J, Kerr RS, Rinkel GJ. Subarachnoid haemorrhage. Lancet. 2007;369(9558): 306–18.
8. Toni D, Fiorelli M, Bastianello S, Sacchetti ML, Sette G, Argentino C, et al. Hemorrhagic transformation of brain infarct: predictability in the first 5 hours from stroke onset and influence on clinical outcome. Neurology. 1996;46(2):341–5.
9. Donnan GA, Fisher M, Macleod M, Davis SM. Stroke. Lancet. 2008;371(9624):1612–23.
10. Berwaerts J, Robb OJ, Dykhuizen RS, Webster J. Course, management and outcome of oral-anticoagulant-related intracranial haemorrhages. Scott Med J. 2000;45(4):105–9.
11. Flaherty ML, Kissela B, Woo D, Kleindorfer D, Alwell K, Sekar P, et al. The increasing incidence of anticoagulant-associated intracerebral hemorrhage. Neurology. 2007;68(2): 116–21.
12. Khatri P, Wechsler LR, Broderick JP. Intracranial hemorrhage associated with revascularization therapies. Stroke. 2007;38(2):431–40.
13. Xi G, Keep RF, Hoff JT. Mechanisms of brain injury after intracerebral haemorrhage. Lancet Neurol. 2006;5(1):53–63.
14. Lyden PD, Zivin JA, Soll M, Sitzer M, Rothrock JF, Alksne J. Intracerebral hemorrhage after experimental embolic infarction. Anticoagulation. Arch Neurol. 1987;44(8):848–50.
15. Lok J, Leung W, Murphy S, Butler W, Noviski N, Lo EH. Intracranial hemorrhage: mechanisms of secondary brain injury. Acta Neurochir Suppl. 2011;111:63–9.
16. Fujimoto S, Katsuki H, Ohnishi M, Takagi M, Kume T, Akaike A. Plasminogen potentiates thrombin cytotoxicity and contributes to pathology of intracerebral hemorrhage in rats. J Cereb Blood Flow Metab. 2008;28(3):506–15.
17. You H, Al-Shahi R. Haemostatic drug therapies for acute primary intracerebral haemorrhage. Cochrane Database Syst Rev. 2006;3:CD005951.
18. Lwaleed BA, Goyal A, Delves GH, Cooper AJ. Seminal hemostatic factors: then and now. Semin Thromb Hemost. 2007;33(1):3–12.
19. Hoots WK. Challenges in the therapeutic use of a "so-called" universal hemostatic agent: recombinant factor VIIa. Hematol Am Soc Hematol Educ Program. See http://www.ncbi.nlm.nih.gov/pubmed/17124094 2006;426–31.

20. Hedner U, Lee CA. First 20 years with recombinant FVIIa (NovoSeven). Haemophilia. 2011;17(1):e172–82.
21. Traynelis SF, Trejo J. Protease-activated receptor signaling: new roles and regulatory mechanisms. Curr Opin Hematol. 2007;14(3):230–5.
22. Coughlin SR. Protease-activated receptors in hemostasis, thrombosis and vascular biology. J Thromb Haemost. 2005;3(8):1800–14.
23. Soh UJ, Dores MR, Chen B, Trejo J. Signal transduction by protease-activated receptors. Br J Pharmacol. 2010;160(2):191–203.
24. Katsuki H. Exploring neuroprotective drug therapies for intracerebral hemorrhage. J Pharmacol Sci. 2010;114(4):366–78.
25. Lee KR, Kawai N, Kim S, Sagher O, Hoff JT. Mechanisms of edema formation after intracerebral hemorrhage: effects of thrombin on cerebral blood flow, blood-brain barrier permeability, and cell survival in a rat model. J Neurosurg. 1997;86(2):272–8.
26. Chen B, Cheng Q, Yang K, Lyden PD. Thrombin mediates severe neurovascular injury during ischemia. Stroke. 2010;41(10):2348–52.
27. Liu DZ, Ander BP, Xu H, Shen Y, Kaur P, Deng W, et al. Blood-brain barrier breakdown and repair by Src after thrombin-induced injury. Ann Neurol. 2010;67(4):526–33.
28. Liu DZ, Tian Y, Ander BP, Xu H, Stamova BS, Zhan X, et al. Brain and blood microRNA expression profiling of ischemic stroke, intracerebral hemorrhage, and kainate seizures. J Cereb Blood Flow Metab. 2010;30(1):92–101.
29. Wang J, Dore S. Inflammation after intracerebral hemorrhage. J Cereb Blood Flow Metab. 2007;27(5):894–908.
30. Wang J. Preclinical and clinical research on inflammation after intracerebral hemorrhage. Prog Neurobiol. 2010;92(4):463–77.
31. Wang J, Tsirka SE. Contribution of extracellular proteolysis and microglia to intracerebral hemorrhage. Neurocrit Care. 2005;3(1):77–85.
32. Stoll G, Schroeter M, Jander S, Siebert H, Wollrath A, Kleinschnitz C, et al. Lesion-associated expression of transforming growth factor-beta-2 in the rat nervous system: evidence for down-regulating the phagocytic activity of microglia and macrophages. Brain Pathol. 2004;14(1):51–8.
33. Gregersen R, Lambertsen K, Finsen B. Microglia and macrophages are the major source of tumor necrosis factor in permanent middle cerebral artery occlusion in mice. J Cereb Blood Flow Metab. 2000;20(1):53–65.
34. Hanisch UK. Microglia as a source and target of cytokines. Glia. 2002;40(2):140–55.
35. Banno M, Mizuno T, Kato H, Zhang G, Kawanokuchi J, Wang J, et al. The radical scavenger edaravone prevents oxidative neurotoxicity induced by peroxynitrite and activated microglia. Neuropharmacology. 2005;48(2):283–90.
36. Min KJ, Yang MS, Kim SU, Jou I, Joe EH. Astrocytes induce hemeoxygenase-1 expression in microglia: a feasible mechanism for preventing excessive brain inflammation. J Neurosci. 2006;26(6):1880–7.
37. Wang J, Tsirka SE. Tuftsin fragment 1–3 is beneficial when delivered after the induction of intracerebral hemorrhage. Stroke. 2005;36(3):613–8.
38. Power C, Henry S, Del Bigio MR, Larsen PH, Corbett D, Imai Y, et al. Intracerebral hemorrhage induces macrophage activation and matrix metalloproteinases. Ann Neurol. 2003;53(6):731–42.
39. Wang J, Rogove AD, Tsirka AE, Tsirka SE. Protective role of tuftsin fragment 1–3 in an animal model of intracerebral hemorrhage. Ann Neurol. 2003;54(5):655–64.
40. Nguyen HX, O'Barr TJ, Anderson AJ. Polymorphonuclear leukocytes promote neurotoxicity through release of matrix metalloproteinases, reactive oxygen species, and TNF-alpha. J Neurochem. 2007;102(3):900–12.
41. Joice SL, Mydeen F, Couraud PO, Weksler BB, Romero IA, Fraser PA, et al. Modulation of blood-brain barrier permeability by neutrophils: in vitro and in vivo studies. Brain Res. 2009;1298:13–23.
42. Barone FC, Feuerstein GZ. Inflammatory mediators and stroke: new opportunities for novel therapeutics. J Cereb Blood Flow Metab. 1999;19(8):819–34.

43. Emsley HC, Tyrrell PJ. Inflammation and infection in clinical stroke. J Cereb Blood Flow Metab. 2002;22(12):1399–419.
44. Aronowski J, Hall CE. New horizons for primary intracerebral hemorrhage treatment: experience from preclinical studies. Neurol Res. 2005;27(3):268–79.
45. Zhang D, Hu X, Qian L, Wilson B, Lee C, Flood P, et al. Prostaglandin E2 released from activated microglia enhances astrocyte proliferation in vitro. Toxicol Appl Pharmacol. 2009;238(1):64–70.
46. Burger D, Dayer JM. Cytokines, acute-phase proteins, and hormones: IL-1 and TNF-alpha production in contact-mediated activation of monocytes by T lymphocytes. Ann N Y Acad Sci. 2002;966:464–73.
47. Hua Y, Wu J, Keep RF, Nakamura T, Hoff JT, Xi G. Tumor necrosis factor-alpha increases in the brain after intracerebral hemorrhage and thrombin stimulation. Neurosurgery. 2006;58(3):542–50; discussion 50.
48. Mayne M, Ni W, Yan HJ, Xue M, Johnston JB, Del Bigio MR, et al. Antisense oligodeoxynucleotide inhibition of tumor necrosis factor-alpha expression is neuroprotective after intracerebral hemorrhage. Stroke. 2001;32(1):240–8.
49. Xi G, Hua Y, Keep RF, Younger JG, Hoff JT. Systemic complement depletion diminishes perihematomal brain edema in rats. Stroke. 2001;32(1):162–7.
50. Lapchak PA. Tumor necrosis factor-alpha is involved in thrombolytic-induced hemorrhage following embolic strokes in rabbits. Brain Res. 2007;1167:123–8.
51. Eigler A, Sinha B, Hartmann G, Endres S. Taming TNF: strategies to restrain this proinflammatory cytokine. Immunol Today. 1997;18(10):487–92.
52. Feuerstein G, Wang X, Barone FC. Cytokines in brain ischemia—the role of TNF alpha. Cell Mol Neurobiol. 1998;18(6):695–701.
53. Wagner KR, Beiler S, Beiler C, Kirkman J, Casey K, Robinson T, et al. Delayed profound local brain hypothermia markedly reduces interleukin-1beta gene expression and vasogenic edema development in a porcine model of intracerebral hemorrhage. Acta Neurochir Suppl. 2006;96:177–82.
54. Lu A, Tang Y, Ran R, Ardizzone TL, Wagner KR, Sharp FR. Brain genomics of intracerebral hemorrhage. J Cereb Blood Flow Metab. 2006;26(2):230–52.
55. Wu J, Yang S, Xi G, Song S, Fu G, Keep RF, et al. Microglial activation and brain injury after intracerebral hemorrhage. Acta Neurochir Suppl. 2008;105:59–65.
56. Zhao X, Zhang Y, Strong R, Grotta JC, Aronowski J. 15d-Prostaglandin J2 activates peroxisome proliferator-activated receptor-gamma, promotes expression of catalase, and reduces inflammation, behavioral dysfunction, and neuronal loss after intracerebral hemorrhage in rats. J Cereb Blood Flow Metab. 2006;26(6):811–20.
57. Allan SM, Tyrrell PJ, Rothwell NJ. Interleukin-1 and neuronal injury. Nat Rev Immunol. 2005;5(8):629–40.
58. Rothwell N. Interleukin-1 and neuronal injury: mechanisms, modification, and therapeutic potential. Brain Behav Immun. 2003;17(3):152–7.
59. Masada T, Hua Y, Xi G, Yang GY, Hoff JT, Keep RF, et al. Overexpression of interleukin-1 receptor antagonist reduces brain edema induced by intracerebral hemorrhage and thrombin. Acta Neurochir Suppl. 2003;86:463–7.
60. Masada T, Hua Y, Xi G, Yang GY, Hoff JT, Keep RF. Attenuation of intracerebral hemorrhage and thrombin-induced brain edema by overexpression of interleukin-1 receptor antagonist. J Neurosurg. 2001;95(4):680–6.
61. Heo JH, Han SW, Lee SK. Free radicals as triggers of brain edema formation after stroke. Free Radic Biol Med. 2005;39(1):51–70.
62. Jian Liu K, Rosenberg GA. Matrix metalloproteinases and free radicals in cerebral ischemia. Free Radic Biol Med. 2005;39(1):71–80.
63. Moskowitz MA, Lo EH, Iadecola C. The science of stroke: mechanisms in search of treatments. Neuron. 2010;67(2):181–98.
64. Kutty RK, Maines MD. Purification and characterization of biliverdin reductase from rat liver. J Biol Chem. 1981;256(8):3956–62.

65. Wu J, Hua Y, Keep RF, Nakamura T, Hoff JT, Xi G. Iron and iron-handling proteins in the brain after intracerebral hemorrhage. Stroke. 2003;34(12):2964–9.
66. Nakamura T, Keep RF, Hua Y, Schallert T, Hoff JT, Xi G. Deferoxamine-induced attenuation of brain edema and neurological deficits in a rat model of intracerebral hemorrhage. J Neurosurg. 2004;100(4):672–8.
67. Huang FP, Xi G, Keep RF, Hua Y, Nemoianu A, Hoff JT. Brain edema after experimental intracerebral hemorrhage: role of hemoglobin degradation products. J Neurosurg. 2002;96(2):287–93.
68. Nakamura T, Xi G, Park JW, Hua Y, Hoff JT, Keep RF. Holo-transferrin and thrombin can interact to cause brain damage. Stroke. 2005;36(2):348–52.
69. Sadrzadeh SM, Anderson DK, Panter SS, Hallaway PE, Eaton JW. Hemoglobin potentiates central nervous system damage. J Clin Invest. 1987;79(2):662–4.
70. Sadrzadeh SM, Eaton JW. Hemoglobin-mediated oxidant damage to the central nervous system requires endogenous ascorbate. J Clin Invest. 1988;82(5):1510–5.
71. Juranek I, Bezek S. Controversy of free radical hypothesis: reactive oxygen species—cause or consequence of tissue injury? Gen Physiol Biophys. 2005;24(3):263–78.
72. Facchinetti F, Dawson VL, Dawson TM. Free radicals as mediators of neuronal injury. Cell Mol Neurobiol. 1998;18(6):667–82.
73. Weiss SJ. Tissue destruction by neutrophils. N Engl J Med. 1989;320(6):365–76.
74. Gu Z, Kaul M, Yan B, Kridel SJ, Cui J, Strongin A, et al. S-nitrosylation of matrix metalloproteinases: signaling pathway to neuronal cell death. Science. 2002;297(5584):1186–90.
75. Huang CY, Fujimura M, Noshita N, Chang YY, Chan PH. SOD1 down-regulates NF-kappaB and c-Myc expression in mice after transient focal cerebral ischemia. J Cereb Blood Flow Metab. 2001;21(2):163–73.
76. Huang CY, Fujimura M, Chang YY, Chan PH. Overexpression of copper-zinc superoxide dismutase attenuates acute activation of activator protein-1 after transient focal cerebral ischemia in mice. Stroke. 2001;32(3):741–7.
77. Wenk J, Brenneisen P, Wlaschek M, Poswig A, Briviba K, Oberley TD, et al. Stable overexpression of manganese superoxide dismutase in mitochondria identifies hydrogen peroxide as a major oxidant in the AP-1-mediated induction of matrix-degrading metalloprotease-1. J Biol Chem. 1999;274(36):25869–76.
78. Gurjar MV, Deleon J, Sharma RV, Bhalla RC. Role of reactive oxygen species in IL-1 beta-stimulated sustained ERK activation and MMP-9 induction. Am J Physiol Heart Circ Physiol. 2001;281(6):H2568–74.
79. Gasche Y, Copin JC, Sugawara T, Fujimura M, Chan PH. Matrix metalloproteinase inhibition prevents oxidative stress-associated blood-brain barrier disruption after transient focal cerebral ischemia. J Cereb Blood Flow Metab. 2001;21(12):1393–400.
80. Chan PH, Schmidley JW, Fishman RA, Longar SM. Brain injury, edema, and vascular permeability changes induced by oxygen-derived free radicals. Neurology. 1984;34(3):315–20.
81. Kim GW, Lewen A, Copin J, Watson BD, Chan PH. The cytosolic antioxidant, copper/zinc superoxide dismutase, attenuates blood-brain barrier disruption and oxidative cellular injury after photothrombotic cortical ischemia in mice. Neuroscience. 2001;105(4):1007–18.
82. Floyd RA. Antioxidants, oxidative stress, and degenerative neurological disorders. Proc Soc Exp Biol Med. 1999;222(3):236–45.
83. Nakashima M, Niwa M, Iwai T, Uematsu T. Involvement of free radicals in cerebral vascular reperfusion injury evaluated in a transient focal cerebral ischemia model of rat. Free Radic Biol Med. 1999;26(5–6):722–9.
84. Cherubini A, Ruggiero C, Polidori MC, Mecocci P. Potential markers of oxidative stress in stroke. Free Radic Biol Med. 2005;39(7):841–52.
85. Lapchak PA, Araujo DM. Development of the nitrone-based spin trap agent NXY-059 to treat acute ischemic stroke. CNS Drug Rev. 2003;9(3):253–62.
86. Siesjo BK, Katsura K, Zhao Q, Folbergrova J, Pahlmark K, Siesjo P, et al. Mechanisms of secondary brain damage in global and focal ischemia: a speculative synthesis. J Neurotrauma. 1995;12(5):943–56.

87. Siesjo BK, Siesjo P. Mechanisms of secondary brain injury. Eur J Anaesthesiol. 1996;13(3): 247–68.

88. Nakamura T, Kuroda Y, Yamashita S, Zhang X, Miyamoto O, Tamiya T, et al. Edaravone attenuates brain edema and neurologic deficits in a rat model of acute intracerebral hemorrhage. Stroke. 2008;39(2):463–9.

89. Edaravone Acute Infarction Study Group. Effect of a novel free radical scavenger, edaravone (MCI-186), on acute brain infarction. Randomized, placebo-controlled, double-blind study at multicenters. Cerebrovasc Dis. 2003;15(3):222–9.

90. Yoshifumi T, editor. Benefits of Pre-treatment with edaravone in tPA intravenous therapy for acute cerebral infarction. In: XXIIIrd international symposium on cerebral blood flow (abstract); 2007; Suppl 1, BP34–06M.

91. Wang J, Zhuang H, Dore S. Heme oxygenase 2 is neuroprotective against intracerebral hemorrhage. Neurobiol Dis. 2006;22(3):473–6.

92. Gong Y, Tian H, Xi G, Keep RF, Hoff JT, Hua Y. Systemic zinc protoporphyrin administration reduces intracerebral hemorrhage-induced brain injury. Acta Neurochir Suppl. 2006;96:232–6.

93. Koeppen AH, Dickson AC, Smith J. Heme oxygenase in experimental intracerebral hemorrhage: the benefit of tin-mesoporphyrin. J Neuropathol Exp Neurol. 2004;63(6):587–97.

94. Wagner KR, Hua Y, de Courten-Myers GM, Broderick JP, Nishimura RN, Lu SY, et al. Tin-mesoporphyrin, a potent heme oxygenase inhibitor, for treatment of intracerebral hemorrhage: in vivo and in vitro studies. Cell Mol Biol (Noisy-le-Grand). 2000;46(3):597–608.

95. Wan S, Hua Y, Keep RF, Hoff JT, Xi G. Deferoxamine reduces CSF free iron levels following intracerebral hemorrhage. Acta Neurochir Suppl. 2006;96:199–202.

96. Lekic T, Hartman R, Rojas H, Manaenko A, Chen W, Ayer R, et al. Protective effect of melatonin upon neuropathology, striatal function, and memory ability after intracerebral hemorrhage in rats. J Neurotrauma. 2010;27(3):627–37.

97. Rosenberg GA. Matrix metalloproteinases in neuroinflammation. Glia. 2002;39(3):279–91.

98. Sternlicht MD, Werb Z. How matrix metalloproteinases regulate cell behavior. Annu Rev Cell Dev Biol. 2001;17:463–516.

99. Nelson AR, Fingleton B, Rothenberg ML, Matrisian LM. Matrix metalloproteinases: biologic activity and clinical implications. J Clin Oncol. 2000;18(5):1135–49.

100. Nagase H. Activation mechanisms of matrix metalloproteinases. Biol Chem. 1997;378(3–4): 151–60.

101. Yong VW, Power C, Forsyth P, Edwards DR. Metalloproteinases in biology and pathology of the nervous system. Nat Rev Neurosci. 2001;2(7):502–11.

102. Mun-Bryce S, Wilkerson A, Pacheco B, Zhang T, Rai S, Wang Y, et al. Depressed cortical excitability and elevated matrix metalloproteinases in remote brain regions following intracerebral hemorrhage. Brain Res. 2004;1026(2):227–34.

103. Rosenberg GA, Navratil M. Metalloproteinase inhibition blocks edema in intracerebral hemorrhage in the rat. Neurology. 1997;48(4):921–6.

104. Tang J, Liu J, Zhou C, Alexander JS, Nanda A, Granger DN, et al. Mmp-9 deficiency enhances collagenase-induced intracerebral hemorrhage and brain injury in mutant mice. J Cereb Blood Flow Metab. 2004;24(10):1133–45.

105. Wang J, Tsirka SE. Neuroprotection by inhibition of matrix metalloproteinases in a mouse model of intracerebral haemorrhage. Brain. 2005;128(Pt 7):1622–33.

106. Abilleira S, Montaner J, Molina CA, Monasterio J, Castillo J, Alvarez-Sabin J. Matrix metalloproteinase-9 concentration after spontaneous intracerebral hemorrhage. J Neurosurg. 2003;99(1):65–70.

107. Alvarez-Sabin J, Delgado P, Abilleira S, Molina CA, Arenillas J, Ribo M, et al. Temporal profile of matrix metalloproteinases and their inhibitors after spontaneous intracerebral hemorrhage: relationship to clinical and radiological outcome. Stroke. 2004;35(6):1316–22.

108. Silva Y, Leira R, Tejada J, Lainez JM, Castillo J, Davalos A. Molecular signatures of vascular injury are associated with early growth of intracerebral hemorrhage. Stroke. 2005;36(1): 86–91.

109. Rosenberg GA. Matrix metalloproteinases and their multiple roles in neurodegenerative diseases. Lancet Neurol. 2009;8(2):205–16.

110. Rosenberg GA, Cunningham LA, Wallace J, Alexander S, Estrada EY, Grossetete M, et al. Immunohistochemistry of matrix metalloproteinases in reperfusion injury to rat brain: activation of MMP-9 linked to stromelysin-1 and microglia in cell cultures. Brain Res. 2001; 893(1–2):104–12.

111. Leib SL, Clements JM, Lindberg RL, Heimgartner C, Loeffler JM, Pfister LA, et al. Inhibition of matrix metalloproteinases and tumour necrosis factor alpha converting enzyme as adjuvant therapy in pneumococcal meningitis. Brain. 2001;124(Pt 9):1734–42.

112. McGeehan GM, Becherer JD, Bast Jr RC, Boyer CM, Champion B, Connolly KM, et al. Regulation of tumour necrosis factor-alpha processing by a metalloproteinase inhibitor. Nature. 1994;370(6490):558–61.

113. Zask A, Levin JI, Killar LM, Skotnicki JS. Inhibition of matrix metalloproteinases: structure based design. Curr Pharm Des. 1996;2:624–61.

114. Yamamoto M, Hirayama R, Naruse K, Yoshino K, Shimada A, Inoue S, et al. Structure-activity relationship of hydroxamate-based inhibitors on membrane-bound Fas ligand and TNF-alpha processing. Drug Des Discov. 1999;16(2):119–30.

115. Black RA, Durie FH, Otten-Evans C, Miller R, Slack JL, Lynch DH, et al. Relaxed specificity of matrix metalloproteinases (MMPS) and TIMP insensitivity of tumor necrosis factor-alpha (TNF-alpha) production suggest the major TNF-alpha converting enzyme is not an MMP. Biochem Biophys Res Commun. 1996;225(2):400–5.

116. Cherney RJ, Wang L, Meyer DT, Xue CB, Arner EC, Copeland RA, et al. Macrocyclic hydroxamate inhibitors of matrix metalloproteinases and TNF-alpha production. Bioorg Med Chem Lett. 1999;9(9):1279–84.

117. Dayer JM, Beutler B, Cerami A. Cachectin/tumor necrosis factor stimulates collagenase and prostaglandin E2 production by human synovial cells and dermal fibroblasts. J Exp Med. 1985;162(6):2163–8.

118. Mun-Bryce S, Rosenberg GA. Matrix metalloproteinases in cerebrovascular disease. J Cereb Blood Flow Metab. 1998;18(11):1163–72.

119. Rosenberg GA. Matrix metalloproteinases in brain injury. J Neurotrauma. 1995;12(5): 833–42.

120. Romanic AM, White RF, Arleth AJ, Ohlstein EH, Barone FC. Matrix metalloproteinase expression increases after cerebral focal ischemia in rats: inhibition of matrix metalloproteinase-9 reduces infarct size. Stroke. 1998;29(5):1020–30.

121. Gearing AJ, Beckett P, Christodoulou M, Churchill M, Clements J, Davidson AH, et al. Processing of tumour necrosis factor-alpha precursor by metalloproteinases. Nature. 1994;370(6490):555–7.

122. Corbel M, Lanchou J, Germain N, Malledant Y, Boichot E, Lagente V. Modulation of airway remodeling-associated mediators by the antifibrotic compound, pirfenidone, and the matrix metalloproteinase inhibitor, batimastat, during acute lung injury in mice. Eur J Pharmacol. 2001;426(1–2):113–21.

123. Falk V, Soccal PM, Grunenfelder J, Hoyt G, Walther T, Robbins RC. Regulation of matrix metalloproteinases and effect of MMP-inhibition in heart transplant related reperfusion injury. Eur J Cardiothorac Surg. 2002;22(1):53–8.

124. Santucci MB, Ciaramella A, Mattei M, Sumerska T, Fraziano M. Batimastat reduces *Mycobacterium tuberculosis*-induced apoptosis in macrophages. Int Immunopharmacol. 2003;3(12):1657–65.

125. Beck G, Bottomley G, Bradshaw D, Brewster M, Broadhurst M, Devos R, et al. (E)-2(R)-[1(S)-(Hydroxycarbamoyl)-4-phenyl-3-butenyl]-2′-isobutyl-2′-(meth anesulfonyl)-4-methylvalerohydrazide (Ro 32-7315), a selective and orally active inhibitor of tumor necrosis factor-alpha convertase. J Pharmacol Exp Ther. 2002;302(1):390–6.

126. Wang X, Feuerstein GZ, Xu L, Wang H, Schumacher WA, Ogletree ML, et al. Inhibition of tumor necrosis factor-alpha-converting enzyme by a selective antagonist protects brain from focal ischemic injury in rats. Mol Pharmacol. 2004;65(4):890–6.

127. Lapchak PA, Chapman DF, Zivin JA. Metalloproteinase inhibition reduces thrombolytic (tissue plasminogen activator)-induced hemorrhage after thromboembolic stroke. Stroke. 2000;31(12):3034–40.

128. Araujo DM, Lapchak PA, editors. Tumor necrosis factor is involved in behavioral deficits, infarct progression and hemorrhage following ischemic strokes: a potential therapeutic target. Washington: SFN; 2004.

129. Guo ZD, Zhang XD, Wu HT, Lin B, Sun XC, Zhang JH. Matrix metalloproteinase 9 inhibition reduces early brain injury in cortex after subarachnoid hemorrhage. Acta Neurochir Suppl. 2011;110(Pt 1):81–4.

130. Guo ZD, Sun XC, Zhang JH. Mechanisms of early brain injury after SAH: matrix metalloproteinase 9. Acta Neurochir Suppl. 2011;110(Pt 1):63–5.

131. Horstmann S, Kalb P, Koziol J, Gardner H, Wagner S. Profiles of matrix metalloproteinases, their inhibitors, and laminin in stroke patients: influence of different therapies. Stroke. 2003;34(9):2165–70.

132. Horstmann S, Su Y, Koziol J, Meyding-Lamade U, Nagel S, Wagner S. MMP-2 and MMP-9 levels in peripheral blood after subarachnoid hemorrhage. J Neurol Sci. 2006;251(1–2): 82–6.

133. Todor DR, Lewis I, Bruno G, Chyatte D. Identification of a serum gelatinase associated with the occurrence of cerebral aneurysms as pro-matrix metalloproteinase-2. Stroke. 1998;29(8): 1580–3.

134. Fujimura M, Watanabe M, Shimizu H, Tominaga T. Expression of matrix metalloproteinases (MMPs) and tissue inhibitor of metalloproteinase (TIMP) in cerebral cavernous malformations: immunohistochemical analysis of MMP-2, -9 and TIMP-2. Acta Neurochir (Wien). 2007;149(2):179–83.

135. Chen CH, Manaenko A, Zhan Y, Liu WW, Ostrowki RP, Tang J, et al. Hydrogen gas reduced acute hyperglycemia-enhanced hemorrhagic transformation in a focal ischemia rat model. Neuroscience. 2010;169(1):402–14.

136. Zhang QL, Du JB, Tang CS. Hydrogen and oxidative stress injury—from an inert gas to a medical gas. Beijing Da Xue Xue Bao. 2011;43(2):315–9.

137. Matchett GA, Martin RD, Zhang JH. Hyperbaric oxygen therapy and cerebral ischemia: neuroprotective mechanisms. Neurol Res. 2009;31(2):114–21.

138. Ostrowski RP, Colohan AR, Zhang JH. Neuroprotective effect of hyperbaric oxygen in a rat model of subarachnoid hemorrhage. Acta Neurochir Suppl. 2006;96:188–93.

139. Ostrowski RP, Tang J, Zhang JH. Hyperbaric oxygen suppresses NADPH oxidase in a rat subarachnoid hemorrhage model. Stroke. 2006;37(5):1314–8.

140. Qin Z, Hua Y, Liu W, Silbergleit R, He Y, Keep RF, et al. Hyperbaric oxygen preconditioning activates ribosomal protein S6 kinases and reduces brain swelling after intracerebral hemorrhage. Acta Neurochir Suppl. 2008;102:317–20.

141. Qin Z, Xi G, Keep RF, Silbergleit R, He Y, Hua Y. Hyperbaric oxygen for experimental intracerebral hemorrhage. Acta Neurochir Suppl. 2008;105:113–7.

142. Zhang JH, Lo T, Mychaskiw G, Colohan A. Mechanisms of hyperbaric oxygen and neuroprotection in stroke. Pathophysiology. 2005;12(1):63–77.

143. Sun L, Zhou W, Mueller C, Sommer C, Heiland S, Bauer AT, et al. Oxygen therapy reduces secondary hemorrhage after thrombolysis in thromboembolic cerebral ischemia. J Cereb Blood Flow Metab. 2010;30(9):1651–60.

144. Huang ZX, Kang ZM, Gu GJ, Peng GN, Yun L, Tao HY, et al. Therapeutic effects of hyperbaric oxygen in a rat model of endothelin-1-induced focal cerebral ischemia. Brain Res. 2007;1153:204–13.

145. Fujiwara N, Mandeville ET, Geng X, Luo Y, Arai K, Wang X, et al. Effect of normobaric oxygen therapy in a rat model of intracerebral hemorrhage. Stroke. 2011;42(5):1469–72.

146. Fujiwara N, Murata Y, Arai K, Egi Y, Lu J, Wu O, et al. Combination therapy with normobaric oxygen (NBO) plus thrombolysis in experimental ischemic stroke. BMC Neurosci. 2009;10:79.

147. Singhal AB, Benner T, Roccatagliata L, Koroshetz WJ, Schaefer PW, Lo EH, et al. A pilot study of normobaric oxygen therapy in acute ischemic stroke. Stroke. 2005;36(4):797–802.

148. Ceulemans AG, Zgavc T, Kooijman R, Hachimi-Idrissi S, Sarre S, Michotte Y. The dual role of the neuroinflammatory response after ischemic stroke: modulatory effects of hypothermia. J Neuroinflammation. 2010;7:74.
149. Froehler MT, Ovbiagele B. Therapeutic hypothermia for acute ischemic stroke. Expert Rev Cardiovasc Ther. 2010;8(4):593–603.
150. van der Worp HB, Macleod MR, Kollmar R. Therapeutic hypothermia for acute ischemic stroke: ready to start large randomized trials? J Cereb Blood Flow Metab. 2010;30(6): 1079–93.
151. Liu L, Yenari MA. Clinical application of therapeutic hypothermia in stroke. Neurol Res. 2009;31(4):331–5.
152. Lyden PD, Krieger D, Yenari M, Dietrich WD. Therapeutic hypothermia for acute stroke. Int J Stroke. 2006;1(1):9–19.
153. Kollmar R, Blank T, Han JL, Georgiadis D, Schwab S. Different degrees of hypothermia after experimental stroke: short- and long-term outcome. Stroke. 2007;38(5):1585–9.
154. Yenari M, Kitagawa K, Lyden P, Perez-Pinzon M. Metabolic downregulation: a key to successful neuroprotection? Stroke. 2008;39(10):2910–7.
155. Lee SM, Zhao H, Maier CM, Steinberg GK. The protective effect of early hypothermia on PTEN phosphorylation correlates with free radical inhibition in rat stroke. J Cereb Blood Flow Metab. 2009;29(9):1589–600.
156. Liu L, Kim JY, Koike MA, Yoon YJ, Tang XN, Ma H, et al. FasL shedding is reduced by hypothermia in experimental stroke. J Neurochem. 2008;106(2):541–50.
157. MacLellan CL, Davies LM, Fingas MS, Colbourne F. The influence of hypothermia on outcome after intracerebral hemorrhage in rats. Stroke. 2006;37(5):1266–70.
158. Kawanishi M, Kawai N, Nakamura T, Luo C, Tamiya T, Nagao S. Effect of delayed mild brain hypothermia on edema formation after intracerebral hemorrhage in rats. J Stroke Cerebrovasc Dis. 2008;17(4):187–95.
159. Hemmen TM, Raman R, Guluma KZ, Meyer BC, Gomes JA, Cruz-Flores S, et al. Intravenous thrombolysis plus hypothermia for acute treatment of ischemic stroke (ICTuS-L): final results. Stroke. 2010;41(10):2265–70.
160. Lapchak PA. Emerging therapies: pleiotropic multi-target drugs to treat stroke victims. Transl Stroke Res. 2011;2(2):129–35.
161. Dirnagl U, Iadecola C, Moskowitz MA. Pathobiology of ischaemic stroke: an integrated view. Trends Neurosci. 1999;22(9):391–7.
162. Zhao BQ, Tejima E, Lo EH. Neurovascular proteases in brain injury, hemorrhage and remodeling after stroke. Stroke. 2007;38(2 Suppl):748–52.
163. White BC, Sullivan JM, DeGracia DJ, O'Neil BJ, Neumar RW, Grossman LI, et al. Brain ischemia and reperfusion: molecular mechanisms of neuronal injury. J Neurol Sci. 2000;179(1–2):1–33.
164. Lapchak PA. A critical assessment of edaravone acute ischemic stroke efficacy trials: is edaravone an effective neuroprotective therapy? Expert Opin Pharmacother. 2010;11(10): 1753–63.
165. Lapchak PA. Neuroprotective and neurotrophic curcuminoids to treat stroke: a translational perspective. Expert Opin Investig Drugs. 2011;20(1):13–22.
166. Christopher D, d'Esterre KMT, Aviv RI, Eisert W, Lee T-Y. Dipyridamole treatment prior to stroke onset: examining post-stroke cerebral circulation and outcome in rabbits. Transl Stroke Res. 2011;2(2):186–94. doi:10.1007/s12975-010-0062-0.
167. Fisher M. New approaches to neuroprotective drug development. Stroke. 2011;42(1 Suppl):S24–7.
168. Woodruff TM, Thundyil J, Tang SC, Sobey CG, Taylor SM, Arumugam TV. Pathophysiology, treatment, and animal and cellular models of human ischemic stroke. Mol Neurodegener. 2011;6(1):11.
169. Tuttolomondo A, Di Sciacca R, Di Raimondo D, Arnao V, Renda C, Pinto A, et al. Neuron protection as a therapeutic target in acute ischemic stroke. Curr Top Med Chem. 2009;9(14):1317–34.

Chapter 3
Experimental Platforms for Assessing White Matter Pathophysiology in Stroke

Ken Arai, Loc-Duyen D. Pham, and Eng H. Lo

Abstract This chapter aims at summarizing current knowledge on experimental systems for analyzing the role of white matter injury relevant to stroke. In this chapter, we will provide a broad but brief survey of existing models at the cell, tissue, and whole-animal levels. Experimental approaches have recently allowed a better understanding of the molecular and cellular pathways underlying oligodendrocyte and oligodendrocyte precursor cell damage and demyelination. Since white matter damage is a clinically important part of stroke, a systematic utilization of these cell/tissue/whole-animal platforms related to white matter pathophysiology may eventually lead us to discover new targets for treating stroke.

1 Introduction

Stroke is the third leading cause of death and a leading cause of adult disability in developed countries. Under stroke conditions, brain function is perturbed due to cerebral ischemia (lack of blood supply to the brain) caused by thrombosis/embolism or hemorrhage. In central areas of ischemic regions, blood flow deficits are severe and brain cells die rapidly. In peripheral areas (the so-called penumbra), blood flow deficits are relatively mild, so that therapeutic salvage is theoretically possible. Thus far, for acute stroke therapy, only thrombolytic therapy with tissue-plasminogen activator has been approved by the US Food and Drug Administration (FDA) to be effective in targeting the salvageable ischemic penumbra. Over the past decade, impressive advances have been made in understanding the basic molecular mechanisms underlying neuronal death. However, clinically effective

K. Arai (✉) • L.-D.D. Pham • E.H. Lo
Neuroprotection Research Laboratory, Massachusetts General Hospital/Harvard Medical School, Charlestown, MA 02129, USA
e-mail: karai@partners.org

P.A. Lapchak and J.H. Zhang (eds.), *Translational Stroke Research*, Springer Series in Translational Stroke Research, DOI 10.1007/978-1-4419-9530-8_3,
© Springer Science+Business Media, LLC 2012

	Models	Advantages	Disadvantages
Cell Platforms	OLG Primary Culture OPC Primary Culture Cell Line	• easy to handle • suitable for intracellular analysis • can assess proliferation/differentiation	• no interaction with other cell types • OLG/OPC exist under artificial conditions
Tissue Platforms	Optic Nerve Brain Slice	• existence of axon-myelin interaction • suitable for electrophysiological analysis	• relatively difficult to handle • relatively artificial conditions • not suitable for proliferation/differentiation assay
Whole Animal Platforms	Mechanical Occlusion Thrombotic Occlusion Endotheli-1 Injection	• OLG/OPC exist in their natural state • can assess numerous biological responses	• difficult to prepare white matter injury models • not suitable for intracellular analysis • relatively large variation

Fig. 3.1 Summary for experimental systems for analyzing the role of white matter injury relevant to stroke

neuroprotectants have not yet been discovered. This has been especially true in stroke, where many drugs targeting excitotoxicity, oxidative stress, and inflammation have all failed [57, 66, 142]. Although there are many difficult reasons for these translational problems [42, 81, 140], one potential issue worth examining further is the lack of emphasis on white matter.

In the central nervous system (CNS), white matter is primarily comprised of axonal bundles ensheathed with myelin. The cells forming these sheaths are the oligodendrocytes (OLGs), which tend to be arranged in rows parallel to axonal tracts. Just before and after birth, OLG precursor cells (OPCs) multiply rapidly, mature into OLGs, and develop processes, which are then involved in the formation of myelin. Damage to OPCs and OLGs causes loss of myelin synthesis and interruption of proper axonal function. Hence, even if we protect neurons in gray matter, loss of myelin and axonal integrity would interfere with neuronal connectivity and function. Neuroprotection cannot be truly attained without oligoprotection. This chapter aims at summarizing current knowledge on experimental systems for analyzing the role of white matter injury relevant to stroke. Cell culture platforms comprise primary cultures of both mature OLGs as well as OPCs. Tissue platforms involve preparations of organotypic slice cultures or optic nerve systems. Whole-animal platforms comprise in vivo models of cerebral ischemia that attempt to target white matter brain areas. While there is no single perfect model system (Fig. 3.1), the collection of these experimental approaches has recently allowed a better understanding of the molecular and cellular pathways underlying OLG/OPC damage and demyelination. Since white matter damage is a clinically important part of stroke, a systematic utilization of these cell/tissue/whole-animal platforms related to white matter pathophysiology may eventually lead us to discover new targets for treating stroke.

2 White Matter Damage in Stroke

Gray matter and white matter are the major two components of the CNS. The ratio between white matter and gray matter in human neocortex is approximately 1. But this ratio is much smaller in rodent neocortex, where white matter only comprises

10–15% of total volume [149]. White matter primarily consists of axonal bundles ensheathed with myelin. Myelin is synthesized by OLGs that tend to be arranged in rows parallel to axonal tracts. OLGs originate from OPCs that multiply rapidly and mature right before and after birth. The OLGs then develop processes that form the myelin sheaths. Although this standard model emphasizes the central role of OLGs, recent data now suggest that even in adult brain, OPCs may also be involved in white matter maintenance. Subpopulations of OPCs persist throughout the adult brain [20, 75, 99, 100]. These OPCs are thought to contribute to myelin mainte-nance and repair by generating new OLGs (see Sect. 5 in detail).

Stroke (also called a brain attack) refers to a heterogeneous spectrum of conditions caused by the occlusion or hemorrhage of blood vessels supplying the brain, and is one of the major causes of death and disability in developed countries [81]. The initial vascular event leads to energy loss, which triggers activation of multiple brain cell death pathways. White matter is especially susceptible to stroke [4, 106, 139]. Because white matter blood flow is lower than in gray matter and there is little collateral blood supply in deep white matter, white matter ischemia is typi-cally severe with rapid cell swelling and tissue edema [81]. Minor white matter strokes often cause extensive neurological deficits by interrupting the passage of large axonal bundles such as those within the internal capsule [81]. Axons contain abundant mitochondria, which is an organelle for a source of reactive oxygen species. In fact, free radical scavenging significantly reduces white matter injury in rodent stroke models [58, 61, 77, 137]. Furthermore, white matter ischemia activates sev-eral kinds of proteases, which weaken the structural integrity of axons and myelin sheath. Neurofilaments are major structural components of white matter axons, and calpains have been demonstrated to be involved in neurofilament degradation under ischemic conditions [132]. Matrix metalloproteinases (MMPs) can directly attack myelin components such as myelin-basic protein [19]. Ischemia-induced degrada-tion of myelin-basic protein is reduced in MMP-9 knockout mice [8]. Importantly, chronic white matter lesions are associated with upregulation of MMPs in autopsied samples from patients with vascular dementia [114].

White matter ischemia is different from gray matter in many ways. For instance, mechanisms responsible for ischemic cell death may be different between white matter and gray matter. Although an increase in intracellular Ca^{2+} is involved in both white matter and gray matter ischemia, the routes of Ca^{2+} entry might differ. In white matter, pathological Ca^{2+} entry occurs in part due to increased intracellular Na^+ and membrane depolarization [59, 131, 133]. Therefore, Na^+ channel blockade has been proposed as a protective strategy in white matter [52, 59, 131, 133]. By contrast, voltage-dependent Ca^{2+} channels and glutamate-activated ionic receptors in gray matter have been traditionally viewed as the primary routes of pathological Ca^{2+} entry [21, 23]. Of course, beyond these differences in calcium handling, there are potentially many other aspects of gray vs. white matter function that differ. Thus far, white matter pathophysiology remains relatively poorly understood compared to gray matter. Because white matter damage is a clinically important part of stroke, we might need to take white matter ischemia more into account to develop effective stroke therapy. From the next section, we will briefly overview current knowledge on exper-imental systems for analyzing the role of white matter injury relevant to stroke.

3 Experimental Platforms for Assessing White Matter Pathophysiology in Stroke

3.1 Cell Culture and Tissue Platforms

In CNS white matter, OLGs form myelin sheaths that encircle axonal bundles. OLGs differentiate from their precursor cells (OPCs). The maturation of OPCs into mature OLGs occurs in a multiple steps defined by the expression of specific cell surface receptors [17, 95, 108]. Overall, there are four stages in this maturation process; oligodendrocyte-type 2 astrocyte (O-2A) cells, pro-OLGs, immature OLGs, and mature OLGs (Fig. 3.2). In the cell culture platform, cultured OLGs can be obtained by differentiating cultured OPCs with CNTF and T3 [84]. Thus far, three methods are mainly used for preparing OPC cultures; Shake-off, immunopanning, and fluorescence-activated cell sorting (FACS). The shake-off method was developed by McCarthy and de Vellis [91]. Because adhesive properties between OPCs and astrocytes/microglia are different, OPCs can be separated from astrocytes/microglia by shaking flasks of mixed glial cultures. Raff and colleagues developed the immunopanning method [12, 112], which can isolate OPCs based on the

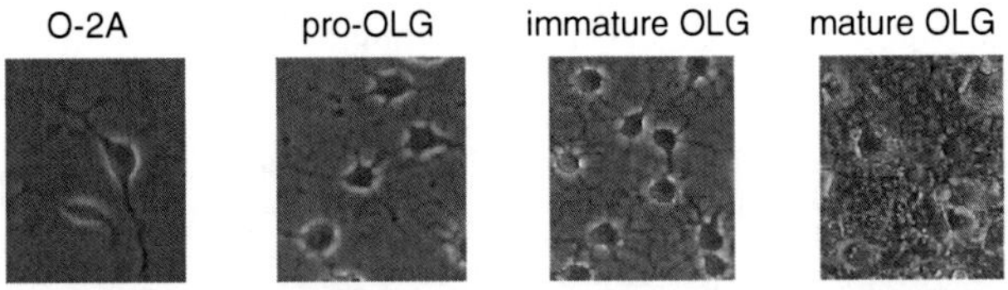

	O-2A	pro-OLG	immature OLG	mature OLG
A2B5	+	+	-	-
PDGF-R	+	+	-	-
NG2	+	+	-	-
O4	-	+	+	+
GalC	-	-	+	+
CNPase	-	-	+	+
MBP	-	-	-	+
PLP	-	-	-	+
MAG	-	-	-	+
MOG	-	-	-	+

Fig. 3.2 Characteristics of OLG maturation. OLG maturation can be defined by the expression of specific cell surface receptors, cell morphology, and proliferative/motility responses. O-2A cells are bipolar cells with high proliferative and motile activities. Pro-OLGs still divide but no longer show motility response. Immature OLGs are one of the two differentiated OLGs, but do not form myelin yet. The other differentiated OLG is mature OLGs, which express the myelin proteins to form the myelin sheath around axons

expression of the A2B5 on their cell surface. The FACS method can separate OPCs that express a fluorescent tag from nonfluorescent OPCs. To data, several transgenic mouse lines that labeled OPCs with fluorescent proteins are available [14, 153], and the FACS method is recently widely used for OPC isolation. In this section, we will introduce main findings in the mechanisms of OLG/OPC damage by stroke-related injury using cell/tissue platforms.

In pathological conditions, excitotoxic cell death is a critical part of neuronal injury, and it has been implicated in acute injury to the CNS and in chronic neuro-degenerative disorders [22, 72, 79]. In addition to neurons, glial cells can also be damaged by excitotoxicity. In the gray matter of the brain, neuronal cell death is often caused by a rise of extracellular glutamate concentration, which activates NMDA receptors and leads to an excessive rise of intracellular Ca^{2+} concentrations. Glutamate can also damage white matter OLGs, in both acute and chronic diseases [90]. Since excessive glutamate release can be seen under brain ischemic conditions, glutamate-induced OLG death has been used as a model for in vitro white matter ischemia to pursue the mechanism of OLG death by excitotoxicity. So far, there are at least three different mechanisms of glutamate-induced OLG death. Using cultured OLGs prepared from newborn rat brain, Oka et al. reported that OLGs in cultures are highly vulnerable to glutamate. They showed that 24-h exposure of glutamate caused OLG death by reversing cystine-glutamate exchange, which induces glutathione depletion [102]. AMPA/kainate receptors have also indicated to be involved in glutamate-induced OLG death. Sanchez-Gomez et al. used primary cultures of OLGs derived from the optic nerves of young adult rat or mouse to examine the involvement of AMPA/kainite receptors in OLG death. In the report, they revealed that excessive activation of AMPA/kainate receptors causes Na^+ and Ca^{2+} influx through the receptor channel complex leading to OLG death [120]. Until recently, it had been thought that NMDA receptor was not involved in OLG death by excitotoxicity. However, three recent studies have demonstrated that OLGs also express NMDA receptors as neurons [64, 94, 119]. NMDA receptors of OLGs are activated by glutamate in white matter ischemia [64], and activation of these receptors leads to rise of intracellular Ca^{2+} concentration [94]. Hence, NMDA receptors might also participate in glutamate-induced OLG damage.

In addition to these primary excitotoxic mechanisms, glutamate can also kill OLGs via immune system-related pathways. Alberdi et al. showed that brief incubation with glutamate followed by exposure to complement is lethal to OLGs in vitro [2]. Thus, even glutamate at nontoxic concentrations can kill OLGs by sensitizing these cells to complement attack. Further, inflammatory cytokines such as TNF-alpha and IL-1beta are also known to be involved in glutamate-induced OLG death. Those cytokines are released by reactive microglia and can impair glutamate uptake and trigger excitotoxic OLG death [134]. Since glutamate is reported to release TNF-alpha from microglia through AMPA/kainate activation, glutamate can also induce OLG death in indirect mechanisms [93].

Although elevations in glutamate certainly occur in stroke, alterations in other extracellular mediators may also be important. In the CNS, extracellular ATP can act as an excitatory neurotransmitter. ATP activates ionotropic P2X receptor and metabotropic P2Y receptor [101, 113]. Both P2X and P2Y receptors are expressed in OLGs. James and Butt used isolated optic nerves to show that ATP increased intracellular Ca^{2+} concentrations in OLGs through P2Y receptors [62]. In addition, they also demonstrated that a P2X receptor agonist evoked a smaller but significant OLG Ca^{2+} signal. In brain ischemia and spinal cord injury models, P2X receptors were reported to mediate signaling cascades leading to neurodegeneration [71, 143]. Also in OLGs, P2X was shown to be involved in cell death. Matute et al. have demonstrated that ATP or P2X agonists, but not P2Y agonists, was toxic to differentiated OLGs in vitro [90].

No matter what the upstream trigger might be, energetic stress should be a common denominator for white matter injury. In this regard, oxygen-glucose deprivation (OGD) is a useful tool for mimicking in vitro ischemia. OGD can induce both OPC as well as OLG death in vitro [6, 30]. Indeed, it appeared that OPCs were more susceptible than OLGs [30]. More recently, using an immortalized mouse OLG cell line, Zhang et al. showed that OGD-induced OLG death was induced by apoptosis via p75 and caspase-3 [148]. Thus, these cell systems can be productively used to dissect the molecular pathways of OGD-induced white matter injury.

Compared to astrocytes, both OLGs and OPCs are relatively weak. To maintain those cells in culture, a wide spectrum of culture supplements such as growth factors are needed. Hence, removing those factors from culture media causes OLG or OPC death. Although this starvation stress does not perfectly reflect in vivo ischemic conditions, the stress can be thought of as one of the in vitro ischemia model and is now well used to induce OLG/OPC damage, especially for OPC cultures. For instance, Cui et al. reported two studies whereby IGF-1 promoted the proliferation of OPCs via PI3K/Akt, MEK/ERK, and src-like tyrosine kinase pathways [27, 28]. Rubio et al. showed that neurotrophin-3 is also important factor for OPC survival [117]. GGF/neuregulin has been also reported to be OPC protective [39]. Taken together, these serum deprivation paradigms provide evidence for the essential nature of trophic coupling for oligoprotection. Indeed, these types of growth factor gain-and-loss experiments may help us test drugs that can be used to salvage OPC and OLG health in a wide range of disorders.

The experimental systems described above are all useful for examining the mechanisms of OLG/OPC death by ischemic stress, and revealed the cellular mechanisms leading to OLG/OPC death under stroke condition (Fig. 3.3). Moreover, the ease of quantitation and the reproducibility of culture platforms might allow for relatively high-throughput screening. However, it must be acknowledged that in vitro systems have some limitations. Cell and tissue systems cannot truly replicate the intercellular interactions and anatomic geometry of in vivo white matter.

3.2   In Vivo Rodent White Matter Ischemia Models

White matter lesions are observed frequently in stroke patients and experimental animal models of cerebral ischemia, and have been thought to contribute to cognitive impairment [35, 92, 105]. Since nonhuman primates have well-developed white

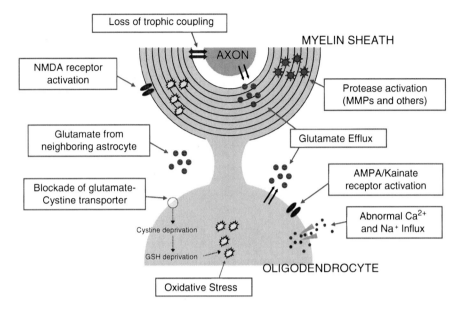

Fig. 3.3 Summary for oligodendrocyte death pathways under white matter stroke. Under ischemic conditions, several deleterious factors/cascades are activated. Glutamate efflux, oxidative stress, and proteinase activation eventually induce oligodendrocyte death. Importantly, deleterious factors secreted by one cell type may affect another cell type

matter and vascular architectures which are similar to those of human brains, it seems to be reasonable to use them for studying the mechanisms of white matter injury. However, from ethical and practical standpoints, there remains an important imperative for developing rodent models. Although basic research of white matter injury via in vivo models is not as well developed as in vitro studies, here we will discuss some useful in vivo white matter stroke models that are widely used in the field.

In the early days of experimental stroke research, invasive surgical techniques predominated, with particular emphasis on transorbital approaches in cats and non-human primates where the middle cerebral artery could be accessed and occluded with aneurysm clips, permanent ligation, or electrocautery. In the early 1980s, Tamura et al. described a new approach whereby a craniotomy was performed next to the zygomatic arch to reveal and occlude proximal portions of the middle cerebral artery as it enters the base of the brain in rats [135]. However, the surgery proved challenging. Subsequently, development of the intraluminal filament model triggered an explosion of experimental work with easier-to-use rodent models [65, 83]. A modified suture could be threaded up the internal carotid artery until the tip occluded the middle cerebral artery, thus inducing focal cerebral ischemia. Additionally, Chopp and colleagues pioneered a technique in which homologous blood clots could be placed in the middle cerebral artery in rats and mice via modified catheters [150, 151].

Data from those rodent middle cerebral artery occlusion (MCAO) models have shown that OLG damage occurs in response to ischemia. Irving et al. observed structural changes of the OLG cytoskeleton by 40 min occlusion of MCA [61]. They assessed these changes by detecting increase of tau immunoreactivity within OLGs. Tau is the microtubule-associated protein, and the increase in immunoreactivity has been found to be a sensitive marker for OLG damage. The MCAO-induced OLG damage assessed by tau immunoreactivity has been confirmed by other groups [58, 138]. Axons and myelin structure have also shown to be damaged in rodent MCAO models [31]. Another report by Irving et al. carefully examined the methods of quantifying white matter injury following prolonged focal cerebral ischemia in the rat. This study showed that myelin basic protein (MBP), Tau 1, and amyloid precursor protein staining can be utilized to assess myelin and axonal integrity in rat MCAO model [60]. Schabitz et al. have also shown the white matter injury in rat stroke model [122]. The use of Luxol fast blue-periodic acid-Schiff and Bielschowsky's silver impregnation allowed the detection of myelin and axons, which could then be semi-quantified as read-outs for specific experiments.

The two-vessel occlusion model by permanent bilateral occlusion of the common carotid arteries has also been used to induce white matter ischemia. This model produces chronic cerebral hypoperfusion in rat and gerbil [38, 121]. These models are characterized by pathological changes in white matter, which appear similar to those in human cerebrovascular white matter lesions [36, 141]. As seen in human white matter ischemia, the rat chronic cerebral hypoperfusion model by the ligation of the bilateral common carotid arteries is accompanied by cognitive impairment [37]. In this model, however, the visual pathway is also injured by the occlusion of the ophthalmic arteries, and thus may affect behavioral assessment [38]. Although rats and gerbils are mostly used for these chronic cerebral hypoperfusion models, mice have also been used recently. With newly designed micro-coils, Shibata et al. developed a new mouse model of chronic cerebral hypoperfusion with relative preservation of the visual pathway [126]. In this model, white matter lesions occurred after 14 days without any gray mater involvement. Another report from Shibata et al. has demonstrated that the mouse hypoperfusion model showed impairment of working memory using the 8-arm radial maze [127]. The importance of the mouse model centers on the ability of future studies to utilize specific knockouts or transgenics for dissecting molecular mechanisms and mediators. For example, the use of knockout mice for MMPs could be anticipated. From a biochemical basis, MMPs can directly attack myelin components [19]. In fact, Nakaji et al. used the mouse chronic cerebral hypoperfusion model to show that gene knockout of MMP-2 reduced the severity of the white matter lesions [97]. Furthermore, MMP-9 knockout mice showed the resistance to MCAO-induced degradation of myelin-basic protein [8]. Depending on the proposed mechanisms, the combination of specific knockout mice with experimental induction of stroke or brain injury should be useful for exploring the underlying mechanisms involved in white matter pathophysiology.

In addition to the methods of occluding blood vessels, the direct injection of endothelin-1 (ET-1) into the neural parenchyma has also been used to induce white matter ischemia [129]. ET-1 is a potent vasoconstrictor peptide, and acts through different receptors called types A and B, which are distributed throughout the CNS [89, 96, 116]. Hughes et al. reported that microinjection of ET-1 into the striatum and cerebral cortex induces focal ischemia with a reduction of 40% of local blood flow in rats [54]. The damage in this model is localized in both gray and white matter without blood–brain barrier breakdown. Frost et al. also used ET-1 microinjection model to induce white matter ischemia [40]. They injected ET-1 into the internal capsule, and tissue necrosis and demyelination in the infarcted white matter were found after a 14-day survival period. Further, infarcts resulted in measurable sensorimotor deficits. Regarding inflammatory response in the ET-1-injection white matter ischemia model, two studies recently came out. They showed that inflammatory response and white matter damage are closely related in the rat models of ET-1 microinjection into the striatum [32, 128].

The optic nerve has been thought of as one of the most useful regions to study OLG functions. As cited in the previous section, numerous groups use OLG cultures from rodent optic nerve to investigate the OLG death by ischemia. Also in vivo system, the mechanisms of OLG damage have been examined in optic nerve. Retinal ganglion cell (RGC) axons comprise the optic nerve. So far, several in vivo models have been created to identify the RGC response to optic nerve damage [25, 63, 70, 124]. A recent effort used photochemically induced thrombosis to evoke anterior ischemic optic neuropathy (AION) in rats and mice [15, 43]. AION may mimic optic nerve strokes, which are among the most common causes of sudden optic nerve-related vision loss. Discrete histopathological changes of OLGs were reported in human AION [68]. Correspondingly, OLG dysfunction was observed in the mouse AION model [43].

Taken together, the emergence of various mechanical, thrombotic and chemical models now provides reasonable ways to move forward in vivo. But it must be acknowledged that white matter volumes in rodents are extremely small compared to human brains. Thus, methods for quantifying outcomes are challenging. Ultimately, comparisons of these rodent outcomes with human stroke patients must be performed carefully since white matter connectivity may be different between these vastly differing levels of brain evolution.

4   Cell–Cell Trophic Coupling in White Matter

As described above, axons and OLG/OPC are vulnerable to damage by excitatory amino acids, oxidative stress, trophic factor deprivation, and activation of apoptotic pathways. Because a single OLG myelinates multiple axons, damage to only one OLG can cause dysfunction in many different neuronal pathways. Therefore, protecting OLGs (and OPCs also) is a reasonable therapeutic approach for white matter stroke treatment. Whereas OPCs and OLGs are clearly important target cells per se, it is likely that cell–cell interactions in the white matter may also be extremely important for white matter protection. The pathophysiology of stroke involves interactions between multiple cell types, and we may need to examine the relevance of dynamic interactions between all the cell types in white matter.

The main cell types comprising white matter are the neuronal axon, OLG (myelin), astrocyte, and endothelial cell. As in gray matter, there is close anatomical and perhaps functional contact between all these cells. Astrocytes are in close apposition to OLGs within the white matter [16], and couple with OLGs through gap junctions to maintain their functions [103]. Furthermore, using astrocyte-OLG coculture system, astrocytes have been shown to promote OLG survival through an alpha6 integrin-laminin-dependent mechanism [26]. Astrocyte-derived soluble factors are also implied to be supportive to both OLGs and OPCs. Conditioned media from astrocyte cultures have been demonstrated to support OLGs/OPCs survival [3, 5, 13, 41, 104, 111, 144]. In turn, recent findings suggest that OLGs not only myelinate axons but also maintain their functional integrity and survival thorough OLG-specific proteins and/or trophic factor release [47, 98, 115]. Thus far, PLP (and its smaller isoform DM20) and CNP are well-documented myelin-associated proteins expressed in OLGs to affect axonal functions. Mice with mutations for those genes revealed that axonal support by OLGs is independent of myelin assembly [47, 98]. Moreover, myelinating OLGs also provide trophic support for axons by secreting soluble factors. To date, brain-derived neurotrophic factor (BDNF), NT3, NGF, and PDGF are reported as OLG-axon signaling mediators [115].

Cerebral endothelial cells also play central and important roles in the cell–cell interactions in both white and gray matters. These cells work with astrocytes to form the blood–brain barrier (BBB) [1, 51, 56]. Compromised BBB function occurs in many CNS diseases. In the context of stroke, BBB damage leads to cerebral edema and hemorrhage [81]. Moreover, beyond outright disruptions of the barrier, damage to cerebral endothelium itself may affect the function and survival of neighboring cells. As discussed previously, under normal conditions, cerebral endothelium secrete trophic factors to support neighboring neurons [34, 48, 49]. This trophic coupling creates a neurovascular niche, wherein cell–cell signaling between cerebral endothelium and neuronal precursor cells help mediate and sustain pockets of ongoing neurogenesis and angiogenesis [24, 45, 55, 85, 147, 154]. Therefore, endothelial dysfunction results in not only lack of blood supply to the brain but also loss of trophic coupling in many aspects. Although the concept of the neurovascular unit is usually used to discuss phenomena in the gray matter [29, 55, 81, 82], it obviously should apply for white matter physiology and pathology as well. Thus far, cell–cell interactions between cerebral endothelium and OLG/OPC are not well understood. However, a recent study suggests that the existence of an "oligovascular niche," whereby cerebral endothelial cells support the survival and proliferation of OPCs [6]. Cross-talk between the vascular and neuronal compartments in the neurovascular niche is mediated by an exchange of soluble signals, and this phenomenon is partly mediated by the ability of cerebral endothelium to secrete a rich repertoire of trophic factors [48, 74, 125]. Similarly, endothelial-derived growth factors such as BDNF and FGF-2 promote OPC proliferation [6]. Importantly, these trophic coupling might be interrupted under pathological conditions. Nonlethal oxidative stress reduces the expression of several growth factors in

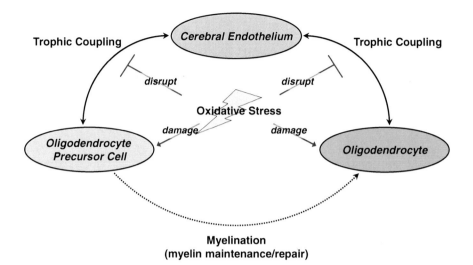

Fig. 3.4 Schematic of the oligovascular signaling in adult white matter. In the oligovascular niche, oligogenesis and angiogenesis might occur to maintain white matter homeostasis. Oligodendrocyte precursor cells are thought to contribute to myelin maintenance and repair by generating new mature oligodendrocytes. In addition, trophic coupling may also exist between cerebral endothelial cells and mature OLGs. Under diseased conditions such as stroke, multiple deleterious factors directly attack OLGs and OPCs, which are vulnerable to oxidative stress. Moreover, oxidative stress can disturb the trophic coupling between cerebral endothelial cells and oligodendrocyte lineage cells, resulting in further white matter damage

cultured cerebral endothelial cells [6, 48]. In stroke patients and rodent stroke models, endothelial dysfunction is often observed by cerebral small-vessel disease in white matter ischemia [50, 67, 123]. Therefore, disruption of endothelium–OPC/OLG coupling should be one of the major causes for the pathogenesis and progression of white matter lesion in CNS diseases including stroke. Thus far, there is no direct evidence showing OPCs support endothelial functions including angiogenesis. Also, endothelium–OLG trophic coupling has not been experimentally proved yet. However, it is well known that both cerebral endothelium and OPC/OLG secrete many kinds of growth factors [6, 33, 34, 48]. Therefore, it is possible that there is two-way trophic coupling between these cells in white matter (Fig. 3.4).

Compared to the mechanisms of cell–cell interaction in gray matter, trophic coupling in white matter remains relatively understudied and poorly understood. For the neurovascular niche, matrix and trophic interactions between endothelial cells and neurons sustain neurogenesis and angiogenesis [74, 85, 125] and may also protect neurons against oxidative and metabolic insults [34, 48]. Analogous interactions within a widely distributed oligovascular niche may provide a similar mechanism for sustaining white matter renewal and integrity. Further studies are warranted to dissect how these mechanisms function in normal brain, and how disruptions in oligovascular signaling may underlie white matter disease and neurodegeneration. Finally, to develop new stroke therapy, one may need to define these mechanisms in aged brains. Aging is the major risk factor for stroke, and it has recently shown that ischemic injury to white matter is an age-dependent process [10, 11]. How these mechanisms of oligovascular signaling are affected by aging and metabolic disease should be extremely important.

5   White Matter Recovery in Rodent Stroke Models

To date, the majority of studies using cell, tissue, and whole-animal models of white matter injury have mostly focused on mechanisms and targets for acute injury. However, it may not be easy to block all multifactorial pathways of brain cell death in stroke patients. Therefore, an emerging emphasis on promoting recovery after stroke is beginning to take shape in our field. In the context of white matter mechanisms, it will be important to carefully define how each of the model platforms can be applied in this regard.

Following the acute phase of stroke injury, the brain gradually transitions into a second phase, characterized by remodeling and repair. The biphasic nature of neurovascular responses represents an endogenous attempt by damaged parenchyma to trigger salvage of the damaged tissue. During the early acute phase of neurovascular injury, BBB perturbations should predominate with key roles for various matrix proteases. During the delayed phase, brain angiogenesis may provide the critical neurovascular substrates for neuronal remodeling. This biphasic phenomenon may in fact allow signals that are deleterious in the acute phase to transition into beneficial effects during stroke recovery. Understanding how neurovascular signals and substrates make the transition from initial injury into angiogenic recovery will be important if we are to find new therapeutic approaches for stroke.

During the acute phase of stroke, the ischemic penumbra suffers milder insults due to residual perfusion from collateral blood vessels compared to the core of the ischemic territory. Over the course of hours to days, the penumbra collapses if therapy is not initiated in time. Besides neuronal death per se, collapse of the acute penumbra can also be viewed in terms of degradation of cell–cell interactions in the neurovascular unit. Loss of signaling between astrocytes and endothelium alters tight junction homeostasis and leads to BBB disruption. Perturbations in neuronal-glial signaling lead to loss of proper neurotransmitter dynamics. And loss of matrix–trophic interactions between the vascular and neuronal elements may trigger parenchymal injury beyond ischemia itself. In the face of this acute neurovascular injury, it is beginning to be recognized that evolution of the penumbra may also mediate recovery as well. The penumbra is not just dying over time. It can also be actively trying to repair itself as endogenous mechanisms of plasticity and remodeling occur over days to weeks after stroke onset [80].

A prominent feature of recovery after stroke may involve neurogenesis and angiogenesis. Neurogenesis has been mainly characterized at sites of ongoing adult neural precursor turnover, comprising subventricular and subgranular zones. Many studies have now found that neurogenesis in these areas appears to be amplified after cerebral ischemia [7, 107]. Newborn neuroblasts stream from these sites toward damaged areas, as the brain attempts to heal itself. Interfering with these neurogenic streams can be deleterious. For example, delayed inhibition of MMPs seems to block these endogenous neurogenic responses and worsen outcomes [73]. In models of transient global brain ischemia, killing off neural precursors in the subventricular zones dramatically worsened spatial memory function over time [110]. Neurogenesis definitely is perturbed as a response to stroke. But whether these endogenous neurogenic alterations are relevant to clinical outcomes in stroke patients remains to be fully understood.

To support normal brain function, neuronal connectivity must be maintained. Hence, remyelination and regenerating axons in the border zone of cerebral infarcts and in secondarily lesioned areas are essential for stroke recovery. Fundamental research into these mechanisms of remyelination after ischemia remains somewhat limited, but a few studies are beginning to lead the way. Mandai et al. reported that myelin repair can occur in peri-infarct areas in mouse MCAO model, as judged by upregulated gene expression of proteolipid protein, one of the major protein components of CNS myelin [87]. Gregersen et al. studied the expression of MBP, another major component of CNS myelin, in peri-infarct areas using rat MCAO model [46]. They revealed that peri-infarct OLGs increased their expression of MBP mRNA from 24 h to maximal levels at day 7. These changes corresponded to the appearance of process-bearing MBP and occasional myelin OLG glycoprotein-immunoreactive OLGs in parallel sections.

As described before, OLGs originate from their precursors OPCs. OPCs are now found not only during development but pockets might also exist throughout the adult brain [20, 75, 99, 100]. Therefore, OPCs may contribute to myelin maintenance and repair by generating new OLGs as a source of remyelination and repair. Experimental evidence is starting to be collected showing that OPCs in adult brain contribute to replenishment of OLGs after ischemic insults. Tanaka et al. examined the alteration of OLGs, OPCs, and myelination in rat MCAO model [136]. They showed a rapid and progressive decrease in the number of OLGs, OPCs, and myelin density after 2 days in the infarct core. By contrast, the peri-infarct area exhibited a moderate reduction in the number of OLGs and the myelin density with a slight increase in OPCs at 2 days after MCAO. Subsequently, a steady increase in the number of OPCs and a gradual recovery of OLGs were found in the peri-infarct area at 2 weeks of MCAO.

If OPCs provide a source of potential white matter repair, what signals and substrates are involved? A recent interesting study by Komitova et al. suggested that enriched environments may enhance the generation of OPCs after focal cortical ischemia. In this study, newly born OPCs were found to be immunoreactive for BDNF [69]. Thus, it is tempting to speculate that BDNF may be related to remyelination by OPC proliferation/differentiation. What remains to be clarified is how this signal may interact with a multitude of other growth factors involved in OPC proliferation/differentiation. Since growth factor expression is upregulated after ischemia in peri-infarct area, a network response of growth factors might have essential roles in remyelination after stroke. These studies have mostly utilized in vivo models.

As we dig down deeper mechanistically, a systems biology approach may ultimately be required as one asks broadly what gene profiles mediate cell–cell interactions in white matter during and after injury.

6   Potential Therapeutic Targets for White Matter Ischemia

The combined use of cell, tissue, and whole-animal platforms discussed here may provide powerful tools for dissecting the pathophysiologic mechanisms of white matter injury in CNS disorders (Fig. 3.5). Of course, no validated drugs yet exit. But several promising leads should be discussed.

Memantine, an uncompetitive NMDA receptor antagonist, has been well examined for the efficacy for preventing white matter from several insults including ischemic stress. Memantine is now licensed for moderate-to-severe Alzheimer's disease in US and EU [78]. Very recently, two studies suggested that memantine may be protective to white matter. Bakiri et al. reported that memantine reduced ischemic damage to mature and precursor OLGs in brain slices assessed by patch-clamp system [9]. Manning et al. showed that memantine attenuated white matter injury in a rat model of periventricular leukomalacia [88]. Therefore, NMDA receptor antagonists might be a good target for white matter injury after stroke.

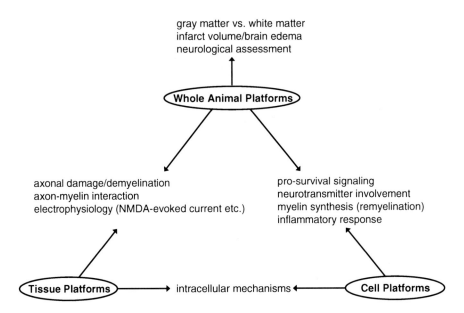

Fig. 3.5 Schematic for experimental systems and their endpoints. All systems have both advantages and disadvantages. A systematic utilization of these systems should enable us to better dissect mechanisms of white matter pathophysiology and help our search for oligoprotectants in stroke and other CNS disorders

Although the radical spin-trap NXY-059 failed in clinical stroke trial recently [109], antioxidant drugs are still potent therapeutic candidates for white matter. Imai et al. used rat transient ischemia models to evaluate the efficacy of ebselen, an antioxidant drug [58]. In this study, they showed that ebselen reduced axonal damage, and that OLG pathology was also reduced. Using rat MCAO model, Irving et al. have demonstrated that a free radical scavenger phenyl-*N*-*tert*-butyl-nitrone (PBN) reduced the number of tau-positive OLGs in the subcortical white matter of the ischemic hemisphere [61]. Lin et al. also examined the efficacy of PBN on white matter injury by hypoxia-ischemia in the neonatal rat brain [77]. In the study, the PBN treatment protected both OLGs and axons from ischemic insults. Subsequently, the same group has demonstrated that PBN also inhibited upregulation of inflammatory cytokines such as IL-1beta, TNF-alpha, and iNOS mRNA expression in the same model [76].

Finally, an ongoing NIH-funded trial is assessing minocycline for acute stroke. Minocycline is a second-generation tetracycline, which can cross the BBB [86, 118]. Minocycline has been shown to be beneficial in a wide range of acute neurological injuries. In rodent brain ischemic models, this drug showed anti-inflammatory effects, based on its ability to inhibit immune mediators such as microglia [145, 146]. Although there is no report that minocycline directly protects OLGs against ischemic stress in adult rodent stroke model, this drug was shown to attenuate hypoxia/ischemia-induced white matter injury in the neonatal rat [18, 130]. Further, Hewlett and Corbett have shown that delayed minocycline treatment reduced long-term functional deficits as well as white matter injury in ET-1-induced rat ischemia model [53]. In spite of these experimental findings, it must be noted that a recent clinical trial using minocycline in ALS patients failed to show efficacy [44]. A potential caveat with this study is the long-term use of minocycline. Among its many actions, minocycline is a powerful metalloproteinase inhibitor. It has been recently suggested that long-term suppression of metalloproteinases may be detrimental for neurovascular homeostasis [152]. In stroke, short-term applications of minocycline may still be possible. Ultimately, however, whether minocycline will be useful for white matter injury in stroke patients will have to be answered in a carefully analyzed randomized trial.

7 Conclusions

OLGs, myelin-forming glial cells in the CNS, are very vulnerable to ischemic stress, resulting in early loss of myelin and white matter dysfunction. Inhibiting axonal damage and/or OLG death, and accelerating the remyelination via OPC proliferation and differentiation may turn out to be critical for preventing acute neuronal disconnections as well as promoting repair and remodeling after stroke. Although there are no clinically validated treatments to date, several promising leads are beginning to be dissected in experimental systems. In this chapter, we tried to provide a broad but brief survey of existing models at the cell, tissue, and whole-animal levels. Many studies have productively used these model systems to dissect pathophysiology as well as assess treatment strategies. As we seek to cross the difficult translational hurdles between basic science and clinical challenges, the combined use of multiple model systems should help.

Acknowledgments and Funding Supported in part by the National Institutes of Health, the American Heart Association, and the Deane Institute. Material including adapted figures for this chapter has been extensively drawn from previously published reviews including: Lo et al., Nat Rev Neurosci 2003; Lo, Nat Med 2008; Arai et al., FEBS J 2009; Arai and Lo, Exp Transl Stroke Med 2009; Arai and Lo, FEBS J 2009.

References

1. Abbott NJ, Ronnback L, Hansson E. Astrocyte-endothelial interactions at the blood–brain barrier. Nat Rev Neurosci. 2006;7(1):41–53.
2. Alberdi E, Sanchez-Gomez MV, Torre I, Domercq M, Perez-Samartin A, Perez-Cerda F, Matute C. Activation of kainate receptors sensitizes oligodendrocytes to complement attack. J Neurosci. 2006;26(12):3220–8.
3. Albrecht PJ, Enterline JC, Cromer J, Levison SW. CNTF-activated astrocytes release a soluble trophic activity for oligodendrocyte progenitors. Neurochem Res. 2007;32(2):263–71.
4. Alix JJ. Recent biochemical advances in white matter ischaemia. Eur Neurol. 2006;56(2):74–7.
5. Arai K, Lo EH. Astrocytes protect oligodendrocyte precursor cells via MEK/ERK and PI3K/Akt signaling. J Neurosci Res. 2010;88(4):758–63.
6. Arai K, Lo EH. An oligovascular niche: cerebral endothelial cells promote the survival and proliferation of oligodendrocyte precursor cells. J Neurosci. 2009;29(14):4351–5.
7. Arvidsson A, Collin T, Kirik D, Kokaia Z, Lindvall O. Neuronal replacement from endogenous precursors in the adult brain after stroke. Nat Med. 2002;8:963–70.
8. Asahi M, Wang X, Mori T, Sumii T, Jung JC, Moskowitz MA, Fini ME, Lo EH. Effects of matrix metalloproteinase-9 gene knock-out on the proteolysis of blood–brain barrier and white matter components after cerebral ischemia. J Neurosci. 2001;21(19):7724–32.
9. Bakiri Y, Hamilton NB, Karadottir R, Attwell D. Testing NMDA receptor block as a therapeutic strategy for reducing ischaemic damage to CNS white matter. Glia. 2008;56(2):233–40.
10. Baltan S. Ischemic injury to white matter: an age-dependent process. Neuroscientist. 2009;15(2):126–33.
11. Baltan S, Besancon EF, Mbow B, Ye Z, Hamner MA, Ransom BR. White matter vulnerability to ischemic injury increases with age because of enhanced excitotoxicity. J Neurosci. 2008;28(6):1479–89.
12. Barres BA, Hart IK, Coles HS, Burne JF, Voyvodic JT, Richardson WD, Raff MC. Cell death and control of cell survival in the oligodendrocyte lineage. Cell. 1992;70(1):31–46.
13. Bartlett WP, Knapp PE, Skoff RP. Glial conditioned medium enables jimpy oligodendrocytes to express properties of normal oligodendrocytes: production of myelin antigens and membranes. Glia. 1988;1(4):253–9.
14. Belachew S, Chittajallu R, Aguirre AA, Yuan X, Kirby M, Anderson S, Gallo V. Postnatal NG2 proteoglycan-expressing progenitor cells are intrinsically multipotent and generate functional neurons. J Cell Biol. 2003;161(1):169–86.
15. Bernstein SL, Guo Y, Kelman SE, Flower RW, Johnson MA. Functional and cellular responses in a novel rodent model of anterior ischemic optic neuropathy. Invest Ophthalmol Vis Sci. 2003;44(10):4153–62.

16. Butt AM, Ibrahim M, Ruge FM, Berry M. Biochemical subtypes of oligodendrocyte in the anterior medullary velum of the rat as revealed by the monoclonal antibody. Rip. Glia. 1995;14(3):185–97.
17. Butts BD, Houde C, Mehmet H. Maturation-dependent sensitivity of oligodendrocyte lineage cells to apoptosis: implications for normal development and disease. Cell Death Differ. 2008;15(7):1178–86.
18. Carty ML, Wixey JA, Colditz PB, Buller KM. Post-insult minocycline treatment attenuates hypoxia-ischemia-induced neuroinflammation and white matter injury in the neonatal rat: a comparison of two different dose regimens. Int J Dev Neurosci. 2008;26(5):477–85.
19. Chandler S, Coates R, Gearing A, Lury J, Wells G, Bone E. Matrix metalloproteinases degrade myelin basic protein. Neurosci Lett. 1995;201(3):223–6.
20. Chang A, Nishiyama A, Peterson J, Prineas J, Trapp BD. NG2-positive oligodendrocyte progenitor cells in adult human brain and multiple sclerosis lesions. J Neurosci. 2000; 20(17):6404–12.
21. Choi DW. Ionic dependence of glutamate neurotoxicity. J Neurosci. 1987;7(2):369–79.
22. Choi DW. Glutamate neurotoxicity and diseases of the nervous system. Neuron. 1988;1(8): 623–34.
23. Choi DW. Cerebral hypoxia: some new approaches and unanswered questions. J Neurosci. 1990;10(8):2493–501.
24. Chopp M, Zhang ZG, Jiang Q. Neurogenesis, angiogenesis, and MRI indices of functional recovery from stroke. Stroke. 2007;38(2 Suppl):827–31.
25. Cioffi GA, Orgul S, Onda E, Bacon DR, Van Buskirk EM. An in vivo model of chronic optic nerve ischemia: the dose-dependent effects of endothelin-1 on the optic nerve microvasculature. Curr Eye Res. 1995;14(12):1147–53.
26. Corley SM, Ladiwala U, Besson A, Yong VW. Astrocytes attenuate oligodendrocyte death in vitro through an alpha(6) integrin-laminin-dependent mechanism. Glia. 2001;36(3): 281–94.
27. Cui QL, Almazan G. IGF-I-induced oligodendrocyte progenitor proliferation requires PI3K/ Akt, MEK/ERK, and Src-like tyrosine kinases. J Neurochem. 2007;100(6):1480–93.
28. Cui QL, Zheng WH, Quirion R, Almazan G. Inhibition of Src-like kinases reveals Akt-dependent and -independent pathways in insulin-like growth factor I-mediated oligodendrocyte progenitor survival. J Biol Chem. 2005;280(10):8918–28.
29. del Zoppo GJ. Stroke and neurovascular protection. N Engl J Med. 2006;354(6):553–5.
30. Deng W, Rosenberg PA, Volpe JJ, Jensen FE. Calcium-permeable AMPA/kainate receptors mediate toxicity and preconditioning by oxygen-glucose deprivation in oligodendrocyte precursors. Proc Natl Acad Sci U S A. 2003;100(11):6801–6.
31. Dewar D, Dawson DA. Changes of cytoskeletal protein immunostaining in myelinated fibre tracts after focal cerebral ischaemia in the rat. Acta Neuropathol. 1997;93(1):71–7.
32. Dos Santos CD, Picanco-Diniz CW, Gomes-Leal W. Differential patterns of inflammatory response, axonal damage and myelin impairment following excitotoxic or ischemic damage to the trigeminal spinal nucleus of adult rats. Brain Res. 2007;1172:130–44.
33. Du Y, Dreyfus CF. Oligodendrocytes as providers of growth factors. J Neurosci Res. 2002; 68(6):647–54.
34. Dugas JC, Mandemakers W, Rogers M, Ibrahim A, Daneman R, Barres BA. A novel purification method for CNS projection neurons leads to the identification of brain vascular cells as a source of trophic support for corticospinal motor neurons. J Neurosci. 2008;28(33): 8294–305.
35. Esiri MM. The interplay between inflammation and neurodegeneration in CNS disease. J Neuroimmunol. 2007;184(1–2):4–16.
36. Farkas E, Donka G, de Vos RA, Mihaly A, Bari F, Luiten PG. Experimental cerebral hypoperfusion induces white matter injury and microglial activation in the rat brain. Acta Neuropathol. 2004;108(1):57–64.
37. Farkas E, Luiten PG. Cerebral microvascular pathology in aging and Alzheimer's disease. Prog Neurobiol. 2001;64(6):575–611.

38. Farkas E, Luiten PG, Bari F. Permanent, bilateral common carotid artery occlusion in the rat: a model for chronic cerebral hypoperfusion-related neurodegenerative diseases. Brain Res Rev. 2007;54(1):162–80.

39. Flores AI, Mallon BS, Matsui T, Ogawa W, Rosenzweig A, Okamoto T, Macklin WB. Akt-mediated survival of oligodendrocytes induced by neuregulins. J Neurosci. 2000;20(20): 7622–30.

40. Frost SB, Barbay S, Mumert ML, Stowe AM, Nudo RJ. An animal model of capsular infarct: endothelin-1 injections in the rat. Behav Brain Res. 2006;169(2):206–11.

41. Gard AL, Burrell MR, Pfeiffer SE, Rudge JS, Williams II WC. Astroglial control of oligo-dendrocyte survival mediated by PDGF and leukemia inhibitory factor-like protein. Development. 1995;121(7):2187–97.

42. Gladstone DJ, Black SE, Hakim AM. Toward wisdom from failure: lessons from neuropro-tective stroke trials and new therapeutic directions. Stroke. 2002;33(8):2123–36.

43. Goldenberg-Cohen N, Guo Y, Margolis F, Cohen Y, Miller NR, Bernstein SL. Oligodendrocyte dysfunction after induction of experimental anterior optic nerve ischemia. Invest Ophthalmol Vis Sci. 2005;46(8):2716–25.

44. Gordon PH, Moore DH, Miller RG, Florence JM, Verheijde JL, Doorish C, Hilton JF, Spitalny GM, MacArthur RB, Mitsumoto H, et al. Efficacy of minocycline in patients with amyo-trophic lateral sclerosis: a phase III randomised trial. Lancet Neurol. 2007;6(12):1045–53.

45. Greenberg DA, Jin K. From angiogenesis to neuropathology. Nature. 2005;438(7070): 954–9.

46. Gregersen R, Christensen T, Lehrmann E, Diemer NH, Finsen B. Focal cerebral ischemia induces increased myelin basic protein and growth-associated protein-43 gene transcription in peri-infarct areas in the rat brain. Exp Brain Res. 2001;138(3):384–92.

47. Griffiths I, Klugmann M, Anderson T, Thomson C, Vouyiouklis D, Nave KA. Current con-cepts of PLP and its role in the nervous system. Microsc Res Tech. 1998;41(5):344–58.

48. Guo S, Kim WJ, Lok J, Lee SR, Besancon E, Luo BH, Stins MF, Wang X, Dedhar S, Lo EH. Neuroprotection via matrix-trophic coupling between cerebral endothelial cells and neurons. Proc Natl Acad Sci U S A. 2008;105(21):7582–7.

49. Guo S, Lo EH. Dysfunctional cell-cell signaling in the neurovascular unit as a paradigm for central nervous system disease. Stroke. 2009;40(3 Suppl):S4–7.

50. Hainsworth AH, Markus HS. Do in vivo experimental models reflect human cerebral small vessel disease? A systematic review. J Cereb Blood Flow Metab. 2008;28(12):1877–91.

51. Hawkins BT, Davis TP. The blood–brain barrier/neurovascular unit in health and disease. Pharmacol Rev. 2005;57(2):173–85.

52. Hewitt KE, Stys PK, Lesiuk HJ. The use-dependent sodium channel blocker mexiletine is neuroprotective against global ischemic injury. Brain Res. 2001;898(2):281–7.

53. Hewlett KA, Corbett D. Delayed minocycline treatment reduces long-term functional deficits and histological injury in a rodent model of focal ischemia. Neuroscience. 2006;141(1): 27–33.

54. Hughes PM, Anthony DC, Ruddin M, Botham MS, Rankine EL, Sablone M, Baumann D, Mir AK, Perry VH. Focal lesions in the rat central nervous system induced by endothelin-1. J Neuropathol Exp Neurol. 2003;62(12):1276–86.

55. Iadecola C. Neurovascular regulation in the normal brain and in Alzheimer's disease. Nat Rev Neurosci. 2004;5(5):347–60.

56. Iadecola C, Nedergaard M. Glial regulation of the cerebral microvasculature. Nat Neurosci. 2007;10(11):1369–76.

57. Ikonomidou C, Turski L. Why did NMDA receptor antagonists fail clinical trials for stroke and traumatic brain injury? Lancet Neurol. 2002;1(6):383–6.

58. Imai H, Masayasu H, Dewar D, Graham DI, Macrae IM. Ebselen protects both gray and white matter in a rodent model of focal cerebral ischemia. Stroke. 2001;32(9):2149–54.

59. Imaizumi T, Kocsis JD, Waxman SG. Anoxic injury in the rat spinal cord: pharmacological evidence for multiple steps in Ca(2+)-dependent injury of the dorsal columns. J Neurotrauma. 1997;14(5):299–311.

60. Irving EA, Bentley DL, Parsons AA. Assessment of white matter injury following prolonged focal cerebral ischaemia in the rat. Acta Neuropathol. 2001;102(6):627–35.
61. Irving EA, Yatsushiro K, McCulloch J, Dewar D. Rapid alteration of tau in oligodendrocytes after focal ischemic injury in the rat: involvement of free radicals. J Cereb Blood Flow Metab. 1997;17(6):612–22.
62. James G, Butt AM. P2X and P2Y purinoreceptors mediate ATP-evoked calcium signalling in optic nerve glia in situ. Cell Calcium. 2001;30(4):251–9.
63. Johnson EC, Deppmeier LM, Wentzien SK, Hsu I, Morrison JC. Chronology of optic nerve head and retinal responses to elevated intraocular pressure. Invest Ophthalmol Vis Sci. 2000;41(2):431–42.
64. Karadottir R, Cavelier P, Bergersen LH, Attwell D. NMDA receptors are expressed in oligo-dendrocytes and activated in ischaemia. Nature. 2005;438(7071):1162–6.
65. Kawamura S, Yasui N, Shirasawa M, Fukasawa H. Rat middle cerebral artery occlusion using an intraluminal thread technique. Acta Neurochir (Wien). 1991;109(3–4):126–32.
66. Kennedy TP, Vinten-Johansen J. A review of the clinical use of anti-inflammatory therapies for reperfusion injury in myocardial infarction and stroke: where do we go from here? Curr Opin Investig Drugs. 2006;7(3):229–42.
67. Knottnerus IL, Ten Cate H, Lodder J, Kessels F, van Oostenbrugge RJ. Endothelial dysfunc-tion in lacunar stroke: a systematic review. Cerebrovasc Dis. 2009;27(5):519–26.
68. Knox DL, Kerrison JB, Green WR. Histopathologic studies of ischemic optic neuropathy. Trans Am Ophthalmol Soc. 2000;98:203–20; discussion 221–2.
69. Komitova M, Perfilieva E, Mattsson B, Eriksson PS, Johansson BB. Enriched environment after focal cortical ischemia enhances the generation of astroglia and NG2 positive polyden-drocytes in adult rat neocortex. Exp Neurol. 2006;199(1):113–21.
70. Krueger-Naug AM, Emsley JG, Myers TL, Currie RW, Clarke DB. Injury to retinal ganglion cells induces expression of the small heat shock protein Hsp27 in the rat visual system. Neuroscience. 2002;110(4):653–65.
71. Le Feuvre RA, Brough D, Touzani O, Rothwell NJ. Role of P2X7 receptors in ischemic and excitotoxic brain injury in vivo. J Cereb Blood Flow Metab. 2003;23(3):381–4.
72. Lee JM, Zipfel GJ, Choi DW. The changing landscape of ischaemic brain injury mechanisms. Nature. 1999;399(6738 Suppl):A7–14.
73. Lee SR, Kim HY, Rogowska J, Zhao BQ, Bhide P, Parent JM, Lo EH. Involvement of matrix metalloproteinase in neuroblast cell migration from the subventricular zone after stroke. J Neurosci. 2006;26(13):3491–5.
74. Leventhal C, Rafii S, Rafii D, Shahar A, Goldman SA. Endothelial trophic support of neu-ronal production and recruitment from the adult mammalian subependyma. Mol Cell Neurosci. 1999;13(6):450–64.
75. Levine JM, Reynolds R, Fawcett JW. The oligodendrocyte precursor cell in health and dis-ease. Trends Neurosci. 2001;24(1):39–47.
76. Lin S, Cox HJ, Rhodes PG, Cai Z. Neuroprotection of alpha-phenyl-n-tert-butyl-nitrone on the neonatal white matter is associated with anti-inflammation. Neurosci Lett. 2006;405(1–2):52–6.
77. Lin S, Rhodes PG, Lei M, Zhang F, Cai Z. Alpha-phenyl-n-tert-butyl-nitrone attenuates hypoxic-ischemic white matter injury in the neonatal rat brain. Brain Res. 2004;1007(1–2):132–41.
78. Lipton SA. NMDA receptors, glial cells, and clinical medicine. Neuron. 2006;50(1):9–11.
79. Lipton SA, Rosenberg PA. Excitatory amino acids as a final common pathway for neurologic disorders. N Engl J Med. 1994;330(9):613–22.
80. Lo EH. A new penumbra: transitioning from injury into repair after stroke. Nat Med. 2008;14(5):497–500.
81. Lo EH, Dalkara T, Moskowitz MA. Mechanisms, challenges and opportunities in stroke. Nat Rev Neurosci. 2003;4(5):399–415.
82. Lok J, Gupta P, Guo S, Kim WJ, Whalen MJ, van Leyen K, Lo EH. Cell-cell signaling in the neurovascular unit. Neurochem Res. 2007;32(12):2032–45.

83. Longa EZ, Weinstein PR, Carlson S, Cummins R. Reversible middle cerebral artery occlusion without craniectomy in rats. Stroke. 1989;20:84–91.
84. Lopes-Cardozo M, Sykes JE, Van der Pal RH, van Golde LM. Development of oligodendrocytes. Studies of rat glial cells cultured in chemically-defined medium. J Dev Physiol. 1989;12(3):117–27.
85. Louissaint Jr A, Rao S, Leventhal C, Goldman SA. Coordinated interaction of neurogenesis and angiogenesis in the adult songbird brain. Neuron. 2002;34(6):945–60.
86. Macdonald H, Kelly RG, Allen ES, Noble JF, Kanegis LA. Pharmacokinetic studies on minocycline in man. Clin Pharmacol Ther. 1973;14(5):852–61.
87. Mandai K, Matsumoto M, Kitagawa K, Matsushita K, Ohtsuki T, Mabuchi T, Colman DR, Kamada T, Yanagihara T. Ischemic damage and subsequent proliferation of oligodendrocytes in focal cerebral ischemia. Neuroscience. 1997;77(3):849–61.
88. Manning SM, Talos DM, Zhou C, Selip DB, Park HK, Park CJ, Volpe JJ, Jensen FE. NMDA receptor blockade with memantine attenuates white matter injury in a rat model of periventricular leukomalacia. J Neurosci. 2008;28(26):6670–8.
89. Masaki T, Yanagisawa M. Endothelins. Essays Biochem. 1992;27:79–89.
90. Matute C, Alberdi E, Domercq M, Sanchez-Gomez MV, Perez-Samartin A, Rodriguez-Antiguedad A, Perez-Cerda F. Excitotoxic damage to white matter. J Anat. 2007;210(6):693–702.
91. McCarthy KD, de Vellis J. Preparation of separate astroglial and oligodendroglial cell cultures from rat cerebral tissue. J Cell Biol. 1980;85(3):890–902.
92. Medana IM, Esiri MM. Axonal damage: a key predictor of outcome in human CNS diseases. Brain. 2003;126(Pt 3):515–30.
93. Merrill JE, Benveniste EN. Cytokines in inflammatory brain lesions: helpful and harmful. Trends Neurosci. 1996;19(8):331–8.
94. Micu I, Jiang Q, Coderre E, Ridsdale A, Zhang L, Woulfe J, Yin X, Trapp BD, McRory JE, Rehak R, et al. NMDA receptors mediate calcium accumulation in myelin during chemical ischaemia. Nature. 2006;439(7079):988–92.
95. Miller RH. Oligodendrocyte origins. Trends Neurosci. 1996;19(3):92–6.
96. Motte S, McEntee K, Naeije R. Endothelin receptor antagonists. Pharmacol Ther. 2006;110(3):386–414.
97. Nakaji K, Ihara M, Takahashi C, Itohara S, Noda M, Takahashi R, Tomimoto H. Matrix metalloproteinase-2 plays a critical role in the pathogenesis of white matter lesions after chronic cerebral hypoperfusion in rodents. Stroke. 2006;37(11):2816–23.
98. Nave KA, Trapp BD. Axon-glial signaling and the glial support of axon function. Annu Rev Neurosci. 2008;31:535–61.
99. Nishiyama A. NG2 cells in the brain: a novel glial cell population. Hum Cell. 2001;14(1):77–82.
100. Nishiyama A, Chang A, Trapp BD. NG2+ glial cells: a novel glial cell population in the adult brain. J Neuropathol Exp Neurol. 1999;58(11):1113–24.
101. North RA. Molecular physiology of P2X receptors. Physiol Rev. 2002;82(4):1013–67.
102. Oka A, Belliveau MJ, Rosenberg PA, Volpe JJ. Vulnerability of oligodendroglia to glutamate: pharmacology, mechanisms, and prevention. J Neurosci. 1993;13(4):1441–53.
103. Orthmann-Murphy JL, Abrams CK, Scherer SS. Gap junctions couple astrocytes and oligodendrocytes. J Mol Neurosci. 2008;35(1):101–16.
104. Pang Y, Cai Z, Rhodes PG. Effects of lipopolysaccharide on oligodendrocyte progenitor cells are mediated by astrocytes and microglia. J Neurosci Res. 2000;62(4):510–20.
105. Pantoni L, Garcia JH. Pathogenesis of leukoaraiosis: a review. Stroke. 1997;28(3):652–9.
106. Pantoni L, Garcia JH, Gutierrez JA. Cerebral white matter is highly vulnerable to ischemia. Stroke. 1996;27(9):1641–6; discussion 1647.
107. Parent JM, Vexler ZS, Gong C, Derugin N, Ferriero DM. Rat forebrain neurogenesis and striatal neuron replacement after focal stroke. Ann Neurol. 2002;52:802–13.
108. Pfeiffer SE, Warrington AE, Bansal R. The oligodendrocyte and its many cellular processes. Trends Cell Biol. 1993;3(6):191–7.
109. Proctor PH, Tamborello LP. SAINT-I worked, but the neuroprotectant is not NXY-059. Stroke. 2007;38(10):e109; author reply e110.

110. Raber J, Fan Y, Matsumori Y, Liu Z, Weinstein PR, Fike JR, Liu J. Irradiation attenuates neurogenesis and exacerbates ischemia-induced deficits. Ann Neurol. 2004;55:381–9.
111. Raff MC, Lillien LE, Richardson WD, Burne JF, Noble MD. Platelet-derived growth factor from astrocytes drives the clock that times oligodendrocyte development in culture. Nature. 1988;333(6173):562–5.
112. Raff MC, Miller RH, Noble M. A glial progenitor cell that develops in vitro into an astrocyte or an oligodendrocyte depending on culture medium. Nature. 1983;303(5916):390–6.
113. Ralevic V, Burnstock G. Receptors for purines and pyrimidines. Pharmacol Rev. 1998; 50(3):413–92.
114. Rosenberg GA, Sullivan N, Esiri MM. White matter damage is associated with matrix metalloproteinases in vascular dementia. Stroke. 2001;32(5):1162–8.
115. Rosenberg SS, Ng BK, Chan JR. The quest for remyelination: a new role for neurotrophins and their receptors. Brain Pathol. 2006;16(4):288–94.
116. Rubanyi GM, Polokoff MA. Endothelins: molecular biology, biochemistry, pharmacology, physiology, and pathophysiology. Pharmacol Rev. 1994;46(3):325–415.
117. Rubio N, Rodriguez R, Arevalo MA. In vitro myelination by oligodendrocyte precursor cells transfected with the neurotrophin-3 gene. Glia. 2004;47(1):78–87.
118. Saivin S, Houin G. Clinical pharmacokinetics of doxycycline and minocycline. Clin Pharmacokinet. 1988;15(6):355–66.
119. Salter MG, Fern R. NMDA receptors are expressed in developing oligodendrocyte processes and mediate injury. Nature. 2005;438(7071):1167–71.
120. Sanchez-Gomez MV, Alberdi E, Ibarretxe G, Torre I, Matute C. Caspase-dependent and caspase-independent oligodendrocyte death mediated by AMPA and kainate receptors. J Neurosci. 2003;23(29):9519–28.
121. Sarti C, Pantoni L, Bartolini L, Inzitari D. Cognitive impairment and chronic cerebral hypoperfusion: what can be learned from experimental models. J Neurol Sci. 2002;203–204:263–6.
122. Schabitz WR, Li F, Fisher M. The N-methyl-D-aspartate antagonist CNS 1102 protects cerebral gray and white matter from ischemic injury following temporary focal ischemia in rats. Stroke. 2000;31(7):1709–14.
123. Schmidt R, Scheltens P, Erkinjuntti T, Pantoni L, Markus HS, Wallin A, Barkhof F, Fazekas F. White matter lesion progression: a surrogate endpoint for trials in cerebral small-vessel disease. Neurology. 2004;63(1):139–44.
124. Selles-Navarro I, Ellezam B, Fajardo R, Latour M, McKerracher L. Retinal ganglion cell and nonneuronal cell responses to a microcrush lesion of adult rat optic nerve. Exp Neurol. 2001;167(2):282–9.
125. Shen Q, Goderie SK, Jin L, Karanth N, Sun Y, Abramova N, Vincent P, Pumiglia K, Temple S. Endothelial cells stimulate self-renewal and expand neurogenesis of neural stem cells. Science. 2004;304(5675):1338–40.
126. Shibata M, Ohtani R, Ihara M, Tomimoto H. White matter lesions and glial activation in a novel mouse model of chronic cerebral hypoperfusion. Stroke. 2004;35(11):2598–603.
127. Shibata M, Yamasaki N, Miyakawa T, Kalaria RN, Fujita Y, Ohtani R, Ihara M, Takahashi R, Tomimoto H. Selective impairment of working memory in a mouse model of chronic cerebral hypoperfusion. Stroke. 2007;38(10):2826–32.
128. Souza-Rodrigues RD, Costa AM, Lima RR, Dos Santos CD, Picanco-Diniz CW, Gomes-Leal W. Inflammatory response and white matter damage after microinjections of endothelin-1 into the rat striatum. Brain Res. 2008;1200:78–88.
129. Sozmen EG, Kolekar A, Havton LA, Carmichael ST. A white matter stroke model in the mouse: axonal damage, progenitor responses and MRI correlates. J Neurosci Methods. 2009;180(2):261–72.
130. Stolp HB, Ek CJ, Johansson PA, Dziegielewska KM, Potter AM, Habgood MD, Saunders NR. Effect of minocycline on inflammation-induced damage to the blood–brain barrier and white matter during development. Eur J Neurosci. 2007;26(12):3465–74.
131. Stys PK. White matter injury mechanisms. Curr Mol Med. 2004;4(2):113–30.
132. Stys PK, Jiang Q. Calpain-dependent neurofilament breakdown in anoxic and ischemic rat central axons. Neurosci Lett. 2002;328(2):150–4.

133. Stys PK, Ransom BR, Waxman SG. Tertiary and quaternary local anesthetics protect CNS white matter from anoxic injury at concentrations that do not block excitability. J Neurophysiol. 1992;67(1):236–40.

134. Takahashi JL, Giuliani F, Power C, Imai Y, Yong VW. Interleukin-1beta promotes oligodendrocyte death through glutamate excitotoxicity. Ann Neurol. 2003;53(5):588–95.

135. Tamura A, Graham DI, McCulloch J, Teasdale GM. Focal cerebral ischemia in the rat: description of technique and early neuropathological consequences following middle cerebral artery occlusion. J Cereb Blood Flow Metab. 1981;1:53–60.

136. Tanaka K, Nogawa S, Suzuki S, Dembo T, Kosakai A. Upregulation of oligodendrocyte progenitor cells associated with restoration of mature oligodendrocytes and myelination in peri-infarct area in the rat brain. Brain Res. 2003;989(2):172–9.

137. Ueno Y, Zhang N, Miyamoto N. Edaravone attenuates white matter lesions through endothelial protection in a rat chronic hypoperfusion model. Neuroscience. 2009;162(2):317–27.

138. Valeriani V, Dewar D, McCulloch J. Quantitative assessment of ischemic pathology in axons, oligodendrocytes, and neurons: attenuation of damage after transient ischemia. J Cereb Blood Flow Metab. 2000;20(5):765–71.

139. Volpe JJ. Cerebral white matter injury of the premature infant-more common than you think. Pediatrics. 2003;112(1 Pt 1):176–80.

140. Wahlgren NG, Ahmed N. Neuroprotection in cerebral ischaemia: facts and fancies—the need for new approaches. Cerebrovasc Dis. 2004;17 Suppl 1:153–66.

141. Wakita H, Tomimoto H, Akiguchi I, Matsuo A, Lin JX, Ihara M, McGeer PL. Axonal damage and demyelination in the white matter after chronic cerebral hypoperfusion in the rat. Brain Res. 2002;924(1):63–70.

142. Wang CX, Shuaib A. Neuroprotective effects of free radical scavengers in stroke. Drugs Aging. 2007;24(7):537–46.

143. Wang X, Arcuino G, Takano T, Lin J, Peng WG, Wan P, Li P, Xu Q, Liu QS, Goldman SA, et al. P2X7 receptor inhibition improves recovery after spinal cord injury. Nat Med. 2004; 10(8):821–7.

144. Yonezawa M, Back SA, Gan X, Rosenberg PA, Volpe JJ. Cystine deprivation induces oligodendroglial death: rescue by free radical scavengers and by a diffusible glial factor. J Neurochem. 1996;67(2):566–73.

145. Yrjanheikki J, Keinanen R, Pellikka M, Hokfelt T, Koistinaho J. Tetracyclines inhibit microglial activation and are neuroprotective in global brain ischemia. Proc Natl Acad Sci U S A. 1998;95(26):15769–74.

146. Yrjanheikki J, Tikka T, Keinanen R, Goldsteins G, Chan PH, Koistinaho J. A tetracycline derivative, minocycline, reduces inflammation and protects against focal cerebral ischemia with a wide therapeutic window. Proc Natl Acad Sci U S A. 1999;96(23):13496–500.

147. Zacchigna S, Lambrechts D, Carmeliet P. Neurovascular signalling defects in neurodegeneration. Nat Rev Neurosci. 2008;9(3):169–81.

148. Zhang J, Li Y, Zheng X, Gao Q, Liu Z, Qu R, Borneman J, Elias SB, Chopp M. Bone marrow stromal cells protect oligodendrocytes from oxygen-glucose deprivation injury. J Neurosci Res. 2008;86(7):1501–10.

149. Zhang K, Sejnowski TJ. A universal scaling law between gray matter and white matter of cerebral cortex. Proc Natl Acad Sci U S A. 2000;97(10):5621–6.

150. Zhang Z, Chopp M, Zhang RL, Goussev A. A mouse model of embolic focal cerebral ischemia. J Cereb Blood Flow Metab. 1997;17:1081–8.

151. Zhang ZG, Zhang RL, Jiang Q, Raman SB, Cantwell L, Chopp M. A new rat model of thrombotic focal cerebral ischemia. J Cereb Blood Flow Metab. 1997;17:123–35.

152. Zhao BQ, Wang S, Kim HY, Storrie H, Rosen BR, Mooney DJ, Wang X, Lo EH. Role of matrix metalloproteinases in delayed cortical responses after stroke. Nat Med. 2006; 12(4):441–5.

153. Zhu X, Bergles DE, Nishiyama A. NG2 cells generate both oligodendrocytes and gray matter astrocytes. Development. 2008;135(1):145–57.

154. Zlokovic BV. The blood–brain barrier in health and chronic neurodegenerative disorders. Neuron. 2008;57(2):178–201.

Chapter 4
Neuroprotection in Stroke

Aarti Sarwal, Muhammad Shazam Hussain, and Ashfaq Shuaib

Abstract Stroke is a devastating disease affecting 15 million people worldwide and is the leading cause of adult disability (Circulation 121(7):e46–e215, 2010). Even where advanced technology and facilities are available, 60% of those who suffer a stroke die or become dependent (N Engl J Med 333(24):1581–1587, 1995). The introduction of thrombolytic agents and evolution of interventional strategies for recanalization have revolutionized the acute management of stroke (N Engl J Med 333(24):1581–1587, 1995). Despite this, a large number of strokes escape any acute intervention due to various contraindications or delay in presentation of the patient (Neurology 56(8):1015–1020, 2001). Even for patients who undergo acute therapy, ongoing secondary injury as a result of natural progression of the disease may cause progressive damage despite acute recanalization. This makes it pertinent to look for strategies to salvage all possible brain tissue irrespective of the use or success of acute stroke intervention. A variety of novel strategies to this effect are being investigated. Most of these originate from new knowledge of mechanisms of neuronal cell death and the new found concept of brain plasticity (Neuron 67(2):181–198, 2010). Once considered exclusively a disorder of blood vessels, growing evidence has led to the realization that the biological processes underlying stroke are driven by the interaction of neurons, glia, vascular cells, and matrix components, which actively participate in mechanisms of tissue injury and repair. Yet despite success in animal and in vitro experiments, none of these strategies have yet been able to be applied to bedside clinical practice (Neuron 67(2):181–198, 2010;

A. Sarwal
Wake Forest School of Medicine, Reynolds M, Medical Center Blvd,
Winston Salem, NC 27157, USA

M.S. Hussain
Cerebrovascular Center, Cleveland Clinic, 9500 Euclid Ave, S80, Cleveland, OH 44106, USA

A. Shuaib (✉)
Division of Neurology, 2E3.13 Walter C. Mackenzie Health Sciences Center,
University of Alberta, 8440-112th Street, Edmonton, AB, Canada T6G 2B7
e-mail: ashfaq.shuaib@ualberta.ca

P.A. Lapchak and J.H. Zhang (eds.), *Translational Stroke Research*, Springer Series
in Translational Stroke Research, DOI 10.1007/978-1-4419-9530-8_4,
© Springer Science+Business Media, LLC 2012

Ann Neurol 59(3):467–477, 2006). This chapter aims to describe the reasons why agents that have shown promise in animal models may have negative results in clinical trials. This is discussed in context of our current understanding of the pathophysiology of cerebral ischemia, the ischemic cascade, and neuroprotection. Finally, the future of neuroprotection in acute ischemic stroke considering the lesions learnt from previous trials will be explored.

1 Cerebral Ischemia: Modes of Cell Injury and the Ischemic Cascade

Our brain is especially vulnerable to ischemic insult because of its high intrinsic metabolic activity and large concentrations of the neurotransmitter excitotoxins such as glutamate [1]. Ischemia develops as a consequence of acute thrombotic or embolic occlusion of a cerebral blood vessel. Within seconds to minutes after the loss of blood flow, a series of biochemical events is initiated that ultimately leads to cell death. Neurological dysfunction occurs within seconds to minutes of vessel occlusion, but the evolution of ischemic injury and cell death continues in stages for minutes, hours, and even days, depending upon the vulnerability of the particular brain region, its cellular constituents, and the extent of residual perfusion [2]. This sequence of events is called the ischemic cascade.

Following blockage of a cerebral blood vessel, two areas of damage are present: the ischemic core and the ischemic penumbra. The ischemic core refers to the irreversibly damaged tissue distal to an occluded blood vessel, characterized by <20% of baseline blood flow levels or below 12 mL/100 g of brain tissue/min [3]. In this region, depleted ATP stores, disruption of ion homeostasis, irreversible failure of energy metabolism, and loss of cellular integrity lead to cell death and necrosis within minutes [4]. A second meta-stable zone, termed the ischemic penumbra or peri-infarct zone [5], is characterized by reduced tissue perfusion (approximately 12–22 mL/100 g of brain tissue/min) [3], barely sufficient to support aerobic metabolism and normal ionic gradients. This zone is heterogeneous, meta-stable, and differentially susceptible to ischemic injury. It denotes the "at risk" region—functionally impaired but potentially salvageable. However, unless perfusion is improved or cells made relatively more resistant to injury, the infarct core expands into the ischemic penumbra, and therapeutic opportunity is lost [2, 4]. The penumbra has been extensively studied in recent years using pathological, electrophysiological, biochemical, radiological, and noninvasive imaging techniques [3]. The ischemic cascade (Fig. 4.1) is the process which occurs in the ischemic penumbra and results in the tissue damage.

Different compartments in brain are affected by ischemia. There are various modes of cell injury which ultimately combine together resulting in neuronal cell death. Within the ischemic penumbra, multiple mechanisms have been identified over the past few decades that irreversibly damage brain tissue. Each of these mechanisms is a potential therapeutic opportunity and treatment target. Ischemic cell

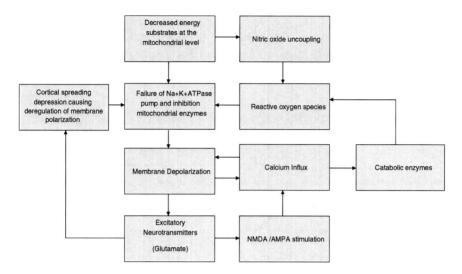

Fig. 4.1 Ischemic cascade: simplified diagram highlighting the intermingling events that ultimately lead to cell death [2]

death develops as a consequence of numerous ionic, biochemical, and cellular events that impose overwhelming stresses upon already compromised tissue [4, 10, 11]. With reduced cerebral blood flow, oxygen and energy substrates decrease, resulting in failure of the sodium-potassium ATPase [4]. This leads to membrane depolarization, with the release of excitatory neurotransmitters, such as glutamate. Glutamate, the most abundant neurotransmitter, accumulates into the extracellular space as a consequence of energy and ion pump failure, as well as failure of reuptake mechanisms [6]. This leads to prolonged stimulation of alpha-amino-3-hydroxy-5-methyl-4-isoxasole-propionate (AMPA) receptors and N-methyl-D-aspartic acid (NMDA) ionotropic receptor subtypes to dramatically enhance the influx of calcium, sodium, and water into neurons. This calcium influx causes further membrane depolarization, leading to further increases in glutamate. Mitochondrial calcium loading plays a critical role in cell viability after initial insult [2]. Excitotoxicity and calcium dysregulation are major factors contributing to the early stages of ischemic cell death [2]. Massive calcium influx activates catabolic processes mediated by proteases, lipases, and nucleases [7].

Activation of several Ca^{2+}-dependent enzymes (like NADPH oxidase) leads to production of NO, arachidonic acid metabolites, and superoxide. These oxidative and nitrosative stresses act as additional triggers of cell death [8, 9]. Oxidative phosphorylation becomes uncoupled, leading to further ATP depletion and reactive oxygen species production with further release of stored Ca^{2+} from mitochondria. These processes further accelerate a series of catastrophic events that lead to acute cell death. Other channels and ion pumps including Na^+/Ca^{2+} exchanger [10], hemichannels [11], acid-sensing ion channels [12], volume-regulated anion channels [13], and TRP channels [14] are activated during ischemia and have been implicated in the Ca^{2+} accumulation. In particular, ASIC1a, activated by ischemia-induced acidosis, is involved in the Ca^{2+} influx [12]. ASICs are stimulated within the pH range commonly found in ischemic brain tissue, thus explaining the well-established link between acidosis and worsening of ischemic outcome in animals and humans. Prostaglandin E2 EP1 receptors have a role in the failure of the Na^+/Ca^{2+} exchanger during ischemia [15].

Cerebral ischemia-reperfusion injury is also associated with an inflammatory response with contributions from various leucocytes and microglia. Focal ischemia evokes a robust inflammatory response that begins within a few hours of onset and typifies the secondary or delayed response to ischemia. Lipopeptides, advanced glycation end-products (AGE), modified lipids, heat shock proteins, hyaluronic acid, and the nuclear protein HMGB1 lead to activation of the innate immune system [16]. Bcl-2 family proteins are key regulators of apoptosis acting to inhibit or promote cell death. Caspases are also mediators of cell death. The net result of these processes is cell death [2].

In addition to excitotoxic events at the molecular and cellular level, massive release of glutamate and ionic imbalance negatively impact the evolution of ischemic injury at the tissue level. Cortical spreading depolarization (CSD), a high energy consuming phenomenon akin to the cortical spreading depression of Leao, is triggered by high levels of extracellular glutamate and K^+ and is characterized by slowly propagating massive depolarization of neurons and astrocytes along with drastic disruption of ionic gradients [17]. In experimental animals, a single CSD expands the area of severe hypoperfusion by more than 20% [18]. The infarct expansion is probably due to mismatch between high metabolic demand to support membrane repolarization and marginal tissue perfusion due to constrained penumbral blood flow [19].

The size of penumbra and its progression to infarction is also linked to the extent of collateral circulation, brain temperature, and blood glucose levels degree of acidosis in the ischemic brain tissue [20]. The fact that ischemic cascade becomes activated soon after an ischemic stroke but cell death occurs in stages over hours to days makes a strong argument in favor of ongoing acute intervention to minimize damage from acute ischemic stroke. Delaying progression of the ischemic cascade and keeping neurons viable are the targets of neuroprotective therapy.

2 Neuroprotection Strategies: Translation of Available Research

Neuroprotection strategies aimed at preserving viable cells in ischemic penumbra have been the target of new research in acute ischemic stroke for the past few years. Limiting the area and impact of injury to neuronal cells in the ischemic penumbra should improve neurological recovery and reduce disability from stroke. Immediate access to acute stroke intervention remains the paramount therapeutic goal, but pairing acute recanalization strategies with neuroprotective treatments acting at one or several steps in the ischemic cascade may have the greatest chance of clinical success. A number of approaches have been taken to neuroprotection. These include calcium channel blockade [21], calcium chelation [22], free-radical trapping, gamma-aminobutyric acid (GABA) and serotonin agonism [23], and AMPA and NMDA receptor antagonism [24]. A large number of neuroprotective agents targeting these mechanisms have been shown to successfully reduce infarct size in animal models of focal and global ischemia [25]. Unfortunately, neuroprotective compounds administered as monotherapy have failed to demonstrate efficacy for the treatment of acute ischemic stroke in Phase III clinical trials in humans. A brief listing of some of the trials highlighting their mechanism of action is charted in Table 4.1.

Table 4.1 Summary of failed trials of neuroprotective agents

Mechanism	Product	Outcome	Comments
Calcium channel blocker	Nimodipine [64]	Mixed effects on outcome. Not approved	Blood pressure affects outcome
Sodium channel blocker	Fosphenytoin [65]	Phase III trial halted due to lack of efficacy	
Potassium channel opener	BMS-204352 [66]	Phase III trial failed. A second trial is being considered	
NMDA antagonist	Selfotel [46]	Preliminary results showed no efficacy in Phase III trials	Poorly tolerated with a potential neurotoxic effect in brain ischemia
NMDA polyamine site blocker	Eliprodil [67]	Phase III trials abandoned	Hinders neuronal survival
NMDA receptor antagonist	Aptiganel [24]	Preliminary results showed no efficacy in Phase III trials	May have detrimental effects in an undifferentiated population of stroke patients
Antagonist at the glycine site of the NMDA receptor	Gavestinel [47]	Preliminary results showed no efficacy in Phase III trials	
Lipid peroxidation inhibitor	Tirilazad [68]	Review of six trials showed that it worsened outcome	Worked in reperfusion models only
Sodium channel blocker and nitric oxide blocker	Lubeluzole [22]	No efficacy in Phase III trial	May be associated with a significant increase in heart conductance disorders
Murine anti-ICAM-1 antibody	Enlimomab [69]	Worsened outcome	Immunogenicity adverse events

(continued)

Table 4.1 (continued)

Mechanism	Product	Outcome	Comments
Neutrophil inhibitory factor	UK-279,276 [70]	Unsuccessful Phase II trials	Worked in reperfusion models only
Growth factor	Trafermin [71]	Phase II/III trials halted due to lack of efficacy	
GABAA receptor modulator	Clomethiazole [72]	Phase III trials halted due to lack of efficacy	Administered up to 12 h following onset of stroke
5-HT1A receptor agonist	Repinotan [23]	Development halted due to disappointing results in Phase IIb trial	
Astrocyte-modulating agent	ONO-2506 [73]	Unfavorable interim analysis of Phase II trial in the US	Phase II/III study currently ongoing in Japan
Stroke-Acute Ischemic NXY Treatment Trial (SAINT I)	NXY-059 [28, 29]	SAINT I—Phase III trial with 1,699 patients; positive result for improvement on mRS	Preclinical data may have overestimated the effect. Positive SAINT I likely a
Stroke-Acute Ischemic NXY Treatment Trial (SAINT II)		SAINT II—larger 3,306 patients; no significant difference to placebo on all primary and secondary measures	chance finding. Very heterogenous population, no selection with imaging
Multifactorial	Albumin [74]	Recruitment into Part 1 of the ALIAS Trial was halted for safety reasons	Second trial underway
NMDA receptor antagonist	Magnesium [49]	Phase III trial underway	
Multifactorial	Hypothermia [54]	Several Phase 1 trials underway	One large trial demonstrated safety and feasibility in acute
Kappa opioid peptide receptor antagonist	Cervene [75]		Lack of efficacy
Cell-membrane stabilizer	Citicholine [50]		ICTUS Trial underway
Chelator of zinc and calcium ions	DP-b99 [22]		Lack of efficacy
Mitochondrial bioenergetization	Transcranial laser therapy (TLT) [51]		Lack of efficacy

Adapted with permission from Beresford et al. [63]

3 Challenges: Why Previous Translational Clinical Trials Have Failed

The field of acute stroke provides several formidable challenges, including the narrow time window of viability of the ischemic brain, complex pathophysiological processes that depend on a variety of underlying mechanisms (e.g., thromboembolism vs. hemodynamic ischemia, white matter vs. gray matter ischemia) and dilemmas in diagnosing a multitude of clinical presentations and the need to differentiate ischemia from hemorrhage [26]. More than 70 neuroprotective agents have been tested in more than 140 randomized, controlled, clinical trials in acute ischemic stroke, enrolling over 25,000 patients, but no agent was unequivocally beneficial in definitive Phase III trials [27]. The recent SAINT trials particularly highlighted this, with reasonable animal data and a Phase 1 clinical trial which was positive, but the larger Phase III clinical trial showed no benefit of the treatment arm over the placebo group [28, 29]. Some of the possible reasons for the lack of translational success are as follows.

3.1 Preclinical Development

Several animal models and clinical trials have been designed and developed over the last 3 decades for the evaluation of neuroprotectant medications in the treatment of acute ischemic stroke. It is clear that preclinical (experimental) studies vary considerably in quality and reliability. A great number of variables come into play in the design and execution of these studies that influence their quality, consistency, and outcome. These variables include animal-related factors (species, strain, age, sex, comorbidities), anesthetic agents and drugs, extent and rigor of physiologic monitoring, animal model-related factors (choice of ischemia model such as temporary or permanent occlusion; presence or absence of reperfusion; duration of ischemia, reperfusion, and survival), modes of outcome assessment, quality of statistical study design, and of course, the characteristics of the neuroprotective agent itself [30]. The mechanisms of cerebral ischemia have been studied in the settings of both transient global (mimicking cardiac arrest and severe systemic hypoxemia) and focal ischemia, and a variety of neuroprotective agents have been found to decrease the size of ischemic lesions [31–33]. However, the majority of these interventions failed to extend benefit from animal models to human patients or affect clinical outcomes significant enough to allow widespread clinical application. A variety of general design features might have accounted for the failure of many potential neuroprotectants during the development process. Majority of preclinical studies have been carried out using animal models that may not extrapolate or translate into metabolic response to ischemia in humans. Rodent models of focal and global ischemia do not represent the mechanisms that lead to cerebral infarction in patients with acute stroke due to inherent but substantial anatomical and physiological differences. Primate models may be more useful as their brain structure more closely resembles the human brain. In animal studies, the duration and location of ischemia and interventions are designed in a homogenous fashion with physiological variables such as temperature and blood pressure very tightly regulated. Acute ischemic stroke in humans, on the other hand, represents culmination of an acute thrombotic or embolic event brewing over an underlying process of ongoing atherosclerosis or chronic disease process. The underlying pathology, acuity of ischemia, its progression, and associated physiological variables are all heterogeneous, adding significant variability to the study populations. Also, preclinical studies utilize young healthy animals, while human stroke populations tend to be elderly, exposed to multiple chronic illnesses, and on medications that potentially affect final outcomes. Efforts to engineer comorbidities in animal models to make them more representative might be helpful.

With regard to the agents themselves, pharmacodynamic and pharmacokinetics differ significantly among humans and animals. Limiting side effects at clinically effective doses may differ significantly among a rodent and a human with ischemic stroke. Higher concentrations may be required for neuroprotective effects to be observed in models of permanent ischemia as compared with transient ischemia. Degree of reperfusion in a stroke may be extremely variable in humans and the same agent or dose may have different effects in patients who completely recanalize an acute ischemic territory vs. another patient that does not. These factors are hard to follow or randomize in a clinical setting. Future trials can be designed to assess degree of perfusion as a marker of separating patients into different target groups.

The Stroke Therapy Academic Industry Roundtable (STAIR) has produced guidelines for the preclinical and clinical evaluation of drugs for stroke treatment with the aim of reducing the number of failed trials in humans in the future [34, 35]. These recommend the modeling of clinical stroke through a greater use of functional tests in animal models and long-term outcome measures (Table 4.2).

3.2   Evolving Understanding of the Ischemic Cascade

Our knowledge and understanding of the ischemic cascade has greatly developed over time. It is now clear that previously there was significant oversimplification of target mechanisms. As an example, it is now known that NMDA receptor location and subunit composition differentially regulate neuronal survival or death, although the strength of the calcium signal may specify the fate of neurons best following NMDA overactivation [36]. With notable exceptions, selective enhancement of synaptic receptors promotes neuronal survival, whereas activating extrasynaptic receptors promote cell death [37]. Under ischemic conditions, the vulnerability to cell death is reduced when postsynaptic scaffolding proteins are modified and PSD-95 binding is prevented [38]. Moreover, successfully targeting the glutamate receptor complex early on would also diminish the opening of downstream channels, e.g., transient receptor potential (TRP) channels and acid-sensing ion channels, as well as suppress the frequency of CSDs (see below), ongoing within the ischemic penumbra.

Table 4.2 Major highlights of updated recommendations of stroke therapy academic industry roundtable (STAIR for preclinical stroke studies STAIR VI [76])

Define minimum effective and maximum tolerated dose with target concentration
Therapeutic window for thrombolytic and neuroprotective drugs should be guided penumbral imaging using perfusion/diffusion MRI
Basic physiological parameters such as blood pressure, temperature, blood gases, and blood glucose should be routinely monitored
Cerebral blood flow using Doppler flow or perfusion imaging should be documented to assess adequate sustained occlusion and to monitor reperfusion in temporary ischemia models
Treatment efficacy should be established in at least two species. Follow-up studies should be conducted in aged animals and animals with comorbidities such as hypertension, diabetes, and hypercholesterolemia. Efficacy studies should be performed in both male and female animals
Interaction studies with medications commonly used in stroke patients should be performed for advanced preclinical drug development candidates
Outcome measures should include histological as well as behavioral measurements in both immediate recovery as well as delayed period after stroke
Relevant biomarker endpoints such as diffusion/perfusion MRI and serum markers of tissue injury should be included that can be also obtained in human trials to indicate that the therapeutic target has been modified
Positive results obtained in 1 laboratory should be replicated in at least 1 independent laboratory before advancing to clinical studies
Planning studies and reporting results should address adequate randomization and eliminating bias, defining inclusion/exclusion criteria and reporting the reasons for excluding animals from the final analysis

Hence, understanding and leveraging the repertoire of diverse responses and mediators at the receptor and second messenger level can only enhance possibilities for successfully targeting excitotoxicity for the treatment of ischemic injury. Similarly, the complexity of Ca^{2+} regulation during ischemia and reperfusion—ineffectiveness of therapies aimed at blocking influx of Ca^{2+}.

Overproduction of reactive nitrogen and oxygen species is surely damaging to brain cells. But at lower, homeostatic levels, radicals are critical signaling molecules participating in normal neuronal and vascular function [39]. Finding novel ways to scavenge or suppress deleterious radicals without interfering with endogenous signaling will be important to design effective therapies. A desirable strategy would be to develop drugs that only become activated by the pathological state intended for inhibition, i.e., during oxidative stress [40]. Another approach would be to test antioxidant molecules with pleiotropic properties, such as activated protein-C (APC). APC blocks ROS generation, suppresses postischemic inflammation, and blocks apoptosis in neurons and endothelium, thereby providing both parenchymal and vascular protection [41]. APC as well as other pleiotropic agents with a broad spectrum of activity in experimental stroke is undergoing clinical trial testing [30].

Taming postischemic inflammation may block secondary events that extend brain injury. However, as noted below, during the later stages in the injury process, inflammation promotes critical events necessary for tissue repair. Therefore, therapeutic interventions targeting postischemic inflammation should be mindful of temporal considerations by minimizing the destructive potential of inflammation in the acute phase while enhancing its beneficial contributions to tissue repair in the late stages of cerebral ischemia.

One aspect of drug selection as an extension to mechanism might be nonselective blockade of all receptor function and subtherapeutic dosing. An alternative viewpoint is that the drugs tested in clinical trials were poorly chosen, because many of them block all receptor functions [40]. As a result, the dosing may have been suboptimal, due to the development of untoward side effects that prevented dose escalation to neuroprotective levels [2].

A new concept of neurovascular unit has recently been proposed. Therapies to target the vascular unit of brain to recanalize, reduce clot propagation, and improve collateralization have revolutionized acute stroke treatment, but therapies targeting the neural elements still await translational success. The concept of targeting a multimodal approach aimed at the neurovascular unit may provide an integrated framework for investigating mechanisms and therapies. Salvaging neurons alone may not be sufficient. One has to protect glial and vascular elements as well. Future investigations might be better served by pursuing targets that are expressed on multiple cell types in brain. Furthermore, prevention of cell death alone may also not be sufficient. In order for therapies to be clinically meaningful, cellular function must be preserved and/or restored. Neurons that are alive but lack proper connectivity or do not express the correct array of transmitter release-reuptake and synaptic activity may not be functional. Astrocytes that survive but cannot provide hemodynamic coupling would be unable to link neuronal activation with the required vascular responses. Oligodendrocytes that are respiring but metabolically impaired might not provide the necessary myelination for axonal conduction. Endothelial cells that survive but do not maintain blood-brain barrier properties would facilitate damage to adjacent parenchyma. Ultimately, any stroke therapy must include both prevention of cell death as well as rescue of integrated neurovascular function.

Another problem facing translation of neuroprotective trials is that there are multiple pathways within the ischemic cascade, yet the majority of neuroprotective agents only focus on single aspects of the cascade. An analysis found that only 3 of the 178 trials targeted a specific stroke mechanism [26]. Targeting only one mechanism may not translate into successful delaying of the ischemic cascade. For example, the glutamate hypothesis with agents targeting NMDA and AMPA receptors may have oversimplified the complexity of the cell death process and underestimated the diversity of expressing cell types as well as the heterogeneity of glutamate responses following receptor overstimulation in stroke [2]. Moreover, it is likely that NMDA–AMPA pathways comprise only a subset of routines disrupting ionic imbalance following glutamate receptor activation. As originally conceived, the NMDA–AMPA model did not consider the delayed and invariably fatal glutamate-independent calcium influx following oxygen-glucose deprivation or the failure of glutamate receptor blockade to halt cell death after very intense oxygen-glucose deprivation [42]. Equally importantly, it could not account in an obvious way for the death of nonneuronal cell types such as astrocytes, vascular cells, or microglia in ischemic tissue. The degree of reperfusion markedly affects penetration and exposure of neuroprotective agents in penumbra. Most trials have used neuroprotective agents as alternative not adjuncts to reperfusion therapy. By slowing the ischemic cascade and sustaining the penumbra temporarily, these agents may also allow for extension of the accepted therapeutic windows for reperfusion.

3.3   Trial Design

Many trials may have been handicapped in their ability to show agent efficacy because of inadequate sample size, inappropriate time windows permitted for patient enrollment, inappropriate choice of outcome time points, or failure to target a specific stroke mechanism.

Another variable that has been addressed in most recent studies relates to the timing of the medication to the arterial occlusion. The complexity of the penumbral area and collateral circulation varies greatly from individual to individual and makes it difficult to determine exact time frames for penumbral survival, both in animals and in humans. Enrollment of patients beyond the critical time window for rescuing salvageable tissue will lead to a dilution of true treatment effects [26]. Compared to animal models, time range of administration of agents with respect to onset of stroke human studies is too broad and late. Few large clinical neuroprotection trials have been conducted within the treatment window in which therapeutic efficacy is considered possible [30]. As an example, NMDA antagonists were effective neuroprotection agents when given within 90 min after occlusion in rat models. However, the majority of patients in the human trials did not receive the medication until 6 h after onset [43]. Newer clinical trials in ischemic stroke have aimed at enrolling patients in a time window less than 6 h from onset of symptoms. Of the 160 clinical trials of neuroprotection for ischemic stroke conducted as of late 2007, only 40 represent larger-phase completed trials, and fully one half of the latter utilized a window to treatment of >6 h, despite strong preclinical evidence that this delay exceeds the likely therapeutic window of efficacy in acute stroke [30]. Hypothermia is one of the interventions where the timing issue has been a constant issue of contention. Timing of induction with respect to onset of symptoms, maintenance, and duration of rewarming all have significant effect on outcomes. Unresolved technical issues surrounding the timing and method for inducing hypothermia and the deleterious effect of rewarming have limited its clinical application.

Stroke severity is another variable difficult to translate from a rodent model to humans. Efforts are now being made to ensure that patients enrolled in clinical studies have moderate to moderately severe deficits on clinical assessment or, with sufficiently large trials, ensure that stratification for severity will adequately ensure groups to be well-balanced.

3.4   Measuring Outcomes

Finally, the outcome measures in animal and human models are quite different. In animal studies, the tools that measure the effects of neuroprotection are limited to assessing change in the volume of infarction by imaging or pathological tests, and a very basic evaluation of motor or cognitive skills. These simple scoring systems are very different than the assessments that are required in human subjects. Most scores (e.g., the modified Rankin Score or Barthel Index) measure more complex human behavior including the ability to walk, take care of body needs, and the ability to go back to activities that were possible prior to the AIS. Unfortunately, these measures do not correlate well, as small volume infarctions in strategic locations may produce severe deficits. An analysis by Kidwell et al. [26] reported the trends on the emergence of validated outcome clinical trial rating scales to assess functional end points. Over time, validated scales assessing the degree of neurological deficit, such as the NIHSS, functional activities of daily living, and global disability, have increasingly displaced the poorly specified global judgments or ad hoc scales used by early investigators [26].

4   Recommendations for Future Trials

Choice of the drugs targeting a single mechanism nonselectively should be revisited. A more suitable approach would be to use therapeutic agents that selectively target excessive channel opening (e.g., uncompetitive inhibitors with a relatively fast off rate, such as memantine [44, 45]), thereby retaining normal receptor functions. Understanding and leveraging the repertoire of diverse responses and mediators at the receptor and second messenger level can only enhance possibilities for successfully targeting excitotoxicity for the treatment of ischemic injury. Treatment using a combination of strategies that work on different pathways might help. These may be combined with tPA or other reperfusion injuries to extend the therapeutic window.

It is clear that there are differences between the rodent brain and that of humans and nonhuman primates, and there are associated difficulties with the scaling up of dosing regimens from rodents to humans. Therefore, it is considered appropriate to include an intermediate step involving large animals such as primates in the translation of neuroprotective agents from the preclinical to the clinical setting.

Time is one of the most important determinants of success in any treatment strategy. In addition, it is important to allow for sufficient number and severity of patients to participate in the study to achieve meaningful results. The sample size in clinical trials should ensure sufficient power to detect small, but clinically significant, differences in treatment effect.

Efforts to improve the quality of data collected as well as outcome measures to determine treatment effects in clinical trials should be made. Recently, there has been considerable interest in the use of surrogate markers to evaluate the drug effect in a smaller number of patients with AIS. The use of MRI technology to measure the volume of infarction or computed tomography utilizing perfusion techniques as a marker of efficacy of therapy is increasingly being tested in smaller number of stroke patients. If such techniques are proven to be predictors of treatment effect, they may prove an excellent method to test drugs in a smaller patient population prior to more extensive and expensive trials. To try to facilitate the analysis and comparison of results from future trials of neuroprotective agents in acute stroke, consistent and standardized outcomes of stroke should be used to assess a spectrum of stroke recovery.

Although the clinical trials involving neuroprotective agents have been negative, they still have provided useful information to try to devise optimal ways to test such agents in AIS.

5   Specific Agents and Therapeutic Targets

Here we discuss a few of these strategies in relation to their major features affecting clinical application.

Glutamate hypothesis and NMDA–AMPA models have led us to try anti-excitotoxic strategies like NMDA and AMPA receptor modulators [30]. Agents like Selfotel [46], Aptiganel [24], Gavestinel [47], and Eliprodil [48] had adverse effects in clinical trials. Magnesium is an NMDA receptor antagonist and limits glutamate-mediated excitotoxicity. While the IMAGES trial (Muir Stroke 1995) failed to show efficacy of magnesium in acute ischemic stroke, another effort known as the FAST-MAG Pilot Trial demonstrated the feasibility, safety, and potential efficacy of paramedic administration of magnesium to protect the brain during transport and extend the time window for revascularization therapy [49]. This is a particularly exciting concept where the loading dose and the infusion are initiated in the ambulance within minutes to hours after onset of symptoms. In the pilot FAST-MAG trial, the study infusion was initiated at a median of 100 min after symptom onset, with those having the infusion initiated by paramedics in the field at 26 min as compared to in-hospital initiated at 139 min. This approach appeared to be safe in both ischemic and hemorrhagic stroke, with no serious adverse events reported [49].

Anti-apoptosis therapy targeting Caspase-3, Caspase-9, Bcl-2/Bax dependent mechanisms—BNP brain-derived neurotrophic growth factors is also being explored. Citicholine (CDP-choline) is an intermediate in the biosynthesis of phosphatidylcholine, a major membrane phospholipid, and is thus an integral part of membrane synthesis. It is believed to work in membrane stabilization, reduction of free radicals, and inhibition of apoptosis. Initial animal trials were promising, but subsequent clinical studies failed to show improved outcome [50]. The International Citicoline Trial on Acute Stroke (ICTUS) is double blinded randomized controlled trial that is presently underway, and with a planned enrollment of 2,600 patients, will be the largest citicholine trial performed.

The NeuroThera Effectiveness and Safety Trials (NESTs) utilized infrared laser therapy for mitochondrial bioenergization to initiate mitochondria to absorb this energy, likely through cytochrome c oxidase, and utilize it for ATP formation. It is premised that increased ATP synthesis may improve energy metabolism in ischemic penumbra and prevent apoptosis [51]. Final results of the NEST-2 randomized trial to treat acute ischemic stroke failed to meet its primary end point of efficacy, although it was safe and well tolerated [52]. It excluded stroke patients who had received thrombolysis or those with evidence of hemorrhagic infarct or high severity [53]. The follow-up international, multicentered NEST-3 study is now underway.

Hypothermia is another clinical strategy that has multiple therapeutic effects at different levels of the ischemic cascade, but has not shown clinical efficacy in trials [54]. Experimental evidence suggests that hypothermia protects the brain through pleiotropic effects acting on multiple cell types within the neurovascular unit. It reduces brain metabolism, maintains the blood-brain barrier, and reduces glutamate release, ROS production, peri-infarct depolarization, inflammatory markers, microglial activation, leucocyte infiltration, and MMP activation [55]. In addition, it suppresses apoptotic cell death and decreases mitochondrial release of cytochrome C and apoptosis-inducing factor [56]. Thus, it is a strategy that targets multiple aspects of the ischemic cascade. Hypothermia has been shown to improve outcomes in other types of ischemic/anoxic injury, e.g., global anoxia following cardiac arrest (hypothermia after cardiac arrest study group *NEJM* 2002) and severe hypoxic encephalopathy in infant [57]. Variability in protocols for inducing and maintaining hypothermia as well cooling strategies like general whole body surface cooling, selective scalp cooling, systemic endovascular cooling [58], and selective regional cooling (i.e., to the ischemic area only) has been a factor in reproducing efficacy of this modality. Two recent randomized trials of hypothermia called Coolaid [58] and (ICTuS-L) [59] did not have an effect on disability outcomes. While both studies addressed the feasibility of hypothermia in clinical studies, none of them were powered for efficacy.

The Safety and Efficacy of NeuroFlo in Acute Ischemic Stroke Trial was a recent randomized trial of the safety and efficacy of treatment of NeuroFlo™ in improving neurological outcome vs. standard medical management. This catheter is based on the observation in animals that partial aortic occlusion of the abdominal aorta resulted in prompt increase in cerebral blood flow and the effect persisted after occlusion was relived. The trial met its primary safety end point, but not its primary efficacy end point [60]. However, certain subgroups did show promising trends and follow-up studies are being planned.

There has been recent interest in exploring other mechanisms in the search for an effective agent for neuroprotection. The revelation of toll-like receptors (TLRs) as key mediators of tissue injury in response to stroke has identified a new target critical to understanding the underlying mechanisms of stroke injury and potential therapies. TLRs have an endogenous ability to self-regulate wherein prior exposure to low level TLR activation induces protection against a subsequent challenge that would otherwise cause damage. Recent studies show that TLR pathways can be reprogrammed via prior exposure to TLR ligands leading to decreased infarct size and improved neurological outcomes in response to ischemic injury [61]. Future possibilities of utilizing neural interface systems to record and modulate neural signals in the acute period are still on the horizon but open to the possibility of a new approach [62].

6 Conclusion

Despite the negative outcome of the trials, it is important to emphasize that failure of these trials has led to an improvement in the methodology of stroke research. Lessons learnt from the shortcoming on these trials can help pave the way to a new approach to find clinical applications of neuroprotection.

We have made tremendous progress over the last few decades in the hyperacute management of ischemic stroke and understanding the physiology of acute ischemic injury leading to neuronal death. Thrombolytic treatment with rt-PA or endovascular recanalization is currently the only therapy available for acute ischemic stroke with restrictions surrounding patient selection and timing of administration. This highlights the unmet need for alternative strategies. Neuroprotection is one such strategy and aims to preserve viable cells in the ischemic penumbra by interfering with the damaging events of the ischemic cascade. By limiting the area and impact of injury to the neuronal cells in the ischemic penumbra, neuroprotective agents could potentially improve recovery from stroke. Interventions targeting a single neurotoxic mechanism alone might not be sufficient to prevent neuronal cell death. We might need interventions that target multiple sites and processed with pleotrophic actions even if polypharmacy is required. However, poor translation from animal models to clinical trial design has accounted for the failure of many potential neuroprotective agents and led to the development of STAIR criteria. These criteria need to be further refined, to try to improve the chances of successful translation and thus provide additional therapeutic options to ischemic stroke patients.

References

1. Choi DW. Excitotoxic cell death. J Neurobiol. 1992;23(9):1261–76.
2. Moskowitz MA, Lo EH, Iadecola C. The science of stroke: mechanisms in search of treatments. Neuron. 2010;67(2):181–98.
3. Heiss WD. Ischemic penumbra: evidence from functional imaging in man. J Cereb Blood Flow Metab. 2000;20(9):1276–93.
4. Lo EH. Experimental models, neurovascular mechanisms and translational issues in stroke research. Br J Pharmacol. 2008;153 Suppl 1:S396–405.
5. Astrup J, Siesjo BK, Symon L. Thresholds in cerebral ischemia—the ischemic penumbra. Stroke. 1981;12(6):723–5.
6. Choi DW, Rothman SM. The role of glutamate neurotoxicity in hypoxic-ischemic neuronal death. Annu Rev Neurosci. 1990;13:171–82.

7. Ankarcrona M, Dypbukt JM, Bonfoco E, Zhivotovsky B, Orrenius S, Lipton SA, et al. Glutamate-induced neuronal death: a succession of necrosis or apoptosis depending on mitochondrial function. Neuron. 1995;15(4):961–73.
8. Dirnagl U, Iadecola C, Moskowitz MA. Pathobiology of ischaemic stroke: an integrated view. Trends Neurosci. 1999;22(9):391–7.
9. Lo EH, Dalkara T, Moskowitz MA. Mechanisms, challenges and opportunities in stroke. Nat Rev Neurosci. 2003;4(5):399–415.
10. Bano D, Nicotera P. Ca^{2+} signals and neuronal death in brain ischemia. Stroke. 2007;38(2 Suppl):674–6.
11. Contreras JE, Sanchez HA, Veliz LP, Bukauskas FF, Bennett MV, Saez JC. Role of connexin-based gap junction channels and hemichannels in ischemia-induced cell death in nervous tissue. Brain Res Brain Res Rev. 2004;47(1–3):290–303.
12. Xiong ZG, Zhu XM, Chu XP, Minami M, Hey J, Wei WL, et al. Neuroprotection in ischemia: blocking calcium-permeable acid-sensing ion channels. Cell. 2004;118(6):687–98.
13. Kimelberg HK, Macvicar BA, Sontheimer H. Anion channels in astrocytes: biophysics, pharmacology, and function. Glia. 2006;54(7):747–57.
14. Aarts MM, Tymianski M. TRPMs and neuronal cell death. Pflugers Arch. 2005;451(1):243–9.
15. Abe T, Kunz A, Shimamura M, Zhou P, Anrather J, Iadecola C. The neuroprotective effect of prostaglandin E2 EP1 receptor inhibition has a wide therapeutic window, is sustained in time and is not sexually dimorphic. J Cereb Blood Flow Metab. 2009;29(1):66–72.
16. Oppenheim JJ, Yang D. Alarmins: chemotactic activators of immune responses. Curr Opin Immunol. 2005;17(4):359–65.
17. Somjen GG. Mechanisms of spreading depression and hypoxic spreading depression-like depolarization. Physiol Rev. 2001;81(3):1065–96.
18. Shin HK, Dunn AK, Jones PB, Boas DA, Moskowitz MA, Ayata C. Vasoconstrictive neurovascular coupling during focal ischemic depolarizations. J Cereb Blood Flow Metab. 2006;26(8):1018–30.
19. Back T, Ginsberg MD, Dietrich WD, Watson BD. Induction of spreading depression in the ischemic hemisphere following experimental middle cerebral artery occlusion: effect on infarct morphology. J Cereb Blood Flow Metab. 1996;16(2):202–13.
20. Simon R, Xiong Z. Acidotoxicity in brain ischaemia. Biochem Soc Trans. 2006;34(Pt 6): 1356–61.
21. Horn H, Federspiel A, Wirth M, Muller TJ, Wiest R, Wang JJ, et al. Structural and metabolic changes in language areas linked to formal thought disorder. Br J Psychiatry. 2009;194(2): 130–8.
22. Diener HC, Schneider D, Lampl Y, Bornstein NM, Kozak A, Rosenberg G. DP-b99, a membrane-activated metal ion chelator, as neuroprotective therapy in ischemic stroke. Stroke. 2008; 39(6):1774–8.
23. Berends AC, Luiten PG, Nyakas C. A review of the neuroprotective properties of the 5-HT1A receptor agonist repinotan HCl (BAYx3702) in ischemic stroke. CNS Drug Rev. 2005;11(4): 379–402.
24. Albers GW, Goldstein LB, Hall D, Lesko LM. Aptiganel Acute Stroke Investigators. Aptiganel hydrochloride in acute ischemic stroke: a randomized controlled trial. JAMA. 2001;286(21): 2673–82.
25. O'Collins VE, Macleod MR, Donnan GA, Horky LL, van der Worp BH, Howells DW. 1,026 experimental treatments in acute stroke. Ann Neurol. 2006;59(3):467–77.
26. Kidwell CS, Liebeskind DS, Starkman S, Saver JL. Trends in acute ischemic stroke trials through the 20th century. Stroke. 2001;32(6):1349–59.
27. Chacon MR, Jensen MB, Sattin JA, Zivin JA. Neuroprotection in cerebral ischemia: emphasis on the SAINT trial. Curr Cardiol Rep. 2008;10(1):37–42.
28. Diener HC, Lees KR, Lyden P, Grotta J, Davalos A, Davis SM, et al. NXY-059 for the treatment of acute stroke: pooled analysis of the SAINT I and II trials. Stroke. 2008;39(6):1751–8.
29. Shuaib A, Lees KR, Lyden P, Grotta J, Davalos A, Davis SM, et al. NXY-059 for the treatment of acute ischemic stroke. N Engl J Med. 2007;357(6):562–71.

30. Ginsberg MD. Current status of neuroprotection for cerebral ischemia: synoptic overview. Stroke. 2009;40(3 Suppl):S111–4.
31. Nakase T, Yoshioka S, Suzuki A. Free radical scavenger, edaravone, reduces the lesion size of lacunar infarction in human brain ischemic stroke. BMC Neurol. 2011;11:39.
32. Jalal FY, Bohlke M, Maher TJ. Acetyl-L-carnitine reduces the infarct size and striatal glutamate outflow following focal cerebral ischemia in rats. Ann N Y Acad Sci. 2010;1199:95–104.
33. Walberer M, Nedelmann M, Ritschel N, Mueller C, Tschernatsch M, Stolz E, et al. Intravenous immunoglobulin reduces infarct volume but not edema formation in acute stroke. Neuroimmunomodulation. 2010;17(2):97–102.
34. Stroke Therapy Academic Industry Roundtable (STAIR). Recommendations for standards regarding preclinical neuroprotective and restorative drug development. Stroke. 1999;30(12): 2752–8.
35. Fisher M. Stroke Therapy Academic Industry Roundtable. Recommendations for advancing development of acute stroke therapies: stroke therapy academic industry roundtable 3. Stroke. 2003;34(6):1539–46.
36. Stanika RI, Pivovarova NB, Brantner CA, Watts CA, Winters CA, Andrews SB. Coupling diverse routes of calcium entry to mitochondrial dysfunction and glutamate excitotoxicity. Proc Natl Acad Sci U S A. 2009;106(24):9854–9.
37. Chen Q, He S, Hu XL, Yu J, Zhou Y, Zheng J, et al. Differential roles of NR2A- and NR2B-containing NMDA receptors in activity-dependent brain-derived neurotrophic factor gene regulation and limbic epileptogenesis. J Neurosci. 2007;27(3):542–52.
38. Cui H, Hayashi A, Sun HS, Belmares MP, Cobey C, Phan T, et al. PDZ protein interactions underlying NMDA receptor-mediated excitotoxicity and neuroprotection by PSD-95 inhibitors. J Neurosci. 2007;27(37):9901–15.
39. Faraci FM. Protecting the brain with eNOS: run for your life. Circ Res. 2006;99(10):1029–30.
40. Lipton SA. Pathologically activated therapeutics for neuroprotection. Nat Rev Neurosci. 2007;8(10):803–8.
41. Yamaji K, Wang Y, Liu Y, Abeyama K, Hashiguchi T, Uchimura T, et al. Activated protein C, a natural anticoagulant protein, has antioxidant properties and inhibits lipid peroxidation and advanced glycation end products formation. Thromb Res. 2005;115(4):319–25.
42. Aarts MM, Arundine M, Tymianski M. Novel concepts in excitotoxic neurodegeneration after stroke. Expert Rev Mol Med. 2003;5(30):1–22.
43. Green AR, Shuaib A. Therapeutic strategies for the treatment of stroke. Drug Discov Today. 2006;11(15–16):681–93.
44. Orgogozo JM, Rigaud AS, Stoffler A, Mobius HJ, Forette F. Efficacy and safety of memantine in patients with mild to moderate vascular dementia: a randomized, placebo-controlled trial (MMM 300). Stroke. 2002;33(7):1834–9.
45. Kavirajan H, Schneider LS. Efficacy and adverse effects of cholinesterase inhibitors and memantine in vascular dementia: a meta-analysis of randomised controlled trials. Lancet Neurol. 2007;6(9):782–92.
46. Davis SM, Albers GW, Diener HC, Lees KR, Norris J. Termination of acute stroke studies involving selfotel treatment. ASSIST steering committed. Lancet. 1997;349(9044):32.
47. Sacco RL, DeRosa JT, Haley Jr EC, Levin B, Ordronneau P, Phillips SJ, et al. Glycine antagonist in neuroprotection for patients with acute stroke: GAIN Americas: a randomized controlled trial. JAMA. 2001;285(13):1719–28.
48. Hogg S, Perron C, Barneoud P, Sanger DJ, Moser PC. Neuroprotective effect of eliprodil: attenuation of a conditioned freezing deficit induced by traumatic injury of the right parietal cortex in the rat. J Neurotrauma. 1998;15(7):545–53.
49. Saver JL, Kidwell C, Eckstein M, Starkman S. FAST-MAG Pilot Trial Investigators. Prehospital neuroprotective therapy for acute stroke: results of the field administration of stroke therapy-magnesium (FAST-MAG) pilot trial. Stroke. 2004;35(5):e106–8.
50. Clark WM, Wechsler LR, Sabounjian LA, Schwiderski UE. Citicoline Stroke Study Group. A phase III randomized efficacy trial of 2000 mg citicoline in acute ischemic stroke patients. Neurology. 2001;57(9):1595–602.

51. Lapchak PA. Taking a light approach to treating acute ischemic stroke patients: transcranial near-infrared laser therapy translational science. Ann Med. 2010;42(8):576–86.
52. Stemer AB, Huisa BN, Zivin JA. The evolution of transcranial laser therapy for acute ischemic stroke, including a pooled analysis of NEST-1 and NEST-2. Curr Cardiol Rep. 2010;12(1): 29–33.
53. Zivin JA, Albers GW, Bornstein N, Chippendale T, Dahlof B, Devlin T, et al. Effectiveness and safety of transcranial laser therapy for acute ischemic stroke. Stroke. 2009;40(4):1359–64.
54. Huh PW, Belayev L, Zhao W, Koch S, Busto R, Ginsberg MD. Comparative neuroprotective efficacy of prolonged moderate intraischemic and postischemic hypothermia in focal cerebral ischemia. J Neurosurg. 2000;92(1):91–9.
55. Sakoh M, Gjedde A. Neuroprotection in hypothermia linked to redistribution of oxygen in brain. Am J Physiol Heart Circ Physiol. 2003;285(1):H17–25.
56. Tang XN, Yenari MA. Hypothermia as a cytoprotective strategy in ischemic tissue injury. Ageing Res Rev. 2010;9(1):61–8.
57. Shankaran S, Laptook AR, Ehrenkranz RA, Tyson JE, McDonald SA, Donovan EF, et al. Whole-body hypothermia for neonates with hypoxic-ischemic encephalopathy. N Engl J Med. 2005;353(15):1574–84.
58. De Georgia MA, Krieger DW, Abou-Chebl A, Devlin TG, Jauss M, Davis SM, et al. Cooling for acute ischemic brain damage (COOL AID): a feasibility trial of endovascular cooling. Neurology. 2004;63(2):312–7.
59. Hemmen TM, Raman R, Guluma KZ, Meyer BC, Gomes JA, Cruz-Flores S, et al. Intravenous thrombolysis plus hypothermia for acute treatment of ischemic stroke (ICTuS-L): final results. Stroke. 2010;41(10):2265–70.
60. Shuaib A, Bornstein NM, Diener HC, Dillon W, Fisher M, Hammer MD, et al. Partial aortic occlusion for cerebral perfusion augmentation: safety and efficacy of NeuroFlo in acute ischemic stroke trial. Stroke. 2011;42(6):1680–90.
61. Marsh BJ, Williams-Karnesky RL, Stenzel-Poore MP. Toll-like receptor signaling in endogenous neuroprotection and stroke. Neuroscience. 2009;158(3):1007–20.
62. Simeral JD, Kim SP, Black MJ, Donoghue JP, Hochberg LR. Neural control of cursor trajectory and click by a human with tetraplegia 1000 days after implant of an intracortical microelectrode array. J Neural Eng. 2011;8(2):025027.
63. Beresford IJ, Parsons AA, Hunter AJ. Treatments for stroke. Expert Opin Emerg Drugs. 2003;8(1):103–22.
64. Horn J, Limburg M. Calcium antagonists for ischemic stroke: a systematic review. Stroke. 2001;32(2):570–6.
65. Sareen D. Neuroprotective agents in acute ischemic stroke. J Assoc Physicians India. 2002; 50:250–8.
66. Jensen BS. BMS-204352: a potassium channel opener developed for the treatment of stroke. CNS Drug Rev. 2002;8(4):353–60.
67. Wood PL, Hawkinson JE. N-methyl-D-aspartate antagonists for stroke and head trauma. Expert Opin Investig Drugs. 1997;6(4):389–97.
68. Haley Jr EC. High-dose tirilazad for acute stroke (RANTTAS II). RANTTAS II investigators. Stroke. 1998;29(6):1256–7.
69. Tuttolomondo A, Di Sciacca R, Di Raimondo D, Renda C, Pinto A, Licata G. Inflammation as a therapeutic target in acute ischemic stroke treatment. Curr Top Med Chem. 2009;9(14):1240–60.
70. Krams M, Lees KR, Hacke W, Grieve AP, Orgogozo JM, Ford GA, et al. Acute stroke therapy by inhibition of neutrophils (ASTIN): an adaptive dose-response study of UK-279,276 in acute ischemic stroke. Stroke. 2003;34(11):2543–8.
71. Bogousslavsky J, Victor SJ, Salinas EO, Pallay A, Donnan GA, Fieschi C, et al. Fiblast (trafermin) in acute stroke: results of the European-Australian phase II/III safety and efficacy trial. Cerebrovasc Dis. 2002;14(3–4):239–51.
72. Lyden P, Wahlgren NG. Mechanisms of action of neuroprotectants in stroke. J Stroke Cerebrovasc Dis. 2000;9(6 Pt 2):9–14.

73. Pettigrew LC, Kasner SE, Albers GW, Gorman M, Grotta JC, Sherman DG, et al. Safety and tolerability of arundic acid in acute ischemic stroke. J Neurol Sci. 2006;251(1–2):50–6.
74. Hill MD, Martin RH, Palesch YY, Tamariz D, Waldman BD, Ryckborst KJ, et al. The albumin in acute stroke part 1 trial: an exploratory efficacy analysis. Stroke. 2011;42(6):1621–5.
75. Clark W, Ertag W, Orecchio E, Raps E. Cervene in acute ischemic stroke: results of a double-blind, placebo-controlled, dose-comparison study. J Stroke Cerebrovasc Dis. 1999; 8(4):224–30.
76. Fisher M, Feuerstein G, Howells DW, Hurn PD, Kent TA, Savitz SI, et al. Update of the stroke therapy academic industry roundtable preclinical recommendations. Stroke. 2009;40(6): 2244–50.
77. Writing Group Members, Lloyd-Jones D, Adams RJ, Brown TM, Carnethon M, Dai S, et al. Heart disease and stroke statistics—2010 update: a report from the American Heart Association. Circulation. 2010;121(7):e46–215.
78. The National Institute of Neurological Disorders and Stroke rt-PA Stroke Study Group. Tissue plasminogen activator for acute ischemic stroke. N Engl J Med. 1995;333(24):1581–7.
79. Barber PA, Zhang J, Demchuk AM, Hill MD, Buchan AM. Why are stroke patients excluded from TPA therapy? An analysis of patient eligibility. Neurology. 2001;56(8):1015–20.
80. Richard Green A, Odergren T, Ashwood T. Animal models of stroke: do they have value for discovering neuroprotective agents? Trends Pharmacol Sci. 2003;24(8):402–8.

Part II
Mechanisms and Targets

Chapter 5
Protein Aggregation and Multiple Organelle Damage After Brain Ischemia

Chunli H. Liu, Fan Zhang, Tibor Krisrian, Brian Polster, Gary M. Fiskum, and Bingren Hu

Abstract Protein aggregation leads to a broad range of conformational diseases. Protein folding during synthesis and assembly to subcellular structures require cooperation of several sophisticated ATP-dependent cellular systems including protein chaperoning, protein trafficking, protein assembling, as well as protein degradation. Emerging evidence strongly suggests that brain ischemia leads to dysfunction of these ATP-dependent protein assembly and quality control systems, resulting in protein misfolding, ubiquitination, and aggregation during the postischemic phase. Misfolded proteins are aggregated with subcellular organelles and eventually damage these organelles, leading to multiple organelle failure and delayed neuronal death after brain ischemia. In this chapter, we will first review the provenance and history of protein aggregation in the brain ischemic field and then update latest concept and mechanisms of protein aggregation and multiple organelle damage after brain ischemia.

Abbreviations

EPTA   Ethanolic phosphotungstic acid
Ubi-proteins Ubiquitin-conjugated proteins
DG   Dentate gyrus

C.H. Liu, MD • F. Zhang, PhD • T. Krisrian, PhD • B. Polster, PhD
• G.M. Fiskum, PhD • B. Hu, PhD (✉)
Department of Anesthesiology, University of Maryland School of Medicine,
Baltimore, MD 21201, USA
e-mail: bhu@anes.umm.edu

P.A. Lapchak and J.H. Zhang (eds.), *Translational Stroke Research*, Springer Series
in Translational Stroke Research, DOI 10.1007/978-1-4419-9530-8_5,
© Springer Science+Business Media, LLC 2012

1 Introduction

Transient brain ischemia leads delayed neuronal death that does not occur immediately but takes place after several days of reperfusion [32]. Such a slowly progressing type of neuronal death is also referred to as the maturation phenomenon [31]. Delayed neuronal death also occurs in the penumbral area and is somewhat different from the cell rupture type of neuronal necrosis in the core area after focal ischemia [49]. In rat and mouse transient ischemia models, delayed neuronal death occurs at 48–72 h of reperfusion after an initial period of ischemia. Hippocampal CA1 pyramidal neurons are more vulnerable to an ischemic episode than neurons in the DG and most neocortical regions (Fig. 5.1). During the delayed period, i.e., from the end of the ischemic episode to 48–72 h of reperfusion, neurons destined to die appear essentially normal under the light microscope (Fig. 5.1). However, disaggregation of ribosomes, deposition of dark substances or materials, abnormalities in the endoplasmic reticulum (ER), Golgi apparatus, mitochondria, and synapses have been observed under electron microscopy [14, 29, 33, 44, 45, 54, 55, 64]. Delayed neuronal death also occurs after other types of brain injuries, such as in several brain areas after hypoglycemia [4, 51].

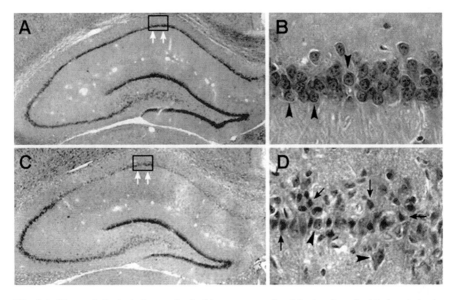

Fig. 5.1 Histopathological changes in the hippocampus after 15 min of cerebral ischemia in the two-vessel occlusion with hypotension ischemia (2VO) model. (**a**) A hippocampal section from a sham-operated rat. (**b**) A higher magnification of the CA1 inset region indicated by *white arrows* in the sham-operated rat (**a**). *Arrowheads* indicate normal neurons. (**c**) A hippocampal section from a rat subjected to 15 min of ischemia followed by 72 h of reperfusion. (**d**) A higher magnification of the CA1 inset region indicated by *white arrows* in the postischemic rat (**c**). *Arrowheads* indicate survival neurons and *black arrows* point to ischemic dead neurons. Dead neurons are identified by cell body and nuclear shrunken as well as acidophilically stained cytoplasm

Several molecular mechanisms underlying delayed neuronal death after brain ischemia have been postulated. Among others, persistent inhibition of protein synthesis [10, 11, 22], increases in ubiquitinated proteins (ubi-proteins) [21, 27, 39–43, 47], and expression of stress proteins [18, 50, 55, 59, 66] have been observed in neurons before delayed neuronal death takes place after brain ischemia. Recent studies support the notion that these intracellular events are commonly associated with protein misfolding and aggregation after brain ischemia [15, 25, 27, 28, 39–42, 51, 68].

2   Historical Review of Protein Aggregation

Electron-dense deposits were clearly observed by electron microscopy (EM) in CA1 pyramidal neurons during the reperfusion phase in a rat transient forebrain ischemia model and were termed "dark substances" [33]. This observation was recapitulated later and described as "electron-dense fluffy dark material" in an ultrastructural study of programmed cell death after brain ischemia [14]. However, at the time, identity and molecular composition of these dark deposits in postischemic neurons were unknown.

Hu and colleagues carried out a series of morphological, biochemical, and molecular studies demonstrating that these dark deposits are, in fact, protein aggregates made of misfolded proteins in postischemic neurons [17, 25, 27–29, 39–42, 68].

During the EM studies of postsynaptic density (PSD) after brain ischemia, Hu and colleagues found that the dark deposits in postischemic neurons can be heavily stained with ethanolic phosphotungstic acid (EPTA) [29, 44]. In those studies, EPTA stains mainly PSD structures and nuclear histones in nonischemic control neurons under EM, but, unexpectedly, it additionally labels intracellular dark deposits in CA1 pyramidal neurons before delayed neuronal death occurs after ischemia. EPTA is known to stain protein complexes under EM [9, 25, 28, 29, 44]. This piece of evidence provided a clue indicating that these dark deposits might be protein aggregates. This view was confirmed by studies showing that the dark deposits were strongly labeled with ubiquitin immunogold under EM, suggesting that ubi-protein conjugates are major components of the dark deposits [25, 28]. EPTA-stained protein aggregates appear as soon as 1–2 h of reperfusion and progressively accumulate until delayed neuronal death occurs at 3 days of reperfusion after transient cerebral ischemia. Figure 5.2a shows that EPTA-stained abnormal protein aggregates are absent in sham-operated nonischemic control neurons (Fig. 5.2a, Sham) and extensively distributed in the cytoplasm of CA1 neurons from a rat subjected to 15 min of ischemia followed by 24 h of reperfusion (Fig. 5.2a, 24 h, arrows). EPTA-stained abnormal protein aggregates are attached to the outer membranes of intracellular organelles such as the ER (arrowheads) and mitochondria (M) at 24 h of reperfusion (Fig. 5.2a, 24 h). Protein aggregates are also attached to the cytoplasmic face of the nuclear membrane (Fig. 5.2a, 24 h, small arrows). Ubiquitin immunogold

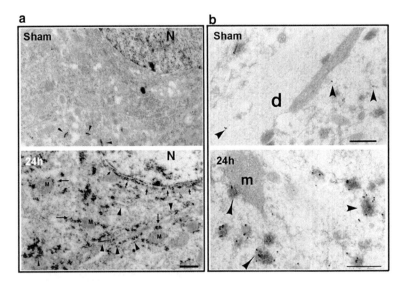

Fig. 5.2 (a) EM micrographs of ethanolic phosphotungstic acid (EPTA) staining of CA1 pyramidal neurons in a sham-operated control rat (Sham) and in a rat subjected to 15 min ischemia followed by 24 h of reperfusion. EPTA-stained materials are extensively distributed in the cytoplasm of 24 h postischemic CA1 neurons but not in control neurons. EPTA-stained protein aggregates are attached to the membranes of intracellular organelles such as the endoplasmic reticulum (ER) and mitochondrial outer membranes, visible in negative contrast (*arrowheads*) at 24 h of reperfusion. EPTA-stained protein aggregates are also attached to the cytoplasmic face of the nuclear membrane (*small arrows*) of CA1 neurons at 24 h of reperfusion. *N* nucleus. Scale bar, 1 µm. (b) Ubiquitin immunogold labeling in the apical dendrites (*d*) of CA1 pyramidal neurons in sham-operated control (Sham) and the postischemic brain at 24 h reperfusion. Immunolabeling in the control brain is usually present in the cytoplasm (*arrowhead* in **b**). Heavy immunolabeling of ubiquitin is colocalized with the protein aggregates (*arrowheads*) at 24 h of reperfusion in CA1 neurons but not in the control. Some protein aggregates are associated with mitochondria (*m* in **b**). *d* dendrites, scale bar, 0.5 µm [28]

labeling in CA1 pyramidal neurons from sham-operated control rat brains is usually present in the cytoplasm (Fig. 5.2b, Sham, arrowhead). Heavy ubiquitin immunolabeling is colocalized with the protein aggregates at 24 h in postischemic CA1 neurons (Fig. 5.2b, 24 h, arrowhead).

Consistent with morphological studies, a host of molecular, biochemical, and confocal microscopic studies revealed that these protein aggregates are composed mainly several categories of proteins: (1) molecular chaperones and folding enzymes; (2) components of protein translation complex including ribosomal proteins, protein synthesis initiation, and elongation factors; (3) components of the protein trafficking and fusion machinery; (4) components of the ubiquitin-proteasomal and autophagy pathways; and (5) several protein kinases such as CaMKII and protein kinase C. Consequently, ischemia-induced protein aggregation damages protein synthesis, protein quality control systems, protein trafficking and membrane fusion, as well as depletion of protein kinases in postischemic neurons [17, 39–42, 68].

3   Mechanisms of Protein Aggregation After Brain Ischemia

Protein aggregation has been observed in neurons virtually of all neurodegenerative disorders, suggesting the protein aggregation may represent a common mechanism leading to neuronal degeneration [3, 6]. Recent studies strongly suggest that protein aggregation also play a critical role in delayed neuronal death after acute brain injury such as brain ischemia and hypoglycemia [17, 25, 27, 28, 41, 42, 51, 68]. However, pathological factors leading to protein aggregation are different among various neurological diseases. Protein aggregation in chronic neurodegenerative diseases is mostly caused by gene mutation and environmental stress, whereas, as described below, ischemia-induced protein aggregation is likely initiated by ATP depletion and damage to several protein quality control systems [38, 41, 42, 68].

There are several different aspects of protein aggregation between brain ischemia and chronic neurodegenerative diseases. Brain ischemia-induced protein aggregates can be observed mainly by transmission EM, but they cannot be seen by conventional light microscopy with a histological staining (compare Figs. 5.1 and 5.2), suggesting that brain ischemia induces formation of "microaggregates" relative to "the large protein aggregates" observable by light microscopy in most chronic neurodegenerative diseases. Ischemic protein aggregates are composed of protein synthesis, trafficking, and quality control components [17, 26, 30, 39–42, 68]. In comparison, chronic neurodegenerative diseases have their own characteristic protein aggregates. In addition, ischemic protein aggregates, together with protein synthesis, protein trafficking and assembly, and protein quality control systems, as well as protein kinase regulatory systems, are clustered with subcellular structures and organelle membranes in postischemic neurons and thus damage to these protein-processing systems and their resided structures (Fig. 5.3; [15, 26, 38–42, 68]). This chapter will focus on protein aggregation and multiple organelle damage after brain ischemia.

Con-Translational Protein Aggregation after Brain Ischemia: Newly synthesized polypeptides are the major sources of unfolded protein in normal cells [7, 20]. Recent studies suggest that newly synthesized polypeptides are aggregated with their parent ribosomal protein synthesis machines, i.e., cotranslational protein aggregation after ischemia. The ribosome is the protein synthesis machine and consists of a 40S small subunit and a 60S large subunit in mammalian cells. The small subunit mediates the interactions with eukaryotic protein synthesis initiation factors (eIFs), transfer RNAs (tRNAs), and messenger RNAs (mRNAs) to determine the sequences of the proteins being made. The large subunit, with assistance by eukaryotic protein synthesis elongation factors (eEFs), catalyzes peptide bond formation.

Translating genetic information encoded in DNA into a linear polypeptide on a ribosome is normally done with fidelity [16]. However, to become a nontoxic functional protein, a newly synthesized polypeptide has to overcome a key rate-limiting hurdle—folding into a unique three-dimensional conformation. Nascent peptide chain folding occurs at the levels of protein domain; folding may not take place until

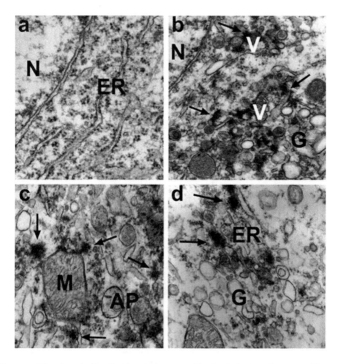

Fig. 5.3 The most dominant ultrastructural changes after brain ischemia are continuous accumulation of intracellular vesicles (V), protein aggregates (*arrows*), multivesicular bodies (MVB), and autophagosomes (AP), as well as aberrant organelles such as dilated endoplasmic reticulum (ER), fragmented Golgi (G), and protein aggregate-associated mitochondria (M). CA1 tissue sections were obtained from rats subjected either to sham surgery or 15 min of ischemia followed by 24 h of reperfusion. (**a**) A sham neuron shows normal nucleus (N) and the ER; (**b–e**) 24-h-reperfused neurons show in (**b**) abnormal clusters of vesicles and fragmented bubble-like Golgi (G), which become the most dominant intraneuronal structures, in (**c**) aggregate-associated mitochondrion (M) and autophagosomes (APs), and in (**d**) protein aggregates (*arrows*), dilated ER, and fragmented Golgi (G). Scale bar = 0.5 μm

a whole protein domain has been synthesized on the ribosome. This folding process during polypeptide elongation on a ribosome is referred to as cotranslational folding [16, 19, 20]. Because ribosomes are highly abundant in cells, many nascent peptide chains are regularly emerging from their parent ribosomes at any given moment. Nascent polypeptide chains expose their hydrophobic segments and are highly prone to intramolecular misfolding and intermolecular aggregation driven by the hydrophobic force [20]. Therefore, cotranslational folding generally requires (1) assistant or protective chaperone proteins, (2) their cochaperones, and (3) cellular energy supply. If a nascent peptide chain cannot correctly reach its final conformation, it is normally degraded immediately by the ubiquitin-proteasomal system to avoid aggregation. Cotranslational folding is a very sophisticated process and must cooperate with protein folding and degradation machinery in the crowded cellular milieu [16]. Any errors either owing to wrong genetic coding (such as in genetic neurodegenerative diseases) or dysfunction of protein folding and degradation machinery (such as ischemia, see below) may lead to cotranslational protein aggregation and delayed cell death.

The assistant or protective proteins for protein folding are generally referred to as molecular chaperones. In general, molecular chaperones (or simply chaperones) are able to recognize and shield exposed hydrophobic segments to avoid protein aggregation. There are several families of chaperones and cochaperones in the cytosol involved in regulation of protein folding, trafficking, and degradation. The best known chaperones and their assistant cochaperones for cotranslational folding are the heat-shock protein 70 (HSP70) family and heat-shock protein 40 (HSP40) family [37, 56]. The HSP70 family in mammalian cytosol consists of a constitutive cognate form of HSC70 and a stress-inducible form of HSP70. HSC70 is the major form for cotranslational folding in normal cells. HSC70 contains an N-terminus ATPase domain and a C-terminus peptide-binding domain and, with assistance by HSP40, carries out cycles of binding and release of nascent polypeptides during cotranslational folding in an ATP-dependent manner. Cotranslational folding of nascent peptide chains usually requires numerous such binding and release cycles. At the end of the each cycle, HSC70 must release ADP for rebinding to ATP in order to start another cycle, which is facilitated by a nucleotide exchange factor (NEF) [1]. Therefore, HSC70, HSP40, NEF, and ATP must cooperatively regulate cotranslational folding in order to avoid nascent peptide chain aggregation during protein synthesis.

If nascent peptide chains cannot be folded successfully as a result of translational errors, they must quickly recruit a ubiquitin ligase such as HSC70-interacting protein (CHIP) for degradation [48]. This interaction results in ubiquitination of chaperone-bound polypeptide substrate and degradation of the misfolded substrate by proteasomes, thus linking the cotranslational folding to the proteasomal pathway.

Brain ischemia disables cotranslational folding and degradation machinery, resulting in abnormal aggregation of cotranslational folding complexes together with their associated ribosomes after ischemia [41, 42, 68]. When cellular ATP is decreased to below 80% of its normal cellular level, all ATP-dependent cellular processes are stopped or markedly slowed down [60]. ATP-dependent cellular processes include maintenance of cellular ionic homeostasis [60], cotranslational folding [19], protein chaperoning [20], ubiquitin-proteasome-mediated protein degradation [17], and the autophagy pathway [38]. The common result of these ischemia-induced cellular alterations can lead to the overload of unprocessed nascent peptide chains on ribosomes after ischemia. The following evidence suggests that overload of misfolded nascent polypeptides on ribosomes leads to their abnormal aggregation and irreversible damage to protein synthesis machinery after brain ischemia: (1) Electron microscopy shows that ribosomes are abnormally aggregated in CA1 vulnerable neurons after ischemia (Fig. 5.4); (2) Ribosomal proteins and cotranslational chaperone-cochaperone complexes are highly accumulated into the detergent-insoluble protein aggregate-containing fraction during the postischemic phase (Fig. 5.5 [68]); (3) Despite the fact that ischemia-depleted ATP is virtually recovered in neurons during reperfusion, the overall rate of protein

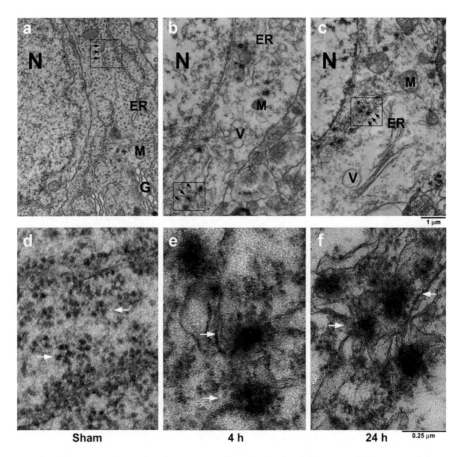

Fig. 5.4 Electron micrographs of osmium-uranium-lead-stained CA1 pyramidal neurons from a sham-operated control rat and rats subjected to 15 min of cerebral ischemia followed by 4 and 24 h of reperfusion. *Upper:* The rough endoplasmic reticulum (ER), mitochondria (M), nucleus (N), Golgi apparatus (G), and ribosomal rosettes (*arrows*) are normally distributed in sham-operated CA1 neurons (**a** Sham). At 4 and 24 h of reperfusion after ischemia, the ER and mitochondria (M) are dilated, membranous vesicles (V) are accumulated, and ribosomes (*arrow*) are clumped into large aggregates (**b, c**). Lower panel: Higher magnification of the ribosomal area (*square*) indicated in the upper panel with three *black arrows* in sham (**a**), 4-h (**b**) and 24-h (**c**) micrographs. Ribosomal rosettes and ER-associated ribosomes are normally distributed in sham-operated control CA1 neurons (**d** *arrows*). After ischemia, ribosomes are abnormally clumped into large aggregates (**e, f** *arrows*). Scale bars: 1 µm in (**a–c**) and 0.25 µm in (**d–f**)

synthesis is irreversibly inhibited in ischemic vulnerable neurons [22, 24]; (4) Persistent inhibition of protein synthesis is a hallmark of delayed neuronal death, i.e., neurons with persistent inhibition of protein synthesis eventually die in a delayed fashion after ischemia [10, 11, 22, 24]. Persistent protein synthesis inhibition after ischemia could not be explained by a brief (about 30 min) eIF2alpha phosphorylation [8, 11–13]. Emerging evidence strongly suggests that irreversible

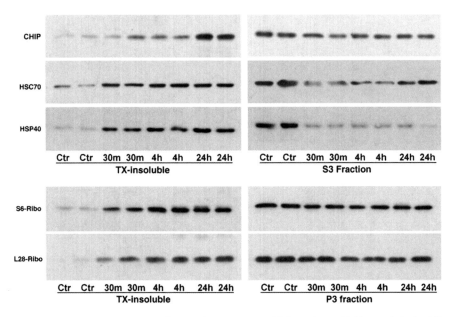

Fig. 5.5 Immunoblots of CHIP, HSC70, HSP40, ribosomal S6 protein, and L28 protein in the 1% Triton X100 and 400 mM KCl detergent/salt-insoluble protein aggregate-containing fraction (*left*) and in either cytosolic S3 fraction or microsomal P3 fraction (*right*) after brain ischemia. Samples were prepared from the dorsolateral neocortical tissues of sham-operated control rats (Ctr) and rats subjected to 20 min of cerebral ischemia followed by 30 min, 4, and 24 h of reperfusion. Each sample was derived from a different rat. Two separate samples in each experimental group were run on the same SDS-PAGE. *Upper immunoblotting panels*: The blots were labeled with antibodies against CHIP, HSC70, HSP40, S6, and L28, and visualized with the ECL system (*upper immunoblot panels*)

inhibition of protein synthesis in vulnerable neurons is caused by protein aggregation after ischemia, and (5) transient cerebral ischemia also leads to irreversible aggregation of components in the ubiquitin-proteasomal system [17].

Cotranslational protein aggregation in ischemic vulnerable neurons explains irreversible inhibition of protein synthesis after an episode of ischemia [41, 42, 68]. As demonstrated in Fig. 5.5, an ischemia-induced cascade of energy failure, intracellular calcium overload, and acidosis cumulatively damages ATP-dependent protein quality control systems for cotranslational folding, cotranslational chaperoning, and translation-coupled protein degradation after brain ischemia [23, 27, 39, 40]. As a result, nascent polypeptide chains cannot fold efficiently. Consequently, ribosomes, together with their associated molecular chaperones, aberrant nascent peptide chains, and protein synthesis initiation and elongation factors, are ubiquitinated and irreversibly aggregated with each other or with other subcellular structures. Cotranslational protein aggregation essentially demolishes protein synthesis machinery in ischemic vulnerable neurons. Cotranslational protein aggregates

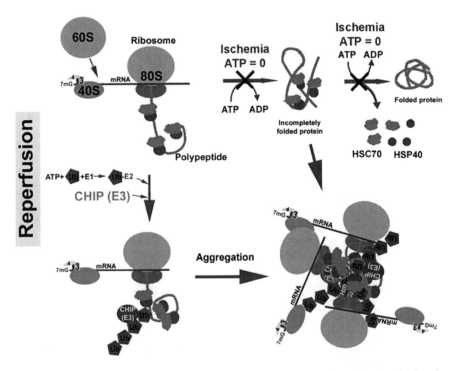

Fig. 5.6 A hypothetical model of cotranslational protein aggregation after brain ischemia. Protein synthesis takes place on translation complexes composed of 40S and 80S ribosomes, messenger RNAs with a cap structure (7mGpppAmpApCpApGp), protein synthesis initiation factors such as 1, 3, and 4, cotranslational chaperones, e.g., HSC70 and HSP40, and translation-coupled protein degradation components including ubiquitin, proteasomes, and ubiquitin ligase CHIP. An ischemia-induced cascade of energy failure, intracellular calcium overload, and acidosis cumulatively damages ATP-dependent protein quality control machinery for cotranslational folding, cotranslational chaperoning, and translation-coupled protein degradation after brain ischemia. As a result, nascent peptide chains cannot fold efficiently. Consequently, ribosomes, together with their associated molecular chaperones, aberrant nascent polypeptides, and protein translational initiation, elongation, and termination factors, are ubiquitinated. Ubiquitinated abnormal proteins are too numerous to be degraded; they irreversibly aggregate with each other or with other subcellular structures. Cotranslational protein aggregates accumulate over time, and demolish protein synthesis machinery and their associated organelles, and eventually contribute to delayed neuronal death in ischemic vulnerable neurons after ischemia

accumulate over time and contribute to delayed neuronal death in ischemic vulnerable neurons after ischemia (Fig. 5.6).

Protein Aggregation Damages ER: The active role of ER damage in ischemic neuronal death has long been suspected, but it was only until recently that signs of ER failure have been gradually revealed [11, 52, 53]. These signs of ER damage include (1) diminished ER stress-induced increase in eukaryotic initiation factor 2alpha (eIF2) phosphorylation and (2) persistent inhibition of protein synthesis, in vulnerable CA1, relative to resistant DG neurons (Fig. 5.7). The signs of ER damage are likely caused by protein aggregation on ER membranes as observed by EM (see Fig. 5.3d).

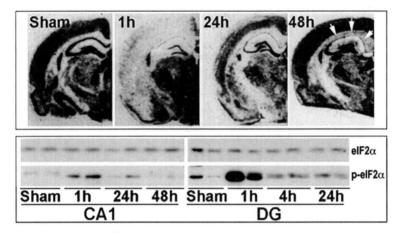

Fig. 5.7 *Upper*: [14C]-leucine incorporation of protein synthesis. Protein synthesis is persistently inhibited in vulnerable CA1 but is lesser affected in the resistant DG neurons after ischemia (the autoradiographs in the upper panel were produced in Dr. Wieloch's laboratory, Lund University, Sweden). *Lower*: The total eIF2a level is unchanged, but the phosphorylated eIF2a (p-eIF2a) is increased only slightly in CA1 region and dramatically in the resistant DG area after ischemia. The data show diminished ER stress response or persistent ER damage in CA1 neurons after ischemia

Protein Aggregation and Lysosomal Damage: Lysosomal rupture can culminate in cell demise. Recent studies show that lysosomal malfunction without structural rupture also leads to delayed neuronal degeneration as seen in a group of inherited lysosomal enzyme-deficiency diseases [5]. Lysosomal enzyme deficiency-induced neuronal changes have many features of autophagy and organelle damage similar to those observed after brain ischemia. Recent studies from Hu's laboratory (unpublished data) further show that lysosomal membrane proteins LAMP1 and LAMP2/LAMP2a dramatically accumulate in CA1 neurons after ischemia (Fig. 5.8), suggesting that lysosomal deficiency and autophagy damage after brain ischemia [38, 58].

Mitochondrial Damage After Brain Ischemia: The role of mitochondrial dysfunction in mechanisms of ischemic brain damage has long been acknowledged. Two decades ago, calcium overload as result of excitotoxicity was considered as a major cause of mitochondrial damage. These conclusions were supported by data showing that transient ischemia is followed by a gradual rise in Ca_i^{2+} [62], by delayed calcium sequestration in mitochondria [67], and by delayed mitochondrial respiratory dysfunction [63]. Another hypothesis is that direct oxidation of mitochondrial proteins by free radicals inhibits mitochondrial respiration [2]. An additional possibility is that the mitochondrial inner membrane permeability transition (MPT) is involved in respiratory dysfunction [34, 36, 61].

The mitochondria damaged by MPT-induced extensive swelling can be removed from the cytosol. For example, ubiquitinated proteins and abnormal protein aggregate associated with mitochondria have been detected in the vulnerable CA1 sector

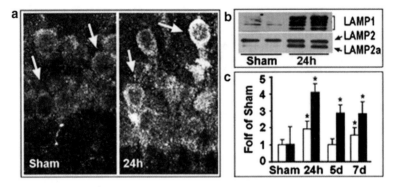

Fig. 5.8 (**a**) LAMP2 immunostaining (*arrows*) is increased at 24 h of reperfusion in CA1 neurons after ischemia. (**b**) Immunoblottings also show that LAMP1 and LAMP2/LAMP2a levels are dramatically increased in the CA1 region after ischemia. (**c**) Quantification of LAMP2 (*white bar*) and LAMP2a (*black bar*) levels in (**b**). Data are expressed as mean ± SD (n = 6), one-way ANOVA, Dunnett's test (*p < 0.01)

of the hippocampus following forebrain ischemia (Fig. 5.3c; [28]). Ultrastructural studies of neuronal mitochondria after ischemic insult revealed several types of changes in mitochondrial morphology. Dispersed around the swollen mitochondria were mitochondria with poorly defined cristae and increased matrix density. When ischemia was followed by 2 h of reperfusion, some mitochondria demonstrated dilated cristae within an electron-dense matrix. These morphological changes are similar to those we observed when isolated mitochondria were exposed to 100 nmol Ca^{2+}/mg protein in the absence of adenine nucleotides [35]. Despite the morphological and functional damage, mitochondria are still capable of producing ATP to keep cell survival for a period of 2–3 days after transient cerebral ischemia [34]. However, without functional intracellular regenerating, repair, and removal systems such as the ER, Golgi, and autophagy, mitochondrial damage continues. Multiple organelle damage develops over time and eventually leads to neuronal death in a delayed fashion after brain ischemia.

4 Concluding Remark

Protein aggregation in ischemic vulnerable neurons should represent a new mechanism underlying delayed neuronal death after brain ischemia. An ischemia-induced cascade of energy failure, intracellular calcium overload, overproduction of ROS, and acidosis cumulatively disables and damages ATP-dependent protein quality control machinery for protein folding, transportation, assembly, and degradation after brain ischemia, leading to abnormal protein aggregation with subcellular organelles (see Fig. 5.3). Protein aggregation-induced organelle damage accumulates over time and, when it reaches a critical degree, may lead to delayed cell death after ischemia. This hypothesis seems consistent with several existing ones: (1) Energy failure and loss of ionic homeostasis seems necessary to induce neuronal damage after ischemia. Both depletion of ATP and loss of ionic homeostasis disable cotranslational folding machinery and cause overproduction of toxic abnormal proteins in cells [20, 52]; (2) Permanent inhibition of protein synthesis is a reliable indicator for neuronal death after ischemia [10, 22, 24, 46]. Abnormal aggregation of ribosomal proteins results in permanent inhibition of protein synthesis after brain ischemia; (3) Free ubiquitin is permanently depleted in ischemic vulnerable neurons [15, 25, 28, 43, 47]. Abnormal proteins accumulating in neurons deplete intracellular free ubiquitin to form ubi-proteins after ischemia; (4) Expression of molecular chaperones before ischemia protects neurons against ischemia [18, 50, 56, 59, 66]. Molecular chaperones may directly neutralize accumulated unfolded proteins and assist in protein degradation. However, molecular chaperones are less effective in dissociation of protein aggregates, and protein aggregation is virtually irreversible [28]; (5) Dark substances or materials were observed previously in CA1 neurons after ischemia [14, 33, 64]. It should be mentioned that it is still a subject of debate whether protein aggregates per se are toxic to cells. However, it is generally held that protein aggregation reflects cellular overload of abnormal proteins that may exert highly toxic effects on neurons in the early stage before aggregation [57]. The studies from the authors' laboratory support the idea that clumping of inactive protein aggregates on cellular machinery such as protein synthesis machinery and multiple organelles in living neurons for an extended period of time will eventually lead to neuronal death after brain ischemia.

References

1. Alberti S, Esser C, Hohfeld J. BAG-1—a nucleotide exchange factor of Hsc70 with multiple cellular functions. Cell Stress Chaperones. 2003;8:225–31.
2. Almeida A, Brooks KJ, Sammut I, Keelan J, Davey GP, Clark JB, Bates TE. Postnatal development of the complexes of the electron transport chain in synaptic mitochondria from rat brain. Dev Neurosci. 1995;17:212–8.
3. Alves-Rodrigues A, Gregori L, Figueiredo-Pereira ME. Ubiquitin, cellular inclusions and their role in neurodegeneration. Trends Neurosci. 1998;21:516–20.
4. Auer RN, Kalimo H, Olsson Y, Siesjo BK. The temporal evolution of hypoglycemic brain damage. II. Light- and electron-microscopic findings in the hippocampal gyrus and subiculum of the rat. Acta Neuropathol. 1985;67:25–36.
5. Bellettato CM, Scarpa M. Pathophysiology of neuropathic lysosomal storage disorders. J Inherit Metab Dis. 2010;33:347–62.
6. Bucciantini M, Giannoni E, Chiti F, Baroni F, Formigli L, Zurdo J, Taddei N, Ramponi G, Dobson CM, Stefani M. Inherent toxicity of aggregates implies a common mechanism for protein misfolding diseases. Nature. 2002;416:507–11.
7. Bukau B, Hesterkamp T, Luirink J. Growing up in a dangerous environment: a network of multiple targeting and folding pathways for nascent polypeptides in the cytosol. Trends Cell Biol. 1996;6:480–6.
8. Burda J, Martin ME, Garcia A, Alcazar A, Fando JL, Salinas M. Phosphorylation of the alpha subunit of initiation factor 2 correlates with the inhibition of translation following transient cerebral ischaemia in the rat. Biochem J. 1994;302:335–8.

9. Burry RW, Lasher RS. A quantitative electron microscopic study of synapse formation in dispersed cell cultures of rat cerebellum stained either by Os-UL or by EPTA. Brain Res. 1978;147:1–15.

10. Cooper HK, Zalewska T, Kawakami S, Hossmann KA, Kleihues P. Delayed inhibition of protein synthesis during recirculation after compression ischemia of the rat brain. Acta Neurol Scand Suppl. 1977;64:130–1.

11. DeGracia DJ, Hu BR. Irreversible translation arrest in the reperfused brain. J Cereb Blood Flow Metab. 2007;27:875–93.

12. DeGracia DJ, Kumar R, Owen CR, Krause GS, White BC. Molecular pathways of protein synthesis inhibition during brain reperfusion: implications for neuronal survival or death. J Cereb Blood Flow Metab. 2002;22:127–41.

13. DeGracia DJ, Neumar RW, White BC, Krause GS. Global brain ischemia and reperfusion: modifications in eukaryotic initiation factors associated with inhibition of translation initiation. J Neurochem. 1996;67:2005–12.

14. Deshpande J, Bergstedt K, Linden T, Kalimo H, Wieloch T. Ultrastructural changes in the hippocampal CA1 region following transient cerebral ischemia: evidence against programmed cell death. Exp Brain Res. 1992;88:91–105.

15. Dewar D, Graham DI, Teasdale GM, McCulloch J. Cerebral ischemia induces alterations in tau and ubiquitin proteins. Dementia. 1994;5:168–73.

16. Frydman J. Folding of newly translated proteins in vivo: the role of molecular chaperones. Annu Rev Biochem. 2001;70:603–47.

17. Ge P, Luo Y, Liu CL, Hu B. Protein aggregation and proteasome dysfunction after brain ischemia. Stroke. 2007;38:3230–6.

18. Giffard RG, Xu L, Zhao H, Carrico W, Ouyang Y, Qiao Y, Sapolsky R, Steinberg G, Hu B, Yenari MA. Chaperones, protein aggregation, and brain protection from hypoxic/ischemic Injury. J Exp Biol. 2004;207:3213–20.

19. Hardesty B, Tsalkova T, Kramer G. Co-translational folding. Curr Opin Struct Biol. 1999;9:111–4.

20. Hartl FU, Hayer-Hartl M. Molecular chaperones in the cytosol: from nascent chain to folded protein. Science. 2002;295:1852–8.

21. Hayashi T, Tanaka J, Kamikubo T, Takada K, Matsuda M. Increase in ubiquitin conjugates dependent on ischemic damage. Brain Res. 1993;620:171–3.

22. Hossmann K-A. Disturbances of cerebral protein synthesis and ischemic cell death. Prog Brain Res. 1993;96:167–77.

23. Hu BR. Co-translational protein folding and aggregation after brain ischemia. In: Chan PH, Lajtha A, editors. Handbook of neurochemistry and molecular neurobiology, Acute ischemic injury and repair in the nervous system, vol. 23. 3rd ed. Berlin: Springer; 2007. p. 109–20.

24. Hu BR, Wieloch T. Stress-induced inhibition of protein synthesis initiation: modulation of initiation factor 2 and guanine nucleotide exchange factor activity following transient cerebral ischemia in the rat. J Neurosci. 1993;13:1830–8.

25. Hu BR, Janelidze S, Ginsberg MD, Busto R, Perez-Pinzon M, Sick TJ, Siesjo BK, Liu CL. Protein aggregation after focal brain ischemia and reperfusion. J Cereb Blood Flow Metab. 2001;21:865–75.

26. Hu BR, Kamme F, Wieloch T. Alterations of Ca2+/calmodulin-dependent protein kinase II and its messenger RNA in the rat hippocampus following normo- and hypothermic ischemia. Neuroscience. 1995;68(4):1003–16.

27. Hu BR, Martone ME, Liu CL. Protein aggregation, unfolded protein response and delayed neuronal death after brain ischemia. In: Buchan VA, Ito U, editors. Maturation phenomenon in cerebral ischemia. Heidelberg: Springer; 2004. p. 225–37.

28. Hu BR, Martone ME, Jones YZ, Liu CL. Protein aggregation after transient cerebral ischemia. J Neurosci. 2000;20(9):3191–9.

29. Hu BR, Park M, Martone ME, Fischer WH, Ellisman MH, Zivin JA. Assembly of proteins to postsynaptic densities after transient cerebral ischemia. J Neurosci. 1998;18:625–33.

30. Hu BR, Wieloch T. Persistent translocation of Ca2+/calmodulin-dependent protein kinase II to synaptic junctions in the vulnerable hippocampal CA1 region following transient ischemia. J Neurochem. 1995;64:277–84.
31. Ito U, Spatz M, Walker Jr JT, Klatzo I. Experimental cerebral ischemia in Mongolian gerbils I. Light microscopic observations. Acta Neuropathol. 1975;32:209–23.
32. Kirino T. Delayed neuronal death in the gerbil hippocampus following ischemia. Brain Res. 1982;239:57–69.
33. Kirino T, Tamura A, Sano K. Delayed neuronal death in the rat hippocampus following transient forebrain ischemia. Acta Neuropathol. 1984;64:139–47.
34. Kristián T, Siesjö BK. Calcium in ischemic cell death. Stroke. 1998; 29:705–18.
35. Kristian T, Bernardi P, Siesjö BK. Acidosis promotes the permeability transition in energized mitochondria: implications for reperfusion injury. J Neurotrauma. 2001;18:1059–74.
36. Kristián T. Metabolic stages, mitochondria and calcium in hypoxic/ischemic brain damage. Cell Calcium. 2004;36:221–33.
37. Li GC, Mivechi NF, Weitzel G. Heat shock proteins, thermotolerance, and their relevance to clinical hyperthermia. Int J Hyperthermia. 1995;11:459–88.
38. Liu C, Gao Y, Barrett J, Hu B. Autophagy and protein aggregation after brain ischemia. J Neurochem. 2010;115:68–78.
39. Liu CL, Hu BR. Protein ubiquitination in postsynaptic densities following transient cerebral ischemia. J Cereb Blood Flow Metab. 2004;24(11):1219–25.
40. Liu CL, Hu BR. Alterations of N-ethylmaleimide-sensitive ATPase following transient cerebral ischemia. Neuroscience. 2004;128(4):767–74.
41. Liu CL, Ge P, Zhang F, Hu BR. Co-translational protein aggregation after transient cerebral ischemia. Neuroscience. 2005a;134:1273–84.
42. Liu C, Chen S, Kamme F, Hu BR. Ischemic preconditioning prevents protein aggregation after transient cerebral ischemia. Neuroscience. 2005b;134:69–80.
43. Magnusson K, Wieloch T. Impairment of protein ubiquitination may cause delayed neuronal death. Neurosci Lett. 1989;96:264–70.
44. Martone ME, Jones YZ, Young SJ, Ellisman MH, Zivin JA, Hu BR. Modification of postsynaptic densities after transient cerebral ischemia: a quantitative and three-dimensional ultrastructural study. J Neurosci. 1999;19:1988–97.
45. Martone ME, Hu BR, Ellisman MH. Alterations of hippocampal postsynaptic densities following transient ischemia. Hippocampus. 2000;10:610–6.
46. Mies G, Ishimaru S, Xie Y, Seo K, Hossmann KA. Ischemic thresholds of cerebral protein synthesis and energy state following middle cerebral artery occlusion in rat. J Cereb Blood Flow Metab. 1991;11:753–61.
47. Morimoto T, Ide T, Ihara Y, Tamura A, Kirino T. Transient ischemia depletes free ubiquitin in the gerbil hippocampal CA1 neurons. Am J Pathol. 1996;148:249–57.
48. Murata S, Chiba T, Tanaka K. CHIP: a quality-control E3 ligase collaborating with molecular chaperones. Int J Biochem Cell Biol. 2003;35:572–8.
49. Nedergaard M. Neuronal injury in the infarct border: a neuropathological study in the rat. Acta Neuropathol. 1987;73:267–74.
50. Nowak Jr TS. Synthesis of a stress protein following transient ischemia in the gerbil. J Neurochem. 1985;45:1635–41.
51. Ouyang YB, Hu BR. Protein ubiquitination in rat brain following hypoglycemic coma. Neurosci Lett. 2001;298:159–62.
52. Paschen W. Role of calcium in neuronal cell injury: which subcellular compartment is involved? Brain Res Bull. 2000;53:409–13.
53. Paschen W. Endoplasmic reticulum dysfunction in brain pathology: critical role of protein synthesis. Curr Neurovasc Res. 2004;1:173–81.
54. Petito CK, Lapinski RL. Postischemic alterations in ultrastructural cytochemistry of neuronal Golgi apparatus. Lab Invest. 1986;55:696–702.

55. Rafols JA, Daya AM, O'Neil BJ, Krause GC, Neumar RW, White BC. Global brain ischemia and reperfusion: Golgi apparatus ultrastructure in neurons selectively vulnerable to death. Acta Neuropathol. 1995;90:17–30.
56. Rajdev S, Hara K, Kokubo Y, Mestril R, Dillmann W, Weinstein PR, Sharp FR. Mice overexpressing rat heat shock protein 70 are protected against cerebral infarction. Ann Neurol. 2000;47:782–91.
57. Ross CA, Wood JD, Schilling G, Peters MF, Nucifora Jr FC, Cooper JK, Sharp AH, Margolis RL, Borchelt DR. Polyglutamine pathogenesis. Philos Trans R Soc Lond B Biol Sci. 1999;354:1005–11.
58. Saftig P, Schröder B, Blanz J. Lysosomal membrane proteins: life between acid and neutral conditions. Biochem Soc Trans. 2010;38:1420–3.
59. Sharp FR, Massa SM, Swanson RA. Heat-shock protein protection. Trends Neurosci. 1999;22:97–9.
60. Siesjö BK, Siesjö P. Mechanisms of secondary brain injury. Eur J Anaesthesiol. 1996;13:247–68.
61. Siesjö BK, Elmér E, Janelidze S, Keep M, Kristián T, Ouyang YB, Uchino H. Role and mechanisms of secondary mitochondrial failure. Acta Neurochir Suppl. 1998;73:7–13.
62. Silver IA, Erecińska M. Ion homeostasis in rat brain in vivo: intra- and extracellular [Ca2+] and [H+] in the hippocampus during recovery from short-term, transient ischemia. J Cereb Blood Flow Metab. 1992;12:759–72.
63. Sims NR, Pulsinelli WA. Altered mitochondrial respiration in selectively vulnerable brain subregions following transient forebrain ischemia in the rat. J Neurochem. 1987;49:1367–74.
64. Tomimoto H, Yanagihara T. Electron microscopic investigation of the cerebral cortex after cerebral ischemia and reperfusion in the gerbil. Brain Res. 1992;598:87–97.
65. Truettner JS, Hu K, Liu CL, Dietrich WD, Hu B. Subcellular stress response and induction of molecular chaperones and folding proteins after transient global ischemia in rats. Brain Res. 2009;1249:9–18.
66. Yenari MA, Dumas TC, Sapolsky RM, Steinberg GK. Gene therapy for treatment of cerebral ischemia using defective herpes simplex viral vectors. Ann N Y Acad Sci. 2001;939:340–57.
67. Zaidan E, Sims NR. The calcium content of mitochondria from brain subregions following short-term forebrain ischemia and recirculation in the rat. J Neurochem. 1994;63:1812–9.
68. Zhang F, Liu CL, Hu BR. Irreversible aggregation of protein synthesis machinery after focal brain ischemia. J Neurochem. 2006;98:102–12.

Chapter 6
Antioxidants and Stroke: Success and Pitfalls

Fernanda M.F. Roleira, Elisiário J. Tavares-da-Silva, Jorge Garrido, and Fernanda Borges

Abstract Stroke is a problem that affects more than 15 million people worldwide. It is the third most common cause of death and is also an important cause of morbidity and of severe long-term disability. In this chapter, stroke mechanisms are outlined and particular attention is given to the oxidative and nitrosative stress, essentially to the role of the reactive oxygen (ROS) and nitrogen (RNS) species in stroke events. Following this approach, the most recent literature relative to success and pitfalls of the use of antioxidants of natural (including endogenous and those from diet) or synthetic origin in stroke is herein reviewed. Antioxidant therapies have enjoyed general success in preclinical studies, across disparate animal models, but little benefit in human intervention studies or clinical trials. This mismatch has been often attributed both to limitations of the animal models and to pitfalls in the clinical trial design. Finally, some ADME/Tox problems related to a number of drawbacks were also discussed.

1 Introduction

1.1 Stroke: The Problem

Stroke is a problem that affects more than 15 million people worldwide. It is the third most common cause of death, after coronary heart disease and cancer, among

F.M.F. Roleira (✉) • E.J. Tavares-da-Silva
CEF/Grupo de Química Farmacêutica, Universidade de Coimbra, Coimbra, Portugal
e-mail: froleira@ff.uc.pt

J. Garrido
CIQUP/Departamento de Engenharia Química, Instituto Superior de Engenharia,
IPP, Porto, Portugal

F. Borges (✉)
CIQUP/Departamento de Química e Bioquímica, Universidade do Porto, Porto, Portugal
e-mail: fborges@fc.up.pt

P.A. Lapchak and J.H. Zhang (eds.), *Translational Stroke Research*, Springer Series
in Translational Stroke Research, DOI 10.1007/978-1-4419-9530-8_6,
© Springer Science+Business Media, LLC 2012

adult population, particularly in the elderly. The overall death rate will increase in the next decade by 12% globally and 20% in low-income families. Stroke is also an important cause of morbidity and of severe long-term disability. In fact, the problems of stroke are not consequences of mortality, but are related to the stroke patients who survive but are physically and mentally disabled by stroke-induced brain damage. More than 30% of stroke survivors will have severe disability and it has been calculated that by 2015 over 50 million healthy life-years will be lost because of stroke [1, 2].

Strokes are usually divided into the following three categories: (a) Ischemic stroke, which is known to have a large number of different causes and/or predisposing factors (e.g., hypertension, diabetes, atherothrombosis, and cigarette smoking); (b) Spontaneous intracerebral haemorrhage that has hypertension, alcoholic beverage, and/or use of anticoagulant treatment as risk factors; (c) Aneurismal subarachnoid haemorrhage that has cigarette smoking as a major risk factor besides the above mentioned [1, 2]. Ischemic stroke accounts for 70–80% of all strokes, while haemorrhagic stroke is responsible for almost 15%. Transient global cerebral ischemia can occur during cardiac arrest, cardiopulmonary bypass surgery, and other situations that deprive the brain of oxygen and glucose. Ischemia rapidly causes a cascade of events in vulnerable regions of the brain, including the hippocampus, striatum, cerebral cortex, and cerebellum that eventually lead to neuronal damage and death. Therefore, the affected tissues can die, often leading to permanent impairment of mental and physical faculties, the shutdown of essential functions, or even death [1, 2].

1.2 Stroke Mechanisms: An Outline

Stroke is a poorly understood multifactorial event and only partially predictable by means of clinical, laboratory, and imaging data. Ischemic stroke is caused by a transient or permanent local reduction of cerebral blood flow, resulting in the deficiency of glucose and oxygen supply to the affected brain region. This usually happens due to vessel occlusion by an embolus or to a primary thrombosis of a major cerebral artery [1, 2]. Starved of oxygen, the brain cells die. However, cell death continues to occur for many hours, even after blood flow is returned to the brain. Many of the cells that are injured, but not killed by oxygen deprivation, die in the hours following the stroke often killed by additional mechanisms.

The mechanisms involved in ischemic injury and subsequent neurological damage are of different aetiology and complexity. Briefly, at least four fundamental mechanisms leading to cell death during ischemic brain injury have been recently proposed: excitotoxicity and ionic imbalance, inflammation, oxidative/nitrosative stress, and apoptotic-like cell death [3–5]. In short, the reduction of blood flow in the brain leads to a decrease of the production of high-energy phosphates. The energy failure causes membrane depolarization and uncontrolled release of excitatory aminoacids, such as glutamate, in the extracellular space (excitotoxicity).

Glutamate acts on various types of receptors causing calcium overload of neuronal cells. In turn, calcium activates proteolytic enzymes that begin to degrade both intracellular and extracellular structures and other enzymes (e.g., phospholipase A2 and cyclooxygenase [COX]). Many of these events are influenced by the generation of reactive species that exert their deleterious effects through lipid peroxidation and membrane damage. Secondary to ischemia proinflammatory genes are expressed and several inflammatory mediators are released. Although excitotoxicity typically leads to necrosis, the occurrence of apoptosis has also been reported to occur after cerebral ischemia [3–5].

In general, all the mentioned mechanisms mediate injury within neurons, glia, and vascular elements, and at the subcellular level, they impact the function of mitochondria, nuclei, cell membranes, endoplasmic reticula, and lysosomes. Therefore, it is believed that therapeutics that simultaneously targets several of these mechanisms could be effective for stroke treatment [6].

2 Stroke and Oxidative/Nitrosative Stress

2.1 General Concepts

Reactive species (RS) of oxygen (ROS) or nitrogen (RNS) are produced during the normal cell metabolic activities. In order to maintain the delicate balance of RS generation and detoxification and to prevent their accumulation in the body, several antioxidant cellular defence systems exist that comprise antioxidant enzymes (e.g., superoxide dismutase, SOD; catalase, CAT; glutathione peroxidase, GPx; glutathione transferase, GST) and non-enzymatically acting compounds such as endogenous antioxidants (e.g., α-tocopherol, vitamin E; ascorbic acid, vitamin C; glutathione, GSH) (Fig. 6.1) [7].

Oxidative and nitrosative stress (O&NS) state occurs when cellular antioxidant defences are insufficient to keep the levels of reactive species below a toxic threshold. These states may be due to excessive production of RS or to the failure of antioxidant defences, or both. In conclusion, O&NS results from an imbalance in redox state, which occurs when the steady-state balance of pro-oxidants to antioxidants is shifted in the direction of the former, creating the potential for oxidative damage [7].

2.2 Oxidative and Nitrosative Stress

O&NS is linked to several pathological processes, namely neurodegenerative and cardiovascular diseases, cancer, diabetes, and may therefore contribute significantly to disease mechanisms. O&NS has been related to inflammation, shock, stroke, and ischemia/reperfusion injury [4, 7–9]. In particular, there is to date a large body of

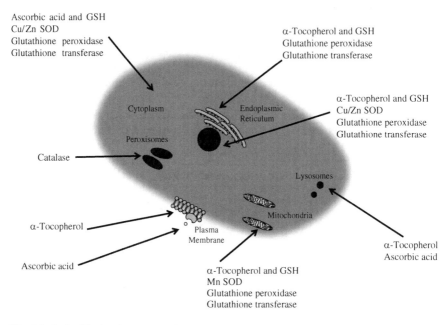

Fig. 6.1 Antioxidant endogenous system

evidence, from experimental as well as from clinical studies, that implicates O&NS in the pathophysiology of acute ischemic stroke (AIS).

Brain is especially prone to O&NS damage since it is characterized by: (a) high-energy requirements and high oxygen consumption; (b) an enrichment in peroxidizable fatty acids; (c) high levels of transition metals, namely iron; (d) a relative deficit of antioxidant defences, such as CAT and GPx when compared with other tissues.

During brain ischemia, a highly complex cascade of metabolic events is initiated, several of which converge in the generation of oxygen and nitrogen reactive species. Three major classes of pro-oxidant enzymes, which lead to the generation of RS, may take part in oxidative stress propagation during cerebral ischemia: (a) nitric oxide synthase (NOS); (b) COX, xanthine oxidase and NADPH oxidase; (c) myeloperoxidase (MPO) and monoamine oxidase (MAO). Free radicals are produced in high concentration from the activation of phospholipase A2 by lipid peroxidation and breakdown of lipid membranes, as well as from the free fatty acid metabolism, via the cyclooxygenase pathway and the metabolism of adenine nucleotides and purines, via the xanthine oxidase pathway.

Reactive species are believed to mediate much of the damage that occurs after transient brain ischemia and in the penumbral region of infarcts caused by permanent ischemia. In these cases, a failure of antioxidant protective system occurs that leads to the accumulation of reactive species, such as superoxide anion and nitric oxide, which leads to the destabilization of cellular membranes, damage of blood–brain barrier, disintegration of DNA, and ultimately, to the neuronal death. During reperfusion, the oxidative and nitrosative burden is so immense that endogenous antioxidant defence system is impaired and the antioxidants present are completely consumed. Upon reoxygenation, O&NS is rapidly built up and numerous non-enzymatic oxidation reactions take place in the cytosol or in organelles. Nevertheless, although studies in animal models of cerebral ischemia have pointed to a major role of O&NS in ischemia/reperfusion injury, evidence of reperfusion damage in human stroke is scant [4, 7–10].

In general, O&NS alter the structural and functional integrity of cells by a variety of mechanisms causing, for example, a loss of cell membrane fluidity; a disturbance of the electrolyte balance, mainly by altering the membrane channels; an interfere with the regulation of vascular smooth muscle cells; and inactivation of a series of enzymes. In addition, O&NS cause structural DNA lesions. As major intracellular O&NS sensitive targets, mitochondria, intracellular signalling cascades, and transcription factors have been identified.

2.2.1   Reactive Species

Reactive Oxygen Species

ROS are continuously generated in the body as a consequence of aerobic metabolism. In general, many ROS also exert vital functions that range from host defence to neuronal signal transduction (e.g., H_2O_2 as inducer of transcription factors). When generated in excess, ROS can have destructive effects in molecular targets (lipids, proteins, nucleic acids, or carbohydrates), modifying their chemical structure and generating oxidation-derived products that are usually assumed as markers of biomolecular oxidative damage.

Several ROS such as superoxide ($O_2^{\cdot-}$), perhydroxyl ($HO_2^{\cdot}$), hydroxyl ($^{\cdot}OH$) radicals, and hydrogen peroxide (H_2O_2) are formed in the brain during the initial reduction of oxygen (Fig. 6.2). Hydroxyl ($^{\cdot}OH$) radicals are potent oxidants, whereas $O_2^{\cdot-}$ radicals are comparatively less potent but have a longer half-life and are capable of generating $^{\cdot}OH$ radicals (Haber–Weiss reaction). This reaction is particularly important in the presence of iron which mediates generation of the $^{\cdot}OH$ radicals from H_2O_2 (Fenton reaction). Another mechanism of $^{\cdot}OH$ production is through $O_2^{\cdot-}$ and nitric oxide (NO), which is constantly produced in the brain by NO synthase (NOS). Superoxide radical ($O_2^{\cdot-}$) can be also produced by glutamate receptor activation or through the activation of endogenous NADPH-dependent oxidase. Glutamate may also directly activate oxidative pathways leading to the intracellular accumulation of peroxides (Fig. 6.2) [7–10].

Recent studies have focused on the role of $O_2^{\cdot-}$ generating systems in immune cells and their consequences on reperfusion injury. In fact, immune cells which infiltrate ischemic tissue or plug ischemic microvasculature can also generate reactive species, through several enzyme systems such as nicotinamide adenine dinucleotide phosphate (NADPH) oxidase originally found on leukocytes, but now recognized in several types of cells in the brain. Inhibition of NADPH oxidase can potentially reduce the amount of $O_2^{\cdot-}$ generated during reperfusion, and thus limit

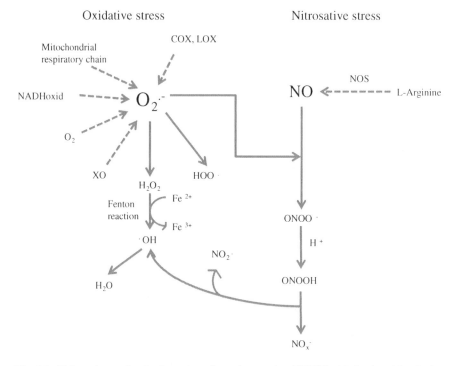

Fig. 6.2 Main pathways for the formation of reactive species. *NADPHoxid* nicotinamide adenine dinucleotide phosphate-oxidase; *XO* xanthine oxidase; *COX* cyclooxygenase; *LOX* 5-lipoxygenase; *NOS* nitric oxide synthases

reperfusion injury. Such a strategy has the potential to treat AIS and reduce complications of recanalizing strategies by using it in combination with thrombolytics or mechanical thrombectomy devices [4, 7–10].

Reactive Nitrogen Species

Nitric oxide (NO) is an inorganic gas that plays a part in the control of cerebral blood flow, thrombogenesis, and modulation of neuronal activity. NO is synthesized by three different isoforms of the enzyme NOS and is produced in the endothelial cells, neurons, glia, and macrophages. NOS mediates the conversion of L-arginine and oxygen to NO and citrulline (Fig. 6.2) [10, 11].

In normal conditions, NO, produced by the constitutive endothelium NOS (eNOS), is the transmitter of vasodilatation signal from endothelial cells to vascular smooth muscles and hence it is neuroprotective. In contrast, NO produced by the neuronal and inducible isoforms of NOS (nNOS, iNOS) can be neurotoxic. This probably occurs through NO-induced formation of peroxynitrite (ONOO⁻) and other toxic free radicals leading to lipid peroxidation damage. NO can also potentiate damage by inhibiting enzymes needed for mitochondrial respiration, glycolysis, and DNA replication. Moreover, NO has been reported to stimulate the release of the neurotransmitter glutamate and could contribute to excitotoxicity. NO outcompetes the SOD reaction using $O_2^{\cdot-}$ to produce $ONOO^-$. The latter reacts rapidly with CO_2 to produce short-lived reaction intermediates that are probably responsible for many of the cytotoxic effects of NO (Fig. 6.2) [10, 11].

Under cerebral ischemia, high concentrations of NO generated by the calcium-dependent activation of the constitutive neuronal NOS (nNOS) and by the activation of the inducible form of NOS (iNOS) in macrophages and other cell types intervene in inflammatory and cytotoxic actions that lead to neuronal death. It is also known that cytotoxicity and inflammatory stimuli (by proinflammatory cytokines) upregulate the genes of NOS that leads to the long-lasted generation of NO and therefore to the activation of enzymatic pathways of necrosis and apoptosis. Consequently, inhibition of NOS and NO production has been considered to be effective therapeutic strategies of stroke treatment. Nevertheless, due to the dual neurotoxic and neuroprotective role of NO in cerebral ischemia, conflicting findings in experimental stroke models have also been generated.

Transition Metals

Iron is the most abundant metal in the brain. The highest concentrations are in the basal ganglia, but an iron staining was also detected in the cerebral cortices. The chief storage sites of iron, oligodendrocytes, are found in close association with blood vessels where they may play a role in transport of iron across the blood–brain barrier. Neurons also show iron staining and neuronal iron stores increase progressively with age [12].

During a transient or a permanent reduction in cerebral blood flow (stroke), there is impaired delivery of oxygen to the tissue. This leads to energy failure and cell death during which cytoplasmic iron is released. Some of this iron is protein bound; however, a series of enzymatic processes and proteolysis liberate the iron. Free iron is particularly dangerous and can be taken up by the surrounding cells in peri-infarctive region. Ionic iron catalyses the generation of ROS via Fenton reaction and there is some indirect evidence that it may exacerbate the damage caused by ischemia (Fig. 6.2) [12].

3   Stroke and Therapy

Stroke continues to kill 5.5 million people each year and the development of safe and effective treatments is a major challenge to experimental and clinical neuroscience. Despite decades of research, no therapies that can prevent the neuronal death and the ensuing neurological deficits after stroke are currently available.

The major therapeutic strategy for treatment of AIS is rapid recanalization, either by pharmacological means through thrombolytic agents or mechanical thrombectomy. However, the time window (up to 3 h after the onset of symptoms) for intervention limits these therapies to a small number of patients, and their inappropriate use can actually worsen outcome, such as worsened brain oedema or symptomatic brain haemorrhage, a phenomenon commonly referred to as "reperfusion injury."

The challenge is now to investigate the various mechanisms underlying and playing a key role in the post-ischemic injury at the molecular and cellular level as well as factors influencing an individual to ischemic injury. Neuroprotective agents with extended therapeutic window that prevent multiple neurochemical cascades are therefore urgently required either for the acute phase of stroke or for the ischemia and reperfusion.

4 Antioxidants and Stroke: Success and Pitfalls

Antioxidants are substances that can protect cells from the damage caused by RS and transition metals. They have been defined as any substance that when present at low concentrations, compared to those of an oxidizable substrate, significantly delays or prevents oxidation of biomolecules. Antioxidants may exert their effects by different mechanisms, such as suppressing the formation of RS (scavenging activity) and/or by sequestering transition metal ions (chelation activity) and/or inhibiting enzymes involved in the production of RS (Fig. 6.3). Briefly, they operate by preventing or slowing the progression of oxidative damage reactions.

Literature review shows that antioxidant therapies for stroke have enjoyed general success in preclinical studies across disparate animal models, but little benefit in human intervention studies or clinical trials. This mismatch has been attributed both to limitations of the animal models, to pitfalls in clinical trial design, and to the absence of satisfactory biomarkers.

Consecutively, the most recent literature relative to success and pitfalls of the use of antioxidants in stroke will be herein reviewed. The most important antioxidants that have been tested in preclinical and clinical trials as potential drugs for stroke treatment are summarized in Table 6.1. They are organized in the following sections according to their origin: natural (including endogenous and those from diet) or synthetic.

4.1 Natural Antioxidants

4.1.1 Vitamins

Several studies have been performed trying to establish a relationship between vitamins intake or their plasma levels and stroke incidence or clinical outcomes, but in general the concluding results are not consistent [13, 14].

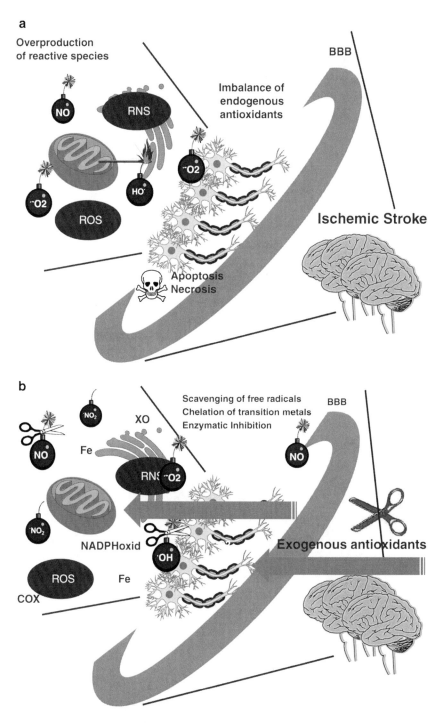

Fig. 6.3 Ischemic stroke and antioxidants. (**a**) Oxidative and nitrosative damage, (**b**) antioxidant benefits

Table 6.1 Some of the antioxidants tested in preclinical and clinical trials for stroke treatment

Chemical structure	Name	References
Natural antioxidants		
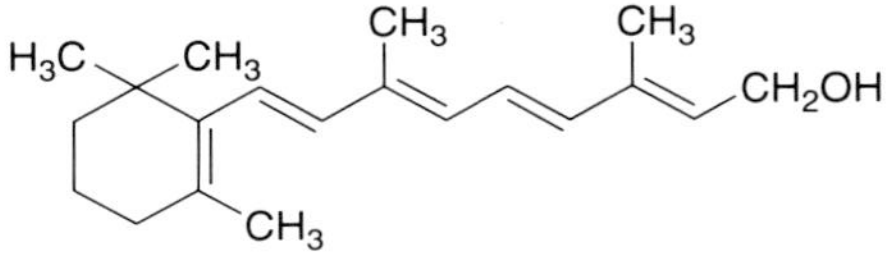	Vitamin A (retinol)	[13, 15]
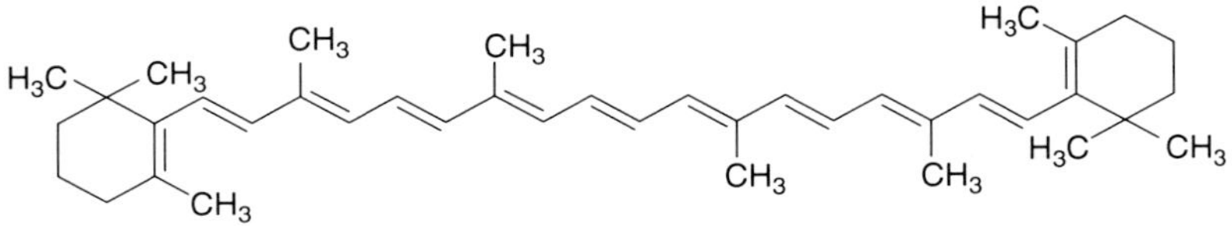	β-Carotene	[13]
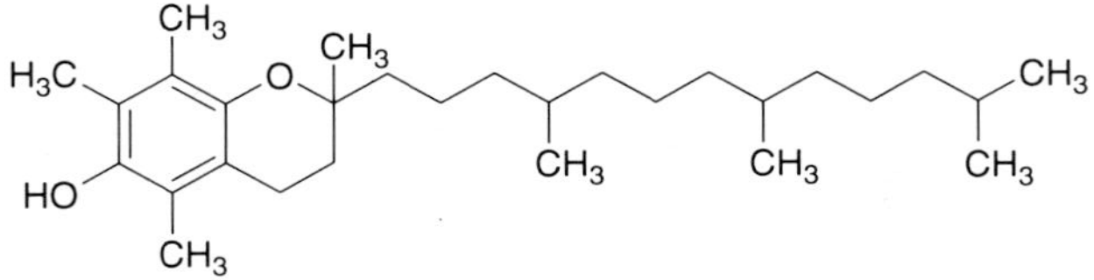	Vitamin E (alpha-tocopherol)	[13, 16]
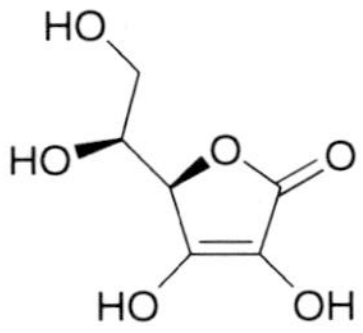 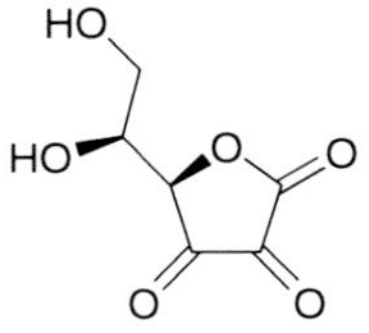	Vitamin C	[13, 17, 18]

Ascorbic acid

Dehydroascorbic acid

[14]

B-group vitamins

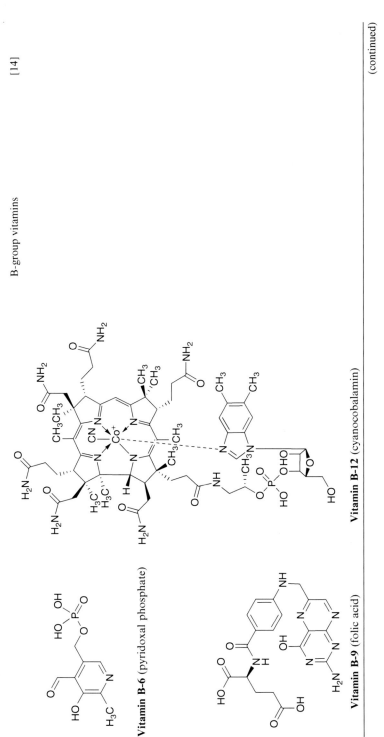

Vitamin B-6 (pyridoxal phosphate)

Vitamin B-9 (folic acid)

Vitamin B-12 (cyanocobalamin)

(continued)

Table 6.1 (continued)

Chemical structure	Name	References
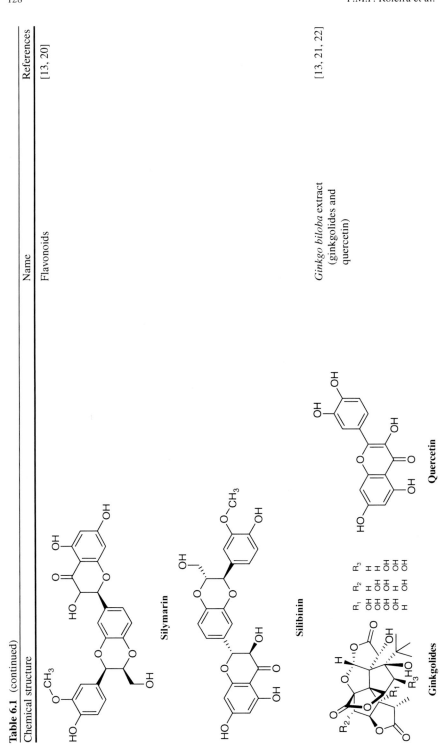	Flavonoids	[13, 20]
	Ginkgo biloba extract (ginkgolides and quercetin)	[13, 21, 22]

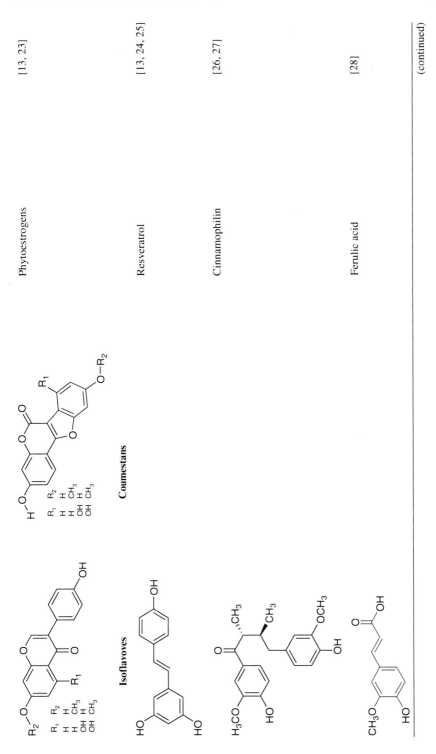

Phytoestrogens [13, 23]

Resveratrol [13, 24, 25]

Cinnamophilin [26, 27]

Ferulic acid [28]

(continued)

Table 6.1 (continued)

Chemical structure	Name	References
	Caffeic acid phenethyl ester	[29]
	Curcumin	[30]
	Uric acid	[31, 32]
	Melatonin	[33, 34]
	Alpha-lipoic acid	[35]

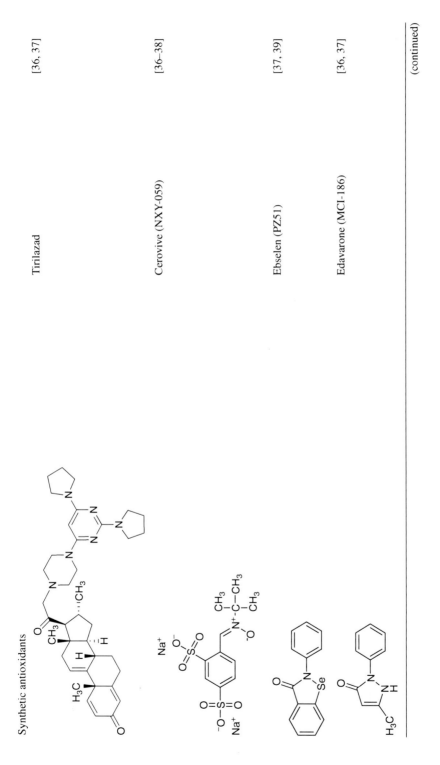

Synthetic antioxidants

Tirilazad   [36, 37]

Cerovive (NXY-059)   [36–38]

Ebselen (PZ51)   [37, 39]

Edavarone (MCI-186)   [36, 37]

(continued)

Table 6.1 (continued)

Chemical structure	Name	References
	Curcumin derivative (CNB-001)	[40]
	Allopurinol	[36, 41, 42]
	Nimesulide	[36, 43]
Nrf2 inducers	tert-Butylhydroquinone (t-BHQ)	[44]

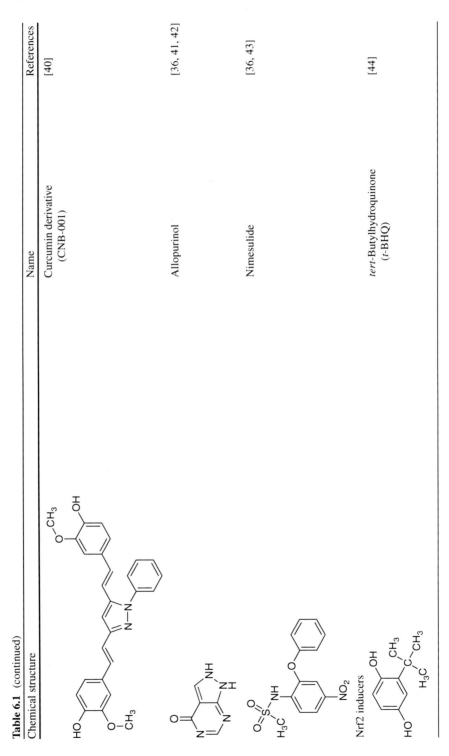

Vitamin A and Carotenoids

There are some observational studies which stressed that the intake of vitamin A and carotenoids was not associated with the risk of stroke, but other divergent results were found in the literature [13]. Neuroprotection against stroke with vitamin A derivatives (all-*trans*-retinol, all-*trans*-retinoic acid, and 9-*cis*-retinoic acid) was recently observed in an animal model of ischemia [15].

Vitamin E

Concerning vitamin E, the majority of clinical trials did not demonstrate that vitamin E supplementation diminishes the incidence of stroke or stroke-related mortality over a follow-up period of 5 years [13]. On the other hand, several studies performed with animal models generate positive evidences to purpose the use of vitamin E for the treatment of oxidative stress-related diseases, such as stroke [16].

Vitamin C

Plasma vitamin C concentrations may be important since they can be used as a biological marker of factors associated with reduced stroke risk and therefore they may be useful in identifying those at high risk of stroke [17]. In a recent study, Lagowska-Lenard et al. [18] look at the influence of vitamin C on markers of oxidative stress in the earliest period of ischemic stroke and conclude that the clinical and functional status of patients after 3 months from the stroke episode was not improved within the administration of vitamin C (500 mg/day, iv). In opposition, in other studies it was concluded that a lower intake of vitamin C, and in turn its plasma levels, is associated with higher risk of ischemic stroke [13]. Yet, no correlation has been found in the most part of the observational studies concerning the intake of vitamin C and the risk of stroke [13].

B-Group Vitamins

A number of scientific data obtained so far do not justify the use of antioxidant vitamin supplements for stroke risk reduction. Yet, in a recent study [14], it was reported that a poor dietary intake of folic acid (vitamin B-9), vitamin B-6, and vitamin B-12 is associated with increased risk of stroke.

During this period, a novel strategy was developed to investigate the potential benefits of dietary antioxidants on stroke risk that consists in the simultaneous supplementation with multiple vitamins. Although explored by different researchers, the results are still controversial [13]. Some studies revealed no evidence to support the use of antioxidant supplements for primary or secondary prevention of various diseases, including stroke [19].

At this time, it seems that the administration of vitamins (either isolated or in combination) does not exert a particular benefit in diminishing ischemic stroke risk, but it also seems unquestionable the existence of a relationship between a highest intake of fruit and vegetables and the lower incidence of stroke and a reduced mortality from stroke. This can be explained by a potential interaction between antioxidants in plants, which can exert synergic effects resulting in benefits for people who have a healthy intake of fruits and vegetables.

4.1.2 Flavonoids

Flavonoids are widely distributed throughout the plant kingdom and have been recognized to have potent antioxidant and free radical scavenging activities. The relationship between flavonoid intake and the risk of stroke has been investigated and no consistent results were obtained [13]. Recently, the results obtained with silymarin and silibinin, phenolic flavonoids derived from *Silybum marianum*, in an animal model disclose their beneficial effects on preventing inflammation-related neurodegenerative diseases as stroke [20].

4.1.3 *Ginkgo biloba* Extract

Ginkgo biloba extract, obtained from the leaves of *Ginkgo* tree, contains several biologically active substances namely the flavonol glycosides of the following aglycones: quercetin, kaempferol and myricetin, and terpenoids (ginkgolides), which have free radical scavenging activity [13]. *G. biloba* extract is widely used in the treatment of AIS in China and occasionally in Europe. Nevertheless, a recent study that evaluated the effects of *G. biloba* extract in patients with AIS did not provide evidence to support its routine use [21]. Yet, other current studies support the use of *G. biloba* extract as a preventive or therapeutic agent in cerebral ischemia [22].

4.1.4 Phytoestrogens

A recent study has shown that soy phytoestrogens act as neuroprotective agents against stroke-like injury in vitro [23]. On the contrary, an observational study, performed with women screened for breast cancer, did not show a direct relationship between dietary intake of phytoestrogens and cerebrovascular diseases [13].

4.1.5 Resveratrol

Resveratrol is a phenolic compound present in grapes and wine which belongs to the stilbene class, with established antioxidant properties. A number of studies have confirmed that resveratrol might reduce cerebral ischemia damage in different animal stroke models [13, 24]. The mechanism of action on the prevention of the deleterious effects triggered by O&NS is proposed to be multifactorial, because it is not limited to antioxidant and anti-inflammatory actions, but also include activation of sirtuin 1 (NAD-dependent deacetylase sirtuin-1) and vitagenes [25].

4.1.6 Cinnamophilin

Cinnamophilin (CINN) is a recently discovered antioxidant and free radical scavenger, isolated from *Cinnamomum philippinense*. CINN penetrates the blood–brain barrier and not only reduces brain infarction and O&NS but also improves behavioural outcome at 24-h post-stroke in an animal model [26]. CINN effectively protects against ischemic brain damage with a therapeutic window up to 6 h in vivo and in vitro [27].

4.1.7 Cinnamic Acids and Derivatives

Ferulic Acid

Ferulic acid (FA), a component of *Anglica sinensis Didl.* and *Ligusticum chuoanx-iong Hort.*, was shown to be a free radical scavenger and to have anti-inflammatory effects in a transient middle cerebral artery occlusion model. These evidences suggest FA as a promising drug for ischemic stroke [28].

Caffeic Acid Phenethyl Ester

Propolis extract has provided an active component identified as caffeic acid phenethyl ester (CAPE), which reduces neurovascular inflammation and protects rat brain, following transient focal cerebral ischemia, due to its antioxidant and anti-inflammatory properties [29]. This observation also suggests CAPE as a potential drug for ischemic stroke treatment.

4.1.8 Curcumin

Curcumin is the principal curcuminoid of the popular Indian spice turmeric, which is a member of the ginger family (Zingiberaceae). The other two identified curcuminoids are desmethoxycurcumin and bis-desmethoxycurcumin. These curcuminoids are natural phenolic compounds with proven antioxidant properties that have shown to be neuroprotective in a variety of preclinical stroke models [30].

4.1.9 Uric Acid

After intense controversy about the function of uric acid, whether it is primarily pro- or antioxidant at physiological conditions, it was finally declared its benefits as a neuroprotector in AIS [31]. In a recent study, the role of elevated serum uric acid (sUA) in AIS among hemodialysis (HD) patients was evaluated. From this study, an inverse association between sUA and the occurrence of AIS in HD patients has been pointed out [32].

4.1.10   Melatonin

Melatonin is a potent-free radical scavenger and has been used as a highly effective treatment in different animal models of excitotoxicity or ischemia/reperfusion injury. Melatonin is regarded as safe and non-toxic for human treatment and its administration via the oral route or intravenous injection is considered to be convenient. Numerous studies have proposed melatonin as a putative drug candidate to be used as a neuroprotective agent for human stroke [33, 34].

4.1.11   Alpha-Lipoic Acid

Alpha-lipoic acid (ALA) is an organosulfur compound derived from octanoic acid. Only the R-(+)-enantiomer (R-LA) exists in nature and is an essential cofactor of various mitochondrial enzyme complexes. Endogenously synthesized R-LA is essential for life and aerobic metabolism. Both R-LA and R/S-LA are available as over-the-counter nutritional supplements and have been used nutritionally and clinically since the 1950s for a number of diseases and conditions. Although pioneer, a recent study suggests the potential use of ALA pretreatment as a neuroprotectant in stroke patients [35].

4.2   Synthetic Antioxidants

4.2.1   Tirilazad

Tirilazad is a 21-aminosterol (lazaroid) with recognized antioxidant activity, namely in preventing damage caused by lipid peroxidation. After phase I and phase II clinical trials [36], it was concluded that tirilazad is not effective for AIS. In fact, tirilazad has been proposed, after studies in animal models, as a drug appropriate for stroke treatment as it protects brain tissue and reduces brain damage. However, when studied in man, tirilazad did not improve the outcome after stroke, but appeared to marginally worsen it [36, 37].

4.2.2   Cerovive

Cerovive (NXY-059) is the disulfonyl derivative of the neuroprotective spin-trap phenylbutylnitrone (PBN). NXY-059 was an antioxidant drug candidate tested in human clinical trials [36, 37]. It really showed benefits in a 1,722 patients phase I trial, but failed in a 3,206 scaled-up phase II trial [37]. In a more recent study the use of preclinical meta-analysis before initiation of future clinical trials with NXY-059 was suggested [38].

4.2.3   Ebselen

Ebselen (PZ51) is a compound that mimics the action of GPx. It has recently been proposed as a potent antioxidant that effectively inhibits both non-enzymatic and enzymatic in vitro lipid peroxidation, with positive results in both experimental and clinical studies of acute cerebral ischemia [39]. It was suggested that the suppression of lipid peroxidation as well as the inhibition of inducible NOS may play a role in the protective effects of PZ51, observed in animal models [39]. However, the human clinical trials with ebselen performed so far did not show clinical improvement at 3 months post-treatment [37].

4.2.4   Edavarone

Edavarone (MCI-186) is a radical scavenging antioxidant introduced in Japanese clinics for treatment of AIS [36, 37]. In a 2003 human trial study, edavarone showed significant benefits in 252 AIS patients [36, 37]. Yet, it will be necessary scaled-up Phase II and III trials in order to attest edavarone as an adequate drug for human ischemia/reperfusion injury.

4.2.5   Curcumin Derivative CNB-001

A novel synthesized multi-target curcuminoid (CNB-001) which is a pyrazole derivative of curcumin has been shown, in animal studies, to improve the behavioural and molecular deficits seen in ischemic stroke and traumatic brain injury with a therapeutic window comparable to tissue plasminogen activator, the only drug approved by the US Food and Drug Administration to treat stroke [40]. On the basis of its in vivo efficacy it was concluded that CNB-001 has a great potential for the treatment of ischemic stroke as well as other central nervous system (CNS) pathologies.

4.2.6   Allopurinol

Several molecules that are able to prevent the formation of free radicals and are active as xanthine oxidase inhibitors, such as allopurinol or its metabolite oxypurinol, have also the capability to reduce cerebral infarct size and oedema in permanent ischemia [36]. In a very recent study, it was postulated that treatment with allopurinol is effective in reducing arterial wave reflection in stroke survivors [41].

In a randomized double-blind placebo-controlled trial, it was concluded that allopurinol yields potentially beneficial effects on inflammatory indices in patients with recent ischemic stroke. The overall results suggest further evaluation of allopurinol as a preventive measure after stroke [42].

4.2.7 Nimesulide

Nimesulide is a COX-2 selective non-steroidal anti-inflammatory drug (NSAID) that has been ascribed to encompass antioxidant properties. Nimesulide was recently shown to have also neuroprotective effects in animal models, particularly in reducing ischemic brain injury. As a result, nimesulide was proposed as a promising neuroprotectant in brain ischemia [36, 43].

4.3 Nrf2 Inducers

Nrf2 is the transcription factor nuclear factor erythroid 2-related factor 2 that coordinates the expression of genes required for free radical scavenging, detoxification of xenobiotics, and maintenance of redox potential. Nrf2 is a master regulator of the antioxidant response. An interesting in vivo study revealed that Nrf2 activation protects the brain from cerebral ischemia [44]. Accordingly, Nrf2 activators could represent a class of drugs potentially useful for the treatment of stroke. There is a wide range of natural and synthetic molecules with remarkable activity as potent inducers of Nrf2 activity [44]. One of the most well-characterized inducers is *tert*-butylhydroquinone (t-BHQ), a metabolite of the widely used food antioxidant butylated hydroxyanisole, which is already approved for human use. These findings open an interesting gateway to develop new drugs for stroke.

5 ADME/Tox of Antioxidants

The enormous difficulty in drug discovery process is to come across the complexity of the human organism. At the molecular level, a synchronized system of transporters, channels, receptors, and enzymes acts as gatekeepers to foreign molecules. These constraints interfere directly with pharmacodynamic and pharmokinetic properties of a drug, namely absorption, distribution, metabolism, excretion, and toxicity (ADME/Tox). Understanding the interactions between small molecules and their molecular targets should improve in a near future the ability to predict the reasons that are responsible for the withdrawal of many marketed drugs as well as late-stage failures of drugs in the development process [45]. It is widely recognized that either predicting or determining the ADME/Tox properties of hits and leads allows to prevent the failure of some drug candidates before they reach the clinical trials.

In this context, when conceiving an antioxidant to be used as a neuroprotective agent in stroke disease, a number of physicochemical factors related to their ADME/Tox properties such as total polar surface area (TPSA), charge state, molecular size, molecular flexibility, number of rotatable bonds, hydrogen-bonding potential, and lipophilicity must be seriously considered and evaluated. In fact, neuroprotective agents must penetrate the blood–brain barrier (BBB) to attain a critical therapeutic concentration within the CNS. Antioxidants that readily pass through the BBB are superior therapeutic candidates for use in neurological disorders.

The transport of molecules across the BBB is a highly restricted and controlled process. Depending on the lipophilicity of the drugs, they can either readily penetrate through the BBB membranes by passive diffusion or require a carrier-mediated transport. The parameters considered optimum for a compound to transport across the BBB are: unionized compounds, $\log P$ value near to 2, molecular weight less than 400 Da, and a cumulative number of hydrogen bonds between 8 and 10 [45].

There are several strategies to enhance drug delivery to the brain with great pharmaceutical interest. Various non-invasive methods that rely on drug chemical manipulation encompassing transformation into lipophilic analogues, prodrugs, carrier-mediated drug delivery, receptor/vector-mediated drug delivery and intranasal drug delivery have been widely used [45]. Concerning the chemical changes, they are usually designed to improve a number of requiring drug physicochemical properties such as membrane permeability or solubility. For example, esterification or amidation of hydroxy-, amino-, or carboxylic acid functions of putative drugs may greatly enhance the lipid solubility, and hence, improve the capacity to penetrate into the brain. One drawback lies in the fact that the increase of lipophilicity of molecules generally increases the volume of distribution, particularly due to plasma protein-binding that affects all other pharmacokinetic parameters [45].

In this context, lipophilic compounds structurally based on the natural antioxidants caffeic, hydrocaffeic, ferulic and hydroferulic acids, but with an additional hexyl chain, linked through an amide or ester bond, were recently synthesized. Their antioxidant activity, partition coefficients, and redox potentials were determined [46]. From the structure–property–activity relationships obtained so far, the existence of a clear correlation between the redox potentials and the antioxidant activity could be concluded. In addition, it was found that the synthesized cinnamic derivatives have a proper lipophilicity to cross the BBB. Their predicted ADME properties are also in accordance with the general requirements for potential CNS drugs. Accordingly, these phenolic compounds have potential to be supplementary studied as antioxidants for tackling the oxidative status linked to the neurodegenerative processes such as stroke. Another example is the natural antioxidant curcumin which has been found to have poor capacity for BBB penetration following acute administration in stroke. Medicinal chemistry artwork has been used to modify curcumin scaffold, resulting in second-generation curcuminoids with enhanced BBB penetration, improved pharmacokinetics, and capable of interacting with multiple viable targets to treat stroke [30, 40].

There are few literature reports of predicted or determined ADME/Tox properties of antioxidants. For instance, antioxidants may be lipid (e.g., vitamin E) or water soluble (e.g., vitamin C) and possess varying degrees of BBB penetrance. For example, ascorbic acid (vitamin C) does not readily cross the BBB, but its oxidized form, dehydroascorbic acid, has the ability to rapidly cross the BBB and accumulate in CNS and consequently exert potent cerebroprotection [13]. Concerning toxicity, vitamin E and vitamin C appear to be safe and free of serious adverse effects when used at high doses in adults [47]. The 21-aminosteroids (lazaroids) as tirilazad are a class of antioxidants with lower BBB penetrance [48]. On the contrary, the synthetic antioxidant ebselen is a relatively small, lipophilic compound, which is rapidly absorbed from the gastrointestinal tract. It maintains a stable plasma concentration and can easily penetrate the BBB [49]. Cerovive (NXY-059) had been shown to be well-tolerated in stroke patients at a plasma unbound concentration of 260 µmol/L [50].

6 Conclusions

In summary, besides the pharmacological properties, ADME/Tox properties have been found to be crucial determinants of the ultimate clinical success of a drug, particularly a drug to be applied in CNS therapeutics.

Concerning antioxidants to treat stroke, their translational disappointment can arise from a combination of various factors including scant studies of the ADME/Tox, failure to understand the drug candidate's mechanism of action in relationship to human disease, and failure to conduct preclinical studies using concentration and time parameters relevant to the clinical setting. In this manner, ADME/Tox screening must be performed during the antioxidant drug discovery process, in an effort to eradicate or improve drugs with problematic ADME/Tox profiles. Further, larger clinical studies in this area are needed to clarify the temporal relationships between antioxidant capacity and oxidative damage following ischemia and reperfusion in man and to form the basis of appropriate antioxidant intervention strategies to minimize long-term brain injury in cerebral ischemia. In addition, the assessment of oxidation by-products and antioxidant status during ischemia/reperfusion may be important in predicting free radical-induced cerebral injury in stroke patients.

Since multiple pathogenetic factors are implicated in stroke, the current hitting-one-target therapeutic strategy could probably be inefficient. New therapeutic strategies for stroke may be fulfilled by combining different drugs or using an alternative approach that aims to attain multiple targets with a single structure.

References

1. Green AR, Shuaib A. Therapeutic strategies for the treatment of stroke. Drug Discov Today. 2006;11:681–93.
2. Hillbom M. Oxidants, antioxidants, alcohol and stroke. Front Biosci. 1999;4:e67–71.
3. Brouns R, De Deyn PP. The complexity of neurobiological processes in acute ischemic stroke. Clin Neurol Neurosurg. 2009;111:483–95.

4. Lo EH, Dalkara T, Moskowitz MA. Mechanisms, challenges and opportunities in stroke. Nat Rev Neurosci. 2003;4:399–415.
5. Doyle KP, Simon RP, Stenzel-Poore MP. Mechanisms of ischemic brain. Neuropharmacology. 2008;55:310–8.
6. Lapchak PA, Araujo DM. Advances in ischemic stroke treatment: neuroprotective and combination therapies. Expert Opin Emerg Drugs. 2007;12:97–112.
7. McCulloch J, Dewar D. A radical approach to stroke therapy. Proc Natl Acad Sci USA. 2001;98:10989–91.
8. Seneş M, Kazan N, Coşkun O, Zengi O, Inan L, Yücel D. Oxidative and nitrosative stress in acute ischaemic stroke. Ann Clin Biochem. 2007;44:43–7.
9. Allen CL, Bayraktutan U. Oxidative stress and its role in the pathogenesis of ischaemic stroke. Int J Stroke. 2009;4:461–70.
10. Behl C, Moosmann B. Oxidative nerve cell death in Alzheimer's disease and stroke: antioxidants as neuroprotective compounds. Biol Chem. 2002;383:521–36.
11. Chen H, Yoshioka H, Kim GS, Jung JE, Okami N, Sakata H, Maier CM, Narasimhan P, Goeders CE, Chan PH. Oxidative stress in ischemic brain damage: mechanisms of cell death and potential molecular targets for neuroprotection. Antioxid Redox Signal. 2011;14:1505–17.
12. Carbonell T, Rama R. Iron, oxidative stress and early neurological deterioration in ischemic stroke. Curr Med Chem. 2007;14:857–74.
13. Cherubini A, Ruggiero C, Morand C, Lattanzio F, Dell'Aquila G, Zuliani G, Di Iorio A, Andres-Lacueva C. Dietary antioxidants as potential pharmacological agents for ischemic stroke. Curr Med Chem. 2008;15:1236–48.
14. Sanchez-Moreno C, Jimenez-Escrig A, Martin A. Stroke: roles of B vitamins, homocysteine and antioxidants. Nutr Res Rev. 2009;22:49–67.
15. Sato Y, Meller R, Yang T, Taki W, Simon RP. Stereo-selective neuroprotection against stroke with vitamin A derivatives. Brain Res. 2008;1241:188–92.
16. Yamagata K, Tagami M, Yamori Y. Neuronal vulnerability of stroke-prone spontaneously hypertensive rats to ischemia and its prevention with antioxidants such as vitamin E. Neuroscience. 2010;170:1–7.
17. Myint PK, Luben RN, Welch AA, Bingham SA, Wareham NJ, Khaw K. Plasma vitamin C concentrations predict risk of incident stroke over 10 y in 20 649 participants of the European Prospective Investigation into Cancer-Norfolk prospective population study. Am J Clin Nutr. 2008;87:64–9.
18. Lagowska-Lenard M, Stelmasiak Z, Bartosik-Psujek H. Influence of vitamin C on markers of oxidative stress in the earliest period of ischemic stroke. Pharmacol Rep. 2010;62:751–6.
19. Bjelakovic G, Nikolova D, Gluud LL, Simonetti RG, Gluud C. Antioxidant supplements for prevention of mortality in healthy participants and patients with various diseases. Cochrane Database Syst Rev. 2009;1(1):1–264.
20. Hou Y, Liou K, Chern C, Wang Y, Liao J, Chang S, Chou Y, Shen Y. Preventive effect of silymarin in cerebral ischemia-reperfusion-induced brain injury in rats possibly through impairing NF-kB and STAT-1 activation. Phytomedicine. 2010;17:963–73.
21. Zeng X, Liu M, Yang Y, Li Y, Asplund K. *Ginkgo biloba* for acute ischaemic stroke. Cochrane Database Syst Rev. 2005;4:CD003691. doi:10.1002/14651858.CD003691.pub2.
22. Saleem S, Zhuang H, Biswal S, Christen Y, Doré S. *Ginkgo biloba* extract neuroprotective action is dependent on heme oxygenase 1 in ischemic reperfusion brain injury. Stroke. 2008;39:3389–96.
23. Schreihofer DA, Redmond L. Soy phytoestrogens are neuroprotective against stroke-like injury in vitro. Neuroscience. 2009;158:602–9.
24. Almeida LMV, Leite MC, Thomazi AP, Battu C, Nardin P, Tortorelli LC, Zanotto C, Posser T, Wofchuk ST, Leal RB, Gonçalves CA, Gottfried C. Resveratrol protects against oxidative injury induced by H_2O_2 in acute hippocampal slice preparations from Wistar rats. Arch Biochem Biophys. 2008;480:27–32.
25. Sun AY, Wang Q, Simonyi A, Sun GY. Resveratrol as a therapeutic agent for neurodegenerative diseases. Mol Neurobiol. 2010;41:375–83.

26. Lee EJ, Chen HY, Lee MY, Chen TY, Hsu YS, Hu YL, Chang GL, Wu TS. Cinnamophilin reduces oxidative damage and protects against transient focal cerebral ischemia in mice. Free Radic Biol Med. 2005;39:495–510.

27. Lee EJ, Chen HY, Hung YC, Chen TY, Lee MY, Yu SC, Chen YH, Chuang IC, Wu TS. Therapeutic window for cinnamophilin following oxygen-glucose deprivation and transient focal cerebral ischemia. Exp Neurol. 2009;217:74–83.

28. Cheng CY, Su SY, Tang NY, Ho TY, Chiang SY, Hsieh CL. Ferulic acid provides neuroprotection against oxidative stress-related apoptosis after cerebral ischemia/reperfusion injury by inhibiting ICAM-1 mRNA expression in rats. Brain Res. 2008;1209:136–50.

29. Khan M, Elango C, Ansari MA, Singh I, Singh AK. Caffeic acid phenethyl ester reduces neurovascular inflammation and protects rat brain following transient focal cerebral ischemia. J Neurochem. 2007;102:365–77.

30. Lapchak PA. Neuroprotective and neurotrophic curcuminoids to treat stroke: a translational perspective. Expert Opin Investig Drugs. 2011;20:13–22.

31. Proctor PH. Uric acid and neuroprotection. Stroke. 2008;39:e126.

32. Chen YM, Ding XQ, Teng J, Zou JZ, Zhong YH, Fang Y, Liu ZH, Xu SW, Wang YM, Shen B. Serum uric acid is inversely related to acute ischemic stroke morbidity in hemodialysis patients. Am J Nephrol. 2011;33:97–104.

33. Macleod MR, Horky LL, Howells DW, Donnan GA. Systematic review and meta-analysis of the efficacy of melatonin in experimental stroke. J Pineal Res. 2005;38:35–41.

34. Cheung RTF, Tipoe GL, Tam S, Ma ESK, Zou LY, Chan PS. Preclinical evaluation of pharmacokinetics and safety of melatonin in propylene glycol for intravenous administration. J Pineal Res. 2006;41:337–43.

35. Connell BJ, Saleh M, Khan BV, Saleh TM. Lipoic acid protects against reperfusion injury in the early stages of cerebral ischemia. Brain Res. 2011;1375:128–36.

36. Margaill I, Plotkine M, Lerouet D. Antioxidant strategies in the treatment of stroke. Free Radic Biol Med. 2005;39:429–43.

37. Kamat CD, Gadal S, Mhatre M, Williamson KS, Pye QN, Hensley K. Antioxidants in central nervous system diseases: preclinical promise and translational changes. J Alzheimers Dis. 2008;15:473–93.

38. Bath PMW, Gray LJ, Bath AJG, Buchan A, Green AR. Effects of NXY-059 in experimental stroke: an individual animal meta-analysis. Br J Pharmacol. 2009;157:1157–71.

39. Sui H, Wang W, Wang PH, Liu LS. Protective effect of antioxidant ebselen (PZ51) on the cerebral cortex of stroke-prone spontaneously hypertensive rats. Hypertens Res. 2005;28:249–54.

40. Lapchak PA, Schubert DR, Maher PA. Delayed treatment with a novel neurotrophic compound reduces behavioral deficits in rabbit ischemic stroke. J Neurochem. 2011;116:122–31.

41. Khan F, George J, Wong K, McSwiggan S, Struthers AD, Belch JF. Allopurinol treatment reduces arterial wave reflection in stroke survivors. Cardiovasc Ther. 2008;26:247–52.

42. Muir SW, Harrow C, Dawson J, Lees KR, Weir CJ, Sattar N, Walters MR. Allopurinol use yields potentially beneficial effects on inflammatory indices in those with recent ischemic stroke: a randomized, double-blind, placebo-controlled trial. Stroke. 2008;39:3303–7.

43. Candelario-Jalil E. Nimesulide as a promising neuroprotectant in brain ischemia: new experimental evidences. Pharmacol Res. 2008;57:266–73.

44. Shih AY, Li P, Murphy TH. A small molecule-inducible Nrf2-mediated antioxidant response provides effective prophylaxis against cerebral ischemia in vivo. J Neurosci. 2005;25:10321–35.

45. Pathan SA, Iqbal Z, Zaidi SM, Talegaonkar S, Vohra D, Jain GK, Azeem A, Jain N, Lalani JR, Khar RK, Ahmad FJ. CNS drug delivery systems: novel approaches. Recent Pat Drug Deliv Formul. 2009;3:71–89.

46. Roleira F, Siquet C, Orru E, Garrido EM, Garrido J, Milhazes N, Podda G, Paiva-Martins F, Reis S, Carvalho RA, Tavares-da-Silva E, Borges F. Lipophilic phenolic antioxidants: correlation between antioxidant profile, partition coefficients and redox properties. Bioorg Med Chem. 2010;18:5816–25.

47. Meyers DG, Maloley PA, Weeks D. Safety of antioxidant vitamins. Arch Intern Med. 1996;156:925–35.
48. Schmid-Elsaesser R, Zausinger S, Hungerhuber E, Plesnila N, Baethmann A, Reulen HJ. Superior neuroprotective efficacy of a novel antioxidant (U-101033E) with improved blood–brain barrier permeability in focal cerebral ischemia. Stroke. 1997;28:2018–24.
49. Ichikawa S, Omura K, Katayama T, Okamura N, Oht-suka T, Ishibashi S, Masayasu H. Inhibition of superoxide anion production in guinea pig polymorphonuclear leukocytes by a seleno-organic compound, ebselen. J Pharmacobiodyn. 1987;10:595–7.
50. Lees KR, Barer D, Ford GA, Hacke W, Kostulas V, Sharma AK, Odergren T. Tolerability of NYX-059 at higher target concentrations in patients with acute stroke. Stroke. 2003;34:482–7.

Chapter 7
Caspase-Independent Stroke Targets

Ruoyang Shi, Jiequn Weng, Paul Szelemej, and Jiming Kong

Abstract Delayed neuronal death in the penumbral region of a stroke is largely responsible for many negative implications seen in stroke victims. This type of neuronal death occurs in many forms, including apoptosis, necrosis, and alternative mechanisms. Although caspases are usually associated with apoptosis, there are several morphologically and biochemically distinct types of cell death that are independent of caspase activation. Downstream effectors and processes of mitochondrial damage, such as AIF, endonuclease G, BNIP3, mitophagy, mitochondrial biogenesis, chaperone-mediated autophagy, reactive oxygen species production as well as parallel endoplasmic reticular stress and lysosomal dysfunction, have all been shown to play a role in post-stroke delayed neuronal cell death. In this chapter, we attempt to summarize these caspase-independent events and their potential therapeutic applications as targets for intervention.

1 Introduction

Caspases (cysteine aspartyl serine proteases) are a unique class of cysteine proteases that play essential roles in programmed cell death (PCD), especially in apoptosis. Although caspases are evolutionarily conserved and frequently activated during cell death [1, 2], it appears that simply equating cell death, apoptosis, and caspase activation to one another does not apply to mammals, as the relationships between these concepts are far more complex than that. There are several reasons for this. First of all, apoptosis is only one of many types of PCD, as mammalian cells die through several biochemically and morphologically distinct pathways. Second, caspase activation does not always necessarily lead to apoptosis [3]. Third, caspase inhibition

R. Shi • J. Weng • P. Szelemej • J. Kong (✉)
Department of Human Anatomy and Cell Science, University of Manitoba,
745 Bannatyne Avenue, Winnipeg, MB, Canada R3E 0J9
e-mail: kongj@cc.umanitoba.ca

P.A. Lapchak and J.H. Zhang (eds.), *Translational Stroke Research*, Springer Series in Translational Stroke Research, DOI 10.1007/978-1-4419-9530-8_7, © Springer Science+Business Media, LLC 2012

often does not prevent cell death but rather causes a shift to caspase-independent self-destruction processes within the cell. The main reason for this is that many stimuli that promote caspase-dependent apoptosis also simultaneously provoke mitochondrial damage, leading to other caspase-independent apoptotic pathways that cannot be reversed by blocking caspase activity. Caspase-dependent self-destruction pathways are only one of many cell death pathways. Hence, in many cases, cytoprotection can only be achieved by targeting caspase-independent processes and molecules that precede activation of caspases in the first place [4].

In addition to caspases [5], emerging evidence from various experimental systems suggests that caspase-independent mechanisms contribute significantly to the overall cell death process [4, 6]. For example, it has been reported that, under cerebral ischemia, neuronal death was found in caspase-3 knockout mice [7] as well as wild-type mice. This indicates that there is a direct link between ischemia and cell death, regardless of whether or not there was caspase activation [8]. Caspase inhibitors, such as the broad, nonspecific caspase inhibitor zVAD-FMK, provide limited neuroprotection in brain ischemia [9] and instead often cause a switch to caspase-independent apoptotic processes. Knockouts of caspases 3 or 9, which play central roles in cell death signaling, did not ultimately alter the number of remaining cells left upon completion of the experiment. This again shows that caspases are not the sole causal factors of apoptosis [4]. In addition, caspase-independent neuronal death was identified in Apaf-1-deficient neurons; Apaf-1 is a key factor downstream in the caspase-dependent apoptotic cascade [10]. MacManus et al. found that, after cerebral ischemia, not only was a small amount of caspase-activated DNase (CAD)-related DNA fragments with a size of 200–1,000 base pairs found, but high molecular weight (on the order of 50 kb pair magnitudes) DNA fragments that may be completely independent of caspase activity also appeared in dying neurons. This indicates that some other process is occurring [11]. These facts collectively support the notion that caspase-independent PCD is distinct from and a viable alternative to the traditional caspase-dependent apoptosis better understood in delayed neuronal cell death. In this chapter, we will critically evaluate the contribution of caspase-independent apoptotic processes in ischemic stroke and discuss the possibility of targeting caspase-independent cell death mechanisms. Inhibition of these caspase-independent processes could potentially lead to cytoprotection of neurons during times of hypoxic stress. This is a quickly developing field with several potential clinical implications. Neurons could potentially be targeted by drugs that could be used therapeutically to stave off premature cell death directly following stroke.

2 Mechanisms of Delayed Neuronal Death During Ischemia

Neurons, like other cells, have death machinery that consists of a set of genes, intracellular signaling and transduction pathways, as well as corresponding transduction factors and enzymes. To date, at least 11 prominent cell death pathways have been identified in various mammalian tissues, among which seven types are observed in the CNS [12]. We will focus on the three major well-known categories that are involved in delayed neuronal death: apoptosis, necrosis, and autophagic cell death. In addition, under certain stimuli like ischemic insults, damaged neurons can shift from one type of cell death to another throughout the course of the chronic degeneration process. The eventual mode of death—apoptotic, necrotic, or autophagic cell death—depends upon a number of parameters. These include metabolic state, energy resources, availability of growth factors, cell maturity, stress stimuli, and several other factors.

2.1   Apoptotic Neuronal Death

Apoptosis is a form of PCD (Type I PCD). In apoptosis, a controlled and regulated sequence of events leads to the organized elimination of cells in an energy-dependent manner without releasing harmful toxins into the surrounding environment. It is widely observed that delayed neuronal death following brain ischemia shows features of apoptosis, which is marked by a well-defined sequence of morphological changes. Cell shrinkage and membrane-bound apoptotic bodies can be found under light microscopy, while nuclear chromatin condensation and fragmentation are observable through electron microscopy in dying, ischemic neurons. DNA fragmentation has also been detected during the processes of delayed neuronal death by terminal dUTP nick-end-labeling staining (TUNEL), in situ end-labeling (ISEL, in the CA1 pyramidal neurons), as well as DNA laddering on gel electrophoresis of extracted nuclear DNA [13]. Further, apoptosis-related caspases such as caspase-3 were activated in the CA1 region following transient brain ischemia [14]. Recent studies have shown that several mitochondrial proteins are released as a result of mitochondrial outer membrane permeabilization (MOMP), including apoptosis-inducing factor (AIF), Omi, and endonuclease G (Endo G), which can all promote atypical apoptotic responses in a caspase-independent fashion [15]. It is clear that AIF and/or Endo G-mediated caspase-independent cell death is characterized by large-scale DNA fragmentation and peripheral chromatin condensation, which are distinct from oligonucleosomal DNA fragmentation and global chromatin condensation in caspase-dependent apoptosis [16]. In later stages of a brief episode of brain ischemia, phagocytosis of fragmented DNA by microglial cells in the CA1 region was also found [13]. Collectively, whether in a caspase-dependent or -independent manner, all of these observations suggest that apoptotic processes are indeed involved in delayed neuronal death due to ischemic attacks.

2.2   Necrotic Neuronal Death

In stark contrast to apoptosis, necrosis is the end result of an uncontrolled and disordered bioenergetic catastrophe resulting from ATP depletion and progressive enzymatic degradation. It is thought to be initiated mainly by toxic insults or physical damage, and is morphologically marked by vacuolation of the cytoplasm, extensive mitochondrial swelling, dilatation of the endoplasmic reticulum (ER), and early plasma membrane damage without major nuclear changes [12, 17, 18]. Although necrosis is more likely to be induced by severe ischemia at an earlier stage of insult, it is also involved in delayed neuronal death. Neurons will lose control of their ionic balance, imbibe water, and lyse eventually [19–21]. A typical necrotic neuronal cell death pathway has been found in acute excitotoxicity, which results from the over-release of neurotransmitters. The engagement of excitatory neurotransmitters and their cell membrane receptors, such as NMDA, kainite, and AMPA, are responsible for the elevated cytosolic Ca^{2+} and subsequent cell death [22, 23]. It has been reported that inhibition of Ca^{2+} uptake by mitochondria can suppress necrotic cell death, as increased Ca^{2+} levels in the cytoplasm lead to more controlled cell death pathways [24]. Moreover, the lysosomal damage and the calpain–cathepsin liberation must play essential roles in necrotic neuronal cell death, as lysosomal protease inhibitors have been shown to protect against delayed neuronal death produced by global ischemia [25, 26]. Under some specific circumstances, necrosis may also be well regulated and is described as being "programmed necrosis" (also known as Type III PCD), a type of PCD. Necrosis is actually usually the opposite of controlled and regulated cell death; that is, spontaneous, quick, and disordered destruction/lysis of the cell [27]. This cell death pathway can be triggered by the DNA damage present after cerebral ischemia, as the DNA repair protein poly (ADP-ribose) polymerase 1(PARP-1) is hyperactivated and actively engaged in the necrotic process, which leads to post-ischemic brain damage [28].

2.3 Autophagic Neuronal Death

Another mechanism participating in delayed ischemic neuronal death, which has attracted attention recently, is autophagy. Autophagy is a highly regulated process that involves the degradation of a cell's own cytoplasmic macromolecules and organelles in mammalian cells via the lysosomal system. It is not only an adaptive response to nutrient limitation but also a mechanism for cell suicide [14, 16–21]. Deregulation of autophagy has been implicated in cell death in neurodegenerative disorders and is known as "autophagic cell death" [29–32] or "type II programmed cell death" (Type II PCD) [30, 33, 34]. "Autophagic cell death," distinct from apoptosis and necrosis, is characterized by extensive autophagic degradation of cellular components prior to nuclear destruction [35, 36]. The most representative morphological feature of autophagy is the formation of numerous autophagosomes in the cytosol with a condensed nucleus [37]. This type of cell death can be inhibited by 3-methyladenine (3-MA) and wortmannin or by downregulation of autophagic proteins such as Beclin 1. This implies that autophagy is a programmed death dependent on genes and is not merely a failing survival attempt. Accumulating evidence suggests that autophagy contributes to the neuronal degeneration following cerebral ischemia. The induction of autophagy has been shown in both neonatal and adult mouse cortices and hippocampi after ischemic injury; increased LC3-II (an autophagosomal marker) levels were detected as early as 8 h and were more pronounced at 24 and 72 h after hypoxic ischemia [38, 39]. Many damaged neurons showed features of autophagic cell death during cerebral hypoxia/ischemia (H/I) in adult mice (e.g., increased lysosomal cysteine proteinases, cytoplasmic autophagic vacuoles, and the induction of GFP-LC3 immunofluorescence) [13, 40]. Furthermore, Puyal et al. provided evidence that inhibiting autophagy provides powerful neuroprotection in a situation where most other pharmacological treatments, including caspase inhibition, are ineffective [41].

2.4   Crosstalk Between Different Types of Neuronal Death

The existing hybrid form of cell death that shares features of various death mechanisms and/or combines different metabolic pathways in dying neurons should not be overlooked. In particular, the crosstalk between autophagous and apoptotic processes needs to be clarified because both autophagy and apoptosis can be triggered by common upstream signals. Sometimes this results in a mixed phenotype of both cell death patterns. In many other instances, the neuron switches between the two responses in a mutually exclusive manner, perhaps as a result of variable thresholds for both processes, or as a result of a cellular "decision" between the two responses. To some extent, apoptosis and autophagic death under ischemic conditions can be described as two sides of the same coin. It appears that simply inhibiting one type of mechanism will switch the cellular response toward the other one, and both catabolic phenomena can inhibit each other. For example, Yu et al. demonstrated that caspase inhibitors may not only arrest apoptosis but also have the unanticipated effect of promoting autophagic cell death [42]. By contrast, Yousefi et al. proved that calpain-mediated cleavage of Atg5 could switch autophagy to apoptosis [43]. Recently, several pathways that link the apoptotic and autophagic machineries have been deciphered at the molecular level [15, 44].

Induction of autophagy may also cause necrotic cell death. Lenardo and colleagues showed that catalase, a key enzyme of the cellular antioxidant defence mechanism, was selectively eliminated during autophagic cell death and catalase depletion caused necrotic cell death, which could be prevented by autophagy inhibition as well as antioxidants [45]. To summarize, we can say that, in delayed ischemic neuronal death, the interactions and connections between multiple mechanisms will jointly seal the fate of the neurons.

3   Caspase-Independent Mechanisms and Mediators in Delayed Neuronal Death

The term delayed neuronal death was first coined by Kirino to describe the selective loss of the hippocampal CA1 neurons that do not become morphologically obvious until 2–3 days following a brief, transient episode of brain ischemia [46]. A wealth of evidence suggests that this second round of neuronal injury, which is referred to as delayed neuronal death in the neighboring areas of the infarct core, occurs hours to days following stroke. Neurons in this penumbral area have impaired function but remain viable for a period of time before gradually succumbing to the insult. These neurons die in a delayed manner, so targeting this damaged but salvageable region of brain tissue will potentially reduce infarct volume and improve the neurological outcome [47].

Mitochondria are essential organelles involved in oxidative phosphorylation, calcium homeostasis, reactive oxygen species (ROS) management, and PCD. Convergence of a number of cell death pathways emanating from membrane receptors, cytosol, the nucleus, lysosome, and endoplasmic reticulum upon the mitochondria results in mitochondrial destabilization [48]. A common consequence of these death pathways is damage to the mitochondria, resulting in MOMP. A number of assays have been developed to measure MOMP as an indicator of cytotoxicity [49]. MOMP is a key event during the cell death process and often defines the point of no return [50–52]. It is lethal because it results in the release of caspase-activating molecules and caspase-independent death effectors, metabolic failure in the mitochondria, or both [52]. The effectors of mitochondria-related PCD can be divided into two categories: (1) downstream mitochondrial death effectors (including cytochrome C, AIF, and others) and (2) agents that directly target and destabilize mitochondrial membranes or through upstream signaling mechanisms [53]. The local regulation and execution of MOMP involve proteins from the Bcl-2 family, mitochondrial lipids, proteins that regulate bioenergetic metabolite flux, and putative components of the permeability transition pore [52]. Drugs designed to suppress excessive MOMP may potentially prevent pathological cell death. Identifying the category of mitochondrial-related PCD is critical for searching targets that will provide optimal protection against brain ischemia. In the following sections, we will discuss the two categories and their therapeutic potential (Fig. 7.1).

3.1   Downstream Effectors of Mitochondrial Damage (MOMP)

3.1.1   The AIF or Endo G-Mediated Caspase-Independent Atypical Apoptosis

AIF is a mitochondrial flavoprotein and normally resides in the intermembrane mitochondrial space, where it performs an oxidoreductase function [54]. Caspase-independent chromatin condensation and DNA degradation [55, 56] are promoted upon cytosolic release and subsequent nuclear translocation of AIF during MOMP (along with AIF's obligate cyclophilin A cofactor) [57, 58]. Although there is substantial variation among the degree to which cell death is affected by AIF (depending upon specific experimental and pathophysiological parameters) [59], this caspase-independent inducer of neuronal death following acute injury has consistently shown itself to be a major determining factor of this process [60, 61]. Indeed,

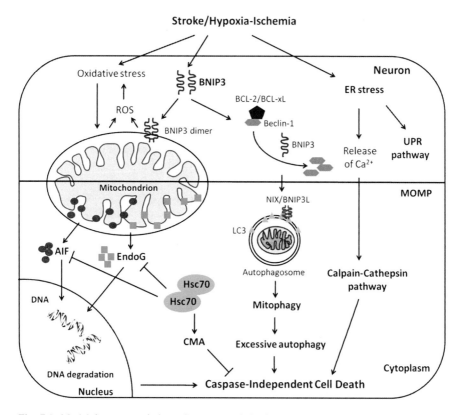

Fig. 7.1 Model for caspase-independent neuronal death pathways in stroke. Hypoxia/ischemia induces HIF-1α-mediated BNIP3 expression, subsequent stable homodimerization, mitochondrial membrane insertion of the BNIP3 dimer, which results in MOMP, and the generation of reactive oxygen species (ROS). Oxidative stress can also be induced directly by H/I injury, which triggers mitochondrial damage and ROS release. As a consequence of the MOMP, neurons undergo different death pathways unrelated to caspase activation, such as the release/nuclear translocation of apoptosis-inducing factor (AIF) and endonuclease G (Endo G), mitophagy, excessive autophagy-induced cell death (by the release of Beclin 1 from the Bcl-2/Bcl-xL complex), endoplasmic reticulum (ER) stress-induced unfolded protein response (UPR), and calpain–cathepsin pathways. For each pathway, please refer to the main text for further details. *LC3* microtubule-associated protein 1 light chain 3; *Hsc70* the heat-shock cognate protein of 70 kDa, in complex with its co-chaperones; *CMA* chaperone-mediated autophagy

the action of AIF cannot be blocked by the broad caspase inhibitor zVAD-FMK [62] and zDEVD-FMK [63]. This is indicative of the fact that this protein is involved in caspase-independent atypical apoptosis. Inhibition of AIF reduced 37–60% of neuronal death induced by glutamate excitotoxicity and oxygen-glucose deprivation-induced neuronal death, two major contributing factors of ischemic insult. Harlequin mutant (Hq) mice, which express about 20% lower AIF levels than wild-type mice, exhibited significantly less brain damage upon MCAO [64]. Similarly, neuroprotective effects caused by AIF deficiency have been shown to be further amplified through the utilization of broad spectrum caspase inhibitors or antioxidants following neonatal stroke [65, 66]. The nuclear translocation of CypA that was typically observed in wild-type neurons did not manifest itself in Hq mice subjected to stroke environments. This is fully compatible with in vitro data suggesting that a proapoptotic DNA degradation complex is formed between AIF and its interacting CypA cofactor [56]. An approximately 50% reduction in infarct volume was observed in the CypA$^{-/-}$ mice, as compared to that of WT mice, which correlates quite well with the suppressed H/I-induced nuclear translocation of AIF [55]. Thus, AIF clearly acts as a lethal, caspase-independent effector of atypical apoptosis in ischemic neuronal death.

Although there is no doubt that AIF plays a critical role in caspase-independent death, the mechanisms responsible for AIF release from the mitochondria still need to be explored further. The Bcl-2 family, which also controls caspase-dependent cell death, regulates AIF efflux and subsequently caspase-independent cell death [67]. It has been found that the anti-apoptotic protein Bcl-xL can prevent AIF translocation in neuronal cultures challenged with transient oxygen-glucose deprivation [63]. Thus, Bcl-xL may account for the acquisition of resistance to neuronal cell death following brief ischemia. Contrarily, Bax expression leads to an increase in AIF efflux from mitochondria in neurons [10]. Truncated Bid (tBid) is also associated with AIF release, and it has been demonstrated that cleavage of Bid occurs coincidently with AIF translocation from mitochondria to the nucleus [64]. Furthermore, incubation of freshly isolated mitochondria with tBid resulted in AIF release as well [68]. Accordingly, the following model of AIF release has been proposed: tBid and Bax interact with mitochondria to open mitochondrial permeability transition (MPT) pores [69] or form a membrane pore where they cleave AIF to thereby dissociate it from the inner mitochondrial membrane [70]. The cleaved AIF then leaves the mitochondria and translocates to the nucleus to execute DNA degradation and cell death [71].

Another mitochondrial protein that potentially contributes to caspase-independent cell death is Endo G [72, 73]. Genetic and biochemical evidence has emerged that Endo G is released from the intermembrane space of mitochondria and translocates to the nucleus to initiate a caspase-independent apoptosis during early embryogenesis [74] as well as in pathological conditions such as transient cerebral ischemia [75] with oxygen and glucose deprivation (OGD) [76]. Once released from mitochondria, Endo G cleaves chromatin DNA into nucleosomal fragments independently of CAD [72, 73]. Like AIF, the release of Endo G is under the control of the Bcl-2 family [70]. The release of Endo G from mitochondria was originally described following treatment of isolated mitochondria with tBid. In the study, mitochondrial efflux of Endo G could be induced by tBid in normal mice but not in Bcl-2 transgenic knockout mice [73]. This indicates that Bcl-2 must be an important bridge between tBid and Endo G. Recently, we found that another BH3-only proapoptotic protein, Bcl2/adenovirus E1B 19-kDa interacting protein 3 (BNIP3), activated a caspase-independent neuronal death pathway mediated by Endo G in hypoxia and stroke (see Sect. 3.1.2) [77].

Human Omi/HtrA2, the stress-regulated endoprotease, has also been implicated in caspase-independent cell death mechanisms that originate from MOMP, due to its ability to promote the cleavage of caspase-unrelated substrates (e.g., cytoskeletal proteins) [78]. However, the contribution of Omi/HtrA2 in acute neuronal injury is poorly characterized. Cytosolic translocation of Omi/HtrA2 and its enhanced inter-action with the X-linked inhibitor of apoptosis protein (XIAP) have been reported to occur in vivo, in mice subjected to transient focal cerebral ischemia (tFCI) [79]. Cytosolic translocation of Omi/HtrA2 and its enhanced interaction with XIAP could not be prevented by Z-VAD-fmk administration, but was significantly reduced in the brain of mice overexpressing the SOD1 gene (superoxide dismutase). Omi/Htra2 is closely resembled by direct IAP-binding protein with a low pI (like DIABLO, whose murine ortholog is known as Smac). Cytosolic translocation of Smac/DIABLO has been shown to be promoted by tFCI in both rats [80] and mice [81, 82], the latter case being through a pathway that can potentially be counteracted by SOD1 overexpression. Both of these results point toward a significant contribu-tion of ROS (but not of caspases) to the molecular pathways leading to an ischemia-induced Omi/HtrA2 and Smac/DIABLO release [83].

3.1.2   The BNIP3-Activated and Endo G-Mediated Neuronal Death Pathway

BNIP3 (formerly NIP3) is a proapoptotic, mitochondrial protein classified in the Bcl-2 family based on limited sequence homology to the Bcl-2 homology 3 (BH3) domain and C-terminal transmembrane (TM) domain. This subfamily includes BNIP3, NIX (also called BNIP3α and BNIP3L), BNIP3h, and a *Caenorhabditis elegans* ortholog (ceBNIP3) [84–88]. When expressed, BNIP3 is able to cause cell death in a variety of cells [86], including neurons [77, 89]. Typically, the BH3 domain of proapoptotic Bcl-2 family members mediates Bcl-2/Bcl-X(L) heterodi-merization and confers proapoptotic activity. Deletion mapping of BNIP3 excluded its BH3-like domain and identified the NH2-terminus (residues 1–49) and TM domain as critical for Bcl-2 heterodimerization, and either region was sufficient for Bcl-X(L) interaction. Additionally, the removal of the BH3-like domain in BNIP3 did not diminish its apoptotic activity. The TM domain of BNIP3 is critical for homodimerization, proapoptotic function, and mitochondrial targeting [90]. Cell transfection studies have shown that the BNIP3-induced cell death is characterized by early plasma membrane permeabilization and mitochondrial damage without cytochrome c release and caspase activation [87, 90]; this suggests that BNIP3 induces cell death through a caspase-unrelated mechanism. Integration of BNIP3 into mitochondrial membranes causes MPT pore opening, suppresses mitochon-drial membrane potential, and increases ROS production [17].

   Under normal physiological conditions, BNIP3 is not detectable in healthy brain neurons. Expression of endogenous BNIP3 is induced in a variety of cells and tis-sues under hypoxic conditions [91–93]. Characterization of the BNIP3 gene revealed that the BNIP3 promoter contained a functional HIF-1-responsive element (HRE) and that expression of BNIP3 could be potently activated by both hypoxia and forced expression of HIF-1α [91]. In our previous studies, we reported that oxidative stress functioned as a redox signal to induce HIF-1α accumulation and subsequent activation of BNIP3 [89]. We have further established a BNIP3-induced and Endo G-mediated neuronal death pathway in stroke [77]. In this pathway, BNIP3 is transcriptionally upregulated by HIF-1 in hypoxia and stroke, causing mitochondrial dysfunction, which results in mitochondrial release of Endo G. Endo G then translocates to the nucleus and cleaves chromatin DNA, which leads to a form of caspase-independent neuronal cell death. The BNIP3 pathway explains well why brain-specific knockout of HIF-1α reduces hypoxic-ischemic damage to the brain, despite the fact that HIF-1 activation is an established adaptive response to hypoxia [94]. Judging from the time-course of BNIP3 expression, it is likely that the BNIP3 pathway contributes greatly to the delayed neuronal death found in stroke.

3.1.3 Mitophagy/Mitochondrial Biogenesis/Chaperone-Mediated Autophagy/ROS Production

Since the proteins released from the MOMP exert pleiotropic effects, ranging from caspase activation to chromatin condensation to DNA strand breakage and generation of ROS [52, 95], successful inhibition of MOMP would be expected to prevent the release and consequent destructive effects of those caspase-dependent and caspase-independent cell death effectors. Also, considering that mitochondria are critical organelles for maintaining the cellular homeostasis and the fact that they are not static according to different physiological and pathological conditions, the mechanisms underlying the dynamic regulation of mitochondrial turnover, content, function, and number in the cells [96] will provide useful insights for therapeutic target-searching. This is especially important for postmitotic cells such as neurons; that is why we would like to introduce the concepts of mitophagy, mitochondrial biogenesis chaperone-mediated autophagy (CMA), and their potential roles in caspase-independent neuronal death here.

Mitophagy, the specific autophagic elimination of mitochondria, has been identified in yeast, mediated by autophagy-related 32 (Atg32), and in mammals during red blood cell differentiation, mediated by NIP-like protein X (NIX; also known as BNIP3L) [97]. Lemasters and colleagues coined the term mitophagy to describe the engulfment of mitochondria into vesicles that are coated with the autophagosome marker MAP1 light chain 3 (LC3), a process that can occur within 5 min [98]. Another study found that, in neurons cultured with caspase inhibitors to prevent cell dissolution, mitochondria were completely removed by mitophagy following apoptosis induction through MOMP [99]. As MOMP occurs upstream of caspase activation, these results suggest that mitochondrial damage can induce mitophagy through an alternative pathway. Mitophagy involves distinct steps to recognize defective or superfluous organelles and to target them to autophagosomes. In yeast, Atg32 seems to recruit redundant and/or damaged mitochondria to autophagosomes [100]. In mammals, NIX mediates developmental removal of mitochondria during erythropoiesis [101, 102], since NIX has a WXXL-like motif, which binds to LC3 on isolation membranes and mediates the binding and sequestration of mitochondria into autophagosomes [103, 104]. Considerable evidence also shows that BNIP3, as well as NIX, can trigger mitochondrial depolarization, and that mitochondrial depolarization is sufficient to cause mitophagy [105, 106]. These early studies indicate that mitophagy regulates mitochondrial number to match metabolic demand and might also be a form of quality control to remove damaged mitochondria.

Oxidative stress triggered by hypoxic-ischemic insult causes extensive mitochondrial fission, an event that precedes neuronal death. Interestingly, abrogation of neuronal death can occur due to overexpression of mitofusin 2, a mitochondrial fusion protein [107]. This suggests that the mitochondrial fission and fusion is tightly regulated, and crucial for neuronal viability. A recent study finally demonstrated that H/I induced mitochondrial biogenesis. After hypoxia, increases are seen in mitochondrial DNA, total mitochondrial number, expression of the mitochondrial transcription factors downstream of PGC-1α (mitochondrial transcription factor A and nuclear respiratory factor 1), and the mitochondrial protein HSP60 [108]. This is an exciting finding that suggests mitochondrial biogenesis is a novel endogenous neuroprotective response. Taken together, after a lethal ischemic insult, selective elimination of mitochondria by autophagy in conjunction with mitochondrial biogenesis regulates the changes in mitochondrial number that are required to meet the metabolic demand. Damaged mitochondria are also selectively removed by mitophagy to maintain quality control. Therefore, effective regulation of mitophagy and/or enhancement of mitochondrial biogenesis are potential neuroprotective strategies.

CMA is a type of autophagy responsible for the degradation of cytosolic proteins bearing a consensus motif, biochemically related to KFERQ, that targets them for lysosomal degradation [109]. This motif is recognizable by the heat-shock cognate protein of 70 kDa (i.e., hsc70), in complex with its co-chaperones [110]. The substrate/chaperone complex binds to the lysosome-associated membrane protein type-2A protein (i.e., LAMP-2A, a CMA receptor) after being delivered to the surface of lysosomes [111]. With the assistance of a resident lysosomal chaperone (lys-hsc70), the substrate protein is translocated across the lysosomal membrane in an ATP-dependent fashion, subsequent to unfolding [112]. Once in the lysosomal lumen, CMA substrates are rapidly degraded (in 5–10 min) by the broad array of lysosomal proteases. CMA is maximally upregulated under stressful conditions, such as prolonged nutrient deprivation (serum removal in cultured cells or starvation in rodents) [113, 114], mild oxidative stress [115], and exposure to toxins [116]. Thus, CMA is part of the cellular quality control systems and is essential for the cellular responses to stress. CMA activity has been detected in several primary cell cultures including astrocytes, dopaminergic neurons, and cortical neurons [117, 118], while impairment of CMA underlies the pathogenesis of certain human pathologies such as neurodegenerative disorders. The Hsc70 plays critical roles in CMA activity, since only a subset of lysosomes containing Hsc70 in their lumen are competent for CMA [119]. Furthermore, upon MOMP, the chaperone Hsp70 has been shown to play a critical role in sequestering AIF within the cytosol, thereby interfering with its proapoptotic potential [120]. Additionally, Hsp70 has been shown to act at another level, by preventing the mitochondrial release of AIF [121], thus having two distinct molecular mechanisms of suppressing AIF-dependent DNA degradation. As could be expected, a greater infarction volume is seen within the cortex of hsp70$^{-/-}$ mice suffering of ischemic insult than that of WT mice [122]. In the contrary scenario, Hsp70 overexpression in a neonatal H/I model has been shown to limit the mitochondrial-nuclear translocation of AIF and consequently provided significant neuroprotection [123]. All of these results suggest that the Hsp70-mediated CMA pathway is closely involved in the caspase-independent neuronal death, which provides us with another promising therapeutic target for stroke treatment.

3.2   Alternative Targets Upstream of Mitochondrial Damage (MOMP)

3.2.1   Endoplasmic Reticulum and Lysosome Targets for Stroke: UPR Pathway/Calpain–Cathepsin Pathway

The endoplasmic reticulum (ER) serves two major functions in the cell. It facilitates the proper folding of newly synthesized proteins destined for secretion, cell surface or intracellular organelles, and it provides the cell with a Ca^{2+} reservoir [124–126]. ER stress occurs in various physiological and pathological conditions, including glucose starvation and hypoxia in stroke, where the capacity of the ER to fold proteins becomes saturated. Accumulation of unfolded proteins triggers ER stress and a subsequent evolutionarily conserved ER-to-nucleus signaling pathway, called the unfolded protein response (UPR), which reduces global protein synthesis and induces the synthesis of chaperones and other proteins that increase the capacity of the ER [127]. The signaling pathway that mediates the UPR in yeast consists of the transmembrane signaling protein inositol-requiring kinase 1 (IRE1), which can be activated by the dissociation of Grp78/BiP (ER-specific member of heat shock protein 70 family) from its ER-sensing domain in response to accumulation of unfolded proteins. The Hac1, a transcription factor, is then expressed and transmits the signal to the nucleus [125, 128, 129]. In mammalian cells, the UPR signaling is more complicated; at least three mechanistically different ER stress transducers—RNA-dependent protein kinase (PKR)-like ER kinase (PERK), ATF6, and IRE1—operate in parallel to mediate the UPR [127]. In addition to activating the UPR, ER stress also leads to a release of Ca^{2+} from the ER into the cytosol, which, in turn, can activate various kinases and proteases possibly involved in autophagy signaling [43, 130, 131]. These activated non-caspase proteases, such as calpains and cathepsins, act as upstream MOMP triggers, which are also potential targets for neuroprotection in brain ischemia.

Calpains, a family of 14Ca^{2+}-activated neutral cysteine proteases, are activated both in physiological states and also during various pathological conditions such as in the presence of free radicals [132, 133], brain ischemia-reperfusion [134, 135], apoptosis [132, 136], and Alzheimer's [137] or Parkinson's [138] diseases. Excessive activation of calpain due to an increase in free Ca^{2+} leads to cytoskeletal protein breakdown, subsequent loss of structural integrity and disturbances of axonal transport, and finally to neuronal death. Recently, Yamashima et al. reported sustained activation of μ-calpain in post-ischemic CA1 neurons, which was shown to cause spillage of cathepsins (a family of hydrolytic proteases) from lysosomes [134].

Lysosomes contain over 80 types of hydrolytic enzymes. In terms of executing neuronal death, two classes of lysosomal hydrolytic enzymes appear to be most active: aspartyl (cathepsin D) and cysteine (cathepsins B, H, L) proteases [139]. Cathepsin D mediates execution of neuronal death induced by aging, transient forebrain ischemia, and excitotoxicity [140], while cathepsins B and L execute hippocampal neuronal death after global ischemia [141]. The spreading of hydrolytic enzymes into the cytoplasm through injury or rupture of the lysosomal membrane was confirmed in ischemic brain injuries. This is the basis of the "calpain–cathepsin hypothesis" [26, 142], a hypothesis encompassing calpain and cathepsin as the key mediators of this process. For example, recent experiments on mice in which the endogenous calpain inhibitor calpastatin has been overexpressed or knocked out underscore the importance of calpains as an activator of lethal MOMP in neuronal cell death [143]. Cathepsin B can also act as a MOMP-inducer, linking lysosomal damage (which causes cathepsin B release) to MOMP [144]. Cathepsin B knockout cells are particularly resistant to induction of apoptosis by TNF-α, and cathepsin B knockout mice show reduced liver damage in response to TNF-α [145] or cholestasis [146]. Moreover, the cathepsin B inhibitor CA-074 can protect neurons from focal cerebral ischemia [147].

3.2.2 Roles of the Bcl-2 Protein Family in Ischemic Delayed Neuronal Death: In Mitochondrial Impairment and Autophagic Neuronal Death

There are three functional groups of proteins comprising the BCL-2 family: the multi-domain proapoptotic effectors, BAX and BAK; the multi-BH domain pro-survival members BCL-2, BCL-xL, BCL-w, MCL-1, and A1; and the proapoptotic BH3-only members BID, BIM, BIK, BAD, NOXA, PUMA, BNIP3, and NIX. These groups of proteins work cooperatively to link the upstream signals to downstream regulators of the pro-survival and pro-death members to determine the final destiny of the cell [148–151]. Presently, Bcl-2 family proteins have been emphasized in two major models that describe the mechanisms of MOMP. In one model, MOMP results from protein-permeable pores formed across the outer mitochondrial membrane by multi-domain proapoptotic members of BAX and BAK [152, 153], which can interact with the proapoptotic BH3-only proteins (e.g., Bid and Bim) that function as intracellular sensors of stress [154, 155]. The other group of Bcl-2 proteins (i.e., the multi-BH domain pro-survival members; e.g., Bcl-2, Bcl-xL) exerted anti-apoptotic functions by sequestering their proapoptotic counterparts into inactive complexes [156] as well as via other mechanisms, such as by modulating Ca^{2+} fluxes at the endoplasmic reticulum [157]. In the second model, MOMP was triggered by an abrupt increase in the permeability of the mitochondrial inner membrane to low-molecular weight solutes, which is known as MPT [158, 159]. The activity of the permeability transition pore complex (PTPC), which responds to proapoptotic signals such as Ca^{2+} overload and oxidative stress, is regulated by both the pro- and anti-apoptotic Bcl-2 proteins [160].

Besides causing mitochondrial impairment, Bcl-2 anti-apoptotic family members also target the lysosomal degradation pathway of autophagy. Bcl-2 has been shown to block caspase-independent cell death and the degradation of mitochondria, which are two processes postulated to involve autophagy. Recently, Pattingre et al. directly demonstrated the role of Bcl-2 in the negative regulation of autophagy. They showed that Bcl-2 inhibits Atg6/Beclin 1 (an autophagy-related protein) as well as the subsequent Beclin 1-dependent autophagic processes. Beclin 1 mutants incapable of binding to Bcl-2 also promote autophagic cell death through induction of autophagy, as if having been displaced by PCD-inducing proteins like BNIP3. These observations led to the hypothesis that Bcl-2 down-regulation of autophagy through its interaction with Beclin 1 may prevent cell death from occurring [161]. This regulatory activity of Bcl-2 on autophagy is specifically attributed to its expression at the ER membrane, indicating that signaling events originating from the ER are crucial for autophagy [162]. Finally, many laboratories have now shown that ER stress triggers autophagy, and this effect is also regulated by UPR stress sensors [163–166]. Stimuli that increase cytosolic calcium could activate ER stress and subsequent autophagy, which can be blocked by Bcl-2 [130].

3.2.3   The BNIP3-Activated and Bcl-2-Beclin 1-Mediated Autophagic Neuronal Death Pathway (Caspase-Independent)

Beclin 1 (also known as Atg6), the first identified mammalian autophagy gene product [167], was originally isolated as a BCL-2-interacting protein [168–170]. A functional BH3-like domain was identified in Beclin 1, and its mutation disrupted the interaction of Beclin 1 with Bcl-xL [171]. Pharmacological BH3 mimetic ABT-737 could competitively inhibit the interaction between Beclin 1 and BCL-2/Bcl-xL, stimulating autophagy [171, 172]. In addition, the BH3-only proteins including BNIP3 have been shown to regulate autophagy under different settings, possibly due to disrupting the interaction between Beclin 1 and Bcl-2/Bcl-xL via their BH3 domains [172, 173]. It appears that prolonged BNIP3 expression or acute overexpression beyond an autophagic survival threshold may result in autophagic cell death. It was recently reported that prolonged exposure of several apoptosis-competent cancer lines to hypoxia induced autophagy and cell death in a BNIP3-dependent manner [174]. These results suggest that liberating Beclin 1 from BCL-2/BCL-xL may be one of the mechanisms by which BH3-only members, including BNIP3, promote autophagy [175]. On the other hand, as loss of MPT appears to induce autophagy, BNIP3 may also induce autophagy indirectly as a consequence of such mitochondrial injury [105].

Recently, we reported that BNIP3 contributes to delayed neuronal death following stroke and provided evidence that BNIP3 is markedly upregulated at 48 and 72 h after cerebral ischemia [77]. It is also known that BNIP3-induced neuronal death is caspase- and cytochrome C-release-independent and characterized by early mitochondrial damage. We have further proposed a BNIP3-activated and Bcl-2/Beclin 1-mediated autophagic neuronal death pathway in stroke. Here are some results:

1. BNIP3 was induced in stroke, both in cultured primary neurons in an OGD model and in a neonatal cerebral H/I animal model. Levels of BNIP3 were low for up to 24 h but started to accumulate after 48 h postoperatively. The expressed BNIP3 was in its active form because it was membrane-bound and localized to mitochondria.
2. An increase of autophagy was observed as determined by the ratio of LC3-II to LC3-I, an autophagy marker protein, both in vitro and in vivo.
3. The time course and expression levels of Beclin 1 (BECN1), an autophagy-regulating protein, correlated positively with the expression of BNIP3. The expression of both proteins was accompanied by an increased autophagic neuronal death rate. This increase could be attenuated by the specific autophagy inhibitor, 3-methyladenine (3-MA).
4. The presence of large autolysosomes and numerous autophagosomes in neurons exposed to OGD injury was confirmed by electron microscopy. Autophagic cell death seemed to contribute to a great portion of the delayed neuronal death after H/I (Shi et al., CNS Neuroscience and Therapeutics, In press) (Fig. 7.2).

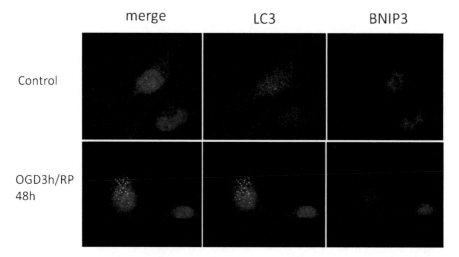

Fig. 7.2 Upregulated BNIP3 expression and increased LC3 translocation after H/I and RP treatment in cortical neurons. Neurons were treated with OGD for 3 h, followed by RP for 48 h. BNIP3 was stained with *red* while LC3 and nuclei were marked by *green* and *blue*, respectively

Our lab has found a unique caspase-independent cell death pathway that features the mitochondrial localization of BNIP3 then Endo G and AIF release from mitochondria and translocation into the nuclei, which results in eventual cell death (unpublished data). It is possible that autophagy also plays a part in this pathway by affecting mitochondrial stabilization. Further studies need to focus on elucidating the interactions between BNIP3, Beclin 1, and other possible intermediate autophagy-related proteins. Furthermore, molecular sequences underlying this pathway also need to be clarified in detail.

4   Therapy

4.1   Potential Therapeutic Applications Targeting on Mitochondrial Damage (MOMP)

In the best case scenarios, patients affected by stroke are treated within dozens of minutes, but usually treatment occurs only within hours, well after the MOMP has been initiated. Novel neuroprotective strategies that actively prevent the acute and delayed loss of neurons are urgently awaited. Such intervention should target the post-mitochondrial effectors and be administered during the first phases of cell death that occur before irreversible catabolic reactions have been ignited. Furthermore, in an early intervention that before affected neurons have undergone MOMP, combination therapies that associate MOMP blockers with post-MOMP effectors' inhibitors might achieve improved neuroprotection by preventing cell death both upstream and downstream of mitochondria [51].

Previous studies in models of delayed neuronal death have demonstrated protective effects of inhibiting caspase activities by various caspase inhibitors or genetic inactivation of distinct caspases in transgenic mice. However, it has been proven that caspase inhibition cannot prevent impairment of the induction of long-term potentiation after ischemia. This suggests that caspase inhibition alone does not preserve functional plasticity of neurons [176]. More recently, therapeutic strategies have been directed toward the prevention of the mitochondrial release of AIF and Endo G due to the discoveries in caspase-independent cell death signaling in delayed neuronal death after cerebral ischemia. Reduction of mitochondrial AIF protein levels has led to a significant decline in neuronal cell death after ischemia. This can be directly or indirectly achieved by RNA interference [64], inhibition of Poly (ADP-ribose) polymerase using 3-aminobenzamide [177], expression of hepatocyte growth factor (HGF) [178], or inhibition of neuronal NO synthase [179]. In more recent work, it has been shown that both the chaperone and ATPase domains of Hsp70 are vital in hindering the release of AIF from mitochondria, while only the latter would be required to cytosolically sequester AIF, thus preventing its nuclear translocation [121]. In addition, data obtained in vivo via an ischemic rat model (which followed the brain-targeted overexpression of different Hsp70 mutants, through plasmid transfection) suggests that the C-terminal portion of Hsp70 is in itself sufficient for neuroprotection [180]. This implies independence of both the functionality as well as the presence of its N-terminal ATPase domain. These experimental results are beginning to provide a solid foundation for the development of neuroprotective strategies through the Hsp70-mediated inhibition of AIF. Like AIF, downregulation of Endo G appears to be another promising method of neuroprotection against delayed neuronal death. Data from our laboratory showed that knockdown of BNIP3 by RNAi inhibited Endo G translocation and protected neurons from hypoxia-induced cell death [77]. Reduction in activity of cps-6, which encodes a homologue of human mitochondrial Endo G, by a genetic mutation or RNA interference, delayed appearance of cell corpses [181]. Transgenic mice with mutant Endo G heterozygous gene were more resistant to neuronal death induced by tumor necrosis factor-α (TNF-α) or staurosporine and had less DNA fragments compared to the wild-type mice [74]. Interestingly, the DNase of Endo G could be inactivated by Hsp70 in an ATP-dependent manner, which was indicated by assays based on purified cellular components [182], backing up the potential value of targeting on Hsp70. In addition to lethal proteins, mitochondria can also generate and release highly toxic ROS, which contribute to cell death. So, ROS scavengers can have cytoprotective effects in vitro and in vivo [183].

Inhibiting MOMP further upstream, at the initiation phase that precedes mitochondrial damage, has achieved significant levels of neuroprotection in vivo. For example, in mice lacking the neuronal nitric oxide synthase (NOS) gene [184], pharmacological [185] and genetic [186] inhibition of the tumor suppressor protein p53 as well as systemic administration of small molecules [187] or peptides [188] designed to block the c-JunN-terminal kinase (JNK) signaling pathway showed neuroprotective effects against ischemia, trauma, and excitotoxicity. Autophagy can serve as an ER-associated degradation system in mammalian cells and play a fundamental role in regulating the accumulation of disease-associated mutant proteins in the ER. Also, non-transformed cells like primary cells may be especially sensitive to ER stress-induced autophagy. Therefore, various ER stressors being potential targets may help prevent the ER-stress-induced excessive autophagy and consequent autophagic cell death. This is based on data showing that the inhibition of autophagy by Atg5 deficiency in MEFs or by 3-methyladenine in colon epithelial cells inhibits cell death induced by ER stressors (Ca^{2+} ionophore, thapsigargin, and tunicamycin) [127, 166].

4.2 Potential Therapeutic Applications Targeting on Bcl-2 Family and Autophagy

It has been reported that some inhibitors of the permeability transition and/or mitochondrial ion channels inhibit cell death in models of stroke [52, 189, 190]. Seeing as BH3-only members of the Bcl-2 family seem to be important instigators of MOMP in many pathways, small-molecule antagonists of these proteins should be effective inhibitors of MOMP-triggered cell death. Pharmacological inhibitors of the pore-forming activity of Bax have been shown to exert neuroprotective effects in an animal model of global brain ischemia [191]. Also, by inhibiting the association of Bax with mitochondria [192], the endogenous bile acid tauroursodeoxycholic acid (TUDCA) protected rats against neurological injury after Tfci [193] and intracerebral hemorrhage [194]. For the anti-apoptotic molecules, a single injection of a DNA plasmid encoding the protective Bcl-2 gene provided neuroprotection of injured neurons in vivo [195]. Mice overexpressing anti-apoptotic members such as Bcl-2 [196, 197] or Bcl-xL [198] displayed decreased tissue damage after permanent focal ischemia. Contrarily, the main strategies for neuroprotection of the proapoptotic members may lie in the downregulation of death gene expression. Plesnila and colleagues have developed low-molecular mass 4-phenylsulfanyl-phenylamine derivatives targeting Bid and have shown that they have the ability to inhibit tBid-induced Samc release, which prevents caspase-3 activation and cell death in isolated mitochondria and in cancer cell lines [199]. The Bid inhibitors preserve mitochondrial integrity and prevent activation of caspase-3 as well as nuclear translocation of AIF and DNA condensation [64]. Neurons from Bid-null mice are resistant to cell death stimuli after oxygen-glucose deprivation and maintain a significantly reduced caspase-3 cleavage. Adult mice lack of Bid exhibit decreased cytochrome c release from mitochondria and reduced infarct volumes after transient focal ischemia as compared to wild-type mice [200, 201]. Similarly, $Bax^{-/-}$ mice had obviously less hippocampal tissue loss or controlled cortical impact [202] upon neonatal H/I [203] than $Bax^{+/-}$ and $Bax^{+/+}$ animals. Stimulation of a BH3-only protein-targeting ubiquitin ligase might be attempted [204].

A death-promoting role of autophagy in cerebral ischemia has already been suggested by many studies. Knockdown of Atg7 can protect hippocampal pyramidal neurons against hypoxia-ischemia [13, 38, 40, 205]. Puyal et al. reported that post-ischemic intracerebroventricular injections of autophagy inhibitor 3-MA strongly reduced the lesion volume (by 46%) even when given >4 h after the beginning of the ischemia, demonstrating for the first time that post-ischemic pharmacological inhibition of autophagy can offer neuroprotection [206]. Moreover, the neuroprotective efficacy of inhibiting autophagy more than 4 h after the onset of ischemia highlighted that autophagy should be a primary target for preventing delayed neuronal death in the ischemic penumbra [41]. This could have clinical potential, as long-term lesions and damage could be minimized despite a stroke having already occurred.

4.3   Other Targets and Novel Techniques

In addition to therapeutic targets discussed above, new techniques have emerged as well. Protein therapeutics combining the super anti-apoptotic factor FNK, generated from anti-apoptotic Bcl-2 gene and the protein transduction domain (PTD) of the HIV Tat protein, have been found effective in preventing delayed neuronal death in the hippocampus caused by transient global ischemia [207]. Delivery of the HGF gene to subarachnoid space prevented delayed neuronal death in gerbil hippocampal CA1 neurons after brain ischemia [208]. The Ca^{2+} channel antagonist, nimodipine, reduced brain injury and improved functional outcome in rodents [209, 210] and nonhuman primate stroke models [211], and may improve outcome in human stroke patients [212, 213]. Dantrolene reduced ischemic injury to neurons in mice [214] and gerbils [215], which suggests an important contribution of Ca^{2+} release from ER stores in the ischemic cell death process. Nitric oxide contributes to ischemia-induced oxidative stress and neuronal death [216]. Drugs that inhibit NOS or scavenge nitric oxide were reported to reduce neuronal damage in rodent stroke models [217, 218]. Several antioxidants have also been reported to be effective in rodent stroke models, including vitamin E [219], lipoate [220], and uric acid [221].

The tetrapeptide inhibitor tyrosine–valine–alanine–aspartate–chloromethyl ketone (Ac-YVAD-cmk) was reported to rescue cultured neurons from cell death due to oxygen/glucose deprivation by targeting lysosomal enzyme cathepsin B [222]. Based on the calpain–cathepsin hypothesis, Yamashima et al. demonstrated in the monkey brain that, other than in the CA1 region, 89.8% of caudate nucleus neurons were free from post-ischemic neuronal death on Day 5 with 4 mg/kg of CA-074 treatment, while 75.0% of the cortical V layer neurons and 91.6% of the cerebellar neurons survived with 4 mg/kg of E-64c treatment. The inhibitory effect of delayed neuronal death by E-64c was overall more remarkable than that of CA-074. This is probably because E-64c can inhibit not only cathepsins B and L but also calpains [139]. Furthermore, other approaches have been found that reduce delayed neuronal death unrelated with caspases include the actin depolymerizing agent, cytochalasin D, which was reported to be effective in reducing focal ischemic brain injury in mice [223], and also dietary folic acid, which decreases the vulnerability of neurons to excitotoxic and oxidative insults by reducing homocysteine levels [224].

5   Conclusion

In conclusion, the delayed neuronal death (penumbral region primarily, as the infarct core undergoes inevitable necrosis) responsible for many adverse and debilitating effects in stroke victims is due to intricate interactions between several signaling and transduction pathways. These are various types of PCD, several of which are independent of caspases and apoptosis. Examples of these are autophagic cell death processes. Understanding the functionalities and interactions of these mediators can help find potential drug targets, ideally in which key PCD pathways could be inhibited or in which at least cascadic amplification could be dampened. It has been shown that caspase inhibition in and of itself is not sufficient for maintenance of neuronal plasticity, despite some success in caspase inhibition maintaining neuronal life. If done early enough, one could target MOMP before its occurrence in ischemia with blockers, such as AIF or Endo G release. Further, the Bcl-2 family of genes has garnered great interest. BH3-only members like Bid and BNIP3 are critical instigators of MOMP, and Bcl-2 itself is an anti-apoptotic agent. By inhibiting release, downregulating or antagonizing pro-PCD intermediates, and/or upregulating expression of anti-PCD intermediates or administering anti-PCD intermediates, one can stave off or even prevent delayed neuronal death until reperfusion ensues. FNK (a derivative of Bcl-2), HGF, Ca^{2+} channel antagonists, NO synthase inhibitors, NO scavengers, cathepsin/calpain inhibitors, actin depolymerizing agents, folic acid, and numerous antioxidants (like Vitamin E, lipoate, and uric acid) have all shown neuroprotective potential in animal models.

The longer the time it takes for a stroke victim to get treatment, the larger the infarct core becomes, as the surrounding penumbral tissue reaches the "point of no return." The reversibly damaged penumbra is often initially far larger than the core (due to collateral capillary networks in the area), so much of the adversity experienced after a stroke is also theoretically reversible if caught early enough. That is, the amount of tissue saved can be amplified with earlier interventions. It is possible that, in the future, we could prophylactically treat people at high risk for stroke before the infarct even occurs (as time-to-treatment equals more neuron death) by administering neuroprotective, therapeutic agents. As PCD pathways would be chronically downregulated, neurons would be protected from the very beginning of infarct, minimizing penumbral cell death and the size of the inevitable infarct core. Furthermore, the administration of therapeutic agents into the infarct region would be facilitated through the deteriorated blood–brain barrier (which would otherwise form a barrier against distributing the agents of interest), as the integrity of the barrier is impaired through ischemic infarction processes. In conclusion, all of the aforementioned research has great potential in changing the course of treatment in stroke and the clinical implications thereof. As research continues looking into different neuroprotective ways in which to inhibit such PCD pathways, one can expect many innovative developments in the near future in this emerging field.

Acknowledgments This work was supported by grants from the Canadian Institutes of Health Research, Canadian Stroke Network and Manitoba Health Research Council (to J. Kong). Dr. Jiming Kong received a salary award from the Heart and Stroke Foundation of Canada. Ms. Ruoyang Shi received a Manitoba Health Research Council/Manitoba Institute of Child Health Graduate Studentship.

References

1. Boyce M, Degterev A, Yuan J. Caspases: an ancient cellular sword of Damocles. Cell Death Differ. 2004;11:29–37.
2. Golstein P, Aubry L, Levraud JP. Cell-death alternative model organisms: why and which? Nat Rev Mol Cell Biol. 2003;4:798–807.
3. Garrido C, Kroemer G. Life's smile, death's grin: vital functions of apoptosis-executing proteins. Curr Opin Cell Biol. 2004;16:639–46.
4. Kroemer G, Martin SJ. Caspase-independent cell death. Nat Med. 2005;11:725–30.
5. Lo EH, Dalkara T, Moskowitz MA. Mechanisms, challenges and opportunities in stroke. Nat Rev Neurosci. 2003;4:399–415.

6. Lang-Rollin IC, Rideout HJ, Noticewala M, Stefanis L. Mechanisms of caspase-independent neuronal death: energy depletion and free radical generation. J Neurosci. 2003;23:11015–25.
7. Le DA, Wu Y, Huang Z, Matsushita K, Plesnila N, Augustinack JC, et al. Caspase activation and neuroprotection in caspase-3-deficient mice after in vivo cerebral ischemia and in vitro oxygen glucose deprivation. Proc Natl Acad Sci U S A. 2002;99:15188–93.
8. Didenko VV, Ngo H, Minchew CL, Boudreaux DJ, Widmayer MA, Baskin DS. Caspase-3-dependent and -independent apoptosis in focal brain ischemia. Mol Med. 2002;8:347–52.
9. Himi T, Ishizaki Y, Murota S. A caspase inhibitor blocks ischaemia-induced delayed neuronal death in the gerbil. Eur J Neurosci. 1998;10:777–81.
10. Cregan SP, Fortin A, MacLaurin JG, Callaghan SM, Cecconi F, Yu SW, et al. Apoptosis-inducing factor is involved in the regulation of caspase-independent neuronal cell death. J Cell Biol. 2002;158:507–17.
11. MacManus JP, Rasquinha I, Tuor U, Preston E. Detection of higher-order 50- and 10-kbp DNA fragments before apoptotic internucleosomal cleavage after transient cerebral ischemia. J Cereb Blood Flow Metab. 1997;17:376–87.
12. Repici M, Mariani J, Borsello T. Neuronal death and neuroprotection: a review. Methods Mol Biol. 2007;399:1–14.
13. Nitatori T, Sato N, Waguri S, Karasawa Y, Araki H, Shibanai K, et al. Delayed neuronal death in the CA1 pyramidal cell layer of the gerbil hippocampus following transient ischemia is apoptosis. J Neurosci. 1995;15:1001–11.
14. Chen J, Nagayama T, Jin K, Stetler RA, Zhu RL, Graham SH, et al. Induction of caspase-3-like protease may mediate delayed neuronal death in the hippocampus after transient cerebral ischemia. J Neurosci. 1998;18:4914–28.
15. Maiuri MC, Zalckvar E, Kimchi A, Kroemer G. Self-eating and self-killing: crosstalk between autophagy and apoptosis. Nat Rev Mol Cell Biol. 2007;8:741–52.
16. Cho BB, Toledo-Pereyra LH. Caspase-independent programmed cell death following ischemic stroke. J Invest Surg. 2008;21:141–7.
17. Vande Velde C, Cizeau J, Dubik D, Alimonti J, Brown T, Israels S, et al. BNIP3 and genetic control of necrosis-like cell death through the mitochondrial permeability transition pore. Mol Cell Biol. 2000;20:5454–68.
18. Kerr JF, Wyllie AH, Currie AR. Apoptosis: a basic biological phenomenon with wide-ranging implications in tissue kinetics. Br J Cancer. 1972;26:239–57.
19. Garcia JH, Liu KF, Ye ZR, Gutierrez JA. Incomplete infarct and delayed neuronal death after transient middle cerebral artery occlusion in rats. Stroke. 1997;28:2303–9; discussion 10.
20. Nedergaard M. Neuronal injury in the infarct border: a neuropathological study in the rat. Acta Neuropathol. 1987;73:267–74.
21. Sairanen T, Karjalainen-Lindsberg ML, Paetau A, Ijas P, Lindsberg PJ. Apoptosis dominant in the periinfarct area of human ischaemic stroke—a possible target of anti-apoptotic treatments. Brain. 2006;129:189–99.
22. Bennett BL, Sasaki DT, Murray BW, O'Leary EC, Sakata ST, Xu W, et al. SP600125, an anthrapyrazolone inhibitor of Jun N-terminal kinase. Proc Natl Acad Sci U S A. 2001;98:13681–6.
23. Sattler R, Tymianski M. Molecular mechanisms of glutamate receptor-mediated excitotoxic neuronal cell death. Mol Neurobiol. 2001;24:107–29.
24. Stout AK, Raphael HM, Kanterewicz BI, Klann E, Reynolds IJ. Glutamate-induced neuron death requires mitochondrial calcium uptake. Nat Neurosci. 1998;1:366–73.
25. Syntichaki P, Xu K, Driscoll M, Tavernarakis N. Specific aspartyl and calpain proteases are required for neurodegeneration in C. elegans. Nature. 2002;419:939–44.
26. Yamashima T, Kohda Y, Tsuchiya K, Ueno T, Yamashita J, Yoshioka T, et al. Inhibition of ischaemic hippocampal neuronal death in primates with cathepsin B inhibitor CA-074: a novel strategy for neuroprotection based on "calpain-cathepsin hypothesis". Eur J Neurosci. 1998;10:1723–33.
27. Yuan J, Lipinski M, Degterev A. Diversity in the mechanisms of neuronal cell death. Neuron. 2003;40:401–13.

28. de Murcia G, Schreiber V, Molinete M, Saulier B, Poch O, Masson M, et al. Structure and function of poly(ADP-ribose) polymerase. Mol Cell Biochem. 1994;138:15–24.
29. Nixon RA. Autophagy in neurodegenerative disease: friend, foe or turncoat? Trends Neurosci. 2006;29:528–35.
30. Qin AP, Liu CF, Qin YY, Hong LZ, Xu M, Yang L, et al. Autophagy was activated in injured astrocytes and mildly decreased cell survival following glucose and oxygen deprivation and focal cerebral ischemia. Autophagy. 2010;6:738–53.
31. Chu CT. Eaten alive: autophagy and neuronal cell death after hypoxia-ischemia. Am J Pathol. 2008;172:284–7.
32. Chu CT, Plowey ED, Dagda RK, Hickey RW, Cherra III SJ, Clark RS. Autophagy in neurite injury and neurodegeneration: in vitro and in vivo models. Methods Enzymol. 2009;453:217–49.
33. Canu N, Tufi R, Serafino AL, Amadoro G, Ciotti MT, Calissano P. Role of the autophagic-lysosomal system on low potassium-induced apoptosis in cultured cerebellar granule cells. J Neurochem. 2005;92:1228–42.
34. Uchiyama Y. Autophagic cell death and its execution by lysosomal cathepsins. Arch Histol Cytol. 2001;64:233–46.
35. Schwartz LM, Smith SW, Jones ME, Osborne BA. Do all programmed cell deaths occur via apoptosis? Proc Natl Acad Sci U S A. 1993;90:980–4.
36. Bursch W, Ellinger A, Gerner C, Frohwein U, Schulte-Hermann R. Programmed cell death (PCD). Apoptosis, autophagic PCD, or others? Ann N Y Acad Sci. 2000;926:1–12.
37. Bursch W, Hochegger K, Torok L, Marian B, Ellinger A, Hermann RS. Autophagic and apoptotic types of programmed cell death exhibit different fates of cytoskeletal filaments. J Cell Sci. 2000;113(Pt 7):1189–98.
38. Koike M, Shibata M, Tadakoshi M, Gotoh K, Komatsu M, Waguri S, et al. Inhibition of autophagy prevents hippocampal pyramidal neuron death after hypoxic-ischemic injury. Am J Pathol. 2008;172:454–69.
39. Zhu C, Wang X, Xu F, Bahr BA, Shibata M, Uchiyama Y, et al. The influence of age on apoptotic and other mechanisms of cell death after cerebral hypoxia-ischemia. Cell Death Differ. 2005;12:162–76.
40. Adhami F, Liao G, Morozov YM, Schloemer A, Schmithorst VJ, Lorenz JN, et al. Cerebral ischemia-hypoxia induces intravascular coagulation and autophagy. Am J Pathol. 2006;169:566–83.
41. Puyal J, Clarke PG. Targeting autophagy to prevent neonatal stroke damage. Autophagy. 2009;5:1060–1.
42. Yu L, Alva A, Su H, Dutt P, Freundt E, Welsh S, et al. Regulation of an ATG7-beclin 1 program of autophagic cell death by caspase-8. Science. 2004;304:1500–2.
43. Yousefi S, Perozzo R, Schmid I, Ziemiecki A, Schaffner T, Scapozza L, et al. Calpain-mediated cleavage of Atg5 switches autophagy to apoptosis. Nat Cell Biol. 2006;8:1124–32.
44. Rubinsztein DC, DiFiglia M, Heintz N, Nixon RA, Qin ZH, Ravikumar B, et al. Autophagy and its possible roles in nervous system diseases, damage and repair. Autophagy. 2005;1:11–22.
45. Yu L, Wan F, Dutta S, Welsh S, Liu Z, Freundt E, et al. Autophagic programmed cell death by selective catalase degradation. Proc Natl Acad Sci U S A. 2006;103:4952–7.
46. Kirino T, Tamura A, Sano K. A reversible type of neuronal injury following ischemia in the gerbil hippocampus. Stroke. 1986;17:455–9.
47. Lo EH. A new penumbra: transitioning from injury into repair after stroke. Nat Med. 2008;14:497–500.
48. Horbinski C, Chu CT. Kinase signaling cascades in the mitochondrion: a matter of life or death. Free Radic Biol Med. 2005;38:2–11.
49. Galluzzi L, Zamzami N, de la Motte Rouge T, Lemaire C, Brenner C, Kroemer G. Methods for the assessment of mitochondrial membrane permeabilization in apoptosis. Apoptosis. 2007;12:803–13.
50. Kroemer G, Reed JC. Mitochondrial control of cell death. Nat Med. 2000;6:513–9.
51. Galluzzi L, Morselli E, Kepp O, Kroemer G. Targeting post-mitochondrial effectors of apoptosis for neuroprotection. Biochim Biophys Acta. 2009;1787:402–13.

52. Green DR, Kroemer G. The pathophysiology of mitochondrial cell death. Science. 2004;
    305:626–9.
53. Vosler PS, Graham SH, Wechsler LR, Chen J. Mitochondrial targets for stroke: focusing
    basic science research toward development of clinically translatable therapeutics. Stroke.
    2009;40:3149–55.
54. Krantic S, Mechawar N, Reix S, Quirion R. Apoptosis-inducing factor: a matter of neuron
    life and death. Prog Neurobiol. 2007;81:179–96.
55. Zhu C, Wang X, Deinum J, Huang Z, Gao J, Modjtahedi N, et al. Cyclophilin A participates
    in the nuclear translocation of apoptosis-inducing factor in neurons after cerebral hypoxia-
    ischemia. J Exp Med. 2007;204:1741–8.
56. Cande C, Vahsen N, Kouranti I, Schmitt E, Daugas E, Spahr C, et al. AIF and cyclophilin A
    cooperate in apoptosis-associated chromatinolysis. Oncogene. 2004;23:1514–21.
57. Lorenzo HK, Susin SA, Penninger J, Kroemer G. Apoptosis inducing factor (AIF): a phyloge-
    netically old, caspase-independent effector of cell death. Cell Death Differ. 1999;6:516–24.
58. Susin SA, Lorenzo HK, Zamzami N, Marzo I, Snow BE, Brothers GM, et al. Molecular
    characterization of mitochondrial apoptosis-inducing factor. Nature. 1999;397:441–6.
59. Penninger JM, Kroemer G. Mitochondria, AIF and caspases—rivaling for cell death execu-
    tion. Nat Cell Biol. 2003;5:97–9.
60. Plesnila N, Zhu C, Culmsee C, Groger M, Moskowitz MA, Blomgren K. Nuclear transloca-
    tion of apoptosis-inducing factor after focal cerebral ischemia. J Cereb Blood Flow Metab.
    2004;24:458–66.
61. Hisatomi T, Sakamoto T, Murata T, Yamanaka I, Oshima Y, Hata Y, et al. Relocalization of
    apoptosis-inducing factor in photoreceptor apoptosis induced by retinal detachment in vivo.
    Am J Pathol. 2001;158:1271–8.
62. Daugas E, Susin SA, Zamzami N, Ferri KF, Irinopoulou T, Larochette N, et al. Mitochondrio-
    nuclear translocation of AIF in apoptosis and necrosis. FASEB J. 2000;14:729–39.
63. Cao G, Clark RS, Pei W, Yin W, Zhang F, Sun FY, et al. Translocation of apoptosis-inducing
    factor in vulnerable neurons after transient cerebral ischemia and in neuronal cultures after
    oxygen-glucose deprivation. J Cereb Blood Flow Metab. 2003;23:1137–50.
64. Culmsee C, Zhu C, Landshamer S, Becattini B, Wagner E, Pellecchia M, et al. Apoptosis-
    inducing factor triggered by poly(ADP-ribose) polymerase and Bid mediates neuronal cell death
    after oxygen-glucose deprivation and focal cerebral ischemia. J Neurosci. 2005;25:10262–72.
65. Zhu C, Qiu L, Wang X, Hallin U, Cande C, Kroemer G, et al. Involvement of apoptosis-
    inducing factor in neuronal death after hypoxia-ischemia in the neonatal rat brain. J
    Neurochem. 2003;86:306–17.
66. Zhu C, Wang X, Huang Z, Qiu L, Xu F, Vahsen N, et al. Apoptosis-inducing factor is a major
    contributor to neuronal loss induced by neonatal cerebral hypoxia-ischemia. Cell Death
    Differ. 2007;14:775–84.
67. Tsujimoto Y. Cell death regulation by the Bcl-2 protein family in the mitochondria. J Cell
    Physiol. 2003;195:158–67.
68. van Loo G, Saelens X, Matthijssens F, Schotte P, Beyaert R, Declercq W, et al. Caspases are
    not localized in mitochondria during life or death. Cell Death Differ. 2002;9:1207–11.
69. Crompton M. The mitochondrial permeability transition pore and its role in cell death.
    Biochem J. 1999;341(Pt 2):233–49.
70. Donovan M, Cotter TG. Control of mitochondrial integrity by Bcl-2 family members and
    caspase-independent cell death. Biochim Biophys Acta. 2004;1644:133–47.
71. Cande C, Cecconi F, Dessen P, Kroemer G. Apoptosis-inducing factor (AIF): key to the con-
    served caspase-independent pathways of cell death? J Cell Sci. 2002;115:4727–34.
72. Li LY, Luo X, Wang X. Endonuclease G is an apoptotic DNase when released from mito-
    chondria. Nature. 2001;412:95–9.
73. van Loo G, Schotte P, van Gurp M, Demol H, Hoorelbeke B, Gevaert K, et al. Endonuclease
    G: a mitochondrial protein released in apoptosis and involved in caspase-independent DNA
    degradation. Cell Death Differ. 2001;8:1136–42.

74. Zhang J, Dong M, Li L, Fan Y, Pathre P, Dong J, et al. Endonuclease G is required for early embryogenesis and normal apoptosis in mice. Proc Natl Acad Sci U S A. 2003;100: 15782–7.

75. Lee BI, Lee DJ, Cho KJ, Kim GW. Early nuclear translocation of endonuclease G and subsequent DNA fragmentation after transient focal cerebral ischemia in mice. Neurosci Lett. 2005;386:23–7.

76. Tanaka S, Takehashi M, Iida S, Kitajima T, Kamanaka Y, Stedeford T, et al. Mitochondrial impairment induced by poly(ADP-ribose) polymerase-1 activation in cortical neurons after oxygen and glucose deprivation. J Neurochem. 2005;95:179–90.

77. Zhang Z, Yang X, Zhang S, Ma X, Kong J. BNIP3 upregulation and EndoG translocation in delayed neuronal death in stroke and in hypoxia. Stroke. 2007;38:1606–13.

78. Vande Walle L, Van Damme P, Lamkanfi M, Saelens X, Vandekerckhove J, Gevaert K, et al. Proteome-wide identification of HtrA2/Omi Substrates. J Proteome Res. 2007;6:1006–15.

79. Saito A, Hayashi T, Okuno S, Nishi T, Chan PH. Modulation of the Omi/HtrA2 signalling pathway after transient focal cerebral ischemia in mouse brains that overexpress SOD1. Brain Res Mol Brain Res. 2004;127:89–95.

80. Siegelin MD, Kossatz LS, Winckler J, Rami A. Regulation of XIAP and Smac/DIABLO in the rat hippocampus following transient forebrain ischemia. Neurochem Int. 2005;46:41–51.

81. Saito A, Hayashi T, Okuno S, Ferrand-Drake M, Chan PH. Interaction between XIAP and Smac/DIABLO in the mouse brain after transient focal cerebral ischemia. J Cereb Blood Flow Metab. 2003;23:1010–9.

82. Shibata M, Hattori H, Sasaki T, Gotoh J, Hamada J, Fukuuchi Y. Subcellular localization of a promoter and an inhibitor of apoptosis (Smac/DIABLO and XIAP) during brain ischemia/reperfusion. Neuroreport. 2002;13:1985–8.

83. Saito A, Hayashi T, Okuno S, Nishi T, Chan PH. Oxidative stress is associated with XIAP and Smac/DIABLO signaling pathways in mouse brains after transient focal cerebral ischemia. Stroke. 2004;35:1443–8.

84. Boyd JM. Adenovirus E1B 19 kDa and Bcl-2 proteins interact with a common set of cellular proteins. Cell. 1994;79:1121.

85. Chen G, Cizeau J, Vande Velde C, Park JH, Bozek G, Bolton J, et al. Nix and Nip3 form a subfamily of pro-apoptotic mitochondrial proteins. J Biol Chem. 1999;274:7–10.

86. Chen G, Ray R, Dubik D, Shi L, Cizeau J, Bleackley RC, et al. The E1B 19K/Bcl-2-binding protein Nip3 is a dimeric mitochondrial protein that activates apoptosis. J Exp Med. 1997; 186:1975–83.

87. Cizeau J, Ray R, Chen G, Gietz RD, Greenberg AH. The C. elegans orthologue ceBNIP3 interacts with CED-9 and CED-3 but kills through a BH3- and caspase-independent mechanism. Oncogene. 2000;19:5453–63.

88. Yasuda M, D'Sa-Eipper C, Gong XL, Chinnadurai G. Regulation of apoptosis by a Caenorhabditis elegans BNIP3 homolog. Oncogene. 1998;17:2525–30.

89. Zhang S, Zhang Z, Sandhu G, Ma X, Yang X, Geiger JD, et al. Evidence of oxidative stress-induced BNIP3 expression in amyloid beta neurotoxicity. Brain Res. 2007;1138:221–30.

90. Ray R, Chen G, Vande Velde C, Cizeau J, Park JH, Reed JC, et al. BNIP3 heterodimerizes with Bcl-2/Bcl-X(L) and induces cell death independent of a Bcl-2 homology 3 (BH3) domain at both mitochondrial and nonmitochondrial sites. J Biol Chem. 2000;275:1439–48.

91. Bruick RK. Expression of the gene encoding the proapoptotic Nip3 protein is induced by hypoxia. Proc Natl Acad Sci U S A. 2000;97:9082–7.

92. Guo K, Searfoss G, Krolikowski D, Pagnoni M, Franks C, Clark K, et al. Hypoxia induces the expression of the pro-apoptotic gene BNIP3. Cell Death Differ. 2001;8:367–76.

93. Sowter HM, Ratcliffe PJ, Watson P, Greenberg AH, Harris AL. HIF-1-dependent regulation of hypoxic induction of the cell death factors BNIP3 and NIX in human tumors. Cancer Res. 2001;61:6669–73.

94. Helton R, Cui J, Scheel JR, Ellison JA, Ames C, Gibson C, et al. Brain-specific knock-out of hypoxia-inducible factor-1alpha reduces rather than increases hypoxic-ischemic damage. J Neurosci. 2005;25:4099–107.

95. Saelens X, Festjens N, Vande Walle L, van Gurp M, van Loo M, Vandenabeele P. Toxic proteins released from mitochondria in cell death. Oncogene. 2004;23:2861–74.
96. Diaz F, Moraes CT. Mitochondrial biogenesis and turnover. Cell Calcium. 2008;44:24–35.
97. Youle RJ, Narendra DP. Mechanisms of mitophagy. Nat Rev Mol Cell Biol. 2011;12:9–14.
98. Kim I, Rodriguez-Enriquez S, Lemasters JJ. Selective degradation of mitochondria by mitophagy. Arch Biochem Biophys. 2007;462:245–53.
99. Tolkovsky AM, Xue L, Fletcher GC, Borutaite V. Mitochondrial disappearance from cells: a clue to the role of autophagy in programmed cell death and disease? Biochimie. 2002;84:233–40.
100. Nowikovsky K, Reipert S, Devenish RJ, Schweyen RJ. Mdm38 protein depletion causes loss of mitochondrial K+/H+ exchange activity, osmotic swelling and mitophagy. Cell Death Differ. 2007;14:1647–56.
101. Schweers RL, Zhang J, Randall MS, Loyd MR, Li W, Dorsey FC, et al. NIX is required for programmed mitochondrial clearance during reticulocyte maturation. Proc Natl Acad Sci U S A. 2007;104:19500–5.
102. Kundu M, Lindsten T, Yang CY, Wu J, Zhao F, Zhang J, et al. Ulk1 plays a critical role in the autophagic clearance of mitochondria and ribosomes during reticulocyte maturation. Blood. 2008;112:1493–502.
103. Schwarten M, Mohrluder J, Ma P, Stoldt M, Thielmann Y, Stangler T, et al. Nix directly binds to GABARAP: a possible crosstalk between apoptosis and autophagy. Autophagy. 2009; 5:690–8.
104. Novak I, Kirkin V, McEwan DG, Zhang J, Wild P, Rozenknop A, et al. Nix is a selective autophagy receptor for mitochondrial clearance. EMBO Rep. 2010;11:45–51.
105. Elmore SP, Qian T, Grissom SF, Lemasters JJ. The mitochondrial permeability transition initiates autophagy in rat hepatocytes. FASEB J. 2001;15:2286–7.
106. Twig G, Elorza A, Molina AJ, Mohamed H, Wikstrom JD, Walzer G, et al. Fission and selective fusion govern mitochondrial segregation and elimination by autophagy. EMBO J. 2008;27:433–46.
107. Jahani-Asl A, Cheung EC, Neuspiel M, MacLaurin JG, Fortin A, Park DS, et al. Mitofusin 2 protects cerebellar granule neurons against injury-induced cell death. J Biol Chem. 2007; 282:23788–98.
108. Yin W, Signore AP, Iwai M, Cao G, Gao Y, Chen J. Rapidly increased neuronal mitochondrial biogenesis after hypoxic-ischemic brain injury. Stroke. 2008;39:3057–63.
109. Dice JF. Peptide sequences that target cytosolic proteins for lysosomal proteolysis. Trends Biochem Sci. 1990;15:305–9.
110. Chiang HL, Terlecky SR, Plant CP, Dice JF. A role for a 70-kilodalton heat shock protein in lysosomal degradation of intracellular proteins. Science. 1989;246:382–5.
111. Cuervo AM, Dice JF. A receptor for the selective uptake and degradation of proteins by lysosomes. Science. 1996;273:501–3.
112. Agarraberes FA, Terlecky SR, Dice JF. An intralysosomal hsp70 is required for a selective pathway of lysosomal protein degradation. J Cell Biol. 1997;137:825–34.
113. Cuervo AM, Knecht E, Terlecky SR, Dice JF. Activation of a selective pathway of lysosomal proteolysis in rat liver by prolonged starvation. Am J Physiol. 1995;269:C1200–8.
114. Wing SS, Chiang HL, Goldberg AL, Dice JF. Proteins containing peptide sequences related to Lys-Phe-Glu-Arg-Gln are selectively depleted in liver and heart, but not skeletal muscle, of fasted rats. Biochem J. 1991;275(Pt 1):165–9.
115. Kiffin R, Christian C, Knecht E, Cuervo AM. Activation of chaperone-mediated autophagy during oxidative stress. Mol Biol Cell. 2004;15:4829–40.
116. Cuervo AM, Hildebrand H, Bomhard EM, Dice JF. Direct lysosomal uptake of alpha 2-microglobulin contributes to chemically induced nephropathy. Kidney Int. 1999;55:529–45.
117. Cuervo AM, Stefanis L, Fredenburg R, Lansbury PT, Sulzer D. Impaired degradation of mutant alpha-synuclein by chaperone-mediated autophagy. Science. 2004;305:1292–5.
118. Martinez-Vicente M, Talloczy Z, Kaushik S, Massey AC, Mazzulli J, Mosharov EV, et al. Dopamine-modified alpha-synuclein blocks chaperone-mediated autophagy. J Clin Invest. 2008;118:777–88.

119. Cuervo AM, Dice JF, Knecht E. A population of rat liver lysosomes responsible for the selective uptake and degradation of cytosolic proteins. J Biol Chem. 1997;272:5606–15.
120. Ravagnan L, Gurbuxani S, Susin SA, Maisse C, Daugas E, Zamzami N, et al. Heat-shock protein 70 antagonizes apoptosis-inducing factor. Nat Cell Biol. 2001;3:839–43.
121. Ruchalski K, Mao H, Li Z, Wang Z, Gillers S, Wang Y, et al. Distinct hsp70 domains mediate apoptosis-inducing factor release and nuclear accumulation. J Biol Chem. 2006; 281: 7873–80.
122. Lee SH, Kwon HM, Kim YJ, Lee KM, Kim M, Yoon BW. Effects of hsp70.1 gene knockout on the mitochondrial apoptotic pathway after focal cerebral ischemia. Stroke. 2004;35:2195–9.
123. Matsumori Y, Hong SM, Aoyama K, Fan Y, Kayama T, Sheldon RA, et al. Hsp70 overexpression sequesters AIF and reduces neonatal hypoxic/ischemic brain injury. J Cereb Blood Flow Metab. 2005;25:899–910.
124. Berridge MJ. The endoplasmic reticulum: a multifunctional signaling organelle. Cell Calcium. 2002;32:235–49.
125. Bernales S, Papa FR, Walter P. Intracellular signaling by the unfolded protein response. Annu Rev Cell Dev Biol. 2006;22:487–508.
126. Momoi T. Conformational diseases and ER stress-mediated cell death: apoptotic cell death and autophagic cell death. Curr Mol Med. 2006;6:111–8.
127. Hoyer-Hansen M, Jaattela M. Connecting endoplasmic reticulum stress to autophagy by unfolded protein response and calcium. Cell Death Differ. 2007;14:1576–82.
128. Bertolotti A, Zhang Y, Hendershot LM, Harding HP, Ron D. Dynamic interaction of BiP and ER stress transducers in the unfolded-protein response. Nat Cell Biol. 2000;2:326–32.
129. Bertolotti A, Ron D. Alterations in an IRE1-RNA complex in the mammalian unfolded protein response. J Cell Sci. 2001;114:3207–12.
130. Hoyer-Hansen M, Bastholm L, Szyniarowski P, Campanella M, Szabadkai G, Farkas T, et al. Control of macroautophagy by calcium, calmodulin-dependent kinase kinase-beta, and Bcl-2. Mol Cell. 2007;25:193–205.
131. Demarchi F, Bertoli C, Copetti T, Tanida I, Brancolini C, Eskelinen EL, et al. Calpain is required for macroautophagy in mammalian cells. J Cell Biol. 2006;175:595–605.
132. Ray SK, Fidan M, Nowak MW, Wilford GG, Hogan EL, Banik NL. Oxidative stress and Ca2+ influx upregulate calpain and induce apoptosis in PC12 cells. Brain Res. 2000;852:326–34.
133. Schoonbroodt S, Ferreira V, Best-Belpomme M, Boelaert JR, Legrand-Poels S, Korner M, et al. Crucial role of the amino-terminal tyrosine residue 42 and the carboxyl-terminal PEST domain of I kappa B alpha in NF-kappa B activation by an oxidative stress. J Immunol. 2000;164:4292–300.
134. Yamashima T, Tonchev AB, Tsukada T, Saido TC, Imajoh-Ohmi S, Momoi T, et al. Sustained calpain activation associated with lysosomal rupture executes necrosis of the postischemic CA1 neurons in primates. Hippocampus. 2003;13:791–800.
135. Yamashima T, Saido TC, Takita M, Miyazawa A, Yamano J, Miyakawa A, et al. Transient brain ischaemia provokes Ca2+, PIP2 and calpain responses prior to delayed neuronal death in monkeys. Eur J Neurosci. 1996;8:1932–44.
136. Ray SK, Wilford GG, Crosby CV, Hogan EL, Banik NL. Diverse stimuli induce calpain overexpression and apoptosis in C6 glioma cells. Brain Res. 1999;829:18–27.
137. Lee MS, Kwon YT, Li M, Peng J, Friedlander RM, Tsai LH. Neurotoxicity induces cleavage of p35 to p25 by calpain. Nature. 2000;405:360–4.
138. Mouatt-Prigent A, Karlsson JO, Agid Y, Hirsch EC. Increased M-calpain expression in the mesencephalon of patients with Parkinson's disease but not in other neurodegenerative disorders involving the mesencephalon: a role in nerve cell death? Neuroscience. 1996;73:979–87.
139. Yamashima T. Ca2+-dependent proteases in ischemic neuronal death: a conserved "calpain-cathepsin cascade" from nematodes to primates. Cell Calcium. 2004;36:285–93.
140. Adamec E, Mohan PS, Cataldo AM, Vonsattel JP, Nixon RA. Up-regulation of the lysosomal system in experimental models of neuronal injury: implications for Alzheimer's disease. Neuroscience. 2000;100:663–75.

141. Yamashima T. Implication of cysteine proteases calpain, cathepsin and caspase in ischemic neuronal death of primates. Prog Neurobiol. 2000;62:273–95.
142. Chan PH. Role of oxidants in ischemic brain damage. Stroke. 1996;27:1124–9.
143. Takano J, Tomioka M, Tsubuki S, Higuchi M, Iwata N, Itohara S, et al. Calpain mediates excitotoxic DNA fragmentation via mitochondrial pathways in adult brains: evidence from calpastatin mutant mice. J Biol Chem. 2005;280:16175–84.
144. Muntener K, Zwicky R, Csucs G, Rohrer J, Baici A. Exon skipping of cathepsin B: mitochondrial targeting of a lysosomal peptidase provokes cell death. J Biol Chem. 2004;279:41012–7.
145. Guicciardi ME, Miyoshi H, Bronk SF, Gores GJ. Cathepsin B knockout mice are resistant to tumor necrosis factor-alpha-mediated hepatocyte apoptosis and liver injury: implications for therapeutic applications. Am J Pathol. 2001;159:2045–54.
146. Canbay A, Guicciardi ME, Higuchi H, Feldstein A, Bronk SF, Rydzewski R, et al. Cathepsin B inactivation attenuates hepatic injury and fibrosis during cholestasis. J Clin Invest. 2003; 112:152–9.
147. Benchoua A, Braudeau J, Reis A, Couriaud C, Onteniente B. Activation of proinflammatory caspases by cathepsin B in focal cerebral ischemia. J Cereb Blood Flow Metab. 2004;24: 1272–9.
148. Danial NN, Korsmeyer SJ. Cell death: critical control points. Cell. 2004;116:205–19.
149. Leber B, Lin J, Andrews DW. Embedded together: the life and death consequences of interaction of the Bcl-2 family with membranes. Apoptosis. 2007;12:897–911.
150. Reed JC. Proapoptotic multidomain Bcl-2/Bax-family proteins: mechanisms, physiological roles, and therapeutic opportunities. Cell Death Differ. 2006;13:1378–86.
151. Youle RJ, Strasser A. The BCL-2 protein family: opposing activities that mediate cell death. Nat Rev Mol Cell Biol. 2008;9:47–59.
152. Lalier L, Cartron PF, Juin P, Nedelkina S, Manon S, Bechinger B, et al. Bax activation and mitochondrial insertion during apoptosis. Apoptosis. 2007;12:887–96.
153. Tajeddine N, Galluzzi L, Kepp O, Hangen E, Morselli E, Senovilla L, et al. Hierarchical involvement of Bak, VDAC1 and Bax in cisplatin-induced cell death. Oncogene. 2008;27:4221–32.
154. Zamzami N, El Hamel C, Maisse C, Brenner C, Munoz-Pinedo C, Belzacq AS, et al. Bid acts on the permeability transition pore complex to induce apoptosis. Oncogene. 2000;19:6342–50.
155. Kuwana T, Bouchier-Hayes L, Chipuk JE, Bonzon C, Sullivan BA, Green DR, et al. BH3 domains of BH3-only proteins differentially regulate Bax-mediated mitochondrial membrane permeabilization both directly and indirectly. Mol Cell. 2005;17:525–35.
156. Kim R. Unknotting the roles of Bcl-2 and Bcl-xL in cell death. Biochem Biophys Res Commun. 2005;333:336–43.
157. Pinton P, Rizzuto R. Bcl-2 and Ca2+ homeostasis in the endoplasmic reticulum. Cell Death Differ. 2006;13:1409–18.
158. Zamzami N, Larochette N, Kroemer G. Mitochondrial permeability transition in apoptosis and necrosis. Cell Death Differ. 2005;12 Suppl 2:1478–80.
159. Zoratti M, Szabo I, De Marchi U. Mitochondrial permeability transitions: how many doors to the house? Biochim Biophys Acta. 2005;1706:40–52.
160. Brenner C, Cadiou H, Vieira HL, Zamzami N, Marzo I, Xie Z, et al. Bcl-2 and Bax regulate the channel activity of the mitochondrial adenine nucleotide translocator. Oncogene. 2000; 19:329–36.
161. Pattingre S, Tassa A, Qu X, Garuti R, Liang XH, Mizushima N, et al. Bcl-2 anti-apoptotic proteins inhibit Beclin 1-dependent autophagy. Cell. 2005;122:927–39.
162. Rodriguez D, Rojas-Rivera D, Hetz C. Integrating stress signals at the endoplasmic reticulum: the BCL-2 protein family rheostat. Biochim Biophys Acta. 2011;1813:564–74.
163. Kouroku Y, Fujita E, Tanida I, Ueno T, Isoai A, Kumagai H, et al. ER stress (PERK/eIF2alpha phosphorylation) mediates the polyglutamine-induced LC3 conversion, an essential step for autophagy formation. Cell Death Differ. 2007;14:230–9.
164. Ogata M, Hino S, Saito A, Morikawa K, Kondo S, Kanemoto S, et al. Autophagy is activated for cell survival after endoplasmic reticulum stress. Mol Cell Biol. 2006;26:9220–31.

165. Yorimitsu T, Nair U, Yang Z, Klionsky DJ. Endoplasmic reticulum stress triggers autophagy. J Biol Chem. 2006;281:30299–304.
166. Ding WX, Ni HM, Gao W, Hou YF, Melan MA, Chen X, et al. Differential effects of endoplasmic reticulum stress-induced autophagy on cell survival. J Biol Chem. 2007;282:4702–10.
167. Aita VM, Liang XH, Murty VV, Pincus DL, Yu W, Cayanis E, et al. Cloning and genomic organization of beclin 1, a candidate tumor suppressor gene on chromosome 17q21. Genomics. 1999;59:59–65.
168. Liang XH, Jackson S, Seaman M, Brown K, Kempkes B, Hibshoosh H, et al. Induction of autophagy and inhibition of tumorigenesis by beclin 1. Nature. 1999;402:672–6.
169. Qu X, Yu J, Bhagat G, Furuya N, Hibshoosh H, Troxel A, et al. Promotion of tumorigenesis by heterozygous disruption of the beclin 1 autophagy gene. J Clin Invest. 2003;112:1809–20.
170. Yue Z, Jin S, Yang C, Levine AJ, Heintz N. Beclin 1, an autophagy gene essential for early embryonic development, is a haploinsufficient tumor suppressor. Proc Natl Acad Sci U S A. 2003;100:15077–82.
171. Maiuri MC, Le Toumelin G, Criollo A, Rain JC, Gautier F, Juin P, et al. Functional and physical interaction between Bcl-X(L) and a BH3-like domain in Beclin-1. EMBO J. 2007;26: 2527–39.
172. Maiuri MC, Criollo A, Tasdemir E, Vicencio JM, Tajeddine N, Hickman JA, et al. BH3-only proteins and BH3 mimetics induce autophagy by competitively disrupting the interaction between Beclin 1 and Bcl-2/Bcl-X(L). Autophagy. 2007;3:374–6.
173. Bellot G, Garcia-Medina R, Gounon P, Chiche J, Roux D, Pouyssegur J, et al. Hypoxia-induced autophagy is mediated through hypoxia-inducible factor induction of BNIP3 and BNIP3L via their BH3 domains. Mol Cell Biol. 2009;29:2570–81.
174. Azad MB, Chen Y, Henson ES, Cizeau J, McMillan-Ward E, Israels SJ, et al. Hypoxia induces autophagic cell death in apoptosis-competent cells through a mechanism involving BNIP3. Autophagy. 2008;4:195–204.
175. Chinnadurai G, Vijayalingam S, Gibson SB. BNIP3 subfamily BH3-only proteins: mitochondrial stress sensors in normal and pathological functions. Oncogene. 2008;27 Suppl 1:S114–27.
176. Gillardon F, Kiprianova I, Sandkuhler J, Hossmann KA, Spranger M. Inhibition of caspases prevents cell death of hippocampal CA1 neurons, but not impairment of hippocampal long-term potentiation following global ischemia. Neuroscience. 1999;93:1219–22.
177. Strosznajder R, Gajkowska B. Effect of 3-aminobenzamide on Bcl-2, Bax and AIF localization in hippocampal neurons altered by ischemia-reperfusion injury. the immunocytochemical study. Acta Neurobiol Exp (Wars). 2006;66:15–22.
178. Niimura M, Takagi N, Takagi K, Mizutani R, Ishihara N, Matsumoto K, et al. Prevention of apoptosis-inducing factor translocation is a possible mechanism for protective effects of hepatocyte growth factor against neuronal cell death in the hippocampus after transient forebrain ischemia. J Cereb Blood Flow Metab. 2006;26:1354–65.
179. Li X, Nemoto M, Xu Z, Yu SW, Shimoji M, Andrabi SA, et al. Influence of duration of focal cerebral ischemia and neuronal nitric oxide synthase on translocation of apoptosis-inducing factor to the nucleus. Neuroscience. 2007;144:56–65.
180. Sun Y, Ouyang YB, Xu L, Chow AM, Anderson R, Hecker JG, et al. The carboxyl-terminal domain of inducible Hsp70 protects from ischemic injury in vivo and in vitro. J Cereb Blood Flow Metab. 2006;26:937–50.
181. Parrish J, Li L, Klotz K, Ledwich D, Wang X, Xue D. Mitochondrial endonuclease G is important for apoptosis in C. elegans. Nature. 2001;412:90–4.
182. Kalinowska M, Garncarz W, Pietrowska M, Garrard WT, Widlak P. Regulation of the human apoptotic DNase/RNase endonuclease G: involvement of Hsp70 and ATP. Apoptosis. 2005;10:821–30.
183. Rustin P. The use of antioxidants in Friedreich's ataxia treatment. Expert Opin Investig Drugs. 2003;12:569–75.
184. Elibol B, Soylemezoglu F, Unal I, Fujii M, Hirt L, Huang PL, et al. Nitric oxide is involved in ischemia-induced apoptosis in brain: a study in neuronal nitric oxide synthase null mice. Neuroscience. 2001;105:79–86.

185. Culmsee C, Zhu X, Yu QS, Chan SL, Camandola S, Guo Z, et al. A synthetic inhibitor of p53 protects neurons against death induced by ischemic and excitotoxic insults, and amyloid beta-peptide. J Neurochem. 2001;77:220–8.
186. Morrison RS, Wenzel HJ, Kinoshita Y, Robbins CA, Donehower LA, Schwartzkroin PA. Loss of the p53 tumor suppressor gene protects neurons from kainate-induced cell death. J Neurosci. 1996;16:1337–45.
187. Gao Y, Signore AP, Yin W, Cao G, Yin XM, Sun F, et al. Neuroprotection against focal ischemic brain injury by inhibition of c-Jun N-terminal kinase and attenuation of the mitochondrial apoptosis-signaling pathway. J Cereb Blood Flow Metab. 2005;25:694–712.
188. Guan QH, Pei DS, Zong YY, Xu TL, Zhang GY. Neuroprotection against ischemic brain injury by a small peptide inhibitor of c-Jun N-terminal kinase (JNK) via nuclear and nonnuclear pathways. Neuroscience. 2006;139:609–27.
189. Mattson MP, Kroemer G. Mitochondria in cell death: novel targets for neuroprotection and cardioprotection. Trends Mol Med. 2003;9:196–205.
190. Stavrovskaya IG, Narayanan MV, Zhang W, Krasnikov BF, Heemskerk J, Young SS, et al. Clinically approved heterocyclics act on a mitochondrial target and reduce stroke-induced pathology. J Exp Med. 2004;200:211–22.
191. Hetz C, Vitte PA, Bombrun A, Rostovtseva TK, Montessuit S, Hiver A, et al. Bax channel inhibitors prevent mitochondrion-mediated apoptosis and protect neurons in a model of global brain ischemia. J Biol Chem. 2005;280:42960–70.
192. Rodrigues CM, Sola S, Sharpe JC, Moura JJ, Steer CJ. Tauroursodeoxycholic acid prevents Bax-induced membrane perturbation and cytochrome C release in isolated mitochondria. Biochemistry. 2003;42:3070–80.
193. Rodrigues CM, Spellman SR, Sola S, Grande AW, Linehan-Stieers C, Low WC, et al. Neuroprotection by a bile acid in an acute stroke model in the rat. J Cereb Blood Flow Metab. 2002;22:463–71.
194. Rodrigues CM, Sola S, Nan Z, Castro RE, Ribeiro PS, Low WC, et al. Tauroursodeoxycholic acid reduces apoptosis and protects against neurological injury after acute hemorrhagic stroke in rats. Proc Natl Acad Sci U S A. 2003;100:6087–92.
195. Saavedra RA, Murray M, de Lacalle S, Tessler A. In vivo neuroprotection of injured CNS neurons by a single injection of a DNA plasmid encoding the Bcl-2 gene. Prog Brain Res. 2000;128:365–72.
196. Martinou JC, Dubois-Dauphin M, Staple JK, Rodriguez I, Frankowski H, Missotten M, et al. Overexpression of BCL-2 in transgenic mice protects neurons from naturally occurring cell death and experimental ischemia. Neuron. 1994;13:1017–30.
197. Zhao H, Yenari MA, Cheng D, Sapolsky RM, Steinberg GK. Bcl-2 overexpression protects against neuron loss within the ischemic margin following experimental stroke and inhibits cytochrome c translocation and caspase-3 activity. J Neurochem. 2003;85:1026–36.
198. Wiessner C, Allegrini PR, Rupalla K, Sauer D, Oltersdorf T, McGregor AL, et al. Neuronspecific transgene expression of Bcl-XL but not Bcl-2 genes reduced lesion size after permanent middle cerebral artery occlusion in mice. Neurosci Lett. 1999;268:119–22.
199. Culmsee C, Plesnila N. Targeting Bid to prevent programmed cell death in neurons. Biochem Soc Trans. 2006;34:1334–40.
200. Plesnila N, Zinkel S, Le DA, Amin-Hanjani S, Wu Y, Qiu J, et al. BID mediates neuronal cell death after oxygen/glucose deprivation and focal cerebral ischemia. Proc Natl Acad Sci U S A. 2001;98:15318–23.
201. Plesnila N, Zinkel S, Amin-Hanjani S, Qiu J, Korsmeyer SJ, Moskowitz MA. Function of BID—a molecule of the bcl-2 family—in ischemic cell death in the brain. Eur Surg Res. 2002;34:37–41.
202. Tehranian R, Rose ME, Vagni V, Pickrell AM, Griffith RP, Liu H, et al. Disruption of Bax protein prevents neuronal cell death but produces cognitive impairment in mice following traumatic brain injury. J Neurotrauma. 2008;25:755–67.
203. Gibson ME, Han BH, Choi J, Knudson CM, Korsmeyer SJ, Parsadanian M, et al. BAX contributes to apoptotic-like death following neonatal hypoxia-ischemia: evidence for distinct apoptosis pathways. Mol Med. 2001;7:644–55.

204. Fischer SF, Vier J, Kirschnek S, Klos A, Hess S, Ying S, et al. Chlamydia inhibit host cell apoptosis by degradation of proapoptotic BH3-only proteins. J Exp Med. 2004;200:905–16.
205. Rami A, Langhagen A, Steiger S. Focal cerebral ischemia induces upregulation of Beclin 1 and autophagy-like cell death. Neurobiol Dis. 2008;29:132–41.
206. Puyal J, Vaslin A, Mottier V, Clarke PG. Postischemic treatment of neonatal cerebral ischemia should target autophagy. Ann Neurol. 2009;66:378–89.
207. Asoh S, Ohsawa I, Mori T, Katsura K, Hiraide T, Katayama Y, et al. Protection against ischemic brain injury by protein therapeutics. Proc Natl Acad Sci U S A. 2002;99:17107–12.
208. Hayashi K, Morishita R, Nakagami H, Yoshimura S, Hara A, Matsumoto K, et al. Gene therapy for preventing neuronal death using hepatocyte growth factor: in vivo gene transfer of HGF to subarachnoid space prevents delayed neuronal death in gerbil hippocampal CA1 neurons. Gene Ther. 2001;8:1167–73.
209. Nuglisch J, Karkoutly C, Mennel HD, Rossberg C, Krieglstein J. Protective effect of nimodipine against ischemic neuronal damage in rat hippocampus without changing postischemic cerebral blood flow. J Cereb Blood Flow Metab. 1990;10:654–9.
210. Mossakowski MJ, Gadamski R. Nimodipine prevents delayed neuronal death of sector CA1 pyramidal cells in short-term forebrain ischemia in Mongolian gerbils. Stroke. 1990;21:IV120–2.
211. Hadley MN, Zabramski JM, Spetzler RF, Rigamonti D, Fifield MS, Johnson PC. The efficacy of intravenous nimodipine in the treatment of focal cerebral ischemia in a primate model. Neurosurgery. 1989;25:63–70.
212. Nag D, Garg RK, Varma M. A randomized double-blind controlled study of nimodipine in acute cerebral ischemic stroke. Indian J Physiol Pharmacol. 1998;42:555–8.
213. Kakarieka A, Schakel EH, Fritze J. Clinical experiences with nimodipine in cerebral ischemia. J Neural Transm Suppl. 1994;43:13–21.
214. Mattson MP, Zhu H, Yu J, Kindy MS. Presenilin-1 mutation increases neuronal vulnerability to focal ischemia in vivo and to hypoxia and glucose deprivation in cell culture: involvement of perturbed calcium homeostasis. J Neurosci. 2000;20:1358–64.
215. Wei H, Perry DC. Dantrolene is cytoprotective in two models of neuronal cell death. J Neurochem. 1996;67:2390–8.
216. Huang Z, Huang PL, Panahian N, Dalkara T, Fishman MC, Moskowitz MA. Effects of cerebral ischemia in mice deficient in neuronal nitric oxide synthase. Science. 1994;265:1883–5.
217. Satoh K, Ikeda Y, Shioda S, Tobe T, Yoshikawa T. Edarabone scavenges nitric oxide. Redox Rep. 2002;7:219–22.
218. Kamii H, Mikawa S, Murakami K, Kinouchi H, Yoshimoto T, Reola L, et al. Effects of nitric oxide synthase inhibition on brain infarction in SOD-1-transgenic mice following transient focal cerebral ischemia. J Cereb Blood Flow Metab. 1996;16:1153–7.
219. Hara H, Nagasawa H, Kogure K. Nimodipine prevents postischemic brain damage in the early phase of focal cerebral ischemia. Stroke. 1990;21:IV102–4.
220. Wolz P, Krieglstein J. Neuroprotective effects of alpha-lipoic acid and its enantiomers demonstrated in rodent models of focal cerebral ischemia. Neuropharmacology. 1996;35:369–75.
221. Yu ZF, Bruce-Keller AJ, Goodman Y, Mattson MP. Uric acid protects neurons against excitotoxic and metabolic insults in cell culture, and against focal ischemic brain injury in vivo. J Neurosci Res. 1998;53:613–25.
222. Gray J, Haran MM, Schneider K, Vesce S, Ray AM, Owen D, et al. Evidence that inhibition of cathepsin-B contributes to the neuroprotective properties of caspase inhibitor Tyr-Val-Ala-Asp-chloromethyl ketone. J Biol Chem. 2001;276:32750–5.
223. Endres M, Fink K, Zhu J, Stagliano NE, Bondada V, Geddes JW, et al. Neuroprotective effects of gelsolin during murine stroke. J Clin Invest. 1999;103:347–54.
224. Kruman II, Culmsee C, Chan SL, Kruman Y, Guo Z, Penix L, et al. Homocysteine elicits a DNA damage response in neurons that promotes apoptosis and hypersensitivity to excitotoxicity. J Neurosci. 2000;20:6920–6.

Chapter 8
Hypoxia-Inducible Factor: A New Hope to Counteract Stroke

Chunhua Chen and Changman Zhou

Abstract Stroke is a major health problem and a leading cause of death and disability. Past research aimed to develop therapeutic strategies that prevent neuronal death and improve recovery. Yet, few successful therapeutic strategies have emerged. Recent findings demonstrated that the induction of the hypoxia signaling pathway of hypoxia-inducible factor (HIF) could mediate neuroprotective events. HIF-1 has been suggested to be an important player in neurological outcomes following ischemic stroke due to the functions of its downstream genes and implicated in both cellular protection and cell death in cerebral ischemia according to the duration and severity of ischemic injury. Therefore, HIF-1 has become an attractive target to counteract stroke. Here, we highlight the multifaceted role of HIF in stroke and some of the recently developed drugs that disrupt the HIF-1 signaling pathway which may provide potent leads for stroke therapy.

1 Introduction

Hypoxia-inducible factors (HIFs) are essential mediators of the cellular oxygen signaling pathway. They are heterodimeric transcription factors consisting of an oxygen-sensitive alpha subunit (HIF-α) and a constitutive beta subunit (HIF-β) (Fig. 8.1). To date, three HIFs (HIF-1, -2, and -3) have been identified to regulate transcriptional process in response to low oxygen levels [38]. However, among the three HIFs, HIF-1 that controlled by the activity of the alpha subunit has been reported to play an important role in stroke. Genes regulated by HIF-1 are involved in angiogenesis, glucose transport and metabolism, erythropoiesis, inflammation,

C. Chen • C. Zhou, MD, PhD (✉)
Department of Anatomy and Embryo, Peking University Health Science Center,
38 Xueyuan Rd, Haidian Qu, Beijing 100191, China
e-mail: cmzhou@bjmu.edu.cn

P.A. Lapchak and J.H. Zhang (eds.), *Translational Stroke Research*, Springer Series
in Translational Stroke Research, DOI 10.1007/978-1-4419-9530-8_8,
© Springer Science+Business Media, LLC 2012

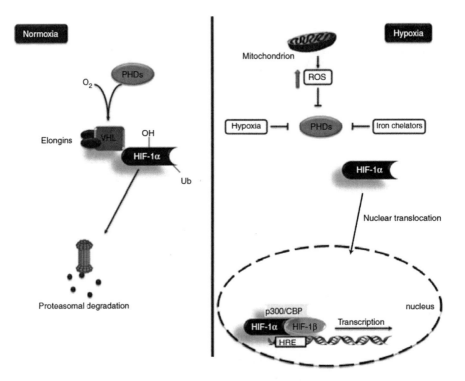

Fig. 8.1 Cells control hypoxia-inducible factor (HIF)-1 in two ways. When oxygen is plentiful, hydroxylase enzymes promote degradation of the HIF-1α subunit, which is aided by the VHL tumor suppressor protein, and also block HIF-1α's ability to bind p300 and other proteins needed for gene transcription. Low oxygen concentrations inhibit those activities of the hydroxylases, thus turning up HIF-1 activity. HIF-1α protein translocates and accumulates in the nucleus, where it dimerizes with HIF-1β subunits and recruits the transcription coactivator p300/CBP, forming the active HIF complex. This complex binds to the hypoxia-responsive element in the promoter and activates transcription of target genes (cited from Correia et al. [12])

apoptosis, and cellular stress. All of these genes may relate to the state of neuronal cells following stroke especially ischemic stroke which makes up 80% of all strokes and results from a thrombotic or embolic occlusion of a major cerebral artery (most often the middle cerebral artery) or its branches. However, whether the effect of HIF-1 is anti- or pro-cell survival is still a matter of debate. It seems that the discrepancy could be attributed to the type, duration, severity, and phase of the insult [11]. On the one hand, early inhibition of HIF-1 reduced ischemic brain damage in rats via inhibition of VEGF, p53, and caspase-3 [9] (Fig. 8.2); on the other hand, upregulation of HIF-1 using deferoxamine (DFO) pretreatment protected against cerebral ischemia [21, 28] and this protective effect was significantly attenuated in neuron-specific HIF-1-deficient mice [2]. Moreover, downregulation of neuron-specific HIF-1 increased brain damage induced by cerebral ischemia [2, 23]. Under mild conditions, hypoxia induces stabilization of the HIF-1 complex, leading to the

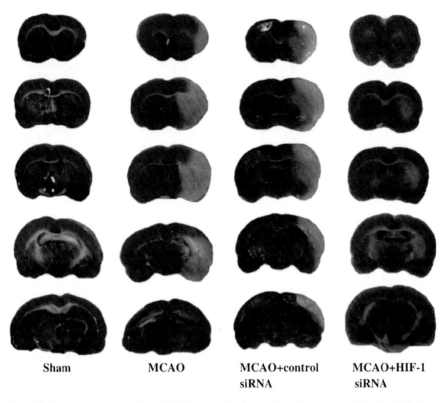

| Sham | MCAO | MCAO+control siRNA | MCAO+HIF-1 siRNA |

Fig. 8.2   Representative samples of TTC-stained brain sections from rats sacrificed at 24 h after ischemia. Severe infarction is shown in MCAO and MCAO + control siRNA rats. The white areas represent the infarct regions in these sections. HIF-1α siRNA treatment reduced the infarct volume sharply, especially in cortex. No ischemic lesion is found in the sham group (cited from Chen et al. [9])

transcriptional activation of adaptive genes. However, under conditions of sustained hypoxia, HIF-1 stabilization increases cellular p53 and promotes the transactivation of pathologic genes like bax [6, 19]. Because of the potentially important and divergent roles of HIF-1 and its downstream genes following ischemia, it has been suggested that regulating HIF-1 induction and accumulation is a highly promising therapeutic approach for stroke.

2   HIF-1 Expression in Hypoxic and Ischemic Brains

The expression of HIF-1 changes in response to different oxygen levels in brains after hypoxic and ischemic injury. It was first reported in 1996 that mRNAs of HIF-1 were induced in the brains of rats or mice when they were exposed to reduced ambient O_2 concentrations for 30–60 min [57]. Systemic hypoxia for the durations of 1, 3, or 6 h rapidly increased the nuclear content of HIF-1α in mouse brains [4]. Bergeron et al. first reported that after focal ischemia in an adult rat brain, mRNA encoding HIF-1α was upregulated in the brain [3]. HIF and its target genes were induced 7.5 h after the onset of ischemia and increased further at 19 and 24 h. Significantly, in the ischemic brain, HIF-1α appeared to be mostly induced in the penumbra, the salvageable tissue. However, Demougeot et al. reported that the HIF-1α level decreased gradually from the ischemic core to more distant regions [13]. A recent study showed a biphasic activation of HIF-1 lasted for up to 10 days in wild-type mice [2].

3   The Multifaceted Role of HIF-1 on Outcomes of Ischemic Stroke In Vitro and Vivo

The role of the transcription factor HIF-1α on outcomes of ischemic stroke is controversial. Yeh et al. provided evidence that time-dependent HIF-1α expression has prodeath and prosurvival effects during ischemic injury and suggested that focal ischemia triggered a robust, biphasic (early and late phase) HIF-1α accumulation that was mainly restricted to the neurons of the ischemic cortex and striatum. Early-phase HIF-1α expression facilitated apoptosis, but late-phase expression increased cell survival. HIF-1α mRNA was not increased in the early phase, but increased in the late phase. HIF-1α accumulation early after MCAO is because of increased protein stability, but that which occurs during the late phase results from augmented transcription [58]. It is noteworthy that HIF-1 regulation and effect of HIF-1 on ischemic outcomes may be affected by ischemic duration such as the results of Baranova's 30-min tMCAO vs. Helton's 75-min bilateral common carotid artery occlusion [2]. Whether short ischemic period may offer limited transcriptional activity in the absence of p300 vs. full transcriptional function in the presence of p300 needs to be considered [10]. The following two points and two tables (Tables 8.1 and 8.2) suggested the multifaceted role of HIF-1 on outcomes of ischemic stroke in vitro and vivo.

Table 8.1 Molecules of HIF target gene involved in the neuroprotection in stroke

Erythropoietin (EPO)	EPO has been found to protect cells from hypoxic/ischemic injury [4, 14, 59]
Vascular endothelial growth factor (VEGF)	VEGF counteracts detrimental ischemic injuries [25, 29]
Cytochrome c	HIF inhibits the release of cytochrome c, caspase activation, and PARP cleavage [42]
p53	HIF suppresses p53 activation and thereby maintains cell survival [27]

Table 8.2   Molecules of HIF target gene involved in the detrimental effect in stroke

p53	Combined with HIF may promote apoptotic cell death in ischemic areas [31, 37]
BNIP3 Nix	HIF regulated proapoptotic protein BNIP3 after global brain ischemia in the rat [44, 51]
iNOS	Upregulated by HIF and inhibition of inducible nitric oxide synthase ameliorates cerebral ischemic damage [24, 30]
Caspase-3	HIF-1 binds to caspase-3 promoter after cerebral ischemia [8, 9, 55]

3.1   Neuroprotective Effect of HIF-1 in Ischemic Stroke

In cerebral ischemic stroke, the reduction in blood supply to brain tissue results in a decrease in the availability of glucose and oxygen, which are necessary for normal brain function. Accumulating evidence indicates HIF-1α protein accumulation triggers the expression of target genes involved in various adaptive responses [45, 48]. Bergeron et al. reported that increased expression of HIF-1 target genes as a result of HIF-1 activation by hypoxia may contribute to tissue viability in the ischemic penumbra by increasing glucose transport and glycolysis [3]. Accordingly, Sharp et al. observed that after focal cerebral ischemia, mRNAs encoding HIF-1α, GLUT-1, and several glycolytic enzymes, including lactate dehydrogenase, were upregulated in the areas around the infarction. Further, preconditioning with the HIF-1 inducers $CoCl_2$ or DFO, 24 h before the stroke, decreased infarction by 75 and 56%, respectively, compared with vehicle injected controls. Thus, HIF-1 activation could contribute to protective brain preconditioning [47]. Baranova et al. observed that neuron-specific knockdown of HIF-1 aggravated brain damage after MCAO and reduced the survival rate of mice. Moreover, the authors also demonstrated that the pharmacologic HIF-1 activators, 3,4-dihydroxybenzoic acid, DFO, and 2,2-dipyridyl significantly reduced ischemic injury in wild-type mice, whereas the effectiveness of these compounds was significantly attenuated in neuron-specific HIF-1α knockdown mice [2]. Consistent with this report, Siddiq et al. showed that hypoxia or hypoxia-mimetic agents confer neuroprotection against oxidative death in vitro and ischemic injury in vivo through the inhibition of PHDs, and consequent HIF-1α stabilization and upregulation of HIF-1-dependent target genes [50]. Another in vitro study demonstrated that DFO exerted protective effect on OGD cultured cortical neurons through the induction of HIF-1α and EPO [28]. Recent findings showed that intranasal administration of DFO, a noninvasive method, prevents and treats stroke damage after MCAO in rats [22]. Zhou et al. evaluated the effect of 2-methoxyestradiol (2ME2), a natural metabolite of estrogen that inhibits HIF-1α in global cerebral ischemia in rats. As expected, the authors observed that 2ME2 treatment significantly reduced HIF-1 levels and lowered the neuronal cell survival, which suggests that 2ME2-induced HIF-1 suppression aggravates outcomes after global ischemia in rats [60].

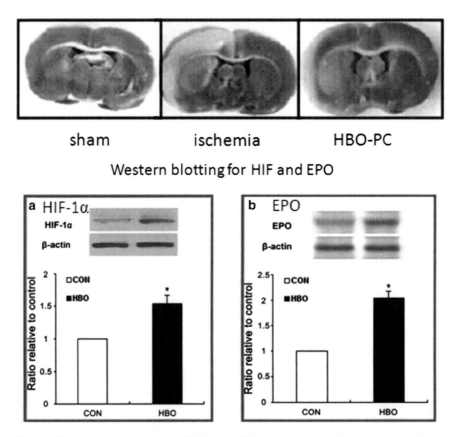

Fig. 8.3 Neuroprotection induced by HBO preconditioning may be mediated by an upregulation of HIF-1 and its target gene EPO [17]

Several mechanisms have been proposed to explain HIF-1-mediated cytoprotective effects. HIF-1-induced EPO expression seems to be a crucial event to protect cells against hypoxic/ischemic injury (Fig. 8.3). In vivo evidence showed that EPO administration directly into the brain reduces neurologic dysfunction in rodent models of stroke [40]. Furthermore, neutralization of endogenous brain EPO potentiates ischemic brain injury, confirming a pivotal role for the endogenous EPO system in neuronal survival after ischemia [41]. A clinical trial conducted in 13 patients who received recombinant human EPO intravenously once daily for the first 3 days after stroke showed a reduction in the infarct size when compared with controls, this effect being associated with an improvement in clinical outcome [14]. In addition, VEGF, another downstream gene of HIF-1, also counteracts detrimental ischemic injuries. Marti et al. demonstrated that the VEGF/VEGF receptor system leads to the growth of new vessels after cerebral ischemia, thereby minimizing the detrimental effects of cerebral ischemia [29]. There is also evidence showing that HIF-1 exerts anti-apoptotic effects through inhibition of cytochrome c release, caspase activation and poly (ADP-ribose) polymerase (PARP) cleavage, and the ability to suppress p53 activation [27, 36, 42]. These results support the idea that HIF-1α could act as a sensor to express therapeutic genes in a cerebral ischemic stroke [49]. Recently, Guo et al. observed that HIF-1 protects cells from ischemic injury by maintaining the cellular redox status and found that HIF-1α knockdown in SH-SY5Y cells downregulates key enzymes in glucose metabolism that are critical in the production of important reducing agents, decreases GSH/GSSG ratio, increases ROS levels, and induces cell death under hypoxic or OGD conditions. Additionally, HIF-1 maintains cellular redox state and prevents ischemic injury mediated by ROS production, possibly through the upregulation of the expression of key enzymes such as GLUT-1, glucose-6-phosphate dehydrogenase, and 6-phosphogluconate dehydrogenase [18]. The IVS9-675C>A polymorphism of the HIF-1α gene was analyzed in patients with acute ischemic stroke and in a control group and found that this polymorphism potentiates the risk of stroke, which reinforce the idea that HIF-1 could be an effective therapeutic target for the development of more effective therapeutic interventions against stroke [53].

3.2   Detrimental Effect of HIF-1 in Ischemic Stroke

It has been reported that HIF-1 may mediate apoptosis during hypoxia/ischemia and contribute to cellular and tissue damage. HIF-1-induced apoptosis has been observed in embryonic stem cells under hypoxic conditions [7]. Halterman et al. indicates that in response to hypoxia, HIF-1α accumulates, stabilizes active wild-type p53, and this increase in p53 protein is responsible for the apoptosis. Similarly, HIF-1α signaling elicits delayed death involving the participation of p53 in ischemic primary cortical neurons in vivo [19] and in vitro [20]. A direct interaction between p53 and HIF-1 is not detectable in vitro, but HIF-1-mediated activation of p53 may result indirectly after HIF-1 binding of MDM2 which results in accumulation of p53. In addition, p53 may repress prosurvival HIF-1 transcriptional activity by binding the limited amounts of the shared coactivator p300 [43]. Other reports indicate that during severe hypoxia, dephosphorylated HIF-1 binds to tumor suppressor p53 to stabilize p53 [52] and activates the expression of various genes including bax, a proapoptotic member of Bcl-2 family proteins [15]. Bax translocates to mitochondria and releases cytochrome c to cytosol to interact with Apaf-1 to activate caspase-9, which in turn activates downstream caspases, such as caspase-3. Another molecule involved in the protection effect is VEGF which is controlled by HIF. The increased expression of VEGF after ischemia may contribute to disruption of the blood–brain barrier and subsequently of brain edema. It is reported that this effect of VEGF may be related to the synthesis/release of nitric oxide and subsequent activation of soluble guanylate cyclase [26], and the antagonists of VEGF reduce brain edema and injury, implicating a pathological role of VEGF in the early stage of stroke [54]. Additionally, it has been suggested that HIF-1 may induce BNIP3,

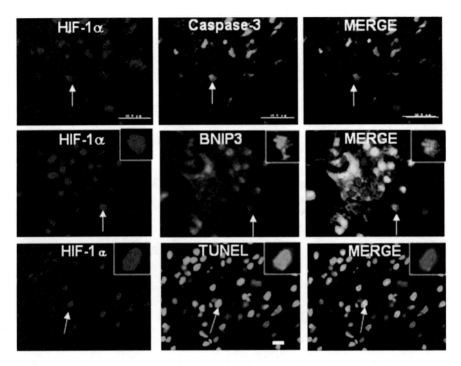

Fig. 8.4 Double-fluorescence labeling shows the colocalization of HIF-1α with Caspase-3 and BNIP3 in the cerebral cortex after 24 h of MCAO: HIF-1α (*red*) is expressed by TRITC, Caspase-3 and BNIP3 (*green*) is expressed by FITC. HIF-1α also colocalizes with TUNEL in the nuclei of neurons (*yellow*). The magnified images of the neurons were shown in *insert panes* (cited from Chen et al. [8])

which can cause mitochondrial dysfunction, membrane depolarization, MPTP opening, and a form of cell death having features of both necrosis and apoptosis [32]. Recently, Chen et al. showed inhibition of HIF-1 by tricyclodecan-9-ylxanthogenate (D609) that was modulated thorough PC-PLC DAG phosphatidic pathway decreased the VEGF expression and BNIP3 and offered protection [8] (Fig. 8.4). BNIP3 influences mitochondrial function in the early apoptotic process and can overcome Bcl-2 suppression of apoptosis [39] or cytochrome c release. In addition, increased nitric oxide from inducible nitric oxide synthase [35], dopamine from tyrosine hydroxylase [33], and lactate from lactate dehydrogenase [46] might also cause potentially detrimental effects on ischemic brain tissue.

4   Targeting HIF-1 Signaling: Implications for Stroke

Recombinant tissue plasminogen activator (r-tPA) is the only approved therapy for selected patients with acute ischemic stroke. However, r-tPA has a very limited treatment time window and may result in bleeding complications [5]. Extensive research has provided evidence that free radicals are involved in ischemic brain damage. The oxidative damage theory has led to numerous experimental and clinical studies using antioxidants as neuroprotectants [16, 56]. Many antioxidants have been tested in phase III clinical trials such as ebselen, tirilazad, edaravone, and NXY-059. There are antioxidants still undergoing phase III clinical trials, such as uric acid [1]. However, except for edaravone, which has been approved as a stroke drug in Japan, most of these completed trials failed due to lack of efficacy. As HIF-1 is a master regulator of a large number of genes that are critical in survival response to hypoxia in diverse cell types, future studies on HIF-1 regulation and its role in brain injury under ischemia will elaborate the fundamental mechanism of neuronal survival in ischemia and will assist in designing novel treatments for human stroke.

Accumulating evidence indicates that ROS may play a critical role in regulating HIF-1 expression in ischemic brain. This novel concept has high potential to be used to screen a broad range of antioxidant drugs. In the past clinical trials, the antioxidant drugs have been focused on their ability to scavenge ROS and to reduce lipid peroxidation. Different from these previous studies on antioxidant drugs for stroke treatments, this new screening would focus on the efficacy of potential antioxidant drugs in preventing neuronal death based on the novel pathway. This will help to find out why the failed drugs did not work, which may be due to their inability in affecting HIF-1α stabilization. Hopefully, this will also help find antioxidants that work for human stroke treatments. It is noteworthy that one caveat for antioxidant treatments is that they are not specific to HIF-1; instead, their effects are much broader. Furthermore, researchers in academic institutions and pharmaceutical companies currently screen small molecules that selectively inhibit HIF-specific prolyl hydroxylases (PHDs) for drug discovery in stroke treatments as increasing HIF-1 stabilization by inhibiting its hydroxylation is believed to provide therapeutic benefits for stroke patients [34]. However, HIF-1α may also be degraded in hypoxia through other mechanisms as discussed above. For example, both ubiquitin-dependent and -independent proteasomal pathways could be involved in the degradation of HIF-1α. Consideration of all these factors may provide more efficient approaches in treating the disease.

Major activators and inhibitors of HIF-1 activity which include oxygen sensing, redox status, a variety of molecular factors, and pharmacological agents are presented in Table 8.3 [11]. Regulation of HIF-1 activity may have potentials in the treatment of acute ischemic cerebral stroke.

Table 8.3 Regulation of HIF-1 activity

Related mechanisms	
HIF-1 activator	
Calcineurin	Promotes HIF-1 expression by dephosphorylating RACK1 and blocking RACK1 dimerization
Cytokines—e.g., IL-1β, TNF-1α	Induce HIF-1α mRNA expression via PI3K pathway
Growth factors, e.g., IGF-1	Prolonged HIF-1 accumulation
Hypoxia	Inhibition of HIF-1α subunit rapid degradation
Iron chelators, e.g., desferrioxamine	Inhibits hydroxylation of proline in ODD domain

(continued)

Table 8.3 (continued)

Related mechanisms	
ROS	Modulating upstream signaling pathways such as hydroxylases or phosphatases, or via NF-kappaB
Thrombin	Redox-sensitive cascade activated by ROS may be involved in this response
Vitamin E	Induced the expression of the α subunit of HIF-1
Sodium orthovanadate	An inhibitor of tyrosine phosphatases increased the basal level of HIF-1 proteins and HIF-1 activity
Mitogen-activated protein kinase (MAPK)	MAPK may mediate HIF-1 activation by various molecules, including nitric oxide
HIF-1 inhibitor	
Guanylate-cyclase activators, e.g., YC-1	YC-1 completely blocks HIF-1α expression at the posttranscriptional level and linked with the oxygen-sensing pathway
HIF prolyl 4-hydroxylases	Belongs to a family of iron- and 2-oxoglutarate-dependent dioxygenases that negatively regulate the stability of HIF-1
HSP90 inhibitor—17-AAG	Inhibition and eventual degradation of Hsp90 thus destabilization of HIF-1
2-Methoxyestradiol (2ME2)	Decreased nuclear HIF-1α-binding activity and affected the expression of downstream genes
Thioredoxin inhibitors	Decrease HIF-1α functional transcriptional activity and DNA binding
Topoisomerase (Topo)-I inhibitors	Repression of HIF-1-dependent induction of gene expression
Tyrosine kinases inhibitor, e.g., genistein	Blocked the synthesis of both HIF-1 subunits as well as HIF-1 DNA-binding activity

Cited from Chen et al. [11]

5 Concluding Remarks

Stroke or brain attack is a sudden problem affecting the blood vessels of the brain. According to the World Health Organization, 15 million people suffer stroke worldwide each year. Of these, five million die and another five million are permanently disabled (http://www.strokecenter.org/patients/stats.htm). HIF-1 has a pivotal role in response to stroke and could be an attractive and feasible target of therapeutic interventions to prevent or mitigate brain injury. Evidence demonstrated that mild hypoxic preconditioning and hypoxia-mimetic agents have beneficial effects in cerebral ischemic stroke. Selective inhibition of early—but not late—expressed HIF-1 is neuroprotective after focal ischemic brain damage. Elucidation of the molecular mechanisms involved in HIF-1 signaling pathway regulation is crucial to develop new pharmacological interventions aimed to minimize or delay ischemic or hemorrhagic brain injury.

References

1. Amaro S, Planas AM, Chamorro A. Uric acid administration in patients with acute stroke: a novel approach to neuroprotection. Expert Rev Neurother. 2008;8:259–70.
2. Baranova O, Miranda LF, Pichiule P, Dragatsis I, Johnson RS, Chavez JC. Neuron-specific inactivation of the hypoxia inducible factor 1 alpha increases brain injury in a mouse model of transient focal cerebral ischemia. J Neurosci. 2007;27:6320–32.
3. Bergeron M, Yu AY, Solway KE, Semenza GL, Sharp FR. Induction of hypoxia-inducible factor-1 (HIF-1) and its target genes following focal ischaemia in rat brain. Eur J Neurosci. 1999;11:4159–70.
4. Bernaudin M, Nedelec AS, Divoux D, MacKenzie ET, Petit E, Schumann-Bard P. Normobaric hypoxia induces tolerance to focal permanent cerebral ischemia in association with an increased expression of hypoxia-inducible factor-1 and its target genes, erythropoietin and VEGF, in the adult mouse brain. J Cereb Blood Flow Metab. 2002;22:393–403.
5. Brown DL, Barsan WG, Lisabeth LD, Gallery ME, Morgenstern LB. Survey of emergency physicians about recombinant tissue plasminogen activator for acute ischemic stroke. Ann Emerg Med. 2005;46:56–60.
6. Bullock JJ, Mehta SL, Lin Y, Lolla P, Li PA. Hyperglycemia-enhanced ischemic brain damage in mutant manganese SOD mice is associated with suppression of HIF-1alpha. Neurosci Lett. 2009;456:89–92.
7. Carmeliet P, Dor Y, Herbert JM, Fukumura D, Brusselmans K, Dewerchin M, Neeman M, Bono F, Abramovitch R, Maxwell P, Koch CJ, Ratcliffe P, Moons L, Jain RK, Collen D, Keshert E. Role of HIF-1alpha in hypoxia-mediated apoptosis, cell proliferation and tumour angiogenesis. Nature. 1998;394:485–90.
8. Chen C, Hu Q, Yan J, Lei J, Qin L, Shi X, Luan L, Yang L, Wang K, Han J, Nanda A, Zhou C. Multiple effects of 2ME2 and D609 on the cortical expression of HIF-1alpha and apoptotic genes in a middle cerebral artery occlusion-induced focal ischemia rat model. J Neurochem. 2007;102:1831–41.
9. Chen C, Hu Q, Yan J, Yang X, Shi X, Lei J, Chen L, Huang H, Han J, Zhang JH, Zhou C. Early inhibition of HIF-1alpha with small interfering RNA reduces ischemic-reperfused brain injury in rats. Neurobiol Dis. 2009;33:509–17.
10. Chen W, Jadhav V, Tang J, Zhang JH. HIF-1alpha inhibition ameliorates neonatal brain injury in a rat pup hypoxic-ischemic model. Neurobiol Dis. 2008;31:433–41.
11. Chen W, Ostrowski RP, Obenaus A, Zhang JH. Prodeath or prosurvival: two facets of hypoxia inducible factor-1 in perinatal brain injury. Exp Neurol. 2009;216:7–15.
12. Correia SC, Moreira PI. Hypoxia-inducible factor 1: a new hope to counteract neurodegeneration? J. Neurochem. 2010;112:1–12.
13. Demougeot C, Van HM, Bertrand N, Prigent-Tessier A, Mossiat C, Beley A, Marie C. Cytoprotective efficacy and mechanisms of the liposoluble iron chelator 2,2′-dipyridyl in the rat photothrombotic ischemic stroke model. J Pharmacol Exp Ther. 2004;311:1080–7.
14. Ehrenreich H, Hasselblatt M, Dembowski C, Cepek L, Lewczuk P, Stiefel M, Rustenbeck HH, Breiter N, Jacob S, Knerlich F, Bohn M, Poser W, Ruther E, Kochen M, Gefeller O, Gleiter C, Wessel TC, De RM, Itri L, Prange H, Cerami A, Brines M, Siren AL. Erythropoietin therapy for acute stroke is both safe and beneficial. Mol Med. 2002;8:495–505.
15. Gibson ME, Han BH, Choi J, Knudson CM, Korsmeyer SJ, Parsadanian M, Holtzman DM. BAX contributes to apoptotic-like death following neonatal hypoxia-ischemia: evidence for distinct apoptosis pathways. Mol Med. 2001;7:644–55.
16. Green AR, Shuaib A. Therapeutic strategies for the treatment of stroke. Drug Discov Today. 2006;11:681–93.
17. Gu G, Li Y, Peng Z, Xu J, Kang Z, Xu W, Tao H, Ostrowski RP, Zhang JH, Sun X. Mechanism of ischemic tolerance induced by hyperbaric oxygen preconditioning involves upregulation of hypoxia-inducible factor-1a and erythropoietin in rats. J Appl Physiol. 2008;104:1185–91.

18. Guo S, Miyake M, Liu KJ, Shi H. Specific inhibition of hypoxia inducible factor 1 exaggerates cell injury induced by in vitro ischemia through deteriorating cellular redox environment. J Neurochem. 2009;108:1309–21.
19. Halterman MW, Federoff HJ. HIF-1alpha and p53 promote hypoxia-induced delayed neuronal death in models of CNS ischemia. Exp Neurol. 1999;159:65–72.
20. Halterman MW, Miller CC, Federoff HJ. Hypoxia-inducible factor-1alpha mediates hypoxia-induced delayed neuronal death that involves p53. J Neurosci. 1999;19:6818–24.
21. Hamrick SE, McQuillen PS, Jiang X, Mu D, Madan A, Ferriero DM. A role for hypoxia-inducible factor-1alpha in desferoxamine neuroprotection. Neurosci Lett. 2005;379:96–100.
22. Hanson LR, Roeytenberg A, Martinez PM, Coppes VG, Sweet DC, Rao RJ, Marti DL, Hoekman JD, Matthews RB, Frey WH, Panter SS. Intranasal deferoxamine provides increased brain exposure and significant protection in rat ischemic stroke. J Pharmacol Exp Ther. 2009;330:679–86.
23. Helton R, Cui J, Scheel JR, Ellison JA, Ames C, Gibson C, Blouw B, Ouyang L, Dragatsis I, Zeitlin S, Johnson RS, Lipton SA, Barlow C. Brain-specific knock-out of hypoxia-inducible factor-1alpha reduces rather than increases hypoxic-ischemic damage. J Neurosci. 2005;25: 4099–107.
24. Iadecola C, Zhang F, Xu X. Inhibition of inducible nitric oxide synthase ameliorates cerebral ischemic damage. Am J Physiol. 1995;268:R286–92.
25. Jin KL, Mao XO, Greenberg DA. Vascular endothelial growth factor: direct neuroprotective effect in in vitro ischemia. Proc Natl Acad Sci U S A. 2000;97:10242–7.
26. Lafuente JV, Bulnes S, Mitre B, Riese HH. Role of VEGF in an experimental model of cortical micronecrosis. Amino Acids. 2002;23:241–5.
27. Li J, Zhang X, Sejas DP, Bagby GC, Pang Q. Hypoxia-induced nucleophosmin protects cell death through inhibition of p53. J Biol Chem. 2004;279:41275–9.
28. Li YX, Ding SJ, Xiao L, Guo W, Zhan Q. Desferoxamine preconditioning protects against cerebral ischemia in rats by inducing expressions of hypoxia inducible factor 1 alpha and erythropoietin. Neurosci Bull. 2008;24:89–95.
29. Marti HJ, Bernaudin M, Bellail A, Schoch H, Euler M, Petit E, Risau W. Hypoxia-induced vascular endothelial growth factor expression precedes neovascularization after cerebral ischemia. Am J Pathol. 2000;156:965–76.
30. Matrone C, Pignataro G, Molinaro P, Irace C, Scorziello A, Di Renzo GF, Annunziato L. HIF-1alpha reveals a binding activity to the promoter of iNOS gene after permanent middle cerebral artery occlusion. J Neurochem. 2004;90:368–78.
31. Melillo G, Musso T, Sica A, Taylor LS, Cox GW, Varesio L. A hypoxia-responsive element mediates a novel pathway of activation of the inducible nitric oxide synthase promoter. J Exp Med. 1995;182:1683–93.
32. Mellor HR, Harris AL. The role of the hypoxia-inducible BH3-only proteins BNIP3 and BNIP3L in cancer. Cancer Metastasis Rev. 2007;26:553–66.
33. Millhorn DE, Raymond R, Conforti L, Zhu W, Beitner-Johnson D, Filisko T, Genter MB, Kobayashi S, Peng M. Regulation of gene expression for tyrosine hydroxylase in oxygen sensitive cells by hypoxia. Kidney Int. 1997;51:527–35.
34. Nangaku M, Izuhara Y, Takizawa S, Yamashita T, Fujii-Kuriyama Y, Ohneda O, Yamamoto M, van Ypersele de Strihou C, Hirayama N, Miyata T. A novel class of prolyl hydroxylase inhibitors induces angiogenesis and exerts organ protection against ischemia. Arterioscler Thromb Vasc Biol. 2007;27:2548–54.
35. Palmer LA, Semenza GL, Stoler MH, Johns RA. Hypoxia induces type II NOS gene expression in pulmonary artery endothelial cells via HIF-1. Am J Physiol. 1998;274:L212–9.
36. Piret JP, Lecocq C, Toffoli S, Ninane N, Raes M, Michiels C. Hypoxia and CoCl$_2$ protect HepG2 cells against serum deprivation- and t-BHP-induced apoptosis: a possible anti-apoptotic role for HIF-1. Exp Cell Res. 2004;295:340–9.
37. Piret JP, Mottet D, Raes M, Michiels C. Is HIF-1alpha a pro- or an anti-apoptotic protein? Biochem Pharmacol. 2002;64:889–92.

38. Rankin EB, Giaccia AJ. The role of hypoxia-inducible factors in tumorigenesis. Cell Death Differ. 2008;15:678–85.
39. Ray R, Chen G, Vande VC, Cizeau J, Park JH, Reed JC, Gietz RD, Greenberg AH. BNIP3 heterodimerizes with Bcl-2/Bcl-X(L) and induces cell death independent of a Bcl-2 homology 3 (BH3) domain at both mitochondrial and nonmitochondrial sites. J Biol Chem. 2000;275:1439–48.
40. Sadamoto Y, Igase K, Sakanaka M, Sato K, Otsuka H, Sakaki S, Masuda S, Sasaki R. Erythropoietin prevents place navigation disability and cortical infarction in rats with permanent occlusion of the middle cerebral artery. Biochem Biophys Res Commun. 1998;253: 26–32.
41. Sakanaka M, Wen TC, Matsuda S, Masuda S, Morishita E, Nagao M, Sasaki R. In vivo evidence that erythropoietin protects neurons from ischemic damage. Proc Natl Acad Sci U S A. 1998;95:4635–40.
42. Sasabe E, Tatemoto Y, Li D, Yamamoto T, Osaki T. Mechanism of HIF-1alpha-dependent suppression of hypoxia-induced apoptosis in squamous cell carcinoma cells. Cancer Sci. 2005;96: 394–402.
43. Schmid T, Zhou J, Kohl R, Brune B. p300 relieves p53-evoked transcriptional repression of hypoxia-inducible factor-1 (HIF-1). Biochem J. 2004;380:289–95.
44. Schmidt-Kastner R, Guirre-Chen C, Kietzmann T, Saul I, Busto R, Ginsberg MD. Nuclear localization of the hypoxia-regulated pro-apoptotic protein BNIP3 after global brain ischemia in the rat hippocampus. Brain Res. 2004;1001(1):133–42.
45. Semenza GL, Agani F, Feldser D, Iyer N, Kotch L, Laughner E, Yu A. Hypoxia, HIF-1, and the pathophysiology of common human diseases. Adv Exp Med Biol. 2000;475:123–30.
46. Semenza GL, Jiang BH, Leung SW, Passantino R, Concordet JP, Maire P, Giallongo A. Hypoxia response elements in the aldolase A, enolase 1, and lactate dehydrogenase A gene promoters contain essential binding sites for hypoxia-inducible factor 1. J Biol Chem. 1996;271:32529–37.
47. Sharp FR, Bergeron M, Bernaudin M. Hypoxia-inducible factor in brain. Adv Exp Med Biol. 2001;502:273–91.
48. Sharp FR, Ran R, Lu A, Tang Y, Strauss KI, Glass T, Ardizzone T, Bernaudin M. Hypoxic preconditioning protects against ischemic brain injury. NeuroRx. 2004;1:26–35.
49. Shi Q, Zhang P, Zhang J, Chen X, Lu H, Tian Y, Parker TL, Liu Y. Adenovirus-mediated brain-derived neurotrophic factor expression regulated by hypoxia response element protects brain from injury of transient middle cerebral artery occlusion in mice. Neurosci Lett. 2009;465: 220–5.
50. Siddiq A, Aminova LR, Troy CM, Suh K, Messer Z, Semenza GL, Ratan RR. Selective inhibition of hypoxia-inducible factor (HIF) prolyl-hydroxylase 1 mediates neuroprotection against normoxic oxidative death via HIF- and CREB-independent pathways. J Neurosci. 2009; 29:8828–38.
51. Sowter HM, Ratcliffe PJ, Watson P, Greenberg AH, Harris AL. HIF-1-dependent regulation of hypoxic induction of the cell death factors BNIP3 and NIX in human tumors. Cancer Res. 2001;61:6669–73.
52. Suzuki H, Tomida A, Tsuruo T. Dephosphorylated hypoxia-inducible factor 1alpha as a mediator of p53-dependent apoptosis during hypoxia. Oncogene. 2001;20:5779–88.
53. Tupitsyna TV, Slominskii PA, Shadrina MI, Shetova IM, Skvortsova VI, Limborskaia SA. Association of the IVS9-675C > A polymorphism of the HIF-1alpha gene with acute ischemic stroke in the Moscow population. Genetika. 2006;42:858–61.
54. Van BN, Thibodeaux H, Palmer JT, Lee WP, Fu L, Cairns B, Tumas D, Gerlai R, Williams SP, van Lookeren CM, Ferrara N. VEGF antagonism reduces edema formation and tissue damage after ischemia/reperfusion injury in the mouse brain. J Clin Invest. 1999;104:1613–20.
55. Van HM, Prigent-Tessier AS, Garnier PE, Bertrand NM, Filomenko R, Bettaieb A, Marie C, Beley AG. Evidence of HIF-1 functional binding activity to caspase-3 promoter after photo-thrombotic cerebral ischemia. Mol Cell Neurosci. 2007;34:40–7.

56. Wang CX, Shuaib A. Neuroprotective effects of free radical scavengers in stroke. Drugs Aging. 2007;24:537–46.
57. Wiener CM, Booth G, Semenza GL. In vivo expression of mRNAs encoding hypoxia-inducible factor 1. Biochem Biophys Res Commun. 1996;225:485–8.
58. Yeh SH, Ou LC, Gean PW, Hung JJ, Chang WC. Selective inhibition of early-but not late-expressed HIF-1alpha is neuroprotective in rats after focal ischemic brain damage. Brain Pathol. 2011;21:249–62.
59. Zaman K, Ryu H, Hall D, O'Donovan K, Lin KI, Miller MP, Marquis JC, Baraban JM, Semenza GL, Ratan RR. Protection from oxidative stress-induced apoptosis in cortical neuronal cultures by iron chelators is associated with enhanced DNA binding of hypoxia-inducible factor-1 and ATF-1/CREB and increased expression of glycolytic enzymes, p21(waf1/cip1), and erythropoietin. J Neurosci. 1999;19:9821–30.
60. Zhou D, Matchett GA, Jadhav V, Dach N, Zhang JH. The effect of 2-methoxyestradiol, a HIF-1 alpha inhibitor, in global cerebral ischemia in rats. Neurol Res. 2008;30:268–71.

Chapter 9
Thrombin in Ischemic Stroke Targeting

Bo Chen

Abstract Accumulating research data have suggested pleiotropic roles of thrombin in brain tissue, including vascular disruption, inflammatory response, oxidative stress, and direct cellular toxicity. As a result, thrombin might contribute to stroke pathology through multiple mechanisms. A better understanding of thrombin toxicity would lead to new therapeutic targets in clinical application. In this brief review, we will discuss the basic biology of thrombin, the mechanism of thrombin toxicity, and the translational aspects of antithrombotic therapy in stroke.

1 Introduction

Stroke refers to the symptoms caused by sudden loss of blood supply to the brain. In principle, any factors that would narrow or block the brain vasculature might result in the reduction of the blood supply and therefore trigger a stroke attack. Thrombin is the key regulator in clot formation and a natural target for stroke prevention and intervention.

Thrombin was generally considered as a clot maker until 1991 when Shaun Coughlin and his team isolated the cellular receptor for thrombin, which is called protease-activated receptor-1 (PAR1) [1]. A small family of receptors (PAR2, PAR3, and PAR4) was subsequently discovered [2–4]. What are other functions of thrombin? What is the signaling pathway for the new receptors? This finding opens up many new avenues of research and brings deeper insight into the role of thrombin in physiology and pathological development of various diseases.

B. Chen, PhD (✉)
Department of Neurology, Cedars-Sinai Medical Center, Davis Research Building,
2094D, 110 N. George Burns Road, Los Angeles, CA 90048, USA
e-mail: Bo.Chen@cshs.org

P.A. Lapchak and J.H. Zhang (eds.), *Translational Stroke Research*, Springer Series
in Translational Stroke Research, DOI 10.1007/978-1-4419-9530-8_9,
© Springer Science+Business Media, LLC 2012

2 Molecular Basis of Thrombin Toxicity

2.1 Coagulation Cascade

Thrombin is probably one of the best studied proteases. The human thrombin is synthesized in the liver, and secreted into the blood stream in the zymogen form, prothrombin. In the event of the vascular damage, prothrombin is cleaved and transformed into thrombin by a process called coagulation cascade [5–7]. In brief, damages on the endothelial cells expose the tissue factor that is normally embedded in the vascular adventitia. Tissue factor then mediates the conversion of factor VII into its active form, and together activate factor X. The latter will assemble with Factor V and calcium into the prothrombinase complex, and cleave prothrombin into thrombin.

Thrombin directs blood clotting process in several ways. First, thrombin cleaves fibrinogen to fibrin, unmasking the cross-linking sites within the molecules and allowing fibrin to form the meshwork of blood clots. The stability of the clots is further enhanced by the action of thrombin on factor XIII and thrombin-activatable fibrinolysis inhibitor (TAFI). Second, thrombin activates platelets through PAR1, leading to platelet aggregation and degranulation. Thrombin also interacts with platelet glycoprotein GPIb independent of its proteolytic activity. Activated platelets express many procoagulant factors, such as phosphatidylserine, that further stimulate the coagulation pathways. Third, thrombin amplifies the coagulant signals through several positive feedback loops. For example, it activates factors VIII and XI, both of which are involved in the activation of factor X, the ultimate protease that catalyzes the conversion of prothrombin to thrombin. It also activates factor V, the cofactor for factor X in the prothrombinase complex. These two positive feedback pathways greatly promote the generation of thrombin after its initial activation.

As thrombin is so powerfully procoagulant, several regulatory mechanisms exist to limit thrombin activity near the injured sites [8]. Antithrombin is one of the major inhibitors in the circulation system. It not only blocks the action of thrombin but also other serine proteases including factor X. On the contrary, protease nexin I (PNI) and heparin cofactor II (HCII) are specific for thrombin. Interestingly, the latter two can also be found in extravascular tissue [9, 10]. In addition to the direct inhibitory mechanism, thrombin can also cleave protein C into activated protein C (APC), and in return, APC will interfere with factor V and VIII, preventing excessive thrombin generation in the long term. APC also exhibits potential neuroprotective effect, which will be discussed in the following passages.

2.2 From Structure to Function

As discussed, thrombin plays a central role in hemostasis regulation by its actions on a variety of substrate (e.g., fibrinogen, protein C, PAR1). Most of the serine proteases cleave the substrate in a relatively indiscriminative manner. For example, trypsin prefers basic residues such as arginine and lysine. Thrombin, however, exhibits a strong specificity in substrate binding and cleavage. What is the structural basis for the substrate specificity?

The crystal structure of thrombin revealed a central groove where the catalytic action on substrate occurs [11, 12]. In the core of the groove is the active site that comprises three amino acid residues, histidine, aspartic acid, and serine. Adjacent to the groove are two surface loops named 60 loop and γ loop (i.e., autolysis loop). Partial deletion of these two loops resulted in enhanced inhibition of thrombin activity by some trypsin inhibitors, suggesting that these loops are involved in restricting the access of the macromolecular substrates to the active site [13, 14]. The substrate specificity is further conferred by two anion-binding exosites residing on opposite ends of the groove. The exosites comprise mostly basic residues, and thus interact with negatively charged amino acids of the substrate by electrostatic force. Exosite I is originally viewed as the fibrinogen binding site, but later found to be attractive to PAR1, GPIb, factor V, and factor VIII as well [15–19]. The leech-derived anticoagulant, hirudin, functions by binding to exosite I and therefore blocking the fibrin conversion and the formation of clot meshwork [20, 21]. On the other end, Exosite II is known as the binding site for heparin, a cofactor for anti-thrombin [22]. Overall, the existence of these surface structures determines the binding specificity of thrombin substrates.

Recent studies by Di Cero group have shed light on the substrate cleavage specificity of thrombin [23]. They replaced each of the surface amino acids of thrombin with alanine and created a pool of engineered mutant thrombins. They found that tryptophan near the active site (W215) was the single most important determinant for thrombin specificity over fibrinogen, PAR1, and protein C. For example, a mutation of tryptophan to glutamic acid (W215E) resulted in a mutant thrombin that shows tenfolds more specificity over protein C. The double mutant with W215E and partial deletion of γ loop (Δ146–149) resulted in a variant with appreciable activity at PAR1 but minimal activity at fibrinogen and protein C. These mutant thrombins not only reveal significant insight into the structural basis of thrombin specificity but also provide a powerful toolset to dissect the function of thrombin in physiologic condition and in disease model such as stroke. Potentially, some of mutants with desirable therapeutic benefits (such as making APC) but without the undesired property (such as making clots) could be used as a therapeutic agent for stroke treatment.

2.3   Activation of PARs

Thrombin not only interacts with blood proteins involved in coagulation pathway but also acts on cellular receptors known as protease-activated receptors (PARs) [24–26]. By far four members of the PAR family have been discovered and they are labeled sequentially as PAR1, PAR2, PAR3, and PAR4. In human platelet cells,

PAR1 is the major mediator for thrombin signal and can be activated in the presence of low concentration of thrombin. PAR4 is also a thrombin receptor but only responds to high-dose thrombin. In rodents, however, the platelets lack PAR1 receptor and instead, PAR3 is responsible for thrombin and transactivates PAR4 and the subsequent cellular pathways. PAR2 is not cleavable by thrombin but trypsin and trypsin-like serine proteases including factor VII and X.

PAR1 is the most well studied among the four. It is widely expressed in neurons, glial cells, platelets (in human), and endothelial cells. In human brain, PAR1 is expressed abundantly in astrocytes and moderately in neurons [27]. Thrombin is considered as the main activator for PAR1, but other proteases, such as factor X, plasmin, APC, and matrix metalloproteinase 1 (MMP1), can also activate PAR1 under certain circumstances [28–31].

PAR1 is a G-protein-coupled receptor. The extracellular domain of PAR1 contains a sequence (DPRSFL in human thrombin) that is recognized and cleaved by thrombin. Next to the cleavage sequence is a hirudin-like domain, which fits in the exosite I of thrombin and enhances the specificity to thrombin. During PAR1 activation, the extracellular "tail" is cleaved by thrombin and unmask the new N-terminus that will be tethered back and act as a ligand to the receptor, leading to intracellular signaling transduction [32]. Activated PAR1 is rapidly desensitized by phosphorylation and arrestin binding; it is then uncoupled from G protein, internalized, and sorted to lysosome for degradation [24–26].

Among the PAR family, PAR1, PAR3, and PAR4 are known to mediate thrombin-induced cellular response. PAR-2 is not generally considered as a thrombin receptor. However, there are several studies showing that PAR-2 might couple with PAR-1 and mediate differential cellular functions [33–35]. For example, cleaved PAR-1 could transactivate PAR-2 and contribute to endothelial response to thrombin stimulation. How different receptors coordinate the thrombin response remains largely unknown. The in vivo roles of PAR-2, PAR-3, and PAR-4 have not been well studied in stroke model. Experimental results from these studies would enrich our understanding about the function of these receptors and suggest potential therapeutic targets.

3   Cellular Mechanism of Thrombin Toxicity in Stroke

3.1   Thrombin and Coagulation

Thrombin plays critical roles in hemostasis regulation. As excessive clot formation leads to ischemia, thrombin has been a primary target for the prevention or treatment of stroke. Numerous drugs have been screened or designed to target functional domain of thrombin (e.g., Dabigatran to the active site, hirudin to the exosite I, heparin to the exosite II), synthesis of thrombin (Warfarin inhibits vitamin K reductase which is involved in prothrombin maturation), activation of thrombin (Rivaroxaban to factor X that converts prothrombin to thrombin), thrombin signaling pathway (Vorapaxar antagonize PAR1 receptor), or other components involved in blood clotting, such as platelet activation and aggregation, as well as thrombolysis and fibrinolysis.

The application of these drugs are not limited to stroke but also other disease that are related to coagulation, such as cardiovascular disease, atherosclerosis, and venous thromboembolism (VTE). So far, tPA remains the only FDA-approved treatment for acute ischemic stroke, and Warfarin as well as Dabigatran is administered to patients with atrial fibrillation for prevention of the recurrent stroke. A lot of reviews have been published to discuss the new therapies with higher efficacy and lower risk profile [36–39]. Readers are referred to these reviews for the recent clinical progress. The discussion below will involve more of thrombin toxicity mechanism other than coagulation.

3.2   Thrombin and Blood–Brain Barrier Damage

Previous studies have indicated that thrombin could act on endothelial cells and disrupt the barrier function. Thrombin activates PAR1 receptors associated with Gα 12/13 and Gα11/q, leading to upregulation of calcium and activation of RhoA/PKC pathways. These actions will affect the remodeling of adherent junction proteins and induce endothelial permeability [40]. In addition, thrombin can participate in direct barrier degradation by activating MMP2 in human microvascular endothelial cells [41]. Most of these investigations on vascular permeability are conducted in cultured endothelial cells and the relevance of these pathways in vivo remains to be clarified.

Recently, our group tested the involvement of thrombin in causing BBB disruption using an in vivo model of focal ischemia [42]. In this experiment, thrombin was delivered to the brain directly via an intra-arterial catheter. Compared to the saline control group, thrombin treatment greatly increased the severe vascular disruption as labeled by 2MDa dextran conjugate with FITC [43]. Coadministration of intravenous argatroban, a direct thrombin inhibitor, reduced the vascular disruption in a dose-dependent manner, suggesting that the observed pathology is specific to thrombin. It is not clear at this point whether or not the increment in vascular damage was caused by increased microthrombosis beyond the initial injured sites, or by direct activation of PAR1. Further studies are needed to identify and dissect the key factors involved in the severe vascular disruption during ischemia.

3.3   Thrombin and Vasospasm

Another line of study on the interaction between thrombin and vasculature concerns the regulation of vascular contractility. In the physiological condition, thrombin modulates the strength and frequency of arterial contractility by activating the PAR1 receptor on smooth muscle and endothelial cells [44, 45]. This could lead to very severe damage in the event of intracerebral hemorrhage (ICH) or subarachnoid hemorrhage (SAH). In these pathological condition, ruptured vasculature releases

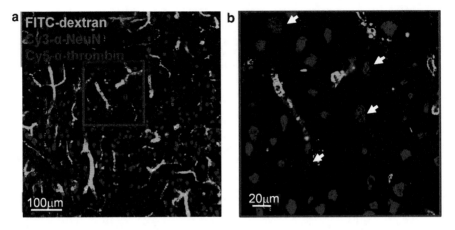

Fig. 9.1 Neuronal specific binding of thrombin in extravascular space. (**a**) Double staining with NeuN revealed that most cellular staining of thrombin was associated with neurons as labeled by the neuronal nuclei marker NeuN. The image was taken at 20× objective of Olympus FV1000 Confocal Microscope. (**b**) Image of high magnification (60×) delineate clear staining of thrombin (*red*) wrapping the neuronal nuclei (*blue*)

excessive amount of thrombin that in turn elicit vasospasm and subsequent ischemic damage to the surrounding tissues. The detailed mechanism is not understood well, and several signaling molecules have been studied in this process, including PAR1. More recently, oxidative stress has been indicated to contribute to the thrombin-mediated vasospasm after SAH [46]. Production of reactive oxygen species (ROS) might be triggered by thrombin in a yet unknown mechanism. ROS then block the desensitization of PAR1, allowing excessive and continuous activation of PAR1 pathway and calcium inflow.

3.4 Thrombin and Neurotoxicity

It might come as a surprise to many when linking thrombin—once thought to participate in stroke damage via coagulation—to neurotoxicity. Several lines of evidence converge in support of the neurotoxicity hypothesis: first, accumulating data from in vitro studies indicate that thrombin induces apoptotic cell death in glia and neurons [47, 48]. Second, several groups investigating the mechanisms of tissue injury around cerebral hematoma prove that leakage of several compounds into surrounding tissue is neurotoxic: oxidized hemoglobin and breakdown products; iron; thrombin; and matrix metalloproteinases [49–54]. Finally, thrombin seems to be particularly toxic to stressed neuronal cells [55].

It became very interesting to us when we stained the ischemic brain sections with a thrombin antibody, and discovered a strong attachment of thrombin to the neuronal cells (Fig. 9.1). We further investigated the presence of PAR1 receptor in the rodent brain, and found abundant PAR1 expression in the neurons. In addition, the neuronal death after ischemia increased as intra-arterial thrombin was infused, and decreased as argatroban was delivered intravenously. These results suggested the possibility that vascular leakage of thrombin contributed to the ischemic neuronal damage.

Thrombin toxicity in the brain is likely mediated by PAR1 signaling in the parenchymal tissue. PAR1 knockout mice are more resistant to short ischemic insult, and intracerebroventricular injection of a PAR1 antagonist reduces the infarct size as well [56]. Many questions remain elusive: what are the pathways underlying thrombin toxicity on neurons? Tyrosine kinase, RhoA, and NMDA pathways have been indicated in a few literatures [47, 57]. Does thrombin act on neurons through direct or indirect mechanism? Is the downstream of thrombin/PAR1 signaling different in neurons from that in platelets and endothelial cells?

3.5   Thrombin and Inflammation

Thrombin interacts with inflammatory cells including microglial cells and leukocytes. Among many cellular signaling pathways induced by PAR1 activation, the mitogen-activated protein kinase (MAPK) appears as an important mediator of thrombin toxicity. Either thrombin or PAR1 agonist could induce a robust MAPK phosphorylation in the cultured cells [58, 59]. In addition, direct injection of thrombin to the striatum causes severe edema and neuronal injury, which could be alleviated by the co-injection of chemical inhibitors targeted to MAPK kinases p38, ERK, and JNK [60]. Follow-up experiment in organotypic slice culture associates reactive microglial cells with the thrombin-induced inflammatory response and subsequent tissue toxicity [61].

Furthermore, thrombin promotes inflammation by stimulating the expression of pro-inflammation factors [25, 62]. The secretion of IL-6 and IL-8 is increased in endothelial cells and monocytes after thrombin activation. Expression of adhesion molecules such as ICAM-1 and VCAM-1 is also augmented by thrombin [63–65]. These molecular events serve to recruit platelets and leukocytes to the injury sites and facilitate the entry of inflammatory cells to the parenchymal tissue. Other coagulation factors including tissue factors, factor VII, factor X, and even fibrin, could also interact with inflammatory response in different mechanisms.

3.6   Thrombin and Oxidative Stress

Thrombin can contribute to oxidative stress through the NADPH oxidase on microglial cells. In organotypic hippocampal slice, the level of inducible nitric oxide synthase (iNOS) is increased 4 hours after thrombin administration, and the expression and translocation of NADPH oxidase is also observed 8 hours after thrombin, leading to the production of ROS and subsequent neuronal death [66]. Several other cytokines involved in inflammatory response, such as IL-4 and IL-13, might also participate in this process in a thrombin-dependent or -independent mechanism [67].

3.7   Thrombin in Angiogenesis and Neurogenesis

The angiogenic potential of thrombin has been explored in several studies in vitro. Thrombin induces capillary tube formation in a dose-dependent manner. Low-dose thrombin promotes angiogenesis. Several mechanisms have been implicated, including the activation of APC and upregulation of the receptor for vascular endothelial growth factor (VEGF) [68, 69]. Whether or not this is a physiologically relevant event in vivo remains unknown. It will be highly interesting if thrombin actually participates in angiogenic response in injured tissue after stroke. In addition to angiogenesis, thrombin might play an important role in vascular development [70]. Mice deficient in thrombin receptor PAR1 have abnormal bleeding and half of them die during embryonic development.

Thrombin might also participate in neurogenesis. When thrombin is directly injected to a hemorrhagic brain, the expression of doublecortin (DCX), a marker for immature neurons, increases [71]. Experiment using other markers for neurogenesis (e.g., BrdU) would be necessary to further examine the concept of thrombin-mediated neurogenesis. It would be of great interest to investigate the potential engagement of thrombin in neurogenesis in post-ischemic brain as well.

4   Translational Challenge for Antithrombotic Therapy in Stroke

As is discussed in the above passages, thrombin seems to mediate pleiotropic effects in the brain (Fig. 9.2). It is like a coin of two sides. On the bad side, it causes blood–brain barrier damage, neurotoxicity, oxidative stress, and inflammatory response, all

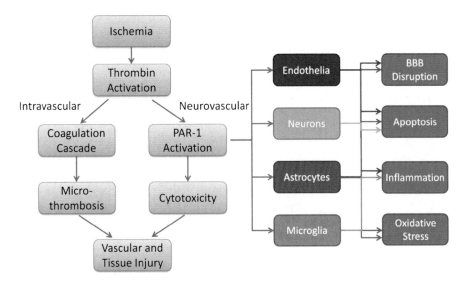

Fig. 9.2 Scheme of thrombin toxicity during ischemic stroke

of which add detrimental effect to the ischemic brain. On the other side, however, there are probably some good reasons why our brains express thrombin receptors. Some studies have associated thrombin with neurogenesis and angiogenesis. So targeting thrombin in stroke is going to be a dilemma. The challenge to both basic and clinical research is to dissect the bad part of thrombin from its good part. In the following passages, I will propose a few perspectives to approach this complicated issue.

4.1   Concentration Matters

Thrombin concentration in the systemic circulation is under delicate regulation. High level of thrombin contributes to the stroke risk with potential high incidence of thrombosis, and low thrombin activity results in intracerebral bleeding complications. So too much thrombin causes ischemia, and too little thrombin leads to hemorrhage. The key is really to keep the concentration of active thrombin within the right range.

Similar finding exists for neurotoxicity of thrombin. Striggow and his colleagues reported the dramatic changes in the ischemic brain when different amount of thrombin was present in the medium [48]. A significant portion of neurons died with thrombin concentration peaked at 10 U/mL, and as thrombin concentration went down the ratio of cell death also decreased. When thrombin concentration was reduced to 0.01 U/mL, however, the cell survival rate is actually higher than the control group without thrombin. This observation supported the speculation that thrombin could be actively engaged in some repair/neurogenesis process in the physiological context. Adding to this fascinating hypothesis is the study showing that low concentration of thrombin (0.12 U/mL) that is tolerable to the neurons in normoglycemia condition could be very toxic to neurons under hypoglycemia [55]. This result suggested that thrombin concentration is not the only factor that modulates cellular response. Other internal factors that are induced by environmental stress also contribute to the regulation of the thrombin signaling.

Another example comes from thrombin effect on BBB damage. Thrombin is generally believed to increase BBB disruption, but low concentration of thrombin could also protect endothelial barrier function in vitro [72]. The result could be complicated when translated in vivo. In our recent study investigating the role of thrombin in BBB disruption, we infused a small amount of thrombin (1 U) to the rat brain over 4 hours, a dose that was previously used to precondition the brain against ischemia [73–75]. Surprisingly, this low-dose thrombin resulted in severe damage to the brain vasculature. How do we interpret the result? One possibility is that the ischemic tissues are more susceptible to thrombin insult, so the low-dose thrombin that is well tolerated by normal brain becomes extremely toxic during ischemia. Another possibility is that the infused thrombin functions as the "seed" and promotes the generation of additional thrombin through the coagulation cascade [5, 76].

4.2 Timing Matters

Timing might be another secret that distinguishes the beneficial and detrimental effects of thrombin. In our in vivo model, thrombin could contribute to neurovascular damage as early as 2 hours after stroke onset. In such an acute phase, the severe disruption on blood–brain barrier allows thrombin, aggregating around injured sites, to enter the parenchyma tissue. The protease activity by thrombin could be overwhelming in a devastating brain and exacerbate the ischemic damage. Over time, however, the excessive thrombin might be gradually removed, and the remaining thrombin would start participating in neurogenesis and angiogenesis, innate mechanisms that are intended to reconstruct the brain tissue.

At this point, this scenario about biphasic roles of thrombin largely stays on the hypothetical level; however, a recent study by Sharp group supported the possibility of this concept [77]. They studied the biphasic role of src kinase in a thrombin-induced hemorrhagic stroke model. They found that src is mostly involved in BBB breakdown and edema formation in the acute phase (1 day after thrombin injection). But in the later phase (2–6 days after), src starts to participate in BBB repair by stimulating microvascular endothelial cells and perivascular astrocyte. The role of thrombin in brain recovery after stroke requires further examination on the function of PAR1 receptor on the neurons and how that is related to tissue regeneration in the context of stroke.

4.3 Cell Matters

The exact mechanism of how thrombin mediates vascular and tissue injury during acute ischemia requires further investigation. It seems that thrombin induces very distinct responses in different cells. A specific pathology as seen in stroke could be the combinatorial results of several cell types. For example, thrombin toxicity on vasculature might be mediated by its direct effect on blood–brain barrier tight junctions, or by affecting neurons and glial cells which disrupt the integrity of neurovascular unit, or by inducing inflammatory response that brings indirect detrimental outcome to the brain.

As most thrombin-related cellular responses are mediated by PAR1 receptor, for future studies cell-specific PAR1 knockout animals would be very helpful to dissect the roles of thrombin in different cell types. Alternatively, virus-based shRNA could be infused to knock down the expression of PAR1 or other downstream genes. Novel markers that are able to detect thrombin activation in vivo are highly desired for research attempt to relate thrombin to pathophysiological changes during stroke.

4.4 Signaling Matters

The major thrombin receptor, PAR1, could also be activated by other proteases and mediate very distinct cellular pathways. For example, Zlokovic group have presented convincing data in support of a neuroprotective role for APC in a PAR1-dependent mechanism [78–81]. APC compared to thrombin activates different gene expression profiles acting through the same PAR1. APC blocks p53-mediated apoptosis in ischemic human brain endothelium and is neuroprotective. Thrombin does just the opposite, increasing p53 expression. It is thus important to understand the differential regulation on PAR1 and the molecular mechanisms that would switch one signaling pathways to the other.

The other example comes from the thrombin preconditioning studies. Several groups have shown that thrombin injected 7 days before stroke onset greatly increases brain tolerance against ischemic attack. However, people would probably concern about the safety issues given the potential toxicity caused by thrombin. Essentially, this dilemma exists for all therapeutic attempts targeting thrombin or similar molecules that have pleiotropic effects. It will be important to investigate what might be the diverging point following thrombin signaling and what the critical and unique pathways are to mimic/reproduce the thrombin preconditioning effect.

Ideally, we would like to target to the level where good and bad pathways diverge. In reality, signaling regulation is often very complicated and the specificity is fulfilled via different mechanisms. The distribution and amount of receptors on the cell surface would affect the sensing of thrombin concentration; pathological changes would usually change the expression profile of the receptor pathway and shift the outcome; in different types of cell, the receptor might be coupled with different sets of g-proteins and the downstream effectors. In a recent report, Trejo group discovered that PAR1 could be clustered with co-receptors such as endothelial protein C receptor (EPCR) on a lipid raft known as caveolae [82]. This finding revealed a new mechanism explaining why the same receptor in the same cell could possibly mediate distinct downstream pathway.

4.5 The Future of Thrombin Targeting Therapy

Based on the above discussion, successful application of thrombin targeting therapy would require accurate understanding of when, where, and to what extend thrombin activity affects brain functioning. Advancement in imaging would certainly be critical. Most current thrombin measurements, such as prothrombin time, give readout of overall coagulation activity throughout the systemic circulation. However, thrombin activation is a dynamic process depending on the timing and local microenvironment. It is especially true for a heterogeneous brain disease like stroke. Recently several innovative imaging techniques show the potential to capture these dynamic features.

Tsien group from UC San Diego has developed a group of probes called activatable cell penetrating peptide (ACPP) that are designed to detect specific protease activity in vivo [83–85]. The probe consists of polyglutamate (negatively charged), a sequence cleavable by the target protease, polyarginine (positively charged), and a fluorophore. In the case of thrombin, the cleavable sequence is DPRSFL, taken from original thrombin cleavage sites of PAR1. Thrombin cleavage of DPRSFL jettisons the polyglutamate, allowing the dye-labeled polyarginine to bind to cells. Uncleaved probes are washed out of the tissue during reperfusion. The considerable strength of this probe is the ability to label thrombin activation in situ at the time window of interest. It preserves the footprint of thrombin activation at the cellular level, and allows the researchers to relate thrombin activity to other molecular events, such as cellular injury. When the fluorophore is replaced with an MRI contrast agent like gadolinium, the ACPP could potentially be applied in clinical imaging and provide molecular information in vivo.

A similar probe was designed by Weissleder group from Massachusetts General Hospital [86]. In principle, the probe consists of a bundle of near infrared (NIR) fluorochromes joined to one delivery vehicle via a thrombin cleavable sequence. In the intact probe, the proximity of individual NIR fluorochrome results in quenching of the fluorescence. In the event of the thrombin cleavage, the NIR fluorochromes are released from the delivery vehicle and the level of fluorescence is significantly increased. This design of probe allows a real-time optical measurement of the thrombin activity. If a similar quenching group could be matched for MRI-based contrast agent such as gadolinium, the probe could be applied for real-time in vivo detection of thrombin activation.

In addition to imaging, advances in new drug design are also anticipated. Given the multiple facets of thrombin toxicity, a drug that could simultaneously antagonize several adverse effects would be highly desired. One promising molecule under investigation is the APC. Recently, the Griffin group has generated several mutant APCs (5A-APC and 3K3A-APC) that have enhanced activity at the vascular and neuronal PAR-1 receptor, but significantly less anticoagulant activity compared to the wildtype [87–90]. This feature will likely reduce the bleeding complication commonly seen in most antithrombotic therapy. The mutant 3K3A-APC is now under further preclinical evaluation in animal models, and Xigris (human recombinant APC) is going for a clinical study as neuroprotection for acute ischemia.

References

1. Vu TK, Hung DT, Wheaton VI, Coughlin SR. Molecular cloning of a functional thrombin receptor reveals a novel proteolytic mechanism of receptor activation. Cell. 1991;64:1057–68.
2. Ishihara H, Connolly AJ, Zeng D, Kahn ML, Zheng YW, Timmons C, Tram T, Coughlin SR. Protease-activated receptor 3 is a second thrombin receptor in humans. Nature. 1997;386:502–6.
3. Nakanishi-Matsui M, Zheng YW, Sulciner DJ, Weiss EJ, Ludeman MJ, Coughlin SR. Par3 is a cofactor for par4 activation by thrombin. Nature. 2000;404:609–13.
4. Coughlin SR. Thrombin signalling and protease-activated receptors. Nature. 2000;407:258–64.
5. Furie B, Furie BC. Mechanisms of thrombus formation. N Engl J Med. 2008;359:938–49.
6. Fenton 2nd JW, Ofosu FA, Moon DG, Maraganore JM. Thrombin structure and function: why thrombin is the primary target for antithrombotics. Blood Coagul Fibrinolysis. 1991;2:69–75.
7. Davie EW, Kulman JD. An overview of the structure and function of thrombin. Semin Thromb Hemost. 2006;32 Suppl 1:3–15.
8. Rau JC, Beaulieu LM, Huntington JA, Church FC. Serpins in thrombosis, hemostasis and fibrinolysis. J Thromb Haemost. 2007;5 Suppl 1:102–15.

9. Choi BH, Suzuki M, Kim T, Wagner SL, Cunningham DD. Protease nexin-1. Localization in the human brain suggests a protective role against extravasated serine proteases. Am J Pathol. 1990;137:741–7.
10. He L, Vicente CP, Westrick RJ, Eitzman DT, Tollefsen DM. Heparin cofactor ii inhibits arterial thrombosis after endothelial injury. J Clin Invest. 2002;109:213–9.
11. Bode W, Mayr I, Baumann U, Huber R, Stone SR, Hofsteenge J. The refined 1.9 a crystal structure of human alpha-thrombin: interaction with d-phe-pro-arg chloromethylketone and significance of the tyr-pro-pro-trp insertion segment. EMBO J. 1989;8:3467–75.
12. Rydel TJ, Tulinsky A, Bode W, Huber R. Refined structure of the hirudin-thrombin complex. J Mol Biol. 1991;221:583–601.
13. Le Bonniec BF, Guinto ER, Esmon CT. Interaction of thrombin des-etw with antithrombin iii, the kunitz inhibitors, thrombomodulin and protein c. Structural link between the autolysis loop and the tyr-pro-pro-trp insertion of thrombin. J Biol Chem. 1992;267:19341–8.
14. Dang QD, Sabetta M, Di Cera E. Selective loss of fibrinogen clotting in a loop-less thrombin. J Biol Chem. 1997;272:19649–51.
15. Tsiang M, Jain AK, Dunn KE, Rojas ME, Leung LL, Gibbs CS. Functional mapping of the surface residues of human thrombin. J Biol Chem. 1995;270:16854–63.
16. Pechik I, Madrazo J, Mosesson MW, Hernandez I, Gilliland GL, Medved L. Crystal structure of the complex between thrombin and the "E" region of fibrin. Proc Natl Acad Sci U S A. 2004;101:2718–23.
17. Mathews II, Padmanabhan KP, Ganesh V, Tulinsky A, Ishii M, Chen J, Turck CW, Coughlin SR, Fenton 2nd JW. Crystallographic structures of thrombin complexed with thrombin receptor peptides: existence of expected and novel binding modes. Biochemistry. 1994;33:3266–79.
18. De Cristofaro R, De Candia E, Landolfi R, Rutella S, Hall SW. Structural and functional mapping of the thrombin domain involved in the binding to the platelet glycoprotein ib. Biochemistry. 2001;40:13268–73.
19. Esmon CT, Lollar P. Involvement of thrombin anion-binding exosites 1 and 2 in the activation of factor v and factor viii. J Biol Chem. 1996;271:13882–7.
20. Stone SR, Braun PJ, Hofsteenge J. Identification of regions of alpha-thrombin involved in its interaction with hirudin. Biochemistry. 1987;26:4617–24.
21. Rydel TJ, Ravichandran KG, Tulinsky A, Bode W, Huber R, Roitsch C, Fenton 2nd JW. The structure of a complex of recombinant hirudin and human alpha-thrombin. Science. 1990;249:277–80.
22. Li W, Johnson DJ, Esmon CT, Huntington JA. Structure of the antithrombin-thrombin-heparin ternary complex reveals the antithrombotic mechanism of heparin. Nat Struct Mol Biol. 2004;11:857–62.
23. Marino F, Pelc LA, Vogt A, Gandhi PS, Di Cera E. Engineering thrombin for selective specificity toward protein c and par1. J Biol Chem. 2010;285:19145–52.
24. Traynelis SF, Trejo J. Protease-activated receptor signaling: new roles and regulatory mechanisms. Curr Opin Hematol. 2007;14:230–5.
25. Coughlin SR. Protease-activated receptors in hemostasis, thrombosis and vascular biology. J Thromb Haemost. 2005;3:1800–14.
26. Soh UJ, Dores MR, Chen B, Trejo J. Signal transduction by protease-activated receptors. Br J Pharmacol. 2010;160:191–203.
27. Junge CE, Lee CJ, Hubbard KB, Zhang Z, Olson JJ, Hepler JR, Brat DJ, Traynelis SF. Protease-activated receptor-1 in human brain: localization and functional expression in astrocytes. Exp Neurol. 2004;188:94–103.
28. Kuliopulos A, Covic L, Seeley SK, Sheridan PJ, Helin J, Costello CE. Plasmin desensitization of the par1 thrombin receptor: kinetics, sites of truncation, and implications for thrombolytic therapy. Biochemistry. 1999;38:4572–85.
29. Trivedi V, Boire A, Tchernychev B, Kaneider NC, Leger AJ, O'Callaghan K, Covic L, Kuliopulos A. Platelet matrix metalloprotease-1 mediates thrombogenesis by activating par1 at a cryptic ligand site. Cell. 2009;137:332–43.

30. Ludeman MJ, Kataoka H, Srinivasan Y, Esmon NL, Esmon CT, Coughlin SR. Par1 cleavage and signaling in response to activated protein c and thrombin. J Biol Chem. 2005;280:13122–8.
31. Nesi A, Fragai M. Substrate specificities of matrix metalloproteinase 1 in par-1 exodomain proteolysis. Chembiochem. 2007;8:1367–9.
32. Rohatgi T, Sedehizade F, Reymann KG, Reiser G. Protease-activated receptors in neuronal development, neurodegeneration, and neuroprotection: thrombin as signaling molecule in the brain. Neuroscientist. 2004;10:501–12.
33. O'Brien PJ, Prevost N, Molino M, Hollinger MK, Woolkalis MJ, Woulfe DS, Brass LF. Thrombin responses in human endothelial cells. Contributions from receptors other than par1 include the transactivation of par2 by thrombin-cleaved par1. J Biol Chem. 2000;275:13502–9.
34. Shi X, Gangadharan B, Brass LF, Ruf W, Mueller BM. Protease-activated receptors (par1 and par2) contribute to tumor cell motility and metastasis. Mol Cancer Res. 2004;2:395–402.
35. McCoy KL, Traynelis SF, Hepler JR. Par1 and par2 couple to overlapping and distinct sets of g proteins and linked signaling pathways to differentially regulate cell physiology. Mol Pharmacol. 2010;77:1005–15.
36. Murray V, Norrving B, Sandercock PA, Terent A, Wardlaw JM, Wester P. The molecular basis of thrombolysis and its clinical application in stroke. J Intern Med. 2010;267:191–208.
37. Eriksson BI, Quinlan DJ, Eikelboom JW. Novel oral factor xa and thrombin inhibitors in the management of thromboembolism. Annu Rev Med. 2011;62:41–57.
38. Garcia D, Libby E, Crowther MA. The new oral anticoagulants. Blood. 2010;115:15–20.
39. Choi J, Kermode JC. New therapeutic approaches to combat arterial thrombosis: better drugs for old targets, novel targets, and future prospects. Mol Interv. 2011;11:111–23.
40. Gavard J, Gutkind JS. Protein kinase c-related kinase and rock are required for thrombin-induced endothelial cell permeability downstream from galpha12/13 and galpha11/q. J Biol Chem. 2008;283:29888–96.
41. Nguyen M, Arkell J, Jackson CJ. Thrombin rapidly and efficiently activates gelatinase a in human microvascular endothelial cells via a mechanism independent of active mt1 matrix metalloproteinase. Lab Invest. 1999;79:467–75.
42. Chen B, Cheng Q, Yang K, Lyden PD. Thrombin mediates severe neurovascular injury during ischemia. Stroke. 2010;41:2348–52.
43. Chen B, Friedman B, Cheng Q, Tsai P, Schim E, Kleinfeld D, Lyden PD. Severe blood-brain barrier disruption and surrounding tissue injury. Stroke. 2009;40:e666–74.
44. Hirano K. The roles of proteinase-activated receptors in the vascular physiology and pathophysiology. Arterioscler Thromb Vasc Biol. 2007;27:27–36.
45. Kai Y, Maeda Y, Sasaki T, Kanaide H, Hirano K. Basic and translational research on proteinase-activated receptors: the role of thrombin receptor in cerebral vasospasm in subarachnoid hemorrhage. J Pharmacol Sci. 2008;108:426–32.
46. Kameda K, Kikkawa Y, Hirano M, Matsuo S, Sasaki T, Hirano K. Combined argatroban and anti-oxidative agents prevents increased vascular contractility to thrombin and other ligands after subarachnoid hemorrhage. Br J Pharmacol. 2012;165(1):106–19.
47. Donovan FM, Pike CJ, Cotman CW, Cunningham DD. Thrombin induces apoptosis in cultured neurons and astrocytes via a pathway requiring tyrosine kinase and RhoA activities. J Neurosci. 1997;17:5316–26.
48. Striggow F, Riek M, Breder J, Henrich-Noack P, Reymann KG, Reiser G. The protease thrombin is an endogenous mediator of hippocampal neuroprotection against ischemia at low concentrations but causes degeneration at high concentrations. Proc Natl Acad Sci U S A. 2000;97:2264–9.
49. Ohnishi M, Katsuki H, Fujimoto S, Takagi M, Kume T, Akaike A. Involvement of thrombin and mitogen-activated protein kinase pathways in hemorrhagic brain injury. Exp Neurol. 2007;206:43–52.
50. Clark JF, Loftspring M, Wurster WL, Beiler S, Beiler C, Wagner KR, Pyne-Geithman GJ. Bilirubin oxidation products, oxidative stress, and intracerebral hemorrhage. Acta Neurochir Suppl. 2008;105:7–12.

51. Qing WG, Dong YQ, Ping TQ, Lai LG, Fang LD, Min HW, Xia L, Heng PY. Brain edema after intracerebral hemorrhage in rats: the role of iron overload and aquaporin 4. J Neurosurg. 2009;110:462–8.
52. Katsu M, Niizuma K, Yoshioka H, Okami N, Sakata H, Chan PH. Hemoglobin-induced oxidative stress contributes to matrix metalloproteinase activation and blood-brain barrier dysfunction in vivo. J Cereb Blood Flow Metab. 2010;30(12):1939–50.
53. Copin JC, Merlani P, Sugawara T, Chan PH, Gasche Y. Delayed matrix metalloproteinase inhibition reduces intracerebral hemorrhage after embolic stroke in rats. Exp Neurol. 2008;213:196–201.
54. Maddahi A, Chen Q, Edvinsson L. Enhanced cerebrovascular expression of matrix metalloproteinase-9 and tissue inhibitor of metalloproteinase-1 via the mek/erk pathway during cerebral ischemia in the rat. BMC Neurosci. 2009;10:56.
55. Weinstein JR, Lau AL, Brass LF, Cunningham DD. Injury-related factors and conditions down-regulate the thrombin receptor (par-1) in a human neuronal cell line. J Neurochem. 1998;71:1034–50.
56. Junge CE, Sugawara T, Mannaioni G, Alagarsamy S, Conn PJ, Brat DJ, Chan PH, Traynelis SF. The contribution of protease-activated receptor 1 to neuronal damage caused by transient focal cerebral ischemia. Proc Natl Acad Sci U S A. 2003;100:13019–24.
57. Gingrich MB, Junge CE, Lyuboslavsky P, Traynelis SF. Potentiation of NMDA receptor function by the serine protease thrombin. J Neurosci. 2000;20:4582–95.
58. Borbiev T, Birukova A, Liu F, Nurmukhambetova S, Gerthoffer WT, Garcia JG, Verin AD. P38 map kinase-dependent regulation of endothelial cell permeability. Am J Physiol Lung Cell Mol Physiol. 2004;287:L911–8.
59. Vandell AG, Larson N, Laxmikanthan G, Panos M, Blaber SI, Blaber M, Scarisbrick IA. Protease-activated receptor dependent and independent signaling by kallikreins 1 and 6 in CNS neuron and astroglial cell lines. J Neurochem. 2008;107:855–70.
60. Fujimoto S, Katsuki H, Kume T, Akaike A. Thrombin-induced delayed injury involves multiple and distinct signaling pathways in the cerebral cortex and the striatum in organotypic slice cultures. Neurobiol Dis. 2006;22:130–42.
61. Fujimoto S, Katsuki H, Ohnishi M, Takagi M, Kume T, Akaike A. Thrombin induces striatal neurotoxicity depending on mitogen-activated protein kinase pathways in vivo. Neuroscience. 2007;144:694–701.
62. Petaja J. Inflammation and coagulation. An overview. Thromb Res. 2011;127 Suppl 2:S34–7.
63. Chi L, Li Y, Stehno-Bittel L, Gao J, Morrison DC, Stechschulte DJ, Dileepan KN. Interleukin-6 production by endothelial cells via stimulation of protease-activated receptors is amplified by endotoxin and tumor necrosis factor-alpha. J Interferon Cytokine Res. 2001;21:231–40.
64. Kaplanski G, Fabrigoule M, Boulay V, Dinarello CA, Bongrand P, Kaplanski S, Farnarier C. Thrombin induces endothelial type ii activation in vitro: Il-1 and tnf-alpha-independent il-8 secretion and e-selectin expression. J Immunol. 1997;158:5435–41.
65. Minami T, Abid MR, Zhang J, King G, Kodama T, Aird WC. Thrombin stimulation of vascular adhesion molecule-1 in endothelial cells is mediated by protein kinase c (PKC)-delta-nf-kappa b and PKC-zeta-gata signaling pathways. J Biol Chem. 2003;278:6976–84.
66. Choi SH, Lee DY, Kim SU, Jin BK. Thrombin-induced oxidative stress contributes to the death of hippocampal neurons in vivo: role of microglial NADPH oxidase. J Neurosci. 2005;25:4082–90.
67. Park KW, Baik HH, Jin BK. Il-13-induced oxidative stress via microglial NADPH oxidase contributes to death of hippocampal neurons in vivo. J Immunol. 2009;183:4666–74.
68. Tsopanoglou NE, Maragoudakis ME. On the mechanism of thrombin-induced angiogenesis. Potentiation of vascular endothelial growth factor activity on endothelial cells by up-regulation of its receptors. J Biol Chem. 1999;274:23969–76.
69. Haralabopoulos GC, Grant DS, Kleinman HK, Maragoudakis ME. Thrombin promotes endothelial cell alignment in matrigel in vitro and angiogenesis in vivo. Am J Physiol. 1997;273:C239–45.

70. Griffin CT, Srinivasan Y, Zheng YW, Huang W, Coughlin SR. A role for thrombin receptor signaling in endothelial cells during embryonic development. Science. 2001;293:1666–70.
71. Yang S, Song S, Hua Y, Nakamura T, Keep RF, Xi G. Effects of thrombin on neurogenesis after intracerebral hemorrhage. Stroke. 2008;39:2079–84.
72. Bae JS, Kim YU, Park MK, Rezaie AR. Concentration dependent dual effect of thrombin in endothelial cells via par-1 and pi3 kinase. J Cell Physiol. 2009;219:744–51.
73. Henrich-Noack P, Striggow F, Reiser G, Reymann KG. Preconditioning with thrombin can be protective or worsen damage after endothelin-1-induced focal ischemia in rats. J Neurosci Res. 2006;83:469–75.
74. Xi G, Keep RF, Hua Y, Xiang J, Hoff JT. Attenuation of thrombin-induced brain edema by cerebral thrombin preconditioning. Stroke. 1999;30:1247–55.
75. Jiang Y, Wu J, Hua Y, Keep RF, Xiang J, Hoff JT, Xi G. Thrombin-receptor activation and thrombin-induced brain tolerance. J Cereb Blood Flow Metab. 2002;22:404–10.
76. Mann KG, Butenas S, Brummel K. The dynamics of thrombin formation. Arterioscler Thromb Vasc Biol. 2003;23:17–25.
77. Liu DZ, Ander BP, Xu H, Shen Y, Kaur P, Deng W, Sharp FR. Blood-brain barrier breakdown and repair by src after thrombin-induced injury. Ann Neurol. 2010;67:526–33.
78. Guo H, Liu D, Gelbard H, Cheng T, Insalaco R, Fernandez JA, Griffin JH, Zlokovic BV. Activated protein c prevents neuronal apoptosis via protease activated receptors 1 and 3. Neuron. 2004;41:563–72.
79. Cheng T, Liu D, Griffin JH, Fernandez JA, Castellino F, Rosen ED, Fukudome K, Zlokovic BV. Activated protein c blocks p53-mediated apoptosis in ischemic human brain endothelium and is neuroprotective. Nat Med. 2003;9:338–42.
80. Cheng T, Petraglia AL, Li Z, Thiyagarajan M, Zhong Z, Wu Z, Liu D, Maggirwar SB, Deane R, Fernandez JA, LaRue B, Griffin JH, Chopp M, Zlokovic BV. Activated protein c inhibits tissue plasminogen activator-induced brain hemorrhage. Nat Med. 2006;12:1278–85.
81. Liu D, Cheng T, Guo H, Fernandez JA, Griffin JH, Song X, Zlokovic BV. Tissue plasminogen activator neurovascular toxicity is controlled by activated protein c. Nat Med. 2004;10: 1379–83.
82. Russo A, Soh UJ, Paing MM, Arora P, Trejo J. Caveolae are required for protease-selective signaling by protease-activated receptor-1. Proc Natl Acad Sci U S A. 2009;106:6393–7.
83. Jiang T, Olson ES, Nguyen QT, Roy M, Jennings PA, Tsien RY. Tumor imaging by means of proteolytic activation of cell-penetrating peptides. Proc Natl Acad Sci U S A. 2004;101: 17867–72.
84. Olson ES, Aguilera TA, Jiang T, Ellies LG, Nguyen QT, Wong EH, Gross LA, Tsien RY. In vivo characterization of activatable cell penetrating peptides for targeting protease activity in cancer. Integr Biol (Camb). 2009;1:382–93.
85. Olson ES, Jiang T, Aguilera TA, Nguyen QT, Ellies LG, Scadeng M, Tsien RY. Activatable cell penetrating peptides linked to nanoparticles as dual probes for in vivo fluorescence and mr imaging of proteases. Proc Natl Acad Sci U S A. 2010;107:4311–6.
86. Jaffer FA, Tung CH, Gerszten RE, Weissleder R. In vivo imaging of thrombin activity in experimental thrombi with thrombin-sensitive near-infrared molecular probe. Arterioscler Thromb Vasc Biol. 2002;22:1929–35.
87. Mosnier LO, Gale AJ, Yegneswaran S, Griffin JH. Activated protein c variants with normal cytoprotective but reduced anticoagulant activity. Blood. 2004;104:1740–4.
88. Wang Y, Thiyagarajan M, Chow N, Singh I, Guo H, Davis TP, Zlokovic BV. Differential neuroprotection and risk for bleeding from activated protein c with varying degrees of anticoagulant activity. Stroke. 2009;40:1864–9.
89. Guo H, Singh I, Wang Y, Deane R, Barrett T, Fernandez JA, Chow N, Griffin JH, Zlokovic BV. Neuroprotective activities of activated protein c mutant with reduced anticoagulant activity. Eur J Neurosci. 2009;29:1119–30.
90. Guo H, Wang Y, Singh I, Liu D, Fernandez JA, Griffin JH, Chow N, Zlokovic BV. Species-dependent neuroprotection by activated protein c mutants with reduced anticoagulant activity. J Neurochem. 2009;109:116–24.

Chapter 10
Toll-Like Receptor Agonists as Antecedent Therapy for Ischemic Brain Injury: Advancing Preclinical Studies to the Nonhuman Primate

Frances Rena Bahjat, Keri B. Vartanian, G. Alexander West, and Mary P. Stenzel-Poore

Abstract Antecedent therapy for ischemic brain injury has the potential to protect a large, high-risk patient population from the devastating effects of cerebral ischemia associated with cardiac surgery. Substantial evidence has shown that preconditioning with a modestly damaging stimulus induces powerful endogenous neuroprotection. Pharmacological agents that stimulate toll-like receptors (TLRs) induce robust neuroprotective effects as preconditioning stimuli against cerebral ischemia in mouse and nonhuman primate models of stroke. Here we describe the progress of our preclinical development of TLR agonists as antecedent therapy against cerebral ischemic injury. The objective was to discuss studies that begin with in vitro validation in cell cultures to in vivo efficacy studies using mouse and nonhuman primate models of stroke, with particular emphasis on the TLR9 agonist CpG oligonucleotide. We provide an in-depth discussion of our novel rhesus macaque stroke model and cover the progress we have made in therapeutic testing and evaluation in these animals. These studies represent a logical path for the

F.R. Bahjat, PhD • K.B. Vartanian, PhD
Department of Molecular Microbiology and Immunology, Oregon Health
and Science University, 3181 Sam Jackson Park Road, Portland, OR 97239, USA

G.A. West, MD, PhD
Colorado Brain and Spine Institute, Neurotrauma Research Laboratory,
Swedish Medical Center, Englewood, CO, USA

M.P. Stenzel-Poore, PhD (✉)
Department of Molecular Microbiology and Immunology, Oregon Health
and Science University, 3181 Sam Jackson Park Road, Portland, OR 97239, USA

Division of Neuroscience, Oregon National Primate Research Center,
Beaverton, OR, USA
e-mail: poorem@ohsu.edu

P.A. Lapchak and J.H. Zhang (eds.), *Translational Stroke Research*, Springer Series
in Translational Stroke Research, DOI 10.1007/978-1-4419-9530-8_10,
© Springer Science+Business Media, LLC 2012

development of TLR agonists as antecedent therapy for the prevention of the damaging neurological complications resulting from cerebral ischemic injury.

1 Introduction

Antecedent therapy for ischemic brain injury targets a specific patient population at a high risk of experiencing ischemia to the brain. To utilize antecedent treatment, there must be prior knowledge that a patient has an increased likelihood of experiencing a brain ischemia within a predictable timeframe. Two major patient populations in this high-risk category are those undergoing major cardiovascular surgical procedures [1, 2] and those with symptoms of a transient ischemic attack (TIA) [3]. Cerebral ischemia due to cardiovascular surgeries, such as carotid artery bypass grafting, occurs at a relatively high rate. One MRI study showed that new postoperative ischemic brain lesions were detectable in 43% of patients who underwent cardiac surgery [1]. Reports indicate that the majority of postoperative strokes occur within 2 days of surgery [2] and result in increased cognitive decline and likelihood of death [1, 2]. TIAs occur in 300,000 patients in the United States each year and research indicates that approximately 10% of TIA patients will suffer from a subsequent cerebral attack [3] within 2 days. Thus, the development of an antecedent therapeutic intervention to induce neuroprotection during this high-risk time window would be extremely beneficial.

Ischemic injury causes tissue damage and cellular death due to reduced vascular perfusion leading to significant pathology. Antecedent stroke therapies function to "precondition" the brain against neuronal damage initiated by an ischemic event. A notable and well-reported preconditioning effect is the application of a brief period of ischemia that does not induce significant damage in the target tissue. Brief ischemia initiated prior to stroke has demonstrated robust neuroprotection against ischemic injury damage in animal models [4]. However, applying a regimen of brief cerebral ischemia lacks clinical applicability as it carries significant risk and currently requires an additional surgical procedure to implement. Importantly, advances in the use of remote peripheral ischemic preconditioning (RIPC), whereby ischemic preconditioning is applied to a nontarget tissue, has shown some success in human studies. Several human clinical trials have been conducted to evaluate RIPC in patients undergoing open abdominal aortic aneurysm repair, coronary artery bypass graft procedures, and pediatric patients undergoing congenital heart defect repair. These studies demonstrated decreased myocardial injury during ischemia/reperfusion (I/R) as a result of RIPC performed before the procedures, as measured by serum cardiac markers. Additionally, renal protection was also observed in one RIPC human study, suggesting that multiple organs may be protected simultaneously in the context of preconditioning. Thus, it has been demonstrated that preconditioning can be successful in protecting various organs in human patients, validating the underlying concepts of antecedent therapy to promote protection against ischemia.

Alternative pharmacological approaches to preconditioning have emerged that also demonstrate robust neuroprotective properties against ischemia with the added benefit of less invasive administration. Among these, a promising role for Toll-like receptor (TLR) agonists has been described. TLRs are a family of evolutionarily conserved receptors that are traditionally associated with regulation of innate immunity. TLRs are activated by their cognate pathogen-associated molecular pattern to induce inflammatory and type I IFN-associated responses. Interestingly, low-dose treatment with the TLR4 agonist, lipopolysaccharide (LPS), acts as a preconditioning agent in the setting of multiple types of injurious events including LPS toxicity [5] and heterologous injury, such as ischemia [6, 7]. LPS was first described as a preconditioning agent in cerebral ischemia by Tasaki et al. wherein studies identified robust neuroprotection in a rat model of cerebral ischemic injury [6]. While the precise molecular mechanisms governing LPS-induced protection are unclear, it is currently believed to represent a form of "tolerance" [8] which is a hyporesponsive state induced by low-level activation of TLR signaling [5]. When tolerance is induced, such as via low-dose treatment with LPS, the response to a subsequent exposure to an injurious signal is dampened. This is most often associated with reduced NFkappaB activity, attenuated secretion of proinflammatory cytokines, and enhanced protective type I IFN signaling [9, 10]. Interestingly, NFkappaB activation and a robust proinflammatory response are triggered by ischemic injury, and as such, are considered major therapeutic targets to attenuate injury. Investigation into the mechanism of LPS-induced neuroprotection reveals a significant shift in gene regulation in the brain following stroke towards a dominant type I IFN response which suggests gene expression may be redirected away from injurious NFkappaB signaling [11]. Thus, preconditioning with LPS may serve as a therapeutic approach by reducing damaging responses, such as NFkappaB activation, induced by ischemia.

Since the identification of LPS as an effective preconditioning agent against subsequent cerebral ischemia, there have been several interesting connections between TLR signaling and the pathophysiology of stroke. Research indicates that mice deficient in either TLR4 or TLR2 have significantly reduced infarct sizes in response to permanent or transient middle cerebral artery occlusion (tMCAO) [12–14]. Additionally, a recent study in humans also linked increased expression levels of TLR2 and TLR4 on hematopoetic cells with worsened outcome in stroke [15]. This suggests that TLR2 and TLR4 are activated by endogenous danger signals, known as damage-associated molecular patterns, which are released during ischemia. Thus, by preconditioning an animal with LPS, it is likely that a state of tolerance is induced that dampens the damaging pathways of TLR4 and TLR2 following stroke. This link emphasizes the therapeutic potential of TLR4 and TLR2 as inducers of tolerance to stroke. However, the toxic and pyrogenic nature of LPS, a cell wall component of gram-negative bacteria, impedes effective clinical translation into humans. Thus, an alternative preconditioning agent must be used to target and redirect damaging TLR4 and TLR2 signaling in response to stroke.

A unique feature of TLR agonists is their ability to induce cross-tolerance, a mechanism whereby the responsiveness of other TLRs are dampened due to prior exposure to a given TLR agonist [9]. This mechanism can be exploited to identify an innocuous TLR agonist able to induce tolerance against the detrimental TLR4 and TLR2 signaling that occurs in response to stroke. Several TLR agonists recognized by TLR2, TLR4, or TLR9 have been shown to protect against ischemic injury in a mouse stroke model [6, 16, 17]. Specifically, preconditioning with these TLR agonists 3 days prior to stroke resulted in reduced infarct volumes. One such promising therapeutic agent is unmethylated CpG oligodeoxynucleotides (CpG ODN), a TLR9 agonist that has demonstrated the ability to cause cross-tolerance of TLR4 and TLR2 which dampens their proinflammatory signaling capacity [9]. This suggests that TLR signaling is a potential target for antecedent stroke therapy.

In this chapter, we will describe the evolution of our preclinical development program involving TLR agonists as antecedent therapy against cerebral ischemic injury or stroke. Our process for the evaluation of TLR agonists extends from in vitro cell-based assays to in vivo animal models of stroke. It is prudent to test a drug candidate in more than one animal model and across multiple species to maximize the predictive value with respect to efficacy and safety, Such considerations are reflected in the set of recommendations by the Stroke Therapy Academic Industry Round Table (STAIR) [18], which included the preclinical efficacy testing of putative neuroprotective agents in nonhuman primates prior to clinical studies in man. Thus, we have developed a unique nonhuman primate (NHP) model of cerebral ischemia to advance our preclinical studies of antecedent treatment with TLR agonists. This research approach sets the stage for the translation of basic science research discoveries to the realm of therapeutic testing and potentially on to clinical applications.

2 In Vitro Target Validation Using a Cell-Based Model of Ischemic Injury

A typical first step in the evaluation of a preclinical therapeutic candidate involves in vitro cell-based assays. In vitro screening assays are developed to predict the effectiveness of a prospective drug on a given biological target and evaluate potential toxicities in mammalian cells.

In vitro evaluation of potential therapies involves the development of a model that closely mimics the in vivo system or a particular cellular response elicited in vivo. Major considerations in creating in vitro models include the selection of appropriate target cells and the creation of physiologically relevant environmental conditions. Modeling stroke in vitro involves isolation and culture of cells of the central nervous system (CNS), which are the ultimate cellular target in stroke. CNS cells are then exposed to an environment that mimics ischemia by depriving the cells of oxygen and glucose. Oxygen and glucose deprivation can be administered prior to or following treatment with potential therapeutics to evaluate their protective potential through the evaluation of cell death.

2.1   Target Cells

Cell-based assays can be implemented using commercially available immortalized cell lines or primary cells isolated directly from the tissue of interest. While both systems are valuable, primary cells more accurately emulate the in vivo system because the cells retain more of the lineage-specific properties associated with the in vivo environment, such as expression of certain receptors, that are lost with the long-term maintenance and passage of cell lines in culture.

Primary CNS cells can be isolated from microdissected brain tissue and cultured for in vitro assays. Primary cortical cells are derived from the cortical tissue of late-stage embryonic rodents, typically rat or mouse. Primary cortical cells consist of multiple cell types within one culture including neurons, astrocytes, and microglia, offering a distinct advantage over cultures of a single cell type because these mixed cultures more accurately reflect the cellular composition of cortical tissue with interactions between cells and their secreted products. This is particularly important in screening a potential therapy as individual cell responses and complex cross-talk between cell types can affect outcome.

2.2   Oxygen-Glucose Deprivation

Blockage of blood flow in ischemia deprives tissue of nutrients and oxygen. To model this in vitro, exclusion of glucose and oxygen in a controlled environment is required. Cell culture media devoid of glucose can be used in combination with oxygen deprivation using an incubation chamber that maintains an anaerobic environment (85% N_2, 10% CO_2, 5% H_2) and physiological temperature (37°C). Thus, by exposing cells to these precisely controlled conditions, it is possible to mimic the in vivo environment of stroke in an in vitro setting.

3   In Vitro Screening of TLR Agonists Against Modeled Ischemia

The efficacy of antecedent treatment of primary cortical cells with TLR agonists was evaluated using the in vitro OGD model. The cell types represented in the primary cortical cell cultures (i.e., astrocytes, neurons, microglia) express TLRs, and thus, all are potential target populations. The TLR expression profile on each cell type is variable; activation of one cell type ultimately affects other cell types in the culture. In our studies, primary cortical cells were treated with a TLR agonist 24 h prior to exposure to 3 h of OGD. Protection against cell death was evaluated by quantifying dead cells based on the ratio of propidium iodide (a fluorescent molecule taken up exclusively by dead or dying cells) to 4′,5′-diamidino-2-phenylindole

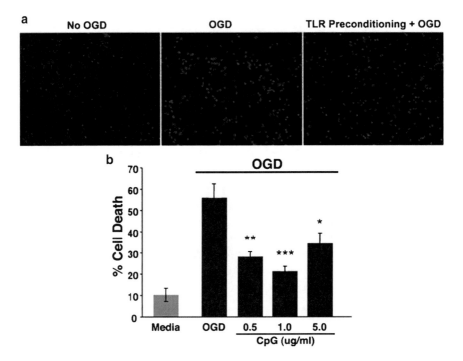

Fig. 10.1 TLR ligands protect CNS cells against ischemic injury in vitro. (**a**) Representative images of primary cortical cells stained with PI (*pink*) and DAPI (*blue*). (**b**) Quantification of the percent of cell death in cortical cells preconditioned by prior exposure to varying doses of CpG (TLR9 agonist) followed by exposure to 3 h OGD (previously published [17]) and subsequent normoxic conditions for 24 h

(DAPI, a nuclear stain) stained cells. The results indicated that pretreatment with TLR agonists effectively protected the primary cortical cells from OGD-induced cell death (Fig. 10.1). To evaluate the potential toxicity of TLR agonists, the highest effective concentration of the TLR agonist can be applied to the primary cortical cells to evaluate cell death. None of the TLR agonists have demonstrated toxicities at the effective preconditioning doses, indicating that TLR agonists are not directly toxic to the representative CNS cell populations at these protective concentrations. These results suggest that preconditioning CNS cells with TLR agonists may provide a potentially safe, effective regimen for antecedent stroke therapy.

3.1   Limitations of In Vitro System

While in vitro testing of cell-based assays is an informative first step, a major limitation of the in vitro paradigm for drug screening is that it is a closed system wherein no other tissues or organs are affected by the administration of the drug. The absence of these important aspects could dramatically alter the therapeutic outcome and could also mask systemic drug toxicities. Therapeutically targeting the CNS presents a major hurdle because of the existence of the blood-brain barrier, an aspect that this in vitro model does not recapitulate. Additionally, although primary culture of neonatal neuronal cells has been widely used to study the mechanisms of hypoxia-induced injury, differences likely exist between neonatal and adult cells. Also, depending upon the location of the specific drug target (e.g., peripheral circulation, CNS, microvascular endothelial cells), in vitro systems may not be practical or informative as to the potential in vivo drug efficacy. In vivo models are exceedingly beneficial as they begin to address these major deficits inherent to in vitro preclinical research.

4   In Vivo Animal Models: An Important Component of Preclinical Development

Many different strategies have been devised to develop novel therapeutics for a host of human conditions. Common among these approaches is their dependence upon animal models of disease to evaluate potential clinical candidates for both safety and efficacy. Importantly, all pivotal preclinical data required by the regulatory agencies involved in drug approvals must be obtained in mammalian systems. As such, animal model studies play a key role in the design of clinical trials and are a crucial component in reducing late-stage attrition due to lack of efficacy of a drug for unforeseen safety issues. Animal testing can be highly informative for risk assessment purposes (i.e., assessing the risk of taking a drug to market for a particular clinical indication) even when using animal models whose clinical predictive power is unproven. This is especially true with respect to safety, as lessons learned from animal studies in multiple species and in disease-specific contexts can dramatically reduce the risk of administering doses that may be unsafe in humans. Animal studies can identify appropriate efficacious dosing strategies in the examined disease state, as well as identify the highest dose resulting in no adverse effects (NOAEL). First-in-humans trials are typically conducted using doses that are projected based upon these types of results from preclinical animal modeling and thus studies of this type can be quite informative.

4.1   Modeling Human Stroke

Many characteristic elements of human ischemic brain injury can be modeled in animals using various experimental approaches (see reviews [19, 20]). The most common type of human stroke is focal ischemia due to occlusion of the middle cerebral artery (MCA) that provides blood flow to the brain [20]. Occlusion of the MCA results in changes in the hemodynamics of the brain whereby perfusion within the vascular territory of the occluded artery varies so that some regions of the brain receive relatively no blood flow. Following artery occlusion and subsequent reperfusion in humans and experimental animals, a progressive ischemia-induced pathology evolves. The early phase involves the development of a core zone of severe ischemia that represents irreversibly injured tissue. Late-phase injury develops in the ischemic penumbra, which is an area sensitive to eventual infarction without restored blood flow. This penumbra region is a major target for stroke therapies as it is considered a salvageable region of brain tissue most likely to be protected by pharmacological, thrombolytic, or mechanical (recanalization) intervention. Thus, it is critical that cerebral ischemia modeled in animals also contain a core and penumbra infarct zone to obtain physiologically relevant data on targeting these distinct features of stroke-induced brain damage.

4.2   In Vivo Modeling of Cerebral Ischemic Injury in the Mouse

Rodent models are often the first system used for establishing in vivo efficacy. These models are readily employed because of their ease of use, affordability, and high-throughput capacity. Most commercially available rodent strains are inbred and thus are essentially genetically identical, a feature that decreases variability in the model that may arise from genetic differences between individual animals. While not representative of the heterogeneity existing in the human population, genetically comparable animals are more efficient and effective for target validation or initial evaluation of candidate therapeutics. An important advantage in the mouse model is the extensive availability of tools for evaluating biological responses following the application of a drug, the implementation of disease model, or other manipulation. Furthermore, the use of genetically altered strains of mice, such as mice that are deficient in a select gene, is invaluable for the further study of disease pathology, drug–target interactions, and identification of therapeutic mechanisms of action.

There are several in vivo mouse models of stroke that have been developed to reflect different types of human stroke. Focal ischemia, the most common type of human stroke, is often modeled using tMCAO, which utilizes the temporary insertion of a silicone-coated surgical suture to unilaterally block blood flow to the MCA. The progression of ischemic brain damage induced through this method involves early- and late-phase infarction that creates a core and penumbra zone, similar to that seen in humans. In mice, the early-phase damage that creates the infarct "core" occurs in the striatum through rapid tissue necrosis [21]. The late-phase infarct that creates the "penumbra" occurs in the cortical tissue primarily due to apoptosis, programmed cell death, and secondary mediators of ischemic damage such as inflammation [21]. Additionally, the transient nature of this model allows for the modeling of reperfusion injury that is common in human stroke.

Control of the duration of ischemia allows the establishment of reproducible infarcts with relatively predictable core and penumbral damage. Furthermore, the tMCAO induces neurological deficits in mice that can be subjectively quantified by a blinded investigator. Using defined graded scales that assign a score for appearance, movement, and behavior based on specific observable features such as body symmetry, cleanliness of fur, and spontaneous activity, observers can evaluate changes in behavioral and motor skills associated with neurological outcome in the presence and absence of drug treatment. This neurological evaluation is often deemed a necessary part of assessing therapeutic efficacy due to its perceived clinical relevance to human stroke patients.

4.3   Evaluation of TLR Agonists in Mouse tMCAO

Key components of in vivo evaluation of a potential stroke therapeutic include determination of the dose, time window of effectiveness, and route of administration. The use of in vivo models for this line of investigation is crucial because the in vivo models incorporate drug behavior and the responses elicited by other tissues. The in vivo system also accounts for the role of the BBB in drug distribution and efficacy. For example, systemic administration of CpG ODN will activate TLR9 receptors found on hematopoietic cells such as neutrophils and dendritic cells (DC). The introduction of these hematopoietic cell responses (i.e., cellular activation or secretion of cytokines) in the context of drug treatment could alter the response to the drug or be responsible for additional efficacy.

Unlike traditional approaches to drug prevention that aim to ensure drug is present at the time of injury, typically preconditioning agents are delivered several days prior to the ischemic event to allow time for the protective effects to develop. Due to the mechanisms involved in preconditioning with TLR agonists, the drug is not required to be present at the time of the insult to elicit neuroprotective effects. To evaluate the effects of antecedent TLR stimulation in mice, TLR agonists were administered 3 days prior to initiating a tMCAO. Evaluation of the infarct was performed 24 h following tMCAO using 2,3,5-triphenyltetrazolium (TTC) staining (which reacts with live tissue and staining a red color). The results showed that preconditioning with a TLR agonist dramatically reduced infarct size (Fig. 10.2a, b). This reduction is dose-dependent when given 1–3 days prior to stroke. The reduction in infarct size correlated with improved behavioral outcome (Fig. 10.2c).

To determine the time window of effectiveness, the lowest effective dose of a TLR agonist is typically administered at varying times prior to tMCAO. The protective time window for TLR agonist administration generally extends from 1 to 3 days, prior to the tMCAO, as demonstrated by reduced infarct sizes in these animals [17]. This is a feasible time window for translation to clinical studies as this protective window falls within the timeframe most likely seen with surgical or TIA patients at risk of exposure to subsequent ischemia.

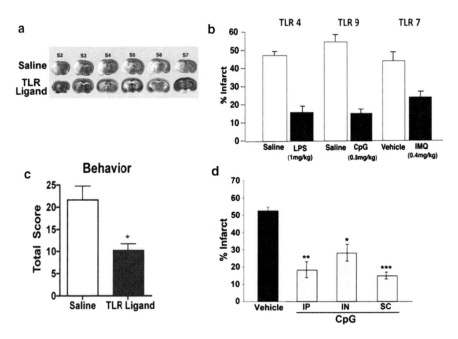

Fig. 10.2 TLR ligands provide protection against ischemic injury in a mouse model of stroke. (**a**) Representative TTC stained coronal brain sections from animals given saline or exposure to a TLR ligand prior to MCAO. (**b**) Quantification of percent infarct for mice preconditioned with TLR ligands for TLR4, TLR9, or TLR7 at 24 h post-MCAO show dramatic reduction in damage compared to controls. (**c**) Total behavioral score for mice subjected to preconditioning with a TLR ligand prior to MCAO is significantly improved compared to controls. (**d**) The TLR9 ligand CpG administered intraperitoneally (IP), intranasally, (IN), or subcutaneously (SC) provides protection against ischemic injury, demonstrated by reduced infarct size

Route of drug administration is always an important consideration for preclinical evaluation of novel therapeutics. The preferred route of administration needs to be convenient and simple to implement with minimal patient discomfort. Intraperitoneal (IP) administration was utilized for the majority of the murine efficacy studies. This is an effective dose route for research purposes, but the IP route is impractical for human studies as it can be cumbersome and painful. Two more practical routes of delivery are intranasal administration and subcutaneous injection. Drugs administered intranasally (IN) are rapidly absorbed and transported through the cell and capillary rich mucosal surface by olfactory neurons and the cerebral spinal fluid [22]. Subcutaneous (SC) injection delivers the drug just beneath the skin and demonstrates robust and predictable absorption through the capillaries. These systems are advantageous due to their ease of use, typically high drug bioavailability, and relatively pain-free application.

To determine the optimal clinically relevant route of administration, the neuroprotective efficacy of a TLR9 agonist CpG ODN delivered intranasally and subcutaneously was evaluated, as compared to IP injection. The results indicated that all drug delivery routes were successful in reducing the mean infarct size; however, the subcutaneous route was more effective than intranasal (Fig. 10.2d). Efficiency in these clinically preferred methods of drug administration is favorable and the resultant data are informative for preclinical evaluation in the nonhuman primate.

4.4 Limitations of the Mouse Model

Mouse models routinely provide investigators with important insight into the understanding of the pathophysiology of stroke. Reproducible rodent models of stroke are instrumental in the discovery and development of novel therapeutics. These models provide a platform for efficient optimization of lead compounds based on overall efficacy and on-target potency, as well as a first look at in vivo pharmacokinetics (PK) and pharmacodynamics (PD). Although the response after cerebral ischemia/ reperfusion (I/R) injury in mice shares many common characteristics with higher mammalian species and is an excellent model for exploration of cellular and molecular mechanisms, there are potential limitations when extrapolating data from mice to humans.

Despite demonstrations of reproducible efficacy in rodent models, many recent efforts to translate preclinical neuroprotective strategies from rodents to humans have been disappointing [23]. The lack of congruency observed between rodent and human therapeutic responses could potentially be attributed to species-specific differences including vast differences in brain tissue structure, composition (such as white matter), and tolerance for ischemia, as well as variation in drug–target interaction, target expression, or target distribution. Additionally, drugs often exhibit species-dependent properties with regard to in vivo PK, PD, or precise mechanism of action. These limitations have prompted the development and implementation of NHP models for evaluation of stroke therapeutics.

5 Modeling Cerebral Ischemic Injury in the Nonhuman Primate

Because of the profound failure of neuroprotective drugs developed in rodents as they are advanced to human trials, there is a new impetus among researchers to include preclinical testing of potential neuroprotective candidate drugs in nonhuman primates. NHP models offer significant advantages for preclinical testing of candidate neuroprotective agents. Old-world monkeys such as macaques and baboons have a close phylogenetic relationship to humans, and therefore, are more likely to share similar endogenous mechanisms of ischemic injury and neuroprotection compared to rodents. NHPs are more closely related to humans physiologically as they have similar neuroanatomy, vasculature [24], and gyroencephalic morphology. Humans and NHPs also have a similar ratio of gray to white matter with a mean GM/WM ratio of ~1.4 depending upon age and sex of the subject [25], which yields similar thresholds to ischemic injury [26]. The complex behavior of these animals allows for a better assessment of neurological or behavioral deficits than that of rodent models. The use of clinically meaningful neurological scales provides a functional outcome measure in NHP models of cerebral ischemia that more accurately reflects the human condition in stroke. Finally, studies performed in NHPs allow for better feasibility assessment of drug development candidates. Studies in NHP can provide PK/PD profiles for drugs, as well as information regarding efficacious and toxic dose ranges, data that can be more applicable to human subjects. As such, NHP studies potentially offer an important translational bridge to clinical studies by providing a pragmatic model for target validation, efficacy testing, and optimization of novel therapeutic strategies prior to clinical studies in man.

5.1 Challenges with Current Surgical NHP Models of Cerebral Ischemia

There are several primate models of ischemic brain injury, most of which involve occlusion of the MCA as in the rodent models. In the NHP, MCAO is accomplished via the use of aneurysm clips, thrombus, or by embolization. Most models of stroke in NHP require significant neurosurgical proficiency to implement. To gain access to the MCA, craniotomy is typically performed, but due to the very thick overlying temporalis muscle and the required brain retraction, this route can cause mechanical injury to the brain and thus greater variance in the overall outcome. Thus, MCAO models in monkeys often produce infarcts that vary in location and size [27, 28]. Further, to induce damage beyond the basal ganglia, extended durations of ischemia are necessary. Unfortunately, durations of occlusion that result in cortical involvement also produce a combination of cortical and subcortical damage often associated with high morbidity and mortality. Collateral flow to the region from the anterior cerebral arteries (ACA) appears to contribute to the resilience of the ipsilateral cortex to damage induced by MCAO in the monkey. In the baboon, occlusion of both ACAs and the internal carotid artery (ICA) does induce cortical damage; however, there is also marked basal ganglia damage, which carries a high risk of mortality and often requires extended postoperative care [29, 30].

5.2 MCA/ACA Occlusion in the Nonhuman Primate: A Practical Model for Preclinical Development of Antecedent Therapies

A model involving a transorbital, reversible, two-vessel occlusion approach to focal ischemia-induced stroke in male rhesus macaques was recently described by West et al. [31]. This approach involves the simultaneous occlusion of the right MCA (distal to the orbitofrontal branch) and both ACA using vascular clips. There are apparent differences between mouse, monkeys, and humans in how occlusion

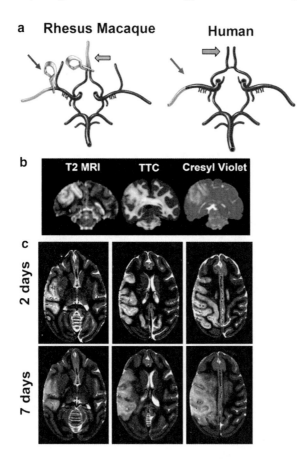

Fig. 10.3 MCA/ACA occlusion model in rhesus macaques. (**a**) Circle of Willis in rhesus macaques and humans. MCA in humans shown with a unilateral blood clot (*purple arrow*). Anterior cerebral arteries in humans (ACA; top of picture, *green arrow*) are separated into left and right branches. The macaque vasculature is similar, but the ACAs are fused (*blue arrow*). Aneurysm clamps are applied unilaterally to the MCA (*red arrow*) and bilaterally across the ACAs to produce the reversible two-vessel occlusion in rhesus macaques. (**b**) Comparison of infarct between T_2 weighted image, TTC staining, and Cresyl Violet staining at 2 days post-MCA/ACA in rhesus macaques. Results indicated comparable infarct sizes between these three measurement techniques. (**c**) T_2 images collected at 2 and 7 days after 60 min of MCA/ACA occlusion. Three different axial slices from the same animal at approximately the same levels reveal the transition from well-defined edema in cortical regions at day 2, progressing to a more diffuse distribution at day 7. Postmortem analysis suggests that the T_2 scan at day 2 accurately assesses the volume and extent of overall damage

affects stroke evolution and functional outcomes based on differences in vascular architecture. In the monkey, bilateral occlusion of the ACA is critical for reducing collateral flow to the ipsilateral cortex, which facilitates marked ischemia-induced focal cortical damage. Specifically, the two proximal ACAs in the rhesus macaque merge to form a single pericallosal artery [24], allowing for collateral flow in ACA and presumably more protection to the cortical region during occlusion (Fig. 10.3a).

The MCA/ACA occlusion approach, rather than the MCAO approach used in mice, aims to eliminate this protective effect and produce a phenotype more similar to human MCA stroke. This injury model affects cortical regions primarily which attempts to mimic the injury observed in a subclass of thrombotic stroke in humans resulting from M1 occlusion of the MCA artery. This model offers a relatively homogeneous I/R injury phenotype for studies of cortical injury and neuroprotection. The major advantages of this approach are that it is less invasive than a craniotomy and results in less residual tissue injury and thus less variability.

Another advantage of the MCA/ACA two-clip model is that subcortical or striatal damage is minimized compared to most other models of MCA occlusion. The overall morbidity and mortality is significantly better than other reported models. Importantly, manipulating animals that are functionally compromised in order to evaluate the extent of motor deficit resulting from MCA/ACA occlusion carries significant risk to those individuals that must handle these animals. Minimizing the damage in these anatomical regions ensures normal mentation in most animals, thus reducing the risk of injury to the observer. This aspect of the model is also important for minimizing postoperative critical care and essential for the performance of practical neurological evaluations with a minimal need for direct hands-on manipulation of animals.

5.3 Quantification of Brain Infarct Volume

Infarct volumes resulting from MCA/ACA occlusion in the rhesus macaque can be quantified using magnetic resonance imaging (MRI) or direct examination of brain tissue via traditional histological staining approaches [31]. Notably, the area of infarct measured by TTC and cresyl violet staining has been shown to closely corroborate the accompanying MRI results, such that either method alone is sufficient (Fig. 10.3b). The staining techniques require tissue sectioning and therefore only allow for a view of damage at one point in time. Tissue that is considered "live" due to staining at a particular time point may, in fact, eventually succumb hours or even days later. Thus, MRI approaches may offer an advantage for identification of infarcted tissue and for evaluation of changes over time.

The volume of infarcted brain tissue typically amounts to ~22% of ipsilateral hemisphere or ~45% of total cortical area in Chinese adult male rhesus macaques (aged 6–12 years and 6–12 kg body weight) given a 60-min MCA/ACA occlusion. T2 MRI scans were collected at 2 and 7 days (Fig. 10.3c) after 60 min of MCA/ACA occlusion revealing well-defined edema at day 2 and more diffuse distribution of edema by day 7 poststroke in NHPs. Postmortem analysis confirmed that the T2 scan at day 2 accurately assesses the volume and extent of overall damage; therefore, day 2 post-stroke T2-weighted scans are typically used to quantify infarct volume using manual tracing of the apparent infarct area. In addition, the volume of infarct closely correlates with the resulting neurological findings described below. While the volume of the infarct typically correlates well with neurological and histopathological outcomes in untreated animals, the possibility exists that these parameters will not always correlate depending upon the drug target.

5.4 Quantification of Focal and Global Neurological Deficits

5.4.1 Neurological Deficits

Neurological deficits observed following MCA/ACA occlusion in the rhesus macaque include deficits in motor activity and changes in behavior. Hemiparesis is commonly observed following stroke, although a wide range of motor deficits can be seen in both humans and experimental animals. Quantification of the extent of motor deficit following stroke can therefore be problematic and has been historically limited to the use of subjective neurological scales. The extent of brain injury induced by a one hour MCA/ACA occlusion typically results in motor deficits that are readily apparent immediately after postoperative recovery in the adult male rhesus macaque. By producing a reliable pattern of damage involving the primary motor cortex, this model also demonstrates readily perceivable focal neurological deficits including paresis or plegia of primarily the upper extremity and face on the contralateral side from the surgical occlusion. Focal deficits typically extend more proximally with increasing infarct severity.

Deficits in the lower limb are more variably induced and typically involve paresis, with full paralysis being infrequently observed only in the most severely affected animals consistent with involvement of the lateral primary motor cortex. Improvements in muscle strength can be observed in proximal upper and lower limbs, but distal function including grip strength or coordination tends not to improve following stroke without therapeutic intervention during 7-day observation. Anecdotal observations during recent drug efficacy studies suggest that improvement over time in proximal and distal limbs is possible in this model, and this may be an important outcome measure for consideration in future studies with longer durations of observation poststroke (i.e., 14–28 days).

Facial deficits are commonly observed in this model and resemble those found in human stroke. Recent anatomical characterization shows that rhesus macaques have nearly identical facial musculature relative to humans, as evidenced by the comparative facial communications [32]. The ability of animals to exert facial expressions in the form of silent bared-teeth or relaxed open-mouth displays are often compromised following occlusion in this model. Such altered facial expressions when compared to assessments made prior to stroke are indicative of weakness or paralysis of facial muscles, a phenotype similar to that observed in many human stroke patients. Thus, assessments of facial movement and function can be used as a measure of focal neurological deficit in the MCA/ACA occlusion NHP model. A unique feature of rhesus macaques and other Old-World monkeys is that they have cheek pouches, a pocket-like fold of skin where food is hoarded for later consumption. The ability to clear food from the cheek pouch requires functional facial muscles, so the assessment of the size of the pouch (i.e., size of a walnut vs. an orange) and ability of the animals to chew and manipulate food with their mouths provides an additional feature to assess facial deficits.

Motor deficits are easily identifiable in the rhesus macaque, whereas changes in alertness or cognition are not as readily ascertained due to the demeanor of this species as compared to baboons or humans. Rhesus macaques exhibit fewer manipulative behaviors in the presence of human observers as compared to baboons, making it difficult to appreciate changes in behavior poststroke [33]. These animals require substantial acclimation to human observers prior to enrollment onto a preclinical study. A specific assessment of individual behaviors prior to stroke is necessary to determine the demeanor and method of assessment that best fits a given animal. Some animals are not motivated by treats or have particular preferences and this information can be useful in order to more accurately assess their behavioral status poststroke.

5.4.2   Implementation of Neurological Scales

The Spetzler scale was developed in 1980 to evaluate neurological deficits following reversible cerebral ischemia in the baboon [34]. The Spetzler scale has since been further validated in stroke studies in other NHP models [35]. Using a similar strategy, the Spetzler neurological scale has been tailored for the examination of rhesus macaques, since this species requires specific manipulations to adequately assess deficits when an observer is present. Prestroke evaluations using positive stimuli in the form of treats are performed to acclimate rhesus macaques to the presence of observers and note any atypical individual behavioral characteristics. Poststroke, we then perform similar direct and indirect stimulations in order to elicit specific motor functions for evaluation. Daily evaluation of NHPs following MCA/ACA occlusion reveals varying degrees of motor and behavioral deficits that correlate with the total volume of infarcted tissue.

Clinically, neurological scales differ in their emphases on specific deficits and often are insensitive to subtle changes in specific motor function [36]. In NHPs, neurological scores assessed using a modified Spetzler scale correlate well with infarct volumes of control animals in this rhesus macaque MCA/ACA occlusion model (Fig. 10.4a) [31]. However, the Spetzler scale is fairly insensitive to small differences in motor function due to the inclusion of limited functional categories. Additionally, the Spetzler scale is designed to assign a single score as a composite score comprising upper and lower limb deficits. Our typical stroke outcomes in preclinical studies testing often include atypical frequencies of upper and lower limb motor deficits. Moreover, varying degrees of functional capabilities are often observed and may be reflective of drug efficacy. These functional differences are not distinguishable using the current iteration of the Spetzler scale in this rhesus macaque model. To accommodate the observed changes in the functional outcomes in our model, we have developed more refined neurological scales including a simplified scale that scores the upper and lower limb independently. A third scale has

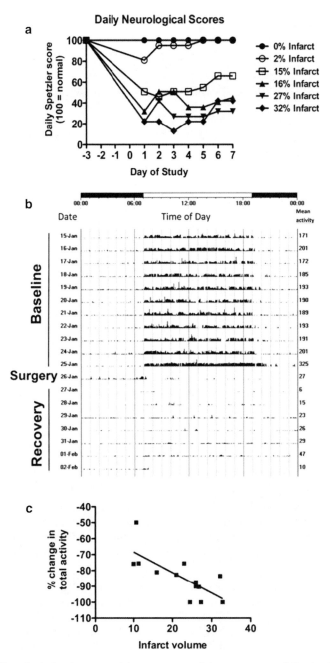

Fig. 10.4 Neurological and motor activity assessment of rhesus macaques following MCA/ACA occlusion. (**a**) Daily Spetzler neurological scores over 7 days following a 60-min MCA/ACA occlusion. Data are shown for animals having a range of infarct volumes. Greater infarct volumes led to greater reductions in motor abilities (lower neurological scores). (**b**) Actograms from a male rhesus macaque. Each row represents 1 day of activity recording over time (*top* to *bottom*; white bar = lights on, black bar = lights off). (**c**) The percent change in total activity compared to prestroke values was calculated using the following equation: (mean poststroke activity$_{\text{Jan 21–Jan 24}}$/mean prestroke activity$_{\text{Jan 29–Feb 1}}$ × 100)−100

been developed that more comprehensively captures behavior and motor function. The later two scales in our model are similar to the widely used NIH Stroke scale used in humans. To date, our experimental results suggest that these novel scales result in similar overall conclusions as results derived using the Spetzler scale. Although the possibility always exists that for a given drug target for stroke, one or more scales may be more effective at capturing efficacy and the use of multiple neurological assessments should be considered.

5.4.3 Actigraphy: An Objective Neurological Assessment

While sophisticated neurological scoring systems have great value, there are obvious limitations providing a discontinuous view of the status of motor function of a subject and require significant training and expertise to implement. As discussed above, subjective neurological scales may obscure drug efficacy and thus limit our ability to translate novel therapies to humans. In light of this, other methods to assess motor deficits may be needed. In the search for objective measurements of motor activity, lightweight miniaturized accelerometer devices offer great promise.

Accelerometers are devices that monitor and record the intensity, amount, and duration of motion and produce a signal that is amplified, digitized, and stored as activity counts. Recordings can be conducted for days, weeks, or months depending upon the chosen frequency with which an activity measurement is logged, referred to as an epoch. When the recording period is complete, the stored data can be transferred to a computer for analysis. Data may be expressed graphically as actograms or reported numerically as total activity counts per epoch, thereby estimating total activity, daytime activity, and nighttime activity. Data derived from miniaturized accelerometers, such as the Actiwatch device, have been validated for use as objective measures of total physical activity in humans [37] and NHPs [38, 39]. Similar devices have been employed for the evaluation of the extent of motor impairment and to quantitatively assess rehabilitation and recovery outcome following stroke in humans. Studies in humans employ devices worn on the wrists of patients suffering from stroke, whereas rhesus macaque studies use accelerometers worn on collars and can only detect total movement rather than comparisons between affected and unaffected limbs. A major advantage of accelerometer activity monitoring is that it enables continuous objective evaluation of motor activity using a noninvasive, safe, and convenient method.

To evaluate activity in the MCA/ACA occlusion model, rhesus macaques are fitted with the Actiwatch Mini TM device (Respironics, Inc., Bend, OR) attached to a lightweight loose-fitting aluminum collar (Primate Products, Inc., Immokalee, FL) approximately 3 weeks prior to surgery. The Actiwatch mini device consists of an ultra lightweight piezoelectric accelerometer with a 32 kb memory capacity (as previously described [39]), which records movement in all directions with a force sensitivity of 0.05 g and a maximum sampling frequency of 32 Hz. Activity is monitored during the acclimation period, as well as after surgical occlusion for 1 week postsurgery. Activity is continually logged at 60-s intervals or epochs generating a wealth of information about minute-by-minute changes in the activity of the monkeys.

Actograms depicting the activity can also be drawn using Actiware Sleep software (version 3.4; Cambridge Neurotechnology Ltd) (Fig. 10.4b). Outcome measures including the mean total daily activity, mean daytime activity (defined as activity during the period between 0700 and 1900 h for our rhesus population), and mean nocturnal activity (activity between 1900 and 0700 h) can be calculated, as well as the ratio of daytime to nocturnal activity. In adult male rhesus macaques (age 6–12), baseline recordings show robust daytime activity with reductions at night. Following 60 min of MCA/ACA occlusion, activity drops dramatically (~80% reduction). Over the next 7 days of recovery following stroke, activity levels gradually increase, do not change, or decrease depending upon the stroke severity; however, activity levels do not typically recover to prestroke levels for affected animals. In general, unmanipulated animals show an average of ~13% change in day-to-day activity level based on our historical data. Therefore, the decrease of ~80% observed in animals subjected to experimental stroke is a significant decrease in activity and these changes correlate well with infarct volumes (Fig. 10.4c, $p < 0.004$ Pearson correlation).

One challenge in using actigraphy in experimental stroke research is that it is difficult to establish testing standards, given the variety of different technologies and algorithms for detecting movement and the absence of standardized units of activity measures across these technologies. Thus, it is not clear how actigraphic information will be consistently employed by stroke researchers for comparison of drug efficacy data or stroke severity in animals across experimental laboratories. We have found recently that significant correlations exist among infarct volume, neurological score, and activity data measured by the Actiwatch accelerometer device in our NHP stroke model. These findings argue that actigraphy may represent a quantitative objective measure of neurological findings that improves our ability to discern drug efficacy in experimental NHP stroke models in addition to the establishment of meaningful functional outcome scales.

6   Target Selection and Validation of TLR Agonists in the NHP

6.1   CpG ODNs: Repositioning an Existing Clinical-Stage Therapeutic

Echoing the words of pharmacologist Sir James Black, "The most fruitful basis for the discovery of a new drug is to start with an old drug." This preclinical development strategy is often referred to as drug repositioning, repurposing, or reprofiling. Therapeutics with attributes appropriate for repurposing typically have already exhibited in human studies the following features: (1) the desired biological activity without apparent toxicity, (2) reasonable pharmacokinetic and ADME attributes, and (3) known dose-limiting toxicities consistent with mechanism of action, when present. A candidate therapeutic of this type for use as antecedent therapy in cerebral ischemic injury is the TLR9 agonist, CpG ODN, which we have established as efficacious in our in vitro OGD model and in vivo mouse MCAO model.

Anti-sense ODN drugs not containing CpG motifs have been approved by the US FDA, establishing a regulatory pathway for this general class of drugs [40]. Clinical-stage CpG ODN therapeutics include ProMune™ (CpG 7909; Class B or K type), ACTILON™ (CpG 10101; Class C), QbG10 (D-type or Class A CpG ODN contained in bacteriophage Qb capsid; Cytos Biotechnology), and HEPLISAV™ (ISS 1018; Class B or K type). These agents have been evaluated for treatment of cancers and viral infection as single agents, in combination with standard of care, and as vaccine adjuvants.

6.2 Species-Specificity of CpG ODN Responsiveness

Mouse studies have revealed great potential for synthetic CpG ODNs to mitigate damage due to ischemic injury. While K-type CpG ODNs represent a class of molecules shown to be protective in our in vitro and in vivo mouse ischemic injury models, data were derived using CpG ODN 1826, a sequence known to preferentially stimulate mouse cells as opposed to primate cells. The relevance of efficacy data derived using this specific CpG ODN to treat humans was not known. Thus, in order to evaluate the clinical potential of this approach for treatment of human ischemic injury, the feasibility of using a specific CpG ODN sequence needed to be evaluated carefully.

Species-specific differences in the immune response to CpG ODNs exist; therefore, it is prudent to choose a CpG ODN that has the greatest clinical potential for translation to humans. The magnitude and nature of the immune response elicited by a given CpG ODN sequence are governed by variations in the distribution of the TLR9 receptor and in sequence recognition. Previous work by Klinman and colleagues revealed that the variation in the individual response of immune cells could be explained by differences in CpG ODN sequence and length [41]. Structural differences between various classes of CpG ODNs have been shown to stimulate distinct cell populations in a species-dependent manner. K-type and D-type ODNs used in our studies predominantly stimulate plasmacytoid dendritic cells (pDC) in primates and humans, while ultimately indirectly affecting additional bystander cell populations (i.e., monocytes, NK cells, and B cells) [42]. This heterogeneity in responsiveness to CpG ODN may affect the magnitude of efficacy in higher primates, which justifies validation of TLR-induced preconditioning in NHPs prior to initiation of human studies.

In light of these factors, we chose to evaluate two different mixtures of CpG ODNs in this NHP stroke model. Three different CpG ODNs of varying lengths and structures are combined to form each of the mixtures (referred to as K mix and D mix) used in our preclinical studies. These mixtures were selected based on their potential to broadly stimulate an immune response in peripheral blood cells from a majority of primates [43].

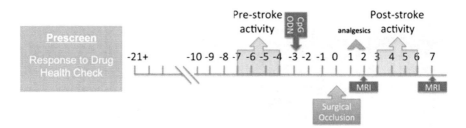

Fig. 10.5 Study timeline. Animals are selected for study based on their ability to respond to the drug in a prescreening test and overall state of health. Physical exams are performed by a veterinarian experienced with NHP physiology and a panel of standard clinical laboratory assays are performed. Animals are acclimated in home cages for ~3 weeks prior to subjecting them to experimental stroke and observers visit the animals frequently. The baseline Actiwatch activity is calculated using data from days -7 to -4. Surgical occlusion is performed on day 0 and animals are given analgesics for 2 days post-op. Daily neurological exams are performed beginning on the day after vascular occlusion and MR imaging is performed on day 2 post-op and at terminal sacrifice on day 7. Poststroke Actiwatch activity is calculated using data from days 3 to 6 to avoid analgesic effects

6.3   Evaluation of CpG as a Preconditioning Agent in the NHP

6.3.1   Selection of NHP: Health and Drug Responsiveness

An important factor in selecting appropriate animals for a study involves evaluating their potential responsiveness to treatment. CpG ODNs have many effector cell populations. In immune cell cultures, initial pDC activation will elicit IFNα-dependent and independent responses in B cells, monocytes, NK cells, T cells, etc. These bystander effects are usually easier to evaluate than direct effects of the drug on pDC that typically comprise less than 1% of circulating peripheral blood mononuclear cells (PBMC). Thus, subjects responsive to CpG ODNs were either selected for study according to induced expression of early activation antigen, CD69 and/or CD86, on CD20+ B cells (K and D mix) or CD69 on CD16+ NK cells (D mix). A functional assay evaluating total CpG ODN-induced IL-6 secretion (K and D mix) at 48 h following in vitro stimulation of PBMC also established responsiveness with similar results. Additionally, physical exams and clinical labs were used to evaluate the general health of each animal prior to inclusion on our studies. The timeline utilized for these studies, including the prescreening, preconditioning, stroke, and poststroke evaluation is diagrammed in Fig. 10.5.

6.3.2   Preclinical Efficacy of CpG in Nonhuman Primates

Our evaluation of a mixture of K-type (a.k.a. B-type) CpG ODNs as a prophylactic neuroprotectant in the MCA/ACA occlusion NHP model successfully showed reduced

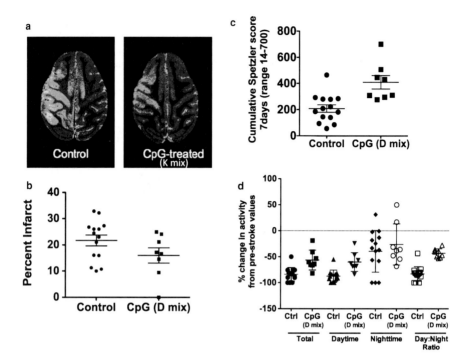

Fig. 10.6 Efficacy of CpG ODN in NHP stroke model. (**a**) Representative pseudo-colored MRI image of infarct from animals given CpG ODN (K mix) or saline prior to exposure to experimental stroke (as published in Bahjat et al. [44]). (**b**) Percent infarct was reduced by preconditioning with K or D mix. Percent infarct data for control and D mix treatment are represented in the figure (bar represents mean ± SEM). (**c**) Treatment with D mix ODNs also resulted in a significant improvement in neurological score, reflective of enhanced motor function as compared to control animals (bar represents mean ± SEM). Normal animals are given a cumulative score of 700 on the Spetzler scale. (**d**) Activities measured by Actiwatch devices in adult male rhesus macaques subjected to MCA/ACA occlusion. Data include activity 4 days prior to surgery compared to 4 days after surgery comprised of days following the cessation of postsurgical administration of analgesics. Symbols reflect the percent change from prestroke values for individual animals (bar represents mean ± SD). Saline controls (Ctrl) are compared to animals treated with D mix CpG ODNs

infarct sizes in response to cerebral ischemic injury [31] (Fig. 10.6a, b). Pretreatment of adult male rhesus macaques with a mixture of K-type CpG ODNs given prior to stroke reduced the mean infarct volume present 48 h following MCA/ACA occlusion [44]. To our knowledge, this is the first demonstration that a pharmacological preconditioning agent can protect against ischemic brain damage in the NHP. Additionally, our ongoing efficacy study testing a mixture of three different D-type ODNs (referred to as D mix) administered 3 days prior to stroke in NHPs shows a significant improvement in neurological scores and activity measurements (Fig. 10.6c, d).

Importantly, this efficacy data generated for CpG-ODN satisfy the Stroke Therapeutics Academic Industry Roundtable (STAIR) recommendations and provide a potentially valuable precursor to the advancement of these compounds to human clinical trials [45]. In addition, as recommended by the STAIR council, this work demonstrates efficacy in stroke models in two different species, mouse and monkey, further strengthening the potential for successful translation in clinical trials.

7   NHP Model Challenges and Additional Applications

7.1   Species Differences in Vascular Biology

There are apparent differences between monkeys and humans in how vascular occlusion affects stroke evolution and functional outcomes. Specifically, the two proximal ACAs in the rhesus macaque merge to form a single pericallosal artery [24], allowing for collateral flow in ACA and presumably more protection to the cortical region during occlusion. The MCA/ACA approach aims to eliminate this protective effect and produce a phenotype more similar to human stroke. In humans, the typical infarct size is small (5–14%) and these sized infarcts can be generated in NHPs. However, the large variability between animals and the limited dynamic range within which to discern efficacy prohibits the assessment of drugs in a setting that more accurately reflects smaller infarcts.

7.2   NHP Species Heterogeneity

Rhesus macaques have the widest population distribution of any nonhuman primate due to a high tolerance for a range of habitats. They are native to regions extending from Afghanistan (west) to China (east) and from Peking (north), through islands of Southeast Asia (south). Intraspecific genetic differences are known to exist among rhesus macaques of different countries of origin [46]. The genomic sequences of Indian and Chinese rhesus macaques are remarkably divergent [47]. The apparent species diversity may be a concern to the outcome of biomedical research, wherein a wide variety of regionally distinct populations are being used collectively as research subjects. It could be important to consider retrospective genetic analyses in the event a lack of congruency is observed between or among datasets. Alternatively, it could be argued that the diversity of rhesus macaques is more similar to that of humans and represents a more clinically relevant model system to test potential therapeutics.

7.3   Rhesus Macaque MCA/ACAO: A Model for Human Cerebral Ischemic Injury

Although there are some important differences between the rhesus macaque NHP ischemic brain injury model and the human phenotype, it is precisely those differences that make the NHP model a useful preclinical tool. Many factors contribute to the overall ischemic injury phenotype including the location of the lesion, the extent and duration of ischemia, the size of the ischemic penumbra, the number and patency of collateral arteries which determines the durability of collateral blood flow, and the degree of recanalization. In human patients, the variability in these factors results in highly diverse clinical outcomes, which can be a challenging population in which to validate a novel therapeutic strategy. While variable ischemic injury phenotypes and magnitudes are observed in human patients, experimental MCA/ACA occlusion in the rhesus macaque results in fairly uniform infarcts of similar size and phenotype, a feature useful for preclinical evaluation of potential neuroprotective therapies.

Models of cerebral ischemia in NHPs also provide an important opportunity to investigate the evolution of cerebral ischemic injury in primates, with respect to vascular pathobiology and extent of tissue injury and recovery. Using this model, reversible ischemia can be induced for varying lengths of time (i.e., 45, 60, or 90 min) and the corresponding infarcts and neurological outcomes can be compared. While infarcts in this model are located predominantly in the cortex, extended durations of occlusion (75–90 min) other brain structures, i.e., subcortical structures, are much more likely to be involved resulting in increased morbidity and mortality.

This model may also allow for the study of neural repair due to the fairly consistent injury to areas of the motor cortex, i.e., the study of subsequent remapping or reorganization and recovery of adjacent tissue near the infarct area. Testing the effects of constraint-induced therapy or other pharmacological or physical modalities in NHP stroke may provide valuable insights into human stroke and provide potential therapeutic avenues to improve long-term outcomes in patients. This model could have important implications for the evaluation of therapies for ischemic brain injury, for the understanding of the pathophysiology involved, and for the discovery of novel targets for neuroprotection.

References

1. Barber PA, Darby DG, Desmond PM, Gerraty RP, Yang Q, Li T, et al. Identification of major ischemic change. Diffusion-weighted imaging versus computed tomography. Stroke. 1999;30(10):2059–65.
2. McKhann GM, Grega MA, Borowicz Jr LM, Baumgartner WA, Selnes OA. Stroke and encephalopathy after cardiac surgery: an update. Stroke. 2006;37(2):562–71.
3. Johnston SC, Gress DR, Browner WS, Sidney S. Short-term prognosis after emergency department diagnosis of TIA. JAMA. 2000;284(22):2901–6.
4. Simon RP, Niiro M, Gwinn R. Prior ischemic stress protects against experimental stroke. Neurosci Lett. 1993;163:135–7.
5. Virca GD, Kim SY, Glaser KB, Ulevitch RJ. Lipopolysaccharide induced hyporesponsiveness to its own action in RAW 264.7 cells. J Biol Chem. 1989;264(36):21951–6.
6. Tasaki K, Ruetzler CA, Ohtsuki T, Martin D, Nawashiro H, Hallenbeck JM. Lipopolysaccharide pre-treatment induces resistance against subsequent focal cerebral ischemic damage in spontaneously hypertensive rats. Brain Res. 1997;748(1–2):267–70.
7. Heemann U, Szabo A, Hamar P, Muller V, Witzke O, Lutz J, et al. Lipopolysaccharide pre-treatment protects from renal ischemia/reperfusion injury: possible connection to an interleukin-6-dependent pathway. Am J Pathol. 2000;156(1):287–93.

8. Vartanian K, Stenzel-Poore M. Toll-like receptor tolerance as a mechanism for neuroprotection. Transl Stroke Res. 2010;1(4):252–60.
9. Broad A, Kirby JA, Jones DE. Toll-like receptor interactions: tolerance of MyD88-dependent cytokines but enhancement of MyD88-independent interferon-beta production. Immunology. 2007;120(1):103–11.
10. Biswas SK, Lopez-Collazo E. Endotoxin tolerance: new mechanisms, molecules and clinical significance. Trends Immunol. 2009;30(10):475–87.
11. Marsh B, Stevens SL, Packard AE, Gopalan B, Hunter B, Leung PY, et al. Systemic lipopoly-saccharide protects the brain from ischemic injury by reprogramming the response of the brain to stroke: a critical role for IRF3. J Neurosci. 2009;29(31):9839–49.
12. Caso JR, Pradillo JM, Hurtado O, Lorenzo P, Moro MA, Lizasoain I. Toll-like receptor 4 is involved in brain damage and inflammation after experimental stroke. Circulation. 2007;115(12):1599–608.
13. Lehnardt S, Lehmann S, Kaul D, Tschimmel K, Hoffmann O, Cho S, et al. Toll-like receptor 2 mediates CNS injury in focal cerebral ischemia. J Neuroimmunol. 2007;190(1–2):28–33.
14. Ziegler G, Harhausen D, Schepers C, Hoffmann O, Rohr C, Prinz V, et al. TLR2 has a detrimental role in mouse transient focal cerebral ischemia. Biochem Biophys Res Commun. 2007;359(3):574–9.
15. Brea D, Blanco M, Ramos-Cabrer P, Moldes O, Arias S, Perez-Mato M, et al. Toll-like receptors 2 and 4 in ischemic stroke: outcome and therapeutic values. J Cereb Blood Flow Metab. 2011;31(6):1424–31.
16. Hua F, Ma J, Ha T, Kelley J, Williams DL, Kao RL, et al. Preconditioning with a TLR2 specific ligand increases resistance to cerebral ischemia/reperfusion injury. J Neuroimmunol. 2008;199(1–2):75–82.
17. Stevens SL, Ciesielski TM, Marsh BJ, Yang T, Homen DS, Boule JL, et al. Toll-like receptor 9: a new target of ischemic preconditioning in the brain. J Cereb Blood Flow Metab. 2008;28(5):1040–7.
18. Stroke Therapy Academic Industry Round Table (Fisher M. Chair). Enhancing the development and approval of acute stroke therapies: stroke therapy academic industry roundtable. Stroke. 2005;36(8):1808–13.
19. Fukuda S, del Zoppo GJ. Models of focal cerebral ischemia in the nonhuman primate. ILAR J. 2003;44(2):96–104.
20. Karpiak SE, Tagliavia A, Wakade CG. Animal models for the study of drugs in ischemic stroke. Annu Rev Pharmacol Toxicol. 1989;29:403–14.
21. Carmichael ST. Rodent models of focal stroke: size, mechanism, and purpose. NeuroRx. 2005;2(3):396–409.
22. American Academy of Pediatrics Committee on Drugs. Alternative routes of drug administration—advantages and disadvantages (subject review). Pediatrics. 1997;100(1):143–52.
23. DeGraba T, Pettigrew L. Why do neuroprotective drugs work in animals but not humans? Neurol Clin. 2000;18:475–93.
24. Kapoor K, Kak VK, Singh B. Morphology and comparative anatomy of circulus arteriosus cerebri in mammals. Anat Histol Embryol. 2003;32(6):347–55.
25. Ge Y, Grossman RI, Babb JS, Rabin ML, Mannon LJ, Kolson DL. Age-related total gray matter and white matter changes in normal adult brain. Part II: quantitative magnetization transfer ratio histogram analysis. AJNR Am J Neuroradiol. 2002;23(8):1334–41.
26. Arakawa S, Wright PM, Koga M, Phan TG, Reutens DC, Lim I, et al. Ischemic thresholds for gray and white matter: a diffusion and perfusion magnetic resonance study. Stroke. 2006;37(5):1211–6.
27. Maeda M, Takamatsu H, Furuichi Y, Noda A, Awaga Y, Tatsumi M, et al. Characterization of a novel thrombotic middle cerebral artery occlusion model in monkeys that exhibits progressive hypoperfusion and robust cortical infarction. J Neurosci Methods. 2005;146(1):106–15.
28. Hirouchi Y, Suzuki E, Mitsuoka C, Jin H, Kitajima S, Kohjimoto Y, et al. Neuroimaging and histopathological evaluation of delayed neurological damage produced by artificial occlusion

of the middle cerebral artery in Cynomolgus monkeys: establishment of a monkey model for delayed cerebral ischemia. Exp Toxicol Pathol. 2007;59(1):9–16.

29. Huang J, Mocco J, Choudhri TF, Poisik A, Popilskis SJ, Emerson R, et al. A modified transorbital baboon model of reperfused stroke. Stroke. 2000;31(12):3054–63.

30. Mack WJ, Komotar RJ, Mocco J, Coon AL, Hoh DJ, King RG, et al. Serial magnetic resonance imaging in experimental primate stroke: validation of MRI for pre-clinical cerebroprotective trials. Neurol Res. 2003;25(8):846–52.

31. West GA, Golshani KJ, Doyle K, Lessov NS, Hobbs TR, Kohama SG, et al. A new model of cortical stroke in the rhesus macaque. J Cereb Blood Flow Metab. 2009;29(6):1175–86.

32. Burrows AM, Waller BM, Parr LA. Facial musculature in the rhesus macaque (*Macaca mulatta*): evolutionary and functional contexts with comparisons to chimpanzees and humans. J Anat. 2009;215(3):320–34.

33. Iredale SK, Nevill CH, Lutz CK. The influence of observer presence on baboon (Papio spp.) and rhesus macaque (*Macaca mulatta*) behavior. Appl Anim Behav Sci. 2010;122(1):53–7.

34. Spetzler RF, Selman WR, Weinstein P, Townsend J, Mehdorn M, Telles D, et al. Chronic reversible cerebral ischemia: evaluation of a new baboon model. Neurosurgery. 1980;7(3): 257–61.

35. Mori E, Ember JA, Copeland BR, Thomas WS, Koziol JA, del Zoppo GJ. Effect of tirilazad mesylate on middle cerebral artery occlusion/reperfusion in nonhuman primates. Cerebrovasc Dis. 1995;5(5):342–9.

36. De Haan R, Horn J, Limburg M, Van Der Meulen J, Bossuyt P. A comparison of five stroke scales with measures of disability, handicap, and quality of life. Stroke. 1993;24(8):1178–81.

37. De Vries SI, Van Hirtum HW, Bakker I, Hopman-Rock M, Hirasing RA, Van Mechelen W. Validity and reproducibility of motion sensors in youth: a systematic update. Med Sci Sports Exerc. 2009;41(4):818–27.

38. Papailiou A, Sullivan E, Cameron JL. Behaviors in rhesus monkeys (*Macaca mulatta*) associated with activity counts measured by accelerometer. Am J Primatol. 2008;70(2):185–90.

39. Mann TM, Williams KE, Pearce PC, Scott EA. A novel method for activity monitoring in small non-human primates. Lab Anim. 2005;39(2):169–77.

40. Krieg AM. Therapeutic potential of Toll-like receptor 9 activation. Nat Rev Drug Discov. 2006;5(6):471–84.

41. Leifer CA, Verthelyi D, Klinman DM. Heterogeneity in the human response to immunostimulatory CpG oligodeoxynucleotides. J Immunother. 2003;26(4):313–9.

42. Gursel M, Verthelyi D, Gursel I, Ishii KJ, Klinman DM. Differential and competitive activation of human immune cells by distinct classes of CpG oligodeoxynucleotide. J Leukoc Biol. 2002;71(5):813–20.

43. Verthelyi D, Kenney RT, Seder RA, Gam AA, Friedag B, Klinman DM. CpG oligodeoxynucleotides as vaccine adjuvants in primates. J Immunol. 2002;168(4):1659–63.

44. Bahjat FR, Williams-Karnesky RL, Kohama SG, West GA, Doyle KP, Spector MD, et al. Proof of concept: pharmacological preconditioning with a Toll-like receptor agonist protects against cerebrovascular injury in a primate model of stroke. J Cereb Blood Flow Metab. 2011;31(5):1229–42.

45. Fisher M, Feuerstein G, Howells DW, Hurn PD, Kent TA, Savitz SI, et al. Update of the stroke therapy academic industry roundtable preclinical recommendations. Stroke. 2009;40(6): 2244–50.

46. Viray J, Rolfs B, Smith DG. Comparison of the frequencies of major histocompatibility (MHC) class-II DQA1 and DQB1 alleles in Indian and Chinese rhesus macaques (*Macaca mulatta*). Comp Med. 2001;51(6):555–61.

47. Ferguson B, Street SL, Wright H, Pearson C, Jia Y, Thompson SL, et al. Single nucleotide polymorphisms (SNPs) distinguish Indian-origin and Chinese-origin rhesus macaques (*Macaca mulatta*). BMC Genomics. 2007;8:43.

Chapter 11
Angiogenesis and Arteriogenesis as Stroke Targets

Jieli Chen and Michael Chopp

Abstract Stroke is the third leading cause of morbidity and long-term disability. Reestablishment of functional microvasculature such as promotion of angiogenesis and arteriogenesis in the ischemic border creates a hospitable microenvironment for neuronal plasticity leading to functional recovery. To capitalize on angiogenesis and arteriogenesis as therapeutic targets for stroke treatment, knowledge of the precise molecular mechanisms which stimulate these vascular processes is necessary. Vascular endothelial growth factor, its receptors, the Angiopoietin-1 (Ang1)/Tie2 system and endothelial nitric oxide synthase, among other angiogenic factors mediate and contribute to post-ischemic angiogenesis and arteriogenesis. This chapter reviews molecular mechanisms which promote angiogenesis and arteriogenesis following cerebral ischemia and the associated vascular remodeling effects of experimental pharmacological (Statins and Niaspan) and cellular (bone marrow stromal cells) approaches for the treatment of stroke.

1 Introduction

Stroke is the third leading cause of morbidity and long-term disability. Treatment of stroke has primarily focused on early neuroprotection, with attempts to therapeutically introduce agents designed to prevent or reduce ischemic cell damage. While promising

J. Chen, MD (✉)
Department of Neurology, Henry Ford Hospital, E & R Building,
Room 3091, Detroit, MI 48202, USA
e-mail: jieli@neuro.hfh.edu

M. Chopp
Department of Neurology, Henry Ford Hospital, E & R Building,
Room 3056, Detroit, MI 48202, USA

Department of Physics, Oakland University, Rochester, MI 48309, USA

P.A. Lapchak and J.H. Zhang (eds.), *Translational Stroke Research*, Springer Series
in Translational Stroke Research, DOI 10.1007/978-1-4419-9530-8_11,
© Springer Science+Business Media, LLC 2012

results with these drugs have been achieved in animal stroke models, all Phase III clinical trials conducted so far indicate that these drugs have failed to live up to their promise. Currently, the only FDA-approved medical therapy for acute ischemic stroke is tissue plasminogen activator (tPA), a thrombolytic agent that targets the thrombus within the blood vessel. The most recent trial published with IV tPA in stroke (European Cooperative Acute Stroke Study (ECASS III)) studied patients between 3 and 4.5 h from symptom onset and found a benefit to treatment in the rate of favorable outcome when compared to placebo, with no difference in mortality [1, 2]. Thrombolytic therapy with t-PA was approved for selected patients with acute ischemic stroke when therapy is initiated within 4.5 h of onset. However, t-PA treatment is limited by a narrow time window and with an increased risk for intracranial hemorrhage, so that only 4–7% of all stroke patients receive thrombolysis [3–5]. Thus, there is a compelling need to develop delayed pharmacological or cell-based therapeutic approaches specifically designed to reduce neurological deficits after stroke beyond the hyper-acute phase of ischemia, and to use multiagent therapy to exploit the capacity of the brain for neuroregeneration and neuroplasticity.

Neurorestorative events include angiogenesis, arteriogenesis, neurogenesis, synaptogenesis, and white matter remodeling. Regulation of cerebral blood flow (CBF) is critical for the maintenance of neural function [6]. Reestablishment of functional microvasculature such as angiogenesis and arteriogenesis in the ischemic border creates a hospitable microenvironment for neuronal plasticity leading to functional recovery [7–9]. Stroke patients with higher density of cerebral blood vessels do better and survive longer than those with lower vascular density [10, 11]. Early clinical improvement after stroke is also linked to the presence of arteriolar collaterals (Arteriogenesis). Absence of significant collateralization increases mortality after stroke [12]. Thrombolysis treatment has a greater clinical impact in patients with better collateral supply after stroke [13]. Therefore, stimulating arteriogenesis and angiogenesis in particular may provide a new treatment strategy for patients with stroke. In this chapter, we discuss vascular remodeling (angiogenesis and arteriogenesis) as a stroke target.

2 Biological Basis of Angiogenesis and Arteriogenesis

2.1 Angiogenesis

Angiogenesis is the physiological process involving the growth of new blood vessels from preexisting vessels. The major physiological stimuli for angiogenesis are tissue ischemia and hypoxia [14]. Angiogenic growth factors bind to specific receptors located on the brain endothelial cells and stimulate endothelial cell proliferation, migration, and sprouting from the existing vessel toward ischemic brain area. Sprouting endothelial cells form blood vessel tube-like structures. During angiogenesis, the initial vascular plexus forms mature vessels by sprouting, branching, pruning, differential growth of endothelial cells, and recruitment of supporting cells, such as pericytes and smooth muscle cells (SMCs) [15, 16]. Many factors regulate angiogenesis and vascular maturation, such as vascular endothelial growth factor (VEGF), its receptors, the Angiopoietin-1 (Ang1)/Tie2 system [17] and endothelial nitric oxide synthase (eNOS), basic fibroblast growth factor (bFGF), platelet-derived growth factor (PDGF) et al. VEGF, the hypoxia-inducible endothelial cell mitogen and vascular permeability factor, appears to play a pivotal role in most of these processes [7].

2.2 *Arteriogenesis*

Arteriogenesis describes the formation of mature arteries from preexistent interconnecting arterioles after an arterial occlusion [18, 19]. It shares some features with angiogenesis, but the pathways leading to arteriogenesis are different. The differences are that collaterals develop from preexisting arterioles and that circulating monocytes adhere to endothelium that had been activated by the high shear stress generated by the large pressure differences between perfusion territories. Much of the growth and remodeling are achieved by attraction, adhesion, activation, and invasion of circulating cells, mostly monocytes, but also T-cells and basophils [18]. Monocytes are the major producers of growth factors and of proteolytic enzymes that enable SMCs to migrate and divide [20]. The arteriogenic response consists of the formation of new arterioles, which presumably occurs when preexisting capillaries acquire smooth muscle coating, and these newly formed and/or preexisting arterioles transform into channels with larger diameters [21, 22]. Arteriogenesis is an important process for adapting preexisting vessels into functional collateral conduits for delivery of oxygen-enriched blood to tissue distal to occlusion of a large, peripheral conduit artery [23].

In contrast to angiogenesis, arteriogenesis is independent of oxygen levels and usually occurs in a normoxic environment. Arteriogenesis is promoted by changes in shear stress forces sensed by the vascular endothelium [24]. The increased fluid shear stress is likely to result in increased expression of endothelial adhesion molecules that would in turn promote recruitment of blood-derived mononuclear cells to sites of arterial formation. Integrins alpha 5 beta 1 and v beta 3 as well as focal adhesion kinase (FAK) signaling axis [25, 26], tyrosine receptor kinases [27], and G protein-coupled receptors [28] have been proposed to act as shear stress sensors on the endothelial cell membrane. The signaling transduction cascades that are initiated as a result of the fluid shear stress lead to the activation of endothelial cells and increase the expression of intercellular adhesion molecule-1(ICAM-1/Mac-1) [29] and vascular cell adhesion molecule-1 (VCAM-1) [30], and increase production of several chemokines such as tumor necrosis factor-alpha (TNF-a) [31], granulocyte-macrophage colony-stimulating factor (GM-CSF) [32, 33], Monocyte chemotactic protein-1 (MCP-1) [29], and nitric oxide (NO) [34]. Cytokines recruit circulating monocytes to the activated endothelium, where they adhere, invade the interstitial space, and mature to macrophages [35]. The macrophages, in addition to other cellular elements, produce abundant cytokines such as VEGF, NO, more MCP-1, FGF-1, and FGF-2. The new milieu leads to endothelial and SMC proliferation, migration, vessel enlargement and maturation, and synthesis of extracellular matrix. Ultimately, the small preexisting arterioles remodel into large functional conducting collaterals [24]. In addition, bone marrow (BM)-derived circulating cells may also be involved in arteriogenesis. The expression of growth factors and cytokines and the cooperation of surrounding and infiltrating cells seem to be essential in orchestrating the complex processes occurring during arteriogenesis.

2.3   Difference Between Angiogenesis and Arteriogenesis

In contrast to angiogenesis, which is dominated by angiogenic factors, arteriogenesis relies on a complex interplay of many growth factors, cytokines, different cell types, a multitude of proteolytic enzymes, and, at least during the initial stages, an environment of inflammation [36]. Tissue ischemia/hypoxia is needed for angiogenesis but not for arteriogenesis, which is governed initially by physical forces that activate the endothelium of preexistent arterioles. Arteriogenesis is potentially able to fully replace an occluded artery whereas angiogenesis cannot. Arteriogenesis proceeds much faster than angiogenesis because of a structural dilatation of preexisting collateral vessels followed by mitosis of all vascular cell types, which restores resting blood flow within 3 days. Recovery of dilatory reserve (maximal flow) takes longer. The slower angiogenesis is unable to significantly restore flow even if angiogenesis reduces the minimal terminal resistance of the entire chain of resistors by new capillaries in parallel [36].

This chapter focuses on discussion of the effects of VEGF/VEGFRs, Ang1, Ang2/Tie2, and eNOS in the regulation of angiogenesis, vascular stabilization, and arteriogenesis.

3   Angiopoietin/Tie2 Effect on Angiogenesis and Arteriogenesis

Angiopoietin-1 (Ang-1) and Ang-2 belong to a family of endothelial growth factors that function as ligands for the endothelial-specific receptor tyrosine kinase, Tie-2. Ang-1 reduces endothelial permeability and enhances vascular stabilization and maturation [37, 38]. The Ang-1/Tie2 signaling pathway controls angiogenesis to form large and small vessels in the mature vascular system and mediates vascular endothelial integrity [39], whereas Ang-2 is thought to be an endogenous antagonist of the action of Ang-1 at Tie-2, and disrupts blood vessel formation [40–42]. Ang2 has no effect on endothelial cell proliferation, but stimulates chemotaxis and tube-like structure formation [43]. Ang2 and VEGF have emerged as potent inducers of blood–brain barrier (BBB) breakdown while Ang1 is a potent anti-leakage factor [44]. Ang1 promotes migration, sprouting, and survival of endothelial cells through activation of different signaling pathways triggered by the receptor Tie-2. Ang-1 also stimulates the formation of pericytes and SMCs with endothelial cells and thus promotes maturity and stability of newly formed blood vessels. Interactions between endothelial cells and mural cells (pericytes and vascular SMCs) in the blood vessel wall have recently come into focus as central processes in the regulation of vascular formation, stabilization, and function. Ang1 expression co-localizes with pericyte reactive cells in the ischemic brain. A pericyte-derived Ang1 multimeric complex induces occludin gene expression in brain capillary endothelial cells through Tie2 activation in vitro [45]. The Ang1/Tie2 system controls pericyte recruitment, endothelial cell survival, and is implicated in blood vessel formation and vascular stabilization [46]. Transgenic overexpression of Ang1 increases vascular stabilization [47], prevents plasma leakage in the ischemic brain, and consequently decreases ischemic lesion volume [48, 49]. A recent study shows that Ang1 also promotes sustained improvement of ventricular perfusion via arteriogenesis in a swine chronic myocardial ischemia model [50]. Ang-1 has both angiogenic and arteriogenic properties when overexpressed alone or in combination with VEGF [51–53].

4   VEGF/VEGFR1 and VEGFR2 Effect on Angiogenesis and Arteriogenesis

VEGF is a potent cytokine and was originally identified as an endothelial cell specific growth factor stimulating angiogenesis and vascular permeability. VEGF binds to two receptor protein tyrosine kinases, VEGFR1 (Flt-1) and VEGFR2 (KDR). Interaction of VEGF with particular subtypes of receptors activates a circuit of signaling pathways, e.g., phosphoinositide 3-kinase (PI3K)/Akt, Ras/Raf-MEK/ Erk (extracellular signal-regulated kinases), and eNOS/NO. These participate in the generation of specific biological responses connected with proliferation, migration, increasing vascular permeability, or promoting endothelial cell survival [54]. VEGFR-1 (ligands include VEGF-A, -B), and placental growth factor (PIGF) and VEGFR-2 (ligands include VEGF-A, -C, and -D) are predominantly expressed on vascular endothelial cells. VEGFR1 signaling is required for endothelial cell survival and mediates cell migration, while activation of VEGFR-2 regulates capillary tube formation and mediates cell proliferation [55, 56] and appears to be both necessary and sufficient to mediate VEGF-dependent angiogenesis and induction of vascular permeability [57]. Neuropilin-1 (NRP-1) expression acts as a cofactor for VEGF (165) enhancing the angiogenic stimulus [58]. VEGF-A 165 (recombinant protein) is capable of inducing migration of VEGFR2-negative human aortic smooth muscle cells (hAOSMCs), and this induction is mediated through a molecular cross-talk of NRP-1, VEGFR1, and PI3K/Akt signaling kinase. VEGF-A 165 induces hAOSMC migration parallel with the induction of NRP-1 and VEGFR1 expressions and their associations along with the activation of PI3K/Akt signaling kinase [59]. However, VEGF is expressed on the transcriptional or translational level in collaterals proper and in the tissue surrounding them [20].

VEGF-A, called VEGF or vascular permeability factor, has emerged as the single most important regulator of blood vessel formation after stroke. However, in most preclinical and clinical studies, the introduction of VEGF did not improve functional outcome after stroke in clinical trial. Early post-ischemic (1 h) administration of rhVEGF (165) to ischemic rats increases BBB leakage and infarction volume in the ischemic brain [60]. By contrast, late (48 h) administration of rhVEGF (165) to ischemic rats enhances angiogenesis in the ischemic penumbra and significantly improves neurological recovery [60]. Drawbacks of administering VEGF directly include potential adverse effects of inducing hemorrhage and phagocytized neuropil and high macrophage density [61]. Therefore, increased angiogenesis induced by VEGF alone is not sufficient for improvement of functional outcome after stroke and induction of brain plasticity. A more reasonable approach by which to capitalize on the potential benefit of VEGF, is to combine it with other agents which decrease BBB leakage and inflammatory effects.

5   Combination of VEGF and Ang1 Effect on Angiogenesis and Arteriogenesis

Angiogenesis is controlled by the net balance between molecules that have positive and negative regulatory activity. This concept had led to the notion of the "angiogenic switch," depending on an increased production of one or more of the positive regulators of angiogenesis. Combination of Ang1 with VEGF treatment decreases BBB leakage and promotes angiogenesis [52]. Ang1/Tie2 cooperates with the VEGF system to establish dynamic blood vessel structures [62] and induces a synergistic angiogenic effect, and promotes the formation of mature neovessels without the side effects on BBB permeability [63]. A combination of the submaximal doses of Ang1 and VEGF enhances angiogenesis and is more potent than the maximal dose of either alone [52]. In addition, combined VEGF/VEGFR-2 and Ang-1/Tie-2 signaling pathways also play an important role in the mobilization and recruitment of hematopoietic stem cells and circulating endothelial progenitor cells (EPCs) [64]. Autologous secretion of Ang-1 by transduced endothelial cell results in Tie-2 activation and in the presence of SMCs expressing VEGF which evokes coordinated sprouting in vitro and increases in flow and number of arteries in vivo [65]. Therefore, combination of Ang1/VEGF increases angiogenesis and vascular stabilization and also promotes arteriogenesis.

6   eNOS Effect on Angiogenesis and Arteriogenesis

eNOS is required for proper endothelial cell migration, proliferation, and differentiation [66]. Enhanced eNOS phosphorylation induces a broad range of effects including the promotion of angiogenesis and mural cell recruitment to immature angiogenic sprouts [67, 68] and regulates vascular tone and vascular remodeling [69, 70]. Endothelium-derived nitric oxide (NO) is a mediator of angiogenesis. In the endothelial cells, NO has a relevant role in the maintenance of the confluent endothelial monolayer by inhibiting apoptotic-related mechanisms [71]. NO upregulates alpha(v)beta(3) integrin ($_v\beta_3$) on endothelial cells, a critical mediator of cell-matrix adhesion and migration and promotes endothelial cell migration and differentiation into capillaries [72]. NO also enhances endothelial cell proliferation, in part by increasing the expression of VEGF or fibroblast growth factor [73, 74]. In addition, NO also has vasodilatory effect. The hemodynamic effects of NO as a potent vasodilator may play a role in its angiogenic effects. The vasodilation effect is mediated by the generation of cyclic GMP (cGMP). cGMP and cyclic AMP are the main second messengers in SMC relaxation. NO binds to a heme-protein, soluble guanylate cyclase that converts GMP to cGMP [75], and promotes vasodilatory effects. eNOS also augments angiogenic properties of ex vivo expanded EPCs. eNOS-modified EPC (eNOS-EPCs) stimulate a significant increase in EPC migration and ability to differentiate into endothelial-like spindle-shaped cells, which therefore promotes angiogenesis [76].

eNOS is not only associated with angiogenesis but also involved in collateral vessel growth (arteriogenesis). L-NAME (NW-nitro-L-arginine methyl ester), an inhibitor of NO synthase, inhibits arteriogenesis more than angiogenesis [77]. Endothelial growth factors such as FGF and VEGF enhance the activity of eNOS. However, the main physiologic factor responsible for eNOS activation is the shearing stress produced by friction of the flowing blood against the immobile vessel wall [75, 78]. Shear stress is an important physiological stimulator of eNOS activity. Shear stress causes rapid NO release and activation of eNOS by Akt-dependent serine phosphorylation [34, 79]. eNOS knockout (eNOS$^{-/-}$) mice have defects in arteriogenesis and functional blood flow reserve after muscle stimulation and pericyte recruitment [67]. Compared to wild-type, eNOS$^{-/-}$ mice evidenced reduced collateral remodeling, angiogenesis, and flow-mediated dilation of the arterial bed supplying the collaterals, resulting in lower perfusion and greater ischemic injury following femoral artery ligation [80]. eNOS overexpression in an ischemic rat hindlimb significantly increased skeletal muscle blood flow, muscle oxygen tension, and collateral arteries [81]. Maintenance or improvement of NO production and signaling, such as with regular exercise, improves endothelial cell function and thus help preserve the arteriogenic potential of preexisting collateral networks [82]. Physical activity improves long-term stroke outcome by eNOS-dependent mechanisms related to improved angiogenesis and arteriogenesis [83].

7 Therapeutic Implications for Stroke

Recently, there is increasing interest in fostering functional restoration after neurological injury by inducing brain plasticity, specifically induction of angiogenesis and arteriogenesis [7, 8, 15, 84–86]. Vascular remodeling promotes neurorestoration and is an important goal for all neurorestorative therapy of ischemic neural tissues [87]. The generation of new blood vessels facilitates highly coupled neurorestorative processes including neurogenesis and synaptogenesis which in turn lead to improved functional recovery. In this chapter, we highlight select pharmacologic (Statins and Niaspan, an agent which increases high-density lipoprotein cholesterol (HDL-C)) and cell-based (bone marrow stromal cells, BMSCs) therapies that enhance the recovery process in the subacute and chronic phase after stroke.

7.1 Statins

Statins (or HMG-CoA reductase inhibitors) are a class of drugs used to lower cholesterol levels by inhibiting the enzyme HMG-CoA reductase, which plays a central role in the production of cholesterol in the liver. Inhibition of HMG-CoA reductase blocks not only the synthesis of cholesterol but also of a number of isoprenoid intermediates that have several biological functions. For example, farnesylpyrophosphate (FPP) and geranylgeranylpyrophosphate (GGPP) serve as important lipid attachments for the posttranslational modification of proteins including heterotrimeric G proteins and small GTP-binding proteins. Therefore, decreasing levels of these isoprenoid intermediates might have several biological consequences.

The efficacy, safety, and tolerability of Statins have been confirmed in randomized, controlled, multicenter trials involving large numbers of patients aged ≥65 years [88]. Treatment of patients within 4 weeks after acute ischemic stroke with Statins significantly increases favorable outcome at 12 weeks [89]. Experimental studies have shown that Statins protect against ischemia-reperfusion injury of the heart, and exert pro-angiogenic effects by stimulating the growth of new blood vessels in ischemic limbs of normocholesterolemic animals [90]. Low-dose Statins administered 24 h after stroke promote angiogenesis and arteriogenesis in the ischemic brain and improve neurological functional outcome after stroke in young and in older retired breeder rats (middle age) [91, 92]. Low-dose Statins were more effective than high-dose Statins in both augmentation of collateral flow recovery and inhibition of atherosclerosis [93]. Higher doses of Statins inhibit angiogenesis, cell growth, and cell migration [94, 95].

The mechanism by which Statins provide benefit against stroke is likely multifactorial, involving reduction of the low-density lipoprotein cholesterol (LDL-C) along with stabilization of vulnerable plaques. In addition, many of the Statin's pleiotropic effects are cholesterol-independent, such as improvement of endothelial function, increased NO bioavailability, antioxidant properties, inhibition of inflammatory responses, immunomodulatory actions, upregulation of eNOS, decrease of platelet activation, and regulation of progenitor cells [96, 97]. Molecular mechanisms underlying the role of Statins in the induction of brain plasticity and subsequent improvement of neurological outcome after treatment of stroke include the Statin-mediated increase of eNOS, VEGF/VEGFR2, PI3K/Akt, and small G proteins in the ischemic brain [98]. These proteins play an important role in regulating vascular effects [99, 100]. Statins increase brain endothelial cell expression of VEGF/VEGFR2 and thereby activate PI3K/Akt, which regulates endothelial cell proliferation and migration and increases angiogenesis [101]. Simvastatin enhanced p-PI3K/Akt substrate eNOS activity, inhibited apoptosis, and accelerated vascular structure formation in vitro and in ischemic limbs of normocholesterolemic rabbits [102]. Simvastatin treatment of stroke also increases Ang1/Tie2 signaling activity and VE-cadherin expression, and thereby reduces BBB leakage and promotes vascular stabilization [103]. In addition, Simvastatin upregulation of Presenilin 1 expression and Notch signaling activity in the ischemic brain may facilitate an increase in arteriogenesis in the ischemic brain after stroke [104]. In the cerebrovascular system, Statin-associated enhancement of collateralization of cerebral arteries leads to a better outcome after acute ischemic stroke [105].

Bone marrow (BM)-derived EPCs contribute to vascular maintenance by participating in angiogenesis, re-endothelialization, and remodeling. Statins also enhance angiogenesis by increasing the mobilization and functionality of EPCs. Atorvastatin treatment of hindlimb ischemia strongly induces angiogenesis with increases in angiogenic cytokines (VEGF, IL-8, Ang-1, Ang-2, eNOS), Heme oxygenase-1 (HO-1), and EPC numbers [106]. Statins increase differentiation of EPCs in vitro and their mobilization in vivo [99, 107] through the PI3K/Akt pathway [108]. Atorvastatin promotes a threefold increase in the number of circulating EPCs in patients with coronary heart disease after 4 weeks of treatment [109]. Therefore, targeting angiogenesis and arteriogenesis by Statin treatment may benefit functional outcome after stroke.

7.2 Increasing HDL: Niacin (Niaspan)

Higher serum HDL–C level is related to better cognitive recovery after stroke at all ages [110]. Low serum HDL-C level is associated with more severe stroke in young stroke patients (≤50 years) [111] and cognitive impairment and dementia [112]. A low level of HDL-C is predictive of mortality from coronary artery disease (CAD) and stroke, and rapidly progressive stroke [113–116]. These findings offer new therapeutic strategies for stroke.

HDL-C binds to the HDL scavenger receptor class B type I (SR-BI) and promotes eNOS expression and activity by maintaining the lipid environment in caveolae where eNOS is colocalized with partner signaling molecules [117]. HDL-C also enhances vasorelaxation, protects endothelial cells from apoptosis, and promotes endothelial cell migration and re-endothelialization by increasing eNOS expression and phosphorylation of eNOS [23, 118–120]. HDL-C causes potent stimulation of eNOS activity and enhances endothelium- and NO-dependent relaxation in the aorta [121]. In addition, HDL also slows down EPC senescence [122] and promotes EPC proliferation, migration, and "tube" formation. PI3K/Akt-dependent cyclin D1 activation plays an essential role in HDL-induced EPC proliferation, migration, and angiogenesis [123]. Reconstituted HDL-C stimulates differentiation of peripheral mononuclear cells to EPCs via the PI3K/Akt pathway and enhances ischemia-induced angiogenesis in a murine ischemic hindlimb model [124].

Niacin (vitamin B$_3$ or nicotinic acid) is the most effective medication in current clinical use for increasing HDL-C, and it substantially lowers triglycerides and LDL cholesterol [125–127]. Niacin binds to adipose nicotinic acid receptors which downregulate free fatty acid mobilization, as well as reduces hepatic very-low-density lipoprotein (VLDL) output [128]. Niacin also accelerates intracellular Apolipoprotein B degradation by inhibiting triacylglycerol synthesis in human hepatoblastoma cells [129]. Niacin increases HDL-C levels through inhibition of the HDL-C Apo-A1 catabolism pathway, blocking an HDL-C holoparticle receptor, thus prolonging HDL-C half-life [130]. Niacin also has multiple lipoprotein and anti-atherothrombosis effects that improve endothelial function, reduce inflammation, increase plaque stability, and diminish thrombosis [131]. Niacin promotes vasodilatation and improves endothelial function by enhancing the function of eNOS to increase plasma and tissue NO levels [118].

Niaspan is a prolonged release formulation of Niacin. Treatment of stroke with Niaspan starting 24 h after MCAo and continuing daily for 14 days increases serum HDL-C level, increases arteriogenesis and angiogenesis after stroke as well as improves functional outcome after stroke [132, 133]. Niaspan restoration of neurological function post stroke is independent of its cardiovascular, anti-inflammatory, anti-oxidative, anti-thrombotic [134, 135] effects, and other traditional therapeutic targets. Niaspan increases the expression of VEGF and Ang1, and phosphorylation of Akt, eNOS, and Tie2 in the ischemic brain [132, 133] thereby promotes angiogenesis in the ischemic brain. In addition, TNF-alpha serves as a pivotal modulator of arteriogenesis [136]. Tumor necrosis factor-alpha-converting enzyme (TACE) is the principal protease involved in the activation of pro-TNF-alpha [137] and regulates the function of several transmembrane proteins and cell adhesion molecular shedding [138]. Niaspan treatment of stroke increases TACE expression and Notch signaling activity and promotes arteriogenesis after stroke. The increased arteriogenesis significantly correlated with the functional outcome after stroke and may play a role in the regulation of functional outcome after stroke [133]. Therefore, targeting vascular remodeling with Niaspan treatment may play important role in improving functional outcome after stroke.

7.3   Bone Marrow Stromal Cells

Bone marrow-derived MSCs have great potential as therapeutic agents, since they are easy to isolate and can be expanded from patients without serious ethical and technical problems for autologous transplantation. The regenerative potential of BMSCs has been measured in myocardial, limb, and brain ischemia [139–143]. BMSCs when administered intravenously selectively migrate to the ischemic area in the brain and increase angiogenesis as well as enhance functional recovery after stroke [144]. Rats receiving BMSC transplantation at 2 h after stroke exhibit reduced scar size, limited apoptosis, and enhanced angiogenic factor expression and vascular density in the ischemic region relative to the control group, as well as significant improvements in functional outcome [145].

The mechanisms of action of BMSCs to treat stroke stand in sharp contrast to those originally targeted for stem and progenitor cells. Stem and progenitor cells, being totipotent cells, when placed in injured brain, are designed to replace dead and injured tissue [146]. BMSCs also likely contain a subpopulation of stem-like cells, which can differentiate into brain cells [147]. However, these cells are a minor subpopulation of the BMSCs and do not contribute to the restoration of function, and only a very small percentage of the BMSCs assume parenchymal cell phenotype [143]. BMSCs express mRNAs for a wide spectrum of angiogenic/arteriogenic cytokines including VEGF, FGF2, placental growth factor, insulin-like growth factor, and Ang1 [148–152]. BMSCs when injected intravenously enter brain and stimulate the local parenchymal cell production of growth factors from the endogenous cells, primarily astrocytes and endothelial cells [9, 153, 154], and promote angiogenesis and vascular stabilization [144], which is partially mediated by VEGF/Flk1 and Ang1/Tie2 [155]. These cytokines and growth factors have both paracrine and autocrine activities [153].

BMSCs have the ability to effectively recruit and participate in angiogenesis and also augment arteriogenesis [9, 142–144, 152, 156, 157]. Media collected from BMSC cultures promote in vitro proliferation and integration of endothelial cells and SMCs, and enhance collateral flow recovery and remodeling when directly injected into a mouse ischemic hindlimb [152]. Using autologous BMSC treatment significantly improved blood flow and increased arteriogenesis in a chronic limb ischemia model [158]. Combination treatment of stroke using BMSCs with statins enhances BMSC migration into the ischemic brain, amplifies arteriogenesis and angiogenesis, and improves functional outcome after stroke [159]. BMSCs selectively target the injury site, participate in arteriogenesis and angiogenesis [157], and induce a neovascular response resulting in a significant increase in blood flow to the ischemic area which aids in repair of injured brain [159]. The BMSC stimulated and angiogenic and arteriogenic vessels also produce trophic and growth factors that contribute to brain plasticity and recovery of neurological function post stroke [9, 155]. Therefore, BMSCs behave as small biochemical and molecular "factories," producing and inducing within vascular and parenchymal cells many cytokines [160–162] and trophic factors that contribute to improvement in neurological function post stroke.

8 Clinical Trials of Stroke Treatment with Statins, Niaspan, and BMSCs

Statins have been successfully tested in clinical trials to prevent recurrent stroke and acute stroke treatment in the ischemic stroke patients in J-STARS (Japan Statin Treatment Against Recurrent Stroke), SPOTRIAS (Translational Research in Acute

Stroke), STARS07 (Stroke Treatment with Acute Reperfusion and Simvastatin), Neu-START (Neuroprotection With Statin Therapy Acute Recovery Trial), and EUREKA (The Effects of Very Early Use of Rosuvastatin in Preventing Recurrence of Ischemic). Niaspan is in clinical trials in Henry Ford Hospital (HFH) to investigate the role of Niaspan in stroke recovery. Niaspan treatment is imitated at 3–7 days after stroke, and patients are evaluated within 90 days after stroke. Transplantation of MSCs from bone marrow is considered safe and has been widely tested in clinical trials of cardiovascular, neurological, and immunological disease with encouraging results [163, 164]. There are 79 registered clinical trial sites for evaluating BMSC therapy throughout the world [165]. The clinical trials of intravenous infusion of autologous BMSCs appear to be a feasible and safe therapy, also showed promising results in improving functional recovery in stroke patients [163, 166, 167]. A clinical trial is being performed at University of California in which autologous MSCs are being used to treat stroke patients.

9   Conclusion

Induction of angiogenesis and arteriogenesis in the ischemic brain creates a hospitable microenvironment for neuronal plasticity leading to functional recovery. Pharmacological (e.g., Statins and Niaspan) and BMSC therapy, and likely other cell-based therapies such as human umbilical cord blood cells [168, 169] and human umbilical tissue-derived cells [170] stimulate and enhance angiogenesis and arteriogenesis after stroke as well as improve functional outcome after stroke. Increasing angiogenesis and arteriogenesis in the ischemic brain may be a target for restorative stroke treatment leading to multifaceted processes that enhance functional recovery after stroke.

Acknowledgments   This work was supported by National Institute of Aging grant RO1 AG301811 (J.C) and R01-AG037506 (M.C).

References

1. Cronin CA. Intravenous tissue plasminogen activator for stroke: a review of the ECASS III results in relation to prior clinical trials. J Emerg Med. 2010;38(1):99–105.
2. Carpenter CR, et al. Thrombolytic therapy for acute ischemic stroke beyond three hours. J Emerg Med. 2011;40(1):82–92.
3. Katzan IL, et al. Utilization of intravenous tissue plasminogen activator for acute ischemic stroke. Arch Neurol. 2004;61(3):346–50.
4. Weimar C, et al. Intravenous thrombolysis in German stroke units before and after regulatory approval of recombinant tissue plasminogen activator. Cerebrovasc Dis. 2006;22(5–6): 429–31.
5. Schwammenthal Y, et al. Trombolysis in acute stroke. Isr Med Assoc J. 2006;8(11):784–7.

6. Pratt PF, Medhora M, Harder DR. Mechanisms regulating cerebral blood flow as therapeutic targets. Curr Opin Investig Drugs. 2004;5(9):952–6.

7. Plate KH. Mechanisms of angiogenesis in the brain. J Neuropathol Exp Neurol. 1999;58(4): 313–20.

8. Renner O, et al. Time- and cell type-specific induction of platelet-derived growth factor receptor-beta during cerebral ischemia. Brain Res Mol Brain Res. 2003;113(1–2):44–51.

9. Chen J, et al. Intravenous administration of human bone marrow stromal cells induces angiogenesis in the ischemic boundary zone after stroke in rats. Circ Res. 2003;92(6):692–9.

10. Krupinski J, et al. Role of angiogenesis in patients with cerebral ischemic stroke. Stroke. 1994;25(9):1794–8.

11. Wei L, et al. Collateral growth and angiogenesis around cortical stroke. Stroke. 2001;32(9): 2179–84.

12. Christoforidis GA, et al. Angiographic assessment of pial collaterals as a prognostic indicator following intra-arterial thrombolysis for acute ischemic stroke. AJNR Am J Neuroradiol. 2005;26(7):1789–97.

13. Liebeskind DS. Collaterals in acute stroke: beyond the clot. Neuroimaging Clin N Am. 2005;15(3):553–73, x.

14. Dor Y, Keshet E. Ischemia-driven angiogenesis. Trends Cardiovasc Med. 1997;7(8):289–94.

15. Risau W. Mechanisms of angiogenesis. Nature. 1997;386(6626):671–4.

16. Folkman J, D'Amore PA. Blood vessel formation: what is its molecular basis? Cell. 1996;87(7):1153–5.

17. Patan S. Vasculogenesis and angiogenesis. Cancer Treat Res. 2004;117:3–32.

18. Schaper W, Buschmann I. Arteriogenesis, the good and bad of it. Eur Heart J. 1999;20(18): 1297–9.

19. Buschmann I, Schaper W. Arteriogenesis versus angiogenesis: two mechanisms of vessel growth. News Physiol Sci. 1999;14:121–5.

20. Scholz D, Cai WJ, Schaper W. Arteriogenesis, a new concept of vascular adaptation in occlusive disease. Angiogenesis. 2001;4(4):247–57.

21. Buschmann I, Schaper W. The pathophysiology of the collateral circulation (arteriogenesis). J Pathol. 2000;190(3):338–42.

22. van Royen N, et al. Stimulation of arteriogenesis; a new concept for the treatment of arterial occlusive disease. Cardiovasc Res. 2001;49(3):543–53.

23. Seetharam D, et al. High-density lipoprotein promotes endothelial cell migration and reendothelialization via scavenger receptor-B type I. Circ Res. 2006;98(1):63–72.

24. Heil M, Schaper W. Influence of mechanical, cellular, and molecular factors on collateral artery growth (arteriogenesis). Circ Res. 2004;95(5):449–58.

25. Jalali S, et al. Integrin-mediated mechanotransduction requires its dynamic interaction with specific extracellular matrix (ECM) ligands. Proc Natl Acad Sci USA. 2001;98(3):1042–6.

26. Cai WJ, et al. Activation of the integrins alpha 5beta 1 and alpha v beta 3 and focal adhesion kinase (FAK) during arteriogenesis. Mol Cell Biochem. 2009;322(1–2):161–9.

27. Chen KD, et al. Mechanotransduction in response to shear stress. Roles of receptor tyrosine kinases, integrins, and Shc. J Biol Chem. 1999;274(26):18393–400.

28. Chachisvilis M, Zhang YL, Frangos JA. G protein-coupled receptors sense fluid shear stress in endothelial cells. Proc Natl Acad Sci USA. 2006;103(42):15463–8.

29. Hoefer IE, et al. Arteriogenesis proceeds via ICAM-1/Mac-1- mediated mechanisms. Circ Res. 2004;94(9):1179–85.

30. Behm CZ, et al. Molecular imaging of endothelial vascular cell adhesion molecule-1 expression and inflammatory cell recruitment during vasculogenesis and ischemia-mediated arteriogenesis. Circulation. 2008;117(22):2902–11.

31. Hoefer IE, et al. Direct evidence for tumor necrosis factor-alpha signaling in arteriogenesis. Circulation. 2002;105(14):1639–41.

32. Kosaki K, et al. Fluid shear stress increases the production of granulocyte-macrophage colony-stimulating factor by endothelial cells via mRNA stabilization. Circ Res. 1998;82(7):794–802.

33. Buschmann IR, et al. GM-CSF: a strong arteriogenic factor acting by amplification of monocyte function. Atherosclerosis. 2001;159(2):343–56.
34. Cai WJ, et al. Expression of endothelial nitric oxide synthase in the vascular wall during arteriogenesis. Mol Cell Biochem. 2004;264(1–2):193–200.
35. Arras M, et al. Monocyte activation in angiogenesis and collateral growth in the rabbit hindlimb. J Clin Invest. 1998;101(1):40–50.
36. Schaper W, Scholz D. Factors regulating arteriogenesis. Arterioscler Thromb Vasc Biol. 2003;23(7):1143–51.
37. Suri C, et al. Requisite role of angiopoietin-1, a ligand for the TIE2 receptor, during embryonic angiogenesis. Cell. 1996;87(7):1171–80.
38. Pfaff D, Fiedler U, Augustin HG. Emerging roles of the Angiopoietin-Tie and the ephrin-Eph systems as regulators of cell trafficking. J Leukoc Biol. 2006;80(4):719–26.
39. Marti HH, Risau W. Angiogenesis in ischemic disease. Thromb Haemost. 1999;82 Suppl 1:44–52.
40. Nourhaghighi N, et al. Altered expression of angiopoietins during blood–brain barrier breakdown and angiogenesis. Lab Invest. 2003;83(8):1211–22.
41. Maisonpierre PC, et al. Angiopoietin-2, a natural antagonist for Tie2 that disrupts in vivo angiogenesis. Science. 1997;277(5322):55–60.
42. Tammela T, et al. Angiopoietin-1 promotes lymphatic sprouting and hyperplasia. Blood. 2005;105(12):4642–8.
43. Mochizuki Y, et al. Angiopoietin 2 stimulates migration and tube-like structure formation of murine brain capillary endothelial cells through c-Fes and c-Fyn. J Cell Sci. 2002;115(Pt 1):175–83.
44. Nag S, Kapadia A, Stewart DJ. Review: molecular pathogenesis of blood–brain barrier breakdown in acute brain injury. Neuropathol Appl Neurobiol. 2011;37(1):3–23.
45. Hori S, et al. A pericyte-derived angiopoietin-1 multimeric complex induces occludin gene expression in brain capillary endothelial cells through Tie-2 activation in vitro. J Neurochem. 2004;89(2):503–13.
46. Iurlaro M, et al. Rat aorta-derived mural precursor cells express the Tie2 receptor and respond directly to stimulation by angiopoietins. J Cell Sci. 2003;116(Pt 17):3635–43.
47. Suri C, et al. Increased vascularization in mice overexpressing angiopoietin-1. Science. 1998;282(5388):468–71.
48. Zhang ZG, et al. Angiopoietin-1 reduces cerebral blood vessel leakage and ischemic lesion volume after focal cerebral embolic ischemia in mice. Neuroscience. 2002;113(3):683–7.
49. Thurston G, et al. Angiopoietin-1 protects the adult vasculature against plasma leakage. Nat Med. 2000;6(4):460–3.
50. Shim WS, et al. Angiopoietin-1 promotes functional neovascularization that relieves ischemia by improving regional reperfusion in a swine chronic myocardial ischemia model. J Biomed Sci. 2006;13(4):579–91.
51. Shyu KG, et al. Direct intramuscular injection of plasmid DNA encoding angiopoietin-1 but not angiopoietin-2 augments revascularization in the rabbit ischemic hindlimb. Circulation. 1998;98(19):2081–7.
52. Chae JK, et al. Coadministration of angiopoietin-1 and vascular endothelial growth factor enhances collateral vascularization. Arterioscler Thromb Vasc Biol. 2000;20(12):2573–8.
53. Siddiqui AJ, et al. Combination of angiopoietin-1 and vascular endothelial growth factor gene therapy enhances arteriogenesis in the ischemic myocardium. Biochem Biophys Res Commun. 2003;310(3):1002–9.
54. Namiecinska M, Marciniak K, Nowak JZ. VEGF as an angiogenic, neurotrophic, and neuroprotective factor. Postepy Hig Med Dosw (Online). 2005;59:573–83.
55. Ortega N, Hutchings H, Plouet J. Signal relays in the VEGF system. Front Biosci. 1999;4:D141–52.
56. Zhang Z, et al. VEGF-dependent tumor angiogenesis requires inverse and reciprocal regulation of VEGFR1 and VEGFR2. Cell Death Differ. 2010;17(3):499–512.
57. Ferrara N, Gerber HP, LeCouter J. The biology of VEGF and its receptors. Nat Med. 2003;9(6):669–76.

58. Hess AP, et al. Expression of the vascular endothelial growth factor receptor neuropilin-1 in the human endometrium. J Reprod Immunol. 2009;79(2):129–36.
59. Banerjee S, et al. VEGF-A165 induces human aortic smooth muscle cell migration by activating neuropilin-1-VEGFR1-PI3K axis. Biochemistry. 2008;47(11):3345–51.
60. Zhang ZG, et al. VEGF enhances angiogenesis and promotes blood–brain barrier leakage in the ischemic brain. J Clin Invest. 2000;106(7):829–38.
61. Manoonkitiwongsa PS, et al. Contraindications of VEGF-based therapeutic angiogenesis: effects on macrophage density and histology of normal and ischemic brains. Vascul Pharmacol. 2006;44(5):316–25.
62. Satchell SC, Anderson KL, Mathieson PW. Angiopoietin 1 and vascular endothelial growth factor modulate human glomerular endothelial cell barrier properties. J Am Soc Nephrol. 2004;15(3):566–74.
63. Valable S, et al. VEGF-induced BBB permeability is associated with an MMP-9 activity increase in cerebral ischemia: both effects decreased by Ang-1. J Cereb Blood Flow Metab. 2005;25(11):1491–504.
64. Hattori K, et al. Vascular endothelial growth factor and angiopoietin-1 stimulate postnatal hematopoiesis by recruitment of vasculogenic and hematopoietic stem cells. J Exp Med. 2001;193(9):1005–14.
65. Gluzman Z, et al. Endothelial cells are activated by angiopoeitin-1 gene transfer and produce coordinated sprouting in vitro and arteriogenesis in vivo. Biochem Biophys Res Commun. 2007;359(2):263–8.
66. Lee PC, et al. Impaired wound healing and angiogenesis in eNOS-deficient mice. Am J Physiol. 1999;277(4 Pt 2):H1600–8.
67. Yu J, et al. Endothelial nitric oxide synthase is critical for ischemic remodeling, mural cell recruitment, and blood flow reserve. Proc Natl Acad Sci USA. 2005;102(31):10999–1004.
68. Jozkowicz A, et al. Genetic augmentation of nitric oxide synthase increases the vascular generation of VEGF. Cardiovasc Res. 2001;51(4):773–83.
69. Rudic RD, et al. Direct evidence for the importance of endothelium-derived nitric oxide in vascular remodeling. J Clin Invest. 1998;101(4):731–6.
70. Murohara T, et al. Nitric oxide synthase modulates angiogenesis in response to tissue ischemia. J Clin Invest. 1998;101(11):2567–78.
71. Lopez-Farre A, et al. Role of nitric oxide in the control of apoptosis in the microvasculature. Int J Biochem Cell Biol. 1998;30(10):1095–106.
72. Lee PC, et al. Nitric oxide induces angiogenesis and upregulates alpha(v)beta(3) integrin expression on endothelial cells. Microvasc Res. 2000;60(3):269–80.
73. Dulak J, et al. Nitric oxide induces the synthesis of vascular endothelial growth factor by rat vascular smooth muscle cells. Arterioscler Thromb Vasc Biol. 2000;20(3):659–66.
74. Ziche M, et al. Nitric oxide promotes proliferation and plasminogen activator production by coronary venular endothelium through endogenous bFGF. Circ Res. 1997;80(6):845–52.
75. Michel JB. Role of endothelial nitric oxide in the regulation of the vasomotor system. Pathol Biol (Paris). 1998;46(3):181–9.
76. Kaur S, et al. Genetic engineering with endothelial nitric oxide synthase improves functional properties of endothelial progenitor cells from patients with coronary artery disease: an in vitro study. Basic Res Cardiol. 2009;104(6):739–49.
77. Yang HT, et al. Prior exercise training produces NO-dependent increases in collateral blood flow after acute arterial occlusion. Am J Physiol Heart Circ Physiol. 2002;282(1):H301–10.
78. Michel JB. Role of endothelial nitric oxide in the regulation of arterial tone. Rev Prat. 1997;47(20):2251–6.
79. Fulton D, et al. Regulation of endothelium-derived nitric oxide production by the protein kinase Akt. Nature. 1999;399(6736):597–601.
80. Dai X, Faber JE. Endothelial nitric oxide synthase deficiency causes collateral vessel rarefaction and impairs activation of a cell cycle gene network during arteriogenesis. Circ Res. 2010;106(12):1870–81.

81. Brevetti LS, et al. Overexpression of endothelial nitric oxide synthase increases skeletal muscle blood flow and oxygenation in severe rat hind limb ischemia. J Vasc Surg. 2003;38(4):820–6.

82. Prior BM, et al. Arteriogenesis: role of nitric oxide. Endothelium. 2003;10(4–5):207–16.

83. Gertz K, et al. Physical activity improves long-term stroke outcome via endothelial nitric oxide synthase-dependent augmentation of neovascularization and cerebral blood flow. Circ Res. 2006;99(10):1132–40.

84. Cramer SC, Chopp M. Recovery recapitulates ontogeny. Trends Neurosci. 2000;23(6):265–71.

85. Landers M. Treatment-induced neuroplasticity following focal injury to the motor cortex. Int J Rehabil Res. 2004;27(1):1–5.

86. Cairns K, Finklestein SP. Growth factors and stem cells as treatments for stroke recovery. Phys Med Rehabil Clin N Am. 2003;14(1 Suppl):S135–42.

87. Hurtado O, et al. Neurorepair versus neuroprotection in stroke. Cerebrovasc Dis. 2006;21 Suppl 2:54–63.

88. Jacobson TA. Overcoming 'ageism' bias in the treatment of hypercholesterolaemia: a review of safety issues with statins in the elderly. Drug Saf. 2006;29(5):421–48.

89. Moonis M, et al. HMG-CoA reductase inhibitors improve acute ischemic stroke outcome. Stroke. 2005;36(6):1298–300.

90. Skaletz-Rorowski A, Walsh K. Statin therapy and angiogenesis. Curr Opin Lipidol. 2003;14(6):599–603.

91. Chen J, et al. Vascular endothelial growth factor mediates atorvastatin-induced mammalian achaete-scute homologue-1 gene expression and neuronal differentiation after stroke in retired breeder rats. Neuroscience. 2006;141(2):737–44.

92. Chen J, et al. Statins induce angiogenesis, neurogenesis, and synaptogenesis after stroke. Ann Neurol. 2003;53(6):743–51.

93. Sata M, et al. Statins augment collateral growth in response to ischemia but they do not promote cancer and atherosclerosis. Hypertension. 2004;43(6):1214–20.

94. Skaletz-Rorowski A, et al. The pro- and antiangiogenic effects of statins. Semin Vasc Med. 2004;4(4):395–400.

95. Urbich C, et al. Double-edged role of statins in angiogenesis signaling. Circ Res. 2002;90(6):737–44.

96. Walter DH, et al. Statin therapy accelerates reendothelialization: a novel effect involving mobilization and incorporation of bone marrow-derived endothelial progenitor cells. Circulation. 2002;105(25):3017–24.

97. Liao JK. Clinical implications for statin pleiotropy. Curr Opin Lipidol. 2005;16(6):624–9.

98. Chen J, et al. Atorvastatin induction of VEGF and BDNF promotes brain plasticity after stroke in mice. J Cereb Blood Flow Metab. 2005;25(2):281–90.

99. Dimmeler S, et al. HMG-CoA reductase inhibitors (statins) increase endothelial progenitor cells via the PI 3-kinase/Akt pathway. J Clin Invest. 2001;108(3):391–7.

100. Maeda T, Kawane T, Horiuchi N. Statins augment vascular endothelial growth factor expression in osteoblastic cells via inhibition of protein prenylation. Endocrinology. 2003;144(2):681–92.

101. Walter DH, Zeiher AM, Dimmeler S. Effects of statins on endothelium and their contribution to neovascularization by mobilization of endothelial progenitor cells. Coron Artery Dis. 2004;15(5):235–42.

102. Kureishi Y, et al. The HMG-CoA reductase inhibitor simvastatin activates the protein kinase Akt and promotes angiogenesis in normocholesterolemic animals. Nat Med. 2000;6(9):1004–10.

103. Khaidakov M, et al. Statins and angiogenesis: is it about connections? Biochem Biophys Res Commun. 2009;387(3):543–7.

104. Zacharek A, et al. Simvastatin increases notch signaling activity and promotes arteriogenesis after stroke. Stroke. 2009;40(1):254–60.

105. Lakhan SE, Bagchi S, Hofer M. Statins and clinical outcome of acute ischemic stroke: a systematic review. Int Arch Med. 2010;3:22.

106. Matsumura M, et al. Effects of atorvastatin on angiogenesis in hindlimb ischemia and endothelial progenitor cell formation in rats. J Atheroscler Thromb. 2009;16(4):319–26.

107. Llevadot J, et al. HMG-CoA reductase inhibitor mobilizes bone marrow—derived endothelial progenitor cells. J Clin Invest. 2001;108(3):399–405.
108. Dimmeler S, Dernbach E, Zeiher AM. Phosphorylation of the endothelial nitric oxide synthase at ser-1177 is required for VEGF-induced endothelial cell migration. FEBS Lett. 2000;477(3):258–62.
109. Vasa M, et al. Increase in circulating endothelial progenitor cells by statin therapy in patients with stable coronary artery disease. Circulation. 2001;103(24):2885–90.
110. Newman GC, et al. Association of diabetes, homocysteine, and HDL with cognition and disability after stroke. Neurology. 2007;69(22):2054–62.
111. Sanossian N, et al. Do high-density lipoprotein cholesterol levels influence stroke severity? J Stroke Cerebrovasc Dis. 2006;15(5):187–9.
112. van Exel E, et al. Association between high-density lipoprotein and cognitive impairment in the oldest old. Ann Neurol. 2002;51(6):716–21.
113. Tanne D, Yaari S, Goldbourt U. High-density lipoprotein cholesterol and risk of ischemic stroke mortality. A 21-year follow-up of 8586 men from the Israeli Ischemic heart disease study. Stroke. 1997;28(1):83–7.
114. Zivkovic SA, et al. Rapidly progressive stroke in a young adult with very low high-density lipoprotein cholesterol. J Neuroimaging. 2000;10(4):233–6.
115. Corti MC, et al. HDL cholesterol predicts coronary heart disease mortality in older persons. JAMA. 1995;274(7):539–44.
116. Weverling-Rijnsburger AW, et al. High-density vs low-density lipoprotein cholesterol as the risk factor for coronary artery disease and stroke in old age. Arch Intern Med. 2003;163(13): 1549–54.
117. Mineo C, et al. Endothelial and antithrombotic actions of HDL. Circ Res. 2006;98(11):1352–64.
118. Kuvin JT, et al. A novel mechanism for the beneficial vascular effects of high-density lipoprotein cholesterol: enhanced vasorelaxation and increased endothelial nitric oxide synthase expression. Am Heart J. 2002;144(1):165–72.
119. Mineo C, et al. High density lipoprotein-induced endothelial nitric-oxide synthase activation is mediated by Akt and MAP kinases. J Biol Chem. 2003;278(11):9142–9.
120. Assanasen C, et al. Cholesterol binding, efflux, and a PDZ-interacting domain of scavenger receptor-BI mediate HDL-initiated signaling. J Clin Invest. 2005;115(4):969–77.
121. Nofer JR, et al. HDL induces NO-dependent vasorelaxation via the lysophospholipid receptor S1P3. J Clin Invest. 2004;113(4):569–81.
122. Pu DR, Liu L. HDL slowing down endothelial progenitor cells senescence: a novel anti-atherogenic property of HDL. Med Hypotheses. 2008;70(2):338–42.
123. Zhang Q, et al. Essential role of HDL on endothelial progenitor cell proliferation with PI3K/Akt/cyclin D1 as the signal pathway. Exp Biol Med (Maywood). 2010;235(9): 1082–92.
124. Sumi M, et al. Reconstituted high-density lipoprotein stimulates differentiation of endothelial progenitor cells and enhances ischemia-induced angiogenesis. Arterioscler Thromb Vasc Biol. 2007;27(4):813–8.
125. Elam MB, et al. Effect of niacin on lipid and lipoprotein levels and glycemic control in patients with diabetes and peripheral arterial disease: the ADMIT study: a randomized trial. Arterial disease multiple intervention trial. JAMA. 2000;284(10):1263–70.
126. Schachter M. Strategies for modifying high-density lipoprotein cholesterol: a role for nicotinic acid. Cardiovasc Drugs Ther. 2005;19(6):415–22.
127. Shepherd J, Betteridge J, Van Gaal L. Nicotinic acid in the management of dyslipidaemia associated with diabetes and metabolic syndrome: a position paper developed by a European Consensus Panel. Curr Med Res Opin. 2005;21(5):665–82.
128. Kamanna VS, Kashyap ML. Mechanism of action of niacin. Am J Cardiol. 2008;101(8A): 20B–6.
129. Jin FY, Kamanna VS, Kashyap ML. Niacin accelerates intracellular ApoB degradation by inhibiting triacylglycerol synthesis in human hepatoblastoma (HepG2) cells. Arterioscler Thromb Vasc Biol. 1999;19(4):1051–9.

130. Jin FY, Kamanna VS, Kashyap ML. Niacin decreases removal of high-density lipoprotein apolipoprotein A-I but not cholesterol ester by Hep G2 cells. Implication for reverse cholesterol transport. Arterioscler Thromb Vasc Biol. 1997;17(10):2020–8.
131. Rosenson RS. Antiatherothrombotic effects of nicotinic acid. Atherosclerosis. 2003; 171(1):87–96.
132. Chen J, et al. Niaspan increases angiogenesis and improves functional recovery after stroke. Ann Neurol. 2007;62(1):49–58.
133. Chen J, et al. Niaspan treatment increases tumor necrosis factor-alpha-converting enzyme and promotes arteriogenesis after stroke. J Cereb Blood Flow Metab. 2009;29(5):911–20.
134. Chapman MJ. Therapeutic elevation of HDL-cholesterol to prevent atherosclerosis and coronary heart disease. Pharmacol Ther. 2006;111(3):893–908.
135. Toth PP. High-density lipoprotein as a therapeutic target: clinical evidence and treatment strategies. Am J Cardiol. 2005;96(9A):50K–8K; discussion 34K–5K.
136. Grundmann S, et al. Anti-tumor necrosis factor-{alpha} therapies attenuate adaptive arteriogenesis in the rabbit. Am J Physiol Heart Circ Physiol. 2005;289(4):H1497–505.
137. Edwards DR, Handsley MM, Pennington CJ. The ADAM metalloproteinases. Mol Aspects Med. 2008;29(5):258–89.
138. Garton KJ, et al. Stimulated shedding of vascular cell adhesion molecule 1 (VCAM-1) is mediated by tumor necrosis factor-alpha-converting enzyme (ADAM 17). J Biol Chem. 2003;278(39):37459–64.
139. Krause DS. Plasticity of marrow-derived stem cells. Gene Ther. 2002;9(11):754–8.
140. Menasche P. Cell transplantation for the treatment of heart failure. Semin Thorac Cardiovasc Surg. 2002;14(2):157–66.
141. Wang JS, et al. Marrow stromal cells for cellular cardiomyoplasty: feasibility and potential clinical advantages. J Thorac Cardiovasc Surg. 2000;120(5):999–1005.
142. Chen J, et al. Therapeutic benefit of intracerebral transplantation of bone marrow stromal cells after cerebral ischemia in rats. J Neurol Sci. 2001;189(1–2):49–57.
143. Li Y, et al. Human marrow stromal cell therapy for stroke in rat: neurotrophins and functional recovery. Neurology. 2002;59(4):514–23.
144. Chen J, et al. Therapeutic benefit of intravenous administration of bone marrow stromal cells after cerebral ischemia in rats. Stroke. 2001;32(4):1005–11.
145. Wu J, et al. Intravenously administered bone marrow cells migrate to damaged brain tissue and improve neural function in ischemic rats. Cell Transplant. 2008;16(10):993–1005.
146. Riess P, et al. Transplanted neural stem cells survive, differentiate, and improve neurological motor function after experimental traumatic brain injury. Neurosurgery. 2002;51(4):1043–52; discussion 1052–4.
147. Sanchez-Ramos J, et al. Adult bone marrow stromal cells differentiate into neural cells in vitro. Exp Neurol. 2000;164(2):247–56.
148. Tang YL, et al. Paracrine action enhances the effects of autologous mesenchymal stem cell transplantation on vascular regeneration in rat model of myocardial infarction. Ann Thorac Surg. 2005;80(1):229–36; discussion 236–7.
149. Kinnaird T, et al. Bone marrow-derived cells for enhancing collateral development: mechanisms, animal data, and initial clinical experiences. Circ Res. 2004;95(4):354–63.
150. Ponte AL, et al. The in vitro migration capacity of human bone marrow mesenchymal stem cells: comparison of chemokine and growth factor chemotactic activities. Stem Cells. 2007;25(7):1737–45.
151. Wu Y, et al. Mesenchymal stem cells enhance wound healing through differentiation and angiogenesis. Stem Cells. 2007;25(10):2648–59.
152. Kinnaird T, et al. Marrow-derived stromal cells express genes encoding a broad spectrum of arteriogenic cytokines and promote in vitro and in vivo arteriogenesis through paracrine mechanisms. Circ Res. 2004;94(5):678–85.
153. Matsuda-Hashii Y, et al. Hepatocyte growth factor plays roles in the induction and autocrine maintenance of bone marrow stromal cell IL-11, SDF-1 alpha, and stem cell factor. Exp Hematol. 2004;32(10):955–61.

154. Annabi B, et al. Hypoxia promotes murine bone-marrow-derived stromal cell migration and tube formation. Stem Cells. 2003;21(3):337–47.
155. Zacharek A, et al. Angiopoietin1/Tie2 and VEGF/Flk1 induced by MSC treatment amplifies angiogenesis and vascular stabilization after stroke. J Cereb Blood Flow Metab. 2007;27(10):1684–91.
156. Chen J, et al. Intravenous bone marrow stromal cell therapy reduces apoptosis and promotes endogenous cell proliferation after stroke in female rat. J Neurosci Res. 2003;73(6):778–86.
157. Al-Khaldi A, et al. Postnatal bone marrow stromal cells elicit a potent VEGF-dependent neoangiogenic response in vivo. Gene Ther. 2003;10(8):621–9.
158. Al-Khaldi A, et al. Therapeutic angiogenesis using autologous bone marrow stromal cells: improved blood flow in a chronic limb ischemia model. Ann Thorac Surg. 2003;75(1):204–9.
159. Cui X, et al. Chemokine, vascular and therapeutic effects of combination Simvastatin and BMSC treatment of stroke. Neurobiol Dis. 2009;36(1):35–41.
160. Eaves CJ, et al. Mechanisms that regulate the cell cycle status of very primitive hematopoietic cells in long-term human marrow cultures. II. Analysis of positive and negative regulators produced by stromal cells within the adherent layer. Blood. 1991;78(1):110–7.
161. Majumdar MK, et al. Phenotypic and functional comparison of cultures of marrow-derived mesenchymal stem cells (MSCs) and stromal cells. J Cell Physiol. 1998;176(1):57–66.
162. Seshi B, Kumar S, Sellers D. Human bone marrow stromal cell: coexpression of markers specific for multiple mesenchymal cell lineages. Blood Cells Mol Dis. 2000;26(3):234–46.
163. Bang OY, et al. Autologous mesenchymal stem cell transplantation in stroke patients. Ann Neurol. 2005;57(6):874–82.
164. Sykova E, et al. Bone marrow stem cells and polymer hydrogels-two strategies for spinal cord injury repair. Cell Mol Neurobiol. 2006;26(7–8):1113–29.
165. Malgieri A, et al. Bone marrow and umbilical cord blood human mesenchymal stem cells: state of the art. Int J Clin Exp Med. 2010;3(4):248–69.
166. Suarez-Monteagudo C, et al. Autologous bone marrow stem cell neurotransplantation in stroke patients. An open study. Restor Neurol Neurosci. 2009;27(3):151–61.
167. Lee JS, et al. A long-term follow-up study of intravenous autologous mesenchymal stem cell transplantation in patients with ischemic stroke. Stem Cells. 2010;28(6):1099–106.
168. Chen J, et al. Intravenous administration of human umbilical cord blood reduces behavioral deficits after stroke in rats. Stroke. 2001;32(11):2682–8.
169. Newcomb JD, et al. Timing of cord blood treatment after experimental stroke determines therapeutic efficacy. Cell Transplant. 2006;15(3):213–23.
170. Zhang L, et al. Delayed administration of human umbilical tissue-derived cells improved neurological functional recovery in a rodent model of focal ischemia. Stroke. 2011;42(5): 1437–44.

Chapter 12
Hematopoietic Growth Factor Family for Stroke Drug Development

Ihsan Solaroglu and Murat Digicaylioglu

Abstract Hematopoietic growth factors (HGFs) are a family of cytokines that play important roles in the survival, proliferation, and differentiation of hematopoietic progenitors, as well as in the functional activation of mature cells. However, there is a growing body evidences have suggested that HGFs have important non-hematopoietic functions in the central nervous system. Preclinical studies have demonstrated significant benefits from the use of these cytokines in the acute and chronic treatment of various neurological disorders. Unfortunately, the efficacy of these candidate neuroprotectants in animal experiments does not reliably predict efficacy in stroke patients. This chapter presents a broad overview of data from preclinical neuroprotection and clinical trial studies.

1 Background

Stroke is a major cause of death and disability worldwide. Each year about 700,000 people experience a new or recurrent stroke in the United States [1]. Disabilities and neurological impairments in stroke patients cause mild to severe levels of loss of quality of life, with severe burden on the patients, their families and society [2]. For the last 2 decades, a great deal of attention has been paid to identify the pathophysiological pathways involved in stroke and to design neuroprotectant agents to modulate

I. Solaroglu, MD(✉)
Department of Neurosurgery, Neuroscience Research Laboratories, Koç Univesity,
School of Medicine, Rumelifeneri Yolu, Sariyer, Istanbul 34450, Turkey
e-mail: isolaroglu@hotmail.com

M. Digicaylioglu, MD, PhD
Departments of Neurosurgery and Physiology, University of Texas, Health Science Center,
Interdisciplinary Graduate Training Program in Neuroscience,
7703 Floyd Curl Drive-7843, San Antonio, TX 78229-3900, USA

P.A. Lapchak and J.H. Zhang (eds.), *Translational Stroke Research*, Springer Series
in Translational Stroke Research, DOI 10.1007/978-1-4419-9530-8_12,
© Springer Science+Business Media, LLC 2012

these pathways. Although numerous agents have been found to reduce infarct size and improve clinical outcome in experimental models, the clinical use of these neuroprotective agents has been hampered by toxicity of the compounds. Recombinant tissue plasminogen activator (rt-PA) remains the only approved agent for acute stroke management in humans which can only be available to a very limited number of patients [1, 3]. Moreover, a narrow therapeutic window and potential side effects limit the efficacy of rt-PA.

Hematopoietic growth factors (HGFs) are a family of cytokines that play important roles in the survival, proliferation, and differentiation of hematopoietic progenitors, as well as in the functional activation of mature cells. However, there is a growing body evidences have suggested that HGFs have important non-hematopoietic functions in the central nervous system (CNS) [2, 4–9]. The expression of HGFs and their receptors are crucial during early embryonic development of the neural system, and structural and functional integrity of the brain. Moreover, a number of growth factors, including granulocyte colony-stimulating factor (G-CSF), granulocyte-macrophage colony-stimulating factor (GM-CSF), and erythropoietin (EPO) have been exert neuroprotective effects both in vitro and in vivo [5, 10–12]. Numerous studies performed to date have demonstrated significant benefits from the use of these cytokines in the acute and chronic treatment of various neurological disorders. Unfortunately, the efficacy of these candidate neuroprotectants in animal experiments does not reliably predict efficacy in stroke patients. This chapter presents a broad overview of the two well-known members of HGFs; G-CSF and EPO, in neuroprotection and clinical trial results of these agents in human stroke.

2   Granulocyte Colony-Stimulating Factor

2.1   The Cytokine G-CSF and Its Receptor

Human G-CSF is encoded by a single gene that is located on chromosome 17 q11–12. Structurally, G-CSF is a glycoprotein consisting of four anti-parallel alpha-helices with a molecular mass of 19 kDa [13]. G-CSF is produced by a variety of cells including bone marrow stromal cells, endothelial cells, macrophages, fibroblasts, and astrocytes in response to specific stimulation [14]. The proliferation and differentiation of neutrophilic progenitor cells is largely dependent on the binding of G-CSF to its specific receptor. The G-CSF receptor (G-CSFR) is a type I membrane protein and has a composite structure consisting of an immunoglobulin-like domain, a cytokine receptor-homologous domain and three fibronectin type III domains in the extracellular region. However, G-CSFR does not have its intracellular kinase domain as do other hematopoietin receptor superfamily members. The cytoplasmic domain of G-CSFR consists of three amino acid sequences (Boxes 1–3) which are critical for the signal transduction of the G-CSFR [15]. G-CSFR is expressed not only on a variety of hematopoietic cells including neutrophils, and their precursors, monocytes, platelets, lymphocytes, and leukemia cells, but also on non-hematopoietic cells such as endothelial cells, neurons and glial cells [7, 16].

2.2 Clinical Applications of G-CSF

Filgrastim is a recombinant human G-CSF was approved by the United States Food and Drug Administration (FDA) for use in chemotherapy induced neutropenia. Many of patients have already been receiving G-CSF as a highly efficacious and safe treatment. G-CSF is also being used clinically to facilitate hematopoietic recovery after bone marrow transplantation, to treat severe congenital neutropenia and to mobilize peripheral blood progenitor cells in healthy donors [17, 18].

2.3 G-CSF and G-CSFR in the Brain

G-CSFR and its ligand have been shown to be expressed by neurons in a variety of brain regions including pyramidal cells in cortical layers (particularly in layers II and V), Purkinje cells in the cerebellum, subventricular zone (SVZ) and cerebellar nuclei in rats. Cells in the CA3 region of the hippocampus, subgranular zone and hilus of the dentate gyrus, entorhinal cortex, and olfactory bulb, positively staining for G-CSF have also been identified previously [16]. Moreover, G-CSFR expression has shown in the frontal cortex of human brain by postmortem studies. Our experiments also showed the localization of G-CSFR on neural cells in rat brain and spinal cord samples [7, 9, 19, 20]. Recently, it has been shown that G-CSF as an essential neurotrophic factor is necessary for the structural and functional integrity of the hippocampal formation [21].

Co-expression and upregulation of G-CSF and its receptor, in neurons after middle cerebral artery occlusion (MCAO) and reperfusion injury has been reported in the rodent CNS, indicating an autocrine protective response of the injured brain [16]. Kleinschnitz et al. showed that 4 h after permanent MCAO, G-CSF mRNA levels massively increases compared to normal cortex and decreases after 2 days to its control levels [22]. The increase in G-CSF mRNA expression was not only seen in the ischemic lesion but also in the nonischemic frontal cortex after focal cerebral ischemia.

2.4 G-CSF as a Novel Neuroprotectant: Preclinical Studies

Recent studies have shown the presence of G-CSF/G-CSFR system in the brain, and their roles in neuroprotection and neural tissue repair as well as improvement in functional recovery. As evidenced by the number of publications and the resulting

Fig. 12.1 The figure shows the neuroprotective actions of G-CSF suggested by the existing literature. G-CSF has various effects including anti-apoptotic, anti-inflammatory, neurotrophic, and
excitoprotective properties that are involved in neuroprotection. In addition, G-CSF induces angiogenesis and neurogenesis after experimental stroke

insights, G-CSF emerges as a novel neuroprotectant with significant clinical potential. The major advantage of G-CSF is its well-described pharmacological behavior
and its clinical use over several years. G-CSF induces multiple mechanisms of
action after stroke, such as induction of neurogenesis and angiogenesis and inhibition of inflammation and apoptosis (Fig. 12.1). Potential additional mechanisms
include the mobilization of stem cells from the bone marrow as well as immunomodulatory effects after stroke [23].

A series of reports have demonstrated that G-CSF displays anti-inflammatory
properties [8, 24, 25]. Inflammation, in response to brain injury, involves infiltration
of inflammatory cells into the injured brain parenchyma and activation of resident
brain cells, which are capable of generating pro-inflammatory cytokines such as
tumor necrosis factor-alpha (TNF-alpha) and interleukin-1-beta (IL-1beta). G-CSF
administration is associated with a marked decrease in the number of infiltrated neutrophils within the hemisphere undergoing infarction experimentally [26, 27]. It is
well known that peripherally derived leukocytes can produce and secrete proteolytic

enzymes and free oxygen radicals which mediate further blood–brain barrier (BBB) injury. One of the possible mechanisms of action of G-CSF in the prevention of BBB breakdown is its ability to attenuate the inflammatory response [26]. Park et al. reported that intra-peritoneal G-CSF administration after intracerebral hemorrhage reduced brain edema and inflammation, and decreased the BBB permeability in the rat [24]. A magnetic resonance imaging (MRI) study showed that G-CSF significantly reduces the amount of edematous brain tissue after cerebral ischemia [28]. In animal models of cerebral ischemia, G-CSF attenuates inflammation by reducing pro-inflammatory cytokine expressions including TNF-alpha, transforming growth factor-beta (TGF-beta), and IL-1beta [8, 25]. Moreover, G-CSF suppresses IL-1beta mRNA upregulation, activation of inducible nitric oxide synthase (iNOS) positive microglia, and decreases iNOS levels [1, 25, 28, 29]. However, a recent study by Taguchi et al. suggested that G-CSF exacerbates the inflammatory response both at the border of the ischemic region and also in non-ischemic brain tissue leading to contradictory results [30].

Matrix metalloproteinases (MMPs), which is a family of zinc-binding proteolytic enzymes involved in the remodeling of the extracellular matrix, have been implicated in the pathogenesis of stroke in both animal and human studies. Activation of MMP-9 increases BBB permeability, inflammatory infiltration, and infarction volume after cerebral ischemia reperfusion injury [31]. The disruption of the basal lamina of the vascular structures by MMP-9 involves in decreased vascular integrity after ischemic insults. In the peri-ischemic cortex of G-CSF deficient mice acute upregulation of MMP-9 occurs following experimental stroke. However, G-CSF substitution suppresses expression of MMP-9. Although further studies are needed to precisely define the pathways, reduction of MMP-9 activity by G-CSF implicates the importance of endogenous G-CSF in brain recovery mechanisms [32].

Another tissue-protective mechanism of G-CSF is its ability to protect cells against apoptosis. Excitatory amino acids, particularly glutamate, have a crucial role in ischemic neuronal death. Ischemic brain injury results in increased extracellular concentrations of glutamate which induces apoptotic cell death. G-CSF was shown to inhibit glutamate release in the ischemic brain regions and exerts an anti-apoptotic effect on neurons both in vivo and in vitro [1, 16, 20, 24, 27, 29, 33]. A role for the Janus kinase [5] 2/signal transducer and activator of transcription (STAT) 3 signaling pathway has been suggested as a major mediator of the anti-apoptotic actions of G-CSF which is mediated via binding to the cell surface receptor G-CSFR in neurons [1, 16, 20, 27, 29] (Figs. 12.2 and 12.3). The PI3K/phosphoinositide-dependent kinase/Akt signaling and extracellular signal-regulated kinase (ERK) family has been also shown to contribute to the anti-apoptotic action of G-CSF. Hence, the tissue-protective molecular pathways that are triggered by G-CSF have similarities to those activated during hematopoiesis. G-CSF displays anti-apoptotic activity not only in neuronal cells but also in glial cells during ischemic injury by effecting different pathways such as cellular inhibitor of apoptosis-2 (cIAP2) [20] (Fig. 12.4). Increased glial survival is important for the support and the production of trophic factors in the CNS and may play a key role in conferring resistance to ischemic damage.

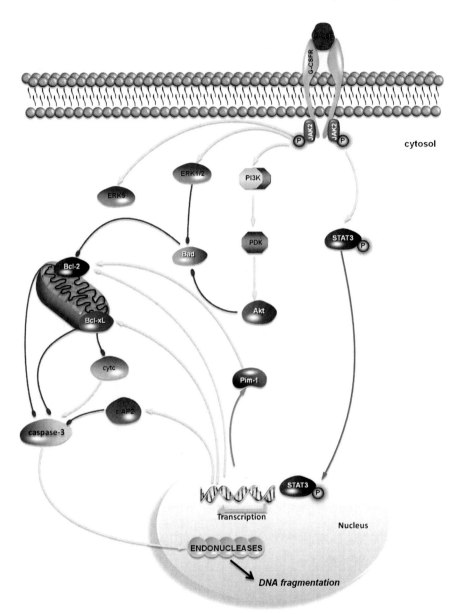

Fig. 12.2 Signaling pathways suggested for G-CSF–mediated neuroprotection. G-CSFR activates the Janus-tyrosine kinase 2, protein kinase B (Akt), and signal transducer and activator of transcription (STAT) proteins. Activation of JAK2/STAT3 leads to increased expression of the anti-apoptotic proteins; cellular inhibitors of apoptotic protein-2, Bcl-xL and bcl-2, which prevents apoptotic cell death by inhibiting activation of the caspases in neurons and glia. Activation of the extracellular signal-related kinases (ERKs) and PI3K/Akt also enhances neuronal survival

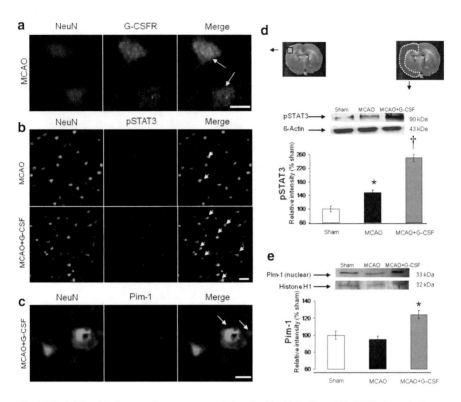

Fig. 12.3 (**a**) Double immunofluorescence staining for NeuN (*red*) and G-CSFR (*green*) shows the expression of G-CSFR in neurons (merge, *yellow*). (**b**) Double immunofluorescence staining for NeuN (*green*) and pSTAT3 (*red*) in the rat brain cortex from the middle cerebral artery occlusion (MCAO) and G-CSF-treated groups. Notice that there is an increased number of pSTAT3 and NeuN colocalized cells (merge, *yellow*) in the G-CSF-treated group. (**c**) Double immunofluorescence staining for NeuN (*green*) and Pim-1 (*red*) shows Pim-1 expression in a neuron from a G-CSF-treated brain. (**d**, **e**) Western blot analysis shows that G-CSF treatment significantly increased pSTAT3 and nuclear Pim-1 levels. (Reprinted from, Solaroglu et al. [20] Copyright (2006), with permission from Elsevier.)

Stem cells can be defined as immature cells with prolonged self-renewal capacity and ability to differentiate into multiple cell types. Regenerative medicine holds the promise of replacing damaged tissues largely by stem cell activation to restore adequate function. Neural stem cells (NSCs) are able to generate other cell types that constitute the CNS, including neurons, oligodendrocytes, and astrocytes. G-CSF modulates the regenerative ability of cells in the brain and enhances their capacity to acquire neuronal characteristics. G-CSF has been shown to have a functional role in the differentiation of adult rat NSCs both in vitro and in vivo [16]. G-CSF increases the number of NSCs expressing mature neural markers from SVZ or hippocampus. Peripheral infusion of G-CSF enhances the recruitment of progenitor cells from the lateral ventricular wall into the ischemic area of the neocortex in the rat. Moreover, G-CSF increases hippocampal neurogenesis not only in ischemic animals but also in the intact, nonischemic rat. The absence of G-CSF results in an

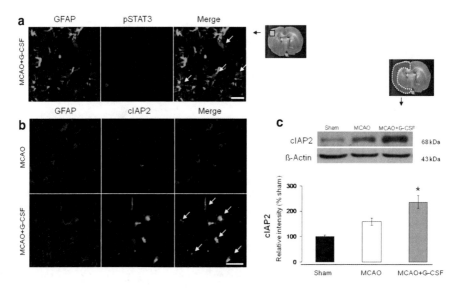

Fig. 12.4 (**a**) Double immunofluorescence staining was performed for GFAP (*green*) and pSTAT3 (*red*). At 24 h, an increased pSTAT3 immunostaining is seen in glia in the rat brain cortex from the G-CSF-treated group. Glia expressing pSTAT3 appeared yellow (*arrows*). (**b**) Double immunofluorescence staining for GFAP (*red*) and cIAP2 (*green*) revealed an extensive expression of cIAP2 in glial (merge, *arrows*) in the G-CSF-treated group. (**c**) Representative Western Blot analysis showing cIAP2 expression in brain tissues from sham, MCAO, and G-CSF-treated groups 24 h after reperfusion. Quantification of WB analysis showed significantly increased cIAP2 levels in the G-CSF group. (Reprinted from, Solaroglu et al. [20], Copyright (2006), with permission from Elsevier.)

impairment of adult neurogenesis and its modulation in the dentate gyrus [21]. G-CSF treatment increases the number of proliferating cells in the dentate gyrus and in the SVZ in also aged rats, which may be important for the further clinical development of the drug in elderly patients with stroke [34]. Investigators have also demonstrated the capability of G-CSF combination therapy with other growths factors to stimulate proliferation and differentiation of neural stem/progenitor cells in the animal brain. G-CSF can also act synergistically with SCF to stimulate angiogenesis and neurogenesis after injury. Administration of G-CSF alone or in combination with stem cell factor (SCF) during chronic stroke exerted trophic effects in the subacute phase of focal cerebral ischemia by synergistically facilitating the proliferation of intrinsic neural stem/progenitor cells in the neurogenic sites [35]. In the chronic phase of stroke administration of both SCF and G-CSF promotes neural progenitor cell proliferation, differentiation, and migration into the infarct area, which participates in infarction size reduction [36]. A combination of G-CSF and EPO could also synergistically promote proliferation of the neural progenitor cells residing in the hippocampus and subventricular region of brain [37].

Whether G-CSF can decrease neuronal injury by mobilization of hematopoietic stem cells into damaged brain areas is currently a matter of discussion. Six et al.

suggested neuroprotective effect of G-CSF to be mediated by the mobilization of autologous HSCs after focal cerebral ischemia in mice [38]. However, this study has no evidence to support the notion that the decrease in infarction volumes is through the mobilization of HSCs from bone marrow to the infarct region in the brain. Similarly, Shyu et al. proposed that the G-CSF treatment may mobilize autologous HSCs into circulation, enhance their translocation into ischemic brain, and thus significantly improve lesion repair [39]. A recent report from Piao et al. showed that SCF and G-CSF treatment can mobilize circulating bone marrow stem cells (BMSCs) into peripheral blood, enhance homing bone marrow-derived (BM-derived) cells into the brain, promote differentiation of BM-derived stem cells into endothelial cells and neuron-like cells, and increase angiogenesis [40]. G-CSF and SCF administration modulates the availability of bone marrow derived cells in the brain and enhances their capacity to acquire neuronal characteristics [19]. Taken together, SCF and G-CSF intervention itself has the capability to stimulate BMSCs to form neurons. However, it has also been suggested that BM-derived cells are not involved in G-CSF-induced reduction of ischemic injury leading to contradictory results [29]. The information currently available seems limited to suggest that the mobilization of BMCs to the damaged brain is central to the neuroprotective capabilities of G-CSF. Further experiments are required to assess whether mobilized stem cells migrate into the brain and whether they involved into the tissue repair. However, regardless of the source of the cells, these findings suggest a new therapeutic strategy to restore functional and structural recovery after stroke.

In addition to the above neuroprotective actions, an additional beneficial action of G-CSF on ischemic tissues has been identified in recent studies. Angiogenesis is critical for the development and repair of the CNS. Although angiogenetic response after an ischemic insult has been reported to affect the overall clinical outcome of stroke, the exact pattern of neovascularization and its relation to restoration of function after stroke are still remains unknown [41]. Many growth factors are involved in the regulation of the angiogenesis both in normal and pathological conditions. Detailed studies of the mechanisms for G-CSF promoted angiogenesis reveal multiple pathways. G-CSF stimulates endothelial proliferation and vascular surface areas in the injured hemisphere, and also increases endothelial nitric oxide synthase (eNOS) and angiopoietin-2 expression [26]. It has been shown that G-CSF enhances collateral artery growth and reduces infarct volume in a mouse model of brain ischemia [42]. G-CSF in combination with EPO could induce angiogenesis through encouraging the homing of BMSCs and their differentiation into vascular-endothelial cells at ischemic sites following experimental cerebral ischemia [37]. Similarly, combination of G-CSF and SCF promotes angiogenesis following permanent MCAO in mice with a 1.5-fold increase in vessel formation, a large percentage of which contain endothelial cells of bone marrow origin [43]. Here activation angiogenetic signaling also seems to play a role in the G-CSF-induced neuroprotection.

2.5 G-CSF for the Treatment of Human Stroke

A great deal of preclinical work has been carried out on G-CSF and a number of mechanisms of its action in the CNS have been identified both in vivo and in vitro. The beneficial actions of G-CSF including neuroprotective and regenerative activity, reducing cerebral edema, improving survival, and enhancing sensorimotor and functional recovery makes it a potential drug for the treatment of human stroke. Most importantly, G-CSF has been shown to be a safe drug, as many patients have received it over the last decade after myelosuppressive treatments.

A series of guidelines have been published by the Stroke Therapy Academic Industry Roundtable (STAIR) concerning the preclinical development of stroke recovery drugs [44]. STAIR suggested that putative stroke recovery drugs should be tested in rodents, recovery of sensorimotor function and monitoring of physiological parameters should be performed, the route and the time window of administration should be carefully considered, and careful toxicological studies should be done in several species, including both intact animals and animals with stroke. Currently, neuroprotective effects of G-CSF have been tested in multiple animal models in different species with different time points, and in different doses and application routes. The current treatment options in ischemic stroke are limited to thrombolytic therapy using intravenously or intraarterially administered rt-PA. Hence, the use of G-CSF in conjunction with rt-PA investigated as suggested by us previously [9]. G-CSF reduces infarct volume and mortality when given prior to delayed rt-PA treatment in experimental model of thromboembolic stroke. Also, early combination with rt-PA does not lead to side effects [45].

The number of randomized controlled trials with G-CSF application in stroke patients is very limited. A very small, nonplacebo-controlled, randomized trial published by Shyu et al. [46]. Seven patients within 7 days of ischemic stroke received subcutaneous G-CSF injections (15 µg/kg per day) for 5 days. G-CSF therapy reported to be well tolerated, safe and feasible and leads to improved neurologic outcomes without adverse effects. Although these results seem promising, the number of patients enrolled to the study was too small. On the other hand, the results of first Phase IIa dose-escalation clinical trial (1, 3 and 10 µg/kg body weight for 1 or 5 days within a time window of between 1 and 4 weeks post-stroke) suggest that G-CSF treatment, at standard hematology doses, mobilizes bone marrow CD34+ stem cells in older patients. More importantly, administration is feasible and appears to be safe and well tolerated [47]. Although the treatment appears to be well tolerated and safe, it is still unknown whether mobilized stem cells migrate into the brain and whether treatment improves functional outcome. A recently published multicenter, randomized, placebo-controlled dose escalation study tested four intravenous dose regimens (total cumulative doses of 30–180 µg/kg over the course of 3 days) of G-CSF in acute ischemic stroke patients. Authors reported that G-CSF is well-tolerated even at high dosages, and that a substantial increase in leukocytes does not lead to an increase in thromboembolic events [48].The first randomized controlled trial that assessed the safety and feasibility of G-CSF in chronic stroke patients showed that subcutaneous G-CSF treatment in elderly chronic stroke patients with concomitant vascular disease is safe and reasonably well tolerated [49]. Although results of clinical trials discussed above are promising, it is too early to know whether CSFs improve functional outcome in human stroke. Hence, further large clinical trials are now required.

3   Erythropoietin

3.1   The Cytokine EPO and Its Receptor

The cytokine EPO, which is produced by all vertebrates investigated so far, is a 30.4 kD glycoprotein which was originally described to stimulate erythropoiesis [10, 50]. The human EPO gene is located on the long arm of chromosome 7 (q11–q22). EPO, mainly produced by fetal liver and adult kidney in response to hypoxia, promotes the maturation and proliferation of erythroid progenitor cells. EPO is essential for regulating circulating levels of red blood cells and increases the number of circulating erythrocytes primarily by preventing apoptosis of erythroid progenitors [1, 50, 51]. EPO acts by binding to its specific transmembrane receptor (EPOR) which belongs to the single-chain cytokine type I receptor family [52]. EPO activates a variety of intracellular cascades, including the Janus kinase 2(JAK)/STAT, the Ras/mitogen-activated protein kinase (MAPK), and phosphatidylinositol 3-kinase (PI3-K)/Akt. Activation of these cascades subsequently activates their downstream substrates and affects target genes which regulate erythropoiesis [10]. Production and secretion of EPO and the expression of EPOR are regulated by the tissue oxygen supply, mostly via the activation of the hypoxia-inducible factor-1 (HIF-1) pathway [53, 54]. Initially, EPO was thought to only be involved in erythropoiesis, but this idea was set aside when it was discovered that the expression of EPO and EPOR is not only limited to the kidney and liver, but rather extends to such organs as the heart, brain and placenta [50, 55].

3.2   Clinical Applications of EPO

In 1989, rhEPO was approved by FDA for the treatment of anemia associated with chronic renal failure. During the last 20 years, recombinant human erythropoietin (rhEPO) has been widely used in clinical practice for the treatment of such conditions as anemia associated with chronic kidney disease, acquired immunodeficiency syndrome, bone marrow transplantation, myelodysplastic syndromes, autoimmune diseases, cancer patients on chemotherapy and surgical patients to avoid allogeneic red blood cell transfusion [50].

3.3   EPO and EPOR in the Brain

Both EPO and the EPOR are functionally expressed in the nervous system of rodents, primates, and humans and the role of EPO in the nervous system extends beyond tissue oxygenation. The expression of EPO and its receptor are crucial during early embryonic development of the neural system. EPO and EPOR are expressed during critical periods of neuronal development, and EPO and EPOR null animals reveal defects in neurogenesis [56]. The major sites of EPO production and secretion are the hippocampus, internal capsule, cortex, astrocytes midbrain, oligodendrocytes and in the endothelial cells of the BBB. The EPOR is expressed by a variety of cells including neurons, microglia, astrocytes, and cerebral endothelial cells and by the myelin sheaths on human peripheral nerves [4, 54, 55]. Endogenous EPO produced in the CNS acts locally in a paracrine or autocrine fashion, but its specific function(s) in the brain remain unclear. However, a hematopoietic function of brain-derived EPO is doubtful due to significantly higher concentrations of EPO in the blood than in the CNS [2].

3.4   Neuroprotective Properties of EPO: Preclinical Studies

During the last 10 years, it has become clear that EPO is neuroprotective both in vivo and in vitro models of cerebral ischemia by inducing a broad range of cellular responses in the brain directed to protect and repair tissue damage (Fig. 12.5).

One of the proposed mechanisms of EPO-induced neuroprotection is its ability to protect cells against oxidative and excitotoxic injury [11, 12]. Glutamate is the most abundant excitatory neurotransmitter in the human CNS, released in response to presynaptic neuronal membrane depolarization. It is well known that N-Methyl-D-aspartate (NMDA) mimics the action of glutamate on the NMDA receptor and the accompanied cell death after exposure [57]. Pretreatment of cultured mouse neocortical neurons with EPO prevents NMDA-induced neuronal death [58]. This finding is in agreement with previous reports showing a neuroprotective effect of EPO on glutamate-mediated and ischemia-induced neuronal cell death [4]. EPO has been shown to reduce kainic acid and glutamate-induced excitotoxicity in cultured neurons [59–61]. Sakanaka et al. suggested that EPO may exert its neuroprotective effect by reducing the nitric oxide (NO)-mediated formation of free radicals or antagonizing their toxicity. EPO also may increase the activities of antioxidant enzymes, such as superoxide dismutase, glutathione peroxidase, and catalase in neurons [60]. Additionally, activation of EPOR may suppress ischemic cell death by inhibiting the exocytosis of glutamate in cultured cerebellar granule neurons [62]. However, EPO-induced neuroprotection against NMDA excitotoxicity depended significantly on TNFRI presence. EPO prevented neuronal damage induced by kainic acid in WT but not TNFRI KO mice, suggesting there is a complex and extensive crosstalk between protective pathways [61].

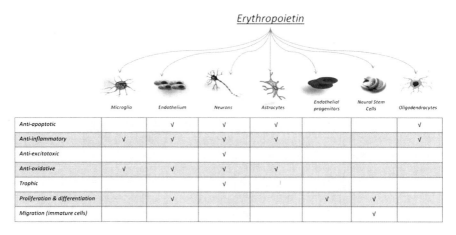

	Microglia	Endothelium	Neurons	Astrocytes	Endothelial progenitors	Neural Stem Cells	Oligodendrocytes
Anti-apoptotic		√	√	√			√
Anti-inflammatory	√	√	√	√			√
Anti-excitotoxic			√				
Anti-oxidative	√	√	√	√			
Trophic			√	❘			
Proliferation & differentiation		√			√	√	
Migration (immature cells)						√	

Fig. 12.5 Multimodal neuroprotective profile of erythropoietin. EPO activates anti-apoptotic, anti-oxidant and anti-inflammatory signaling pathways in neurons, glial and cerebrovascular endothelial cells, and stimulates angiogenesis and neurogenesis

Anti-apoptotic effect of EPO has been reported to be a fundamental mechanism of neuroprotection and tissue preservation following ischemic neuronal injury (Fig. 12.6). Activation of neuronal EPORs prevents apoptosis induced by NMDA or NO by triggering cross-talk between the signaling pathways of JAK2 and nuclear factor-B (NF-$_\kappa$B) in primary cultures of cerebrocortical neurons [5]. After the activation of JAK2, EPO modulates neuronal survival and apoptosis through a variety of downstream signals that range from the level of STAT proteins to caspases [55]. EPO also exerts a profound stimulatory action on rat microglial cell viability in vitro through a mechanism involving simultaneously cell proliferation and the inhibition of apoptosis. However, this effect is much weaker on cultured astrocytes [63]. Several additional downstream pathways also have been shown to be linked intimately to the anti-apoptotic mechanisms of EPO. EPO may prevent apoptosis through the PI3-K/Akt pathway [64, 65], which maintains the mitochondrial membrane potential, thereby preventing the subsequent release of cytochrome c, and modulates caspase activity [54, 66]. Both Bcl-2 and Bcl-x$_L$ block neuronal apoptosis by preventing mitochondrial cytochrome c release as well as modulating intracellular calcium levels. In an in vivo model of focal cerebral ischemia, EPO has been shown to exert anti-apoptotic effect by also elevating ERK-1/-2, Akt and c-Jun-amino terminal kinase (JNK)-1/-2 phosphorylation and by stimulating Bcl-x$_L$ expression [67]. EPO treatment consistently reduces infarct volume in focal cerebral ischemia in animal models [58, 67, 69]. Systemic administration of EPO abolished the appearance of TUNEL-positive cells and prevented histological damage of the penumbra after middle-cerebral artery occlusion in rats, findings consistent with the inhibition of neuronal apoptosis by EPO [68]. By anti-apoptotic mechanisms discussed above, EPO protects neurons from cell death, which in turn, results in a much

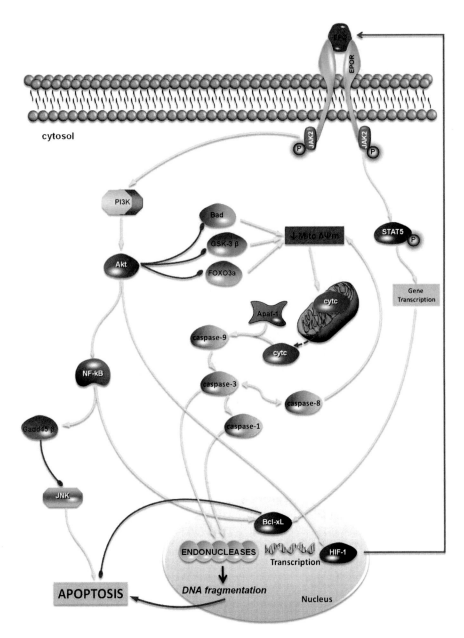

Fig. 12.6 Erythropoietin mediated signaling pathways. EPO and the EPO receptor (EPOR) can increase cell survival through pathways that involve the Janus-tyrosine kinase 2 (Jak2) proteins, protein kinase B (Akt), and STAT proteins. A series of interconnected cellular pathways such as forkhead family member FOXO3a, glycogen synthase kinase-3β (GSK-3β), nuclear factor-κB (NF-κB), mitochondrial membrane potential ($\Delta\Psi_m$), cytochrome c (cytc), and caspases play role in EPO-mediated neuroprotection

smaller volume of injury. In contrast to its anti-apoptotic effect, EPO has not shown any activity in preventing necrosis, it might do so indirectly by preserving vascular integrity and the induction of angiogenesis which will be discussed in detail [10, 69].

Additional potential protective mechanism of EPO might also include anti-inflammatory effects. It is clear that, inflammation is an important pathogenic component in cerebral ischemia, induced by production of cytokines and chemokines, followed by leukocyte infiltration and glial activation [70]. EPO suppresses the inflammatory mononuclear cell infiltration and the burst of pro-inflammatory cytokines including TNF, IL-6, and monocyte chemoattractant protein 1 (MCP-1) in an animal model of stroke [70]. Similarly, administration of exogenous rhEPO showed significant brain-protective activity in the neonatal rat by preventing the secondary, delayed rise in IL-1beta and attenuating the infiltration of leukocytes into the damaged brain [71]. However, it has been suggested that rhEPO attenuates inflammation by reducing neuronal apoptosis and increasing resistance to inflammatory injury rather than by a direct inhibition of cytokine release [72]. Generation of molecular signals by injured neurons such as the appearance of phosphatidylserine (PS) within the cell membrane, known to be a critical signal for the attraction and activation of microglia and astrocytes [71]. The exposure of membrane PS residues in neurons can precipitate a latent cellular inflammation and microglial phagocytosis of viable neurons [73]. These studies provided convincing evidence that the anti-inflammatory action of rhEPO is secondary to its neuroprotective activity. However, EPO has also potent anti-inflammatory effects in animal models that do not involve extensive apoptosis [65, 74]. Hence, further experiments are needed to explore precise mechanisms of anti-inflammatory effects of EPO.

It is well established that disrupted endothelium leads to a breakdown of the BBB, hemorrhage, and edema, which increases cerebral infarction volume after ischemic insult. EPO plays a dual role in vascular protection by preserving endothelial cell integrity and by promoting angiogenesis [10]. Apoptosis is also implicated in endothelial cell death after focal cerebral ischemia, and this may lead to increased BBB permeability [31, 75, 76]. EPO acts as cytoprotective during ischemic vascular injury through direct modulation of Akt1 phosphorylation, mitochondrial membrane potential, and caspase activity in endothelial cells [54]. By activating an EPOR-dependent intracellular signaling pathway, EPO inhibits eNOS expression and stabilize junctions by regulating expression of junction proteins including ZO-1, VE-cadherin and occludin [75].

Experiments in stroke models have shown a multifaceted role of EPO on angiogenesis. It has been shown that EPO enhances angiogenesis both in vivo and in vitro via regulation of vascular endothelial growth factor (VEGF). EPO significantly increases the numbers of capillary segments in the boundary regions of ischemia, and induces capillary-like tube formation on cerebral endothelial cells [76]. Treatment with rhEPO promotes early and higher angiogenic activity in the penumbra region and upregulates the expression of major angiogenic factors Angiopoietin-2, Tie-2, and VEGF after ischemic stroke. Moreover, EPO treatment not only protects vascular endothelial cells and stimulates angiogenic activity, but also promotes local cerebral blood flow recovery after permanent cerebral ischemia in mice [77]. A very recent study suggested that interaction of TNF-alpha with TNFR1 sensitizes cerebral endothelial cells for EPO-induced angiogenesis by upregulation of EPOR [78].

Angiogenesis and neurogenesis are coupled processes. Accelerated new vessel formation seems to be essential for enhancing endogenous neurogenesis and improving functional recovery [79]. Increased number of neurons in close proximity to angiogenic microvessels observed in rhEPO-treated stroke animals, which may support the idea that an angiogenic response results in the establishment of a favorable environment for the regeneration of neurons [80]. Exogenous EPO can promote neurogenesis both in vitro and in vivo, and EPOR is critical for adult and post-stroke neurogenesis [81–83]. EPO infusion into the adult lateral ventricles resulted in a decrease in the numbers of NSCs in the SVZ, an increase in newly generated cells migrating to the olfactory bulb EPO appears to act directly on NSCs, promoting the production of neuronal progenitors at the expense of multipotent progenitors [56]. Endothelial cells activated by EPO promote the migration of neural progenitor cells by secreting MMP2 and MMP9 via the PI3K/Akt and ERK1/2 signaling pathways, and provide insight into the molecular mechanisms underlying neurogenesis [82]. EPO is not critical only for post-stroke neurogenesis but also for maintenance of neural progenitor cell proliferation independent of insult, injury, or ischemia. Neural progenitor cells express EPOR at higher levels compared with mature neurons, and EPO stimulates proliferation of embryonic neural progenitor cells [84].

Substantial evidence also indicates that EPO mediates protective effects by maintaining normal vascular autoregulation [85]. Application of systemic rhEPO following experimental subarachnoid hemorrhage restores the autoregulation of cerebral blood flow, reverses basilar artery vasoconstriction, and enhances neuronal survival and functional recovery [10, 86, 87].

3.5 EPO Derivates for Neuroprotection

Because of the adverse effects associated with long-term rhEPO administration, various engineered forms of EPO were generated and discussed as a non-hematopoietic alternative for treatment of stroke. Among the EPO derivatives are most notably the low-sialic acid EPO (Neuro-EPO), desialylated EPO (asialoEPO) and carbamylated EPO (CEPO) [88]. AsialoEPO, which has the same EPOR affinity, possesses a very short plasma half-life and is fully neuroprotective [2]. Both CEPO and Neuro-EPO are also conferring neuroprotection similar to rhEPO [89, 90, 93]. However, when administered systemically, their passage through the BBB is restricted due to their molecular weight. Hence, the intranasal administration route is likely to offer a desirable and efficient alternative for the transport of these EPO analogs to the CNS [91].

3.6 EPO for the Treatment of Human Stroke

The cumulative preclinical data from numerous translational studies statistically confirmed the efficacy of EPO as a treatment option in ischemic stroke [2]. The efficacy of EPO treatment was measured by examining the stroke volume, behavioral testing and animal survival. Signaling pathways induced by EPO were also studied extensively. The meta-analysis of Jerndal et al. identified a significant improvement in experimental stroke after EPO treatment. However, authors emphasized that these results must be interpreted with caution due to potential sources of bias, such as the lack of allocation concealment and randomization to group of some of the included studies [92]. Efficacy of EPO and also EPO analogues in neuroprotection after stroke was confirmed in a meta-analysis of experimental studies, by Jerndal et al. [93].

Based on a wealth of preclinical data that shows exogenously administered EPO and its hematopoietic and non-hematopoietic analogs are potent neuroprotectants in animal models of ischemic stroke, a limited clinical trial in stroke patients was initiated [94]. In this study, a small number of patients received daily doses of intravenously administered EPO for the first 3 days after stroke. The results showed a trend towards reduced infarct sizes as well as improved neurological scores. These results further encouraged the randomized phase II/III German Multicenter EPO Stroke Trial, of which the results were published in 2009. This trial was so far the largest clinical EPO stroke study that included over 533 patients [95]. Unfortunately, the results of this trial were negative and unable to confirm previous results from preclinical trials. In addition, they reported increased mortality in patients treated with EPO and strongly cautioned against combined treatment of rt-PA and EPO.

In 2011, Yip et al. proposed that endothelial progenitor cell (EPC) mobilization from the bone marrow of patients treated with EPO contributes to the improved outcome after ischemic stroke [96]. In this clinical study 167 patients suffering from ischemic stroke were randomized and received either placebo or 5,000 IU of EPO subcutaneously 48 and 72 h after the onset of stroke symptoms. The reported 90-day clinical outcomes based on National Institutes of Health Stroke Scale (NIHSS) were promising. The authors correlated this beneficial outcome of EPO treatment with increase of EPCs found in the blood of treated patients. Furthermore they proposed the increase in EPC release might contribute to the repair of damaged brain vasculature and cerebral vasculature function by restoring the endothelium, thereby playing a significant role in damage reduction after ischemic stroke. EPC release from the bone marrow into the blood by EPO is known [97]. Furthermore EPCs contribute to neurorestorative repair, such as neovascularization and angiogenesis [98–100]. The notion that the number of EPCs in the peripheral blood can indicate stroke severity has been discussed by other authors [98] who remained skeptical about the suitability of systemically administered EPO for stroke treatment. This discussion revealed that the negative outcome of the Multicenter EPO Trail created new concerns and cautions against the use of EPO in stroke therapy, but has also encouraged other groups to initiate other clinical trials. This failed Multicenter

EPO Trial raised questions about the efficacy of systemically administered EPO in patients. The very limited access of intravascular or subcutaneously administered EPO through the BBB into the brain is the main concern. The intact BBB severely limits the access of blood born small peptides and larger molecules to the brain [101, 102]. Only 0.008–00.1% in mice and 0.05–0.1% in rats respectively of systemically applied EPO reach the brain [103]. In a rat model of embolic stroke, 0.02–0.07% of exogenous EPO was detected in the brain when administered intravenously. Even under neuropathological conditions such as ischemic stroke when the BBB's tight control is compromised, EPO's access to the brain and its diffusion through the parenchyma to the affected cells is limited and time consuming. In order to circumvent the documented inefficiency of EPO to reach the brain tissue due to the BBB, scientists have suggested using very high doses of systemic administered EPO. However, overdosing with EPO is likely to trigger unwanted side effects and can potentially cause more harm than benefit [93]. Using non-hematopoietic EPO derivates may reduce these side effects, however, the BBB's impermeability to these molecules still remains a factor that significantly reduces their timely availability in therapeutic doses in brain tissue.

The enormous human and socio-economic burden of stroke and the promising neuroprotective potential of EPO in preclinical trials demand further consideration of this therapeutic under more defined conditions for clinical use. Albert Einstein famously defined insanity as "doing the same thing over and over again and expecting different results." In order to avoid falling within the limits of this definition, alternative administration methods for EPO must be considered [2].

One of these options is the intranasal administration of drugs, among them EPO. Intranasal administration of a variety of drugs has been documented in the last 2 decades [2]. The efficacy of this method makes it suitable for a wide variety of neuroprotective drug candidates into the brain and is associated with beneficial effects [33, 104–122]. In a recent comparison, intranasally administration of radiolabelled EPO resulted in 7–31 times higher peak concentrations when compared to systemic delivery [123]. Moreover, 20% of the original EPO dose was measured in normal brain tissue within 20 min and 33% within 120 min following intranasal administration in post-stroke animal. In contrast less than 0.5% in control and 1.75% in injured mice of systemic administered EPO reached the brain tissue within 480 min after injection. Similar results were obtained by other groups who were able to demonstrate independently the efficacy of intranasal administration of therapeutic drugs in ischemic stroke [121, 124]. Two access path into the brain that connect the nasal cavity to the brain are discussed currently: the trigeminal nerve [125] and the rostral migratory stream [125]. It is entirely possible that both routes are used to transfer intranasaly administered drugs to different regions of the brain and potentially the spinal cord. Identification of the access route can provide the information necessary to tailor therapeutics that will suit the characteristics of the access route. However, before entering a clinical trial using the intranasal delivery of neuroprotective drugs limitations of this administration method should be carefully evaluated and the differences in human and rodent anatomy considered [126].

As described above the preclinical trials using EPO as a neuroprotectant are heterogenous due to a wide variety of protocols used in the treatment time points, assessment of neurological outcome and the histological evaluation of variations in neuropathological effects of ischemia on brain tissue. In order to overcome heterogeneity of experimental protocols for testing candidate drugs for ischemic stroke, the STAIR has published guidelines (STAIR I-IV) [44, 127–131] that present the limitations of current treatment options and provide recommendations for the preclinical testing of novel treatments of drug candidates, such as recombinant EPO. Furthermore, before further clinical trials will be performed, the necessity of the experimental studies that investigate the efficacy of EPO in animal stroke models with comorbidity and the safety of a combination with thrombolysis is emphasized [132].

References

1. Tissue plasminogen activator for acute ischemic stroke. The national institute of neurological disorders and stroke rt-pa stroke study group. N Engl J Med. 1995;333:1581–7.
2. Digicaylioglu M. Erythropoietin in stroke: quo vadis. Expert Opin Biol Ther. 2010; 10:937–49.
3. Hacke W, Donnan G, Fieschi C, Kaste M, von Kummer R, Broderick JP, Brott T, Frankel M, Grotta JC, Haley Jr EC, Kwiatkowski T, Levine SR, Lewandowski C, Lu M, Lyden P, Marler JR, Patel S, Tilley BC, Albers G, Bluhmki E, Wilhelm M, Hamilton S. Association of outcome with early stroke treatment: pooled analysis of atlantis, ecass, and ninds rt-pa stroke trials. Lancet. 2004;363:768–74.
4. Digicaylioglu M, Bichet S, Marti HH, Wenger RH, Rivas LA, Bauer C, Gassmann M. Localization of specific erythropoietin binding sites in defined areas of the mouse brain. Proc Natl Acad Sci USA. 1995;92:3717–20.
5. Digicaylioglu M, Lipton SA. Erythropoietin-mediated neuroprotection involves cross-talk between jak2 and nf-kappab signalling cascades. Nature. 2001;412:641–7.
6. Marti HH, Wenger RH, Rivas LA, Straumann U, Digicaylioglu M, Henn V, Yonekawa Y, Bauer C, Gassmann M. Erythropoietin gene expression in human, monkey and murine brain. Eur J Neurosci. 1996;8:666–76.
7. Solaroglu I, Cahill J, Jadhav V, Zhang JH. A novel neuroprotectant granulocyte-colony stimulating factor. Stroke. 2006;37:1123–8.
8. Solaroglu I, Cahill J, Tsubokawa T, Beskonakli E, Zhang JH. Granulocyte colony-stimulating factor protects the brain against experimental stroke via inhibition of apoptosis and inflammation. Neurol Res. 2009;31:167–72.
9. Solaroglu I, Jadhav V, Zhang JH. Neuroprotective effect of granulocyte-colony stimulating factor. Front Biosci. 2007;12:712–24.
10. Brines M, Cerami A. Emerging biological roles for erythropoietin in the nervous system. Nat Rev Neurosci. 2005;6:484–94.
11. Kaptanoglu E, Solaroglu I, Okutan O, Surucu HS, Akbiyik F, Beskonakli E. Erythropoietin exerts neuroprotection after acute spinal cord injury in rats: effect on lipid peroxidation and early ultrastructural findings. Neurosurg Rev. 2004;27:113–20.
12. Solaroglu I, Solaroglu A, Kaptanoglu E, Dede S, Haberal A, Beskonakli E, Kilinc K. Erythropoietin prevents ischemia-reperfusion from inducing oxidative damage in fetal rat brain. Childs Nerv Syst. 2003;19:19–22.
13. Wells JA, de Vos AM. Hematopoietic receptor complexes. Annu Rev Biochem. 1996;65: 609–34.

14. Demetri GD, Griffin JD. Granulocyte colony-stimulating factor and its receptor. Blood. 1991;78:2791–808.
15. Fukunaga R, Ishizaka-Ikeda E, Pan CX, Seto Y, Nagata S. Functional domains of the granulocyte colony-stimulating factor receptor. EMBO J. 1991;10:2855–65.
16. Schneider A, Kruger C, Steigleder T, Weber D, Pitzer C, Laage R, Aronowski J, Maurer MH, Gassler N, Mier W, Hasselblatt M, Kollmar R, Schwab S, Sommer C, Bach A, Kuhn HG, Schabitz WR. The hematopoietic factor g-csf is a neuronal ligand that counteracts programmed cell death and drives neurogenesis. J Clin Invest. 2005;115:2083–98.
17. Bensinger WI, Weaver CH, Appelbaum FR, Rowley S, Demirer T, Sanders J, Storb R, Buckner CD. Transplantation of allogeneic peripheral blood stem cells mobilized by recombinant human granulocyte colony-stimulating factor. Blood. 1995;85:1655–8.
18. Welte K, Gabrilove J, Bronchud MH, Platzer E, Morstyn G. Filgrastim (r-methug-csf): the first 10 years. Blood. 1996;88:1907–29.
19. Corti S, Locatelli F, Strazzer S, Salani S, Del Bo R, Soligo D, Bossolasco P, Bresolin N, Scarlato G, Comi GP. Modulated generation of neuronal cells from bone marrow by expansion and mobilization of circulating stem cells with in vivo cytokine treatment. Exp Neurol. 2002;177:443–52.
20. Solaroglu I, Tsubokawa T, Cahill J, Zhang JH. Anti-apoptotic effect of granulocyte-colony stimulating factor after focal cerebral ischemia in the rat. Neuroscience. 2006;143:965–74.
21. Diederich K, Sevimli S, Dorr H, Kosters E, Hoppen M, Lewejohann L, Klocke R, Minnerup J, Knecht S, Nikol S, Sachser N, Schneider A, Gorji A, Sommer C, Schabitz WR. The role of granulocyte-colony stimulating factor (G-CSF) in the healthy brain: a characterization of G-CSF-deficient mice. J Neurosci. 2009;29:11572–81.
22. Kleinschnitz C, Schroeter M, Jander S, Stoll G. Induction of granulocyte colony-stimulating factor mrna by focal cerebral ischemia and cortical spreading depression. Brain Res Mol Brain Res. 2004;131:73–8.
23. Schabitz WR, Schneider A. New targets for established proteins: exploring G-CSF for the treatment of stroke. Trends Pharmacol Sci. 2007;28:157–61.
24. Park HK, Chu K, Lee ST, Jung KH, Kim EH, Lee KB, Song YM, Jeong SW, Kim M, Roh JK. Granulocyte colony-stimulating factor induces sensorimotor recovery in intracerebral hemorrhage. Brain Res. 2005;1041:125–31.
25. Sehara Y, Hayashi T, Deguchi K, Zhang H, Tsuchiya A, Yamashita T, Lukic V, Nagai M, Kamiya T, Abe K. Decreased focal inflammatory response by G-CSF may improve stroke outcome after transient middle cerebral artery occlusion in rats. J Neurosci Res. 2007;85:2167–74.
26. Lee ST, Chu K, Jung KH, Ko SY, Kim EH, Sinn DI, Lee YS, Lo EH, Kim M, Roh JK. Granulocyte colony-stimulating factor enhances angiogenesis after focal cerebral ischemia. Brain Res. 2005;1058:120–8.
27. Schabitz W-R, Kollmar R, Schwaninger M, Juettler E, Bardutzky J, Scholzke MN, Sommer C, Schwab S. Neuroprotective effect of granulocyte colony-stimulating factor after focal cerebral ischemia. Stroke. 2003;34:745–51.
28. Gibson CL, Bath PM, Murphy SP. G-CSF reduces infarct volume and improves functional outcome after transient focal cerebral ischemia in mice. J Cereb Blood Flow Metab. 2005;25:431–9.
29. Komine-Kobayashi M, Zhang N, Liu M, Tanaka R, Hara H, Osaka A, Mochizuki H, Mizuno Y, Urabe T. Neuroprotective effect of recombinant human granulocyte colony-stimulating factor in transient focal ischemia of mice. J Cereb Blood Flow Metab. 2006;26:402–13.
30. Taguchi A, Wen Z, Myojin K, Yoshihara T, Nakagomi T, Nakayama D, Tanaka H, Soma T, Stern DM, Naritomi H, Matsuyama T. Granulocyte colony-stimulating factor has a negative effect on stroke outcome in a murine model. Eur J Neurosci. 2007;26:126–33.
31. Tsubokawa T, Solaroglu I, Yatsushige H, Cahill J, Yata K, Zhang JH. Cathepsin and calpain inhibitor e64d attenuates matrix metalloproteinase-9 activity after focal cerebral ischemia in rats. Stroke. 2006;37:1888–94.

32. Sevimli S, Diederich K, Strecker JK, Schilling M, Klocke R, Nikol S, Kirsch F, Schneider A, Schabitz WR. Endogenous brain protection by granulocyte-colony stimulating factor after ischemic stroke. Exp Neurol. 2009;217:328–35.

33. Hanson LR, Frey 2nd WH. Intranasal delivery bypasses the blood-brain barrier to target therapeutic agents to the central nervous system and treat neurodegenerative disease. BMC Neurosci. 2008;9 Suppl 3:S5.

34. Popa-Wagner A, Stocker K, Balseanu AT, Rogalewski A, Diederich K, Minnerup J, Margaritescu C, Schabitz WR. Effects of granulocyte-colony stimulating factor after stroke in aged rats. Stroke. 2010;41:1027–31.

35. Kawada H, Takizawa S, Takanashi T, Morita Y, Fujita J, Fukuda K, Takagi S, Okano H, Ando K, Hotta T. Administration of hematopoietic cytokines in the subacute phase after cerebral infarction is effective for functional recovery facilitating proliferation of intrinsic neural stem/progenitor cells and transition of bone marrow-derived neuronal cells. Circulation. 2006;113:701–10.

36. Zhao LR, Berra HH, Duan WM, Singhal S, Mehta J, Apkarian AV, Kessler JA. Beneficial effects of hematopoietic growth factor therapy in chronic ischemic stroke in rats. Stroke. 2007;38:2804–11.

37. Liu SP, Lee SD, Lee HT, Liu DD, Wang HJ, Liu RS, Lin SZ, Shyu WC. Granulocyte colony-stimulating factor activating hif-1alpha acts synergistically with erythropoietin to promote tissue plasticity. PLoS One. 2010;5:e10093.

38. Six I, Gasan G, Mura E, Bordet R. Beneficial effect of pharmacological mobilization of bone marrow in experimental cerebral ischemia. Eur J Pharmacol. 2003;458:327–8.

39. Shyu WC, Lin SZ, Yang HI, Tzeng YS, Pang CY, Yen PS, Li H. Functional recovery of stroke rats induced by granulocyte colony-stimulating factor-stimulated stem cells. Circulation. 2004;110:1847–54.

40. Piao CS, Gonzalez-Toledo ME, Xue YQ, Duan WM, Terao S, Granger DN, Kelley RE, Zhao LR. The role of stem cell factor and granulocyte-colony stimulating factor in brain repair during chronic stroke. J Cereb Blood Flow Metab. 2009;29:759–70.

41. Arai K, Jin G, Navaratna D, Lo EH. Brain angiogenesis in developmental and pathological processes: neurovascular injury and angiogenic recovery after stroke. FEBS J. 2009;276:4644–52.

42. Sugiyama Y, Yagita Y, Oyama N, Terasaki Y, Omura-Matsuoka E, Sasaki T, Kitagawa K. Granulocyte colony-stimulating factor enhances arteriogenesis and ameliorates cerebral damage in a mouse model of ischemic stroke. Stroke. 2011;42:770–5.

43. Toth ZE, Leker RR, Shahar T, Pastorino S, Szalayova I, Asemenew B, Key S, Parmelee A, Mayer B, Nemeth K, Bratincsak A, Mezey E. The combination of granulocyte colony-stimulating factor and stem cell factor significantly increases the number of bone marrow-derived endothelial cells in brains of mice following cerebral ischemia. Blood. 2008;111:5544–52.

44. Recommendations for standards regarding preclinical neuroprotective and restorative drug development. Stroke. 1999;30:2752–8.

45. Kollmar R, Henninger N, Urbanek C, Schwab S. G-CSF, rt-PA and combination therapy after experimental thromboembolic stroke. Exp Transl Stroke Med. 2010;2:9.

46. Shyu WC, Lin SZ, Lee CC, Liu DD, Li H. Granulocyte colony-stimulating factor for acute ischemic stroke: a randomized controlled trial. CMAJ. 2006;174:927–33.

47. Sprigg N, Bath PM, Zhao L, Willmot MR, Gray LJ, Walker MF, Dennis MS, Russell N. Granulocyte-colony stimulating factor mobilizes bone marrow stem cells in patients with subacute ischemic stroke: the stem cell trial of recovery enhancement after stroke (stems) pilot randomized, controlled trial (ISRCTN 16784092). Stroke. 2006;37:2979–83.

48. Schabitz WR, Laage R, Vogt G, Koch W, Kollmar R, Schwab S, Schneider D, Hamann GF, Rosenkranz M, Veltkamp R, Fiebach JB, Hacke W, Grotta JC, Fisher M, Schneider A. Axis: a trial of intravenous granulocyte colony-stimulating factor in acute ischemic stroke. Stroke. 2010;41:2545–51.

49. Floel A, Warnecke T, Duning T, Lating Y, Uhlenbrock J, Schneider A, Vogt G, Laage R, Koch W, Knecht S, Schabitz WR. Granulocyte-colony stimulating factor (G-CSF) in stroke patients with concomitant vascular disease-a randomized controlled trial. PLoS One. 2011;6:e19767.

50. Jelkmann W. Erythropoietin after a century of research: younger than ever. Eur J Haematol. 2007;78:183–205.

51. Ghezzi P, Brines M. Erythropoietin as an antiapoptotic, tissue-protective cytokine. Cell Death Differ. 2004;11:S37–44.

52. Byts N, Siren AL. Erythropoietin: a multimodal neuroprotective agent. Exp Transl Stroke Med. 2009;1:4.

53. Joyeux-Faure M. Cellular protection by erythropoietin: new therapeutic implications? J Pharmacol Exp Ther. 2007;323:759–62.

54. Maiese K, Li F, Chong ZZ. New avenues of exploration for erythropoietin. JAMA. 2005;293:90–5.

55. Chong ZZ, Kang JQ, Maiese K. Hematopoietic factor erythropoietin fosters neuroprotection through novel signal transduction cascades. J Cereb Blood Flow Metab. 2002;22:503–14.

56. Tsai PT, Ohab JJ, Kertesz N, Groszer M, Matter C, Gao J, Liu X, Wu H, Carmichael ST. A critical role of erythropoietin receptor in neurogenesis and post-stroke recovery. J Neurosci. 2006;26:1269–74.

57. van der Kooij MA, Groenendaal F, Kavelaars A, Heijnen CJ, van Bel F. Neuroprotective properties and mechanisms of erythropoietin in in vitro and in vivo experimental models for hypoxia/ischemia. Brain Res Rev. 2008;59:22–33.

58. Bernaudin M, Marti HH, Roussel S, Divoux D, Nouvelot A, MacKenzie ET, Petit E. A potential role for erythropoietin in focal permanent cerebral ischemia in mice. J Cereb Blood Flow Metab. 1999;19:643–51.

59. Morishita E, Masuda S, Nagao M, Yasuda Y, Sasaki R. Erythropoietin receptor is expressed in rat hippocampal and cerebral cortical neurons, and erythropoietin prevents in vitro glutamate-induced neuronal death. Neuroscience. 1996;76:105–16.

60. Sakanaka M, Wen T-C, Matsuda S, Masuda S, Morishita E, Nagao M, Sasaki R. In vivo evidence that erythropoietin protects neurons from ischemic damage. Proc Natl Acad Sci USA. 1998;95:4635–40.

61. Taoufik E, Petit E, Divoux D, Tseveleki V, Mengozzi M, Roberts ML, Valable S, Ghezzi P, Quackenbush J, Brines M, Cerami A, Probert L. TNF receptor I sensitizes neurons to erythropoietin- and VEGF-mediated neuroprotection after ischemic and excitotoxic injury. Proc Natl Acad Sci. 2008;105:6185–90.

62. Kawakami M, Sekiguchi M, Sato K, Kozaki S, Takahashi M. Erythropoietin receptor-mediated inhibition of exocytotic glutamate release confers neuroprotection during chemical ischemia. J Biol Chem. 2001;276:39469–75.

63. Vairano M, Russo CD, Pozzoli G, Battaglia A, Scambia G, Tringali G, Aloe-Spiriti MA, Preziosi P, Navarra P. Erythropoietin exerts anti-apoptotic effects on rat microglial cells in vitro. Eur J Neurosci. 2002;16:584–92.

64. Ruscher K, Freyer D, Karsch M, Isaev N, Megow D, Sawitzki B, Priller J, Dirnagl U, Meisel A. Erythropoietin is a paracrine mediator of ischemic tolerance in the brain: evidence from an in vitro model. J Neurosci. 2002;22:10291–301.

65. Chong ZZ, Kang JQ, Maiese K. Erythropoietin fosters both intrinsic and extrinsic neuronal protection through modulation of microglia, akt1, bad, and caspase-mediated pathways. Br J Pharmacol. 2003;138:1107–18.

66. Chong ZZ, Kang JQ, Maiese K. Apaf-1, Bcl-xL, cytochrome c, and caspase-9 form the critical elements for cerebral vascular protection by erythropoietin. J Cereb Blood Flow Metab. 2003;23:320–30.

67. Lee ST, Chu K, Sinn DI, Jung KH, Kim EH, Kim SJ, Kim JM, Ko SY, Kim M, Roh JK. Erythropoietin reduces perihematomal inflammation and cell death with eNOS and STAT3 activations in experimental intracerebral hemorrhage. J Neurochem. 2006;96:1728–39.

68. Kilic E, Kilic Ü, Soliz J, Bassetti CL, Gassmann M, Hermann DM. Brain-derived erythropoietin protects from focal cerebral ischemia by dual activation of erk-1/-2 and akt pathways. FASEB J. 2005;19:2026–8.

69. Sirén A-L, Fratelli M, Brines M, Goemans C, Casagrande S, Lewczuk P, Keenan S, Gleiter C, Pasquali C, Capobianco A, Mennini T, Heumann R, Cerami A, Ehrenreich H, Ghezzi P. Erythropoietin prevents neuronal apoptosis after cerebral ischemia and metabolic stress. Proc Natl Acad Sci. 2001;98:4044–9.

70. Maiese K, Li F, Chong ZZ. Erythropoietin in the brain: can the promise to protect be fulfilled? Trends Pharmacol Sci. 2004;25:577–83.

71. Villa P, Bigini P, Mennini T, Agnello D, Laragione T, Cagnotto A, Viviani B, Marinovich M, Cerami A, Coleman TR, Brines M, Ghezzi P. Erythropoietin selectively attenuates cytokine production and inflammation in cerebral ischemia by targeting neuronal apoptosis. J Exp Med. 2003;198:971–5.

72. Sun Y, Calvert JW, Zhang JH. Neonatal hypoxia/ischemia is associated with decreased inflammatory mediators after erythropoietin administration. Stroke. 2005;36:1672–8.

73. Henson PM, Bratton DL, Fadok VA. The phosphatidylserine receptor: a crucial molecular switch? Nat Rev Mol Cell Biol. 2001;2:627–33.

74. Agnello D, Bigini P, Villa P, Mennini T, Cerami A, Brines ML, Ghezzi P. Erythropoietin exerts an anti-inflammatory effect on the cns in a model of experimental autoimmune encephalomyelitis. Brain Res. 2002;952:128–34.

75. Chong ZZ, Kang JQ, Maiese K. Erythropoietin is a novel vascular protectant through activation of akt1 and mitochondrial modulation of cysteine proteases. Circulation. 2002; 106:2973–9.

76. Martínez-Estrada OM, Rodríguez-Millán E, González-de Vicente E, Reina M, Vilaró S, Fabre M. Erythropoietin protects the in vitro blood–brain barrier against vegf-induced permeability. Eur J Neurosci. 2003;18:2538–44.

77. Wang L, Zhang Z, Wang Y, Zhang R, Chopp M. Treatment of stroke with erythropoietin enhances neurogenesis and angiogenesis and improves neurological function in rats. Stroke. 2004;35:1732–7.

78. Li Y, Lu Z, Keogh CL, Yu SP, Wei L. Erythropoietin-induced neurovascular protection, angiogenesis, and cerebral blood flow restoration after focal ischemia in mice. J Cereb Blood Flow Metab. 2006;27:1043–54.

79. Wang L, Chopp M, Teng H, Bolz M, Francisco MA, Aluigi DM, Wang XL, Zhang RL, Chrsitensen S, Sager TN, Szalad A, Zhang ZG. Tumor necrosis factor α primes cerebral endothelial cells for erythropoietin-induced angiogenesis. J Cereb Blood Flow Metab. 2011;31:640–7.

80. Taguchi A, Soma T, Tanaka H, Kanda T, Nishimura H, Yoshikawa H, Tsukamoto Y, Iso H, Fujimori Y, Stern DM, Naritomi H, Matsuyama T. Administration of cd34+ cells after stroke enhances neurogenesis via angiogenesisin a mouse model. J Clin Invest. 2004;114:330–8.

81. Keogh CL, Yu SP, Wei L. The effect of recombinant human erythropoietin on neurovasculature repair after focal ischemic stroke in neonatal rats. J Pharmacol Exp Ther. 2007; 322:521–8.

82. Shingo T, Sorokan ST, Shimazaki T, Weiss S. Erythropoietin regulates the in vitro and in vivo production of neuronal progenitors by mammalian forebrain neural stem cells. J Neurosci. 2001;21:9733–43.

83. Studer L, Csete M, Lee S-H, Kabbani N, Walikonis J, Wold B, McKay R. Enhanced proliferation, survival, and dopaminergic differentiation of CNS precursors in lowered oxygen. J Neurosci. 2000;20:7377–83.

84. Wang L, Zhang ZG, Zhang RL, Gregg SR, Hozeska-Solgot A, LeTourneau Y, Wang Y, Chopp M. Matrix metalloproteinase 2 (mmp2) and mmp9 secreted by erythropoietin-activated endothelial cells promote neural progenitor cell migration. J Neurosci. 2006;26: 5996–6003.

85. Chen Z-Y, Asavaritikrai P, Prchal JT, Noguchi CT. Endogenous erythropoietin signaling is required for normal neural progenitor cell proliferation. J Biol Chem. 2007;282: 25875–83.

86. Alafaci C, Salpietro F, Grasso G, Sfacteria A, Passalacqua M, Morabito A, Tripodo E, Calapai G, Buemi M, Tomasello F. Effect of recombinant human erythropoietin on cerebral ischemia following experimental subarachnoid hemorrhage. Eur J Pharmacol. 2000;406:219–25.

87. Grasso G. Neuroprotective effect of recombinant human erythropoietin in experimental subarachnoid hemorrhage. J Neurosurg Sci. 2001;45:7–14.

88. Santhanam AVR, Smith LA, Akiyama M, Rosales AG, Bailey KR, Katusic ZS. Role of endothelial no synthase phosphorylation in cerebrovascular protective effect of recombinant erythropoietin during subarachnoid hemorrhage- induced cerebral vasospasm. Stroke. 2005; 36:2731–7.

89. Erbayraktar S, Grasso G, Sfacteria A, Xie QW, Coleman T, Kreilgaard M, Torup L, Sager T, Erbayraktar Z, Gokmen N, Yilmaz O, Ghezzi P, Villa P, Fratelli M, Casagrande S, Leist M, Helboe L, Gerwein J, Christensen S, Geist MA, Pedersen LØ, Cerami-Hand C, Wuerth J-P, Cerami A, Brines M. Asialoerythropoietin is a nonerythropoietic cytokine with broad neuroprotective activity in vivo. Proc Natl Acad Sci USA. 2003;100:6741–6.

90. Montero M, Poulsen FR, Noraberg J, Kirkeby A, van Beek J, Leist M, Zimmer J. Comparison of neuroprotective effects of erythropoietin (EPO) and carbamylerythropoietin (CEPO) against ischemia-like oxygen-glucose deprivation (OGD) and NMDA excitotoxicity in mouse hippocampal slice cultures. Exp Neurol. 2007;204:106–17.

91. Rodriguez Cruz Y, Mengana Tamos Y, Munoz Cernuda A, Subiros Martines N, Gonzalez-Quevedo A, Sosa Teste I, Garcia Rodriguez JC. Treatment with nasal neuro-EPO improves the neurological, cognitive, and histological state in a gerbil model of focal ischemia. Scientific World J. 2010;10:2288–300.

92. Lapchak PA. Erythropoietin molecules to treat acute ischemic stroke: a translational dilemma! Expert Opin Investig Drugs. 2010;19:1179–86.

93. Jerndal M, Forsberg K, Sena ES, Macleod MR, O'Collins VE, Linden T, Nilsson M, Howells DW. A systematic review and meta-analysis of erythropoietin in experimental stroke. J Cereb Blood Flow Metab. 2010;30:961–8.

94. Minnerup J, Heidrich J, Rogalewski A, Schabitz W-R, Wellmann J. The efficacy of erythropoietin and its analogues in animal stroke models: a meta-analysis. Stroke. 2009;40: 3113–20.

95. Ehrenreich H, Hasselblatt M, Dembowski C, Cepek L, Lewczuk P, Stiefel M, Rustenbeck HH, Breiter N, Jacob S, Knerlich F, Bohn M, Poser W, Ruther E, Kochen M, Gefeller O, Gleiter C, Wessel TC, De Ryck M, Itri L, Prange H, Cerami A, Brines M, Siren AL. Erythropoietin therapy for acute stroke is both safe and beneficial. Mol Med. 2002;8:495–505.

96. Ehrenreich H, Weissenborn K, Prange H, Schneider D, Weimar C, Wartenberg K, Schellinger PD, Bohn M, Becker H, Wegrzyn M, Jahnig P, Herrmann M, Knauth M, Bahr M, Heide W, Wagner A, Schwab S, Reichmann H, Schwendemann G, Dengler R, Kastrup A, Bartels C. Recombinant human erythropoietin in the treatment of acute ischemic stroke. Stroke. 2009;40:e647–56.

97. Yip HK, Tsai TH, Lin HS, Chen SF, Sun CK, Leu S, Yuen CM, Tan TY, Lan MY, Liou CW, Lu CH, Chang WN. Effect of erythropoietin on level of circulating endothelial progenitor cells and outcome in patients after acute ischemic stroke. Crit Care. 2011;15:R40.

98. Bahlmann FH, DeGroot K, Duckert T, Niemczyk E, Bahlmann E, Boehm SM, Haller H, Fliser D. Endothelial progenitor cell proliferation and differentiation is regulated by erythropoietin. Kidney Int. 2003;64:1648–52.

99. Hohenstein B, Kuo MC, Addabbo F, Yasuda K, Ratliff B, Schwarzenberger C, Eckardt KU, Hugo CP, Goligorsky MS. Enhanced progenitor cell recruitment and endothelial repair after selective endothelial injury of the mouse kidney. Am J Physiol Renal Physiol. 2010; 298:F1504–14.

100. Hirata A, Minamino T, Asanuma H, Fujita M, Wakeno M, Myoishi M, Tsukamoto O, Okada K, Koyama H, Komamura K, Takashima S, Shinozaki Y, Mori H, Shiraga M, Kitakaze M, Hori M. Erythropoietin enhances neovascularization of ischemic myocardium and improves left ventricular dysfunction after myocardial infarction in dogs. J Am Coll Cardiol. 2006;48:176–84.

101. Minnerup J, Wersching H, Schabitz WR. Erythropoietin for stroke treatment: dead or alive? Crit Care. 2011;15:129.

102. Pardridge WM. Drug targeting to the brain. Pharm Res. 2007;24:1733–44.
103. Pardridge WM. Blood–brain barrier delivery. Drug Discov Today. 2007;12:54–61.
104. Freychet L, Rizkalla SW, Desplanque N, Basdevant A, Zirinis P, Tchobroutsky G, Slama G. Effect of intranasal glucagon on blood glucose levels in healthy subjects and hypoglycaemic patients with insulin-dependent diabetes. Lancet. 1988;1:1364–6.
105. Liu XF, Fawcett JR, Thorne RG, DeFor TA, Frey 2nd WH. Intranasal administration of insulin-like growth factor-i bypasses the blood-brain barrier and protects against focal cerebral ischemic damage. J Neurol Sci. 2001;187:91–7.
106. Liu XF, Fawcett JR, Thorne RG, Frey 2nd WH. Non-invasive intranasal insulin-like growth factor-i reduces infarct volume and improves neurologic function in rats following middle cerebral artery occlusion. Neurosci Lett. 2001;308:91–4.
107. Ross TM, Martinez PM, Renner JC, Thorne RG, Hanson LR, Frey 2nd WH. Intranasal administration of interferon beta bypasses the blood-brain barrier to target the central nervous system and cervical lymph nodes: a non-invasive treatment strategy for multiple sclerosis. J Neuroimmunol. 2004;151:66–77.
108. Thorne RG, Pronk GJ, Padmanabhan V, Frey 2nd WH. Delivery of insulin-like growth factor-i to the rat brain and spinal cord along olfactory and trigeminal pathways following intranasal administration. Neuroscience. 2004;127:481–96.
109. Yu YP, Xu QQ, Zhang Q, Zhang WP, Zhang LH, Wei EQ. Intranasal recombinant human erythropoietin protects rats against focal cerebral ischemia. Neurosci Lett. 2005;387:5–10.
110. Reger MA, Watson GS, Frey 2nd WH, Baker LD, Cholerton B, Keeling ML, Belongia DA, Fishel MA, Plymate SR, Schellenberg GD, Cherrier MM, Craft S. Effects of intranasal insulin on cognition in memory-impaired older adults: modulation by apoe genotype. Neurobiol Aging. 2006;27:451–8.
111. Thorne RG, Hanson LR, Ross TM, Tung D, Frey 2nd WH. Delivery of interferon-beta to the monkey nervous system following intranasal administration. Neuroscience. 2008;152:785–97.
112. Francis GJ, Martinez JA, Liu WQ, Xu K, Ayer A, Fine J, Tuor UI, Glazner G, Hanson LR, Frey 2nd WH, Toth C. Intranasal insulin prevents cognitive decline, cerebral atrophy and white matter changes in murine type i diabetic encephalopathy. Brain. 2008;131:3311–34.
113. Ma M, Ma Y, Yi X, Guo R, Zhu W, Fan X, Xu G, Frey 2nd WH, Liu X. Intranasal delivery of transforming growth factor-beta1 in mice after stroke reduces infarct volume and increases neurogenesis in the subventricular zone. BMC Neurosci. 2008;9:117.
114. Martinez JA, Francis GJ, Liu WQ, Pradzinsky N, Fine J, Wilson M, Hanson LR, Frey 2nd WH, Zochodne D, Gordon T, Toth C. Intranasal delivery of insulin and a nitric oxide synthase inhibitor in an experimental model of amyotrophic lateral sclerosis. Neuroscience. 2008; 157:908–25.
115. Reger MA, Watson GS, Green PS, Baker LD, Cholerton B, Fishel MA, Plymate SR, Cherrier MM, Schellenberg GD, Frey 2nd WH, Craft S. Intranasal insulin administration dose-dependently modulates verbal memory and plasma amyloid-beta in memory-impaired older adults. J Alzheimers Dis. 2008;13:323–31.
116. Ross TM, Zuckermann RN, Reinhard C, Frey 2nd WH. Intranasal administration delivers peptoids to the rat central nervous system. Neurosci Lett. 2008;439:30–3.
117. Novakovic ZM, Leinung MC, Lee DW, Grasso P. Intranasal administration of mouse [d-leu-4] ob3, a synthetic peptide amide with leptin-like activity, enhances total uptake and bioavailability in swiss webster mice when compared to intraperitoneal, subcutaneous, and intramuscular delivery systems. Regul Pept. 2009;154:107–11.
118. Garcia-Rodriguez JC, Sosa-Teste I. The nasal route as a potential pathway for delivery of erythropoietin in the treatment of acute ischemic stroke in humans. Scientific World J. 2009;9:970–81.
119. Hanson L, Roeytenberg A, Martinez PM, Coppes VG, Sweet DC, Rao RJ, Marti DL, Hoekman JD, Matthews RB, Frey WH, Panter SS. Intranasal deferoxamine provides increased brain exposure and significant protection in rat ischemic stroke. J Pharmacol Exp Ther. 2009;330(3):679–86.

120. Dhuria SV, Hanson LR, Frey 2nd WH. Intranasal drug targeting of hypocretin-1 (orexin-a) to the central nervous system. J Pharm Sci. 2009;98:2501–15.
121. Fletcher L, Kohli S, Sprague SM, Scranton RA, Lipton SA, Parra A, Jimenez DF, Digicaylioglu M. Intranasal delivery of erythropoietin plus insulin-like growth factor-i for acute neuroprotection in stroke. J Neurosurg. 2009;111(1):164–70.
122. Alcala-Barraza SR, Lee MS, Hanson LR, McDonald AA, Frey WH, McLoon LK. Intranasal delivery of neurotrophic factors BDNF, CNTF, EPO, and NT-4 to the CNS. J Drug Target. 2009;18(3):179–90.
123. Yu H, Kim K. Direct nose-to-brain transfer of a growth hormone releasing neuropeptide, hexarelin after intranasal administration to rabbits. Int J Pharm. 2009;378:73–9.
124. Alcala-Barraza SR, Lee MS, Hanson LR, McDonald AA, Frey 2nd WH, McLoon LK. Intranasal delivery of neurotrophic factors BDNF, CNTF, EPO, and NT-4 to the CNS. J Drug Target. 2010;18:179–90.
125. Johnson NJ, Hanson LR, Frey WH. Trigeminal pathways deliver a low molecular weight drug from the nose to the brain and orofacial structures. Mol Pharm. 2010;7:884–93.
126. Scranton RA, Fletcher L, Sprague S, Jimenez DF, Digicaylioglu M. The rostral migratory stream plays a key role in intranasal delivery of drugs into the CNS. PLoS One. 2011;6: e18711.
127. Hermann DM. Enhancing the delivery of erythropoietin and its variants into the ischemic brain. Scientific World J. 2009;9:967–9.
128. Recommendations for clinical trial evaluation of acute stroke therapies. Stroke. 2001; 32:1598–1606.
129. Fisher M. Recommendations for advancing development of acute stroke therapies: stroke therapy academic industry roundtable 3. Stroke. 2003;34:1539–46.
130. Fisher M, Albers GW, Donnan GA, Furlan AJ, Grotta JC, Kidwell CS, Sacco RL, Wechsler LR. Enhancing the development and approval of acute stroke therapies: stroke therapy academic industry roundtable. Stroke. 2005;36:1808–13.
131. Fisher M, Hanley DF, Howard G, Jauch EC, Warach S. Recommendations from the stair v meeting on acute stroke trials, technology and outcomes. Stroke. 2007;38:245–8.
132. Fisher M, Feuerstein G, Howells DW, Hurn PD, Kent TA, Savitz SI, Lo EH. Update of the stroke therapy academic industry roundtable preclinical recommendations. Stroke. 2009; 40:2244–50.

Chapter 13
Soluble Epoxide Hydrolase as a Stroke Target

Jonathan W. Nelson and Nabil J. Alkayed

Abstract Soluble epoxide hydrolase (sEH) is a promising therapeutic target for stroke. Both pharmacological inhibition and genetic knockout of sEH have shown protection in experimental models of cerebral ischemia. Additionally, human single-nucleotide polymorphisms in the gene that encodes for sEH, designated *ephx2*, have been correlated with stroke incidence and outcome. This chapter starts out by introducing sEH in the context of the five other mammalian epoxide hydrolases before delving specifically into sEH biology. Up-to-date research into sEH's protein structure, role in metabolizing epoxyeicosatrienoic acids, regional localization in brain, as well as subcellular localization are discussed in relation to brain function and disease. The chapter concludes by evaluating the prospect of using sEH inhibitors in clinical trials for the treatment of stroke based on the Stroke Therapy Academic and Industry Roundtable criteria.

1 Epoxide Hydrolases

Epoxide hydrolases (EHs, EC 3.3.2.3) are a large family of enzymes that catalyze the conversion of an epoxide group to a corresponding vicinal diol. Found ubiquitously in nature, they play a key role in the degradation of both environmental and

J.W. Nelson
Department of Anesthesiology and Perioperative Medicine, Oregon Health
& Science University, 3181 S.W. Sam Jackson Park Road, Portland, OR 97239-3098, USA

Department of Molecular and Medical Genetics, Oregon Health & Science University,
3181 S.W. Sam Jackson Park Road, Portland, OR 97239-3098, USA

N.J. Alkayed (✉)
Department of Anesthesiology and Perioperative Medicine, Oregon Health
& Science University, 3181 S.W. Sam Jackson Park Road, Portland, OR 97239-3098, USA
e-mail: alkayedn@ohsu.edu

P.A. Lapchak and J.H. Zhang (eds.), *Translational Stroke Research*, Springer Series
in Translational Stroke Research, DOI 10.1007/978-1-4419-9530-8_13,
© Springer Science+Business Media, LLC 2012

metabolic epoxides into water-soluble products to facilitate their clearance. Through gene duplication and divergent evolution, EHs became specialized to perform vital functions beyond metabolic conversion of xenobiotics [1]. Concomitant with EH evolution, epoxide-containing molecules have transitioned beyond metabolic by-products into important mediators of signal transduction [2]. Thus, EHs play a large variety of roles ranging from cytoprotection to cell signaling, making them attractive therapeutic targets [3].

In mammals, five distinct EH enzymes have been identified: microsomal epoxide hydrolase (mEH), cholesterol epoxide hydrolase (ChEH), hepoxilin hydrolase (HH), leukotriene A_4 hydrolase (LTA$_4$H), and soluble epoxide hydrolase (sEH) [3]. The first EH purified from mammals was mEH: a 50 kD protein containing a hydrophobic amino terminus which, as its name implies, acts as a membrane anchor to microsomes. [4, 5]. The major function of mEH in the cell is the detoxification of metabolically derived arene epoxides [6]. In contrast to mEH, relatively little is known about ChEH. While its activity and endogenous substrate are known, the gene encoding ChEH has yet to be identified [7]. HH also has yet to have its coding gene identified. A defining feature of HH is its high substrate specificity for hepoxilins which play key roles in inflammation, smooth muscle tone, and carbohydrate metabolism [3]. Interestingly, a recent article suggests that HH and sEH are actually the same enzyme [8]. LTA$_4$H, in contrast to the other EHs, is a bifunctional zinc metaloprotease with both EH and aminopeptidase activities [9]. Interestingly, the two catalytic domains share a common carboxyl recognition site and, therefore, a substrate for one catalytic domain inhibits the catalytic activity of the other domain [10].

Knowledge of other EHs is important in not only providing a perspective on the final EH, sEH, but it also may be useful in solving mysteries that remain unanswered about the enzyme. This is particularly important because sEH, the focus of this chapter, is a potential therapeutic target for the treatment of stroke.

1.1 Soluble Epoxide Hydrolase Protein

sEH is a 62.5 kD homodimeric protein encoded by the gene EPHX2 [11]. Similar to LTA$_4$H, sEH is a bifunctional enzyme with both EH and lipid phosphatase (LP) activity [12]. Both the amino-terminal LP domain and carboxyl-terminal EH domain contain a canonical α/β-hydrolase fold; however, based on sequence analysis, the domains evolved from different ancestral genes. The EH domain most likely evolved from haloalkane dehalogenase while the LP domain evolved from haloacid dehalogenase [13]. The fused sEH gene has been identified in the purple sea urchin (*Strongylocentrotus purpuratus*) but not in the nematode *Caenorhabditis elegans*, hinting at the timeline of the sEH gene fusion event [14]. Interestingly, more recent sEH orthologs in frog and chicken contain an LP domain lacking phosphatase activity, suggesting both that (1) sEH has only recently become a functional LP and (2) the evolutionary driving force for the gene fusion event was not enzymatic but possibly structural [15].

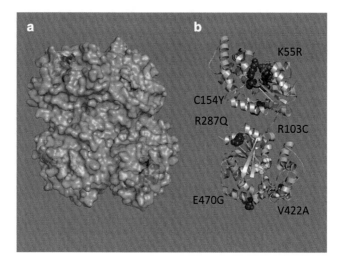

Fig. 13.1 Crystal structure of sEH and localized human polymorphisms. (**a**) sEH forms a domain swapped dimer between two monomers (*tan* and *gray*). The epoxide hydrolase domain (*yellow*) binds to the lipid phosphatase domain (*purple*) of the opposite monomer. (**b**) Localization of human polymorphisms (*blue*) on sEH monomer

The structure of both murine [16] and human [17] sEH proteins has been solved through protein crystallization. Unlike LTA$_4$H, the two catalytic domains do not share any residues; they are separated by a proline-rich linker suggesting that the activity of the two domains is independent of the other [17]. However, the domain-swapped quaternary architecture of sEH has fueled speculation that there may be communication between the two domains [12] (Fig. 13.1a). In support of this theory, the EH enzymatic activity of the sEH enzyme in which the LP has been deleted is reduced compared to the complete sEH protein [18]. Additionally, mutations in LP domain of sEH affect EH enzymatic activity [19] and, conversely, mutations in the EH affect LP activity [20]. However, in support of the other hypothesis in which there is no communication between EH and LP domains, Cronin et al. demonstrated that the enzymatic activity of each domain is unaffected by the presence of a substrate for the opposite domain. Furthermore, they showed that the LP activity of sEH is unchanged in the presence of EH inhibitors [21]. More work needs to be done in order to understand the interaction between these two catalytic domains. A major hurdle still remaining is the identification of an endogenous substrate for the sEH LP domain.

While the catalytic mechanism of the LP domain has been described, the endogenous substrate for this domain remains elusive [22]. It was demonstrated that sEH is capable of metabolizing an intermediate in cholesterol metabolism, isoprenoid phosphate [20]. Further work has shown that both sEH domains may play a direct role in the regulation of cholesterol [23]. Despite this, there is no consensus as to what the endogenous substrate of the LP domain is. Given that the majority of known phosphatases are involved in cell signaling, it is likely that the LP domain of sEH plays an important role in the regulation of a physiological function [24]. Therefore, until this endogenous substrate has been identified, there is a fundamental gap in the sEH field for multiple reasons. Without an identified substrate for the LP domain, it is impossible to (1) correctly interpret sEH knockout (sEHKO) phenotypes, (2) determine the interaction between the LP domain and EH domain, and (3) understand how inhibitors of EH activity affect levels of the LP substrate.

sEH is dually localized to both the cytosol and peroxisomes. This localization pattern is primarily thought to be the result of an impaired peroxisome targeting signal 1 (PTS1) on the carboxyl terminus of sEH. Instead of the canonical Ser–Lys–Leu, human sEH contains Ser–Lys–Met [25]. The substitution of a methionine for a leucine for the final residue in human PTS1 reduces the affinity of sEH to the peroxisome transport protein Pex5, resulting in poor transport and thus cytosolic localization. Interestingly, the mouse and rat PTS1 signal, Ser–Lys–Ile, is even more impaired than the human PTS1 [26]. There may be other factors, however, that influence the subcellular localization of sEH [25, 26].

The purpose for sEHs' localization pattern in peroxisomes is another unresolved issue in the sEH field. The following evidence supports a functional role for sEH in peroxisomes. Plant orthologues of sEH are dually localized to both the cytosol and glyoxysomes (specialized plant peroxisomes) [3]. Additionally, treating mice with clofibrate (a peroxisome proliferation-inducing agent) increased sEH activity, supporting a functional link between the number of peroxisomes and sEH [27]. However, a substrate for either the EH or LP domain of the sEH contained within peroxisomes remains unknown.

While there is agreement that sEH is dually localized to both the cytosol and peroxisomes, exactly which cells express sEH is currently under dispute. sEH is undeniably expressed in the liver, kidney, and heart among other human tissues tested for immunoreactivity [25, 28]. sEH is also expressed in the brain, which is of particular interest to stroke researchers. Its expression in human brain has been validated by Western blotting, immunohistochemistry, and an enzymatic activity assay [29]. Expression of sEH in mouse brain has also been validated by liquid chromatography tandem mass spectroscopy (LC-MS/MS) [30], Western blotting [31], and immunohistochemistry. Interestingly, expression of sEH in the brain has been shown to be sexually dimorphic with higher sEH protein expression in male mice [32]. This finding is supported by further data indicating that estrogen is a negative regulator sEH expression in the brain [33]. This would suggest that sEH may play a larger role in the pathology of male mice than female mice. Indeed, targeted disruption of sEH resulted in improvements in the hypertension of male but not female knockout animals [34].

Multiple studies have attempted to identify which cells in the brain express sEH; however, some of the data is conflicting in this regard. To date, there has only been one study using human tissue to localize sEH. This study localized sEH to multiple cell types in the brain, including neurons, glia, and endothelia [29]. Most studies agree that sEH is expressed in the vasculature [35–37]. However, expression of sEH in neurons is under dispute. While an additional study clearly demonstrates expression in neurons [38], others directly contradict this finding and report that sEH is primarily expressed in glia with only sparse neuronal expression in the medulla [36] or amygdale [39]. Data about the cell-specific expression of sEH from cell culture models suggests that both rat primary cultured astrocytes [40] and neurons are immunoreactive for sEH antibodies [35].

Regardless of which cells in the brain express sEH, its abundance strongly suggests that it plays a critical role in central nervous system physiology. Based on our current understanding, the most likely role is the regulation of the eicosanoid-signaling molecule epoxyeicosatrienoic acids (EETs): the endogenous substrate for its EH domain.

2   sEH and Epoxyeicosatrienoic Acids

EETs are synthesized from arachidonic acid (AA) by cytochrome P450 epoxyge-nases (CYP450). The CYP450 pathway has been described as the third arm of AA metabolism: the first two pathways being lipoxygenase (LOX) and cycloxygenase (COX) pathways which produce leukotrienes and prostaglandins, respectively. AA metabolites are known therapeutic targets to commonly proscribed drugs, including aspirin and Singulair [41].

Epoxygenation of AA, particularly by CYP450 2C or 2J isoforms, can occur on any one of the four double bonds present, resulting in four regioisomers, 5,6-, 8,9-, 11,12-, or 14,15-EET [42]. In addition to regioisomer specificity, EETs can also either be R,S or S,R enantiomers resulting in eight chemically distinct EET [43]. Both the regioisomer and stereoisomer profile of EETs varies between CYP450 synthesizing enzyme isoform [44]. This is particularly important because the physi-ological function varies between each EET isoform. Although often lumped together, it is important to note that different EETs have different functions [45].

EETs, acting through multiple mechanisms, have beneficial effects on the outcome of ischemic brain injury. The effects of EETs have been attributed to their contribu-tion to the endothelial derived hyperpolarizing factor (EDHF) response [46]. By contributing to the EDHF response, EETs maintain vascular tone, thus improving hemodynamics during ischemic events. It is becoming increasingly clear that the EDHF response plays a key role in parenchymal arterioles, rather than in larger cerebral arteries [47].

Recent studies, however, have shown that the influence of EETs extends beyond acting as an EDHF [48]. In particular, EETs have been shown to play a role in neuroprotection, promotion of angiogenesis, as well as suppression of platelet aggregation, oxidative stress, and postischemic inflammation [49]. This broad profile of effects has been comprehensively reviewed [49].

Despite the identification of a broad spectrum of effects, the exact mechanism of action of EETs remains unknown. The most plausible explanation is that EETs mediate their actions through binding to an EET-specific receptor [43]. In line with this thinking, EETs have been shown to interact with large-conductance calcium-activated potassium channels (BK_{Ca}) [50], transient receptor potential cation channels, subfamily V, member 4 (TRPV4) [51], thrombaxane receptors [52], and prostaglandin EP_2 receptors [53].

In addition to interactions with membrane-bound receptors, EETs or EET metabolites are peroxisome proliferator-activated receptor (PPAR) ligands. Through binding of the fatty acid-binding domain of PPAR alpha (PPARα) [54, 55] or PPARγ [56], EETs turn on the transcriptional activity of PPARs. The anti-inflammatory effects of EETs in particular may be explained through activation of PPARγ [57]. Furthermore, the colocalization of EET-synthesizing enzymes and PPARs strengthens the hypothesis of an interaction between the two [58].

While there are many beneficial actions of EETs, they are attenuated by the enzyme sEH which mitigates their biological activity by conversion of an epoxide to a vicinal diol creating dihydroxyeicosatrienoic acids (DHETs) [59]. While EETs are regulated through multiple mechanisms including degradation through β-oxidation and incorporation into membrane lipids, metabolism through the sEH pathway has been shown to have the greatest effect on EETs' bioavailability [60]. Particularly pertinent to this chapter, 14,15-EET is the preferred substrate metabolized by sEH and therefore most likely to be affected by sEH inhibition or gene deletion [59].

Given that sEH is the most potent regulator of EETs' diverse biological actions, it is no surprise that mutations in the gene encoding for sEH, EPHX2, have been linked to multiple pathological conditions including ischemia.

3   sEH Single-Nucleotide Polymorphisms

EPHX2 comprises 19 exons spanning 45 kb on the short arm of chromosome 8 [61]. The large number of exons as well as the large distance between them make it exceptionally cost-inefficient to sequence the entire EPHX2 gene. Consequently, the majority of genetic association studies have sought associations with previously identified polymorphisms from past studies or community resources, such as HapMap (www.hapmap.org), rather than seeking to identify novel mutations [62].

Of the mutations and polymorphisms identified in the EPHX2 gene, arguably the most interesting are the missense mutations that change the amino acid sequence of sEH. Multiple missense mutations have been identified that are located within both the LP and EH domains (Fig. 13.1b). Of these, six have been extensively examined for their effect on sEH properties, such as EH activity [19], LP activity [63], and subcellular localization [26]. As noted earlier, some of the most interesting findings have been that mutations in one domain of sEH affect the enzymatic activity of the other. However, because the EH activity of sEH is most understood, the effect of each polymorphism is often interpreted solely in light of its effect on EETs' metabolism.

EPHX2 polymorphisms have been linked to familial hypercholesterolemia [64], subclinical atherosclerosis [65], coronary heart disease [66], subclinical cardiovascular disease [67], and vasodilatory response [68]. Of these, perhaps the most interesting was the association of the K55R mutation with increased risk of cardiovascular heart disease (CHD) within the study population [66]. The K55R mutation is known to increase EH activity which, therefore, would lead to a decrease in cardioprotective EETs' levels. It, therefore, follows that the K55R mutation would be associated with an increased risk of CHD.

Table 13.1 Stroke genomic association studies

Ethnicity	Population number	Main finding	References
ARIC	315 cases, 1,021 controls	Haplotypes can increase or decrease the incidence of stroke	[69]
Chinese	200 cases, 350 controls	The R287Q polymorphism was protective from stroke	[71]
Europeans	601 cases, 736 controls	One intronic polymorphism was significantly associated with stroke	[62]
Danish	1,430 cases, 37,159 controls	No association with EPHX2 polymorphisms and stroke	[72]
Swedish	197 cases, 5,560 controls	The K55R polymorphism increased the risk of stroke in males	[70]

More than any other disease, EPHX2 polymorphisms have been studied for their link to ischemic protection or risk (Table 13.1). One of the first studies to associate EPHX2 polymorphisms with cerebral ischemia found that haplotypes within EPHX2 can either increase or decrease an individual's risk of stroke [69]. This is a very interesting finding because it demonstrates that the same protein, sEH, can either be advantageous or disadvantageous depending on its polymorphism status. This is consistent with previous work demonstrating that EPHX2 polymorphisms can either increase or decrease EH activity.

More specific than a haplotype, the same mutation that increased CHD risk, K55R, was shown to increase the stroke risk for Swedish males [70]. Conversely, another study directly linked the R287Q polymorphism of sEH to protection from ischemic stroke [71]. Previous work demonstrated that the R287Q polymorphism affects sEH by decreasing its EH activity. Therefore, this polymorphism should result in increased levels of EETs, in particular during ischemia, resulting in protection. Further linking the R287Q polymorphism specifically to neuroprotection, it was shown that introducing sEH harboring the R287Q polymorphism to primary cultured rat neurons results in decreased cell death when the neurons were subjected to an in vitro model of stroke, oxygen–glucose deprivation [35].

Because of the insight gained from biochemical studies characterizing the effect of sEH polymorphisms on enzymatic activity, investigators have made specific hypotheses about which polymorphisms would be associated with protection. For instance, one study only genotyped polymorphisms of sEH shown to reduce enzymatic activity and looked for protection from both cerebral and myocardial ischemia [72]. Unfortunately, this study was unable to find any statistically significant associations. It is important to note that the protective association between the R287Q polymorphism and ischemia has not always been replicated [62]. However, this is probably due to the fact that stroke is a complex disorder with multiple factors playing crucial roles in patient outcomes.

In addition to sEH polymorphisms being associated with human cardiovascular disease, mutations in the EPHX2 gene in animal models of cardiovascular disease have been identified that affect activity and expression level. Genetic differences between spontaneously hypertensive rats (SHR) and Wistar-Kyoto rats (WKY) from Charles River were shown to be responsible for increased levels of sEH expression in SHR rats. However, the inverse expression pattern was found true for SHR and WKY rats from the Heidelberg SP substrain suggesting an association of these polymorphisms with sEH expression but not hypertension [73]. Another study found that a variant of EPHX2 that increased transcription, protein expression, and activity was associated with spontaneously hypertensive heart failure (SHHF) rats [74]. Further linking sEH to stroke, single-nucleotide polymorphisms (SNPs) contained within the promoter of EPHX2 are also thought to be responsible for differences in the levels of sEH expression and activity determined in SHR/A3 (stroke prone) and SHR/N (stroke resistant) rats [75].

Clearly, these studies identifying a link between EPHX2 polymorphism status and ischemic risk support a role for sEH in ischemic outcome. Investigations into how genetic knockout of sEH affects ischemic outcomes have further supported a link between EPHX2 and ischemia.

4   sEH Knockout Mice

sEHKO mice were first generated in 2000 for the purpose of studying the effect of sEH on hypertension [34]. Since then, sEHKO mice have been an invaluable tool for studying the effect of sEH on cardiac arrest [76], inflammation [77], cholesterol regulation [23], atherosclerosis [78], vascular remodeling [79], hyperlipidemia [80], hyperglycemia [81], as well as myocardial and cerebral ischemia.

sEHKO mice have been shown to be protected from experimental cerebral ischemia, further identifying sEH as a risk factor for stroke [31]. Consistent with previous research showing that sEH is key in regulating EETs' levels, sEHKO mice were shown to have increased levels of EETs, specifically in brain extracts, compared to wild-type mice. This increase in EETs' levels presumably leads to the increased blood flow observed in sEHKO mice during middle cerebral artery occlusion (MCAO), resulting in a decreased infarct size.

In addition to maintaining blood flow during ischemia, sEH gene deletion also attenuates the inflammatory response to ischemia [33]. While expression of inflammatory cytokines tumor necrosis factor alpha (TNF-alpha), interleukin 6 (IL-6), interferon gamma (INF-gamma), and interleukin beta (IL-beta) was upregulated in the ipsilateral hemisphere (stroke side) compared to the contralateral hemisphere, this induction was significantly reduced in sEHKO mice compared to wild-type mice.

While sEHKO mice were shown to be protected from experimental ischemia, they had reduced survival after cardiac arrest and cardiopulmonary resuscitation [76]. This finding was unexpected, but may be explained by EETs' potent effect as a vasodilator. Because sEHKO mice have increased levels of EETs and thus dilated vessels, they were unable to restore sufficient blood upon resuscitation. It is unlikely that sEHKO mice have a cardiac dysfunction given that they had indistinguishable cardiac mass and myocardial function compared to wild-type mice.

Interpreting the phenotypes of knockout animals is difficult, and sEHKO mice in particular have multiple confounders. As noted earlier, while sEHKO mouse phenotypes are often interpreted in light of their effect on EETs, sEHKO mice also lack the LP domain. Another confounder of sEHKO mice involves physiologic compensation. Because sEHKO mice lack sEH throughout development, this allows time for alternative mechanisms to develop to make up for the lack of sEH activity. This was observed in one study which showed that sEHKO mice compensate for increased levels of vasodilating EETs with increased levels of vasoconstricting 20-HETE [82].

Knockout confounders notwithstanding, sEHKO mice support a protective role for inhibiting the activity of sEH during ischemia. However, in order to translate this basic science finding into the clinic, specific pharmacological agents needed to be developed that inhibit sEH.

5 Development of sEH Inhibitors

Because the main function of sEH is thought to be the regulation of EETs through its EH domain, potent and selective inhibitors for this domain have been developed for research and clinical use. Even though these inhibitors have been shown not to influence sEH's other enzymatic activity, LP, they are still termed sEH inhibitors. On a side note, there are a few sEH lipid phosphate inhibitors available, although they have not been as extensively developed or studied as EH inhibitors [18].

Based on the crystal structure of sEH and the catalytic mechanism epoxide hydrolysis, it was noted that amides or urea groups fit exceptionally well into the EH catalytic pocket. Specifically, the carbonyl oxygen of either an amide or urea would interact with Tyrosine 381 and Tyrosine 465 while the N–H group would act as a hydrogen donor to Aspartate 333, the nucleophile that attacks epoxide bonds [83]. In addition to ureas and amides, chalcone oxides, carbamates, acyl hydrazones, *trans*-3-phenylglycidols, and aminoheterocycles have been developed as sEH inhibitors. These inhibitors are, however, not commonly used [83]. Currently, the best sEH inhibitors are 1,3-disubstituted ureas, amides, and carbamates with IC_{50} values in the low nanomolar range [84]. The most currently used inhibitors are selective; however, it was observed that the sEH inhibitor AUDA activates PPARα [85].

Supporting previous work demonstrating that elevated EETs' levels are beneficial against a number of pathological conditions, sEH inhibitors have multiple effects. They are anti-inflammatory, antialgesic, antiatherosclerotic, antihypertensive, as well as renal, neuronal, and myocardial protective [84]. This wide profile of effects has made the development of sEH inhibitors an attractive target for both basic research and pharmacological companies. Indeed, University of California, Arete, Boehringer Ingelheim, Merk, and GlaxoSmithKline have all filed patents regarding the development and application of sEH inhibitors [83].

6   sEH Inhibitors and Ischemia

The multiple effects of increasing EETs with sEH inhibitors make testing their effect on ischemic brain damage a natural fit. Specifically, decreasing inflammation, apoptosis, and thrombosis all should have a beneficial effect on improving stroke outcomes [49]. In fact, multiple sEH inhibitors have been used in several animal models which have all shown protection from cerebral ischemia (Table 13.2).

Alkayed and colleagues identified sEH inhibition as a viable therapeutic strategy for the treatment of stroke [38]. They found that giving mice the sEH inhibitor AUDA-BE, either 30 min before a 2-h MCAO or at the time of reperfusion, significantly decreased infarct volume compared to vehicle-treated mice. Unexpectedly and in contrast to findings with sEHKO mice, they did not observe any difference in blood flow as measured by iodoantipyrine autoradiography; however, they were able to link the protection to EETs. When they treated mice both with AUDA-BE and MS-PPOH, an inhibitor of EETs' synthesis, the ischemic protection of the sEH inhibitor was lost.

Another research group showed that chronic treatment with a slightly different sEH inhibitor, AUDA, improved stroke outcome in stroke-prone spontaneously hypertensive rats (SHRSP) [86]. AUDA was given to SHRSP for 6 weeks before receiving an MCAO. Similar to the results found by Alkayed and colleagues, AUDA-treated SHRSP mice had a smaller infarct volume. This effect was found to be independent of changes in blood pressure or vascular structure even though they observed an increase in the passive compliance of cerebral vessels with AUDA treatment.

In another study, the sEH inhibitor t-AUCB was shown to protect SHRSP rats from ischemic brain damage [48]. In this study, SHRSP rats were treated for 1 week with t-AUCB before being subjected to MCAO. t-AUCB is a significantly different inhibitor than AUDA or AUDA-BE in that it does not have the long fatty acid chain thought to be responsible for AUDA's PPARα activation. Therefore, the ischemic protection observed in t-AUCB-treated mice suggests that AUDA's effect was largely mediated by its effect on EETs' stabilization and not on PPARα activation.

Additionally, this study observed transcriptional changes of apoptotic genes in brains from AUDA-treated rats. Interestingly, it found a reduction in proapoptotic transcription factors in both SHRSP and WYK rats compared to vehicle-treated rats. This change in proapoptotic transcription factors supports a role for neural in

Table 13.2 Ischemia sEH inhibitor studies

Ischemia sEH inhibitor studies					
Inhibitor	Model	Dose	Administration	Duration	References
AUDA	Rats	25 mg/L	Drinking water	6 weeks	[86]
AUDA-BE	Mice	10 mg/kg	Intraperitoneal	0, 30 min	[38]
4-PCO	Primary neurons	1 μM	Culture media	60 min	[35]
AUDA	Rats	2 mg/day	Drinking water	6 weeks	[48]
t-AUCB	Rats	2 mg/day	Drinking water	1 week	[48]

addition to vascular protection by treatment with sEH inhibitors. Further supporting a role for sEH inhibitors as neuroprotective agents, a study using the sEH inhibitor 4-PCO found that primary cultured rat neurons overexpressing sEH were protected from oxygen–glucose deprivation-induced cell death [35].

These studies all agree that the use of sEH inhibitor is protective against stroke. In fact, they suggest that sEH inhibitors may be protective against the effects of cerebral ischemia by multiple mechanisms making them even a more attractive therapeutic option than previous stroke therapeutics that target a single mechanism. Therefore, the question should be asked: Are we ready for sEH inhibitor clinical trials for stroke?

7   Are We Ready for an sEH Inhibitor Clinical Trial for Stroke?

Within this chapter, multiple lines of evidence have been presented that suggest that sEH is a novel and exciting target for the treatment of stroke. From human and rodent genetic polymorphisms that are linked to stroke risk to the protective effects against ischemia of both sEH gene deletion and pharmacological inhibition on ischemia, study after study supports the protective effect of increasing EETs' levels through the inhibition of sEH. Furthermore, the availability of an sEH inhibitor which has successfully completed phase I clinical trials makes moving forward with sEH clinical trials for the treatment of stroke even more tempting. But, are we ready?

The Stroke Therapy Academic and Industry Roundtable (STAIR) preclinical criteria were generated to aid in making this exact judgment [87]. The STAIR was formed in response to the repeated failure of clinical trials for the treatment of stroke. In particular, strategies that block calcium channels, scavenge-free radicals (e.g., NXY-059), and glutamate receptor antagonists were moved through clinical trials without success, despite laboratory evidence to the contrary.

To better judge whether a neuroprotective strategy has sufficient supporting evidence to warrant a clinical trial, the STAIR panel produced a set of preclinical research criteria. These criteria were designed to enhance the likelihood of a favorable clinical trial outcome if utilized by investigators.

The STAIR preclinical criteria have been summarized into ten key points [88]. As can be seen in Table 13.3, sEH inhibitors meet six of the ten criteria. These include evidence of ischemic protection from more than one laboratory. Animal testing of sEH inhibitors has also shown them to be protective in more than one species, including a human disease model. Furthermore, sEH inhibitors can be delivered in a feasible manner, such as 1 h after the time of occlusion. However, there are STAIR preclinical criteria that sEH inhibitors have not yet met. These include testing in both male and female mice, as well as testing at two different doses of inhibitor. Furthermore, behavioral measurements have not been made on animals treated with sEH inhibitors, in particular long-term behavioral studies are lacking. Therefore, based on the STAIR preclinical recommendations, more work is needed before sEH inhibitors are considered for clinical trials.

Table 13.3 sEH and the initial STAIR criteria

Item	Criteria	Description	sEH status	References
1.	Laboratory validation	Focal model tested in two or more laboratories	+	[38, 86]
2.	Animal species	Focal model tested in two or more species	+	[38, 86]
3.	Health of animals	Focal model tested in old or diseased animals	+	[86]
4.	Sex of animals	Focal model tested in male and female animals	−	
5.	Reperfusion	Tested in temporary and permanent models of focal ischemia	+	[38, 86]
6.	Time window	Drug administered at least 1 h after occlusion in focal model	+	[38]
7.	Dose response	Drug administered using at least two doses in focal model	−	
8.	Route of delivery	Tested using a feasible mode of delivery	+	[36, 48]
9.	End point	Both behavioral and histological outcomes measured	−	
10.	Long-term effect	Outcome measured at 4 or more weeks after occlusion in focal models	−	

Unfortunately, publication of the STAIR preclinical criteria in 1999 did not improve the effectiveness of subsequent clinical trials for stroke. Therefore, in 2009, the STAIR preclinical criteria were expanded. In addition to emphasizing the importance of the initial criteria, further recommendations were made. These include suggestions that laboratory stroke studies should be designed to ensure that they are sufficiently powered to support their conclusion. They also recommend that it is essential that stroke studies are randomized and blinded. Furthermore, they emphasize the importance of preclinical experiments on aged and diseased animals, especially animals with comorbid conditions, such as hypertension, diabetes, and hypercholesterolemia. Finally, they recommend collecting a biomarker in the animal studies that can also be obtained from patients during the clinical trial to indicate that the therapeutic treatment is working in humans as it did in preclinical animal studies.

Currently, sEH inhibitors have not been reported to have met any of these amended STAIR criteria. Therefore, they should act as a guide for current investigators to focus their sEH ischemia studies in addition to the missing components of the initial STAIR criteria. A retrospective analysis of stroke clinical trials up to 2006 showed that the effectiveness of treatment strategies in clinical trials was not different than those reported in preclinical experiments, emphasizing the importance of thorough preclinical studies [88]. Therefore, a more complete preclinical evaluation of sEH inhibitors is essential for making the decision to move them forward through clinical trials.

8   Concluding Comments and Future Directions

sEH inhibition is a promising target for the treatment of stroke. Where past failed therapeutic strategies have focused on protection through a single pathway, sEH inhibition results in protection through multiple mechanisms, including vascular and neuronal components [49]. Furthermore, substantial efforts have been made to create potent and selective sEH inhibitors, culminating in the creation of AR9281 which successfully completed phase 1 clinical trials [89]. However, according to consensus guidelines created by the STAIR panel, many studies still remain to be completed before a judgment can be made on whether sEH inhibitors are ready to move through clinical trials [87].

Both basic science and translational research questions still remain in the field of sEH ischemia research. In terms of basic science research, these questions include (1) what is the endogenous substrate for the LP domain of sEH; (2) is there an interaction between the sEH LP and EH catalytic domains; (3) what cell types in the brain express sEH; and (4) what is the role for sEH in peroxisomes. Answers to these questions will increase our understanding of sEH and also may lead to novel and, perhaps, more effective sEH-based stroke therapies.

In terms of translational research, the STAIR preclinical criteria should be used by investigators as a guide to what research still remains to be carried out. While all the STAIR criteria should be addressed, perhaps the most important criterion to be addressed is the effectiveness of sEH inhibitors in female animals. Indeed, because sEH has been shown to be more highly expressed in the brain of male mice, the effectiveness of sEH inhibition for the treatment of stroke in female mice may be attenuated [32]. Such a sexual dimorphism should not be viewed as a setback to sEH research, but rather as an important insight into how future clinical trials should be designed.

Research into sEH is a dynamic field full of talented investigators with a broad range of expertise. In the future, there are sure to be novel and important insights into the role and function of sEH. Research into the interaction between sEH and ischemia will play a vital role in moving the sEH field forward. In particular, experiments which move sEH inhibitors closer to clinical trials for the treatment of stroke will be important in translating the vast amount of laboratory-based research into use in the clinic.

References

1. Holmquist M. Alpha/beta-hydrolase fold enzymes: structures, functions and mechanisms. Curr Protein Pept Sci. 2000;1:209–35.
2. Fleming I. Epoxyeicosatrienoic acids, cell signaling and angiogenesis. Prostaglandins Other Lipid Mediat. 2007;82:60–7.
3. Newman JW, Morisseau C, Hammock BD. Epoxide hydrolases: their roles and interactions with lipid metabolism. Prog Lipid Res. 2005;44:1–51.

4. Oesch F. Purification and specificity of a human microsomal epoxide hydratase. Biochem J. 1974;139:77–88.
5. Friedberg T, Löllmann B, Becker R, Holler R, Oesch F. The microsomal epoxide hydrolase has a single membrane signal anchor sequence which is dispensable for the catalytic activity of this protein. Biochem J. 1994;303(Pt 3):967–72.
6. Lu AY, Miwa GT. Molecular properties and biological functions of microsomal epoxide hydrase. Annu Rev Pharmacol Toxicol. 1980;20:513–31.
7. de Medina P, Paillasse MR, Segala G, Poirot M, Silvente-Poirot S. Identification and pharmacological characterization of cholesterol-5,6-epoxide hydrolase as a target for tamoxifen and AEBS ligands. Proc Natl Acad Sci U S A. 2010;107:13520–5.
8. Cronin A, Decker M, Arand M. Mammalian soluble epoxide hydrolase is identical to liver hepoxilin hydrolase. J Lipid Res. 2011;52(4):712–9.
9. Orning L, Gierse JK, Fitzpatrick FA. The bifunctional enzyme leukotriene-A4 hydrolase is an arginine aminopeptidase of high efficiency and specificity. J Biol Chem. 1994;269:11269–73.
10. Rudberg PC, Tholander F, Andberg M, Thunnissen MMGM, Haeggström JZ. Leukotriene A4 hydrolase: identification of a common carboxylate recognition site for the epoxide hydrolase and aminopeptidase substrates. J Biol Chem. 2004;279:27376–82.
11. Knehr M, Thomas H, Arand M, Gebel T, Zeller HD, Oesch F. Isolation and characterization of a cDNA encoding rat liver cytosolic epoxide hydrolase and its functional expression in *Escherichia coli*. J Biol Chem. 1993;268:17623–7.
12. Newman JW, Morisseau C, Harris TR, Hammock BD. The soluble epoxide hydrolase encoded by EPXH2 is a bifunctional enzyme with novel lipid phosphate phosphatase activity. Proc Natl Acad Sci U S A. 2003;100:1558–63.
13. Beetham JK, Grant D, Arand M, Garbarino J, Kiyosue T, Pinot F, et al. Gene evolution of epoxide hydrolases and recommended nomenclature. DNA Cell Biol. 1995;14:61–71.
14. Harris TR, Aronov PA, Hammock BD. Soluble epoxide hydrolase homologs in *Strongylocentrotus purpuratus* suggest a gene duplication event and subsequent divergence. DNA Cell Biol. 2008;27:467–77.
15. Harris TR, Morisseau C, Walzem RL, Ma SJ, Hammock BD. The cloning and characterization of a soluble epoxide hydrolase in chicken. Poult Sci. 2006;85:278–87.
16. Argiriadi MA, Morisseau C, Hammock BD, Christianson DW. Detoxification of environmental mutagens and carcinogens: structure, mechanism, and evolution of liver epoxide hydrolase. Proc Natl Acad Sci U S A. 1999;96:10637–42.
17. Gomez GA, Morisseau C, Hammock BD, Christianson DW. Structure of human epoxide hydrolase reveals mechanistic inferences on bifunctional catalysis in epoxide and phosphate ester hydrolysis. Biochemistry. 2004;43:4716–23.
18. Tran KL, Aronov PA, Tanaka H, Newman JW, Hammock BD, Morisseau C. Lipid sulfates and sulfonates are allosteric competitive inhibitors of the N-terminal phosphatase activity of the mammalian soluble epoxide hydrolase. Biochemistry. 2005;44:12179–87.
19. Przybyla-Zawislak BD, Srivastava PK, Vazquez-Matias J, Mohrenweiser HW, Maxwell JE, Hammock BD, et al. Polymorphisms in human soluble epoxide hydrolase. Mol Pharmacol. 2003;64:482–90.
20. Enayetallah AE, Grant DF. Effects of human soluble epoxide hydrolase polymorphisms on isoprenoid phosphate hydrolysis. Biochem Biophys Res Commun. 2006;341:254–60.
21. Cronin A, Mowbray S, Dürk H, Homburg S, Fleming I, Fisslthaler B, et al. The N-terminal domain of mammalian soluble epoxide hydrolase is a phosphatase. Proc Natl Acad Sci U S A. 2003;100:1552–7.
22. Cronin A, Homburg S, Dürk H, Richter I, Adamska M, Frère F, et al. Insights into the catalytic mechanism of human sEH phosphatase by site-directed mutagenesis and LC-MS/MS analysis. J Mol Biol. 2008;383:627–40.
23. EnayetAllah AE, Luria A, Luo B, Tsai H, Sura P, Hammock BD, et al. Opposite regulation of cholesterol levels by the phosphatase and hydrolase domains of soluble epoxide hydrolase. J Biol Chem. 2008;283:36592–8.

24. Arand M, Cronin A, Oesch F, Mowbray SL, Jones TA. The telltale structures of epoxide hydrolases. Drug Metab Rev. 2003;35:365–83.
25. Enayetallah AE, French RA, Barber M, Grant DF. Cell-specific subcellular localization of soluble epoxide hydrolase in human tissues. J Histochem Cytochem. 2006;54:329–35.
26. Luo B, Norris C, Bolstad ESD, Knecht DA, Grant DF. Protein quaternary structure and expression levels contribute to peroxisomal-targeting-sequence-1-mediated peroxisomal import of human soluble epoxide hydrolase. J Mol Biol. 2008;380:31–41.
27. Lundgren B, DePierre JW. Proliferation of peroxisomes and induction of cytosolic and microsomal epoxide hydrolases in different strains of mice and rats after dietary treatment with clofibrate. Xenobiotica. 1989;19:867–81.
28. Enayetallah AE, French RA, Thibodeau MS, Grant DF. Distribution of soluble epoxide hydrolase and of cytochrome P450 2C8, 2C9, and 2J2 in human tissues. J Histochem Cytochem. 2004;52:447–54.
29. Sura P, Sura R, Enayetallah AE, Grant DF. Distribution and expression of soluble epoxide hydrolase in human brain. J Histochem Cytochem. 2008;56:551–9.
30. Shin J, Engidawork E, Delabar J, Lubec G. Identification and characterisation of soluble epoxide hydrolase in mouse brain by a robust protein biochemical method. Amino Acids. 2005;28:63–9.
31. Zhang W, Otsuka T, Sugo N, Ardeshiri A, Alhadid YK, Iliff JJ, et al. Soluble epoxide hydrolase gene deletion is protective against experimental cerebral ischemia. Stroke. 2008;39:2073–8.
32. Zhang W, Iliff JJ, Campbell CJ, Wang RK, Hurn PD, Alkayed NJ. Role of soluble epoxide hydrolase in the sex-specific vascular response to cerebral ischemia. J Cereb Blood Flow Metab. 2009;29:1475–81.
33. Koerner IP, Zhang W, Cheng J, Parker S, Hurn PD, Alkayed NJ. Soluble epoxide hydrolase: regulation by estrogen and role in the inflammatory response to cerebral ischemia. Front Biosci. 2008;13:2833–41.
34. Sinal CJ, Miyata M, Tohkin M, Nagata K, Bend JR, Gonzalez FJ. Targeted disruption of soluble epoxide hydrolase reveals a role in blood pressure regulation. J Biol Chem. 2000;275:40504–10.
35. Koerner IP, Jacks R, DeBarber AE, Koop D, Mao P, Grant DF, et al. Polymorphisms in the human soluble epoxide hydrolase gene EPHX2 linked to neuronal survival after ischemic injury. J Neurosci. 2007;27:4642–9.
36. Bianco RA, Agassandian K, Cassell MD, Spector AA, Sigmund CD. Characterization of transgenic mice with neuron-specific expression of soluble epoxide hydrolase. Brain Res. 2009;1291:60–72.
37. Iliff JJ, Close LN, Selden NR, Alkayed NJ. A novel role for P450 eicosanoids in the neurogenic control of cerebral blood flow in the rat. Exp Physiol. 2007;92:653–8.
38. Zhang W, Koerner IP, Noppens R, Grafe M, Tsai H, Morisseau C, et al. Soluble epoxide hydrolase: a novel therapeutic target in stroke. J Cereb Blood Flow Metab. 2007;27:1931–40.
39. Marowsky A, Burgener J, Falck JR, Fritschy J, Arand M. Distribution of soluble and microsomal epoxide hydrolase in the mouse brain and its contribution to cerebral epoxyeicosatrienoic acid metabolism. Neuroscience. 2009;163:646–61.
40. Rawal S, Morisseau C, Hammock BD, Shivachar AC. Differential subcellular distribution and colocalization of the microsomal and soluble epoxide hydrolases in cultured neonatal rat brain cortical astrocytes. J Neurosci Res. 2009;87:218–27.
41. Imig JD, Hammock BD. Soluble epoxide hydrolase as a therapeutic target for cardiovascular diseases. Nat Rev Drug Discov. 2009;8:794–805.
42. Capdevila JH, Falck JR, Harris RC. Cytochrome P450 and arachidonic acid bioactivation. Molecular and functional properties of the arachidonate monooxygenase. J Lipid Res. 2000;41:163–81.
43. Spector AA, Norris AW. Action of epoxyeicosatrienoic acids on cellular function. Am J Physiol Cell Physiol. 2007;292:C996–1012.

44. Daikh BE, Lasker JM, Raucy JL, Koop DR. Regio- and stereoselective epoxidation of arachidonic acid by human cytochromes P450 2C8 and 2C9. J Pharmacol Exp Ther. 1994;271:1427–33.
45. Gross GJ, Hsu A, Falck JR, Nithipatikom K. Mechanisms by which epoxyeicosatrienoic acids (EETs) elicit cardioprotection in rat hearts. J Mol Cell Cardiol. 2007;42:687–91.
46. Campbell WB, Fleming I. Epoxyeicosatrienoic acids and endothelium-dependent responses. Pflugers Arch. 2010;459:881–95.
47. Cipolla MJ, Smith J, Kohlmeyer MM, Godfrey JA. SKCa and IKCa channels, myogenic tone, and vasodilator responses in middle cerebral arteries and parenchymal arterioles: effect of ischemia and reperfusion. Stroke. 2009;40:1451–7.
48. Simpkins AN, Rudic RD, Schreihofer DA, Roy S, Manhiani M, Tsai H, et al. Soluble epoxide inhibition is protective against cerebral ischemia via vascular and neural protection. Am J Pathol. 2009;174:2086–95.
49. Iliff JJ, Alkayed NJ. Soluble epoxide hydrolase inhibition: targeting multiple mechanisms of ischemic brain injury with a single agent. Future Neurol. 2009;4:179–99.
50. Larsen BT, Miura H, Hatoum OA, Campbell WB, Hammock BD, Zeldin DC, et al. Epoxyeicosatrienoic and dihydroxyeicosatrienoic acids dilate human coronary arterioles via BK(Ca) channels: implications for soluble epoxide hydrolase inhibition. Am J Physiol Heart Circ Physiol. 2006;290:H491–9.
51. Watanabe H, Vriens J, Prenen J, Droogmans G, Voets T, Nilius B. Anandamide and arachidonic acid use epoxyeicosatrienoic acids to activate TRPV4 channels. Nature. 2003;424:434–8.
52. Behm DJ, Ogbonna A, Wu C, Burns-Kurtis CL, Douglas SA. Epoxyeicosatrienoic acids function as selective, endogenous antagonists of native thromboxane receptors: identification of a novel mechanism of vasodilation. J Pharmacol Exp Ther. 2009;328:231–9.
53. Yang C, Kwan Y, Au AL, Poon CC, Zhang Q, Chan S, et al. 14,15-Epoxyeicosatrienoic acid induces vasorelaxation through the prostaglandin EP(2) receptors in rat mesenteric artery. Prostaglandins Other Lipid Mediat. 2010;93:44–51.
54. Cowart LA, Wei S, Hsu M, Johnson EF, Krishna MU, Falck JR, et al. The CYP4A isoforms hydroxylate epoxyeicosatrienoic acids to form high affinity peroxisome proliferator-activated receptor ligands. J Biol Chem. 2002;277:35105–12.
55. Fang X, Hu S, Xu B, Snyder GD, Harmon S, Yao J, et al. 14,15-Dihydroxyeicosatrienoic acid activates peroxisome proliferator-activated receptor-alpha. Am J Physiol Heart Circ Physiol. 2006;290:H55–63.
56. Liu Y, Zhang Y, Schmelzer K, Lee T, Fang X, Zhu Y, et al. The antiinflammatory effect of laminar flow: the role of PPARgamma, epoxyeicosatrienoic acids, and soluble epoxide hydrolase. Proc Natl Acad Sci U S A. 2005;102:16747–52.
57. Ricote M, Li AC, Willson TM, Kelly CJ, Glass CK. The peroxisome proliferator-activated receptor-gamma is a negative regulator of macrophage activation. Nature. 1998;391:79–82.
58. Wray J, Bishop-Bailey D. Epoxygenases and peroxisome proliferator-activated receptors in mammalian vascular biology. Exp Physiol. 2008;93:148–54.
59. Yu Z, Xu F, Huse LM, Morisseau C, Draper AJ, Newman JW, et al. Soluble epoxide hydrolase regulates hydrolysis of vasoactive epoxyeicosatrienoic acids. Circ Res. 2000;87:992–8.
60. Seubert JM, Sinal CJ, Graves J, DeGraff LM, Bradbury JA, Lee CR, et al. Role of soluble epoxide hydrolase in postischemic recovery of heart contractile function. Circ Res. 2006;99:442–50.
61. Sandberg M, Meijer J. Structural characterization of the human soluble epoxide hydrolase gene (EPHX2). Biochem Biophys Res Commun. 1996;221:333–9.
62. Gschwendtner A, Ripke S, Freilinger T, Lichtner P, Müller-Myhsok B, Wichmann H, et al. Genetic variation in soluble epoxide hydrolase (EPHX2) is associated with an increased risk of ischemic stroke in white Europeans. Stroke. 2008;39:1593–6.
63. Srivastava PK, Sharma VK, Kalonia DS, Grant DF. Polymorphisms in human soluble epoxide hydrolase: effects on enzyme activity, enzyme stability, and quaternary structure. Arch Biochem Biophys. 2004;427:164–9.
64. Sato K, Emi M, Ezura Y, Fujita Y, Takada D, Ishigami T, et al. Soluble epoxide hydrolase variant (Glu287Arg) modifies plasma total cholesterol and triglyceride phenotype in familial

hypercholesterolemia: intrafamilial association study in an eight-generation hyperlipidemic kindred. J Hum Genet. 2004;49:29–34.

65. Wei Q, Doris PA, Pollizotto MV, Boerwinkle E, Jacobs DRJ, Siscovick DS, et al. Sequence variation in the soluble epoxide hydrolase gene and subclinical coronary atherosclerosis: interaction with cigarette smoking. Atherosclerosis. 2007;190:26–34.

66. Lee CR, North KE, Bray MS, Fornage M, Seubert JM, Newman JW, et al. Genetic variation in soluble epoxide hydrolase (EPHX2) and risk of coronary heart disease: The Atherosclerosis Risk in Communities (ARIC) study. Hum Mol Genet. 2006;15:1640–9.

67. Burdon KP, Lehtinen AB, Langefeld CD, Carr JJ, Rich SS, Freedman BI, et al. Genetic analysis of the soluble epoxide hydrolase gene, EPHX2, in subclinical cardiovascular disease in the Diabetes Heart Study. Diab Vasc Dis Res. 2008;5:128–34.

68. Lee CR, Pretorius M, Schuck RN, Burch LH, Bartlett J, Williams SM, et al. Genetic variation in soluble epoxide hydrolase (EPHX2) is associated with forearm vasodilator responses in humans. Hypertension. 2011;57:116–22.

69. Fornage M, Lee CR, Doris PA, Bray MS, Heiss G, Zeldin DC, et al. The soluble epoxide hydrolase gene harbors sequence variation associated with susceptibility to and protection from incident ischemic stroke. Hum Mol Genet. 2005;14:2829–37.

70. Fava C, Montagnana M, Danese E, Almgren P, Hedblad B, Engström G, et al. Homozygosity for the EPHX2 K55R polymorphism increases the long-term risk of ischemic stroke in men: a study in Swedes. Pharmacogenet Genomics. 2010;20:94–103.

71. Zhang L, Ding H, Yan J, Hui R, Wang W, Kissling GE, et al. Genetic variation in cytochrome P450 2J2 and soluble epoxide hydrolase and risk of ischemic stroke in a Chinese population. Pharmacogenet Genomics. 2008;18:45–51.

72. Lee J, Dahl M, Grande P, Tybjaerg-Hansen A, Nordestgaard BG. Genetically reduced soluble epoxide hydrolase activity and risk of stroke and other cardiovascular disease. Stroke. 2010;41:27–33.

73. Fornage M, Hinojos CA, Nurowska BW, Boerwinkle E, Hammock BD, Morisseau CHP, et al. Polymorphism in soluble epoxide hydrolase and blood pressure in spontaneously hypertensive rats. Hypertension. 2002;40:485–90.

74. Monti J, Fischer J, Paskas S, Heinig M, Schulz H, Gösele C, et al. Soluble epoxide hydrolase is a susceptibility factor for heart failure in a rat model of human disease. Nat Genet. 2008;40:529–37.

75. Corenblum MJ, Wise VE, Georgi K, Hammock BD, Doris PA, Fornage M. Altered soluble epoxide hydrolase gene expression and function and vascular disease risk in the stroke-prone spontaneously hypertensive rat. Hypertension. 2008;51:567–73.

76. Hutchens MP, Nakano T, Dunlap J, Traystman RJ, Hurn PD, Alkayed NJ. Soluble epoxide hydrolase gene deletion reduces survival after cardiac arrest and cardiopulmonary resuscitation. Resuscitation. 2008;76:89–94.

77. Manhiani M, Quigley JE, Knight SF, Tasoobshirazi S, Moore T, Brands MW, et al. Soluble epoxide hydrolase gene deletion attenuates renal injury and inflammation with DOCA-salt hypertension. Am J Physiol Renal Physiol. 2009;297:F740–8.

78. Zhang L, Vincelette J, Cheng Y, Mehra U, Chen D, Anandan S, et al. Inhibition of soluble epoxide hydrolase attenuated atherosclerosis, abdominal aortic aneurysm formation, and dyslipidemia. Arterioscler Thromb Vasc Biol. 2009;29:1265–70.

79. Simpkins AN, Rudic RD, Roy S, Tsai HJ, Hammock BD, Imig JD. Soluble epoxide hydrolase inhibition modulates vascular remodeling. Am J Physiol Heart Circ Physiol. 2010;298:H795–806.

80. Revermann M, Schloss M, Barbosa-Sicard E, Mieth A, Liebner S, Morisseau C, et al. Soluble epoxide hydrolase deficiency attenuates neointima formation in the femoral cuff model of hyperlipidemic mice. Arterioscler Thromb Vasc Biol. 2010;30:909–14.

81. Luo P, Chang H, Zhou Y, Zhang S, Hwang SH, Morisseau C, et al. Inhibition or deletion of soluble epoxide hydrolase prevents hyperglycemia, promotes insulin secretion, and reduces islet apoptosis. J Pharmacol Exp Ther. 2010;334:430–8.

82. Luria A, Weldon SM, Kabcenell AK, Ingraham RH, Matera D, Jiang H, et al. Compensatory mechanism for homeostatic blood pressure regulation in Ephx2 gene-disrupted mice. J Biol Chem. 2007;282:2891–8.

83. Shen HC. Soluble epoxide hydrolase inhibitors: a patent review. Expert Opin Ther Pat. 2010;20:941–56.
84. Revermann M. Pharmacological inhibition of the soluble epoxide hydrolase-from mouse to man. Curr Opin Pharmacol. 2010;10:173–8.
85. Fang X, Hu S, Watanabe T, Weintraub NL, Snyder GD, Yao J, et al. Activation of peroxisome proliferator-activated receptor alpha by substituted urea-derived soluble epoxide hydrolase inhibitors. J Pharmacol Exp Ther. 2005;314:260–70.
86. Dorrance AM, Rupp N, Pollock DM, Newman JW, Hammock BD, Imig JD. An epoxide hydrolase inhibitor, 12-(3-adamantan-1-yl-ureido)dodecanoic acid (AUDA), reduces ischemic cerebral infarct size in stroke-prone spontaneously hypertensive rats. J Cardiovasc Pharmacol. 2005;46:842–8.
87. Fisher M, Feuerstein G, Howells DW, Hurn PD, Kent TA, Savitz SI, et al. Update of the stroke therapy academic industry roundtable preclinical recommendations. Stroke. 2009;40:2244–50.
88. O'Collins VE, Macleod MR, Donnan GA, Horky LL, van der Worp BH, Howells DW. 1,026 experimental treatments in acute stroke. Ann Neurol. 2006;59:467–77.
89. Anandan S, Webb HK, Chen D, Wang YJ, Aavula BR, Cases S, et al. 1-(1-acetyl-piperidin-4-yl)-3-adamantan-1-yl-urea (AR9281) as a potent, selective, and orally available soluble epoxide hydrolase inhibitor with efficacy in rodent models of hypertension and dysglycemia. Bioorg Med Chem Lett. 2011;21:983–8.

Chapter 14
Membrane Potential as Stroke Target

Jens P. Dreier, Maren Winkler, Dirk Wiesenthal, Michael Scheel, and Clemens Reiffurth

Abstract All neurons in the mammalian brain develop a sustained depolarization in the absence of oxygen. In many structures of the grey matter including the brain cortex and the basal ganglia such sustained depolarizations develop abruptly in a large population of neurons and propagate in the tissue. Therefore, they are often referred to as spreading depolarizations. Spreading depolarizations seem to facilitate neuronal death and have now been demonstrated in the human brain in patients with aneurismal subarachnoid hemorrhage, delayed ischemic stroke after subarachnoid hemorrhage, and malignant hemispheric stroke. Therapies that target spreading depolarizations may potentially treat these conditions. These tsunami-like spreading depolarizations in the diseased brain are distinguished from the brief depolarizations that convey the flow of information in the healthy brain. However, understanding the latter is a prerequisite to develop an understanding of the former.

1 Basics of the Neuronal Membrane Potential

In the central nervous system, information is transferred within and between neurons by electrical and chemical signals. Electrical signals are receptor, synaptic, and action potentials. They result from transient alterations in current flow into and out

J.P. Dreier (✉) • M. Winkler • D. Wiesenthal • C. Reiffurth
Center for Stroke Research Berlin, Translation in Stroke Research,
Charité University Medicine Berlin, Charitéplatz 1, 10117 Berlin, Germany
e-mail: jens.dreier@charite.de

M. Scheel
Department of Neuroradiology, Charité University Medicine Berlin,
Charitéplatz 1, 10117 Berlin, Germany

P.A. Lapchak and J.H. Zhang (eds.), *Translational Stroke Research*, Springer Series
in Translational Stroke Research, DOI 10.1007/978-1-4419-9530-8_14,
© Springer Science+Business Media, LLC 2012

of the neuron that drives the membrane potential away from its resting potential. The resting potential is the electrical potential across the membrane in the absence of signaling activity while the term membrane potential refers to the electrical potential difference across the membrane at any moment in time.

Anions are negatively charged ions, and cations are positively charged ions. They carry the electrical current in an ionic solution. The direction of current flow is defined as the direction of net movement of positive charge. Thus, the term influx describes the net current flow of positive charge into the cell, and the term efflux that of net positive charge out of the cell.

Ion channels embedded in the cell membrane control the in- and efflux of ions. They can be gated or non-gated. Non-gated channels essentially maintain the resting membrane potential. They are not significantly affected by extrinsic factors and always remain open while gated channels open and close. At rest, gated channels are mostly closed. Their opening probability is essentially controlled by three extrinsic factors: stretch of the membrane, change in membrane potential (= voltage), or ligand binding.

The membrane potential is possible since the lipid bilayer of the membrane acts as a capacitor. It separates the ions on both sides of the membrane. At rest, an excess of positive charges is found on the outside of the membrane and an excess of negative charges on the inside. This charge separation across the membrane causes the resting membrane potential. The resting potential ranges typically between -60 and -70 mV in neurons. The net flux of charge across the membrane is zero at rest.

2 Neuronal Depolarization vs. Hyperpolarization

All electrical signaling is due to brief alterations away from the resting potential. Decline in charge separation and thus a shift of the membrane potential toward a less negative potential is called depolarization. Depolarization is caused by the opening of gated ion channels that allow positive charges to enter the cell (= net influx). An increase in charge separation due to net efflux, in contrast, is called hyperpolarization. The term electrotonic potential describes a passive de- or hyperpolarization of the membrane. Hyperpolarization is always passive. Whereas small depolarizations are also passive, at a critical level of depolarization, termed threshold, the neuron responds actively with an all-or-none action potential since many voltage-gated ion channels open suddenly and augment the depolarizing response explosively.

The driving force for electrical potential changes away from the resting potential comes from the electrochemical gradients across the membrane. The chemical driving force for a particular ion is due to its concentration gradient while the electrical driving force results from the potential difference between intra- and extracellular space. Chemical and electrical driving force for a given ion can act synergistically or can oppose each other. The flux of a given ion is then the product of the cell membrane conductance to that ion and the sum of the chemical and electrical driving forces that are summed up as the electrochemical driving force.

3 The Neuronal Resting Membrane Potential Is Between the Potassium and the Sodium Equilibrium Potential but Closer to the Potassium Equilibrium Potential

Potassium and sodium are the most prevalent cation species in the nervous system. Under physiological conditions, the potassium concentration is high in the intra- but low in the extracellular space since the concentration of large, impermeable anions in the form of amino acids and proteins is much higher in the intra- than in the extracellular space. Those large, impermeable anions constantly attract small cations such as potassium to enter the cells. If the membrane were exclusively permeable to potassium at rest, the resting potential would equal the potassium equilibrium potential at around -100 mV which can be calculated from the intracellular potassium concentration of $\sim$135 mM and the extracellular potassium concentration of $\sim$3 mM using the Nernst equation. In equilibrium the chemical driving force due to the potassium concentration gradient from the intra- to the extracellular space would equal the opposing electrical driving force due to the excess of negative charge in the intra- compared to the extracellular space.

The permeability of the membrane is in fact relatively high for potassium at rest because of a relatively high number of non-gated potassium channels but the membrane also has low permeability to sodium, calcium, and chloride. Therefore, the membrane potential is not identical to the potassium equilibrium potential.

Under physiological conditions, the sodium concentration is low in the intra- but high in the extracellular space. If the membrane were exclusively permeable to sodium, the resting potential would equal the sodium equilibrium potential at around $+70$ mV which can be calculated from the intracellular sodium concentration of $\sim$10 mM and the extracellular sodium concentration of $\sim$155 mM using the Nernst equation. However, the sodium permeability due to the few non-gated sodium channels is low compared to the potassium permeability due to the many non-gated potassium channels. Therefore, the resting potential is between the potassium and sodium equilibrium potentials but much closer to the potassium equilibrium potential. On the other hand, this means that, at rest, there is a small driving force for potassium to leave the cells due to the small difference between resting membrane potential and potassium equilibrium potential, and a large driving force for sodium to enter the cells because of the large difference between resting membrane potential and sodium equilibrium potential. Similar considerations as for sodium apply to calcium but the intra- and extracellular calcium concentrations are of magnitudes smaller than those of sodium. Therefore, calcium's direct contributions to changes in the membrane potential are almost negligible compared to those of sodium.

4 The Na, K-ATPase Maintains the Neuronal Steady-State Equilibrium

Importantly, in such a system there must be a constant leak of sodium and calcium into the cells and of potassium out of the cells due to all the electrochemical gradients that are not in equilibrium at rest and the presence of open, non-gated channels that allow for the flux of ions across the membrane. Hence, resting conditions do not mean that the system is in a stable equilibrium but they imply a dynamic or steady-state equilibrium that requires energy to be maintained. Without external energy, the steady state must eventually break down.

The energy to maintain the steady state comes from the hydrolysis of ATP. The neuron invests the energy from this chemical reaction into the transport of sodium, calcium, and potassium against the electrochemical driving forces. Hence, the steady influx of sodium and calcium and outflux of potassium through the non-gated ion channels are compensated by an equal steady outflux of sodium and calcium and influx of potassium through specialized, ATP-consuming membrane pumps. The most important of those membrane pumps is the Na, K-ATPase which consumes about 50% of the brain's energy for this purpose [31]. The brain only weighs 2% of the whole body weight but consumes ~20% of the whole body energy at rest [3]. Thus, at rest, about 10% of the energy consumed by the whole body serves the purpose to maintain the steady-state ionic equilibrium of the brain with its large electrochemical driving forces that provide the basis for the information processing.

The Na, K-ATPase also contributes directly to the membrane potential. It tends to hyperpolarize the membrane to a somewhat more negative potential than would be achieved by simple passive diffusion since it hydrolyzes one molecule of ATP to extrude three sodium ions and bring in two potassium ions with each cycle. This unequal cation flux produces a net efflux of positive charge and results in the typical neuronal resting potential between -60 and -70 mV.

5 The Action Potential

Action potentials are electrical impulses that are particularly important for the rapid flow of information over long distances in the central nervous system. At onset of an action potential, sodium channels open and sodium influx starts to exceed potassium efflux. Hence, the membrane depolarizes, i.e., the membrane potential is caused to shift away from the potassium equilibrium potential to approach the sodium equilibrium potential. When the neuronal membrane is depolarized past the threshold, voltage-gated sodium channels open rapidly producing additional influx of positive charge. This opens more and more voltage-gated sodium channels and accelerates membrane depolarization explosively. At its peak, the membrane potential almost reaches the sodium equilibrium potential. However, voltage-gated sodium channels close due to inactivation and voltage-gated potassium channels open. This causes a net efflux of positive charge, and the membrane potential re-approaches the resting potential, i.e., the membrane repolarizes. The action potential is very fast. It takes about 1 ms.

Importantly, the flows of sodium and potassium during an action potential do not lead to any appreciable change in their concentration gradients across the membrane. They only suffice to alter the charge separation between intra- and extracellular space so that the membrane is briefly dis- and then recharged [2]. The preservation of the ion concentration gradients during action potentials implies that sending those information impulses will not cause the system to lose significant amounts of energy under physiological conditions since the vast bulk of electrochemical energy is contained in the ion concentration gradients rather than the electrical potential difference across the membrane. This is related to the fact that the bulks of cytoplasm and extracellular fluid are electroneutral; the excess of positive and negative charges separated by the membrane is only ~1/200,000 of the total number of ions within the cytoplasm [14].

Hence, the elegance of the brain is that it operates in a steady state on a very high level of free energy, but the actual energy consumption during information processing is very low in relation to its enormous information processing power. In fact, the human brain consumes only 20–40 W for all the tasks it has to perform [22]. However, if the steep ion concentration gradients between intra- and extracellular space break down, an enormous amount of chemical energy is then required for the sodium and calcium pumps to reestablish the very high level of free energy that characterizes the normal steady state. This ionic breakdown only happens under pathological conditions such as migraine aura, intoxication with chemicals such as potassium, pharmacological inhibition of the sodium pump or the respiratory chain, or massive release of transmitters like glutamate and acetylcholine, during hypoxia, hypoglycemia, brain trauma, or stroke. The hallmark of this process is near-complete sustained neuronal depolarization to ~-10 mV [11].

6 Near-Complete Sustained Depolarization

When the physiological, steep ion concentration gradients breaks down, the critical question is whether this was due to signaling overload of the system in relation to the normal membrane pump activity or due to break down of pump function. In the former condition, the ionic breakdown is usually relatively harmless. Additional pump activity will be recruited, and this will reestablish the normal steady state without any tissue damage [15, 18]. Under this condition, the process will take about 1 min but even then, tissue ATP will fall transiently by about 50% [16].

However, in the latter condition of breakdown in pump function, the pumps cannot reestablish the normal steady state. This situation is critical since the ionic breakdown, associated with intracellular overload of calcium and sodium as well as loss of intracellular potassium, will persist which interferes in many ways with the

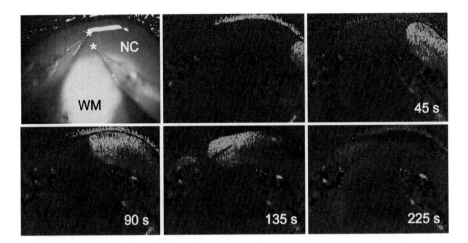

Fig. 14.1 Imaging of spreading depolarization in a human neocortical brain slice obtained from a patient with refractory epilepsy. Two ion-selective/reference microelectrodes (*asterisks*) are placed in layer II/III and V of the neocortex (NC) with the underlying white matter (WM). Pseudocolored subtraction images of the transilluminated tissue reveal spreading depolarization evoked by raising the extracellular potassium concentration. Note propagation of the event in the tissue

structural metabolism of the neurons. If the neurons remain in this unphysiological state for too long, they will eventually die. In animals, neurons can die within this state or they can show a transient electrical recovery after being in this state for a prolonged period and then die later on [11]. How long this pathological state will be tolerated depends presumably on the type of neuron, the developmental state, and the noxious condition that led to the ionic breakdown.

Such near-complete sustained depolarizations are observed in various animals from locusts and cockroaches to monkeys [21, 24, 26]. Therefore, it is not surprising that they also occur in the human brain. In patients with brain trauma, intracerebral and subarachnoid hemorrhage as well as malignant hemispheric stroke, near-complete sustained depolarizations have now been demonstrated in abundance [6, 8, 12]. Moreover, recordings with novel technology for invasive full-band direct current signals have evidenced that a similar spectrum from short-lasting to very prolonged sustained depolarizations occurs in time and space in the injured human brain as in animals [10, 19].

Near-complete sustained depolarizations last locally for at least 30 s up to hours. A characteristic feature of the near-complete sustained depolarizations is that they propagate in the tissue at a rate of about 3 mm/min (Fig. 14.1). Therefore, they are often referred to as spreading depolarizations. The near-complete breakdown of the ion concentration gradients is associated with water movement into the cells. Therefore, spreading depolarization leads to cytotoxic edema with cell swelling and distortion of dendritic spines that can be visualized with dual-photon microscopy [20, 29].

7 Spreading Depolarization and Cell Death

In animals, the role of spreading depolarizations for cell death was studied intensely in the ischemic penumbra after middle cerebral artery occlusion where their occurrence and cumulative duration correlated with infarct size [5, 17] and infarct growth [13]. The classic experimental approach to demonstrate their deleterious effect exploited the feature that they propagate in the tissue. In those experiments, a harmless, rapidly reversible spreading depolarization was ignited chemically outside of the ischemic penumbra. The depolarization wave then propagated into the penumbra where it partially changed its biophysical features. With each incoming depolarization wave, the ischemic core enlarged as evaluated by continuous apparent diffusion coefficient (ADC) of water mapping [4, 28]. In addition, using the endothelin-1 (ET-1) model in rats, direct histological evidence was provided that spreading depolarization has a deleterious effect in low-flow regions. In this model, the vasoconstrictive effect of ET-1 can be titrated. A concentration was applied brain topically where ET-1 induced a local low-flow region giving rise to spontaneous depolarization waves in about half of the animals. Subsequent histological analysis revealed focal necrosis restricted to the area exposed to ET-1. In the remaining half of the animals, the same ET-1 concentration led to a low-flow region but neither depolarization waves nor focal necrosis encountered (= nonresponders). However, when a depolarization wave was then triggered chemically in nonresponders and propagated into the ET-1 exposed cortex, those animals also developed focal necrosis in the ET-1-exposed cortex. The findings suggested that the depolarization waves initiated the cellular damage in the low-perfusion region independently of whether the depolarization wave started in the healthy surrounding cortex or in the low-perfusion region [9].

8 Spreading Depolarization as a Target for Treatment

Therapeutically, it is tempting to block the occurrence of spreading depolarization in order to conserve the intracellular environment of the neurons when the system is deprived of ATP. Despite the strong electrochemical driving forces, if the membrane would have no ion conductance, the neurons would be protected from loss of potassium, overload with sodium and calcium as well as the numerous deleterious changes associated therewith, including glutamate transporter reversal with massive glutamate release, etc. However, there is currently no drug that would achieve a complete electrical isolation between intra- and extraneuronal space. Moreover, given the multitude of different ion channels in the neuronal membrane, this is obviously a challenging if not impossible target for stroke therapy.

At least, it could be possible to slow down the depolarization process if those ion channels are inhibited that are responsible for the fast, initial component of spreading depolarization. However, under energy deprivation, even this component is not mediated by one cation channel but a number of different channels including at least N-methyl-D-aspartate, AMPA/kainate and GABA receptor controlled channels [23], and possibly more such as acid-sensing [33] and transient receptor potential channels [1] as well as pannexin hemichannels [30].

However, spreading depolarization also leads to secondary hemodynamic changes that can be either beneficial, if arterioles in the microcirculation dilate and blood flow increases (= spreading hyperemia), or detrimental, if arterioles constrict and blood flow declines severely (= spreading ischemia). The latter could be an important mechanism by which the depolarization waves recruit further tissue into ischemia and subsequent necrosis. Hence, the neurovascular response to spreading depolarization might be another, more modest but perhaps more realistic target for novel treatment strategies [11]. The role of the hemodynamic response to spreading depolarization for lesion progression is currently under intense experimental [7, 25, 27, 32] and clinical [10] investigation.

References

1. Aarts MM, Tymianski M. TRPMs and neuronal cell death. Pflugers Arch. 2005;451:243–9.
2. Alle H, Roth A, Geiger JR. Energy-efficient action potentials in hippocampal mossy fibers. Science. 2009;325:1405–8.
3. Attwell D, Laughlin SB. An energy budget for signaling in the grey matter of the brain. J Cereb Blood Flow Metab. 2001;21:1133–45.
4. Busch E, Gyngell ML, Eis M, Hoehn-Berlage M, Hossmann KA. Potassium-induced cortical spreading depressions during focal cerebral ischemia in rats: contribution to lesion growth assessed by diffusion-weighted NMR and biochemical imaging. J Cereb Blood Flow Metab. 1996;16:1090–9.
5. Dijkhuizen RM, Beekwilder JP, van der Worp HB, Berkelbach van der Sprenkel JW, Tulleken KA, Nicolay K. Correlation between tissue depolarizations and damage in focal ischemic rat brain. Brain Res. 1999;840:194–205.
6. Dohmen C, Sakowitz OW, Fabricius M, Bosche B, Reithmeier T, Ernestus RI, Brinker G, Dreier JP, Woitzik J, Strong AJ, Graf R. Spreading depolarizations occur in human ischemic stroke with high incidence. Ann Neurol. 2008;63:720–8.
7. Dreier JP, Korner K, Ebert N, Gorner A, Rubin I, Back T, Lindauer U, Wolf T, Villringer A, Einhaupl KM, Lauritzen M, Dirnagl U. Nitric oxide scavenging by hemoglobin or nitric oxide synthase inhibition by N-nitro-L-arginine induces cortical spreading ischemia when K+ is increased in the subarachnoid space. J Cereb Blood Flow Metab. 1998;18:978–90.
8. Dreier JP, Woitzik J, Fabricius M, Bhatia R, Major S, Drenckhahn C, Lehmann TN, Sarrafzadeh A, Willumsen L, Hartings JA, Sakowitz OW, Seemann JH, Thieme A, Lauritzen M, Strong AJ. Delayed ischaemic neurological deficits after subarachnoid haemorrhage are associated with clusters of spreading depolarizations. Brain. 2006;129:3224–37.
9. Dreier JP, Kleeberg J, Alam M, Major S, Kohl-Bareis M, Petzold GC, Victorov I, Dirnagl U, Obrenovitch TP, Priller J. Endothelin-1-induced spreading depression in rats is associated with a microarea of selective neuronal necrosis. Exp Biol Med (Maywood). 2007;232:204–13.
10. Dreier JP, Major S, Manning A, Woitzik J, Drenckhahn C, Steinbrink J, Tolias C, Oliveira-Ferreira AI, Fabricius M, Hartings JA, Vajkoczy P, Lauritzen M, Dirnagl U, Bohner G, Strong AJ. Cortical spreading ischaemia is a novel process involved in ischaemic damage in patients with aneurysmal subarachnoid haemorrhage. Brain. 2009;132:1866–81.
11. Dreier JP. The role of spreading depression, spreading depolarization and spreading ischemia in neurological disease. Nat Med. 2011;17(4):439–47.
12. Fabricius M, Fuhr S, Bhatia R, Boutelle M, Hashemi P, Strong AJ, Lauritzen M. Cortical spreading depression and peri-infarct depolarization in acutely injured human cerebral cortex. Brain. 2006;129:778–90.

13. Hartings JA, Rolli ML, Lu XC, Tortella FC. Delayed secondary phase of peri-infarct depolarizations after focal cerebral ischemia: relation to infarct growth and neuroprotection. J Neurosci. 2003;23:11602–10.
14. Koester J. Membrane potential. In: Kandel ER, Schwartz JH, Jessell TM, editors. Principles of neural science. 3rd ed. New York: McGraw-Hill; 1993. p. 81–94.
15. LaManna JC, Rosenthal M. Effect of ouabain and phenobarbital on oxidative metabolic activity associated with spreading cortical depression in cats. Brain Res. 1975;88:145–9.
16. Mies G, Paschen W. Regional changes of blood flow, glucose, and ATP content determined on brain sections during a single passage of spreading depression in rat brain cortex. Exp Neurol. 1984;84:249–58.
17. Mies G, Iijima T, Hossmann KA. Correlation between peri-infarct DC shifts and ischaemic neuronal damage in rat. Neuroreport. 1993;4:709–11.
18. Nedergaard M, Hansen AJ. Spreading depression is not associated with neuronal injury in the normal brain. Brain Res. 1988;449:395–8.
19. Oliveira-Ferreira AI, Milakara D, Alam M, Jorks D, Major S, Hartings JA, Lückl J, Martus P, Graf R, Dohmen C, Bohner G, Woitzik J, Dreier JP. Experimental and preliminary clinical evidence of an ischemic zone with prolonged negative DC shifts surrounded by a normally perfused tissue belt with persistent electrocorticographic depression. J Cereb Blood Flow Metab. 2010;30(8):1504–19.
20. Risher WC, Ard D, Yuan J, Kirov SA. Recurrent spontaneous spreading depolarizations facilitate acute dendritic injury in the ischemic penumbra. J Neurosci. 2010;30:9859–68.
21. Rodgers CI, Armstrong GA, Shoemaker KL, LaBrie JD, Moyes CD, Robertson RM. Stress preconditioning of spreading depression in the locust CNS. PLoS One. 2007;2:e1366.
22. Rolfe DF, Brown GC. Cellular energy utilization and molecular origin of standard metabolic rate in mammals. Physiol Rev. 1997;77:731–58.
23. Rossi DJ, Oshima T, Attwell D. Glutamate release in severe brain ischaemia is mainly by reversed uptake. Nature. 2000;403:316–21.
24. Rounds HD. KCl-induced 'spreading depression' in the cockroach. J Insect Physiol. 1967;13: 869–72.
25. Shin HK, Dunn AK, Jones PB, Boas DA, Moskowitz MA, Ayata C. Vasoconstrictive neurovascular coupling during focal ischemic depolarizations. J Cereb Blood Flow Metab. 2006;26: 1018–30.
26. Strong AJ, Smith SE, Whittington DJ, Meldrum BS, Parsons AA, Krupinski J, Hunter AJ, Patel S, Robertson C. Factors influencing the frequency of fluorescence transients as markers of peri-infarct depolarizations in focal cerebral ischemia. Stroke. 2000;31:214–22.
27. Strong AJ, Anderson PJ, Watts HR, Virley DJ, Lloyd A, Irving EA, Nagafuji T, Ninomiya M, Nakamura H, Dunn AK, Graf R. Peri-infarct depolarizations lead to loss of perfusion in ischaemic gyrencephalic cerebral cortex. Brain. 2007;130:995–1008.
28. Takano K, Latour LL, Formato JE, Carano RA, Helmer KG, Hasegawa Y, Sotak CH, Fisher M. The role of spreading depression in focal ischemia evaluated by diffusion mapping. Ann Neurol. 1996;39:308–18.
29. Takano T, Tian GF, Peng W, Lou N, Lovatt D, Hansen AJ, Kasischke KA, Nedergaard M. Cortical spreading depression causes and coincides with tissue hypoxia. Nat Neurosci. 2007;10:754–62.
30. Thompson RJ, Zhou N, MacVicar BA. Ischemia opens neuronal gap junction hemichannels. Science. 2006;312:924–7.
31. Tidow H, Aperia A, Nissen P. How are ion pumps and agrin signaling integrated? Trends Biochem Sci. 2010;35(12):653–9.
32. Windmuller O, Lindauer U, Foddis M, Einhaupl KM, Dirnagl U, Heinemann U, Dreier JP. Ion changes in spreading ischaemia induce rat middle cerebral artery constriction in the absence of NO. Brain. 2005;128:2042–51.
33. Xiong ZG, Zhu XM, Chu XP, Minami M, Hey J, Wei WL, MacDonald JF, Wemmie JA, Price MP, Welsh MJ, Simon RP. Neuroprotection in ischemia: blocking calcium-permeable acid-sensing ion channels. Cell. 2004;118:687–98.

Chapter 15
Hypothermia to Identify Therapeutic Targets for Stroke Treatment

Masaaki Hokari and Midori A. Yenari

Abstract Hypothermia has emerged as a viable neuroprotectant at the clinical level. The reasons for its protective effect are likely due to its ability to affect multiple facets of ischemic brain injury. While at times difficult to implement in humans due to various comorbidities, hypothermia can also be viewed as a tool by which neuroprotective targets may be identified. In this review, we discuss optimal conditions for therapeutic hypothermia, as evidenced by laboratory studies, as well as many of the effects of cooling on several cell death and cell survival pathways. The collective scientific literature indicates that temperature need only be decreased by a few degrees in order to confer protection, but early cooling and cooling of somewhat long duration (12–24 h) seem to be the more critical factors that determine success. Hypothermia seems to halt many damaging processes that lead to brain tissue injury, while upregulating factors that aid in its recovery. However, it should be noted that not all forms of brain injury benefit from therapeutic cooling. While global and focal cerebral ischemia appears to benefit from hypothermia, it is less clear whether brain hemorrhage responds in the same way. Thus, the laboratory literature also emphasizes the importance of careful preclinical studies prior to applying such concepts to humans.

1 Introduction

Hypothermia is recognized as one of the most robust neuroprotectants studied in the laboratory. Its therapeutic effect may be due to multimodal effects, rather than one specific mechanism. In the laboratory, it has been shown to affect multiple

M. Hokari, MD, PhD • M.A. Yenari, MD (✉)
Department of Neurology, University of California, San Francisco and the San Francisco
Veterans Affairs Medical Center, 4150 Clement Street, San Francisco, CA 94121, USA
e-mail: Yenari@alum.mit.edu

P.A. Lapchak and J.H. Zhang (eds.), *Translational Stroke Research*, Springer Series in Translational Stroke Research, DOI 10.1007/978-1-4419-9530-8_15, © Springer Science+Business Media, LLC 2012

aspects of cerebral injury, including reduction in metabolic activity, glutamate release, inflammation, production of reactive oxygen species, and mitochondrial cytochrome c release. Although stroke models vary in methodology, several laboratories have consistently demonstrated that hypothermia ameliorates the extent of brain injuries and improves neurologic function.

At the clinical level, a few prospective randomized studies have now shown that therapeutic cooling improves neurological outcome from cardiac arrest [1, 2] and neonatal hypoxia-ischemia [3–6]. There are fewer randomized multicenter trials for stroke patients, but a few studies have demonstrated its feasibility [7–9] and safety [10, 11]. Clinical studies in stroke seem to be hampered by effective cooling strategies in awake patients and the potential for adverse cardiac and infectious events in this older patient population. Thus, hypothermia research in stroke and related conditions might be used as a model of neuroprotection, by which other, perhaps less cumbersome and risky therapies could be developed.

2 Optimal Conditions for Hypothermia

Laboratory studies have systematically examined the depth, duration, and timing of cooling, to the extent that some of these parameters may be applied clinically. While some of these parameters are still in need of refinement, they may also offer insight into how cooling is protective.

2.1 Optimal Target Temperature

Generally, hypothermia can be classified by the depth of cooling from a normal body temperature of 37–38°C: mild hypothermia (32–35°C), moderate hypothermia (28–32°C), and deep hypothermia (−28°C). Because of the numerous complications of deep hypothermia and the difficulty in achieving and maintaining these temperatures, mild to moderate hypothermia is becoming more attractive alternatives [12]. For example, cardiac arrhythmias become more frequent with temperatures below 30°C, platelet dysfunction occurs when temperatures decrease below 35°C, and coagulation is reduced with temperatures below 33°C [13]. The scientific literature indicates that brain temperature lowered to the mild to moderate range is similarly protective as deep hypothermia [14, 15].

In a meta-analysis of animal studies [16], the benefit of hypothermia was inversely related to the temperature achieved and hypothermia reduced infarct size by >40% with temperatures of 34°C or below. However, specific studies that directly compared different temperatures failed to show dose-dependent neuroprotection comparing temperatures of 30 vs. 33°C [17] and 27 vs. 32°C [18]. A more recent experiment not included in their review compared cooling with 36, 35, 34, 33, and

32°C for 4 h, starting 90 min after middle cerebral artery (MCA) occlusion in rats. Cooling to 35°C failed to improve outcome and the largest benefit was obtained at 34°C [19]. Most clinical studies of hypothermia in acute ischemic stroke [7, 8, 11, 20, 21] as well as in post-anoxic encephalopathy after cardiac arrest [1, 2], perinatal hypoxic-ischemic encephalopathy (HIE) [22], and traumatic brain injury [23, 24] have studied cooling to levels of 32–34°C. Unfortunately, discomfort and shivering increase with lower temperatures in awake stroke patients, and cooling to these levels therefore generally requires sedation, mechanical ventilation, and admission to an intensive care unit (ICU) [13]. On the other hand, cooling patients with severe head injury to 35°C appeared to be as beneficial as 33°C while causing fewer complications [25]. In addition, temperature reductions to 35.5 or 35°C via surface cooling have been shown to be feasible and safe in stroke patients without sedation, provided pethidine (meperidine) is given to prevent shivering [26]. Using a similar strategy, the NoCSS (Nordic Cooling Stroke Study) was a randomized trial which tested the effect of temperature reduction to 35°C in awake patients with surface cooling for 9 h, started within 6 h of ischemic stroke onset, but unfortunately the trial was terminated because of slow recruitment.

Thus, cooling to 32–34°C may be optimum target temperature, and while cooling to 35°C may be safe but its neuroprotective potential is less clear.

2.2 Timing and Duration of Cooling

From numerous laboratory studies, it is clear that cooling is a remarkable neuroprotectant when applied during ischemia. Therefore, hypothermia should be initiated as soon as possible to achieve its optimal beneficial effect. But because many patients do not present to the emergency room immediately after symptom onset, a critical question is how long after stroke cooling can be applied and still be effective. In our prior review [27], we reported that reduction of infarct size is commonly observed when cooling is begun within 60 min of stroke onset in permanent [28] and 180 min of stroke onset in temporary MCA occlusion models. In global cerebral ischemia, one study showed that prolonged cooling delayed 6 h resulted in marked neuroprotection, but similar delays have not been shown for focal cerebral ischemia [29, 30].

A meta-analysis of animal studies could not determine a clear temporal therapeutic window due to the small numbers of published reports, but the authors found that the best protection was observed in the temporary occlusion models where cooling was begun during ischemia and continued for some time into the reperfusion period [31]. In the European Hypothermia After Cardiac Arrest Study, cooling improved functional outcome despite median delay of 8 h after the restoration of spontaneous circulation and attainment of the target temperature [2]. All things considered, 3 h may be the maximum time to start cooling in focal ischemia and about 6 h in global ischemia. Although longer delays may still protect if cooling is prolonged, this has not been adequately studied in focal ischemia.

The optimal duration of hypothermia is also unclear. Some groups have used brief durations of hypothermia (0.5–5 h), whereas others longer periods (12–48 h). In a few studies of focal cerebral ischemia where the duration of intraischemic hypothermia was compared directly, durations of 1–3 h appeared effective, whereas 0.5–1 h were not [17, 32]. In global cerebral ischemia, intraischemic hypothermia (rectal temperature 28–32°C) completely prevented hippocampal cell damage if continued for 4 or 6 h, whereas 2 h of hypothermia protected less well, and 1 or 0.5 h did not protect at all [33]. Longer durations may be necessary especially when the initiation of cooling is delayed. Some studies showed that postischemic hypothermia merely delayed the onset of irreversible neuronal injury, unless combined with a second neuroprotectant [34, 35]. However, these latter studies only applied hypothermia for 3 h. Other investigators have shown that prolonged hypothermia initiated 4–6 h after forebrain ischemia for 24 h can provide sustained functional and histological neuroprotection as far as 6-month postischemic onset [36]. Thus, rodent data indicated that prolonged reduction in temperature robust neuroprotection when hypothermia is delayed by several hours, provided cooling is maintained for more than 24 h [37]. Thus, the extent of a neuroprotection is influenced by the length of the delay and the duration of hypothermia.

In clinical stroke, hypothermia may be a more effective neuroprotection strategy if applied for a long duration after the ischemic event, as most patients do not present until hours after the onset of stroke [38]. Although a long cooling time seems attractive, this may be offset by an increased risk of complications.

3 Mechanism of Hypothermia

Hypothermia has the potential to affect multiple aspects of cell physiology to confer neuroprotection, although no data exist to elucidate the relative importance of these actions individually. Possible neuroprotective mechanisms associated with hypothermia are to lower metabolic rate, to reduce glutamate release, to limit calcium influx after glutamate, to reduce inflammatory response to insult, to limit BBB disruption, to suppress formation of reactive oxygen species, and to upregulate antiapoptotic or neurotrophic factors.

3.1 Effects on Brain Metabolism

Temperature has a profound effect on brain metabolism during ischemia [39]. Hypothermia has long been thought to protect the brain by reducing cerebral metabolism during conditions of reduced substrate. Hypothermia lowers the cerebral metabolic rates of glucose (CMRglu) and oxygen (CMRO2) and slows ATP breakdown [40]. On average, hypothermia reduces brain oxygen consumption by approximately 5%/°C fall in body temperature in the range of 22–37°C [41, 42], and in anesthetized animals CMRO2 decreases linearly between 37 and 8°C [43]. However, the Q10 for brain metabolism does not correlate to the remarkable reduction in lesion size. Thus, the widely held notion that hypothermia protects because lower temperature slows metabolism does not completely explain the neuroprotective effect of cooling [44].

3.2 Effects on Neurotransmitters and Intracellular Calcium Accumulation

Several investigators showed that hypothermia significantly decreased glutamate release due to cerebral ischemia [45–47], and excessive levels of glutamate are well known to potentiate ischemic injury. Hypothermia also suppresses ischemic down-regulation of the α-smino-3-hydroxy-5-methyl-4-isoxazole-propionic acid (AMPA) receptor submit GluR [48]. Thus, a potentially protective effect of hypothermia is that it could salvage neurons from delayed calcium entry via AMPA receptors [49].

Hypothermia also prevented the increase in intracellular calcium associated with ischemia [50–52]. The mechanism of this effect is uncertain yet, although it could be linked to the above reduction in glutamate release, as well as functions of other calcium entry channels. Recent work from our group has shown that hypothermia prevents ischemia-induced upregulation of a newly characterized calcium-sensing receptor (CaSR). CaSR is thought to sense changes in extracellular calcium levels, but also appears to reciprocally downregulate GABA receptors, thereby decreasing inhibitory tone. Cerebral ischemia upregulates CaSR while downregulating GABA-B R1, a phenomenon that therapeutic hypothermia suppresses [53]. Thus, CaSR might be studied as a potential target for treatment of stroke and related conditions.

Excitotoxic neurotransmitters are released early after ischemia onset, and it is well known that glutamate antagonists have a rather narrow temporal therapeutic window of 1–2 h even in models of temporary MCAO. Since earlier cooling is superior to delayed cooling, this may explain some of the protective effect of hypothermia. However, cooling even after glutamate release has occurred is still protective [54].

3.3 Effects on Inflammation

Hypothermia markedly suppresses various aspects of the postischemic inflammatory response. Mild and moderate hypothermia reduced accumulation of granulocytes and upregulation of the intercellular adhesion molecule ICAM-1 [55–57]. Hypothermia also suppressed microglial production of toxic mediators, such as iNOS, nitric oxide, and peroxynitrite [58]. Other studies showed that hypothermia also suppresses the inflammatory immune transcription factor NFκB [59, 60] with downstream suppression of NFκB-regulated immune factors. The finding that hypothermia suppresses NFκB also contributes to scientific knowledge regarding the role of NFκB in brain ischemia. Conflicting studies using genetic mutant models or pharmacologic inhibitors of NFκB suggested that NFκB was damaging to the brain during ischemia because of its pro-immune properties, but its ability to upregulate trophic factors also suggested that it might also have beneficial roles. Studying the patterns of NFκB activation under the robust neuroprotective conditions of hypothermia showed that cooling overall suppressed NFκB, and might suggest that inhibiting NFκB might be a more prudent therapeutic strategy.

Nevertheless, the immune reactions following stroke are activated somewhat early (within hours), but activated microglia and infiltrated immunes are present in the brain for months, although their functions may change depending on the phase of stroke. This suppression of inflammation by cooling might also explain, in part, why longer durations of cooling are more effective, and why it may be essential to cool for longer durations when cooling is delayed.

3.4 Effects on BBB Disruption

Small differences in intraischemic or postischemic temperature have been reported to influence early blood–brain barrier (BBB) breakdown. While mild hypothermia (34°C) reduced the extravasation of a BBB tracer (horseradish peroxidase), hyperthermia (39°C) aggravated BBB leakage compared with normothermia [61]. Matrix metalloproteinases (MMPs) are recognized to contribute significantly to the destruction of the BBB and extracellular matrix. A few studies have now shown that therapeutic mild hypothermia not only reduces BBB disruption and edema, but also reduces generation of MMPs and preserves various matrix proteins [62–66].

3.5 Effects on Reactive Oxygen Species and Apoptosis

Several studies have now shown that hypothermia attenuates free radical formation and thus provides protection [67, 68]. The generation of reactive oxygen species (ROS) leads to oxidative stress which leads to a variety of damaging consequences to the ischemic brain. ROS can lead to lipid peroxidation, direct DNA damage, and can trigger other cell death pathways such as apoptosis. Hypothermia, by preventing the generation of ROS, has been shown to inhibit all of these downstream processes including the prevention of cytochrome c release, downregulation of pro-apoptotic factors, and the prevention of caspase activation (reviewed previously [54, 69]).

3.6 Upregulation of Endogenous Protective Factors

While hypothermia is largely associated with the downregulation or suppression of various damaging metabolic processes, hypothermia is also associated with the upregulation of factors known to reduce tissue injury. Other data suggest that hypothermia could have other protective effects, including the upregulation of anti-apoptotic gene Bcl-2, several trophic factors, and Akt (reviewed previously [70]). The precise reasons for this differential effect of cooling on the expression of various genes are unclear, but it is known that there are a few temperature-dependent transcription factors and a family of cold shock proteins have been described in other model systems [71]. The study of such temperature dependent molecular mechanisms has not yet been studied in brain injury models. Regardless, these observations indicate that organisms have in place a complex response system to temperature changes that could be potentially explored as therapeutic targets.

4 Hypothermia in Neurological Disorders

4.1 Cerebral Ischemia

4.1.1 Ischemic Stroke (Focal Cerebral Ischemia)

Abundant experimental studies have shown the neuroprotective effects of mild or moderate hypothermia when cooling was initiated within a few hours of focal ischemia onset. The benefit of hypothermia was greater in temporary, compared to permanent occlusion models. A few clinical studies of hypothermia in acute ischemic stroke have been published or are ongoing [12, 72]. These studies have collectively shown that mild hypothermia is feasible, though not completely without complications. A significant challenge in applying hypothermia to ischemic stroke patients is that they are generally awake and do not tolerate cooling, unlike other neurological conditions, such as cardiac arrest and severe brain injury.

There are a few small clinical studies of hypothermia in stroke. Using intravascular cooling devices to cool acute ischemic stroke patients, a few studies have shown feasibility and tolerability, as well as regimens to prevent or reduce shivering in the awake stroke patient [8, 9]. In a recent randomized multicenter study, an endovascular cooling device was used in combination with rt-PA administration in acute stroke patients (ICTuS-L). Here, patients could be treated within 0–6 h of symptom onset followed by endovascular cooling to 33°C for 24 h. While the study was not powered to study efficacy, this regimen appeared well tolerated, although cooled patients tended to experience pneumonia more often [10].

4.1.2 Cerebral Hypoxia/Anoxia (Global Cerebral Ischemia)

Several experimental studies have demonstrated the neuroprotective effects of mild or moderate hypothermia for global ischemia. Experimental studies have established the durability of this protective effect and have defined a temporal therapeutic window which can be lengthened provided cooling is prolonged [73]. The clinical benefit of hypothermia has also been demonstrated in two multicenter clinical trials [1, 2]. Thus, therapeutic cooling has become increasingly embraced by not only tertiary medical centers but also community hospitals as well [11].

Therapeutic cooling has also been shown to be effective in preventing perinatal brain injury from HIE. There have been four large studies of newborns with HIE [3–6]. These studies have shown benefit in infants with moderate and severe HIE; however, long-term, lifelong benefits are especially key in pediatric populations, and there are no reports of outcomes beyond 21 months of age. This condition has also been studied in the laboratory, although not as extensively as in adult models. However, hypothermic protection in neonatal animal models have shown similar associations such as reduced excitatory amino acid accumulation, preservation of metabolic substrates, and inhibition of caspase activation [74, 75].

4.2 Hemorrhagic Stroke

4.2.1 Intracerebral Hemorrhage

While less studied compared to ischemic stroke, recent experimental studies have begun to explore the role of hypothermia in intracerebral hemorrhage (ICH). A few reports have shown that hypothermia, used in a similar manner as that described above for ischemia, can reduce brain edema, inflammation, and BBB disruption [76, 77]. One study demonstrated a favorable effect of cooling on lesion size and functional deficits [78], but this was not as consistent a finding across labs. In fact, some labs failed to find improvements in histological endpoints and neurological function in models of ICH [76, 77] and one report described increased bleeding in the brain among cooled animals [79].

The optimal hypothermic conditions may also be different from that for cerebral ischemia. Although earlier hypothermia is better in ischemia, the study that demonstrated increased bleeding with cooling actually found some protection provided cooling was delayed 12 h, but worsened outcome if cooling began earlier [79]. The authors speculated that cooling could affect critical coagulant and thrombolytic systems in the acute period, or that cooling exacerbated complications of the initial increased blood pressure observed in this model. These findings bear further studies to clarify the reasons for worsening in certain scenarios, and whether cooling might be detrimental if not applied in an optimal manner.

At the clinical level, Kollmar et al. [80] recently reported that 12 patients with large ICH were treated with hypothermia to 35°C for 10 days (initiated 3–12 h after symptoms onset) and these patients were compared to data from a local hemorrhage data bank. In the hypothermia group, edema volume remained stable during 14 days, whereas edema significantly increased in the control group. However, larger controlled clinical trials of hypothermia in ICH are lacking.

4.2.2 Subarachnoid Hemorrhage

Subarachnoid hemorrhage (SAH) is often due to aneurysmal rupture, and hypothermia is often used intraoperatively during aneurysm repair. There are two animal studies of SAH that hypothermia exhibited neuroprotection. Torok et al. [81] applied mild hypothermia for 2 h and reported reduced intracranial pressure and improved post-hemorrhagic neurological deficits and postoperative weight gain by days 1–7 if applied up to 3 h after SAH. Schubert et al. [82] examined the effects of moderate hypothermia on the acute changes after massive experimental SAH as evaluated by DWI and magnetic resonance spectroscopy (MRS). They concluded that hypothermia ameliorated early development of cytotoxic edema, lactate accumulation, and a general metabolic stress response after SAH in rat. The mechanisms underlying this protective effect have not been explored as extensively as in brain ischemia models, but one study showed that cooling led to suppression of the stress response [83]. At the clinical level, Muroi et al. assessed the effect of the hypothermia and high-dose barbiturate combined therapy on the inflammatory response in seven patients compared to eight patients who received no such intervention. Hypothermia decreased systemic and cerebrospinal fluid levels of interleukin (IL)-6, IL-1β, and leukocyte counts compared to untreated patients, although cooling increased tumor necrosis factor alpha (TNF-α) [84].

However, in a large multicenter randomized study, mild intraoperative hypothermia during surgery for intracranial aneurysm turned out not to improve neurologic outcomes among favorable-grade patients with aneurysmal SAH (WFNS grade 1–3) [85]. Therefore, recent studies have been focusing on therapeutic hypothermia for patients with poor grade SAH. Seule et al. evaluated the feasibility and safety of mild hypothermia treatment in patients with aneurysmal poor grade SAH who are experiencing intracranial hypertension and/or cerebral vasospasm [86]. They concluded prolonged systemic hypothermia may be considered as a last-resort option for a carefully selected group of younger SAH patients with resistant intracranial hypertension or cerebral vasospasm although severe side effects occurred in many patients treated with long-term hypothermia. Thus, the clinical effectiveness of therapeutic cooling for SAH remains unclear, probably because many factors including fever [87], vasospasm [88], and excessive decompression by skull removal affect the prognosis of SAH.

5 Identification of Therapeutic Targets Through Hypothermia

While hypothermia is the most robust neuroprotectant studied in the laboratory to date, and one that has now been shown to have benefit in certain patient populations, it is not always practical or feasible to cool many patients with acute neurological injuries. Thus, hypothermia might be viewed as a "model of neuroprotection" where therapeutic strategies could be designed taking into consideration some of the benefits of hypothermia without the risks or challenges.

5.1   Therapies That Target Multiple Facets of Ischemic Injury

Hypothermia appears to affect multiple aspects of ischemia pathogenesis. Thus, it may behoove investigators to identify drugs that may also have similar multifaceted properties. One such multifaceted drug is minocycline [89]. Like hypothermia, minocycline appears to inhibit inflammation and apoptosis and preserve BBB integrity. It can be given in an oral form, and there is already abundant clinical experience. A small study of minocycline in stroke patients was shown to improve neurological outcome [90], and there is an ongoing trial of minocycline plus rt-PA in acute stroke [91].

5.2   Hypothermia to Identify New Potential Therapeutic Targets

With increasing availability of gene profiling technologies, it is possible to identify large numbers of genes and changes in those genes in a relatively short period of time. In work by Kobayashi et al. [92], rats were subjected to global cerebral ischemia under normothermic, hypothermic, or hyperthermic conditions. Brain samples were subjected to gene profiling which identified 33 genes that were temperature-sensitive. Many of these genes and their gene products had been studied in the scientific literature previously, and cooling seemed to lead to the suppression of genes, most of which have been described as being neurotoxic. However, cooling also led to the upregulation of several cytoprotective genes. Interestingly, many of the altered genes by hypothermia were similarly altered by preconditioning, an endogenous defense mechanism. Thus, future research may focus on the significance of these genes in ischemia pathogenesis, and whether their modulation represents a potential therapeutic target.

In our own lab, we recently reported the upregulation of CaSR, a protein normally found in the parathyroid gland to sense subtle changes in extracellular calcium. While not previously recognized to exist in the brain, it was robustly upregulated by ischemia. By applying therapeutic cooling in a forebrain ischemia model, we found that hypothermia reduced CaSR while simultaneously upregulating GABA-B-R1, a receptor of the inhibitory neurotransmitter, GABA [53]. Thus, the identification that hypothermia suppresses ischemia-induced CaSR might also stimulate research in CaSR inhibitors for treatment of stroke.

5.3   Hypothermia to Clarify the Role of Putative Targets

A few proteins and signaling pathways have met with varied results in the scientific literature. NFκB inhibitors or its deficiency have been reported to both protect and exacerbate ischemic brain injury. These conflicting observations could be due to off-target effects of pharmacological inhibitors, or unanticipated biological changes in lifelong NFκB deficiency. The finding that hypothermia suppresses NFκB might, for instance, encourage the development of safer or more specific NFκB inhibitors. However, not all protein and gene changes observed by cooling translate into an instant therapeutic target. One example of this is the 70 kDa heat shock protein (HSP70). While HSP70 is upregulated in the ischemic brain and is known to protect the brain from stroke and related injuries [93], hypothermia downregulates its expression [94]. Thus, validation of the significance of observed changes by hypothermia would be obviously needed.

6 Pharmacologic Cooling Methods

While small animals can be easily cooled in the laboratory, cooling patients is a significant clinical challenge, especially awake stroke patients. There have been substantial technical advances in mechanical cooling methods [12], but there is also a growing body of literature of pharmacological approaches to cooling and bear mention. Such approaches include neurotension [95–97], 3-iodothyronamine (T1AM) [98], and hydrogen sulfide (H2S) [99]. Neurotension is an endogenous peptide involved in circadian temperature regulation, and analogues have been developed that penetrate BBB. Single intraperitoneal injections to rats decreased body temperatures by about 5°C within 1 h, and cooling could be maintained for about 7 h [95–97]. However, these approaches have not yet been studied clinically.

7 Conclusion

Hypothermia has long been known to be a potent putative neuroprotectant. Experimental evidence and clinical experience show that hypothermia protects the brain from cerebral injury in multiple ways. There appears to be a bright future for the application of therapeutic hypothermia in acute stroke brain injury. It is likely that benefit will be greatest when treatment is initiated very early, within several hours of symptom onset. Hypothermia may be able to extend the therapeutic window for other neuroprotective therapies.

However, many questions still need to be answered regarding the use of therapeutic hypothermia in clinical practice, such as optimal target temperature and duration, the therapeutic window in humans, cost-effective. In addition, sufficient numbers of attempts at hypothermia mono-therapy have failed to support this notion. Therefore, despite the complexity, it seems wise to approach stroke therapy with combination therapies with neuroprotective, anti-inflammatory, and thrombolytic agents are likely to be investigated in the clinical setting in the future.

Acknowledgments This work was supported by the Department of Veterans Affairs, grants to MAY from the NIH R01 NS 40156, P50 NS014543, Department of Veterans Affairs Merit Review Award I01BX007080, and the Department of Defense DAMD17-03-1-0532. Grants were administered by the Northern California Institute for Research and Education, and supported by resources of the San Francisco Veterans Affairs Medical Center.

References

1. Bernard SA, Gray TW, Buist MD, et al. Treatment of comatose survivors of out-of-hospital cardiac arrest with induced hypothermia. N Engl J Med. 2002;346:557–63.
2. HACA. Mild therapeutic hypothermia to improve the neurologic outcome after cardiac arrest. N Engl J Med. 2002;346:549–56.
3. Gluckman PD, Wyatt JS, Azzopardi D, et al. Selective head cooling with mild systemic hypothermia after neonatal encephalopathy: multicentre randomised trial. Lancet. 2005;365:663–70.
4. Shankaran S, Laptook AR, Ehrenkranz RA, et al. Whole-body hypothermia for neonates with hypoxic-ischemic encephalopathy. N Engl J Med. 2005;353:1574–84.
5. Azzopardi D, Strohm B, Edwards AD, et al. Moderate hypothermia to treat perinatal asphyxial encephalopathy. N Engl J Med. 2009;361:1349–58.
6. Simbruner G, Mittal RA, Rohlmann F, Muche R. Systemic hypothermia after neonatal encephalopathy: outcomes of neo.nEURO.network RCT. Pediatrics. 2010;126:e771–8.
7. Schwab S, Schwarz S, Spranger M, et al. Moderate hypothermia in the treatment of patients with severe middle cerebral artery infarction. Stroke. 1998;29:2461–6.
8. De Georgia MA, Krieger DW, Abou-Chebl A, et al. Cooling for Acute Ischemic Brain Damage (COOL AID): a feasibility trial of endovascular cooling. Neurology. 2004;63:312–7.
9. Lyden PD, Allgren RL, Ng K, et al. Intravascular cooling in the treatment of stroke (ICTuS): early clinical experience. J Stroke Cerebrovasc Dis. 2005;14:107–14.
10. Hemmen TM, Raman R, Guluma KZ, et al. Intravenous thrombolysis plus hypothermia for acute treatment of ischemic stroke (ICTuS-L): final results. Stroke. 2010;41:2265–70.
11. Schwab S, Georgiadis D, Berrouschot J, et al. Feasibility and safety of moderate hypothermia after massive hemispheric infarction. Stroke. 2001;32:2033–5.
12. Lyden PD, Krieger D, Yenari M, Dietrich WD. Therapeutic hypothermia for acute stroke. Int J Stroke. 2006;1:9–19.
13. Polderman KH, Herold I. Therapeutic hypothermia and controlled normothermia in the intensive care unit: practical considerations, side effects, and cooling methods. Crit Care Med. 2009;37:1101–20.
14. Busto R, Dietrich WD, Globus MY, Ginsberg MD. The importance of brain temperature in cerebral ischemic injury. Stroke. 1989;20:1113–4.
15. Busto R, Dietrich WD, Globus MY, et al. Small differences in intraischemic brain temperature critically determine the extent of ischemic neuronal injury. J Cereb Blood Flow Metab. 1987;7:729–38.
16. van der Worp HB, Macleod MR, Kollmar R. Therapeutic hypothermia for acute ischemic stroke: ready to start large randomized trials? J Cereb Blood Flow Metab. 2010;30:1079–93.
17. Maier CM, Ahern K, Cheng ML, et al. Optimal depth and duration of mild hypothermia in a focal model of transient cerebral ischemia: effects on neurologic outcome, infarct size, apoptosis, and inflammation. Stroke. 1998;29:2171–80.
18. Huh PW, Belayev L, Zhao W, et al. Comparative neuroprotective efficacy of prolonged moderate intraischemic and postischemic hypothermia in focal cerebral ischemia. J Neurosurg. 2000;92:91–9.
19. Kollmar R, Blank T, Han JL, et al. Different degrees of hypothermia after experimental stroke: short- and long-term outcome. Stroke. 2007;38:1585–9.

20. Georgiadis D, Schwarz S, Kollmar R, Schwab S. Endovascular cooling for moderate hypothermia in patients with acute stroke: first results of a novel approach. Stroke. 2001;32:2550–3.
21. Krieger DW, De Georgia MA, Abou-Chebl A, et al. Cooling for acute ischemic brain damage (cool aid): an open pilot study of induced hypothermia in acute ischemic stroke. Stroke. 2001;32:1847–54.
22. Jacobs S, Hunt R, Tarnow-Mordi W, et al. Cooling for newborns with hypoxic ischaemic encephalopathy. Cochrane Database Syst Rev. 2007:CD003311.
23. Hutchison JS, Ward RE, Lacroix J, et al. Hypothermia therapy after traumatic brain injury in children. N Engl J Med. 2008;358:2447–56.
24. Peterson K, Carson S, Carney N. Hypothermia treatment for traumatic brain injury: a systematic review and meta-analysis. J Neurotrauma. 2008;25:62–71.
25. Tokutomi T, Miyagi T, Takeuchi Y, et al. Effect of 35 degrees C hypothermia on intracranial pressure and clinical outcome in patients with severe traumatic brain injury. J Trauma. 2009;66:166–73.
26. Kammersgaard LP, Rasmussen BH, Jorgensen HS, et al. Feasibility and safety of inducing modest hypothermia in awake patients with acute stroke through surface cooling: a case-control study: the Copenhagen Stroke Study. Stroke. 2000;31:2251–6.
27. Krieger DW, Yenari MA. Therapeutic hypothermia for acute ischemic stroke: what do laboratory studies teach us? Stroke. 2004;35:1482–9.
28. Clark DL, Penner M, Orellana-Jordan IM, Colbourne F. Comparison of 12, 24 and 48 h of systemic hypothermia on outcome after permanent focal ischemia in rat. Exp Neurol. 2008;212:386–92.
29. Colbourne F, Corbett D. Delayed and prolonged post-ischemic hypothermia is neuroprotective in the gerbil. Brain Res. 1994;654:265–72.
30. Colbourne F, Corbett D, Zhao Z, et al. Prolonged but delayed postischemic hypothermia: a long-term outcome study in the rat middle cerebral artery occlusion model. J Cereb Blood Flow Metab. 2000;20:1702–8.
31. van der Worp HB, Sena ES, Donnan GA, et al. Hypothermia in animal models of acute ischaemic stroke: a systematic review and meta-analysis. Brain. 2007;130:3063–74.
32. Zhang ZG, Chopp M, Chen H. Duration dependent post-ischemic hypothermia alleviates cortical damage after transient middle cerebral artery occlusion in the rat. J Neurol Sci. 1993;117:240–4.
33. Carroll M, Beek O. Protection against hippocampal CA1 cell loss by post-ischemic hypothermia is dependent on delay of initiation and duration. Metab Brain Dis. 1992;7:45–50.
34. Dietrich WD, Busto R, Alonso O, et al. Intraischemic but not postischemic brain hypothermia protects chronically following global forebrain ischemia in rats. J Cereb Blood Flow Metab. 1993;13:541–9.
35. Shuaib A, Waqar T, Wishart T, Kanthan R. Post-ischemic therapy with CGS-19755 (alone or in combination with hypothermia) in gerbils. Neurosci Lett. 1995;191:87–90.
36. Colbourne F, Corbett D. Delayed postischemic hypothermia: a six month survival study using behavioral and histological assessments of neuroprotection. J Neurosci. 1995;15:7250–60.
37. Colbourne F, Li H, Buchan AM. Indefatigable CA1 sector neuroprotection with mild hypothermia induced 6 hours after severe forebrain ischemia in rats. J Cereb Blood Flow Metab. 1999;19:742–9.
38. Olsen TS, Weber UJ, Kammersgaard LP. Therapeutic hypothermia for acute stroke. Lancet Neurol. 2003;2:410–6.
39. Shackelford RT, Hegedus SA. Factors affecting cerebral blood flow—experimental review: sympathectomy, hypothermia, CO_2 inhalation and pavarine. Ann Surg. 1966;163:771–7.
40. Erecinska M, Thoresen M, Silver IA. Effects of hypothermia on energy metabolism in mammalian central nervous system. J Cereb Blood Flow Metab. 2003;23:513–30.
41. Hagerdal M, Harp J, Nilsson L, Siesjo BK. The effect of induced hypothermia upon oxygen consumption in the rat brain. J Neurochem. 1975;24:311–6.

42. Hagerdal M, Harp J, Siesjo BK. Effect of hypothermia upon organic phosphates, glycolytic metabolites, citric acid cycle intermediates and associated amino acids in rat cerebral cortex. J Neurochem. 1975;24:743–8.
43. Ehrlich MP, McCullough JN, Zhang N, et al. Effect of hypothermia on cerebral blood flow and metabolism in the pig. Ann Thorac Surg. 2002;73:191–7.
44. Yenari M, Wijman C, Steinberg G. Effects of hypothermia on cerebral metabolism, blood flow, and autoregulation. New York: Marcel Dekker; 2004. p. 141–78.
45. Busto R, Globus MY, Dietrich WD, et al. Effect of mild hypothermia on ischemia-induced release of neurotransmitters and free fatty acids in rat brain. Stroke. 1989;20:904–10.
46. Matsumoto M, Scheller MS, Zornow MH, Strnat MA. Effect of S-emopamil, nimodipine, and mild hypothermia on hippocampal glutamate concentrations after repeated cerebral ischemia in rabbits. Stroke. 1993;24:1228–34.
47. Mitani A, Kataoka K. Critical levels of extracellular glutamate mediating gerbil hippocampal delayed neuronal death during hypothermia: brain microdialysis study. Neuroscience. 1991;42:661–70.
48. Young RS, Zalneraitis EL, Dooling EC. Neurological outcome in cold water drowning. JAMA. 1980;244:1233–5.
49. Colbourne F, Grooms SY, Zukin RS, et al. Hypothermia rescues hippocampal CA1 neurons and attenuates down-regulation of the AMPA receptor GluR2 subunit after forebrain ischemia. Proc Natl Acad Sci U S A. 2003;100:2906–10.
50. Hu BR, Kamme F, Wieloch T. Alterations of Ca2+/calmodulin-dependent protein kinase II and its messenger RNA in the rat hippocampus following normo- and hypothermic ischemia. Neuroscience. 1995;68:1003–16.
51. Takata T, Nabetani M, Okada Y. Effects of hypothermia on the neuronal activity, [Ca2+]i accumulation and ATP levels during oxygen and/or glucose deprivation in hippocampal slices of guinea pigs. Neurosci Lett. 1997;227:41–4.
52. Taylor CP, Burke SP, Weber ML. Hippocampal slices: glutamate overflow and cellular damage from ischemia are reduced by sodium-channel blockade. J Neurosci Methods. 1995;59:121–8.
53. Kim JY, Kim N, Chang W, Yenari MA. Mild hypothermia suppresses calcium sensing receptor (CaSR) induction following forebrain ischemia while increasing GABA-B receptor1 (GABA-B-R1) expression. Transl Stroke Res. 2011;2(2):195–201.
54. Liu L, Yenari MA. Therapeutic hypothermia: neuroprotective mechanisms. Front Biosci. 2007;12:816–25.
55. Inamasu J, Suga S, Sato S, et al. Intra-ischemic hypothermia attenuates intercellular adhesion molecule-1 (ICAM-1) and migration of neutrophil. Neurol Res. 2001;23:105–11.
56. Kawai N, Okauchi M, Morisaki K, Nagao S. Effects of delayed intraischemic and postischemic hypothermia on a focal model of transient cerebral ischemia in rats. Stroke. 2000;31:1982–9. discussion 1989.
57. Wang GJ, Deng HY, Maier CM, et al. Mild hypothermia reduces ICAM-1 expression, neutrophil infiltration and microglia/monocyte accumulation following experimental stroke. Neuroscience. 2002;114:1081–90.
58. Han HS, Qiao Y, Karabiyikoglu M, et al. Influence of mild hypothermia on inducible nitric oxide synthase expression and reactive nitrogen production in experimental stroke and inflammation. J Neurosci. 2002;22:3921–8.
59. Yenari MA, Han HS. Influence of hypothermia on post-ischemic inflammation: role of nuclear factor kappa B (NFkappaB). Neurochem Int. 2006;49:164–9.
60. Webster CM, Kelly S, Koike MA, et al. Inflammation and NFkappaB activation is decreased by hypothermia following global cerebral ischemia. Neurobiol Dis. 2009;33:301–12.
61. Dietrich WD, Busto R, Halley M, Valdes I. The importance of brain temperature in alterations of the blood–brain barrier following cerebral ischemia. J Neuropathol Exp Neurol. 1990;49:486–97.
62. Burk J, Burggraf D, Vosko M, et al. Protection of cerebral microvasculature after moderate hypothermia following experimental focal cerebral ischemia in mice. Brain Res. 2008;1226:248–55.

63. Hamann GF, Burggraf D, Martens HK, et al. Mild to moderate hypothermia prevents microvascular basal lamina antigen loss in experimental focal cerebral ischemia. Stroke. 2004;35:764–9.

64. Lee JE, Yoon YJ, Moseley ME, Yenari MA. Reduction in levels of matrix metalloproteinases and increased expression of tissue inhibitor of metalloproteinase-2 in response to mild hypothermia therapy in experimental stroke. J Neurosurg. 2005;103:289–97.

65. Truettner JS, Alonso OF, Dalton Dietrich W. Influence of therapeutic hypothermia on matrix metalloproteinase activity after traumatic brain injury in rats. J Cereb Blood Flow Metab. 2005;25:1505–16.

66. Wagner S, Nagel S, Kluge B, et al. Topographically graded postischemic presence of metalloproteinases is inhibited by hypothermia. Brain Res. 2003;984:63–75.

67. Globus MY, Busto R, Lin B, et al. Detection of free radical activity during transient global ischemia and recirculation: effects of intraischemic brain temperature modulation. J Neurochem. 1995;65:1250–6.

68. Maier CM, Sun GH, Cheng D, et al. Effects of mild hypothermia on superoxide anion production, superoxide dismutase expression, and activity following transient focal cerebral ischemia. Neurobiol Dis. 2002;11:28–42.

69. Yenari MA. Heat shock proteins and neuroprotection. Adv Exp Med Biol. 2002;513:281–99.

70. Yenari M, Kitagawa K, Lyden P, Perez-Pinzon M. Metabolic downregulation: a key to successful neuroprotection? Stroke. 2008;39:2910–7.

71. Han HS, Yenari MA. Effect of gene expression by therapeutic hypothermia in cerebral ischemia. Future Neurol. 2007;2:435–40.

72. Hemmen TM, Lyden PD. Induced hypothermia for acute stroke. Stroke. 2007;38:794–9.

73. Behringer W. Global brain ischemia: animal studies. In: Tisherman SA, Sterz F, editors. Therapeutic hypothermia. New York: Springer; 2005. p. 1–10.

74. Laptook AR. Use of therapeutic hypothermia for term infants with hypoxic-ischemic encephalopathy. Pediatr Clin North Am. 2009;56:601–16.

75. Tang XN, Yenari MA. Hypothermia as a cytoprotective strategy in ischemic tissue injury. Ageing Res Rev. 2010;9:61–8.

76. Fingas M, Clark DL, Colbourne F. The effects of selective brain hypothermia on intracerebral hemorrhage in rats. Exp Neurol. 2007;208:277–84.

77. MacLellan CL, Davies LM, Fingas MS, Colbourne F. The influence of hypothermia on outcome after intracerebral hemorrhage in rats. Stroke. 2006;37:1266–70.

78. Kawanishi M, Kawai N, Nakamura T, et al. Effect of delayed mild brain hypothermia on edema formation after intracerebral hemorrhage in rats. J Stroke Cerebrovasc Dis. 2008;17:187–95.

79. MacLellan CL, Girgis J, Colbourne F. Delayed onset of prolonged hypothermia improves outcome after intracerebral hemorrhage in rats. J Cereb Blood Flow Metab. 2004;24:432–40.

80. Kollmar R, Staykov D, Dorfler A, et al. Hypothermia reduces perihemorrhagic edema after intracerebral hemorrhage. Stroke. 2010;41:1684–9.

81. Torok E, Klopotowski M, Trabold R, et al. Mild hypothermia (33 degrees C) reduces intracranial hypertension and improves functional outcome after subarachnoid hemorrhage in rats. Neurosurgery. 2009;65:352–9.

82. Schubert GA, Poli S, Schilling L, et al. Hypothermia reduces cytotoxic edema and metabolic alterations during the acute phase of massive SAH: a diffusion-weighted imaging and spectroscopy study in rats. J Neurotrauma. 2008;25:841–52.

83. Kawamura Y, Yamada K, Masago A, et al. Hypothermia modulates induction of hsp70 and c-jun mRNA in the rat brain after subarachnoid hemorrhage. J Neurotrauma. 2000;17:243–50.

84. Muroi C, Frei K, El Beltagy M, et al. Combined therapeutic hypothermia and barbiturate coma reduces interleukin-6 in the cerebrospinal fluid after aneurysmal subarachnoid hemorrhage. J Neurosurg Anesthesiol. 2008;20:193–8.

85. Todd MM, Hindman BJ, Clarke WR, Torner JC. Mild intraoperative hypothermia during surgery for intracranial aneurysm. N Engl J Med. 2005;352:135–45.

86. Seule MA, Muroi C, Mink S, et al. Therapeutic hypothermia in patients with aneurysmal subarachnoid hemorrhage, refractory intracranial hypertension, or cerebral vasospasm. Neurosurgery. 2009;64:86–92. discussion 83–92.

87. Fernandez A, Schmidt JM, Claassen J, et al. Fever after subarachnoid hemorrhage: risk factors and impact on outcome. Neurology. 2007;68:1013–9.

88. Proust F, Hannequin D, Langlois O, et al. Causes of morbidity and mortality after ruptured aneurysm surgery in a series of 230 patients. The importance of control angiography. Stroke. 1995;26:1553–7.

89. Elewa HF, Hilali H, Hess DC, et al. Minocycline for short-term neuroprotection. Pharmacotherapy. 2006;26:515–21.

90. Lampl Y, Boaz M, Gilad R, et al. Minocycline treatment in acute stroke: an open-label, evaluator-blinded study. Neurology. 2007;69:1404–10.

91. Fagan SC, Waller JL, Nichols FT, et al. Minocycline to improve neurologic outcome in stroke (MINOS): a dose-finding study. Stroke. 2010;41:2283–7.

92. Kobayashi MS, Asai S, Ishikawa K, et al. Global profiling of influence of intra-ischemic brain temperature on gene expression in rat brain. Brain Res Rev. 2008;58:171–91.

93. Kelly S, Yenari MA. Neuroprotection: heat shock proteins. Curr Med Res Opin. 2002;18 Suppl 2:s55–60.

94. Kumar K, Wu X, Evans AT, Marcoux F. The effect of hypothermia on induction of heat shock protein (HSP)-72 in ischemic brain. Metab Brain Dis. 1995;10:283–91.

95. Gordon CJ, McMahon B, Richelson E, et al. Neurotensin analog NT77 induces regulated hypothermia in the rat. Life Sci. 2003;73:2611–23.

96. Katz LM, Young A, Frank JE, et al. Neurotensin-induced hypothermia improves neurologic outcome after hypoxic-ischemia. Crit Care Med. 2004;32:806–10.

97. Tyler-McMahon BM, Stewart JA, Farinas F, et al. Highly potent neurotensin analog that causes hypothermia and antinociception. Eur J Pharmacol. 2000;390:107–11.

98. Scanlan TS, Suchland KL, Hart ME, et al. 3-Iodothyronamine is an endogenous and rapid-acting derivative of thyroid hormone. Nat Med. 2004;10:638–42.

99. Blackstone E, Roth MB. Suspended animation-like state protects mice from lethal hypoxia. Shock. 2007;27:370–2.

Chapter 16
Stroke Preconditioning to Identify Endogenous Protective or Regenerative Mechanisms

Liren Qian, Prativa Sherchan, and Xuejun Sun

Abstract Many neuroprotectants have shown effectiveness by reducing infarction and improving neurologic functions in animal models of stroke, but few of these neuroprotectants have been successful in clinical scenario. In clinical trials, pharmacological agents have shown to be either ineffective or have potential adverse effects. Consequently, efforts have been directed toward understanding and enhancing the endogenous protective mechanisms by which the brain protects itself against noxious stimuli in an attempt to recover from the damage encountered. Preconditioning-induced ischemic tolerance is an effective approach to understand how the brain protects itself. In this chapter, we summarize the development of preconditioning followed by discussion of various stimuli that can induce brain preconditioning and the downstream signaling pathways involved in preconditioning-induced protection. Specifically, we discuss the potential clinical application of preconditioning for brain injuries such as stroke.

1 Introduction

Many neuroprotectants have been shown to reduce infarction and improve neurologic functions in animal models of stroke, but few neuroprotectants have offered protective effects ranging from basic research to clinical applications [1]. The death

L. Qian (✉)
Department of Hematology, Navy General Hospital, Fucheng Road, Beijng, China

P. Sherchan
Division of Physiology, School of Medicine, Loma Linda University, Loma Linda, CA 92350, USA

X. Sun (✉)
Department of Hematology, Navy General Hospital, Fucheng Road, Beijng 100048, China

Department of Diving Medicine, Second Military Medical University, Shanghai 200433, China
e-mail: sunxjk@hotmail.com

P.A. Lapchak and J.H. Zhang (eds.), *Translational Stroke Research*, Springer Series
in Translational Stroke Research, DOI 10.1007/978-1-4419-9530-8_16,
© Springer Science+Business Media, LLC 2012

of neurons represents a catastrophic event, because dead neurons are not replaced by division of surviving neurons. Preserving the viability of neurocytes therefore has been recognized as a major therapeutic target. Drug treatments are either ineffective or confounded by adverse effects. This necessitates the exploration of novel therapeutic approaches, such as ischemic preconditioning.

Preconditioning is a procedure by which a noxious stimulus near to but below the threshold of damage is applied to the tissue. Groups of investigators from diverse fields have used different approaches to show that stimuli can promote preconditioning-dependent protective responses. Preconditioning is an attractive experimental strategy to identify endogenous protective or regenerative mechanisms. Also, by inducing tolerance in individuals in whom ischemic events are anticipated, such as transient ischemic attack (TIA) or high-risk surgical cohorts, preconditioning could be used as a therapeutic technique.

Thus, in this chapter, we firstly summarize the development of preconditioning. Next, various stimuli that can induce brain preconditioning are discussed in another section along with a brief discussion of the downstream signaling pathways that converge on some common fundamental mechanisms. Lastly, we discuss the potential clinical application of preconditioning on brain injuries such as stroke.

2   History of Preconditioning

Preconditioning is a general concept in which an entity is exposed to some stimulus in order to prepare that entity to be more resilient against the stimulus when and if the stimulus is encountered in the future. The neuroprotective concept of preconditioning is based on the phenomenon in which the brain protects itself against future injury by adapting to low doses of noxious insults [2] (Fig. 16.1).

This phenomenon first drew attention in an animal laboratory in 1971. Maroko et al. [3] observed that interventions at the time of coronary occlusion could reduce the size of the resulting infarct in the open chest dog. At that time, a large number of drugs were used to limit infarct size in the setting of ischemia/reperfusion, but none of these studies could be consistently reproduced in laboratories. Many factors affecting infarct size such as temperature [4], risk zone size [5], and collateral flow [6] were seldom adequately controlled. It was not even known whether therapeutically limiting infarct size was possible until 1986. In 1986, Murry et al. [7] preconditioned a group of dogs with four 5 min circumflex occlusions, each separated by 5 min of reperfusion, followed by a sustained 40 min occlusion and 4 days of reperfusion thereafter. They found that multiple brief ischemic episodes indeed protected the heart from a subsequent sustained ischemic insult. Infarct size in control animals averaged 29.4% of the area at risk. Infarct size in control hearts averaged only 7.3% of the area at risk. Therefore, reduction of infarct size has been firmly established as the "gold standard" of ischemic preconditioning. Based on these findings, in 1986 Murry et al. put forward the concept of ischemic preconditioning, namely, a phenomenon in which tissue is rendered resistant to the deleterious effects of prolonged

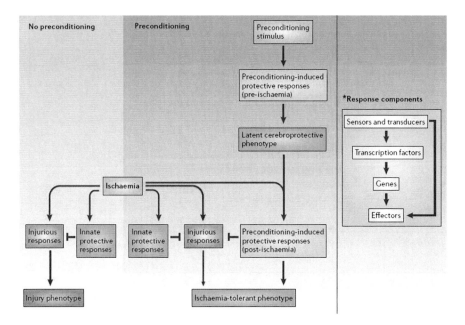

Fig. 16.1 Cerebroprotection by preconditioning. This figure was published in ref. [2]

ischemia and reperfusion by prior exposure to brief periods of vascular occlusion. This powerful cardioprotective effect appears to be a ubiquitous endogenous response to an episode of acute ischemia-reperfusion injury.

The preconditioning and subsequent ischemic tolerance were demonstrated in brain in 1990 by Kitagawa et al. [8]. In gerbils, prior exposure to brief periods of sublethal ischemia reduced hippocampal neuronal injury significantly after a more substantial ischemic insult. The interval between the preconditioning ischemia and the subsequent lethal ischemia relate to the extent of neuronal injury. This phenomenon has drawn attention of scientists in different fields, and the beneficial effects of preconditioning have been reproduced in various species tested including humans and a variety of organs other than the heart and brain including the kidney, liver, and so on [9].

3   Stimuli

It is generally accepted that preconditioning requires small doses of harmful stimulus to induce protection against subsequent injurious challenge. Ischemia was initially demonstrated as the preconditioning stimulus. However, numerous other stimuli that can evoke brain preconditioning been discovered (Table 16.1). Preconditioning stimuli include but are not limited to transient global and focal ischemia, various chemical and biological agents, hypoxia, oxygen–glucose deprivation,

Table 16.1 Stimuli that induce brain preconditioning

Global and focal ischemia
Chemical and biological agents
Endotoxin
Free radicals
Anesthetic agents
Lipopolysaccharide
Adenosine
Thrombin
N-methyl-D-aspartate (NMDA) agonists
3-nitropropionic acid (3NP)
Cyanide
Na^+/K^+-ATPase inhibitors
Physiologic factors
Hypoxia
Oxygen–glucose deprivation
Metabolic inhibitors
Diethyldithiocarbamate
Cortical spreading depression
Brief episodes of seizure
Hypothermia/hyperthermia
Traumatic injury
Hyperbaric oxygenation

diethyldithiocarbamate, cortical spreading depression, brief episodes of seizure, hypothermia, and hyperthermia.

Different kinds of preconditioning stimuli can protect against diverse types of injury, and this phenomenon has been named "cross-tolerance." Accordingly, a stimulus that promotes tolerance in the brain can promote tolerance in other organs as well. Furthermore, exogeneously delivered stimuli, such as inflammatory cytokines and metabolic inhibitors, can also induce ischemic tolerance which makes it possible to use these agents as drugs after stroke. Moreover, because inducers and mechanisms of tolerance might have similar features, the induction of tolerance in one organ can spread to other organs. This phenomenon, known as "remote preconditioning," has been described recently.

Whether a particular stimulus is too weak to elicit a response, of sufficient intensity to serve as a preconditioning trigger, or too robust to be harmful is determined by the intensity, duration, and /or frequency of the stimulus.

4 Sensors and Transducers

Over the past decade, research on preconditioning has resulted in various promising strategies for the treatment of acute brain injury. Depending on the specific preconditioning stimulus, a state of neuronal tolerance can be established in at least two temporal profiles: one in which the trigger induces protection within minutes (rapid or acute tolerance) [10] and the other kind in which the protected state develops after a delay of several hours to days (delayed tolerance) [11]. Some stimuli induce both phases of ischemic tolerance, while others can induce only the acute phase or only the delayed phase [12]. Rapid ischemic tolerance produces protection within minutes and does not require new protein synthesis [13]. Delayed tolerance develops over time, with protection against subsequent injury being present at 24–48 h after the preconditioning event, peaking at around 3 days and slowly disappearing over a week period. And this type of delayed tolerance is dependent on protein synthesis [14].

4.1 NMDA Receptor Activation and Excitotoxicity Protection

Cellular events such as cellular depolarization and Ca^{2+} influx can be triggered by cerebral ischemia, resulting in excitotoxic cell death. The activation of the N-methyl-D-aspartate (NMDA) glutamate receptors through increases in intracellular calcium mediates ischemic tolerance [15]. There is convincing evidence that NMDA receptor activation is critical to the induction of ischemic tolerance in neurons [16]. Exposure to sublethal NMDA concentrations provided neuroprotection against lethal death stimuli applied at later time points [17]. And the development of ischemic tolerance induced by brief repeated ischemic insults in brains can be altered by NMDA receptor antagonists [18]. It was reported [19] that pretreatment with NMDA receptor antagonists prevented OGD-induced neurons death while α-amino-3-hydroxy-5-methyl-4-isoxa-zolep-propionate (AMPA) receptor and voltage-dependent Ca^{2+} channel blockers did not have protective effects. Glutamate preconditioning exhibited neuron protection against OGD-induced damage. The protective efficacy could be blocked by NMDA or AMPA receptor antagonists. The protection efficacy depends on the duration of preconditioning exposure and the interval between preconditioning exposure and test challenge.

4.2 PKC and MAPK

Activation of kinase pathways is thought to play a critical role in the development of the brain tolerance. Recently, many studies have shown that phospholipase C, protein kinase C (PKC), and mitogen-activated protein kinase (MAPK) were critical in ischemic tolerance in the brain [20–22]. The early phase of ischemic preconditioning is mediated by rapid posttranslational modification of preexisting proteins through signaling pathways that involve PKC and MAPK. The late preconditioning is mediated by protective gene expression and by the synthesis of new protective proteins. This mechanism involves redox-sensitive activation of transcriptional factors through PKC and tyrosine kinase signaling pathways.

In general, PKC is regulated by two sequential mechanisms: phosphorylation triggered by the 3-phosphoinositide-dependent kinase (PDK)-1 and binding to DAG. In the last decade, researchers have discovered another function of PKC. Mochly-Rosen et al. discovered that each isoform of PKC dock on a unique binding protein called receptor for activated C kinase (RACK) when activated [23]. These RACKS are located only on certain cell organs to take the isoform to a specific substrate protein. Binding to the RACK completes the activation and causes the isoform to phosphorylate any nearby substrate. It is thought that only certain isoforms participate in preconditioning. Among the different PKC isozymes, the ϵPKC has been proposed to be a key player in the induction of brain preconditioning. And recently, it was demonstrated that after lethal ischemia, δPKC is translocated and involved in the cell death pathway, which may also participate in preconditioning.

MAPKs, a serine/threonine protein kinase family, play a crucial role in triggering the intracellular events leading to the activation of the adaptive response observed in ischemic preconditioning, both in the heart [24] and in the brain [25]. Shamloo et al. [25] had investigated the role of MEK1/2 and ERK1/2 in ischemic preconditioning. They found that 3 min of ischemic preconditioning increased the expression of phosphorylated MEK and of phosphorylated ERK in the rat hippocampus. In addition, 9 min of lethal ischemia increased phosphorylated ERK in both nonpreconditioned and preconditioned animals. However, the levels of phosphorylated ERK had returned to normal in the tolerant hippocampi whereas it stayed elevated in the damaged hippocampi 24 h after ischemic damage. ERK5 activation was also demonstrated to participate in the neuroprotection of ischemic preconditioning [26].

4.3 The Akt Pathway

The protooncogene Akt is a central component of the phosphatidylinositol 3-kinase (PI 3-kinase) signal transduction pathway. Akt promotes cell survival by phosphorylating several proapoptotic proteins, including caspase-9, Bad, forkhead transcription factors (FKHR), and glycogen synthase kinase 3 (GSK3). Noshita et al. found that after 4 h of middle cerebral artery occlusion (MCAO) in mice, phosphorylated Akt was decreased in the ischemic core but increased in cortical peri-infarct region. And double staining showed different cellular distributions for phosphorylated Akt and DNA fragmentation, and pretreatment with a PI 3-kinase inhibitor prevented the increased Akt phosphorylation and promoted DNA fragmentation [27]. Following sublethal cerebral ischemia in gerbils, Akt phosphorylation was persistently stimulated in the hippocampal CA1 region, but in nonpreconditioned animals, Akt phosphorylation showed no obvious decrease after the subsequent lethal ischemia [28]. Preconditioning, induced by transient focal ischemia in rats, reduced neuronal death produced by subsequent lethal MCAO in the penumbra. This neuroprotective effect was associated with persistent Akt activation in this region [29]. Furthermore, several studies have now identified Akt, which is phosphorylated by phosphatidylinositol-3-kinase, as crucial to establishing the tolerant phenotype, secondary to an Akt-mediated phosphorylation of mixed lineage kinase 3 [30].

4.4   Toll-Like Receptor

It is important to note that recent studies have suggested that Toll-like receptors (TLR), well-known components of the innate immune system, might play a role in preconditioning [31]. LPS preconditioning in the brain shares several characteristics with ischemic preconditioning in the brain [32], such as the delayed induction of tolerance following preconditioning and a dependency on de novo protein synthesis. Importantly, similar to ischemic preconditioning that exposure to brief ischemia does not induce brain damage, a preconditioning dose of LPS does not cause brain injury. Ischemic tolerance induced by LPS appears to occur through stimulation of TLR4 (Toll-like receptor 4), a kind of molecules involved in sensing the presence of pathogens. The activation of inflammatory pathways plays an important role in LPS-induced ischemic tolerance. Neuroprotection can also conferred by a preconditioning dose of CpG ODNs, activator of TLR9, administered in advance of ischemia. It was reported that TNFα, IFNα/β, and its downstream signaling mediator are critical effectors of LPS and CPG ODN-induced neuroprotection [33–35].

4.5   Transcription Factors

Several transcription factors are known to be associated with ischemia and probably participate in brain preconditioning, including activating protein 1 (AP1), CREB, hypoxia-inducible factor (HIF), NF-κB, early growth response 1, and the redox-regulated transcriptional activator SP1 [36]. The function of HIFs on brain preconditioning has been extensively studied. The HIF isoforms participate in regulating genes that are involved in adaptation to ischemic injuries by mediating the transactivation of adaptive, pro-survival genes, particularly those involved in glucose metabolism and angiogenesis. HIF-1α can dimerize with HIF-1β and translocate to the nucleus to promote the transcription of genes that enhance hypoxic resistance.

The HIF2α isoform is regulated in a similar way, but details regarding transcriptional regulation by HIF2α are still unclear.

AP-1 complexes participate in regulating multiple genes to control cellular proliferation, differentiation, and death. AP-1 transcription factors include c-fos, fra-1, fra-2, fosB, c-jun, junB, and junD. Ischemic injury is associated with significant changes in the expression of these factors. Ischemic injury is also associated with phosphorylation of CREB in the brain. CREB participates in regulating multiple genes to control cellular proliferation, differentiation, and survival. CREB is important in stimulus-transcription coupling and in receptor-mediated changes in gene expression. These changes are probably mediated via stimulation of glutamate receptors and subsequent increases in cytosolic calcium. This pathway is thought to be involved in the manifestation of ischemic preconditioning.

Caspases might be essential brain tolerance induction catalysts, given that cyclic AMP responsive element-binding protein (CREB), the p50 and p65 subunits of nuclear factor-κB (NF-κB), and PKC and other kinases are caspase substrates.

4.6 Role of Adenosine and ATP-Sensitive K+ (K_{ATP})

Adenosine, a prototypical paracrine mediator and "retaliatory metabolite," the production of which is linked to ATP degradation, leads to a cascade of signaling events including K_{ATP} channels. This cascade results in increased resistance to subsequent ischemic damage. The role of K_{ATP} channels in preconditioning was demonstrated in a rat delayed ischemic tolerance model. In rats, ischemic preconditioning increased A1 receptor immunoreactivity in the hippocampal CA1 at days 1, 3, and 7 after preconditioning induction [37]. But such a change was not found in mice after sublethal 3-nitropropionate treatment. Interestingly, early ischemic tolerance is also blocked by pharmacological inhibition of KATP channels in vitro [13].

4.7 Role of Free Radicals in Brain Tolerance

Short exposure to hypoxia/ischemia may produce free radicals which may be involved in ischemic tolerance in the brain [38]. Free radicals increase oxidation of lipids, proteins, and DNA. Free radicals like superoxide may be derived from the actions of xanthine oxidase [39] and nitric oxide (NO) can be generated by the action of nitric oxide synthases [40]. NO has been linked to the trigger and end-effector phases of delayed preconditioning. NO and superoxide can interact to form the toxic peroxynitrite [40]. Excessive NO formation may also contribute to ischemic brain injury through the formation of reactive nitrogen species such as peroxynitrite. It was demonstrated that administrating diethyldithiocarbamate which elicits ROS production and inhibits superoxide dismutase exerted neuroprotective effects after global ischemia was induced 2–4 days later in gerbils [41]. Treatment with a free radical scavenger can raise the threshold of preconditioning, and a free radical generator can trigger a preconditioned state. And preconditioning may exert its protective effects by preventing the formation of free radicals or by reducing their toxic effects on the brain [42]. Ischemic preconditioning which reduced the volume of cerebral infarction also caused significant increases in the activity of the antioxidant superoxide dismutase enzyme [43]. Hydrogen peroxide could induce Heme oxygenase-1(HO-1) which has been shown to exert protective activity against ischemic damage [44].

Additionally, ROS can modulate cell-signaling pathway by increasing the activity of protein tyrosine kinases, such as the MAPKs and c-Jun N-terminal kinase. Free radicals can also modulate cell-signaling proteins by acting on some transcription factors and protein kinase, such as NF-κB and AP-1. These transcription factors that regulate genes play a critical role in cell survival and in antioxidant system.

Another free radicals modulating cell-signaling pathway involves the tyrosine kinases by autophosphorylation of tyrosines. Phosphorylation of tyrosine kinases was associated with activating phospholipases such as C and D.

5 Effectors

The genomic expression pattern in response to ischemia is unique in a preconditioned animal, and differs considerably from the pattern activated by either preconditioning or ischemia in a nonpreconditioned animal. And changes in gene transcription after preconditioning, or after ischemia in a preconditioned brain, are also different. Some genes, such as adenosine A receptor and vascular endothelial growth factor (VEGF), are affected within minutes or hours after preconditioning. However, some genes such as β-actin, serine/threonine protein kinase, arachidonate 12-lipoxygenase, calretinin, the S100A5 calcium-binding protein, dihydropyrimidine dehydrogenase, and the zinc transporter ZnT1 are affected days later after preconditioning. Some change transiently (CCAAT/enhancer-binding protein-related transcription factor, adenosine A2a receptor18, and metallothionein II), whereas others change for a protracted time (heat shock proteins, BCL2, p38-MAPK, TGFβ1, Iκ-Bα, glial fibrillary acidic protein, and β-tubulin).

5.1 Heat Shock Proteins

Heat shock proteins (HSPs) are found in all living cells, where they act as molecular chaperones, carrying out multiple functions that are essential to protein housekeeping.

Many studies have found the potential role of HSPs in both in vitro and in vivo models of the preconditioned response. It has been found that Heat shock protein 70 (HSP70), 27-kDa heat shock protein (HSP27), HSP40, HSP90, and small heat shock protein HSPB2 were correlated with preconditioning-induced protection in the brain [45–49]. And the changes in HSP110/105 expression after ischemia were found similar to those of inducible HSP70, which was previously reported to be involved in preconditioning-related protection, suggesting that these two chaperones might work in concert to protect neurons against subsequent lethal damage. Although lots of studies have implicated the role of heat shock proteins in neuroprotection, Abe et al. found that ischemic tolerance can be produced without heat shock protein expression, suggesting that they are not entirely critical for ischemic tolerance [50].

5.2 Trophic Factors

Trophic factors such as nerve growth factor (NGF) and brain-derived neurotrophic factor (BDNF) are involved in ischemic tolerance in brain. Following cerebral ischemia or hypoglycemic coma, the levels of TrkA (TrkA for NGF) and BDNF are decreased in hippocampal neurons. However, only the expression of TrkB (TrkB for BDNF) is increased in the hippocampal formation. It was found that NGF and BDNF can protect hippocampal neurons against ischemic cell damage [51–53]. Also, the levels of NGF and BDNF increase within the first 6 h after ischemic preconditioning [54]. Preconditioning applied 3 days earlier was able to attenuate increases in the levels of BDNF mRNA observed at 12 h after reperfusion using a rat model. NGF, BDNF, and their receptors recover better in tolerant hippocampal neurons. Because trophic factors are so important during development and participate so actively in neuroplastic and neuroprotective mechanisms in the brain, more investigations on their roles in the tolerant brain are warranted.

6   Clinic Application

Studies of preconditioning and ischemic are coming in great numbers. These studies help us understand its mechanistic basis and its clinical potential. Preconditioning-derived strategies include hematopoietic cytokines, immunological tolerance, and physical measures such as remote ischemic preconditioning. Some of these interventions seem to be safe and effective in protecting the ischemic brain, and they are now being tested in randomized clinical trials to protect the brain. However, there are still many uncertainties about this therapeutic approach, including which doses of preconditioning are safe and effective, and whether TIAs are an ischemic preconditioning equivalent in human beings. Preconditioning strategies have been applied on clinic more than a decade in heart. And there is a little clinical evidence for the use of preconditioning to protect the brain.

Weih et al. found that comparing with stroke patients without TIA preceding, preceding TIA was to be associated with less-severe stroke on admission and improved outcome on follow-up [55].

Wegener et al. [56] found that ischemic lesions and final infarct volumes were smaller in stroke patients with prior TIA than in those without in an MRI study. Although these findings strongly suggest TIA as the clinical correlate of preconditioning, other explanations for milder strokes after preceding TIA must be considered. In a recent review article on cerebral IT [57], mediators of IT could be used as biochemical markers of IT in stroke patients. Castillo et al. [58] tested this hypothesis by evaluating blood levels of TNF-α and IL-6 in acute stroke patients with or without prior ipsilateral TIA. Better outcome was found in patients with TIA, who showed high plasma concentrations of TNF-α and low concentrations of IL-6. Hence, authors proposed the index of TNF-α/IL-6 as a marker of IT phenomenon in humans. With further research, we proposed that preconditioning must be a necessary method in stroke therapy.

7 Conclusion

It is important to develop strategies to protect the brain either prior to vascular surgeries or in patients at high risk of stroke. While it would be dangerous and impractical to precondition at-risk patients with ischemia, the identification of underlying preconditioning mechanisms may lead to safer therapeutic factors that can be administered before brain injury. The most direct and significant application of understanding the mechanism of ischemic tolerance is therapeutic access to this protective state. In addition, narrow safety margin of precondition may prove a limiting factor of the therapeutic utility of precondition in clinics. Several approaches, including remote-PC by limb ischemia, pharmacological-PC with nitroglycerine, and anesthetic-PC, are tested in clinical trials to protect the heart from cardiovascular interventions with high risk of cardiac ischemic event. Results are promising and give hope for clinical trials of PC to protect brain.

References

1. Hoyte L, Kaur J, Buchan AM. Lost in translation: taking neuroprotection from animal models to clinical trials. Exp Neurol. 2004;188:200–4.
2. Gidday JM. Cerebral preconditioning and ischaemic tolerance. Nat Rev Neurosci. 2006;7: 437–48.
3. Maroko PR, Kjekshus JK, Sobel BE, Watanabe T, Covell JW, Ross Jr J, Braunwald E. Factors influencing infarct size following experimental coronary artery occlusio. Circulation. 1971;43:67–82.
4. Chien GL, Wolff RA, Davis RF, Vanwinkle DM. "Normothermic-range" temperature affects myocardial infarct size. Cardiovasc Res. 1994;28:1014–7.
5. Ytrehus K, Liu Y, Tsuchida A, Miura T, Liu GS, Yang X-M, Herbert D, Cohen MV, Downey JM. Rat and rabbit heart infarction: effects of anesthesia, perfusate, risk zone, and method of infarct sizing. Am J Physiol Heart Circ Physiol. 1994;267:H2383–90.
6. Reimer KA, Jennings RB. The "wavefront phenomenon" of myocardial ischemic cell death. II. Transmural progression of necrosis within the framework of ischemic bed size (myocardium at risk) and collateral flow. Lab Invest. 1979;40:633–44.
7. Murry CE, Jennings RB, Reimer KA. Preconditioning with ischemia: a delay of lethal cell injury in ischemic myocardium. Circulation. 1986;74:1124–36.
8. Kitagawa K, Matsumoto M, Tagaya M, Hata R, Ueda H, Niinobe M, Handa N, Fukunaga R, Kimura K, Mikoshiba K. "Ischemic tolerance" phenomenon found in the brain. Brain Res. 1990;528:21–4.
9. Yellon DM, Downey JM. Preconditioning the myocardium: from cellular physiology to clinical cardiology. Physiol Rev. 2003;83:1113–51.
10. Perez-Pinzon MA, Xu GP, Dietrich WD, Rosenthal M, Sick TJ. Rapid preconditioning protects rats against ischemic neuronal damage after 3 but not 7 days of reperfusion following global cerebral ischemia. J Cerebr Blood Flow Metab. 1997;17(2):175–82.
11. Kitagawa K, Matsumoto M, Kuwabara K, Tagaya M, Ohtsuki T, Hata R, Ueda H, Handa N, Kimura K, Kamada T. 'Ischemic tolerance' phenomenon found in the brain. Brain Res. 1990;528:21–4.
12. Stagliano NE, Perez-Pinzon MA, Moskowitz MA, Huang PL. Focal ischemic preconditioning induces rapid tolerance to middle cerebral artery occlusion in mice. J Cerebr Blood Flow Metab. 1999;19:757–61.

13. Perez-Pinzon MA, Born JG. Rapid preconditioning neuroprotection following anoxia in hippocampal slices: role of the K+ ATP channel and protein kinase C. Neuroscience. 1999; 89:453–9.

14. Kirino T. Ischemic tolerance. J Cerebr Blood Flow Metab. 2002;22:1283–96.

15. Grabb MC, Choi DW. Ischemic tolerance in murine cortical cell culture: critical role for NMDA receptors. J Neurosci. 1999;19:1657–62.

16. Jiang X, Tian F, Mearow K, Okagaki P, Lipsky RH, Marini AM. The excitoprotective effect of N-methyl-D-aspartate receptors is mediated by a brain-derived neurotrophic factor autocrine loop in cultured hippocampal neurons. J Neurochem. 2005;94:713–22.

17. Bhave SV, Ghoda L, Hoffman PL. Brain-derived neurotrophic factor mediates the anti-apoptotic effect of NMDA in cerebellar granule neurons: signal transduction cascades and site of ethanol action. J Neurosci. 1999;19:3277–86.

18. Kato H, Liu Y, Araki T, Kogure K. MK-801, but not anisomycin, inhibits the induction of tolerance to ischemia in the gerbil hippocampus. Neurosci Lett. 1992;139:118–21.

19. Lin CH, Chen PS, Gean PW. Glutamate preconditioning prevents neuronal death induced by combined oxygen-glucose deprivation in cultured cortical neurons. Eur J Pharmacol. 2008;589:85–93.

20. Lange-Asschenfeldt C, Raval AP, Dave KR, Mochly-Rosen D, Sick TJ, Perez-Pinzon MA. Epsilon protein kinase C mediated ischemic tolerance requires activation of the extracellular regulated kinase pathway in the organotypic hippocampal slice. J Cereb Blood Flow Metab. 2004;24:636–45.

21. Perez-Pinzon MA. Neuroprotective effects of ischemic preconditioning in brain mitochondria following cerebral ischemia. J Bioenerg Biomembr. 2004;36:323–7.

22. Raval AP, Dave K, Mochly-Rosen D, Sick T, Perez-Pinzon M. εPKC is required for the induction of tolerance by ischemic and NMDA-mediated preconditioning in the organotypic hippocampal slice. J Neurosci. 2003;23:384–91.

23. Johnson JA, Gray MO, Chen C-H, Mochly-Rosen D. A protein kinase C translocation inhibitor as an isozyme-selective antagonist of cardiac function. J Biol Chem. 1996;271:24962–6.

24. Maulik N, Watanabe M, Zu YL, et al. Ischemic preconditioning triggers the activation of MAP kinases and MAPKAP kinase 2 in rat hearts. FEBS Lett. 1996;396:233–7.

25. Shamloo M, Wieloch T. Changes in protein tyrosine phosphorylation in the rat brain after cerebral ischemia in a model of ischemic tolerance. J Cereb Blood Flow Metab. 1999;19:173–83.

26. Wang RM, Zhang QG, Li CH, Zhang GY. Activation of extracellular signal-regulated kinase 5 may play a neuroprotective role in hippocampal CA3/DG region after cerebral ischemia. J Neurosci Res. 2005;80:391–9.

27. Noshita N, Lewen A, Sugawara T, Chan PH. Evidence of phosphorylation of Akt and neuronal survival after transient focal cerebral ischemia in mice. J Cereb Blood Flow Metab. 2001;21:1442–50.

28. Yano S, Morioka M, Fukunaga K, Kawano T, Hara T, Kai Y, Hamada J, Miyamoto E, Ushio Y. Activation of Akt/protein kinase B contributes to induction of ischemic tolerance in the CA1 subfield of gerbil hippocampus. J Cereb Blood Flow Metab. 2001;21:351–60.

29. Nakajima T, Iwabuchi S, Miyazaki H, Okuma Y, Kuwabara M, Nomura Y, Kawahara K. Preconditioning prevents ischemia-induced neuronal death through persistent Akt activation in the penumbra region of the rat brain. J Vet Med Sci. 2004;66:521–7.

30. Miao B, Yin XH, Pei DS, Zhang QG, Zhang GY. Neuroprotective effects of preconditioning ischemia on ischemic brain injury through down-regulating activation of JNK1/2 via N-methyl-D-aspartate receptor-mediated Akt1 activation. J Biol Chem. 2005;280:21693–9.

31. Atkinson TJ. Toll-like receptors, transduction-effector pathways, and disease diversity: evidence of an immunobiological paradigm explaining all human illness? Int Rev Immunol. 2008;27:255–81.

32. Tasaki K, Ruetzler CA, Ohtsuki T, et al. Lipopolysaccharide pre-treatment induces resistance against subsequent focal cerebral ischemic damage in spontaneously hypertensive rats. Brain Res. 1997;748:267–70.

33. Furuya K, Ginis I, Takeda H, Chen Y, Hallenbeck J. Cell permeable exogenous ceramide reduces infarct size in spontaneously hypertensive rats supporting in vitro studies that have implicated ceramide in induction of tolerance to ischemia. J Cereb Blood Flow Metab. 2001;21:226–32.
34. Marshall JD, Heeke DS, Abbate C, et al. Induction of interferon-gamma from natural killer cells by immunostimulatory CpG DNA is mediated through plasmacytoid-endritic-cell-produced interferon-alpha and tumour necrosis factor-alpha. Immunology 2006;117:38–46.
35. Katakura K, Lee J, Rachmilewitz D, Li G, Eckmann L, Raz E. Toll-like receptor 9-induced type I IFN protects mice from experimental colitis. J Clin Invest. 2005;115:695–702.
36. Ryu H, et al. Sp1 and Sp3 are oxidative stressinducible, antideath transcription factors in cortical neurons. J Neurosci. 2003;23:3597–606.
37. Zhou AM, Li WB, Li QJ, Liu HQ, Feng RF, Zhao HG. A short cerebral ischemic preconditioning up-regulates adenosine receptors in the hippocampal CA1 region of rats. Neurosci Res. 2004;48:397–404.
38. Carbonell T, Rama R. Iron, oxidative stress and early neurological deterioration in ischemic stroke. Curr Med Chem. 2007;14:857–74.
39. Cadet JL, Brannock C. Free radicals and the pathobiology of brain dopamine systems. Neurochem Int. 1998;32:117–31.
40. Guix FX, Uribesalgo I, Coma M, Munoz FJ. The physiology and pathophysiology of nitric oxide in the brain. Prog Neurobiol. 2005;76:126–52.
41. Ohtsuki T, Matsumoto M, Kuwabara K, Kitagawa K, Suzuki K, Taniguchi N, Kamada T. Influence of oxidative stress on induced tolerance to ischemia in gerbil hippocampal neurons. Brain Res. 1992;599:246–52.
42. Busija DW, Gaspar T, Domoki F, Katakam PV, Bari F. Mitochondrial-mediated suppression of ROS production upon exposure of neurons to lethal stress: mitochondrial targeted preconditioning. Adv Drug Deliv Rev. 2008;60:1471–7.
43. Toyoda T, Kassell NF, Lee KS. Induction of ischemic tolerance and antioxidant activity by brief focal ischemia. Neuroreport. 1997;8:847–51.
44. Panahian N, Yoshiura M, Maines MD. Overexpression of heme oxygenase-1 is neuroprotective in a model of permanent middle cerebral artery occlusion in transgenic mice. J Neurochem. 1999;72:1187–203.
45. Kato H, Liu Y, Kogure K, Kato K. Induction of 27-kDa heat shock protein following cerebral ischemia in a rat model of ischemic tolerance. Brain Res. 1994;634:235–44.
46. Dhodda VK, Sailor KA, Bowen KK, Vemuganti R. Putative endogenous mediators of preconditioning-induced ischemic tolerance in rat brain identified by genomic and proteomic analysis. J Neurochem. 2004;89:73–89.
47. Liu Y, Kato H, Nakata N, Kogure K. Temporal profile of heat shock protein 70 synthesis in ischemic tolerance induced by preconditioning ischemia in rat hippocampus. Neuroscience. 1993;56:921–7.
48. Chen J, Graham SH, Zhu RL, Simon RP. Stress proteins and tolerance to focal cerebral ischemia. J Cereb Blood Flow Metab. 1996;16:566–77.
49. Tanaka S, Kitagawa K, Ohtsuki T, Yagita Y, Takasawa K, Hori M, Matsumoto M. Synergistic induction of HSP40 and HSC70 in the mouse hippocampal neurons after cerebral ischemia and ischemic tolerance in gerbil hippocampus. J Neurosci Res. 2002;67:37–47.
50. Abe H, Nowak TS. Postischemic temperature as a modulator of the stress response in brain: dissociation of heat shock protein 72 induction from ischemic tolerance after bilateral carotid artery occlusion in the gerbil. Neurosci Lett. 2000;295:54–8.
51. Beck T, Lindholm D, Castren E, Wree A. Brain-derived neurotrophic factor protects against ischemic cell damage in rat hippocampus. J Cereb Blood Flow Metab. 1994;14:689–92.
52. Shigeno T, Mima T, Takakura K, Graham DI, Kato G, Hashimoto Y, Furukawa S. Amelioration of delayed neuronal death in the hippocampus by nerve growth factor. J Neurosci. 1991;11:2914–9.
53. Tanaka K, Tsukahara T, Hashimoto N, Ogata N, Yonekawa Y, Kimura T, Taniguchi T. Effect of nerve growth factor on delayed neuronal death after cerebral ischaemia. Acta Neurochir (Wien). 1994;129:64–71.

54. Truettner J, Busto R, Zhao W, Ginsberg MD, Perez-Pinzon MA. Effect of ischemic preconditioning on the expression of putative neuroprotective genes in the rat brain. Brain Res Mol Brain Res. 2002;103:106–15.

55. Weih M, Kallenberg K, Bergk A, Dirnagl U, Harms L, Wernecke KD, Einhaupl KM. Attenuated stroke severity after prodromal TIA: a role for ischemic tolerance in the brain? Stroke. 1999;30:1851–4.

56. Wegener S, Gottschalk B, Jovanovic V, Knab R, Fiebach JB, Schellinger PD, Kucinski T, Jungehulsing GJ, Brunecker P, Muller B, Banasik A, Amberger N, Wernecke KD, Siebler M, Röther J, Villringer A, Weih M. MRI in Acute Stroke Study Group of the German Competence Network Stroke: transient ischemic attacks before ischemic stroke: preconditioning the human brain? A multicenter magnetic resonance imaging study. Stroke. 2004;35:616–21.

57. Dirnagl U, Becker K, Meisel A. Preconditioning and tolerance against cerebral ischaemia: from experimental strategies to clinical use. Lancet Neurol. 2009;8:398–412.

58. Castillo J, Moro MA, Blanco M, Leira R, Serena J, Lizasoain I, Davalos A. The release of tumor necrosis factor-alpha is associated with ischemic tolerance in human stroke. Ann Neurol. 2003;54:811–9.

Chapter 17
microRNAs in Ischemic Brain: The Fine-Tuning Specialists and Novel Therapeutic Targets

Ashutosh Dharap, Venkata P. Nakka, and Raghu Vemuganti

Abstract CNS injuries are associated with extensive spatio-temporal alterations in gene expression that dictate the extent of brain damage and subsequent plasticity. Studies in CNS injury models show that microRNAs (miRNAs) modulate gene expression, which contributes to development of the injury. The expression levels of brain miRNAs show diverse spatio-temporal profiles in the healthy brain and are altered significantly over a sustained period of time in injured states. Apart from the conventional mechanisms of translational inhibition of mRNAs, recent studies have revealed that miRNAs can also target gene promoters to induce or repress their expression, adding another layer to the already complex network of miRNA–gene interactions. The extent of the influence of miRNAs in the CNS is only beginning to be elucidated. miRNAs play a major role in the etiology of neurodegenerative diseases as well as in brain injuries such as traumatic brain injury, spinal cord injury, and ischemia. Studies have shown that they considerably influence the outcome of the ischemic pathophysiology. However, miRNAs have also been observed to play a role in endogenous (preconditioning) and exogenous neuroprotective treatments, indicating their role in inducing ischemic tolerance. Lately, miRNAs have shown promise for serving as clinical biomarkers of stroke and other CNS disorders.

1 Introduction

A noncoding RNA (ncRNA) is a functional RNA molecule that is not translated into a protein. The ncRNAs include transfer RNA (tRNA) and ribosomal RNA (rRNA) that comprise the support machinery for translation in the cytoplasm, as well as

A. Dharap • V.P. Nakka • R. Vemuganti, PhD (✉)
Department of Neurological Surgery and Neuroscience Training Program,
University of Wisconsin, 600 Highland Ave, Madison, WI 53792, USA
e-mail: vemuganti@neurosurgery.wisc.edu

P.A. Lapchak and J.H. Zhang (eds.), *Translational Stroke Research*, Springer Series
in Translational Stroke Research, DOI 10.1007/978-1-4419-9530-8_17,
© Springer Science+Business Media, LLC 2012

small nucleolar RNA (snoRNA), piwi-interacting RNA (piRNA), long ncRNA (lncRNA), short interfering RNA (siRNA), microRNA (miRNA), small Cajal body-specific RNA (scaRNA), centromere-associated RNA (crasiRNA), telomere small RNA (tel-sRNA), CRISPR RNA (crRNA), repeat-associated siRNA (rasiRNA), transacting siRNA (tasiRNA), promoter-associated RNA (PAR), X-inactivation RNA (xiRNA), and tRNA-derived RNA [49, 108, 128]. All these ncRNAs are thought to control gene expression at either the transcriptional [95] or the translational level [5, 23, 30].

The first indication of inhibitory regulation of gene expression by nucleic acid sequences was observed in 1990 when exogenous sequences introduced to overexpress the chalcone synthase (CHS) gene unexpectedly caused a 50-fold reduction in the expression of the endogenous CHS gene in the petunia plants [84, 118]. Soon after, Lee et al. [59] observed that in *Ceanorhabditis elegans* (*C. elegans*), an endogenous ncRNA called lin-4 specifically downregulates the expression of the Lin-14 protein. This study further observed that the 3' untranslated region (3'-UTR) of Lin-14 mRNA contains sequence repeats that are complementary to regions in the lin-4. This suggested that lin-4 may be a regulator of Lin-14 translation by interacting at the 3'-UTR. A subsequent study identified that a ncRNA conserved in the animal kingdom known as let-7 can bind to the 3'-UTR of half a dozen mRNAs to repress them [91, 99]. In 1998, a collaborative effort led by Andrew Fire and Craig Mello demonstrated potent and specific translational repression by exogenously introduced double-stranded RNAs in *C. elegans* that repressed the translation of several target genes with shared sequence homologies. They termed this process as "RNA interference" (RNAi) [30]. For this work, Fire and Mello were awarded the Nobel Prize in Physiology or Medicine in 2006. Subsequently, several studies tested and confirmed the ability of ncRNAs to selectively target mRNAs and inhibit their translation, and a trio of publications in the journal Science in 2001 established the term "microRNA" for these yet uncategorized ncRNAs [55–58].

The miRNA field began developing rapidly in the early 2000s. A uniform system for miRNA annotation was laid out by Victor Ambros and David Bartel that enabled the discovery of hundreds of miRNAs in several species [5]. The establishment of "The microRNA Registry" in 2004 and "miRBase" in 2006 by the Wellcome Trust Sanger Institute (Hinxton, UK) provided further impetus for miRNA research by providing a one-stop, systematically categorized, interactive repository for miRNAs that includes miRNA targets, genomic coordinates, cluster information, and tissue expression [35, 36]. The current version 16 of miRBase has 15,172 entries representing hairpin precursor miRNAs expressing 17,341 mature miRNA products in 142 species (http://www.miRbase.org).

2   MiRNA Biogenesis

Different miRNAs are transcribed from either independent miRNA encoding genes [56] or from introns of the protein-coding genes [101]. In many cases, multiple miRNAs are transcribed as a polycistronic cluster from a common promoter [4, 60].

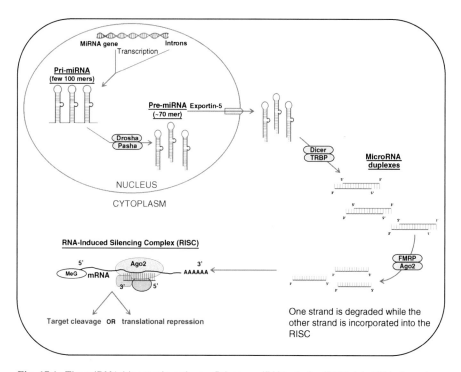

Fig. 17.1 The miRNA biogenesis pathway. Primary miRNAs (pri-miRNAs) (>100 bp) are transcribed from dedicated genes or introns. In the nucleus, pri-miRNAs are cleaved by the RNAse Drosha to form ~70 bp stem-loop structures called pre-miRNAs. The pre-miRNAs are exported into the cytosol by exportin-5 where RNAse Dicer cleaves them into 20–25 nucleotide long duplexes. The miRNA duplexes are recruited by Ago2, which degrades one strand and retains the other to form mature miRNA-containing RISC. This complex selectively binds to mRNAs that are complementary to the miRNA strand loaded into the RISC and either degrade the mRNAs or repress their translation. *TRBP* transactivating response RNA-binding protein; *FMRP* fragile X mental retardation protein; *MeG* methylguanine

The miRNAs that are derived from such clusters may not necessarily have common targets. The miRNA genes are transcribed by RNA Polymerase II [60, 136] to form a primary transcript (pri-miRNA) that can be several hundreds to thousands of nucleotides long and contains one or more stem-loop structures. The pri-miRNA is recognized by a nuclear protein called DiGeorge Syndrome Critical Region 8 (DGCR8), which pairs with the RNAse Drosha to cleave the pri-miRNA into a ~70 nucleotide hairpin loop called the pre-miRNA with a two-nucleotide 3′ overhang [34]. The pre-miRNA is transported to the cytoplasm by Exportin 5, which recognizes the 3′ overhang. In the cytoplasm, RNAse Dicer cleaves the pre-miRNA into an imperfect miRNA/miRNA* duplex of ~22–25 nucleotides (Fig. 17.1). Although both strands may form functional, mature miRNAs, typically one strand is degraded and the other is bound to the Dicer and Argonaute (Ago) proteins to form the RNA-induced Silencing Complex (RISC) [116], which identifies and interacts with its targets via complementary binding [107, 117]. Ago is thought to preferentially recruit those strands that have many targets as opposed to ones that have fewer or no targets, leaving the nonrecruits for degradation [46].

3   microRNA-Mediated Repression

A miRNA-containing RISC selectively binds to specific mRNA targets by forming Watson-Crick complementary base pairing with high specificity in the 3' UTR region of the mRNAs [41, 62, 94]. A specific group of nucleotides located in the 5' end of the miRNA sequence called the "seed sequence" is thought to be the main determinant for targeting specific mRNAs [33]. However, Ago proteins in the RISC can also target transcripts directly [31, 97]. Posttranscriptional gene silencing by a miRNA primarily occurs by the prevention of translation of the complementary mRNA [6, 32, 94]. In mammals, for translation to initiate, it is necessary for the eukaryotic initiation factor 4E to interact with the 7-methylguanosine (m7G) cap on the mRNA [32, 100]. However, when the RISC binds to the mRNA, the Ago protein interacts with the m7G cap rendering it inaccessible to the initiation factor and hence the mRNA will be unavailable for translation [50]. RISC can also directly degrade the mRNA targets resulting in translational arrest [72]. How the RISC determines whether to repress translation or degrade the mRNA is not yet conclusively known. Once RISC binds to a target, the RISC-mRNA assembly can be stored in cytoplasmic processing bodies called "P-bodies" or "GW bodies," which contain factors such as decapping enzymes, decapping activators, and 5'–3' exonucleases that play a role in translational repression and mRNA decay [10, 90, 93]. The mRNA-RISC complexes may be temporarily stored in P-bodies or they may be permanently degraded. This process has been well documented in studies with the CAT-1 mRNA (a target of the liver-specific miR-122), which localized to the P-bodies explicitly in the presence of miR-122 even in cells that did not originally contain miR-122 (but were provided with exogenous miR-122). This localization to the P-bodies was absent when miR-122 was reduced due to stress [10]. Interestingly, in neurons many mRNAs were shown to be transported to dendritic spines in the form of "repressed messenger ribonucleoproteins," which will be released and translated following synaptic activity [106]. This seems to suggest that miRNAs not only repress mRNA translation, but also mediate the selective packing and transport of mRNAs from one area of a cell where they are not needed (or detrimental) to another area of the cell that requires them. Such compartmentalized regulation of transcription and translation in the dendrites is thought to be necessary for processes such as memory, learning, and long-term potentiation [40, 89, 112]. Recent studies showed that miRNAs can also bind to the 5'-UTR or open reading frame (ORF) of a mRNA to repress translation [31, 79, 104]. A recent study also demonstrated that certain miRNAs can repress gene expression by targeting the promoters that contain miRNA-binding sites [134]. For example, the promoter of progesterone receptor (PR) was shown to be silenced by four exogenously administered miRNAs (miR-372, miR-373, miR-520c-3p, and miR-423-5p) [134].

This study suggested decreased RNA polymerase II occupancy and increased H3 lysine 9 dimethylation (H3K9me2) at the PR promoter indicating chromatin-level silencing as the mechanism for miRNA-mediated gene silencing [134]. The miR-423-5p mimic was also shown to repress the immunoglobulin superfamily member 1 gene that contained the target site for miR-423-5p [134]. Interestingly, endogenous miR-423-5p had no effect on PR expression suggesting that for promoter silencing a miRNA is needed at a particular threshold that can be achieved only by exogenous administration of a miRNA mimic. A recent study also showed that certain miRNAs target both 5′-UTR and 3′-UTR of a miRNA [57]. In this model, the 5′ end of the miRNA targets 3′-UTR of a mRNA, while the 3′ end of the miRNA targets the 5′-UTR of the same mRNA. Thus, a miRNA redundantly controls a mRNA in a fail-safe mechanism.

4   microRNA-Mediated Activation

In the first few years after their discovery, the excitement around miRNAs was focused solely around their ability to repress the translation of mRNAs. Subsequently, Vasudevan et al. [119] observed that upon cell cycle arrest, the AU-rich elements of the TNFα mRNA act as an activation signal to recruit miR-369-3 that upregulated rather than suppressed the TNFα translation. Likewise, let-7a and a synthetic miRNA called miRcxcr4 induced translation of their target mRNAs during cell cycle arrest, as opposed to their repressive actions in proliferating cells [119]. This data was supported by a subsequent study that showed that when NIH3T3 cells entered cell cycle arrest upon contact inhibition, exogenously added miRcxcr4 induced the miRcxcr4-responsive CX reporter and let-7 induced the let-7-responsive HMGA2 reporter [120]. This indicates that the cell cycle-dependent activation of genes by miRNAs is a widespread phenomenon. Orom et al. [88] confirmed the miRNA-mediated gene induction by showing that miR-10a induced the translation of the ribosomal proteins Rps6, Rps16, and Rps19 by binding to the 5′-UTR of their mRNAs in response to anisomycin. Significantly, miR-10a is highly conserved evolutionarily with respect to its sequence as well as localization to the Hox cluster of developmental genes [68] and is observed to preferentially bind to a region immediately downstream to the 5′-TOP motif in mRNAs that contain a 5′-TOP motif [88]. 5′-TOP motif is known to confer sensitivity to mitogens such as anisomycin and regulates the translation of ribosomal proteins [75].

miRNAs were also shown to induce gene expression by sequence-specific binding to promoters. Place et al. [95] showed that transfection of miR-373 into PC3 cells induced the transcription of the E-cadherin and cold shock domain-containing protein C2 (CSDC-C2) genes, which contained miR-373 target sites within their promoter. Mismatch mutations in the miR-373 binding sites of the promoters as well as treatment with antimiR-373 prevented this induction indicating that this miR-373 binding to promoters specifically mediates E-cadherin and CSDC-C2 gene expression.

5   microRNAs in Brain Development and Plasticity

Krichevsky et al. [51] showed that many miRNAs are developmentally regulated in the mammalian brain. This study also demonstrated that certain miRNAs like miR-9 and miR-131 are essential for normal brain development and their dysfunction leads to developmental defects. Subsequently, several studies showed the temporal profiles of miRNAs in brain development, neural differentiation, and the association of brain developmental disorders with abnormal expression of specific miRNAs [77, 102, 110]. During different periods of mammalian development, there is a constant change in the spatio-temporal profiles of gene expression. The miRNA profiles also showed a characteristic signature at different stages of development in various tissues of mammals indicating a dynamic relationship between miRNAs and mRNAs that is needed for normal development [113]. Importantly, miR-92b is restricted to proliferating cells, miR-124 and miR-138 are expressed in differentiating cells of the CNS, and miR-9, miR-135c, miR-153a, miR-219 and let-7a, 7b, and 7c are expressed in both proliferating and differentiating embryonic CNS cells [113]. Moreover, miRNA expression is cell type-specific and several miRNAs are found to be spatially confined in their expression [47]. A good example is miR-181 which strongly expresses in the cells of the visual system, pretectum, tectum, central pallium, and medulla oblongata [47]. While miR-222 expression is confined to the forebrain and midbrain, miR-34 expression is present only in the caudal, ventral, and lateral isthmus and hindbrain [47]. Furthermore, in rodents miR-124 is enriched in the CNS, while miR-1 is enriched in the heart and muscle [76]. This was supported by experiments in which transfection of miR-1 and miR-124 in HeLa cells shifted the expression profiles of those cells toward that of muscle cells and neural cells, respectively [63].

In addition to directing growth and differentiation of cells and tissues, miRNAs were found to finely control the morphology and functionality of neural cell processes. A brain-specific miRNA miR-134 negatively regulates the size of dendritic spines in rat hippocampal neurons by targeting the mRNA of Limk1, a protein kinase that controls spine development [106]. This has implications in embryonic development as well as memory and learning in adults. MiR-132 is an activity-dependent rapid response miRNA regulated by the cAMP response element-binding protein (CREB) pathway that controls the activity-dependent refinement of neuronal circuitry [67, 129]. It is known that neuronal activity inhibits translation of the p250GAP mRNA leading to dendrite growth [82] and p250GAP mRNA is a target of miR-132 [125]. Furthermore, introduction of miR-132 enhances dendrite morphogenesis of hippocampal neurons, while treatment with antagomiR-132 reversed this effect [67, 125]. Studies with dominant-interfering mutants suggested that the miR-132-p250GAP pathway acts via the downstream Rac signaling pathway to modulate activity-dependent synaptic plasticity [125]. Pietrzykowski et al. [92] demonstrated the role of miRNAs in drug adaptation via molecular plasticity. They observed that in the mammalian brain, alcohol intake upregulates miR-9, which targets the large-conductance calcium- and voltage-activated potassium channel

(BK; a well-established alcohol target) to produce its splice variants, which might contribute to alcohol tolerance. Chronic cocaine administration to rats was shown to significantly downregulate the miRNAs let-7d and miR-124, which are upstream to several genes induced by cocaine [13]. The prominent targets of miR-124 and let-7d are the brain-derived neurotrophic factor (BDNF) and dopamine receptor D3R (DA D3R), respectively. While BDNF has been known to positively regulate dendritic spine size and modulate cocaine-induced plasticity in reward and memory, D3R up-regulation is implicated in synaptic plasticity and behavioral changes induced by chronic cocaine administration [8]. Cocaine administration was also shown to upregulate miR-181a which targets several genes suppressed in the mesolimbic dopaminergic system [13]. Additionally, miR-181 which is upregulated following cocaine administration targets the period circadian protein homolog 2 (Per2) mRNA, the down-regulation of which is associated with the development of association cues related to cocaine administration [1]. Thus, the changes in miR-124, let-7d, and miR-181 and their interaction with genes involved in the drug-reward circuitry indicate a crucial role for miRNAs in this system.

6   microRNAs in Ischemic Brain Damage

The stroke pathophysiology has been well documented by several human and animal studies that have dissected the roles of many genes in various stages of disease progression [48, 64, 87, 108, 122, 131]. Like other areas of biology, the discovery of miRNAs has changed the way we look at stroke. We now know that miRNAs are important in all aspects of stroke from etiology to pathology to therapeutic prevention. Atherosclerotic plaques and hypertension are two major factors that predispose humans to stroke [22]. A recent study showed that induction of miR-21 by shear stress might be a promoter of atherosclerotic plaque rupture leading to stroke [126]. Interestingly, miR-21 induction following angioplasty was found to increase Bcl-2 expression resulting in a better survival of vascular smooth muscle cells and its depletion is associated with plaque formation [45]. Furthermore, vascular cell adhesion molecule-1 (VCAM1) induced by atherosclerosis binds to integrins on macrophages to mediate their adhesion [81, 86]. It was shown that up-regulation of VCAM1 might be mediated by deceased levels of its upstream miRNA miR-126, leading to enhanced leukocyte adherence to the endothelium [37]. Furthermore, deletion of miR-126 led to the loss of endothelial cell migration and negatively affected angiogenesis resulting in leaky vessels [123]. More recently, an endothelial cell-specific loss of miR-126 was demonstrated in type-2 diabetic patients, indicating its role in the increased susceptibility of diabetics to stroke [135]. The miRNA miR-210 was also implicated in ischemic pathophysiology as inhibition of miR-210 reduces endothelial cell migration and angiogenesis which are implicated in hypoxia-induced angiogenesis and plaque rupture [28, 29].

Recently, a polymorphism in the 3'-UTR of the angiotensin II type 1 receptor (AT1R) mRNA known as A1166C was shown to be associated with hypertension which is a risk factor for stroke [20, 74]. Subsequent studies revealed that this polymorphism disrupts the miR-155-binding site in the AT1R 3'-UTR resulting in increased levels of AT1R protein in patients homozygous for the A1166C allele [12, 71]. Furthermore, transforming growth factor beta 1 (TGF-B1) was negatively correlated with miR-155, but positively correlated with the AT1R expression in A166C homozygotes suggesting a role for TGF-B1 in mediating the interplay between miR-155 and AT1R [11]. In line with this observation, miR-155 was found to be significantly lower in the aorta of adult spontaneously hypertensive rats compared to age-matched Wistar-Kyoto rats, and this was correlated with increasing blood pressure [130].

Several groups recently profiled the postischemic changes in the cerebral miR-NAome [23, 44, 65, 115]. These studies showed that the cerebral miRNAome alters rapidly following ischemia and the changes sustain through the progression of the brain damage. Pathophysiological studies from our laboratory showed that 11 miR-NAs were altered as early as 3 h of reperfusion, and this number increases with the progression of reperfusion with 24 miRNAs upregulated and 22 miRNAs down-regulated at 3 days of reperfusion [23]. Several miRNAs were observed to sustain their altered state of expression from 3 h to 3 days of reperfusion, indicating that they are crucial in mediating the mechanism of acute stage ischemic pathophysiology [23]. Recent studies also evaluated the functional significance of some stroke-responsive miRNAs. Bioinformatics showed that the antioxidant enzyme superoxide dismutase 2 (SOD2) mRNA is a major target of miR-145, a miRNA that was observed to be upregulated following focal ischemia [23]. Furthermore, knockdown of miR-145 led to a significant induction of SOD2 and concomitant neuroprotection after focal ischemia [23]. A subsequent study demonstrated that miR-145 and several other miRNAs (let-7b, miR-26a, and miR-34) could silence the anti-inflammatory cytokine interferon-beta (IFN-β) in monocyte-derived macrophages [127]. IFN-β is known to have neuroprotective effects and delivery of IFN-β in a cerebral ischemia model attenuated the infiltration of neutrophils and monocytes into brain and a 70% reduction of the infarct volume [121]. Thus, induction of miR-145 after focal ischemia appears to contribute to neuronal death by curtailing the ability of the brain to fight oxidative stress and inflammation. Additionally, miR-497, which was also observed to be induced in the postischemic brain, seems to play a role in modulating apoptosis [132, 133]. Bioinformatics showed that miR-497 targets the 3'-UTR of the anti-apoptotic gene Bcl2 mRNA and treatment of mice with antagomiR-497 leads to significant induction of Bcl2 in neurons, reduced apoptosis, and smaller infarcts after focal ischemia [132, 133]. The role of miRNAs in controlling Bcl2 in the postischemic brain was also demonstrated by Yin et al. [132, 133]. When mouse cerebrovascular endothelial cells (CECs) were subjected to oxygen-glucose deprivation (OGD), expression of the anti-apoptotic transcription factor PPAR-δ was observed to be significantly reduced; and PPAR-δ overexpression suppressed caspase-3 activity via increased bcl-2 protein levels without affecting the bcl-2 mRNA levels. This treatment led to protection of CECs from cell death [132, 133]. Interestingly, the bcl-2 mRNA is a target of miR-15a, and when mice were subjected to focal ischemia following PPAR-δ overexpression, miR-15a levels decreased leading to increased bcl-2 protein that curtailed the blood–brain barrier disruption and infarct development [132, 133]. Thus, miR-497 and miR-15a seem to modulate Bcl2 protein levels in different cell types in the ischemic brain, demonstrating the highly specific but redundant target selection process of conserved miRNAs. Such characteristics could be exploited for developing future cell type-specific therapies for stroke. Interestingly, miR-21 which was implicated in Bcl2 regulation and plaque formation received attention due to its anti-apoptotic nature. The expression of miR-21 was found to be increased in the neurons of the "ischemic boundary zone" when rats were subjected to focal ischemia, and this up-regulation lasted for up to a week after the stroke onset [11]. The ischemic boundary zone is the area around the core of infarct that contains viable cells and can be rescued with proper intervention. When synthetic miR-21 was exogenously transfected into cultured cortical neurons subjected to OGD, there was a significant reduction in the levels of its target Fas ligand (a member of the TNF superfamily; induces cell death) and this was associated with a substantial decrease in OGD-induced apoptosis [11]. Thus, the presence or absence of miR-21 and the magnitude of its expression seem to be important determinants of cell survival in stroke.

In a recent study exploring 3′-UTR variants of Angiopoietin-1 (Angpt1) in humans, Chen et al. [15] discovered a variant called "rs2507800" containing a SNP with an "A" to "T" change. They found that the "A allele" suppressed the translation of Angpt1 via miR-211 binding to the 3′-UTR, but the suppression was lost in the "T allele." Thus, the subjects carrying the TT genotype had higher plasma levels of Angpt1 than those with an AA or AT genotype. Angpt1 is a known protector of vascular inflammation and leakage [25, 42]. When the association of the rs2507800 variant was tested in stroke patients, those with the TT genotype demonstrated a significant reduction in overall stroke risk as compared to those with the AA or AT genotype [15].

Human pluripotent stem cells (hPSCs) offer a promising potential for cell-based therapies in brain injury patients [21]. Human neural progenitor cells (hNPCs) derived from hPSCs are shown to proliferate and migrate towards the region of injury when transplanted into the ischemic mouse brain [21]. This study observed that the brain-specific miRNA miR-9 was upregulated in the proliferative phase, but downregulated during the migratory phase. It was also observed that miR-9 reduces the levels of stathmin (which increases microtubule instability) to promote early hNPC proliferation but delayed migration, and when miR-9 was lost, hNPCs switched to an enhanced migratory mode [21].

7   microRNAs as Biomarkers of Ischemia

Many studies showed that blood contains significant amounts of miRNAs and they change in response to CNS injuries and disorders [19, 43, 105, 108]. Blood miRNA profiles were also shown to be reliable markers of myocardial infarction

[2, 70, 124]. Furthermore, miRNAs in blood were shown to be very stable over days indicating their suitability as biomarkers of disease progression [126]. Jeyaseelan et al. [44] showed that following focal ischemia in rodents, the miRNA levels change significantly in blood. In young stroke patients, miRNAs that are implicated in stroke-related events such as endothelial dysfunction, angiogenesis, and erythropoiesis were observed in peripheral blood samples at levels significantly different from the healthy controls and were detectable for several months after stroke [115]. Blood miRNAs were also shown to be significantly altered after stroke, seizures, and brain hemorrhage, and a combination of injuries led to more significant changes [65]. These studies demonstrate the potential of blood miRNA profiles to evaluate brain injury after stroke.

8 microRNAs in Neuroprotection and Ischemic Tolerance

A brief, nonlethal ischemic attack induces tolerance against a future massive ischemic attack. This phenomenon is known as preconditioning (PC)-induced ischemic tolerance [80]. PC is associated with new protein synthesis [9] as well as transient repression of gene expression [111]. Recent studies showed that PC significantly alters the miRNA expression profiles in rodent brain [24, 66]. The miRNAs reacted quickly to PC, resulting in 26 upregulated and 25 downregulated miRNAs by 6 h of reperfusion [24]. As in the case after focal ischemia, several of these miRNAs showed sustained alteration up to 3 days after PC. Pathways analysis revealed that MAP kinase (MAPK) signaling and mTOR signaling are the top targets of the miRNAs induced after PC [24]. The targeting of mTOR is particularly interesting as mTOR is a transcription and translation regulator that is negatively controlled in response to cellular energy stress [96]. Interestingly, inhibition of mTOR has been shown to increase the lifespan of mammals [7, 18, 38]. Future studies might show the importance of miRNAs in controlling the lifespan. More than a dozen members of the MAPK family mediate variety of signals [18]. Although MAPK signaling is known to play a significant role in cerebral ischemia [73, 85], its role in PC-induced ischemic tolerance is not known. Growing evidence suggests that some members of the MAPK family such as p38 and ERK are involved in promoting neuronal cell death [61, 85, 114]. Attenuation of the p38 MAPK activation in heart ischemia leads to a reduction in the infarct size [69]. However, sustained high levels of p38 MAPK following cardiac ischemia were associated with increased infarction and poor recovery [69, 78]. Interestingly, activation of p38 MAPK was shown to curtail the PC-induced ischemic tolerance in heart [103]. Thus, altered miRNAs during PC might influence MAPK and its downstream effects.

   Another prominent target of the miRNAs altered after ischemic PC is the methyl CpG-binding protein 2 (MeCP2), which is a global transcriptional activator/repressor [14, 24, 66, 83]. Cerebral MeCP2 protein levels were observed to increase rapidly after PC without any changes in its mRNA levels; and MeCP2 knock-out mice showed increased susceptibility to focal ischemia [66]. Conservation of energy by repression of nonessential gene expression and acceleration of critical protein expression is a feature of tolerance. Hence, the down-regulation of MeCP2-targeting miRNAs following PC might be a finely tuned early response of neuronal cells to allow MeCP2 protein to be rapidly translated from its mRNA to serve essential downstream functions that contribute to tolerance to future ischemic events. Consistent with the transient nature of PC, MeCP2 expression returns to baseline in the absence of further ischemic insults. Overall, these observations support a role for miRNAs in PC-induced ischemic tolerance.

9   Therapeutic Potential of microRNAs

Despite hundreds of studies that documented the relationship between miRNAs and gene regulation in various pathologies, there is very little evidence of the therapeutic potential of miRNAs. The therapeutic application of miRNAs can involve two strategies. One is to use antagomiRs against miRNAs that are upstream to neuroprotective pathways so that the beneficial proteins will be formed at higher levels. The second strategy is to use premiRs that can release miRNAs upstream to neurotoxic pathways so that these pathways will be suppressed leading to cell protection. Both antagomiRs and premiRs showed substantial promise in experimental studies [23, 52, 54, 132, 133]. However, there are several roadblocks for formulating miRNA therapeutics that include the following: (1) The premiRs or antagomiRs are susceptible to degradation inside cells, (2) they might not reach the organ of interest when administered systemically, (3) they might induce unwanted side effects due to the ability to target multiple mRNAs, (4) targeted delivery of premiRs and antagomiRs to specific cells or tissues is technically challenging, and (5) not enough is known about the regulatory loops that govern miRNA–mRNA interactions. Ongoing research will clarify some of these issues. For example, miRNAs are being polyaminated or encapsulated in microvesicles to escape the action of nucleases [3, 98]. Furthermore, their structure is modified for better sequence recognition [27] and they are combined with synthetic molecules that improve delivery and target specificity in vivo [16]. Si et al. [109] reduced breast tumor growth by targeting miR-21 with $2'$-O-methyl-modified antagomiR-21. Krutzfeldt et al. [53] showed that a single injection of cholesterol-conjugated antagomiR-122 silences miR-122 in the liver for 23 days and this treatment decreases the growth of hepatocellular carcinoma. Hepatic miR-122 was also implicated in increased plasma cholesterol levels. Elmen et al. [27] showed that systemically administered locked nucleic acid (LNA)-modified antagomiR-122 accumulates in the cytosol of hepatocytes leading to silencing of miR-122 and decreased plasma cholesterol in nonhuman primates. The LNA modification involves formation of a methylene bridge that connects the $2'$-O-oxygen with the $4'$-C atom of the ribose ring resulting in increased thermal stability of the duplexes, decreased degradation, improved target specificity, and reduced toxicity. The LNA-modified antagomiR-122 passed the Phase I clinical trials [39].

Certain miRNAs are formed as polycistronic clusters and Ebert et al. [26] showed that miRNA sponges (vectors containing promoters that encode transcripts that have multiple, tandem binding sites for all the miRNAs of a cluster) can be used to simultaneously silence a complete cluster. This approach is ideal to silence oncomiR-17-92 cluster that is implicated in various cancers [26]. Despite the promising results of these experiments, it is important to note that systemically administered compounds reach peripheral organs including liver, but transfer to CNS is still challenging due to active blood–brain barrier. Modifications like LNA or O-methylation increase the size of antagomiRs and premiRs making them difficult to reach CNS. Apart from delivery constraints, as modulating a miRNA can change hundreds of targets, premiRs and antagomiRs might alter the translation of nonspecific mRNAs leading to undesired side effects. However, if properly controlled and dosed, miRNA therapies might help to prevent pathological changes associated with acute conditions like stroke and traumatic CNS injury.

Acknowledgments Supported by NIH grant 061071 and NIH grant 074444.

References

1. Abarca C, Albrecht U, Spanagel R. Cocaine sensitization and reward are under the influence of circadian genes and rhythm. Proc Natl Acad Sci USA. 2002;99:9026–30.
2. Adachi T, Nakanishi M, Otsuka Y, Nishimura K, Hirokawa G, Goto Y, Nonogi H, Iwai N. Plasma microRNA 499 as a biomarker of acute myocardial infarction. Clin Chem. 2010;56:1183–5.
3. Akao Y, Iio A, Itoh T, Noguchi S, Itoh Y, Ohtsuki Y, Naoe T. Microvesicle-mediated RNA molecule delivery system using monocytes/macrophages. Mol Ther. 2011;19:395–9.
4. Altuvia Y, Landgraf P, Lithwick G, Elefant N, Pfeffer S, Aravin A, Brownstein MJ, Tuschl T, Margalit H. Clustering and conservation patterns of human microRNAs. Nucl Acid Res. 2005;33:2697–706.
5. Ambros V, Bartel B, Bartel DP, Burge CB, Carrington JC, Chen X, Dreyfuss G, Eddy SR, Griffiths-Jones S, Marshall M, Matzke M, Ruvkun G, Tuschl T. A uniform system for microRNA annotation. RNA. 2003;9:277–9.
6. Ambros V. The functions of animal microRNAs. Nature. 2004;431:350–55. Review.
7. Anisimov VN, Zabezhinski MA, Popovich IG, Piskunova TS, Semenchenko AV, Tyndyk ML, Yurova MN, Antoch MP, Blagosklonny MV. Rapamycin extends maximal lifespan in cancer-prone mice. Am J Pathol. 2010;176:2092–7.
8. Bahi A, Boyer F, Bussard G, Dreyer JL. Silencing dopamine D3-receptors in the nucleus accumbens shell in vivo induces changes in cocaine-induced hyperlocomotion. Eur J Neurosci. 2005;21:3415–26.
9. Barone FC, White RF, Spera PA, Ellison J, Currie RW, Wang X, Feuerstein GZ. Ischemic preconditioning and brain tolerance: temporal histological and functional outcomes, protein synthesis requirement, and interleukin-1 receptor antagonist and early gene expression. Stroke. 1998;29:1937–50.
10. Bhattacharyya SN, et al. Relief of microRNA-mediated translational repression in human cells subjected to stress. Cell. 2006;125:1111–24.
11. Buller B, Liu X, Wang X, Zhang RL, Zhang L, Hozeska-Solgot A, Chopp M, Zhang ZG. MicroRNA-21 protects neurons from ischemic death. FEBS J. 2010;277:4299–307.

12. Ceolotto G, Papparella I, Bortoluzzi A, Strapazzon G, Ragazzo F, Bratti P, Fabricio AS, Squarcina E, Gion M, Palatini P, Semplicini A. Interplay between miR-155, AT1R A1166C polymorphism, and AT1R expression in young untreated hypertensives. Am J Hypertens. 2011;24:241–6.
13. Chandrasekar V, Dreyer JL. MicroRNAs miR-124, let-7d and miR-181a regulate cocaine-induced plasticity. Mol Cell Neurosci. 2009;42:350–62.
14. Chahrour M, Jung SY, Shaw C, Zhou X, Wong ST, Qin J, Zoghbi HY. MeCP2, a key contributor to neurological disease, activates and represses transcription. Science. 2008; 320:1224–9.
15. Chen J, Yang T, Yu H, Sun K, Shi Y, Song W, Bai Y, Wang X, Lou K, Song Y, Zhang Y, Hui R. A functional variant in the 3'-UTR of angiopoietin-1 might reduce stroke risk by interfering with the binding efficiency of microRNA 211. Hum Mol Genet. 2010;19:2524–33.
16. Chen Y, Zhu X, Zhang X, Liu B, Huang L. Nanoparticles modified with tumor-targeting scFv deliver siRNA and miRNA for cancer therapy. Mol Ther. 2010;18:1650–6.
17. Cobb MH. MAP kinase pathways. Prog Biophys Mol Biol. 1999;71:479–500.
18. Cox LS, Mattison JA. Increasing longevity through caloric restriction or rapamycin feeding in mammals: common mechanisms for common outcomes? Aging Cell. 2009;8:607–13.
19. Cox MB, Cairns MJ, Gandhi KS, Carroll AP, Moscovis S, Stewart GJ, Broadley S, Scott RJ, Booth DR, Lechner-Scott J. MicroRNAs miR-17 and miR-20a inhibit T cell activation genes and are under-expressed in MS whole blood. PLoS One. 2010;5:e12132.
20. De Oliveira-Sales EB, Nishi EE, Boim MA, Dolnikoff MS, Bergamas-chi CT, Campos RR. Up-regulation of AT1R and iNOS in the rostral ventrolateral medulla (RVLM) is essential for the sympathetic hyperac- tivity and hypertension in the 2 K-1C Wistar rat model. Am J Hypertens. 2010;23:708–15.
21. Delaloy C, Liu L, Lee JA, Su H, Shen F, Yang GY, Young WL, Ivey KN, Gao FB. MicroRNA-9 coordinates proliferation and migration of human embryonic stem cell-derived neural pro-genitors. Cell Stem Cell. 2010;6:323–35.
22. Dempsey RJ, Vemuganti R, Varghese T, Hermann BP. A review of carotid atherosclerosis and vascular cognitive decline: a new understanding of the keys to symptomology. Neurosurgery. 2010;67:484–93.
23. Dharap A, Bowen K, Place R, Li LC, Vemuganti R. Transient focal ischemia induces extensive temporal changes in rat cerebral microRNAome. J Cereb Blood Flow Metab. 2009;29: 675–87.
24. Dharap A, Vemuganti R. Ischemic pre-conditioning alters cerebral microRNAs that are upstream to neuroprotective signaling pathways. J Neurochem. 2010;113:1685–91.
25. Ding YH, Luan XD, Li J, Rafols JA, Guthinkonda M, Diaz FG, Ding Y. Exercise-induced overexpression of angiogenic factors and reduction of ischemia/reperfusion injury in stroke. Curr Neurovasc Res. 2004;1:411–20.
26. Ebert MS, Neilson JR, Sharp PA. MicroRNA sponges: competitive inhibitors of small RNAs in mammalian cells. Nat Methods. 2007;4:721–6.
27. Elmen J, Lindow M, Schutz S, Lawrence M, Petri A, Obad S, Lindholm M, Hedtjarn M, Hansen HF, Berger U, Gullans S, Kearney P, Sarnow P, Straarup EM, Kauppinen S. LNA-mediated microRNA silencing in non-human primates. Nature. 2008;452:896–9.
28. Fasanaro P, D'Alessandra Y, Di Stefano V, Melchionna R, Romani S, Pompilio G, Capogrossi MC, Martelli F. MicroRNA-210 modulates endothelial cell response to hypoxia and inhibits the receptor tyrosine kinase ligand Ephrin-A3. J Biol Chem. 2008;283:15878–83.
29. Fasanaro P, Greco S, Lorenzi M, Pescatori M, Brioschi M, Kulshreshtha R, Banfi C, Stubbs A, Calin GA, Ivan M, Capogrossi MC, Martelli F. An integrated approach for experimental target identification of hypoxia-induced miR-210. J Biol Chem. 2009;284:35134–43.
30. Fire A, Xu S, Montgomery MK, Kostas SA, Driver SE, Mello CC. Potent and specific genetic interference by double-stranded RNA in *Caenorhabditis elegans*. Nature. 1998;391:806–11.
31. Forman JJ, Coller HA. The code within the code: microRNAs target coding regions. Cell Cycle. 2010;9:1533–41.

32. Gebauer F, Hentze MW. Molecular mechanisms of translational control. Nat Rev Mol Cell Biol. 2004;5:827–35.
33. Ghildiyal M, Zamore PD. Small silencing RNAs: an expanding universe. Nat Rev Genet. 2009;10:94–108.
34. Gregory R, Chendrimada T, Shiekhattar R. MicroRNA biogenesis: isolation and characterization of the microprocessor complex. Methods Mol Biol. 2006;342:33–47.
35. Griffiths-Jones S. The microRNA Registry. Nucl Acid Res. 2004;32:D109–11.
36. Griffiths-Jones S, Grocock RJ, van Dongen S, Bateman A, Enright AJ. miRBase: microRNA sequences, targets and gene nomenclature. Nucl Acid Res. 2006;34:D140–4.
37. Harris TA, Yamakuchi M, Ferlito M, Mendell JT, Lowenstein CJ. MicroRNA-126 regulates endothelial expression of vascular cell adhesion molecule 1. Proc Natl Acad Sci USA. 2008;105:1516–21.
38. Harrison DE, Strong R, Sharp ZD, Nelson JF, Astle CM, Flurkey K, Nadon NL, Wilkinson JE, Frenkel K, Carter CS, Pahor M, Javors MA, Fernandez E, Miller RA. Rapamycin fed late in life extends lifespan in genetically heterogeneous mice. Nature. 2009;460:392–5.
39. Haussecker D, Kay MA. MiR-122 continues to blaze the trail for microRNA therapeutics. Mol Ther. 2010;18:240–2.
40. Huang F, Chotiner JK, Steward O. The mRNA for elongation factor 1alpha is localized in dendrites and translated in response to treatments that induce long-term depression. J Neurosci. 2005;25:7199–209.
41. Jackson RJ, Standart N. How do microRNAs regulate gene expression? Sci STKE 2007; 367:re1.
42. Jeansson M, Gawlik A, Anderson G, Li C, Kerjaschki D, Henkelman M, Quaggin SE. Angiopoietin-1 is essential in mouse vasculature during development and in response to injury. J Clin Invest. 2011;121:2278–89.
43. Jensen MB, Chacon MR, Sattin JA, Levine RL, Vemuganti R. Potential biomarkers for the diagnosis of stroke. Expert Rev Cardiovasc Ther. 2009;7:389–93.
44. Jeyaseelan K, Lim KY, Armugam A. MicroRNA expression in the blood and brain of rats subjected to transient focal ischemia by middle cerebral artery occlusion. Stroke. 2008;39:959–66.
45. Ji R, Cheng Y, Yue J, Yang J, Liu X, Chen H, Dean DB, Zhang C. MicroRNA expression signature and antisense-mediated depletion reveal an essential role of MicroRNA in vascular neointimal lesion formation. Circ Res. 2007;100:1579–88.
46. Kai Z, Pasquinelli A. MicroRNA assassins: factors that regulate the disappearance of miR-NAs. Nat Struct Mol Biol. 2010;17:5–10.
47. Kapsimali M, Kloosterman WP, de Bruijn E, Rosa F, Plasterk RH, Wilson SW. MicroRNAs show a wide diversity of expression profiles in the developing and mature central nervous system. Genome Biol. 2007;8:R173.
48. Kapadia R, Yi JH, Vemuganti R. Mechanisms of anti-inflammatory and neuroprotective actions of PPAR-gamma agonists. Front Biosci. 2008;13:1813–26.
49. Ketting RF. The many faces of RNAi. Dev Cell. 2011;20:148–61.
50. Kiriakidou M, Tan GS, Lamprinaki S, De Planell-Saguer M, Nelson PT, Mourelatos Z. An mRNA m7G cap binding-like motif within human Ago2 represses translation. Cell. 2007;129:1141–51.
51. Krichevsky AM, King KS, Donahue CP, Khrapko K, Kosik KS. A microRNA array reveals extensive regulation of microRNAs during brain development. RNA. 2003;9: 1274–781.
52. Krutzfeldt J, Kuwajima S, Braich R, Rajeev KG, Pena J, Tuschl T, Manoharan M, Stoffel M. Specificity, duplex degradation and subcellular localization of antagomirs. Nucl Acid Res. 2007;35:2885–92.
53. Krutzfeldt J, Rajewsky N, Braich R, Rajeev KG, Tuschl T, Manoharan M, Stoffel M. Silencing of microRNAs in vivo with 'antagomirs'. Nature. 2005;438:685–9.
54. Kuhn DE, Nuovo GJ, Terry Jr AV, Martin MM, Malana GE, Sansom SE, Pleister AP, Beck WD, Head E, Feldman DS, Elton TS. Chromosome 21-derived microRNAs provide an etio-

logical basis for aberrant protein expression in human Down syndrome brains. J Biol Chem. 2010;285:1529–43.

55. Lagos-Quintana M, Rauhut R, Lendeckel W, Tuschl T. Identification of novel genes coding for small expressed RNAs. Science. 2001;294:853–8.

56. Lau NC, Lim LP, Weinstein EG, Bartel DP. An abundant class of tiny RNAs with probable regulatory roles in *Caenorhabditis elegans*. Science. 2001;294:858–62.

57. Lee I, Ajay SS, Yook JI, Kim HS, Hong SH, Kim NH, Dhanasekaran SM, Chinnaiyan AM, Athey BD. New class of microRNA targets containing simultaneous 5'-UTR and 3'-UTR interaction sites. Genome Res. 2009;19:1175–83.

58. Lee RC, Ambros V. An extensive class of small RNAs in *Caenorhabditis elegans*. Science. 2001;294:862–4.

59. Lee RC, Feinbaum RL, Ambros V. The *C. elegans* heterochronic gene lin-4 encodes small RNAs with antisense complementarity to lin-14. Cell. 1993;75:843–54.

60. Lee Y, Kim M, Han J, Yeom KH, Lee S, Baek SH, Kim VN. MicroRNA genes are transcribed by RNA polymerase II. EMBO J. 2004;23:4051–160.

61. Lennmyr F, Karlsson S, Gerwins P, Ata KA, Terént A. Activation of mitogen-activated protein kinases in experimental cerebral ischemia. Acta Neurol Scand. 2002;106:333–40.

62. Lewis BP, Burge CB, Bartel DP. Conserved seed pairing, often flanked by adenosines, indicates that thousands of human genes are microRNA targets. Cell. 2005;120:15–20.

63. Lim LP, Lau NC, Garrett-Engele P, Grimson A, Schelter JM, Castle J, Bartel DP, Linsley PS, Johnson JM. Microarray analysis shows that some microRNAs downregulate large numbers of target mRNAs. Nature. 2005;433:769–73.

64. Lipton P. Ischemic cell death in brain neurons. Physiol Rev. 1999;79:1431–568.

65. Liu DZ, Tian Y, Ander BP, Xu H, Stamova BS, Zhan X, Turner RJ, Jickling G, Sharp FR. Brain and blood microRNA expression profiling of ischemic stroke, intracerebral hemorrhage, and kainate seizures. J Cereb Blood Flow Metab. 2010;30:92–101.

66. Lusardi TA, Farr CD, Faulkner CL, Pignataro G, Yang T, Lan J, Simon RP, Saugstad JA. Ischemic preconditioning regulates expression of microRNAs and a predicted target, MeCP2, in mouse cortex. J Cereb Blood Flow Metab. 2010;30:744–56.

67. Magill ST, Cambronne XA, Luikart BW, Lioy DT, Leighton BH, Westbrook GL, Mandel G, Goodman RH. microRNA-132 regulates dendritic growth and arborization of newborn neurons in the adult hippocampus. Proc Natl Acad Sci USA. 2010;107:20382–7.

68. Mansfield JH, Harfe BD, Nissen R, Obenauer J, Srineel J, Chaudhuri A, Farzan-Kashani R, Zuker M, Pasquinelli AE, Ruvkun G, Sharp PA, Tabin CJ, McManus MT. MicroRNA-responsive 'sensor' transgenes uncover Hox-like and other developmentally regulated patterns of vertebrate microRNA expression. Nat Genet. 2004;36:1079–83.

69. Marais E, Genade S, Huisamen B, Strijdom JG, Moolman JA, Lochner A. Activation of p38 MAPK induced by a multi-cycle ischaemic preconditioning protocol is associated with attenuated p38 MAPK activity during sustained ischaemia and reperfusion. J Mol Cell Cardiol. 2001;33:769–78.

70. Margulies KB. MicroRNAs as novel myocardial biomarkers. Clin Chem. 2009;55:1897–99.

71. Martin MM, Buckenberger JA, Jiang J, Malana GE, Nuovo GJ, Chotani M, Feldman DS, Schmittgen TD, Elton TS. The human angiotensin II type 1 receptor 1166 A/C polymorphism attenuates microrna-155 binding. J Biol Chem. 2007;282:24262–9.

72. Mathonnet G, Fabian MR, Svitkin YV, Parsyan A, Huck L, Murata T, Biffo S, Merrick WC, Darzynkiewicz E, Pillai RS, Filipowicz W, Duchaine TF, Sonenberg N. MicroRNA inhibition of translation initiation in vitro by targeting the cap-binding complex eIF4F. Science. 2007;317:1764–7.

73. Mehta SL, Manhas N, Raghubir R. Molecular targets in cerebral ischemia for developing novel therapeutics. Brain Res Rev. 2007;54:34–66.

74. Mettimano M, Romano-Spica V, Ianni A, Specchia M, Migneco A, Savi L. AGT and AT1R gene polymorphism in hypertensive heart disease. Int J Clin Pract. 2002;56:574–7.

75. Meyuhas O. Synthesis of the translational apparatus is regulated at the translational level. Eur J Biochem. 2000;267:6321–30.

76. Mishima T, Mizuguchi Y, Kawahigashi Y, Takizawa T, Takizawa T. RT-PCR-based analysis of microRNA (miR-1 and −124) expression in mouse CNS. Brain Res. 2007;1131:37–43.

77. Miska EA, Alvarez-Saavedra E, Townsend M, Yoshii A, Sestan N, Rakic P, Constantine-Paton M, Horvitz HR. Microarray analysis of microRNA expression in the developing mammalian brain. Genome Biol. 2004;5:R68.

78. Moolman JA, Hartley S, van Wyk J, Marais E, Lochner A. Inhibition of myocardial apoptosis by ischaemic and beta-adrenergic preconditioning is dependent on p38 MAPK. Cardiovasc Drugs Ther. 2006;20:13–25.

79. Moretti F, Thermann R, Hentze MW. Mechanism of translational regulation by miR-2 from sites in the 5′ untranslated region or the open reading frame. RNA. 2010;16:2493–502.

80. Murry CE, Jennings RB, Reimer KA. Preconditioning with ischemia: a delay of lethal cell injury in ischemic myocardium. Circulation. 1986;74:1124–36.

81. Nakashima Y, Raines EW, Plump AS, Breslow JL, Ross R. Upregulation of VCAM-1 and ICAM-1 at atherosclerosis-prone sites on the endothelium in the ApoE-deficient mouse. Arterioscler Thromb Vasc Biol. 1998;18:842–51.

82. Nakazawa T, Kuriu T, Tezuka T, Umemori H, Okabe S, Yamamoto T. Regulation of dendritic spine morphology by an NMDA receptor-associated Rho GTPase-activating protein, p250GAP. J Neurochem. 2008;105:1384–93.

83. Nan X, Ng HH, Johnson CA, Laherty CD, Turner BM, Eisenman RN, Bird A. Transcriptional repression by the methyl-CpG-binding protein MeCP2 involves a histone deacetylase complex. Nature. 1998;393:386–9.

84. Napoli C, Lemieux C, Jorgensen R. Introduction of a chimeric chalcone synthase gene into *Petunia* results in reversible co-suppression of homologous genes in trans. Plant Cell. 1990;2:279–89.

85. Nozaki K, Nishimura M, Hashimoto N. Mitogen-activated protein kinases and cerebral ischemia. Mol Neurobiol. 2001;23:1–19.

86. O'Brien KD, Allen MD, McDonald TO, Chait A, Harlan JM, Fishbein D, McCarty J, Ferguson M, Hudkins K, Benjamin CD. Vascular cell adhesion molecule-1 is expressed in human coronary atherosclerotic plaques. Implications for the mode of progression of advanced coronary atherosclerosis. J Clin Invest. 1993;92:945–51.

87. Onteniente B, Rasika S, Benchoua A, Guégan C. Molecular pathways in cerebral ischemia: cues to novel therapeutic strategies. Mol Neurobiol. 2003;27:33–72.

88. Orom UA, Nielsen FC, Lund AH. MicroRNA-10a binds the 5′UTR of ribosomal protein mRNAs and enhances their translation. Mol Cell. 2008;30:460–71.

89. Ostroff LE, Fiala JC, Allwardt B, Harris KM. Polyribosomes redistribute from dendritic shafts into spines with enlarged synapses during LTP in developing rat hippocampal slices. Neuron. 2002;35:535–45.

90. Parker R, Sheth U. P bodies and the control of mRNA translation and degradation. Mol Cell. 2007;25:635–46.

91. Pasquinelli AE, Reinhart BJ, Slack F, Martindale MQ, Kuroda MI, Maller B, Hayward DC, Ball EE, Degnan B, Müller P, Spring J, Srinivasan A, Fishman M, Finnerty J, Corbo J, Levine M, Leahy P, Davidson E, Ruvkun G. Conservation of the sequence and temporal expression of let-7 heterochronic regulatory RNA. Nature. 2000;408:86–9.

92. Pietrzykowski AZ, Friesen RM, Martin GE, Puig SI, Nowak CL, Wynne PM, Siegelmann HT, Treistman SN. Posttranscriptional regulation of BK channel splice variant stability by miR-9 underlies neuroadaptation to alcohol. Neuron. 2008;59:274–87.

93. Pillai RS, Bhattacharyya SN, Artus CG, Zoller T, Cougot N, Basyuk E, Bertrand E, Filipowicz W. Inhibition of translational initiation by let-7 microRNA in human cells. Science. 2005;309:1573–6.

94. Pillai RS, Bhattacharyya SN, Filipowicz W. Repression of protein synthesis by miRNAs: how many mechanisms? Trends Cell Biol. 2007;17:118–26.

95. Place RF, Li LC, Pookot D, Noonan EJ, Dahiya R. MicroRNA-373 induces expression of genes with complementary promoter sequences. Proc Natl Acad Sci USA. 2008; 105:1608–13.

96. Potter WB, O'Riordan KJ, Barnett D, Osting SM, Wagoner M, Burger C, Roopra A. Metabolic regulation of neuronal plasticity by the energy sensor AMPK. PLoS One. 2010; 5:e8996.

97. Pratt A, MacRae I. The RNA-induced silencing complex: a versatile gene-silencing machine. J Biol Chem. 2009;284:17897–901.

98. Rahbek UL, Nielsen AF, Dong M, You Y, Chauchereau A, Oupicky D, Besenbacher F, Kjems J, Howard KA. Bioresponsive hyperbranched polymers for siRNA and miRNA delivery. J Drug Target. 2010;18:812–20.

99. Reinhart BJ, Slack FJ, Basson M, Pasquinelli AE, Bettinger JC, Rougvie AE, Horvitz HR, Ruvkun G. The 21-nucleotide let-7 RNA regulates developmental timing in *Caenorhabditis elegans*. Nature. 2000;403:901–6.

100. Richter JD, Sonenberg N. Regulation of cap-dependent translation by eIF4E inhibitory proteins. Nature. 2005;433:477–80.

101. Rodriguez A, Griffiths-Jones S, Ashurst JL, Bradley A. Identification of mammalian microRNA host genes and transcription units. Genome Res. 2004;14:1902–10.

102. Rogaev EI. Small RNAs in human brain development and disorders. Biochemistry. 2005;70:1404–7.

103. Rose BA, Force T, Wang Y. Mitogen-activated protein kinase signaling in the heart: angels versus demons in a heart-breaking tale. Physiol Rev. 2010;90:1507–46.

104. Schnall-Levin M, Zhao Y, Perrimon N, Berger B. Conserved microRNA targeting in *Drosophila* is as widespread in coding regions as in 3'UTRs. Proc Natl Acad Sci USA. 2010;107:15751–6.

105. Scholer N, Langer C, Döhner H, Buske C, Kuchenbauer F. Serum microRNAs as a novel class of biomarkers: a comprehensive review of the literature. Exp Hematol. 2010;38: 1126–30.

106. Schratt GM, Tuebing F, Nigh EA, Kane CG, Sabatini ME, Kiebler M, Greenberg ME. A brain-specific microRNA regulates dendritic spine development. Nature. 2006;439:283–9.

107. Schwarz D, Zamore P. Why do miRNAs live in the miRNP? Genes Dev. 2002;16:1025–31.

108. Sharp FR, Jickling GC, Stamova B, Tian Y, Zhan X, Liu D, Kuczynski B, Cox CD, Ander BP. Molecular markers and mechanisms of stroke: RNA studies of blood in animals and humans. J Cereb Blood Flow Metab. 2011;31:1513–31.

109. Si LM, Zhu S, Wu H, Lu Z, Wu F, Mo YY. miR-21-mediated tumor growth. Oncogene. 2007;26:2799–803.

110. Smirnova L, Gräfe A, Seiler A, Schumacher S, Nitsch R, Wulczyn FG. Regulation of miRNA expression during neural cell specification. Eur J Neurosci. 2005;21:1469–77.

111. Stenzel-Poore MP, Stevens SL, Xiong Z, Lessov NS, Harrington CA, Mori M, Meller R, Rosenzweig HL, Tobar E, Shaw TE, Chu X, Simon RP. Effect of ischaemic preconditioning on genomic response to cerebral ischaemia: similarity to neuroprotective strategies in hibernation and hypoxia-tolerant animals. Lancet. 2003;362:1028–37.

112. Steward O, Worley P. Local synthesis of proteins at synaptic sites on dendrites: role in synaptic plasticity and memory consolidation? Neurobiol Learn Mem. 2002;78:508–27.

113. Strauss WM, Chen C, Lee CT, Ridzon D. Nonrestrictive developmental regulation of microRNA gene expression. Mamm Genome. 2006;17:833–40.

114. Subramaniam S, Unsicker K. ERK and cell death: ERK1/2 in neuronal death. FEBS J. 2010;277:22–9.

115. Tan KS, Armugam A, Sepramaniam S, Lim KY, Setyowati KD, Wang CW, Jeyaseelan K. Expression profile of MicroRNAs in young stroke patients. PLoS One. 2009;4:e7689.

116. Tang F, Hajkova P, O'Carroll D, Lee C, Tarakhovsky A, Lao K, Surani MA. MicroRNAs are tightly associated with RNA-induced gene silencing complexes in vivo. Biochem Biophys Res Commun. 2008;372:24–9.

117. Van den Berg A, Mols J, Han J. RISC-target interaction: cleavage and translational suppression. Biochim Biophys Acta. 2008;1779:668–77.
118. Van der Krol AR, Mur LA, de Lange P, Mol JN, Stuitje AR. Inhibition of flower pigmentation by antisense CHS genes: promoter and minimal sequence requirements for the antisense effect. Plant Mol Biol. 1990;14:457–66.
119. Vasudevan S, Tong Y, Steitz JA. Switching from repression to activation: microRNAs can up-regulate translation. Science. 2007;318:1931–4.
120. Vasudevan S, Tong Y, Steitz JA. Cell-cycle control of microRNA-mediated translation regulation. Cell Cycle. 2008;7:1545–9.
121. Veldhuis WB, Derksen JW, Floris S, Van Der Meide PH, De Vries HE, Schepers J, Vos IM, Dijkstra CD, Kappelle LJ, Nicolay K, Bar PR. Interferon-beta blocks infiltration of inflammatory cells and reduces infarct volume after ischemic stroke in the rat. J Cereb Blood Flow Metab. 2003;23:1029–39.
122. Vemuganti R, Dempsey RJ. Carotid atherosclerotic plaques from symptomatic stroke patients share the molecular fingerprints to develop in a neoplastic fashion: a microarray analysis study. Neuroscience. 2005;131:359–74.
123. Wang S, Aurora AB, Johnson BA, Qi X, McAnally J, Hill JA, Richardson JA, Bassel-Duby R, Olson EN. The endothelial-specific microRNA miR-126 governs vascular integrity and angiogenesis. Dev Cell. 2008;15:261–71.
124. Wang GK, Zhu JQ, Zhang JT, Li Q, Li Y, He J, Qin YW, Jing Q. Circulating microRNA: a novel potential biomarker for early diagnosis of acute myocardial infarction in humans. Eur Heart J. 2010;31:659–66.
125. Wayman GA, Davare M, Ando H, Fortin D, Varlamova O, Cheng HY, Marks D, Obrietan K, Soderling TR, Goodman RH, Impey S. An activity-regulated microRNA controls dendritic plasticity by down-regulating p250GAP. Proc Natl Acad Sci USA. 2008;105:9093–8.
126. Weber M, Baker MB, Moore JP, Searles CD. MiR-21 is induced in endothelial cells by shear stress and modulates apoptosis and eNOS activity. Biochem Biophys Res Commun. 2010;393:643–8.
127. Witwer KW, Sisk JM, Gama L, Clements JE. MicroRNA regulation of IFN-beta protein expression: rapid and sensitive modulation of the innate immune response. J Immunol. 2010; 184:2369–76.
128. Wright MW, Bruford EA. Naming 'junk': human non-protein coding RNA (ncRNA) gene nomenclature. Hum Genomics. 2011;5:90–8.
129. Wu J, Xie X. Comparative sequence analysis reveals an intricate network among REST, CREB and miRNA in mediating neuronal gene expression. Genome Biol. 2006;7:R85.
130. Xu CC, Han WQ, Xiao B, Li NN, Zhu DL, Gao PJ. Differential expression of microRNAs in the aorta of spontaneously hypertensive rats. Sheng Li Xue Bao. 2008;60:553–60.
131. Yi JH, Park SW, Kapadia R, Vemuganti R. Role of transcription factors in mediating post-ischemic cerebral inflammation and brain damage. Neurochem Int. 2007;50:1014–27.
132. Yin KJ, Deng Z, Hamblin M, Xiang Y, Huang H, Zhang J, Jiang X, Wang Y, Chen YE. Peroxisome proliferator-activated receptor delta regulation of miR-15a in ischemia-induced cerebral vascular endothelial injury. J Neurosci. 2010;30:6398–408.
133. Yin KJ, Deng Z, Huang H, Hamblin M, Xie C, Zhang J, Chen YE. miR-497 regulates neuronal death in mouse brain after transient focal cerebral ischemia. Neurobiol Dis. 2010;38:17–26.
134. Younger ST, Corey DR. Transcriptional gene silencing in mammalian cells by miRNA mimics that target gene promoters. Nucl Acid Res. 2011;39:5682–91.
135. Zampetaki A, Kiechl S, Drozdov I, Willeit P, Mayr U, Prokopi M, Mayr A, Weger S, Oberhollenzer F, Bonora E, Shah A, Willeit J, Mayr M. Plasma microRNA profiling reveals loss of endothelial MiR-126 and other microRNAs in type 2 diabetes. Circ Res. 2010;107:810–7.
136. Zhou X, Ruan J, Wang G, Zhang W. Characterization and identification of microRNA core promoters in four model species. PLoS Comput Biol. 2007;3:e37.

Chapter 18
Neuroglobin: A Novel Target for Endogenous Neuroprotection

Zhanyang Yu, Ning Liu, and Xiaoying Wang

Abstract Augmentation of endogenous protective mechanisms has been thought to be promising strategies to develop new therapies against stroke. Neuroglobin (Ngb) is a tissue oxygen-binding globin that is highly and specifically expressed in brain neurons. Accumulating evidences have proved Ngb is a unique endogenous neuroprotective molecule against hypoxic/ischemic insults in cultured neurons and in stroke animals, which strongly suggest that development of pharmacological strategies that upregulate endogenous Ngb expression may lead to a novel therapeutic approach for stroke intervention. In this chapter, recent experimental findings from our laboratory and others in understanding Ngb's biological function, gene expression regulation, and neuroprotective mechanisms are summarized. We also propose strategies to identify small molecules that upregulate endogenous Ngb for neuroprotection against stroke and related neurological disorders. Briefly, the strategies comprise two translational features. First, we will establish both mouse and human Ngb gene activation reporter stable cell lines to ensure that only compounds capable of activating both mouse and human Ngb promoters will be selected for further testing. Second, validation is the key component of any single step in drug screening and therapeutic development processes.

1 Introduction

In the past 20 years, more than one hundred clinical trials for neuroprotectants against stroke by inhibiting or blocking a specific step in the ischemia response cascade have failed. However, enormous knowledge has been learnt from previous investigations. In particular, preconditioning studies have demonstrated that

Z. Yu • N. Liu • X. Wang, MD, PhD (✉)
Neuroprotection Research Laboratory, Departments of Neurology and Radiology,
Massachusetts General Hospital, Harvard Medical School,
149 13th Street, Room 2411A, Charlestown, MA 02129, USA
e-mail: wangxi@helix.mgh.harvard.edu

P.A. Lapchak and J.H. Zhang (eds.), *Translational Stroke Research*, Springer Series
in Translational Stroke Research, DOI 10.1007/978-1-4419-9530-8_18,
© Springer Science+Business Media, LLC 2012

activation of endogenous protective mechanisms can prevent or limit brain damage. Augmentation of these endogenous protective mechanisms could be more promising strategies to develop new therapies against stroke. Neuroglobin (Ngb) is a tissue oxygen-binding globin that is highly and specifically expressed in brain neurons [1]. Accumulating evidences have proved Ngb is an endogenous neuroprotective molecule against hypoxic/ischemic insults in cultured neurons and in stroke animals. Enhanced Ngb gene expression inversely correlates with the severity of histological and functional deficits after ischemic stroke [2–5]. Additionally, Ngb knockdown deteriorates outcome of hypoxic/ischemic brain injuries [6]. These experimental findings strongly suggest that development of pharmacological strategies that upregulate endogenous Ngb expression may lead to a novel therapeutic approach for stroke intervention. In this chapter, we will review recent experimental findings from our laboratory and others in understanding Ngb's biological function, gene expression regulation, and neuroprotective mechanisms. Additionally, developing strategies to identify small molecules that upregulate endogenous Ngb for neuroprotection against stroke and related neurological disorders are also discussed.

2 Part 1: Neuroprotective Roles and Mechanisms of Neuroglobin

Globins are oxygen-binding proteins that widely exist and play important physiological roles in many taxa including bacteria, plant, fungi, and animal. To date, four members of globin family have been identified, including hemoglobin, myoglobin, cytoglobin, and Ngb. Among them, hemoglobin and myoglobin are the best-known globin proteins. In 2000, Ngb was for the first time identified as a new globin family member that is highly expressed in brain neurons [1]. Ngb is a 151 amino acid protein in both mouse and human with protein sequence identity of 94%, but they share less than 25% identity with other vertebrate globins. Besides brain neurons, Ngb is also highly expressed in peripheral nervous system, endocrine tissues, and retina [7]. In the past 11 years, a large array of experimental studies has extensively investigated the functions and mechanisms of Ngb, which are summarized in several nice review articles [8–13], though our knowledge on this regard remains limited. In Part 1 of this chapter, we will first briefly introduce Ngb gene and mainly focus on reviewing the emerging experimental findings of Ngb's neuroprotective effects and mechanisms against stroke and other neurodegenerative disorders.

2.1 Tissue-Specific Expression of Neuroglobin Gene

The tissue-specific expression pattern of Ngb gene is indicative of its physiological function. In situ hybridization showed that Ngb mRNA was widely distributed throughout the adult rat brain, including cerebral cortex, hippocampus, and subcortical structures such as thalamus, hypothalamus, olfactory bulb, and cerebellum

[14–16]. The distribution of Ngb protein is consistent with its mRNA localization, and the subcellular immunoreactivity is restricted to the cytoplasm.

Among all Ngb-expressing cells, the highest expression is seen in retina, with the estimated concentration about 100-fold higher than in the brain [17]. Ngb mRNA was detected in the perikarya of the nuclear and ganglion layers of the neuronal retina, whereas the protein was present mainly in the plexiform layers and in the ellipsoid region of photoreceptor inner segment [18]. The distribution of Ngb correlates with the subcellular localization of mitochondria and with the relative oxygen demands. These findings suggest that Ngb supplies oxygen to the retina, similar to myoglobin in the myocardium and the skeletal muscle. Although Ngb concentration in the brain is relatively lower than in the retina, considering neuron is the major cell type in the brain specifically expressing Ngb, thus Ngb may be a unique molecule that plays certain roles in maintaining normal neuron function and protectively responding to pathological insults.

2.2 Regulation of Neuroglobin Gene Under Pathological Conditions

Since its discovery over a decade ago, Ngb's structure, biochemical properties, and its expression in cerebral neurons suggested it plays a role in providing oxygen to the brain. Its neuronal expression is upregulated under hypoxic/ischemic conditions both in cultured cells [19, 20] and in stroke animal brains [2, 6, 21, 22]. Ngb is also upregulated in the cerebellum of mouse pups in response to hypoxic-ischemic insults caused by maternal seizures during intrauterine life [23]. More importantly, a recent report showed that Ngb expression is increased in the cortical peri-infarct region in stroke patients, suggesting its clinical relevance for endogenous neuroprotection [24]. In contrast to acute hypoxic conditions, a chronic hypoxia (10% oxygen for 14 days) did not increase Ngb gene expression in mRNA or protein level in mouse [25]. Moreover, 2-h exposure of mice to 7.6% oxygen did not upregulate brain Ngb either [18]. However, others reported opposite results that housing rats in 10% oxygen for up to 14 days upregulated Ngb mRNA in the rat brain [26]. Thus, there might also be species-dependent differences for neurglobin's responses to hypoxic conditions. In addition to hypoxia/ischemia, aging might also be a factor influencing Ngb's expression. Sun et al. have demonstrated that Ngb's expression level was decreased to about a half in aged rats (24 months) compared to young ones (3, 12 months) in various regions of brain, thus implying the pathophysiological importance of Ngb in aging-related neurodegenerative diseases [27].

2.3 Neuroprotective Roles of Neuroglobin Against Hypoxic/Ischemic and Oxidative Stresses

The oxygen-binding and neuron-specific expression properties make Ngb a new target molecule for study of neuronal hypoxia/ischemia pathophysiology and development of new neuroprotective agent. Gene expression alteration approaches were applied to address the neuroprotective effect of Ngb. The first report in this category by Sun et al. showed that antisense-mediated knockdown of Ngb rendered cortical neurons more vulnerable to hypoxia, whereas Ngb overexpression conferred protection of cultured neurons against hypoxia [2]. Similar effect was observed in neuroblastoma cell line SH-SY5Y that Ngb overexpression enhanced cell survival under conditions of anoxia or oxygen and glucose deprivation (OGD) [28]. In animal stroke models, intracerebral administration of a Ngb-overexpressing adeno-associated virus vector significantly reduced infarct size in rats following middle cerebral artery occlusion (MCAO), and the outcome was reversed when Ngb antisense oligonucleotide was applied [6]. Using Ngb-overexpressing transgenic (Ngb-Tg) mice, Khan et al. found that the cerebral infarct size after MCAO was reduced by approximately 30% compared to wild type [29]. Our lab also generated a Ngb-overexpressing transgenic mouse line and tested its neuroprotective effects in transient focal cerebral ischemia [5]. Our results were broadly consistent with an earlier report from Dr. Greenberg's group [29], and further documented that reduction of brain infarction in Ngb-overexpressing transgenic mice can be sustained up to 14 days after ischemia compared to wild-type controls, suggesting that Ngb overexpression is neuroprotective against transient focal cerebral ischemia, although the involved mechanisms need to be further characterized. We need to emphasize that above experiments with nontissue-specific Ngb transgenic approaches are "outcome" studies; all findings are very informative for the effects of Ngb upregulation in stroke, but have limitations to fundamentally define or interpret the role of endogenous Ngb, therefore a neuron-specific and inducible Ngb knockdown approach would be very useful in further investigations of Ngb's function in normal vs. ischemic brain.

Previous studies on the neuroprotective effects of Ngb are mostly based on models using transgenic overexpression or Ngb-viral delivery. However, for stroke therapy development, it would be more practical and easily applicable if high concentration of neuroglobin protein can be delivered into brain tissue after stroke. A very recent report by Cai et al. [30] presented new evidences toward this direction. They delivered Ngb protein into mouse brain tissue using the 11-amino-acid human immunodeficiency virus transactivator of transcription (TAT) protein transduction domain. The results showed that brain damages were protected from ischemia stroke by both pretreatment of TAT-Ngb or posttreatment right after reperfusion onset in a mild MCAO mouse model, but no beneficial outcome was observed if the TAT-Ngb was administered 2 h after ischemia onset. This study suggests that Ngb overexpression might be beneficial for early stroke treatment, as well as for stroke intervention to individuals with higher stroke risk.

2.4 Neuroprotective Effects of Neuroglobin Against Other Neurological Disorders

As a brain-specific oxygen-binding protein, it is not surprising that Ngb can also be protective against other models for neurodegenerative disorders. For example, Ngb overexpression was shown to be protective against A-beta and NMDA toxicity in both cultured neurons and in Alzheimer's disease (AD) transgenic mice in [31, 32].

2.5 Molecular Mechanisms of Neuroglobin's Neuroprotection

Accumulating experimental studies have demonstrated that Ngb is protective against hypoxic/ischemic brain injury [5, 29]. Although the underlying mechanisms remain poorly defined, initial evidences suggest that the neuroprotective effect of Ngb may be largely linked to its structural features in terms of O_2 and NO binding. Furthermore, putative signal transduction and mitochondrial function preservation may also be involved in the protective mechanisms.

2.5.1 Oxygen Sensing and ROS Scavenging by Neuroglobin: Related to Its Structural Features and Ligand-Binding Properties

Ngb protein in human or mouse exists as a monomer, which is distinct from the heterotetrameric hemoglobins. The 3D structure of human [33] or mouse Ngb [34] has been solved, showing that the heme is inserted into the protein in two different orientations. The lack of orientation selectivity is possibly related to the presence of a large cavity lining the heme and to the increased mobility of heme contacts [34]. Spectroscopic and kinetic experiments indicate that human Ngb displays a typical globin fold and the heme-iron is hexacoordinate [14], with proximal HisF8 and distal HisE7 that provide the two axial coordination bonds. Binding of exogenous ligand such as O_2 and CO displaces the endogenous HisE7 heme distal ligand. An elongated protein matrix cavity in the 3D structure would facilitate O_2 diffusion to the heme [35]. Ngb was originally thought to function in O_2 storage and transportation similarly to hemoglobin; however, the fact that Ngb exerts a high O_2 binding rate and low O_2 dissociation rate, plus the relatively low Ngb protein concentration (~1 μM) in the brain [9], has cast doubt on these function in general, but instead it may function in O_2 sensing [36, 37].

A number of studies have indicated that specific binding property of Ngb makes this molecule neuroprotective against hypoxic or ischemic injury via scavenging reactive species, because Ngb has long been found to bind to NO directly with high intrinsic affinity and low dissociation rate [38]. In support of this function, a high degree of colocalization of neuronal nitric oxide synthase (nNOS) and Ngb has been detected within anterior basomedial amygdala (BMA), lateral hypothalamus, and laterodorsal tegmental nucleus (LDTg) [39], implying that in these neurons NO could be the endogenous ligand for Ngb. Furthermore, Brunori et al. [40] found that the oxygenated derivative of Ngb, Ngb-O_2, reacts with NO rapidly to produce NO_3^- and met-Ngb. This pathway would dispose of NO by means of a rapid reaction with Ngb-O_2, which may in turn protect cellular respiration jeopardized by the inhibitory effect of NO on cytochrome c oxidase activity [41, 42]. Another recent study by

Tsio et al. [43] revealed that human Ngb can function as a nitrite reductase, therefore being able to inhibit cellular respiration via NO binding to cytochrome c oxidase. Additional studies showed that the six-to-five-coordinate status of Ngb regulates intracellular hypoxic NO-signaling pathways. Overall, those studies suggest that Ngb may function as a physiological oxidative stress sensor and a posttranslationally redox-regulated nitrite reductase that generates NO under six-to-five-coordinate heme pocket control.

Dr. Greenberg's group has extensively studied the effects of different oxidative challenges on cultured neurons with and without Ngb overexpression. Ngb-transfected HN33 neuroblastoma cells were less sensitive to NO-induced cell death compared to wild-type cells, suggesting the ability of Ngb for neutralizing the neurotoxic effects of reactive nitrogen species [44]. Ngb overexpression conferred protection in SH-SY5Y cells directly injured by H_2O_2 [45]. Beta-amyloid-induced cytotoxicity to PC12 cells, marked by reactive oxygen species production and lipids peroxidation, was ameliorated by Ngb overexpression [27]. Findings from above experiments suggest that Ngb may have the function of reactive oxygen species scavenging.

2.5.2 Regulation of Signal Transduction

In addition to the structural features and ligand-binding properties described above, Ngb has also been hypothesized to act as a signal transducer. Dr. Morishima's lab found that ferric human Ngb (met-Ngb) binds to the GDP-bound state of the α subunit of the heterotrimeric G protein (Gα) and exerts guanine-nucleotide dissociation inhibitor (GDI) activity [46]. The ferric Ngb inhibits the exchange of GDP for GTP, thus prevents the Gα subunit from binding to the G$\beta\gamma$ complex and activates the downstream signal transduction pathway, which is protective against oxidative stress [47]. In a continuous study, the binding sites between Gα subunit and Ngb were identified [48]. This hypothesis was additionally supported by the observation that the GDI activity of human Ngb is required for its neuroprotection for PC12 cells under oxidative stress conditions. In this study, the human Ngb that retained GDI activities rescued PC12 cell death caused by hypoxia/reoxygenation. In contrast, zebrafish Ngb and human Ngb mutants, which did not function as GDI, did not rescue cell death, suggesting the importance of the GDI activity in Ngb's neuroprotection [49].

Khan et al. recently showed evidence that Ngb binds two members of the Rho GTPase family, Rac1 and Rho A, as well as the Pak1 kinase, a key regulator of actin assembly and Rho-GDI-GTPase signaling complex activity under hypoxia. They thereafter hypothesized that Ngb may play a neuroprotective role by inhibiting the dissociation of the GTPase Rac-1 from its endogeneous GDI and resulting in preventing the hypoxia-induced actin polymerization and microdomain aggregation [50]. Moreover, Ngb was also found to be interacting with other targets such as flotillin-1 (a lipid raft microdomain-associated protein) [51] and the cysteine protease inhibitor cystatin C [52], suggesting a possibility that Ngb modulates the intracellular transport of cystatin C to protect against neuronal death caused by oxidative stress.

Based on those findings and together with the kinetics study of Ngb reaction with O_2 and NO, Brunori et al. proposed that Ngb might function as a sensor of the relative O_2 and NO concentrations in the tissue [40]. A supportive evidence is that Ngb oxygenation is quickly reversible, and the oxygenated derivative, Ngb-O_2, reacts rapidly with NO to produce NO_3^- and met-Ngb. This process competes effectively with direct formation of Ngb-NO, which excludes the production of met-Ngb and its protective signaling function as a GDI [46].

The proper function of Ngb requires a met-Ngb reductase to maintain the balance between redox and oxidized Ngb. Complying with this scenario, Dr. Brunori's lab recently opened another direction in exploring Ngb's functioning mechanism based on findings that NADH:flavorubredoxin oxidoreductase (FlRd-red) from *E. coli* is able to slowly reduce Ngb at catalytic concentrations [53]. More interestingly, a BLAST search of the human genome showed that *E. coli* NADH:flavorubredoxin oxidoreductase shares significant similarity with apoptosis-inducing factor (AIF), the principal mediator of the so-called caspase-independent programmed cell death [54]. In healthy cells, AIF is located within the mitochondrion, but upon permeabilization of mitochondrial outer membrane, it translocates first to the cytosol and then to the nucleus, where it triggers chromatin condensation followed by massive DNA fragmentation [54]. It is therefore possible that AIF may reduce cytoplasmic met-Ngb on its way from the mitochondrion to nucleus, and depending on O_2 tension, the reduced Ngb can either interfere with classical apoptotic pathway by reducing ferric Cytochrome c (Cyt c) [55] or involve in NO scavenging [40].

Previous studies have suggested Ngb is very likely to function in multiple pathways leading to its neuroprotection role, as described above. Based on this, our laboratory has hypothesized that Ngb may also play a role in regulating genes' expression in response to hypoxia/OGD; we therefore performed a microarray screening to examine the effect of Ngb overexpression on the expression of hypoxic-response genes in mouse cortical neurons. We showed that 20 genes were down-regulated at early phase of OGD/Reoxygenaton in wild-type neurons, while 12 of them were no longer significantly changed in Ngb-overexpressing neurons [56]. These genes are broadly involved in neuronal function and survival, indicating that Ngb may play roles in multiple cell survival signaling pathways.

Another example of Ngb's involvement in signal transduction was found in animal model of Alzheimer Disease (AD). Previous study has shown that Ngb is able to attenuate neurotoxicity induced by beta-amyloid [32]. Recently, Chen et al. [57] looked into the molecular mechanisms involved in this effect. They found that Ngb overexpression attenuates tau hyperphosphorylation, a characterized pathological hallmark of AD brains, probably through activating Akt signaling pathway. One should notice that almost all of these tests are "correlation" studies; the impact of Ngb on these signaling pathways might be direct or indirect. Direct and causative evidence is still largely lacking. Nonetheless, Ngb has been demonstrated to be protective in AD and could serve as a therapeutic target.

2.5.3   Maintenance of Mitochondrial Function

It has been well documented that Ngb expression is confined to metabolically active, oxygen-consuming cell types [10]. At the subcellular level, Ngb is associated with mitochondria and thus linked to the oxidative metabolism [58]. Mitochondria plays key roles in energy production, ROS homeostasis, and cell death signaling. Mitochondria responds to various insults to cells and its dysfunction is associated with a large variety of clinical phenotypes. It has been demonstrated that mitochondria comprises a central locus for energetic perturbations and oxidative stress in hypoxia/ischemia [59, 60]. Experimental studies have shown that overexpression of Ngb promotes cell survival of PC12 cells against beta-amyloid toxicity and attenuates beta-amyloid-induced mitochondrial dysfunction [61] and eliminates hypoxia-induced mitochondrial aggregation and neuron death [50]. Our lab has also demonstrated that Ngb overexpression improves mitochondrial function and reduces oxidative stress in Ngb-overexpressing neurons after hypoxic insult in cultured mouse cortical neurons [62]. We found that at earlier time points after hypoxia/reoxygenation, no difference in neurotoxicity can be detected between Ngb-overexpressing and wild-type control neurons, whereas the rates of decline of several mitochondria functions biomarkers, including ATP levels, 3-(4,5-dimethyl-2-thiazolyl)-2,5-diphenyl-2H-tetrazolium bromide (MTT) reduction, and mitochondrial membrane potential, were significantly ameliorated in Ngb-transgenic neurons compared to wild-type neurons. But at late time point, there was a significant reduction of neurotoxicity in Ngb-overexpressing neurons. Furthermore, Ngb overexpression reduced superoxide anion generation after hypoxia/reoxygenation, but glutathione levels were significantly improved compared to wild-type controls. Our data here suggested that Ngb might affect both mitochondrial function and free radical generation as its potential neuroprotective mechanisms. However, there are multiple and probably inextricable feedback loops between preservation of mitochondrial energetics vs. direct radical scavenging [60, 63–65]. We acknowledge that it will likely be impossible to unequivocally separate mitochondrial effects vs. reactive oxygen species' effects of Ngb.

Hypoxia and OGD result in mitochondrial depolarization [66]. The mitochondrial permeability transition pore (mPTP) is a protein pore formed across the inner and outer membrane of the mitochondria under pathological conditions such as stroke. In response to hypoxia/ischemia, mPTP opening cause release of Cyt c from mitochondria to cytosol [67], followed by activation of caspase-dependent or -independent apoptosis pathways [68–70]. Studies in our lab have shown that Ngb overexpression is correlated with reduced mPTP opening and decreased Cyt c release as well (unpublished data). This suggests an inhibitory role of Ngb in OGD-induced mPTP opening, which has been thought to be one of the major causes of cell death in a variety of tissue ischemic damage, as occurs in heart attack and stroke, thus Ngb's inhibitory effect in mPTP opening may be an important mechanism of Ngb's neuroprotection.

To further dissect the molecular mechanisms of Ngb's neuroprotection, our laboratory recently investigated protein interaction partners of mouse Ngb by performing yeast two-hybrid assay using a mouse brain cDNA library. We identified several candidates that bind to Ngb, including Na/K ATPase beta 1, cytochrome c1, ubiquitin C, voltage-dependant anion channel (VDAC), and a few more (unpublished data). Among these Ngb-binding protein candidates, some of them are biologically

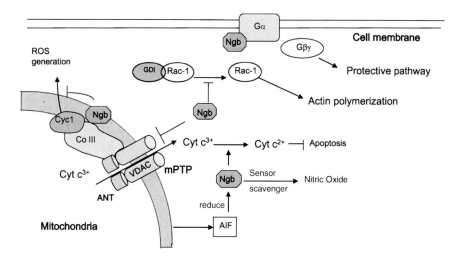

Fig. 18.1 Potential molecular mechanisms of Ngb's neuroprotection by modulating cell death/survival signaling pathways. Ngb may function as an O_2 and NO sensor. Ngb has guanine-nucleotide dissociation inhibitor (GDI) activity and can prevent $G\alpha$ from binding to $G\beta\gamma$ complex and activates the downstream signaling pathway. Ngb may inhibit the dissociation of the GTPase Rac-1 from its endogeneous GDI, thus preventing the hypoxia-induced actin polymerization and microdomain aggregation. Ngb may interact with VDAC and inhibit hypoxia/OGD-induced mPTP opening and Cyt c release from mitochondria. Ngb could convert Cyt c^{3+} to Cyt c^{2+}, and subsequently interfere apoptotic signaling cascades or scavenge nitric oxide. Ngb may interact with Cyc1 and inhibit hypoxia/OGD-induced ROS generation by mitochondria complex III

important for neuronal function and survival. For example, cytochrome c1 is a subunit of the cytochrome bc1 complex (mitochondria complex III), which plays an important role in mitochondria function for energy transduction and generation of superoxide anion [71], and it also plays pathological roles in response to oxidative stress [72, 73] and regulates hypoxia-inducible-factor-1 activation induced by hypoxia [74–76]. Cytochrome c1 is localized in the intermembrane space between the outer and inner membrane of mitochondria [77]. The mitochondrial outer membrane contains the protein "porin," which forms an aqueous channel through which proteins up to 10 kd can pass and go into the intermembrane space. It has been known that hypoxia-induced superoxide, as well as apoptotic signaling molecules such as Bax, may cause permeabilization of mitochondrial outer membrane [78, 79]. Thus, as a 16 kd monomer, Ngb might be able to pass the outer membrane to bind and affect the function of cytochrome c1. Additionally, results from our laboratory showed that Ngb overexpression is able to decrease OGD-induced mitochondria permeability transition pore (mPTP) opening and Cyt c release from mitochondria. The interaction between Ngb and VDAC, an mPTP component, is supportive of Ngb's effect in mPTP. However, the phenomenon of Ngb binding to other proteins and binding status-correlated alteration of associated cell signaling and mitochondrial function requires further investigation. The major hypotheses about Ngb-involved signaling pathways are summarized in Fig. 18.1.

In summary, Ngb is a newly discovered globin member, but its functions in both physiological and pathological conditions remain poorly understood. Emerging experimental studies suggest that Ngb is neuroprotective against hypoxic/ischemic neuronal and brain injuries, but the mechanisms have not been clearly defined. As an O_2-binding globin protein, the traditional O_2 storage and transport activity seem not realistic for Ngb. Instead, Ngb is more likely to act as a scavenger of reactive species, as well as a signal transduction molecule involved in neuroprotective pathways. Ngb might also be closely related to mitochondria function. There are a number of proposed theoretical concepts based on the molecular property of Ngb and experimental data obtained from cell-free systems, but they have rarely been tested in cell cultures and animal models. Further investigations in Ngb's biological functions and neuroprotective mechanisms are highly warranted. Most importantly, emerging data have clearly documented that upregulating or forcing Ngb expression is neuroprotective against hypoxic/ischemic brain injury [5, 29]. Therefore, targeting Ngb for endogenous neuroprotection would be translationally significant. We need to seek strategies of identifying small molecules that elevate endogenous Ngb, which would help us in the development of new intervention and therapy approaches for stroke and other related neurological disorders.

3   Part 2: Targeting Neuroglobin for Endogenous Neuroprotection

3.1   Targeting Endogenous Neuroprotective Molecules

Endogenous neuroprotective molecule or transcription factors in response to stroke damage or related disorders have been thought as new therapeutic targets. The most common approach is cell-based high-throughtput small molecule screening for discovering gene regulation-related neuroprotective drugs. For example, a large-scale chemical screen for regulators of the arginase 1 promoter identified the soy isoflavone daidzein as a neuroprotectant [80]. A cell-based HIF ODD-luciferase reporter system that monitored stability of the overexpressed luciferase-labeled HIF PHD substrate was used for screening hypoxic-adapted drugs [81]. One more example, Neh2-luciferase reporter system was recently developed for high-throughout screening Nrf2 activators that may be beneficial for treatment and prevention of chronic neurodegenerative diseases [82]. Activation of NF-κB was also targeted in a cell-based high-throughput screening to identify neuroprotectants [83]. A number of identified small molecules are being tested in preclinical animal studies, some of them initially showed promising results [80, 84, 85].

As reviewed in Part 1 of this chapter, Ngb has been thought to be a novel endogenous neuroprotective molecule [2, 11, 86]. Since Ngb is an intracellular protein, delivery of exogenous Ngb protein is generally considered unfeasible as a therapy, especially for treating CNS disorders such as stroke. Thus, seeking small molecules capable of upregulating endogenous Ngb may lead to development of new approaches for intervention and treatment of stroke and related neurological disorders. Indeed, recent reports showed that Ngb can be upregulated by certain neuroprotective compounds, such as Valproic acid, Neuro-EPO, and 17β-estradiol [87–89], but the causality of upregulated Ngb to their neuroprotective properties was undefined. However, those studies at least indicate that strategies for seeking small molecules for upregulating Ngb are feasible. Our laboratory is now working on this direction to establish a cell-based screening system for identifying small molecules capable of upregulating cellular Ngb expression, and thereafter to develop a new therapeutic intervention to stroke and related neurological disorders. In the following, we will discuss strategies for Ngb-upregulating compound screening, validation, and preclinical evaluation.

3.2 Identifying Compounds That Upregulate Neuroglobin for Endogenous Neuroprotection

Any identified molecules that increase Ngb expression may potentially enhance endogenous neuroprotective mechanisms, and these compounds might potentially be further developed to be novel therapeutic agents for intervention of ischemia stroke. Since cell-based high-throughput drug screening has become a highly promising method for drug discoveries [90], it would be a useful tool to achieve our goals.

Cell-based high-throughput drug screening is a relatively cost- and time-effective method that allows simultaneous screening of hundreds to thousands of small molecule compounds; its powerful and convenient features have been highly recommended in primary screening of compounds or drug candidates. A typical high-throughput screening program integrates several critical discovery or expertise groups: target identification (genomics and molecular biology groups); reagent preparation (compound preparation and purification groups); and compound management, assay development, and high-throughput library screening (discovery groups) [91].

Cell-based high-throughput screening has been recently used and successfully discovered compounds for upregulating intracellular proteins. For example, Thong C. Ma et al. used a cell-based high-throughput screening to identify activators of the arginase 1 promoter and found soy isoflavone daidzein as a potent neuroprotectant [80]. Another example is that Jill Jarecki et al. developed a high-throughput screening strategy to identify small molecules that increase intracellular protein SMN2 promoter activity for treating genetic disease spinal muscular atrophy (SMA) [92]. Generally, the common high-throughput screening strategy may have two key steps or components: (1) Establishing biological detection system; (2) Screening and validating compounds. In real experiments, the screening strategy would be slightly varied in some involved technologies based on different biological characterizations of targeting molecules [93]. Below we briefly discuss our screening strategies for identifying Ngb activators.

3.2.1 Characterization of Neuroglobin Promoter and Development of Reporter Constructs

Cell-based reporter assay for high-throughtput screening promoter activators may enable us to rapidly and efficiently identify promising compound candidates. For the first step, development of a reporter constructs becomes a major task, to which promoter analysis of targeting gene would be an important research basis for constructing this reporter system. Experimental reports that selected and utilized promoters for successfully screening targeted arginase 1 and SMA genes could serve as examples for others to follow [94–97]. Regarding Ngb gene, a recent study identified a 2.0 kb promoter region for human Ngb [98]. However, functional analysis and validation on this human Ngb promoter under physiological condition are yet to be well clarified. Importantly, mouse Ngb promoter analysis has not been studied yet. Thus, based on available knowledge, to identify the promoter region for constructing the reporter system of cell-based assay, functional analysis of Ngb promoter is necessary and critical. However, we have not established specific and commonly accepted criteria for selecting the gene promoter. Published studies usually used around 2 kb or longer fragment upstream of the targeting gene transcription site as promoter region. For example, a 3.4 kb DNA fragment from the *SMN2* gene was used to identify small molecule compounds that upregulates intracellular SMN2 protein [92]; a 1,959-bp fragment of the human KLF5 promoter upstream of the ATG start codon was cloned to identify novel small molecule compounds that inhibit the proproliferative Kruppel-like factor 5 expression [99]; a 4.8 kb of DNA upstream of the Arg1 transcription site was selected to identify neuroprotective drugs that increase Arginase 1 expression [80]. According to the above studies, it seems that the promoter may need to meet the following requirements: (1) Contains core promoter region; (2) Selected gene promoter activity in response to known positive stimulations must be similar to Ngb gene expression of cells or tissues and validated by RT-PCR and western blot analyses.

Accordingly, we have proposed working protocol to identify Ngb gene promoter. Briefly, a series of 5′-deletion fragments of both mouse and human Ngb promoter upstreams of the transcription start sites will be constructed and their promoter activities will be further examined upon reported known Ngb gene activators, such as hypoxia/ischemia, hemin [2, 100], and compounds such as Valproic acid, 17β-estradiol, and Neuro-EPO [87–89]. These series of examinations will help us to select the most suitable mouse or human Ngb promoter. Then the selected promoter region will be genetically engineered. In brief, mouse or human Ngb promoters will be constructed into a commercially available reporter plasmid which has already incorporated with a luciferase gene; verification of above proposed constructs can be checked and confirmed by direct sequencing.

3.2.2 Establishment and Validation of Both Mouse and Human Stable Cell Lines

The standard drug development procedure is cell assay—animal model—human clinical trial. From a translational perspective, an ideal compound would be able to induce or boost endogenous neuroprotective molecules in both animal and human neurons and brains. Therefore, we plan to establish both mouse and human cell-based reporter systems for Ngb up-regulation compound screening. The first step is to construct Ngb reporter plasmids, in which the luciferase activity is controlled by mouse or human Ngb promoter and is quantitatively measurable. Then the next step is to establish two separate stable cell lines for mouse Ngb promoter activation reporter and human Ngb promoter activation reporter systems, respectively. As non-dividing primary neurons are not suitable for establishing stable cell lines, another important issue is how to select proper human and mouse neuron-like cell lines. The most commonly used cell lines on this regard are human neuroblastoma cell line SH-SY5Y and mouse neuroblastoma cell line N2a. This is because neuroblastoma cell lines could be differentiated with morphological and biochemical characteristics of mature neurons, thus would be suitable as in vitro models for neuroscience studies [101]. Indeed, they are wildly used for establishing stable cell lines [102, 103]. To further determine whether SHSY5Y and N2a cells are ideal models for establishing stable cell lines, both cell lines need to be validated by comparing with primary neurons for Ngb expression patterns in response to the reported known positive stimulations. Next, mouse and human Ngb promoter constructs will be stably integrated into the genome of mouse neuroblastoma or human neuroblastoma cell lines by following standard protocols [80]. To characterize and validate the reporter clones, standard PCR will be performed to confirm presence of an integrated reporter construct. Then, selected stably transfected clones will be tested in basal luciferase expression and inducing capability with the reported known positive stimulations for validation.

3.2.3 Compound Screening and Pharmacological Testing

Upon validation of the selected mouse and human Ngb up-regulation reporter cell lines, compound screening can be performed on a cell-based high-throughput assay system. Briefly, to increase assay throughput, it has been recommended to use 96 or 384-well plate format [104]. To evaluate the quality or performance of our systems, a small subset screen will be performed. Data is initially generated using a positive and negative control. The assay conditions would be optimized in order to meet the assay validation of $Z' < 1$ ($Z' = 1 - [3(\text{standard deviation of signal}) + 3(\text{standard deviation of the baseline})/\text{signal} - \text{baseline}])$, upon testing stimulations of reported known Ngb activators [105]. After optimizing screening condition, the next step is to identify drugs that can induce Ngb promoter activity in our stable cell lines.

Once certain compound candidates are identified from the cell-based high-throughput assay systems, the following validation assessments need to be considered. (1) Only the identified compounds that can upregulate both mouse and human Ngb gene promoter activities will be selected for further investigations; (2) To estimate both the potency and efficacy of selected compounds, EC50 (effective concentration) should be tested [106]; (3) To determine whether changes of Ngb reporter activities truly reflect endogenous Ngb mRNA or protein expression alterations,

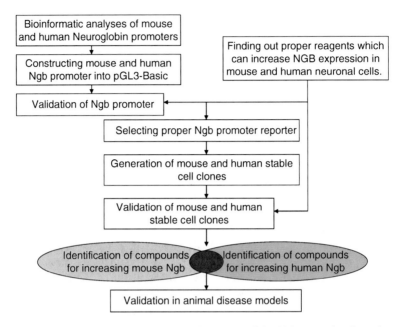

Fig. 18.2 Strategies of identifying compounds that boost cellular Ngb expression for endogenous protection. The following schematic outline summarizes our developmental strategies that comprise two translational features. First, we will establish both mouse and human Ngb gene activation reporter stable cell lines to ensure that only compounds capable in activating both mouse and human Ngb promoters will be selected for further testing. Second, validation is the key component of any single step in drug screening and therapeutic development processes

effects of selected compounds in Ngb mRNA or protein levels of cultured cell lines and primary neurons need to be examined by quantitative RT-PCR and western blot for validation; (4) To test whether the upregulating Ngb capability of selected compounds is truly neuroprotective to hypoxic/ischemic insults, we need to validate neuroprotective effects of those compounds against neuronal injury induced by OGD (oxygen-glucose deprivation) or other insult models in neuron cultures; (5) Lastly, we need to validate that the selected compounds are capable of increasing Ngb expression in the brain neurons and protect against cerebral ischemia in mouse models. We are aware that all efficacy and safety aspects to a new drug need to be fully evaluated before a clinical test can be applied [107].

In summary, the strategies we have proposed comprise two translational features. First, we will establish both mouse and human Ngb gene activation reporter stable cell lines to ensure that only compounds capable of activating both mouse and human Ngb promoters will be selected for further testing. Second, validation is the key component of any single step in drug screening and therapeutic development processes. A schematic outline of Fig. 18.2 summarizes our developmental strategies for identifying compounds that upregulate Ngb expression for endogenous neuroprotection.

References

1. Burmester T, Weich B, Reinhardt S, Hankeln T. A vertebrate globin expressed in the brain. Nature. 2000;407:520–3.
2. Sun Y, Jin K, Mao XO, Zhu Y, Greenberg DA. Ngb is up-regulated by and protects neurons from hypoxic-ischemic injury. Proc Natl Acad Sci USA. 2001;98:15306–11.
3. Peroni D, Negro A, Bahr M, Dietz GP. Intracellular delivery of Ngb using HIV-1 tat protein transduction domain fails to protect against oxygen and glucose deprivation. Neurosci Lett. 2007;421:110–4.
4. Hundahl C, Kelsen J, Kjaer K, Ronn LC, Weber RE, Geuens E, Hay-Schmidt A, Nyengaard JR. Does Ngb protect neurons from ischemic insult? A quantitative investigation of Ngb expression following transient mcao in spontaneously hypertensive rats. Brain Res. 2006;1085:19–27.
5. Wang X, Liu J, Zhu H, Tejima E, Tsuji K, Murata Y, Atochin DN, Huang PL, Zhang C, Lo EH. Effects of Ngb overexpression on acute brain injury and long-term outcomes after focal cerebral ischemia. Stroke. 2008;39:1869–74.
6. Sun Y, Jin K, Peel A, Mao XO, Xie L, Greenberg DA. Ngb protects the brain from experimental stroke in vivo. Proc Natl Acad Sci USA. 2003;100:3497–500.
7. Reuss S, Saaler-Reinhardt S, Weich B, Wystub S, Reuss MH, Burmester T, Hankeln T. Expression analysis of Ngb mRNA in rodent tissues. Neuroscience. 2002;115:645–56.
8. Nienhaus K, Nienhaus GU. Searching for Ngb's role in the brain. IUBMB Life. 2007;59:490–7.
9. Brunori M, Vallone B. A globin for the brain. FASEB J. 2006;20:2192–7.
10. Burmester T, Hankeln T. Ngb: a respiratory protein of the nervous system. News Physiol Sci. 2004;19:110–3.
11. Greenberg DA, Jin K, Khan AA. Ngb: an endogenous neuroprotectant. Curr Opin Pharmacol. 2008;8:20–4.
12. Giuffre A, Moschetti T, Vallone B, Brunori M. Is Ngb a signal transducer? IUBMB Life. 2008;60:410–3.
13. Brunori M, Vallone B. Ngb, seven years after. Cell Mol Life Sci. 2007;64:1259–68.
14. Wystub S, Laufs T, Schmidt M, Burmester T, Maas U, Saaler-Reinhardt S, Hankeln T, Reuss S. Localization of Ngb protein in the mouse brain. Neurosci Lett. 2003;346:114–6.
15. Zhang C, Wang C, Deng M, Li L, Wang H, Fan M, Xu W, Meng F, Qian L, He F. Full-length cDNA cloning of human Ngb and tissue expression of rat Ngb. Biochem Biophys Res Commun. 2002;290:1411–9.
16. Geuens E, Brouns I, Flamez D, Dewilde S, Timmermans JP, Moens L. A globin in the nucleus! J Biol Chem. 2003;278:30417–20.
17. Schmidt M, Giessl A, Laufs T, Hankeln T, Wolfrum U, Burmester T. How does the eye breathe? Evidence for Ngb-mediated oxygen supply in the mammalian retina. J Biol Chem. 2003;278:1932–5.
18. Hundahl C, Stoltenberg M, Fago A, Weber RE, Dewilde S, Fordel E, Danscher G. Effects of short-term hypoxia on Ngb levels and localization in mouse brain tissues. Neuropathol Appl Neurobiol. 2005;31:610–7.
19. Schmidt-Kastner R, Haberkamp M, Schmitz C, Hankeln T, Burmester T. Ngb mRNA expression after transient global brain ischemia and prolonged hypoxia in cell culture. Brain Res. 2006;1103:173–80.
20. Shao G, Gong KR, Li J, Xu XJ, Gao CY, Zeng XZ, Lu GW, Huo X. Antihypoxic effects of Ngb in hypoxia-preconditioned mice and SH-SY5Y cells. Neurosignals. 2009;17:196–202.
21. Shang A, Zhou D, Wang L, Gao Y, Fan M, Wang X, Zhou R, Zhang C. Increased Ngb levels in the cerebral cortex and serum after ischemia-reperfusion insults. Brain Research. 2006;1078:219–26.
22. Fordel E, Thijs L, Moens L, Dewilde S. Ngb and cytoglobin expression in mice. Evidence for a correlation with reactive oxygen species scavenging. FEBS J. 2007;274:1312–7.

23. Lima DC, Cossa AC, Perosa SR, de Oliveira EM, da Silva JAJ, da Silva Fernandes MJ, et al. Ngb is up-regulated in the cerebellum of pups exposed to maternal epileptic seizures. Int J Dev Neurosci. 2011;29(8):891–7.

24. Jin K, Mao Y, Mao X, Xie L, Greenberg DA. Ngb expression in ischemic stroke. Stroke. 2010;41:557–9.

25. Mammen PP, Shelton JM, Goetsch SC, Williams SC, Richardson JA, Garry MG, Garry DJ. Ngb, a novel member of the globin family, is expressed in focal regions of the brain. J Histochem Cytochem. 2002;50:1591–8.

26. Li RC, Lee SK, Pouranfar F, Brittian KR, Clair HB, Row BW, Wang Y, Gozal D. Hypoxia differentially regulates the expression of Ngb and cytoglobin in rat brain. Brain Res. 2006;1096:173–9.

27. Sun Y, Jin K, Mao XO, Xie L, Peel A, Childs JT, Logvinova A, Wang X, Greenberg DA. Effect of aging on Ngb expression in rodent brain. Neurobiol Aging. 2005;26:275–8.

28. Fordel E, Thijs L, Martinet W, Schrijvers D, Moens L, Dewilde S. Anoxia or oxygen and glucose deprivation in SH-SY5Y cells: a step closer to the unraveling of Ngb and cytoglobin functions. Gene. 2007;398:114–22.

29. Khan AA, Wang Y, Sun Y, Mao XO, Xie L, Miles E, Graboski J, Chen S, Ellerby LM, Jin K, Greenberg DA. Ngb-overexpressing transgenic mice are resistant to cerebral and myocardial ischemia. Proc Natl Acad Sci USA. 2006;103:17944–8.

30. Cai B, Lin Y, Xue XH, Fang L, Wang N, Wu ZY. Tat-mediated delivery of Ngb protects against focal cerebral ischemia in mice. Exp Neurol. 2011;227:224–31.

31. Li RC, Pouranfar F, Lee SK, Morris MW, Wang Y, Gozal D. Ngb protects PC12 cells against beta-amyloid-induced cell injury. Neurobiol Aging. 2008;29:1815–22.

32. Khan AA, Mao XO, Banwait S, Jin K, Greenberg DA. Ngb attenuates beta-amyloid neurotoxicity in vitro and transgenic alzheimer phenotype in vivo. Proc Natl Acad Sci USA. 2007;104:19114–9.

33. Pesce A, Dewilde S, Nardini M, Moens L, Ascenzi P, Hankeln T, Burmester T, Bolognesi M. Human brain Ngb structure reveals a distinct mode of controlling oxygen affinity. Structure. 2003;11:1087–95.

34. Vallone B, Nienhaus K, Brunori M, Nienhaus GU. The structure of murine Ngb: novel pathways for ligand migration and binding. Proteins. 2004;56:85–92.

35. Pesce A, Dewilde S, Nardini M, Moens L, Ascenzi P, Hankeln T, Burmester T, Bolognesi M. The human brain hexacoordinated Ngb three-dimensional structure. Micron. 2004;35:63–5.

36. Kriegl JM, Bhattacharyya AJ, Nienhaus K, Deng P, Minkow O, Nienhaus GU. Ligand binding and protein dynamics in Ngb. Proc Natl Acad Sci USA. 2002;99:7992–7.

37. Fago A, Hundahl C, Dewilde S, Gilany K, Moens L, Weber RE. Allosteric regulation and temperature dependence of oxygen binding in human Ngb and cytoglobin. Molecular mechanisms and physiological significance. J Biol Chem. 2004;279:44417–26.

38. Van Doorslaer S, Dewilde S, Kiger L, Nistor SV, Goovaerts E, Marden MC, Moens L. Nitric oxide binding properties of Ngb. A characterization by EPR and flash photolysis. J Biol Chem. 2003;278:4919–25.

39. Hundahl CA, Kelsen J, Dewilde S, Hay-Schmidt A. Ngb in the rat brain (ii): co-localisation with neurotransmitters. Neuroendocrinology. 2008;88:183–98.

40. Brunori M, Giuffre A, Nienhaus K, Nienhaus GU, Scandurra FM, Vallone B. Ngb, nitric oxide, and oxygen: functional pathways and conformational changes. Proc Natl Acad Sci USA. 2005;102:8483–8.

41. Moncada S, Erusalimsky JD. Does nitric oxide modulate mitochondrial energy generation and apoptosis? Nat Rev Mol Cell Biol. 2002;3:214–20.

42. Brunori M, Giuffre A, Forte E, Mastronicola D, Barone MC, Sarti P. Control of cytochrome c oxidase activity by nitric oxide. Biochim Biophys Acta. 2004;1655:365–71.

43. Tiso M, Tejero J, Basu S, Azarov I, Wang X, Simplaceanu V, Frizzell S, Jayaraman T, Geary L, Shapiro C, Ho C, Shiva S, Kim-Shapiro DB, Gladwin MT. Human Ngb functions as a redox-regulated nitrite reductase. J Biol Chem. 2011;286:18277–89.

44. Jin K, Mao XO, Xie L, Khan AA, Greenberg DA. Ngb protects against nitric oxide toxicity. Neurosci Lett. 2008;430:135–7.
45. Fordel E, Thijs L, Martinet W, Lenjou M, Laufs T, Van Bockstaele D, Moens L, Dewilde S. Ngb and cytoglobin overexpression protects human SH-SY5Y neuroblastoma cells against oxidative stress-induced cell death. Neurosci Lett. 2006;410:146–51.
46. Wakasugi K, Nakano T, Morishima I. Oxidized human Ngb acts as a heterotrimeric galpha protein guanine nucleotide dissociation inhibitor. J Biol Chem. 2003;278:36505–12.
47. Schwindinger WF, Robishaw JD. Heterotrimeric g-protein betagamma-dimers in growth and differentiation. Oncogene. 2001;20:1653–60.
48. Kitatsuji C, Kurogochi M, Nishimura S, Ishimori K, Wakasugi K. Molecular basis of guanine nucleotide dissociation inhibitor activity of human Ngb by chemical cross-linking and mass spectrometry. J Mol Biol. 2007;368:150–60.
49. Watanabe S, Wakasugi K. Neuroprotective function of human Ngb is correlated with its guanine nucleotide dissociation inhibitor activity. Biochem Biophys Res Commun. 2008;369:695–700.
50. Khan AA, Mao XO, Banwait S, DerMardirossian CM, Bokoch GM, Jin K, Greenberg DA. Regulation of hypoxic neuronal death signaling by Ngb. FASEB J. 2008;22:1737–47.
51. Wakasugi K, Nakano T, Kitatsuji C, Morishima I. Human Ngb interacts with flotillin-1, a lipid raft microdomain-associated protein. Biochem Biophys Res Commun. 2004;318:453–60.
52. Wakasugi K, Nakano T, Morishima I. Association of human Ngb with cystatin c, a cysteine proteinase inhibitor. Biochemistry. 2004;43:5119–25.
53. Giuffre A, Moschetti T, Vallone B, Brunori M. Ngb: enzymatic reduction and oxygen affinity. Biochem Biophys Res Commun. 2008;367:893–8.
54. Modjtahedi N, Giordanetto F, Madeo F, Kroemer G. Apoptosis-inducing factor: vital and lethal. Trends Cell Biol. 2006;16:264–72.
55. Fago A, Mathews AJ, Moens L, Dewilde S, Brittain T. The reaction of Ngb with potential redox protein partners cytochrome b5 and cytochrome c. FEBS Lett. 2006;580:4884–8.
56. Yu Z, Liu J, Guo S, Xing C, Fan X, Ning M, Yuan JC, Lo EH, Wang X. Ngb-overexpression alters hypoxic response gene expression in primary neuron culture following oxygen glucose deprivation. Neuroscience. 2009;162:396–403.
57. Chen LM, Xiong YS, Kong FL, Qu M, Wang Q, Chen XQ, et al. Ngb attenuates Alzheimer-like tau hyperphosphorylation by activating akt signaling. J Neurochem. 2012;120(1):157–64.
58. Burmester T, Gerlach F, Hankeln T. Regulation and role of Ngb and cytoglobin under hypoxia. Adv Exp Med Biol. 2007;618:169–80.
59. Nicholls DG, Budd SL. Mitochondria and neuronal survival. Physiol Rev. 2000;80:315–60.
60. Sims NR, Anderson MF. Mitochondrial contributions to tissue damage in stroke. Neurochem Int. 2002;40:511–26.
61. Li RC, Pouranfar F, Lee SK, Morris MW, Wang Y, Gozal D. Ngb protects PC12 cells against beta-amyloid-induced cell injury. Neurobiol Aging. 2007;29:1815–22.
62. Liu J, Yu Z, Guo S, Lee SR, Xing C, Zhang C, Gao Y, Nicholls DG, Lo EH, Wang X. Effects of Ngb overexpression on mitochondrial function and oxidative stress following hypoxia/reoxygenation in cultured neurons. J Neurosci Res. 2009;87:164–70.
63. Saito A, Maier CM, Narasimhan P, Nishi T, Song YS, Yu F, Liu J, Lee YS, Nito C, Kamada H, Dodd RL, Hsieh LB, Hassid B, Kim EE, Gonzalez M, Chan PH. Oxidative stress and neuronal death/survival signaling in cerebral ischemia. Mol Neurobiol. 2005;31:105–16.
64. Chan PH. Reactive oxygen radicals in signaling and damage in the ischemic brain. J Cereb Blood Flow Metab. 2001;21:2–14.
65. Perez-Pinzon MA, Dave KR, Raval AP. Role of reactive oxygen species and protein kinase c in ischemic tolerance in the brain. Antioxid Redox Signal. 2005;7:1150–7.
66. Larsen GA, Skjellegrind HK, Berg-Johnsen J, Moe MC, Vinje ML. Depolarization of mitochondria in isolated ca1 neurons during hypoxia, glucose deprivation and glutamate excitotoxicity. Brain Res. 2006;1077:153–60.

67. Zhang WH, Wang H, Wang X, Narayanan MV, Stavrovskaya IG, Kristal BS, Friedlander RM. Nortriptyline protects mitochondria and reduces cerebral ischemia/hypoxia injury. Stroke. 2008;39:455–62.

68. Zhu S, Stavrovskaya IG, Drozda M, Kim BY, Ona V, Li M, Sarang S, Liu AS, Hartley DM, Wu DC, Gullans S, Ferrante RJ, Przedborski S, Kristal BS, Friedlander RM. Minocycline inhibits cytochrome c release and delays progression of amyotrophic lateral sclerosis in mice. Nature. 2002;417:74–8.

69. Susin SA, Zamzami N, Castedo M, Hirsch T, Marchetti P, Macho A, Daugas E, Geuskens M, Kroemer G. Bcl-2 inhibits the mitochondrial release of an apoptogenic protease. J Exp Med. 1996;184:1331–41.

70. Petit PX, Goubern M, Diolez P, Susin SA, Zamzami N, Kroemer G. Disruption of the outer mitochondrial membrane as a result of large amplitude swelling: the impact of irreversible permeability transition. FEBS Lett. 1998;426:111–6.

71. Sun J, Trumpower BL. Superoxide anion generation by the cytochrome bc1 complex. Arch Biochem Biophys. 2003;419:198–206.

72. Berry EA, Guergova-Kuras M, Huang LS, Crofts AR. Structure and function of cytochrome bc complexes. Annu Rev Biochem. 2000;69:1005–75.

73. Hunte C, Palsdottir H, Trumpower BL. Protonmotive pathways and mechanisms in the cytochrome bc1 complex. FEBS Lett. 2003;545:39–46.

74. Klimova T, Chandel NS. Mitochondrial complex III regulates hypoxic activation of HIF. Cell Death Differ. 2008;15:660–6.

75. Guzy RD, Hoyos B, Robin E, Chen H, Liu L, Mansfield KD, Simon MC, Hammerling U, Schumacker PT. Mitochondrial complex III is required for hypoxia-induced ROS production and cellular oxygen sensing. Cell Metab. 2005;1:401–8.

76. Brunelle JK, Bell EL, Quesada NM, Vercauteren K, Tiranti V, Zeviani M, Scarpulla RC, Chandel NS. Oxygen sensing requires mitochondrial ROS but not oxidative phosphorylation. Cell Metab. 2005;1:409–14.

77. Iwata S, Lee JW, Okada K, Lee JK, Iwata M, Rasmussen B, Link TA, Ramaswamy S, Jap BK. Complete structure of the 11-subunit bovine mitochondrial cytochrome bc1 complex. Science. 1998;281:64–71.

78. Madesh M, Hajnoczky G. VDAC-dependent permeabilization of the outer mitochondrial membrane by superoxide induces rapid and massive cytochrome c release. J Cell Biol. 2001;155:1003–15.

79. Billen LP, Kokoski CL, Lovell JF, Leber B, Andrews DW. Bcl-XL inhibits membrane permeabilization by competing with BAX. PLoS Biol. 2008;6:e147.

80. Ma TC, Campana A, Lange PS, Lee HH, Banerjee K, Bryson JB, Mahishi L, Alam S, Giger RJ, Barnes S, Morris Jr SM, Willis DE, Twiss JL, Filbin MT, Ratan RR. A large-scale chemical screen for regulators of the arginase 1 promoter identifies the soy isoflavone daidzein as a clinically approved small molecule that can promote neuronal protection or regeneration via a camp-independent pathway. J Neurosci. 2010;30:739–48.

81. Smirnova NA, Rakhman I, Moroz N, Basso M, Payappilly J, Kazakov S, Hernandez-Guzman F, Gaisina IN, Kozikowski AP, Ratan RR, Gazaryan IG. Utilization of an in vivo reporter for high throughput identification of branched small molecule regulators of hypoxic adaptation. Chem Biol. 2010;17:380–91.

82. Smirnova NA, Haskew-Layton RE, Basso M, Hushpulian DM, Payappilly JB, Speer RE, Ahn YH, Rakhman I, Cole PA, Pinto JT, Ratan RR, Gazaryan IG. Development of Neh2-luciferase reporter and its application for high throughput screening and real-time monitoring of Nrf2 activators. Chem Biol. 2011;18:752–65.

83. Manuvakhova MS, Johnson GG, White MC, Ananthan S, Sosa M, Maddox C, McKellip S, Rasmussen L, Wennerberg K, Hobrath JV, White EL, Maddry JA, Grimaldi M. Identification of novel small molecule activators of nuclear factor-κb with neuroprotective action via high-throughput screening. J Neurosci Res. 2011;89:58–72.

84. Signore AP, Zhang F, Weng Z, Gao Y, Chen J. Leptin neuroprotection in the CNS: mechanisms and therapeutic potentials. J Neurochem. 2008;106:1977–90.

85. Ehrenreich H, Aust C, Krampe H, Jahn H, Jacob S, Herrmann M, Sirén AL. Erythropoietin: novel approaches to neuroprotection in human brain disease. Metab Brain Dis. 2004;19: 195–206.
86. Garry DJ, Mammen PP. Neuroprotection and the role of Ngb. Lancet. 2003;362:342–3.
87. Jin K, Mao XO, Xie L, John V, Greenberg DA. Pharmacological induction of Ngb expression. Pharmacology. 2011;87:81–4.
88. De Marinis E, Ascenzi P, Pellegrini M, Galluzzo P, Bulzomi P, Arevalo MA, Garcia-Segura LM, Marino M. 17β-estradiol—a new modulator of Ngb levels in neurons: role in neuroprotection against H_2O_2-induced toxicity. Neurosignals. 2010;18:223–35.
89. Gao Y, Mengana Y, Cruz YR, Muñoz A, Testé IS, García JD, Wu Y, Rodríguez JC, Zhang C. Different expression patterns of ngb and epor in the cerebral cortex and hippocampus revealed distinctive therapeutic effects of intranasal delivery of neuro-epo for ischemic insults to the gerbil brain. J Histochem Cytochem. 2011;59:214–27.
90. Sittampalam GS, Kahl SD, Janzen WP. High-throughput screening: advances in assay technologies. Curr Opin Chem Biol. 1997;1:384–91.
91. Landro JA, Taylor IC, Stirtan WG, Osterman DG, Kristie J, Hunnicutt EJ, Rae PM, Sweetnam PM. HTS in the new millennium: the role of pharmacology and flexibility. J Pharmacol Toxicol Methods. 2000;44:273–89.
92. Jarecki J, Chen X, Bernardino A, Coovert DD, Whitney M, Burghes A, Stack J, Pollok BA. Diverse small-molecule modulators of SMN expression found by high-throughput compound screening: early leads towards a therapeutic for spinal muscular atrophy. Hum Mol Genet. 2005;14:2003–18.
93. Croston GE. Functional cell-based uHTS in chemical genomic drug discovery. Trends Biotechnol. 2002;20:110–5.
94. Gray MJ, Poljakovic M, Kepka-Lenhart D, Morris Jr SM. Induction of arginase I transcription by IL-4 requires a composite DNA response element for STAT6 and C/EBPbeta. Gene. 2005;353:98–106.
95. Echaniz-Laguna A, Miniou P, Bartholdi D, Melki J. The promoters of the survival motor neuron gene (SMN) and its copy (SMNc) share common regulatory elements. Am J Hum Genet. 1999;64:1365–70.
96. DiDonato CJ, Brun T, Simard LR. Complete nucleotide sequence, genomic organization, and promoter analysis of the murine survival motor neuron gene (SMN). Mamm Genome. 1999;10:638–41.
97. Majumder S, Varadharaj S, Ghoshal K, Monani U, Burghes AH, Jacob ST. Identification of a novel cyclic amp-response element (CRE-II) and the role of CREB-1 in the camp-induced expression of the survival motor neuron (SMN) gene. J Biol Chem. 2004;279:14803–11.
98. Zhang W, Tian Z, Sha S, Cheng LY, Philipsen S, Tan-Un KC. Functional and sequence analysis of human Ngb gene promoter region. Biochim Biophys Acta. 2011;1809:236–44.
99. Bialkowska AB, Du Y, Fu H, Yang VW. Identification of novel small-molecule compounds that inhibit the proproliferative kruppel-like factor 5 in colorectal cancer cells by high-throughput screening. Mol Cancer Ther. 2009;8:563–70.
100. Zhu Y, Sun Y, Jin K, Greenberg DA. Hemin induces Ngb expression in neural cells. Blood. 2002;100:2494–8.
101. Agholme L, Lindstrom T, Kågedal K, Marcusson J, Hallbeck M. An in vitro model for neuroscience: differentiation of SH-SY5Y cells into cells with morphological and biochemical characteristics of mature neurons. J Alzheimers Dis. 2010;20:1069–82.
102. Sarang SS, Yoshida T, Cadet R, Valeras AS, Jensen RV, Gullans SR. Discovery of molecular mechanisms of neuroprotection using cell-based bioassays and oligonucleotide arrays. Physiol Genomics. 2002;11:45–52.
103. Bertsch U, Winklhofer K, Hirschberger T, Bieschke J, Weber P, Hartl FU, Tavan P, Tatzelt J, Kretzschmar HA, Giese A. Systematic identification of antiprion drugs by high-throughput screening based on scanning for intensely fluorescent targets. J Virol. 2005;79:7785–91.
104. Iljin K, Ketola K, Vainio P, Halonen P, Kohonen P, Fey V, Grafström RC, Perälä M, Kallioniemi O. High-throughput cell-based screening of 4910 known drugs and drug-like

small molecules identifies disulfiram as an inhibitor of prostate cancer cell growth. Clin Cancer Res. 2009;15:6070–8.

105. Zhang JH, Chung TD, Oldenburg KR. A simple statistical parameter for use in evaluation and validation of high throughput screening assays. J Biomol Screen. 1999;4:67–73.

106. Papke RL. Estimation of both the potency and efficacy of alpha7 nAChR agonists from single-concentration responses. Life Sci. 2006;78:2812–9.

107. Berg EL, Hytopoulos E, Plavec I, Kunkel EJ. Approaches to the analysis of cell signaling networks and their application in drug discovery. Curr Opin Drug Discov Devel. 2005;8:107–14.

Chapter 19
Characterization of Novel Neuroprotective Lipid Analogues for the Treatment of Stroke

Pamela Maher, Alain César Biraboneye, and Jean-Louis Kraus

Abstract Epidemiological, clinical, and biochemical studies have shown that various types of dietary fatty acids can modify the risk of stroke suggesting that fatty acids might be a good starting point for the development of new neuroprotective compounds for the treatment of stroke. We took two distinct approaches to the modification of fatty acids for their use as neuroprotective compounds. In the first approach, we made simplified forms of the ganglioside GM1 that contain a saturated, unsaturated, or cyclic fatty acid linked via an amide bond to a hydrophilic moiety. In the second approach, we directly linked fatty acids to the carboxylic acid bioisostere 1,2,4-oxadiazolidine-3,5-dione. Bioisosteres act as biomimetics of the parent group with respect to key physicochemical properties but provide additional, potentially more beneficial properties. We tested these new compounds in two distinct models of nerve cell death that mimic several different aspects of ischemic injury and death. We found that only a small subset of the tested compounds were effective in both neuroprotection assays, suggesting that these may be the best leads for novel therapeutic compounds for stroke. Interestingly, their neuroprotective effects appear to be mediated by distinct mechanisms.

P. Maher, PhD (✉)
The Salk Institute for Biological Studies, 10010 North Torrey Pines Road,
La Jolla, CA 92037, USA
e-mail: pmaher@salk.edu

A.C. Biraboneye • J.-L. Kraus
Laboratoire de Chimie Biomoleculaire, CNRS, IBDML-UMR-6216,
Campus de Luminy Case 907 13288, Marseille Cedex 09, France

P.A. Lapchak and J.H. Zhang (eds.), *Translational Stroke Research*, Springer Series   373
in Translational Stroke Research, DOI 10.1007/978-1-4419-9530-8_19,
© Springer Science+Business Media, LLC 2012

1   Introduction

Stroke due to occlusion of a cerebral artery is one of the most common neurological diseases and has a major impact on mortality and morbidity in the elderly population of industrialized countries [1]. The nerve cell death associated with cerebral ischemia is due to multiple factors resulting from the lack of oxygen, including the loss of ATP, excitotoxicity, oxidative stress, reduced neurotrophic support, and multiple other metabolic stresses [2]. Therefore, a drug directed against a single molecular target may not be effective in treating the nerve cell death associated with stroke. In addition, drugs that inhibit a single CNS activity with high potency are likely to be toxic because the activity is probably also required for normal brain function. Indeed, there is no effective treatment for stroke that is approved by the FDA except for recombinant tissue-type plasminogen activator (rt-PA). It follows that combinations of drugs or individual drugs that are broadly neuroprotective may be required.

Epidemiological, clinical, and biochemical studies have shown that various types of dietary fatty acids can modify the risks of stroke [3–5]. These observations have led to an increasing focus on the potential neuroprotective activities of free fatty acids in pharmacological research. Besides being key biomolecules in metabolic processes, free fatty acids serve as the substrates for cell membrane biogenesis (glyco- and phospholipids) and the precursors of intracellular signaling molecules such as prostaglandins, leukotrienes, thromboxanes, and platelet activating factor. Polyunsaturated fatty acids have been implicated in the prevention of various human diseases, including obesity, diabetes, coronary artery disease, stroke, and inflammatory and neurological diseases [6]. Saturated fats are usually regarded as unhealthy, but nutritionists believe that the type of saturated fat is also important. Stearic acid is biochemically classified as a saturated fatty acid, both for the purposes of food labeling and dietary recommendations. Stearic acid, one of the most common fatty acids in brain phospholipids, originates in the circulation and is sequestered from blood by the brain along with precursor fatty acids [7]. The neuroprotective effects of stearic acid against the toxicity of oxygen–glucose deprivation and glutamate on rat cortical or hippocampal slices have been already reported [8]. The behavior of stearic acid is especially unique with respect to its effects on serum cholesterol levels. A beneficial effect of stearic acid on clotting factors can result in a less thrombogenic state. Together, these results suggest that fatty acids might be a good starting point for the development of new neuroprotective compounds for the treatment of stroke.

We took two very different approaches to the modification of fatty acids for their use as neuroprotective drugs. In the first approach, we made simplified forms of the ganglioside GM1 that has previously been shown to have neuroprotective properties [9]. These compounds contain a saturated, unsaturated, or cyclic fatty acid linked via an amide bond to a hydrophilic moiety including a carboxylic acid, a phosphonic acid, a tetrazole, or an ascorbic acid moiety [10]. In the second approach, we directly linked fatty acids to the carboxylic acid bioisostere 1,2,4-oxadiazolidine-3,5-dione [11]. Bioisosterism provides an approach for the rational modification of lead compounds into safer and more clinically effective agents. Bioisosteres act as biomimetics of the parent group with respect to key physicochemical properties (e.g., pK_a, $clog P$) but provide additional, potentially more beneficial properties.

We tested these new compounds in two distinct models of nerve cell death that mimic several different aspects of the insults that are implicated in nerve cell injury and death following ischemia [12]. We found that only a small subset of the tested compounds are effective in both neuroprotection assays suggesting that these may be the best leads for novel therapeutic compounds for stroke. Interestingly, their neuroprotective effects appear to be mediated by distinct mechanisms.

2 Results

To date, most of the compounds that have been tested in clinical trials for the treatment of ischemic stroke were chosen based on specific targets that were proposed to be involved in the nerve cell death in stroke [13]. In order to identify novel compounds that might be effective for the treatment of stroke, a new approach is needed. Rather than taking a target-based approach, we have chosen to utilize two cell culture-based assays with death being an endpoint as a screening tool. In this way, we can identify potential neuroprotective compounds that act at multiple sites in two distinct cell death pathways and therefore could provide a better chance at protecting the cells.

The first model is oxidative glutamate toxicity or oxytosis. Oxidative glutamate toxicity is a mechanism by which glutamate kills nerve cells due to an impairment of GSH synthesis [14]. High levels of glutamate inhibit cystine uptake by the cystine/glutamate exchanger system x_c^-, thereby leading to GSH depletion. Experimentally, this mechanism can be investigated in isolation from excitotoxicity in neuronal cell lines [15–18] or immature primary cultured neurons still devoid of NMDA receptors [18, 19]. Using the murine hippocampal cell line HT22, it was shown that after GSH depletion, a massive increase in the production of reactive oxygen species (ROS) ensues [20], followed by activation of a cGMP-gated calcium channel leading to calcium influx [21], which initiates a form of programmed cell death with features of both apoptosis and necrosis [22]. All of these changes are associated with the nerve cell death following stroke [12] and neuroprotective compounds can act at any or all of these steps to prevent cell death [14]. This is a simple, very reproducible model of neuronal cell death that has allowed the identification of number of neuroprotective pathways [17, 18, 23–28], and an impressive majority of these results were replicated in primary neuronal cultures. Furthermore, oxidative glutamate toxicity can be partly responsible for nerve cell death in excitotoxic paradigms [29].

The second model is in vitro ischemia. Most cell culture models of ischemia utilize primary cortical cultures that are exposed to some form of oxygen and glucose deprivation. While these assays may come closer to mimicking the in vivo conditions than assays employing nerve cell lines, they have a number of problems that make them difficult to use for the routine screening of potential neuroprotective compounds. The most serious problem is that the conditions needed to kill a fixed percentage of cells are highly variable from experiment to experiment, making it

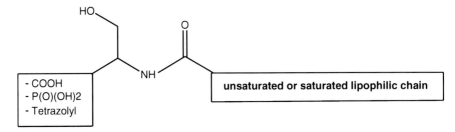

Fig. 19.1 General simplified structure of GM1-like analogues

very difficult to get accurate values for neuroprotective activity. In order to circumvent this and other problems, we decided to use the HT22 nerve cell line in combination with chemical ischemia as a screen for potential neuroprotective drugs for this type of cell damage. In order to induce ischemia in the HT22 cells, we used the compound iodoacetic acid (IAA), a well-known, irreversible inhibitor of the glycolytic enzyme glyceraldehyde 3-phosphate dehydrogenase [30]. IAA has been used in a number of other studies to induce ischemia in nerve cells [31–35]. The changes following IAA treatment of neural cells are very similar to changes which have been seen in animal models of CNS trauma [36] and ischemia [12]. These include alterations in membrane potential, breakdown of phospholipids [37], loss of ATP, and an increase in ROS. A 2-h treatment of the HT22 cells with IAA showed a dose-dependent increase in cell death 20 h later with <5% survival at 20 μM [38]. Furthermore, we have recently shown that success in this assay is predictive of the ability to work in the rabbit stroke/ischemia model [38–40].

For the compounds based on GM1, two series of analogues, whose general structure is outlined in Fig. 19.1, were synthesized. In the first series, various lipophilic saturated (stearyl), unsaturated (linoleyl, linolenyl, and arachidonyl), or cyclic polyunsaturated (retinoyl, abietyl, and abscisyl) moieties were introduced, while in the second series, the carboxylic function was replaced by different hydrophilic groups which could be either carboxylic bioisostere groups such as phosphonic and tetrazole groups or an ascorbic acid moiety. Introduction of ascorbic acid was supported by several reports which have shown that ascorbic acid conjugates can improve blood–brain barrier (BBB) permeation properties [41–43].

All of the GM1-like analogues were screened for their neuroprotective effects in the two cell culture-based neuroprotection assays. Only two of the 21 analogues were very effective in both assays; the abietyl analogue with the carboxylic function protected by an ethyl ester and the stearyl analogue coupled to an ascorbic acid moiety through a ω-aminohexanoic acid linker (Figs. 19.2 and 19.3; Table 19.1). While these two compounds did show both the highest potency and efficacy in the in vitro ischemia assay, other compounds that were just slightly less potent (e.g. EC_{50} 2.6 vs. 1.8) and efficacious (83% maximal protection vs. 92% maximal protection) were completely ineffective in the oxidative glutamate toxicity assay. Furthermore, the linoleyl analogue coupled to an ascorbic acid moiety through a

Structures of Neuroprotective Fatty Acid Analogues

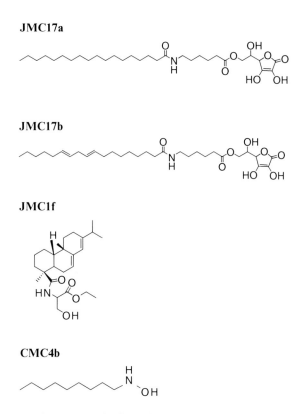

Fig. 19.2 Structures of neuroprotective fatty acid analogues

ω-aminohexanoic acid linker also showed some activity in the oxidative glutamate toxicity assay despite having a relatively high EC_{50} (6.4 μM) in the in vitro ischemia assay. Thus, these results strongly suggest that the two neuroprotection assays do indeed measure different parameters and support the idea of using multiple neuroprotection assays to identify the best compounds.

Three alkyl chain lengths (hexyl, decyl, and stearyl) were selected for study for the compounds based on linking a fatty acid to the carboxylic acid bioisostere 1,2,4-oxadiazolidine-3,5-dione. For these compounds, we tested both the final derivatives as well as their synthetic intermediates in both of the cell culture-based neuroprotection assays. Of the 12 compounds tested, only one was effective in both neuroprotection assays. This was an intermediate which contains a hydroxylamine function linked to a linear decyl chain (Figs. 19.2 and 19.3; Table 19.1).

Having identified several new fatty acid-related compounds with neuroprotective effects, the question of their mechanism of action was an important issue.

Effects of Fatty Acid Analogues on Cell Survival

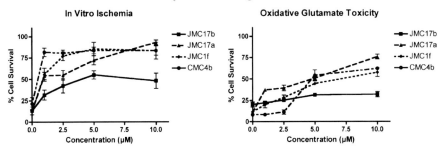

Fig. 19.3 Dose–response curves for neuroprotection in the in vitro ischemia and oxidative gluta-mate toxicity assays. For in vitro ischemia, HT22 cells were treated with 20 μM IAA for 2 h alone or in the presence of increasing doses of the fatty acid analogues. The same concentrations of the fatty acid analogues were also included in the fresh medium added after the 2-h treatment with IAA. Percentage survival was measured after 24 h by the MTT assay. Similar results were obtained in three to five independent experiments. For oxidative glutamate toxicity, HT22 cells were treated with 5 mM glutamate for 24 h alone or in the presence of increasing doses of the fatty acid ana-logues. Percentage survival was measured after 24 h by the MTT assay. Similar results were obtained in two to three independent experiments

Previously, it was shown that some compounds that protect from IAA toxicity do so by preventing the loss of GSH and/or ATP [38], while compounds that protect from oxidative glutamate toxicity do so by preventing the loss of GSH and/or the increase in ROS and/or the increase in intracellular calcium [23]. However, none of the neuroprotective fatty acid analogues had any effect on either GSH or ATP levels in the in vitro ischemia assay nor did they prevent the loss of GSH in the oxidative glutamate toxicity assay. Surprisingly, despite the observation that all of the fatty acid analogues lacked direct antioxidant activity as determined by the TEAC assay, a test tube measure of antioxidant activity (Table 19.1), they all had some efficacy in reducing ROS accumulation in response to glutamate (Fig. 19.4). Fisetin, a neu-roprotective flavonoid, was used as a positive control [23]. However, this effect did not appear sufficient to fully explain their neuroprotective effects.

We next investigated the p38 MAPK and JNK pathways because these kinase pathways can be activated during both in vitro ischemia [40] and oxidative gluta-mate toxicity [25] in the HT22 cells as well as in other in vitro and animal models of ischemia and their activation can contribute to nerve cell death [44]. Thus, it was possible that either or both of these pathways could be specifically altered by the fatty acid analogues, thereby reducing the observed cell death. Using specific anti-bodies to the phosphorylated forms of p38 MAPK and JNK as well as their specific substrates, MAPKAP kinase 2 and c-jun, respectively, the effects of the fatty acid analogues on these pathways were examined. All of the GM-1 analogues with EC_{50}s below 10 μM and most of the fatty acid bioisostere analogues and their synthetic intermediates were tested in these assays. From these studies, it was found that only

Table 19.1 Protection of HT22 cells by the fatty acid analogues and their effects on key signaling pathways

Compound	EC_{50} HT22/IAA (μM)[a]	% Maximum protection[b]	EC_{50} HT22/ glutamate (μM)[c]	% Maximum protection[d]	JNK/p38 activation[e]	ERK activation[f]	Akt activation[g]	TEAC[h]
JMC17b	6.4	55	–	32	No effect	No effect	No effect	0
JMC17a	1.8	92	4.6	76	No effect	No effect	No effect	0
JMC1f	1.05	76	9	55	+	+	No effect	0
CMC4b	1.0	83	5	62	No effect	+	+	0

[a] Half maximal effective concentrations (EC_{50}) for protection from IAA toxicity were determined by exposing HT22 cells to different doses of each analogue in the presence of 20 μM IAA for 2 h (HT22/IAA). Cell viability was determined after 24 h by the MTT assay

[b] The maximal percent of survival at the most effective dose (which varied from 1 to 10 μM depending on the analogue) is also indicated. The average % survival in cells treated with IAA alone was $15 \pm 4\%$

[c] Half maximal effective concentrations (EC_{50}) for protection from oxidative glutamate toxicity were determined by exposing the HT22 cells to different doses of each analogue in the presence of 5 mM glutamate for 24 h. Cell viability was determined after 24 h by the MTT assay

[d] The maximal percent of survival at the most effective dose (which varied from 1 to 10 μM depending on the analogue) was determined. The average % survival in cells treated with glutamate alone was $15 \pm 5\%$

[e] The effect of the analogues on JNK and p38 MAPK activation by IAA treatment was determined by SDS-PAGE and Western blotting with phospho-specific antibodies as described [10]

[f] The effect of the analogues on ERK activation in the presence of IAA was determined by SDS-PAGE and Western blotting with phospho-specific antibodies as described [10]

[g] The effect of the analogues on Akt activation in the presence of IAA was determined by SDS-PAGE and Western blotting with phospho-specific antibodies as described [40]

[h] The TEAC value, a test tube measure of antioxidant activity, was determined as described [56]

Effects of Fatty Acid Analogues on ROS Production

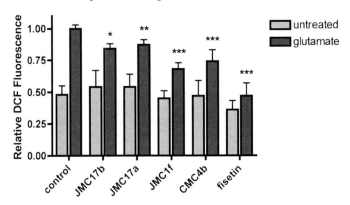

Fig. 19.4 Effect of fatty acid analogues on glutamate-induced ROS production. HT22 cells were treated with 5 mM glutamate for 16 h alone or in the presence of increasing doses of the fatty acid analogues. The medium was then replaced with Hank's buffered salt solution containing 10 μM CM-H$_2$DCFDA to detect ROS for 30 min. The fluorescence ($\lambda_{excitation}$ =495 nm; $\lambda_{emission}$ =525 nm) was read on a Gemini fluorescent plate reader [55] and the results normalized to the value for cells treated with glutamate alone. Fisetin was used as a positive control [23]

one compound, JMC1f, reduced both p38 MAPK and JNK activation and completely inhibited phosphorylation of their respective substrates MAPKAP kinase 2 and c-jun [10]. In contrast, the other neuroprotective GM1-like analogues (JMC17a and JMC17b) had no effect on the activation of these kinases or on the phosphorylation of their substrates. In addition, the neuroprotective fatty acid bioisostere synthetic intermediate CMC4b also had no effect on the activation of these kinases or on the phosphorylation of their substrates [11].

We next examined the Ras-ERK cascade which has been implicated in nerve cell survival [25, 45–47] and might be activated by the new fatty acid analogues. For these studies, we again tested all of the GM1-like analogues with EC$_{50}$s less than 10 μM and most of the fatty acid bioisostere analogues and their synthetic intermediates using SDS-PAGE and immunoblotting in combination with antibodies to the phosphorylated and therefore activated form of ERK. Both the GM1 analogue JMC1f and the fatty acid bioisostere synthetic intermediate CMC4b enhanced ERK activation.

Finally, we investigated the phosphoinositide-3-kinase (PI3 kinase)-Akt signaling pathway because this pathway is also implicated in nerve cell survival both in vitro and in vivo [48, 49]. Consistent with these observations, phosphorylation of Akt on Ser473 was largely eliminated following treatment with IAA but was restored by the fatty acid bioisostere synthetic intermediate CMC4b but not by any of the GM1 analogues (Fig. 19.5). CNB001 was used as a positive control [40]. Together these results suggest that these fatty acid analogues exert their neuroprotective effects by distinct mechanisms.

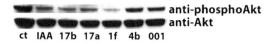
anti-phosphoAkt
anti-Akt
ct IAA 17b 17a 1f 4b 001

Fig. 19.5 Effect of the fatty acid analogues on the inhibition of Akt activation during in vitro ischemia. HT22 cells were treated 20 μM IAA for 2 h alone or in the presence of increasing doses of the fatty acid analogues. The same concentrations of the fatty acid analogues were also included in the fresh medium added after the 2-h treatment with IAA. Cells were harvested 2 h later and analyzed for Akt phosphorylation by SDS-PAGE and Western blotting as described [40]

3 Discussion

There are no drugs that prevent nerve cell death associated with CNS stroke and ischemia despite billions spent by the pharmaceutical industry. This failure may be a result of the current approach to drug development in which a potential drug target is identified and a high-affinity ligand is made. However, when the ligand/drug is tested in clinical trials for stroke, it is often ineffective or toxic, perhaps because there are multiple pathways to cell death, or the target is required for normal brain function. While the single-target paradigm for drug development is clearly effective for some indications, it is not likely to work for CNS ischemia where there are multiple cell types and physiological parameters involved in the pathology. An alternative approach is first to identify potential neuroprotective compounds and then refine them on the basis of assays that reflect multiple biological activities relevant to a wide range of CNS insults.

Oxidative glutamate toxicity and in vitro ischemia are two cell culture assays employing the hippocampal nerve cell line HT22 that were used to determine if the fatty acid analogues had the potential to be effective in stroke. Both assays have been used to screen a large number of flavones to predict potential efficacy for stroke in animals [38], and both utilize neurotoxic conditions that are similar to those observed in ischemic stroke [14, 34]. In the oxidative glutamate toxicity assay, glutamate blocks cystine uptake, leading to glutathione (GSH) depletion, lipoxygenase activation, ROS production, and cGMP-dependent calcium influx resulting in cell death. We have previously shown that three compounds that block this pathway by maintaining GSH levels (fisetin), inhibiting lipoxygenases (baicalein), or modulating kinase activation (CNB001) are effective in the rabbit stroke model [38–40].

In the chemical ischemia assay, IAA inhibits the glycolytic enzyme GAPDH, resulting in a loss of Akt and ERK phosphorylation, an induction of p38 MAP kinase and JNK activation, ATP depletion, and oxidative stress as determined by increased ROS and lipid peroxidation [10, 37, 38, 40]. The loss of ATP in ischemia results in cell death by necrosis [50]. This assay gives a toxicity profile very similar to oxygen–glucose deprivation [37] but is much more reproducible for pathway analysis and drug screening.

Only three of the GM1 analogues were effective in the two assays: JMC1f ($EC_{50} = 1.05$ μM), JMC17a ($EC_{50} = 1.8$ μM) and JMC17b ($EC_{50} = 6.4$ μM). JMC1f contains in its structure an abietyl lipophilic moiety, while JMC17a and JMC17b contain stearyl and linoleyl moieties, respectively, coupled to an ascorbic acid moiety through a ω-aminohexanoic acid linker. Based on the results with the entire group of compounds, it was observed that both lipophilic and hydrophilic moieties influenced their neuroprotective effects. These studies showed differing neuroprotective effects for analogues bearing saturated or unsaturated side chains. Despite the recent results from Strokin et al. [51] showing that polyunsaturated fatty acids such as docosahexanoic acid provide potent protection against neurodegeneration after hypoxia/hypoglycemia, our results clearly show that compounds including a saturated stearyl chain are as or more active than their corresponding unsaturated analogues. For example, among the lipophilic moieties, both the long-chain polyunsaturated linolenyl and the cyclic polyunsaturated abietyl moieties appeared to significantly reduce neuroprotection in the in vitro ischemia assay. Surprisingly, the protection of the carboxylic function by an ethyl ester not only significantly enhanced the activity of the compound with the abietyl moiety in the in vitro ischemia assay but also endowed it with activity in the oxidative glutamate toxicity assay. However, overall, the protection of the carboxylic function by an ethyl ester had divergent effects on the neuroprotective activity, enhancing it in some cases (e.g., with an abietyl or linolenyl lipophilic group) and reducing it in others (e.g., with a stearyl, retinoyl, and abscisyl lipophilic group). The replacement of the carboxylic function with different hydrophilic groups also had mixed effects. While the substitution of the carboxylic function with a phosphonic group had little effect on activity, one of the analogues in which the serine moiety was replaced by an ascorbic acid (JMC17a) was found to be highly active in both neuroprotection assays. The other ascorbic acid-substituted analogue (JMC17b) was less potent in the in vitro ischemia assay ($EC_{50} = 6.4$ vs. 2.9 for the analogue with a carboxylic acid group) but gained measurable activity in the oxidative glutamate toxicity assay. This result is quite encouraging since, as already mentioned in the Introduction section, ascorbic acid could help anti-ischemic agents reach the brain after BBB permeation via active transport. Ascorbic acid is metabolized in vivo into its oxidized form (dihydroascorbic acid [DHAA]) which is then transported across the BBB via either the glucose transporter GLUT1 [52] or the Na^+-dependent ascorbate transporter SVTC2 [53]. Thus, it can be concluded that both lipoyl and hydrophilic moieties induce or modulate the observed neuroprotective effects.

Only one (CMC4b) of the 1,2,4-oxadiazolidine-3,5-dione carboxylic acid bioisostere fatty acid analogues was effective in both neuroprotection assays. Based on the results with the entire group of compounds, including the bioisoteres as well as their synthetic intermediates, the replacement of the carboxylic acid function by a hydroxylamine group appeared to greatly enhance the observed neuroprotective effect. We emphasize this point because, to our knowledge, the neuroprotective effects of N-alkylhydroxylamine compounds have never been reported. Because N-hydroxylamine compounds are known to be relatively unstable and sensitive to oxidation, their potential use as neuroprotective agents could be limited. However, it is possible to stabilize such compounds through the formation of their corresponding acid salts [54]. In contrast to the results with the GM1-like analogues, the stearyl alkyl chain was not consistent with a strong neuroprotective effect since only the intermediate with a decanoyl chain was effective in both neuroprotection assays.

We also found that the acidic properties of the hydrophilic groups weakly influence and the electronic parameters of the hydrophilic groups strongly influence the observed neuroprotective effects in the in vitro ischemia assay.

Surprisingly, the pathway analysis results for all four of the fatty acid analogues that are effective in both neuroprotection assays suggest that they act through distinct pathways. While JMC1f appears to inhibit the activation of multiple, pro-death kinase pathways, the other three compounds do not. Furthermore, while the induction of ERK activation by both CMC4b and JMC1f may contribute to their neuroprotective effects, this is not the case for either JMC17a or JMC17b. Interestingly, in contrast to the other fatty acid analogues, CMC4b also maintains activation of the PI3K-Akt pathway. Indeed, its effects on the different signaling pathways are quite similar to another compound, CNB-001, recently described by one of us [40], which is extremely effective in the rabbit stroke model. While we know that these compounds are not antioxidants, they can reduce ROS production in response to glutamate treatment, suggesting that they may also have effects on the mitochondria. Thus, further investigation is needed to clearly define their mode of action. Nevertheless, given the potent effects of these fatty acid analogues in two distinct cell-based neuroprotection assays, exploration of their neuroprotective effects in animal models of ischemia is clearly warranted.

References

1. Ingall T. Stroke-incidence, mortality and risk. J Insur Med. 2004;36:143–52.
2. Dirnagl U, Iadecola C, Moskowitz MA. Pathobiology of ischaemic stroke: an integrated view. Trends Neurosci. 1999;22:391–7.
3. McArthur MJ, Atshaves BP, Frolov A, Foxworth WD, Kier AB, Schroeder F. Cellular uptake and intracellular trafficking of long chain fatty acids. J Lipid Res. 1999;40:1371–83.
4. Belayev L, Khoutorova L, Atkins KD, Bazan NG. Robust docosahexaenoic acid-mediated neuroprotection in a rat model of transient, focal cerebral ischemia. Stroke. 2009;40:3121–6.
5. Nguemeni C, Delplanque B, Rovere C, Simon-Rousseau N, Gandin C, Agnani G, Nahon JL, Heurteaux C, Blondeau N. Dietary supplementation of alpha-linolenic acid in an enriched rapeseed oil diet protects from stroke. Pharmacol Res. 2010;61:226–33.
6. Bazan NG. Lipid signaling in neural plasticity, brain repair and neuroprotection. Mol Neurobiol. 2005;32:89–103.
7. Tholstrup T, Marckmann P, Jespersen J, Sandstrom B. Fat high in stearic acid favorably affects blood lipids and factor VII coagulant activity in comparison with fats high in palmitic acid or high in myristic and lauric acids. Am J Clin Nutr. 1994;59:371–7.
8. Wang Z-J, Li G-M, Tang W-L, Yin M. Neuroprotective effects of stearic acid against toxicity of oxygen/glucose deprivation or glutamate on rat cortical or hippocampal slices. Acta Pharmacol Sin. 2006;27:145–50.
9. Hadjiconstantinou M, Neff NH. GM1 ganglioside: in vivo and in vitro trophic actions on central neurotransmitter systems. J Neurochem. 1998;70:1335–45.
10. Biraboneye AC, Madonna S, Laras Y, Krantic S, Maher P, Kraus J-L. Potential neuroprotective drugs in cerebral ischemia: new saturated and polyunsaturated lipids coupled to hydrophilic moieties: synthesis and biological activity. J Med Chem. 2009;52:4358–69.
11. Biraboneye AC, Madonna S, Maher P, Kraus J-L. Neuroprotective effects of N-alkyl-1,2,4-oxadiazolidine-3,5-diones and their corresponding synthetic intermediates

N-alkylhydroxylamines and N-1-alkyl-3-carbonyl-1-hydroxyureas against in vitro cerebral ischemia. ChemMedChem. 2010;5:79–85.

12. Lipton P. Ischemic death in brain neurons. Physiol Rev. 1999;79:1431–568.

13. Green AR, Shuaib A. Therapeutic strategies for the treatment of stroke. Drug Discov Today. 2006;11:681–93.

14. Tan S, Schubert D, Maher P. Oxytosis: a novel form of programmed cell death. Curr Top Med Chem. 2001;1:497–506.

15. Miyamoto M, Murphy TH, Schnaar RL, Coyle JT. Antioxidants protect against glutamate-induced cytotoxicity in a neuronal cell line. J Pharmacol Exp Ther. 1989;250:1132–40.

16. Schubert D, Kimura H, Maher P. Growth factors and vitamin E modify neuronal glutamate toxicity. Proc Natl Acad Sci U S A. 1992;89:8264–7.

17. Davis JB, Maher P. Protein kinase C activation inhibits glutamate-induced cytotoxicity in a neuronal cell lines. Brain Res. 1994;652:169–73.

18. Maher P, Davis JB. The role of monoamine metabolism in oxidative glutamate toxicity. J Neurosci. 1996;16:6394–401.

19. Murphy TH, Baraban JM. Glutamate toxicity in immature cortical neurons precedes development of glutamate receptor currents. Brain Res. 1990;57:146–50.

20. Tan S, Sagara Y, Liu Y, Maher P, Schubert D. The regulation of reactive oxygen species production during programmed cell death. J Cell Biol. 1998;141:1423–32.

21. Li Y, Maher P, Schubert D. Requirement for cGMP in nerve cell death caused by glutathione depletion. J Cell Biol. 1997;139:1317–24.

22. Tan S, Wood M, Maher P. Oxidative stress in nerve cells induces a form of cell death with characteristics of both apoptosis and necrosis. J Neurochem. 1998;71:95–105.

23. Ishige K, Schubert D, Sagara Y. Flavonoids protect neuronal cells from oxidative stress by three distinct mechanisms. Free Radic Biol Med. 2001;30:433–46.

24. Ishige K, Chen Q, Sagara Y, Schubert D. The activation of dopamine D4 receptors inhibits oxidative stress-induced nerve cell death. J Neurosci. 2001;21:6069–76.

25. Maher P. How protein kinase C activation protects nerve cells from oxidative stress-induced cell death. J Neurosci. 2001;21:2929–38.

26. Sagara Y, Ishige K, Tsai C, Maher P. Tyrphostins protect neuronal cells from oxidative stress. J Biol Chem. 2002;277:36204–15.

27. Lewerenz J, Letz J, Methner A. Activation of stimulatory heteromeric G proteins increases glutathione and protects neuronal cells against oxidative stress. J Neurochem. 2003;87:522–31.

28. Sahin M, Saxena A, Jost P, Lewerenz J, Methner A. Induction of Bcl-2 by functional regulation of G-protein coupled receptors protects from oxidative glutamate toxicity. Free Radic Res. 2006;40:1113–23.

29. Schubert D, Piasecki D. Oxidative glutamate toxicity can be a part of the excitotoxicity cascade. J Neurosci. 2001;21:7455–62.

30. Winkler BS, Sauer MW, Starnes CA. Modulation of the Pasteur effect in retinal cells: implications for understanding compensatory metabolic mechanisms. Exp Eye Res. 2003;76:715–23.

31. Reshef A, Sperling O, Zoref-Shani E. Activation and inhibition of protein kinase C protect rat neuronal cultures against ischemia-reperfusion insult. Neurosci Lett. 1997;238:37–40.

32. Sperling O, Bromberg Y, Oelsner H, Zoref-Shani E. Reactive oxygen species play an important role in iodoacetate-induced neurotoxicity in primary rat neuronal cultures and in differentiated PC12 cells. Neurosci Lett. 2003;351:137–40.

33. Rego AC, Areias FM, Santos MS, Oliveira CR. Distinct glycolysis inhibitors determine retinal cell sensitivity to glutamate-mediated injury. Neurochem Res. 1999;24:351–8.

34. Sigalov E, Fridkin M, Brenneman DE, Gozes I. VIP-related protection against iodoacetate toxicity in pheochromocytoma (PC12) cells: a model for ischemic/hypoxic injury. J Mol Neurosci. 2000;15:147–54.

35. Reiner PB, Laycock AG, Doll CJ. A pharmacological model of ischemia in the hippocampal slice. Neurosci Lett. 1990;119:175–8.

36. Kokiko ON, Hamm RJ. A review of pharmacological treatments used in experimental models of traumatic brain injury. Brain Inj. 2007;21:259–74.

37. Taylor BM, Fleming WE, Benjamin CW, Wu Y, Mathews R, Sun FF. The mechanism of cytoprotective action of lazaroids I: inhibition of reactive oxygen species formation and lethal cell injury during periods of energy depletion. J Pharmacol Exp Ther. 1996;276:1224–31.
38. Maher P, Salgado KF, Zivin JA, Lapchak PA. A novel approach to screening for new neuroprotective compounds for the treatment of stroke. Brain Res. 2007;1173:117–25.
39. Lapchak PA, Maher P, Schubert D, Zivin JA. Baicalein, an antioxidant 12/15-lipoxygenase inhibitor improves clinical rating scores following multiple infarct embolic strokes. Neuroscience. 2007;150:585–91.
40. Lapchak PA, Schubert DR, Maher PA. Delayed treatment with a novel neurotrophic compound reduces behavioral deficits in rabbit ischemic stroke. J Neurochem. 2011;116:122–31.
41. Manfredini S, Pavan B, Ventuani S, Scaglianti M, Compagnone D, Biodi C, Scatturin A, Tanganelli S, Ferraro L, Prasad P, Dalpiaz A. Design, synthesis and activity of ascorbic acid prodrugs of nipecotic, kynurenic and diclophenamic acids, liable to increase neurotropic activity. J Med Chem. 2002;45:559–62.
42. Laras Y, Quelever G, Carino C, Pietrancosta N, Sheha M, Bihel F, Wolfe MS, Kraus J-L. Substituted thiazolamide coupled to a redox delivery system: a new gamma-secretase inhibitor with enhanced pharmacokinetic profile. Org Biomol Chem. 2005;3:612–8.
43. Laras Y, Sheha M, Pietrancosta N, Kraus J-L. Thiazolamide-ascorbic acid conjugate: a gamma secretase inhibitor with enhanced blood-brain barrier permeation. Aust J Chem. 2007;60:128–32.
44. Mehta SL, Manhas N, Rahubir R. Molecular targets in cerebral ischemia for developing novel therapeutics. Brain Res Rev. 2007;54:34–66.
45. Dugan LL, Creedon DJ, Johnson EM, Holtzman DM. Rapid suppression of free radical formation by nerve growth factor involves the mitogen-activated protein kinase pathway. Proc Natl Acad Sci U S A. 1997;94:4086–91.
46. Bonni A, Brunet A, West AE, Datta SR, Takasu MA, Greenberg ME. Cell survival promoted by the Ras-MAPK signaling pathway by transcription-dependent and transcription-independent mechanisms. Science. 1999;286:1358–62.
47. Hetman M, Xia Z. Signaling pathways mediating anti-apoptotic action of neurotrophins. Acta Neurobiol Exp (Wars). 2000;60:531–45.
48. Zhao H, Sapolsky RM, Steinberg GK. Phosphoinositide-3-kinase/Akt survival signal pathways are implicated in neuronal survival after stroke. Mol Neurobiol. 2006;34:249–70.
49. Liu C, Wu J, Xu K, Cai F, Gu J, Ma L, Chen J. Neuroprotection by baicalein in ischemic brain injury involves PTEN/AKT pathway. J Neurochem. 2010;112:1500–12.
50. Vanlangenakker N, Vanden Berghe T, Krysko DV, Festjens N, Vandenbeele P. Molecular mechanisms and pathophysiology of necrotic cell death. Curr Mol Med. 2008;8:207–20.
51. Strokin M, Chechneva O, Reymann KG, Reiser G. Neuroprotection of rat hippocampal slices exposed to oxygen-glucose deprivation by enrichment with docosahexaenoic acid and by inhibition of hydrolysis of docosahexaenoic acid-containing phospholipids by calcium independent phospholipase A2. Neuroscience. 2006;140:547–53.
52. Agus DB, Gambhir SS, Pardridge WM, Spielholz C, Baselga J, Vera JC, Golde DW. Vitamin C crosses the blood-brain barrier in the oxidized form through glucose transporters. J Clin Invest. 1997;100:2842–8.
53. Tsukaguchi H, Tokui T, Mackenzie B, Berger UV, Chen XZ, Wang YX, Brubaker RF, Hediger MA. A family of mammalian Na^+-dependent L-ascorbic acid transporters. Nature. 1999;399:70–5.
54. Blixt J. Patent application 09/380458; 1999.
55. Lewerenz J, Albrecht P, Tien ML, Henke N, Karumbayaram S, Kornblum HI, Wiedua-Pazos M, Schubert D, Maher P, Methner A. Induction of Nrf2 and xCT and involved in the action of the neuroprotective antibiotic ceftriazone in vitro. J Neurochem. 2009;111:332–43.
56. Maher P. A comparison of the neurotrophic activities of the flavonoid fisetin and some of its derivatives. Free Radic Res. 2006;40:1105–11.

Chapter 20
Na⁺/H⁺ Exchangers as Therapeutic Targets for Cerebral Ischemia

Yejie Shi and Dandan Sun

Abstract The Na$^+$/H$^+$ exchangers (NHE) are a family of membrane transporters that catalyzes the exchange of intracellular H$^+$ with extracellular Na$^+$ and plays a role in regulating intracellular pH and cell volume. Following cerebral ischemia, the "housekeeping" NHE isoform 1 (NHE-1) is stimulated by intracellular acidosis to remove excess H$^+$. Overstimulation of NHE-1 causes accumulation of Na$^+$ and Ca^{2+} inside the cell through the reversal mode of Na$^+$/Ca^{2+} exchange (NCX) and eventually contributes to cell death. Pharmacological inhibition or genetic knockdown of NHE-1 is neuroprotective in both in vitro and in vivo ischemia models as shown by reduced neuronal death and blockade of intracellular Ca^{2+} and Na$^+$ accumulation. Inhibition of NHE-1 not only reduces brain infarct volume but also improves long-term neurological functions. Inhibition of NHE-1 also has a profound effect on neuroinflammation and edema formation, providing a longer treatment time window for stroke therapy. Therefore, NHE-1 merges as an important target for developing new therapeutics for stroke treatment.

Y. Shi
Neuroscience Training Program, University of Wisconsin, Madison, WI 53705, USA

Department of Neurology, University of Pittsburgh, Pittsburgh, PA 15213, USA

D. Sun, MD, PhD (✉)
Neuroscience Training Program, University of Wisconsin, Madison, WI 53705, USA

Department of Neurological Surgery, University of Wisconsin, Madison, WI 53705, USA

Department of Neurology, University of Pittsburgh, Pittsburgh, PA 15213, USA
e-mail: sund@upmc.edu

P.A. Lapchak and J.H. Zhang (eds.), *Translational Stroke Research*, Springer Series
in Translational Stroke Research, DOI 10.1007/978-1-4419-9530-8_20,
© Springer Science+Business Media, LLC 2012

1 Introduction

Stoke is the third leading cause of death and the leading cause of serious, long-term disabilities in the USA. Stroke occurs when the blood vessel that supplies oxygen and nutrient to the brain is interrupted; it can be classified into two major categories: ischemic and hemorrhagic. In an ischemic stroke, a blood vessel becomes occluded, and the blood supply to part of the brain is totally or partially blocked, causing brain injury. Hemorrhagic stroke results from a blood vessel rupturing and prevents blood flow to the brain. About 87% of strokes are due to cerebral ischemia (AHA 2011). Currently, the only treatment for ischemic stroke approved by US Food and Drug Administration is the clot-dissolving drug, tissue plasminogen activator (t-PA). However, due to the short therapeutic time window (less than 4 h), only 3–5% of the stroke patients are candidates for t-PA treatment. Therefore, more therapeutic targets are needed for improving stroke treatment.

The development of ischemic brain injury is a complex process, and its underlying mechanisms are not well understood. Ischemia deprives the cell of energy required to maintain intracellular ionic homeostasis. One characteristic of the ionic dysregulation is intracellular acidosis, as a result of a shift from aerobic to anaerobic glycolysis, failure to eliminate the metabolic waste CO_2, and altered function of acid-based transporters [63]. The Na^+/H^+ exchanger, which catalyzes the electroneutral secondary active transport of $1Na^+$ with $1H^+$, plays an important role in regulating intracellular pH (pH_i), Na^+ concentration ($[Na^+]_i$), and cell volume. Following cerebral ischemia, NHE is stimulated by intracellular acidosis to remove excess H^+. During this process, Na^+ accumulates in the cell and stimulates the reversal mode of Na^+/Ca^{2+} exchange (NCX) (Fig. 20.1). As a result, Ca^{2+} accumulates in the cell and eventually contributes to cell death. Thus, NHE presents a target in the acute stage of cerebral ischemia in reducing ionic dysregulation and ischemic cell death.

In this chapter, we reviewed recent evidences that overstimulation of NHE plays a role in ischemic brain injury and the underlying mechanisms. A better understanding of the function and regulation of NHE following cerebral ischemia will provide insight into developing more effective neuroprotective agents for stroke treatment.

2 The Na^+/H^+ Exchanger Isoform 1

2.1 Structural Characteristics of NHE-1

NHEs are a family of membrane transport proteins which catalyze the secondary active electroneutral exchange of one Na^+ for one H^+. To date, nine NHE isoforms (NHE1–9) have been cloned in mammalian tissues, and NHE-1 is the most abundantly expressed isoform in brains [48]. All characterized NHE isoforms consist of about

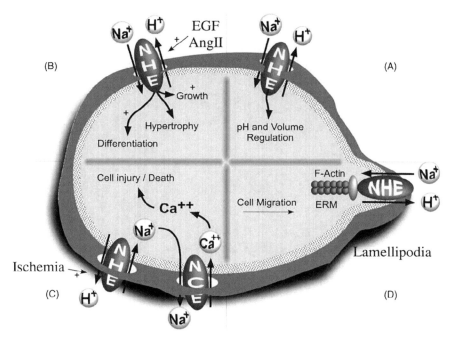

Fig. 20.1 Physiological functions of NHE-1. (**a**) Steady-state pH regulation by NHE-1. (**b**) Hormones such as epidermal growth factor (EGF) and angiotensin II (AngII) activate NHE-1. This leads to increased cell growth and cell differentiation. In the myocardium, this can lead to hypertrophy. (**c**) Activation of NHE-1 during ischemia and reperfusion leads to increased intracellular Na⁺ that results in increased intracellular Ca²⁺ through the Na⁺/Ca²⁺ exchanger (NCE) and ultimately cell injury and cell death. (**d**) In some cells, NHE-1 is found in lamellipodia, where it binds to cytoskeletal proteins through ERM proteins, and its activity are important in cell migration (Adapted from Fliegel [17])

600–900 amino acids with approximately 40% amino acid homology. As shown in Fig. 20.2, NHE-1 is made of 815 amino acids with both N- and C-terminal in the cytosol, forming 12 transmembrane (TM) domains which are highly conserved in most NHE isoforms. The highly conserved N-terminal domain is responsible for cation translocation. TM 6 and TM 7 share 95% homology, and both play a key role in the translocation of Na⁺ and H⁺ [54]. The less conserved C-terminal domain is the main regulatory site for NHE-1 activity and comprised of distinct subdomains which can be modified by phosphorylation or binding to other regulatory factors. In response to environmental signals, diverse signaling pathways act on the C-terminal regulatory domain. The distal C-terminal tail of NHE-1 contains a number of serine and threonine residues that can be phosphorylated by several protein kinases, including extracellular signal-related kinase (ERK1/2), p90 ribosomal S kinase (p90^RSK) [45], and p38 mitogen-activated protein kinase (MAPK) [30].

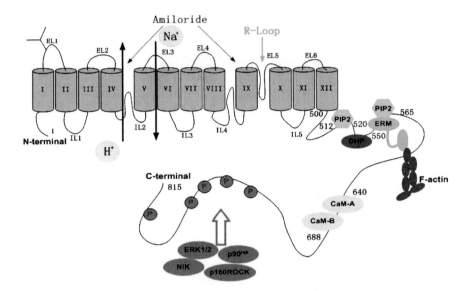

Fig. 20.2 Basic structure and regulatory elements of NHE-1. The topology of the protein illustrates regions of the cytoplasmic domain which are important in regulation or protein–protein interaction, the positions of reentrant loops, and the membrane-associated segments. The associated factors PIP2, ERM, CaM, and CHP are also shown in their approximate known binding sites. "P" indicates the approximate site for phosphorylation of the cytosolic tail of the protein (Adapted from Luo and Sun [40])

2.2 NHE-1 Inhibitors

The major inhibitors of NHE-1 activity fall into two categories: the first consists of amiloride and its analogues, such as ethylisopropylamiloride (EIPA) and dimethylamiloride (DMA). Another class contains the benzoylguanidines and derivatives, such as HOE 694, HOE 642 (cariporide), and EMD-85131 (eniporide) [42]. The NHE isoforms show different sensitivities to inhibition by amiloride: both NHE-1 and NHE-2 appear very sensitive to amiloride and its derivatives, while other isoforms, such as NHE-3 and NHE-4, are insensitive. Moreover, amiloride has been shown to inhibit NHEs in a nonspecific manner, whereas guanidine derivatives are more efficient and selective inhibitors of NHE-1 [42]. HOE compounds have been tested for the treatment of NHE-1-mediated cardiac ischemia-reperfusion injury [44]. Of these, HOE 642 is the most potent, being $\sim 10^4$ to 10^5 times more specific for NHE-1 than NHE-3 and NHE-5 [54].

2.3 Expression and Functions of NHE-1

The isoform NHE-1 is found in the plasma membrane of most mammalian cells and normally described as the housekeeping isoform [17]. Other isoforms have a more restricted tissue distribution and appear to regulate more specialized functions: NHE-2 and NHE-3 are expressed predominantly in the kidney and gastrointestinal tract, while NHE-5 is expressed mainly in the brain [67]. NHE-6 and NHE-7 are exclusively localized in intracellular organelles such as mitochondrial and trans-Golgi; these isoforms are expressed in tissues with high metabolism rates such as heart, brain, and skeletal muscle [7,47]. NHE-8 and NHE-9 are also found to distribute in kidney, stomach, and intestine [12].

Of all the isoforms, NHE-1 is the most extensively characterized member. It resides exclusively on the cell surface but is also present in discrete microdomains of the plasma membrane in different types of cells [48]. For example, NHE-1 is concentrated along the border of lamellipodia in fibroblasts [19], the basolateral membrane of epithelia [6], and the intercalated disks and t-tubules of cardiac myocytes [52]. NHE-1 is believed to exert two fundamental functions. First, it serves as the principal alkalinizing mechanism in many cell types against the damaging effects of excess intracellular acidification. Together with bicarbonate transporting systems, NHE-1 plays a crucial role in maintaining cytoplasmic acid–base balance. Second, it provides a major resource for Na⁺ influx, coupled to Cl⁻ and H_2O uptake, which is required to restore cell volume to steady-state levels following cell shrinkage induced by external osmolality [56]. The cell type-specific localization of NHE-1 in distinct subdomains of the plasma membrane also suggests that it may play more subtle, specialized roles in cell function. For example, NHE-1 expression may be a significant factor in regulating cell morphology, adhesion, and migration (Fig. 20.1). It has been reported that NHE-1 plays an important role in remodeling the cortical actin cytoskeleton and cell shape of fibroblasts through its association with the cytoskeletal associated proteins ezrin, radixin, and moesin (ERMs) [14]. Both the cation translocation and anchorage to the cytoskeleton are required for remodeling focal adhesions at the front and trailing edges of the cell necessary for guided movement [13].

2.4   NHE-1 and Cerebral Ischemia

Neurons are susceptible to injury from acidosis due to their high metabolic rate. In addition, intracellular acidosis can affect neuronal excitability by modulating ion channel gating [60]. Thus, efficient acid-extrusion mechanisms are critical for neuronal function. NHE-1 activity has been found in almost all neuronal and glial cell types. Primary cortical astrocytes and neurons from NHE-1 null mice exhibit reduced steady-state pH_i, as well as absent H⁺ extrusion after intracellular acidification [31,38]. Similar results have also been reported in acutely dissociated CA1 pyramidal neurons from NHE-1 null mice [64]. A recent study shows that NHE-1 is an important regulator of microglia pH_i and plays an essential role in microglial activation [37]. These studies demonstrate that NHE-1 plays a key role in pH_i regulation of neurons and glial cells under steady-state conditions and after intracellular acidification.

NHE-1 is important for neuronal and glial function under physiological conditions. But, overstimulation of NHE-1 has a deleterious effect following cerebral ischemia. During ischemia/hypoxia, NHE-1 is activated by intracellular acidosis, causing an increase in $[Na^+]_i$. The excessive accumulation of Na^+ then leads to an increase in intracellular Ca^{2+} ($[Ca^{2+}]_i$) via NCX, which accelerates the Ca^{2+}-mediated signaling cascade of deleterious events [38]. NHE-1 inhibition has been demonstrated to have a protective effect against Ca^{2+}-mediated damage in different tissues following ischemia/hypoxia, including the endothelial and smooth muscle cells [5] and myocytes [1].

3   NHE-1 and Ischemic Neuronal Death

3.1   Regulation of NHE-1 Following Ischemia

NHE-1 activity is stimulated after ischemia [63]. Diverse stimuli may enhance NHE-1 activity via activation of numerous receptor tyrosine kinases and G protein-coupled receptors. Many growth factors and peptide hormones that activate NHE-1 are thought to transduce their signals through a common MAPK pathway involving mitogen-activated, extracellular signal-related kinase (MEK-ERK)-p90[RSK]. Many factors released after ischemia may stimulate members of MAPK pathway, such as ERK1/2, cJun N-terminal kinase (c-JNK), and p38 MAPK. Among these, the ERK signaling cascade is emerging as an important regulator of neuronal responses to the pathologic stimuli [10]. After activation, ERK1/2 phosphorylates several down-stream elements, including Elk-1 or p90[RSK]. It functions as an important intermediate in signal transduction pathways to transmit extracellular signals into key regulatory membrane, cytoplasmic, and/or nuclear targets. Luo et al. demonstrated that ERK1/2 play a role in stimulation of neuronal NHE-1 following in vitro ischemia [39]. NHE-1 activity was significantly increased in pure cortical neuron cultures during 10–60-min reoxygenation (REOX) after 2-h oxygen and glucose deprivation/reoxygenation (OGD). At 10-min REOX, phosphorylated NHE-1 was increased with a concurrent elevation of phosphorylation of p90[RSK] [39]. Activation of ERK1/2–p90[RSK] pathways following in vitro ischemia phosphorylates NHE-1 and increases its activity, which subsequently contributes to neuronal damages [3]. This was confirmed later in an in vivo ischemia study [41] where ERK and p90[RSK] was stimulated at 3 min after 60-min middle cerebral artery occlusion (MCAO), contributing to phosphorylation of NHE-1. Furthermore, both ERK inhibitor U0126 and p90[RSK] inhibitor FMK reduced brain infarct volume after MCAO [41,46]. This may in part result from blocking NHE-1 activity. However, ERK plays a role in Ca^{2+}-dependent glutamate release via phosphorylation of a synaptic vesicle protein synapsin I [28]. Therefore, it is possible that the neuroprotective effects observed by ERK and p90[RSK] inhibition may also result from attenuating glutamate release in ischemic brains.

3.2 Inhibition of NHE-1 Reduces Neuronal Death Following In Vitro Hypoxia–Ischemia

NHE-1 activity is overstimulated in neurons after OGD to correct the intracellular acidosis. Overstimulation of NHE-1 activity also disrupts Na$^+$ and Ca^{2+} homeostasis and thus contributes to ischemic neuronal damage. Luo et al. reported that 3-h OGD followed by 60-min REOX triggered a sevenfold increase in [Na$^+$]$_i$ and a 1.5-fold increase in [Ca^{2+}]$_i$ in neurons. Twenty-one hours of REOX led to $68 \pm 10\%$ cell death [38]. Inhibition of NHE-1 with the potent inhibitor HOE 642 or genetic ablation of NHE-1 reduced OGD-induced cell death by ~40–50% and attenuates intracellular Na$^+$ and Ca^{2+} accumulation [38]. NHE-1 inhibition was also shown to be protective against glutamate-mediated neurotoxicity. Inhibition of NHE-1 with KR-33028 significantly attenuated glutamate-induced LDH release in cortical neurons in vitro [36]. KR-33028 also had an antiapoptotic effect as shown by TUNEL positivity and caspase-3 activity. Another nonspecific NHE inhibitor SM-20220 dose-dependently attenuated glutamate-induced neuronal death and inhibited the acute cellular swelling following glutamate exposure [43]. SM-20220 also suppresses the persistent [Ca^{2+}]$_i$ increase and intracellular acidification following glutamate exposure. In conclusion, NHE-1 inhibition suppressed neuronal death and cellular swelling induced by glutamate or hypoxia/ischemia through inhibition of both Ca^{2+} influx and acidification in the neurons. These studies imply that activation of the NHE system may exacerbate the progress of cerebral damage and edema after cerebral ischemia.

3.3 Inhibition of NHE-1 Is Neuroprotective Following Cerebral Ischemia

The role of NHE-1 has been investigated following cerebral ischemia in different model systems including focal cerebral ischemia, global cerebral ischemia, and neonatal hypoxia–ischemia (HI). Luo et al. reported a decrease in brain infarct volume with HOE 642 treatment 5 min before 2-h MCAO and 24-h reperfusion [38]. Genetic ablation of NHE-1 reduced mitochondrial cytochrome c release and apoptosis following 30-min MCAO and 24-h reperfusion [62]. Nuclear translocation of apoptosis-inducing factor (AIF), activation of caspase-3, and TUNEL-positive staining and chromatin condensation was significantly in NHE-1$^{+/-}$ mice [62]. Pretreatment of adult gerbils with the amiloride derivative EIPA, a nonselective NHE inhibitor, significantly reduces the extent of CA1 pyramidal neuronal loss following global ischemia [25].

Neuroprotective effect of HOE 642 has also been investigated in HI. HI is a common cause of brain injury in neonates [16]. Disruption of ionic homeostasis is an important consequence of HI and may contribute to brain injury. Brain intracellular alkalosis has been shown to correlate with the severity of brain injury in term infants with neonatal HI [55]. The infants with the most alkaline brain pH$_i$ demonstrated more severe brain injury in the first 2 weeks after birth and worse neurodevelopmental outcome at 1 year of age [49]. This persistent brain intracellular alkalosis is thought to result from excessive activation of NHE. Administration of the nonselective NHE inhibitor N-methyl-isobutyl-amiloride (MIA) ameliorates neonatal brain injury in a mouse HI model [29]. In a recent study, inhibition of NHE-1 with HOE 642 is found to be neuroprotective in neonatal HI brain injury. HOE 642 preserves hippocampal structures and reduces neurodegeneration induced by HI. Inhibition of NHE-1 not only reduces neurodegeneration during the acute stage of HI but also improves the striatum-dependent motor learning and spatial learning at 8 weeks of age after HI [9]. These findings suggest that NHE-1-mediated disruption of ionic homeostasis contributes to striatal and CA1 pyramidal neuronal injury after neonatal HI.

4 NHE-1 in Glial Function and Neuroinflammation

Besides the early responses starting from minutes to hours after stroke onset, there are also delayed inflammatory responses happening from days of stroke onset and can further last for weeks (Fig. 20.3), including microglia/macrophage activation, astrogliosis, neutrophil infiltration, and cytokine/chemokine release. Although some of these responses are found to be helpful for tissue repair processes, they can also be cytotoxic and contribute to cell death. Since these inflammatory responses last for a longer time, they may provide potential therapeutic targets in a prolonged

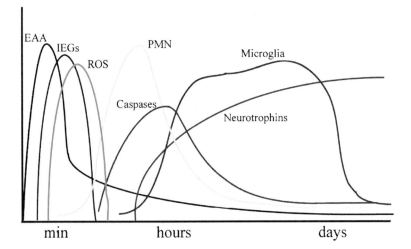

Fig. 20.3 Cellular events of cerebral ischemia. Approximate time frames of selected cellular events in the ischemic brain that may be relevant to therapeutic time windows of neuroprotective strategies. *EAA* excitatory amino acids; *IEGs* immediate early genes; *ROS* reactive oxygen species; *PMN* polymorphonuclear leukocytes (Adapted from Hsu et al. [23])

treatment time window for stroke. NHE-1 is abundantly expressed in glial and microglial cells. One would expect NHE-1 may be involved in glial cell function and thus contribute to delayed inflammatory responses.

4.1 Inhibition of NHE-1 Is Neuroprotective to Astrocytes Following In Vitro Ischemia

Acidosis-induced swelling of glial cells has been reported in C6 glioma cells [26] and astrocytes [53]. This acidosis-induced damage is associated with NHE activity. It suggests that under ischemia-induced acidosis condition, NHE activity appears to be detrimental instead of defending the normal pH_i. Intracellular acidosis has also been reported to be associated with glutamate uptake activity in astrocytes after oxidative stress, in which EIPA-sensitive NHE-1 activity may play a role in the maintenance of pH_i [11]. Kintner et al. showed that NHE-1 is the major pH_i regulatory mechanism in cortical astrocytes, and ablation of NHE-1 in astrocytes attenuates ischemia-induced disruption of ionic regulation and swelling [31]. NHE-1 activity is significantly elevated after in vitro ischemia in mouse cortical astrocytes and leads to overload of intracellular Na^+ and Ca^{2+} [31]. NHE-1$^{-/-}$ astrocytes exhibited lower resting pH_i levels compared to NHE-1$^{+/+}$ astrocytes. And HOE 642 significantly reduced the resting level of pH_i in NHE-1$^{+/+}$ astrocytes. Moreover, NHE-1$^{+/+}$ astrocytes exhibited a rapid pH_i recovery after NH_4Cl prepulse acid load. This recovery was inhibited by HOE 642 or removal of extracellular Na^+. In NHE-1$^{-/-}$ astrocytes, the pH recovery after acidification was impaired. Two hours of OGD leads to an ~80% increase in pH_i recovery rate in NHE-1$^{+/+}$ astrocytes. OGD induced a fivefold rise in intracellular Na^+ and 26% swelling after OGD. With NHE-1 inhibition or genetic ablation, the $[Na^+]_i$ rise and cell swelling are significantly reduced [31].

The role of NHE-1 in astrocyte function has also been investigated in in vivo ischemia. It has been reported that following an ischemic insult, the surviving astrocytes adjacent to the injured tissue undergo hypertrophy, referred to as reactive astrogliosis [58]. Astrocytes with slight cytoplasmic hypertrophy were detected in ischemic noninfarct area as early as 1–3 h postischemia [51]. Eventually, glial cytoplasmic processes create meshwork around the area of necrosis and form the glial scar. It has been reported that NHE-1 was upregulated in hippocampal astrocytes following global cerebral ischemia [9,25]. Hwang et al. reported that inhibition of NHE-1 reduced astrocyte activation gerbil hippocampal CA1 region following global ischemia [25]. However, in mouse brain following either global or focal cerebral ischemia, no effects of NHE-1 inhibition with HOE 642 were detected on astrogliosis [9,57]. Despite the minor effect of HOE 642 treatment on hippocampal astrogliosis, it significantly decreases the hippocampal pyramidal neuronal injury and improves the spatial learning after neonatal HI. Cengiz et al. reported that HOE 642-treated mice had an increased time spent in the training quadrant in the hidden water maze test [9]. More studies are needed to further elucidate the role of NHE-1 in astrogliosis following in vivo ischemia.

4.2 NHE-1 Is Required for Microglial Activation and Proinflammatory Responses After Ischemia

Microglia are resident macrophages ubiquitously distributed throughout the CNS and serve as neurological sensors. They can be rapidly activated under many pathological conditions, including neurodegenerative disease, brain tumor, trauma, infection, and stroke [18,27,61]. Activated microglia undergo a series of transformations including morphological change, proliferation, migration, and upregulation of surface markers such as CD45, CD4, and MHC class I molecules [61]. Activated microglia release biologically active substances, including reactive oxygen species (ROS), nitrogen species, cytokines, and growth factors [21,65]. Although microglia can contribute to tissue repair by cleaning up the cell debris and secretion of neurotrophic factors [34], activated microglia can also exacerbate brain injury by cytotoxic proinflammatory responses [65,66], including the production of many factors such as IL-1b, TNF-a, ROS, and NO [27,32].

One important component for microglial activation and function is the NADPH oxidase (NOX). NOX catalyzes the reduction reaction of molecular oxygen to superoxide anion using NADPH as an electron donor and is the major source of ROS production in microglia [4,21]. In this process, H^+ accumulates inside microglia as a by-product, causing depolarization and cytoplasmic acidification [12]. NOX is sensitive to intracellular pH and has an optimal pH_i of 7.2 [22]. Intracellular acidosis may impair NOX function. Thus, NHE-1 is required to maintain an optimal pH_i and sustain microglial respiratory burst. In cultured microglia, activation by several stimuli depends on NHE-1-mediated H^+ homeostasis [37]. Inhibition of NHE-1 with HOE 642 impaired pH_i regulation in microglia under basal conditions. HOE 642 also reduced the production of superoxide anion as well as proinflammatory cytokines IL-6, IL-1β, and TNF-α induced by LPS or in vitro ischemia [37]. These results were supported by evidence from in vivo ischemia. In our recent study using the mouse MCAO model, activation of microglia is significantly reduced with NHE-1 inhibition or genetic knockdown [57]. Inhibition of NHE-1 also reduces proinflammatory cytokine production and phagocytosis. Since microglia activation is found to last for days after stroke onset, the effect of NHE-1 provides a promising therapeutic target with prolonger time window.

4.3 NHE-1 and Blood–Brain Barrier Dysfunction

The blood–brain barrier (BBB) is a physical and metabolic barrier between blood vessels and surrounding parenchyma tissues and is vital for the homeostasis and normal functions of the CNS. Cerebral microvascular endothelial cells form the anatomical basis of the BBB, and tight junctions (TJs) in the BBB play essential roles in the maintenance of barrier functions [20]. It has been well established that the disruption of TJ proteins in the BBB is a critical event during cerebral ischemia and that this disruption is followed by the passive diffusion of water into the brain, which is responsible for vasogenic edema and secondary brain damage [24]. Cerebral vasogenic edema is a major fatal complication of acute ischemic stroke [2]. Although neuronal cells along with glial cells have long been a prime therapeutic target against cerebral ischemia injury, BBB protection can thus be seen as one aspect of rescuing the neurovascular unit, an important objective in neuroprotection [15].

NHE inhibitors have been reported to ameliorate brain edema after ischemic insult [33,59]; however, its relevance to BBB endothelial cells was not well investigated until recently. NHE-1 and NHE-2 are present in BBB endothelial cells. Hypoxia or aglycemia significantly increases cerebral microvascular endothelial NHE activity [35], suggesting that BBB NHE may be stimulated during ischemia. In a recent study, effects of a specific NHE-1 inhibitor sabiporide on the disruption of TJ proteins in BBB endothelial cells have been investigated after cerebral ischemia [50]. NHE-1 inhibition by sabiporide attenuates ischemia/aglycemic hypoxia-induced BBB hyperpermeability and TJ protein disruption. In BBB endothelial cells, abnormally high levels of $[Ca^{2+}]_i$ during ischemia/hypoxia can impair barrier function since Ca^{2+} plays a critical role in the maintenance of TJ integrity and normal barrier function [8]. Sabiporide was shown to inhibit the Ca^{2+} overload induced by hypoxia in endothelial cells [50]. The preservation of the functional and structural integrity of the BBB endothelial cells is likely to be a crucial issue for any therapeutic approach to the amelioration of ischemic brain damage, and it reveals another mechanism on how NHE-1 contributes to ischemic brain injury.

5 Conclusions

In this chapter, we reviewed recent experimental findings on NHE-1 in cerebral ischemia. NHE-1 plays an important role in maintaining ionic homeostasis under normal physiological conditions. However, excessive stimulation of NHE-1 appears to be a major contributor to the cellular damage after ischemia or hypoxia. This involves both neurons and glial cells and different stages of ischemia (acute and chronic phases after stroke onset), and affects long-term neurological functions. Both pharmacological inhibition and transgenic knockdown of NHE-1 show protections in different cerebral ischemic models (both in vitro and in vivo). Inhibition of NHE-1 also has a profound effect on neuroinflammation and edema formation, providing a longer treatment time window for stroke therapy. Therefore, NHE-1 emerges as an important target for developing new therapeutics for stroke treatment.

Acknowledgments This work was supported by NIH grants R01NS 48216 and R01NS 38118 (D. Sun).

References

1. An J, Varadarajan SG, Camara A, Chen Q, Novalija E, Gross GJ, Stowe DF. Blocking Na(+)/H(+) exchange reduces [Na(+)](i) and [Ca(2+)](i) load after ischemia and improves function in intact hearts. Am J Physiol Heart Circ Physiol. 2001;281:H2398–409.
2. Ayata C, Ropper AH. Ischaemic brain oedema. J Clin Neurosci. 2002;9:113–24.
3. Back SA, Luo NL, Borenstein NS, Levine JM, Volpe JJ, Kinney HC. Late oligodendrocyte progenitors coincide with the developmental window of vulnerability for human perinatal white matter injury. J Neurosci. 2001;21:1302–12.
4. Bedard K, Krause KH. The NOX family of ROS-generating NADPH oxidases: physiology and pathophysiology. Physiol Rev. 2007;87(1):245–313.
5. Besse S, Tanguy S, Boucher F, Huraux C, Riou B, Swynghedauw B, de Leiris J. Protection of endothelial-derived vasorelaxation with cariporide, a sodium-proton exchanger inhibitor, after prolonged hypoxia and hypoxia-reoxygenation: effect of age. Eur J Pharmacol. 2006;531:187–93.
6. Biemesderfer D, Reilly RF, Exner M, Igarashi P, Aronson PS. Immunocytochemical characterization of Na(+)-H+ exchanger isoform NHE-1 in rabbit kidney. Am J Physiol. 1992;263:F833–40.
7. Brett CL, Wei Y, Donowitz M, Rao R. Human Na^+/H^+ exchanger isoform 6 is found in recycling endosomes of cells, not in mitochondria. Am J Physiol Cell Physiol. 2002;282:C1031–41.
8. Brown RC, Davis TP. Calcium modulation of adherens and tight junction function: a potential mechanism for blood-brain barrier disruption after stroke. Stroke. 2002;33:1706–11.
9. Cengiz P, Kleman N, Uluc K, Kendigelen P, Hagemann T, Akture E, Messing A, Ferrazzano P, Sun D. Inhibition of Na+/H+ exchanger isoform 1 is neuroprotective in neonatal hypoxic ischemic brain injury. Antioxid Redox Signal. 2011;14:1803–13.
10. Chu CT, Levinthal DJ, Kulich SM, Chalovich EM, DeFranco DB. Oxidative neuronal injury. The dark side of ERK1/2. Eur J Biochem. 2004;271:2060–6.
11. Daskalopoulos R, Korcok J, Farhangkhgooee P, Karmazyn M, Gelb AW, Wilson JX. Propofol protection of sodium-hydrogen exchange activity sustains glutamate uptake during oxidative stress. Anesth Analg. 2001;93:1199–204.
12. De Vito P. The sodium/hydrogen exchanger: a possible mediator of immunity. Cell Immunol. 2006;240:69–85.
13. Denker SP, Barber DL. Cell migration requires both ion translocation and cytoskeletal anchoring by the Na-H exchanger NHE1. J Cell Biol. 2002;159(6):1087–96.
14. Denker SP, Huang DC, Orlowski J, Furthmayr H, Barber DL. Direct binding of the Na–H exchanger NHE1 to ERM proteins regulates the cortical cytoskeleton and cell shape independently of H+ translocation. Mol Cell. 2000;6:1425–36.
15. Fagan SC, Hess DC, Hohnadel EJ, Pollock DM, Ergul A. Targets for vascular protection after acute ischemic stroke. Stroke. 2004;35:2220–5.
16. Ferriero DM. Neonatal brain injury. N Engl J Med. 2004;351:1985–95.
17. Fliegel L. The Na+/H+ exchanger isoform 1. Int J Biochem Cell Biol. 2005;37:33–7.
18. Graeber MB, Streit WJ. Microglia: biology and pathology. Acta Neuropathol. 2010; 119(1):89–105.
19. Grinstein S, Woodside M, Waddell TK, Downey GP, Orlowski J, Pouyssegur J, Wong DC, Foskett JK. Focal localization of the NHE-1 isoform of the Na+/H+ antiport: assessment of effects on intracellular pH. EMBO J. 1993;12:5209–18.
20. Harhaj NS, Antonetti DA. Regulation of tight junctions and loss of barrier function in pathophysiology. Int J Biochem Cell Biol. 2004;36:1206–37.
21. Harrigan TJ, Abdullaev IF, Jourd'heuil D, Mongin AA. Activation of microglia with zymosan promotes excitatory amino acid release via volume-regulated anion channels: the role of NADPH oxidases. J Neurochem. 2008;106:2449–62.
22. Henderson LM, Chappell JB, Jones OT. Internal pH changes associated with the activity of NADPH oxidase of human neutrophils. Further evidence for the presence of an H+ conducting channel. Biochem J. 1988;251:563–7.

23. Hsu CY, Ahmed SH, Lees KR. The therapeutic time window—theoretical and practical considerations. J Stroke Cerebrovasc Dis. 2000;9(6 Pt 2):24–31.

24. Huber JD, Egleton RD, Davis TP. Molecular physiology and pathophysiology of tight junctions in the blood-brain barrier. Trends Neurosci. 2001;24:719–25.

25. Hwang IK, Yoo KY, An SJ, Li H, Lee CH, Choi JH, Lee JY, Lee BH, Kim YM, Kwon YG, Won MH. Late expression of Na⁺/H⁺ exchanger 1 (NHE1) and neuroprotective effects of NHE inhibitor in the gerbil hippocampal CA1 region induced by transient ischemia. Exp Neurol. 2008;212:314–23.

26. Jakubovicz DE, Klip A. Lactic acid-induced swelling in C6 glial cells via Na⁺/H⁺ exchange. Brain Res. 1989;485:215–24.

27. Jin R, Yang G, Li G. Inflammatory mechanisms in ischemic stroke: role of inflammatory cells. J Leukoc Biol. 2010;87(5):779–89.

28. Jovanovic JN, Sihra TS, Nairn AC, Hemmings Jr HC, Greengard P, Czernik AJ. Opposing changes in phosphorylation of specific sites in synapsin I during Ca2+-dependent glutamate release in isolated nerve terminals. J Neurosci. 2001;21:7944–53.

29. Kendall GS, Robertson NJ, Iwata O, Peebles D, Raivich G. N-methyl-isobutyl-amiloride ameliorates brain injury when commenced before hypoxia ischemia in neonatal mice. Pediatr Res. 2006;59:227–31.

30. Khaled AR, Moor AN, Li A, Kim K, Ferris DK, Muegge K, Fisher RJ, Fliegel L, Durum SK. Trophic factor withdrawal: p38 mitogen-activated protein kinase activates NHE1, which induces intracellular alkalinization. Mol Cell Biol. 2001;21:7545–57.

31. Kintner DB, Su G, Lenart B, Ballard AJ, Meyer JW, Ng LL, Shull GE, Sun D. Increased tolerance to oxygen and glucose deprivation in astrocytes from Na⁺/H⁺ exchanger isoform 1 null mice. Am J Physiol Cell Physiol. 2004;287:C12–21.

32. Kreutzberg GW. Microglia: a sensor for pathological events in the CNS. Trends Neurosci. 1996;19:312–8.

33. Kuribayashi Y, Itoh N, Kitano M, Ohashi N. Cerebroprotective properties of SM-20220, a potent Na⁺/H⁺ exchange inhibitor, in transient cerebral ischemia in rats. Eur J Pharmacol. 1999;383:163–8.

34. Lalancette-Hebert M, Gowing G, Simard A, Weng YC, Kriz J. Selective ablation of proliferating microglial cells exacerbates ischemic injury in the brain. J Neurosci. 2007;27(10):2596–605.

35. Lam TI, Wise PM, O'Donnell ME. Cerebral microvascular endothelial cell Na/H exchange: evidence for the presence of NHE1 and NHE2 isoforms and regulation by arginine vasopressin. Am J Physiol Cell Physiol. 2009;297:C278–89.

36. Lee BK, Lee DH, Park S, Park SL, Yoon JS, Lee MG, Lee S, Yi KY, Yoo SE, Lee KH, Kim YS, Lee SH, Baik EJ, Moon CH, Jung YS. Effects of KR-33028, a novel Na+/H+ exchanger-1 inhibitor, on glutamate-induced neuronal cell death and ischemia-induced cerebral infarct. Brain Res. 2009;1248:22–30.

37. Liu Y, Kintner DB, Chanana V, Algharabli J, Chen X, Gao Y, Chen J, Ferrazzano P, Olson JK, Sun D. Activation of microglia depends on Na+/H+ exchange-mediated H+ homeostasis. J Neurosci. 2010;30:15210–20.

38. Luo J, Chen H, Kintner DB, Shull GE, Sun D. Decreased neuronal death in Na⁺/H⁺ exchanger isoform 1-null mice after in vitro and in vivo ischemia. J Neurosci. 2005;25:11256–68.

39. Luo J, Kintner DB, Shull GE, Sun D. ERK1/2-p90RSK-mediated phosphorylation of Na(+)/H(+) exchanger isoform 1. A role in ischemic neuronal death. J Biol Chem. 2007;282:28274–84.

40. Luo J, Sun D. Physiology and pathophysiology of Na(+)/H(+) exchange isoform 1 in the central nervous system. Curr Neurovasc Res. 2007;4:205–15.

41. Manhas N, Shi Y, Taunton J, Sun D. p90 activation contributes to cerebral ischemic damage via phosphorylation of Na+/H+ exchanger isoform 1. J Neurochem. 2010;114:1476–86.

42. Masereel B, Pochet L, Laeckmann D. An overview of inhibitors of Na(+)/H(+) exchanger. Eur J Med Chem. 2003;38:547–54.

43. Matsumoto Y, Yamamoto S, Suzuki Y, Tsuboi T, Terakawa S, Ohashi N, Umemura K. Na$^+$/H$^+$ exchanger inhibitor, SM-20220, is protective against excitotoxicity in cultured cortical neurons. Stroke. 2004;35:185–90.
44. Mentzer Jr RM, Bartels C, Bolli R, Boyce S, Buckberg GD, Chaitman B, Haverich A, Knight J, Menasche P, Myers ML, Nicolau J, Simoons M, Thulin L, Weisel RD. Sodium-hydrogen exchange inhibition by cariporide to reduce the risk of ischemic cardiac events in patients undergoing coronary artery bypass grafting: results of the EXPEDITION study. Ann Thorac Surg. 2008;85:1261–70.
45. Moor AN, Gan XT, Karmazyn M, Fliegel L. Activation of Na$^+$/H$^+$ exchanger-directed protein kinases in the ischemic and ischemic-reperfused rat myocardium. J Biol Chem. 2001;276:16113–22.
46. Namura S, Iihara K, Takami S, Nagata I, Kikuchi H, Matsushita K, Moskowitz MA, Bonventre JV, Alessandrini A. Intravenous administration of MEK inhibitor U0126 affords brain protection against forebrain ischemia and focal cerebral ischemia. Proc Natl Acad Sci U S A. 2001;98:11569–74.
47. Numata M, Orlowski J. Molecular cloning and characterization of a novel (Na+, K+)/H+ exchanger localized to the trans-Golgi network. J Biol Chem. 2001;276:17387–94.
48. Orlowski J, Grinstein S. Diversity of the mammalian sodium/proton exchanger SLC9 gene family. Pflugers Arch. 2004;447:549–65.
49. Ott M, Robertson JD, Gogvadze V, Zhivotovsky B, Orrenius S. Cytochrome c release from mitochondria proceeds by a two-step process. Proc Natl Acad Sci U S A. 2002;99:1259–63.
50. Park SL, Lee DH, Yoo SE, Jung YS. The effect of Na(+)/H(+) exchanger-1 inhibition by sabiporide on blood-brain barrier dysfunction after ischemia/hypoxia in vivo and in vitro. Brain Res. 2010;1366:189–96.
51. Petito CK, Babiak T. Early proliferative changes in astrocytes in postischemic noninfarcted rat brain. Ann Neurol. 1982;11:510–8.
52. Petrecca K, Atanasiu R, Grinstein S, Orlowski J, Shrier A. Subcellular localization of the Na$^+$/H$^+$ exchanger NHE1 in rat myocardium. Am J Physiol. 1999;276:H709–17.
53. Plesnila N, Haberstok J, Peters J, Kolbl I, Baethmann A, Staub F. Effect of lactacidosis on cell volume and intracellular pH of astrocytes. J Neurotrauma. 1999;16:831–41.
54. Putney LK, Denker SP, Barber DL. The changing face of the Na$^+$/H$^+$ exchanger, NHE1: structure, regulation, and cellular actions. Annu Rev Pharmacol Toxicol. 2002;42:527–52.
55. Robertson NJ, Cowan FM, Cox IJ, Edwards AD. Brain alkaline intracellular pH after neonatal encephalopathy. Ann Neurol. 2002;52:732–42.
56. Rotin D, Grinstein S. Impaired cell volume regulation in Na(+)-H+ exchange-deficient mutants. Am J Physiol. 1989;257:C1158–65.
57. Shi Y, Chanana V, Watters JJ, Ferrazzano P, Sun D. Role of sodium/hydrogen exchanger isoform 1 in microglial activation and proinflammatory responses in ischemic brains. J Neurochem. 2011;119(1):124–35.
58. Sofroniew MV, Vinters HV. Astrocytes: biology and pathology. Acta Neuropathol. 2010;119:7–35.
59. Suzuki Y, Matsumoto Y, Ikeda Y, Kondo K, Ohashi N, Umemura K. SM-20220, a Na(+)/H(+) exchanger inhibitor: effects on ischemic brain damage through edema and neutrophil accumulation in a rat middle cerebral artery occlusion model. Brain Res. 2002;945:242–8.
60. Takahashi KI, Copenhagen DR. Modulation of neuronal function by intracellular pH. Neurosci Res. 1996;24:109–16.
61. Tambuyzer BR, Ponsaerts P, Nouwen EJ. Microglia: gatekeepers of central nervous system immunology. J Leukoc Biol. 2009;85:352–70.
62. Wang Y, Luo J, Chen X, Chen H, Cramer SW, Sun D. Gene inactivation of Na+/H+ exchanger isoform 1 attenuates apoptosis and mitochondrial damage following transient focal cerebral ischemia. Eur J Neurosci. 2008;28:51–61.
63. Yao H, Haddad GG. Calcium and pH homeostasis in neurons during hypoxia and ischemia. Cell Calcium. 2004;36:247–55.

64. Yao H, Ma E, Gu XQ, Haddad GG. Intracellular pH regulation of CA1 neurons in Na(⁺)/H(⁺) isoform 1 mutant mice. J Clin Invest. 1999;104:637–45.
65. Yenari MA, Kauppinen TM, Swanson RA. Microglial activation in stroke: therapeutic targets. Neurotherapeutics. 2010;7(4):378–91.
66. Yoshioka H, Niizuma K, Katsu M, Okami N, Sakata H, Kim GS, Narasimhan P, Chan PH. NADPH oxidase mediates striatal neuronal injury after transient global cerebral ischemia. J Cereb Blood Flow Metab. 2010;31(3):868–80.
67. Zachos NC, Tse M, Donowitz M. Molecular physiology of intestinal Na+/H+ exchange. Annu Rev Physiol. 2005;67:411–43.

Chapter 21
Iron as a Therapeutic Target in Intracerebral Hemorrhage: Preclinical Testing of Deferoxamine

Ya Hua, Richard F. Keep, Yuxiang Gu, and Guohua Xi

Abstract Intracerebral hemorrhage (ICH) is a subtype of stroke with high mortality. Experimental studies have found that brain iron overload occurs after ICH and iron has a key role in ICH-induced brain injiury. Deferoxamine, an iron chelator, reduces hematoma-induced brain edema, neuronal death, brain atrophy and neurological deficits. Iron chelation with deferoxamine could be a new therapy for ICH.

1 Introduction

Each year, approximately 720,000 people suffer a stroke in the USA. The causes of stroke are, in general, either hemorrhagic or nonhemorrhagic. Intracerebral hemorrhage is a common and often fatal stroke subtype [1, 2]. About 15% of patients (30,000 annually) die from spontaneous ICH.

At present, there is no effective treatment that attenuates brain edema and improves long-term outcome in ICH. Our previous studies have shown that iron plays an important role in edema formation, brain atrophy, and behavioral deficits. Iron chelation with deferoxamine could be a new therapy for ICH. Such a therapy would be of great benefit not only to the patients but also their caregivers and to society in general by reducing the cost of care for hemorrhagic stroke patients.

Y. Hua, MD • R.F. Keep, PhD • Y. Gu, MD, PhD • G. Xi, MD (✉)
Department of Neurosurgery, University of Michigan,
5018 BSRB, 109 Zina Pitcher Pl, Ann Arbor, MI 48109-2200, USA
e-mail: guohuaxi@umich.edu

P.A. Lapchak and J.H. Zhang (eds.), *Translational Stroke Research*, Springer Series
in Translational Stroke Research, DOI 10.1007/978-1-4419-9530-8_21,
© Springer Science+Business Media, LLC 2012

2 Brain Iron Overload After ICH

Iron is essential for normal brain function, but iron overload may have devastating effects [3]. Iron overload contributes to many kinds of brain injury including ICH, Alzheimer's disease, and Parkinson's disease [4–6]. In ICH, the clot lyses and iron is released from heme within the first week, and this appears to contribute to acute brain edema formation [7]. However, brain atrophy occurs several weeks later, suggesting that iron exposure causes cell damage that results in delayed cell death (e.g., iron may cause sufficient damage that the cell cannot repair itself). If such death occurs in cells that are storing iron released from the hematoma, the new iron release may affect nearby cells, leading to amplification of the lesion. Alternately, the iron stored in cells after resolution of the clot may be naturally released for clearance across the blood–brain barrier or through the cerebrospinal fluid, but with potential for causing further tissue damage.

After erythrocyte lysis, iron concentrations in the brain can reach very high levels. Our data have showed a threefold increase of brain nonheme iron after intracerebral hemorrhage in rats, and it remains high for at least 28 days [5]. Intracerebral infusion of iron causes brain edema and an iron chelator reduces hematoma- and hemoglobin-induced edema, suggesting that iron plays an important role in edema formation after ICH [8, 9].

3 Deferoxamine

Deferoxamine (DFX), an iron chelator, is an FDA-approved drug for the treatment of acute iron intoxication and of chronic iron overload due to transfusion-dependent anemias. It has a molecular weight of 657. Deferoxamine can rapidly penetrate the blood–brain barrier and accumulate in the brain tissue at a significant concentration after systemic administration [10, 11]. The terminal half-life of DFX after intravenous infusion is 3.05 h [12]. Deferoxamine chelates iron by forming a stable complex that prevents the iron from entering into further chemical reactions. It readily chelates iron from ferritin and hemosiderin but not readily from transferrin.

Deferoxamine binds ferric iron and prevents the formation of hydroxyl radical via the Fenton/Haber-Weiss reaction. Deferoxamine reduces hemoglobin-induced brain Na^+/K^+ ATPase inhibition and neuronal toxicity [13–15]. Favorable effects of iron chelator therapy have been reported in various cerebral ischemia models [16, 17].

Although DFX is an iron chelator, it can have other effects. Thus, it can act as a direct free radical scavenger [16, 17], and it can induce ischemic tolerance in the brain [18]. The latter has been demonstrated in vivo and in vitro, and it may be related to a DFX induction of hypoxia-inducible transcription factor 1 binding to DNA [18].

4 Animal Models

We have used two animal models for our DFX testing. The first is a rat ICH model. Experimental models of ICH have been available since the 1960s and commonly involve the injection of autologous blood into the frontal lobe of dogs, cats, pigs, or monkeys [19–22]. Now, rodents have been found to provide equally convenient and suitable models [7, 23–26]. They have the advantages of a lower cost (for animals and procedures), a relative homogeneity within strains owing to inbreeding, a close resemblance of the cerebrovascular anatomy and physiology to that of higher species, and a small brain size well suited to immunohistochemical and biochemical studies. We chose to inject fresh autologous blood (usually 100 μL) into the caudate nucleus at a rate of 10 μL/min because it produces a reproducible lesion that lends itself to quantitative measurement. We found that a more rapid injection rate resulted in a variable reflux of blood along the needle track and poorly reproducible lesions. The selection of the caudate as the site of infusion was based on considerable research indicating that behavioral deficits and treatment effects can be assessed chronically with high sensitivity and reliability. There are other models of ICH. In particular, a collagenase model has been used extensively [27, 28]. However, the collagenase-induced widespread disruption of the extracellular matrix (including the endothelial basement membrane) is not found in human ICH. It also appears to induce areas of ischemia that we do not find after injection of blood [23, 29]. Recent human studies suggest that ischemia is not a major component of ICH-induced injury [30].

ICH causes greater neurological deficits, more severe brain swelling, greater induction of heat shock proteins, and enhances microglial activation in aged rats compared to young rats [31]. These results suggest that age is a significant factor in determining brain injury after ICH. In addition, behavioral data show that the temporal profiles of recovery in aged and young rats are identical, suggesting that it is differences in acute injury that cause the greater brain swelling and neurological deficits in aged rats rather than less plasticity. These age-dependent changes in ICH-induced brain injury have led us to use aged rats for our rat experiments.

The second ICH model that we have used to study DFX is the pig. The well-developed white matter of the porcine brain (in contrast to the rat) makes it very useful for studying white matter injury after ICH. We have used the porcine ICH model to examine whether systemic DFX treatment reduces brain edema and brain atrophy after ICH. These experiments address three issues. The first is whether the protective effects of DFX occur in more than one species, thereby increasing confidence that the drug would work in humans. The second relates to clot size. The duration over which clot lysis and iron release are likely to be dependent on clot size (i.e., 3 days for a 100 μL clot in rats and 1 week for a 2.5 mL clot in pigs). Thus, it is possible that there may be a longer therapeutic time window for DFX in pigs (hematoma 2.5 mL) than rats (hematoma 100 μL), i.e., there may be a beneficial effect at 24 or 48 h in the pig. If so, it may suggest that there would be a longer time window in humans as well. The third is to answer whether DFX protects white matter injury or gray matter injury or both.

5   Major Endpoints

The three major endpoints in our studies are brain edema, brain atrophy, and neu-rological scores. Perihematomal edema is thought by many, but not all, to be a major cause of death and disability following ICH, particularly in relation to her-niation [32–35]. In combination with the presence of the hematoma, further mass effect due to edema formation can result in a midline shift and herniation. Perihematomal edema, as seen on CT scan, can result from clot retraction. However, that represents a redistribution of fluid between hematoma and brain. A progressive mass effect can only result from a movement of fluid from the blood to brain, either in the form of hematoma enlargement and/or progressive perihematomal edema. Although all agree that the latter does occur, its extent/prevalence has been debated [34]. It is easy to examine where perihematomal tissue can be sampled and water content determined directly in animal models. In animals, there is substantial evi-dence for progressive edema formation [7, 22, 23, 36, 37]. It should be noted that apart from its direct potential importance as a clinical endpoint, brain edema is also an easily quantifiable marker of brain injury. Therapeutic agents that reduce perihe-matomal cellular injury will almost certainly reduce edema formation (although the converse may not necessarily be true). Our brain edema measurements are done in concert with Na^+ and K^+ content measurements. ICH-induced edema is associated with an increase in brain Na^+, while K^+ (a predominantly intracellular cation) loss from the brain is an indirect measure of brain injury.

Brain atrophy has been found in rats, pigs, and humans after ICH [38–40]. The underlying cause(s) of this atrophy is, however, unknown. Our recent studies sug-gested iron overload is associated with brain atrophy after ICH, and deferoxamine reduces ICH-induced brain atrophy in rats. In rat brain sections, hematoxylin and eosin staining was used to examine brain atrophy [41]. Coronal sections from 1 mm posterior to the blood injection site were used [39]. The caudate, cortex, and lateral ventricular areas on both sides (ipsilateral and contralateral) were measured using NIH image program. Brain atrophy was determined at day 56 because our previous study demonstrates that brain atrophy develops gradually and peaks between 1 and 2 months after ICH in the rat [41].

We also use behavioral outcomes as an endpoint. Three sensorimotor behavioral tests, forelimb placing, forelimb use asymmetry, and corner turn test, are sensitive enough to detect behavioral deficits in the first day after ICH in rats [42]. We have found that forelimb placing and forelimb use asymmetry tests can detect even mild neurological deficits such as that found after 15 min of middle cerebral artery occlu-sion with reperfusion that does not cause detectable tissue damage but which ren-ders the brain less vulnerable to a second, longer occlusion that otherwise causes extensive cell death (i.e., ischemic tolerance—the shorter duration of occlusion may transiently traumatize cells which preconditions them so that they are resistant to later, longer occlusions) [43]. All three tests appear to be well suited to models of unilateral brain injury because they measure asymmetry. Thus, they can factor out confounding variables for behavioral tests such as decreased overall activity following surgery. These sensorimotor tests are also not altered by repeated testing, and they do not require special training or food deprivation [44].

In contrast to the rat, little work has been done developing behavioral tests for pig ICH. Such tests would enhance the field.

6 Optimal Deferoxamine Dose Determination

The optimal DFX dose was examined in an ICH model in aged rats [45]. Fischer 344 rats (18 months old) had an intracaudate injection of 100-μL autologous whole blood and were treated with different doses of DFX (10, 50, and 100 mg/kg, i.m.) or vehicle 2 and 6 h post-ICH and then every 12 h up to 7 days. Rats were euthanized at day 3 for brain edema determination and day 56 for brain atrophy measurement. Behavioral tests were performed during the experiments. The maximal dose, 100 mg/kg, was chosen based on our previous studies which indicated this dose was effective in reducing ICH-induced brain injury [9, 41]. The lower doses were chosen based on normal dosing in humans for other conditions. Deferoxamine is normally given as a 1,000-mg dose followed by 500 mg every 4–12 h if needed. The maximal recommended daily dose is 6,000 mg. Assuming a body weight of 80 kg, the initial human dose would be 12.5 mg/kg, hence, our lower dose of 10 mg/kg. An intermediate dose of 50 mg/kg was also chosen because of the possibility that ICH may require a higher therapeutic dose than a systemic disease. In this study, we found that systemic treatment of 50 or 100 mg/kg DFX significantly reduced ICH-induced perihematomal brain edema at 3 days after ICH (Fig. 21.1) and reduced ipsilateral ventricle enlargement and caudate atrophy 2 months after ICH (Fig. 21.2) compared to the vehicle-treated rats with ICH. DFX also improved behavioral outcomes after ICH (Fig. 21.3). On the other hand, while ICH rats treated with 10 mg/kg DFX showed significant brain edema reduction and attenuation of ipsilateral ventricle enlargement compared to vehicle-treated ICH rats, caudate atrophy was not significantly reduced, and residual neurological deficits were present at 56 days after ICH in the corner turn test. These results indicate that a dose higher than 10 mg/kg is the optimal dose of DFX in rat ICH model.

7 Therapeutic Window and Optimal Durations
   for Deferoxamine Treatment

The therapeutic time window and optimal duration for DFX treatment were examined in a rat model of ICH [46]. Fischer 344 rats (18 months old) had an intracaudate injection of 100-μL autologous whole blood, followed by intramuscular DFX or vehicle beginning at different time points or continuing for different durations. Subgroups of rats were euthanized at day 3 for brain edema measurement and day

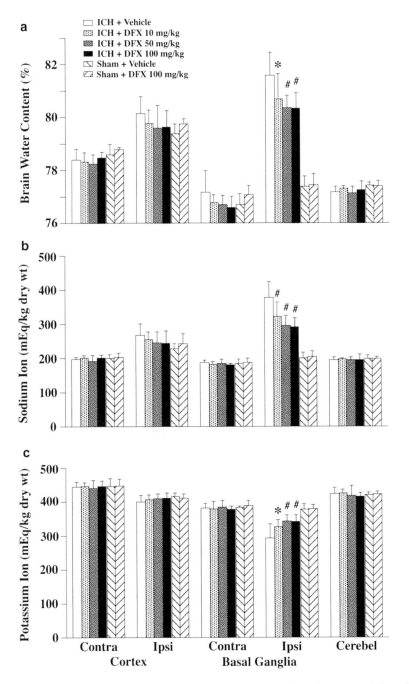

Fig. 21.1 Bar graph showing brain water (**a**), sodium (**b**), and potassium (**c**) content at 3 days after ICH. Values are expressed as the means ± SD. *Contra*, contralateral; *Ipsi*, ipsilateral; *Cerebel*, cerebellum. *$p<0.05$, #$p<0.01$ vs. ICH + Vehicle group. Figure reprinted with permission from Okauchi et al. [45]

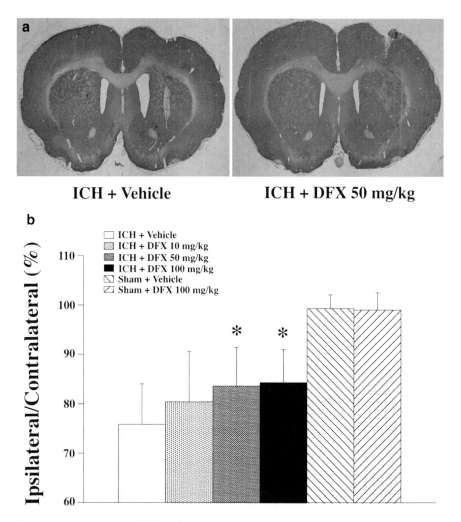

Fig. 21.2 (**a**) Coronal gross H&E sections 8 weeks after ICH treated with vehicle or DFX 50 mg/kg. (**b**) Bar graph demonstrating caudate size expressed as a percentage of the contralateral side. Values are expressed as the means ± SD. *$p < 0.05$, #$p < 0.01$ vs. ICH + Vehicle group. Figure reprinted with permission from Okauchi et al. [45]

56 for brain atrophy determination. Behavioral tests were carried out on days 1, 28, and 56 post-ICH. We found that systemic administration of DFX, when begun within 12 h after ICH, reduced brain edema. DFX treatment started 2 h after ICH and administered for 7 days or more attenuated ICH-induced ventricle enlargement, caudate atrophy, and neurological deficits. DFX attenuated ICH-induced brain atrophy and neurological deficits without detectable side effects when started within 24 h and administered for 7 days. However, DFX treatment with a 48 h delay failed to reduce brain damage. These results suggest the therapeutic time window for brain atrophy and functional outcome is 24 h in aged rats (Fig. 21.4).

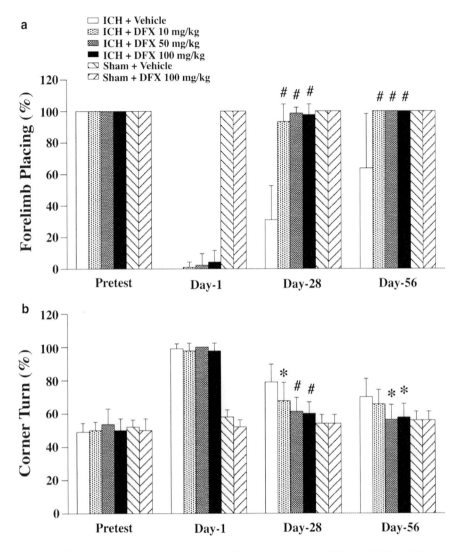

Fig. 21.3 Forelimb placing (**a**) and corner turn (**b**) test scores prior to ICH, and 1, 28, and 56 days after ICH. Values are expressed as the means ± SD. *$p < 0.05$, #$p < 0.01$ vs. ICH + Vehicle group. Figure reprinted with permission from Okauchi et al. [45]

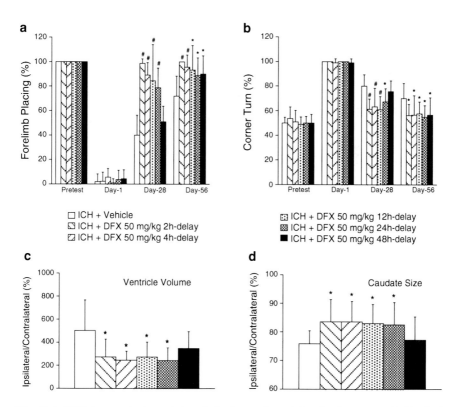

Fig. 21.4 DFX therapeutic time window for brain atrophy and functional outcome. (**a**) Forelimb placing test, (**b**) corner turn test, (**c**) ventricle volume expressed as a percentage of the contralateral side at 8 weeks after ICH, and (**d**) caudate size expressed as a percentage of the contralateral side at 8 weeks after ICH. Values are expressed as the means ± SD. $*p < 0.05$, $\#p < 0.01$ vs. ICH + Vehicle group. Figure reprinted with permission from Okauchi et al. [46]

8 Deferoxamine Studies in Pigs

The effects of DFX on ICH-induced brain injury were also examined in pigs [47]. Pigs received an injection of autologous blood into the right frontal lobe. Deferoxamine (50 mg/kg, i.m.) or vehicle was administered 2 h after ICH and then every 12 h up to 7 days. Animals were euthanized 3 or 7 days later to examine iron accumulation, white matter injury, and neuronal death.

In the pig, ICH resulted in development of a reddish perihematomal zone and iron accumulation, ferritin upregulation, and neuronal death within that zone. Deferoxamine reduced the perihematomal reddish zone (Fig. 21.5), white matter injury, and the number of Perls', ferritin, and Fluoro-Jade C-positive cells (Fig. 21.6).

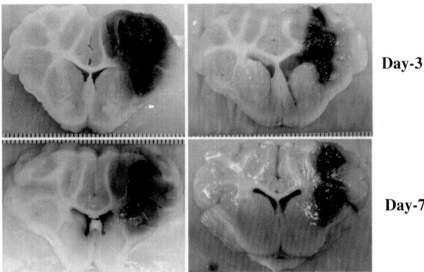

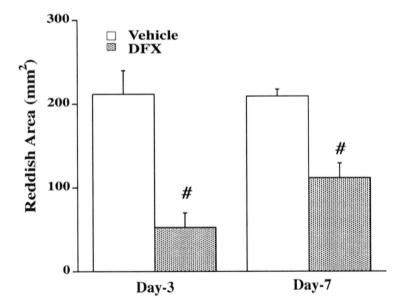

Fig. 21.5 Deferoxamine reduces reddish zone around hematoma at day 3 and day 7 in a pig ICH model. Values are means ± SD, $n=4$, $^{\#}p<0.01$ vs. vehicle. Figure reprinted with permission from Gu et al. [47]

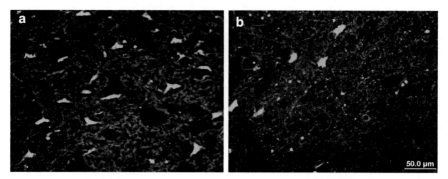

Vehicle   DFX

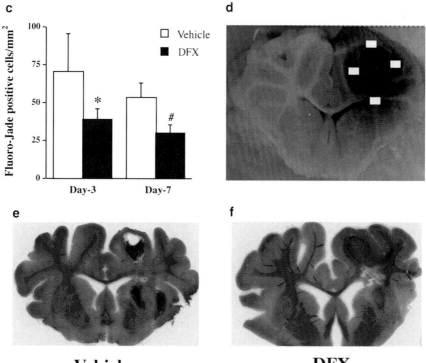

Vehicle   DFX

Fig. 21.6 Fluoro-Jade C-positive cells in the perihematomal area (**a–c**) and Luxol fast blue staining (**e, f**) after ICH. (**d**) Shows four sampled fields for Fluoro-Jade C cell counting. Pigs had ICH and were treated with either vehicle or deferoxamine. Values are means $\pm$ SD, $n=4$, $*p<0.05$, $^{\#}p<0.01$ vs. vehicle. (**a, b**) Scale bar $= 50$ μm. Figure reprinted with permission from Gu et al. [47]

9 Summary

The iron chelator, deferoxamine, reduces ICH-induced brain edema, brain atrophy, and neurological deficits in aged rats and pigs. Our preclinical data have led to a phase I clinical trial, "Safety and tolerability of deferoxamine in acute cerebral hemorrhage." That phase I trial has been completed, and deferoxamine is now going on to phase II clinical trial [48].

References

1. Kase CS, Caplan LR. Intracerebral hemorrhage. Boston: Butterworth-Heinemann; 1994.
2. Broderick JP, Brott T, Tomsick T, Miller R, Huster G. Intracerebral hemorrhage more than twice as common as subarachnoid hemorrhage. J Neurosurg. 1993;78:188–91.
3. Chiueh CC. Iron overload, oxidative stress, and axonal dystrophy in brain disorders. Pediatr Neurol. 2001;25:138–47.
4. Thompson KJ, Shoham S, Connor JR. Iron and neurodegenerative disorders. Brain Res Bull. 2001;55:155–64.
5. Wu J, Hua Y, Keep RF, Nakamura T, Hoff JT, Xi G. Iron and iron-handling proteins in the brain after intracerebral hemorrhage. Stroke. 2003;34:2964–9.
6. Xi G, Keep RF, Hoff JT. Mechanisms of brain injury after intracerebral hemorrhage. Lancet Neurol. 2006;5:53–63.
7. Xi G, Keep RF, Hoff JT. Erythrocytes and delayed brain edema formation following intracerebral hemorrhage in rats. J Neurosurg. 1998;89:991–6.
8. Huang F, Xi G, Keep RF, Hua Y, Nemoianu A, Hoff JT. Brain edema after experimental intracerebral hemorrhage: Role of hemoglobin degradation products. J Neurosurg. 2002;96:287–93.
9. Nakamura T, Keep R, Hua Y, Schallert T, Hoff J, Xi G. Deferoxamine-induced attenuation of brain edema and neurological deficits in a rat model of intracerebral hemorrhage. J Neurosurg. 2004;100:672–8.
10. Palmer C, Roberts RL, Bero C. Deferoxamine posttreatment reduces ischemic brain injury in neonatal rats. Stroke. 1994;25:1039–45.
11. Keberle H. The biochemistry of desferrioxamine and its relation to iron metabolism. Ann NY Acad Sci. 1964;119:758–68.
12. Porter JB. Deferoxamine pharmacokinetics. Semin Hematol. 2001;38 Suppl 1:63–8.
13. Sadrzadeh SM, Anderson DK, Panter SS, Hallaway PE, Eaton JW. Hemoglobin potentiates central nervous system damage. J Clin Invest. 1987;79:662–4.
14. Regan RF, Panter SS. Neurotoxicity of hemoglobin in cortical cell culture. Neurosci Lett. 1993; 153:219–22.
15. Guo Y, Regan RF. Delayed therapy of hemoglobin neurotoxicity. Acad Emerg Med. 2001; 8:510.
16. Hurn PD, Koehler RC, Blizzard KK, Traystman RJ. Deferoxamine reduces early metabolic failure associated with severe cerebral ischemic acidosis in dogs. Stroke. 1995;26:688–94. discussion 694–685.
17. Liachenko S, Tang P, Xu Y. Deferoxamine improves early postresuscitation reperfusion after prolonged cardiac arrest in rats. J Cereb Blood Flow Metab. 2003;23:574–81.
18. Prass K, Ruscher K, Karsch M, Isaev N, Megow D, Priller J, Scharff A, Dirnagl U, Meisel A. Desferrioxamine induces delayed tolerance against cerebral ischemia in vivo and in vitro. J Cereb Blood Flow Metab. 2002;22:520–5.
19. Whisnant JP, Sayer GP, Millikan CH. Experimental intracerebral hematoma. Arch Neurol. 1963;9:586–92.

20. Sussman BJ, Barber JB, Goald H. Experimental intracerebral hematoma. Reduction of oxygen tension in brain and cerebrospinal fluid. J Neurosurg. 1974;41:177–86.
21. Takasugi S, Ueda S, Matsumoto K. Chronological changes in spontaneous intracerebral hematoma–an experimental and clinical study. Stroke. 1985;16:651–8.
22. Wagner KR, Xi G, Hua Y, Kleinholz M, de Courten-Myers GM, Myers RE, Broderick JP, Brott TG. Lobar intracerebral hemorrhage model in pigs: Rapid edema development in perihematomal white matter. Stroke. 1996;27:490–7.
23. Yang GY, Betz AL, Chenevert TL, Brunberg JA, Hoff JT. Experimental intracerebral hemorrhage: Relationship between brain edema, blood flow, and blood–brain barrier permeability in rats. J Neurosurg. 1994;81:93–102.
24. Xi G, Wagner KR, Keep RF, Hua Y, de Courten-Myers GM, Broderick JP, Brott TG, Hoff JT. The role of blood clot formation on early edema development following experimental intracerebral hemorrhage. Stroke. 1998;29:2580–6.
25. Xi G, Hua Y, Keep RF, Younger JG, Hoff JT. Systemic complement depletion diminishes perihematomal brain edema. Stroke. 2001;32:162–7.
26. Hua Y, Xi G, Keep RF, Hoff JT. Complement activation in the brain after experimental intracerebral hemorrhage. J Neurosurg. 2000;92:1016–22.
27. Rosenberg GA, Mun-Bryce S, Wesley M, Kornfeld M. Collagenase-induced intracerebral hemorrhage in rats. Stroke. 1990;21:801–7.
28. Rosenberg GA, Navratil M. Metalloproteinase inhibition blocks edema in intracerebral hemorrhage in the rat. Neurology. 1997;48:921–6.
29. Xi G, Hua Y, Bhasin RR, Ennis SR, Keep RF, Hoff JT. Mechanisms of edema formation after intracerebral hemorrhage: effects of extravasated red blood cells on blood flow and blood–brain barrier integrity. Stroke. 2001;32:2932–8.
30. Zazulia AR, Diringer MN, Videen TO, Adams RE, Tundt K, Aiyagari V, Grubb Jr RL, Powers WJ. Hypoperfusion without ischemia surrounding acute intracerebral hemorrhage. J Cereb Blood Flow Metab. 2001;21:804–10.
31. Gong Y, Hua Y, Keep RF, Hoff JT, Xi G. Intracerebral hemorrhage: effects of aging on brain edema and neurological deficits. Stroke. 2004;35:2571–5.
32. Ropper AH. Lateral displacement of the brain and level of consciousness in patients with an acute hemispheral mass. N Engl J Med. 1986;314:953–8.
33. Clasen RA, Huckman MS, Von Roenn KA, Pandolfi S, Laing I, Clasen JR. Time course of cerebral swelling in stroke: a correlative autopsy and ct study. Adv Neurol. 1980;28:395–412.
34. Zazulia AR, Diringer MN, Derdeyn CP, Powers WJ. Progression of mass effect after intracerebral hemorrhage. Stroke. 1999;30:1167–73.
35. Xi G, Keep RF, Hoff JT. Pathophysiology of brain edema formation. Neurosurg Clin N Am. 2002;13:371–83.
36. Enzmann DR, Britt RH, Lyons BE, Buxton JL, Wilson DA. Natural history of experimental intracerebral hemorrhage: sonography, computed tomography and neuropathology. AJNR Am J Neuroradiol. 1981;2:517–26.
37. Tomita H, Ito U, Ohno K, Hirakawa K. Chronological changes in brain edema induced by experimental intracerebral hematoma in cats. Acta Neurochir Suppl. 1994;60:558–60.
38. Skriver EB, Olsen TS. Tissue damage at computed tomography following resolution of intracerebral hematomas. Acta Radiol Diagn (Stockh). 1986;27:495–500.
39. Felberg RA, Grotta JC, Shirzadi AL, Strong R, Narayana P, Hill-Felberg SJ, Aronowski J. Cell death in experimental intracerebral hemorrhage: the "black hole" model of hemorrhagic damage. Ann Neurol. 2002;51:517–24.
40. Xi G, Fewel ME, Hua Y, Thompson BG, Hoff J, Keep R. Intracerebral hemorrhage: pathophysiology and therapy. Neurocrit Care. 2004;1:5–18.
41. Hua Y, Nakamura T, Keep RF, Wu J, Schallert T, Hoff JT, Xi G. Long-term effects of experimental intracerebral hemorrhage: the role of iron. J Neurosurg. 2006;104:305–12.
42. Hua Y, Schallert T, Keep RF, Wu J, Hoff JT, Xi G. Behavioral tests after intracerebral hemorrhage in the rat. Stroke. 2002;33:2478–84.

43. Hua Y, Wu J, Pecina S, Yang S, Schallert T, Keep R, Xi G. Ischemic preconditioning procedure induces behavioral deficits in the absence of brain injury? Neurol Res. 2005;27:261–7.

44. Schallert T, Fleming SM, Leasure JL, Tillerson JL, Bland ST. Cns plasticity and assessment of forelimb sensorimotor outcome in unilateral rat models of stroke, cortical ablation, parkinsonism and spinal cord injury. Neuropharmacology. 2000;39:777–87.

45. Okauchi M, Hua Y, Keep RF, Morgenstern LB, Xi G. Effects of deferoxamine on intracerebral hemorrhage-induced brain injury in aged rats. Stroke. 2009;40:1858–63.

46. Okauchi M, Hua Y, Keep RF, Morgenstern LB, Schallert T, Xi G. Deferoxamine treatment for intracerebral hemorrhage in aged rats: therapeutic time window and optimal duration. Stroke. 2010;41:375–82.

47. Gu Y, Hua Y, Keep RF, Morgenstern LB, Xi G. Deferoxamine reduces intracerebral hematoma-induced iron accumulation and neuronal death in piglets. Stroke. 2009;40:2241–3.

48. Selim M. Deferoxamine mesylate: a new hope for intracerebral hemorrhage: from bench to clinical trials. Stroke. 2009;40:S90–1.

Chapter 22
Potential Therapeutic Targets for Cerebral Resuscitation After Global Ischemia

Yan Xu

Abstract Global cerebral ischemia is caused by the disruption of blood supply to the entire brain, often as a consequence of a cardiac arrest. In humans, 10-min global cerebral ischemia is traditionally considered lethal and irreversible. Despite the continuous development of basic and advanced life support protocols for cardiac resuscitation, statistics on long-term outcome after cardiac arrest have not significantly improved for several decades. The primary clinical challenge is the lack of an effective treatment strategy for cerebral resuscitation. Popular animal models of incomplete or complete forebrain ischemia by two-vessel or four-vessel occlusion are irrelevant to clinical situations, and treatments developed based on these models have thus far not yielded any useful clinical applications. This chapter will review the recent advancements in clinically relevant global ischemia models, the characteristics and mechanisms of brain injuries after global ischemia, and the therapeutic targets that can potentially alter the molecular and cellular cascades for an improved cerebral recovery.

1 Introduction

Cardiovascular disease, which often leads to cardiac arrest as its most deadly manifestation, remains a leading cause of mortality in the world [1]. In the United States alone, approximately 1,000 people die of cardiac arrest each day. This amounts to over 350,000 people, who would otherwise be physically competent to live on average a decade longer, to collectively lose 3–5 million years of potential life each year [2]. The current treatment strategy for cardiac arrest is grossly inadequate.

Y. Xu, PhD (✉)
Departments of Anesthesiology, Pharmacology and Chemical Biology, Computational Biology, and Structural Biology, University of Pittsburgh School of Medicine, 2048 Biomedical Science Tower 3, 3501 Fifth Avenue, Pittsburgh, PA 15260, USA
e-mail: xuy@anes.upmc.edu

P.A. Lapchak and J.H. Zhang (eds.), *Translational Stroke Research*, Springer Series in Translational Stroke Research, DOI 10.1007/978-1-4419-9530-8_22,
© Springer Science+Business Media, LLC 2012

The statistics suggests that the survival rate of cardiac arrest is <1% worldwide and ~5% in developed countries [1]. Among ~70,000 patients each year in the USA who are successfully resuscitated by cardiopulmonary resuscitation (CPR), 60% die later in hospitals due to brain damage; less than 10% of these patients have a chance to resume their former life activities.

The number of studies dealing with global cerebral ischemia has been significantly less than those dealing with focal vascular obstruction. The slower advances in global ischemia research, especially in relation to mechanisms of long-term post-resuscitation neuronal injury and death, are largely due to (a) the complexity of pathophysiologic changes after cardiac arrest and difficulties in dealing with multiple time points—from acute (minutes) to chronic (days to months, if long-term survival is the objective of an investigation); (b) the lack of a reproducible and *clinically relevant* animal model; and (c) the lack of experimental modalities for longitudinal follow-up studies of long-term outcome. In addition, as with reductionist approaches in other disciplines of science, the desire for a "clean" primary insult in order to isolate the injury pathways has driven ischemia research to localized injury models. Indeed, the majority of investigations into the mechanisms of neuronal death have largely focused on models of focal or incomplete forebrain ischemia. Most of these models do not produce a primary insult to the brainstem or cerebellum, and are often devoid of the secondary derangement of the so-called post-resuscitation syndrome [3, 4]. This latter aspect has been shown recently to be the major cause of poor long-term outcome after even a brief period of global ischemia.

More recent animal studies have shown that the brain's ability to withstand severe insults from global cerebral ischemia is much greater than what has been traditionally realized [5]. Promise of a significantly better outcome in the future for cardiac arrest patients is supported by a recent case report [6] documenting the full recovery of a 54-year-old patient without apparent neurological deficits after 96 min of recorded no-pulse due to an out-of-hospital cardiac arrest. To channel more research efforts to advancing both the science and the treatment of global cerebral ischemia, we will in this chapter compare the frequently used animal models, summarize the manifestation of injury and associated mechanisms, review recent research advances, discuss new clinical treatment strategies and breakthroughs, and present the exciting outlook for future directions.

2 Animal Models

Since the development of CPR in the 1950s and its promotion to the general public in the 1970s, the American Heart Association has periodically updated the practical guidelines for administering CPR [7–22]. Many of the changes in the guidelines are based on new knowledge gained through laboratory research using animal models on the injury and recovery processes, particularly those related to cerebral resuscitation.

Various animal models have been developed to mimic human ischemic response to global cessation of blood flow and oxygen supply to the brain. Large animals such as pigs, cats, and dogs have the advantage of more closely resembling human physiology, but high experimental cost, poor reproducibility, and ethical concerns have limited their use. The most popular animal models for global cerebral ischemia studies are those using rodents, particularly the two-vessel occlusion (2VO) model, four-vessel occlusion (4VO) model, and cardiac arrest and resuscitation models.

2.1 Two-Vessel Occlusion

The 2VO model was developed in 1972 in rats [23] and further modified in 1984 [24]. It is probably the simplest and most popular global ischemia model because it involves minimal surgical preparation. A typical protocol is summarized in Table 22.1. Blood flow to the brain is stopped by reversible bilateral occlusion of

Table 22.1 Two-vessel and four-vessel occlusion protocols in rats or mice

Preparation
1. Induce anesthesia with 3–4% isoflurane in a chamber until unconscious
2. Intubate transorally using 14G (rat) or 20G (mouse) cannula. Maintain ventilation with 30%:70% O_2:N_2O (or air) mixture containing 1.5–2% isoflurane
3. Shave and prepare surgical sites (neck, inguinal region) with 1% povidone iodine or comparable antiseptic. Place rodent in supine position
4. Paralyze the rodent with 2 mg/kg pancuronium subcutaneously. Add half a dose every hour
5. Make 1 cm incision in inguinal region and expose femoral/iliac artery and vein without disturbing the femoral/iliac nerve
6. Ligate distal end of the artery and cannulate the proximal end of femoral artery and vein using RenaPulse and micro-Renathane tubing (Braintree Scientific Inc.), respectively. Secure vessel to tube with a single tie of a suture
7. Cover surgical site with sterile saline-soaked gauze

Post-op care
1. After ischemia, carefully remove cannula from vein then artery, making sure to use a second suture to tie off the vessel proximally before tube is fully removed
2. Close wound and apply prophylaxis (e.g., lidocaine/bupivacaine) to incisions
3. Wean rodent off the ventilator, extubate after spontaneous breathing is regained
4. Return animal to a clean cage and monitor closely for at least 3 days for signs of pain, weight loss, and seizure activity

Two-vessel occlusion (rats or mice) [24]
1. Follow preparation procedures
2. Make a midline neck incision to expose carotid arteries without separating them from adjacent muscle bundles
3. Loop a 20-cm long Renasil SIL-037 (Braintree Scientific) behind each carotid artery. Pass each tube through the anterolateral neck tissue and skin. Close the skin with nonabsorbable sutures
4. Sample arterial blood to confirm that blood gases are in normal range ($PaO_2 > 80$ mmHg, $PaCO_2 = 35$–41 mmHg, and pH = 7.35–7.45); body temperature is 36–37°C

(continued)

Table 22.1 (continued)

---

5. Administer Labetalol (0.25 mg/kg) intravenously to target arterial blood pressure of 50 ± 5 mmHg

6. (Optional) Remove venous blood into a heparinized syringe to further reduce blood pressure as needed

7. Just before ischemia, turn off anesthesia but maintain ventilation

8. Pull both Renasil tubes laterally so that the end is 15–20 cm from the midline

9. Maintain tension for the predetermined ischemia duration (optional: Test arterial blood gases at 10 min into ischemia)

10. Release tension, and remove the Renasil tubing by gently pulling one end

11. Reinfuse venous blood if previously drawn

12. Reinstate anesthesia and paralytics as needed (judging by heart rate change)

13. Inject 0.05 mEq/mL sodium bicarbonate intravenously and measure blood gases after 30 min

14. Follow post-op care procedures

Four-vessel occlusion (two-stage method) [161]

1. Anesthetize rodent using 2% isoflurane through a nose cone

2. Position rodent in stereotaxic apparatus and prepare dorsal surface of neck for surgery (shave and apply antiseptic)

3. Make an incision behind occipital bone over first two cervical vertebra

4. Separate paraspinal muscles from the midline and expose the right and left alar foramina of the first cervical vertebra

5. Use a 0.5-mm diameter electrocautery needle to cauterize both vertebral arteries permanently by pushing cautery tip in and down centrally 1–3 mm

6. Clean blood and close incision

7. Remove rodent from stereotaxic apparatus and place in supine position, while maintaining general anesthesia

8. Follow steps 2 and 3 in two-vessel occlusion procedures

9. Allow rodent to wake up and recover for at least 24 h

10. On the day of the ischemia, hold the awake animal in one hand and pull the tubing to occlude common carotid arteries (CCA). Animal should lose consciousness rapidly

11. Maintain tube tension and body temperature at 37°C throughout the ischemia period

12. Follow post-op care procedures

Four-vessel occlusion (one-stage method) [29]

1. Anesthetize rat using 2% isoflurane through a nose cone

2. Make a 2.5 cm midline incision to neck and gently retract subcutaneous connective tissue and muscles to expose the trachea and thyroid

3. Isolate the bilateral CCA and loosely place 2-0 silk sutures as landmarks for placing microvascular clips later

4. Expose the longus colli muscle by retracting the trachea and esophagus to the right side; identify the anterior tubercle of the atlas

5. Expose the cervical vertebral bodies from the atlas to the upper half of the fourth cervical vertebral body to visualize the bilateral vertebral arteries (VA) between the second and third transverse processes

6. Isolate the vertebral arteries without interrupting blood flow

7. Occlude in the following order the right CCA, left CCA, right VA, and left VA with microvascular clips

8. Release clips in reverse order after predetermined ischemia duration

9. Verify reperfusion and close wound

10. Follow post-op care procedures

---

both common carotid arteries (CCA), combined with moderate to severe hypotension (arterial blood pressure ≤50 mmHg). There are several procedural variations to this model, including induction of hypotension by phlebotomy and the use of different hypotension drugs [25–27]. Strictly speaking, 2VO is a forebrain, incomplete ischemia model, not a global ischemia model, because collateral flow due to blood supply from the vertebral arteries can still provide a certain degree of cerebral perfusion. To minimize collateral flow, gerbils are preferred over rats and mice in the 2VO model. Gerbils do not have posterior communicating arteries necessary for completing the circle of Willis, through which the collateral flow is established.

Reproducibility of the primary ischemic insults to the brain in the 2VO model is poor, particularly in rats and mice. Variability arises from different degrees of collateral circulation (hence different severities of global ischemia) and variations in physiological response to hypotension. Thus, comparison of results among different research groups, or different animals or even different strains of same animals, is difficult if not impossible. This variability also makes it difficult to evaluate the effects of any treatment.

2.2   Four-Vessel Occlusion

The 4VO model was developed in 1979 [28] to address the issue of collateral circulation due to blood flow through vertebral arteries. A detailed protocol for 4VO is also included in Table 22.1. Because of the complex preparation, the procedure is typically performed on 2 separate days. Animals are first prepared under general anesthesia for electrocauterization of both vertebral arteries, followed 24 h later with reversible bilateral occlusion of CCA. If the experimenters are skillful, 4VO is considered more reproducible than 2VO and thus is widely used as a model for complete forebrain ischemia. However, various complications from the procedures do occur frequently, and the model is not clinically relevant to any global ischemia conditions in humans. Two separate surgeries not only are cumbersome, the preparatory procedure also creates preconditioning effects [29] that can alter, and even complicate, the injury development and outcome of subsequent ischemia. Moreover, permanently occluding the vertebral arteries can have extraneous effects, obscuring the brain's responses to the primary ischemic insults [30]. Despite the popularity of this laboratory model, its clinical relevance is always questionable.

2.3   Cardiac Arrest and Resuscitation

Because 2VO and 4VO limit the ischemic insult only to the forebrain without directly affecting the brain stem, cerebellum, spinal cord, and other organs, postischemic care for long-term survival is possible in these models. However, neither model provides clinically relevant settings to mimic global ischemia in humans.

Most therapeutic treatments developed using these models have failed in human clinical trials. It has long been recognized that developing intact animal models that closely resemble the clinical situations of cardiac arrest and resuscitation is of crucial importance for evaluating effective post-resuscitation therapies and outcome [31]. Earlier development of cardiac arrest and resuscitation models involved intracardiac injection of potassium chloride (KCl) [32] and asphyxia-induced cardiac arrest [33]. The former model mimics KCl overdose and is more suitable for studying the dying instead of recovery processes, unless the arrest time is very transient. The early asphyxial cardiac arrest model creates an anoxic condition to a beating heart to exhaust the heart's energy reserves, resulting in myocardium damage and cardiac arrest. Although it mimics the clinical situations of suffocation (such as in severe asthma attack) and drowning, the early asphyxia model causes too severe damage to the heart and other organs to be suitable for long-term outcome studies. As the original authors pointed out [33], post-insult recovery is practical only for up to 6 h after 7–10 min of asphyxia in rats. Because ischemic injuries in brain tissues usually take 3 days to fully develop, a global ischemia model that allows for only a few hours of post-resuscitation evaluation is inadequate.

A highly reproducible outcome model of cardiac arrest and resuscitation in rats was developed in the late 1990s [31, 34]. The model involves minimal surgical preparation (comparable to 2VO) and thus can be easily implemented for long-term survival studies. A schematic of the model is shown in Fig. 22.1, with a detailed protocol given in Table 22.2. The model combines two clinical procedures to achieve cardiac arrest and resuscitation. The first procedure is to use an ultra-short-acting $\beta 1$ adrenergic blocker such as esmolol to transiently stop the heart from beating, as in the case of cardiopulmonary bypass and coronary artery surgeries or as in the preservation of donor hearts in heart transplantations. This procedure stops the heart within ~20 s, thus protecting the myocardium from hypoxia injury [35]. The esmolol dose used in the model is comparable to the clinical concentration for reversible cardioplegia [36–38]. Resuscitation is achieved with the second clinical procedure whereby oxygenated blood is retro-gradely infused into the heart through the femoral artery and abdominal aorta, comparable to the clinical procedure of cardiopulmonary support [39]. Figure 22.2 depicts typical arterial blood pressure changes before, during, and after a 12-min cardiac arrest and resuscitation using this model. The model has several advantages over other global ischemia models. First, it is highly reproducible with minimal procedural variations. The effect of short-acting $\beta 1$-blockers is predictable and nearly independent of animal body weight because the heart size is relatively invariant in adult rats. Within approximately 20 s of the esmolol injection, electro-mechanical dissociation is achieved (see insert to Fig. 22.2). This dissociation is maintained as long as oxygen is absent. Upon infusion of an oxygenated blood and epinephrine mixture in conjunction with mechanical ventilation, heartbeat is resumed within 1–2 min. Second, because the heart is preserved during the onset of cardiac arrest, the ischemia are well tolerated by the myocardium and other organs, thereby providing a relatively "clean" global ischemic insult to the brain.

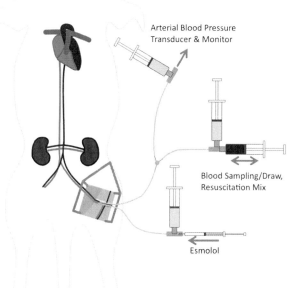

Arterial Blood Pressure
Transducer & Monitor

Blood Sampling/Draw,
Resuscitation Mix

Esmolol

Fig. 22.1 A schematic representation of a highly reproducible cardiac arrest and resuscitation model in rats. This model uses a clinical cardioplegia procedure combined with apnea to induce cardiac arrest, and a cardiopulmonary support procedure for resuscitation. Because the effects of the short acting β-adrenergic blocker and the resuscitation mixture are relatively invariant, the injury in this model is highly reproducible. Minimal surgical preparation makes this clinically relevant model ideally suited for outcome studies after global ischemia

Table 22.2 Cardiac arrest and resuscitation protocols in rats [31, 34]

1. Follow preparation procedures in Table 22.1
2. Sample arterial blood to confirm that blood gases are in normal range ($PaO_2 > 80$ mmHg, $PaCO_2 = 35$–41 mmHg, and pH = 7.35–7.45); tympanic temperature is 36–37°C
3. Draw 1 mL/kg of 10× concentrated resuscitation mixture into a 3-mL syringe. (10× resuscitation mixture is 80 µg/mL epinephrine, 0.5 mEq/mL sodium bicarbonate, and 50 U/mL heparin mixed in sterile saline)
4. Slowly withdraw 2.0–2.5 mL arterial blood into the syringe with resuscitation mixture. Keep it warm at 37°C
5. Administer 1 mg/kg vecuronium intravenously. Wait 5 min
6. Discontinue anesthesia initiate cardiac arrest 30 s later by a bolus injection of 6.7–7.5 mg esmolol and turning off the ventilator at the same time
7. Wait for the predetermined cardiac arrest duration
8. Before resuscitation, mix blood and resuscitation mixture, remove all air bubbles
9. To resuscitate
   (a) Restart ventilation with 100% O_2 and no anesthetic
   (b) Steadily infuse oxygenated blood + resuscitation mixture through femoral artery so that blood pressure is maintained at 50 mmHg for 50 s
   (c) Transiently and periodically relax the infusion pressure to check for spontaneous cardiac activities. As soon as heartbeat is detected, stop infusion and discard the remaining blood and resuscitation mixture
10. Ventilate the animal for 2 h with 100% O_2 initially and gradually change to 30%/70% O_2/air mixture. Reinstate anesthesia as needed
11. Follow post-op care procedures in Table 22.1

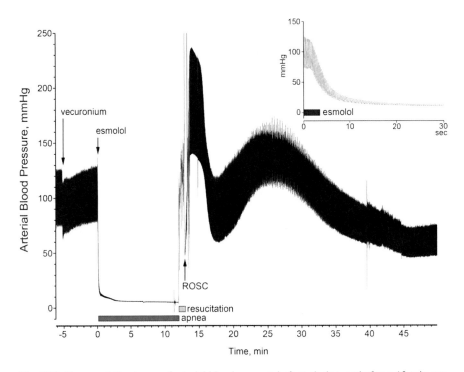

Fig. 22.2 Representative traces of arterial blood pressure before, during, and after a 12-min cardiac arrest. Rapid onset of cardiac arrest was initiated by a bolus intravenous injection of ~7 mg esmolol, followed by 12-min apnea (*dark gray horizontal bar*). The first 30 s of detailed recording during and immediately after esmolol injection (*black bar*) are shown in the insert. Notice the rapid onset of cardiac arrest within 20 s, and controlled resuscitation (*light gray bar*) within ~1 min to achieve return of spontaneous circulation (ROSC)

Third, because this model uses minimal surgical procedures, animals can be used for long-term survival studies, including neurological deficit evaluation and behavior testing. Fourth, because both cardiac arrest and resuscitation can be remotely controlled, the model is best suited for experiments utilizing state-of-the-art instruments such as *in vivo* magnetic resonance spectroscopy (MRS) and imaging (MRI). For example, *in vivo* quantitative perfusion- and diffusion-weight MRI has been performed without interruption before, during, and after the entire cardiac arrest and resuscitation procedure, such that the cerebral perfusion during the most critical period of reperfusion can be quantified noninvasively [5, 31, 34, 40, 41]. The clinical relevance of the model to long-term recovery after global ischemia permits investigators to shift the focus from chasing the injury and dying processes to studying more critical cellular events that control the brain's revitalization and recovery.

3 Manifestation of Injuries

3.1 Selective Cell Vulnerability to Global Ischemia

During the past 4 decades, a massive amount of information was generated from intensive research into the cellular and molecular cascades in the brain parenchyma after global ischemia. Although many injury pathways are shared by focal and global ischemia, manifestation of these injuries differs significantly. In focal ischemia, there is a gradation of blood flow reduction: in the ischemic core, the perfusion is greatly reduced but does not necessarily reach zero, whereas in the peripheral of the ischemia-affected territory (the so-called ischemic penumbra), the perfusion is decreased to a varying degree with moderate to severe reduction in glucose and oxygen availability. Because of the trickle-flow effects that mobilize ischemia-induced toxins, particularly the reactive oxygen species (ROS) discussed below, focal ischemia is characterized by a pan-necrosis of all cell types [42]. In global ischemia, in contrast, perfusion completely stops. This usually initiates the conditions for subsequent injuries upon reperfusion, but injury progression is limited during the period of no-flow. As a consequence, global ischemia has a unique histology pattern of selective cell death—often injured and dead cells are found side by side with apparently healthy cells, as shown in Fig. 22.3. Moreover, severe ischemic injuries are found in

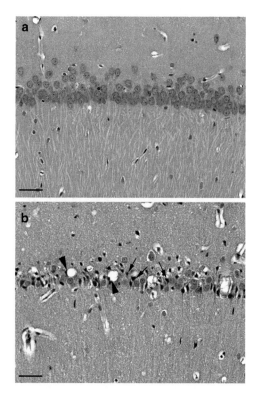

Fig. 22.3 Typical histology changes after global cerebral ischemia. Hematoxylin and eosin (H&E) stained sections of paraffin-embedded tissues showing the CA1 regions of hippocampus from rats 5 days after 0 (**a**, sham control) and (**b**) 12 min of cardiac arrest. Neuronal loss is indicated by the nuclear pyknosis (*arrows*) and vacuolization of the neuronal cytoplasm (*arrowheads*). Bar = 50 μm

the so-called selectively vulnerable regions of the brain, particularly the pyramidal cell layer of the hippocampal CA1 and CA3 regions, hilar cells of the dentate gyrus, and certain regions in the cortex. Surprisingly, a large percentage of neurons in the hippocampal CA2 region and in the striatum (with the exception of medium-sized striatum neurons) are spared after moderate global ischemia. Thus, if the mechanisms of delayed cell injuries in the selective vulnerable regions are understood and prevented, there is a large time window for potential treatment.

3.2   *Protracted Post-Resuscitation Hypoperfusion*

One of the profound pathophysiological changes after resuscitation from cardiac arrest is severe and protracted hypoperfusion after an initial and transient phase of hyperemia. The characteristics of this change are accurately reproduced in the cardiac arrest and resuscitation model discussed above. Since the model is fully compatible with noninvasive MRI measurements, the cerebral perfusion in rats can be followed with high temporal and spatial resolutions during the cardiac arrest and resuscitation, as well as over the long-term recovery period. In conventional MRI, the measured signals arise from the protons in the water molecules in the blood and tissue. These signals can be modulated by using various types of externally applied radiofrequency pulses. In the so-called arterial spin-labeling (ASL) technique, a physical property of the water protons in the arterial blood is tagged by a radiofrequency pulse when the blood flows through the CCA in the neck region. These tagged protons, while perfusing through the vessels and microvasculature of the brain tissue, behave differently from the untagged protons. High-resolution difference images obtained with and without the ASL produce regional maps of cerebral perfusion. Figure 22.4 shows examples of MRI perfusion maps in the transverse sections of rat brains. The color of the maps indicates the regional cerebral perfusion in units of mL blood per 100 g of brain tissue per minute. Notice that arterial blood pressure normalizes around 15–20 min after resuscitation, but cerebral perfusion, after a transient hyperemic phase, is severely depressed to as low as 20% of the normal level in some regions. In rats, this protracted hypoperfusion lasts at least 48 h. The underlying mechanisms of hypoperfusion remain unclear and are the focus of many experimental studies. Severe hypoperfusion for a long duration can cause secondary ischemic injuries, exacerbating the injury conditions already created by the primary insult from the global ischemia. It is believed that the generally poor long-term outcome after cardiac arrest and resuscitation, even after a short period of no-flow, is caused by hypoperfusion. Thus, treating the underlying causes of hypoperfusion will likely be beneficial for long-term outcome.

Very few current treatments have proven effective in reversing post-resuscitation hypoperfusion. In a recent study [41] with transgenic rats that overexpress cytosolic superoxide dismutase (SOD-1), it was discovered that cerebral perfusion recovered significantly faster in the SOD-1 transgenic rats than in the wild-type control animals, albeit the initial phase of hypoperfusion was similar. This result suggests a

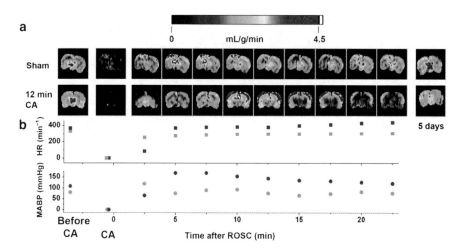

Fig. 22.4 Noninvasive magnetic resonance imaging measurements of cerebral perfusion before and during cardiac arrest, and after resuscitation. (**a**) Representative perfusion images of rat brains were acquired before and during cardiac arrest (CA), immediately after the return of spontaneous circulation (ROSC), and 5 days after CA. Image intensity is color-coded using the color scale above the images in units of mL/g/min. (**b**) Corresponding mean arterial blood pressure (*circles*) and heart rate (*squares*) are plotted as a function of time. The point of ROSC is arbitrarily designated as time 0. *Green* sham-operated group; *Blue* 12-min cardiac arrest group

link between hypoperfusion and the formation of free radicals, particularly reactive oxygen species (ROS), and how effectively the ROS are removed.

3.3 Reactive Astrocytosis and Glial Scaring

While most investigations in the past have centered on neuronal cell injury and death after global ischemia, several recent studies have turned the attention to the role of glial cells in brain recovery after global ischemia [43–45]. Changes in astrocyte morphology are distinctive in the hippocampus after global ischemia. Reactive astrocytosis, in response to global ischemia, is characterized by a hypertrophy of astrocytic processes, dedifferentiation of glia, and upregulation of intermediate filament expression. Three intermediate filament proteins, glial fibrillary acidic protein (GFAP), vimentin (VIM), and nestin can express in astrocytes in different developmental stages. GFAP is a marker for mature astrocytes, whereas VIM is a marker for precursor astrocytes, radial glia, reactive astrocytes, and activated microglia. After global ischemia, a large number of GFAP positive astrocytes become VIM positive as well (Fig. 22.5). The processes of these cells become thicker and denser after reperfusion. The formation of a dense network of astrocytic processes can create physical barriers (scars) to hinder neuronal regrowth. Recent studies also suggest that the selective vulnerability of the CA1 pyramidal neurons is related to the

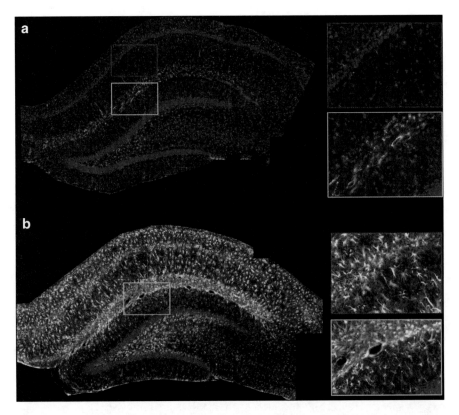

Fig. 22.5 Astrocytosis after cardiac arrest and reperfusion. Micrographs of the hippocampus from rat brains without (**a**) and after 12-min cardiac arrest (**b**) were acquired after 7 days of reperfusion. Comparison of immunochemical staining for glial fibrillary acidic protein (GFAP, *red*) and vimentin (VIM, *green*) and nucleus staining with 4',6-diamidino-2-phenylindole (DAPI, *blue*) between (**a**) and (**b**) shows hypertrophy of astroglial processes, reactive astrocytosis, and neuronal loss in the CA1 region. *Inserts* to the *right* are from the boxed regions in the CA1 (*red box*) and molecular layer (*green box*) of the hippocampus

selective vulnerability of the surrounding astrocytes. CA1 astrocytes were more susceptible to injury, showing the loss of glutamate transport activities and rapid increases in mitochondrial free radical formation [43].

4   Mechanisms of Injuries and Repair After Global Ischemia and Reperfusion

Because of the multifactorial nature of the injury cascade in global ischemia, the discussions here will be limited only to the key events that have been considered as targets for potential therapies. More in-depth reviews of the molecular mechanisms of brain injury after global ischemia and reperfusion can be found elsewhere (see, for example, [46]).

4.1   Cellular Cascade of Neuronal Injuries

4.1.1   Rapid Intracellular Energy Depletion

The brain is highly dependent on a minute-to-minute blood supply of glucose and oxygen as an energy source. The brain's first response to global ischemia is the rapid depletion of high-energy metabolites. Figure 22.6 shows an example from noninvasive measurements of brain energetics using *in vivo* MRS before and during cardiac arrest, and after resuscitation. Energy production in the brain primarily involves the aerobic catabolism of glucose and adenylate kinase reaction. It has been estimated that the brain's normal ATP reserve at any given time is about 4.5 µmoles/g. During complete global ischemia, the energy reserve is depleted at a rate of ~0.5 µmoles/g/s [47]. Thus, the energy reserve will last only about 9 s. Phosphocreatine (PCr), in the presence of creatine kinase, can regenerate an additional 5 µmoles/g of ATP, or equivalently, provide an additional 10 s before ATP is completely depleted.

Energy failure is believed to cause neuronal overexcitation, which in turn accelerates energy depletion. Many of the injury mechanisms discussed below are to some extent associated with energy failure.

4.1.2   Excitotoxicity

Because of the rapid energy depletion in global ischemia, injuries are initiated, but not amplified or fully expressed. Initiation involves a process called excitotoxicity, through which an excessive amount of excitatory amino acid neurotransmitters, primarily glutamate and aspartate, are released during ischemia-induced depolarization. The released glutamate binds to two classes of glutamate receptors: the ligand-gated inotropic receptor channels and the G protein-coupled metabotropic receptors. There are three types of inotropic receptors, which are activated by NMDA, AMPA, or kainic acid. Sustained activation of these receptors leads to a large influx of extracellular cations. Na^+ ions can enter through the activated ionophores of all three receptor types, bringing with them the passive influx of Cl^- and water, thereby causing cytosolic edema. The NMDA receptor channels are special in admitting substantial Ca^{2+}; upon ischemic insult, the acute influx of Ca^{2+} is primarily through this receptor. Although the AMPA-regulated ionophore is also Ca^{2+} permeable [48–50], the Ca^{2+} entry by activation of the AMPA/kainate receptor is largely through indirect pathways, such as voltage-gated Ca^{2+} channels [51–53] or reversal of the $3Na^+/Ca^{2+}$ exchanger that normally extrudes Ca^{2+} [54]. Metabotropic glutamate receptor activation has been shown to couple to multiple second messenger systems, including, most importantly, the liberation of inositol 1,4,5-tris phosphate (IP_3), which mediates Ca^{2+} release from intracellular stores in the endoplasmic reticulum and "calcisomes" [55, 56], and the liberation of diacylglycerol (DAG), which activates the protein kinase C (PKC) in the presence of free Ca^{2+} and phosphatidylserine.

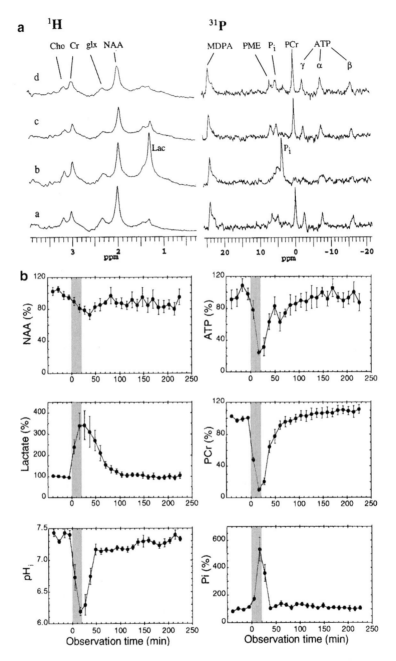

Fig. 22.6 Interleaved *in vivo* NMR measurements of brain energy metabolism. (**a**) Representative interleaved ¹H/³¹P NMR spectra of a rat brain before (a) and during (b) a 20-min cardiac arrest and 1 h (c) and 3 h (d) after resuscitation. (**b**) Time course of brain metabolic changes before, during, and after 20-min cardiac arrest and resuscitation (*n* = 6). *PME* phosphomonoester; *Cho* choline; *Cr* creatine; *glx* glutamine/glutamine; *Lac* lactate

The substantial accumulation of intracellular free Ca^{2+}, irrespective of the pathways, is believed to set the stage for several destructive cascades, leading ultimately to neuronal degeneration. Among the most important cascades are the activation of catabolic enzymes and generation of free radicals (see below), cytoskeleton breakdown via proteolysis, and protein misfolding and aggregation. In addition, the activation of PKC, which translocates to the inner cell membrane and catalyses the phosphorylation of many substrate proteins, causes sustained modifications in neuronal excitability and behavior [57, 58]. Left untreated, these processes eventually lead to *necrotic cell death.*

4.1.3 Burst Production of Reactive Oxygen Species upon Reperfusion

It is now well established that after ischemic insult, ROS are formed, especially the superoxide anion ($O_2{}^{\cdot-}$). Several mechanisms are considered significant or likely relevant in the brain [59–61], including (i) prostaglandin (PG) synthesis, i.e., conversion of arachidonic acid (AA) to PGG_2 and subsequent peroxidized conversion to PGH_2 with concomitant release of $O_2{}^{\cdot-}$, [62]; (ii) activation of neuronal nitric oxide synthase (nNOS) [63]; (iii) purine catabolism, i.e., $O_2{}^{\cdot-}$ releases in the successive reactions of hypoxanthine to xanthine and xanthine to urea catabolized by the xanthine oxidase [64]; (iv) neutrophil activation [65, 66]; and (v) respiration in the mitochondrial respiratory chain.

The association of excitotoxicity with pathway (i) and (ii) has been studied extensively. Elevation of cytosolic Ca^{2+} activates phospholipases and nNOS. The phospholipases are a family of ubiquitous enzymes that hydrolyze functional groups of phospholipids [67, 68]. In particular, the Ca^{2+}-dependent phospholipase A_2 cleaves fatty acyl chains from the β-position of phospholipids. AA in mammalian cells is esterified almost exclusively at the β-position. Thus, lipolysis during ischemia can cause accumulation of free fatty acids, especially AA. Further metabolism of AA by pathway (i) leads to the amplified production of $O_2{}^{\cdot-}$ and further to hydroxyl radical ($\cdot$OH) via the superoxide-driven, Fe^{2+}-catalyzed, Haber–Weiss reaction [69]. The activation of nNOS produces NO that reacts with $O_2{}^{\cdot-}$ to generate peroxynitrite $ONOO^-$. In the presence of O_2, such as during reperfusion, $\cdot$OH and $ONOO^-$ lead to protein and membrane peroxidation [69, 70].

Pathways (iii), (iv), and (v) in ROS production are highly relevant to the vascular response to ischemia and reperfusion. Two major areas have gathered much attention: (a) pro-inflammatory stimulation and polymorphonuclear neutrophil (PMN) infiltration, and (b) endothelial cell reactions. Oxidants derived from xanthine oxidase contribute to the formation of platelet-activating factor and leukotrienes, the inactivation of anti-adhesive molecules, and the expression of endothelial cell adhesion molecules (e.g., P-selectin and ICAM-1). All of these set the stage for the respiratory burst of ROS production upon reperfusion by NADH/NADPH oxidase activities in both neutrophils and endothelial cells. This burst is believed to be the determinant factor for the onset of post-resuscitation hypoperfusion and the delayed loss of autoregulation in the brain.

ROS is also the primary inducer of mitochondrial injury via the activation and membrane translocation of BAX, a proapoptotic protein. Activation and translocation of BAX result in mitochondrial release of cytochrome c, which activates proteolytic enzyme caspases. Caspases are responsible for DNA fragmentations and eventual *apoptotic neuronal death*.

Under normal circumstances, the cellular defense against the burden of $O_2^{\cdot-}$ and H_2O_2 is through superoxide dismutases (SOD) (22.1), catalases (22.2), and nonspecific peroxidases (22.3).

$$2O_2^{\cdot-} + 2H^+ \rightarrow H_2O_2 + O_2 \tag{22.1}$$

$$2H_2O_2 \rightarrow 2H_2O + O_2 \tag{22.2}$$

$$H_2O_2 + RH_2 \rightarrow 2H_2O + R \tag{22.3}$$

When this defense system is overwhelmed by the rapid production of ROS during the onset of reperfusion, catalysis through SOD becomes a critical step for the clearance of $O_2^{\cdot-}$. SODs are Mn based in the mitochondria and Cu and Zn based in the cytosol or on the extracellular surfaces. One of the possible ways to enhance the capacities of SOD is to overexpress cytosolic CuZn SOD (SOD-1) for the purpose of improving the recovery of cerebral reperfusion after resuscitation [41].

4.2 Post-Resuscitation Neural Repairs

4.2.1 Coupling of Angiogenesis and Neurogenesis

Recent research advances have led to the belief that cellular repairs in the brain after ischemia and reperfusion are possible via causal coupling between angiogenesis and neurogenesis. Generalizing the current theory of neovascularization and neuronal recruitment in the brain of adult songbirds [71], one can hypothesize that the cardinal events in the sequence of cellular processes after global ischemia and reperfusion are the induction of vascular endothelial growth factor (VEGF), subsequent VEGF receptor activation, and the upregulation of neurotrophic factors such as brain-derived neurotrophic factor (BDNF) in the VEGF-expanded neo-vasculature. More specifically, global ischemia activates the hypoxia-inducing factor-1a (HIF-1a), which binds to the promoter of the VEGF gene to trigger the overproduction of VEGF. The surge of VEGF upregulates the expression of VEGF receptor 1 (VEGFR-1, flt-1) and VEGFR-2 (KDR/flk-1). The VEGF-expanded vasculature leads to the upregulation of both BDNF mRNA and protein levels, resulting in a BDNF gradient in the brain parenchyma for neuron recruitment into ischemia-damaged region. Ample evidence from *in vitro* studies suggests that the messengers for the intra- and inter-cellular signaling among these events after hypoxia and ischemia are, paradoxically, the reactive oxygen species, particularly $O_2^{\cdot-}$ and H_2O_2

generated upon activation of NADPH oxidase-like membrane complexes. In the untreated brain, these events are not orchestrated: (1) the burst of ROS production upon reperfusion exerts more cytotoxic effects than mitigating effects of signaling vascular regeneration; (2) the endothelial cells are injured to such an extent that expression of VEGF and VEGF receptors is impaired, failing to respond to the need for neovascularization; and (3) many neurons become sick after ischemia and thus cannot be recruited for neuronal repair. Future treatment can be developed to create intraparenchymal conditions that promote neuronal cell survival (e.g., by overexpression of BDNF), neuronal cell differentiation from quiescent adult stem cells, and neovascular generation (e.g., by overexpression of VEGF).

4.2.2 VEGF's Role in Reperfusion

VEGF has several different splice variants, including VEGF-121, -165, -189, and -206. VEGF121 is weakly acidic and does not bind to heparin, whereas VEGF165 is a basic, heparin-binding protein. VEGF189 and VEGF206 are more basic and bind to heparin with even greater affinity than VEGF165. VEGF121 is freely diffusible; VEGF165 is also secreted, although a significant fraction remains bound to the cell surface and the extracellular matrix (ECM). In contrast, VEGF189 and VEGF206 are almost completely sequestered in the ECM, but may be released in soluble form by heparin and heparinase. The long forms may also be released by plasmin following cleavage at the COOH terminus. Hence, VEGF proteins are available to endothelial cells either as freely diffusible proteins (VEGF121, VEGF165) or from protease activation and cleavage of the longer isoforms.

Different VEGF splice variants have distinct roles. Many studies have shown that VEGF121 and VEGF165 are essential for new vessel formation [72] and thus can be used for therapeutic angiogenesis in ischemic and reperfusion injuries [73–75]. Therapeutic angiogenesis against ischemia and reperfusion injury in brain tissue by overexpression of VEGF is indicated by clinical studies that show higher microvessel density in the ischemic hemisphere compared to the contralateral hemisphere, and a significant correlation between blood vessel density and survival of stroke patients. Experimental studies have also shown that VEGF can induce a robust angiogenic response in the adult rat brain, and that application of exogenous VEGF reduces infarct volume and diminishes neuronal damage in a rat focal ischemia model.

There are three signaling tyrosine-kinase receptors in the VEGF family: VEGFR-1 (flt-1), VEGFR-2 (KDR/flk-1), and VEGFR-3 (flt-4). The dominant isoforms of VEGF in the brain, VEGF121 and VEGF165, bind only to VEGFR-1 and VEGFR-2. It is believed that VEGFR-1 is responsible for endothelial cell migration but plays no role in cell proliferation, whereas VEGFR-2 is critical for cell proliferation and migration [76]. The transcription of VEGFR-1, but not VEGFR-2, is enhanced by hypoxic conditions [77]. Because VEGFR-2 activates downstream cell proliferation and migration, it is conceivable that therapeutic overexpression of VEGF will upregulate the production of VEGFR-2 at the posttranscription level.

Two other receptors, neuropilin-1 and neuropilin-2, are genetically related. VEGF165, but not VEGF121, can bind to neuropilin-1 and -2, suggesting that VEGF165 may also activate neuropilin receptors of neuronal cells. Neuropilin-1 is known to work synergistically with VEGF for vascular development.

4.2.3   ROS as an Intracellular Messenger

Although reactive oxygen species are commonly viewed in reperfusion injury as bad molecular entities due to their cytotoxic effects outlined above, many recent studies have concluded that ROS play a pivotal role in signaling the downstream events for endothelial and neural cell mitosis [71, 78–85]. Of particular significance in relation to reperfusion injury is the cascade of events associated with the induction of VEGF mRNA expression and activation of VEGFR-2 (KDR/flk-1), leading to post-resuscitation neovascularization and angiogenesis. It has been shown that the small GTPase Rac1-regulated and gp91^{phox}-containing NADPH oxidase is critically important in VEGF-induced signaling and angiogenesis [79, 86], especially in mitogenic and chemotactic effects of VEGF on endothelial cells mediated by the activation of KDR. ROS is also required to mediate the activation of transcription factors such as the nuclear factor-κB (NF-κB), which plays an important role in many medically significant biological processes, including immune, inflammatory, and antiapoptotic responses. Thus, on the one hand, it is desirable to increase the scavenging capability of the neuronal cells to protect cells from ROS-related injuries (e.g., overexpression of SOD); and on the other hand, it is also desirable to maximize the benefit of ROS for angiogenesis and neurogenesis. The delicate balance between angiogenesis and antioxidation requires fine tuning.

5   Dependence of Outcome on Resuscitation Efficacy

It is generally believed that longer duration of cardiac arrest is always associated with poorer outcome. In humans, after 5 min of out-of-hospital cardiac arrest, every additional minute of no-flow decreases the survival rate by ~10%. However, the true limit for which the brain can withstand cardiac arrest-induced global ischemia is unknown. A recent report of successful resuscitation and recovery after more than one and a half hours of complete no-flow in man [6] offers hope that the intrinsic ability of the brain to survive global ischemia is higher than expected. Other factors may contribute to a poor outcome after resuscitation. In animal experiments [5], it was demonstrated that the long-term outcome is more dependent on the resuscitation efficacy than on the duration of no-flow. Rats with 8 min of complete no-flow plus 1 min of effective resuscitation ("8 + 1" group) had a much better survival rate than rats with only 5 min of no-flow plus 4 min ("5 + 4") of ineffective, but continuous, resuscitation. Likewise, 12 min of cardiac arrest plus 1 min of resuscitation ("12 + 1") led to a significantly better outcome than 8 min of cardiac arrest plus

5 min of resuscitation ("8 + 5"). More dramatically, the animals in the "12 + 1" group performed better than the "5 + 4" group. Notice that the total time for cardiac arrest and resuscitation in the "5 + 4" group (total 9 min) is shorter than the complete no-flow time (12 min) in the "12 + 1" group. Thus, not only did the no-flow time fail to predict outcome, the animal data seemed to also suggest that resuscitation efficacy is the major determinant of outcome within a certain time limit of no-flow. Indeed, retrospective analyses of witnessed, index cases of in-hospital cardiac arrest in the National Registry of Cardiopulmonary Resuscitation showed that the arrest interval is not a predictor of survival to hospital discharge [87].

The post-resuscitation physiology in humans is significantly more complicated, and the results from rodent studies cannot be directly translated into clinical practice. However, this animal study does raise an important question: is it the long duration of cardiac arrest *per se* or the intrinsic difficulty of CPR after a long cardiac arrest (and hence the resulting ineffective and long resuscitation effort) that worsens the outcome in human cardiac arrest patients? Understandably, an increase in no-flow time in humans is often associated with increasing difficulties to resuscitate by means of conventional CPR. Thus, as pointed out by Xu *et al.* [5], a suboptimal application of CPR may have detrimental effects on cardiac arrest patients, even when the no-flow time is short. It is reasonable to ask whether the benefit of waiting for more effective resuscitation—especially if such waiting time can be predetermined—will outweigh the risk of premature and suboptimal administration of CPR by untrained bystanders. Both retrospective evaluation of existing human data and prospective controlled studies are needed to answer this perplexing question.

When prior cardiac arrest and resuscitation clinical trials are viewed from this perspective, the raw data seem to be consistent with the conclusions from the animal studies. In the First Brain Resuscitation Clinical Trial [88], a subset of patients with resuscitation time less than 15 min had a 42% rate of recovery, significantly better than the 19% for those with CPR administered for longer than 15 min. Likewise, in the Second Brain Resuscitation Clinical Trial [89], a 31% or 14% rate of good cerebral recovery was achieved among patients who had a resuscitation time shorter or longer than 20 min, respectively.

More recently, two independent clinical trials [90, 91] investigated the effects of mild hypothermia on clinical outcomes after cardiac arrest. It is worth noting, however, that neither study reports the duration from the first resuscitation attempt to the restoration of spontaneous circulation (ROSC). One trial recorded the time from the first direct current (DC) shock to ROSC, but the time elapsed from the initial bystander resuscitation efforts was not counted.

Both trials had more bystander-performed basic life supports in the normothermic group than in the hypothermic group. In one trial, the difference was significant. Because the clinical outcome for the hypothermic group was better than the normothermic group, the conventional prediction was that if the hypothermic group had had more bystander resuscitations, the benefit of hypothermia would have been more pronounced. In light of the results from precisely controlled animal studies, a counterargument would question whether the worse outcome in the normothermic groups in the two trials was actually exacerbated, not only by normothermia, but also by the *inadequate* resuscitation by the bystanders. It would be interesting to determine whether a correlation exists between longer resuscitation efforts by the bystanders and the worse long-term outcome in these two clinical trials, independent of hypothermia or normothermia. The potential benefit of therapeutic hypothermia can only be reassured after the confounding variable of resuscitation efficacy is taken into consideration.

6 Clinical Management of Post-Resuscitation Syndromes

Effective post-cardiac arrest care, requiring services from a multidisciplinary team with complementary expertise, is a determining factor for long-term outcome. Recommendations for contemporary approaches [92] include at least three parallel strategies to treat (a) myocardial dysfunction, (b) post-resuscitation systemic perfusion disorders, and (c) ischemic cellular injuries of the brain. Standardized and goal-directed protocols [93, 94] have been introduced and implemented. It should be noted, however, that some of the recommendations, such as oxygenation management, are mere guidelines based on data from animal studies. Prospective clinical trials are still needed to further improve these protocols.

6.1 Treatment of Myocardial Dysfunction

The heart is at the core of immediate post-resuscitation care. Establishing circulatory stability is essential to achieve a better prognosis. Fluid therapy and inotropic drug therapy are usually considered first in order to maintain a moderately elevated mean arterial blood pressure (65–100 mmHg). If such a pressure and circulatory stability cannot be obtained, mechanical supports are indicated, including intra-aortic balloon pump (IABP), percutaneous cardiopulmonary support (PCPS), extracorporeal membrane oxygenation (ECMO), and left ventricular assist device (LVAD). Not all devices are equally effective in all circumstances and the selection should depend on the nature and severity of myocardial dysfunction. For example, IABP is beneficial for acute coronary syndrome complicated by cardiogenic shock [95] but not for ST-elevation myocardial infarction (STEMI) [96]. In severe myocardial dysfunction patients, PCPS and ECMO should be considered. Coronary revascularization should be attempted in all STEMI cases.

6.2 Treatment of Reperfusion Disorder

The so-called no-reflow phenomenon and protracted hypoperfusion are known factors for poor long-term prognosis. There are currently no proven methods that are universally applicable. Treatment with high perfusion pressure to flush through the vascular system and therapeutic thrombolysis have been proposed on the basis of a large body of literature from animal studies. Earlier small-scale clinical trials in humans showed promising positive results of thrombolysis therapy with recombinant tissue plasminogen activators (rt-PA), particularly for patients with initially unsuccessful out-of-hospital CPR [97, 98]. The prospective, double-blinded, randomized, and placebo-controlled multicenter clinical trial on thrombolysis in cardiac arrest (TROICA) [99, 100] was halted because preliminary results from the trial showed that the benefits of thrombolysis therapy in cardiac arrest were not statistically significant over the placebo in terms of return of spontaneous circulation, 24-h and 30-day survival, and hospital admission. Hemorrhage is a major concern of thrombolytic therapy, but the TROICA trial showed no significant difference between treatment and placebo in symptomatic intracranial hemorrhage and major bleeding. It is still debatable whether in certain subsets of the TROICA trial, thrombolysis is actually beneficial. For example, in confirmed cases of pulmonary embolism, treatment with rt-PA can significantly improve survival rate and long-term outcome [101].

6.3 Treatment of Brain Injuries After Reperfusion

Close to 70% of cardiac arrest deaths after successful resuscitation are caused by ischemic brain injuries [102]. Currently, the only clinical intervention that proves efficacious for neuroprotection after cardiac arrest is therapeutic hypothermia. While the cellular and molecular mechanisms of hypothermic protection is the subject of many ongoing investigations, the major contributing factors likely include the preservation of energy metabolism, reduction of cerebral oxygen demand, and the resulting decrease in both the rate and amount of ROS production. Therapeutic hypothermia is particularly indicated if the initial cardiac rhythm is ventricular fibrillation or ventricular tachycardia (VF/VT). An initial rhythm of asystole or pulseless electric activity is usually associated with significantly poorer outcome. However, even in these patients, therapeutic hypothermia might also be beneficial.

The current recommendation is that therapeutic hypothermia be implemented as soon as possible. However, the optimal procedure for implementation is not yet fully established and requires prospective, randomized, and resuscitation-efficacy-controlled studies in humans. Important parameters to be determined include optimal temperatures at different stages, method and rate of cooling, the proper assessment of the core temperature, duration of hypothermia, and rewarming steps as well as their speed and timing. Two recent clinical trials [90, 91] used 33°C within 2 h of cardiac arrest for 12 h, or 32–34°C for 24 h, respectively. Rewarming is recommended at a rate of 0.25–0.5°C/h [103], with hemodynamics being closely monitored.

Therapeutic hypothermia is likely to reduce the rate of ROS production, thereby decreasing the amount of oxidative damage after the reestablishment of cerebral perfusion. In conjunction with therapeutic hypothermia, oxygenation after resuscitation should also be carefully controlled. Mechanical ventilation with 100% oxygen, while commonly used during resuscitation, is not recommended for post-resuscitation management. Arterial hyperoxia, with $P_aO_2 > 300$ mmHg, is potentially harmful. Thus, there is a delicate balance between hyperoxia and hypoxia; neither is beneficial for long-term recovery. Moreover, carefully controlled ventilation is important to avoid respiratory acidosis and to maintain P_aCO_2 in the desirable physiological range.

Control of post-resuscitation seizures is another important component of neuro-protective strategies, as seizures are related to worse neurological outcomes [104]. Early preventative treatment with anticonvulsant drugs, such as phenytoin, leveti-racetam, and valproic acid, has been advocated [104].

7 Future Directions

Exciting new advances in neuroscience provide hope for improved prognosis of cardiac arrest patients. The goal to achieve a significantly better long-term outcome is both realistic and attainable in the foreseeable future. New research frontiers are being advanced in the areas of gene therapy, stem cell therapy, and nanomedicine. Translational research is urgently needed to bring the mechanistic insights from the bench-top to the bedside.

7.1 Gene Therapy for Global Ischemia

At the dawn of the gene therapy era, a review of ~100 early gene therapy trials concluded that "human gene transfer is feasible, can evoke biologic responses that are relevant to human disease, and can provide important insights into human biology" [105]. Since then, significant advancements have been made in the development of novel gene delivery systems that are organ-, tissue-, or disease-specific. Common gene carriers suitable for clinical application for localized and systemic delivery include replication-deficient herpes simplex viruses (HSV), recombinant adeno-associated viruses (rAAV), liposomes, and nanoparticles.

The central idea of gene therapy is to use genetic methods to manipulate the expression of gene products, such that the pro-injury genes are suppressed and pro-tective genes are promoted. Thus, most gene therapies are designed based on a known molecular mechanism or pathway. As our knowledge of injury mechanisms evolves, better and more effective gene targets will be identified and used. For example, early gene therapy attempts targeted the mismatch between energy pro-duction and utilization during ischemia by overexpression of glucose transport [106, 107], with the hope that readily available intracellular high-energy substrates would correct and reverse the injury pathways associated with energy depletion.

Similarly, when excitotoxicity is targeted, one of the possible ways to correct the intracellular calcium overload is to overexpress the calcium-buffering protein calbindin D28K [108, 109]. Inhibition of the apoptotic pathway can be achieved by the overexpression of the antiapoptotic protein BCL-2 [110–116], a 26-kD membrane-bound protein capable of blocking cytosolic translocation of mitochondrial cytochrome c. However, inhibiting the late-stage molecular events in the injury cascade is not always effective [117].

Given their roles in angiogenesis and neurogenesis, VEGF and BDNF are logical candidates for gene manipulation. In rats, widespread BDNF overexpression in the hippocampus can be achieved by intraventricular microinjection of rAAV vectors encoding BDNF, as exemplified in Fig. 22.7. VEGF overexpression can be achieved in a similar way [118]. Critical issues associated with moving this technology to pre-clinical trials have been carefully considered [119]. A key element to consider in gene therapy design is how to control the gene dose. Uncontrolled overexpression of VEGF can cause elevated intracranial pressure, tissue edema, and worse injuries [120].

Many newly discovered molecular processes associated with reperfusion injuries could be gene therapy targets. Of particular note are the brain parenchyma immune and inflammatory responses to ischemic injuries. Microglia become activated and can contribute to endogenous cerebral defense. Expansion of processes from microglia and reactive astrocytes, and the secretion of cytokines from these cells, can play a biphasic role in both injurious and protective pathways. Timed and conditioned manipulations of pro-inflammatory and anti-inflammatory signals in the brain may prove to be an effective strategy to control late-phase recovery after reperfusion. In this regard, gene silencing by RNA interference technology is a versatile approach that is likely to find many fruitful clinical applications in the near future.

7.2 Stem Cell Therapy for Global Cerebral Ischemia

The realization that neurons can regenerate in the adult brain revolutionizes the thinking about potential therapies for neurodegenerative diseases, as well as ischemic and hypoxic brain injuries. Stem cell therapy holds tremendous promise in repairing the damaged neurons and glia after global ischemia to ensure a progressive long-term brain recovery. The most fundamental property of stem cells is their ability to self-renew indefinitely and to differentiate into multiple cell or tissue types. This ability is well documented for embryonic stem (ES) cells. Recent studies suggest that non-ES cells are more plastic than previously realized. For example, transplantation of bone marrow or enriched hematopoietic stem cells leads to detection of various cell types of donor origin, including skeletal and cardiac myoblasts [121–124], endothelium [125, 126], epithelium [127, 128], and neuroectoderm [129–134]. It was also demonstrated that mesenchymal stem cells from adult marrow are pluripotent; they are capable of differentiating at the single cell level, not only into mesenchymal cells, but also into cells with visceral mesoderm,

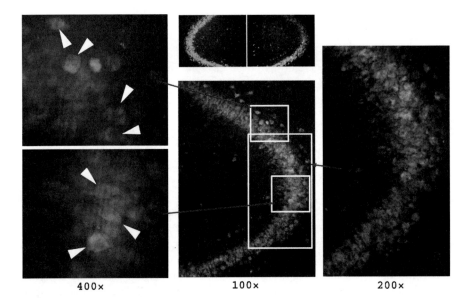

Fig. 22.7 Gene therapy for global cerebral ischemia. Exogenous brain-derived neurotrophic factor with a hemagglutinin tag (BDNF-HA) was overexpressed in the CA3 region of hippocampus 35 days after rAAV5-BDNF-HA microinjection into the left lateral ventricle. The *top* section of the *middle column* shows that the HA-tag staining (*red*, for exogenous BDNF) is mostly in the ipsilateral side of the injection. The contralateral side at this level has no HA-positive staining and can thus serve as control. At 200× (*right column*) and 400× (*left column*) magnifications, the neuronal nuclei (NeuN staining in *green*) can be seen clearly inside neurons that are positively stained for BDNF-HA (*white arrowheads*)

neuroectoderm, and endoderm characteristics [135, 136]. Other sources of postnatal stem cells have been identified in recent years. Myofibroblasts are important cells in growth, development, and repair. They are found throughout the body and include bone marrow stromal cells, astrocytes, and pericytes [137, 138]. Wharton's jelly is the gelatinous connective tissue from the umbilical cord and is composed of myofibroblast-like stromal cells, collagen fibers, and proteoglycans [139]. Stable cell lines from umbilical cord matrix cells have been established [140, 141]. These cells are developmentally between embryonic and adult stem cells and share many of the properties of ES cells. Umbilical cord matrix stem cells are easily attainable and have been shown to be multipotent. More recently, inducible pluripotent stem (iPS) cells [142–146] showed tremendous promise for future clinical application of transcriptional modifications for the purpose of inducing pluripotency in adult tissue repair.

Stem cell therapy for global ischemia remains an uncharted territory; only a few studies have been carried out to date [44, 141]. For some neurodegenerative diseases, such as Parkinson's disease, a viable treatment option is to transplant stem cells preconditioned for dopaminergic neuron differentiation into substantia nigra.

However, global ischemia is characterized by disseminated neuronal loss throughout the brain. Thus, localized stem cell transplantation with the goal of neuronal repairs by the classical mechanisms of *trans*-differentiation or stem cell fusion with the host cells is unlikely to work. Moreover, even if the dissemination of exogenous stem cells by transplantation can be achieved, there is still an insurmountable mismatch between the huge number of neurons to be replaced or repaired after global ischemia and the relatively small number of stem cells that can be practically transplanted. However, using the well-controlled cardiac arrest and resuscitation model discussed above, it was discovered unexpectedly that Oct4+ umbilical cord matrix stem cells transplanted into the left hemisphere of a rat brain could provide a nearly equal level of protection in both hemispheres [141]. Follow-up studies, including sham operation, defined medium injection, controls, and three-dimensional stem cell tracking of transplanted stem cells, ruled out distant cell migration or complications such as preconditioning to microinjection. It was concluded that the protection seen in the contralateral side was indeed the result of stem cell protection. The degree of protection found in these animals could not be explained by *trans*-differentiation or cell fusion. Thus, the results from stem cell treatment of global ischemia suggested the possibility of a third novel mechanism of stem cell repair— one that elicits one or multiple synergistic extracellular signaling pathways. This possibility is strongly supported by recent studies where intravenous injection of human umbilical cord blood (HUCB) cells into rats was shown to reduce brain injury during a 1-h middle cerebral artery occlusion [147, 148]. These focal ischemia studies unequivocally demonstrated that cell entry into the central nervous system is not absolutely required for neuroprotection by the peripherally injected HUCB cells. The secretion of the "therapeutic molecules" (including the neurotrophic factors) and the nonimmune anti-inflammatory effects are the two necessary components of the observed HUCB cell neuroprotection.

The prospect of stem cell signaling as the underlying mechanism of protection and repair after global ischemia is exciting, suggesting that therapeutic conditions can be created by mimicking those produced by the stem cells. It can be speculated that the presence of Oct-4+ stem cells during ischemia and reperfusion activates and accelerates the proliferation and recruitment of the endogenous neural stem cells, including reentry of quiescent stem cells, into the rescue effort. Other possibilities include the creation of an extracellular milieu that enhances and restores the intrinsic ability of the brain tissue to self-repair [149]. For example, the very presence of transplanted stem cells may serve as a first responder to ischemic stress signals by priming the activation and reintegration of developmental signaling cascades [150], thereby encouraging adult brain tissue to reenter a youthful state [151, 152] in the recovery stage. Retrograde signaling from long-distance connections might also play an important role in determining the fate of neurons after ischemia and reperfusion injuries. Future studies will fully explore the underlying processes of stem cell signaling. The possibilities to find and use the signaling molecules for therapeutic purposes are endless.

7.3 Nanotechnology and Nanomedicine

In addition to the use of nanoparticles as gene delivery vehicles, nanomedicine holds previously unimaginable promise for both diagnosis and treatment of neurological conditions [153]. Specifically for long-term repair of injured brain after cardiac arrest and resuscitation, nanomaterials can be engineered to carry out diverse functions ranging from cellular scaffolds for axonal regrowth to scavengers to sweep up ROS during reperfusion. A unique feature of nanoparticles is that they are amenable to various chemical modifications by adding biologically active epitopes such that they possess the desired properties of targeting a particular organ or a certain population of cells. For example, in one study a small peptide derived from the rabies viral glycoprotein, which can easily pass the blood–brain barrier, is hijacked in the engineering of nanoparticles for systemic delivery of short interfering RNA into the brain [154]. Several different formulations of nanoparticles have been developed recently [155–160], and they can be adopted for use to treat global ischemia and reperfusion injuries.

8 Conclusion

The brain's ability to withstand global ischemia is significantly higher than previously appreciated. Poor prognosis after cardiac arrest and resuscitation is likely related to secondary injuries from the post-resuscitation syndrome. Protracted hypoperfusion can exacerbate cellular injuries from the primary ischemic insult by amplifying the oxidative damages due to the delivery of oxygen and the circulation of reactive oxygen species. Contemporary therapeutic strategies combine three parallel approaches to treat myocardial dysfunction for the maintenance of circulatory stability, to correct systemic perfusion abnormalities by thrombolysis in certain cases, and to treat underlying causes of cellular injuries. Therapeutic hypothermia is currently the only proven method to improve long-term outcome and should be implemented as soon as possible during and after resuscitation. Recent advances in neurosciences have enabled many new technologies to be developed for clinical applications, including gene therapy, stem cell therapy, and nanomedicine.

Acknowledgments The author would like to thank Professor Pei Tang for many stimulating discussions and Dr. Serguei Liachenko, Dr. Sachiko Jomura, Dr. Renee Dallasen, Mr. Marc Uy, and Ms. Nicole Brandon for their contributions to the research projects in the author's laboratory. This work was supported in part by a grant from the United States National Institutes of Health (R01NS/HL036124).

References

1. Mehra R. Global public health problem of sudden cardiac death. J Electrocardiol. 2007; 40(6 Suppl):S118–22.
2. Becker L. The epidemiology of sudden death. In: Paradis NA, Nowak RM, Halperin HR, editors. Cardiac arrest: the science and practice of resuscitation medicine. Baltimore: Williams & Wilkins; 1996. p. 28–47.
3. Safar P. Effects of the postresuscitation syndrome on cerebral recovery from cardiac arrest. Crit Care Med. 1985;13(11):932–5.
4. Safar P. Cerebral resuscitation after cardiac arrest: a review. Circulation. 1986;74(6 Pt 2): IV138–53.
5. Xu Y, Liachenko S, Tang P. Dependence of early cerebral reperfusion and long-term outcome on resuscitation efficiency after cardiac arrest in rats. Stroke. 2002;33(3):837–43.
6. White RD, Goodman BW, Svoboda MA. Neurologic recovery following prolonged out-of-hospital cardiac arrest with resuscitation guided by continuous capnography. Mayo Clin Proc. 2011;86(6):544–8.
7. Berg MD, Schexnayder SM, Chameides L, Terry M, Donoghue A, Hickey RW, et al. Part 13: Pediatric basic life support. Circulation. 2010;122(18 Suppl 3):S862–75.
8. Berg RA, Hemphill R, Abella BS, Aufderheide TP, Cave DM, Hazinski MF, et al. Part 5: Adult basic life support. Circulation. 2010;122(18 Suppl 3):S685–705.
9. Bhanji F, Mancini ME, Sinz E, Rodgers DL, McNeil MA, Hoadley TA, et al. Part 16: Education, implementation, and teams. Circulation. 2010;122(18 Suppl 3):S920–33.
10. Cave DM, Gazmuri RJ, Otto CW, Nadkarni VM, Cheng A, Brooks SC, et al. Part 7: CPR techniques and devices. Circulation. 2010;122(18 Suppl 3):S720–8.
11. Field JM, Hazinski MF, Sayre MR, Chameides L, Schexnayder SM, Hemphill R, et al. Part 1: Executive summary. Circulation. 2010;122(18 Suppl 3):S640–56.
12. Jauch EC, Cucchiara B, Adeoye O, Meurer W, Brice J, Chan Y, et al. Part 11: Adult stroke. Circulation. 2010;122(18 Suppl 3):S818–28.
13. Kattwinkel J, Perlman JM, Aziz K, Colby C, Fairchild K, Gallagher J, et al. Part 15: Neonatal resuscitation. Circulation. 2010;122(18 Suppl 3):S909.
14. Kleinman ME, Chameides L, Schexnayder SM, Samson RA, Hazinski MF, Atkins DL, et al. Part 14: Pediatric advanced life support. Circulation. 2010;122(18 Suppl 3):S876–908.
15. Link MS, Atkins DL, Passman RS, Halperin HR, Samson RA, White RD, et al. Part 6: Electrical therapies. Circulation. 2010;122(18 Suppl 3):S706–19.
16. Morrison LJ, Kierzek G, Diekema DS, Sayre MR, Silvers SM, Idris AH, et al. Part 3: Ethics. Circulation. 2010;122(18 Suppl 3):S665–75.
17. Neumar RW, Otto CW, Link MS, Kronick SL, Shuster M, Callaway CW, et al. Part 8: Adult advanced cardiovascular life support. Circulation. 2010;122(18 Suppl 3):S729–67.
18. O'Connor RE, Brady W, Brooks SC, Diercks D, Egan J, Ghaemmaghami C, et al. Part 10: Acute coronary syndromes. Circulation. 2010;122(18 Suppl 3):S787–817.
19. Peberdy MA, Callaway CW, Neumar RW, Geocadin RG, Zimmerman JL, Donnino M, et al. Part 9: Post‚ÄìCardiac arrest care. Circulation. 2010;122(18 Suppl 3):S768–86.
20. Sayre MR, O'Connor RE, Atkins DL, Billi JE, Callaway CW, Shuster M, et al. Part 2: Evidence evaluation and management of potential or perceived conflicts of interest. Circulation. 2010;122(18 Suppl 3):S657–64.
21. Travers AH, Rea TD, Bobrow BJ, Edelson DP, Berg RA, Sayre MR, et al. Part 4: CPR overview. Circulation. 2010;122(18 Suppl 3):S676–84.
22. Vanden Hoek TL, Morrison LJ, Shuster M, Donnino M, Sinz E, Lavonas EJ, et al. Part 12: Cardiac arrest in special situations. Circulation. 2010;122(18 Suppl 3):S829–61.
23. Eklof B, Siesjo BK. The effect of bilateral carotid artery ligation upon the blood flow and the energy state of the rat brain. Acta Physiol Scand. 1972;86(2):155–65.

24. Smith ML, Bendek G, Dahlgren N, Rosen I, Wieloch T, Siesjo BK. Models for studying long-term recovery following forebrain ischemia in the rat. 2. A 2-vessel occlusion model. Acta Neurol Scand. 1984;69(6):385–401.
25. Harukuni I, Bhardwaj A. Mechanisms of brain injury after global cerebral ischemia. Neurol Clin. 2006;24(1):1–21.
26. Pulsinelli WA, Levy DE, Duffy TE. Cerebral blood flow in the four-vessel occlusion rat model. Stroke. 1983;14(5):832–4.
27. Kagstrom E, Smith ML, Siesjo BK. Local cerebral blood flow in the recovery period following complete cerebral ischemia in the rat. J Cereb Blood Flow Metab. 1983;3(2):170–82.
28. Pulsinelli WA, Brierley JB. A new model of bilateral hemispheric ischemia in the unanesthetized rat. Stroke. 1979;10(3):267–72.
29. Yamaguchi M, Calvert JW, Kusaka G, Zhang JH. One-stage anterior approach for four-vessel occlusion in rat. Stroke. 2005;36(10):2212–4.
30. McBean DE, Kelly PA. Rodent models of global cerebral ischemia: a comparison of two-vessel occlusion and four-vessel occlusion. Gen Pharmacol. 1998;30(4):431–4.
31. Liachenko S, Tang P, Hamilton RL, Xu Y. A reproducible model of circulatory arrest and remote resuscitation in rats for NMR investigation. Stroke. 1998;29(6):1229–38. discussion 38–9.
32. Blomqvist P, Wieloch T. Ischemic brain damage in rats following cardiac arrest using a long-term recovery model. J Cereb Blood Flow Metab. 1985;5(3):420–31.
33. Hendrickx HH, Rao GR, Safar P, Gisvold SE. Asphyxia, cardiac arrest and resuscitation in rats. I. Short term recovery. Resuscitation. 1984;12(2):97–116.
34. Liachenko S, Tang P, Hamilton RL, Xu Y. Regional dependence of cerebral reperfusion after circulatory arrest in rats. J Cereb Blood Flow Metab. 2001;21(11):1320–9.
35. Geissler HJ, Davis KL, Laine GA, Ostrin EJ, Mehlhorn U, Hekmat K, et al. Myocardial protection with high-dose beta-blockade in acute myocardial ischemia. Eur J Cardiothorac Surg. 2000;17(1):63–70.
36. Ede M, Ye J, Gregorash L, Summers R, Pargaonkar S, LeHouerou D, et al. Beyond hyperkalemia: beta-blocker-induced cardiac arrest for normothermic cardiac operations. Ann Thorac Surg. 1997;63(3):721–7.
37. Mehlhorn U, Sauer H, Kuhn-Regnier F, Sudkamp M, Dhein S, Eberhardt F, et al. Myocardial beta-blockade as an alternative to cardioplegic arrest during coronary artery surgery. Cardiovasc Surg. 1999;7(5):549–57.
38. Warters RD, Allen SJ, Davis KL, Geissler HJ, Bischoff I, Mutschler E, et al. Beta-blockade as an alternative to cardioplegic arrest during cardiopulmonary bypass. Ann Thorac Surg. 1998;65(4):961–6.
39. Urbanek P, Bock H, Vicol C. Percutaneous cardiopulmonary support (PCPS). Cardiology. 1994;84(3):216–21.
40. Liachenko S, Tang P, Xu Y. Deferoxamine improves early postresuscitation reperfusion after prolonged cardiac arrest in rats. J Cereb Blood Flow Metab. 2003;23(5):574–81.
41. Xu Y, Liachenko SM, Tang P, Chan PH. Faster recovery of cerebral perfusion in SOD1-overexpressed rats after cardiac arrest and resuscitation. Stroke. 2009;40(7):2512–8.
42. Endres M, Dirnagl U. Neuroprotective strategies in animal and in vitro-models of neuronal damage: ischemia and stroke. In: Alzheimer C, editor. Molecular and cellular biology of neuroprotection in the CNS. New York: Springer; 2002.
43. Ouyang YB, Voloboueva LA, Xu LJ, Giffard RG. Selective dysfunction of hippocampal CA1 astrocytes contributes to delayed neuronal damage after transient forebrain ischemia. J Neurosci. 2007;27(16):4253–60.
44. Hirko AC, Dallasen R, Jomura S, Xu Y. Modulation of inflammatory responses after global ischemia by transplanted umbilical cord matrix stem cells. Stem Cells. 2008;26(11):2893–901.
45. Kubo K, Nakao S, Jomura S, Sakamoto S, Miyamoto E, Xu Y, et al. Edaravone, a free radical scavenger, mitigates both gray and white matter damages after global cerebral ischemia in rats. Brain Res. 2009;1279:139–46.

46. White BC, Sullivan JM, DeGracia DJ, O'Neil BJ, Neumar RW, Grossman LI, et al. Brain ischemia and reperfusion: molecular mechanisms of neuronal injury. J Neurol Sci. 2000;179(S1–2):1–33.

47. Whittingham TS. Aspects of brain energy metabolism and cerebral ischemia. In: Schurr A, Rigor BM, editors. Cerebral ischemia and resuscitation. Boston: CRC Press; 1990. p. 101–22.

48. Gilbertson TA, Scobey R, Wilson M. Permeation of calcium ions through non-NMDA glutamate channels in retinal bipolar cells. Science. 1991;251(5001):1613–5.

49. Hollmann M, Hartley M, Heinemann S. Ca2+ permeability of KA-AMPA–gated glutamate receptor channels depends on subunit composition. Science. 1991;252(5007):851–3.

50. Iino M, Ozawa S, Tsuzuki K. Permeation of calcium through excitatory amino acid receptor channels in cultured rat hippocampal neurones. J Physiol (Lond). 1990;424:151–65.

51. Michaels RL, Rothman SM. Glutamate neurotoxicity in vitro: antagonist pharmacology and intracellular calcium concentrations. J Neurosci. 1990;10(1):283–92.

52. Treanor J, Kawaoka Y, Miller R, Webster RG, Murphy B. Nucleotide sequence of the avian influenza A/Mallard/NY/6750/78 virus polymerase genes. Virus Res. 1989;14(3):257–69.

53. Murphy SN, Miller RJ. Regulation of Ca++ influx into striatal neurons by kainic acid. J Pharmacol Exp Ther. 1989;249(1):184–93.

54. O'Neil BJ, Krause GS, Grossman LI, Grunberger G, Rafols JA, DeGracia DJ, et al. Global brain ischemia and reperfusion by cardiac arrest and resuscitation. In: Paradis NA, Halperin HR, Nowak RM, editors. Cardiac arrest: the science and practice of resuscitation medicine. Baltimore: Williams & Wilkins; 1996. p. 84–112.

55. Meldolesi J, Volpe P, Pozzan T. The intracellular distribution of calcium. Trends Neurosci. 1988;11(10):449–52.

56. Pozzan T, Volpe P, Zorzato F, Bravin M, Krause KH, Lew DP, et al. The Ins(1,4,5)P3-sensitive Ca2+ store of non-muscle cells: endoplasmic reticulum or calciosomes? J Exp Biol. 1988;139:181–93.

57. Nathanson JA, Scavone C, Scanlon C, McKee M. The cellular Na+ pump as a site of action for carbon monoxide and glutamate: a mechanism for long-term modulation of cellular activity. Neuron. 1995;14(4):781–94.

58. Kaczmarek LK. The role of protein kinase C in the regulation of ion channels and neurotransmitter release. Trends Neurosci. 1987;10:30–4.

59. Ikeda Y, Long DM. The molecular basis of brain injury and brain edema: the role of oxygen free radicals. Neurosurgery. 1990;27(1):1–11.

60. Schmidley JW. Free radicals in central nervous system ischemia. Stroke. 1990;21(7): 1086–90.

61. Kontos HA. Oxygen radicals in CNS damage. Chem Biol Interact. 1989;72(3):229–55.

62. Kukreja RC, Kontos HA, Hess ML, Ellis EF. PGH synthase and lipoxygenase generate superoxide in the presence of NADH or NADPH. Circ Res. 1986;59(6):612–9.

63. Safar P. Prevention and therapy of postresuscitation neurologic dysfunction and injury. In: Paradis NA, Halperin HR, Nowak RM, editors. Cardiac arrest: the science and practice of resuscitation medicine. Baltimore: Williams & Wilkins; 1996. p. 859–87.

64. McCord JM. Oxygen-derived free radicals in postischemic tissue injury. N Engl J Med. 1985;312(3):159–63.

65. Weiss SJ. Tissue destruction by neutrophils [see comments]. N Engl J Med. 1989;320(6): 365–76.

66. McCord JM. Oxygen-derived radicals: a link between reperfusion injury and inflammation. Fed Proc. 1987;46(7):2402–6.

67. Irvine RF. How is the level of free arachidonic acid controlled in mammalian cells? Biochem J. 1982;204(1):3–16.

68. van den Bosch H. Intracellular phospholipases A. Biochim Biophys Acta. 1980;604(2): 191–246.

69. Watson BD. Evaluation of the concomitance of lipid peroxidation in experimental models of cerebral ischemia and stroke. Prog Brain Res. 1993;96:69–95.

70. Chan PH, Kinouchi H, Epstein CJ, Carlson E, Chen SF, Imaizumi S, et al. Role of superoxide dismutase in ischemic brain injury: reduction of edema and infarction in transgenic mice following focal cerebral ischemia. Prog Brain Res. 1993;96:97–104.

71. Louissaint Jr A, Rao S, Leventhal C, Goldman SA. Coordinated interaction of neurogenesis and angiogenesis in the adult songbird brain. Neuron. 2002;34(6):945–60.

72. Harrigan MR, Ennis SR, Masada T, Keep RF. Intraventricular infusion of vascular endothelial growth factor promotes cerebral angiogenesis with minimal brain edema. Neurosurgery. 2002;50(3):589–98.

73. Hayashi T, Abe K, Itoyama Y. Reduction of ischemic damage by application of vascular endothelial growth factor in rat brain after transient ischemia. J Cereb Blood Flow Metab. 1998;18(8):887–95.

74. Hayashi T, Abe K, Suzuki H, Itoyama Y. Rapid induction of vascular endothelial growth factor gene expression after transient middle cerebral artery occlusion in rats. Stroke. 1997;28(10):2039–44.

75. Marti HJ, Bernaudin M, Bellail A, Schoch H, Euler M, Petit E, et al. Hypoxia-induced vascular endothelial growth factor expression precedes neovascularization after cerebral ischemia. Am J Pathol. 2000;156(3):965–76.

76. Neufeld G, Cohen T, Gengrinovitch S, Poltorak Z. Vascular endothelial growth factor (VEGF) and its receptors. FASEB J. 1999;13(1):9–22.

77. Gerber HP, Condorelli F, Park J, Ferrara N. Differential transcriptional regulation of the two vascular endothelial growth factor receptor genes. Flt-1, but not Flk-1/KDR, is up-regulated by hypoxia. J Biol Chem. 1997;272(38):23659–67.

78. Lee TH, Avraham H, Lee SH, Avraham S. Vascular endothelial growth factor modulates neutrophil transendothelial migration via up-regulation of interleukin-8 in human brain microvascular endothelial cells. J Biol Chem. 2002;277(12):10445–51.

79. Ushio-Fukai M, Tang Y, Fukai T, Dikalov SI, Ma Y, Fujimoto M, et al. Novel role of gp91(phox)-containing NAD(P)H oxidase in vascular endothelial growth factor-induced signaling and angiogenesis. Circ Res. 2002;91(12):1160–7.

80. Wang Z, Castresana MR, Newman WH. Reactive oxygen and NF-kappaB in VEGF-induced migration of human vascular smooth muscle cells. Biochem Biophys Res Commun. 2001;285(3):669–74.

81. Page EL, Robitaille GA, Pouyssegur J, Richard DE. Induction of hypoxia-inducible factor-1alpha by transcriptional and translational mechanisms. J Biol Chem. 2002;277(50): 48403–9.

82. Brandes RP, Miller FJ, Beer S, Haendeler J, Hoffmann J, Ha T, et al. The vascular NADPH oxidase subunit p47phox is involved in redox-mediated gene expression. Free Radic Biol Med. 2002;32(11):1116–22.

83. Colavitti R, Pani G, Bedogni B, Anzevino R, Borrello S, Waltenberger J, et al. Reactive oxygen species as downstream mediators of angiogenic signaling by vascular endothelial growth factor receptor-2/KDR. J Biol Chem. 2002;277(5):3101–8.

84. Kim SH, Won SJ, Sohn S, Kwon HJ, Lee JY, Park JH, et al. Brain-derived neurotrophic factor can act as a pronecrotic factor through transcriptional and translational activation of NADPH oxidase. J Cell Biol. 2002;159(5):821–31.

85. Han BH, Holtzman DM. BDNF protects the neonatal brain from hypoxic-ischemic injury *in vivo* via the ERK pathway. J Neurosci. 2000;20(15):5775–81.

86. Abid MR, Tsai JC, Spokes KC, Deshpande SS, Irani K, Aird WC. Vascular endothelial growth factor induces manganese-superoxide dismutase expression in endothelial cells by a Rac1-regulated NADPH oxidase-dependent mechanism. Faseb J. 2001;15(13):2548–50.

87. Scouras NE, Bircher NG, Jordan D, Xu Y. Retrospective evaluation of resuscitation intervals and efficacy on mortality. Circulation. 2010;122(122):A139.

88. Group BRCTIS. Randomized clinical study of thiopental loading in comatose survivors of cardiac arrest. Brain Resuscitation Clinical Trial I Study Group. N Engl J Med. 1986;314(7): 397–403.

89. Group BRCTIS. A randomized clinical study of a calcium-entry blocker (lidoflazine) in the treatment of comatose survivors of cardiac arrest. Brain Resuscitation Clinical Trial II Study Group. N Engl J Med. 1991;324(18):1225–31.
90. Hypothermia after Cardiac Arrest Study Group. Mild therapeutic hypothermia to improve the neurologic outcome after cardiac arrest. N Engl J Med. 2002;346(8):549–56.
91. Bernard SA, Gray TW, Buist MD, Jones BM, Silvester W, Gutteridge G, et al. Treatment of comatose survivors of out-of-hospital cardiac arrest with induced hypothermia. N Engl J Med. 2002;346(8):557–63.
92. Stub D, Bernard S, Duffy SJ, Kaye DM. Post cardiac arrest syndrome: a review of therapeutic strategies. Circulation. 2011;123(13):1428–35.
93. Gaieski DF, Band RA, Abella BS, Neumar RW, Fuchs BD, Kolansky DM, et al. Early goal-directed hemodynamic optimization combined with therapeutic hypothermia in comatose survivors of out-of-hospital cardiac arrest. Resuscitation. 2009;80(4):418–24.
94. Sunde K, Pytte M, Jacobsen D, Mangschau A, Jensen LP, Smedsrud C, et al. Implementation of a standardised treatment protocol for post resuscitation care after out-of-hospital cardiac arrest. Resuscitation. 2007;73(1):29–39.
95. Cheng JM, Valk SD, den Uil CA, van der Ent M, Lagrand WK, van de Sande M, et al. Usefulness of intra-aortic balloon pump counterpulsation in patients with cardiogenic shock from acute myocardial infarction. Am J Cardiol. 2009;104(3):327–32.
96. Sjauw KD, Engstrom AE, Vis MM, van der Schaaf RJ, Baan Jr J, Koch KT, et al. A systematic review and meta-analysis of intra-aortic balloon pump therapy in ST-elevation myocardial infarction: should we change the guidelines? Eur Heart J. 2009;30(4):459–68.
97. Bottiger BW, Bode C, Kern S, Gries A, Gust R, Glatzer R, et al. Efficacy and safety of thrombolytic therapy after initially unsuccessful cardiopulmonary resuscitation: a prospective clinical trial. Lancet. 2001;357(9268):1583–5.
98. Lederer W, Lichtenberger C, Pechlaner C, Kroesen G, Baubin M. Recombinant tissue plasminogen activator during cardiopulmonary resuscitation in 108 patients with out-of-hospital cardiac arrest. Resuscitation. 2001;50(1):71–6.
99. Spohr F, Arntz HR, Bluhmki E, Bode C, Carli P, Chamberlain D, et al. International multicentre trial protocol to assess the efficacy and safety of tenecteplase during cardiopulmonary resuscitation in patients with out-of-hospital cardiac arrest: the Thrombolysis in Cardiac Arrest (TROICA) Study. Eur J Clin Invest. 2005;35(5):315–23.
100. Bottiger BW, Arntz HR, Chamberlain DA, Bluhmki E, Belmans A, Danays T, et al. Thrombolysis during resuscitation for out-of-hospital cardiac arrest. N Engl J Med. 2008;359(25): 2651–62.
101. Perrott J, Henneberry RJ, Zed PJ. Thrombolytics for cardiac arrest: case report and systematic review of controlled trials. Ann Pharmacother. 2010;44(12):2007–13.
102. Laver S, Farrow C, Turner D, Nolan J. Mode of death after admission to an intensive care unit following cardiac arrest. Intensive Care Med. 2004;30(11):2126–8.
103. Arrich J. Clinical application of mild therapeutic hypothermia after cardiac arrest. Crit Care Med. 2007;35(4):1041–7.
104. Cokkinos P. Post-resuscitation care: current therapeutic concepts. Acute Card Care. 2009; 11(3):131–7.
105. Crystal RG. Transfer of genes to humans: early lessons and obstacles to success. Science. 1995;270(5235):404–10.
106. Lawrence MS, Sun GH, Kunis DM, Saydam TC, Dash R, Ho DY, et al. Overexpression of the glucose transporter gene with a herpes simplex viral vector protects striatal neurons against stroke. J Cereb Blood Flow Metab. 1996;16(2):181–5.
107. Ho DY, Mocarski ES, Sapolsky RM. Altering central nervous system physiology with a defective herpes simplex virus vector expressing the glucose transporter gene. Proc Natl Acad Sci U S A. 1993;90(8):3655–9.
108. Phillips RG, Meier TJ, Giuli LC, McLaughlin JR, Ho DY, Sapolsky RM. Calbindin D28K gene transfer via herpes simplex virus amplicon vector decreases hippocampal damage in vivo following neurotoxic insults. J Neurochem. 1999;73(3):1200–5.

109. Yenari MA, Minami M, Sun GH, Meier TJ, Kunis DM, McLaughlin JR, et al. Calbindin d28k overexpression protects striatal neurons from transient focal cerebral ischemia. Stroke. 2001;32(4):1028–35.

110. Xu L, Koumenis IL, Tilly JL, Giffard RG. Overexpression of bcl-xL protects astrocytes from glucose deprivation and is associated with higher glutathione, ferritin, and iron levels. Anesthesiology. 1999;91(4):1036–46.

111. Xu L, Lee JE, Giffard RG. Overexpression of bcl-2, bcl-XL or hsp70 in murine cortical astrocytes reduces injury of co-cultured neurons. Neurosci Lett. 1999;277(3):193–7.

112. Tamatani M, Mitsuda N, Matsuzaki H, Okado H, Miyake S, Vitek MP, et al. A pathway of neuronal apoptosis induced by hypoxia/reoxygenation: roles of nuclear factor-kappaB and Bcl-2. J Neurochem. 2000;75(2):683–93.

113. Lawrence MS, Ho DY, Sun GH, Steinberg GK, Sapolsky RM. Overexpression of Bcl-2 with herpes simplex virus vectors protects CNS neurons against neurological insults in vitro and *in vivo*. J Neurosci. 1996;16(2):486–96.

114. Dubois-Dauphin M, Pfister Y, Vallet PG, Savioz A. Prevention of apoptotic neuronal death by controlling procaspases? A point of view. Brain Res Brain Res Rev. 2001;36(2–3):196–203.

115. Nakamura M, Raghupathi R, Merry DE, Scherbel U, Saatman KE, McIntosh TK. Overexpression of Bcl-2 is neuroprotective after experimental brain injury in transgenic mice. J Comp Neurol. 1999;412(4):681–92.

116. Wang HD, Fukuda T, Suzuki T, Hashimoto K, Liou SY, Momoi T, et al. Differential effects of Bcl-2 overexpression on hippocampal CA1 neurons and dentate granule cells following hypoxic ischemia in adult mice. J Neurosci Res. 1999;57(1):1–12.

117. Phillips RG, Lawrence MS, Ho DY, Sapolsky RM. Limitations in the neuroprotective potential of gene therapy with Bcl-2. Brain Res. 2000;859(2):202–6.

118. Li SF, Meng QH, Yao WC, Hu GJ, Li GL, Li ZJ, et al. Recombinant AAV1 mediated vascular endothelial growth factor gene expression promotes angiogenesis and improves neural function: experiment with rats. Zhonghua Yi Xue Za Zhi. 2009;89(3):167–70.

119. Manoonkitiwongsa PS. Critical questions for preclinical trials on safety and efficacy of vascular endothelial growth factor-based therapeutic angiogenesis for ischemic stroke. CNS Neurol Disord Drug Targets. 2010;10(2):215–34.

120. Li Z, Wang R, Li S, Wei J, Zhang Z, Li G, et al. Intraventricular pre-treatment with rAAV-VEGF induces intracranial hypertension and aggravates ischemic injury at the early stage of transient focal cerebral ischemia in rats. Neurol Res. 2008;30(8):868–75.

121. Ferrari G, Cusella-De Angelis G, Coletta M, Paolucci E, Stornaiuolo A, Cossu G, et al. Muscle regeneration by bone marrow-derived myogenic progenitors. Science. 1998;279(5356):1528–30.

122. Gussoni E, Soneoka Y, Strickland CD, Buzney EA, Khan MK, Flint AF, et al. Dystrophin expression in the mdx mouse restored by stem cell transplantation. Nature. 1999;401(6751):390–4.

123. Orlic D, Kajstura J, Chimenti S, Jakoniuk I, Anderson SM, Li B, et al. Bone marrow cells regenerate infarcted myocardium. Nature. 2001;410(6829):701–5.

124. Jackson KA, Majka SM, Wang H, Pocius J, Hartley CJ, Majesky MW, et al. Regeneration of ischemic cardiac muscle and vascular endothelium by adult stem cells. J Clin Invest. 2001;107(11):1395–402.

125. Isner JM, Kalka C, Kawamoto A, Asahara T. Bone marrow as a source of endothelial cells for natural and iatrogenic vascular repair. Ann N Y Acad Sci. 2001;953:75–84.

126. Asahara T, Masuda H, Takahashi T, Kalka C, Pastore C, Silver M, et al. Bone marrow origin of endothelial progenitor cells responsible for postnatal vasculogenesis in physiological and pathological neovascularization. Circ Res. 1999;85(3):221–8.

127. Petersen BE, Bowen WC, Patrene KD, Mars WM, Sullivan AK, Murase N, et al. Bone marrow as a potential source of hepatic oval cells. Science. 1999;284(5417):1168–70.

128. Krause DS, Theise ND, Collector MI, Henegariu O, Hwang S, Gardner R, et al. Multi-organ, multi-lineage engraftment by a single bone marrow-derived stem cell. Cell. 2001;105(3):369–77.

129. Prockop DJ, Azizi SA, Colter D, Digirolamo C, Kopen G, Phinney DG. Potential use of stem cells from bone marrow to repair the extracellular matrix and the central nervous system. Biochem Soc Trans. 2000;28(4):341–5.

130. Prockop DJ, Azizi SA, Phinney DG, Kopen GC, Schwarz EJ. Potential use of marrow stromal cells as therapeutic vectors for diseases of the central nervous system. Prog Brain Res. 2000;128:293–7.

131. Phinney DG, Kopen G, Isaacson RL, Prockop DJ. Plastic adherent stromal cells from the bone marrow of commonly used strains of inbred mice: variations in yield, growth, and differentiation. J Cell Biochem. 1999;72(4):570–85.

132. Phinney DG, Kopen G, Righter W, Webster S, Tremain N, Prockop DJ. Donor variation in the growth properties and osteogenic potential of human marrow stromal cells. J Cell Biochem. 1999;75(3):424–36.

133. Brazelton TR, Rossi FM, Keshet GI, Blau HM. From marrow to brain: expression of neuronal phenotypes in adult mice. Science. 2000;290(5497):1775–9.

134. Mezey E, Chandross KJ, Harta G, Maki RA, McKercher SR. Turning blood into brain: cells bearing neuronal antigens generated *in vivo* from bone marrow. Science. 2000;290(5497):1779–82.

135. Jiang Y, Jahagirdar BN, Reinhardt RL, Schwartz RE, Keene CD, Ortiz-Gonzalez XR, et al. Pluripotency of mesenchymal stem cells derived from adult marrow. Nature. 2007;447:880–1. doi:10.1038/nature05812.

136. Schwartz RE, Reyes M, Koodie L, Jiang Y, Blackstad M, Lund T, et al. Multipotent adult progenitor cells from bone marrow differentiate into functional hepatocyte-like cells. J Clin Invest. 2002;109(10):1291–302.

137. Powell DW, Mifflin RC, Valentich JD, Crowe SE, Saada JI, West AB. Myofibroblasts. II. Intestinal subepithelial myofibroblasts. Am J Physiol. 1999;277(2 Pt 1):C183–201.

138. Powell DW, Mifflin RC, Valentich JD, Crowe SE, Saada JI, West AB. Myofibroblasts. I. Paracrine cells important in health and disease. Am J Physiol. 1999;277(1 Pt 1):C1–9.

139. Kobayashi K, Kubota T, Aso T. Study on myofibroblast differentiation in the stromal cells of Wharton's jelly: expression and localization of alpha-smooth muscle actin. Early Hum Dev. 1998;51(3):223–33.

140. Mitchell KE, Weiss ML, Mitchell BM, Martin P, Davis D, Morales L, et al. Matrix cells from Wharton's jelly form neurons and glia. Stem Cells. 2003;21(1):50–60.

141. Jomura S, Uy M, Mitchell K, Dallasen R, Bode CJ, Xu Y. Potential treatment of cerebral global ischemia with Oct-4+ umbilical cord matrix cells. Stem Cells. 2007;25(1):98–106.

142. Okita K, Ichisaka T, Yamanaka S. Generation of germline-competent induced pluripotent stem cells. Nature. 2007;448(7151):313–7.

143. Park IH, Zhao R, West JA, Yabuuchi A, Huo H, Ince TA, et al. Reprogramming of human somatic cells to pluripotency with defined factors. Nature. 2008;451(7175):141–6.

144. Kim JB, Zaehres H, Wu G, Gentile L, Ko K, Sebastiano V, et al. Pluripotent stem cells induced from adult neural stem cells by reprogramming with two factors. Nature. 2008;454(7204):646–50.

145. Kaji K, Norrby K, Paca A, Mileikovsky M, Mohseni P, Woltjen K. Virus-free induction of pluripotency and subsequent excision of reprogramming factors. Nature. 2009;458(7239):771–5.

146. Kim JB, Greber B, Arauzo-Bravo MJ, Meyer J, Park KI, Zaehres H, et al. Direct reprogramming of human neural stem cells by OCT4. Nature. 2009;461(7264):649–53.

147. Borlongan CV, Hadman M, Sanberg CD, Sanberg PR. Central nervous system entry of peripherally injected umbilical cord blood cells is not required for neuroprotection in stroke. Stroke. 2004;35(10):2385–9.

148. Vendrame M, Gemma C, de Mesquita D, Collier L, Bickford PC, Sanberg CD, et al. Anti-inflammatory effects of human cord blood cells in a rat model of stroke. Stem Cells Dev. 2005;14(5):595–604.

149. Prockop DJ, Gregory CA, Spees JL. One strategy for cell and gene therapy: harnessing the power of adult stem cells to repair tissues. Proc Natl Acad Sci U S A. 2003;100 Suppl 1:11917–23.

150. Duncan AW, Rattis FM, DiMascio LN, Congdon KL, Pazianos G, Zhao C, et al. Integration of Notch and Wnt signaling in hematopoietic stem cell maintenance. Nat Immunol. 2005;6(3):314–22.

151. Conboy IM, Rando TA. Aging, stem cells and tissue regeneration: lessons from muscle. Cell Cycle. 2005;4(3):407–10.

152. Conboy IM, Conboy MJ, Wagers AJ, Girma ER, Weissman IL, Rando TA. Rejuvenation of aged progenitor cells by exposure to a young systemic environment. Nature. 2005;433(7027): 760–4.

153. Gilmore JL, Yi X, Quan L, Kabanov AV. Novel nanomaterials for clinical neuroscience. J Neuroimmune Pharmacol. 2008;3(2):83–94.

154. Alvarez-Erviti L, Seow Y, Yin H, Betts C, Lakhal S, Wood MJ. Delivery of siRNA to the mouse brain by systemic injection of targeted exosomes. Nat Biotechnol. 2011;29(4):341–5.

155. Nowacek AS, Miller RL, McMillan J, Kanmogne G, Kanmogne M, Mosley RL, et al. NanoART synthesis, characterization, uptake, release and toxicology for human monocyte-macrophage drug delivery. Nanomedicine (Lond). 2009;4(8):903–17.

156. Kingsley JD, Dou H, Morehead J, Rabinow B, Gendelman HE, Destache CJ. Nanotechnology: a focus on nanoparticles as a drug delivery system. J Neuroimmune Pharmacol. 2006;1(3): 340–50.

157. Gorantla S, Dou H, Boska M, Destache CJ, Nelson J, Poluektova L, et al. Quantitative magnetic resonance and SPECT imaging for macrophage tissue migration and nanoformulated drug delivery. J Leukoc Biol. 2006;80(5):1165–74.

158. Dou H, Morehead J, Destache CJ, Kingsley JD, Shlyakhtenko L, Zhou Y, et al. Laboratory investigations for the morphologic, pharmacokinetic, and anti-retroviral properties of indinavir nanoparticles in human monocyte-derived macrophages. Virology. 2007;358(1):148–58.

159. Dou H, Grotepas CB, McMillan JM, Destache CJ, Chaubal M, Werling J, et al. Macrophage delivery of nanoformulated antiretroviral drug to the brain in a murine model of neuroAIDS. J Immunol. 2009;183(1):661–9.

160. Dou H, Destache CJ, Morehead JR, Mosley RL, Boska MD, Kingsley J, et al. Development of a macrophage-based nanoparticle platform for antiretroviral drug delivery. Blood. 2006; 108(8):2827–35.

161. O'Neill MJ, Clemens JA. Rodent models of global cerebral ischemia. Curr Protoc Neurosci. 2001;Chapter 9:9.5.1–9.5.25.

Chapter 23
The Splenic Response to Ischemic Stroke: Neuroinflammation, Immune Cell Migration, and Experimental Approaches to Defining Cellular Mechanisms

Christopher C. Leonardo, Hilary Seifert, and Keith R. Pennypacker

1 Introduction

The neural sequelae resulting from ischemic stroke involve multiple physiological systems, acute injury and delayed cellular degeneration. Clinical studies targeting early neural injury uncovered in animal models of stroke have resulted in poor outcomes. The limited success of these approaches has broadened investigation into the delayed phase of infarct expansion, whereby the penumbral tissue surrounding the initial core infarct can be rescued prior to irreversible injury. Of particular relevance are the neuroimmune mechanisms that are activated from several hours to days following ischemic stroke and eventually lead to permanent infarction that cannot be rescued. New findings from animal studies suggest that the therapeutic window is wider than previously thought, thus allowing for selective targeting of inflammatory processes prior to irreversible injury.

Stem cell therapies in rodent models are perhaps most promising of the experimental therapeutics. In particular, neural stem cells (NSCs) and human umbilical cord blood (HUCB) cells have proven effective in minimizing neural injury and enhancing behavioral recovery. Despite their success in experimental models, however, stem cell therapies, unlike chemical-based pharmaceuticals, present the problem of clinical standardization. Thus, it is crucial to elucidate the precise mechanisms by which these cell therapies confer protection in order to develop novel therapeutics. Interestingly, an increasing number of studies show that a peripheral immune response originating from the spleen represents a potential target for stroke therapy. In fact, the cells administered as therapies migrate to the spleen to alter peripheral signaling from this organ. These recent data highlight the need for additional studies

C.C. Leonardo • H. Seifert • K.R. Pennypacker (✉)
University of South Florida, College of Medicine, Tampa, FL, USA
e-mail: kpennypa@health.usf.edu

P.A. Lapchak and J.H. Zhang (eds.), *Translational Stroke Research*, Springer Series
in Translational Stroke Research, DOI 10.1007/978-1-4419-9530-8_23,
© Springer Science+Business Media, LLC 2012

into the role of the peripheral immune response emanating from the spleen, as splenic signaling appears to be a pharmaceutical target not only for the treatment of stroke, but other tissues and organs undergoing ischemia-reperfusion injury.

2   The Temporal Response: Acute Injury and Delayed Neuroinflammation

The acute response to embolic stroke has been extensively characterized. Energy failure occurs within minutes of stroke onset and triggers a cellular switch to anaerobic respiration. The acute cellular injury that follows is a result of several processes including necrosis from cellular edema, plasma membrane degradation from the production of free oxygen and nitrogen radicals, and excitotoxic neuronal death resulting from excessive synaptic glutamate released from astrocytes and compromised neurons. Moreover, excessive calcium uptake through glutamate receptor activation and release of intracellular stores facilitates cellular edema and triggers apoptotic signaling cascades. All of these mechanisms work in concert to destroy the umbra that is directly perfused by the occluded artery (for a comprehensive review, see [1]).

The field of stroke research has gained considerable knowledge regarding the early response to ischemia. Nevertheless, therapeutic approaches targeting acute injury mechanisms have failed to provide efficacy in reducing neural injury or improving functional outcomes. A clear example of these failures concerns selective targeting of glutamate signaling. Despite success in animal models, targeting glutamatergic neurotransmission through receptor antagonism was ineffective [2] and produced adverse side effects [3] in clinical trials. While a variety of factors may account for the failure of glutamatergic blockade, it is now clear that the delayed neuroinflammatory response is an instrumental component of infarct expansion and must be targeted to improve neurological and functional outcomes.

The phrase "neuroinflammatory response" has been widely used in the description of various pathologies involving the interruption of blood and oxygen flow to the brain. While this phrase generally refers to responses elicited by activated microglia, reactive astrocytes, and peripheral immune cells that include the production and secretion of proinflammatory cytokines, chemokines, and proteases, the precise manner in which this response is regulated and contributes to infarct expansion is not well understood. Despite the gaps in knowledge, however, the temporal sequence of neurodegeneration provides some clues. Expansion of the core infarct to the adjacent, viable penumbra occurs from hours to days after the initial insult [4]. In the rat model of permanent middle cerebral artery occlusion (pMCAO), a standard model used to recapitulate ischemic stroke injury, the infarct stabilizes at 96 h post insult [5]. These data also suggest a therapeutic window of 48 h in which to improve outcomes for those afflicted by ischemic stroke, providing that the appropriate mechanisms can be targeted. Increased expression of inflammatory mediators, blood–brain barrier (BBB) degradation, and subsequent infiltration of peripheral immune cells are believed to exacerbate injury during this period of infarct expansion [6]. Although there is an abundance of data characterizing alterations in brain-derived cytokine, chemokine, and matrix metalloproteinase (MMP) expression [7–10], these data have not translated into promising strategies for targeting inflammatory signaling within the ischemic brain. Furthermore, extravasation of peripheral immune cells into the brain parenchyma adds another layer of complexity that requires consideration of multiple cell types and signaling pathways that synergize to promote neural injury.

In regard to this latter point concerning peripheral immune cell signaling, there is now evidence that modulating the splenic immune response is a promising approach to attenuate infarct expansion, as the spleen appears to mediate proinflammatory signaling and contribute to infarct expansion. This chapter summarizes recent data from experimental rodent models that have provided insights into promising avenues to treat the delayed inflammatory component of core expansion.

3   Microglia and Peripheral Macrophages

While the details regarding the pathological events that constitute the delayed neuroinflammatory response are still being elucidated, it has become evident that both resident microglia and peripheral immune cells play important roles in injury progression. Microglia have been shown to play a dual role in stroke pathology, secreting both proinflammatory [4, 7] and neuroprotective [11, 12] molecules. In fact, there has been much debate concerning the specific time point(s) at which microglial activation should be attenuated to combat stroke injury, as ablation of these cells prior to stroke/reperfusion exacerbates injury in mice [13]. Despite this controversy, it has been noted that while microglia are beneficial in response to milder forms of injury, extensive injury associated with BBB breakdown appears to induce an activated, proinflammatory phenotype that releases various cytotoxins [14].

Consistent with this notion, activated microglia/macrophages are present throughout the course of infarct expansion [15, 16] and generate free radicals and nitric oxide (NO) [17]. The leaky BBB permits entry of peripheral immune cells that contribute to neural injury. One study revealed that a population of amoeboid, proinflammatory (identified with isolectin IB4 staining) microglia/macrophages was first detected at 51 h, was seated in and around the point of MCAO, and had migrated to the corpus striatum by 72 h post stroke. The temporal and spatial distribution of these cells suggested that they were not resident cells of the brain, but instead represented peripheral monocytes/macrophages that had infiltrated into the brain parenchyma [18].

Numerous other studies have demonstrated infiltration of neutrophils [19, 20], T cells, and IB4-positive microglia/macrophages [21] using experimental rodent models. Additionally, reduced numbers of each of these cell types have been associated with improved outcomes following experimental treatments that alter the splenic inflammatory profile (these studies will be discussed in greater detail within the context of splenic signaling).

The robust infiltration of peripheral monocytes poses another problem that has hindered progress in defining the precise role of microglia: the considerable overlap of cell surface antigens between resident microglia and peripheral monocytes. Because cell surface antigens exploited for immunohistological identification are highly conserved across the monocyte lineage, commonly used cellular labeling and subfractionation techniques cannot be used to differentiate activated microglia from peripheral macrophages. For example, IB4 binds activated microglia and peripheral macrophages with high affinity. CD11b-expressing cells also bind IB4 [18], and this surface protein has been detected in both spleen-derived macrophages [22, 23] and cells seated within the brain parenchyma [18, 23, 24]. Microglia/macrophages also upregulate Iba-1 upon activation, and increased expression was noted in rat brain after transient MCAO (tMCAO) [25] though it is also expressed in peripheral tissues, including spleen. Although discriminating between resident microglia and peripheral macrophages has proven difficult, flow cytometric methodologies that incorporate CD45 gating techniques now offer an additional method of differentiating between activated and resting microglia/macrophage cells [26]. Thus, advances in technology and a greater understanding of the various immune cell phenotypes may expand our knowledge base regarding the precise roles of these specific immune cell populations.

4   The Spleen as a Therapeutic Target in Acute Neural Injury

The spleen is an immune organ that contains vast quantities of peripheral immune cells, including macrophages/monocytes, B cells, T cells, NK cells, dendritic cells, and neutrophils. The spleen is also highly vascularized and contains a store of erythrocytes. The role of the spleen in promoting ischemic injury has been well documented in peripheral organs. Studies of liver showed that reactive oxygen species, TNF-α, and nitric oxide production increased in both kupffer cells and neutrophils after ischemia/reperfusion [27], leading to damage of these and other tissues throughout the body [28]. The profound effects of spleen-derived inflammatory signaling were further demonstrated by experiments in which the spleens were removed from animals prior to insult. In these studies, splenectomy reduced ischemia-induced hepatic injury [29], leukocyte infiltration, and TNF-α release [30]. Splenectomy also provided protection from intestinal ischemia/reperfusion and the resulting inflammation that damages other organs [31].

Consistent with studies demonstrating a role for the spleen in ischemia of peripheral tissues, numerous reports have now indicated that the spleen contributes to inflammation and tissue damage in several models of brain injury involving ischemia. In particular, two seminal studies showed that the spleen decreases in size after permanent cerebral ischemia in rat [32] and ischemia/reperfusion injury in mouse [33]. Importantly, splenectomy prior to pMCAO reduced microglia/macrophage activation, neutrophil migration, and total infarct volume in rat [32]. Similar protection was demonstrated in a rat model of intracerebral hemorrhage, where removal of the spleen prior to insult reduced brain water content, numbers of myeloperoxidase- (MPO-) positive neutrophils, and CD11b-positive immune cells [23]. In addition to traditional rodent stroke models, the protective effects of splenectomy were recently extended to a rat model of traumatic brain injury (TBI). These data showed that removal of the spleen immediately following TBI not only decreased mortality, vasogenic edema, and cytokine secretion (TNF-α, IL-1β and IL-6), but also reduced deficits in spatial memory as measured by the Morris Water Maze [34].

Although these data provide convincing evidence that the spleen exacerbates ischemic injury to both neural and peripheral tissues, many questions remain as to the nature of the spleen size reduction after experimental stroke and the mechanisms mediating the protective effects of splenectomy in acute brain injury models. The fact that the spleen is composed of various immune cell types suggests that the release of these cells accounts, at least in part, for the size reduction observed in response to stroke. This hypothesis is consistent with the efficacy of splenectomy in reducing neuroinflammatory signaling, immune cell activation, and migration to the ischemic brain.

Another explanation that would account for reduced splenic mass is the release of erythrocytes into the systemic circulation. This would presumably occur as a compensatory response initiated by the reduction in blood flow to the brain, although this contention is pure speculation and has not been tested experimentally. To date, standard methods of immunological detection and cell tracing lack the sensitivity and precision required to definitively determine if either of these mechanisms is responsible for the observed effect. Nonetheless, other approaches to identifying splenic signals have offered a glimpse into potential mechanisms that involve splenic leukocyte populations.

5   Splenic Leukocytes, Macrophages, and the "Catecholamine Surge"

While it is not feasible to label the entire population of splenocytes for tracing studies, other experimental approaches have provided insights into the cause of spleen size reduction following stroke and the manner in which splenic immune cells respond to ischemic insult. Technological advances have permitted investigations of specific splenocyte subpopulations by exploiting the expression of cell surface antigens unique to B cells, neutrophils, NK cells, and several T cell phenotypes. In particular, flow cytometry employing fluorescence-activated cell sorting (FACS) is a powerful tool for quantifying peripheral leukocytes and allowing for identification of specific immune cell populations.

In general, spleens removed from rodents subjected to experimental stroke show reduced numbers of leukocytes. Flow cytometric analyses showed that total numbers of spleen-derived leukocytes were reduced at 96 h [24] and 3 days [35] after tMCAO in mice, while similar reductions were reported at selected time points ranging from 2 to 28 days after tMCAO in rat [36]. Cell death in response to the ischemic insult itself would be a simple explanation for the observed reductions in leukocyte numbers. Braun et al. demonstrated that caspase-3 immunoreactivity was detected in spleen 24 h after reperfusion and caspase activity was increased at 2, 6, and 24 h [35]. Other experiments using the same rodent model showed TUNEL staining in the spleen that was accompanied by reductions in spleen size and spleen-derived B cells [33]. Thus, if leukocyte depletion is the cause of the spleen size reduction, apoptotic cell death may be the mechanism responsible for the loss of these cells. Although these data provide some support for this mechanism, the loss of selective populations of splenic immune cells is not consistent with the magnitude of total size reduction reported by these groups.

Another possible explanation for the observed leukocyte reduction concerns the stroke-induced "catecholamine surge." Ischemia induces a massive release of nor-epinephrine and epinephrine into the systemic circulation, specifically in strokes involving damage to the insular cortical region that is mainly perfused by the middle cerebral artery [37]. Both norepinephrine and epinephrine can cause splenic atrophy [38]. The majority of splenic nerve fibers innervate the white pulp and modulate activation of lymphocytes, eosinophils, mast cells, and macrophages [39]. Sympathetic activation also causes contraction of smooth muscle via α1-adrenergic receptor activation, while the β2-adrenergic receptor is expressed on innate and adaptive immune cells [40]. The predicted effects of this systemic elevation in catecholamines would include contraction of splenic smooth muscle, potentially leading to the release of immune cells and/or erythrocytes. Another possibility would be activation of β2-adrenergic receptors on NK and B cells that could cause their migration from the spleen. Interestingly, the proposed mechanisms suggest that these events could occur even in the absence of chemotactic signals, and β2-adernergic receptor activation on most immune cells results in their inactivation or the initiation of an anti-inflammatory response early following an ischemic insult. Therefore, although it is well documented that peripheral immune cells are present in the brain following stroke, further studies are needed to elucidate the signaling mechanisms responsible for migration of these cells from the periphery into the ischemic brain and whether this cellular response is beneficial or detrimental.

Indeed, evidence from rodent models indicates that circulating catecholamines are elevated in response to cerebral ischemia and may mediate the splenic response. Increased catecholamine levels have been detected in plasma as early as 2 h after reperfusion in rat [36]. In the mouse model, serum catecholamines were elevated at 6, 24, and 96 h after tMCAO [41], and increased numbers of circulating macrophages have also been documented [33]. Additionally, prophylactic treatment with the pan adrenergic antagonist carvedilol diminished neural injury and prevented spleen size reduction in rats subjected to pMCAO. However, neither denervation of the splenic nerve nor selective β2 blockade reduced infarct volume or blocked the spleen size reduction, while selective α1 blockade prevented spleen size reduction but did not reduce neural injury [42]. These data, while supporting a role for adrenergic receptor activation in the splenic response, did not provide a clear picture of the precise mechanisms by which endogenous elevations of catecholamines modulate this response. Although these data clearly illustrate the profound effects of focal cerebral ischemia on the peripheral immune response, more investigations are necessary to gain an understanding of the complex interplay between adrenergic receptor activation, splenocyte signaling, and immune cell activation after stroke.

6   Cytokine and Chemokine Profiles After Experimental Stroke

The spleen reacts to and plays a role in the pathogenesis of ischemic brain injury. These data have prompted investigations into alterations in the respective inflammatory signaling profiles. Many studies have shown that experimental ischemic injury results in an increase in the expression of proinflammatory cytokines and chemokines, and that the induction of these signaling molecules is not neural-specific injury. While a great deal of research has examined cytokine profiles after stroke, no definitive signaling signature emanating from the spleen or other immune organs has been deciphered that is unique to stroke pathology.

Most studies examining cytokines, chemokines, and their associated receptors in the ischemic brain and spleen have relied on techniques that quantify mRNA content after a few days of the experimental stroke procedure. Among most assayed, the proinflammatory cytokines TNF-α and IL-1β are elevated in ischemic brain at both early (6 h) and extended (22 and 96 h) time points after reperfusion in the tMCAO mouse model. Levels of mRNA for IL-4, -6, -10, and -13 are also enhanced at 6 and 22 h, and IL-4 and -10 mRNA remains elevated at 96 h [10, 22, 41]. IFN-γ mRNA is elevated in brain starting at 2 days post-tMCAO and remains elevated out to 6 days [43]. A parallel expression pattern is observed in spleen, such that TNF-α, IL-1β, IL-4, -6, and -10 message are increased by 22 h post following reperfusion [22, 23]. Additionally, TNF-α and IL-1β gene transcripts were elevated at 48 h [21]. Also, IFN-γ mRNA was increased in spleen at 22 h after insult [22]. However, TNF-α and IFN-γ were elevated in blood 24 h after tMCAO, but returned to levels in sham-operated mice by 96 h [41]. Due to the known proinflammatory action of IFN-γ production and the mRNA expression in the brain and the spleen, future studies aimed at inhibiting the effects of peripheral IFN-γ appear to be a potential avenue for the treatment of stroke.

The expression of chemokine message has been examined in detail in the transient mouse model that disclosed alterations in chemokine receptor expression, specifically CCR -1, -2, -4, -5, -6, -7, and -8, as well as chemokine ligands MIP-2, IP-10, and RANTES [10]. Interestingly, RANTES, a potent chemotaxic molecule, binds CCR -1, -2, and -5. Each of these mRNAs is upregulated in the brain following stroke, suggesting a role for RANTES in the immune cell migration to the brain that is associated with the inflammatory process. Among the chemokine receptors measured, only CCR6 did not exhibit increased mRNA expression in brain at 22 h. Both CCR3 and CCR8 may be involved in early core infarct expansion since these showed increased mRNA expression at 6 h. Similarly, elevations in IP-10, MIP-2,

BRAIN

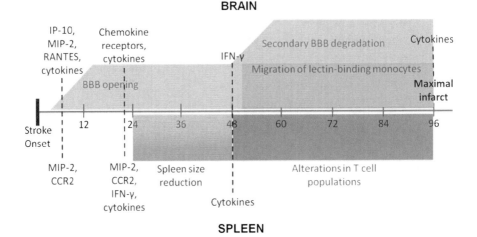

SPLEEN

Fig. 23.1 Cytokines, chemokines, and the temporal response to ischemic stroke. Schematic depicts the profile of upregulated proinflammatory signals after cerebral ischemia and their relation to key events that are associated with infarct expansion

and RANTES mRNA at 6 h suggest a role in the early immune response to ischemic injury (Fig. 23.1). Of the chemokine ligands and receptors in spleen, MIP-2 and CCR2 mRNA were consistently increased between the 6 and 22 h time points, but only CCR2 protein expression has been confirmed at 48 h after reperfusion [21].

As expected, the chemokine profile of the splenic response to stroke is distinct from that of brain, which may provide insight into the pathogenesis of the peripheral immune response and how it influences the neural microenvironment. However, fluctuating cytokine profiles have provided little insight into the progression of stroke injury and there is much to be determined regarding their physiological effects in the injury process. No signaling pattern has been elucidated to discern a clear protective or degenerative pathway in regard to infarct expansion, although the alterations observed suggest many possibilities. For example, increased IL-4 expression would be expected to induce a Th2 anti-inflammatory response by affecting the differentiation of naïve T cells. However, expansion and differentiation into either the Th1 or Th2 phenotype are highly regulated by a host of signaling molecules [44]. Importantly, the net effects of cytokine signaling alone on a shift to Th1 or Th2 immunity are currently unknown. Developing targeted therapies against these molecules will be difficult when considering the host of cytokine/chemokine ligands and receptors that are involved in cellular chemotaxis and signaling.

In light of this caveat, one approach that could prove successful is selective targeting of the peripheral immune cells that secrete and respond to these signals. Many investigators have now expanded focus to the various immune cell populations contained within the spleen in hopes that determining the ways in which these cells respond to ischemic injury will reveal a selective target to dampen the peripheral inflammatory response, thereby blocking penumbral expansion and improving the outcome.

7   Spleen-Derived Lymphocyte Profiles After Experimental Stroke

A great deal of research has been devoted to identifying specific splenocyte sub-populations and their responses to ischemia. In particular, T cells have emerged as a principal candidate for targeted therapy aimed at dampening the proinflammatory peripheral immune response. In general, the majority of research has utilized FACS to label multiple cell surface glycoproteins associated with pro- or anti-inflammatory phenotypes. CD3 is most often used as a pan T cell marker, while CD8+ cells are commonly classified as cytotoxic T cells. Labeling of CD4 is used in conjunction with other markers, particularly CD8, to signify whether the net T cell response is pro- or anti-inflammatory.

Thus, an increased CD4/CD8 ratio may indicate several possible CD4 T cell responses: a Treg response (with coexpression of FoxP3), an anti-inflammatory Th2 response, or a proinflammatory Th1 response. Despite the variety of responses that could result in an increased ratio, the general consensus is that a decreased ratio demonstrates a proinflammatory response. In rat, ischemic spleens harvested 48 h after tMCAO showed a reduction in CD8+ cells that was accompanied by an increased CD4/CD8 ratio, which may reflect the loss of CD8+ cells. Additionally, T and B cells from these spleens showed a reduced proliferative capacity in response to mitogen activation. Interestingly, there was no change in ED1+ macrophages [36]. One interpretation of these data is that the injury resulting from stroke cannot be accounted for by splenic macrophages or cytotoxic T cells. However, it is noteworthy that plasma norepinephrine levels were elevated 2 h after insult and splenocyte counts were performed only at the 48 h time point. Upon consideration of the catecholamine elevations and the time point at which the spleen was assessed, it is also possible that release of T cells accounts for the splenic reduction observed at 48 h. Indeed, increased responses to mitogen activation have also been demonstrated using splenocytes harvested from ischemic rat spleens [45]. Regardless of the mechanisms involved, the critical point is that a cautionary interpretation is necessary when evaluating most studies since it is rarely feasible to investigate all time points necessary to encompass the extended temporal response to cerebral ischemia.

To date, the majority of work devoted to assessing T cell responses has been performed in the transient mouse model. Despite the growing number of investigations, the data obtained thus far have not been particularly helpful in establishing a consistent response profile upon which to develop therapies. In contrast to the CD4/CD8 ratio observed in rat, this ratio was unchanged at 6 and 24 h, but decreased by 96 h after tMCAO (120 min) in mouse. However, Treg cells were elevated in blood at 24 and 96 h, and these elevations occurred concomitantly with increased serum catecholamine levels that persisted up to 96 h [41]. Other data showed that spleens and blood harvested from mice 72 h after tMCAO (45 min) contained fewer CD3+ T cells, CD45R + B cells, and DX5+ NK cells relative to sham controls. No results were reported for CD11b + macrophages, although this surface marker was also labeled [35]. Because lymphocytes expressing these surface antigens were not quantified in brain, these data leave an open question as to whether the observed reductions in presumptive CD4+/FoxP3- T cells [41], CD3+ T cells, B cells, and NK cells [35] were due to die off or migration from the spleen to the cerebral infarct or other lymph organs.

To confuse the issue further, a subsequent study using the same model found no significant differences in CD3, CD4, or CD8+ T cells from spleens or blood harvested 96 h after tMCAO (60 min) [24]. One possible explanation for these inconsistencies could be differences in the duration of occlusion. However, the prevailing theories regarding immunosuppression and splenic apoptosis would not predict such outcomes. For example, if cellular apoptosis accounts for reductions in splenic leukocytes, one would expect a positive correlation between prolonged duration of occlusion and reductions in cell counts. However, this relationship does not appear to exist based on the data obtained thus far.

Because the alterations in the splenic leukocyte profile likely involve additional mechanisms that have yet to be determined, more thorough investigations are required prior to any meaningful interpretation of these changes. For example, studies aimed at quantifying the various leukocyte populations in both spleen and brain at early, intermediate, and delayed time points after each duration of occlusion would provide critical insights necessary for the field to progress toward novel therapies. Furthermore, identification of the signaling pathways induced upon activation of these cells in vivo will be crucial in developing therapies to selectively block the spleen-derived substrates that contribute to infarct expansion.

8 Elucidating Spleen-Derived Lymphocyte Signaling Ex Vivo

The ability to define and target splenic proinflammatory signaling is well recognized. In spite of the technological advancements and considerable expansion of our knowledge base, elucidating the specific cell types and signaling mechanisms in vivo has remained troublesome. Although it is widely accepted that brain-spleen signaling involves immune cell activation and subsequent cytokine and chemokine production, practical issues such as the number of animals required, rapid response of cytokine/chemokine release, cost of Golgi sequestering reagents, and caveats associated with administration of such reagents have impeded efforts directed at measuring protein expression.

As a result, a reductionist approach in which spleens from stroked animals are harvested and cells are subjected to in vitro leukocyte activation assays has provided important clues regarding the mechanisms that may operate within the intact animal. Splenocytes from mice subjected to tMCAO (90 min) were activated using an enzyme-linked immunoabsorbance assay (ELISA) coated with CD3/CD28 antibodies to induce cytokine and chemokine expression. Data showed that splenocytes from mice euthanized at 6 and 22 h after reperfusion upregulated TNF-α, IFN-γ, IL-2, and MCP-1 at both time points relative to sham-operated controls, while IL-6 and IL-10 elevations were detected only at the 22 h time point [10]. Similar experiments using the same model (120 min tMCAO) and assay later replicated the

TNF-α, IFN-γ, and IL-10 findings at 6 and 24 h postreperfusion [41]. This latter study also extended the splenocyte time course to 96 h and showed that while all cytokine levels remained elevated compared to sham controls, they were reduced relative to 24 h tMCAO.

These experiments demonstrated that splenocytes from animals subjected to experimental stroke can be induced to secrete proinflammatory cytokines and chemokines, thus reinforcing the notion that splenic signaling contributes to ischemic pathology. Artificial activation conditions undeniably present unique caveats in terms of mimicking the conditions present in vivo. However, consistencies with reported expression levels of mRNA and differences in expression capacity at 96 h relative to earlier time points are encouraging. Although a cautionary interpretation of these data is appropriate, this ex vivo approach should shed some light on potential mechanisms to be tested in vivo.

9 Knockouts and Transgenics: A Role for T Cells and Interferon Gamma

Contrary to ex vivo approaches where mechanisms are defined and subsequently targeted in vivo, knockout mouse models have been useful tools in investigating the role of peripheral immune cells and their signaling mechanisms. As previously described, converging lines of evidence indicate that IFN-γ production is involved in ischemic injury and may represent a prime target for therapeutic intervention. In an elegant series of experiments, Yilmaz et al. utilized several strains of knockout mice deficient in distinct lymphocyte populations or IFN-γ production to assess their relative contributions to tMCAO injury (60 min). In addition to measuring infarct volume and neurological scores, these experiments assessed the effects of various lymphocyte populations on leukocyte adherence in cortical venules using live, intravital video microscopy. Data showed reduced infarct volume in CD4+ T cell−/−, CD8+ T cell−/−, Rag1−/− (lacking T and B cells), and IFN-γ−/−, but not B cell−/− mice at 24 h postreperfusion. Addition of whole splenocytes from wild-type mice restored the infarct in Rag1−/− mice, whereas only partial restoration of neural injury occurred in Rag1−/− mice that received whole splenocytes from IFN-γ−/− mice. Furthermore, addition of splenocytes from wild-type or IFN-γ−/− mice blocked the reductions in leukocyte adherence observed at 4 and 24 h in Rag1−/−, CD4+ T cell−/−, and CD8+ T cell−/− mice. Despite the reductions in infarct volume and leukocyte adherence in animals deficient in T cells or IFN-γ, none of the knockout mice showed improvements in neurological scores relative to wild-type tMCAO controls [46]. These experiments provided the first thorough examination using knockouts to elucidate a deleterious role for both T cells and IFN-γ signaling in the pathogenesis of stroke. Although the various T cell populations were not quantified in brain, these results demonstrate a role for the spleen in leukocyte adherence to cortical venules and provide evidence that the spleen contributes to mechanisms required for extravasation of peripheral immune cells into the brain parenchyma.

The fact that splenocyte-derived IFN-γ was not required for leukocyte adherence, but was required to achieve total infarction, also indicates a role for other signaling molecules. A subsequent study utilized SCID mice, a transgenic line similar to Rag1$^{-/-}$ in T and B cell deficiency, to determine the contribution of T and B lymphocytes to cytokine and chemokine production after tMCAO (90 min). In these experiments, mice lacking T and B cells showed reduced cortical injury at 22 h postreperfusion, while the striatal infarct was unchanged from wild-type controls. Although total splenocyte counts were reduced in SCID mice at 22 h, an increase in spleen-derived CD11b$^+$/VLA-4$^-$ macrophages was accompanied by elevations in splenic mRNA encoding IFN-γ and MIP-2. With the exception of IL-1β, ischemic brains of SCID mice downregulated message for the cytokines TNF-α, IL-6, IL-10, as well as chemokine receptors CCR -1, -2, -3, and -5 [22].

Taken together, these data raise several implications. The role of IFN-γ in promoting ischemic injury is becoming increasingly clear and has now been reported by several laboratories. In particular, Toll-like receptors (TLR) seated on the cell surfaces of peripheral macrophages and microglia bind specific pathogen-associated molecular patterns (PAMP). However, there are endogenous cellular proteins that can activate these receptors triggering an inflammatory response, specifically proteins that are released from neurons and other neural cell types in response to ischemia-induced necrosis. Some examples are heat shock proteins 60/70, heparin sulfate, and HMGB1 that activate TLR -2 and -4, while DNA and mRNA can activate TLR -3 and -9, respectively [47]. Upon activation, these macrophages induce IFN-γ production by T cells and NK cells through secretion of IFN-α, IL-12, and IL-12/IL-18 synergy [48]. Thus, TLR-mediated signaling in microglia and macrophages may contribute to ischemic brain injury through a feed-forward mechanism that results in the production of IFN-γ-inducible proteins such as IP-10, MIG, I-TAC which help sustain the proinflammatory response.

Interferon inducible protein (IP-10) expression is induced in cells that are activated by INF-γ. IP-10 and the other IFN-γ inducible proteins play a critical role in sustaining the proinflammatory Th1 response. These chemokines belong to the CXC class of chemokines and all three bind the CXCR3 receptor, which is expressed on Th1 cells. The activation of the CXCR3 receptor leads to further amplification of the Th1 response. Interestingly, these chemokines also bind the CCR3 receptor, which is found on Th2 cells, and act as antagonists at this receptor to inhibit the binding of the CC chemokines that bind the CCR3 receptor but do not initiate signal transduction at the CCR3 receptor. This antagonism prevents the activation of Th2 cells through the CCR3 receptor and can result in suppression of the anti-inflammatory response [49] (Fig. 23.2). Taken together with the finding that mRNA for IP-10 is upregulated at 22 h post-tMCAO implicates IP-10 as a potential therapeutic option in treating stroke. However, more research into IP-10 expression is needed to substantiate it as a therapeutic target.

Thus, the selective targeting of these cellular signals may represent an important intervention to rescue the brain from delayed neuroinflammatory injury, though the clinical relevance of findings from rodent models remains uncertain. Because IFN-γ does not appear to mediate all processes known to promote infarct expansion, as demonstrated by the knockout study, it is unclear whether selective inhibition of this

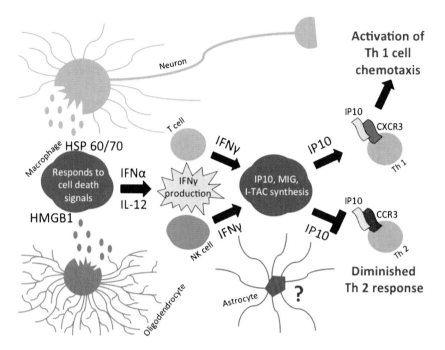

Fig. 23.2 Immune cell crosstalk promotes inflammation. Schematic depicts the potential mechanism by which resident and peripheral immune cells synergize to promote neural injury after ischemia. Endogenous proteins released from necrotic and apoptotic cells activate TLRs seated on cell surfaces of microglia and macrophages. TLR activation leads to the production of IFN-α and IL-12 by these cells and subsequent activation of T cells and NK cells. IFN-γ secreted by T cells and NK cells further activates microglia and macrophages, facilitating the expression of IFN-γ-inducible proteins such as IP-10. The net effect of these actions is activation of Th1 cell chemotaxis and suppression of Th2 cell responsiveness

signal alone will be sufficient to improve outcomes. Interestingly, splenic macrophages have also emerged as a likely candidate. In spite of the absence of VLA-4 at the plasma membrane, increased numbers of macrophages were present in spleen at the time when the BBB is compromised. This suggests that the splenic macrophages might be activating naïve T cells to respond to brain antigens that have leaked out through the compromised BBB. Nonetheless, much remains to be determined regarding the potential for selective targeting of IFN-γ-expressing cell populations and other cell populations that exhibit unique proinflammatory responses to ischemia.

10 Cell Therapies and Neuroprotection: Insights into the Splenic Response

For all of the experimental methodologies that have been tried and tested to date, it is perhaps not surprising that administration of potential therapeutics into the intact animal has provided the greatest insights into the splenic response to cerebral ischemia.

Cell therapies, in particular, have been a popular avenue for treating ischemic injury in experimental rodent models. In recent years, the field of stroke research has been forced to reconsider the proposed mechanism responsible for the efficacy of cell therapies. The original notion was that stem cells would replace dead and dying neural cells, thus aiding in plasticity and repair of compromised tissue within the brain parenchyma. However, increasing evidence suggests that various cell therapies likely exert their protective effects, at least in part, through direct actions at the level of the spleen. Intravenous administration of bone marrow-derived stem cells (BMSC) 2 h after tMCAO resulted in greater than 95% of cells traveling to the spleen, while few were present in brain [25]. These data are consistent with other reports using HUCB cells, which contain a population of stem cells in addition to various mononuclear cell populations, and NSC therapies. In these experiments, intravenous administration resulted in an accumulation of cells in the spleen but not the brain after tMCAO [50], and predominantly in the splenic marginal zone after intracerebral hemorrhage [23]. Hematopoietic stem cell (HSC) therapy also resulted in homing of these cells to the spleen after tMCAO in mouse, although this group reported detection of HSCs in brain at 48 and 72 h postreperfusion [21].

In addition to accumulating primarily in the spleen following systemic administration, these cell therapies have also been shown to alter inflammatory cytokine profiles and reduce activated responses of immune cells. Administration of HSCs 24 h post-tMCAO (45 min) reduced cerebral infarction, TUNEL-stained apoptotic neurons, CD3[+] T cells, and IB4[+] microglia/macrophages at 96 h. HSC treatment also blocked the induction of TNF-α, IL-1β, CCR2, and CX3CR1 gene transcripts in spleen at 48 h [21]. In a subsequent rat study, a mixed population of HSCs and MSCs (BMSCs) administered 2 h after tMCAO (60 min) demonstrated similar efficacy in decreasing infarct volume and caspase-3 immunoreactivity after stroke. Additionally, the vast majority of implanted cells accumulated in the spleen shortly after injection, and detection of BMSCs in brain began to decline by 24 h [25].

In agreement with these data, experiments by Lee et al. demonstrated neuroprotection by NSCs in rat and extended these findings to show a definitive role for the spleen in the pathogenesis of hemorrhagic stroke. While i.c. administration of NSCs was ineffective, i.v. administration 2 h after ICH decreased brain water content, neutrophils, microglia/macrophages, and improved forelimb placement. Splenectomized animals with or without NSCs received similar benefits, yet reductions did not reach levels achieved by NSC administration in nonsplenectomized animals. Importantly, NSCs blocked stroke-induced elevations in splenic TNF-α and IL-6 at 24 h and were found to be in direct contact with CD11b[+] splenic macrophages at 3 days. Consistent with these findings, NSCs blocked LPS-stimulated TNF-α production by cultured peritoneal macrophages when added 2 h after stimulation, and using transwell inserts to remove cellular contact abolished these effects [23]. These data showed that NSCs act at the level of the spleen, possibly through direct actions on splenic macrophages. Importantly, the fact that i.c. administration did not improve any outcome measures further supports the notion that the splenic response is instrumental in stroke neuropathology.

It is noteworthy that in addition to providing neuroprotection, HUCB cell therapy has been repeatedly shown to promote functional recovery in rats subjected to experimental stroke [51–53] and is more efficacious when administered systemically [54]. These cells exert direct effects on splenic immune cells. Specifically, intravenous administration of HUCB cells 24 h after pMCAO reduced the CD8/CD4 ratio in ischemic spleens harvested at 48 h. Additionally, HUCB cell treatment reduced splenocyte proliferation and production of the proinflammatory cytokines TNF-α and IFN-γ upon stimulation with concovalin A [45].

While studies have demonstrated a link between the protective effects of cell therapies and splenic signaling, several reports have implicated the spleen in the neuroprotective actions of other experimental therapeutics in mice. For example, administration of the broad spectrum caspase inhibitor Q-VD-OPH immediately and 3 h after tMCAO (45 min) decreased TUNEL staining at 2 days in spleen and diminished infarct volume, TUNEL staining, and caspase-3 immunoreactivity at 2 days in brain. This compound also attenuated the stroke-induced reduction in splenic NK cells, but not T or B cells [35]. Similarly, administration of recombinant T cell receptor ligands after tMCAO (60 min) reduced infarct volume and microglia/macrophages in brain. Despite a partial restoration of total splenocyte numbers, treatment did not alter the relative abundance of CD3+, CD4+, or CD8+ T cells [24]. In other experiments, cocaine- and amphetamine-regulated transcript (CART) reduced infarct volume at 6, 24, and 96 h after tMCAO when administered at the time of reperfusion. CART administration also reduced CD4/CD8 ratios in spleen at 96 h. Furthermore, CART blocked stroke-induced TNF-α and IFN-γ elevations in blood at 24 h, and splenocytes harvested from 6 to 24 h CART-treated mice failed to produce either cytokine when artificially activated [41].

11   Priming the Adaptive Immune Response

A hallmark of ischemic stroke is opening of the BBB. In experimental autoimmune encephalitis (EAE), neural injury results from a pathogenic immune response to the myelin basic protein (MBP) antigen. The Becker laboratory has pioneered research taking the general concept of MBP tolerization from EAE and applying it to the ischemic stroke model as means to investigate the role of lymphocytes in ischemic injury. Data from these experiments showed that infarct volumes were reduced in rats tolerized to MBP relative to OVA-tolerized controls. Interestingly, adoptive transfer of splenocytes from tolerized rats into naïve rats immediately after tMCAO also reduced neural injury, and neuroprotection was accompanied by reduced numbers of total mononuclear cells, but increased numbers of TGF-β1-expressing mononuclear cells in brain [55]. These results demonstrated that MBP tolerization exerts profound effects on spleen-derived lymphocytes and suggest that altering the net immune response, if possible, is a promising avenue for therapeutic intervention. However, because prophylactic priming is not an option to treat stroke patients, the question remains as to whether the peripheral immune response that is dampened through tolerization can be successfully targeted and inhibited once it is activated by ischemic insult.

References

1. Lipton P. Ischemic cell death in brain neurons. Physiol Rev. 1999;79:1431–568.
2. Citron M, Westaway D, Xia W, Carlson G, Diehl T, Levesque G, et al. Mutant presenilins of Alzheimer's disease increase production of 42 residue amyloid B-protein in both transfected cells and transgenic mice. Nature Med. 1997;3:67–72.
3. Davis SM, Lees KR, Albers GW, Diener HC, Markabi S, Karlsson G, et al. Selfotel in acute ischemic stroke: possible neurotoxic effects of an NMDA antagonist. Stroke. 2000;31(2):347–54.
4. Dirnagl U, Iadecola C, Moskowitz MA. Pathobiology of ischaemic stroke: an integrated view. Trends Neurosci. 1999;22(9):391–7.
5. Newcomb JD, Ajmo CT, Sanberg CD, Sanberg PR, Pennypacker KR, Willing AE. Timing of cord blood treatment after experimental stroke determine therapeutic efficacy. Cell Transplant. 2006;15(3):213–23.
6. Danton GH, Dietrich WD. Inflammatory mechanisms after ischemia and stroke. J Neuropathol Exp Neurol. 2003;62(2):127–36.
7. Allan SM, Rothwell NJ. Cytokines and acute neurodegeneration. Nat Rev Neurosci. 2001;2(10):734–44.
8. Furuya K, Takeda H, Azhar S, McCarron RM, Chen Y, Ruetzler CA, et al. Examination of several potential mechanisms for the negative outcome in a clinical stroke trial of enlimomab, a murine anti-human intercellular adhesion molecule-1 antibody: a bedside-to-bench study. Stroke. 2001;32(11):2665–74.
9. Cuadrado E, Ortega L, Hernandez-Guillamon M, Penalba A, Fernandez-Cadenas I, Rosell A, et al. Tissue plasminogen activator (t-PA) promotes neutrophil degranulation and MMP-9 release. J Leukoc Biol. 2008;84(1):207–14.
10. Offner H, Subramanian S, Parker SM, Afentoulis ME, Vandenbark AA, Hurn PD. Experimental stroke induces massive, rapid activation of the peripheral immune system. J Cereb Blood Flow Metab. 2006;26(5):654–65.
11. Coull JA, Beggs S, Boudreau D, Boivin D, Tsuda M, Inoue K, et al. BDNF from microglia causes the shift in neuronal anion gradient underlying neuropathic pain. Nature. 2005;438(7070):1017–21.
12. da Cunha A, Jefferson JJ, Tyor WR, Glass JD, Jannotta FS, Cottrell JR, et al. Transforming growth factor-beta1 in adult human microglia and its stimulated production by interleukin-1. J Interferon Cytokine Res. 1997;17(11):655–64.
13. Lalancette-Hebert M, Gowing G, Simard A, Weng YC, Kriz J. Selective ablation of proliferating microglial cells exacerbates ischemic injury in the brain. J Neurosci. 2007;27(10): 2596–605.
14. Streit WJ, Mrak RE, Griffin WS. Microglia and neuroinflammation: a pathological perspective. J Neuroinflammation. 2004;1(1):14.
15. Iadecola C. Bright and dark sides of nitric oxide in ischemic brain injury. Trends Neurosci. 1997;20(3):132–9.
16. Vendrame M, Gemma C, De Mesquita D, Collier L, Bickford PC, Sanberg CD, et al. Anti-inflammatory effects of human cord blood cells in a rat model of stroke. Stem Cells Dev. 2005;14:595–604.
17. Giulian D. Immune responses and dementia. Ann N Y Acad Sci. 1997;835:91–110.
18. Leonardo CC, Hall AA, Collier LA, Ajmo Jr CT, Willing AE, Pennypacker KR. Human umbilical cord blood cell therapy blocks the morphological change and recruitment of

CD11b-expressing, isolectin-binding proinflammatory cells after middle cerebral artery occlusion. J Neurosci Res. 2010;88(6):1213–22.

19. Jean WC, Spellman SR, Nussbaum ES, Low WC. Reperfusion injury after focal cerebral ischemia: the role of inflammation and the therapeutic horizon. Neurosurgery. 1998;43(6):1382–96; discussion 96–7.

20. Matsuo Y, Onodera H, Shiga Y, Nakamura M, Ninomiya M, Kihara T, et al. Correlation between myeloperoxidase-quantified neutrophil accumulation and ischemic brain injury in the rat. Effects of neutrophil depletion. Stroke. 1994;25(7):1469–75.

21. Schwarting S, Litwak S, Hao W, Bahr M, Weise J, Neumann H. Hematopoietic stem cells reduce postischemic inflammation and ameliorate ischemic brain injury. Stroke. 2008;39(10): 2867–75.

22. Hurn PD, Subramanian S, Parker SM, Afentoulis ME, Kaler LJ, Vandenbark AA, et al. T- and B-cell-deficient mice with experimental stroke have reduced lesion size and inflammation. J Cereb Blood Flow Metab. 2007;27(11):1798–805.

23. Lee ST, Chu K, Jung KH, Kim SJ, Kim DH, Kang KM, et al. Anti-inflammatory mechanism of intravascular neural stem cell transplantation in haemorrhagic stroke. Brain. 2008;131(Pt 3): 616–29.

24. Subramanian S, Zhang B, Kosaka Y, Burrows GG, Grafe MR, Vandenbark AA, et al. Recombinant T cell receptor ligand treats experimental stroke. Stroke. 2009;40(7):2539–45.

25. Keimpema E, Fokkens MR, Nagy Z, Agoston V, Luiten PG, Nyakas C, et al. Early transient presence of implanted bone marrow stem cells reduces lesion size after cerebral ischaemia in adult rats. Neuropathol Appl Neurobiol. 2009;35(1):89–102.

26. Stevens SL, Bao J, Hollis J, Lessov NS, Clark WM, Stenzel-Poore MP. The use of flow cytometry to evaluate temporal changes in inflammatory cells following focal cerebral ischemia in mice. Brain Res. 2002;932(1–2):110–9.

27. Jaeschke H. Reactive oxygen and mechanisms of inflammatory liver injury. J Gastroenterol Hepatol. 2000;15(7):718–24.

28. Fan C, Zwacka RM, Engelhardt JF. Therapeutic approaches for ischemia/reperfusion injury in the liver. J Mol Med. 1999;77(8):577–92.

29. Okuaki Y, Miyazaki H, Zeniya M, Ishikawa T, Ohkawa Y, Tsuno S, et al. Splenectomy-reduced hepatic injury induced by ischemia/reperfusion in the rat. Liver. 1996;16(3):188–94.

30. Jiang H, Meng F, Li W, Tong L, Qiao H, Sun X. Splenectomy ameliorates acute multiple organ damage induced by liver warm ischemia reperfusion in rats. Surgery. 2007;141(1):32–40.

31. Savas MC, Ozguner M, Ozguner IF, Delibas N. Splenectomy attenuates intestinal ischemia-reperfusion-induced acute lung injury. J Pediatr Surg. 2003;38(10):1465–70.

32. Ajmo Jr CT, Vernon DO, Collier L, Hall AA, Garbuzova-Davis S, Willing A, et al. The spleen contributes to stroke-induced neurodegeneration. J Neurosci Res. 2008;86(10):2227–34.

33. Offner H, Subramanian S, Parker SM, Wang C, Afentoulis ME, Lewis A, et al. Splenic atrophy in experimental stroke is accompanied by increased regulatory T cells and circulating macrophages. J Immunol. 2006;176(11):6523–31.

34. Li M, Li F, Luo C, Shan Y, Zhang L, Qian Z, et al. Immediate splenectomy decreases mortality and improves cognitive function of rats after severe traumatic brain injury. J Trauma. 2011; 71(1):141–7.

35. Braun JS, Prass K, Dirnagl U, Meisel A, Meisel C. Protection from brain damage and bacterial infection in murine stroke by the novel caspase-inhibitor Q-VD-OPH. Exp Neurol. 2007;206(2):183–91.

36. Gendron A, Teitelbaum J, Cossette C, Nuara S, Dumont M, Geadah D, et al. Temporal effects of left versus right middle cerebral artery occlusion on spleen lymphocyte subsets and mitogenic response in Wistar rats. Brain Res. 2002;955(1–2):85–97.

37. Meyer S, Strittmatter M, Fischer C, Georg T, Schmitz B. Lateralization in autonomic dysfunction in ischemic stroke involving the insular cortex. Neuroreport. 2004;15(2):357–61.

38. Mignini F, Streccioni V, Amenta F. Autonomic innervation of immune organs and neuroimmune modulation. Auton Autacoid Pharmacol. 2003;23(1):1–25.

39. Felten DL, Felten SY, Carlson SL, Olschowka JA, Livnat S. Noradrenergic and peptidergic innervation of lymphoid tissue. J Immunol. 1985;135(2 Suppl):755s–65.
40. Kin NW, Sanders VM. It takes nerve to tell T and B cells what to do. J Leukoc Biol. 2006;79(6):1093–104.
41. Chang L, Chen Y, Li J, Liu Z, Wang Z, Chen J, et al. Cocaine-and amphetamine-regulated transcript modulates peripheral immunity and protects against brain injury in experimental stroke. Brain Behav Immun. 2010;25(2):260–9.
42. Ajmo Jr CT, Collier LA, Leonardo CC, Hall AA, Green SM, Womble TA, et al. Blockade of adrenoreceptors inhibits the splenic response to stroke. Exp Neurol. 2009;218(1):47–55.
43. Li HL, Kostulas N, Huang YM, Xiao BG, van der Meide P, Kostulas V, et al. IL-17 and IFN-gamma mRNA expression is increased in the brain and systemically after permanent middle cerebral artery occlusion in the rat. J Neuroimmunol. 2001;116(1):5–14.
44. Zhu J, Paul WE. Peripheral CD4+ T-cell differentiation regulated by networks of cytokines and transcription factors. Immunol Rev. 2010;238(1):247–62.
45. Vendrame M, Gemma C, Pennypacker KR, Bickford PC, Davis Sanberg C, Sanberg PR, et al. Cord blood rescues stroke-induced changes in splenocyte phenotype and function. Exp Neurol. 2006;199(1):191–200.
46. Yilmaz G, Arumugam TV, Stokes KY, Granger DN. Role of T lymphocytes and interferon-gamma in ischemic stroke. Circulation. 2006;113(17):2105–12.
47. Marsh BJ, Williams-Karnesky RL, Stenzel-Poore MP. Toll-like receptor signaling in endogenous neuroprotection and stroke. Neuroscience. 2009;158(3):1007–20.
48. Filen S, Ylikoski E, Tripathi S, West A, Bjorkman M, Nystrom J, et al. Activating transcription factor 3 is a positive regulator of human IFNG gene expression. J Immunol. 2010;184(9):4990–9.
49. Loetscher P, Pellegrino A, Gong JH, Mattioli I, Loetscher M, Bardi G, et al. The ligands of CXC chemokine receptor 3, I-TAC, Mig, and IP10, are natural antagonists for CCR3. J Biol Chem. 2001;276(5):2986–91.
50. Makinen S, Kekarainen T, Nystedt J, Liimatainen T, Huhtala T, Narvanen A, et al. Human umbilical cord blood cells do not improve sensorimotor or cognitive outcome following transient middle cerebral artery occlusion in rats. Brain Res. 2006;1123(1):207–15.
51. Chen J, Sanberg PR, Li Y, Wang L, Lu M, Willing AE, et al. Intravenous administration of human umbilical cord blood reduces behavioral deficits after stroke in rats. Stroke. 2001;32(11):2682–8.
52. Chen SH, Chang FM, Tsai YC, Huang KF, Lin CL, Lin MT. Infusion of human umbilical cord blood cells protect against cerebral ischemia and damage during heatstroke in the rat. Exp Neurol. 2006;199(1):67–76.
53. Vendrame M, Cassady J, Newcomb J, Butler T, Pennypacker KR, Zigova T, et al. Infusion of human umbilical cord blood cells in a rat model of stroke dose-dependently rescues behavioral deficits and reduces infarct volume. Stroke. 2004;35(10):2390–5.
54. Willing AE, Lixian J, Milliken M, Poulos S, Zigova T, Song S, et al. Intravenous versus intrastriatal cord blood administration in a rodent model of stroke. J Neurosci Res. 2003;73(3):296–307.
55. Kees F, Jehkul A, Bucher M, Mair G, Kiermaier J, Grobecker H. Bioavailability of opipramol from a film-coated tablet, a sugar-coated tablet and an aqueous solution in healthy volunteers. Arzneimittelforschung. 2003;53(2):87–92.

Part III
Translational Models

Chapter 24
Overcoming Barriers to Translation from Experimental Stroke Models

Ludmila Belayev

Abstract The current understanding of stroke pathophysiology and the recognition of advantages and disadvantages of animal models are providing a renewed incentive for translational research. However, when neuroprotective drugs that work in animal models for adult stroke fail when tested in humans, this can result in major setbacks for developing stroke therapy. In this review, we will define key challenges and complexities in translational stroke research. In part 1, we focus on how to choose an appropriate experimental stroke model, role of species, strain, sex and age of animals, morphological and functional differences between the brain of humans and animals, and protection of cerebral gray matter (instead of both gray and white matter). In part 2, we discuss preclinical pitfalls, including physiological monitoring, role of anesthesia, dosing and side effects, therapeutic window, and outcome measures. In part 3, we focus on the ischemic penumbra as a target of neuroprotection in experimental and clinical stroke. By recognizing the advances in stroke basic science, translational research is moving forward in the development of new effective acute stroke treatments. Success in this endeavor requires increased interactions and cooperation between basic scientists and clinical researchers.

1 Introduction

Despite tremendous efforts in stroke research and significant improvements in stroke care, therapy is still insufficient. Therapeutic options in the acute phase of a stroke are limited despite a great number of neuroprotective drugs that have been developed in the past few decades [1, 2]. The only approved treatment for ischemic stroke is intravenous recombinant tissue plasminogen activator (tPA), which open

L. Belayev, MD (✉)
Neuroscience Center of Excellence, Louisiana State University Health Sciences Center,
2020 Gravier Street, Suite 9B4, Room 946A, New Orleans, LA 70112, USA
e-mail: lbelay@lsuhsc.edu

P.A. Lapchak and J.H. Zhang (eds.), *Translational Stroke Research*, Springer Series
in Translational Stroke Research, DOI 10.1007/978-1-4419-9530-8_24,
© Springer Science+Business Media, LLC 2012

channels for blood flow through the occluded thrombi and improves neurological outcome [3]. However, only 3–5% of stroke patients are eligible for tPA therapy, which must be initiated within 3 h of stroke onset—otherwise effectiveness is limited and there is potential for hemorrhagic side effects [4].

A major question is why translation to the clinical setting of promising basic neuroprotection studies has not been attained. Research in animal models has made important contributions to stroke pathophysiology and clinical management of stroke. Animal models try to recreate disease in an animal to study the progress and treatment of the disease in a highly controlled fashion. There are, without a doubt, many problems with this approach to clinical research. Even the best animal models cannot hope to mimic all the facets of human disease. A variety of animal models have been developed for modeling ischemic stroke. However, after numerous failed clinical trials of drugs showing preclinical promise, questions have been raised about the relevance of current experimental models for human brain. Specifically, there have been ongoing debates about the usefulness of animal models in the stroke field for at least 20 years. In this chapter, factors that adversely affect the ability to successfully translate data obtained in experimental stroke models to clinical trials are discussed. In addition, the significance of enhanced understanding of the ischemic penumbra as an opportunity for neuroprotection is highlighted.

2 Stroke Modeling and Clinical Relevance of Stroke Models

2.1 Global vs. Focal Cerebral Ischemia

A key bridge for the translation from a basic science advance into a potential therapy for acute ischemic stroke is finding appropriate stroke models. This is because novel therapies can be evaluated more readily to help determine their safety and efficacy profiles. One of the primary advantages of rodent models of stroke is that the consequences of the ischemic insult closely replicate the pathobiology observed in the human brain, making these models very much clinically relevant. Highly reproducible stroke models with excellent outcome consistency are essential for obtaining useful data from preclinical stroke trials as well as for improving interlab comparability.

A variety of animal models have been developed to mimic different stroke subtypes or pathological mechanisms and used for various aspects of stroke research (Table 24.1). They can be generally classified into two categories: focal and global cerebral ischemia models [5]. Global ischemia models mimic the clinical conditions of brain ischemia following cardiac arrest or profound systemic hypotension; focal models represent ischemic stroke, the most common clinical stroke subtype. Some additional stroke models involve special mechanisms to induce focal cerebral ischemia, such as the thromboembolic, endothelin, and photochemical models.

Table 24.1 Animal models of cerebral ischemia

Type of model	Type of ischemia	Representative models	Animal species available
Global ischemia	Complete ischemia	Cardiac arrest Decapitation Neck tourniquet Aortic occlusion	Dog, pig, rodents
	Incomplete ischemia	2-Vessel occlusion plus hypotension 4-Vessel occlusion Hypoxia-ischemia Unilateral common carotid occlusion (CCAo)	Dog, pig, rodents
Focal ischemia	Middle cerebral artery occlusion	Proximal middle cerebral artery occlusion (MCAo) • Permanent by coagulation or intraluminal thread • Temporary by intraluminal thread, clips, or snare	Dog, cat, rabbit, rodents
		Distal MCAo • Permanent or transient by coagulation, clip, or snare • Photochemical distal MCAo by laser • Vasoconstrictive MCAo by local endothelin-1 application	Nonhuman primate, dog, cat, pig, rabbit, rodents
	Miscellaneous models of focal cerebral ischemia	Focal ischemia in spontaneously hypertensive rats	SHR, SP-SHR rats
		Photochemically induced focal cortical thrombosis	Normally in rodents
		Models of cerebral embolism • Blood clot embolization • Microspheres embolization • Photochemically initiated thromboembolism	Normally in rodents
Models of cerebral ischemia in gerbils		Unilateral CCAo Bilateral CCAo	Gerbils

It must be acknowledged that none of these animal stroke models mimics all the complexities of human stroke and the model used should be relevant to the specific experimental questions to be answered. This question is often a novel mechanisms being expounded.

2.1.1 Global Ischemia

The first example of experimental and clinical correlation comes from the idea of selective neuronal vulnerability described in rodent models of transient global ischemia. Global brain ischemia occurs when cerebral blood flow (CBF) is reduced throughout most or all of the brain. It occurs in patients after cardiac arrest or systemic circulatory collapse. With complete ischemia, global blood flow has ceased completely; whereas with incomplete ischemia, global flow is severely reduced, but the amount of flow is insufficient to maintain adequate cerebral metabolism and function [6]. Global ischemia is produced in experimental models by occluding the major arterial supply to the forebrain (Table 24.1). Models of transient global cerebral ischemia in gerbils, rats, and mice all show selective and delayed neuronal death in the "selectively vulnerable" brain regions, while tending to spare glia and vascular elements. The pyramidal neurons of sector CA1 of the hippocampus are the most vulnerable to global ischemia in both animal models [7] and humans [8]. These selective neuronal responses reflect the basic science concept that active cell death mechanisms become triggered after an ischemic insult and molecular pathways can be manipulated and studied [9].

2.1.2 Focal Cerebral Ischemia

The knowledge from clinical practice has not effectively been transferred into the modeling of experimental focal cerebral ischemia. In human ischemic stroke, blood supply to part of the brain is decreased, leading to dysfunction of the brain tissue in that area. There are four reasons why this might happen: thrombosis, embolism, systemic hypoperfusion, or venous thrombosis (Table 24.2) [10]. These four entities predict the extent of the stroke, the area of the brain affected, the underlying cause, and the prognosis [11]. Clearly, there is an urgent need for an appropriate animal stroke model which can mimic common forms of ischemic stroke in the clinical population (Table 24.2).

Most focal cerebral ischemia models, characterized human thrombotic or embolic stroke, are produced in the laboratory by temporary or permanent occlusion of a major cerebral artery (usually, the middle cerebral artery (MCA)). Middle cerebral artery occlusion (MCAo) results in a reduction of CBF in both the striatum and cortex, but the degree and distribution of blood flow reduction depend on the duration of MCAo, the site of occlusion along the MCA, and the amount of collateral blood flow into the MCA territory. Several different types of MCAo models exist, and for the most part, they are either a permanent or temporary (reperfusion) occlusion with MCAo at either the proximal or distal part of the vessel. Procedures that induce vessel occlusion might include clot (autologous blood) placement or mechanical (ligation or "suture" procedures) occlusion.

The model with the best face validity is the thromboembolic model in which microspheres/macrospheres or clotted blood is injected [12]. The microsphere/macrosphere technique involves injection of different sizes of spheres (50 or 300–400 μm

Table 24.2 Common forms of ischemic stroke in humans and corresponding animal models

Type of stroke	Description	Affected vasculature	Causes	Animal models
Thrombotic	Obstruction of a cerebral blood vessel by a blood clot forming locally	Large vessels: internal carotids, vertebral, and the Circle of Willis	Atherosclerosis Vasoconstriction (tightening of the artery) carotid or vertebral artery dissection Inflammatory diseases of the blood vessel wall (Takayasu arteritis, giant cell arteritis, vasculitis)	Permanent or transient MCAo by intraluminal suture, thromboembolism, or electrocoagulation
			Noninflammatory vasculopathy Moyamoya disease Fibromuscular dysplasia	
		Small vessels: branches of the Circle of Willis, MCA, stem, basilar artery	Lipohyalinosis (build-up of fatty hyaline matter in the blood vessel as a result of high blood pressure and aging) Fibrinoid degeneration (stroke involving these vessels are known as lacunar infarcts) Macroatheroma (small atherosclerotic plaques)	
Embolic	Blockage of an artery by an arterial embolus (blood clot, fat, air, cancer cells, bacteria)	Large and small cerebral vessels	Atrial fibrillation Rheumatic disease of the mitral or aortic valve Artificial heart valves Myocardial infarction Cardiomyopathy Infective endocarditis	Thromboembolic MCAo
Systemic hypoperfusion	Reduction of blood flow to all parts of the body	"Watershed" areas: border zone regions supplied by the major cerebral arteries	Cardiac arrest Arrhythmia Myocardial infarction Hemorrhagic shock	Global cerebral ischemia models
Venous thrombosis	Obstruction of the venous sinuses by thrombus	Dural venous sinuses	Blood disorders Chronic inflammatory diseases Trauma	Cerebral venous thrombosis models

diameter) into the internal carotid artery (ICA) or MCA [13]. Thromboembolic clots use either spontaneously formed clots from autologous blood placed into the MCA, or thrombin-induced clots in the MCA, to directly model the clot-induced human stroke [14]. Although these models are relatively easy to produce, the main disadvantage is that the location of infarction(s) is not consistent and infarction size varies, which makes analysis of neuroprotective treatments difficult. However, it provides an excellent system to evaluate new thrombolytic therapies.

Another widely used technique of MCAo involves cauterization of the MCA via a craniotomy [15]; this technique is invasive and does not permit reperfusion. In human ischemic stroke, however, recirculation occurs frequently after focal ischemia, particularly in the case of cerebral embolism. This model results in an infarction of the cortex and caudoputamen areas. Occlusion of the MCA at its origin, *proximal* MCAo, disrupts blood flow to the entire vascular territory of the MCA and can be induced by placement of a vascular clip on the proximal MCA branch (Table 24.1). Occlusion of the *distal* MCA typically spares the striatum and involves primarily the cortex. It is usually produced by transcranial placement of a vascular clip on a distal MCA branch or by electrocauterization. In the rat, mechanical clipping of the MCA and photothrombotic occlusion of the vessel are in common use, but these techniques involve craniotomy [7].

In order to conduct preclinical assessment of potential drugs for acute ischemic stroke, rat intraluminal suture models are the most widely employed [16–18]. This model involves inserting a nylon suture into the external carotid artery of rats and then advancing the thread cranially to block the MCA. The severity of ischemia can be modulated by varying the duration of time (often between 60 and 120 min) the filament is left in place; the suture is then withdrawn to allow for tissue reperfusion. This model has the advantage of not requiring craniotomy with its associated operative trauma and it permits subsequent reperfusion of the occluded MCA. However, brain injury produced by filament MCAo in rodents varies considerably in size and distribution. This variability in infarct volume from animal to animal necessitates the use of large numbers of animals to discern statistical significance in drug testing. Unsuccessful outcomes consist of animals without neurological deficits (insufficient occlusion) and subarachnoid hemorrhage (SAH) resulting from rupture of the intracranial ICA. Several modifications of the initial model have been developed by introducing different filaments and coating techniques of the filament, for example, with silicone, glue, or poly-L-lysine to increase reproducibility of infarct size and to reduce the incidence of SAH or premature reperfusion [18–20].

To summarize, there are two main factors that influence the selection of in vivo stroke models for preclinical studies. (1) Mechanism of the neuroprotective candidates: if the candidate is predicted to reduce ischemic lesion by attenuating brain edema after thrombolytic therapy, the thromboembolic model should be used; if the predicted neuroprotection is associated with a particular brain cortex region, the distal MCAo or photochemical model is recommended because these models are able to produce a cortical lesion. (2) Model quality within a particular lab setting: if the predicted protection mechanism of a drug candidate is shared by several stroke models, the selection of the preferred model could be determined by success rate and outcome consistency. In most cases, the choice is between the intraluminal filament and the Tamura cauterization model.

2.2 Permanent vs. Transient Ischemia

Models resulting in permanent ischemia mimic clinical stroke without reperfusion, whereas transient models reflect human stroke with therapy-induced or spontaneous reperfusion. Permanent MCAo models belong to the most frequently used procedures in stroke research. Occlusion is usually induced by cauterization of the MCA via a craniotomy or advancing a suture into the ICA to occlude the MCA at its origin from the circle of Willis. If the suture is removed after a certain interval, reperfusion is achieved (transient MCAo); if the filament is left in place, the procedure is also suitable as a model of permanent MCAo. This procedure has the advantage (when the same occlusion time is performed) that it is highly reproducible. It does not require craniotomy (avoiding its complications) and produces focal damage similar to that seen in embolic human stroke [21].

As recommended by Stroke Therapy Academic Industry Roundtable (STAIR), permanent MCAo models should be studied first, followed by transient (reperfusion) models [22]. An exception to this recommendation exists for drugs with mechanisms of action requiring successful reperfusion. However, because of greater mortality with permanent occlusion models, most preclinical studies use transient MCAo models. In the case of some nonhuman primates, permanent occlusion of the proximal MCA may be accompanied by significant mortality rates, similar to those seen in humans. In these cases, permanent MCAo may not be suitable.

To summarize, permanent then transient MCAo animal model studies should be completed before beginning clinical trials in humans.

2.3 Significance of the Selection of Species, Strain, Sex, and Age of Animals

While the MCAo models provide stroke researchers with an excellent platform to investigate this disease, controversial or even contradictory results are occasionally seen in the literature utilizing these models. Various factors exert important effects on the outcome in these stroke models, including species, strain, sex, and age of the animal examined.

2.3.1 Role of Species

Several ischemic stroke models have been developed in a variety of species, including rodents, canines, rabbits, cats, and even nonhuman primates [6]. Different species also vary in their susceptibility to the various types of ischemic insults. Large animal species such as cats, dogs, and rabbits have a greater similarity to humans because of similar brain size and structure, but such animals show more individual anatomical variability. Thus, reproducibility of infarction is very low for these animals. In addition, they are very expensive compared to rodents. Larger animals including dogs, some primates, and humans have highly folded and convoluted brains, i.e., gyrencephalic brains, while rodents have smooth, lissencephalic, brains. Although the nonhuman primate brain is most similar to the human brain, the positive effects observed in the rat or mouse may be lost in other larger animals such as the dog, which has an anatomically distinct gyrencephalic brain [23]. The use of a nonprimate large animal may be useful for studying therapeutic efficacy and yet still be more cost-effective than primate studies. The main benefit of the primates is that they can better model complex behaviors and recovery, although the primate model should be chosen with care because not all primates have a gyrencephalic brain and the high cost associated with primate studies.

The most commonly used animals in stroke research are small animal species, rodents in particular. They are inexpensive, reproducibility of infarction is very good (small standard deviation), they have close cerebrovascular anatomy and physiology between different strains, and they have a small brain size that is well-suited to different procedures. They are also easy to maintain and easy to manipulate genetically. Transgenic and knock-out mice have been introduced into ischemia research to permit the study of molecular mechanisms. All together, rodent models are ideal for investigating the mechanisms underlying injury after ischemic stroke as well as for developing effective therapeutic approaches to the disease.

To summarize, neuroprotective agents should be tested in two or more species because different species vary in their susceptibility to and recovery after ischemic insult.

2.3.2 Role of Strain

Strain variability in both cerebral vasculature and neuronal vulnerability can contribute to variances in the outcome of ischemic injury and, hence, drug effect. Differences in genetic background can influence the outcome of cerebral ischemia [24]. For example, in genetically hypertensive strains of rats, e.g., spontaneously hypertensive rats (SHR) and stroke-prone spontaneously hypertensive rats (SP-SHR), MCAo usually results in larger infarct volume and lower infarct size variability than in normotensive strains [25]. Several studies have reported differences in the extent of ischemic lesion volumes among rat strains [25, 26]. Duverger and MacKenzie [25] found that there is interstrain variation in the volume of cerebral infarction after standardized MCAo in rats. They reported that the variability in infarction size in Wistar rats was extreme in comparison to that in Sprague–Dawley, F344, SHR, and stroke-prone SHR rats. Markgraf et al. [26] reported that photothrombotic occlusion of the distal MCA, in conjunction with common carotid occlusion (CCAo), produces large and consistent infarcts of the cerebral neocortex of Sprague–Dawley rats. By contrast, the same procedure, when carried out in Wistar rats, yields substantially smaller and more variable zones of complete infarction but significant volumes of incomplete infarction with selective neuronal necrosis.

Several authors have previously reported that mouse strains differ in their susceptibility to focal and global cerebral ischemia. Hara et al. [19] observed that C57Black/6 mice developed larger lesions than SV-129 mice during focal ischemia/reperfusion, although such differences were not found after permanent focal ischemia [27]. C57Black/6 mice were significantly more susceptible to global ischemic injury and the resistance could not be accounted for by differences in blood pressure or choice of anesthesia. This vulnerability was probably related to the presence of a circle of Willis lacking vascular connections between the anterior and posterior circulations and poor collateral blood flow during bilateral common carotid artery occlusion [28]. Variations in genetic background can lead to erroneous conclusions about the importance of an identified phenotype to a specifically targeted gene. It is important to use proper controls when studying ischemia and cerebrovascular regulation in genetically engineered mice derived from more than a single parent strain.

To summarize, Wistar rats have extensive collateral circulation and are suitable for global cerebral ischemia. SD rats, however, have less collateral vessels and are the species of choice for producing focal cerebral ischemia in rats. The ability of SHR to develop large infarcts following MCAo makes this strain attractive for experimental research, but the relatively lower effectiveness of collateral circulation in this strain may limit efforts to achieve neuroprotection.

2.3.3 Role of Sex on Specific Injury Response

Stroke is a sexually dimorphic disease with differences between males and females observed both clinically and in the laboratory. While males have a higher incidence of stroke throughout much of their lifespan, aged females have a higher burden of stroke [29]. Sex differences in stroke result from a combination of factors including elements intrinsic to the sex chromosomes as well as the effects of exposure to sex hormones throughout the lifespan. The different effects of aspirin on men and women in the primary prevention of cardiovascular disease and stroke provide a striking example of how influential these responses may be. In one study, it was found that aspirin substantially reduced the risk of cardiovascular disease, but not stroke, in men; however in the another study, aspirin reduced the risk of stroke by 24%, but had no effect on cardiovascular disease, in women [29].

Preclinical studies in animal models confirm that the effects of stroke vary based on sex and age. Decreased ischemic damage for equal insults occurs in adult female vs. male rodents in models of global [30] and focal [31] cerebral ischemia. Interestingly, just as in humans, with increasing age, female mice (i.e., 16 months; average lifespan of a mouse is 2–3 years) become more susceptible to stroke damage than males, an effect opposite to that observed in younger mice (aged 2–3 months) [32]. Acute administration of estrogen after stroke reduces ischemic damage in both male and female rodents [33]. Unfortunately, experimental studies frequently use only male animals to reduce variability between mixing sexes and to avoid accounting for the estrus cycle in experimental design.

To summarize, putative neuroprotective agents should be evaluated in both males and females. Based on sex-specific pathways involved in ischemic cell death, agents that target male-specific parts of the pathway may be ineffective in females, and vice versa.

2.3.4   Role of Age and Associated Illnesses

Age is the most important independent risk factor for stroke. Both the incidence of stroke and level of negative outcomes associated with stroke (in terms of disability and mortality rate) increase with age. With the growth of the global elderly population, the prevalence of stroke will also rise. Reports indicate that 75–89% of strokes occur in individuals aged >65 years [34]. Of these strokes, 50% occur in people who are aged ≥70 years and nearly 25% occur in individuals who are aged >85 years. Risk factor profiles and mechanisms of ischemic injury vary between young and old patients with stroke. It is well known that elderly patients tend to have a worse outcome than younger patients because they have comorbidities that heavily affect outcome. In addition, elderly patients often receive less-effective treatment than younger individuals. Although age is one of the most significant prognostic markers for poor outcome, very few studies have been performed in aged animals, especially in animals over 15 months of age. Studies on aged animals (e.g., rodents) can closely mimic the clinical features of older patients with stroke. Aged animals show greater infarction volumes and a higher mortality rate than young animals following MCAo [35]. While animals of all ages have shown neurological deficits after MCAo, aged animals exhibited poorer performances than young animals in functional tests [32]. Aged mice of both sexes showed considerably reduced stroke-induced edema compared with young animals. This finding is consistent with the clinical observation that young patients with stroke are more likely to develop fatal brain edema than older patients [36].

Unfortunately, most experimental studies are conducted on healthy young animals under rigorously controlled laboratory conditions. However, the typical stroke patient is elderly with numerous risk factors and complicating diseases (e.g., diabetes, hypertension, and heart diseases). Stroke researchers frequently avoid using aging animals for MCAo due to the more complex surgical procedure and high cost of purchasing and raising animals. Very few experimental studies exist in the literature that have examined middle aged and aging animals, and these have led to somewhat inconsistent results [37]. This might partly explain the translational failure of neuroprotective drugs in humans.

To summarize, in order to provide an adequate characterization of therapeutic potential, novel neuroprotective agents should be tested in young, old, healthy, and diseased animal models before they are used for human clinical trials.

2.4 Brain Anatomical and Physiological Characteristics of Importance in the Selection of Stroke Animal Models

Although the basic mechanisms of stroke are identical between humans and other mammals, there are differences in brain structure, function, and vascular anatomy. In general, larger animals, especially primates, have a greater similarity to humans because of similar brain size and structure, but such animals show more individual anatomical variability. There are many advantages of using large animals in stroke research. It is easier to perform physiological monitoring (e.g., arterial blood gases, blood pressure, plasma glucose, hematocrit measurements, and electroencephalography). The measurements can be coupled with magnetic resonance imaging (MRI), neurobehavior, neurochemistry, and neuropathology, all performed on the same animal. In addition, measurements of CBF and metabolism can be made easily in large animal models. Finally, the brains of large animals are gyrencepahlic, like humans, which may make them closer to the human brain in structure and function [6]. There are several disadvantages to using large animals; they involve the use of invasive surgery for monitoring and producing cerebral ischemia, less consistent infarct sizes, increased physiological variability, and significant mortality.

In contrast, small-brained animals, such as rats and gerbils, suffer proportionately larger infarcts than do larger animals. These smaller animals, however, have lissencephalic, i.e., smooth, brains and thus may be quite different in anatomy and functional aspects from the human brain. Physiological monitoring is also considerably more challenging in small animals, such as gerbils and mice, and concurrent measurements over time are limited or may not be possible. Of all the rodents, the rat is one of the most suitable species because of the similarities between the anatomy of the circulation of rat and human, particularly when compared to such animals as the gerbil, cat, and dog. Also, physiological variables can be easily controlled to ensure that the resulting injury is reproducible and variability is as limited as possible.

Cerebral energy metabolism and blood flow in mammals is inversely related to their body weight. For example, in the rat, glucose and oxygen metabolism, as well as blood flow, are three times higher than humans. Neuronal and glial densities are also quantitatively different in various mammals. Furthermore, the human brain is gyrated and larger. There are also large differences in the gross cerebral vascular anatomy. Some rodents do not have a complete circle of Willis (gerbils), while others can have even more effective collaterals between large cerebral vessels than humans (e.g., rats). Accordingly, the size, anatomical distribution, and temporal evolution of the infarct differ among species.

The penumbra zone, although well established in rodents, is less well characterized in humans and might be considerably smaller. The effect of neuroprotective treatments is also species-dependent and rodents seem to be more amenable to neuroprotection than higher mammals. Therefore, greater emphasis should be placed on studying experimental stroke and neuroprotection in species that are phylogenetically closer to humans.

To summarize, MCAo model studies should be initially completed in rats, then possibly in cats or primates, before beginning human clinical trials.

2.5   Assessment of Ischemia-Reperfusion Injury in White and Gray Matter

Neuroprotective therapies that work on gray matter do not necessarily have the same effect on white matter, which makes up a large portion of human brain. The cerebral cortex is the part of the brain that most strongly distinguishes mammals from other vertebrates, primates from other mammals, and humans from other primates. The human brain contains a greater proportion of white matter compared with the rat brain. The rodent brain is not gyrencephalic like the human brain. Preclinical studies have largely studied neuroprotection of gray matter since rats have a higher proportion of gray to white matter than humans [38]. The effects of neuroprotective therapy on rodent cerebral white matter tracts are largely unknown. The failure of some neuroprotective trials may be due to an inability of certain agents to protect against axonal damage. The pathophysiology of ischemic injury in white matter is different than in gray matter, and treatment targets likely differ as well. *N*-methyl-D-aspartate (NMDA) receptors, for example, are preferentially located at synapses rather than along axons [39]. For example, MK-801, the NMDA antagonist, was neuroprotective in photothrombotic focal cerebral ischemia [40], but did not attenuate axonal injury in focal cerebral ischemia in cats [41]. Interestingly, however, a recent study showed that axonal and myelin damage could be reduced in rats with the NMDA antagonist CNS 1102 (Cerestat) [42].

To summarize, neuroprotective bioactivity in white and gray matter should be defined before progressing to clinical trials.

3   Preclinical Pitfalls

3.1   Physiological Monitoring

Complete physiological monitoring is required to ensure an interpretable outcome in experimental models of cerebral ischemia. Basic physiological parameters such as body and brain temperatures, blood gases, plasma glucose, and arterial blood pressure should be routinely monitored for as long as possible [5]. The outcome of experimental ischemic insult is markedly influenced by even small variations in brain and rectal temperature. Because brain temperature in animals, particularly during and after insult, cannot be predicted from knowledge of rectal temperature alone, it must be independently monitored. This can be measured indirectly in rats and mice with a temporal muscle temperature probe. It is also recommended to use fasted and intubated animals in cerebral ischemia models to avoid hyperglycemia and hypoxia. Hyperglycemia increases both the frequency with which the animals die of hemispheric edema secondary to infarction and the infarct size in animals that survive [43]. It is important to monitor CBF using Doppler flow or perfusion imaging to document adequate sustained occlusion and to monitor reperfusion in temporary cerebral ischemia models [44]. A lack of scrupulous monitoring of the above variables has been recognized as a major flaw in neuroprotection studies, which might have contributed significantly to the failure of compounds in human clinical trials.

To summarize, full physiological monitoring and meticulous control of blood pressure, arterial blood gases, and rectal and cranial temperatures are powerful modulators of ischemic outcome.

3.2   Role of Anesthesia

The majority of patients with ischemic stroke are not anesthetized. When designing a preclinical neuroprotective study, the protection provided by an anesthetic should be taken into account. Numerous studies have demonstrated that anesthetic agents afford a degree of protection from cerebral ischemia [45]. These anesthetics include isoflurane, halothane, xenon, nitrous oxide, barbiturates, propofol, ketamine, and the local anesthetic lidocaine [46]. Interestingly, the neuroprotective effect of anesthetics depends on severity of insult. In the case of gas anesthetics agents and propofol, neuroprotection may be sustained if the ischemic insult is relatively mild; however, with moderate-to-severe insults, this neuronal protection is not sustained after a prolonged recovery period [47]. In addition, the working mechanism of anesthetics needs to be considered during the experimental design of neuroprotection studies. For example, when an expected neuroprotection is likely via opening of K channels, gas anesthetics may have a compounding effect and should be avoided when possible [46].

Animal studies using both anesthesia and neuroprotective agents are actually studying a combination of two therapies. It will be ideal to produce experimental stroke in conscious animals to avoid the effect of anesthesia on stroke outcome. There are animal models of stroke that are able to induce an ischemic stroke in conscious animals [48, 49]. Such models are likely to model human stroke more closely than those models using anesthesia during the induction of stroke. Specifically, the rabbit blood clot embolism model [48] involves the preparation of animals under anesthesia, and then later the stroke is induced in conscious, unanesthetized animals. Even light surgical anesthesia may substantially reduce infarct size following stroke [50].

To summarize, potential neuroprotection requires careful consideration. The working mechanism of anesthetics needs to be considered and, ideally, the confounding effects of anesthesia must be eliminated from animal stroke models.

3.3   Dosing of Experimental Compounds and Side Effects

An extended dose–response strategy is absolutely critical. The minimum effective and maximum tolerated dose should be defined. Different species can have different thresholds for pharmacological agents, and simple extrapolation from one species to another can be misleading. A drug dose effective in the mouse or rat may not be effective in large animals and/or humans. Small animal studies rarely yield information on harmful side effects because small animals tolerate drugs very well and they are not routinely examined for evidence of side effects. In addition, they are frequently sacrificed in the early hours or days after treatment, so long-term consequences are unknown. Pharmacokinetics and pharmacodynamics may vary considerably among species. As stated in STAIR recommendation [22], dose–response curve should be performed in models of both small and large animals. Larger animals might be more prone to similar side effects as human [51]. In addition, the dose–response curve may be excessively high in stroke efficacy studies. The blood–brain barrier, low CBF (i.e., the drug delivery system to the tissue at risk), and plasma protein-binding may present impediments for determination of appropriate dosing regimens for use in humans [22]. Also, outcomes can vary at different dose ranges; for example, histological protection might be less sensitive than a functional outcome. It is also crucial that at all dosing regimens, plasma and brain drug levels are determined.

Many highly potent neuroprotective drugs display side effects that inhibit the application of effective doses in patients (e.g., MK-801, phencyclidine, and others) [52]. Most anti-excitotoxic compounds cause psychomimetic or cardiovascular side effects, or both [53]. This severely limits the tolerated dose insofar as levels in humans reach only a fifth of the effective concentrations observed in rodent models. It is therefore important to design drugs with better safety profiles and more favorable pharmacokinetics with fewer side effects.

To summarize, the minimally and maximally effective dose and careful detection of side effects should be defined in experimental stroke models of both small and large animals. In addition, it should be documented which drugs in these ranges can access the target organ.

3.4   Therapeutic Window

There is debate about the relevance of a therapeutic window in animals for acute clinical stroke. The window of therapeutic opportunity in animal models is not necessarily predictive of the time window in humans, but determination of relative windows is useful. Whereas in animal studies the time of incidence onset is known and therapy can be started early, patients often present with delay and unclear time of symptom onset. For example, the onset of symptoms might not coincide with the onset of cerebral ischemia, or there might be a delay before the patients become aware of these symptoms, as in strokes at night or stroke syndromes characterized by unawareness of deficits [54]. Thus, it is difficult to define accurately the time window in which a certain drug might be effective in each patient.

Experiments with animal models often begin with a pretreatment protocol. If the therapy works, it is tested at different time points after stroke onset. Only treatment initiated after onset of ischemia qualifies. Most effective neuroprotective agents work best within 60–180 min; rarely are they effective more than 4 h after stroke [38]. It may be inappropriate to perform expensive clinical trials in which the drug is administered, for example, at 3 h after occlusion, because the only preclinical animal study available shows the drug to be effective when given at the time of arterial occlusion or shortly after reperfusion. Prior to designing human clinical trials, it would be helpful to know the effective therapeutic window after stroke onset and also if the time window was defined in both small and large animal models.

It must be noted that the therapeutic window in animal models might not predict the human therapeutic window and can be used only as a general guideline. For example, positron emission tomography (PET) studies suggest that the window of opportunity may be extended in some patients. About one third of patients still had evidence of penumbra (ischemic tissue that is functionally impaired but whose damage is potentially reversible) when assessed at 5–18 h (mean, 10 h) after stroke onset [38].

To summarize, neuroprotective agents should demonstrate efficacy when administered several hours after ischemic onset. Prospective neuroprotective drugs should not be advanced into clinical stroke trials until preclinical studies have investigated their effects when administered many hours, not minutes, after ischemia.

3.5 Outcome Measures

Traditionally, animal studies have relied on reduction in infarct size within the first few hours after stroke as the primary measure of therapeutic efficacy. In contrast, clinical trials judge efficacy by using neurological outcomes, not infarct volume, most often at 3 months after stroke [55]. Thus, therapies might reduce the size of cerebral lesions found in animals, but not the functional impairment when tested in patients. In humans, the size of the lesion does not always correlate with functional impairment, although this correlation has been shown to be more robust with diffusion-perfusion (MRI). Therefore, more emphasis must be placed on function over histology in preclinical studies. As per STAIR recommendation, preclinical outcome measures should include neurobehavior, infarct volume, immunohistochemical analysis, neuropathology, somatosensory-evoked potentials, and electroencephalography [22].

In experimental stroke models, at least two outcome measures should be considered mandatory: functional responses and infarct volume. Most animal studies measure infarct size, but even this is not standardized. Many experiments employ 2,3,5-triphenyltetrazolium chloride (TTC) staining of metabolically active cells. TTC is fast, reproducible, inexpensive, but it is not sensitive enough at early and late time points and is difficult to analyze [56]. In contrast, Fluoro-Jade histochemistry can be used to highlight specific cells and regions that are compromised as early as 1 h after stroke onset. Nissl or hematoxylin and eosin staining also may be challenging at early time points after stroke, but are effective at longer survival times. None of these techniques are clinically relevant in human patients.

A key element of testing potential therapeutics in stroke is the ability to assess functional recovery in addition to the more common assessment of infarct size. Diminished infarct size does not always correlate with functional improvement and improved function is the relevant measure of clinical outcome [56]. Assessment of therapeutic efficacy in preclinical studies should require demonstration of the benefits of functional measures of motor, sensory, or cognitive deficits [57]. For this reason, it is better to use a battery of appropriate tests rather than a single measure. Examples include tests of limb placing, beam walking, grid walking, rotorod performance, balance beam, cognition (e.g., Morris water maze, radial maze, T-maze retention test), and many others [18, 38].

It should be emphasized that some behaviors initially impaired by the stroke recover naturally in rodent models. Thus, it is especially important to be able to distinguish between a test compound's ability to speed normal recovery vs. its ability to produce recovery of functions otherwise lost. Several studies have now demonstrated that it is a necessity to follow animals for much longer time periods. There are examples of compounds shown to be initially effective after a short survival time, but those initial beneficial effects were lost over time [58]. Furthermore, replication of improved functional outcome in at least a second species is likely to optimize the chance of success in large-scale clinical trials. Evaluation in larger species, such as cats or primates, is desirable rather than in rodents only.

Functional scores (National Institutes of Health stroke scale, Barthel index, etc.) are typically used in humans and are assessed 3 or 6 months after stroke. These scores are of great importance because they reflect the extent of neurological deficits. However, functional scores are less amenable to statistical evaluation and might be less sensitive than markers of infarct volume or histopathology. Therefore, objective measures of ischemic damage should be developed and tested soon after ischemic insult and at a later stage (e.g., diffusion-weighted and flow-sensitive imaging, functional and spectroscopical imaging, etc.) [59].

To summarize, multiple endpoints are important and both histological and behavioral outcomes should be assessed at least 2–3 weeks or longer after stroke onset. Replication of improved functional outcome should be confirmed in at least a second species.

4 The Focal Ischemic Penumbra as a Target of Neuroprotection

Acute ischemic stroke produces a brain lesion that encompasses an irreversibly injured core and a peripheral zone (penumbra), where tissue is damaged but potentially salvageable [54]. The purpose of acute stroke treatment is to salvage penumbral tissue and to improve brain function. The penumbra has a limited life span and appears to undergo irreversible damage within a few hours unless reperfusion is initiated and/or neuroprotective therapy is administered [9, 60].The ischemic penumbra (defined as local cerebral blood flow (LCBF) of 20–40% of control) forms an irregular rim around the ischemic core and tends to be greatest in the frontal and occipital cortex [61, 62]. Although reduced LCBF is the major factor responsible for necrotic injury, other factors, including lipid peroxidation, inflammatory responses, and development of brain edema, may also contribute to either the severity or progression of penumbral injury [63, 64]. The rate of progression of the penumbra from reversible to irreversible ischemic injury is also dependent on other factors and may be accelerated in the presence of poor collateral circulation, hyperglycemia, and other exacerbating factors [65].

The ischemic core region is not the target of acute stroke therapy. Reperfusion cannot be implemented quickly enough to reverse the cellular consequences of the ischemic cascade induced by the severe interruption of blood flow to salvage much tissue, nor can purported neuroprotective agents reach this region in sufficient concentration to ameliorate these harmful cellular events. The penumbral zone is the conceptual basis for therapeutic intervention in the acute stroke phase, either by using thrombolytic or neuroprotective drugs. Both interventions are, in fact, focused on rescuing the penumbral zone from transforming into the core of the ischemic insult [66].

4.1 Penumbra in Stroke Animals

The stroke penumbra is transient and yet an important area to rescue due to the fact that the damage is not yet irreversible. Therefore, penumbral changes and infarct expansion can be defined in experimental stroke, in which the timing of MCAo and reperfusion is accurately controlled. LCBF assessed autoradiographically after 2 h of MCAo depicts the penumbra as a region of intermediate CBF depression (20–40% of control) surrounding the ischemic core (5–20% of control) and comprising one half of the entire lesion [67]. Local glucose metabolic rate in the acute penumbra is not reduced despite the critical CBF reduction, so that the penumbral metabolism/blood flow ratio is markedly elevated. In contrast, following 1 h of recirculation, glucose metabolism throughout the initially ischemic hemisphere becomes markedly depressed, and the metabolism/flow ratio appears to have returned to basal levels. The transition of penumbra from a salvageable to irreversibly damaged state is driven by the magnitude and extent of the initial CBF reduction [68]. Thus, brain infarction is a highly deterministic event, precisely predictable on the basis of local perfusion at the time of MCAo. In a similar fashion, these studies also reveal that the early postischemic decline of glucose utilization observed 1 h after recirculation is also a predictor of infarction [69]. Taken together, these data show that the acute penumbra is, in fact, evanescent and that it progressively deteriorates over a few hours after the onset of focal ischemia, eventually becoming an extension of the ischemic core.

4.2 Penumbra in Stroke Patients

The goal of treating ischemic stroke patients is to salvage the penumbra as much and as early as possible. It has been reported that half of acute ischemic patients show the presence of the penumbra in MRI scans [70], and as, such the penumbra is potentially treatable. Unfortunately, the majority of potentially treatable patients are not treated. The use of imaging techniques to detect the existence of the penumbra in stroke patients has opened up a new avenue in patient management. Imaging the penumbra is a new strategy for detecting tissue at risk [71]. Up to 44% of patients may still have penumbral tissue 18 h after stroke [72]. The penumbra can be imaged using different technologies, such as MRI, computer tomography, PET, and single-photon emission computed tomography. In patients with acute ischemic stroke, PET scanning permits measurements of regional blood flow and metabolism in a fashion similar to the autoradiographic animal studies presented above. However, relatively few sequential correlative PET studies have been carried out in a series of patients with acute stroke, owing chiefly to the formidable logistic challenges posed by studying these acutely ill patients over many hours.

For targeting the penumbra in stroke patients, penumbral imaging is necessary for monitoring treatment response as well as for patient screening. The "mismatch" of perfusion-weighted and diffusion-weighted images (PWI-DWI mismatch) is the most commonly used method for imaging the penumbra and may serve as an effective screening method [72]. Previous neuroprotective trials have not been selective enough in targeting patients with cortical stroke and providing evidence for the presence of an ischemic penumbra. Patient selection should be "penumbra-specific" for acute treatment interventions [38]. It is anticipated that future treatments will more appropriately be customized according to an individual's "penumbra window" rather than a rigid time window. Recently, many clinical trials and stroke centers have incorporated penumbral imaging into their protocols despite the lack of consensus on the appropriate way to image the ischemic penumbra and the role for its use in patient selection [73].

To summarize, protection of the penumbra after stroke is a logical approach to preventing extensive irreversible brain damage. Imaging the penumbra is necessary both for monitoring treatment response as well as for patient screening.

5 Conclusions

Stroke research in animal models has made useful contributions to our understanding of the disease, but animal research and clinical practice still seem to be distant from each other. The clinical failure is mostly caused by the choice of drug and its pharmacokinetics, its preclinical evaluation, or clinical trial design and analysis, as well as to reflect on the appropriateness of the therapeutic target. Crucial issues remain unresolved regarding the translation of preclinical developments to the bedside.

The lack of translation between the animal work and clinical benefit is not because the animal models are not useful; instead, we must carefully choose the most appropriate stroke model and continue to improve the quality of our animal experiments. More effective effort to bring basic scientists and clinicians together will provide the stepping stone for overcoming the challenges of translational research.

References

1. Ginsberg MD. Current status of neuroprotection for cerebral ischemia: synoptic overview. Stroke. 2009;40:S111–4.
2. O'Collins VE, Macleod MR, Donnan GA, Horky LL, van der Worp BH, Howells DW. 1,026 experimental treatments in acute stroke. Ann Neurol. 2006;59:467–77.
3. The National Institute of Neurological Disorders and Stroke rt-PA Stroke Study Group. Tissue plasminogen activator for acute ischemic stroke. N Engl J Med. 1995;333:1581–7.
4. Fisher M, Bastan B. Treating acute ischemic stroke. Curr Opin Drug Discov Devel. 2008;11:626–32.
5. Ginsberg MD, Busto R. Small-animal models of global and focal cerebral ischemia. In: Ginsberg MD and Bogousslavsky J, eds. Cerebrovascular Disease: Pathophysiology, Diagnosis and Management. Blackwell Science. 1998;1:14–35.
6. Traystman RJ. Animal models of focal and global cerebral ischemia. ILAR J. 2003;44:85–95.
7. Ginsberg MD, Busto R. Rodent models of cerebral ischemia. Stroke. 1989;20:1627–42.
8. Petito CK, Feldmann E, Pulsinelli WA, Plum F. Delayed hippocampal damage in humans following cardiorespiratory arrest. Neurology. 1987;37:1281–6.
9. Lo EH. Experimental models, neurovascular mechanisms and translational issues in stroke research. Br J Pharmacol. 2008;153:S396–405.
10. Donnan GA, Fisher M, Macleod M, Davis SM. Stroke. Lancet. 2008;371:1612–23.
11. Kidwell CS, Warach S. Acute ischemic cerebrovascular syndrome: diagnostic criteria. Stroke. 2003;34:2995–8.
12. Carmichael ST. Rodent models of focal stroke. NeuroRx. 2005;2:396–409.
13. Gerriets T, Li F, Silva MD, Meng X, Brevard M, Sotak CH, et al. The macrosphere model: evaluation of a new stroke model for permanent middle cerebral artery occlusion in rats. J Neurosci Methods. 2003;122:201–11.
14. Zhang Z, Zhang RL, Jiang Q, Raman SB, Cantwell L, Chopp M. A new rat model of thrombotic focal cerebral ischemia. J Cereb Blood Flow Metab. 1997;17:123–35.
15. Tamura A, Graham DI, McCulloch J, Teasdale GM. Focal cerebral ischaemia in the rat: 1. Description of technique and early neuropathological consequences following middle cerebral artery occlusion. J Cereb Blood Flow Metab. 1981;1:53–60.
16. Koizumi J, Yoshida Y, Nakazawa T. Experimental studies of ischemic brain edema.1. A new experimental model of cerebral embolism in rats in which recirculation can be introduced in the ischemic area. Jpn J Stroke. 1986;8:1–8.
17. Longa EZ, Weinstein PR, Carlson S, Cummins R. Reversible middle cerebral artery occlusion without craniectomy in rats. Stroke. 1989;20:84–91.
18. Belayev L, Alonso OF, Busto R, Zhao W, Ginsberg MD, Belayev L, Alonso OF, Busto R, Zhao W, Ginsberg MD. Middle cerebral artery occlusion in the rat by intraluminal suture. Neurological and pathological evaluation of an improved model. Stroke. 1996;27:1616–22.
19. Hara H, Huang PL, Panahian N, Fishman MC, Moskowitz MA. Reduced brain edema and infarction volume in mice lacking neuronal isoform of nitric oxide synthase after transient MCA occlusion. J Cereb Blood Flow Metab. 1996;16:605–11.

20. Shah ZA, Namiranian K, Klaus J, Kibler K, Doré S. Use of an optimized transient occlusion of the middle cerebral artery protocol for the mouse stroke model. J Stroke Cerebrovasc Dis. 2006;15:133–8.
21. Menzies SA, Hoff JT, Betz AL. Middle cerebral artery occlusion in rats: a neurological and pathological evaluation of a reproducible model. Neurosurgery. 1992;31:100–6.
22. Stroke Therapy Academic Industry Roundtable. Recommendations for standards regarding preclinical neuroprotective and restorative drug development. Stroke. 1999;30:2752–8.
23. D'Alecy LG, Lundy EF, Barton KJ, Zelenock GB. Dextrose containing intravenous fluid impairs outcome and increases death after eight minutes of cardiac arrest and resuscitation in dogs. Surgery. 1986;100:505–11.
24. Oliff HS, Weber E, Eilon E, Marek P, Oliff HS, Weber E, Eilon E, Marek P, Oliff HS, Weber E, Eilon E, Marek P. The role of strain/vendor differences on the outcome of focal ischemia induced by intraluminal middle cerebral artery occlusion in the rat. Brain Res. 1995;675: 20–6.
25. Duverger D, MacKenzie ET. The quantification of cerebral infarction following focal ischemia in the rat: influence of strain, arterial pressure, blood glucose concentration, and age. J Cereb Blood Flow Metab. 1988;8:449–61.
26. Markgraf CG, Kraydieh S, Prado R, Watson BD, Dietrich WD, Ginsberg MD. Comparative histopathologic consequences of photothrombotic occlusion of the distal middle cerebral artery in Sprague–Dawley and Wistar rats. Stroke. 1993;24:286–92.
27. Huang Z, Huang PL, Panahian N, Dalkara T, Fishman MC, Moskowitz MA. Effects of cerebral ischemia in mice deficient in neuronal nitric oxide synthase. Science. 1994;265:1883–5.
28. Fujii M, Hara H, Meng W, Vonsattel JP, Huang Z, Moskowitz MA. Strain-related differences in susceptibility to transient forebrain ischemia in SV-129 and C57black/6 mice. Stroke. 1997;28:1805–10.
29. Turtzo LC, McCullough LD. Sex-specific responses to stroke. Future Neurol. 2010;5:47–59.
30. Hall ED, Pazara KE, Linseman KL. Sex differences in postischemic neuronal necrosis in gerbils. J Cereb Blood Flow Metab. 1991;11:292–8.
31. Alkayed NJ, Harukuni I, Kimes AS, London ED, Traystman RJ, Hurn PD. Gender-linked brain injury in experimental stroke. Stroke. 1998;29:159–65.
32. Liu F, Yuan R, Benashski SE, McCullough LD. Changes in experimental stroke outcome across the life span. J Cereb Blood Flow Metab. 2009;29:792–802.
33. McCullough LD, Alkayed NJ, Traystman RJ, Williams MJ, Hurn PD. Postischemic estrogen reduces hypoperfusion and secondary ischemia after experimental stroke. Stroke. 2001;32: 796–802.
34. Feigin VL, Lawes CM, Bennett DA, Anderson CS. Stroke epidemiology: a review of population-based studies of incidence, prevalence, and case-fatality in the late 20th century. Lancet Neurol. 2003;2:43–53.
35. Chen RL, Balami JS, Esiri MM, Chen LK, Buchan AM. Ischemic stroke in the elderly: an overview of evidence. Nat Rev Neurol. 2010;6:256–65.
36. Wagner JC, Lutsep HL. Thrombolysis in young adults. J Thromb Thrombolysis. 2005;20: 133–6.
37. Liu F, McCullough LD. Middle cerebral artery occlusion model in rodents: methods and potential pitfalls. J Biomed Biotechnol. 2011;2011:464701.
38. Gladstone DJ, Black SE, Hakim AM. Toward wisdom from failure: lessons from neuroprotective stroke trials and new therapeutic directions. Stroke. 2002;33:2123–36.
39. Jones KA, Baughman RW. Both NMDA and non-NMDA subtypes of glutamate receptors are concentrated at synapses on cerebral cortical neurons in culture. Neuron. 1991;7:593–603.
40. Yao H, Markgraf CG, Dietrich WD, Prado R, Watson BD, Ginsberg MD. Glutamate antagonist MK-801 attenuates incomplete but not complete infarction in thrombotic distal middle cerebral artery occlusion in Wistar rats. Brain Res. 1994;642:117–22.
41. Yam PS, Dunn LT, Graham DI, Dewar D, McCulloch J. NMDA receptor blockade fails to alter axonal injury in focal cerebral ischemia. J Cereb Blood Flow Metab. 2000;20:772–9.

42. Schäbitz WR, Li F, Fisher M. The *N*-methyl-D-aspartate antagonist CNS 1102 protects cerebral gray and white matter from ischemic injury following temporary focal ischemia in rats. Stroke. 2000;31:1709–14.
43. Martín A, Rojas S, Chamorro A, Falcón C, Bargalló N, Planas AM. Why does acute hyperglycemia worsen the outcome of transient focal cerebral ischemia? Role of corticosteroids, inflammation, and protein *O*-glycosylation. Stroke. 2006;37:1288–95.
44. Fisher M, Feuerstein G, Howells DW, Hurn PD, Kent TA, Savitz SI, et al. Update of the stroke therapy academic industry roundtable preclinical recommendations. Stroke. 2009;40:2244–50.
45. Head BP, Patel P. Anesthetics and brain protection. Curr Opin Anaesthesiol. 2007;20: 395–409.
46. Liu S, Zhen G, Meloni BP, Campbell K, Winn HR. Rodent stroke model guidelines for preclinical stroke trials (1st edition). J Exp Stroke Transl Med. 2009;2:2–27.
47. Kawaguchi M, Furuya H, Patel PM. Neuroprotective effects of anesthetic agents. J Anesth. 2005;19:150–6.
48. Zivin JA, Fisher M, DeGirolami U, Hemenway CC, Stashak JA. Tissue plasminogen activator reduces neurological damage after cerebral embolism. Science. 1985;230:1289–92.
49. Callaway JK, Knight MJ, Watkins DJ, Beart PM, Jarrott B. Delayed treatment with AM-36, a novel neuroprotective agent, reduces neuronal damage after endothelin-1-induced middle cerebral artery occlusion in conscious rats. Stroke. 1999;30:2704–12.
50. Gelb AW, Bayona NA, Wilson JX, Cechetto DF. Propofol anesthesia compared to awake reduces infarct size in rats. Anesthesiology. 2002;96:1183–90.
51. Danton GH, Dietrich WD. The search for neuroprotective strategies in stroke. AJNR Am J Neuroradiol. 2004;25:181–94.
52. Kornhuber J, Weller M. Psychotogenicity and *N*-methyl-D-aspartate receptor antagonism: implications for neuroprotective pharmacotherapy. Biol Psychiatry. 1997;41:135–44.
53. Gerriets T, Stolz E, Walberer M, Kaps M, Bachmann G, Fisher M. Neuroprotective effects of MK-801 in different rat stroke models for permanent middle cerebral artery occlusion: adverse effects of hypothalamic damage and strategies for its avoidance. Stroke. 2003;34:2234–9.
54. Dirnagl U, Iadecola C, Moskowitz MA. Pathobiology of ischaemic stroke: an integrated view. Trends Neurosci. 1999;22:391–7.
55. Duncan PW, Jorgensen HS, Wade DT. Outcome measures in acute stroke trials: a systematic review and some recommendations to improve practice. Stroke. 2000;31:1429–38.
56. Willing AE. Experimental models: help or hindrance. Stroke. 2009;40:S152–4.
57. Hunter AJ, Mackay KB, Rogers DC. To what extent have functional studies of ischaemia in animals been useful in the assessment of potential neuroprotective agents? Trends Pharmacol Sci. 1998;19:59–66.
58. Valtysson J, Hillered L, Andiné P, Hagberg H, Persson L. Neuropathological endpoints in experimental stroke pharmacotherapy: the importance of both early and late evaluation. Acta Neurochir (Wien). 1994;129:58–63.
59. Heiss WD, Podreka I. Role of PET and SPECT in the assessment of ischemic cerebrovascular disease. Cerebrovasc Brain Metab Rev. 1993;5:235–63.
60. Pulsinelli W. Pathophysiology of acute ischaemic stroke. Lancet. 1992;339:533–6.
61. Ginsberg MD, Belayev L, Zhao W, Huh PW, Busto R. The acute ischemic penumbra: topography, life span, and therapeutic response. Acta Neurochir Suppl. 1999;73:45–50.
62. Hossmann KA. Pathophysiology and therapy of experimental stroke. Cell Mol Neurobiol. 2006;26:1057–83.
63. van der Worp HB, van Gijn J. Clinical practice. Acute ischemic stroke. N Engl J Med. 2007;357:572–9.
64. Lee JM, Grabb MC, Zipfel GJ, Choi DW. Brain tissue responses to ischemia. J Clin Invest. 2000;106:723–31.
65. Fisher M. Characterizing the target of acute stroke therapy. Stroke. 1997;28:866–72.
66. Bacigaluppi M, Hermann DM. New targets of neuroprotection in ischemic stroke. Scientific World J. 2008;8:698–712.

67. Belayev L, Zhao W, Busto R, Ginsberg MD. Transient middle cerebral artery occlusion by intraluminal suture, I: three-dimensional autoradiographic image-analysis of local cerebral glucose metabolism-blood flow interrelationships during ischemia and early recirculation. J Cereb Blood Flow Metab. 1997;17:1266–80.

68. Zhao W, Belayev L, Ginsberg MD. Transient middle cerebral artery occlusion by intraluminal suture, II. Neurological deficits, and pixel-based correlation of histopathology with local blood flow and glucose utilization. J Cereb Blood Flow Metab. 1997;17:1281–90.

69. Ginsberg MD. Adventures in the pathophysiology of brain ischemia: penumbra, gene expression, neuroprotection: the 2002 Thomas Willis Lecture. Stroke. 2003;34:214–23.

70. Liu S, Levine SR, Winn HR. Targeting ischemic penumbra: part I—from pathophysiology to therapeutic strategy. J Exp Stroke Transl Med. 2010;3:47–55.

71. Donnan GA, Baron JC, Ma H, Davis SM. Penumbral selection of patients for trials of acute stroke therapy. Lancet Neurol. 2009;8:261–9.

72. Ebinger M, De Silva DA, Christensen S, Parsons MW, Markus R, Donnan GA, et al. Imaging the penumbra—strategies to detect tissue at risk after ischemic stroke. J Clin Neurosci. 2009;16:178–87.

73. Wintermark M, Albers GW, Alexandrov AV, Alger JR, Bammer R, Baron JC, et al. Acute stroke imaging research roadmap. AJNR Am J Neuroradiol. 2008;29:e23–30.

Chapter 25
Animal Models of Stroke for Preclinical Drug Development: A Comparative Study of Flavonols for Cytoprotection

Carli L. Roulston, Sarah McCann, Robert M. Weston, and Bevyn Jarrott

Abstract Laboratory animals have been used extensively to mimic human ischemic stroke with the aim of discovering drugs that minimize damage and reduce loss of motor and cognitive functions. To date, this has not led to successful clinical translation. Here we consider the available animal models of stroke with discussion focusing on their strengths and limitations for assessing neuroprotective actions. For assessment of drugs outside of thrombolytic therapies, we propose the use of the endothelin-1 (ET-1) model of stroke in conscious animals as a better model than the more common procedure of inducing focal stroke in anesthetized rats with a vascular occluder. The ET-1 model of ischemic stroke allows for stratification of animals to drug trial based on predictive outcomes assessed during stroke induction, mimicking the scenario of stroke patients who are assessed for functional outcome and survival for clinical management and to correctly stratify treatment groups for clinical trials. We highlight the value of applying a predictive outcome score to animals prior to treatment stratification using the water-soluble derivative of the flavonol, 3′,4′-dihydroxyflavonol: an effective neuroprotective drug in rats with modest focal strokes but not in rats with severe strokes. Finally, we discuss the use of flavonols

C.L. Roulston • S. McCann
Cytoprotection Pharmacology Program, O'Brien Institute,
University of Melbourne, Parkville, VIC 3065, Australia

R.M. Weston
Cytoprotection Pharmacology Program, O'Brien Institute,
University of Melbourne, Parkville, VIC 3065, Australia

Brain Injury and Repair Program, Howard Florey Institute,
University of Melbourne, Parkville, VIC 3010, Australia

B. Jarrott (✉)
Brain Injury and Repair Program, Howard Florey Institute,
University of Melbourne, Parkville, VIC 3010, Australia
e-mail: bevyn.jarrott@florey.edu.au

P.A. Lapchak and J.H. Zhang (eds.), *Translational Stroke Research*, Springer Series
in Translational Stroke Research, DOI 10.1007/978-1-4419-9530-8_25,
© Springer Science+Business Media, LLC 2012

as potential neuroprotectants and propose future ideas as to how the positive outcomes observed with flavonols in preclinical stroke models might indeed translate to clinical success.

1 Cerebral Ischemia/Stroke

A cerebral vascular incident in the brain, more commonly referred to as stroke, can result in severe, long-lasting alterations of body function due to an insufficient supply of blood to part of the brain. It can strike without warning with debilitating consequences. On average, stroke is the second leading cause of death above the age of 60 years and the principal cause of adult disability worldwide [110]. Despite the enormous burden that stroke places on individuals and the community, few therapeutic interventions have achieved success in either preventing the spread of damage to the brain after stroke, or in the restoration of brain function after damage has ensued. In individuals who survive a stroke, the resulting brain damage can lead to permanent cognitive and functional disabilities. Furthermore, a stroke can also lead to emotional disorders, the most common of which is depression, which occurs in up to 40% of stroke patients [87]. With improved methods of brain imaging for rapidly and accurately identifying the incidence of a stroke [112], combined with better management and treatment of risk factors, the likelihood of surviving a stroke has increased. However, there continues to be a risk of stroke with age that will progressively increase with an aging population [110].

1.1 Current Treatments

Thrombolysis with intravenous alteplase, a recombinant tissue plasminogen activator (rtPA), that led to restoration of blood flow during a stroke was the first major breakthrough for the medical treatment of ischemic stroke. The excitement of this new treatment that could reverse the effects of stroke through its clot busting actions was short-lived, however, due to its limited therapeutic window. Currently, thrombolytic therapy is suitable for fewer than 5% of stroke patients who are diagnosed within 4.5 h after ischemic thrombotic stroke onset [23]. This low patient assignment to treatment is due to time taken to diagnose a thrombotic stroke and the feared complications of an increased incidence of intracerebral hemorrhage that has been reported outside this treatment window. Although new studies suggest that this therapeutic window might be extended [23], there is still a compelling need to develop safe and effective treatment strategies that target the majority of patients.

Although no other drugs have received FDA approval in the USA [36], a limited number of neuroprotective compounds are known to be in clinical use for the treatment of ischemic stroke outside the USA. Included in these is the antioxidant edaravone, which has been in clinical use in Japan since 2001, with significant reductions in stroke-related disabilities demonstrated at 3 months, when edaravone treatment was commenced within 24 h of stroke onset [70].

1.2 Neuroprotective Compounds with a Singular Protective Function

Current knowledge regarding the pathophysiology of cerebral ischemia indicates multiple mechanisms that contribute to loss of cellular integrity and tissue destruction. These mechanisms include excitotoxicity, overproduction of free radicals, inflammation, and apoptosis [21]. Cerebral ischemia leads to metabolic stress, ionic perturbations, and a complex cascade of biochemical and molecular events ultimately causing death of neurons and capillaries. Numerous drugs have been examined in animal models of stroke, targeting all aspects of the secondary biochemical cascade of events; with the most promising drug candidates having been tested in human clinical trials [36]. Yet the failure of more than 50 promising neuroprotective compounds to progress beyond phase III of clinical trials has been attributed to many varying factors, including differences in the temporal and spatial progression of the ischemic injury between animal models and humans; the therapeutic time window; dosage; duration of therapy; and a limited ability to significantly modify the progression of the injury using drugs that target only a single mode of action [36]. The majority of the drugs designed to reduce damage following stroke have all specifically targeted one aspect of the ischemic cascade. Given that the pathophysiology of cerebral ischemia is multifaceted, it is possible to assume that neuroprotection will require multiple drugs in combination or a drug that possesses multiple neuroprotective properties [7].

1.3 Treatments that Target Reperfusion

The identification of deleterious cellular reactions that occur secondary to cerebral ischemia has led investigators to search for neuroprotection strategies to complement reperfusion. Despite the obvious need to reestablish blood flow to ischemic brain tissue, reperfusion itself leads to further injury if blood reflow is not initiated within the first 1–2 h [76]. Of the many biochemical events that occur in the brain during an ischemic episode, spontaneous or thrombolytic reperfusion provides oxygen as a substrate for numerous enzymatic oxidation reactions turned on during ischemia, several of which involve the excessive generation of toxic reactive oxygen and nitrogen species (RONS) such as superoxide anion and nitric oxide [76, 91]. Cell death occurs through multiple pathways when production of RONS exceeds elimination by antioxidant systems, or when these defense systems are damaged or impaired [17]. Indeed in humans, studies have shown that after acute ischemic stroke, endogenous antioxidant activity is lower than in control subjects [20]. If not scavenged, these highly reactive RONS modify macromolecules through breakdown of lipid membranes, oxidation of DNA, and denaturation of enzyme proteins, and interfere with signal transduction and initiate cell death programs in the form of apoptosis or necrosis within hours after stroke [12]. Therefore, it is not surprising that in recent studies we and others have found that delayed treatment with free radical-scavenging compounds is effective in experimental animal models of ischemic stroke [8, 89]. Moreover, novel free radical-scavenging compounds are reported to have multiple neuroprotective properties targeting more than one mechanism involved in the ischemic cascade. The arylalkylpiperazine compound AM-36 (1-(2-(4-chlorophenyl)-2-hydroxy)-ethyl-4-(3,5-bis(1,1-dimethylethyl)-4-hydroxyphenyl) methylpiperazine) has been shown to have Na^+ channel blocking activity, inhibit neuronal apoptosis, inhibit neuronal NOS (nNOS) and iNOS through sigma receptor activity, and reduce dopamine release and RONS formation following focal cerebral ischemia [9]. Recently, AM-36 administration has also been reported to modulate the neutrophil inflammatory response and reduce blood–brain barrier breakdown following middle cerebral artery occlusion (MCAo) of rats [118]. These effects of AM-36 have all been reported using the endothelin-1 (ET-1) model of ischemic stroke in rats and it is yet to be seen if AM-36 can produce the same or similar effects in other models of ischemic stroke.

1.4 STAIR Guidelines for Preclinical Drug Testing

Overall, the translation to positive outcomes in human clinical trials of drugs, which showed promise in preclinical evaluation using animal models, has been a litany of disappointment. For this reason, publications arising from the STAIR-I [106] and STAIR-II [107] conferences, a collaboration of academic and industry leaders, provided new guidelines for the preclinical and clinical testing of neuroprotective drugs for ischemic stroke treatment. Among the recommendations were: evaluation of combined therapies, dose–response effects, therapeutic window, and achieving reproducible treatment effects in both permanent and transient ischemic animal models that should be realized across several laboratories (Table 25.1). Importantly, identification of the target population is vital to ensure that the animal model used has relevance to both human stroke outcome as well as the treatment being investigated. To this effect, current research into the development of thrombolytics requires a model of stroke that incorporates introduction or induction of a thrombus or embolus to cause stroke.

AstraZeneca's nitrone drug, NXY-059 (Disulfenton, Cerivive™), was the first neuroprotective drug developed under strict adherence to the STAIR criteria before entering Phase III trials [26]. This nitrone compound with free radical-trapping properties generated considerable optimism after a successful SAINT I (Stroke-Acute Ischemic NXY Treatment I) trial. It was shown to be effective in neuroprotection after ischemic stroke when administered to rats and primates where the stroke model studied incorporated reperfusion [57]. Unfortunately, the second large-scale clinical trial (SAINT-II) failed, and despite following the STAIR criteria closely, NXY-059 antioxidant therapy did not achieve clinical success [26]. Several papers have

Table 25.1 Recommendations for preclinical stroke drug development *STAIR* [106, 107]

Evaluate an appropriate dose–response effect across multiple species in both small and large animal models

Establish the period of time after an ischemic event for therapeutic efficacy including the optimal duration of treatment

Both permanent and transient ischemic animal models should be studied in small and large animals that permit extended recovery

Appropriate physiological monitoring with at least two outcome measures considered: functional response and infarct volume. Studies should be blinded and randomized

Identify the target population to ensure that the animal model used has relevance to human stroke: treatment effects should be confirmed in both sexes, in aged animals with consideration to relevant disease such as diabetes and cardiovascular disease

Treatment effects should be replicated in several laboratories, and data both positive and negative should be published

The use of combined therapies that target more than one mechanism of neuronal injury should be considered particularly in relation to current TPA treatment

Careful toxicological studies in several species

Establish baseline information for every patient using functional scales so that each patient serves as his or her own control

Correct assignment of stroke patients to the appropriate trial

Recognition and assessment of spontaneous recovery

since been published to assess the failure of this promising drug, with unspecified methodological problems often quoted and reference to the "unacceptably short," shelf life of parenteral solutions of NXY-059, which hydrolyzes a few percent per year. However, foremost to this was the critical need for more rigorous testing of neuroprotective agents in animal models of stroke with attention to treatment windows and inclusion of diverse stroke outcomes [26].

Given the failures to translate results from animal models of stroke into clinical trials, it is now more important than ever to reconsider the available models and to use only those that mimic more closely the human condition. The most common type of stroke in humans is embolic, usually due to a clot or vascular stenosis from plaque formation that has become lodged within a major artery blocking blood supply to or within the brain itself, accounting for over 88% of cases. Although with a higher mortality outcome, vascular bleeds within the brain occur less frequently with intracerebral hemorrhage accounting for 9% of stroke cases, and the remaining 3% of clinical strokes are due to subarachnoid hemorrhage. While animal models of stroke have been described for intracerebral and subarachnoid hemorrhage, most animal models of stroke are based on ischemic stroke. For this reason, this chapter will focus on ischemic stroke models with discussion focusing on their strengths and limitations when assessing the potential neuroprotective actions of a class of drugs known as flavonols. Finally, based on the preclinical evaluation of flavonols to date, we suggest ideas as to how the positive outcomes observed with the use of flavonols in preclinical stroke models might indeed translate to clinical success.

2 Animal Models of Stroke

Successful preclinical assessment of any therapeutic regimen depends on the reliability and reproducibility of the experimental animal models used. It is now more important than ever to reevaluate the available animal models of ischemic stroke: to use those that mimic most closely the human condition and to use a model that incorporates a relevant clinical target for the treatment under investigation. Methods exist to model both permanent and transient (involving reperfusion) ischemia; however, since most human strokes involve some degree of reperfusion, models that incorporate this element are most relevant. The most widely modeled is occlusion of the middle cerebral artery (MCA) as this is commonly seen in clinical stroke.

When testing therapeutic compounds, a reproducible model with small variation in infarct size has been used in animal models for evaluation of treatment effects, which must be assessed with appropriate outcome measures. Infarct volume, determined histologically or more recently by CT or MRI, has long been the standard outcome measure for assessing the severity of stroke. However, in human stroke a larger infarct volume does not always indicate a worse outcome. Beyond survival, protection of cerebral function constitutes the principal goal. Clinical trials use broad functional outcomes at extended durations of months poststroke to measure success. This is filtering through to animal models where the importance of neurological and behavioral outcome measures is being realized and its emphasis increased in these studies. Most important is the need to extend preclinical assessments in animal models into long-term survival studies to ensure any neuroprotective effects observed short-term while testing compounds are not transient and also to enable investigation of endogenous repair mechanisms that may be enhanced with treatment.

There is no single ideal animal model of stroke as all models have advantages and disadvantages and the model, the species, and strain of animal used must be carefully selected to provide answers to the specific experimental question posed. Reviewed here is a selection of the most commonly used models of transient focal ischemic stroke.

2.1 Rodent Models of Stroke

Rodents are the most commonly used laboratory animals to model stroke. Their widespread availability and low cost, along with the ease with which they can be handled and minimal space requirements, make them ideally suited to the task. High numbers of rodents can be used without cost and time becoming prohibitive, and furthermore, aged as well as hypertensive rats can also be used with relative ease. The rat cerebral vasculature and physiology closely resemble those of humans, despite their small size. Additionally, the ability to genetically alter mice and their increasing availability have led to the widespread use of mice in stroke research, where numerous knockout and transgenic strains are highly applicable. Functional outcome can be measured in rodents using a variety of methods. Methods to assess behavior in mice were initially adapted from those used for rats; however, more specific tests are becoming widespread. Tests with higher sensitivity are also being validated as extended recovery times are being recognized as important, since rodents tend to show good recovery of function despite large infarcts in some cases.

2.1.1 Intraluminal Filament Middle Cerebral Artery Occlusion Model

The intraluminal filament model of intraluminal filament MCA occlusion (MCAo) is the most commonly used method for inducing stroke in rats and mice. It involves endovascular occlusion of the proximal MCA with a suture or filament via neck surgery in an anesthetized animal. The method was introduced by Koizumi et al. [53], who described insertion of the suture via the common carotid artery (CCA) in rats. A modification by Longa et al. [62] introduced insertion of the suture via the external carotid artery (ECA), allowing the CCA to remain patent. Briefly, a midline neck incision is made before isolation of the CCA and ECA. The ECA is transected and the suture advanced, via an incision in the ECA stump, through the internal carotid artery, into the Circle of Willis until it occludes the MCA at its origin. The suture is secured and left in place for a defined period of time (most commonly 60–180 min) before retraction to allow reperfusion. The suture can also be left in place, without retraction to model permanent occlusion of the MCA. A distinct advantage of this model is that vascular occlusion and reperfusion can be monitored during stroke using laser Doppler flowmetry over the affected cortex, since the animal is anesthetized during the procedure.

The application of intraluminal suture MCAo to the mouse translated to increasing interest in genetically modified strains in stroke studies, allowing for characterization of important ischemic cascades that contribute to the evolution of damage after stroke. Minor adaptations to the method were necessary, including changing the occluding filament diameter and length. However in both mice and rats, the resulting ischemic core remains the same, located in the striatum and the cortex which are affected with longer durations of occlusion [31]. Despite the reproducibility of this model of the intraluminal MCAo, a drawback is that branches of the internal carotid artery other than the MCA are often occluded, resulting in injury to areas outside the MCA territory not typically affected in human stroke [48]. As a result of this, the degree of ischemia and the infarct volume are variable (both between and within strains) and diverse structures including the hypothalamus, thalamus, and hippocampus can be affected. Damage to the hypothalamus due to interruption of blood supply from the hypothalamic artery has been linked to thermal dysregulation. Post-MCAo hyperthermia has been reported in the rat, dependent upon ischemic duration, but not rat strain nor the type of occluding filament used [122]. This complication has been shown to increase infarct size in the rat and influence results when testing for neuroprotection [33]. In contrast to the rat model, intraluminal MCAo in the mouse leads to hypothermia, possibly due to its large surface area/volume. This spontaneous hypothermia has been shown to reduce infarct volume and functional deficits in the mouse [3], highlighting the need for strict temperature control poststroke. Hence, changes in thermoregulation must be strictly monitored and taken into account when using this model in preclinical testing.

The variability in infarct size often noted in this model may be partially attributed to the use of different occluding sutures. Uncoated sutures have been reported to lead to incomplete occlusion and a high incidence of subarachnoid hemorrhage, a complication of the model resulting from perforation of the vessel wall [92]. Poly-L-lysine-coated sutures result in more successful MCAo and larger infarcts than uncoated sutures; however, their adhesive quality can lead to vessel rupture upon reperfusion due to tearing of the luminal surface [104]. Silicone-coated sutures have been reported to provide better occlusion and lower rates of subarachnoid hemorrhages than both uncoated and Poly-L-lysine-coated sutures [92, 104]. Matching the size of the suture to the body weight of the animal, or using animals within a narrow weight range also helps reduce infarct variability [2]. Damage to the endothelium caused by insertion of the occluding filament leading to thrombosis during or after MCAo and resulting in incomplete reperfusion has also been a concern with this model, leading to the administration of heparin in some studies [72].

A further complication of intraluminal MCAo is that the surgery involves transection of the ECA which supplies the muscles of mastication, sometimes leading to feeding difficulties and loss of body weight that can affect recovery and behavioral analysis [22]. In the mouse, visual dysfunction due to retinal injury caused by ophthalmic artery interruption can affect neurological and behavioral outcome measures [105]. Given that extended recovery times and neurological assessments are an important feature of a stroke model, high mortality rates in this model, particularly with longer ischemic duration, can make long-term studies difficult, especially in mice which develop poststroke infections due to immunodepression by this surgical procedure [83].

Despite the above complications associated with the intraluminal filament, this MCAo model can be conducted efficiently with high throughput and has the important advantage of precise control over the ischemic period. This is particularly important when investigating compounds that might give rise to complications associated with delayed treatment from stroke onset as those reported with tPA. Additionally, this method is minimally invasive and negates the need for craniectomy and resulting complications that were encountered with the use of earlier models developed using vascular clips on the MCA.

2.1.2 Thromboembolic Models

Thromboembolic models, through introduction or induction of an embolus or thrombus into the cerebrovasculature, can closely replicate the human stroke scenario. Thromboembolic stroke can be induced using methods that employ synthetic material to occlude a vessel or vessels and include both macrosphere [32] and microsphere [71] models. Although initial thromboembolic models mimicked the effects of a clot lodged within a major vessel of the brain, they did not incorporate reperfusion. However, the recent development of thromboembolic models that induce stroke by the introduction or induction of a nonsynthetic blood clot in the cerebrovasculature that can be thrombolyzed does allow for reperfusion. They also have the advantage of producing graded reperfusion, in contrast to the intraluminal filament model. The immediate occlusion and reperfusion that occur in intraluminal MCAo are not representative of most human stroke and surge reperfusion may influence pathophysiological cascades. Obviously, thromboembolic clot models are a methodological requirement for investigating thrombolytic therapies.

Thromboembolic Clot Occlusion of the MCA

Early models of thromboembolic stroke used a suspension of blood clot fragments injected into the carotid arteries of rats by the same surgical techniques as those described for intraluminal filament MCAo. These models were used to assess the effects of tPA; however, they suffered from variability in infarct size. The size of the clot fragments was not easy to control and fragments sometimes traveled to distal arterial branches and even to the hemisphere contralateral to injection [54]. To overcome this problem, models utilizing a single large clot or a specific number of slightly smaller clots were developed. As with intraluminal filament MCAo, embolic models were introduced in mice to take advantage of genetically modified strains and follow the same methods as those used for the rat.

Numerous methods for the preparation of in situ or preformed clots have been described. Preforming clots in vitro and allowing them to stabilize for several hours or overnight results in more durable clots. These clots are usually formed using autologous or homologous blood, allowed to clot spontaneously [125] or with the addition of thrombin [113]. After stabilization, the fibrin-rich segments of the clot can be selected for embolization and may be either injected whole or washed free of red blood cells, creating a "white" clot. Addition of exogenous thrombin induces clots with a higher density of fibrin meshwork designed to strengthen the clot and inflict more reproducible infarcts. However, these clots are also less physiologically relevant and impair the outcome of tPA treatment [73]. Clot preparation therefore influences how clinically relevant the model is. A study of mechanically retrieved thromboemboli from patients suffering ischemic stroke allowed characterization of fresh human clots [67]. The majority of these had a layered fibrin:platelet configuration with interspersed nucleated cells and few red blood cells. Fibrin-rich clots were rare, accounting for only 12% of collected clots. "Red" clots that contained a mass of red blood cells were also uncommon and thought to reflect postocclusion clot formation due to blood stasis.

Recently, a new model of thrombotic stroke in mice has been described by Orset et al. [79], which does not require arterial catheterization and thus circumvents bleeding complications. In this new model, intraluminal injection of thrombin into the MCA of anesthetized mice leads to in situ clot formation and ischemic infarction. A fibrin-rich thrombus is formed which results in a decrease in blood velocity for at least 60 min and which can be successfully lysed with tPA treatment. This model leads to accurate clot placement, reproducible infarct, and no mortality. It does, however, require craniectomy and the application of exogenous thrombin may affect processes outside of thrombus formation [73].

Despite the aforementioned improvements, inherent variation in thromboembolic models remains due to autolysis and clot placement, affecting infarct size and location. They can also result in high levels of subarachnoid hemorrhage, hemorrhagic transformation, and high mortality rates [127]. However, thromboembolic clot models remain the most relevant to human stroke and are essential for testing clot lysing and combination therapy involving lysis and neuroprotective strategies. Moreover, sustained functional deficits can be induced with thromboembolic stroke and long-term recovery can be assessed in both rats [127] and mice [126].

Photothrombotic Stroke

Photothrombotic stroke was first described in the rat by Watson et al. [117] as a model that closely mimics human stroke by activating endogenous thrombosis cascades. This technique is based on the interaction between a systemically administered photosensitive dye and light source directed at the region of the brain to be affected. Most widely used to induce small cortical lesions, photothrombosis can also be used to produce occlusion of individual vessels such as the MCA [14]. Unfortunately, these methods are relatively invasive and not as commonly employed and a cortical lesion induced by photothrombosis is virtually noninvasive. Irradiation with the appropriate wavelength light from arc lamp or laser immediately thereafter intravenous infusion of a fluorescein derivative leads to formation of singlet molecular oxygen which acutely damages the endothelium and results in the formation of platelet-rich thrombi throughout the cortical microvessels irradiated [117]. However, the role of these platelet-rich thrombi in producing the lesion induced by photochemical stroke has been questioned. It has been shown that mice with impaired platelet function or blocked coagulation cascades suffered the same level of damage following photothrombosis as controls [52]. Early BBB disruption and strong cytotoxic and vasogenic edema occur in this model which is not representative of human stroke and may contribute to lesion development in the absence of platelet function. It has therefore been suggested that the use of this model to study pathogenesis of ischemia-induced stroke, or for testing antithrombotic therapies, be limited [52].

Further criticisms of the model include prominent microvessel injury and induction of end-arterial thrombosis in the distal MCA and anterior carotid artery branches, vessels not generally involved in clinical stroke. Additionally, the lesion evolves quickly in a discreet region and involves at best only a small rim of what could be classed as penumbral tissue. This last point was addressed by the introduction of a photothrombotic "ring" model. By altering a laser light source to produce a thin ring of light, a subsequent ring of infarcted tissue is produced which encircles a region at risk that evolves to infarction over time, modeling the penumbral zone, albeit in reverse. Further modification of the laser beam irradiation intensity has resulted in late spontaneous reperfusion (at 72 h postirradiation) and tissue recovery in the inner "penumbral" region in rats [43] and mice [46].

Despite the aforementioned drawbacks, this technique is effective in both rats and mice and results in highly reproducible and defined infarction of the preselected cortical region under investigation. Unlike other stroke models, the size and position of the lesion are independent of variations in the cerebral vasculature, especially useful when using mice which are prone to such variations, impacting on lesion reproducibility [46]. The lesion size, location, and severity can be manipulated by changing the wavelength, intensity, beam shape, and duration of light exposure or dye concentration [117]. In mice, the lesion can be extended to underlying subcortical regions using a higher level of irradiation and in rats the striatum can be lesioned independently, although this does require more invasive surgery. Although this model has little clinical precedent, pathological hallmarks of ischemic stroke including necrotic and apoptotic cell death [44] and recovery mechanisms including angiogenesis [39] and neurogenesis [38] have been documented. Additionally, photochemical stroke induction does not impede long-term recovery and produces enduring sensorimotor deficits [61].

2.1.3 ET-1-Induced Stroke

ET-1 is a long-acting potent venous and arterial constrictor generated predominantly in endothelial cells. It exerts its effects through two classes of receptors, endothelin-A receptors (ETA-R) and endothelin-B receptors (ETB-R) [40]. ET-1 confers constrictor activity through ETA-R and, to a lesser extent, ETB-R, although dilatory actions through ETB-R have been reported that are linked to the production of NO and prostacyclin [40]. Despite this, the overall effect of ET-1 receptor activation is usually to increase vascular tone.

Direct application of ET-1 onto the exposed MCA was first described in anesthetized rats in 1990, resulting in marked reductions in cerebral blood flow beyond the threshold for ischemic damage in the dorsolateral caudate nucleus and sensorimotor cortex without effects in distant areas [88]. ET-1 has since been used to induce stroke by intraparenchymal injection or topical application in rats and in mice during anesthesia. For intraparenchymal administration, stereotaxic injections of ET-1 into predefined cortical or striatal regions are carried out through small burr holes in the skull. Injection of ET-1 into the subcortical white matter in rats or mice can also be used to model white matter ischemia, a type of stroke for which few appropriate models exist [101].

Topical application of ET-1 requires a craniectomy over the cortical region of interest and removal of the dura. Craniectomy exposes the brain to the atmosphere and can affect intracranial pressure, temperature, and blood–brain barrier function [77]. Despite this, the model offers a highly localized dose-dependent reduction in cerebral blood flow and a resulting infarct volume with a large region of penumbral tissue [30], making it a relevant model to investigate the ischemic penumbra and mechanisms of gradually developing ischemia. Reperfusion occurs gradually over several hours in these models, as opposed to the immediate response achieved with removal of an intraluminal filament, and for this reason, more closely mimics the human condition where clots can fragment or lyse either spontaneously or upon exposure to tPA. A further benefit of these methods is the ability to induce an accurately placed lesion, for example, in the forelimb region of the motor and somatosensory cortices, modeling upper extremity impairment in humans, a common functional deficit [34]. Lasting functional deficits are also observed in long-term recovery studies [108, 121]. Another advantage of the ET-1 model is that it does not damage mechanically the cerebral vasculature [68], which could increase adhesion molecule expression and artificially disrupt the blood–brain barrier. Intracortical ET-1-induced vasoconstriction has also been shown to evoke neurogenesis in the subventricular zone of mice in response to ischemic stroke damage similar to that reported using the intraluminal filament model [116].

Of particular concern in all experimental models of stroke discussed above is the use of general anesthesia during induction of stroke, for humans are not usually under anesthesia when a stroke occurs. It has been reported that the use of barbiturate and inhalational anesthetics can: (1) confound experimental findings due to their own neuroprotective effects [6], (2) affect reactive oxygen species (ROS) production [6], (3) cause long-lasting depression of protein synthesis [28], and (4) induce ischemic tolerance [49]. Although contradictory results exist regarding neuroprotective and neurotoxic effects of both inhalation and nonvolatile anesthetics, it is clear that all have substantial direct effects on neuronal death and survival [50]. Anesthetics may also alter cellular disturbances after stroke such as reducing neutrophil activation and passage into tissue [75], which has been shown to occur rapidly in the brain after human stroke [47]. Most obvious though is that anesthetics have the potential to interact with therapeutic agents in an unpredictable manner while altering physiological parameters: blood pressure, pH, blood glucose, and blood gases all critically determine lesion size after focal ischemic stroke [124]. Spontaneously breathing anesthetized rats exhibit altered blood gases and decreased blood pressure, resulting in increased infarct size [124]. Mechanical ventilation reduces the side effects associated with anesthesia [124]; however, it also inhibits the normal physiological response to stroke. In addition, vastly different anesthetic regimens are applied across different laboratories, making it difficult to compare the results of therapeutic interventions. A model of stroke that is induced in the absence of anesthesia, and which affords spontaneous reperfusion, is most desirable for clinical translation of drugs with neuroprotective potential.

ET-1 MCA Constriction in Conscious Rats

Human stroke incidence rarely occurs under anesthesia. The only model of stroke in experimental rodents that mimics the human scenario and avoids the complications associated with anesthesia is the ET-1 model. Sharkey et al. [95] first described stereotaxic injection of ET-1 adjacent to the MCA in conscious rats that resulted in constriction of the MCA followed by gradual reperfusion. Briefly, under anesthesia a guide cannula is stereotaxically implanted into the piriform cortex, adjacent to the MCA. After the cannula is secured, the animal is allowed to recover fully from the surgery and effects of anesthesia (usually a number of days). In the conscious rat, ET-1 is then delivered locally through the cannula to induce stroke [95] (Fig. 25.1).

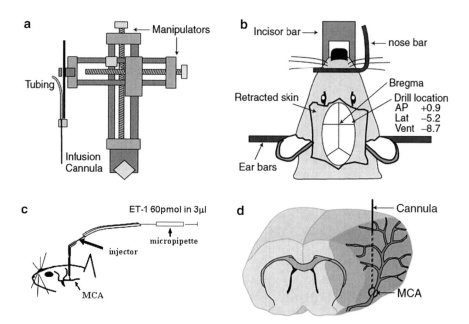

Fig. 25.1 The endothelin-1 model of middle cerebral artery occlusion. (**a**) The infusion cannula and the manipulator on the stereotaxic frame. (**b**) Rat head in stereotaxic frame—the bregma and site of cannula implant are marked. (**c**) Injector for infusion of ET-1 to induce stroke positioned in the guide cannula in a conscious rat free to move. (**d**) Diagram of brain showing areas of damage after infusion of ET-1 adjacent to middle cerebral artery. (Modified diagram from O'Neill and Clemens [78]—this material is reproduced with permission of John Wiley & Sons, Inc.)

Importantly, application of ET-1 to the MCA produced dose-dependent ischemic brain damage with volume of damage of 65 ± 34 mm^3 at 10^{-5} M compared to 0.22 ± 0.57 mm^3 for vehicle ($P < 0.01$) [63, 96]. Hence, the magnitude and duration of cerebral blood flow reduction and the resulting infarct size can be altered by changing the volume or concentration of ET-1 used. These data demonstrate that ET-1 is capable of reducing cerebral blood flow to pathologically low levels and provides a model of controlled focal ischemia that incorporates gradual reperfusion in conscious rats. Recently, concurrent observation of behavioral changes that occur during ET-1 stroke induction, which include symptomatic grooming, lifting the contralateral forepaw, and spontaneous circling, allows for the level of stroke intensity to be ranked that reliably correlates with resulting histological infarct volume and neurological deficits [89].

Prognostic clinical approaches such as the use of the Scandinavian Stroke Scale allow prediction of functional outcome and survival in stroke patients in order to support clinical management and to correctly stratify treatment groups in clinical trials [16]. The use of a similar approach in experimental rodents had not previously been attempted until recently [89]. During ET-1-induced stroke, counter-clockwise circling, clenching, and dragging of the contralateral forepaw are observed [97] and we have recently shown that stroke outcome can accurately be predicted based on these observed changes which correlate strongly with both histological damage and neurological deficits [89]. The use of such a predictive outcome model enables stratification of rats into treatment groups ensuring that stroke severity is evenly represented across all groups [89]. This is particularly important since human stroke sufferers are not preassigned to treatment groups prior to having a stroke, but rather patients are assessed after stroke using various neurological scales before assignment to treatment [16]. This poststroke stratification is rarely reported using other models of stroke for preclinical drug assessments since animals are randomized into treatment groups prior to stroke and behavioral changes cannot be observed during stroke induction if the animal is anesthetized. A further advantage of this model relates to the degree of variability in human stroke damage which can also be accounted for using the ET-1 model since all degrees of stroke severity can be assessed across the various treatment groups. This enables complete characterization of differential effects of the treatment on histological and functional outcome in rats with differing stroke severity [89].

One criticism of this model is that it does not produce total occlusion of the MCA, as seen in the human condition and modeled by the suture and embolic methods, but rather constricts the artery resulting in reduced cerebral blood flow. Therefore, it is not suitable for testing thrombolytic therapies. Surgical implantation of a guide cannula or needle also causes a discreet track of cortical damage, although this does not result in functional deficits [4]. Of some concern is the reported effect that endothelins act as signal mediators between neuronal and glial cells in response to nerve injury and raises the question as to whether topical ET-1 can evoke additional responses outside of changes in vascular tone that might play a role in inducing brain injury. However, while it is possible that the local cellular responses to injury and repair may be altered by ET-1 receptor-mediated neuronal or glial effects, application of ET-1 to neuron-enriched or glia-enriched cultures fails to induce cell toxicity [74]. Additionally, coadministration of a vasodilator with ET-1 has been found to block the generation of ischemic lesions [29]. It is also important to remember that only miniscule volumes of ET-1 are infused into the MCA territory in order to produce ischemic lesions, and recently, it has been shown that only localized immunoreactivity to ET-1 is found in the sensorimotor cortex surrounding the MCA of rats 30 min after application of 120 pmol in 1.5 μL saline [1]. Given that ET-1 is such a powerful vasoconstrictor agent, together with the above data, it is highly unlikely that the damage observed following ET-1 infusion into the MCA territory is due to activation of other brain cells by ET-1, but rather it is the direct result of MCA constriction that results in ischemic injury.

Unfortunately, successful use of ET-1 constriction of the MCA in conscious mice has not been documented. The ability of ET-1 to produce vasoconstriction in the rat is significantly greater than that in mice. This may be due to differences in expression of ETA-R and ETB-R on the MCA between the species; the mouse brain has double the number of the vasodilatory ETB-R compared with the vasoconstricting ETA-R [119]. While intraparenchymal injection of ET-1 has been successfully applied to mice, significantly less damage is seen with higher doses of ET-1 than those used in rats [108, 116]. For these reasons, the use of the conscious ET-1 model is restricted to nongenetically modified animals and mechanisms associated with the ischemic cascade investigated using genetically modified mice still require anesthesia for stroke induction.

The sustained reduction in cerebral blood flow and gradual reperfusion observed during ET-1-induced stroke more closely resemble human stroke than the intraluminal filament model where occlusion and reperfusion are immediate [5]. The low mortality rate and lack of postsurgical complications despite the presence of a considerable cortico-striatal infarct make the ET-1 model ideal for long-term recovery studies [4, 41]. Moreover, ET-1 application induces long-term sensorimotor deficits that are responsive to treatment [4] and this is particularly important to ensure that neuroprotective effects observed while testing compounds are not transient. It also enables investigation of endogenous repair mechanisms that may be enhanced with neuroprotective treatments.

2.2   Nonrodent Models of Stroke

Despite the many advantages of rodent models of stroke, there are obvious limitations to their applicability to humans. Anatomically, the human brain is gyrencephalic, exhibiting a complex structure of cortical gyri and sulci with intercalating grey and white matter. In contrast, the rodent cortex is smooth with a higher proportion of grey matter and clear demarcation between grey and white matter regions. Although early axonal responses to ischemia have been demonstrated in rat models [37], outcome measures predominantly focus on grey matter and neuronal perikarya when assessing neuroprotective agents in rodents. The attention to grey matter damage in rodent models may account for the failure of translation to positive outcomes in human clinical trials of neuroprotective drugs that showed promise from preclinical evaluation in rodents. In addition to gross anatomical differences between the rodent and human brain, physiological and metabolic differences also need to be taken into account which may influence translational significance of study results. Other species currently used to model ischemic stroke include rabbits, cats, dogs, and nonhuman primates that can address some of these differences.

2.2.1   Rabbit

The use of thromboembolic clot models, as described previously, using both small microclots and larger clots has been used extensively in rabbits to develop tPA treatment strategies [42]. Rabbits share the lissencephalic brain structure of rats and mice; however, vascular abnormalities including MCA duplication may be encountered and thus the use of the thread occlusion model is problematic. The use of the rabbit thromboembolic clot models has been proposed as an effective translational tool for incrementally advancing clinical development of potential therapeutics after rodent studies [56]. The rabbit small clot embolic stroke model is produced by the injection of blood clots into the cerebral vasculature, resulting in multiple infarcts in cortical and subcortical brain structures including the hippocampus, thalamus, and caudate putamen [58, 59]. Stroke may be induced in the conscious animal which does not require artificial respiration or other external support during clot infusion. A wide range of clot doses are injected in order to generate both normal and abnormal animals. Behavioral deficits are observed using a dichotomous rating system based on a range of behavioral changes during and after stroke such as circling, loss of balance, and loss of limb/facial sensation, similar to those described in rodent models. Analysis for potential neuroprotectants relies on identifying the quantity of microclots (mg) that produce neurological dysfunction in 50% of a group of animals (P_{50}), with successful intervention considered if the P_{50} is increased beyond controls. One of the major advantages of this model is the inclusion of the percentage of dead animals in all groups which may be particularly important when designing preclinical stroke studies since death is an outcome measured in human studies [56]. In addition, this model may also be effective for quantifying therapeutic windows during translational research to determine if compounds should be further developed or whether they should be abandoned.

2.2.2 Nonhuman Primate

Nonhuman primate models of stroke provide the most clinically relevant platform to investigate ischemic stroke. The nonhuman primate brain is closer to the human anatomically and has an analogous cerebrovasculature and more similar hemostatic components than other species [27]. In 1999, the STAIR group recommended that potential neuroprotective compounds be tested in nonhuman primates before progressing to clinical trial [106]. Despite this recommendation, nonhuman primate studies remain uncommon. The models are costly, labor intensive, require specialist facilities and surgical expertise, and provoke greater public animal welfare concerns than the use of lower order animals. Primate models should therefore only be used to evaluate neuroprotective agents when positive results in rodent models indicate potential advancement to clinical trials.

A standardized model of stroke in nonhuman primates is lacking; however, permanent or transient surgical occlusion of the MCA leading to focal stroke can be induced in a variety of ways in both the anesthetized and awake primate [27]. Appropriate development of such models requires neurosurgical expertise to produce single or multiple vascular occlusions that often involve intracranial approaches to access the proximal portion of the MCA with subsequent temporary application of surgical clips, ligatures, or inflation of cuffs in large, old world monkeys (baboons and macaques) [27]. A commonly used approach with clinical relevance is the transorbital surgical approach to the MCA that allows rapid recovery of the animal with full function with approximately 40–60% of animals displaying both cortical and subcortical injuries [103]. This model incorporates an implantable, inflatable balloon cuff surgically attached to the MCA. Postoperative recovery occurs in a timely fashion, and once awake and fully active, the balloon is inflated in the unanesthetized baboon resulting in consistent, reliable occlusion of the MCA that demonstrates a "recruitment response" of increasingly persistent deficit with repeated occlusion [102]. This model allows for controlled evaluation of the effects of stroke using a device that is self-contained requiring no maintenance after placement. Most importantly, the animals are not constrained and stroke can be induced in the awake, unanesthetized state which is more applicable to the human condition.

There are limitations associated with transorbital approaches to stroke in nonhuman primates including the requirement of experienced neurosurgical teams and invasive techniques that require intensive care unit management within the postoperative period. Removal of the eye for access to the MCA can also confound functional measurements for identifying visuospatial neglect, a key impairment associated with human stroke. Extracranial endeavors to induce autologous clot embolism techniques are in development that involve injection of a large, single nonfragmented blood clot into the left internal carotid artery of macaques that results in robust occlusion of the proximal MCA leading to contralateral motor and sensory deficits that correlate with 24 h endpoint infarct volumes [51, 55]. This procedure is hindered, however, by the lack of spontaneous recanalization observed with this model and the high degree of postoperative care and mortality making long-term assessment beyond 24 h problematic.

2.2.3 New World Monkeys

To avoid the costly, labor-intensive, and animal welfare concerns associated with the use of nonhuman primates for experimental stroke, alternative approaches using small, laboratory-bred common marmosets have been explored. The marmoset is small and relatively easy to handle, and despite having a nongyrencephalic brain, the marmaoset has a brain size up to eight times, with a greater white matter to grey matter ratio than that of rodents. The cerebrovascular anatomy, neuroanatomy, and behavioral repertoire are viewed to be more advanced than rodents and from a functional perspective, marmosets offer the advantage for assessment of hand movement and hand-to-eye coordination. Therefore, this places the marmoset in an ideal intermediate position between rodents and humans when assessing stroke pathophysiology, functional outcome, and the efficacy of novel neuroprotectants.

The use of the intravascular approach by thread occlusion of the MCA for both temporary and permanent occlusion has been investigated in the common marmoset. Through the use of multiple analyses with MRI, behavioral tests, histology, and immunochemistry, intraluminal occlusion of the MCA results in widespread brain damage and long-lasting functional deficits that are reduced by reperfusion. This makes this model suitable to test new therapies against stroke. However, use of this model requires thorough examination of the cerebral vasculature prior to surgical intervention to consider the highly tortuous and variable kinking of the internal carotid artery seen in higher primates. For this reason, topical administration of ET-1 to the proximal or distal branch of the MCA has also been investigated in the common marmoset, which results in a robust infarct to the MCA territory that correlates with contralateral motor and sensory/neglect impairments [111]. The development of the ET-1 model in the marmoset is yet to progress to conscious animals, but initial studies to induce stroke with ET-1 suggest that this model also has the potential to assess long-term function and efficacy of novel therapeutic strategies for clinical stroke. Additionally, the recent success in generating transgenic marmosets has reinforced the value of using this species in modeling human diseases [90].

2.3   Animal Models of Stroke for Neuroprotection Studies

There is not one definitive answer as to why the neuroprotective agents that appear to work in animals fail in human trials. One important aspect should be considered: animal studies are conducted to obtain a consistent homogenous result and thus comparisons between animals are easily made. Frequently, human clinical trials often include all types of stroke with varying degrees of damage and functional loss, whereas the animal studies usually assess one particular type of stroke, with small variations in lesion size and absence of long-term functional assessment. Ideally, neuroprotective drug strategies could be more effectively translated from animal to human trials if they were able to demonstrate their effectiveness in varying degrees of stroke damage, in two different animal species, and in both transient and permanent focal ischemia, to include both long term histological and functional outcome measures, with both immediate and delayed administration times after stroke [106]. To this extent, the quest for developing future neuroprotective strategies in animal models for treatment of human stroke is not futile if all the above parameters are taken into consideration prior to human clinical trial.

3   Flavonols for Future Stroke Therapy

In spite of the many lines of evidence linking oxidative stress to the clinical development of brain injury following stroke, treatment with external antioxidants to regain oxidative equilibrium and to control the evolution of injury has not resulted in successful clinical management of ischemic stroke in human trials [26]. Extensive experimental and epidemiological evidence suggests that intake of food rich in antioxidants is associated with improvement of vascular endothelial function and better outcomes against cardiovascular diseases. In particular, polyphenol compounds which are abundant in the human diet have created much interest in recent years for their presumed role in modulating intracellular signals that promote cellular survival. The most important polyphenol compounds are the flavonoids (both flavonols and flavanols) which are a large group of plant-derived secondary metabolites that give rise to pigmentation for flower coloration and yellow or red/blue pigmentation in fruits and vegetables. The main groups of flavonoids are (1) flavonols which are found in onions, leeks, and broccoli; (2) flavones which are found in parsley and

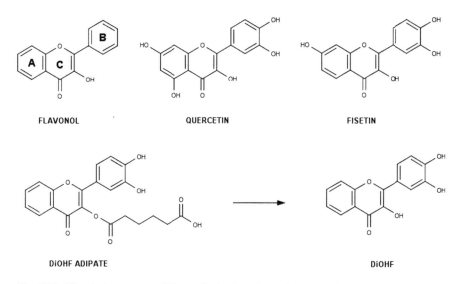

Fig. 25.2 Chemical structures of flavonols that have been shown to achieve neuroprotection in animal models of focal stroke

celery; (3) isoflavones, which are found in soy products; (4) flavanones which are found in citrus fruit and tomatoes; (5) flavanols which are abundant in green tea, red wine, and chocolate; and (6) anthocyanidins whose sources include red wine and berries [81]. Flavonoids are characterized by two benzene rings (A and B) which are connected by an oxygen-containing pyran ring (C) (Fig. 25.2). The three rings are planar and the molecule is relatively polarized. Three intermolecular hydrogen bonds are observed: two with the carbonyl group and the other between the hydroxyl groups in ring B. Within each group, there exists variation in the number and arrangement of hydroxyl groups, as well as the nature and extent of alkylation and/ or glycosylation of these groups.

Because of their wide distribution and relative low toxicity compared to other active plant compounds, flavonoids are ingested in significant quantities in the human diet. Due to associations between cultural dietary patterns and reduced risk of cardiovascular disease, naturally occurring compounds such as flavonoids have been widely researched. Consumption of fruits, red wine, and vegetables associated with a Mediterranean diet for example has been shown to increase the total antioxidant capacity of blood [19]. It is now known that flavonoids exhibit multiple biological effects, including lowering plasma levels of low-density lipoproteins, inhibiting platelet aggregation, reducing cell proliferation, altering allergens and inflammation, and importantly, scavenging free radicals [69]. Additionally, by targeting proinflammatory cell signaling pathways, flavonoids have been widely explored for their anti-cancer activity [82].

Most of the known actions of flavonoids are related to their antioxidant properties including suppression of ROS formation, scavenging of ROS, and up-regulation of antioxidant defenses. For these reasons, the use of flavonoids to treat ischemic stroke has received much attention in recent years with several studies across multiple animal models reviving hope for the translation of neuroprotection seen in animal models to clinical success. In particular, compounds that feature 3′,4′-dihydroxycatechol structure in the B ring, presence of 2,3 unsaturation, together with an oxo function in position 4 of the C ring, offer increased cytoprotection with potent antioxidant activity in both in vitro and in vivo studies [19, 84]. As a general rule, as the number of hydroxyl groups increases, flavonoids have a tendency to show better scavenging effects, with the addition that two neighboring hydroxyl groups have even more potent effects [45]. For this reason, we will focus on flavonoids that confer the above-mentioned structural characteristics for discussion, highlighting three potential flavonols that have recently been studied for preclinical evaluation using in vivo models of stroke (Fig. 25.2).

3.1   Quercetin

One of the original bioflavonoids investigated for treatment of stroke was quercetin, 2-(3,4-dihydroxyphenyl)-3,5,7-trihydroxy-4H-chromen-4-one. It is one of the most common flavonoids ubiquitously found in plants particularly tea, onions, and apple skins, and following ingestion of 100 mg quercetin, it has a half life range within plasma of 31–50 h [13]. Dietary quercetin is mostly present as its glycoside form in which one or more sugar groups are bound to phenolic groups by glycosidic linkage with its water solubility increasing with increasing number of sugar groups. The physical properties of quercetin are determined by its chemical structure which is characteristic of an antioxidant: an ortho-dihydroxy or catechol group in ring B; a 2, 3-double bond; as well as 3- and 5-OH groups with the 4-oxo group [100]. Dietary quercetin absorbed through the intestinal lumen is converted to conjugated metabolites such as quercetin 3′-O-β-D-glucuronide and quercetin 4′-O-β-D-glucuronide, both of which possess considerable activity with long half-lives [13]. For these reasons, quercetin is considered an excellent candidate as an in vivo antioxidant.

The first studies to investigate the effects of quercetin in ischemic reperfusion injury in the rat brain utilized the four vessel occlusion model of global ischemia [99]. It was reported that quercetin (5 mg/kg i.v.), by scavenging superoxide during reperfusion after a period of global forebrain ischemia, partly restored the fall in nitric oxide levels [99]. From these studies, it was proposed that this natural flavonoid could influence the balance of nitric oxide and superoxide that ultimately lowered peroxynitrite concentrations, indicating a potentially protective role of quercetin in various cerebral ischemias. However, the study failed to report any direct effects on neuronal loss in this model of ischemia.

Initial studies investigating the effects of quercetin on neurodegeneration were not encouraging. Using the 6-hydroxy-dopamine lesion model of Parkinson's disease, the in vitro antioxidant activity of several flavonoids was assessed and compared with their in vivo activity [18]. Despite its potent antioxidant activity in vitro, intraperitoneal administration of quercetin failed to protect substantia nigra neurons in vivo, which was attributed to a limited ability of quercetin to cross the blood–brain barrier. Nevertheless, the same group later confirmed the neuroprotective capacity of quercetin following oxidative insult with hydrogen peroxide in PC12 cells in culture [19]. To increase transfer across the blood–brain barrier, quercetin was combined with lecithin, generating a liposomal preparation for transport across lipid membranes. Given 30 min after permanent MCA vessel occlusion using a monofilament suture in rats, a single intraperitoneal dose of lecithin/quercetin preparation (30 mg/kg) resulted in a 56% reduction in lesion volume 24 h after occlusion [19]. In contrast to the lack of effects in the oxidative lesion of experimental Parkinsonism, the results following permanent stroke in rats demonstrated the importance of drug delivery to ensure access to brain nuclei existing within the blood–brain barrier. This was later confirmed when aqueous solutions of quercetin with varying concentrations were not detected by HPLC in the brain up to 4 h after IP administration [85]. In contrast, quercetin was detected by HPLC in the brain only when administered as a liposomal preparation, reaching highest concentrations 1 h after IP administration that was again effective in attenuating neuronal injury following permanent MCAo in rats [85]. Although these data confirmed the potential for the use of quercetin in stroke recovery, it was likely that the interaction of quercetin with lecithin increased the possibilities to provide protection, and it remained to be seen if quercetin could offer neuroprotection following stroke with reperfusion and an extended recovery past 24 h. This is most important given the dramatic increases in free radicals generated during recanalization in the days after an ischemic event in the brain.

In contrast to the above reports suggesting a limited ability for quercetin to cross the blood–brain barrier, neuroprotective effects have been demonstrated with quercetin in a variety of CNS injury models showing improvement of morphology and functional outcomes without the use of liposomal preparations [15, 93, 94]. Furthermore, Youdim et al. [123] demonstrated that quercetin can pass the blood–brain barrier using an in situ rat model where a measurable rate of quercetin uptake into the right cerebral hemisphere of 0.019 mL min^{-1} was reported. This was also demonstrated using an in vitro model of blood–brain barrier permeability where quercetin was shown to pass across RBE-4 cells (rat cerebral capillary endothelial cells) in a time-dependant manner that was not influenced by the presence of 0.1% ethanol [24]. In support of these findings, quercetin (50 mg/kg) dissolved in saline and administered IP 30 min prior to global ischemia in mice and again immediately after reperfusion significantly decreased delayed neuronal damage over a 72 h recovery period demonstrating that quercetin alone could indeed prevent neuronal loss associated with reperfusion injury [15].

In global ischemia, damage results in selective neuronal injury typically in the hippocampus and striatum where cell death is delayed for 3–5 days after ischemic insult. Various manifestations of fundamental neuronal death pathways have been characterized in this model including excitotoxicity, free radical stress, and apoptotic-like mechanisms that are not dissimilar to those which occur following focal ischemia. It is therefore not surprising that quercetin (25 μmol/kg in saline IP) was later shown to attenuate ischemic reperfusion injury following transient focal ischemia induced by photothrombosis, where treatment was delayed, commencing 1 h after injury with continued treatment at 12 hourly intervals over 3 days [60]. This delayed treatment window has since been confirmed where quercetin, given 2 and 4 h after reperfusion in a rat MCAo intraluminal filament model, resulted in marked reduction in cerebral infarct by 48 and 25%, respectively [80]. Most importantly, however, was that both the above-mentioned studies reported for the first time that delayed treatment with quercetin significantly improved functional recovery in both neurological deficit score [80] and performance on the accelerating rota rod [60]. The positive effects of quercetin on functional recovery have also been reported in other models of CNS injury. Administration of 25 μmol/kg quercetin aglycone twice daily, commencing 1 h after injury, resulted in significantly improved recovery of motor function in paraplegic animals following acute spinal cord injury [93].

While most of the physiological benefits of flavonoids are generally thought to be due to their antioxidant and free radical-scavenging effects, quercetin and structurally related molecules increase cell survival where other nonrelated antioxidants fail to have effect [19]. This would suggest that compounds such as quercetin have additional cytoprotective properties that may be linked to their ability to activate intracellular molecules that in turn mediate intracellular events that promote cell survival. Emerging evidence now supports this hypothesis and the mechanism of action for flavanoids such as quercetin might go well beyond free radical-scavenging: quercetin increases endogenous brain glutathione levels following permanent focal stroke in rats, boosting the brains defense resulting in reduced lesion volumes [86]; quercetin markedly reduced ischemia-induced up-regulation of metalloproteinase 9 (MMP-9) and MMP-9 activity [15, 60] as well as postischemic blood–brain barrier permeability and brain edema in a rat model of photothrombosis [60]; emerging evidence suggests that astrocytes are potential targets for the actions of flavonoids and quercetin has been shown to inhibit synthesis of heat shock protein and glial fibrillary acid protein [35]; quercetin is also reported to be an effective Fe^{2+} chelator and able to prevent the Fenton reaction thereby preventing iron-mediated lipid peroxidation [25].

Given the reported actions of quercetin following stroke, together with its well known anti-platelet aggregation and anti-inflammatory activity, this common flavonoid shows great promise as a new drug for treatment of stroke with multiple neuroprotective properties. Additionally, preclinical studies in animal models using quercetin have so far addressed essential criteria set down by STAIR [106, 107]: effective in varying degrees of stroke damage; effective in more than one animal species in both transient and permanent focal ischemia; effective in reducing histological and functional outcome measures; effective with both immediate and delayed administration times after stroke. To date, reports regarding the effects of quercetin following ischemic reperfusion injury in the brain have been limited to anesthetized rodent stroke models and it remains to be seen if the effects of quercetin can be observed in conscious stroke models as well as in nonrodent models that incorporate long-term survival studies.

3.2   Fisetin

Similar to quercetin, fisetin (3,7,3',4'-tetrahydroxy flavone) is also a naturally occurring flavonoid commonly found in strawberries and other fruits and vegetables. Its structural characteristic is similar to quercetin apart from the absence of one hydroxyl group at the 5' position on the A ring (Fig. 25.2). Following both oral and intravenous administration, the levels of free fisetin in serum decline rapidly while the levels of sulfated/glucuronidated fisetin increase which also show significant antioxidant activity [98]. Furthermore, circulating flavonids sulfates/glucuronides can be cleaved to the free form in a tissue-specific manner if there is a local release of β-glucouronidases and/or sulfatases.

Studies have demonstrated that fisetin exhibits a wide variety of activities in addition to its direct antioxidant activity. Fisetin can increase intracellular glutathione levels; has anti-inflammatory activity against microglial cells; inhibits 5-lipoxygenase and thus reduces lipid peroxides and their proinflammatory by-products; can act as a neurotrophic factor promoting nerve cell survival by enhanced proteasome activity; and several studies have shown that fisetin can enhance cognitive function, in particular facilitate long-term learning and memory [64, 65]. Treatment with fisetin markedly suppresses the production of tumor necrosis factor-α, NO, and prostaglandin E_2 in lipopolysaccharide-stimulated BV-2 microglia cells, all of which are important mediators of injury following stroke [128]. Together with the above actions, fisetin would make an excellent candidate for treatment of brain injury, yet only a small number of studies have examined its effects in vivo following stroke.

One of the few reports to investigate the use of fisetin in recovery from stroke in vivo also compared the effects of several flavonoids, including quercetin, all of which were administered in liposomal preparations 30 min after permanent MCA occlusion in rats using the filament model [85]. Where the flavonoid, catechin, failed to have any effect on stroke in comparison to quercetin, fisetin (50 mg/kg) significantly reduced histological damage and infarct volume to levels equivalent to that reported for quercetin [85]. Interestingly, the same group reported that brain concentrations of fisetin remained higher for longer periods. This may be attributed to prolonged serum levels of fisetin sulfate/glucuronides in comparison to other flavonoids [98]. Alternatively, fisetin has also been shown to exhibit a higher brain uptake potential in an in vitro model of blood–brain barrier penetration [115].

In support of the above findings, benefits of fisetin following stroke have also been reported using the rabbit small clot model where a range of flavonoids underwent prescreening in vitro prior to in vivo investigations [66]. Here fisetin was reported to be the most potent flavonoid for protecting HT22 cells from chemical-induced toxicity by increasing glutathione levels, ATP generation, and promoting long-term induction of antioxidant response element-specific transcription factor and HO-1, a phase II detoxification enzyme. Following small clot embolic stroke in the rabbits, fisetin (50 mg/kg) dissolved in 90% β-hydroxypropyl cyclodextrin in PBS containing 10% dimethylsulfoxide, given 5 min after embolization, significantly reduced stroke-induced behavioral deficits and doubled the P_{50} value in comparison to vehicle-treated rabbits, with an increase in the number of behaviorally normal animals in the fisetin group [66]. These investigations highlighted the importance of prescreening compounds and generating data prior to in vivo testing that further validated the neuroprotective activity observed. Additionally, given that death is one of the outcome measures included in the small clot embolic stroke model, the fact that ficetin resulted in a greater survival rate also further supports its clinical potential. To date, however, fisetin has only been reported to have beneficial effects when administered up to 30 min after stroke and this treatment window must be extended in order to truly show benefit for the vast majority of stroke sufferers.

3.3   3′,4′-Dihydroxyflavonol

3,3′,4′-Trihydroxyflavone, more commonly known as 3′,4′-dihydroxyflavonol (DiOHF), is a synthetic flavonol that lacks substitution on the A phenyl ring (Fig. 25.2). DiOHF has structural features that are reported to maximize its radical-scavenging activity: a catechol group in the B ring, a 3-OH group on the C ring, and a C2–C3 double bond. The actions of DiOHF have been studied most extensively in the vascular system. Many reports support its potent vasorelaxant activity through scavenging of superoxide anions to increase the bioavailability of nitric oxide, effects that were significantly more potent than a number of naturally occurring flavonols such as quercetin and fisetin [10]. Similarly, DiOHF has been shown to be a potent inhibitor of superoxide radical generation in the presence of both the free radical generating enzymes, xanthine oxidase and NADPH oxidase [11]. In the same study, the authors also reported for the first time that DiOHF might also be important in preventing ischemic reperfusion injury using a model of rat-perfused hindquarter injury: DiOHF given prior to the onset of ischemia reduced the degree of impairment and vascular injury following reperfusion [11]. Following these studies, DiOHF was then reported to be effective in reducing infarct size following myocardical ischemia and reperfusion injury in anesthetized sheep through attenuation of superoxide production in the postischemic myocardium and by improving coronary blood flow and preserving nitric oxide metabolites in venous outflow from the ischemic zone [114].

Based on the above results, we therefore set out to assess the neuroprotective potential of DiOHF in the ET-1 rat model of stroke using behavioral responses to stratify rats into groups based on stroke severity prior to assessment of recovery of neurological functions and morphologically mapped brain damage [89]. We found that delayed treatment with DiOHF (10 mg/kg i.v. given 3, 24 and 48 h poststroke) significantly reduced infarct volume and restored neurological function in rats with modest stroke ratings ($P < 0.01$), but not in rats with high stroke ratings. An important finding from this study was that although delayed treatment with DiOHF reduced damage in the cortex across all stroke severities, this was not the case in the striatum and there was no significant effect on neurological outcomes where all stroke severities were grouped together. However, in a subgroup of rats with moderate stroke severity, DiOHF markedly reduced damage to the striatum, as well as cortex, and importantly this effect translated to improvement in neurological function. Hence, by using a reliable rating scale, such as that described for the ET-1 model in conscious rats [89], rats can be stratified into groups of mild, moderate, or severe stroke, allowing potential neuroprotective agents to be reliably tested across all stroke severities, at the same time enabling further subgroup assessment. This has particular importance when assessing studies in humans, since the severity of stroke can have a marked impact on the efficacy of treatment [109]. Often large-scale clinical trials have not accounted for this and although potential treatments might be effective in smaller to moderate strokes, their use in the clinic is lost due to an overall failure to reach significance when all subjects are included in the final analysis of effect.

3.3.1 3′,4′-Dihydroxyflavonol-3-Hemiadipate

Poor aqueous solubility of DiOHF required it to be injected in a solvent mixture containing 20% dimethylsulfoxide, 40% polyethyleneglycol, and 40% water [89], which would not be suitable for rapid intravenous injection into patients with an acute stroke. We therefore applied a prodrug strategy by introducing an ionizable hemiadipate group on the hydroxyl group at position 3 of ring C to form a compound 3′,4′-dihydroxyflavonol-3-hemiadipate (diOHF-adipate) (Fig. 25.2) that was soluble at millimolar concentrations in isotonic sodium carbonate solutions [120]. This soluble compound was shown to be hydrolyzed back to DiOHF within 2 min by esterases in tissue and blood upon injection into laboratory animals as well as hydrolyzed in vitro by solutions of butyrylcholinesterase. Furthermore, parenteral injection of this prodrug gave significant cardioprotection to sheep subjected to surgically induced myocardial ischemia/reperfusion [120] and also neuroprotection in the conscious rat ET-1 focal stroke model as judged by behavioral and morphological criteria (Roulston, unpublished data).

4 Conclusion

Flavonols, in addition to their potent antioxidant activity, possess multiple neuro-protective properties, and even when administered hours after an ischemic stroke, provide significant levels of protection. The naturally derived and synthetically generated polyphenolic compounds have the potential to be administered before or together with the current thrombolytic treatment strategy. However, despite the promising studies reported to date, few of the above-mentioned flavonols have been studied extensively enough in animal models of stroke to warrant progression to clinical trial. If further studies can include investigation into dose–response effects, with results replicated in multiple animal models to incorporate both conscious stroke studies and higher order species, as well as studying the effect in combination with current tPA strategies, there is still promise for achieving neuroprotection and recovery from ischemic stroke for many more sufferers. Most importantly, all investigations in the development of potential neuroprotective agents must incorporate studies that include longer survival times after stroke in order to mimic more closely the human scenario and to ensure that any positive effects observed are not merely transient. Thus, any effects on endogenous recovery mechanisms associated with brain repair are yet to be explored using many neuroprotective agents that target the hours after stroke, an effect that may be vital to the future of stroke treatment where neuroprotection strategies might precede cell-based therapies that promote regeneration.

References

1. Adkins DL, Voorhies AC, Jones TA. Behavioral and neuroplastic effects of focal endothelin-1 induced sensorimotor cortex lesions. Neuroscience. 2004;128:473–86.
2. Ardehali MR, Rondouin G. Microsurgical intraluminal middle cerebral artery occlusion model in rodents. Acta Neurol Scand. 2003;107:267–75.
3. Barber PA, Hoyte L, Colbourne F, Buchan AM. Temperature-regulated model of focal ischemia in the mouse: a study with histopathological and behavioral outcomes. Stroke. 2004;35:1720–5.
4. Biernaskie J, Corbett D. Enriched rehabilitative training promotes improved forelimb motor function and enhanced dendritic growth after focal ischemic injury. J Neurosci. 2001;21:5272–80.
5. Biernaskie J, Corbett D, Peeling J, Wells J, Lei H. A serial MR study of cerebral blood flow changes and lesion development following endothelin-1-induced ischemia in rats. Magn Reson Med. 2001;46:827–30.
6. Bhardwaj A, Castro AFI, Alkayed NJ, Hurn PD, Kirsch JR. Anesthetic choice of halothane versus propofol: impact on experimental perioperative stroke. Stroke. 2001;32:1920–5.
7. Callaway JK. Acute stroke therapy: combination drugs and multifunctional neuroprotectants. Curr Neuropharmacol. 2004;2:277–94.
8. Callaway JK, Knight MJ, Watkins DJ, Beart PM, Jarrott B. Delayed treatment with AM-36, a novel neuroprotective agent, reduces neuronal damage after endothelin-1-induced middle cerebral artery occlusion in conscious rats. Stroke. 1999;30:2704–12.
9. Callaway JK, Castillo-Melendez M, Giardina SF, Krstew EK, Beart PM, Jarrott B. Sodium channel blocking activity of AM-36 and sipatrigine (BW619C89): in vitro and in vivo evidence. Neuropharmacology. 2004;47:146–55.
10. Chan EC, Pannangpetch P, Woodman OL. Relaxation to flavones and flavonols in rat isolated thoracic aorta: mechanism of action and structure-activity relationships. J Cardiovasc Pharmacol. 2000;35:326–33.
11. Chan EC, Drummond GR, Woodman OL. 3′,4′-Dihydroxyflavonol enhances nitric oxide bioavailability and improves vascular function after ischemia and reperfusion injury in the rat. J Cardiovasc Pharmacol. 2003;42:727–35.
12. Chan PH. Reactive oxygen radicals in signaling and damage in the ischemic brain. J Cereb Blood Flow Metab. 2001;21:2–14.
13. Chen C, Zhou J, Ji C. Quercetin: a potential drug to reverse multidrug resistance. Life Sci. 2010;87:333–8.
14. Chen F, Suzuki Y, Nagai N, Jin L, Yu J, Wang H, Marchal G, Ni Y. Rodent stroke induced by photochemical occlusion of proximal middle cerebral artery: evolution monitored with MR imaging and histopathology. Eur J Radiol. 2007;63:68–75.
15. Cho JY, Kim IS, Jang YH, Kim AR, Lee SR. Protective effect of quercetin, a natural flavonoid against neuronal damage after transient global cerebral ischemia. Neurosci Lett. 2006; 40:330–5.

16. Counsell C, Dennis M, McDowall M, Warlow C. Predicting outcome after acute and subacute stroke: development and validation of new prognostic models. Stroke. 2002;33:1041–7.

17. Crack PJ, Taylor JM, Flentjar NJ, de Haan J, Hertzog P, Iannello RC, Kola I. Increased infarct size and exacerbated apoptosis in the glutathione peroxidase-1 (Gpx-1) knockout mouse brain in response to ischemia/reperfusion injury. J Neurochem. 2001;78:1389–99.

18. Dajas F, Costa G, Abin-Carriquiry JA, Echeverry C, Martínez-Borges A, Dajas-Bailador F. Antioxidant and cholinergic neuroprotective mechanisms in experimental parkinsonism. Funct Neurol. 2002;17:37–44.

19. Dajas F, Rivera F, Blasina F, Arredondo F, Echeverry C, Lafon L, Morquio A, Heizen H. Cell culture protection and in vivo neuroprotective capacity of flavonoids. Neurotox Res. 2003;5:342–425.

20. Demirkaya S, Topcuoglu MA, Aydin A, Ulas UH, Isimer AI, Vural O. Malondialdehyde, glutathione peroxidase and superoxide dismutase in peripheral blood erythrocytes of patients with acute cerebral ischemia. Eur J Neurol. 2001;8:43–51.

21. Dirnagl U, Iadecola C, Moskowitz MA. Pathobiology of ischaemic stroke: an integrated view. Trends Neurosci. 1999;22:391–7.

22. Dittmar M, Spruss T, Schuierer G, Horn M. External carotid artery territory ischemia impairs outcome in the endovascular filament model of middle cerebral artery occlusion in rats. Stroke. 2003;34:2252–7.

23. Donnan G, Davis S. Breaking the 3 h barrier for treatment of acute ischaemic stroke. Lancet Neurol. 2008;7:981–2.

24. Faria A, Pestana D, Teixeira D, Azevedo J, De Freitas V, Mateus N, Calhau C. Flavonoid transport across RBE4 cells: a blood–brain barrier model. Cell Mol Biol Lett. 2010;15:234–41.

25. Ferrali M, Signorini C, Caciotti B, Sugherini L, Ciccoli L, Giachetti D, Comporti M. Protection against oxidative damage of erythrocyte membrane by the flavonoid quercetin and its relation to iron chelating activity. FEBS Lett. 1997;416:123–9.

26. Feuerstein GZ, Zaleska MM, Krams M, Wang X, Day M, Rutkowski JL, Finklestein SP, Pangalos MN, Poole M, Stiles GL, Ruffolo RR, Walsh FL. Missing steps in the STAIR case: a translational medicine perspective on the development of NXY-059 for treatment of acute ischemic stroke. J Cereb Blood Flow Metab. 2008;28:217–9.

27. Fukuda S, del Zoppo GJ. Models of focal cerebral ischemia in the nonhuman primate. ILAR J. 2003;44:96–104.

28. Fütterer CD, Maurer MH, Schmitt A, Feldmann REJ, Kuschinsky W, Waschke KF. Alterations in rat brain proteins after desflurane anaesthesia. Anaesthesiology. 2004;100:302–8.

29. Fuxe K, Kurosawa N, Cintra A, Hallstrom A, Goiny M, Rosen L, Agnati LF, Ungerstedt U. Involvement of local ischemia in endothelin-1 induced lesions of the neostriatum of the anaesthetized rat. Exp Brain Res. 1992;88:131–9.

30. Fuxe K, Bjelke B, Andbjer B, Grahn H, Rimondini R, Agnati LF. Endothelin-1 induced lesions of the frontoparietal cortex of the rat. A possible model of focal cortical ischemia. Neuroreport. 1997;8:2623–9.

31. Garcia JH, Liu KF, Ho KL. Neuronal necrosis after middle cerebral artery occlusion in Wistar rats progresses at different time intervals in the caudoputamen and the cortex. Stroke. 1995;26:636–42.

32. Gerriets T, Li F, Silva MD, Meng X, Brevard M, Sotak CH, Fisher M. The macrosphere model: evaluation of a new stroke model for permanent middle cerebral artery occlusion in rats. J Neurosci Methods. 2003;122:201–11.

33. Gerriets T, Stolz E, Walberer M, Kaps M, Bachmann G, Fisher M. Neuroprotective effects of MK-801 in different rat stroke models for permanent middle cerebral artery occlusion: adverse effects of hypothalamic damage and strategies for its avoidance. Stroke. 2003;34:2234–9.

34. Gilmour G, Iversen SD, O'Neill MF, Bannerman DM. The effects of intracortical endothelin-1 injections on skilled forelimb use: implications for modelling recovery of function after stroke. Behav Brain Res. 2004;150:171–83.

35. Gitika B, Sai Ram M, Sharma SK, Ilavazhagan G, Banerjee PK. Quercetin protects C6 glial cells from oxidative stress induced by tertiary-butylhydroperoxide. Free Radic Res. 2006;40:95–102.

36. Green AR, Shuaib A. Therapeutic strategies for the treatment of stroke. Drug Discov Today. 2006;11:681–93.

37. Gresle MM, Jarrott B, Jones NM, Callaway JK. Injury to axons and oligodendrocytes following endothelin-1-induced middle cerebral artery occlusion in conscious rats. Brain Res. 2006;1110:13–22.

38. Gu W, Brannstrom T, Wester P. Cortical neurogenesis in adult rats after reversible photo-thrombotic stroke. J Cereb Blood Flow Metab. 2000;20:1166–73.

39. Gu W, Brannstrom T, Jiang W, Bergh A, Wester P. Vascular endothelial growth factor-A and -C protein up-regulation and early angiogenesis in a rat photothrombotic ring stroke model with spontaneous reperfusion. Acta Neuropathol. 2001;102:216–26.

40. Haynes WG, Webb DJ. Endothelin as a regulator of cardiovascular function in health and disease. J Hypertens. 1998;16:1081–98.

41. Hicks AU, Hewlett K, Windle V, Chernenko G, Ploughman M, Jolkkonen J, Weiss S, Corbett D. Enriched environment enhances transplanted subventricular zone stem cell migration and functional recovery after stroke. Neuroscience. 2007;146:31–40.

42. Hoyte L, Kaur J, Buchan AM. Lost in translation: taking neuroprotection from animal models to clinical trials. Exp Neurol. 2004;188:200–4.

43. Hu X, Wester P, Brannstrom T, Watson BD, Gu W. Progressive and reproducible focal cortical ischemia with or without late spontaneous reperfusion generated by a ring-shaped, laser-driven photothrombotic lesion in rats. Brain Res Brain Res Protoc. 2001;7:76–85.

44. Hu XL, Olsson T, Johansson IM, Brannstrom T, Wester P. Dynamic changes of the anti- and pro-apoptotic proteins Bcl-w, Bcl-2, and Bax with Smac/Diablo mitochondrial release after photothrombotic ring stroke in rats. Eur J Neurosci. 2004;20:1177–88.

45. Hyun J, Woo Y, Hwang DS, Jo G, Eom S, Lee Y, Park JC, Lim Y. Relationships between structures of hydroxyflavones and their antioxidative effects. Bioorg Med Chem Lett. 2010;20:5510–3.

46. Jiang W, Gu W, Hossmann KA, Mies G, Wester P. Establishing a photothrombotic 'ring' stroke model in adult mice with late spontaneous reperfusion: quantitative measurements of cerebral blood flow and cerebral protein synthesis. J Cereb Blood Flow Metab. 2006;26:927–36.

47. Jin R, Yang G, Li G. Inflammatory mechanisms in ischemic stroke: role of inflammatory cells. J Leukocyte Biol. 2010;87:779–89.

48. Kanemitsu H, Nakagomi T, Tamura A, Tsuchiya T, Kono G, Sano K. Differences in the extent of primary ischemic damage between middle cerebral artery coagulation and intraluminal occlusion models. J Cereb Blood Flow Metab. 2002;22:1196–204.

49. Kapinya KJ, Löwl D, Fütterer C, Maurer M, Waschke KF, Isaev NK, Dirnagl U. Tolerance against ischemic neuronal injury can be induced by volatile anesthetics and is inducible NO synthase dependent. Stroke. 2002;33:1889–98.

50. Karmarkar SW, Bottum KM, Tischkau SA. Considerations for the use of anesthetics in neu-rotoxicity studies. Comp Med. 2010;60:256–62.

51. Kito G, Nishimura A, Susumu T, Nagata R, Kuge Y, Yokota C, Minematsu K. Experimental thromboembolic stroke in cynomolgus monkey. J Neurosci Methods. 2001;105:45–53.

52. Kleinschnitz C, Braeuninger S, Pham M, Austinat M, Nolte I, Renne T, Nieswandt B, Bendszus M, Stoll G. Blocking of platelets or intrinsic coagulation pathway-driven thrombosis does not prevent cerebral infarctions induced by photothrombosis. Stroke. 2008;39:1262–8.

53. Koizumi J, Yoshida Y, Nazakawa T, Ooneda G. Experimental studies of brain edema, 1: a new experimental model of cerebral embolism in rats in which recirculation can be introduced in the ischemic area. Jpn J Stroke. 1986;8:1–8.

54. Kudo M, Aoyama A, Ichimori S, Fukunaga N. An animal model of cerebral infarction. Homologous blood clot emboli in rats. Stroke. 1982;13:505–8.

55. Kuge Y, Yokota C, Tagaya M, Hasegawa Y, Nishimura A, Kito G, Tamaki N, Hashimoto N, Yamaguchi T, Minematsu K. Serial changes in cerebral blood flow and flow-metabolism uncoupling in primates with acute thromboembolic stroke. J Cereb Blood Flow Metab. 2001;21:202–10.

56. Lapchak PA. Translational stroke research using a rabbit embolic stroke model: a correlative analysis hypothesis for novel therapy development. Transl Stroke Res. 2010;1:96–107.

57. Lapchak PA, Araujo DM. Development of the nitrone-based spin trap agent NXY-059 to treat acute ischemic stroke. CNS Drug Rev. 2003;9:253–62.

58. Lapchak PA, Araujo DM, Song D, Wei J, Zivin JA. Neuroprotective effects of the spin trap agent disodium-[(tert-butylimino)methyl]benzene-1,3-disulfonate N-oxide (generic NXY-059) in a rabbit small clot embolic stroke model: combination studies with the thrombolytic tissue plasminogen activator. Stroke. 2002;33:1411–5.

59. Lapchak PA, Araujo DM, Zivin JA. Comparison of tenecteplase with alteplase on clinical rating scores following small clot embolic strokes in rabbits. Exp Neurol. 2004;185:154–9.

60. Lee JK, Kwak HJ, Piao MS, Jang JW, Kim SH, Kim HS. Quercetin reduces the elevated matrix metalloproteinases-9 level and improves functional outcome after cerebral focal ischemia in rats. Acta Neurochir (Wien). 2011;153:1321–9.

61. Lee JK, Park MS, Kim YS, Moon KS, Joo SP, Kim TS, Kim JH, Kim SH. Photochemically induced cerebral ischemia in a mouse model. Surg Neurol. 2007;67:620–5.

62. Longa EZ, Weinstein PR, Carlson S, Cummins R. Reversible middle cerebral artery occlusion without craniectomy in rats. Stroke. 1989;20:84–91.

63. Macrae IM, Robinson MJ, Graham DI, Reid JL, McCulloch J. Endothelin-1-induced reductions in cerebral blood flow: dose dependency, time course, and neuropathological consequences. J Cereb Blood Flow Metab. 1993;13:276–84.

64. Maher P. The flavonoid fisetin promotes nerve cell survival from trophic factor withdrawal by enhancement of proteasome activity. Arch Biochem Biophys. 2008;476:139–44.

65. Maher P. Modulation of multiple pathways involved in the maintenance of neuronal function during aging by fisetin. Genes Nutr. 2009;4:297–307.

66. Maher P, Salgado KF, Zivin JA, Lapchak PA. A novel approach to screening for new neuroprotective compounds for the treatment of stroke. Brain Res. 2007;1173:117–25.

67. Marder VJ, Chute DJ, Starkman S, Abolian AM, Kidwell C, Liebeskind D, Ovbiagele B, Vinuela F, Duckwiler G, Jahan R, Vespa PM, Selco S, Rajajee V, Kim D, Sanossian N, Saver JL. Analysis of thrombi retrieved from cerebral arteries of patients with acute ischemic stroke. Stroke. 2006;37:2086–93.

68. McAuley MA. Rodent models of focal ischemia. Cerebrovasc Brain Metab Rev. 1995;7:153–80.

69. Middleton Jr E, Kandaswami C, Theoharides TC. The effects of plant flavonoids on mammalian cells: implications for inflammation, heart disease, and cancer. Pharmacol Rev. 2000;52:673–751.

70. Mishina M, Komaba Y, Kobayashi S, Tanaka N, Kominami S, Fukuchi T, Mizunari T, Hamamoto M, Teramoto A, Katayama Y. Efficacy of edaravone, a free radical scavenger, for the treatment of acute lacunar infarction. Neurol Med Chir (Tokyo). 2005;45:344–8.

71. Miyake K, Takeo S, Kaijihara H. Sustained decrease in brain regional blood flow after microsphere embolism in rats. Stroke. 1993;24:415–20.

72. Muller TB, Haraldseth O, Jones RA, Sebastiani G, Lindboe CF, Unsgard G, Oksendal AN. Perfusion and diffusion-weighted MR imaging for in vivo evaluation of treatment with U74389G in a rat stroke model. Stroke. 1995;26:1453–8.

73. Niessen F, Hilger T, Hoehn M, Hossmann KA. Differences in clot preparation determine outcome of recombinant tissue plasminogen activator treatment in experimental thromboembolic stroke. Stroke. 2003;34:2019–24.

74. Nikolov R, Rami A, Krieglstein J. Endothelin-1 exacerbates focal cerebral ischemia without exerting neurotoxic action in vitro. Eur J Pharmacol. 1993;248:205–8.

75. Nishina K, Akamatsu H, Mikawa K, Shiga M, Maekawa N, Obara H, Niwa Y. The effects of clonidine and dexmedetomidine on human neutrophil functions. Anesth Analg. 1999;88:452–8.

76. Nita DA, Nita V, Spulber S, Moldovan M, Popa DP, Zagrean AM, Zagrean L. Oxidative damage following cerebral ischemia depends on reperfusion—a biochemical study in rat. J Cell Mol Med. 2001;52:163–70.

77. Olesen SP. Leakiness of rat brain microvessels to fluorescent probes following craniotomy. Acta Physiol Scand. 1987;130:63–8.
78. O'Neill MJ, Clemens JA. Rodent models of focal cerebral ischemia. Curr Protoc Neurosci. 2001;9.6.1–9.6.32.
79. Orset C, Macrez R, Young AR, Panthou D, Angles-Cano E, Maubert E, Agin V, Vivien D. Mouse model of in situ thromboembolic stroke and reperfusion. Stroke. 2007;38:2771–8.
80. Pandey AK, Hazari PP, Patnaik R, Mishra AK. The role of ASIC1a in neuroprotection elicited by quercetin in focal cerebral ischemia. Brain Res. 2011;1383:289–99.
81. Pietta PG. Flavonoids as antioxidants. J Nat Prod. 2000;63:1035–42.
82. Prasad S, Phromnoi K, Yadav VR, Chaturvedi MM, Aggarwal BB. Targeting inflammatory pathways by flavonoids for prevention and treatment of cancer. Planta Med. 2010;76:1044–63.
83. Prass K, Meisel C, Hoflich C, Braun J, Halle E, Wolf T, Ruscher K, Victorov IV, Priller J, Dirnagl U, Volk HD, Meisel A. Stroke-induced immunodeficiency promotes spontaneous bacterial infections and is mediated by sympathetic activation reversal by poststroke T helper cell type 1-like immunostimulation. J Exp Med. 2003;198:725–36.
84. Rice-Evans C. Flavonoid antioxidants. Curr Med Chem. 2001;8:797–807.
85. Rivera F, Urbanavicius J, Gervaz E, Morquio A, Dajas F. Some aspects of the in vivo neuroprotective capacity of flavonoids: bioavailability and structure-activity relationship. Neurotox Res. 2004;6:543–53.
86. Rivera F, Costa G, Abin A, Urbanavicius J, Arruti C, Casanova G, Dajas F. Reduction of ischemic brain damage and increase of glutathione by a liposomal preparation of quercetin in permanent focal ischemia in rats. Neurotox Res. 2008;13:105–14.
87. Robinson RG. Neuropsychiatric consequences of stroke. Annu Rev Med. 1997;48:217–29.
88. Robinson MJ, Macrae IM, Todd M, Reid JL, McCulloch J. Reduction of local cerebral blood flow to pathological levels by endothelin-1 applied to the middle cerebral artery in the rat. Neurosci Lett. 1990;118:269–72.
89. Roulston CL, Callaway JK, Jarrott B, Woodman OL, Dusting GJ. Using behaviour to predict stroke severity in conscious rats: post-stroke treatment with 3′,4′-dihydroxyflavonol improves recovery. Eur J Pharmacol. 2008;584:100–10.
90. Sasaki E, Suemizu H, Shimada A, Hanazawa K, Oiwa R, Kamioka M, Tomioka I, Sotomaru Y, Hirakawa R, Eto T, Shiozawa S, Maeda T, Ito M, Ito R, Kito C, Yagihashi C, Kawai K, Miyoshi H, Tanioka Y, Tamaoki N, Habu S, Okano H, Nomura T. Generation of transgenic non-human primates with germline transmission. Nature. 2009;459:523–7.
91. Schaller B. Prospects for the future: the role of free radicals in the treatment of stroke. Free Radic Biol Med. 2005;38:411–25.
92. Schmid-Elsaesser R, Zausinger S, Hungerhuber E, Baethmann A, Reulen HJ. A critical reevaluation of the intraluminal thread model of focal cerebral ischemia: evidence of inadvertent premature reperfusion and subarachnoid hemorrhage in rats by laser-doppler flowmetry. Stroke. 1998;29:2162–70.
93. Schültke E, Kendall E, Kamencic H, Ghong Z, Griebel RW, Juurlink BH. Quercetin promotes functional recovery following acute spinal cord injury. J Neurotrauma. 2003;20:583–91.
94. Schültke E, Kamencic H, Zhao M, Tian GF, Baker AJ, Griebel RW, Juurlink BH. Neuroprotection following fluid percussion brain trauma: a pilot study using quercetin. J Neurotrauma. 2005;22:1475–84.
95. Sharkey J, Ritchie IM, Kelly PA. Perivascular microapplication of endothelin-1: a new model of focal cerebral ischaemia in rats. J Cereb Blood Flow Metab. 1993;13:865–71.
96. Sharkey J, Butcher SP, Kelly JS. Endothelin-1 induced middle cerebral artery occlusion: pathological consequences and neuroprotective effects of MK801. J Auton Nerv Syst. 1994;49 Suppl:S177–85.
97. Sharkey J, Butcher SP. Characterisation of an experimental model of stroke produced by intracerebral microinjection of endothelin-1 adjacent to the rat middle cerebral artery. J Neurosci Methods. 1995;60:125–31.

98. Shia CS, Tsai SY, Kuo SC, Hou YC, Chao PD. Metabolism and pharmacokinetics of 3,3′,4′,7-tetrahydroxyflavone (fisetin), 5-hydroxyflavone, and 7-hydroxyflavone and anti-hemolysis effects of fisetin and its serum metabolites. J Agric Food Chem. 2009;57:83–9.

99. Shutenko Z, Henry Y, Pinard E, Seylaz J, Potier P, Berthet F, Girard P, Sercombe R. Influence of the antioxidant quercetin *in vivo* on the level of nitric oxide determined by electron paramagnetic resonance in rat brain during global ischemia and reperfusion. Biochem Pharmacol. 1999;57:199–208.

100. Silva MM, Santos MR, Caroço G, Rocha R, Justino G, Mira L. Structure-antioxidant activity relationships of flavonoids: a re-examination. Free Radic Res. 2002;36:1219–27.

101. Sozmen EG, Kolekar A, Havton LA, Carmichael ST. A white matter stroke model in the mouse: axonal damage, progenitor responses and MRI correlates. J Neurosci Methods. 2009;180:261–72.

102. Spetzler RF, Zabramski JM, Kaufman B, Yeung HN. Acute NMR changes during MCA occlusion: a preliminary study in primates. Stroke. 1983;14:185–91.

103. Spetzler RF, Selman WR, Weinstein P, Townsend J, Mehdorn M, Telles D, Crumrine RC, Macko R. Chronic reversible cerebral ischemia: evaluation of a new baboon model. Neurosurgery. 1980;7:257–61.

104. Spratt NJ, Fernandez J, Chen M, Rewell S, Cox S, van Raay L, Hogan L, Howells DW. Modification of the method of thread manufacture improves stroke induction rate and reduces mortality after thread-occlusion of the middle cerebral artery in young or aged rats. J Neurosci Methods. 2006;155:285–90.

105. Steele Jr EC, Guo Q, Namura S. Filamentous middle cerebral artery occlusion causes ischemic damage to the retina in mice. Stroke. 2008;39:2099–104.

106. Stroke Therapy Academic Industry Roundtable I (STAIR-I). Recommendations for standards regarding preclinical neuroprotective and restorative drug development. Stroke. 1999;30:2752–8.

107. Stroke Therapy Academic Industry Roundtable II (STAIR-II). Recommendations for clinical trial evaluation of acute stroke therapies. Stroke. 2001;32:1598–606.

108. Tennant KA, Jones TA. Sensorimotor behavioral effects of endothelin-1 induced small cortical infarcts in C57BL/6 mice. J Neurosci Methods. 2009;181:18–26.

109. The National Institute of Neurological Disorders and Stroke rt-PA Stroke Study Group. Tissue plasminogen activator for acute ischemic stroke. N Engl J Med. 1995;333:1581–7.

110. Thom T, Haase N, Rosamond W, Howard VJ, Rumsfeld J, Manolio T, et al. Heart disease and stroke statistics—2006 update. A report from the American heart association statistics committee and stroke statistics subcommittee. Circulation. 2006;113:e85–151.

111. Virley D, Hadingham SJ, Roberts JC, Farnfield B, Elliott H, Whelan G, Golder J, David C, Parsons AA, Hunter AJ. A new primate model of focal stroke: endothelin-1-induced middle cerebral artery occlusion and reperfusion in the common marmoset. J Cereb Blood Flow Metab. 2004;24:24–41.

112. Vuadens P, Bogousslavsky J. Diagnosis as a guide to stroke therapy. Lancet. 1998;352 Suppl 3:SIII5–9.

113. Wang CX, Yang T, Shuaib A. An improved version of embolic model of brain ischemic injury in the rat. J Neurosci Methods. 2001;109:147–51.

114. Wang S, Dusting GJ, May CN, Woodman OL. 3′,4′-Dihydroxyflavonol reduces infarct size and injury associated with myocardial ischaemia and reperfusion in sheep. Br J Pharmacol. 2004;142:443–52.

115. Wang Q, Rager JD, Weinstein K, Kardos PS, Dobson GL, Li J, Hidalgo IJ. Evaluation of the MDR-MDCK cell line as a permeability screen for the blood–brain barrier. Int J Pharm. 2005;288:349–59.

116. Wang Y, Jin K, Greenberg DA. Neurogenesis associated with endothelin-induced cortical infarction in the mouse. Brain Res. 2007;1167:118–22.

117. Watson BD, Dietrich WD, Busto R, Wachtel MS, Ginsberg MD. Induction of reproducible brain infarction by photochemically initiated thrombosis. Ann Neurol. 1985;17:497–504.

118. Weston RM, Jarrott B, Ishizuka Y, Callaway JK. AM-36 modulates the neutrophil inflammatory response and reduces breakdown of the blood brain barrier after endothelin-1 induced focal brain ischaemia. Br J Pharmacol. 2006;149:712–23.
119. Wiley KE, Davenport AP. Endothelin receptor pharmacology and function in the mouse: comparison with rat and man. J Cardiovasc Pharmacol. 2004;44 Suppl 1:S4–6.
120. Williams SJ, Thomas CJ, Boujaoude M, Gannon CT, Zanatta SD, Jarrott B, May CN, Woodman OL. Water soluble flavonol prodrugs that protect against ischaemia-reperfusion injury in rat hindlimb and sheep heart. Med Chem Comm. 2011;2:321–4.
121. Windle V, Corbett D. Fluoxetine and recovery of motor function after focal ischemia in rats. Brain Res. 2005;1044:25–32.
122. Woitzik J, Schneider UC, Thome C, Schroeck H, Schilling L. Comparison of different intravascular thread occlusion models for experimental stroke in rats. J Neurosci Methods. 2006;151:224–31.
123. Youdim KA, Qaiser MZ, Begley DJ, Rice-Evans CA, Abbott NJ. Flavonoid permeability across an in situ model of the blood–brain barrier. Free Radic Biol Med. 2004;36:592–604.
124. Zausinger S, Baethmann A, Schmid-Elsaesser R. Anesthetic methods in rats determine outcome after experimental focal cerebral ischemia: mechanical ventilation is required to obtain controlled experimental conditions. Brain Res Brain Res Protoc. 2002;9:112–21.
125. Zhang RL, Chopp M, Zhang ZG, Jiang Q, Ewing JR. A rat model of focal embolic cerebral ischemia. Brain Res. 1997;766:83–92.
126. Zhang L, Schallert T, Zhang ZG, Jiang Q, Arniego P, Li Q, Lu M, Chopp M. A test for detecting long-term sensorimotor dysfunction in the mouse after focal cerebral ischemia. J Neurosci Methods. 2002;117:207–14.
127. Zhang L, Zhang RL, Wang Y, Zhang C, Zhang ZG, Meng H, Chopp M. Functional recovery in aged and young rats after embolic stroke: treatment with a phosphodiesterase type 5 inhibitor. Stroke. 2005;36:847–52.
128. Zheng LT, Ock J, Kwon BM, Suk K. Suppressive effects of flavonoid fisetin on lipopolysaccharide-induced microglial activation and neurotoxicity. Int Immunopharmacol. 2008;8:484–94.

Chapter 26
Clinical Relevance in a Translational Rodent Model of Acute Ischemic Stroke: Incorporating the Biological Variability of Spontaneous Recanalization

Kama Guluma

Abstract Acute ischemic stroke, in which a clot of blood manages to obstruct an artery supplying blood to a part of the brain and leads to brain damage, ranks as the third leading cause of death and is the leading cause of serious long-term disability in the United States. At present, the administration of the clot-dissolving drug, tissue-plasminogen activator, remains the only FDA-approved reperfusion therapy for acute ischemic stroke. Unfortunately, relatively few people (4–14%, depending on the study) access medical care in time or are eligible enough for treatment to benefit [8, 10, 24]. There is a need to develop additional therapies. The occlusion that causes a stroke is a dynamic one that is prone to spontaneous re-canalization and varying degrees of spontaneous reperfusion. The incorporation of this tendency into a translational stroke model will maximize clinical relevance.

1 A Disconnect Between Preclinical Stroke Models and Clinical Stroke; Spontaneous Recanalization

Overviews of the nexus between preclinical work and clinical trial results reveal that there has thus far, unfortunately, been only marginal progress in the translation between seemingly positive results in animal studies to comparable efficacy in human clinical trials [34, 47]. The reason for this is probably multifactorial, but one possibility, as outlined by the Stroke Therapy Academic Industry Roundtable (STAIR) group, is that currently used models of acute ischemic stroke do not precisely mimic clinical stroke, "in that truly permanent occlusion rarely occurs in humans because of spontaneous recanalization" [17].

K. Guluma, MD (✉)
Department of Emergency Medicine, University of California, San Diego, CA, USA
e-mail: kguluma@ucsd.edu

P.A. Lapchak and J.H. Zhang (eds.), *Translational Stroke Research*, Springer Series in Translational Stroke Research, DOI 10.1007/978-1-4419-9530-8_26, © Springer Science+Business Media, LLC 2012

1.1 Spontaneous Reperfusion in Humans

The acute phase of human ischemic stroke is a dynamic process characterized by varying degrees of spontaneous reperfusion [2, 5, 23, 43, 56] and spontaneous variations in early outcome ranging from complete improvement to acute worsening [28, 30, 46]. Physiological extremes of blood pressure and body temperature have been shown to be generally factorial in outcome, from a neuroprotection standpoint [11, 22, 32, 48, 51, 58, 62], but spontaneous reperfusion has a pronounced effect on the overall variation in outcome noted clinically [36, 43, 52, 54]. Even in a conservative analysis, the rate of spontaneous recanalization of cerebral artery occlusions in humans is at 17% by 6–8 h from onset [23]. It is really all that stands between (a) a transient ischemic attack (or TIA)—in which a patient may walk out of the hospital the next day with no deficits—and (b) a malignant stroke—in which the patient may end up in an intensive care unit, critically ill with brain swelling. Of all the experimental treatments that have been tried for stroke—almost all vetted in animals—only emergent reperfusion with tissue-plasminogen activator (tPA) has proven to have any significant efficacy in humans [47], underlining the profound and direct effect that reperfusion has on outcome [60, 63]. Modern imaging modalities now enable us to directly visualize the amount of clot occluding the problematic cerebral artery in humans with acute stroke, and two things are now clear: (1) the thrombus burden is a dynamic entity, and (2) the amount of thrombus burden in the occlusion, and how quickly it is cleared (either spontaneously or via active thrombolysis), has a direct effect on outcome [6, 19, 27, 36, 50].

1.2 What Is a Thromboembolic Stroke Made of?

The inciting event in an acute ischemic stroke is almost always from an acute intraluminal thromboembolic occlusion in the index artery, although other types of emboli (such as fat, air, and infectious debris) and other mechanisms (such as vasospasm and vascular dissection) can be causative on a rare basis. "Red" (erythrocyte and fibrin) clots have been said to develop in areas of blood stasis or reduced flow, such as atrial fibrillation, cardiomyopathies, ventricular aneurysms, and akinetic regions, or in areas of severe arterial stenosis and occlusion, and on the surface of existent thrombi. "White" platelet-fibrin thrombi are said to form on irregular surfaces such as in fast-moving streams or on an irregular plaque [9]. The advent of neurothrombectomy devices is now affording us direct and previously unavailable, real-time insight into the actual makeup of the clots that cause strokes. Marder performed a histopathological study of the first 25 consecutive thromboemboli retrieved, using mechanical embolectomy devices, from a cerebral artery from patients with acute ischemic stroke, and found common component parts of fibrin, platelets, nucleated cells, and "red" clot [37]. Three quarters of the clots showed a complex pattern of platelet-fibrin areas, interspersed with linear or broad deposits of nucleated cells, often with intervening collections of erythrocytes, albeit with a marked diversity in the histological pattern seen. "Red" clots that contained a mass of erythrocytes with evenly dispersed nucleated cells were frequently enclosed within thromboemboli, but in three subjects, were extracted as whole, intact clots. The same pattern of clots was seen in those retrieved from patients with both cardioembolic and arteriopathic etiology, suggesting to the authors that there is an underlying commonality of histological component structure of thromboemboli from both cardiac and arterial sources of clot [37].

A revealing limitation of the study is that, in actuality, another 29 patients (i.e., over 50% of all the patients undergoing mechanical embolectomy in this consecutive series) could not have their clots successfully extracted, and the authors indicated that it might to be related to friability of the clots involved. We cannot be sure as to the makeup of the clots of these 29 patients, but a clue is the fact that in the three "extractible clot" subjects from whom whole "red" clots were extracted, complete extraction and vascular recanalization were incomplete due to the friability of the clots. The fact that the whole red clots that were obtained were difficult to extract in toto due to their friability and that a slight majority of all patients undergoing mechanical embolectomy could not have a successful extraction (also presumably due to clot friability) brings up the distinct (even if arguable) possibility that a significant proportion of stroke patients in their series (3 of the 25 "extractible" patients, and possibly a majority of the 29 "unextractible" patients) were occluded with soft, friable, difficult-to-extract "red" clot. Marder noted that their study did not include patients with distal, superficial branch occlusions, deep penetrator occlusions, or thrombus forming in situ on intracranial atherosclerotic lesions [37]. In another study, Liebeskind noted that in the clots extracted out of patients presenting with a hyperdense middle cerebral artery sign on CT or blooming artifact on MRI (both of which tend to represent a high-grade and proximal type of vessel occlusion), RBCs featured prominently; specifically, he noted that an RBC-dominant occlusion was noted in every case in which these radiological findings were identified [33]. A more comprehensive assessment of the histology of all clots in stroke—not just those that are firm and proximal enough to extract mechanically—would of course need to somehow be performed, but the findings and limitations of the study above suggest that "red" clot features prominently in acute thrombotic stroke. This would certainly reflect what occurs in a comparable event in the human heart, an acute myocardial infarction from an acute coronary artery occlusion, in which "red" clot is the predominant culprit found on aspiration of the occlusion on angiography [41, 42, 57]. An acute cerebrovascular occlusion appears to be a dynamic, chaotic interaction between fibrin, platelets, and erythrocytes, leading to architecturally complex mixtures of "white" and "red" clot, with "red" clot either primarily occluding a vessel or acting as an extension from a "white" clot to do the same [37]. This may be why spontaneous recanalization is a prominent feature of human stroke.

2 Incorporating Natural Spontaneous Reperfusion into a Clinically Relevant Translational Model of Acute Ischemic Stroke

2.1 Previously Utilized Animal Models of Stroke

If human strokes are "spontaneous reperfusion-sensitive," then translational models designed to evaluate the acute phase of stroke should probably be so as well. Previously published literature describing animal models of ischemic stroke has outlined techniques that include craniectomy with direct manual occlusion of a cerebral vessel [7, 25, 55]; photocoagulative induction of in situ thrombosis, either by direct endothelial injury or via secondary endothelial injury from activation of a photosensitive dye [14, 18, 38, 39]; direct application of electric current to the vessel wall to induce vascular injury and thrombosis [20]; embolization of nonbiological material such as latex microspheres, polymers, or magnet-guided ferromagnetic particles [1, 61]; the transluminal insertion of a suture ligature into a target cerebral vessel, which affords some controlled reversibility of the ischemic occlusion [35]; and the embolization of aged, hardened, or processed clot [29, 59, 65, 66].

These techniques are parts of models that have significant utility in answering focused biological questions regarding the fate of neuronal populations downstream from a vascular occlusion, but they present a limitation in that they are designed to optimize the uniformity of the final results obtained and use a relatively hardy occlusion to stabilize the initial ischemic insult over the duration of assessment and minimize any variability. In so doing, they lose the capacity through which to clearly study potential effects on vascular thrombomodulation and autolysis that may underlie the natural variation and outcome in human ischemic stroke. The clinical reality of stroke is that the initial ischemic occlusion is an inherently unstable one, susceptible to biological variation and spontaneous reperfusion, and also progression.

The above techniques understandably cannot be used to study mechanisms of spontaneous reperfusion and thrombolysis, but their use also brings up a more empiric vulnerability when used for translational research. O'Collins noted that a failing in the bench-to-bedside translation of the development of therapies for ischemic stroke may be tendency to exclusively frame drug activity in terms of the dominant schema of stroke pathophysiology, coupled with the sometimes arbitrary attribution of a drug mechanism [47]. To best avoid this, a translational model being used to screen for therapies, such as new drugs for example, would best be reflective of the totality of the disease, otherwise—if designed exclusively with a particular paradigm in mind—it might never be able to detect potentially unknown alternate therapeutic effect or mechanism. Having an incomplete understanding of therapeutic mechanism is an inherent reality of translational research. Take statins as a treatment in stroke, for example: It was assumed for decades that the beneficial effects of statins in acute cardiovascular disease were exclusively based on a cholesterol-lowering effect; only in more recent years has it been discovered that the beneficial effects on stroke may be due to a pleiotropic set of vascular mechanisms in which cholesterol lowering may be only a part [13]. Statins have now been shown to, among other things, decrease platelet activation and upregulate nitric oxide synthase in thrombocytes [31], upregulate endothelial nitric oxide synthase (eNOS) [3, 4, 53], and increase tPA activity [4, 16]. The vascular, anti-inflammatory, and antithrombotic effects of statins are numerous [12], and each of these "noncholesterol" effects has obvious relevance to an acute thrombotic cerebrovascular occlusion but would not be incorporated by a string, cautery, clip, microsphere, or other nonbiological occlusion model of acute cerebral ischemia. The same considerations can be applied to safety, on the flip side of the coin; a drug with potential therapeutic effects, but hidden prothrombotic effects, may perform very favorably in a model that does not account for spontaneous recanalization or thrombomodulation but proves ineffective or even detrimental in human clinical trials where its prothrombotic effects can affect biology [26].

2.2 Options for a "Natural" Spontaneous Reperfusion-Sensitive Thromboembolism in a Stroke Model

The three major perspectives of stroke treatment—and therefore translational stroke research—rest in the chronologically salient concepts of (1) reperfusion, (2) neuroprotection, and (3) rehabilitation/remodeling. Almost every conceived model of acute ischemic stroke has utility with regard to answering a specific scientific question within each of these realms, but when it comes to incorporating effects on the dynamic stability of an acute cerebrovascular occlusion in the first critical hours of stroke (in the realm of reperfusion and all things related) and maximizing fidelity to human disease, very specific considerations come into play.

An available option to place an acute thrombotic clot within a vessel in an animal model of stroke is to induce the formation of a clot in situ. Published techniques include direct endothelial injury from photocoagulation from activation of a photosensitive dye [14, 18, 38, 39] or direct application of electric current to the vessel wall to induce vascular injury and thrombosis [20]. The downside to these types of techniques is that, while they induce a clot, the scenario is hardly reflective of the overall clinical situation, in that one is left with a clot adherent to an acutely traumatized endothelium and damaged adjacent blood vessel wall, whereas in "real life," this is hardly the case. Another option that has been described involves the withdrawal of blood into a catheter containing thrombin and then reinjection of that thrombin-induced blood clot back into the intracranial circulation [65, 66]. However, thrombin-generated clots are significantly more resistant to thrombolysis with tPA than are fresh, spontaneously formed clots [45], and so this approach has a limitation insofar as a fidelity to the human condition is concerned.

A perhaps more realistic approach is the embolization of a blood clot, since this is actually the manner in which a stroke is propagated in humans, either as an embolization from the heart, or an intravascular thrombosis and/or embolization from an atheromatous or hypercoagulable intra-arterial site. This approach has been described, using embolization of aged, hardened, or processed clot [29, 49, 57, 65, 66], all designed toward an embolization with a "white" clot, again, so as to optimize the stability of the insult and subsequent reproducibility of an untreated endpoint. However, given the prevalence of red clot noted in human clot extractions and the real possibility that it may factor prominently in most strokes, an animal model with an occlusion with a "red" clot may be the most clinically relevant. The incidence of spontaneous recanalization is higher when "red" emboli are used [15], but "red" clots and spontaneous recanalization are, as noted above, a definite clinical reality.

3   A Clinically Relevant Translational Model of Acute Ischemic Stroke

An animal model with a thromboembolic stroke with "red" clot may be a useful tool. What follows below is a description of the methods, afforded outcome measures, and performance of such a model, as used by this author.

3.1   *Preparation of Clot*

In the hours prior to the planned embolization of a clot, a donor animal (the author uses male Fisher 344 rats weighing from 300 to 350 g) is placed under general inhalational anesthesia using an admixture of isoflurane and 2 L/min of oxygen. Anesthesia induction was carried out with 5% isoflurane and maintained with 2.5% halothane. After sterile preparation of the right neck, approximately 1 cc of venous blood is obtained via transcutaneous venipuncture of the right internal jugular vein with a 23-gauge needle and syringe. The blood is immediately injected into a length of polyethylene catheter (PE-50) so as to fill the lumen completely. This filled catheter is allowed to sit for approximately 1 min on a level surface, so as to ensure uniform initial stabilization of clot along the length of the catheter, and was then placed in a 37°C water bath. In the meantime, the donor rat is taken off anesthesia and monitored to full recovery.

The clot is allowed to incubate at 37°C from anywhere from 1 to 6 h, depending on the desired age of clot to be used at embolization (which will be performed the same day). At the end of this time period, the PE-50 catheter is removed from the bath and cut into 40-mm segments. A special clot-delivery catheter is fashioned out of a regular PE-50 catheter such that it has a tapered and narrow end capable of navigating the cerebral circulation of a rat and is prefilled with a pH-neutral, physiological electrolyte solution (author uses Plasmalyte™, from Baxter Health Care Corporation, Deerfield IL). The 40-mm clot, bathed in the same solution, is "loaded" into this catheter by ejecting it out of its host catheter and into the special catheter, using a dedicated syringe similarly filled with Plasmalyte (Fig. 26.1). The clot is not

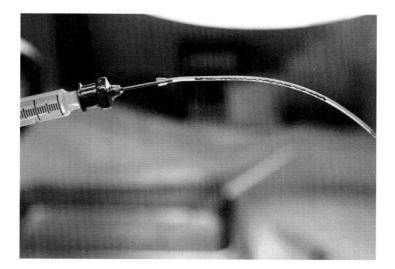

Fig. 26.1 A 40-mm segment of blood clot from the original PE-50 catheter, now extracted and loaded into a Plasmalyte-filled embolization catheter

"washed" of erythrocytes, not refrigerated, and not purposefully manipulated so as to change its natural consistency. When it is ready to be embolized, it is only a few hours old.

3.2   Surgical Technique and Embolization

A rat is placed under general anesthesia with isoflurane as performed above and kept euthermic ($37 \pm 1°C$) using heating pads during the ensuing procedure. After a shave and sterile prep, a midline neck incision is made, and the right common carotid artery (CCA), external carotid artery (ECA), and internal carotid artery (ICA), as well as the right pterygopalatine artery are exposed. The pterygopalatine artery is tied off with suture ligature. The right ECA is then tied off and divided, thereby releasing a stump through which the right ICA can be accessed. A suture ligature is placed around the ECA stump and readied for later use. The right ICA and CCA are then transiently clamped, and an arteriotomy is made in the right ECA stump, through which the previously assembled special PE-50 catheter containing the string of clot is inserted. The suture ligature is then gently tightened around the catheter, just enough to prevent backflow leakage of blood from around the catheter. The ICA is then unclamped, and the catheter is gently threaded into the intracranial portion of the ICA, to a depth of approximately 16 mm from the origin of the ECA, or just until a sudden slight resistance is felt and the catheter feels like it is starting to "seat" (meaning it is starting to occlude the vessel distally). Insertion of a catheter to around this depth results in the tip of the catheter being adjacent to or close to the origin of the middle cerebral artery (MCA) (Fig. 26.2).

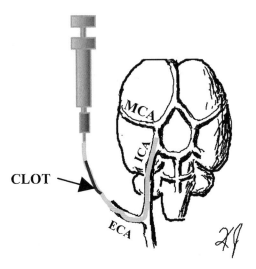

Fig. 26.2 Cannulation of the intracranial ICA with a modified catheter containing fresh clot, with the tip of the catheter just proximal to the origin of the MCA. *MCA* middle cerebral artery; *ICA* internal carotid artery; *ECA* external carotid artery

The CCA is then unclamped, and the clot is gently injected together with 0.2 cc of Plasmalyte (so to ensure good seating of the clot). The catheter is then withdrawn, and the ECA arteriotomy closed completely with the suture ligature. The operative incision is then closed. The animal is subsequently monitored to recovery off anesthesia.

3.3 Neurological Scoring and Model Performance

Upon full recovery from anesthesia and again at defined intervals, the degree of neurological deficit in the animal is assessed. This author uses a multielement neurological deficit scoring (neuroscore) paradigm that assesses both motor and nonmotor deficits. The paradigm is a modification of that used by Nedelman [44], which he described as an evolutionary development from prior work [7, 35, 40, 64], and found to correlate with infarct volume on hematoxylin and eosin staining. To determine neuroscore, nine items are separately assessed and scored, and then the scores totaled to derive the overall neuroscore as outlined in the Table 26.1.

This is an easy-to-use evaluation paradigm in which each item is scored in a relatively binary fashion (e.g., present/absent, normal/abnormal) and assigned a score of 0 or 10 with 10 denoting an abnormal finding, and totaled to a maximum total neuroscore of 90, with a higher neuroscore denoting a more severe neurological deficit. In using this neuroscore paradigm to assess the performance of the model, two things are evident: (1) animals are left with a very stereotypical deficit that approximates an MCA-distribution stroke deficit, with left forepaw weakness, sensory loss, and a visual field deficit predominating, and (2) this deficit profile is very consistent (reproducible) between experiments (Fig. 26.3).

Under controlled experimental conditions (including controlled body temperature and regulated inhalational anesthesia), with the same operator performing

Table 26.1 Items comprising the neuroscore

Item	Scoring (maximum = 90)
Left forelimb extension: hold the animal gently by the tail, suspended meter above the floor, and observe for forelimb flexion. Normal rats extend both forelimbs toward the floor	0: No impairment 10: Impairment; unable to fully extend left forelimb
Instability to lateral push from right: hold the animal gently by the tail, head away, on a graspable surface, while pushing laterally at the level of the shoulders	0: No impairment 10: Impairment
Torso twisting: hold the animal gently by the tail	0: No impairment 10: Impairment; flexing of body to the left and remaining in that position in three consecutive attempts
Walking on ground	0: Normal gait 5: Gait preference/circling to the left 10: Unable to walk on ground
Whisker movements on left side	0: Present 10: Absent
Consciousness	0: Normal 10: No reaction to stimuli
Hearing	0: Normal 10: No reaction to stimuli
Sensory: left-sided touch or prick	0: Normal 10: No reaction to stimuli
Left-sided hemianopia: reaction to visual stimuli approaching from left	0: Normal 10: Repeatedly absent reaction, or circling to the right

procedures and performing blinded assessments, this embolization results in a relatively consistent and stereotyped initial neurological deficit at 1 h after embolization. At 24 h, however, outcome in terms of neurological deficit (and associated lesion volume and clot burden) is notably more variable [21], in a fashion comparable to the clinical evolution of acute ischemic stroke humans in the first hours to days. Some animals end up with a persistent if not worsened neurological deficit, whereas others may have improved, at times to the point of having no detectable deficit, and as a group, there is a transition from having a relatively closely clustered range of neurological deficits on recovery to having a noticeably wider range of deficits at 24 h (Fig. 26.4), as can be expected for a "red" clot model. This is not correlated to any discernable technical factor in which there is a potential for influential variation, such as day of surgery, order of surgery, minor difference in exact age of the clot (in hours), or minor difference in exact body weight of the animal. Operator dependence and differences in clot consistency are not an issue because the variability occurs from animal-to-animal, on the same day, at the hands of a single operator, using clots which are precisely sectioned from the same singular length of clot. Administration of tPA at a time point 1 h after embolization in this model results a significant reduction in neurological deficit and associated lesion volume at 24 h postembolization [21].

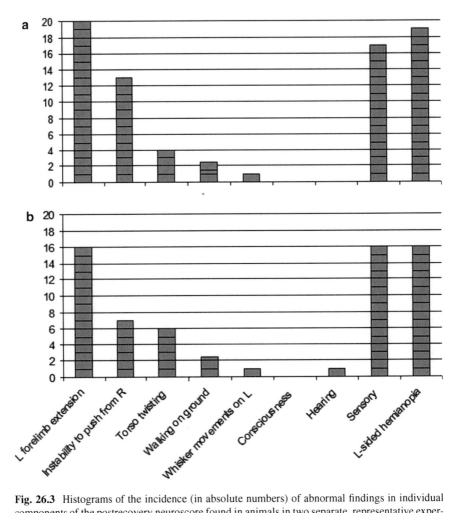

Fig. 26.3 Histograms of the incidence (in absolute numbers) of abnormal findings in individual components of the postrecovery neuroscore found in animals in two separate, representative experiments performed several months apart, 1 experiment with 21 animals (**a**) and another experiment with 18 animals (**b**). The relative incidence profile of abnormalities is very similar between experiments and correlates with a right MCA-distribution stroke. All procedures and assessments were performed by the same investigator

3.4   Assessment of Clot Burden

At 24 h after clot embolization (or at any time point desired), the rat is euthanized by transcardiac exsanguination of up to 10 mL of blood while under general inhalational anesthesia. This technique is chosen so that euthanasia can be rapid (essentially instantaneous) and allow staining of optimally oxygenated tissue. The animal is decapitated using a surgical scalpel blade so as to minimize any pressure on the neck region that might result in a significantly supraphysiological rise in the intraluminal

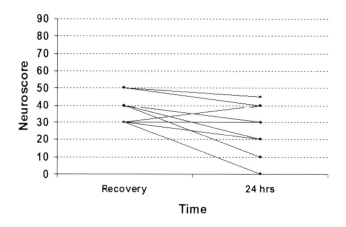

Fig. 26.4 Neurological deficit (neuroscore) data from a group of ten identical animals given an experimental thromboembolic stroke in the model. A higher neuroscore represents a more severe neurological deficit, and each line corresponds to a single animal and represents a connection between the neuroscore upon recovery (about an hour) after an experimental thromboembolic occlusion and that noted in the same animal at 24 h (there are nine lines because two animals had identical recovery and 24-h neuroscores—40 and 30 respectively—and their lines overlap). All animals were males of similar weight, and of the same species (Fisher 344) from the same breeder, and were treated and evaluated identically by the same operator

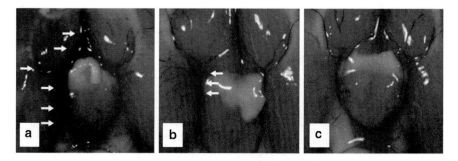

Fig. 26.5 Varying degrees of recanalization and autolysis of clot 24 h later in three rats embolized with identical amounts of fresh clot. (**a**) A clot in situ in the MCA and adjacent circle of Willis, typical of a large residual clot burden (*white arrows*); (**b**) a partially recanalized vessel with residual clot (*white arrows*); and (**c**) no residual clot

pressure of the intracranial vascular tree (thereby minimizing any risk of intravascular clot dislodgement prior to inspection of the vasculature). The brain with cerebral vessels is removed in toto and examined under a dissecting microscope. The position, length, and conformation of any visible intravascular clot are noted. At 24 h, there is a variable amount of residual thrombus left (Fig. 26.5) that coarsely correlates with outcome, in that persistence of a residual occlusive clot at 24 h is uniformly associated with a substantial deficit and lesion, and—on the other hand—a clot is never seen in the complete absence of a deficit or lesion.

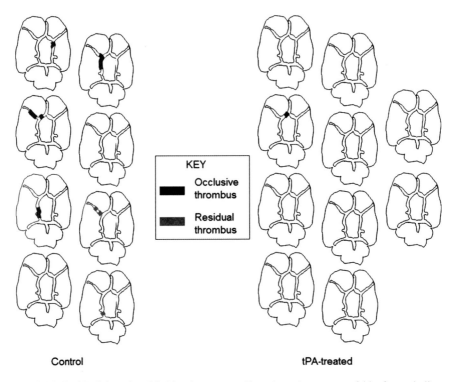

Fig. 26.6 Residual thrombus (clot) burden seen on dissecting microscopy at 24 h after emboliza-tion in control animals (*left*) and tissue-plasminogen activator (tPA)-treated animals (*right*). For each animal, the location, conformation, and length of residual clot and/or residual thrombotic material were entered into a structured data collection form. Treatment with tPA at 1 h postembo-lization resulted in a significant reduction in the amount of clot seen at 24 h, compared to that seen in control animals. In this study, the control group had a mean (±SD) clot burden (measured in total mm) of 1.0 ± 1.3 mm, whereas the tPA group had a mean of 0.1 ± 0.3 mm ($p=0.010$), and only 1/8 tPA-treated animals, compared with 6/8 control animals, had residual clot. The assessments were made in a consistent fashion, by a single investigator who was blinded to treatment assignment

The amount of remaining clot burden assessed this way can be evaluated as a therapeutic endpoint. In this model, an intravenous administration of 10 mg/kg of tPA (20% given as an initial bolus, and 80% given as an infusion over the remaining 30 min) an hour after stroke results in a significant reduction in clot burden at 24 h, compared with control animals (Fig. 26.6) [21].

4   Conclusions

There are a host of preclinical models of acute ischemic stroke, some utilizing non-biological occlusions, others utilizing biological occlusions that have been hard-ened against the natural variability inherent to biological systems. The uniformity of outcome that these models bring to the table is very useful in answering specific scientific questions, especially with regard to effects on neuroprotection and regeneration. There is a need, however, for a clinically relevant model to concurrently screen for effects on the spontaneous recanalization, clot autolysis, thrombolysis, and reperfusion that appear to have a profound effect on immediate outcome in stroke. Furthermore, a new agent with one presumed therapeutic mechanism of action may actually have other unknown, unrelated, and perhaps even more important mechanisms. A comprehensively reflective stroke model may afford a benefit as a screening tool for the unknowns. The aforedescribed model represents a useful addition to the armamentarium.

References

1. Akai F, Maeda M, Hashimoto S, Taneda M, Takagi H. A new animal model of cerebral infarction: magnetic embolization with carbonyl iron particles. Neurosci Lett. 1995;194:139–41.
2. Alexandrov AV, Grotta JC. Arterial reocclusion in stroke patients treated with intravenous tissue plasminogen activator. Neurology. 2002;59(6):862–7.
3. Amin-Hanjani S, Stagliano NE, Yamada M, Huang PL, Liao JK, Moskowitz MA. Mevastatin, an HMG-CoA reductase inhibitor, reduces stroke damage and upregulates endothelial nitric oxide synthase in mice. Stroke. 2001;32(4):980–6.
4. Asahi M, Huang Z, Thomas S, Yoshimura S, Sumii T, Mori T, Qiu J, Amin-Hanjani S, Huang PL, Liao JK, Lo EH, Moskowitz MA. Protective effects of statins involving both eNOS and tPA in focal cerebral ischemia. J Cereb Blood Flow Metab. 2005;25(6):722–9.
5. Baracchini C, Manara R, Ermani M, Meneghetti G. The quest for early predictors of stroke evolution: can TCD be a guiding light? Stroke. 2000;31(12):2942–7.
6. Barreto AD, Albright KC, Hallevi H, Grotta JC, Noser EA, Khaja AM, Shaltoni HM, Gonzales NR, Illoh K, Martin-Schild S, Campbell 3rd MS, Weir RU, Savitz SI. Thrombus burden is associated with clinical outcome after intra-arterial therapy for acute ischemic stroke. Stroke. 2008;39(12):3231–5.
7. Bederson JB, Pitts LH, Tsuji M, Nishimura MC, Davis RL, Bartkowski H. Rat middle cerebral artery occlusion: evaluation of the model and development of a neurologic examination. Stroke. 1986;17(3):472–6.
8. California Acute Stroke Pilot Registry (CASPR) Investigators. Prioritizing interventions to improve rates of thrombolysis for ischemic stroke. Neurology. 2005;64(4):654–9.
9. Caplan LR. Brain embolism, revisited. Neurology. 1993;43(7):1281–7.
10. Caplan LR. The treatment of acute stroke; still struggling. JAMA. 2004. 292:1183–1185.
11. Castillo J, Leira R, Garcia MM, Serena J, Blanco M, Davalos A. Blood pressure decrease during the acute phase of ischemic stroke is associated with brain injury and poor stroke outcome. Stroke. 2004;35(2):520–6.
12. Cimino M, Gelosa P, Gianella A, Nobili E, Tremoli E, Sironi L. Statins: multiple mechanisms of action in the ischemic brain. Neuroscientist. 2007;13(3):208–13.
13. Di Napoli P, Taccardi AA, Oliver M, De Caterina R. Statins and stroke: evidence for cholesterol-independent effects. Eur Heart J. 2002;23(24):1908–21.
14. Dietrich WD, Prado R, Halley M, Watson BD. Microvascular and neuronal consequences of common carotid artery thrombosis and platelet embolization in rats. J Neuropathol Exp Neurol. 1993;52:351–60.
15. Durukan A, Strbian D, Tatlisumak T. Rodent models of ischemic stroke: a useful tool for stroke drug development. Curr Pharm Des. 2008;14(4):359–70.

16. Essig M, Nguyen G, Prie D, Escoubet B, Sraer J-D, Friedlander G. 3-Hydroxy-3-methylglutaryl coenzyme A reductase inhibitors increase fibrinolytic activity in rat aortic endothelial cells. Circ Res. 1998;83:683–9.

17. Fisher M, Feuerstein G, Howells DW, Hurn PD, Kent TA, Savitz SI, Lo EH, STAIR Group. Update of the stroke therapy academic industry roundtable preclinical recommendations. Stroke. 2009;40(6):2244–50.

18. Futrell N, Millikan C, Watson BD, Dietrich WD, Ginsberg MD. Embolic stroke from a carotid arterial source in the rat: pathology and clinical implications. Neurology. 1989;39:1050–6.

19. Gadda D, Vannucchi L, Niccolai F, Neri AT, Carmignani L, Pacini P. Multidetector computed tomography of the head in acute stroke: predictive value of different patterns of the dense artery sign revealed by maximum intensity projection reformations for location and extent of the infarcted area. Eur Radiol. 2005;15(12):2387–95.

20. Guarini S. A highly reproducible model of arterial thrombosis in rats. J Pharmacol Toxicol Methods. 1996;35(2):101–5.

21. Guluma KZ, Lapchak PA. Comparison of the post-embolization effects of tissue-plasminogen activator and simvastatin on neurological outcome in a clinically relevant rat model of acute ischemic stroke. Brain Res. 2010;1354:206–16.

22. Kammersgaard LP, Jørgensen HS, Rungby JA, Reith J, Nakayama H, Weber UJ, Houth J, Olsen TS. Admission body temperature predicts long-term mortality after acute stroke: the copenhagen stroke study. Stroke. 2002;33(7):1759–62.

23. Kassem-Moussa H, Graffagnino C. Nonocclusion and spontaneous recanalization rates in acute ischemic stroke: a review of cerebral angiography studies. Arch Neurol. 2002; 59(12):1870–3.

24. Katzan IL, Hammer MD, Hixson ED, Furlan AJ, Abou-Chebl A, Nadzam DM. Cleveland Clinic Health System Stroke Quality Improvement Team. Utilization of intravenous tissue plasminogen activator for acute ischemic stroke. Arch Neurol. 2004;61(3):346–50.

25. Kawamata T, Dietrich WD, Schallert T, Gotts JE, Cocke RR, Benowitz LI, Finklestein SP. Intracisternal basic fibroblast growth factor enhances functional recovery and up-regulates the expression of a molecular marker of neuronal sprouting following focal cerebral infarction. Proc Natl Acad Sci. 1997;94:8179–84.

26. Kestenbaum B. Reevaluating erythropoiesis-stimulating agents. N Engl J Med. 2010; 362(18):1742 (author reply 1743–4).

27. Kharitonova T, Thorén M, Ahmed N, Wardlaw JM, von Kummer R, Thomassen L, Wahlgren N, SITS investigators. Disappearing hyperdense middle cerebral artery sign in ischaemic stroke patients treated with intravenous thrombolysis: clinical course and prognostic significance. J Neurol Neurosurg Psychiatry. 2009;80(3):273–8.

28. Kimura K, Minematsu K, Yasaka M, Wada K, Yamaguchi T. The duration of symptoms in transient ischemic attack. Neurology. 1999;52:976–80.

29. Kilic E, Hermann DM, Hossman K-A. A reproducible model of thromboembolic stroke in mice. Neuroreport. 1998;9:2967–70.

30. Koudstaal PJ, van Gijn J, Lodder J, Frenken WG, Vermeulen M, Franke CL, Hijdra A, Bulens C. Transient ischemic attacks with and without a relevant infarct on computed tomographic scans cannot be distinguished clinically. Dutch Transient Ischemic Attack Study Group. Arch Neurol. 1991;48:916–20.

31. Laufs U, Gertz K, Huang P, Nickenig G, Böhm M, Dirnagl U, Endres M. Atorvastatin upregulates type III nitric oxide synthase in thrombocytes, decreases platelet activation, and protects from cerebral ischemia in normocholesterolemic mice. Stroke. 2000;31(10):2442–9.

32. Leonardi-Bee J, Bath PM, Phillips SJ, Sandercock PA, IST Collaborative Group. Blood pressure and clinical outcomes in the international stroke trial. Stroke. 2002;33(5):1315–20.

33. Liebeskind DS, Sanossian N, Yong WH, Starkman S, Tsang MP, Moya AL, Zheng DD, Abolian AM, Kim D, Ali LK, Shah SH, Towfighi A, Ovbiagele B, Kidwell CS, Tateshima S, Jahan R, Duckwiler GR, Viñuela F, Salamon N, Villablanca JP, Vinters HV, Marder VJ, Saver JL. CT and MRI Early Vessel Signs Reflect Clot Composition in Acute Stroke. Stroke. 2011;42(5):1237–43.

34. Lo EH. Experimental models, neurovascular mechanisms and translational issues in stroke research. Br J Pharmacol. 2008;153 Suppl 1:S396–405.
35. Longa EZ, Weinstein PR, Carlson S, Cummins R. Reversible middle cerebral artery occlusion without craniectomy in rats. Stroke. 1989;20(1):84–91.
36. Mandava P, Kent TA. Reversal of dense signs predicts recovery in acute ischemic stroke. Stroke. 2005;36(11):2490–2.
37. Marder VJ, Chute DJ, Starkman S, Abolian AM, Kidwell C, Liebeskind D, Ovbiagele B, Vinuela F, Duckwiler G, Jahan R, Vespa PM, Selco S, Rajajee V, Kim D, Sanossian N, Saver JL. Analysis of thrombi retrieved from cerebral arteries of patients with acute ischemic stroke. Stroke. 2006;37(8):2086–93.
38. Markgraf CG, Kraydieh S, Prado R, Watson BD, Dietrich WD, Ginsberg MD. Comparative histopathologic consequences of photothrombotic occlusion of the distal middle cerebral artery in Sprague-Dawley and Wistar rats. Stroke. 1993;24:286–93.
39. Matsuno H, Uematsu T, Umemura K, Takiguchi Y, Asai Y, Muranaka Y, Nakashima M. A simple and reproducible cerebral thrombosis model in rats induced by a photochemical reaction and the effect of a plasminogen-plasminogen activator chimera in this model. J Pharmacol Toxicol Methods. 1993;29:165–73.
40. Menzies SA, Hoff JT, Betz AL. Middle cerebral artery occlusion in rats: a neurological and pathological evaluation of a reproducible model. Neurosurgery. 1992;31(1):100–6; discussion 106–7.
41. Mizuno K, Satomura K, Miyamoto A, Arakawa K, Shibuya T, Arai T, Kurita A, Nakamura H, Ambrose JA. Angioscopic evaluation of coronary-artery thrombi in acute coronary syndromes. N Engl J Med. 1992;326(5):287–91.
42. Mizuno K, Arakawa K, Isojima K, Shibuya T, Satomura K, Kurita A, Nakamura H, Arai T, Kikuchi M. Angioscopy, coronary thrombi and acute coronary syndromes. Biomed Pharmacother. 1993;47(5):187–91.
43. Molina CA, Montaner J, Abilleira S, Ibarra B, Romero F, Arenillas JF, Alvarez-Sabín J. Timing of spontaneous recanalization and risk of hemorrhagic transformation in acute cardioembolic stroke. Stroke. 2001;32(5):1079–84.
44. Nedelmann M, Wilhelm-Schwenkmezger T, Alessandri B, Heimann A, Schneider F, Eicke BM, Dieterich M, Kempski O. Cerebral embolic ischemia in rats: correlation of stroke severity and functional deficit as important outcome parameter. Brain Res. 2007;1130(1):188–96.
45. Niessen F, Hilger T, Hoehn M, Hossmann KA. Differences in clot preparation determine outcome of recombinant tissue plasminogen activator treatment in experimental thromboembolic stroke. Stroke. 2003;34(8):2019–24.
46. NINDS rt-PA Study Group. Tissue plasminogen activator for acute ischemic stroke. N Engl J Med. 1995;333:1581–7.
47. O'Collins VE, Macleod MR, Donnan GA, Horky LL, van der Worp BH, Howells DW. 1,026 experimental treatments in acute stroke. Ann Neurol. 2006;59(3):467–77.
48. Okumura K, Ohya Y, Maehara A, Wakugami K, Iseki K, Takishita S. Effects of blood pressure levels on case fatality after acute stroke. J Hypertens. 2005;23:1217–23.
49. Overgaard K, Sereghy T, Boysen G, Pedersen H, Høyer S, Diemer NH. A rat model of reproducible cerebral infarction using thrombotic blood clot emboli. J Cereb Blood Flow Metab. 1992;12(3):484–90.
50. Puetz V, Dzialowski I, Hill MD, Subramaniam S, Sylaja PN, Krol A, O'Reilly C, Hudon ME, Hu WY, Coutts SB, Barber PA, Watson T, Roy J, Demchuk AM, Calgary CTA Study Group. Intracranial thrombus extent predicts clinical outcome, final infarct size and hemorrhagic transformation in ischemic stroke: the clot burden score. Int J Stroke. 2008;3(4):230–6.
51. Reith J, Jørgensen HS, Pedersen PM, Nakayama H, Raaschou HO, Jeppesen LL, Olsen TS. Body temperature in acute stroke: relation to stroke severity, infarct size, mortality, and outcome. Lancet. 1996;347(8999):422–5.
52. Rha JH, Saver JL. The impact of recanalization on ischemic stroke outcome: a meta-analysis. Stroke. 2007;38(3):967–73.

53. Sironi L, Cimino M, Guerrini U, Calvio AM, Lodetti B, Asdente M, Balduini W, Paoletti R, Tremoli E. Treatment with statins after induction of focal ischemia in rats reduces the extent of brain damage. Arterioscler Thromb Vasc Biol. 2003;23(2):322–7.

54. Stolz E, Cioli F, Allendoerfer J, Gerriets T, Del Sette M, Kaps M. Can early neurosonology predict outcome in acute stroke? A metaanalysis of prognostic clinical effect sizes related to the vascular status. Stroke. 2008;39(12):3255–61.

55. Tamura A, Graham DI, McCulloch J, Teasdale GM. Focal cerebral ischaemia in the rat: description of technique and early neuropathological consequences following middle cerebral artery occlusion. J Cereb Blood Flow Metab. 1981;1:53–60.

56. Toni D, Fiorelli M, Zanette EM, Sacchetti ML, Salerno A, Argentino C, Solaro M, Fieschi C. Early spontaneous improvement and deterioration of ischemic stroke patients. A serial study with transcranial Doppler ultrasonography. Stroke. 1998;29(6):1144–8.

57. Vlaar PJ, Svilaas T, Vogelzang M, Diercks GF, de Smet BJ, van den Heuvel AF, Anthonio RL, Jessurun GA, Tan E, Suurmeijer AJ, Zijlstra F. A comparison of 2 thrombus aspiration devices with histopathological analysis of retrieved material in patients presenting with ST-segment elevation myocardial infarction. JACC Cardiovasc Interv. 2008;1(3):258–64.

58. Vemmos KN, Tsivgoulis G, Spengos K, Zakopoulos N, Synetos A, Manios E, Konstantopoulou P, Mavrikakais M. U-shaped relationship between mortality and admission blood pressure in patients with acute stroke. J Intern Med. 2004;255:257–65.

59. Wang CX, Todd KG, Yang Y, Gordon T, Shuaib A. Patency of cerebral microvessels after focal embolic stroke in the rat. J Cereb Blood Flow Metab. 2001;21:413–21.

60. Wunderlich MT, Goertler M, Postert T, Schmitt E, Seidel G, Gahn G, Samii C, Stolz E, Duplex Sonography in Acute Stroke (DIAS) Study Group. Competence network stroke. Recanalization after intravenous thrombolysis: does a recanalization time window exist? Neurology. 2007;68(17):1364–8.

61. Yang Y, Tao Yang T, Li Q, Wang CX, Shuaib A. A new reproducible focal cerebral ischemia model by introduction of polyvinylsiloxane into the middle cerebral artery: a comparison study. J Neurosci Meth. 2002;118:199–206.

62. Yong M, Diener HC, Kaste M, Mau J. Characteristics of blood pressure profiles as predictors of long-term outcome after acute ischemic stroke. Stroke. 2005;36(12):2619–25.

63. Zaidat OO, Suarez JI, Sunshine JL, Tarr RW, Alexander MJ, Smith TP, Enterline DS, Selman WR, Landis DM. Thrombolytic therapy of acute ischemic stroke: correlation of angiographic recanalization with clinical outcome. AJNR Am J Neuroradiol. 2005;26(4):880–4.

64. Zausinger S, Hungerhuber E, Baethmann A, Reulen H, Schmid-Elsaesser R. Neurological impairment in rats after transient middle cerebral artery occlusion: a comparative study under various treatment paradigms. Brain Res. 2000;863(1–2):94–105.

65. Zhang RL, Chopp M, Zhang ZG, Jiang Q, Ewing JR. A rat model of focal embolic cerebral ischemia. Brain Res. 1997;766:83–92.

66. Zhang Z, Zhang RL, Jiang Q, Raman SBK, Cantwell L, Chopp M. A new rat model of thrombotic focal cerebral ischemia. J Cereb Blood Flow Metab. 1997;17:123–35.

Chapter 27
A Clinically Relevant Rabbit Embolic Stroke Model for Acute Ischemic Stroke Therapy Development: Mechanisms and Targets

Paul A. Lapchak

Abstract Alteplase (tissue plasminogen activator, tPA) is currently the only FDA-approved treatment that can be given to acute ischemic stroke (AIS) patients, if patients present within 3 h of an ischemic stroke. Recent clinical trial evidence now suggests that the therapeutic treatment window for tPA can be expanded 4.5 h, but this is not formally approved by the FDA. Even though there remains a significant risk of intracerebral hemorrhage (ICH) associated with alteplase administration, there is an increased chance of favorable outcome with tPA treatment.

Over the last 30 years, significant progress in the understanding of mechanisms involved in stroke damage have resulted from the sue of a series of in vivo stroke models. The use of preclinical models has also assisted with the identification of new treatments strategies, but the new strategies have not been easily translated from the laboratory animal into the stroke patient. Current research trends emphasize the development of new and potentially useful thrombolytics, neuroprotective agents and devices, which are also being tested for efficacy in preclinical and clinical trials. We have used the rabbit small clot embolic stroke model (RSCEM) to optimize treatment strategies prior to the development of clinical trials. Originally, the RSCEM was used to develop tPA for efficacy, and it remains the only preclinical model used to gain FDA approval of a therapeutic agent for stroke. This chapter will focus on recent studies of new therapeutic approaches developed using the RSCEM. Analysis from existing preclinical and clinical trials indicates that the RSCEM can be used as an effective translational tool to gauge the clinical potential of new treatments.

P.A. Lapchak, PhD, FAHA (✉)
Department of Neurology, Cedars-Sinai Medical Center,
Davis Research Building, D-2091, 110 N. George Burns Road,
Los Angeles, CA 90048, USA
e-mail: Paul.Lapchak@cshs.org

P.A. Lapchak and J.H. Zhang (eds.), *Translational Stroke Research*, Springer Series
in Translational Stroke Research, DOI 10.1007/978-1-4419-9530-8_27,
© Springer Science+Business Media, LLC 2012

1 Introduction

According to the USA stroke statistics, each year, approximately 795,000 people suffer a stroke, 75% of which are first strokes. Eighteen percent of stroke victims outright die from the brain attack. Acute ischemic stroke (AIS) is the third leading cause of death and the leading cause of adult disability in the USA with an estimated cost of $68.9 billion annually. Currently, stroke is treated using the thrombolytic, tissue plasminogen activator (tPA), a drug that promotes recanalization [1–4]. tPA is effective if given up to 4.5 h after a stroke [5, 6]. Although thrombolysis is now widely accepted as a standard of care, only 2.4% of AIS patients are being treated with tPA in the USA [7]. Even though there is no doubt that tPA is quite useful, there are important shortcomings of the drug including the fact that tPA is not neuroprotective. With tPA treatment, there is the risk of hemorrhagic transformation (HT) or intracerebral hemorrhage (ICH) in approximately 3–6% of patients treated within 4.5 h of a stroke [8], but the odds ratio for mortality rate increases after 4 h [8]. Nevertheless, the finding that at least one FDA-approved drug that was developed using the RSCEM is effective at reducing clinical and neurological deficits is an important proof of concept [9, 10].

The NINDS rt-PA trial [11] used the National Institutes of Neurological Disorders and Stroke (NINDS) classification described in detail in reviews by Petty et al. [12, 13]. AIS is subdivided into three main recognizable types with the following population incidence levels: cardioembolic stroke, in which the embolism arises from a cardiac source, is the most predominant subtype in AIS patients making up approximately 29% of the population; atheroembolic (which is associated with the narrowing of a cervicocephalic artery) makes up 16% of the population; and small vessel lacunar stroke (i.e., thrombosis in one of the deep penetrating branches from larger cerebral arteries) represents 16% of the population. In the majority of ischemic strokes, 39% have no identified known cause.

2 The Stroke Cascade

There is one underlying commonality between all types of strokes in humans: a "blood clot" is responsible for blocking cerebral blood flow [1, 14, 15] resulting in ischemia and activation of the ischemic cascade. This topic has been extensively reviewed in the literature [14–17]. Key components of the cascade include reduced tissue metabolism, depletion of energy stores, and, depending upon the duration of the initial insult, triggering of a cascade of excitotoxicity, free radical formation, inflammation, blood–brain barrier (BBB) and vascular injury, and programmed cell death [14].

Reduced blood flow and severe oxygen deficiency lead to an ischemic brain area, comprised of a central core of severely ischemic tissue that will die, surrounded by a tissue zone consisting of moderate ischemic tissue with preserved cellular metabolism and viability [18–20]. For a yet undefined period of time after a stroke, there appears to be a region of salvageable tissue commonly referred to as a "penumbra" to target with novel therapies in order to improve cellular functions.

There is a distinct temporal profile of cascade activation following an ischemic stroke, which is the target for almost all neuroprotective agents being developed [18, 20, 21]. Since there are parallel and simultaneous activation of pathways deleterious to tissues at risk, neuronal, glial, and vascular in nature, it has been hypothesized that no single compound will have maximal efficacy and that a pleiotropic multitarget compound or combination therapy approach may be needed to promote significant neuroprotection and recovery of function [16, 22–29].

3   Current State-of-the-Art Treatment

The only Food and Drug Administration (FDA)-approved treatment for AIS is a plasminogen activator that promotes thrombolysis by activating the endogenous fibrinolytic system [1–4]. The thrombolytic, tPA catalyzes the conversion of plasminogen to form plasmin, which in turn degrades fibrin and leads to clot lysis and recanalization or cerebral reperfusion. tPA has been shown to be effective up to 4.5 h after a stroke [5]. However, tPA therapy has been linked to serious side effects [11], primarily symptomatic ICH. It is interesting to note that patient mortality rate at 90-day posttreatment is similar in tPA and placebo groups; however, the short-term incidence of symptomatic intracerebral hemorrhage (sICH) is significantly higher in tPA-treated patients. In the NINDS tPA trial [11], sICH occurred in 6.4% of tPA-treated patients within 36 h of stroke onset, a significant increase compared to patients receiving placebo [11]. There are also reports in the literature indicating that ICH rates are between 0% and 10% [7, 30]. For over 15 years, clinical trials with tPA have continued. In the most recent ECASS III trial [5], the incidence of ICH (27.0% vs. 17.6%) and sICH (2.4% vs. 0.2%) was higher with tPA treatment than with placebo. However, there were no statistically significant differences in mortality rate (7.7% vs. 8.4%). Thus, even though tPA is effective at causing clot lysis, at least to some extent, with an expanded treatment window, transient or permanent reocclusion was observed in about 35% of cases [11], and these patients typically exhibited worse clinical outcome than those with stable tPA-induced recanalization.

With only one effective FDA-approved stroke treatment, there remains a need to develop safe and effective treatments for stroke. This chapter will specifically focus on the use of a rabbit small clot embolic stroke model (RSCEM) as a gold standard functional assay to develop thrombolytics, neuroprotective agents, and devices to treat stroke with the goal of behavioral improvement. Detailed procedural information is supplied throughout the chapter to encourage other stroke investigators to adopt the model so that we can realize the goal of effective stroke therapy.

4 State-of-the-Art Translational Research

4.1 Treatment Windows

There are two main therapeutic windows for possible successful intervention in stroke patients:

1. **Neuroprotection.** First, one can attempt to intervene at the level of one or more mediators of the ischemic cascade [15, 16]. However, due to therapeutic window limitations, many stroke victims will not receive adequate treatment [31].
2. **Neuronal Repair.** The second opportunity is when neuroprotection is no longer an option. Thus, there is a need for an intervention that can promote neuronal repair and enhance both neuronal and clinical functions [16].

In this chapter, attempts specific to point (1) above will be discussed in detail. We are currently actively pursuing (2) with novel neuroprotective and neurotrophic drugs (see [22, 25, 32]).

4.2 Choice of a Translational Stroke Model

It has long been recognized by the original STAIR committee that a predictable animal model of stroke should be used to develop the next effective stroke therapy. In accordance with original STAIR criteria and recommendations [33], a therapeutic agent being developed for the treatment of AIS should be tested in an animal model that mimics stroke in humans and uses functional recovery as an outcome measure because functional recovery is a major end point in clinical trials [5, 11, 34–36]. In AIS patients, a treatment is and can only be considered successful if it results in a positive clinical outcome, an outcome which is evaluated using one of two standard clinical outcome measures: the National Institutes of Health Stroke Scales (NIHSS) and/or the modified Rankin Scale (mRS), both of which have been extensively reviewed in the literature [36–40].

There has long been debate concerning the use of stroke models and appropriate end points to be used to develop novel forms of therapy for stroke. These have recently been discussed by numerous experts in the field [9, 41–44]. While there is continuing dispute on which animal should be used for therapy development and there is no consensus in the stroke community, there is one animal model of embolic stroke that in many ways parallels stroke in patients. The RSCEM that was used to demonstrate that tPA could effectively improve clinical function [45] is the only currently utilized standard model that has resulted in an FDA-approved stroke treatment. There are pros and cons of all small animal models used for therapy development.

The RSCEM has evolved since the first publications on the topic and has been characterized as a reproducible model where injection of emboli results in reduced cerebral blood flow, cell death, and decreased cortical energy [46–48]. The model is based upon the insertion of a catheter into the common carotid artery (CCA) using the following procedures.

4.3 Characterization of the RSCEM

4.3.1 Surgical Procedure

Extreme care prior to and postsurgery is used to minimize any adverse effects in study rabbits. All studies are institutionally IACUC approved and follow extensive guidelines, including aseptic surgical techniques. Briefly, rabbits are anesthetized with isoflurane (or halothane, where available) via a face mask, 5% in 3–4 L/min at induction and 2–3% in 3–4 L/min as a maintenance dose and placed on a Quantum-Heat™ heat pack. The right internal carotid artery (ICA) is exposed, and both the external carotid artery (ECA) and the CCA are ligated. A Becton, Dickinson (B-D) and Company plastic catheter oriented toward the brain was inserted into the CCA positioned toward the brain and secured with ligatures. The incision around the catheter is closed so that the distal end is accessible outside. The catheter is filled with 0.2 mL of heparinized saline (33 units/mL) and plugged with an injection cap. The animals are allowed to recover from inhalation anesthesia for 2–3 h before embolization to ensure the absence of anesthetic during the process.

Importantly, during or after surgery we also do not use analgesics and/or anti-inflammatory compounds, since all known compounds have neuroprotective effects that can interfere with the study of novel therapies to treat stroke [49–52].

4.3.2 Emboli Preparation

Blood is drawn from one or more donor rabbits and allowed to clot for 3 h at 37°C. The single resulting large blood clot is then suspended in phosphate-buffered saline (PBS) pH 7.4 with 0.1% bovine serum albumin (BSA) and large Polytron-generated clot fragments are sequentially passed through sizing screens to result in a suspension of small-sized blood clots (100–250 μm). The small screen-sized are then suspended in PBS containing 0.1% BSA, and the clot suspension is then spiked with ^{57}Co-labeled microspheres (NEN-Perkin-Elmer Inc.) to allow for tracking into the brain. At this point, an aliquot of the solution is removed for the determination of specific activity (i.e., ^{57}Co/mg clot). This will aid in determination of *clot burden* in brain following embolization. (Caution: Co^{57} is a gamma emitting radioactive. Please use all necessary precautions when handling, injecting, and disposing the isotope and tissue containing isotope.)

4.3.3 Embolization

For embolization, rabbits are placed in a Plas-Labs™ acrylic restrainer, and 1 mL of a clot particle suspension is injected through the carotid catheter into the brain via the CCA, which is then flushed with 5 mL of sterile saline. Rabbits are fully awake during the embolization procedure and they are self-maintaining (i.e., they do not require artificial respiration or other external support). This allows for immediate observation of the effects of embolization on behavior at the time of embolus injection and thereafter. After the embolization process is completed, the catheter is ligated close to the neck, leaving a small portion protruding from the skin.

4.3.4 Behavioral Quantal Analysis

For the RSCEM [9, 32, 53], we use a clinically relevant behavioral end point in combination with a powerful statistical quantal analysis technique [54] to determine how a heterogeneous stroke population will respond to a treatment. The use of clinical rating scores is a desirable primary end point to use when a novel therapeutic is being tested for further development and to support a clinical trial [55], which uses the NIHSS and/or mRS [56–59] as a clinically relevant and reproducible end point. Clinical scores in combination with quantal analysis is a sophisticated statistical analysis method to determine how a large heterogeneous population of stroke "patients," in this case, rabbits, will respond to a treatment. To evaluate the quantitative relationship between clot dose lodged in brain (clot burden) and behavioral deficits or clinical scores, logistic sigmoidal (S-shaped) quantal analysis curves are fit to dose-response data using equations originally described by Waud [60]. To construct quantal analysis curve, a wide range of clot doses (3–7 mg) are injected into the brains vasculature system via the indwelling carotid catheter. The injection of a range of clot "doses" produces a spectrum of behaviorally normal to abnormal animals, which includes death on the continuum of embolization-induced effects [55]. Figure 27.1a shows the progression of quantal analysis from raw data to a statistically fit curve. The representative sigmoidal dose-response curve is derived from a group of embolized rabbits treated with a common control vehicle, DMSO.

In most studies, less than 5% of animals die due to embolization, but in the event that embolization does result in a rapid death, there was a positive correlation with a high clot dose measured in brain, and the animal was represented as an abnormal (or dead) on the quantal analysis curve if the rabbit received treatment. Embolized rabbits are scored as abnormal if they have one or more of the following symptoms: ataxia, leaning, circling, lethargy, nystagmus, loss of balance, loss of limb/facial sensation, and, occasionally, hind-limb paraplegia. Using a simple dichotomous rating system, with a reproducible composite result and low interrater variability (<5%), each rabbit is rated as either behaviorally normal or abnormal by an observer naïve to the study design and/or treatment groups. It is essential that all studies be blinded and randomized to one or more levels (see [9, 33, 61]). For construction of quantal analysis curves, clot burden is plotted against behavior in order to define the

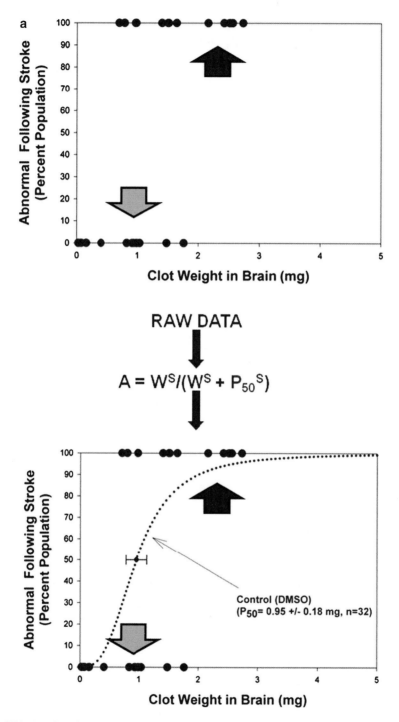

Fig. 27.1 (continued)

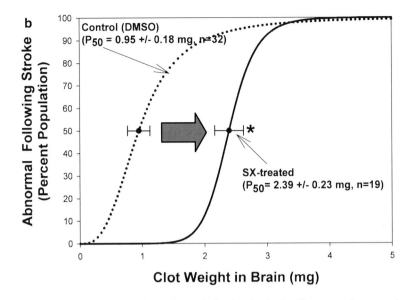

Fig. 27.1 (a) Behavioral analysis following embolization: basic plot. Representative quantal analysis process. The *top* graph shows a raw data plot of behavioral performance, abnormal (or dead). Following stroke vs. clot burden, clot weight in brain (mg), which allows for the calculation of a P_{50} value (mg), which is a numerical representation of the clot dose or the amount of blood clots in brain that produce neurologic dysfunction in 50% of a group of animals. This statistical calculation using the WAUD logistic equation [54] in the *middle panel* allows for the construction of a quantal analysis curve, which represents the overall response of a heterogeneous population of embolized rabbits to a substance such as DMSO. (b) Quantal curve shift analysis interpretation. Representative quantal analysis interpretation graph. The graph shows a plot of behavioral performance, abnormal (or dead). Following stroke vs. clot burden, clot weight in brain (mg), with P_{50} values (mg), calculated for the control group and the SX-treated group. With SX treatment, there is a significant (*$p < 0.05$) increase in P_{50} to 2.39 ± 0.23 mg ($n = 19$) from 0.95 ± 0.18 mg ($n = 32$) in the control group. The *curves* demonstrate a beneficial response to SX, an increase in P_{50}, and an overall shift to the *right* (*green arrow*). Thus, there was a beneficial response in the heterogeneous population of embolized rabbits treated with SX

P_{50} value (described in detail below). In the absence of a "beneficial" treatment regimen, small numbers of microclots lodged in the brain vasculature cause no grossly apparent neurologic dysfunction. However, when large numbers of microclots become lodged in the vasculature, they invariably cause encephalopathy due to ischemia, neuronal degeneration, depletion of ATP, and cell death [47].

Interpretation of Quantal Data: Shift Analysis

A separate quantal curve is generated for each treatment condition, and a statistically significant increase in the P_{50} value (or the amount of blood clots in brain that produce neurologic dysfunction in 50% of a group of animals) compared to a control curve is indicative of neuroprotection or a behavioral improvement in the study population [47, 62–64].

When one looks at a pair of quantal curves from a control group and a treated group, Substance X (SX) for instance, the first observation is that the P_{50} value for the SX-treated group is 100–250% higher than the P_{50} value for the vehicle control group (see, e.g., [62]). Thus, the most straightforward interpretation is that SX exerts *beneficial effects* in the model; the mechanism(s) involved in the beneficial effect is not important for this discussion. Thus, the group treated with SX during an effective therapeutic window will have a subpopulation of rabbits receiving a higher clot burden than their counterparts in the parallel-run control group, and that subpopulation will be functionally or behaviorally normal when they are rated, compared to abnormal (or dead) in the control group receiving the same corresponding clot burden. For example, from the curves, a control rabbit with a 0.95-mg clot burden has a 50% probability of being abnormal, whereas an SX-treated rabbit with a 2.39-mg clot burden has a 50% probability of being abnormal. There is a 2.51-fold shift to the right as shown by the green arrow. Thus, shift analysis indicates that an increase to the right (a higher P_{50}) in a treated group is indicative of neuroprotection, attenuation/blockade of portions of the ischemic cascade, increased cerebral perfusion and recanalization in the case of tPA (and similar thrombolytics), or possibly increased synaptic efficacy in remaining neuronal networks.

Why Should a Heterogeneous Animal Model of Stroke Be Used?

The answer is actually quite straightforward and can be derived from almost any randomized and blinded clinical trial such as the original NINDS tPA clinical trial report [11]. If one looks at enrollment in the trial, the patient population included in the placebo control group had NIHSS scores of 1–32, and the active drug group ranged from 1 to 37 [11]. Thus, the clinical trial encompassed a wide range of NIHSS scores, which is a 1–42 point scale [11, 36], indicative of great heterogeneity within each group. As one can see from Fig. 27.1, in the representative control group identified, the rabbit model also encompasses a wide range of stroke severity, including animals that die following embolization, but this population is kept at a minimum (<5%) by embolizing with clot doses that will not cause excessive death, but may cause deficits.

4.3.5   Characterization of Brain Deficits Following Embolization

Regional Cerebral Blood Flow

In one experiment, regional cerebral blood flow (RCBF) was measured in embolized rabbits using a laser Doppler perfusion monitor (Moor Instruments, UK). Rabbits were maintained on halothane (2%) for the duration of the measurement, after anesthesia induction using 5% halothane. This IACUC-approved procedure is

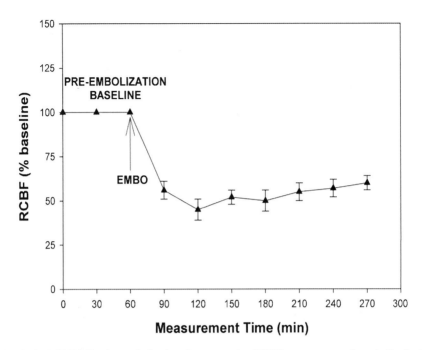

Fig. 27.2 RCBF following embolization. Representative RCBF measurements in embolized rabbits. The panel shows that Doppler analysis is useful in demonstrating a significant reduction in RCBF in embolized rabbits when measurement is done on the surface of the parietal/occipital cortex. After achieving a stable baseline preembolization, embolization (EMBO) results in a large significant decrease in RCBF for hours following embolization

necessary for an invasive measurement such as RCBF but is not the norm when measuring the therapeutic benefit of a treatment on behavior. A burr hole of 2 mm in diameter was created by a twist drill at 1 mm posterior and 5 mm lateral to the bregma on the right side skull. A P10 fiber probe (Moor Instruments, UK) was placed on the surface of the right cerebral cortex in order to monitor blood flow through the cerebral vessels. The probe was immobilized to the skull with bone wax. CBF was continuously monitored to establish a baseline, measure the effect of embolization, and determine the effect of REP. Data was collected and calculated by Moorlab software for windows (Moor Instruments, UK). All data is expressed as percent of baseline for each of the individual group of rabbits [46].

Figure 27.2 is a representative composite panel showing the effects of embolization on CBF in rabbits. Using a laser Doppler probe that was placed on the surface of the parietal/occipital cortex just beneath the skull, a baseline CSF was achieved. After embolization, we continued to measure RCBF in embolized rabbits. As demonstrated, after embolization, RCBF was reduced to 48–55% of preembolization baseline, which is maintained throughout the measurement period.

Physiological Measurements

In various pharmacological studies, we used physiological measurements to assess the effects of embolization on physiological parameters including temperature, blood glucose levels, and blood nitrite levels [65, 66]. Within 5 min of embolization, there was a significant 40–75% increase in blood glucose levels that was maintained for the first 2 h; blood glucose levels returned to control levels by 24 h regardless of the extent of stroke-induced behavioral deficits. In rabbits, embolization did not significantly affect body temperature. However, it is interesting to note that in a proof-of-concept study, Silver and Lapchak [66] used a novel instrumentation to compare the effects of embolization on nitrite levels measured in blood that was continuously obtained from a jugular venous catheter. In the assay, collected blood was passed through an ultrafiltration filter, and the filtrate was chemically reduced to convert free nitrite to nitric oxide (NO), which could be easily measured using a NO-specific electrode. The study showed that there was a significantly greater increase in nitrite levels in embolized rabbits as compared to controls ($p = 0.017$). Using a statistical calculation method described previously [66], the study showed that increased nitrite levels reached statistical significance ($p < 0.05$) within 3 min of embolization. While there was large variability in the measure that was related to the heterogeneous stroke population used in the study, NO_2 levels increased by $424 \pm 256\%$ compared to baseline. This study provides evidence for the feasibility of using nitrite as a biomarker marker of ischemic stroke and demonstrated that nitrite can be measured immediately following a stroke and that there is adequate sensitivity independent of the extent of the stroke.

Infarct Distribution

In another experiment, rabbits were embolized with a small-sized clot suspension and, 24 h later, were euthanized for infarct distribution using a standard 2,3,5-triphenyltetrazolium chloride (TTC) staining technique [67, 68]. Following euthanasia, brains are removed from the skull within 1 min, and coronal slices are immersed in 2% TTC dissolved in saline at 37°C. Cells that are metabolically active (i.e., functionally alive) convert TTC into a red stain, whereas damaged tissue both neuronal and glial, the area of which correlates with infarct volume, is pale in color. After complete development of the red color, each block is digitally photographed. (Caution: TTC is dangerous and all users should wear gloves and a face mask.)

As shown in Fig. 27.3, there is a heterogeneous distribution of infarcts throughout brain following small clot embolism. This heterogeneous distribution makes the definitive calculation of an infarct volume quite difficult using the model. However, it is possible with large experimental groups. Moreover, as described above (Sects. 4.3.2 and 4.3.3), Co^{57} tracer is used to determine clot burden. Thus, the investigator must be aware of the hazards of radioisotope and TTC if infarct volumes are to be considered.

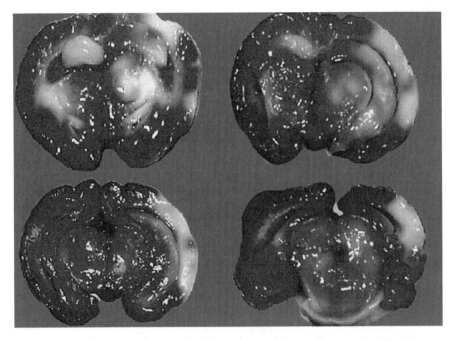

Fig. 27.3 Heterogeneous infarct distribution following embolization. Representative TTC-stained brain sections. Viable tissue is *pink/red*, and ischemic tissue is pale *beige/white*. *Upper left* and *right*: ischemic areas in the cerebral cortex and hippocampus. *Upper right*: an infarct in the right thalamus. *Bottom left* shows a cortical and hippocampal infarct, and the bottom right panel shows a cortical infarct

Cortical Energy Deficits Following Embolization

For this embolization, fully awake rabbits are placed in restrainers, and 1 mL suspension of nonradioactive blood clots was injected through an indwelling carotid catheter positioned toward the brain [47]. The catheter is then flushed with up to 5 mL of sterile saline.

Three hours postembolization, rabbits were euthanized with 1–1.5 mL of Beuthanasia-D via the marginal ear vein. The brain was rapidly removed from the skull within 30 s, and the parietal/occipital cortex was dissected on wet ice and then quick frozen using liquid nitrogen. Frozen cortical (parietal/occipital) tissue was then briefly placed in preweighed centrifuge tubes and quickly weighed so that ATP content could be normalized to tissue wet weight (mg). ATP was extracted from brain tissue using Polytron homogenization with 10% trichloroacetic acid (TCA) at a ratio of 100 mg cortical tissue/mL TCA. TCA has the ability to precipitate all proteins/lipids and release the content into the supernatant once centrifuged. Thus, following centrifugation at 13,200 ×g at 4°C to pellet the tissue, the supernatant was used to measure ATP content. ATP analysis was done using a previously published method luciferin/luciferase luminescence assay [69, 70].

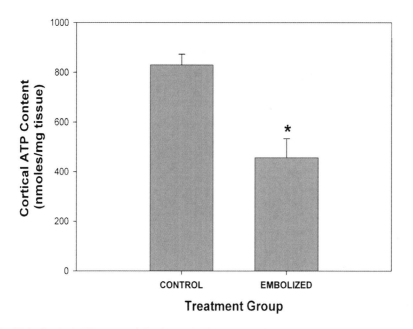

Fig. 27.4 Cortical ATP content following embolization. The data set shows that embolization not only results in infarcts in brain but also significantly (*$p < 0.05$) reduces cortical ATP content. This is indicative of tissue death and dysfunctional mitochondria in cortex following embolic strokes

As shown in Fig. 27.4, when using a high-dose clot suspension for embolization, there is a reproducible decrease in cortical ATP content that correlates with both decreased cortical function (RCBF) and infarcts (TTC) within cortical structures.

Cortical Kinase Signaling Pathways Modified by Embolization

The basic procedures for the study of pathways altered following embolic strokes using the RSCEM have recently been published [32]. Proteins are separated on 10% SDS–polyacrylamide gels and transferred to nitrocellulose. Equal loading and transfer of the samples was confirmed by staining the nitrocellulose with Ponceau S. Transfers were blocked for 1 h at room temperature with 5% nonfat milk in TBS/0.1% Tween 20 and then incubated overnight at 4°C in the primary antibody diluted in 5% BSA in TBS/0.05% Tween 20. The primary antibodies used were phospho-p44/42 MAP kinase antibody (#9101 and #9106, 1/1,000) and phospho-Akt ser473 (#9271, #4051, 1/1,000) from Cell Signaling (Beverly, MA), ORP150 antibody (#10301, 1/100,000) from IBL (Gunma, Japan), anti-CaMKIIα (Zymed, Invitrogen #137300), actin antibody (#A5441, 1/800,000) from Sigma (St. Louis, MO), anti-Akt (#05-591, 1/1,000) from Millipore (Bedford, MA), and pan ERK antibody (#E17120, 1/1,000) from Transduction Laboratories (San Diego, CA). The transfers were rinsed with TBS/0.05% Tween 20 and incubated for 1 h at room temperature in horseradish peroxidase-goat anti-rabbit or goat anti-mouse (Biorad, Hercules,

CA) diluted 1/5,000 in 5% nonfat milk in TBS/0.1% Tween 20. The immunoblots were developed with the Super Signal reagent (Pierce, Rockford, IL).

To assay kinase pathways in cortical tissues, animals were embolized with a suspension of nonradioactive blood clots, and 6 h following embolization, the rabbits were euthanized with 1–1.5 mL of Beuthanasia-D via the marginal ear vein and the brain was rapidly removed from the skull within 30 s, and the complete thickness of the parietal/occipital cortex was dissected on ice and then quick frozen using liquid nitrogen as previously described [47]. The identical area was taken from naive control animals to determine the effect of embolization on the signaling pathways. Brain samples were sonicated in PBS-containing protease (CompleteMini, Roche, Mannheim, Germany) and phosphatase inhibitors (Phosphatase Inhibitor Cocktail 2, Sigma, St. Louis, MO). Equal amounts of protein were solubilized in SDS sample buffer and analyzed as for the cell extracts.

To obtain an initial understanding of changes following embolization, we examined a variety of signaling pathways previously implicated in stroke and cell survival processes including the phosphatidylinositol 3-kinase (PI3K) pathway, the MEK1/2 (MAPK/ERK) pathway, and the c-Raf/ERK1/2 (extracellular-signal-regulated kinase, ERK) pathway [71–76]. Ras-dependent activation of PI3K is important for cell survival [77] via activation of *Akt* (i.e., protein kinase B (PKB), a serine/threonine kinase). Moreover, activation of the ERK and Akt pathways [77] regulates the expression of CaMKIIα [78] and regulates the ER-associated chaperone ORP150 via PI3K/Akt [79]. Studies have also shown that synaptic efficacy may be facilitated by the activation of CaMKIV (calcium/calmodulin-dependent protein kinase IV) and the activation of *CaMKIIα* [80, 81]. Lastly, since Tamatani et al. [79] have reported that ORP150 is part of a cytoprotective pathway that may be involved in cell survival, this was an interesting protein to assess.

The pathways that we identified as being of interest are *Akt* (*PKB*), *ERK*, and *CaM kinase IIα* (*CAMKIIα*); in addition, we studied *oxygen-regulated protein 150* (*ORP150*) [32].

Since the Ras–ERK cascade has been implicated in nerve cell survival in ischemia [82], ERK was measured. There was a significant decrease in ERK phosphorylation in the embolized animals, relative to control animals. PI3K-Akt signaling, which is also implicated in nerve cell survival both in vitro and in vivo [83, 84], was measured. We noted that phosphorylation of Akt on Ser^{473} was also largely eliminated in the embolized animals. When we assayed CaMKIIα protein expression in embolized rabbit cortex and control cortex, we found that within 6 h of a stroke, there was a significant reduction in CaMKIIα expression. Finally, the ER-associated chaperone ORP150 has been linked to the PI3K/Akt and BDNF signaling pathways [79], and the overexpression of ORP150 reduces infarct volume in a mouse stroke model [79]. ORP150 expression was slightly but not significantly decreased in embolized rabbit cortex (Fig. 27.5).

In summary, the RSCEM is a valuable embolic stroke model that allows for the study of a heterogeneous population of stroke "patients." With the use of a sophisticated statistical analysis method, i.e., quantal analysis, an investigator is capable of determining how a diverse population will respond to a specific treatment, whether it is a drug or a novel device that is being tested. The calculation of a

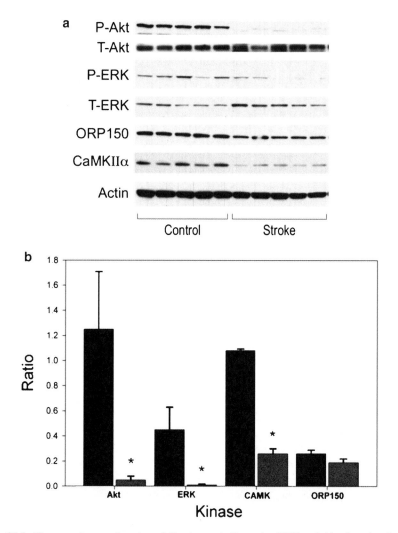

Fig. 27.5 Kinase pathways alterations following embolic stroke. ERK and Akt phosphorylation and ORP150 and CaMKIIa expression in embolized rabbits. Infarcted cortex removed 6 hrs after embolization was analyzed for proteins and phosphoproteins by Western blotting(6). Control cortex from the same region was used for comparison. $n = 5$ per group. Data is presented as the ratio of phosphoprotein to total protein (Akt and ERK) or ORP150 and CaMKII protein to actin. Representative western blots are shown above (a) the quantitative representation of the data (b), presented as ratio of the phosphoblots to total protein (Akt and ERK) or ORP150 and CaMKII a protein to actin. The data for each protein of interest were analyzed by ANOVA followed by Bonferroni's test ($*p < 0.05$; $**p < 0.01$; $***p < 0.001$).

P_{50} value provides a numerical measure of "actual" behavioral or clinical benefit in the model and is used in graphical shift analysis. Moreover, histochemical or biochemical measures can be used to supplement the behavioral data. Signaling pathway data suggests that many different pathways are affected by embolization and the subsequent process of cell death. It is foreseeable that some of the pathways may be useful targets to promote cell survival and improve clinical function.

5   Validation of the RSCEM

An important aspect of the use of an animal model is whether the model has been reasonably validated as a useful model. A true validation for a model would be the development of a drug in the model and, subsequently, the successful verification of efficacy in a randomized and blinded clinical trial. To date, the RSCEM remains the only validated stroke model for drug development [11].

5.1   Profile of tPA Efficacy in the RSCEM

In the RSCEM, tPA has consistently been shown to effectively improve behavioral scores when given 1–1.5 h following embolization [45, 62, 85, 86]. Figure 27.6 provided a diagrammatic synthesis of therapeutic windows for tPA and a free radical scavenger, NXY-059. First, in rabbits using the RSCEM, when a 3.3 mg/kg dose of tPA is administered, efficacy can be demonstrated out to 1.5 h postembolization; tPA is ineffective at promoting behavioral improvement when given 3 h postembolization.

5.2   Profile of tPA Efficacy in Stroke Patients

As all stroke researchers and clinicians know, the NINDS study [11] was the first study to establish that intravenous tPA improved neurological outcome in approximately 160 out of 1,000 treated stroke patients, and the most pronounced efficacy was when tPA was administered within 3 h of stroke onset. However, after delaying treatment 4 to 6 h, tPA was ineffective, and neurological outcome was not different compared to the placebo group. Follow-up clinical studies continue to emphasize the need for rapid treatment with tPA after AIS onset [1, 5, 34, 87–93]. Most recently, tPA has been shown to be effective stroke treatment up to 4.5 h after a stroke [5, 6]. In the more recent ECASS publication, mortality was also not different between groups and only accounted for 7.7% of patients in the tPA group and 8.4% in the placebo group [5]. While there are clear differences between the mortality rates in the original NINDS rt-PA trial and the recent ECASS tPA trial, there were no differences in each of the individual trials. It is important to note that the inclusion of mortality rate or death in rabbits during drug development is extremely important, and the inclusion may point to, at least in part, why the RSCEM was predictive of tPA efficacy in humans. There are substantial benefits to the patient should they elect to receive tPA, but overall, as a drug that is now widely accepted as a standard of care, it is drastically underutilized since recent estimates indicate that only 2–4% of stroke patients are being treated with tPA in the USA [7, 94].

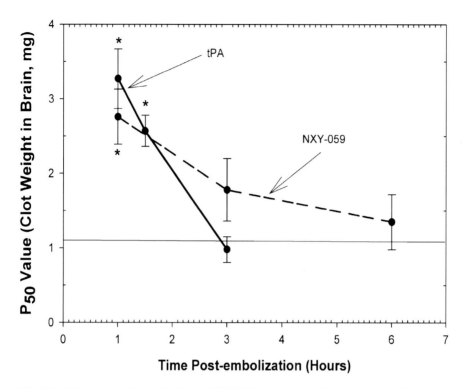

Fig. 27.6 Therapeutic window for tPA and NXY-059. Comparison of the therapeutic windows for tPA and NXY-059. Composite figure showing that tPA (*solid line*) administered up to 1.5 h postembolization results in a significant improvement of behavior represented by an increase (*$p < 0.05$) in P_{50}. The P_{50} measured 3 h following embolization is not significantly different from control ($p > 0.05$). NXY-059 (*dashed line*) effectively increased P_{50} values when given 5 min and 1 h postembolization but did not significantly ($p > 0.05$) improve behavior when given at either 3 or 6 h postembolization ($n = 16$–24). The *horizontal line* represents the mean P_{50} (mg) for the vehicle-treated group

5.2.1   Do All Stroke Patients Benefit from tPA?

One of the unexpected successes of the initial NINDS tPA trial in human stroke was that tPA was effective not only for patients with cardioembolic stroke but also for patients with either atheroembolic or small vessel lacunar stroke [95]. Indeed, the NINDS rt-PA trial publication [95] indicated that 62% of subjects improved with tPA compared with 41.5% with placebo following lacunar stroke, and 38% of subjects improved with tPA compared to 29% with placebo following cardioembolic stroke [95]. This data did not provoke changes in thinking about the pathogenesis of different types of stroke and may have been de-emphasized because of uncertainties about criteria for lacunar stroke. The efficacy of tPA in lacunar stroke likely had a role in accounting for the fact that infarct volume was not predictive of outcome in the NINDS tPA stroke trial [95]. Of note, the increases of endogenous blood levels of tPA and PAI are similar in cardioembolic, thromboembolic, and lacunar stroke [96]. Finally, a recent study using standard TOAST criteria for the three major types of stroke examined 90 consecutive patients and demonstrated that the efficacy of tPA in cardioembolic, in thromboembolic large vessel, and in lacunar stroke was comparable [35].

The importance of tPA efficacy in large vessel thromboembolic stroke and small vessel lacunar stroke has not been addressed by the stroke community. Clinically, this data suggests that clot formation plays a key pathoetiological role in lacunar stroke and that in situ thrombosis associated with atheroembolic stroke may also be very important in stroke pathogenesis [97]. The efficacy of tPA in cardioembolic, large vessel atheroembolic, and small vessel lacunar stroke has important implications for the predictive value of animal models of stroke. Everyone seems to agree that embolism with an autologous clot likely mimics cardioembolic stroke; there is no accepted model with a well-defined clinically relevant behavioral end point that specifically mimics large vessel cardioembolic stroke in animals. However, since tPA improves outcome of all three of these stroke types in patients, it is reasonable to predict that animal models in which tPA is effective might be considered as potential models of important aspects of the pathophysiology of all three major types of stroke in humans. The RSCEM is an animal model with heterogeneous infarct type, size, and distribution, which may closely parallel human stroke. Thus, as proposed by Turner et al. [44] and Lapchak [9], the RSCEM should be included as a predictive animal model when either drugs or devices are being developed.

5.3 Evaluation of Neuroprotective Agents

Table 27.1 presents a comprehensive, but not all inclusive, survey of drug classes and devices that have been studied in the RSCEM. Many compounds that were found to be ineffective after initial screens up to 1 h postembolization were not reported in the peer-reviewed literature due to the difficulty associated with the publication of negative data [61].

Figure 27.7 provides structures for the majority of drugs to be discussed in this section. The drugs chosen for many of the studies target specific mechanisms involved in the stroke cascade [15] such as anti-inflammatory drugs, free radical scavengers, and glutamate or NMDA antagonists. More recently, with the realization that monotherapy will probably not be optimally effective in stroke patients, pleiotropic compounds have been assessed for efficacy. Table 27.1 lists the drugs, effective dose, study, and therapeutic window of the specific drug. This section will provide a detailed analysis of possible strategies that will be useful to treat stroke based upon pharmacological efficacy in the RSCEM.

5.3.1 Anti-inflammatory Drugs

To the best of my knowledge, only two attempts to directly affect an inflammation process have been studied in the RSCEM. Broadly defined as an inflammatory

Table 27.1 Cumulative therapeutic window and dose-response curve data for the RSCEM: A historical overview

Therapy investigated		Effective dose	Therapeutic window	References
Anti-inflammatory	Anti-ICAM-1	1.0 mg/kg	5 min	[98, 99]
	Anti-CD18	1.0 mg/kg	5 min	[99]
Antioxidant	Baicalein	100 mg/kg	1 h	[210]
	Chlorogenic acid	50 mg/kg	1 h	[117]
	Ebselen	20 mg/kg	5 min	[119]
	Fisetin	100 mg/kg	<1 h	[126]
	NXY-059	0.1–100 mg/kg	1 h	[85]
	Radicut	100 mg/kg	3 h	[53]
Device	NILT	10–25 mW/cm^2	6 h	[64]
Miscellaneous	Caffeinol	5 mg/kg Caffeine 0.32 g/kg Ethanol	Ineffective	[144]
	TSC	0.25 mg/kg	1 h	[149]
NMDA antagonist	ABHS	25 mg/kg	5 min	[157]
	Memantine	25 mg/kg	1 h	[164]
Pleiotropic	CEPO	200 µg/kg	3 h	[177]
	CNB-001	100 mg/kg	1 h	[32]
	Simvastatin	20 mg/kg	1 h	[192]
Thrombolytic	tPA	3.3 mg/kg	1.5 h	[62, 86]
	TNK	1.0 mg/kg	3 h	[62]
	Microplasmin	4 mg/kg	1 h	[204]

process, leukocyte adhesion and transendothelial migration are regulated by cell-surface molecules, including the CD18 complex located on leukocytes and its counterreceptor, intercellular adhesion molecule-1 (ICAM-1), expressed on vascular endothelium and other cells. A publication from Bowes et al. [98] evaluated the effect of a polyclonal antibody directed against ICAM-1. In the study, the anti-ICAM antibody, which was given 5 min after embolization, did improve behavior. A second paper from the same author [99] tested anti-CD18 and also found that efficacy was limited to 5 min postembolization. The lack of delayed efficacy may be related to the single-dose regimens studied. Clearly, since inflammation is an ongoing process, this must be addressed long term and not acutely.

5.3.2 Antioxidants-Free Radical Scavengers

Using the RSCEM, many novel structurally independent drugs have been tested for efficacy. As described below, some of the drugs were ineffective, while others promoted significant behavioral improvement with relatively long therapeutic windows.

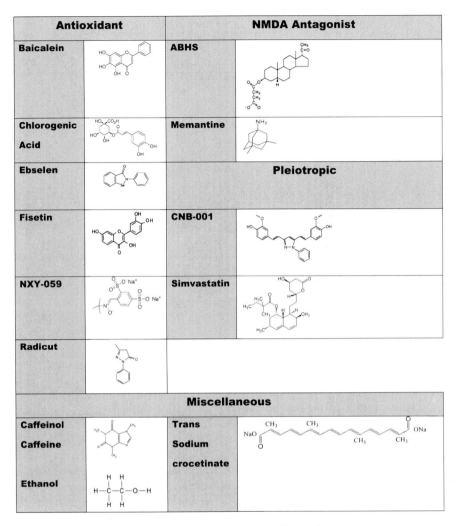

Fig. 27.7 Drug structures (alphabetically according to classification)

Baicalein

Baicalein, (5,6,7-trihydroxyflavone), is a polyphenolic antioxidant 12/15-lipoxyge-
nase inhibitor that has been studied in vitro and in vivo. Baicalein is a polyphenolic
flavonoid compound that is found in natural products such as in the skullcap or
golden root (*Scutellaria baicalensis* Georgi) and fruits of *Oroxylum indicum* [100–102].
Baicalein is a potent antioxidant, free radical scavenger and xanthine oxidase inhibitor [103–105], and it also has antithrombotic, antiproliferative, and antimitogenic
effect [103–105]. Using HT22 cell in vitro, baicalein significantly promotes cell
survival following incubation in the presence of iodoacetic acid (IAA), an irreversible

inhibitor of the glycolytic pathway that results in free radical production, lipid peroxidation, and cell death. In vivo, using the RSCEM, we found that baicalein has a therapeutic window of 1 h [106].

Chlorogenic Acid

Chlorogenic acid (CGA) [1,3,4,5-tetrahydroxycyclohexanecarboxylic acid 3-(3,4-dihydroxycinnamate)] is an ester formed between caffeic acid and (L)-quinic acid (1L-1(OH), 3,4/5-tetrahydroxycyclohexanecarboxylic acid). The phenylpropanoid micronutrient is found in tea, coffee, and fruits and vegetables and their products such as red wine and olive oil [107–112]. CGA has long been known to have multiple mechanisms of action [113–115] including being an antioxidant and anti-inflammatory [115], and it has been described as a high-affinity metalloproteinase-9 (MMP-9) inhibitor [116]. When studied as a possible stroke treatment in the RSCEM, CGA attenuated behavioral deficits associated with embolic strokes when administered up to 1 h after embolization. More importantly, the therapeutic window for a standard effective dose of tPA (3.3 mg/kg) could be increased by administration of CGA, suggesting that CGA may be useful as a cotherapy with a standard thrombolytic treatment regimen [117].

Ebselen

Ebselen (2-phenyl-1,2-benziselenazol-3(2H)-one) is a selenium-based antioxidant compound with documented neuroprotective effects mediated by reduced inflammation, scavenging of reactive nitrogen species including peroxynitrite, and inhibition of apoptotic mechanisms [118–121]. Ebselen as a monotherapy was quite disappointing when studied in the RSCEM since it was only effective when administered 5 min after embolization and was quite toxic [119]. There was evidence that ebselen could be applied with tPA to produce a synergistic behavioral improvement.

Fisetin

Fisetin is polyphenol flavonoid (3,7,3',4'-tetrahydroxyflavone) compound found in edible vegetables, fruits, and wine that has been shown to be neuroprotective in vitro and in vivo [122–125]. Like ebselen, fisetin monotherapy was not an optimal compound to promote behavioral improvement in the RSCEM past a 5-min administration time [126].

NXY-059

NXY-059 otherwise known as disodium-[(tert-butylimino) methyl]benzene-1,3-disulfonate-N-oxide is a water-soluble nitrone that has been tested in the RSCEM and in stroke patients. NXY-059 is a poor choice of compound to treat stroke for a number of reasons: the addition of two sulfonyl groups to the base compound phenyl butyl nitrone (PBN) significantly reduces the ability of NXY-059 to cross the BBB [127, 128], which limits its activity to the vascular compartment [127].

NXY-059 was tested as a possible treatment using the RSCEM. It must be noted that NXY-059 used in the RSCEM studies was synthesized and validated by Dr. Robert Purdy (VASDHS, San Diego & Scripps Institute, La Jolla, CA). Thus, it will be referred to as generic NXY-059 (NXY-059g). The drug was shown to reduce embolism-induced behavioral dysfunction when administered up to 1 h after the insult but not thereafter [85, 129] (see Fig. 27.6).

NXY-059 was studied in the Stroke Acute Ischemic NXY-059 Treatment (SAINT I and II) trials conducted by Astra-Zeneca. In the SAINT I trial [130], they received a 72-h infusion of placebo or NXY-059 initiated within 6 h of stroke onset [130]. The primary outcome, defined as disability at 90 days and assessed according to the mRS ($p=0.038$) for disability, was modestly improved, but function measured by the NIHSS was not [130]. In the SAINT II trial, the same dosing regimen of NXY-059 describe above was used [131], and the mean time to treatment was 3 h and 46 min, similar to that of SAINT I [130, 131]. SAINT II was a complete failure since NXY-059 did not meet its primary outcome assessed using mRS ($p=0.33$). In the SAINT trials, time to treatment (4–6 h) was outside of that predicted efficacy window by the RSCEM (see Table 27.1). Thus, it is likely that the failure of NXY-059 was due to the combination of an inferior agent, poor BBB penetration and distribution, and a short therapeutic window (see [132]).

Radicut

Radicut otherwise known as 3-methyl-1-phenyl-2-pyrazolin-5-one, edaravone, or MCI-186 is a low-molecular-weight free radical scavenger that readily crosses the BBB [133, 134]; thus, Radicut is able to scavenge free radicals outside of the vascular compartment. There is a wealth of scientific information available on Radicut because the drug was first approved for the treatment of stroke by the Japanese Ministry of Health and Welfare (see [24]) in 2001.

In the RSCEM, Radicut administered subcutaneously decreased behavioral deficits for the treated population when administered up to 3 h postembolization [53]. The study using the RSCEM indicates that edaravone may have substantial therapeutic benefit for the treatment of stroke, and the 3-h therapeutic window may allow for the treatment of many stroke patients (Fig. 27.8).

In Japan, Radicut is available in two formulations: ampoules that must be reconstituted or as a solution in an i.v. bag (dose: 30 mg b.i.d., i.v., 14 days). Cumulative data on Radicut suggests that efficacy is quite variable and ranges form from large clinical improvement to smaller improvement in overall clinical function measured using standard stroke scales such as NIHSS and mRS. The use of Radicut to treat stroke outside of Japan is still controversial mainly because there have not been randomized double blind international trials to support the use of the drug.

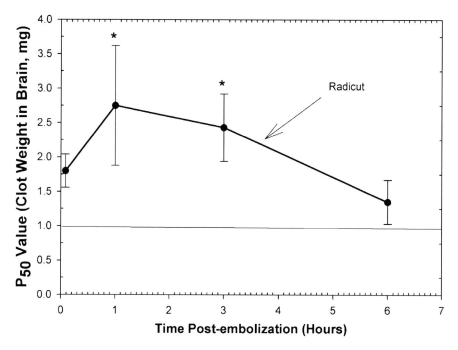

Fig. 27.8 Therapeutic window for Radicut. The figure is showing that Radicut (*solid line*) administered up to 3 h postembolization results in a significant improvement of behavior represented by an increase (*$p < 0.05$) in P_{50}. The *horizontal line* represents the mean P_{50} (mg) for the vehicle-treated group used in this study

To summarize this section on the development of free radical scavengers, studies to date suggest that pleiotropic lipophilic compounds have better efficacy profiles and therapeutic windows than water-soluble compounds. An emphasis should be placed on developing new compounds that readily cross the blood–brain barrier. Moreover, the compounds, although being labeled as antioxidants, should be able to attenuate other pathways in the stroke cascade to reduce damage to neurons, glial cells, and vascular endothelial cells.

5.3.3 Devices: Near-Infrared Laser Therapy (Target Unknown)

Transcranial near-infrared laser therapy (NILT) is a novel technology to produce functional recovery following a stroke. An in depth review of the application of NILT to teat stroke was recently published by Lapchak [135]. NILT produces significant photobiostimulation effects in vitro and in vivo [136–138]. In vivo, NILT with specific infrared wavelengths, specifically in the range of 790–810 nm penetrates deep into the brain [137] and can stimulate the mitochondrial chromophore cytochrome c oxidase (COX) to enhance adenosine triphosphate (ATP) production [47]. Although specific mechanism involved in NILT remains to be elucidated,

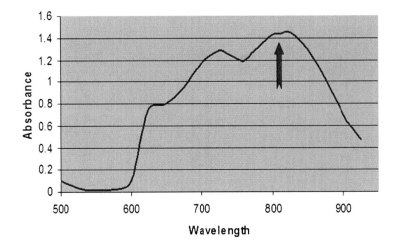

Fig. 27.9 NILT absorption spectrum is a representative graph demonstrating the wavelength absorption spectrum (absorbance) for tissues. Note that peak wavelength (*red arrow*) is consistent with 800–830-nm activation of the mitochondrial COX enzyme complex involved in the synthesis of ATP

NILT is postulated to improve energy metabolism and potentially enhance cell viability [48].

NILT development for the treatment of stroke using the RSCEM began in 2002. The laser probe used in the studies was coupled to a female SMA-905-adapted 2-cm diameter probe via an OZ Optics Ltd. fiber optic patch cord (step index fiber with a 550-μm core diameter and a numerical Aperture of 0.22). The probe utilized specially designed optics to generate a divergent, diffused 5-mm diameter beam ($1/e^2$). Transmission studies in this animal model have demonstrated that when the laser probe is placed on the skin surface posterior to bregma on the midline, the laser beam diffused by the scalp, skull, and dura covers the entire cortical surface of the animal.

Figure 27.9 is a representative graph demonstrating the wavelength absorption spectrum for tissues. Note that the peak wavelength is consistent with the activation of the COX enzyme complex, which contains two copper centers, Cu_A and Cu_B. The Cu_A center has a broad wavelength absorption peak between 800 and 830 nm in its oxidized form. NILT when applied through the skull or embolized rabbits was shown to promote significant behavioral improvement when treatment was initiated within 6 h of small clot embolization [64], see Fig. 27.10. The study also showed that a single NILT treatment provided durable and significant behavioral improvement when measured 3 weeks after embolization.

NILT has been studied in two randomized clinical trials: the Neurothera Effectiveness and Safety trial (NEST-I) [139] and the NEST-2 trial [140]. The NEST-1 trial with a median time to treatment of 18 h provided evidence that NILT could improve clinical function using the NIHSS ($p = 0.035$). However, the NEST-2 trial with a median time to treatment of 14.6 h was not an overall positive trial ($p = 0.094$). When a post hoc analysis of patients with a baseline NIHSS of <16 was

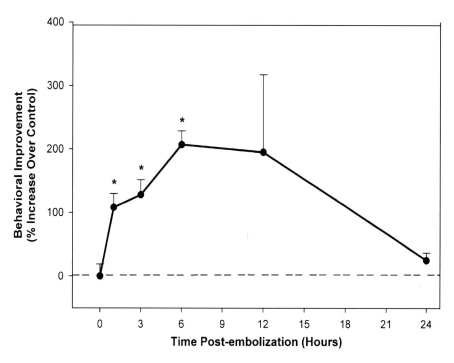

Fig. 27.10 Therapeutic window for NILT. Results are shown as behavioral improvement (mean ± SEM) for NILT treatment (10–25 mW/cm^2) initiated 1, 3, 6, 12, or 24 h following embolization as shown on the *x* axis. The beneficial effect of laser treatment is only observed up to 6 h following an embolic stroke (*$p < 0.05$ compared to control)

completed, there was evidence for a favorable outcome at 90 days on the primary end point ($p < 0.044$). The NEST-3 trial is currently underway [141].

There are two lessons to be learned from the study of NILT in rabbits and patients. First, a therapy with a relatively unknown mechanism of action can be developed and approved for a clinical trial by the FDA. Second, the RSCEM is an animal model that can be used to effectively develop novel therapies, including nonpharmacological treatments.

5.3.4   Miscellaneous

Caffeinol

Caffeinol (0.65 g/kg ethanol and 10 mg/kg caffeine) was tested using the RSCEM based upon preliminary promising data in rodents [142]. The Grotta group has found that caffeinol can reduce infarct volume when caffeinol was administered up to 2 h following ischemia [142]. The authors showed that there was a transient effect of caffeinol on behavioral improvement in the rat that was correlated with cortical but not subcortical structure survival [143]. In the RSCEM, caffeinol was without effects when administered 15 min following embolization either by infusion or bolus injection [144]. Moreover, in our experience, caffeinol also increased ICH incidence in the present of tPA administration. The clinical trial of caffeinol has been canceled due to lack of efficacy.

Trans-sodium Crocetinate

Trans-sodium crocetinate (TSC) is a carotenoid antioxidant that has the ability to enhance oxygen diffusion to hypoxic tissue [145]. A rodent study by Manabe, which is reviewed in Gainer [145], indicates that TSC can reduce infarct volume following ischemia and also increase PO_2 levels. Okonkwo et al. [146] also studied the pharmacology of TSC in rats by measuring PO_2 levels in brain tissue and concluded that TSC significantly increased brain tissue oxygen delivery. Thus, the main mechanism of action of TSC appears to be related to its ability to increase O_2 delivery to brain by increasing the diffusivity of oxygen through plasma. However, carotenoids such as TSC also have the ability to be free radical scavengers [145, 147] and to decrease the production of cytokines such as TNFα and interleukin-10 [148].

When tested in the RSCEM for effects on behavior, TSC improved behavior when given 1 h following embolization and produced synergy with tPA up to 3 h following embolization [149]. This novel form of therapy may have great applicability with other neuroprotectives or devices.

5.3.5   NMDA Antagonists

3α-ol-5-β-Pregnan-20-One Hemisuccinate

CNS active steroids referred to as "neurosteroids" or "neuroactive steroids" have received considerable attention because of their pharmacological activities related to the modulation of NMDA and GABA receptors [150, 151]. In rodents, neurosteroids can protect subcortical and cortical neurons following cerebral ischemia induced by middle cerebral artery occlusion (MCAO) [152, 153]. We have shown that 3α-ol-5-β-pregnan-20-one hemisuccinate (ABHS), a neurosteroid that regulates NMDA-receptor-mediated neurotransmission, is neuroprotective following reversible spinal cord ischemia [154] and can afford neuroprotection even if administered following a long delay after the initiation of ischemia. Moreover, ABHS reduces MCAO-induced infarct volume [155] in rodents. ABHS is a synthetic analog of the endogenous steroid 3α-5β-pregnan-20-one sulfate, which appears to have selectivity for NMDA receptors containing the presence of specific receptor subunits. The replacement of the sulfate by hemisuccinate in ABHS makes the molecule more stable in vivo and resistant to cleavage by endogenous sulfatases. ABHS has been described as a negative modulator of NMDA receptors that specifically contain either the NR2C or NR2D subunits [156]. Thus, it suggests the possibility that ABHS interacts with NMDA receptors of specific compositions to confer neuroprotection and inhibit the deleterious effects of glutamate. This hypothesis was studied in the RSCEM [157]. Unfortunately, even with a targeted strategy, ABHS (25 mg/kg) was only effective at improving behavior when injected 5 min postembolization. However, ABHS did significantly enhance neuroprotection when a low-dose tPA (0.9 mg/kg) regimen, which by itself does not improve behavioral function, was used, and ABHS increased the therapeutic window for tPA, suggesting that it may be most useful as a cotherapy with thrombolytics for stroke.

Memantine

Memantine is an uncompetitive open-channel-blocking NMDA receptor antagonist that binds to the MK-801 recognition site in the NMDA channel with an inhibitory constant (K_i) of approximately 1 μM and antagonizes NMDA-receptor-mediated inward currents in vitro [158–161]. Electrophysiological studies have shown that memantine has similar potency at NR1A/2B, 2C, and 2D receptors and is less effective at the NR1A/2A receptor. Under normal physiological conditions, NMDA receptors are transiently activated by micromolar concentrations of glutamate following depolarization of the postsynaptic membrane, whereas during pathological activation, NMDA receptors are activated by lower concentrations of glutamate but for a longer duration [160, 162, 163]. Thus, with this pharmacological profile, it appeared that memantine would be an excellent candidate for further development using the RSCEM. Memantine administered up to 3 h following embolization improved behavior, but the effect was only significant up to 1 h following embolization [164].

In a subsequent study, we used 2D gels to identify possible mediators or mechanisms involved in stroke-induced neurodegeneration as well as identifying potential new targets. For these studies, we harvested cortical tissue from an untreated control rabbit, an untreated embolized rabbit, and an embolized rabbit treated with 25 mg/kg memantine injected 5 min following embolization. For embolized rabbits, tissues were removed 2 h following stroke onset. Figure 27.11 shows a classical 2D acrylamide gel of proteins from an embolized rabbit brain. The gel shows that there are several hundred of proteins that are separated using the 2D gel system. A protein that is reproducibly more abundant in ischemic tissue than in controls is highlighted in the smaller box and marked with an arrow.

Figure 27.12 shows a composite panel from the three experimental groups. The panel on the left shows that there is little expression of the protein of interest that we have identified as aldolase reductase (AR) using mass spectroscopy. The expression of AR was significantly increased in ischemic cortical tissue (middle panel). In tissue harvested from an embolic stroke rabbit following memantine treatment, we found that there was attenuation of the levels of AR. These data show that high quality 2D gels can readily be made from rabbit brain, and differentially expressed proteins identified in the CNS. Figure 27.13 shows a western blot confirmation of AR and quantitation in embolized and memantine-treated embolized rabbits.

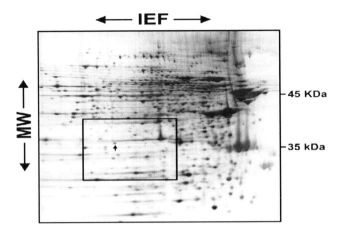

Fig. 27.11 2D Gel electrophoresis of cortical extracts. Total cortical tissue lysates of ischemic brain tissue were focused on pH 4–7 on isoelectric focusing strips and separated in the second dimension in 10% SDS–PAGE gels and were stained with silver. The *arrow* within the small interior box points to a protein of interest identified using differential analysis of protein gels from naïve control rabbits, untreated embolized rabbits, and embolized rabbit treated with memantine

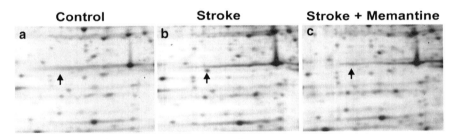

Fig. 27.12 Comparison of 2D gels from three different groups. Total cortical tissue lysates of control brain tissue were focused on pH 4–7 on isoelectric focusing strips and separated in the second dimension in 10% SDS–PAGE gels and were stained with silver. The *arrow* points to the AR protein in (**a**) naive control rabbit, (**b**) an untreated embolized rabbit and, (**c**) an embolized rabbit treated with memantine. Although there is little variation between gels and even between gels from different animals, any result from 2D gels must be confirmed by western blotting

What is the role of AR? AR is a monomeric, nicotinamide adenine dinucleotide phosphate (NADPH)-dependent enzyme that is a member of the aldo–keto reductase family that catalyzes the reduction of aldo sugars such as glucose and other saturated and unsaturated aldehydes. High levels of glucose in tissues lead to the accumulation of sorbitol via the (AR) pathway or polyol pathway. In this pathway, glucose is reduced to sorbitol by AR, and sorbitol is then oxidized by sorbitol dehydrogenase (SDH) to fructose. The enzymatic reaction using AR requires NADPH, whereas SDH requires nicotinamide adenine dinucleotide (NAD^+). AR constitutes the first and the rate-limiting step of the polyol pathway [165, 166]. Recent studies

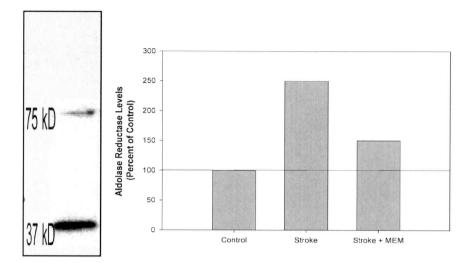

Fig. 27.13 A RAGE pathway mediates memantine neuroprotection. Quantitation of aldolase reductase levels using western blot analysis. Cortical samples from the identical brain areas of naïve control, stroke, and stroke plus memantine were run on standard SDS acrylamide gels and immunoblotted with antisera against aldolase reductase (Immunogen Inc., Norwood, MA). Embolization increased AR cortical levels, which was attenuated by memantine administration

have indicated that generation of advanced glycation end products (AGEs) is facilitated by the AR enzymatic pathway. Identification of glucose metabolism via the polyol or the AR pathway has generated interest in the use of AR pathway inhibitors as a novel metabolic therapy to protect from ischemia-induced cell death because an increase in glucose metabolism via the polyol pathway has been demonstrated in ischemic myocardial tissue [167]. Because recent studies have shown that AR activity increases under ischemic conditions, some attention has been focused on studying the benefits of AR inhibition [167]. Experimental studies have demonstrated that AR inhibition reduced ischemic injury and was associated with attenuation of the rise in cytosolic redox (NADH/NAD$^+$ ratio) and improved glycolysis in ischemic myocardium. AR inhibition also enhanced myocardial glucose oxidation and normalized ATP and ion homeostasis [168–172].

When uncontrolled, the increased production of AGEs forms the basis for nonenzymatic glycation resulting in tissue damage. Nonenzymatic glycation can be mediated by a large variety of carbohydrates, including glucose, methylglyoxal (MG), and 3-deoxyglucosone (3-DG). The identification of this specific pathway is extremely important to the understanding of mechanism involved in NMDA-mediated neurodegeneration and the neuroprotection observed in the presence of memantine. While preliminary, this data is an important step in identifying mechanisms involved in NMDA-induced neurodegeneration and possible protein that can be new drug targets. RAGE represents a suitable target for intervention to reduce ischemia-induced neurodegeneration and behavioral deficits.

The development of NMDA antagonists, independent of their pharmacological properties or specific mechanisms, remains disappointing as an avenue to treat stroke. Since high- and medium-affinity antagonists have very limited therapeutic windows, they will not be useful to large numbers of patients in a clinical situation, unless patients are enrolled during the hyperacute phase following a stroke [31].

5.3.6 Pleiotropic

Carbamylated Erythropoietin

Carbamylated erythropoietin (CEPO) is an erythropoietin (EPO) derivative which, unlike EPO, does not bind to the EPO receptor and does not stimulate the production of red blood cells nor does it possess procoagulation properties [173], see Lapchak [174–176] for reviews on EPO and CEPO. In an RSCEM study [177], CEPO significantly improved clinical rating scores when assessed 48 h following embolic strokes in rabbits. Based upon therapeutic window analysis, it appears that CEPO administration effectively improves motor function when given up to 3 h following a stroke. Because of recent detrimental effects of EPO in stroke patients, CEPO development has been halted due to safety concerns [178].

CNB-001 (Curcuminoid)

Recently, we discovered a novel series of multitarget curcumin-based compounds that have both neurotrophic and neuroprotective properties [179]; the parent drug is a pyrazole derivative of curcumin identified as CNB-001 that lacks the labile dicarbonyl group of curcumin. Evidence from three laboratories suggest that CNB-001 has the unique ability to be neuroprotective and block neuronal degeneration induced by a variety of insults, act as a neurotrophic factor mimetic in vitro [179], and either enhance the production of brain-derived neurotrophic factor (BDNF) [180], normalize BDNF levels after injury [180], or regulate BDNF signaling pathways in vivo [32]. Thus, this novel compound has the potential to block components of the ischemic cascade and possibly restore neuronal function long after a stroke. We recently showed that CNB-001 can significantly improve behavior in the RSCEM up to an hour following stroke and that protection is at least partially mediated via the maintenance of the MAP and PI3K-Akt kinase pathways and elevated CaMKIIα [32].

Simvastatin

Statins, 3-hydroxy-3-methylglutaryl coenzyme A reductase (HMG-CoA reductase) inhibitors, were initially designed as cholesterol-lowering drugs [181–183]. Because statins have been shown to be neuroprotective in rodent stroke models [184–187], we tested simvastatin (20 mg/kg, bolus subcutaneous injection) using the RSCEM. Simvastatin significantly improved clinical function when administered 1 h following embolization but was ineffective at 3 h. In combination studies with tPA, using a standard intravenous dose of 3.3 mg/kg (20% bolus, 80% infused), we found that simvastatin could be safely administered with tPA to improve clinical scores; however, the maximum behavioral improvement with the combination treatment was similar to either monotherapy alone. It has been suggested that simvastatin neuroprotection may be related to a variety of signaling pathways including Rho-kinase (ROCK), Janus kinases (JAKs)/signal transducers and activators of transcription (STAT) (JAK–STAT), or phosphatidylinositol-3-kinase (P13K)/PKB (P13K/Akt) [188–191]. We recently found that simvastatin treatment may produce behavioral improvement in embolized rabbits via a ROCK signaling pathway [192].

5.3.7   Thrombolytic Therapy

Three thrombolytics have been studied using the RSCEM. The original thrombolytic, tPA, is FDA-approved for the treatment of stroke. Since the therapeutic window for tPA in the RSCEM was previously discussed in Sects. 5.1 and 5.2, this section will review the second-generation tPA-like thrombolytics, tenecteplase and the plasminogen activator microplasmin, both of which are in clinical trial development [193, 194].

Tenecteplase

Tenecteplase (TNK) is a second-generation genetically modified form of wild-type tPA [195], a plasminogen activator engineered to have increased fibrin specificity and a longer half-life in plasma compared to tPA. TNK is produced by multiple point mutations designed to prolong the biological half-life from 4 min (tPA) to approximately 18 min and to increase its specificity for fibrin so that thrombolysis is enhanced [196–198]. By inserting mutations at amino acids 103 and 117, TNK has a reduced clearance rate and is 80-fold more resistant to PA inhibitor-1 (PA-I-1) because of the substitution of alanine residues at amino acid positions 296–299, compared to tPA.

We determined both the efficacy and safety profiles of TNK using rabbit embolic stroke models [62, 199]. In the RSCEM, TNK was directly compared to tPA using dose-response analysis and therapeutic window analysis. For dose-response curves, we found that the maximally effective dose of tPA was 3.3 mg/kg, and higher doses could not be administered due to excessive peripheral bleeding [62], whereas TNK was effective with doses in the range of 0.9–3.3 mg/kg, without excessive peripheral bleeding. The therapeutic window for tPA was limited to dosing up to 1 h following embolization, whereas we found that TNK (0.9 mg/kg) was effective up to 3 h following embolization [46]. In the RSCEM, there was a low rate of hemorrhage with TNK administration [46]. Using a variation of the RSCEM where large clots are injected through the carotid cannula into the cerebral vasculature (RLCEM), we found that both thrombolytics, tPA and TNK, produced similar levels of ICH, when tested at 3.3 and 0.6 or 0.9 mg/kg, respectively [199].

Microplasmin

Plasminogen activators like tPA are specific proteolytic enzymes that convert the inactive proenzyme plasminogen to plasmin, which is a potent, nonspecific protease that cleaves blood fibrin clots and several other extracellular proteins. Efforts have been made to develop a fragment of plasmin that can be biologically active as a possible approach to enhance clot lysis. Microplasmin is one approach that has been attempted [200]. Microplasmin is a disulfide-bonded molecule consisting of two polypeptide chains [201, 202]. One of the polypeptides is the 230 amino acid B chain of plasmin, and the other is a 31 amino acid residue of the carboxy-terminal A chain of plasmin. Together, microplasmin is a 28.6-kD positively charged hydrophobic molecule that has less affinity for fibrin and reacts slowly with α2 antiplasmin (α2-AP), which results in a half-life in blood of approximately 4 s [203].

Microplasmin administration was compared with the biological efficacy and safety profile of both tPA and TNK, although not in parallel in the same study [62, 199, 204, 205]. Dose-response analysis of microplasmin showed that doses in the same range previously described for tPA or TNK were required to produce behavioral improvements in embolized rabbits; the maximally effective dose being 4 mg/kg. However, the microplasmin dose-response curve is an inverted U-shaped curve with a narrow efficacy range [204]. Complete therapeutic window analysis was not conducted in the RSCEM; data shows that microplasmin is effective when given i.v. 1 h after embolization. When microplasmin was tested for safety, experiments showed that microplasmin did not increase ICH rate [204].

Taken together, data from the RSCEM shows that thrombolytics, independent of their mode of action or specificity, can be effectively developed as drugs to increase recanalization and cerebral reperfusion and improve behavior. Moreover, the model has adequate sensitivity to distinguish between effective thrombolytic doses and can also be useful to develop therapeutic efficacy window profiles.

6 A Promising Future

Clinical translation of experimental stroke data remains a daunting task. There is a need to have a firm basis for clinical development, a basis that cannot come from any animal model that has not been sufficiently validated. Based upon the comprehensive data in this chapter, there is "A Window for Effective Translation" if one considers the wealth of information provided regarding effective therapies and the time frame when they can be used to promote functional recovery.

Table 27.2 presents some of the novel treatments that have been tested in the RSCEM for efficacy and in clinical trials. As discussed earlier, in the RSCEM, tPA improves behavioral scores when given 1–1.5 h following embolization [62]. In the stroke patient population, the newly expanded therapeutic window is 4.5 h after stroke onset [5]. Using correlative analysis between humans and rabbits, including the extended therapeutic window for tPA in stroke patients, there is a two- to threefold

Table 27.2 Correlative data for clinical and experimental RSCEM studies

Therapy	Effective window in RSCEM[a]	Effect in AIS patients[b]	TW (human/rabbit) ratio AIS/RSCEM
Caffeinol	Ineffective [144]	Clinical development stopped [211]	Cannot be calculated
Ebselen	Improved behavior TW of 5 min [119]	Ongoing trial recruitment [212]	Cannot be calculated
Microplasmin	Improved behavior TW-undefined Effective at 1 h [204]	Ongoing clinical trial [193]	Cannot be calculated
NILT	Improved behavior [64] TW of 6 h	Improved NIHSS scores in patients enrolled with an NIHSS <16. TT of 14.6 h [140] (for details, see [135])	14.6/6 = ratio 2.43
NXY-059	Improved behavior [85, 129] TW of 1 h	No efficacy when studied with a TT of 3.76 h [130, 131]	No efficacy/1 = ratio 0
Radicut	Improved behavior [53] TW of 3 h	Improved NIHSS and/or mRS scores with a TW of 24–72 h (for details, see [24])	24–72/3 = ratio 8–24
tPA	Improved behavior [45, 62] TW of 1–1.5 h	Improved clinical scores with a TW of 3–4.5 h [5, 11]	3–4.5/1–1.5 = ratio 3
TNK	Improved behavior [62] TW of 3 h	Clinical trial stopped [194]	Cannot be calculated

TT, time to treatment; *TW*, therapeutic window

[a]RSCEM end point: behavior using the analysis using quantal analysis

[b]AIS patient end point: clinical rating scores using NIHSS and/or mRS

time differential between the RSCEM model and patients, i.e., when using the RSCEM, it is important to note that a drug that is found to be effective at 1.5 h following embolization may be effective in stroke victims within 3–4.5 h after stroke. This important correlative ratio is now the standard for all future development using the RSCEM.

Of the few treatments that have been tested and successfully developed using the RSCEM, there are two other examples that provide additional insight into requirements for future drug/device development. For example, Radicut, which is effective up to 3 h in the RSCEM [53], has now been shown to be effective when administered 24–72 h after stroke [206–209]. Moreover, NILT was shown to be effective up to 6 h following a stroke [64] and has been shown to have some level of efficacy in patients with a time to treatment of 14.6 h [139, 140], but this seems to depend somewhat on the extent of stroke severity based upon NIHSS scores. The last example is NXY-059, which was effective up to 1 h following embolization in the RSCEM

[1, 129] but was ineffective in the SAINT II clinical trial [131], with a mean time to treatment of 3.6 h.

As shown in Table 27.2, there are three other treatment strategies currently in clinical trials. For ebselen, a 5-min therapeutic window would suggest no efficacy in patients. However, both microplasmin and TNK have effective windows of 1 and 3 h, respectively, in the RSCEM, suggesting that both treatments may show benefit in patients, unless dose-limiting side effects are observed.

7   Summary and Conclusions

The RSCEM is a useful translational animal model to predict the efficacy of treatments that should be advanced from the laboratory into the clinic. The strength behind the RSCEM is directly associated with use of a powerful statistical method to quantitate the effects of a treatment on a heterogeneous population of stroke victims, quite similar to that used in a clinical trial setting, where a standard NIHSS/mRS scale is used.

In addition, this chapter describes the utility of the RSCEM to quantitate physiological changes such as RCBF, histochemical changes using TTC staining, and biochemical changes (i.e., mitochondrial function by way of ATP measurements). Moreover, as described, we have recently used the RSCEM to investigate extracellular and intracellular mechanisms involved in cell survival and signaling. The studies have resulted in the identification of pathways that may be new targets to attenuate stroke-induced cell death (i.e., Akt, ERK, CAMK, ORP150), stroke-induced behavioral deficits (i.e., ROCK), and glutamate-induced cellular dysfunction (i.e., RAGE).

Based upon comprehensive scientific data presented in this chapter, the RSCEM should continually be used as a potential **"gold standard"** model and translational tool in order to develop therapies to attenuate embolism-induced behavioral deficits and possibly promote recovery of function and improve the well-being of stroke victims. The model is well suited for the discovery and testing of all types of pharmacological agents and devices, without limitation.

Acknowledgments This article was supported by an NINDS Translational Research grant U01 NS60685 and NIH American Recovery and Reinvestment Act R01 grant NS060864.

References

1. Lapchak PA. Development of thrombolytic therapy for stroke: a perspective. Expert Opin Investig Drugs. 2002;11(11):1623–32.
2. Schellinger PD, Fiebach JB, Mohr A, Ringleb PA, Jansen O, Hacke W. Thrombolytic therapy for ischemic stroke—a review. Part II—Intra-arterial thrombolysis, vertebrobasilar stroke, phase IV trials, and stroke imaging. Crit Care Med. 2001;29(9):1819–25.

3. Schellinger PD, Fiebach JB, Mohr A, Ringleb PA, Jansen O, Hacke W. Thrombolytic therapy for ischemic stroke—a review. Part I—intravenous thrombolysis. Crit Care Med. 2001;29(9): 1812–8.

4. Verstraete M. Newer thrombolytic agents. Ann Acad Med Singapore. 1999;28(3):424–33.

5. Hacke W, Kaste M, Bluhmki E, Brozman M, Davalos A, Guidetti D, et al. Thrombolysis with alteplase 3 to 4.5 hours after acute ischemic stroke. N Engl J Med. 2008;359(13):1317–29.

6. Lansberg MG, Bluhmki E, Thijs VN. Efficacy and safety of tissue plasminogen activator 3 to 4.5 hours after acute ischemic stroke: a metaanalysis. Stroke. 2009;40(7):2438–41.

7. Chernyshev OY, Martin-Schild S, Albright KC, Barreto A, Misra V, Acosta I, et al. Safety of tPA in stroke mimics and neuroimaging-negative cerebral ischemia. Neurology. 2010;74(17): 1340–5.

8. Lees KR, Bluhmki E, von Kummer R, Brott TG, Toni D, Grotta JC, et al. Time to treatment with intravenous alteplase and outcome in stroke: an updated pooled analysis of ECASS, ATLANTIS, NINDS, and EPITHET trials. Lancet. 2010;375(9727):1695–703.

9. Lapchak PA. Translational stroke research using a rabbit embolic stroke model: a correlative analysis hypothesis for novel therapy development. Transl Stroke Res. 2010;1(2):96–107.

10. Saver JL, Albers GW, Dunn B, Johnston KC, Fisher M. Stroke Therapy Academic Industry Roundtable (STAIR) recommendations for extended window acute stroke therapy trials. Stroke. 2009;40(7):2594–600.

11. NINDS. Tissue plasminogen activator for acute ischemic stroke. The National Institute of Neurological Disorders and Stroke rt-PA Stroke Study Group. N Engl J Med. 1995;333(24): 1581–7.

12. Petty GW, Brown Jr RD, Whisnant JP, Sicks JD, O'Fallon WM, Wiebers DO. Ischemic stroke subtypes: a population-based study of functional outcome, survival, and recurrence. Stroke. 2000;31(5):1062–8.

13. Petty GW, Brown Jr RD, Whisnant JP, Sicks JD, O'Fallon WM, Wiebers DO. Ischemic stroke subtypes: a population-based study of incidence and risk factors. Stroke. 1999;30(12):2513–6.

14. Lapchak PA, Araujo DM. Advances in ischemic stroke treatment: neuroprotective and combination therapies. Expert Opin Emerg Drugs. 2007;12(1):97–112.

15. Dirnagl U, Iadecola C, Moskowitz MA. Pathobiology of ischaemic stroke: an integrated view. Trends Neurosci. 1999;22(9):391–7.

16. Moskowitz MA, Lo EH, Iadecola C. The science of stroke: mechanisms in search of treatments. Neuron. 2010;67(2):181–98.

17. Lipton P. Ischemic cell death in brain neurons. Physiol Rev. 1999;79(4):1431–568.

18. Michel P, Bogousslavsky J. Penumbra is brain: no excuse not to perfuse. Ann Neurol. 2005;58(5):661–3.

19. Moustafa RR, Baron JC. Imaging the penumbra in acute stroke. Curr Atheroscler Rep. 2006; 8(4):281–9.

20. Fisher M. The ischemic penumbra: identification, evolution and treatment concepts. Cerebrovasc Dis. 2004;17:1–6.

21. Muir KW. Heterogeneity of stroke pathophysiology and neuroprotective clinical trial design. Stroke. 2002;33(6):1545–50.

22. Lapchak PA. Emerging therapies: pleiotropic multi-target drugs to treat stroke victims. Transl Stroke Res. 2011;2(2):129–35.

23. White BC, Sullivan JM, DeGracia DJ, O'Neil BJ, Neumar RW, Grossman LI, et al. Brain ischemia and reperfusion: molecular mechanisms of neuronal injury. J Neurol Sci. 2000;179 (S 1–2):1–33.

24. Lapchak PA. A critical assessment of edaravone acute ischemic stroke efficacy trials: is edaravone an effective neuroprotective therapy? Expert Opin Pharmacother. 2010;11(10):1753–63.

25. Lapchak PA. Neuroprotective and neurotrophic curcuminoids to treat stroke: a translational perspective. Expert Opin Investig Drugs. 2011;20(1):13–22.

26. Ginsberg MD. Neuroprotection for ischemic stroke: past, present and future. Neuropharmacology. 2008;55(3):363–89.

27. Fisher M. New approaches to neuroprotective drug development. Stroke. 2011;42(1 Suppl): S24–7.
28. Woodruff TM, Thundyil J, Tang SC, Sobey CG, Taylor SM, Arumugam TV. Pathophysiology, treatment, and animal and cellular models of human ischemic stroke. Mol Neurodegener. 2011;6(1):11.
29. Tuttolomondo A, Di Sciacca R, Di Raimondo D, Arnao V, Renda C, Pinto A, et al. Neuron protection as a therapeutic target in acute ischemic stroke. Curr Top Med Chem. 2009; 9(14):1317–34.
30. Scott PA, Frederiksen SM, Kalbfleisch JD, Xu Z, Meurer WJ, Caveney AF, et al. Safety of intravenous thrombolytic use in four emergency departments without acute stroke teams. Acad Emerg Med. 2010;17(10):1062–71.
31. Ferguson KN, Kidwell CS, Starkman S, Saver JL. Hyperacute treatment initiation in neuroprotective agent stroke trials. J Stroke Cerebrovasc Dis. 2004;13(3):109–12.
32. Lapchak PA, Schubert DR, Maher PA. Delayed treatment with a novel neurotrophic compound reduces behavioral deficits in rabbit ischemic stroke. J Neurochem. 2011;116(1):122–31.
33. STAIR. Recommendations for standards regarding preclinical neuroprotective and restorative drug development. Stroke. 1999;30(12):2752–8.
34. Hacke W, Brott T, Caplan L, Meier D, Fieschi C, von Kummer R, et al. Thrombolysis in acute ischemic stroke: controlled trials and clinical experience. Neurology. 1999;53(7):S3–14.
35. Hsia AW, Sachdev HS, Tomlinson J, Hamilton SA, Tong DC. Efficacy of IV tissue plasminogen activator in acute stroke: does stroke subtype really matter? Neurology. 2003;61(1):71–5.
36. Lyden P, Lu M, Jackson C, Marler J, Kothari R, Brott T, et al. Underlying Structure of the National Institutes of Health Stroke Scale: results of a factor analysis. Stroke. 1999;30:2347.
37. Lyden P, Raman R, Liu L, Grotta J, Broderick J, Olson S, et al. NIHSS training and certification using a new digital video disk is reliable. Stroke. 2005;36(11):2446–9.
38. Sulter G, Steen C, De Keyser J. Use of the Barthel Index and modified Rankin Scale in acute stroke trials. Stroke. 1999;30:1538.
39. Wilson JT, Hareendran A, Grant M, Baird T, Schultz UGR, Muir KW, et al. Improving the assessment of outcomes in strokes. Stroke. 2002;33:2243.
40. Young FB, Lees KR, Weir CJ. Strengthening acute stroke trials through optimal use of disability end points. Stroke. 2003;34(11):2676–80.
41. Ginsberg MD. The validity of rodent brain-ischemia models is self-evident. Arch Neurol. 1996;53(10):1065–7; discussion 70.
42. Ginsberg MD. Life after cerovive: a personal perspective on ischemic neuroprotection in the post-NXY-059 era. Stroke. 2007;38(6):1967–72.
43. Hoyte L, Kaur J, Buchan AM. Lost in translation: taking neuroprotection from animal models to clinical trials. Exp Neurol. 2004;188(2):200–4.
44. Turner R, Jickling G, Sharp F. Are underlying assumptions of current animal models of human stroke correct: from STAIRS to high hurdles? Transl Stroke Res. 2011;2(2):138–43.
45. Zivin JA, Fisher M, DeGirolami U, Hemenway CC, Stashak JA. Tissue plasminogen activator reduces neurological damage after cerebral embolism. Science. 1985;230(4731):1289.
46. Lapchak PA. Effect of internal carotid artery reperfusion in combination with Tenecteplase on clinical scores and hemorrhage in a rabbit embolic stroke model. Brain Res. 2009;1294: 211–7.
47. Lapchak PA, De Taboada L. Transcranial near infrared laser treatment (NILT) increases cortical adenosine-5'-triphosphate (ATP) content following embolic strokes in rabbits. Brain Res. 2010;1306:100–5.
48. Lapchak PA, Streeter J, DeTaboada L. Transcranial near infrared laser therapy (NILT) to treat acute ischemic stroke: a review of efficacy, safety and possible mechanism of action derived from rabbit embolic stroke studies SPIE Proceedings. 7552R:7552–31.
49. Kawaguchi M, Furuya H, Patel PM. Neuroprotective effects of anesthetic agents. J Anesth. 2005;19(2):150–6.
50. Koerner IP, Brambrink AM. Brain protection by anesthetic agents. Curr Opin Anaesthesiol. 2006;19(5):481–6.

51. Matchett GA, Allard MW, Martin RD, Zhang JH. Neuroprotective effect of volatile anesthetic agents: molecular mechanisms. Neurol Res. 2009;31(2):128–34.
52. Nishikawa K, MacIver MB. Excitatory synaptic transmission mediated by NMDA receptors is more sensitive to isoflurane than are non-NMDA receptor-mediated responses. Anesthesiology. 2000;92(1):228–36.
53. Lapchak PA, Zivin JA. The lipophilic multifunctional antioxidant edaravone (radicut) improves behavior following embolic strokes in rabbits: a combination therapy study with tissue plasminogen activator. Exp Neurol. 2009;215(1):95–100.
54. Waud DR. On biological assays involving quantal responses. J Pharmacol Exp Ther. 1972;183:577–607.
55. Lapchak PA. Translational stroke research using a rabbit embolic stroke model: a correlative analysis hypothesis for novel therapy development. Trans Stroke Res. 2010;1:96–107.
56. Kasner SE. Clinical interpretation and use of stroke scales. Lancet Neurol. 2006;5(7):603–12.
57. Lindsell CJ, Alwell K, Moomaw CJ, Kleindorfer DO, Woo D, Flaherty ML, et al. Validity of a retrospective National Institutes of Health Stroke Scale scoring methodology in patients with severe stroke. J Stroke Cerebrovasc Dis. 2005;14(6):281–3.
58. Lyden P, Brott T, Tilley B, Welch KM, Mascha EJ, Levine S, et al. Improved reliability of the NIH Stroke Scale using video training. NINDS TPA Stroke Study Group. Stroke. 1994;25(11):2220–6.
59. Lyden P, Lu M, Jackson C, Marler J, Kothari R, Brott T, et al. Underlying structure of the National Institutes of Health Stroke Scale: results of a factor analysis. NINDS tPA Stroke Trial Investigators. Stroke. 1999;30(11):2347–54.
60. Waud DR. On biological assays involving quantal responses. J Pharmacol Exp Ther. 1972;183(3):577–607.
61. Lapchak PA, Zhang JH. Resolving the negative data publication dilemma in translational stroke research. Transl Stroke Res. 2011;2(1):1–6.
62. Lapchak PA, Araujo DM, Zivin JA. Comparison of tenecteplase with alteplase on clinical rating scores following small clot embolic strokes in rabbits. Exp Neurol. 2004;185(1):154–9.
63. Lapchak PA, Salgado KF, Chao CH, Zivin JA. Transcranial near-infrared light therapy improves motor function following embolic strokes in rabbits: an extended therapeutic window study using continuous and pulse frequency delivery modes. Neuroscience. 2007;148(4):907–14.
64. Lapchak PA, Wei J, Zivin JA. Transcranial infrared laser therapy improves clinical rating scores after embolic strokes in rabbits. Stroke. 2004;35(8):1985–8.
65. Lapchak PA, Araujo DM, Song D, Wei J, Purdy R, Zivin JA. Effects of the spin trap agent disodium-[(tert-butylimino)methyl]benzene-1,3-disulfonate N-oxide (generic NXY-059) on intracerebral hemorrhage in a rabbit large clot embolic stroke model: combination studies with tissue plasminogen activator. Stroke. 2002;33(6):1665–70.
66. Silver JH, Lapchak PA. Continuous monitoring of changes in plasma nitrite following cerebral ischemia in a rabbit embolic stroke model. Transl Stroke Res. 2011;2(2):218–26.
67. Bederson JB, Pitts LH, Germano SM, Nishimura MC, Davis RL, Bartkowski HM. Evaluation of 2,3,5-triphenyltetrazolium chloride as a stain for detection and quantification of experimental cerebral infarction in rats. Stroke. 1986;17(6):1304–8.
68. Guluma KZ, Lapchak PA. Comparison of the post-embolization effects of tissue-plasminogen activator and simvastatin on neurological outcome in a clinically relevant rat model of acute ischemic stroke. Brain Res. 2010;1354:206–16.
69. Kricka LJ. Clinical and biochemical applications of luciferases and luciferins. Anal Biochem. 1988;175:14–21.
70. Fisher RL, Gandolfi AJ, Brendel K. Human liver quality is a dominant factor in the outcome of in vitro studies. Cell Biol Toxicol. 2001;17(3):179–89.
71. Eide FF, Lowenstein DH, Reichardt LF. Neurotrophins and their receptors–current concepts and implications for neurologic disease. Exp Neurol. 1993;121(2):200–14.
72. Olson L, Backman L, Ebendal T, Eriksdotter-Jonhagen M, Hoffer B, Humpel C, et al. Role of growth factors in degeneration and regeneration in the central nervous system; clinical

experiences with NGF in Parkinson's and Alzheimer's diseases. J Neurol. 1994;242 (1 Suppl 1):S12–5.

73. Patapoutian A, Reichardt LF. Trk receptors: mediators of neurotrophin action. Curr Opin Neurobiol. 2001;11(3):272–80.

74. Reichardt LF. Neurotrophin-regulated signalling pathways. Philos Trans R Soc Lond B Biol Sci. 2006;361(1473):1545–64.

75. Wu D. Neuroprotection in experimental stroke with targeted neurotrophins. NeuroRx. 2005;2(1):120–8.

76. Yano H, Chao MV. Neurotrophin receptor structure and interactions. Pharm Acta Helv. 2000; 74(2–3):253–60.

77. Rossler OG, Giehl KM, Thiel G. Neuroprotection of immortalized hippocampal neurons by brain-derived neurotrophic factor and Raf-1 protein kinase: role of extracellular signal-regulated protein kinase and phosphatidylinositol 3-kinase. J Neurochem. 2004;88(5):1240–52.

78. Schratt GM, Nigh EA, Chen WG, Hu L, Greenberg ME. BDNF regulates the translation of a select group of mRNAs by a mammalian target of rapamycin-phosphatidylinositol 3-kinase-dependent pathway during neuronal development. J Neurosci. 2004;24(33):7366–77.

79. Tamatani M, Matsuyama T, Yamaguchi A, Mitsuda N, Tsukamoto Y, Taniguchi M, et al. ORP150 protects against hypoxia/ischemia-induced neuronal death. Nat Med. 2001;7(3): 317–23.

80. Ying SW, Futter M, Rosenblum K, Webber MJ, Hunt SP, Bliss TV, et al. Brain-derived neurotrophic factor induces long-term potentiation in intact adult hippocampus: requirement for ERK activation coupled to CREB and upregulation of Arc synthesis. J Neurosci. 2002;22(5): 1532–40.

81. Yin Y, Edelman GM, Vanderklish PW. The brain-derived neurotrophic factor enhances synthesis of Arc in synaptoneurosomes. Proc Natl Acad Sci U S A. 2002;99(4):2368–73.

82. Mehta SL, Manhas N, Raghubir R. Molecular targets in cerebral ischemia for developing novel therapeutics. Brain Res Rev. 2007;54(1):34–66.

83. Zhao H, Sapolsky RM, Steinberg GK. Phosphoinositide-3-kinase/akt survival signal pathways are implicated in neuronal survival after stroke. Mol Neurobiol. 2006;34(3):249–70.

84. Liu C, Wu J, Xu K, Cai F, Gu J, Ma L, et al. Neuroprotection by baicalein in ischemic brain injury involves PTEN/AKT pathway. J Neurochem. 2010;112:1500–12.

85. Lapchak PA, Araujo DM, Song D, Wei J, Zivin JA. Neuroprotective effects of the spin trap agent disodium-[(tert-butylimino)methyl]benzene-1,3-disulfonate N-oxide (generic NXY-059) in a rabbit small clot embolic stroke model: combination studies with the thrombolytic tissue plasminogen activator. Stroke. 2002;33(5):1411–5.

86. Zivin JA, Lyden PD, DeGirolami U, Kochhar A, Mazzarella V, Hemenway CC, et al. Tissue plasminogen activator. Reduction of neurologic damage after experimental embolic stroke. Arch Neurol. 1988;45(4):387–91.

87. Albers GW, Bates VE, Clark WM, Bell R, Verro P, Hamilton SA. Intravenous tissue-type plasminogen activator for treatment of acute stroke: the Standard Treatment with Alteplase to Reverse Stroke (STARS) study. JAMA. 2000;283(9):1145–50.

88. Alberts MJ. tPA in acute ischemic stroke: United States experience and issues for the future. Neurology. 1998;51(3 Suppl 3):S53–5.

89. Christou I, Alexandrov AV, Burgin WS, Wojner AW, Felberg RA, Malkoff M, et al. Timing of recanalization after tissue plasminogen activator therapy determined by transcranial Doppler correlates with clinical recovery from ischemic stroke. Stroke. 2000;31(8):1812–6.

90. Clark WM, Albers GW, Madden KP, Hamilton S. The rtPA (alteplase) 0- to 6-hour acute stroke trial, part A (A0276g): results of a double-blind, placebo-controlled, multicenter study. Thrombolytic therapy in acute ischemic stroke study investigators. Stroke. 2000;31(4):811–6.

91. Grotta JC, Alexandrov AV. tPA-associated reperfusion after acute stroke demonstrated by SPECT. Stroke. 1998;29(2):429–32.

92. Grotta JC, Burgin WS, El-Mitwalli A, Long M, Campbell M, Morgenstern LB, et al. Intravenous tissue-type plasminogen activator therapy for ischemic stroke: Houston experience 1996 to 2000. Arch Neurol. 2001;58(12):2009–13.

93. Hacke W, Kaste M, Fieschi C, Toni D, Lesaffre E, von Kummer R, et al. Intravenous thrombolysis with recombinant tissue plasminogen activator for acute hemispheric stroke. The European Cooperative Acute Stroke Study (ECASS). JAMA. 1995;274(13):1017–25.

94. Reeves MJ, Arora S, Broderick JP, Frankel M, Heinrich JP, Hickenbottom S, et al. Acute stroke care in the US: results from 4 pilot prototypes of the Paul Coverdell National Acute Stroke Registry. Stroke. 2005;36(6):1232–40.

95. The National Institute of Neurological Disorders and Stroke rt-PA Stroke Study Group. Tissue plasminogen activator for acute ischemic stroke. N Engl J Med. 1995;333(24):1581–7.

96. Zunker P, Schick A, Padro T, Kienast J, Phillips A, Ringelstein EB. Tissue plasminogen activator and plasminogen activator inhibitor in patients with acute ischemic stroke: relation to stroke etiology. Neurol Res. 1999;21(8):727–32.

97. del Zoppo G. Thrombolytic therapy in cerebrovascular disease. Curr Concepts Cerebrovasc Dis. 1988;23:7.

98. Bowes MP, Zivin JA, Rothlein R. Monoclonal antibody to the ICAM-1 adhesion site reduces neurological damage in a rabbit cerebral embolism stroke model. Exp Neurol. 1993;119(2): 215–9.

99. Bowes MP, Rothlein R, Fagan SC, Zivin JA. Monoclonal antibodies preventing leukocyte activation reduce experimental neurologic injury and enhance efficacy of thrombolytic therapy. Neurology. 1995;45(4):815–9.

100. Bousova I, Martin J, Jahodar L, Dusek J, Palicka V, Drsata J. Evaluation of in vitro effects of natural substances of plant origin using a model of protein glycoxidation. J Pharm Biomed Anal. 2005;37(5):957–62.

101. Martin J, Dusek J. The Baikal scullcap (*Scutellaria baicalensis* Georgi)—a potential source of new drugs. Ceska Slov Farm. 2002;51(6):277–83.

102. Roy MK, Nakahara K, Na TV, Trakoontivakorn G, Takenaka M, Isobe S, et al. Baicalein, a flavonoid extracted from a methanolic extract of *Oroxylum indicum* inhibits proliferation of a cancer cell line in vitro via induction of apoptosis. Die Pharmazie. 2007;62(2):149–53.

103. Huang Y, Tsang SY, Yao X, Chen ZY. Biological properties of baicalein in cardiovascular system. Curr Drug Targets. 2005;5(2):177–84.

104. Huang WH, Lee AR, Chien PY, Chou TC. Synthesis of baicalein derivatives as potential anti-aggregatory and anti-inflammatory agents. J Pharm Pharmacol. 2005;57(2):219–25.

105. Ma Z, Otsuyama K, Liu S, Abroun S, Ishikawa H, Tsuyama N, et al. Baicalein, a component of *Scutellaria radix* from Huang-Lian-Jie-Du-Tang (HLJDT), leads to suppression of proliferation and induction of apoptosis in human myeloma cells. Blood. 2005;105(8):3312–8.

106. Lapchak PA, Maher P, Schubert D, Zivin JA. Baicalein, an antioxidant 12/15 lipoxygenase inhibitor improves clinical rating scores following multiple infarct embolic strokes. Neuroscience. 2007;150(3):585–91.

107. Goldfinger TM. Beyond the French paradox: the impact of moderate beverage alcohol and wine consumption in the prevention of cardiovascular disease. Cardiol Clin. 2003;21(3):449–57.

108. Kar P, Laight D, Shaw KM, Cummings MH. Flavonoid-rich grapeseed extracts: a new approach in high cardiovascular risk patients? Int J Clin Pract. 2006;60(11):1484–92.

109. Nijveldt RJ, van Nood E, van Hoorn DE, Boelens PG, van Norren K, van Leeuwen PA. Flavonoids: a review of probable mechanisms of action and potential applications. Am J Clin Nutr. 2001;74(4):418–25.

110. Renaud S, Ruf JC. The French paradox: vegetables or wine. Circulation. 1994;90(6):3118–9.

111. Manach C, Scalbert A, Morand C, Remesy C, Jimenez L. Polyphenols: food sources and bioavailability. Am J Clin Nutr. 2004;79(5):727–47.

112. Scalbert A, Manach C, Morand C, Remesy C, Jimenez L. Dietary polyphenols and the prevention of diseases. Crit Rev Food Sci Nutr. 2005;45(4):287–306.

113. Chassevent F. Chlorogenic acid, physiological and pharmacological activity. Ann Nutr Aliment. 1969;23 Suppl 1:1–14.

114. Jung HA, Park JC, Chung HY, Kim J, Choi JS. Antioxidant flavonoids and chlorogenic acid from the leaves of *Eriobotrya japonica*. Arch Pharm Res. 1999;22(2):213–8.

115. dos Santos MD, Almeida MC, Lopes NP, de Souza GE. Evaluation of the anti-inflammatory, analgesic and antipyretic activities of the natural polyphenol chlorogenic acid. Biol Pharm Bull. 2006;29(11):2236–40.

116. Jin UH, Lee JY, Kang SK, Kim JK, Park WH, Kim JG, et al. A phenolic compound, 5-caffeoylquinic acid (chlorogenic acid), is a new type and strong matrix metalloproteinase-9 inhibitor: isolation and identification from methanol extract of *Euonymus alatus*. Life Sci. 2005;77(22):2760–9.

117. Lapchak PA. The phenylpropanoid micronutrient chlorogenic acid improves clinical rating scores in rabbits following multiple infarct ischemic strokes: synergism with tissue plasminogen activator. Exp Neurol. 2007;205(2):407–13.

118. Parnham M, Sies H. Ebselen: prospective therapy for cerebral ischaemia. Expert Opin Investig Drugs. 2000;9(3):607–19.

119. Lapchak PA, Zivin JA. Ebselen, a seleno-organic antioxidant, is neuroprotective after embolic strokes in rabbits: synergism with low-dose tissue plasminogen activator. Stroke. 2003;34(8):2013–8.

120. Seo JY, Lee CH, Cho JH, Choi JH, Yoo KY, Kim DW, et al. Neuroprotection of ebselen against ischemia/reperfusion injury involves GABA shunt enzymes. J Neurol Sci. 2009; 285(1–2):88–94.

121. Yamagata K, Ichinose S, Miyashita A, Tagami M. Protective effects of ebselen, a seleno-organic antioxidant on neurodegeneration induced by hypoxia and reperfusion in stroke-prone spontaneously hypertensive rat. Neuroscience. 2008;153(2):428–35.

122. Dajas F, Rivera-Megret F, Blasina F, Arredondo F, Abin-Carriquiry JA, Costa G, et al. Neuroprotection by flavonoids. Braz J Med Biol Res. 2003;36(12):1613–20.

123. Maher P. A comparison of the neurotrophic activities of the flavonoid fisetin and some of its derivatives. Free Radic Res. 2006;40(10):1105–11.

124. Maher P. The flavonoid fisetin promotes nerve cell survival from trophic factor withdrawal by enhancement of proteasome activity. Arch Biochem Biophys. 2008;476(2):139–44.

125. Rivera F, Urbanavicius J, Gervaz E, Morquio A, Dajas F. Some aspects of the in vivo neuroprotective capacity of flavonoids: bioavailability and structure-activity relationship. Neurotox Res. 2004;6(7–8):543–53.

126. Maher P, Salgado KF, Zivin JA, Lapchak PA. A novel approach to screening for new neuroprotective compounds for the treatment of stroke. Brain Res. 2007;1173:117–25.

127. Kuroda S, Tsuchidate R, Smith ML, Maples KR, Siesjo BK. Neuroprotective effects of a novel nitrone, NXY-059, after transient focal cerebral ischemia in the rat. J Cereb Blood Flow Metab. 1999;19(7):778–87.

128. Green AR, Lanbeck-Vallen K, Ashwood T, Lundquist S, Lindstrom Boo E, Jonasson H, et al. Brain penetration of the novel free radical trapping neuroprotectant NXY-059 in rats subjected to permanent focal ischemia. Brain Res. 2006;1072(1):224–6.

129. Lapchak PA, Song D, Wei J, Zivin JA. Coadministration of NXY-059 and tenecteplase six hours following embolic strokes in rabbits improves clinical rating scores. Exp Neurol. 2004;188(2):279–85.

130. Lees KR, Zivin JA, Ashwood T, Davalos A, Davis SM, Diener HC, et al. NXY-059 for acute ischemic stroke. N Engl J Med. 2006;354(6):588–600.

131. Shuaib A, Lees KR, Lyden P, Grotta J, Davalos A, Davis SM, et al. NXY-059 for the treatment of acute ischemic stroke. N Engl J Med. 2007;357(6):562–71.

132. Bath PM, Gray LJ, Bath AJ, Buchan A, Miyata T, Green AR. Effects of NXY-059 in experimental stroke: an individual animal meta-analysis. Br J Pharmacol. 2009;157(7):1157–71.

133. Watanabe T, Tahara M, Todo S. The novel antioxidant edaravone: from bench to bedside. Cardiovasc Ther. 2008;26(2):101–14.

134. Yoshida H, Yanai H, Namiki Y, Fukatsu-Sasaki K, Furutani N, Tada N. Neuroprotective effects of edaravone: a novel free radical scavenger in cerebrovascular injury. CNS Drug Rev. 2006;12(1):9–20.

135. Lapchak PA. Taking a light approach to treating acute ischemic stroke patients: transcranial near-infrared laser therapy translational science. Ann Med. 2010;42(8):576–86.

136. Detaboada L, Ilic S, Leichliter-Martha S, Oron U, Oron A, Streeter J. Transcranial application of low-energy laser irradiation improves neurological deficits in rats following acute stroke. Lasers Surg Med. 2006;38(1):70–3.

137. Ilic S, Leichliter S, Streeter J, Oron A, DeTaboada L, Oron U. Effects of power densities, continuous and pulse frequencies, and number of sessions of low-level laser therapy on intact rat brain. Photomed Laser Surg. 2006;24(4):458–66.

138. Oron A, Oron U, Chen J, Eilam A, Zhang C, Sadeh M, et al. Low-level laser therapy applied transcranially to rats after induction of stroke significantly reduces long-term neurological deficits. Stroke. 2006;37(10):2620–4.

139. Lampl Y, Zivin JA, Fisher M, Lew R, Welin L, Dahlof B, Andersson B, Perez J, Caparo C, Ilic S, Oron U. Infrared laser therapy for ischemic stroke—a new treatment strategy: results of the neuorthera effectiveness and safety Trial-1 (NEST-1). Stroke. 2007;38(6):1843–9.

140. Zivin JA, Albers GW, Bornstein N, Chippendale T, Dahlof B, Devlin T, et al. Effectiveness and safety of transcranial laser therapy for acute ischemic stroke. Stroke. 2009;40(4):1359–64.

141. NEST-3. http://clinicaltrials.gov/ct2/show/NCT01120301; 2012. Accessed Jan 11, 2012.

142. Aronowski J, Strong R, Shirzadi A, Grotta JC. Ethanol plus caffeine (caffeinol) for treatment of ischemic stroke: preclinical experience. Stroke. 2003;34(5):1246–51.

143. Belayev L, Khoutorova L, Zhang Y, Belayev A, Zhao W, Busto R, et al. Caffeinol confers cortical but not subcortical neuroprotection after transient focal cerebral ischemia in rats. Brain Res. 2004;1008(2):278–83.

144. Lapchak PA, Song D, Wei J, Zivin JA. Pharmacology of caffeinol in embolized rabbits: clinical rating scores and intracerebral hemorrhage incidence. Exp Neurol. 2004;188(2):286–91.

145. Gainer JL. Trans-sodium crocetinate for treating hypoxia/ischemia. Expert Opin Investig Drugs. 2008;17(6):917–24.

146. Okonkwo DO, Wagner J, Melon DE, Alden T, Stone JR, Helm GA, et al. Trans-sodium crocetinate increases oxygen delivery to brain parenchyma in rats on oxygen supplementation. Neurosci Lett. 2003;352(2):97–100.

147. Giaccio M. Crocetin from saffron: an active component of an ancient spice. Crit Rev Food Sci Nutr. 2004;44(3):155–72.

148. Stennett AK, Gainer JL. TSC for hemorrhagic shock: effects on cytokines and blood pressure. Shock. 2004;22(6):569–74.

149. Lapchak PA. Efficacy and safety profile of the carotenoid trans sodium crocetinate administered to rabbits following multiple infarct ischemic strokes: a combination therapy study with tissue plasminogen activator. Brain Res. 2010;1309:136–45.

150. Mellon SH, Griffin LD. Neurosteroids: biochemistry and clinical significance. Trends Endocrinol Metab. 2002;13(1):35.

151. Lapchak PA, Araujo DM. Preclinical development of neurosteroids as neuroprotective agents for the treatment of neurodegenerative diseases. Int Rev Neurobiol. 2001;46:379–97.

152. Jiang N, Chopp M, Stein D, Feit H. Progesterone is neuroprotective after transient middle cerebral artery occlusion in male rats. Brain Res. 1996;735(1):101–7.

153. Chen J, Chopp M, Li Y. Neuroprotective effects of progesterone after transient middle cerebral artery occlusion in rat. J Neurol Sci. 1999;171(1):24–30.

154. Lapchak PA. The neuroactive steroid 3-alpha-ol-5-beta-pregnan-20-one hemisuccinate, a selective NMDA receptor antagonist improves behavioral performance following spinal cord ischemia. Brain Res. 2004;997:152–8.

155. Weaver Jr CE, Marek P, Park-Chung M, Tam SW, Farb DH. Neuroprotective activity of a new class of steroidal inhibitors of the N-methyl-D-aspartate receptor. Proc Natl Acad Sci U S A. 1997;94(19):10450–4.

156. Malayev A, Gibbs TT, Farb DH. Inhibition of the NMDA response by pregnenolone sulphate reveals subtype selective modulation of NMDA receptors by sulphated steroids. Br J Pharmacol. 2002;135(4):901.

157. Lapchak PA. 3alpha-OL-5-beta-pregnan-20-one hemisuccinate, a steroidal low-affinity NMDA receptor antagonist improves clinical rating scores in a rabbit multiple infarct ischemia model: synergism with tissue plasminogen activator. Exp Neurol. 2006;197(2):531–7.

158. Chen HS, Lipton SA. Pharmacological implications of two distinct mechanisms of interaction of memantine with N-methyl-D-aspartate-gated channels. J Pharmacol Exp Ther. 2005;314(3):961–71.

159. Lipton SA. Pathologically-activated therapeutics for neuroprotection: mechanism of NMDA receptor block by memantine and S-nitrosylation. Curr Drug Targets. 2007;8(5):621–32.

160. Lipton SA, Chen HS. Paradigm shift in neuroprotective drug development: clinically tolerated NMDA receptor inhibition by memantine. Cell Death Differ. 2004;11(1):18–20.

161. Xia P, Chen HS, Zhang D, Lipton SA. Memantine preferentially blocks extrasynaptic over synaptic NMDA receptor currents in hippocampal autapses. J Neurosci. 2010;30(33): 11246–50.

162. Lipton SA. Failures and successes of NMDA receptor antagonists: molecular basis for the use of open-channel blockers like memantine in the treatment of acute and chronic neurologic insults. NeuroRx. 2004;1(1):101–10.

163. Chen HS, Pellegrini JW, Aggarwal SK, Lei SZ, Warach S, Jensen FE, et al. Open-channel block of N-methyl-D-aspartate (NMDA) responses by memantine: therapeutic advantage against NMDA receptor-mediated neurotoxicity. J Neurosci. 1992;12(11):4427–36.

164. Lapchak PA. Memantine, an uncompetitive low affinity NMDA open-channel antagonist improves clinical rating scores in a multiple infarct embolic stroke model in rabbits. Brain Res. 2006;1088(1):141–7.

165. Poulton KR, Rossi ML. Peripheral nerve protein glycation and muscle fructolysis: evidence of abnormal carbohydrate metabolism in ALS. Funct Neurol. 1993;8(1):33–42.

166. Winegrad AI, Morrison AD, Clements Jr RS. Polyol pathway activity in aorta. Adv Metab Disord. 1973;2 Suppl 2:117–27.

167. Kaneko M, Bucciarelli L, Hwang YC, Lee L, Yan SF, Schmidt AM, et al. Aldose reductase and AGE-RAGE pathways: key players in myocardial ischemic injury. Ann N Y Acad Sci. 2005;1043:702–9.

168. Hwang YC, Shaw S, Kaneko M, Redd H, Marrero MB, Ramasamy R. Aldose reductase pathway mediates JAK-STAT signaling: a novel axis in myocardial ischemic injury. FASEB J. 2005;19(7):795–7.

169. Iwata K, Matsuno K, Nishinaka T, Persson C, Yabe-Nishimura C. Aldose reductase inhibitors improve myocardial reperfusion injury in mice by a dual mechanism. J Pharmacol Sci. 2006;102(1):37–46.

170. Ramasamy R. Aldose reductase: a novel target for cardioprotective interventions. Curr Drug Targets. 2003;4(8):625–32.

171. Dan Q, Wong R, Chung SK, Chung SS, Lam KS. Interaction between the polyol pathway and non-enzymatic glycation on aortic smooth muscle cell migration and monocyte adhesion. Life Sci. 2004;76(4):445–59.

172. Wirasathien L, Pengsuparp T, Suttisri R, Ueda H, Moriyasu M, Kawanishi K. Inhibitors of aldose reductase and advanced glycation end-products formation from the leaves of *Stelechocarpus cauliflorus* R.E. Fr. Phytomedicine. 2006;14(7–8):546–50.

173. Leist M, Ghezzi P, Grasso G, Bianchi R, Villa P, Fratelli M, et al. Derivatives of erythropoietin that are tissue protective but not erythropoietic. Science. 2004;305(5681):239–42.

174. Lapchak PA. Erythropoietin molecules to treat acute ischemic stroke: a translational dilemma! Expert opinion investigational. Drugs. 2010;19(10):1179–86.

175. Lapchak PA. Carbamylated erythropoietin to treat neuronal injury: new development strategies. Expert Opin Investig Drugs. 2008;17(8):1175–86.

176. Lapchak PA. The many faces of erythropoietin: from erythropoiesis to a rational neuroprotective strategy—correspondence. Expert Opin Investig Drugs. 2008;17(10):1615–6.

177. Lapchak PA, Kirkeby A, Zivin JA, Sager TN. Therapeutic window for nonerythropoietic carbamylated-erythropoietin to improve motor function following multiple infarct ischemic strokes in New Zealand white rabbits. Brain Res. 2008;1238:208–14.

178. Ehrenreich H, Weissenborn K, Prange H, Schneider D, Weimar C, Wartenberg K, et al. Recombinant human erythropoietin in the treatment of acute ischemic stroke. Stroke. 2009; 40(12):e647–56.

179. Liu Y, Dargusch R, Maher P, Schubert D. A broadly neuroprotective derivative of curcumin. J Neurochem. 2008;105(4):1336–45.
180. Wu A, Ying Z, Schubert D, Gomez-Pinilla F. Brain and spinal cord interaction: a dietary curcumin derivative counteracts locomotor and cognitive deficits after brain trauma. Neurorehabil Neural Repair. 2011;25(4):332–42.
181. Delanty N, Vaughan CJ. Vascular effects of statins in stroke. Stroke. 1997;28(11):2315–20.
182. Hebert PR, Gaziano JM, Chan KS, Hennekens CH. Cholesterol lowering with statin drugs, risk of stroke, and total mortality. An overview of randomized trials. JAMA. 1997;278(4):313–21.
183. Kashyap ML. Cholesterol and atherosclerosis: a contemporary perspective. Ann Acad Med Singapore. 1997;26(4):517–23.
184. Laufs U, Gertz K, Dirnagl U, Bohm M, Nickenig G, Endres M. Rosuvastatin, a new HMG-CoA reductase inhibitor, upregulates endothelial nitric oxide synthase and protects from ischemic stroke in mice. Brain Res. 2002;942(1–2):23–30.
185. Seyfried D, Han Y, Lu D, Chen J, Bydon A, Chopp M. Improvement in neurological outcome after administration of atorvastatin following experimental intracerebral hemorrhage in rats. J Neurosurg. 2004;101(1):104–7.
186. Prinz V, Laufs U, Gertz K, Kronenberg G, Balkaya M, Leithner C, et al. Intravenous rosuvastatin for acute stroke treatment: an animal study. Stroke. 2008;39(2):433–8.
187. Shimamura M, Sato N, Sata M, Kurinami H, Takeuchi D, Wakayama K, et al. Delayed postischemic treatment with fluvastatin improved cognitive impairment after stroke in rats. Stroke. 2007;38(12):3251–8.
188. Wang S, Lee SR, Guo SZ, Kim WJ, Montaner J, Wang X, et al. Reduction of tissue plasminogen activator-induced matrix metalloproteinase-9 by simvastatin in astrocytes. Stroke. 2006;37(7):1910–2.
189. Ma T, Zhao Y, Kwak YD, Yang Z, Thompson R, Luo Z, et al. Statin's excitoprotection is mediated by sAPP and the subsequent attenuation of calpain-induced truncation events, likely via rho-ROCK signaling. J Neurosci. 2009;29(36):11226–36.
190. Sugawara T, Jadhav V, Ayer R, Zhang J. Simvastatin attenuates cerebral vasospasm and improves outcomes by upregulation of PI3K/Akt pathway in a rat model of subarachnoid hemorrhage. Acta Neurochir Suppl. 2008;102:391–4.
191. Wu L, Zhao L, Zheng Q, Shang F, Wang X, Wang L, et al. Simvastatin attenuates hypertrophic responses induced by cardiotrophin-1 via JAK-STAT pathway in cultured cardiomyocytes. Mol Cell Biochem. 2006;284(1–2):65–71.
192. Lapchak PA, Han MK. Simvastatin improves clinical scores in a rabbit multiple infarct ischemic stroke model: synergism with a rock inhibitor, but not the thrombolytic tissue plasminogen activator. Brain Res. 2010;18(1344):217–25.
193. MICROPLASMIN. http://www.strokecenter.org/trials/TrialDetail.aspx?tid=523; 2008. Accessed Jan 11, 2012.
194. Haley Jr EC, Thompson JL, Grotta JC, Lyden PD, Hemmen TG, Brown DL, et al. Phase IIB/III trial of tenecteplase in acute ischemic stroke: results of a prematurely terminated randomized clinical trial. Stroke. 2010;41(4):707–11.
195. Davydov L, Cheng JW. Tenecteplase: a review. Clin Ther. 2001;23(7):982–97. discussion 1.
196. Melandri G, Vagnarelli F, Calabrese D, Semprini F, Nanni S, Branzi A. Review of tenecteplase (TNKase) in the treatment of acute myocardial infarction. Vasc Health Risk Manag. 2009;5(1):249–56.
197. Hefer DV, Munir A, Khouli H. Low-dose tenecteplase during cardiopulmonary resuscitation due to massive pulmonary embolism: a case report and review of previously reported cases. Blood Coagul Fibrinolysis. 2007;18(7):691–4.
198. Dunn CJ, Goa KL. Tenecteplase: a review of its pharmacology and therapeutic efficacy in patients with acute myocardial infarction. Am J Cardiovasc Drugs. 2001;1(1):51–66.
199. Chapman DF, Lyden P, Lapchak PA, Nunez S, Thibodeaux H, Zivin J. Comparison of TNK with wild-type tissue plasminogen activator in a rabbit embolic stroke model. Stroke. 2001;32(3):748–52.

200. Nagai N, Demarsin E, Van Hoef B, Wouters S, Cingolani D, Laroche Y, et al. Recombinant human microplasmin: production and potential therapeutic properties. J Thromb Haemost. 2003;1(2):307–13.
201. Wu HL, Shi GY, Wohl RC, Bender ML. Structure and formation of microplasmin. Proc Natl Acad Sci U S A. 1987;84(24):8793–5.
202. Wu HL, Shi GY, Bender ML. Preparation and purification of microplasmin. Proc Natl Acad Sci U S A. 1987;84(23):8292–5.
203. Collen D. Revival of plasmin as a therapeutic agent? Thromb Haemost. 2001;86(3):731–2.
204. Lapchak PA, Araujo DM, Pakola S, Song D, Wei J, Zivin JA. Microplasmin: a novel thrombolytic that improves behavioral outcome after embolic strokes in rabbits. Stroke. 2002;33(9):2279–84.
205. Lapchak PA, Chapman DF, Zivin JA. Metalloproteinase inhibition reduces thrombolytic (tissue plasminogen activator)-induced hemorrhage after thromboembolic stroke. Stroke. 2000;31(12):3034–40.
206. EAISG. Effect of a novel free radical scavenger, edaravone (MCI-186), on acute brain infarction. Randomized, placebo-controlled, double-blind study at multicenters. Cerebrovasc Dis. 2003;15(3):222–9.
207. Inatomi Y, Takita T, Yonehara T, Fujioka S, Hashimoto Y, Hirano T, et al. Efficacy of edaravone in cardioembolic stroke. Intern Med. 2006;45(5):253–7.
208. Kitagawa Y. Edaravone in acute ischemic stroke. Intern Med. 2006;45(5):225–6.
209. Sinha M, Anuradha H, Juyal R, Shukla R, Garg R, Kar A. Edaravone in acute ischemic stroke, an Indian experience. Neurology Asia. 2009;14:7–10.
210. Lapchak PA, Maher P, Schubert D, Zivin JA. Baicalein, an antioxidant 12/15 lipoxygenase inhibitor improves clinical rating scores following multiple infarct embolic strokes. Neuroscience. 2007;150(3):585–91.
211. Caffeinol. http://www.strokecenter.org/trials/InterventionDetail.aspx?tid=249; 2008. Accessed Jan 11, 2012.
212. Ebselen. http://www.strokecenter.org/trials/TrialDetail.aspx?tid=298; 2011. Accessed Jan 11, 2012.

Chapter 28
Animal Models of Intracranial Aneurysms

Elena I. Liang, Hiroshi Makino, Yoshiteru Tada,
Kosuke Wada, and Tomoki Hashimoto

Abstract We established a new mouse model of intracranial aneurysms that yields large aneurysms within a relatively short time frame. Aneurysms are induced by a combination of pharmacological hypertension and a single injection of elastase into the cerebrospinal fluid. The model gives us a unique opportunity to study and understand the mechanisms for not only aneurysm formation but also the rupture of intracranial aneurysms. Using this model, we can study the effects of various pharmacological agents on aneurysmal rupture in mice and translate our findings toward the clinical setting.

1 Introduction

Intracranial aneurysms are common—as much as 1–7% of the general population may be harboring intracranial aneurysms [1]. Subarachnoid hemorrhage from aneurysmal rupture results in severe morbidity and high mortality [2]. In the U.S., approximately 27,000 people are estimated to suffer from subarachnoid hemorrhage each year [1]. The management of unruptured intracranial aneurysms remains controversial. In some patients, the risks of surgical treatment may be higher than the risks for rupture [3]. In addition, there are limited treatment options for patients with giant aneurysms. Pharmacological stabilization of aneurysms for the prevention of aneurysmal growth and rupture may be an attractive alternative approach.

E.I. Liang, BS • H. Makino, MD • Y. Tada, MD, PhD • K. Wada, MD • T. Hashimoto, MD (✉)
Department of Anesthesia and Perioperative Care, Center for Cerebrovascular Research,
University of California, 1001 Potrero Avenue, No. 3C-38, San Francisco, CA 94110, USA
e-mail: hashimot@anesthesia.ucsf.edu

P.A. Lapchak and J.H. Zhang (eds.), *Translational Stroke Research*, Springer Series
in Translational Stroke Research, DOI 10.1007/978-1-4419-9530-8_28,
© Springer Science+Business Media, LLC 2012

2 Formation, Growth, and Rupture of Intracranial Aneurysms

Conceptually, there are six stages in the natural course of intracranial aneurysms. During stage 1 (aneurysm formation), small aneurysms or preaneurysmal lesions can be formed. Intracranial aneurysm formation may start as an otherwise normal outward vascular remodeling at bifurcations or in curved blood vessels. However, because of the particular geometry of affected blood vessels, outward vascular remodeling becomes asymmetric and focalized, causing preaneurysmal changes that lead to small aneurysm formation—a small bulging of the blood vessel. During stage 2 (aneurysm growth), part of the aneurysm will experience sustained high shear stress as the small aneurysm grows [4].

The development of an aneurysm may be seen as a continuum. In the area exposed to high shear stress, endothelial cells are kept activated or sometimes damaged, leading to infiltration by inflammatory cells. Focalized vascular remodeling continues with increasing inflammation. Abnormal shear stress coupled with sustained inflammation may result in the growth of aneurysms. Once aneurysms are fully formed, in some aneurysms, remodeling processes may lead to the adaptation of aneurysms. These "stabilized" aneurysms remain as unruptured aneurysms (stage 3: stabilization). However, in some aneurysms, focalized outward vascular remodeling coupled with inflammation continues, leading to destabilization of the aneurysmal wall (stage 4: destabilization). In destabilized aneurysms, the vascular wall will be weakened or damaged as a result of maladaptive remodeling, eventually resulting in aneurysmal rupture (stage 5: aneurysmal rupture).

Pharmacological stabilization of aneurysms is intended to change the natural course of aneurysms from destabilization (stage 4) to stabilization (stage 3) and prevent aneurysmal rupture. Such therapies may be in the form of chronic treatment, or only a short course of the treatment may be necessary to change the natural course.

To understand the pathophysiology of intracranial aneurysms and develop pharmacological therapies for the prevention of aneurysmal growth, an animal model of intracranial aneurysms that recapitulates key features of human intracranial aneurysms is needed. An ideal aneurysm model would yield a high rate of large and easily distinguishable aneurysms within a relatively short incubation period.

There are several proposed animal models for studying intracranial aneurysms. Each one mimics different aspects of the pathophysiology of intracranial aneurysms. Aneurysms created in the extracranial arteries have been successfully used to study (1) efficacy of endovascular treatments and (2) flow-dynamic profiles. These extracranial models include surgically created vein graft aneurysms [5–10] and endovascularly or surgically created carotid stump treated with elastase [11–14]. While these models are appropriate for those particular studies, there remains a need for a more suitable model for studying the mechanisms behind spontaneous aneurysm development, growth, and rupture in an intracranial setting.

Considering the availability of techniques for genetic manipulations and the lower cost of treatments with experimental drugs, mice are an attractive species for the development of the animal model of intracranial aneurysms. There are a few different strategies that can be utilized to develop an animal model of intracranial aneurysms in mice. A genetic approach is useful when disease causative genes are identified. Although multiple gene mutations appear to be associated with intracranial aneurysms and aneurysmal subarachnoid hemorrhage, the associations are generally weak [15, 16]. Disease-causing genes have also not yet been identified for intracranial aneurysms. Pharmacological approach is useful when the disease causing genes are unidentified. This approach can be easily applied to various knockout or transgenic mice to study underlying molecular mechanisms. In the pharmacological approach, pharmacological manipulations are used to reproduce physiological or pathological changes that are associated with the human disease. When the final phenotypes are similar between the human disease and the animal model, it is assumed that the human disease and the animal model, at least, partly share common downstream pathways that lead to the final phenotype. In some cases, a combination of pharmacological approach, genetic approach, and surgical approach may be combined to create the disease phenotype. However, as the number of manipulations increases, it becomes more difficult to assess relative contributions of each manipulation to the phenotype. More importantly, when an experimental drug is tested for the treatment of the disease in such complex model, the response to drug of each component needs to be carefully interrogated. Therefore, fewer manipulations are desirable for developing animal models of intracranial aneurysms to study its pathophysiology and develop pharmacological treatments for the prevention of aneurysmal rupture.

3   Kyoto Model

Nobuo Hashimoto et al. at Kyoto University pioneered the development of an elaborate intracranial aneurysm model that combines surgical and pharmacological manipulations [17–20]. The researchers combined (1) renovascular hypertension induced by ligation of posterior branches of the bilateral renal arteries and loading with high salt diet, (2) continuous administration of a lathyrogen, beta-aminopropionitrile, and (3) unilateral ligation of common carotid artery. A combination of unilateral carotid ligation and hypertension was thought to increase blood flow and arterial pressure in particular segments of the Circle of Willis [18]. When used in rats and nonhuman primates, this model resulted in aneurysm formations (1) at the bifurcation between the anterior cerebral artery and the olfactory artery contralateral to the ligated common carotid artery, and (2) at the posterior communicating arteries, locations with abnormal hemodynamic stresses [21]. There appears to be a dose relationship between an increase in blood flow (increase in wall shear stress) and the incidence of aneurysms. Rats that underwent bilateral common carotid ligation displayed a higher incidence of aneurysms in the posterior circulation than those that underwent unilateral ligation [21]. Compared to rats that underwent unilateral ligation, rats that underwent bilateral ligation presumably developed higher

Fig. 28.1 A representative aneurysm from "Kyoto model" in a rat. We reproduced "Kyoto model" in rats. Four months after unilateral carotid artery ligation, induction of renovascular hypertension, and continuous treatment with an irreversible lysyl oxidase inhibitor (beta-aminopropionitrile), an aneurysm formation was found in the left posterior communicating artery

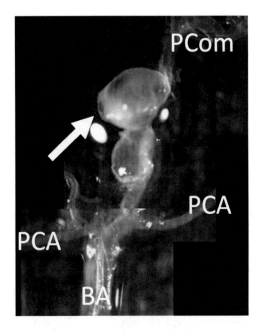

blood flow through the basilar artery since bilateral ligation had rerouted blood flow to the brain almost entirely through the basilar artery. Figure 28.1 shows a representative intracranial aneurysm in a rat from the experiment that was performed in our laboratory to reproduce the "Kyoto model" in rats.

While this model appears to create easily distinguishable aneurysms in rats and nonhuman primates [17–19], the size of aneurysms induced in this model in mice appears to be disproportionally small, and the aneurysms exhibited very subtle histological changes in the elastic lamina without any apparent bulging of the arterial wall [20]. This may be due to fundamentally different effects of unilateral carotid ligation on the blood flow patterns in the Circle of Willis between mice and rats since there are anatomical differences of the cerebral arteries [22, 23]. Unlike rats or primates, mice, especially the C57BL/6 strain, have underdeveloped anastomosis between the anterior and the posterior cerebral circulation through the posterior communicating arteries [22–25]. In C57BL/6 strain, the posterior communicating artery is often absent or undetectable [22, 24, 25]. This complex model in mice is reported to induce discontinuation of elastic lamina (termed "preaneurysmal changes") or microscopic aneurysm formations (termed "advanced aneurysms") at the bifurcation of the olfactory artery and the anterior cerebral artery after a relatively long incubation period (3–5 months). The incidence of "advanced aneurysms" was only 30–43%, and "advanced aneurysms" and "preaneurysmal changes" had to be combined to detect statistically significant effects of drug treatments or gene deletions [26, 27]. Because of the long incubation period, the smaller size of aneurysms, the low incidence of aneurysms, and the complexity of model, the "Kyoto model" in mice may not be suitable for high-throughput studies to dissect various molecular pathways that may play roles in the pathophysiology of intracranial aneurysms.

4   Elastase-Induced Intracranial Aneurysm Model in Mice

To study intracranial aneurysms, we have developed a new mouse model of intracranial aneurysms that yields large aneurysms within a relatively short time frame [28–30]. To induce intracranial aneurysm formation in mice, we combined two well-known factors associated with human intracranial aneurysms—hypertension and disruption of elastic lamina [1, 20, 31–33].

In C57BL/6J male mice (8–10 weeks old), hypertension was induced by continuous infusion of angiotensin II (1,000 ng/kg/min). Disruption of elastic lamina was induced by a single injection of elastase into the cerebrospinal fluid at the right basal cistern using a stereotaxic method. Aneurysms were defined as a localized outward bulging of the vascular wall whose diameter is greater than 150% of the parent artery diameter [28, 29]. Seventy to eighty percent of the mice developed intracranial aneurysms over a 4-week period [28, 29]. Figure 28.2 shows representative aneurysms from this model. Large intracranial aneurysms were mostly found along the right half of the Circle of Willis and its major branches. Histologically, intracranial aneurysms in this model closely resemble human intracranial aneurysms [28, #606] Details of immunohistochemical findings were presented in our published papers [28, 29].

We performed further studies to characterize aneurysmal subarachnoid hemorrhage in this model. We used DOCA-salt hypertension (DOCA: deoxycorticosterone acetate) instead of angiotensin II-induced hypertension in this series. Because of the slower time course for the increase in blood pressure by DOCA-salt hypertension compared to angiotensin II-induced hypertension, a combination of elastase injection and DOCA-salt hypertension generally results in a lower early mortality rate than our original combination with angiotensin II.

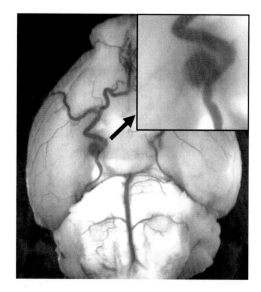

Fig. 28.2 A mouse model of intracranial aneurysm. A combination of systemic hypertension and a single injection of elastase into the cerebrospinal fluid space results in aneurysm formation in mice. The figure shows a representative aneurysm in the right internal carotid artery of a mouse

As a part of well-established method of DOCA-salt hypertension, unilateral nephrectomy was performed [34, 35]. One week later, mice were subjected to a single injection of elastase into the cerebrospinal fluid and implantation of the DOCA pellet. Administration of high-salt drinking water (0.9% NaCl) was started on the same day.

We performed daily neurological examinations using a previously described method with minor modifications [36–38]. Neurological signs were scored as follows: 0, normal function; 1, reduced eating or drinking activity demonstrated by the weight loss greater than 2 g of body weight (approximately 10% weight loss) over 24 h; 2, flexion of torso and forelimb upon lifting of the whole animal by the tail; 3, circling to one side but normal posture at rest; 4, leaning to one side at rest; and 5, no spontaneous activity. Mice were sacrificed when the neurological score was 1 through 5.

Neurological symptoms started developing 1 week after aneurysm induction. When symptomatic mice were sacrificed, all symptomatic mice had ruptured aneurysms with subarachnoid hemorrhage. When all remaining asymptomatic mice were sacrificed after 4 weeks, less than 10% of the asymptomatic animals had blood clots from subarachnoid hemorrhage.

In this model, aneurysmal subarachnoid hemorrhages (i.e., ruptured aneurysms) were found in 50–60% of the mice. Unruptured aneurysms were found in 20–25% of the mice. The incidence of aneurysms, including both ruptured and unruptured aneurysms, was 70–80%, an incidence that was similar to the incidence in angiotensin II-induced hypertension. The rupture rate (mice with ruptured aneurysms/mice with ruptured or unruptured aneurysms) was 70–80%. These observations indicate that the daily neurological exam using this scoring system was sensitive and specific enough to detect aneurysmal subarachnoid hemorrhage in this model.

In this model, aneurysmal rupture occurs between day 7 and day 28. The period between day 7 and day 28 appears to represent a time window during which pharmacological interventions for the prevention of aneurysmal rupture can be tested. Pharmacological interventions are started one day before this time period to ensure the effectiveness of treatment. By treating animals with experimental agents from day 6 to day 28, we would be able to test whether the experimental agents can prevent aneurysmal rupture (i.e., aneurysmal subarachnoid hemorrhage) after aneurysmal formation.

Based on these observations, we developed the experimental protocol to test experimental agents for prevention of aneurysmal rupture. Aneurysms will be induced by a combination of pharmacological hypertension and a single injection of elastase into the cerebrospinal fluid. Hypertension can be induced by DOCA-salt hypertension or angiotensin II-induced hypertension. Six days after aneurysm induction, treatment with experimental agents will begin so that the treatment affects the processes leading to aneurysmal rupture without affecting aneurysmal formation.

This model provides us with a unique opportunity to study mechanisms for the rupture of intracranial aneurysms and conduct preclinical studies for identifying the therapeutic targets for the prevention of aneurysmal rupture.

5 Summary

To understand the pathophysiology of intracranial aneurysms, an animal model yielding aneurysms that resemble key features of human intracranial aneurysms is needed. An ideal animal model of intracranial aneurysms would yield a high rate of large and easily distinguishable aneurysms within a relatively short incubation period. In addition, fewer manipulations are desirable for developing animal models of intracranial aneurysms to study the pathophysiology of aneurysms and develop pharmacological treatments for the prevention of aneurysmal rupture. Several models have been proposed, including the model established at Kyoto University. However, this model was difficult to use in mice and only resulted in very small, undistinguishable aneurysms.

As a result, we established a new mouse model of intracranial aneurysms yields large aneurysms within a relatively short time frame. Aneurysms are induced by a combination of pharmacological hypertension and a single injection of elastase into the cerebrospinal fluid. The model gives us a unique opportunity to study and understand the mechanisms for not only aneurysm formation but also the rupture of intracranial aneurysms. Using this model, we can study the effects of various pharmacological agents on aneurysmal rupture in mice and translate our findings toward the clinical setting.

Acknowledgment This study was funded by the National Institutes of Health (R01NS055876 P01NS04415, both TH), the American Heart Association (11GRNT6380003, TH), and The Brain Aneurysm Foundation (TH).

References

1. Schievink WI. Intracranial aneurysms. N Engl J Med. 1997;336:28–40.
2. Juvela S. Treatment options of unruptured intracranial aneurysms. Stroke. 2004;35:372–4.
3. Wiebers DO, Whisnant JP, Huston III J, Meissner I, Brown Jr RD, Piepgras DG, Forbes GS, Thielen K, Nichols D, O'Fallon WM, Peacock J, Jaeger L, Kassell NF, Kongable-Beckman GL, Torner JC. Unruptured intracranial aneurysms: Natural history, clinical outcome, and risks of surgical and endovascular treatment. Lancet. 2003;362:103–10.
4. Hoi Y, Meng H, Woodward SH, Bendok BR, Hanel RA, Guterman LR, Hopkins LN. Effects of arterial geometry on aneurysm growth: three-dimensional computational fluid dynamics study. J Neurosurg. 2004;101:676–81.
5. German WJ, Black SP. Experimental production of carotid aneurysms. N Engl J Med. 1954;250:104–6.
6. Guglielmi G, Ji C, Massoud TF, Kurata A, Lownie SP, Vinuela F, Robert J. Experimental saccular aneurysms. II. A new model in swine. Neuroradiology. 1994;36:547–50.
7. Massoud TF, Ji C, Guglielmi G, Vinuela F, Robert J. Experimental models of bifurcation and terminal aneurysms: construction techniques in swine. AJNR Am J Neuroradiol. 1994;15: 938–44.
8. Graves VB, Ahuja A, Strother CM, Rappe AH. Canine model of terminal arterial aneurysm. AJNR Am J Neuroradiol. 1993;14:801–3.

9. Macdonald RL, Mojtahedi S, Johns L, Kowalczuk A. Randomized comparison of guglielmi detachable coils and cellulose acetate polymer for treatment of aneurysms in dogs. Stroke. 1998;29:478–85; discussion 485–86.

10. Forrest MD, O'Reilly GV. Production of experimental aneurysms at a surgically created arterial bifurcation. AJNR Am J Neuroradiol. 1989;10:400–2.

11. Cloft HJ, Altes TA, Marx WF, Raible RJ, Hudson SB, Helm GA, Mandell JW, Jensen ME, Dion JE, Kallmes DF. Endovascular creation of an in vivo bifurcation aneurysm model in rabbits. Radiology. 1999;213:223–8.

12. Altes TA, Cloft HJ, Short JG, DeGast A, Do HM, Helm GA, et al. 1999 arrs executive council award. Creation of saccular aneurysms in the rabbit: a model suitable for testing endovascular devices. American Roentgen Ray Society. AJR Am J Roentgenol. 2000;174:349–54.

13. Cawley CM, Dawson RC, Shengelaia G, Bonner G, Barrow DL, Colohan AR. Arterial saccular aneurysm model in the rabbit. AJNR Am J Neuroradiol. 1996;17:1761–6.

14. Kallmes DF, Fujiwara NH, Berr SS, Helm GA, Cloft HJ. Elastase-induced saccular aneurysms in rabbits: a dose-escalation study. AJNR Am J Neuroradiol. 2002;23:295–8.

15. Yasuno K, Bilguvar K, Bijlenga P, Low SK, Krischek B, Auburger G, Simon M, Krex D, Arlier Z, Nayak N, Ruigrok YM, Niemela M, Tajima A, Von und zu Fraunberg M, Doczi T, Wirjatijasa F, Hata A, Blasco J, Oszvald A, Kasuya H, Zilani G, Schoch B, Singh P, Stuer C, Risselada R, Beck J, Sola T, Ricciardi F, Aromaa A, Illig T, Schreiber S, van Duijn CM, Van den Berg LH, Perret C, Proust C, Roder C, Ozturk AK, Gaal E, Berg D, Geisen C, Friedrich CM, Summers P, Frangi AF, State MW, Wichmann HE, Breteler MM, Wijmenga C, Mane S, Peltonen L, Elio V, Sturkenboom MC, Lawford P, Byrne J, Macho J, Sandalcioglu EI, Meyer B, Raabe A, Steinmetz H, Rufenacht D, Jaaskelainen JE, Hernesniemi J, Rinkel GJ, Zembutsu H, Inoue I, Palotie A, Cambien F, Nakamura Y, Lifton RP, Gunel M. Genome-wide association study of intracranial aneurysm identifies three new risk loci. Nat Genet. 2010;42:420–5.

16. Nahed BV, Bydon M, Ozturk AK, Bilguvar K, Bayrakli F, Gunel M. Genetics of intracranial aneurysms. Neurosurgery. 2007;60:213–25; discussion 225–26.

17. Hashimoto N, Handa H, Hazama F. Experimentally induced cerebral aneurysms in rats. Surg Neurol. 1978;10:3–8.

18. Hashimoto N, Handa H, Hazama F. Experimentally induced cerebral aneurysms in rats: part II. Surg Neurol. 1979;11:243–6.

19. Hashimoto N, Kim C, Kikuchi H, Kojima M, Kang Y, Hazama F. Experimental induction of cerebral aneurysms in monkeys. J Neurosurg. 1987;67:903–5.

20. Morimoto M, Miyamoto S, Mizoguchi A, Kume N, Kita T, Hashimoto N. Mouse model of cerebral aneurysm: experimental induction by renal hypertension and local hemodynamic changes. Stroke. 2002;33:1911–5.

21. Hashimoto N, Handa H, Nagata I, Hazama F. Experimentally induced cerebral aneurysms in rats: Part V. Relation of hemodynamics in the circle of Willis to formation of aneurysms. Surg Neurol. 1980;13:41–5.

22. Yang G, Kitagawa K, Matsushita K, Mabuchi T, Yagita Y, Yanagihara T, Matsumoto M. C57bl/6 strain is most susceptible to cerebral ischemia following bilateral common carotid occlusion among seven mouse strains: selective neuronal death in the murine transient forebrain ischemia. Brain Res. 1997;752:209–18.

23. Okuyama S, Okuyama J, Tamatsu Y, Shimada K, Hoshi H, Iwai J. The arterial circle of Willis of the mouse helps to decipher secrets of cerebral vascular accidents in the human. Med Hypotheses. 2004;63:997–1009.

24. Beckmann N. High resolution magnetic resonance angiography non-invasively reveals mouse strain differences in the cerebrovascular anatomy in vivo. Magn Reson Med. 2000;44:252–8.

25. McColl BW, Carswell HV, McCulloch J, Horsburgh K. Extension of cerebral hypoperfusion and ischaemic pathology beyond mca territory after intraluminal filament occlusion in C57Bl/6J mice. Brain Res. 2004;997:15–23.

26. Aoki T, Kataoka H, Ishibashi R, Nozaki K, Egashira K, Hashimoto N. Impact of monocyte chemoattractant protein-1 deficiency on cerebral aneurysm formation. Stroke. 2009;40:942–51.

27. Moriwaki T, Takagi Y, Sadamasa N, Aoki T, Nozaki K, Hashimoto N. Impaired progression of cerebral aneurysms in interleukin-1beta-deficient mice. Stroke. 2006;37:900–5.
28. Nuki Y, Tsou TL, Kurihara C, Kanematsu M, Kanematsu Y, Hashimoto T. Elastase-induced intracranial aneurysms in hypertensive mice. Hypertension. 2009;54:1337–44.
29. Kanematsu Y, Kanematsu M, Kurihara C, Tada Y, Tsou TL, van Rooijen N, Lawton MT, Young WL, Liang EI, Nuki Y, Hashimoto T. Critical roles of macrophages in the formation of intracranial aneurysm. Stroke. 2011;42:173–8.
30. Tada Y, Kanematsu Y, Kanematsu M, Nuki Y, Liang EI, Wada K, et al. A mouse model of intracranial aneurysm: technical considerations. Acta Neurochir Suppl. 2011;111:31–5.
31. Connolly Jr ES, Choudhri TF, Mack WJ, Mocco J, Spinks TJ, Slosberg J, Lin T, Huang J, Solomon RA. Influence of smoking, hypertension, and sex on the phenotypic expression of familial intracranial aneurysms in siblings. Neurosurgery. 2001;48:64–8.
32. Bonita R. Cigarette smoking, hypertension and the risk of subarachnoid hemorrhage: a population-based case–control study. Stroke. 1986;17:831–5.
33. Cajander S, Hassler O. Enzymatic destruction of the elastic lamella at the mouth of cerebral berry aneurysm? An ultrastructural study with special regard to the elastic tissue. Acta Neurol Scand. 1976;53:171–81.
34. Weiss D, Taylor WR. Deoxycorticosterone acetate salt hypertension in apolipoprotein E–/– mice results in accelerated atherosclerosis: the role of angiotensin II. Hypertension. 2008;51:218–24.
35. Kanematsu Y, Kanematsu M, Kurihara C, Tsou TL, Nuki Y, Liang EI, Makino H, Hashimoto T. Pharmacologically induced thoracic and abdominal aortic aneurysms in mice. Hypertension. 2010;55:1267–74.
36. Huang Z, Huang PL, Panahian N, Dalkara T, Fishman MC, Moskowitz MA. Effects of cerebral ischemia in mice deficient in neuronal nitric oxide synthase. Science. 1994;265:1883–5.
37. Yang G, Chan PH, Chen J, Carlson E, Chen SF, Weinstein P, Epstein CJ, Kamii H. Human copper-zinc superoxide dismutase transgenic mice are highly resistant to reperfusion injury after focal cerebral ischemia. Stroke. 1994;25:165–70.
38. Chan PH. Oxygen radicals in focal cerebral ischemia. Brain Pathol. 1994;4:59–65.

Chapter 29
Animal Models of SAH and Their Translation to Clinical SAH

Tommaso Zoerle and R. Loch Macdonald

Abstract Animal models of stroke may be useful for elucidating mechanisms of disease, but they have arguably not been particularly successful at predicting what treatments will be successful for ischemic stroke in humans. Animal models of subarachnoid hemorrhage also have been developed in rodents, dogs, and nonhuman primates. These models mimic angiographic vasospasm and some aspects of subarachnoid hemorrhage such as the transient global ischemia that sometimes occurs at the time of rupture of an aneurysm. Since the detailed acute and delayed pathologic effects of subarachnoid hemorrhage on human brain are not well delineated, how the animal models replicate this is unknown. Nevertheless, meta-analysis of the literature suggests that clinical trials of drugs for angiographic vasospasm in humans have been effective, and that some animal models accurately reflect what the effects of drugs are in humans. Analysis of animal models and comparison of drug effects on angiographic vasospasm in humans and animals suggest injection of autologous blood into the basal cisterns; assessment of vasospasm more than 3 days after the injection and intrathecal delivery of drugs may be better ways to study drugs in animals, in terms of translation to success in humans.

T. Zoerle, MD
Department of Anesthesia and Critical Care Medicine, University of Milano,
Neurosurgical Intensive Care Unit, Fondazione IRCCS Ca' Granda—Ospedale Maggiore,
Policlinico, Via F. Sforza 35, 20122 Milan, Italy

R.L. Macdonald, MD, PhD (✉)
Division of Neurosurgery, Labatt Family Centre of Excellence in Brain Injury and Trauma
Research, Keenan Research Centre of the Li Ka Shing Knowledge Institute of St. Michael's
Hospital, and Department of Surgery, University of Toronto, Toronto, ON, Canada
e-mail: macdonaldlo@smh.ca

P.A. Lapchak and J.H. Zhang (eds.), *Translational Stroke Research*, Springer Series
in Translational Stroke Research, DOI 10.1007/978-1-4419-9530-8_29,
© Springer Science+Business Media, LLC 2012

1 Introduction

Animal models of stroke may be useful for elucidating mechanisms of disease, but they have been poor predictors of successful treatment of stroke in humans [96]. About 500 neuroprotective strategies reduced infarct size in animal models of ischemic stroke, but only acetylsalicylic acid and tissue plasminogen activator have been successful so far in humans. This has led to recommendations for design of animal studies, which are reviewed in other chapters and should be adhered to [93].

Animal models of SAH have faired better. Subarachnoid hemorrhage (SAH) may be traumatic or spontaneous, with 80% of spontaneous cases due to ruptured intracranial aneurysms. Most studies have targeted spontaneous SAH, from which the outcome is poor. One systematic review found overall mortality was 32–67% [27] with about two thirds of survivors suffering long-term physical and neurobehavioral sequelae. Efforts to improve the outcome traditionally focused on delayed effects of SAH, primarily angiographic vasospasm and its sequela, delayed cerebral ischemia (DCI). More recent investigations have examined early brain injury after SAH.

Most work has relied on experiments in vivo, as opposed to purely in vitro studies. Arguments against the applicability of studies in vitro to human SAH are that angiographic vasospasm and DCI occur 3–14 days after the SAH so that acute effects of blood products may not be relevant. Excised arteries are denervated and not subject to pulsatile pressure and flow. Endothelial function may be difficult to preserve and remodeling responses due to incubation in vitro in smooth muscle cells can begin within hours. In most systems, blood products and drugs are applied to both luminal and extraluminal surfaces, which is not what occurs after SAH. Long-term pathological changes in vasospastic arteries are not produced in vitro. On the other hand, it is interesting that in retrospect, nimodipine, the only drug that has an indication in North America for patients with SAH, was tested clinically based on potent vasodilatory properties that were obvious in studies in vitro, and the vasodilatory effects may not even be the only or most important mechanism by which it improved outcome.

This chapter reviews animal models of SAH. We focus on models using whole blood, not blood fractions or other substances, and on more commonly used models and discuss potential translation of results from the models to humans.

2 What Effects of Subarachnoid Hemorrhage to Model

Rupture of an intracranial aneurysm causes SAH and varying degrees of other pathologic effects including increased intracranial pressure (ICP) with transient global ischemia, intracerebral and intraventricular hemorrhage, brain shifts, and early brain injury. There also is the rupture in the wall of the aneurysm. The researcher needs to decide, depending on the goal of the investigation, whether to model SAH alone without increased ICP or SAH plus increased ICP.

The contribution of increased ICP to the effects of SAH requires discussion. Prognostic factors thought to be most important for outcome after SAH include initial neurological condition, age, and DCI [70]. The lower the level of consciousness after rescusitation and treatment of acute hydrocephalus and increased ICP shortly after the SAH, the worse the outcome. This level of consciousness or initial neurological condition contributes to the largest component of variation in outcome. It is assumed that worse neurologic function means there is early brain injury and thus that early brain injury contributes substantially to morbidity and mortality in patients with aneurysmal SAH [5]. One question is how much early brain injury is due to SAH and how much is due to transient global ischemia. This has not been well studied. Models of SAH with increased ICP produce different pathology than transient global ischemia alone [5].

Animals do not develop cerebral aneurysms, so the only method to include vessel rupture in SAH models has been to puncture a subarachnoid artery by open craniotomy or endovascular means. When intracranial aneurysms are induced in rodents by various methods (induced hypertension, lathryism, subarachnoid elastase injection, and others), they seldom rupture spontaneously [58]. The role of arterial or aneurysm rupture in the pathophysiology of SAH has been difficult to clarify. A direct comparison of the effects of equal volumes of blood injected into the basal cisterns or delivered there by arterial rupture while producing equivalent acute alterations in ICP has not been carried out and would be difficult to do because of the problems quantifying the volume of hemorrhage with different models of SAH.

Investigators have focused on vasospasm of the large conducting cerebral arteries, but the importance of this phenomenon has been questioned because pharmacologic prevention of angiographic vasospasm in humans has not been associated with improved clinical outcome [16]. A full discussion of this controversy is beyond the scope of this chapter and can be found elsewhere [65]. Explanations include inadequate powering of clinical trials, insensitivity of the outcome measures used to meaningful improvements in outcome, beneficial effects of rescue therapies used in the placebo groups that balance the benefits of drug treatment, detrimental off-target drug effects and drug side effects, and that other processes cause DCI and poor outcome in addition to angiographic vasospasm.

Other processes suggested to contribute to DCI include microthromboemboli, cortical spreading ischemia, and microcirculatory dysfunction [11, 88]. Microthromboemboli have been observed in the anterior circulation prechiasmatic injection rat and mouse models of SAH [73] and humans [100] and cortical spreading ischemia has thus far only been described in humans [11, 12]. In the absence of effective treatments, it is not possible to determine how these processes contribute to outcome after SAH.

Can DCI be studied in animal models. DCI may be defined as neurological deterioration in the absence of identifiable factors such as increased ICP, seizures, brain shift, and systemic factors. This is difficult to and has seldom been studied in animals. Neurologic exam scales are used in animal models of ischemic stroke, but focal infarctions of the type seen in association with angiographic vasospasm in humans are rare after SAH in animals. Jeon et al. reviewed animal studies of

SAH that reported neurological outcomes [33]. Examination of the time course of neurological and behavior changes in these studies showed that almost in every study these changes improve from a nadir that occurs at the first hours to day after SAH which is usually the first day tested. Many studies document delayed angiographic vasospasm with a time course similar to that occurring in humans; neurologic and behavior changes associated with it are rarely seen. A study of rats with SAH created by endovascular perforation showed progressive cerebral ischemia and vasospasm over the first 2 days after SAH [95]. The authors concluded this was the equivalent of DCI in humans. Guresir et al. studied the time course of vasospasm, neuron cell counts, cerebral blood flow (CBF), and neurological examination in the double hemorrhage rat model [24]. All parameters were worse 5 compared to 3 days after SAH, which the authors considered to be evidence of DCI.

Takata et al. documented delayed deterioration in neurological function days after SAH induced by two injections of blood into the cisterna magna of rats [92]. The ICP was not measured during SAH, but the blood was injected slowly, so the authors suggested the effects were due solely to SAH. CBF was reduced for 14 days after SAH. Neurological function was worse 72 compared to 24 h after SAH induced by endovascular perforation in rats in only one study [89].

In rats and rabbits, delayed neurological deterioration could be produced by bilateral carotid ligation and two injections of blood into the cisterna magna [14, 71] or trends to worse neurological function by two blood injections alone [109]. Another study correlated neurological function with degree of angiographic vasospasm days after SAH in the single injection model in rabbits [40]. Similar prolonged abnormalities in neurological function are documented in the dog double hemorrhage model of SAH [107, 108], but distinct deterioration from stable function, as occurs in humans, has been difficult to document.

Prunell et al. reported TUNEL-positive neurons in brains of rats 2 and 7 days after SAH produced by endovascular perforation or chiasmatic blood injection [69]. If the delayed neuronal death causes symptoms, this could be a pathophysiological mechanism of DCI.

3   Species Differences

The anatomy of the cerebral circulation of species used in SAH models varies and is suggested to contribute to why DCI rarely develops despite severe angiographic vasospasm that can be produced in some models. The pharmacological and molecular biological differences between the species that may influence SAH, vasospasm, DCI, and effects of treatments are unknown.

Vasospasm after a single cisternal blood injection in mice, rats, rabbits, dogs, and monkeys tends to have a rapid onset and a shorter duration than vasospasm in man [45]. The immediate onset of vasospasm may be due to hemolysis caused by injecting the blood through a narrow gauge needle. The short duration is probably because the erythrocytes are cleared rapidly from the subarachnoid space and there is only minimal hemolysis. A second cisternal injection of blood can be given 48 h after the first injection. This prolongs vasospasm and increases its severity. A meta-analysis of animal models of SAH to humans showed assessment of vasospasm more than 3 days after SAH was associated with translation of results to humans [110]. When investigating mechanisms of vasospasm, the severity and duration of vasospasm may be important since remodeling and structural changes in the arteries take days to develop [45].

Models of SAH where clotted blood is surgically placed into the basal cisterns have been reported in dogs [83], pigs [48], and monkeys [15]. The advantage of these models is that vasospasm tends to be more severe and prolonged than after single cisternal injections. The disadvantages are that the models are more technically difficult to create, there is no rupture of an aneurysm (or in most models, a different phenomenon that is rupture of a cerebral artery), and there are no acute ICP changes.

4   Creation of SAH

SAH has been created in animals by injection of autologous blood into the subarachnoid space, deliberate rupture of an intracranial artery, or placement of a blood clot in the subarachnoid space. Criteria for an ideal model of SAH were constant amount of blood in the subarachnoid space in each animal in the same distribution as occurs after aneurysm rupture and by a mechanism that simulates aneurysm rupture [80]. The model should be simple, inexpensive, and reproduce features of human SAH such as that the major cerebral arteries become narrowed in a delayed fashion days after the SAH and remain narrowed for days, that there are histopathological signs of smooth muscle and endothelial cell damage, often associated with intimal hyperplasia towards the end of and following the phase of arterial narrowing [19], and the vasospastic arteries are resistant to dilation with vasodilatory drugs such as tolerable oral or intravenous nimodipine and papaverine [102]. Other pathological features such as cortical spreading depolarizations and microthromboemboli and clinical and radiological features of DCI would theoretically be desirable. At present, there is no model that fulfills all these criteria. The best model, therefore, will depend in part on the question being asked.

5   Specific Animal Models

5.1   *Mouse*

An endovascular perforation model was described by Kamii et al. who advanced a 5-0 monofilament nylon suture with a blunt end up the internal carotid artery of the mouse until resistance was felt [35]. SAH was created by advancing it 5 mm further to perforate the anterior cerebral artery, and then withdrawing the suture. They reported that acute mortality was 28% and there was significant vasospasm of the middle cerebral artery 3, but not 1 or 7, days after SAH. Middle cerebral artery diameter was reduced about 20%.

Parra et al. noted that in the mouse endovascular perforation model, the degree of SAH was variable, which has been noted with all SAH models that induce SAH by rupturing an intracranial artery [63]. Also, the degree of vasospasm depended on the pressure the animals were perfused at, with less vasospasm at higher perfusion pressures. Neurological assessments have been done and show acute changes that resolve over time. Most work on this model has focused on vasospasm. Brain edema [20] and various molecular changes in the brain [87, 103] have been studied days after SAH, and platelet-leucocyte adhesion and microcirculatory changes hours after SAH [29, 82]. One report also documented neuronal loss in the hippocampus but not the cerebral cortex or striatum and impaired neurological function 7 days, in addition to brain edema 3 days after SAH [17]. Vasospasm was not examined. Review of seven studies using this model found 20–62% vasospasm 3 days after SAH [46]. Thus, there are many similarities to SAH in humans, although phenomena not documented include microthromboemboli, cortical spreading depolarizations, pharmacological/structural changes in the large cerebral arteries, and territorial infarctions.

Altay et al. reported inducing SAH in mice by exposing the posterior atlanto-occipital membrane and transecting a cisternal vein with microforceps [2]. Vasospasm was observed 24 h after SAH. There was no mortality.

Cisternal blood injections also are reported in mice. Lin et al. injected 60:1 autologous blood into the cisterna magna of C57 black mice weighing 27–32 g [42]. There was reduction in anterior circulation artery diameters within hours of SAH that persisted for up to 3 days. Mortality was 3%. The prechiasmatic cistern anterior circulation model of SAH first reported in rats was modified for use in mice by injecting 100:1 autologous blood [74]. There was an acute increase in ICP, suggesting there was transient global ischemia. Vasospasm was detected after 7 days and there was brain injury shown by TUNEL- and fluorojade-positive neurons throughout the brain. Microthromboemboli also occur in this model [72]. Thus, the cisternal injection methods have the limitations of the endovascular perforation model listed above, except that microthromboemboli have been documented. They differ in that the volume of SAH is more controllable and the mortality is lower than human SAH. Neurologic deficits are reported in mouse models up to 3 days after SAH, and the scales used have in some cases detected treatment effects [33].

5.2   Rat

Human creativity has resulted in innumerable methods for inducing SAH in rats, but the most common are single or double injections into the cisterna magna, prechiasmatic cistern injection, or endovascular perforation.

Barry and coworkers were among the first to use rats for SAH [3]. They exposed the basilar artery transclivally and punctured it with a microelectrode. Overall mortality was 26%. The technical difficulties and lack of advantages over other models probably contributed to the limited use of this model.

Single injections of autologous blood into the cisterna magna are commonly done in rats, initially through a cannula or needle inserted through a parietal burr hole, or more recently by direct or percutaneous needle puncture of the cisterna magna [46, 86]. Volumes injected range from 0.07 to 0.8 mL, with most using 300–500:1. This causes 19–29% reduction in basilar artery diameter 3–5 days after SAH [46]. Reduction in CBF was reported 24–48 h after SAH by some, but not all investigators [30, 106]. Myonecrosis was reported in smooth muscle cells in the basilar artery at 48 h, although there were no significant changes in the intima or perivascular nerves. Others did not observe histopathologic changes in the vasospastic arteries [7]. This model has been used to examine vasospasm, blood–brain barrier permeability [21], as well as the microcirculation within hours of SAH.

The double injection model adds a second injection of blood usually 48 h after the first. The severity of vasospasm was 32–47% 5–7 days after the first injection [50, 98]. Mortality was 47% within 5 days in one study, and less than 10% in another that injected less blood [94]. The injections usually are given directly into the cisterna magna. Another method is to insert a catheter into the cisterna magna via a parietal burr hole [41, 86]. The model is widely used to study vasospasm. Changes in receptor function and in genes that suggest remodeling may be occurring are beginning to be described [25, 97, 101]. In addition, the brain has been examined for edema, blood–brain barrier permeability, and changes in gene expression [34]. Slow deterioration in neurological function up to 5 days after SAH was taken as evidence of DCI in this model [24]. Most studies have not measured ICP, so independent effects of SAH and increased ICP are unclear.

Ryba et al. reproduced in rats a model established in rabbits by Endo and coworkers [14, 71]. Both common carotid arteries were ligated in Wistar rats. Two weeks later, SAH was induced by injections of 0.1 mL autologous arterial blood 48 h apart. Mortality was 25% and myonecrosis was not observed in the basilar arteries. Measurement of vasospasm was not reported.

Okada et al. invented a model where blood is placed in silicone cuffs around the femoral arteries of rats [60]. The time course of vasospasm is similar to that in humans and the arteries develop pathological changes. Advantages of this model are that there is a contralateral femoral artery that is an internal control. Disadvantages are the use of a systemic rather than cerebral artery, inflammation is qualitatively greater, and obviously there is no relationship to the brain.

Many methods have been reported for rupturing cerebral arteries in rats in order to produce SAH, but endovascular perforation methods have been most popular. Bederson et al. created SAH by threading a 3-0 monofilament suture up the extracranial internal carotid artery until it perforated through the internal carotid or anterior cerebral artery. SAH developed in 89% and intracerebral hemorrhage in 11%. Mortality was 50% within 24 h of SAH. The Sheffield model is very similar [99]. This model produces an acute rise in ICP and reduction in CBF on the side of the hemorrhage that persisted for some time. The model is most used to study acute changes in the microcirculation and acute and more delayed changes in the brain. The degree of SAH is variable, pathological changes in the cerebral arteries are not reported, pharmacological features have been investigated only within hours of SAH [47] and the time course and severity of vasospasm are less well characterized. Comparison of single cisterna magna injection of 300:1 autologous arterial blood to SAH induced by endovascular perforation with a 4-0 or 3-0 suture showed less SAH and lower peak ICP with the smaller suture [80].

Various other technical complications can occur with this model, such as middle cerebral artery occlusion with infarction and subdural bleeding. A method to modify the end of the filament to reduce the incidence of subdural hemorrhage and increase the incidence of SAH was reported [61]. Another method to adjust for the variability of SAH in the endovascular perforation model was to grade the amount of SAH observed at gross pathology when the animals were euthanized 24 h after SAH [90]. This can then be used to adjust the groups for any other parameters measured. A limitation is that there is no control injection of saline group and mortality is high (generally about 50%).

Prunell et al. injected blood into the prechiasmatic cistern of rats [67]. This model produces vasospasm that lasts, with a mortality that varies depending on the volume of blood injected [68]. There is brain injury, microthromboembolism, and learning deficits in the Morris water maze [32, 69]. The model has advantages in that the SAH is in the anterior circulation like that in humans.

Several studies compared the rat single and double cisterna magna injection and endovascular perforation models [23, 66]. The endovascular perforation model has high mortality (44–57%) and substantial variability in the amount of SAH produced. Vasospasm was most severe with the double hemorrhage model and mortality was 9%. In another study, vasospasm was not examined, but it was found that the endovascular perforation, double cisterna magna, and prechiasmatic cistern injections all produce neuronal injury hours to days after SAH [62, 68]. Some limitations of rat models are that severe (usually defined as >50% reduction in arterial diameter), prolonged vasospasm has not been reported. Histological changes in the cerebral arteries have not been reproducible and the other pharmacological features such as reduced contractility and compliance and resistance to reversal with vasodilators are not documented to occur. ICP is seldom monitored in the injection models, so differentiation of the effects of SAH from increased ICP could but generally has not been made. Many studies demonstrated reduced neurologic function in rats with SAH and shown ability of some scales to detect treatment effects [33]. The rat is the only species in which cognitive testing has been done [32, 84, 92]. Results have differed with prolonged, progressive deficits detected in the double hemorrhage model [92] and some deficits in the prechiasmatic injection and endovascular perforation models [32, 84].

5.3 Rabbit

Models in rabbits include single or multiple percutaneous injections into the cisterna magna, transorbital injection, arterial rupture via craniotomy, injection through the exposed dura, and placement of blood in silicone cuffs around the cervical common carotid arteries. Single and double cisterna magna injection models are by far the most frequently used in rabbits. This is the smallest species where vasospasm-related artery changes are well described.

Single injections of autologous arterial blood have ranged from 0.5 to 5 mL or 0.5 to 3 mL/kg have been used, with the most common being 3 mL or 1 mL/kg [46]. Reductions in basilar artery diameter have ranged from 19 to 55%, usually 48–72 h after SAH. There are conflicting reports as to whether histopathological changes occur although a pharmacologically resistant phase of vasospasm is documented, similar to that occurring in large animals and presumably in humans [102]. Rabbits are inexpensive and their size makes the model technically simpler than in rats and mice. There are generally fewer biological reagents specific for rabbits than for rats and mice and less knowledge of the genome. Mortality is very low. Many experiments have studied drug treatments to prevent or reverse vasospasm. Grasso et al. reported vasospasm and cortical neuronal necrosis in a rabbit single injection model 72 h after injection [22]. Necrotic and apoptotic neuronal injury was observed 72 h after SAH in this model [8, 81]. ICP was not measured in these studies, so the contribution of increased ICP to the changes cannot be discerned. Functional and molecular changes in smaller, conducting arteries and in the brain are reported [28, 54].

Two injections of blood into the cisterna magna increase the severity of vasospasm and prolong its duration, so the maximum is 4–6 days after the first injection [56, 112]. Comparison of 1 and 2 injections in rabbits showed no mortality and maximum vasospasm 3 days after a single injection and 6% mortality and maximum vasospasm 5 days after two injections spaced 48 h apart [109].

Endo et al. occluded both common carotid arteries of rabbits [14]. Two weeks later, they injected autologous blood (2.5 and 1.5 mL in 3–3.5 kg rabbits) 48 h apart. There was 23% reduction in basilar artery diameter 3–6 days later and cerebral infarction was observed pathologically in 2 of 13 (15%) animals.

Various methods to perforate the basilar artery and creation of vasospasm by placement of blood in cuffs around the cervical common carotid arteries are infrequently used [45, 46]. No studies examined microthromboemboli or cortical spreading depolarizations. Neurologic scales have been devised to detect effects of SAH and treatment in rabbits [33].

5.4 Dog

Methods reported for induction of SAH in dogs include injections into the cisterna magna and chiasmatic cistern, rupture of arteries, and placement of clot against the middle cerebral artery. Over 90% of reported studies use the double-injection model.

Lougheed and Tom used a transoral, transclival route to induce SAH in 50 dogs [44]. They injected 5 ml arterial blood, but SAH was produced in only 42% with technical failures in the remainder. Other approaches that are seldom reported are percutaneous transorbital access to the chiasmatic cistern via the optic foramen [49, 105] and femoral artery to subarachnoid space shunts [45]. Brawley et al. were probably the first to describe the biphasic time course of vasospasm in animals [4]. SAH was induced by avulsing the anterior cerebral artery. Vasospasm was assessed by a strain gauge around the intracranial internal carotid artery, which showed acute constriction 5 min after SAH followed by return to baseline at 1 h and then reconstriction 4–6 days after SAH.

A method similar to the rat femoral artery model was reported in dogs where blood-filled silicone cuffs were placed around the extracranial carotid arteries of dogs after microsurgical removal of some of the arterial adventitia [51]. Craniotomy and placement of autologous clot against the middle cerebral artery caused 50% reduction in diameter of this artery 7 days later [83].

A single autologous blood injection into the cisterna magna Kuwayama et al. injected 2 mL fresh autologous arterial blood (0.13–0.29 mL/kg) into the cisterna magna of dogs [39]. Angiography showed acute spasm that reduced the basilar artery diameter an average of 37% at 30 min. There was slight dilation and then recurrence of an identical degree of narrowing 2 days after the SAH. Important modifications added by two groups were careful standardization of angiography and maintenance of Trendelenberg position for 15 min after injection to promote pooling of blood in the basal cisterns [1, 38]. This model was used with little modification for numerous studies documenting at 2 days post-SAH a relatively constant 25–35% reduction in basilar artery diameter [37]. Numerous biochemical measurements were carried out upon which theories of the pathogenesis of VSP were based [57, 77]. This model does not produce severe VSP. Pathological changes in the spastic arteries do not occur [13] or at least are not marked [78] after single blood injections and other markers of VSP, such as pharmacological changes of decreased contractility and compliance and resistance to vasodilator drugs, are not found [55]. CBF is not affected. Interpretation of many of these early studies is difficult because they often utilized small numbers of animals, qualitative or unstandardized assessment of angiographic VSP, and they assessed drug effect on acute but not delayed vasospasm.

Eldevik et al. gave multiple injections of blood into the cisterna magna of dogs in an effort to increase the severity and duration of vasospasm [13], but it was Liszczak et al. [43] who developed the now popular dog double hemorrhage model. Maintaining the animal in the 30° head-down position for 15 min after injection facilitates pooling of blood in the basal cisterns and increases the severity of vasospasm. Review of studies using this model shows wide ranges of blood injected (1.25–11 mL per injection or 0.1–0.8 mL/kg) with most investigators using two 4–5 mL injections or two 0.4–0.5 mL/kg injections. Ninety-five percent of studies reported this caused 45–66% reduction in basilar artery diameter 5–7 days after the first injection. The injections are usually 48 h apart. This model has been used extensively to study effects of drugs on vasospasm as well as biochemical and molecular biological changes in the large cerebral arteries [45].

Studies of CBF in the dog double hemorrhage model are conflicting with many showing no effect of SAH on CBF. Even double SAH plus bilateral common carotid artery ligation did not affect CBF or pressure autoregulation 7 days later, although CO_2 reactivity was impaired in some areas [53]. The penetrating arteries were also constricted and remodeled in this model [59, 111] in some studies, but not in other [64]. No neuronal damage occurs in the pons [64]. Magnetic resonance imaging showed decreased T2 signal in gray matter 2 and 7 days and increased T2 signal in white matter 2 days after SAH [31]. The apparent diffusion coefficient increased at both times in white and gray matter. Neuronal damage has been detected in the cortex [31, 75].

The dog double hemorrhage model is well characterized with vasospasm shown to have a time course similar to that in man. Severe vasospasm (>50% diameter reduction) reliably occurs, there are histopathological changes in the cerebral arteries, and vasospasm is resistant to pharmacological vasodilators. Disadvantages include cost, low mortality, use of two injections of blood which some opine creates different pathophysiology than a single SAH, and limited genetic knowledge of the dog. Also, only primitive neurologic assessments have been done in dogs despite the availability of more sophisticated neurologic and cognitive tests [33].

5.5 Nonhuman Primate

Models of SAH in nonhuman primates include percutaneous injection of blood into the cisterna magna, percutaneous or open transorbital injection into the chiasmatic cistern, arterial rupture by removal of a suture placed via craniotomy into an intracranial artery, and craniotomy with clot placement. The clot placement model was used in 72% of studies reported in the last decade [46].

Simeone et al. injected blood into the subarachnoid space of Rhesus monkeys (Macaca mulatta) or created SAH by avulsing a needle placed previously into the internal carotid artery [85]. Mortality was high and vasospasm was relatively transient, which is consistent with the findings of other investigators using single cisternal injections into nonhuman primates and with effects of single cisternal blood injections in dogs [18, 45, 104]. In another study, weekly subarachnoid injections of 4 mL (0.6–1.5 mL/kg) of autologous blood to Rhesus monkeys was associated with mortality of 6% after the first injection and 37% after the fourth. Vasospasm was worst 15 min after SAH, and there was only a 6% reduction in artery diameters after 7 days [104]. Prior investigators were interested in the theory that arterial wall rupture was required to produce structural changes in the arteries and has some role in vasospasm [6]. Comparison of the effect of cisternal injection of 3 mL autologous nonheparinized blood to that created by avulsing the middle cerebral artery by pulling on a ligature on it showed more vasospasm with the latter method, but it is likely that some of the difference was due to a greater volume of SAH produced by the avulsion. The vessel rupture models increase ICP and lower CBF, at least acutely, and produce brain edema, but mortality is high and they don't offer much advantage over the same models in lower species, so they are not used commonly [94]. Single injection models in nonhuman primates also seem to have largely been abandoned.

Multiple cisternal injections also were reported in nonhuman primates. Svendgaard et al. gave two injections of blood 24 h apart into the cisterna magna of baboons and a third 24 h later into the chiasmatic cistern. A total of 14–33 mL of blood (1.6 mL/kg) was injected [91]. Mortality was 32% and average reductions in arterial diameter were 10–20% 7 days later. Increased sensitivity to several agonists in pial arteries was noted 7 days after SAH [76]. The large cerebral arteries tend to have reduced sensitivity and contractility in other reports and in dogs [36]. In another multiple injection model, the injections were given through several different catheters into squirrel monkeys (Saimiri sciureus) [10]. The model was associated with 44–80% constriction of the most severely affected artery 6 days after SAH. Mortality was 17%. There was reduced CBF and increased deoxyglucose uptake during vasospasm 6 days after SAH [9]. The complexity of the model and perhaps lack of advantages over other models have contributed to lack of any further studies with injection models in nonhuman primates.

A seminal advance was the unilateral clot placement model developed by Weir [15]. This involves a frontotemporal craniectomy with opening of the arachnoid membrane over the internal carotid and anterior and middle cerebral arteries. Clotted arterial blood is placed against the exposed cerebral arteries. Acute mortality is about 10% and essentially all animals develop some vasospasm 7 days after SAH. About a quarter develop >50% reduction in arterial diameters, and the time course of vasospasm is the same as in humans. Advantages are the ability to produce prolonged, severe vasospasm with all of the features of the condition in humans. Focal neurologic deficits were produced in early reports, but are very uncommon in recent studies [45]. Pathological changes are observed in the arteries, CBF and autoregulation are impaired, the vasospastic arteries are resistant to dilation with intravenous nimodipine, and they show other pharmacological characteristics of VSP such as decreased compliance and contractility [45]. The contralateral arteries are internal controls. The disadvantages are cost and the technical nature of the procedures. The model also produces very focal SAH, the CSF is clear, and since the cranium is open when the clot is placed, there is no increase in ICP.

Schatlo et al. modified the monkey clot placement by removing more arachnoid from the brain adjacent to the Sylvian fissure [79]. This was associated with cortical laminar necrosis and focal infarctions in the brain under the clots. Otherwise, few studies have examined the brain in this model. Nothing has been done in terms of neurologic and cognitive testing in nonhuman primate models of SAH.

6   Translation to Humans

The experimental models of SAH described above have been used to test the efficacy of various drugs against vasospasm. The results could differ because of different species used, different methods of induction of SAH, as well as in quality of design (i.e., randomization, blinding). We reviewed pharmacologic treatments that have been tested in clinical trials of patients with SAH and compared their effects on angiographic vasospasm in the trials with the effects of the same drugs in the animal studies. The goals were to determine if there is evidence that any animal model of SAH can inform clinical trials and to determine what specific model or study characteristics, if any, more accurately reflect results reported in humans. A systematic review and meta-analysis was conducted according to methods recommended for assessing health technologies [26, 52, 110].

Animal models where SAH was induced by endovascular puncture of a cerebral artery, subarachnoid clot placement, or injection of blood into the subarachnoid space were included. The animal study had to test a drug that had also been tested in randomized clinical trials in humans. The drugs were those included in the systematic review of drug treatments for humans with SAH conducted by Etminan et al. [16]. Since most of the animal models assess vasospasm, the comparison between animal and human study results was based on large-artery vasospasm measured by angiography of some type or transcranial Doppler ultrasound.

Overall, there were 70 papers describing animal studies with data suitable for comparison to the human trials. Two different drugs were tested in 4 papers, leading to 74 comparisons for analysis. All of the species described above were represented, and the most common methods to induce SAH were two injections of blood into the cisterna magna (dog, rabbit, and rat studies), a single injection of blood into the cisterna magna (rabbit and rat studies), and craniotomy followed by clot placement around the cerebral arteries (all monkey and one dog study). Endovascular perforation in rats and mice and prechiasmatic cistern blood injection in rats were used in only a few studies. The drugs tested were tirilazad, tissue plasminogen activator, calcium channel antagonists (nimodipine or nicardipine), erythropoietin, statins, fasudil, endothelin antagonists or endothelin converting enzyme inhibitors, magnesium, and in one study, tissue plasminogen activator plus an endothelin antagonist. We defined successful translation as either a drug effectively decreasing vasospasm in both animals and humans or as not having any effect in both animals and humans.

The key findings were that pharmacologic prevention of vasospasm was significantly effective in animals and humans. Subgroup analysis by drug and species showed that all drugs except magnesium were effective and that prevention was effective in all species. Comparing animal and human studies, fresh blood used to simulate SAH (vs. clot placement) and the evaluation of vasospasm more than 3 days after SAH were independently associated with successful translation. These results suggest that at least in terms of large-artery vasospasm, multiple animal models seem to be useful. Injection of fresh blood rather than clot placement and evaluation of vasospasm more than 3 days after SAH may be preferable for testing such drugs. The limitations include that other clinical endpoints, such as DCI, and overall outcome may be more relevant, but these are not well described in most animal models. Another problem was that the methodological quality of many of the animal studies is poor and this could affect translation of the results. Drugs also may be ineffective in both animals and humans because the dose, timing, and route of administration are not optimal or they could be effective in one situation or the other because these factors are not fully evaluated.

References

1. Allen GS, Bahr AL. Cerebral arterial spasm: part 10. Reversal of acute and chronic spasm in dogs with orally administered nifedipine. Neurosurgery. 1979;4:43–7.
2. Altay T, Smithason S, Volokh N, et al. A novel method for subarachnoid hemorrhage to induce vasospasm in mice. J Neurosci Methods. 2009;183:136–40.
3. Barry KJ, Gogjian MA, Stein BM. Small animal model for investigation of subarachnoid hemorrhage and cerebral vasospasm. Stroke. 1979;10:538–41.
4. Brawley BW, Strandness DEJ, Kelly WA. The biphasic response of cerebral vasospasm in experimental subarachnoid hemorrhage. J Neurosurg. 1968;28:1–8.
5. Cahill J, Calvert JW, Zhang JH. Mechanisms of early brain injury after subarachnoid hemorrhage. J Cereb Blood Flow Metab. 2006;26:1341–53.
6. Clower BR, Smith RR, Haining JL, et al. Constrictive endarteropathy following experimental subarachnoid hemorrhage. Stroke. 1981;12:501–8.
7. Clower BR, Yamamoto Y, Cain L, et al. Endothelial injury following experimental subarachnoid hemorrhage in rats: effects on brain blood flow. Anat Rec. 1994;240:104–14.
8. Cosar M, Eser O, Fidan H, et al. The neuroprotective effect of dexmedetomidine in the hippocampus of rabbits after subarachnoid hemorrhage. Surg Neurol. 2009;71:54–9.
9. Delgado-Zygmunt T, Arbab MA, Shiokawa Y, et al. Cerebral blood flow and glucose metabolism in the squirrel monkey during the late phase of cerebral vasospasm. Acta Neurochir. 1993;121:166–73.
10. Delgado-Zygmunt TJ, Arbab MA, Shiokawa Y, et al. A primate model for acute and late cerebral vasospasm: angiographic findings. Acta Neurochir. 1992;118:130–6.
11. Dreier JP, Major S, Manning A, et al. Cortical spreading ischaemia is a novel process involved in ischaemic damage in patients with aneurysmal subarachnoid haemorrhage. Brain. 2009;132:1866–81.
12. Dreier JP, Windmuller O, Petzold G, et al. Ischemia triggered by red blood cell products in the subarachnoid space is inhibited by nimodipine administration or moderate volume expansion/hemodilution in rats. Neurosurgery. 2002;51:1457–65.
13. Eldevik OP, Kristiansen K, Torvik A. Subarachnoid hemorrhage and cerebrovascular spasm. Morphological study of intracranial arteries based on animal experiments and human autopsies. J Neurosurg. 1981;55:869–76.
14. Endo S, Branson PJ, Alksne JF. Experimental model of symptomatic vasospasm in rabbits. Stroke. 1988;19:1420–5.
15. Espinosa F, Weir B, Shnitka T, et al. A randomized placebo-controlled double-blind trial of nimodipine after SAH in monkeys. Part 2: pathological findings. J Neurosurg. 1984;60:1176–85.
16. Etminan N, Vergouwen MD, Ilodigwe D, et al. Effect of pharmaceutical treatment on vasospasm, delayed cerebral ischemia, and clinical outcome in patients with aneurysmal subarachnoid hemorrhage: a systematic review and meta-analysis. J Cereb Blood Flow Metab. 2011;31:1443–51.
17. Feiler S, Friedrich B, Scholler K, et al. Standardized induction of subarachnoid hemorrhage in mice by intracranial pressure monitoring. J Neurosci Methods. 2010;190:164–70.
18. Fein JM, Flor WJ, Cohan SL, et al. Sequential changes of vascular ultrastructure in experimental cerebral vasospasm. Myonecrosis of subarachnoid arteries. J Neurosurg. 1974;41:49–58.
19. Findlay JM, Weir BK, Kanamaru K, et al. Arterial wall changes in cerebral vasospasm. Neurosurgery. 1989;25:736–45.
20. Gao J, Wang H, Sheng H, et al. A novel apoE-derived therapeutic reduces vasospasm and improves outcome in a murine model of subarachnoid hemorrhage. Neurocrit Care. 2006;4:25–31.
21. Germano A, Costa C, DeFord SM, et al. Systemic administration of a calpain inhibitor reduces behavioral deficits and blood-brain barrier permeability changes after experimental subarachnoid hemorrhage in the rat. J Neurotrauma. 2002;19:887–96.

22. Grasso G, Buemi M, Alafaci C, et al. Beneficial effects of systemic administration of recombinant human erythropoietin in rabbits subjected to subarachnoid hemorrhage. Proc Natl Acad Sci USA. 2002;99:5627–31.

23. Gules I, Satoh M, Clower BR, et al. Comparison of three rat models of cerebral vasospasm. Am J Physiol Heart Circ Physiol. 2002;283:H2551–9.

24. Guresir E, Raabe A, Jaiimsin A, et al. Histological evidence of delayed ischemic brain tissue damage in the rat double-hemorrhage model. J Neurol Sci. 2010;293:18–22.

25. Hansen-Schwartz J, Hoel NL, Xu CB, et al. Subarachnoid hemorrhage-induced upregulation of the 5-HT1B receptor in cerebral arteries in rats. J Neurosurg. 2003;99:115–20.

26. Higgins J, Green S. Cochrane handbook for systematic reviews of interventions. In: The cochrane collaboration, 2008. Available at www.cochrane-handbook.org. Accessed Aug 8, 2011

27. Hop JW, Rinkel GJ, Algra A, et al. Case-fatality rates and functional outcome after subarachnoid hemorrhage: a systematic review. Stroke. 1997;28:660–4.

28. Ishiguro M, Puryear CB, Bisson E, et al. Enhanced myogenic tone in cerebral arteries from a rabbit model of subarachnoid hemorrhage. Am J Physiol Heart Circ Physiol. 2002;283:H2217–25.

29. Ishikawa M, Kusaka G, Yamaguchi N, et al. Platelet and leukocyte adhesion in the microvasculature at the cerebral surface immediately after subarachnoid hemorrhage. Neurosurgery. 2009;64:546–53.

30. Jackowski A, Crockard A, Burnstock G, et al. The time course of intracranial pathophysiological changes following experimental subarachnoid haemorrhage in the rat. J Cereb Blood Flow Metab. 1990;10:835–49.

31. Jadhav V, Sugawara T, Zhang J, et al. Magnetic resonance imaging detects and predicts early brain injury after subarachnoid hemorrhage in a canine experimental model. J Neurotrauma. 2008;25:1099–106.

32. Jeon H, Ai J, Sabri M, et al. Learning deficits after experimental subarachnoid hemorrhage in rats. Neuroscience. 2010;169:1805–14.

33. Jeon H, Ai J, Sabri M, et al. Neurological and neurobehavioral assessment of experimental subarachnoid hemorrhage. BMC Neurosci. 2009;10:103.

34. Josko J, Gwozdz B, Hendryk S, et al. Expression of vascular endothelial growth factor (VEGF) in rat brain after subarachnoid haemorrhage and endothelin receptor blockage with BQ-123. Folia Neuropathol. 2001;39:243–51.

35. Kamii H, Kato I, Kinouchi H, et al. Amelioration of vasospasm after subarachnoid hemorrhage in transgenic mice overexpressing CuZn-superoxide dismutase. Stroke. 1999;30:867–71.

36. Kanamaru K, Weir BK, Findlay JM, et al. Pharmacological studies on relaxation of spastic primate cerebral arteries in subarachnoid hemorrhage. J Neurosurg. 1989;71:909–15.

37. Kaoutzanis M, Yokota M, Sibilia R, et al. Neurologic evaluation in a canine model of single and double subarachnoid hemorrhage. J Neurosci Methods. 1993;50:301–7.

38. Kistler JP, Lees RS, Candia G, et al. Intravenous nitroglycerin in experimental cerebral vasospasm. A preliminary report. Stroke. 1979;10:26–9.

39. Kuwayama A, Zervas NT, Belson R, et al. A model for experimental cerebral arterial spasm. Stroke. 1972;3:49–56.

40. Laslo AM, Eastwood JD, Pakkiri P, et al. CT perfusion-derived mean transit time predicts early mortality and delayed vasospasm after experimental subarachnoid hemorrhage. AJNR Am J Neuroradiol. 2008;29:79–85.

41. Lee JY, Huang DL, Keep R, et al. Characterization of an improved double hemorrhage rat model for the study of delayed cerebral vasospasm. J Neurosci Methods. 2008;168:358–66.

42. Lin CL, Calisaneller T, Ukita N, et al. A murine model of subarachnoid hemorrhage-induced cerebral vasospasm. J Neurosci Methods. 2003;123:89–97.

43. Liszczak TM, Varsos VG, Black PM, et al. Cerebral arterial constriction after experimental subarachnoid hemorrhage is associated with blood components within the arterial wall. J Neurosurg. 1983;58:18–26.

44. Lougheed WM, Tom M. A method of introducing blood into the subarachnoid space in the region of the circle of Willis in dogs. Can J Surg. 1961;4:329–37.

45. Macdonald RL, Weir B. Cerebral vasospasm. San Diego: Academic; 2001.

46. Marbacher S, Fandino J, Kitchen ND. Standard intracranial in vivo animal models of delayed cerebral vasospasm. Br J Neurosurg. 2010;24:415–34.

47. Marshman LA, Morice AH, Thompson JS. Increased efficacy of sodium nitroprusside in middle cerebral arteries following acute subarachnoid hemorrhage: indications for its use after rupture. J Neurosurg Anesthesiol. 1998;10:171–7.

48. Mayberg MR, Okada T, Bark DH. Morphologic changes in cerebral arteries after subarachnoid hemorrhage. Neurosurg Clin N Am. 1990;1:417–32.

49. McQueen JD, Jeanes LD. Influence of hypothermia on intracranial hypertension. J Neurosurg. 1962;19:277–88.

50. Meguro T, Chen B, Lancon J, et al. Oxyhemoglobin induces caspase-mediated cell death in cerebral endothelial cells. J Neurochem. 2001;77:1128–35.

51. Megyesi JF, Findlay JM, Vollrath B, et al. In vivo angioplasty prevents the development of vasospasm in canine carotid arteries. Pharmacological and morphological analyses. Stroke. 1997;28:1216–24.

52. Mignini LE, Khan KS. Methodological quality of systematic reviews of animal studies: a survey of reviews of basic research. BMC Med Res Methodol. 2006;6:10.

53. Miyamoto Y, Matsuda M. Cerebral blood flow and somatosensory evoked potentials in dogs with experimental vasospasm caused by double injection. Archiv Jpn Chirgurie. 1991;60: 289–98.

54. Murakami K, Koide M, Dumont TM, et al. Subarachnoid hemorrhage induces gliosis and increased expression of the pro-inflammatory cytokine high mobility group box 1 protein. Transl Stroke Res. 2011;2:72–9.

55. Nagasawa S, Handa H, Naruo Y, et al. Experimental cerebral vasospasm arterial wall mechanics and connective tissue composition. Stroke. 1982;13:595–600.

56. Nakagomi T, Kassell NF, Sasaki T, et al. Impairment of endothelium-dependent vasodilation induced by acetylcholine and adenosine triphosphate following experimental subarachnoid hemorrhage. Stroke. 1987;18:482–9.

57. Nozaki K, Okamoto S, Uemura Y, et al. Changes of glycogen and ATP contents of the major cerebral arteries after experimentally produced subarachnoid haemorrhage in the dog. Acta Neurochir. 1990;104:38–41.

58. Nuki Y, Tsou TL, Kurihara C, et al. Elastase-induced intracranial aneurysms in hypertensive mice. Hypertension. 2009;54:1337–44.

59. Ohkuma H, Suzuki S, Ogane K. Phenotypic modulation of smooth muscle cells and vascular remodeling in intraparenchymal small cerebral arteries after canine experimental subarachnoid hemorrhage. Neurosci Lett. 2003;344:193–6.

60. Okada T, Harada T, Bark DH, et al. A rat femoral artery model for vasospasm. Neurosurgery. 1990;27:349–56.

61. Park IS, Meno JR, Witt CE, et al. Subarachnoid hemorrhage model in the rat: modification of the endovascular filament model. J Neurosci Methods. 2008;172:195–200.

62. Park S, Yamaguchi M, Zhou C, et al. Neurovascular protection reduces early brain injury after subarachnoid hemorrhage. Stroke. 2004;35:2412–7.

63. Parra A, McGirt MJ, Sheng H, et al. Mouse model of subarachnoid hemorrhage associated cerebral vasospasm: methodological analysis. Neurol Res. 2002;24:510–6.

64. Perkins E, Kimura H, Parent AD, et al. Evaluation of the microvasculature and cerebral ischemia after experimental subarachnoid hemorrhage in dogs. J Neurosurg. 2002;97:896–904.

65. Pluta RM, Hansen-Schwartz J, Dreier J, et al. Cerebral vasospasm following subarachnoid hemorrhage: time for a new world of thought. Neurol Res. 2009;31:151–8.

66. Prunell GF, Mathiesen T, Diemer NH, et al. Experimental subarachnoid hemorrhage: subarachnoid blood volume, mortality rate, neuronal death, cerebral blood flow, and perfusion pressure in three different rat models. Neurosurgery. 2003;52:165–75.

67. Prunell GF, Mathiesen T, Svendgaard NA. A new experimental model in rats for study of the pathophysiology of subarachnoid hemorrhage. Neuroreport. 2002;13:2553–6.
68. Prunell GF, Mathiesen T, Svendgaard NA. Experimental subarachnoid hemorrhage: cerebral blood flow and brain metabolism during the acute phase in three different models in the rat. Neurosurgery. 2004;54:426–36.
69. Prunell GF, Svendgaard NA, Alkass K, et al. Delayed cell death related to acute cerebral blood flow changes following subarachnoid hemorrhage in the rat brain. J Neurosurg. 2005;102:1046–54.
70. Rosengart AJ, Schultheiss KE, Tolentino J, et al. Prognostic factors for outcome in patients with aneurysmal subarachnoid hemorrhage. Stroke. 2007;38:2315–21.
71. Ryba MS, Gordon-Krajcer W, Walski M, et al. Hydroxylamine attenuates the effects of simulated subarachnoid hemorrhage in the rat brain and improves neurological outcome. Brain Res. 1999;850:225–33.
72. Sabri M, Ai J, Macdonald RL. Dissociation of vasospasm and secondary effects of experimental subarachnoid hemorrhage by clazosentan. Stroke. 2011;42:1454–60.
73. Sabri M, Ai J, Marsden PA, et al. Simvastatin re-couples dysfunctional endothelial nitric oxide synthase in experimental subarachnoid hemorrhage. PLoS One. 2011;6:e17062.
74. Sabri M, Jeon H, Ai J, et al. Anterior circulation mouse model of subarachnoid hemorrhage. Brain Res. 2009;1295:179–85.
75. Sabri M, Kawashima A, Ai J, et al. Neuronal and astrocytic apoptosis after subarachnoid hemorrhage: a possible cause for poor prognosis. Brain Res. 2008;1238:163–71.
76. Sahlin C, Owman C, Chang JY, et al. Changes in contractile response and effect of a calcium antagonist, nimodipine, in isolated intracranial arteries of baboon following experimental subarachnoid hemorrhage. Brain Res Bull. 1990;24:355–61.
77. Sasaki T, Murota SI, Wakai S, et al. Evaluation of prostaglandin biosynthetic activity in canine basilar artery following subarachnoid injection of blood. J Neurosurg. 1981;55:771–8.
78. Sasaki T, Wakai S, Asano T, et al. Prevention of cerebral vasospasm after SAH with a thromboxane synthetase inhibitor, OKY-1581. J Neurosurg. 1982;57:74–82.
79. Schatlo B, Dreier JP, Glasker S, et al. Report of selective cortical infarcts in the primate clot model of vasospasm after subarachnoid hemorrhage. Neurosurgery. 2010;67:721–8.
80. Schwartz AY, Masago A, Sehba FA, et al. Experimental models of subarachnoid hemorrhage in the rat: a refinement of the endovascular filament model. J Neurosci Methods. 2000;96:161–7.
81. Seckin H, Simsek S, Ozturk E, et al. Topiramate attenuates hippocampal injury after experimental subarachnoid hemorrhage in rabbits. Neurol Res. 2009;31:490–5.
82. Sehba FA, Flores R, Muller A, et al. Adenosine A(2A) receptors in early ischemic vascular injury after subarachnoid hemorrhage. Laboratory investigation. J Neurosurg. 2010;113:826–34.
83. Shiokawa K, Kasuya H, Miyajima M, et al. Prophylactic effect of papaverine prolonged-release pellets on cerebral vasospasm in dogs. Neurosurgery. 1998;42:109–15.
84. Silasi G, Colbourne F. Long-term assessment of motor and cognitive behaviours in the intraluminal perforation model of subarachnoid hemorrhage in rats. Behav Brain Res. 2009;198:380–7.
85. Simeone FA, Ryan KG, Cotter JR. Prolonged experimental cerebral vasospasm. J Neurosurg. 1968;29:357–66.
86. Solomon RA, Antunes JL, Chen RYZ, et al. Decrease in cerebral blood flow in rats after experimental subarachnoid hemorrhage: a new animal model. Stroke. 1985;16:58–64.
87. Sozen T, Tsuchiyama R, Hasegawa Y, et al. Role of interleukin-1beta in early brain injury after subarachnoid hemorrhage in mice. Stroke. 2009;40:2519–25.
88. Stein SC, Browne KD, Chen XH, et al. Thromboembolism and delayed cerebral ischemia after subarachnoid hemorrhage: an autopsy study. Neurosurgery. 2006;59:781–7.

89. Sugawara T, Ayer R, Jadhav V, et al. Simvastatin attenuation of cerebral vasospasm after subarachnoid hemorrhage in rats via increased phosphorylation of Akt and endothelial nitric oxide synthase. J Neurosci Res. 2008;86:3635–43.

90. Sugawara T, Ayer R, Jadhav V, et al. A new grading system evaluating bleeding scale in filament perforation subarachnoid hemorrhage rat model. J Neurosci Methods. 2008;167:327–34.

91. Svendgaard NA, Brismar J, Delgado T, et al. Late cerebral arterial spasm: the cerebrovascular response to hypercapnia, induced hypertension and the effect of nimodipine on blood flow autoregulation in experimental subarachnoid hemorrhage in primates. Gen Pharmacol. 1983;14:167–72.

92. Takata K, Sheng H, Borel CO, et al. Long-term cognitive dysfunction following experimental subarachnoid hemorrhage: new perspectives. Exp Neurol. 2008;213:336–44.

93. The Stroke Therapy Academic Industry Round Table (STAIR). Recommendations for standards regarding preclinical neuroprotective and restorative drug development. Stroke. 1999;30:2752–8.

94. Titova E, Ostrowski RP, Zhang JH, et al. Experimental models of subarachnoid hemorrhage for studies of cerebral vasospasm. Neurol Res. 2009;31(6):568–81.

95. van den Bergh WM, Schepers J, Veldhuis WB, et al. Magnetic resonance imaging in experimental subarachnoid haemorrhage. Acta Neurochir (Wien). 2005;147:977–83.

96. van der Worp HB, Howells DW, Sena ES, et al. Can animal models of disease reliably inform human studies? PLoS Med. 2010;7:e1000245.

97. Vatter H, Konczalla J, Weidauer S, et al. Characterization of the endothelin-B receptor expression and vasomotor function during experimental cerebral vasospasm. Neurosurgery. 2007;60:1100–8.

98. Vatter H, Weidauer S, Konczalla J, et al. Time course in the development of cerebral vasospasm after experimental subarachnoid hemorrhage: clinical and neuroradiological assessment of the rat double hemorrhage model. Neurosurgery. 2006;58:1190–7.

99. Veelken JA, Laing RJ, Jakubowski J. The Sheffield model of subarachnoid hemorrhage in rats. Stroke. 1995;26:1279–83.

100. Vergouwen MD, Vermeulen M, Coert BA, et al. Microthrombosis after aneurysmal subarachnoid hemorrhage: an additional explanation for delayed cerebral ischemia. J Cereb Blood Flow Metab. 2008;28:1761–70.

101. Vikman P, Beg S, Khurana T, et al. Gene expression and molecular changes in cerebral arteries following subarachnoid hemorrhage in the rat. J Neurosurg. 2006;105:438–44.

102. Vorkapic P, Bevan RD, Bevan JA. Longitudinal time course of reversible and irreversible components of chronic cerebrovasospasm of the rabbit basilar artery. J Neurosurg. 1991;74:951–5.

103. Wakade C, King MD, Laird MD, et al. Curcumin attenuates vascular inflammation and cerebral vasospasm after subarachnoid hemorrhage in mice. Antioxid Redox Signal. 2009;11:35–45.

104. Weir B, Erasmo R, Miller J, et al. Vasospasm in response to repeated subarachnoid hemorrhages in the monkey. J Neurosurg. 1970;33:395–406.

105. Wilkins RH, Levitt P. Intracranial arterial spasm in the dog. A chronic experimental model. J Neurosurg. 1970;33:260–9.

106. Yamamoto S, Teng W, Kakiuchi T, et al. Disturbance of cerebral blood flow autoregulation in hypertension is attributable to ischaemia following subarachnoid haemorrhage in rats: a PET study. Acta Neurochir. 1999;141:1213–9.

107. Yatsushige H, Yamaguchi M, Zhou C, et al. Role of c-Jun N-terminal kinase in cerebral vasospasm after experimental subarachnoid hemorrhage. Stroke. 2005;36:1538–43.

108. Zhou C, Yamaguchi M, Kusaka G, et al. Caspase inhibitors prevent endothelial apoptosis and cerebral vasospasm in dog model of experimental subarachnoid hemorrhage. J Cereb Blood Flow Metab. 2004;24:419–31.

109. Zhou ML, Shi JX, Zhu JQ, et al. Comparison between one- and two-hemorrhage models of cerebral vasospasm in rabbits. J Neurosci Methods. 2007;159:318–24.

110. Zoerle T, Ilodigwe D, Wan H, et al. Pharmacologic prevention of cerebral vasospasm in experimental subarachnoid hemorrhage: systematic review and meta-analysis. Submitted 2011.
111. Zubkov AY, Aoki K, Parent AD, et al. Preliminary study of the effects of caspase inhibitors on vasospasm in dog penetrating arteries. Life Sci. 2002;70:3007–18.
112. Zuccarello M, Boccaletti R, Tosun M, et al. Role of extracellular Ca2+ in subarachnoid hemorrhage-induced spasm of the rabbit basilar artery. Stroke. 1996;27:1896–902.

Part IV
De-Risking of Drug Candidates

Chapter 30
ADME (Absorption, Distribution, Metabolism, Excretion): The Real Meaning—Avoiding Disaster and Maintaining Efficacy for Preclinical Candidates

Katya Tsaioun and Steven A. Kates

Abstract The purpose of preclinical Absorption, Distribution, Metabolism, Excretion, and Toxicity also referred to as early drug metabolism and pharmacokinetics is to reduce the risk of drug development to avoid spending scarce resources on weak lead candidates and expensive R&D programs and clinical trials. This allows drug-development resources to be focused on fewer, but more-likely-to-succeed drug candidates. This chapter provides the rationale for effective preclinical drug development using a systematic approach to limit failure.

1 Introduction

Drug attrition that occurs in late clinical development or during postmarketing is a serious economic problem in the pharmaceutical industry [1]. The cost for drug approvals is approaching $1 billion USD, and the cost of advancing a compound to Phase 1 trials can reach up to $100 million USD according to the Tufts Center for the Study of Drug Development, Tufts University School of Medicine [2]. The study also estimates a $37,000 USD direct out-of-pocket cost for each day a drug is in the development stage and opportunity costs of $1.1 million USD in lost revenue [2]. Given these huge expenditures, substantial savings can accrue from early recognition of problems that would demonstrate a compound's potential to succeed in development [3].

The costs associated with withdrawing a drug from the market are even greater. For example, terfenadine is both a potent hERG cardial channel ligand and is metabolized by the liver enzyme Cytochrome P450 3A4 (CYP3A4). Terfenadine was

K. Tsaioun (✉)
Apredica, a Cyprotex Company, 313, Pleasant Street, Watertown, MA, USA
e-mail: k.tsaioun@Cyprotex.com

S.A. Kates
Ischemix LLC, 63 Great Road, Maynard, MA 01754, USA

P.A. Lapchak and J.H. Zhang (eds.), *Translational Stroke Research*, Springer Series in Translational Stroke Research, DOI 10.1007/978-1-4419-9530-8_30, © Springer Science+Business Media, LLC 2012

frequently coadministered with Cyp3A4 inhibitors ketoconazole or erythromycin [4]. The consequent overload resulted in increases in plasma terfenadine to levels that caused cardiac toxicity [5], resulting in the drug to be withdrawn from the market [6] at an estimated cost of $6 billion USD. Another example is the broad-spectrum antibiotic trovafloxacin, which was introduced in 1997 and soon became Pfizer's top seller. The drug was metabolically activated in vivo and formed a highly reactive metabolite causing severe drug-induced hepatotoxicity [7]. Trovafloxacin was black-labeled in 1998 [8], costing Pfizer $8.5 billion USD in lawsuits [9]. With the new ability to measure hERG and other important ADMET (Absorption, Distribution, Metabolism, Excretion, and Toxicity) parameters early in the discovery and development process, such liabilities are now recognized earlier allowing for safer analogs to be advanced to more expensive formal preclinical and clinical stages.

The purpose of preclinical ADMET also referred to as early DMPK (drug metabolism and pharmacokinetics) is to reduce the risk similar to above and avoid spending scarce resources on weak lead candidates and programs. This allows drug-development resources to be focused on fewer, but more-likely-to-succeed drug candidates. In 1993, 40% of drugs failed in clinical trials because of pharmacokinetic (PK) and bioavailability problems [10]. Since then, major technological advances have occurred in molecular biology and screening to allow major aspects of ADMET to be assessed earlier during the lead optimization stage. By the late 1990s, the pharmaceutical industry recognized the value of early ADMET assessment and began routinely employing it with noticeable results. ADME and DMPK problems decreased from 40 to 11% [4]. Presently, a lack of efficacy and human toxicity are the primary reasons for failure [11].

The terms "drugability" and "druglikeness" were first described by Dr. Christopher Lipinski, who proposed "Lipinski's Rule of 5" due to the frequent appearance of a number "5" in the rules [12]. The Rule of 5 has come to be a compass for the drug discovery industry [13]. It stipulates that small-molecule drug candidates should possess:

- A molecular weight less than 500 g/mol
- A partition coefficient (log P—a measure of hydrophobicity) less than 5
- No more than five hydrogen bond donors
- No more than ten hydrogen bond acceptors

A compound with fewer than three of these properties is unlikely to become a successful orally bioavailable drug. There are exceptions to Lipinski's Rule of 5 that have become marketed drugs, such as those taken up by active transport mechanisms, natural compounds, oligonucleotides, and proteins.

The drug discovery industry is experiencing dramatic structural change and is no longer just the domain of traditional large pharmaceutical companies. Now venture-capital-funded startups, governments, venture philanthropy, and other nonprofit and academic organizations are important participants in the search for new drug targets, pathways, and molecules. These organizations frequently form partnerships, sharing resources, capabilities, risks, and rewards of drug discovery. Thus, it is becoming increasingly important to ensure that investors, donors, and taxpayers' money is efficiently used so that new safe drugs for unmet medical needs may be delivered to the public. ADMET profiling has been proven to remove poor drug candidates from development and accelerate the discovery process.

Although lack of efficacy and unexpected toxicity are the major causes of drug failure in clinical trials, a prime determinant is the ability of a drug to penetrate biological barriers such as cell membranes, intestinal walls, or the blood–brain barrier (BBB). For drugs that target the Central Nervous System (CNS) such as stroke, in vitro efficacy combined with the inability to penetrate the BBB typically results in poor in vivo efficacy in patients. The delivery of systemically administered drugs to the brain of mammals is limited by the BBB as it effectively isolates the brain from the blood because of the presence of tight junctions connecting the endothelial cells of the brain vessels. In addition, specific metabolizing enzymes and efflux pumps such as P-glycoprotein (P-gp) and the multidrug-resistance protein (MRP), located within the endothelial cells of the BBB, actively pump exogenous molecules out of the brain [14, 15]. This is one of the reasons for CNS drugs having a notoriously high failure rate [16]. In recent years, 9% of compounds that entered Phase 1 survived to launch and only 3–5% of CNS drugs were commercialized [16]. Greater than 50% of this attrition resulted from failure to demonstrate efficacy in Phase 2 studies. Over the last decade, Phase 2 failures have increased by 15%. Compounds with demonstrated efficacy against a target in vitro and in animal models frequently proved to be ineffective in humans. Many of these failures occur due to the inability to reach the CNS targets such as in stroke due to lack of BBB permeability. For drugs targeted to reduce damage from a stroke, the delivery method, BBB permeability, and drug metabolism and clearance can provide life or death to a patient if the drug is not delivered to the target tissue in its active form in a matter of hours from the event.

Due to the extraordinary cost of drug development, it is highly desirable to have effective, cost-efficient and high-throughput tools to measure BBB permeability before proceeding to expensive and time-consuming animal BBB permeability studies or human clinical trials. With in vitro tools available, promising drug candidates with ineffective BBB penetration may be improved by removing structural components that mediate interaction(s) with efflux proteins and/or lowering binding to brain tissue at earlier stages of development to increase intrinsic permeability [17].

The development of drugs targeting CNS requires precise knowledge of the drug's brain penetration. Ideally, this information would be obtained as early as possible to focus resources on compounds most likely to reach the target organ. The physical transport and metabolic composition of the BBB is highly complex. Numerous in vitro models have been designed to study kinetic parameters in the CNS, including noncerebral peripheral endothelial cell lines, immortalized rat brain endothelial cells, primary cultured bovine, porcine or rat brain capillary endothelial cells, and cocultures of primary brain capillary cells with astrocytes [18–20]. In vitro BBB models must be carefully assessed for their capacity to reflect accurately the passage of drugs into the CNS in vivo.

Alternatively, several in vivo techniques have been used to estimate BBB passage of drugs directly in laboratory animals. In vivo transport across the BBB was first studied in the 1960s using the early indicator diffusion method (IDM) of Crone [21]. Other in vivo techniques were later proposed including brain uptake index (BUI) measurement [22], in situ brain perfusion method [23, 24], autoradiography, and intracerebral microdialysis [25]. Unfortunately, these methods have limitations including sophisticated equipment, technical expertise, mathematical modeling, species differences, invasiveness, and low throughput, which render them unsuitable for use during early stages of drug discovery and development.

Hence, in vitro and in vivo models remain mere approximations of the complex BBB and their relevance to human pharmacology must be carefully considered. The most appropriate method to conduct controlled experiments is to cross-compare the BBB passage of a series of compounds evaluated with both in vitro and in vivo models. This enables cross-correlations of pharmacokinetic data and the assessment of the predictive power of in vitro and in vivo tests.

2   The Evolving Science of ADMET

Regulatory authorities have relied upon in vivo testing to predict the behavior of new molecules in the human body since the 1950s. Bioavailability, tissue distribution, pharmacokinetics, metabolism, and toxicity are assessed typically in one rodent and one nonrodent species prior to administering a drug to a human to evaluate safety in a clinical trial (Phase 1). Biodistribution is assessed using radioactively labeled compounds later in development because it is expensive both in terms of synthesizing sufficient amounts of radioactively labeled compound and for performing the animal experiments [22].

Pharmacodynamic (PD) effectiveness of test compounds is typically assessed initially through in vitro models such as receptor binding, followed by confirmation through in vivo efficacy models in mice or rats. The predictive ability depends on the therapeutic area and the animal model. Infectious disease models are considered to have the best predictive ability, whereas CNS and oncology animal models are generally the least predictive of human efficacy. Understanding the PK/PD relationship is crucial in determining the mechanism of action and metabolic stability of the molecule which can explain and support efficacy results. In vivo pharmacokinetic (PK) studies in a variety of animal models are routinely used for lead optimization to assess drug metabolism and absorption. There are significant differences in absorption and metabolism among species from animal studies, which may cause conflicting predictions of degradation pathways of new chemical entities (NCEs).

Toxicity and safety studies are performed in models that are relevant to the NCE's mode of action and therapeutic area. In vivo toxicity models are required for IND (Investigational New Drug Application) to the US Food and Drug Administration (US FDA), but have substantial predictive weaknesses. In a retrospective study of 150 compounds from 12 large pharmaceutical companies, the combined animal toxicity study of rodents and nonrodents accurately predicted only 50% of the human hepatotoxicity. This poor level of accuracy in animal toxicity studies caused large numbers of compounds to be removed from development without proceeding into clinical trials with the potential of demonstrating safety in human subjects [26]. The other ~50% whose toxicity could not be predicted was attributed to "idiosyncratic human hepatotoxicity that cannot be detected by conventional animal toxicity studies." Although it is widely recognized that mechanisms for toxicity are frequently quite different between species, animal testing remains the "gold standard" for required regulatory and historical data reasons. The US FDA and other regulatory agencies are in the process of evaluating alternatives to animal testing, with the aim of developing models that are truly predictive of human mechanisms of toxicity and limiting in vivo toxicology testing.

3   The ADMET Feedback Loop

As discussed above, historically ADMET studies were focused on in vivo assays. These are time- and resource-intensive and generally low-throughput assays resulting in their implementation later in the development process, when more resources are released to study the few molecules that have advanced to this stage. With the advent of in vitro high-throughput screening, molecular biology, and miniaturization technologies in the 1990s, early ADMET assays were developed to predict in vivo animal and human results, at a level of speed and cost-effectiveness appropriate for the discovery stage. This produced a major advance in the science of ADMET and has created a new paradigm that drug discovery programs follow in advancing compounds from hit to lead, from lead to advanced lead, and on to nominated clinical candidates. Now, early in the discovery phase, using human enzymes and human-origin cells, drug discovery programs are able to obtain highly actionable information about the drug-likeness of new molecules, the potential to reach target organ, and early indications of known human mechanisms of toxicities. ADMET assessment of varying complexity is currently routinely performed on compounds that have shown in vitro efficacy and in conjunction with or just prior to demonstrating early proof of principle in vivo.

The application of early ADMET is unique to each drug discovery program. The development path from discovery to IND is not straightforward and is dependent on the therapeutic area, route of administration, chemical series, and other parameters. Correspondingly, the importance of the various ADMET assays is based upon the specifics of the drug discovery program. ADMET assays can also be categorized into those that are routine and those reserved for more advanced profiling. This division is also a function of cost-effectiveness and the need for the specific information. For instance, data regarding induction of human liver enzymes and transporters are not relevant during the hit-to-lead phase and are normally obtained for fewer more advanced candidates.

In some cases, the FDA requests data from in vitro ADMET assays. For example, in vitro drug–drug interaction (DDI) studies may now be conducted under the guidance from FDA dated September 2006. The guidance document precisely outlines methods to conduct CYP450 inhibition and induction and P-gp interaction studies [27]. This package is now typically included in an IND submission.

How should a discovery team employ early ADMET? The answer is not simple and formulaic—it is a process. It is useful to start from the ultimate goal and work backwards towards discovery. The drug discovery and development team should first define the target product profile (TPP), which includes indication, intended patient population, route of administration, acceptable toxicities, and ultimately will define the drug label. The TPP invariably will evolve during the life of the project, but having major parameters of TPP established initially maintains a collaboration and focus between disciplines such as biology and chemistry, discovery and development, and preclinical and clinical groups. Once the TPP is identified, then major design elements of the Phase 2 and 3 clinical trials can be outlined leading to questions about the tolerability, toxicity, and safety of the molecule. These parameters will then define the GLP toxicity studies in animals, which will guide the team to the discovery and preclinical development data to be addressed in an early ADMET program.

How is this information implemented in the discovery phase? If a compound has high target receptor binding and biological activity in cells and in relevant in vivo animal models, what are the chances of it becoming a successful drug? A molecule needs to cross many barriers to reach its biological target. In order to obtain this goal, a molecule must be in solution and thus the first step is typically to assess the solubility of a compound. A solubility screen provides information about the NCE's solubility in fluids compatible with administration to humans. Chemical and metabolic stability is a further extension of the intrinsic properties of a molecule. Chemical stability in buffers, simulated gastric and intestinal fluids, and metabolic stability in plasma, hepatocytes, or liver microsomes of different species can be measured to predict the rate of decay of a compound in the different environments encountered in the human body.

The second step is to define the absorption properties and the bioavailability of a molecule. Measurement of permeability across Caco-2 cell monolayers is a good predictor of human oral bioavailability. For CNS drugs, assessment of BBB penetration would be performed at this stage and is usually a key component of lead optimization campaigns. Passive BBB permeability may be assessed using BBB-PAMPA assays, whereas potential for active uptake or efflux may be determined using in vivo models or cell lines naturally expressing endogenous human intestinal or BBB transporters (such as CaCo-2 cell line) or cell lines overexpressing specific transporters (such as MDCK-MDR1).

Measurement of binding to plasma proteins indicates the degree of availability of the free compound in the blood circulation. This is critical as only unbound drugs are able to reach the target and exert their pharmacologic effects. Metabolism and DDI issues are discovered by screening for inhibition of cytochrome P450 liver enzymes (CYP450). All these assays allow chemists and biologists to obtain actionable information and provide a link between structure-activity (SAR) and structure-properties (SPR) relationships that drive decisions on selection of chemical series and molecules.

The next step is the involvement of DDIs and is required for advanced lead optimization. The effect of drug transporters on permeability and the effect of drugs on transporter activity can be measured in Caco-2, MDCK-MDR1, or other models. P-gp interactions are particularly important for CNS drugs due to high expression of these efflux transporters in the human BBB. Early knowledge about these interactions is instrumental to the medicinal chemistry strategy and helps drive lead optimization.

The effect of a compound on CYP450 metabolism can be identified by determining the 50% inhibitory concentration (IC_{50}) for each CYP450. These relationships between the NCE and metabolizing enzymes need to be evaluated in the context of the human effective dose and maximum effective plasma concentrations. These human data are not normally available at early stages of discovery, but could be extrapolated from animal PK/PD results for compounds in more advanced stages of development. It is important to understand these transporter and CYP450 relationships for the following:

1. The compound may affect the effective plasma concentrations of other concomitantly administered drugs if metabolized by the same CYPs (i.e., terfenadine).
2. If the parent drug is a CYP inducer, it may increase the clearance rate of concomitantly administered drugs which are metabolized by these CYPs. This may result in a decrease in these drugs' effective plasma concentrations, thus decreasing their pharmacologic effect.
3. Metabolites formed *via* CYP metabolism may be responsible for undesirable side effects such as organ toxicity.
4. The metabolite of a compound may actually be responsible for compound's efficacy, and not the parent compound. The metabolite may even have a better efficacy, safety, and pharmacokinetics profile than its parent. As a result, metabolism can be exploited to produce a better drug which will impact the medicinal chemistry strategy.
5. The identification of drug-metabolizing enzymes involved in the major metabolic pathways of a compound assists to predict the probable DDIs in humans. This information also may be used to design human clinical trials to detect unnecessary DDI.

ADMET is a tool that supports overall program goals. Similar to the Rule of 5 that requires only 3 of the 4 conditions to be met, seldom will negative results from a single ADMET assay terminate a compound's development or the overall program. The results are more likely to alter the medicinal chemistry direction.

After assessing compounds in a few simple mechanistic systems such as plasma and liver microsomal stability screens in relevant species, lead optimization phase is started that includes assays which identify potential liabilities. Finally, at the stage of advanced lead optimization and development, more complex systems are used to more thoroughly understand a compound's metabolic fate and absorption mechanism to drive efficient development. As ADMET roadblocks are discovered, the cycle is repeated until a clear path is found (Fig. 30.1).

ADMET Feedback Loop

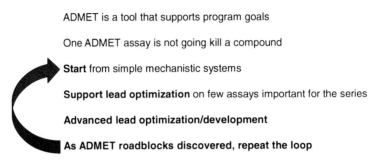

ADMET is a tool that supports program goals

One ADMET assay is not going kill a compound

Start from simple mechanistic systems

Support lead optimization on few assays important for the series

Advanced lead optimization/development

As ADMET roadblocks discovered, repeat the loop

Fig. 30.1 ADMET feedback loop

4   Impact of ADMET

Early ADMET provides the data necessary for selecting preclinical candidates by providing crucial information to medicinal chemists and accelerates the timelines for IND and subsequently NDA submission which translates to lengthier commercialization under patent protection and greater profits. For investors, this is a major parameter. For philanthropic organizations and from standpoint of public policy, it means increasing the time of clinical benefit to the public. Data compiled by the Tufts Center for Drug Discovery have identified that for a typical, moderately successful proprietary drug ($350 million USD annual sales), each day's delay equates to $1.1 million USD in lost patent-protected revenues that provide the return on investment needed to fund drug discovery [3]. Further, shorter discovery and development timelines provide faster liquidity events for venture capital and angel investors. As drug discovery requires a longer commercialization than any other form of product development, its slowness to produce returns is a major impediment for obtaining investment. Accelerating drug discovery and development should attract more investment in drug discovery research.

ADMET technologies remain an active area of research. There are many challenges in accurately measuring BBB penetration which may be one of the reasons for poor human efficacy of CNS drug candidates. Another challenge is detection of all mechanisms of human idiosyncratic toxicity. These mechanisms cause the most expensive, harmful, and disheartening form of drug attrition—postcommercialization toxicity. Many idiosyncratic drug reactions are due to formation of short-lived reactive metabolites that bind covalently to cell proteins [28]. The extent to which a compound will generate these metabolites can now be detected before a compound is administered to humans' signifying progress. Other mechanisms of human toxicity can be observed early in discovery and are briefly described in the following section.

5   New ADMET Tools

Penetrating the BBB is a challenge particular to CNS drug discovery. Another obstacle caused by BBB permeability is that many drugs not intended as CNS therapeutics cause neurotoxicity. Artificial membrane permeability assays (PAMPA and BBB-PAMPA) offer a cost-effective and high-throughput method of screening for passively absorbed compounds, but do not predict active transport in or out of the brain.

5.1   In Vitro Model of Human Adult Blood–Brain Barrier

Many new drugs designed for CNS may show exceptional therapeutic promise due to their high potency at the target site, but lack general efficacy when administered systemically. In many cases, the problem is due to lack of penetration of the BBB and this has become a major problem that has impeded the discovery and development of active CNS drugs. CEA Technologies previously reported the development of a new coculture-based model of human BBB able to predict passive and active transport of molecules into the CNS [29]. This new model consists of primary cultures of human brain capillary endothelial cells cocultured with primary human glial cells [18, 29]. The advantages of this system include:

1. Made of human primary culture cells.
2. Avoids species, age, and interindividual differences since the two cell types are removed from the same person.
3. Expresses functional efflux transporters such as P-gp, MRP-1, 4,5, and BCRP.

This model has potential for assessment of permeability of drug and specific transport mechanisms, which is not possible in PAMPA or other cell models due to incomplete expression of active transporters.

One important step in development of any in vitro model is to cross-correlate in vitro and in vivo data in order to validate experimental models and to assess the predictive power of the techniques [30]. The human BBB model was validated against a "gold standard" in vivo model and has shown an excellent in vitro–in vivo correlation [29, 31]. In this carefully designed in vivo–in vitro correlation study, the authors reported the evaluation of the BBB permeabilities for a series of compounds studied correlatively in vitro using a human BBB model and in vivo with quantitative PET imaging [29]. Six clinical PET tracers with different molecular size ranges (Fig. 30.2) and degree of BBB penetration were used including [^{18}F]-FDOPA and [^{18}F]-FDG, ligands of amino acid, and glucose transporters, respectively. The findings demonstrate that the in vitro coculture model of human BBB has important features of the BBB in vivo including low paracellular permeability, well-developed tight junctions, functional expression of important known efflux transporters, and is suitable for discriminating between CNS and non-CNS compounds. To further

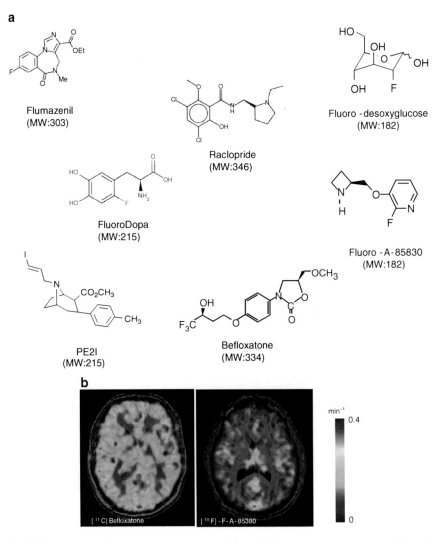

Fig. 30.2 In vitro–in vivo drug transport correlation. (**a**) Chemical structures of radioligands investigated and used clinically. (**b**) Typical imaging data. Coregistered PET-MRI images representing the k1 obtained in human after intravenous injection of [^{11}C]-befloxatone (*left*) and [^{18}F]-F-A-85380 (*right*). The PET images representing the k1 are as follows: PET image obtained at 1 min postinjection (mean value between 30 and 90 s) is considered as independent to the receptor binding. This image (in Bq/mL) is corrected from the vascular fraction (Fv in Bq/mL, considered as 4% of the total blood concentration at 1 min) and divided by the arterial plasma input function (AUC$_{0-1}$ min of the plasma concentration, in Bq*min/mL). The resulting parametric image, expressed in min^{-1}, represents an index of the k1 parameter of the radiotracer. (**c**) In vivo distribution volume (DV) as function of the in vitro P$_e$-out/P$_e$-in ratio (Q). Regression line was calculated, and correlation was estimated by the two-tailed nonparametric Spearman test. [^{11}C] PE2I radioligand was not plotted in the figure since the in vivo K1/k2 parameter in human is not available

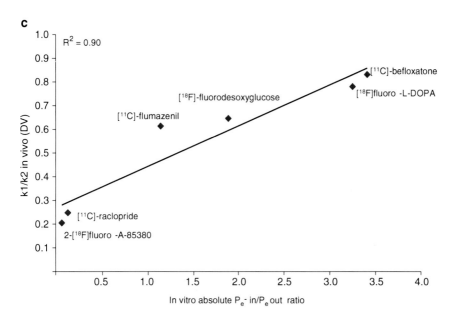

Fig. 30.2 (continued)

demonstrate the relevance of the in vitro human system, drug permeation into the human brain was evaluated using PET imaging in parallel to the assessment of drug permeability across the in vitro model of the human BBB. In vivo plasma–brain exchange parameters used for comparison were determined previously in humans by PET using a kinetic analysis of the radiotracer binding. 2-[^{18}F]Fluoro-A-85380 and [^{11}C]-raclopride show absent or low cerebral uptake with the distribution volume under 0.6. [^{11}C]-Flumazenil, [^{11}C]-befloxatone, [^{18}F]-FDOPA, and [^{18}F]-FDG show a cerebral uptake with the distribution volume above 0.6. The in vitro human BBB model discriminates compounds similar to in vivo human brain PET imaging analysis. This data illustrates the close relationship between in vitro and in vivo pharmacokinetic data (r^2 0.90, $p < 0.001$) (Fig. 30.2). Past in vivo–in vitro studies often did not have good correlations for substances transported into or out of the brain *via* active transport. Presumably, this is due to experiments being performed either with models that did not have adequate expression of active human transporters (such as PAMPA or MDCK cells) or using too high concentrations of compounds in vitro, which are known to saturate the transporters. Using the radioactive-labeled probes and the small amounts of compounds avoids these issues.

In conclusion, this in vitro human BBB model offers great potential for both being developed into a reproducible screen for passive BBB permeability and determining active transport mechanisms. Due to its high-throughput potential, the model may provide testing large numbers of compounds of pharmaceutical importance for CNS diseases. Validation work is in progress in which activity of transporters that are important in CNS BBB is being assessed in a functional assay and compared between CaCo-2 and hBBB models [31].

5.2   Mechanisms of Human Toxicity

Idiosyncratic hepatotoxicity or drug-induced liver injury (DILI) occurs in only one out of about 10,000 patients and is usually statistically impossible to discover during clinical trials. In spite of its name which means "rare event with undefined mechanism," some mechanisms have now been defined including mitochondrial toxicity and the formation of reactive metabolites. Another mechanism of human toxicity that is not limited to the liver but may also affect lung, spleen, and heart tissues is phospholipidosis.

5.2.1   Mitochondrial Toxicity

Mitochondrial toxicity is increasingly implicated in drug-induced idiosyncratic toxicity. Many of the drugs that have been withdrawn from the market due to organ toxicity have been found to be mitochondrial toxicants [32]. Mitochondrial toxicants injure mitochondria by inhibiting respiratory complexes of the electron chain, inhibiting or uncoupling oxidative phosphorylation, inducing mitochondrial oxidative stress, or inhibiting DNA replication, transcription, or translation [33].

Toxicity testing of drug candidates is usually performed in immortalized cell lines that have been adapted for rapid growth in a reduced-oxygen atmosphere. Their metabolism is often anaerobic by glycolysis despite having functional mitochondria and an adequate oxygen supply. Alternatively, normal cells generate ATP for energy consumption aerobically by mitochondrial oxidative phosphorylation. The anaerobic metabolism of transformed cell lines is less sensitive to mitochondrial toxicants causing systematically underreporting in toxicity testing [33, 34]. To address this issue, HepG2 and NIH/3T3 cells can be grown in media in which glucose is replaced by galactose [32]. The change in sugar results in the metabolism of the cell to possess a respiratory substrate that is both more similar to normal cells and sensitive to mitochondrial toxicants without reducing sensitivity to nonmitochondrial toxicants (Fig. 30.3).

5.2.2   Reactive Metabolites Formation

Another property of compounds that can cause idiosyncratic toxicity is their ability to form reactive intermediates [35]. Formation of short-lived reactive metabolites is known to be the mechanism of toxicity of some compounds such as acetaminophen [36]. The formation of reactive metabolites can be identified by incubating test compounds with liver microsomes and adding glutathione to trap the reactive intermediates which are then identified by LC/MS/MS (Fig. 30.4). Conversion of more than 10% of the test agent to reactive intermediates indicates that the compound may be implicated in idiosyncratic toxicity.

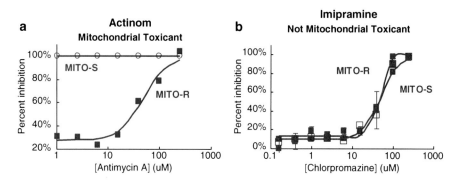

Fig. 30.3 Effect of antimycin A, a compound known to be toxic to mitochondria (**a**) and imipramine (**b**) on parent HepG2 cells (Mito-R—*blue*) and a HepG2 cell line that has been developed to become sensitive to mitochondrial toxicants (Mito-S—*red*)

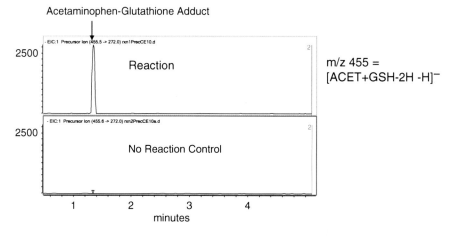

Fig. 30.4 Formation of reactive metabolites of acetaminophen. Acetaminophen was incubated with microsomes and glutathione in the presence and absence of NADPH. An adduct of glutathione with acetaminophen was formed in the presence of NADPH. When NADPH was absent (no reaction control), no adduct was formed

5.2.3 Phospholipidosis

Phospholipidosis is a lysosomal storage disorder and can be caused by drugs that are cationic amphiphiles [37]. The disorder is considered to be mild and often can self-resolve. However, drugs that cause phospholipidosis can also produce organ damage, and thus this disorder is a concern to the regulatory agencies [37]. A cell-based assay for phospholipidosis has been developed [38], which involves accumulation of a fluorescent phospholipid resulting in an increase of fluorescence in the lysosomes of cells that have been treated with drugs that cause phospholipidosis (Fig. 30.5). If phospholipidosis is absent, the phospholipid is degraded and fluorescence does

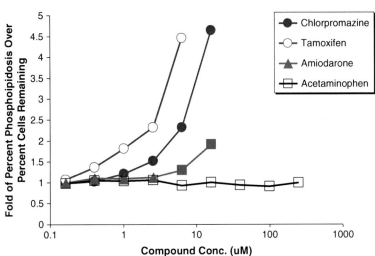

Fig. 30.5 Drug-induced phospholipidosis (PLD) is determined by measuring the accumulation of a fluorescent phospholipid in cells treated with increasing drug concentrations. Fluorescence is measured and normalized to cell number. Fluorescence is increased in cells treated with compounds that are known to cause PLD (chlorpromazine, tamoxifen, amiodarone), but it is not increased in cells treated with a compound that is known not to cause PLD (acetaminophen)

not increase. Increases in fluorescence are normalized to cell numbers since many of these drugs are also cytotoxic (Fig. 30.5).

5.2.4 High-Content Toxicology: The Present and Future of Predictive Toxicology

Drug safety is a major concern for the pharmaceutical industry with greater than 30% of drug candidates failing in clinical trials as a result of toxicity [3, 10]. Furthermore, there are numerous examples of drugs which have been withdrawn from the market or given black box warnings as a result of side effects not identified in clinical trials. Developing and commercializing a drug is a large financial commitment and failure at this stage can be catastrophic for a company. To address this problem, there has been a significant drive to incorporate toxicity assessment at much earlier stages in the drug discovery and development process.

It is well recognized that animal models are often not reflective of human toxicity. This is corroborated by a large percentage of drugs failing in the clinic through toxicity despite having progressed through costly preclinical animal studies. Human hepatotoxicity, as well as hypersensitivity and cutaneous reactions, is particularly difficult to identify during regulatory-based animal studies. Only 50% of drugs found to be hepatotoxic in clinical studies showed concordance with animal toxicity results [39, 40]. In addition, there are profound ethical issues associated with the widespread use of animals for this purpose. Initiatives such as ECVAM, ICCVAM, and NC3Rs are currently addressing this problem by identifying alternatives to animal safety testing. The cosmetics industry is at the forefront, and starting in 2013, there is an anticipated total EU ban of the sale of cosmetics tested on animals.

The introduction of more relevant and sophisticated in vitro human systems is essential to overcome these issues and will enable higher-throughput screening assays to be implemented earlier and more cost-effectively. The widespread use of in vitro methods has to some extent been hampered by their relatively poor predictive capability as traditionally only single markers of toxicity have been investigated. However, in many cases, drug toxicity is a highly complex process which can manifest itself *via* multiple different mechanisms. Predictions of toxicity can only be improved by investigating a broad panel of markers and their relationship to each other.

6   The Power of High-Content Screening

Technologies such as high content screening (HCS) have transformed cell biology and enabled subtle changes in multiple cellular processes to be tracked within the same population of cells. The technique uses either fluorescently labeled antibodies or dyes to stain specific areas of the cell which have critical roles in cell health or the maintenance of cellular function. The impact of concentration-dependent and time-dependent drug exposure on these cellular processes can be investigated and related to specific toxicological or efficacious responses. The ability to analyze multiple endpoints simultaneously, yet selectively, is a major advantage, and as well as being more sensitive, allows greater predictivity and an improved mechanistic understanding over traditional single endpoint measurements [41].

The power of HCS in toxicity assessment was illustrated in two key papers authored by scientists at Pfizer [41, 42], where a panel of six to eight key toxicity markers were identified and used to predict human hepatotoxicity. The articles highlight the improved predictive power of HCS over existing conventional in vitro toxicity assays and over traditional preclinical animal tests. HCS technology is now routinely used for in vitro toxicity assessment in large pharmaceutical companies.

7   CellCiphr™: Bridging the Gap Between In Vitro and In Vivo

It is critical to link the in vitro HCS information to animal or human pathology and establish a relationship to the in vivo data. The patterns observed in the key toxicity markers are often characteristic of specific mechanisms of known pathology.

The CellCiphr™ system utilizes the powerful technique of HCS and combines this with a novel classifier system which is underpinned by a large database of information for drugs with known toxicological profiles. Using this system, the toxicological profiles generated by HCS can be compared with known drugs for which animal or clinical data are available. Specific changes in cellular function observed for particular mechanisms of toxicity can then be recognized and used to predict for unknown NCEs.

It is well recognized that toxicological events can be organ-specific, time-dependent, and concentration-dependent. The CellCiphr™ system investigates three different panels which represent general cytotoxicity (using HepG2 cells) or organ-specific toxicity (using primary rat hepatocytes as a model for hepatotoxicity or H9c2 cardiomyocytes as a model for cardiotoxicity). The panels have been validated for the most relevant parameters for each particular cell type. Depending on the panel, these include the following cell health parameters: cell cycle arrest, nuclear size, oxidative stress, stress kinase activation, DNA damage response, DNA fragmentation, mitochondrial potential and mass, mitosis marker, cytoskeletal disruption, apoptosis, steatosis, phospholipidosis, ROS generation, hypertrophy, and general cell loss. To assess early- and late-stage toxic responses, CellCiphr™ investigates exposure at three different time points. Dose-dependent effects are investigated by exposing the cells to ten different concentrations of the compound.

Data are represented as AC_{50} (concentration at which average response is 50% of control activity) for each cell health parameter, and the collection of AC_{50} values over the entire cell feature set comprises the response profile. Proprietary visual and quantitative data mining tools including CellCiphr™ Classifiers, correlation analysis, and cluster analysis are used to analyze the profiles (Fig. 30.6). Using the CellCiphr™ approach, there are a number of different ways by which the data can be interpreted.

1. Similarity profile plots can identify potential mechanisms of actions by correlating unknown test compound response with known control compounds where the mechanisms of action are already known.
2. The relative toxicity of compounds in a series can be predicted by the CellCiphr™ Classifier and used to rank compounds for prioritization of the most promising candidates.
3. The most potent or earliest cellular response can be extracted from the data which may highlight the optimal endpoint(s) for designing higher-throughput systems to investigate SAR within a series.

Detailed mechanistic data can be generated for specific compounds. In the case of nimesulide which has been withdrawn from the market for severe hepatotoxicity, the CellCiphr™ Hepatotoxicity Profiling Panel scored this drug as the most toxic of the nonsteroidal anti-inflammatory drugs (NSAIDS). The toxicity was associated with a specific mechanistic profile characterized by an early oxidative stress response captured as a decreased mitochondrial membrane potential after 1 h of exposure. This insult drives the development of an apoptotic response at subsequent times points measured in the release of cytochrome C from the mitochondria and activation of the DNA damage response. Finally, prolonged exposure to nimesulide is marked by the accumulation of lipids in lysosomes and other vesicles (Fig. 30.7). The early effect on cell loss may also indicate a necrotic response in addition to apoptosis.

HCS Imaging

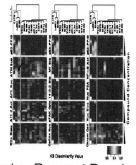

Biomarker Reagent Panels &

Relevant
cells/tissues

Informatics & Classifiers

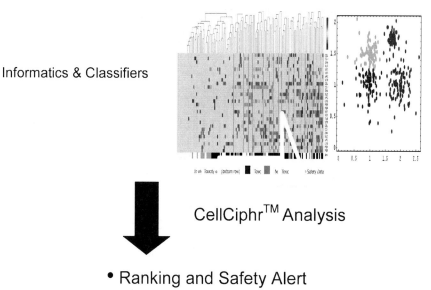

CellCiphr™ Analysis

- Ranking and Safety Alert
- Indices to Guide Directions

Fig. 30.6 The CellCiphr™ system uses HCS imaging platforms to identify specific patterns of biomarker response following exposure to multiple concentrations of compound at multiple time points. Proprietary visual and quantitative data mining tools have been developed to analyze the profiles and compare unknown compound response against known in vivo pathology

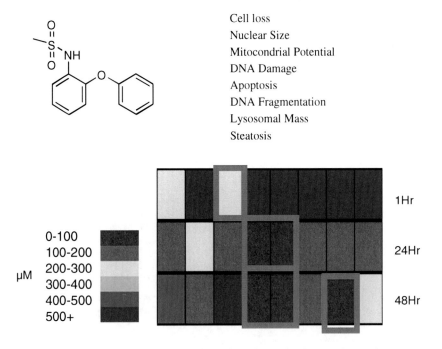

Cell loss
Nuclear Size
Mitocondrial Potential
DNA Damage
Apoptosis
DNA Fragmentation
Lysosomal Mass
Steatosis

Fig. 30.7 Nimesulide, which has been withdrawn from the market in several countries over concerns of severe hepatotoxicity, was scored as the most toxic of the NSAIDS by the CellCiphr™ hepatotoxicity profiling panel. The toxicity is associated with a specific mechanistic profile at sublethal doses

7.1 Future Strategies

In summary, CellCiphr™ is shown here as an example of a novel approach which identifies time-dependent, sublethal effects on cell health and function. The system illustrates significant improvements over existing single endpoint assays and has the ability to predict mechanistic outcomes by correlating with known compound profiles and pathology. By expanding the CellCiphr™ database, improving bioinformatics platform, and increasing the number of panels to cover new organ-specific cells, it will continue to improve the reliability of the classification.

Toxicological response is influenced by many factors including dose administration (including tissue exposure levels), time of exposure, and/or accumulation in specific cells. Many of these factors are influenced by the pharmacokinetics of the drug administered and its effect on absorption, distribution, metabolism, and excretion. Considering ADME data in conjunction with multiparametric measurements of in vitro toxicology is likely to be an important consideration in the future direction of predictive toxicology. Incorporation of human PK parameters prediction in models such as CellCiphr to ensure that cytotoxicity is relevant to projected tissue exposure is actively being pursued.

8 Genotoxicity

Genotoxicity of drugs is an important concern to the regulatory authorities. The FDA recommends a number of in vitro and in vivo tests to measure the mutagenic potential of chemical compounds, including the Ames test in *Salmonella typhimurium* [43]. GreenScreen GC, a new, high-throughput assay that links the regulation of the human GADD45a gene to the production of Green Fluorescent Protein (GFP), has become available. The assay relies on the DNA damage-induced up-regulation of the RAD54 gene in yeast measured using a promoter-GFP fusion reporter [44]. The test is more specific and sensitive for genotoxicity than those currently recommended by the FDA such as the Ames and mouse lymphoma tests.

9 Current Challenges and Future Directions

A large amount of progress in the field of ADMET profiling has occurred in the last 15 years. This progress has decreased the proportion of drug candidates failing in clinical trials for ADME reasons providing optimism in an otherwise declining productivity in drug discovery. The principal barrier now is the toxicity portion of ADMET. The prediction of human-specific toxicology must be improved.

Cell-based assays using established cell lines and cocultures have been used to determine toxicity to various organs, but many of these cell lines have lost some of the physiological activities present in normal cells. HepG2 cells, for instance, have greatly reduced levels of metabolic enzymes. Primary human hepatocytes can be used, but are expensive, suffer from high donor-to-donor variability, and maintain their characteristics for only a short time. Three-dimensional models have been developed for cell-based therapies including micropatterned cocultures of human liver cells that maintain the phenotypic functions of the human liver for several weeks [45]. This development should provide more accurate information about toxicity when used in ADMET screening and could be extended to other organ-specific cells leading to integrated tissue models in the "human on a chip" [46]. The potential of stem cells to differentiate into cell lines of many different lineages may be exploited to develop human and animal stem cell-derived systems for major organ systems [47].

HCS has been used for early cytotoxicity measurement since 2006 and provides great optimism [41]. This method has been optimized for hepatocytes and is more predictive of hepatotoxicity than other currently available methods and in the future could be applied to cells of other organs.

Molecular profiling is another alternative and is defined as any combination or individual application of mRNA expression, proteomic, toxicogenomic, or metabolomic measurements that characterize the state of a tissue [48]. This approach has been applied in an attempt to develop profiles or signatures of certain toxicities. Molecular profiles, in conjunction with agents that specifically perturb cellular systems, have been used to identify patterns of changes in gene expression and other parameters at subtoxic drug concentrations that might be predictive of hepatotoxicity including idiosyncratic hepatotoxicity [49]. In the future, larger data sets, high-throughput gene disruptions, and more-diverse profiling data will lead to more detailed knowledge of disease pathways and will facilitate in target selection and the construction of detailed models of cellular systems for use in ADMET screening to identify toxic compounds early in the discovery process. The combination of in silico, in vitro, and in vivo methods and models into multiple content databases, data mining, and predictive modeling algorithms, visualization tools, and high-throughput data analysis solutions can be integrated to predict systems' ADMET properties. Such models are starting to be built and should be widely available within 10 years [50]. The use of these tools will lead to a greater understanding of the interactions of drugs with their targets and predict their toxicities.

To conclude, the future should provide a decrease in late-stage development failures and withdrawals of marketed drugs, faster timelines from discovery to market, and reduced development costs through the reduction of late-stage failures.

References

1. Kaitin KI. Obstacles and opportunities in new drug development. Clin Pharmacol Ther. 2008;83:210–2.
2. DiMasi JA, Hansen RW, Grabowski HG. The price of innovation: new estimates of drug development costs. J Health Econ. 2003;22:151–85.
3. Kola I, Landis J. Can the pharmaceutical industry reduce attrition rates? Nat Rev Drug Discov. 2004;3:711–5.
4. Honig PK, Woosley RL, Zamani K, Conner DP, Cantilena Jr LR. Changes in the pharmacokinetics and electrocardiographic pharmacodynamics of terfenadine with concomitant administration of erythromycin. Clin Pharmacol Ther. 1992;52:231–8.
5. Honig PK, Wortham DC, Zamani K, Conner DP, Mullin JC, Cantilena LR. Terfenadine-ketoconazole interaction. Pharmacokinetic and electrocardiographic consequences. J Am Med Assoc. 1993;269:1513–8.
6. Thompson D, Oster G. Use of terfenadine and contraindicated drugs. J Am Med Assoc. 1996;275(17):1339–41.
7. Ball P, Mandell L, Niki Y, Tillotson G. Comparative tolerability of the newer fluoroquinolone antibacterials. Drug Saf. 1999;21:407–21.
8. US Food and Drug Administration. Public Health Advisory. Trovan (trovafloxacin/alatroflocacin mesylate). Available at http://www.fda.gov/Drugs/DrugSafety/PostmarketDrugSafety InformationforPatientsandProviders/DrugSafetyInformationforHeathcareProfessionals/Public HealthAdvisories/ucm052276.htm. Accessed 2 Feb 2012.
9. Stephens J. Panel faults Pfizer in '96 clinical trial in Nigeria. The Washington Post. A01. Available at http://www.washingtonpost.com/wp-dyn/content/article/2006/05/06/AR2006050601338.html. Accessed 2 Feb 2012.
10. Kubinyi H. Drug research: myths, hype and reality. Nat Rev Drug Discov. 2003;2:665–8.
11. Schuster D, Laggner C, Langer T. Why drugs fail—a study on side effects in new chemical entities. Curr Pharm Des. 2005;11:3545–59.
12. Lipinski CA, Lombardo F, Dominy BW, Feeney PJ. Experimental and computational approaches to estimate solubility and permeability in drug discovery and development settings. Adv Drug Deliv Rev. 1997;23:3–25.

13. Lipinski CA. Drug-like properties and the causes of poor solubility and poor permeability. J Pharmacol Toxicol Methods. 2000;44:235–49.

14. Schinkel AH, Smit JJ, van Tellingen O, Beijnen JH, Wagenaar E, van Deemter L, Mol CA, van der Valk MA, Robanus-Maandag EC, te Riele HP, Berns AJM, Borst P. Disruption of the mouse mdr1a P-glycoprotein gene leads to a deficiency in the blood–brain barrier and to increased sensitivity to drugs. Cell. 1994;77(4):491–502.

15. Schinkel RS, Minn A. Drug metabolizing enzymes in cerebrovascular endothelial cells afford a metabolic protection to the brain. Cell Mol Biol. 1999;45:15–23.

16. Hurko O, Ryan JL. Translational research in central nervous system drug discovery. NeuroRx. 2005;2(4):671–82.

17. Liu X, Chen C. Strategies to optimize brain penetration in drug discovery. Curr Opin Drug Discov Devel. 2005;8(4):505–12.

18. Megard I, Garrigues A, Orlowski S, Jorajuria S, Clayette P, Ezan E, Mabondzo A. A co-culture-based model of human blood–brain barrier: application to active transport of indinavir and in vivo-in vitro correlation. Brain Res. 2002;927(2):153–67.

19. Begley DJ, Lechardeur D, Chen ZD, Rollinson C, Bardoul M, Roux F, Scherman D, Abbott NJ. Functional expression of P-glycoprotein in an immortalised cell line of rat brain endothelial cells, RBE4. J Neurochem. 1996;67:988–95.

20. Deli MA, Abraham CS, Kataoka Y, Niwa M. Permeability studies on in vitro blood–brain barrier models: physiology, pathology and pharmacology. Cell Mol Neurobiol. 2005;25:59–120.

21. Crone C. The permeability of capillaries in various organs as determined by use of the 'indicator diffusion' method. Acta Physiol Scand. 1963;58:292–305.

22. Oldendorf WH. Measurement of brain uptake of radiolabeled substances using a tritiated water internal standard. Brain Res. 1970;24(2):372–6.

23. Takasato Y, Rapoport SI, Smith QR. An in situ brain perfusion technique to study cerebrovascular transport in the rat. Am J Physiol. 1984;247(3 Pt 2):H484–93.

24. Kakee A, Terasaki T, Sugiyama Y. Brain efflux index as a novel method of analyzing efflux transport at the blood–brain barrier. J Pharmacol Exp Ther. 1996;277(3):1550–9.

25. Elmquist WF, Sawchuk RJ. Application of microdialysis in pharmacokinetic studies. Pharm Res. 1997;14(3):267–88.

26. Olson H, Betton G, Robinson D, Thomas K, Monro A, Kolaja G, Lilly P, Sanders J, Sipes G, Bracken W, Dorato M, Van Deun K, Smith BPB, Heller A. Concordance of the toxicity of pharmaceuticals in humans and in animals. Regul Toxicol Pharmacol. 2000;32:56–67.

27. US Food and Drug Administration Website. Guidance for industry drug interaction studies: study design, data analysis, and implications for dosing and labeling. Available at http://www.fda.gov/Drugs/DevelopmentApprovalProcess/DevelopmentResources/DrugInteractionsLabeling/ucm093606.htm. Accessed 2 Feb 2012.

28. Uetrecht J. Idiosyncratic drug reactions: past, present, and future. Chem Res Toxicol. 2008;21:84–92.

29. Josserand V, Pélerin H, de Bruin B, Jego B, Kuhnast B, Hinnen F, Ducongé F, Boisgard R, Beuvon F, Chassoux F, Daumas-Duport C, Ezan E, Dollé F, Mabondzo A, Tavitian B. Evaluation of drug penetration into the brain: a double study by in vivo imaging with positron emission tomography and using an in vitro model of the human blood–brain barrier. J Pharmacol Exp Ther. 2006;316(1):79–86.

30. Pardridge WM, Triguero D, Yang J, Cancilla PA. Comparison of in vitro and in vivo models of drug transcytosis through the blood–brain barrier. J Pharmacol Exp Ther. 1990;253(2):884–91.

31. Jacewicz M, Guyot A-C, Annand R, Gilbert J, Mabondzo A, Tsaioun K. Contribution of transporters to permeability across cell monolayers. Comparison of Three Models. Apredica, CEA. American Association of Pharmaceutical Scientists, poster presentation, Atlanta, GA, November 2008.

32. Dykens JA, Will Y. The significance of mitochondrial testing in drug development. Drug Discov Today. 2007;12:777–85.

33. Rodríguez-Enríquez S, Juárez O, Rodríguez-Zavala JS, Moreno-Sánchez R. Multisite control of the Crabtree effect in ascites hepatoma cells. Eur J Biochem. 2001;268:2512–9.

34. Marroquin LD, Hynes J, Dykens JA, Jamieson JD, Will Y. Circumventing the Crabtree effect: replacing media glucose with galactose increases susceptibility of HepG2 cells to mitochondrial toxicants. Toxicol Sci. 2007;97:539–47.

35. Williams DP, Park BK. Idiosyncratic toxicity: the role of toxicophores and bioactivation. Drug Discov Today. 2003;8:1044–50.

36. Liu ZX, Kaplowitz N. Role of innate immunity in acetaminophen-induced hepatotoxicity. Expert Opin Drug Metab Toxicol. 2006;2:493–503.

37. Ademuyiwa O, Agarwal R, Chandra R, Behari JR. Lead-induced phospholipidosis and cholesterogenesis in rat tissues. Chem Biol Interact. 2009;179:314–20.

38. Natalie M, Margino S, Erik H, Annelieke P, Geert V, Philippe V. A 96-well flow cytometric screening assay for detecting in vitro phospholipidosis-induction in the drug discovery phase. Toxicol In Vitro. 2009;23:217–26.

39. Olson H, Betton G, Stritar J, Robinson D. The predictivity of the toxicity of pharmaceuticals in humans from animal data—an interim assessment. Toxicol Lett. 1998;102–103:535–8.

40. Olson H, Betton G, Robinson D, Thomas K, Monro A, Kolaja G, Lilly P, Sanders J, Sipes G, Bracken W, Doratoi M, Van Deun K, Smith P, Berger B, Heller A. Concordance of toxicity of pharmaceuticals in humans and animals. Regul Toxicol Pharmacol. 2000;32:56–67.

41. O'Brien PJ, Irwin W, Diaz D, Howard-Cofield E, Krejsa CM, Slaughter MR, Gao B, Kaludercic N, Angeline A, Bernardi P, Brain P, Hougham C. High concordance of drug-induced human hepatotoxicity with in vitro cytotoxicity measured in a novel cell-based model using high content screening. Arch Toxicol. 2006;80:580–604.

42. Xu JJ, Henstock PV, Dunn MC, Smith AR, Chabot JR, de Graaf D. Cellular imaging predictions of clinical drug-induced liver injury. Toxicol Sci. 2008;105(1):97–105.

43. US Food and Drug Administration Guidance for Industry. S2B genotoxicity: a standard battery for genotoxicity testing of pharmaceuticals. Available at http://www.fda.gov/downloads/Drugs/GuidanceComplianceRegulatoryInformation/Guidances/UCM074929.pdf. Accessed 2 Feb 2012.

44. Hastwell PW, Chai L-L, Roberts KJ, Webster TW, Harvey JS, Rees RW, Walmsley RM. High-specificity and high-sensitivity genotoxicity assessment in a human cell line: validation of the GreenScreen HC GADD45a-GFP genotoxicity assay. Mutat Res. 2006;607:160–75.

45. Khetani SR, Bhatia SN. Microscale culture of human liver cells for drug development. Nat Biotechnol. 2008;26:120–6.

46. Viravaidya K, Shuler ML. Incorporation of 3T3-L1 cells to mimic bioaccumulation in a microscale cell culture analog device for toxicity studies. Biotechnol Prog. 2004;20:590–7.

47. Yang L, Soonpaa MH, Adler ED, Roepke TK, Kattman SJ, Kennedy M, Henckaerts E, Bonham K, Abbott GW, Linden RM, Field LJ, Keller GM. Human cardiovascular progenitor cells develop from a KDR$^+$ embryonic-stem-cell-derived population. Nature. 2008;453:524–9.

48. Stoughton RB, Friend SH. How molecular profiling could revolutionize drug discovery. Nat Rev Drug Discov. 2005;4:345–50.

49. Kaplowitz N. Idiosyncratic drug hepatotoxicity. Nat Rev Drug Discov. 2005;4:489–99.

50. Ekins S, Nikolsky Y, Nikolskaya T. Techniques: application of systems biology to absorption, distribution, metabolism, excretion and toxicity. Trends Pharmacol Sci. 2005;26:202–9.

Chapter 31
CeeTox Analysis to De-risk Drug Development: The Three Antioxidants (NXY-059, Radicut, and STAZN)

Paul A. Lapchak

Abstract Translational stroke research is a multistep process where a drug candidate is synthesized, characterized using a series of in vitro assays and efficacy is then assessed in vivo in animal models representative of the clinical disease. Progress with the process is usually quite slow producing thousands to hundreds of thousands of candidates to screen. However, few candidates are ever fully developed into a randomized and double-blind clinical trial. To date, it is estimated that less than 40 small-molecule neuroprotective agents have been systematically studied in both preclinical and clinical trials, and those that have been tested in a randomized double-blind clinical trial have not proven to be positive. There are many points in the drug development process where a drug can fail to meet established criteria for further development. The critical points include lack of efficacy that could be established using a series of translational models for effective screening or toxicity that may be observed using in vitro or in vivo screening techniques. Since compound failure in the clinical situation is often attributed to toxicity, it is useful to introduce a de-risking step in the early preclinical development of novel small-molecule neuroprotective or neurotrophic compounds. This chapter reviews the CeeTox analysis panel that we have used in the development of small-molecule antioxidants such as NXY-059, Radicut, and STAZN.

1 Antioxidants of Interest

Free radical species appear to be important mediators in the progression of the stroke ischemic cascade, resulting in cell death and clinical deficits following acute ischemic stroke (AIS) [1–7]. Free radicals are chemical compounds having

P.A. Lapchak, PhD, FAHA (✉)
Department of Neurology, Cedars-Sinai Medical Center, Davis Research Building,
D-2091, 110 N. George Burns Road, Los Angeles, CA 90048, USA
e-mail: Paul.Lapchak@cshs.org

P.A. Lapchak and J.H. Zhang (eds.), *Translational Stroke Research*, Springer Series
in Translational Stroke Research, DOI 10.1007/978-1-4419-9530-8_31,
© Springer Science+Business Media, LLC 2012

one or more unpaired electrons, which makes them highly reactive with a variety of brain substrates. Free radicals can be classified by their core reactive species, namely, oxygen, nitrogen, or sulfur. Reactive oxygen species (ROS) usually refers to oxygen-based molecules such as superoxide, hydrogen peroxide (H_2O_2), hydroxyl radical, singlet oxygen, whereas a reactive nitrogen species can include nitric oxide (NO) and peroxynitrite. Sulfur free radicals may take the form of GS, which are generated from glutathione (GSH), hydrated sulfur dioxide, or sulfur trioxide anion radicals $^.SO^{3-}$ [8, 9]. Increased levels of oxygen, nitrogen, or sulfur-based free radicals can cause damage to virtually all cellular components, including membranes, DNA, lipid bilayers, and proteins, which are components of all cell types in brain (i.e., neurons, glial cells, vasculature). Free radicals can exert effects directly on cellular components and regulate cellular molecular pathways which contributes to the development of brain edema and cell death [10].

Many attempts have been made to develop chemically and pharmacologically diverse molecules to attenuate or reverse the effects of free radical generation following a stroke. Three molecules are the focus of this de-risking chapter, two of which have already been extensively studied in animals and patients and the third that is being developed for clinical trials. The three molecules of interest with their estimated log octanol–water partition coefficient (ClogP) are discussed below [11, 12]. A higher ClogP value indicates increased lipophilicity, hydrophobicity, and possible improved blood–brain barrier (BBB) penetration, although there are caveats to interpretation of ClogP. Most important is the observation that high values may indicate that membranes or membrane components may be a sink for the drug and it will not cross the BBB due to nonspecific "binding."

The hydrophilic nitrone spin trap agent NXY-059 (*disodium-[(tert-butylimino) methyl]benzene-1,3-disulfonate N-oxide*) (ClogP 0.95) was systematically tested in stroke patients in clinical trials and failed to show efficacy [13]. NXY-059 is a highly water-soluble molecule that does not readily cross the BBB, thus cannot effectively scavenge free radicals produced following an ischemic event [5, 14–17]. It is now well accepted that NXY-059 was a poor choice of compound to develop to treat stroke and that preclinical rodent and marmoset study data overestimated the efficacy of the drug [15, 17–20]. The lack of efficacy of NXY-059 was predicted by Lapchak et al. [15, 20] using the rabbit small clot embolic stroke model (RSCEM). Nevertheless, it is important to use NXY-059 as a "control" compound in de-risking analysis of prospective antioxidants because it was "safe" in patients [13, 21, 22].

The hydrophobic pyrazoline based free radical scavenger edaravone or Radicut® (*5-methyl-2-phenyl-2,4-dihydro-3H-pyrazol-3-one*) (ClogP 1.66) has not been systematically tested in large numbers of patients, but has shown significant efficacy in various small patient trials when administered 24–72 h after a stroke [23]. Radicut is a free radical scavenger approved by the Japanese Ministry of Health and Welfare and marketed in Japan by Mitsubishi Tanabe Pharma Corporation to treat

AIS patients presenting within 24 h of the attack. Radicut was first approved on May 23, 2001 and supplied in injectable edaravone ampoules. The Radicut BAG was subsequently approved by the Japanese Ministry of Health and Welfare on January 19, 2010, as a new innovative method to supply the drug for clinical use [23]. Radicut was studied using the CeeTox assay system because it has an excellent pharmacological profile in stroke models [16, 24] and good, but not perfect safety profile in patients [23].

The hydrophilic nitrone spin trap agent STAZN (*Stilbazulenyl nitrone*) is a second-generation nitrone-based antioxidant that has been studied preclinically in rodent stroke models, where the drug has shown significant efficacy [24–28]. STAZN may be an optimal choice of drug in this class of nitrone-based drugs for further testing and evaluation, especially during the de-risking phase when safety is a concern. Unlike NXY-059, STAZN has a low oxidation potential and is 300 times more potent at inhibiting free radical induced peroxidation [26]. STAZN reduces infarct volume in a rodent stroke model in a dose- and time-dependent manner [25, 29], the effect is durable and neuroprotection is observed with doses 300–600 times lower than NXY-059 [25, 30]. Third, due to the high lipophilicity of STAZN (ClogP 11.0), it readily crosses the BBB in normal rats following peripheral injection and has a long circulating half-life [29]. Thus, STAZN is superior to NXY-059 in many ways and may approach the pharmacological efficacy of Radicut, which is used extensively in Japanese AIS patients [23].

2 Drug Synthesis and Preparation for CeeTox Analysis

Disodium-[(tert-butylimino) methyl]benzene-1,3-disulfonate N-oxide (NXY-059; CAS 168021-79-2) was synthesized and purified according to a previously published procedure and pharmacological testing has previously been reported [31–33]. *5-methyl-2-phenyl-2,4-dihydro-3H-pyrazol-3-one* (Radicut; CAS 89-25-8) was purchased from Sigma Inc. (St. Louis, MO). Pharmacological testing of Radicut has previously been reported [31, 34]. STAZN (*Stilbazulenyl nitrone*) was synthesized and characterized according to the scientific [26] and patent literature [35]. STAZN was synthesized and purified to >97.5% purity by Combi-Blocks (San Diego, CA). NMR analysis in CDCL3 confirmed the structure of STAZN and HPLC analysis using a methanol–water gradient was used to determine the purity of the compound. Pharmacological testing of STAZN has previously been reported [24, 26, 36].

Since Radicut and STAZN are lipophilic compounds, dimethylsulfoxide (DMSO) was used to prepare 20 mM stock solutions that was then further diluted in DMSO to prepare 2 mM stock solutions. NXY-059 is water soluble and was made up in purified, sterile water at concentrations of 20 mM and 2 mM [24, 37].

3   CeeTox Assay Criteria

CeeTox analysis is an effective in vitro predictive toxicity screening assay to determine the toxicity profile of drugs being developed for human use [38, 39] so that drug development can be de-risked during early development stages. The CeeTox assay uses rat hepatoma derived H4IIE cells as a test system [24, 37] to determine the pharmacological effects of small molecules on general cellular toxicity, mitochondrial function, oxidative stress, apoptosis, protein binding, solubility, and microsomal metabolic stability [38]. Rat hepatoma derived H4IIE cells were used as the test system because the cell have a rapid doubling time in culture (i.e., 22 h) [38]. The culture medium used for these cells was Eagles Minimum Essential Medium (MEM) with 10% bovine serum and 10% calf serum. H4IIE cells were seeded into 96-well plates and allowed to equilibrate for approximately 48 h before drug assay to allow cells to move into a stable growth phase prior to treatment. Following the equilibration period the cells were exposed to NXY-059, Radicut, or STAZN at concentrations of 1–300 µM. For all CeeTox studies, drug analysis was done blinded, and two different positive controls were included in the study to assure assay effectiveness.

3.1   Drug Solubility

For solubility, the test compounds were prepared in DMSO or water and the appropriate amounts were then added to complete medium containing 10% bovine serum and 10% calf serum at 37°C. The samples were evaluated using light scattering with a Nepheloskan instrument. A reading that was greater than or equal to three times background was considered the limit of solubility [38]. Solubility was determined by Nephelometry techniques immediately after dosing and prior to harvesting the cells at 6 or 24 h.

3.2   Cell Mass

Cell mass in each well was measured with a modified propidium iodide (PI) [40], a specific nucleic acid binding dye that fluoresces when intercalated within the nucleic acids. The 15 nm shift enhances PI fluorescence approximately 20 times while the excitation maxima are shifted 30–40 nm. Triton-X-100 was used to permeabilize the H4IIE cells, thereby allowing the PI access to intracellular RNA and DNA. Fluorescence was measured using a Packard Fusion plate reader at 540 nm excitation and 610 nm emission [38]. Data is collected as relative fluorescent units (RFU) and expressed as percent change relative to control.

3.3 Membrane Toxicity

The presence of α-glutathione *S*-transferase (α-GST), an enzyme leakage marker, was measured in the culture medium using an ELISA assay purchased from Argutus Medical [38, 41]. At the end of the exposure period, the medium covering the cells in each well was removed and stored at 80°C until assayed. Absorbance values were measured with a Packard SpectraCount™ reader at 450 nm and reference absorbance at 650 nm. Leakage of α-GST from the cell into the culture medium was determined by collecting the culture medium at the end of the exposure period. Thus, the values measured represent total enzyme leakage lost over the exposure period. *3-[4,5-dimethylthiazol-2-yl] 2,5-diphenyltetrazolium bromide (MTT)*: After the medium was removed from a plate for α-GST analysis, the cells remaining in each well were evaluated for their ability to reduce soluble-MTT (yellow) to formazan-MTT (purple) [38, 42, 43]. An MTT stock solution was prepared in complete medium just prior to use and warmed to 37°C. Once the medium was removed from all wells, MTT solution was added to each well and the plate was allowed to incubate at 37°C for 3–4 h. Following incubation, all medium was removed and the purple formazan product was extracted using anhydrous isopropanol. Sample absorbance was read at 570 nm and reference absorbance at 650 nm with a Packard Fusion reader. The control for 100% dead or maximum enzyme release was based on cells treated with 1 mM digitonin at the time of dosing. Percent dead cells relative to digitonin treated cells was determined and then subtracted from 100% to yield the percent live cells.

3.4 Cellular ATP Content

Adenosine triphosphate (ATP) content was determined using a modification of a standard luciferin/luciferase luminescence assay [44] based on a reaction between ATP + D-luciferin + oxygen catalyzed by luciferase to yield oxyluciferin + AMP + PPi + CO_2 + light. The emitted light is proportional to the amount of ATP present [38]. At the end of the 24-h exposure period the medium was removed from the cells and the ATP cell lysis buffer added to each well. Plates were analyzed immediately or stored at −20°C until needed. On the day of analysis, the plates were thawed and calibration curve prepared with ATP in the same liquid matrix as samples. ATP was quantified by adding ATP substrate solution and then reading luminescence on a Packard Fusion Luminescence reader. ATP levels (pmoles ATP/million cells) in treated cells was extrapolated using the regression coefficients obtained from the linear regression analysis of the calibration curve. Background corrected luminescence was used to determine percent change relative to controls by dividing treated values by control values and multiplying by 100.

3.5   Oxidative Stress: 2 Measures

1. *Intracellular glutathione (GSH) levels*: Intracellular glutathione levels were determined using a modification of the procedure published by Griffith [45]. Briefly, the sulfhydryl group of GSH reacts with DTNB (5,5'-dithio-bis-2-nitrobenzoic acid, Ellman's reagent) and produces a yellow colored 5-thio-2-nitrobenzoic acid (TNB). The mixed disulfide, GSTNB (between GSH and TNB) that is concomitantly produced, is reduced by glutathione reductase to recycle the GSH and produce more TNB. The rate of TNB production is directly proportional to the concentration of GSH in the sample. Measurement of the absorbance of TNB at 405 or 412 nm provides an accurate estimation of GSH in the sample. At the end of the exposure period, the medium was removed from the cells and metaphosphoric acid (MPA) was added to each well. Plates were then shaken for 5 min at room temperature and stored at −20°C until needed. The sample plates were thawed just prior to analysis and centrifuged at >2,000×g for 2 min. Sample aliquots were removed and transferred to a clean 96-well plate along with appropriate standard curve controls. Sample pH was neutralized just prior to analysis and each well received an aliquot of PBS reaction buffer containing Ellman's reagent, NADPH, and glutathione reductase. The plates were shaken for 15–30 min at room temperature and glutathione content was determined colorimetrically with a Packard Fusion reader at 415 nm. The assay is based on the concept that all GSH is oxidized to GSSG by DTNB reagent. Two molecules of GSH are required to make one molecule of GSSG (oxidized glutathione). Total GSH was determined by reducing GSSG to 2GSH with glutathione reductase. A standard curve was prepared with GSSG over a wide range of concentrations. These concentrations are then converted to glutathione equivalents (GSX) essentially by multiplying the GSSG standard concentrations by 2. The amount of GSX expressed as pmoles/well was determined using the standard curve and regression analysis and are expressed as percent of control.

2. *Lipid peroxidation measured as 8-isoprostane (8-ISO or 8-epi PGF2α)*: 8-ISO levels were determined using an ELISA (Cayman Chemical, Inc.). 8-ISO is a member of a family of eicosanoids produced nonenzymatically by random oxidation of tissue phospholipids by oxygen radicals. Therefore, an increase in 8-ISO is an indirect measure of increased lipid peroxidation [46]. At the end of the exposure period, plates were either analyzed immediately or stored at −80°C until needed for analysis. Color development, which is indirectly proportional to the amount of 8-ISO present in the sample, was read on a Packard Fusion or equivalent plate reader at 415 nm [38]. Background absorbance produced from Ellman's reagent is subtracted from all wells. Nonspecific binding is subtracted from the maximum binding wells to give a corrected maximum binding expressed as B_o. The percent of bound (B) relative to maximum binding capacity (B_o) for all unknown samples and for standards was determined an expressed as (%B/B_o). The %B/B_o for standards was plotted against the log of 8-ISO added to yield the final standard curve. This curve was used to convert %B/B_o to pg 8-ISO/mL of sample.

3.6   Apoptosis

Caspase-3 activity was determined using a caspase substrate (DEVD, Asp-Glu-Val-Asp) labeled with a fluorescent molecule, 7-Amino-4-methylcoumarin (AMC). Caspase-3 cleaves the tetrapeptide between D and AMC, thus releasing the fluorogenic green AMC [38]. Following the test article exposure to cells in 96-well plates, medium was aspirated from the plates and PBS added to each well. The plates were stored at −80°C to lyse cells and store samples until further analysis. On the day of analysis, the plates were removed from freezer and thawed. Caspase buffer with fluorescent substrate was added to each well and incubated at room temperature for 1 h. AMC release was measured in a spectrofluorometer at an excitation wavelength of 360 nm and an emission wavelength of 460 nm. Values are expressed as RFU. After the sample plates were completely thawed, the caspase substrate buffer mix was added to each plate. The plates were incubated at room temperature for 1 h, shielded from light. The plates were read using a spectrofluorometer at an excitation wavelength of 360 nm and an emission wavelength of 460 nm. Values were expressed as RFU.

3.7   P-glycoprotein Binding Using the MTT Assay

The H4IIE cells possess high levels of P-glycoprotein (PgP) protein in the outer membrane and be used effectively for evaluation of drug binding to PgP [38, 47–49]. For this assay, the cells are incubated with and without cyclosporin A (CSA) (a PgP inhibitor) at a single exposure concentration (50 μM) and the difference in toxicity determined with the MTT assay. Compounds with increased toxicity in the presence of CSA have a high probability of binding to PgP proteins. However, compounds of low toxicity will typically not show a difference relative to the addition of CSA, regardless of whether they bind to PgP. At the end of the 24-h exposure period, the culture medium was removed and the remaining attached cells were assayed for their ability to reduce MTT. Viable cells will have the greatest amount of MTT reduction and highest absorbance values. Percent control values were determined by dividing the mean absorbance/fluorescence of the treatment group by the mean absorbance of the control group and multiplying by 100.

3.8   Metabolic Stability

Metabolic stability was conducted using pooled microsomes from noninduced male Sprague-Dawley rats [38]. The test compounds were incubated for 30 min at 37°C at concentrations of 1 μM. Subsequent HPLC analysis using a Waters Alliance 2795 in combination with a Waters Quattro Premier mass spectrometer measured disappearance of the parent molecule. The compounds were run on a Waters X-BridgeC18

(186003021) 50×2.1 mm column with 3.5 μm particle packing at a flow rate of
1 mL/min and with the temperature maintained at 50°C. Solvent A was water with
0.1% formic acid. Solvent B was acetonitrile with 0.07% formic acid. The data for
three replicates are expressed as percent of parent remaining. Metabolism rankings
are based on extent of metabolism with a percent remaining breakdown of:
100–65% = low; 65–45% = moderate; <45% = high. In metabolic stability assays,
two positive controls were also assayed: midazolam and terfenadine. Both are
highly metabolized with 20–30% remaining.

4   Results

1. *Drug solubility*: Both NXY-059 and Radicut were soluble up to and including
   300 μM in our system. STAZN was soluble up to and including 100 μM in the
   culture medium system, but was not completely soluble at 300 μM. It is extremely
   important to make sure that the assays are conducted within the solubility range
   of all test compounds. Otherwise, insolubility, especially at high concentrations,
   will confound the analysis of results.
2. *Effects on cellular toxicity*: To validate the CeeTox assay, rotenone [50, 51], a
   mitochondrial inhibitor that directly interferes with electron transport chain to
   inhibit ATP synthesis was used as well as camptothecin [52–54], a cytotoxic
   quinoline alkaloid which inhibits the DNA enzyme topoisomerase I were used
   (see Fig. 31.1 for structures). Table 31.1 and Fig. 31.2 provide detailed analysis
   of the two positive controls. Both compounds were highly cytotoxic and reduced
   all five primary CeeTox parameters (cell mass, membrane toxicity, MTT, ATP,
   and GSH).

   In contrast to the cellular toxicity demonstrated with the positive controls, as
   shown in Fig. 31.3, the cellular toxicity profiles of all three antioxidant compounds are quite different from control using 24 h in vitro analysis. For all three
   antioxidants, NXY-059, Radicut, and STAZN (structures shown in Fig. 31.1),
   cell death measured using the release of α-GST into the culture medium was
   only observed at the highest exposure concentration (300 μM).

   Neither NXY-059 (Fig. 31.3a) nor Radicut nor STAZN was acutely toxic, and
   none of the compounds had significant effects on the subcellular markers of
   acute toxicity across the exposure range tested up to 300 μM. In all five assays
   described above, the TC_{50} values were >300 μM. For Radicut, there was a 30%
   decrease in cell proliferation observed at the 300 μM exposure (Fig. 31.3b) and
   for STAZN (Fig. 31.3c) there was a 35% decrease in ATP content at the highest
   exposure concentration in the absence of cell death, indicating membrane leak-
   age-induced by STAZN.
3. *Effects on oxidative stress markers*: Tissue damage due to production of ROS
   occurs when highly reactive chemical species with unpaired electrons are generated both endogenously and by metabolism of parent chemicals. The result of
   these interactions is membrane damage, enzyme malfunction, and hydroxylation

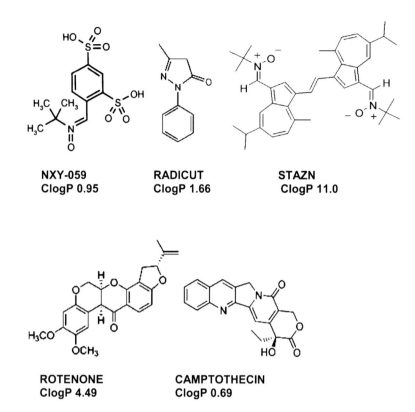

NXY-059
ClogP 0.95

RADICUT
ClogP 1.66

STAZN
ClogP 11.0

ROTENONE
ClogP 4.49

CAMPTOTHECIN
ClogP 0.69

Fig. 31.1 The chemical structure of the three study compounds and two positive controls. The three antioxidant molecules of interest with their estimated log octanol–water partition coefficient (ClogP). A higher ClogP value indicates increased lipophilicity and improved blood–brain barrier (BBB) penetration. The two positive control molecules with their respective ClogP values

Table 31.1 Summary of general cytotoxicity

Compound	Cell number TC$_{50}$ (µM)	MemTox TC$_{50}$ (µM)	MTT TC$_{50}$ (µM)	ATP TC$_{50}$ (µM)	Predicted C_{tox} (µM)
NXY-059	>300	>300	>300	>300	>300
Radicut	>300	>300	>300	>300	>300
STAZN	>300	>300	>300	>300	260
Rotenone	0.04	0.93	0.07	0.05	Control (~0.03)
Camptothecin	0.6	>300	0.5	0.3	Control (~0.1)

TC_{50} concentration that produced a half-maximal response, which was extracted from the graphs in Fig. 31.1; *MemTox* membrane toxicity measured using α-glutathione S-transferase (GST) leakage (*n* = 3 replicates); *MTT* 3-[4,5-dimethylthiazol-2-yl]-2,5-diphenyltetrazolium bromide; *ATP* adenosine triphosphate content
Cell number (or mass) in Fig. 31.1 was measured using propidium iodide

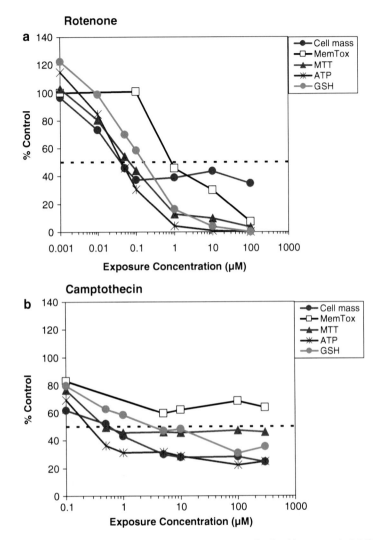

Fig. 31.2 Effect of rotenone and camptothecin on H4IIE cells: Positive controls (**a**) Rotenone (**b**) Camptothecin

of DNA, which can lead to mutagenesis. Oxidative stress can also occur when a chemical either is a direct electrophile or is metabolized to an electrophilic entity. Electrophiles can produce oxidative damage indirectly by depleting cellular antioxidants such as glutathione (GSH). Once depleted, the cell would be considerably more susceptible to oxidant damage from endogenously produced ROS. The potential clinical consequences of increased oxidative stress underscore the importance of evaluating changes in cellular oxidative stress using key biomarkers. The antioxidant defense mechanism of healthy cells can be divided into two primary categories: (1) Enzymatic antioxidant systems which would include superoxide

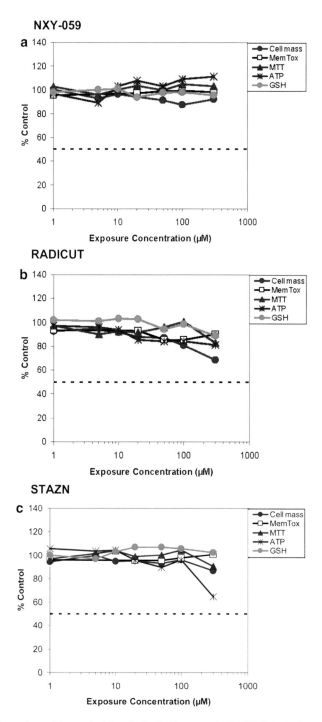

Fig. 31.3 Comparison of three antioxidants in the CeeTox assay (**a**) NXY-059 (**b**) Radicut (**c**) STAZN

dismutase, catalase, peroxidases, and glutathione reductase and (2) nonenzymatic oxidants such as glutathione, vitamin E, and vitamin A. Changes in ATP can also be indicative of oxidative or metabolic stress.

The CeeTox toxicity panel provides information on ATP, GSH/GSSG, and membrane lipid peroxidation. Table 31.2 provides data for a direct comparison of the three antioxidants on two cellular markers, which were used to determine the effects of the drugs on markers of oxidative stress. For this measure, we used intracellular GSH content and lipid peroxidation measured as 8-isoprostane [38]. There were no effects of the antioxidants on either marker. As positive controls in the assay, rotenone and camptothecin were used. Both compounds produced oxidative stress measured as a decrease in GSH and an increase in 8-isoprostane (Fig. 31.4).

4. *Effect on apoptosis or caspase-3 activity*: Apoptosis is a mode of cell death by which a cell can control its own fate [55, 56]. Apoptotic processes occur in development, differentiation, tumor deletion, and in response to exogenous stimuli. There are multiple pathways that can initiate the process of apoptosis; one well-characterized and committed step is activation of caspase-3 [38]. Therefore, caspase-3 activation has been included in the CeeTox general health panel of assays as a marker for initiation of apoptosis. As a measure of the effects of the study drugs on apoptotic mechanisms, we used caspase-3 activity [38], which is a key mediator of apoptosis in neuronal cells, but also has nonapoptotic functions [57]. None of the antioxidants had an effect on caspase-3 activity (Table 31.2).

5. *Drug PgP binding*: Table 31.3 shows that all three compounds showed low interaction with the Permeability-glycoprotein when cells are incubated with and without CSA, a PgP inhibitor at a single exposure concentration (50 μM). Compounds with increased toxicity in the presence of CSA have a high probability of binding to PgP proteins. However, compounds of low toxicity will typically not show a difference relative to the addition of CSA, regardless of whether they bind to PgP.

6. *Metabolic stability studies*: As shown in Table 31.4, NXY-059, Radicut, and STAZN were metabolically stable, that is they were not highly metabolized when incubated with rat microsomes (phase 1 metabolism) for 30 min at 37°C. There are caveats to the use of a rat microsomal preparation for stability studies. As we have recently shown [24], when STAZN was studied in rat microsomes, there was little metabolism. However, STAZN was highly metabolized by mouse microsomes and to a lesser extent by dog and human microsomes [24]. Thus, a four species microsomal screen may be quite useful to determine the metabolic stability of a novel test compound.

7. C_{Tox} *ranking*: For each compound studies, a C_{Tox} value was generated by CeeTox Inc., using a patented proprietary algorithm [58]. Figure 31.5 provides the C_{Tox} values generated using the TC_{50} values from Fig. 31.1. The C_{Tox} ranking, which is an estimate of a sustained concentration expected or necessary to produce toxicity, in a rat 14 day repeat dose study was 260 μM for STAZN, whereas the C_{Tox} ranking for NXY-059 and Radicut were both >300 μM.

Table 31.2 Drug effects on measures of oxidative stress and apoptosis

Compound	Total GSH TC$_{50}$ (μM)	Percent change in total GSH	8-Isoprostane membrane lipid peroxidation	Caspase 3 activity
NXY-059	>300	NC	NC	NC
Radicut	>300	NC	NC	NC
STAZN	>300	NC	NC	NC
Rotenone	0.39	−99	1	NC
Camptothecin	13	−72	0	3/100

TC$_{50}$ = concentration that produced a half-maximal response, which was extracted from the graphs presented in Fig. 31.2

GSH data: decrease in total GSH indicated by (−)

Membrane lipid peroxidation (8-ISO): *NC* no change

Membrane lipid peroxidation data: 0 = no change; 1 = modest increase with maximum values <15 pg/mL; 2 = concentration related increase with maximum values >15 pg/mL; 3 = concentration related increase with a maximum value >30 pg/mL

Caspase 3 data: 0–200 = 1; 200–400 = 2; 400–600 = 3; 600–800 = 4; 800–1,000 = 5; 1,000–1,200 = 6; 1,200–1,400 = 7; 1,400–1,600 = 8; 1,600–1,800 = 9; 1,800–2,000 = 10; *NC* No change

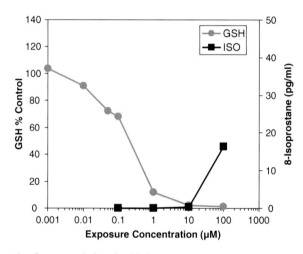

Fig. 31.4 Example of rotenone-induced oxidative stress

Table 31.3 Drug effects on P-glycoprotein (PgP) binding

Compound	% Control (compound)	% Control (compound + CSA)	% Difference
NXY-059	100.0	93.6	NC
Radicut	95.9	94.8	NC
STAZN	100.3	83.4	16.9

PgP interaction ranking (based on % difference in the absence and presence of CSA): *NC* no change

Table 31.4 Summary of metabolic stability

Metabolic stability in vitro		
Compound (1 μM)	% Remaining	Metabolism ranking
NXY-059	93	Low
Radicut	28	High
STAZN	98	Low

Metabolism rankings are based on extent of metabolism with a percent remaining breakdown as follows: 100–65 = low; 65–45 = moderate; <45 = high. % Remaining indicates % of parent compound remaining as measured using HPLC/MS. Metabolites were not assayed

C_{TOX} Ranking (μM) – Probability of *in vivo* Effects		
1 High 20	21 Moderate 50	51 Low 300
ROTENONE (0.03)	-----	RADICUT (> 300)
CAMPTOTHECIN (0.1)	-----	NXY-059 (> 300)
-----	-----	STAZN (260)

Fig. 31.5 C_{Tox} ranking profile of three antioxidants

5   Conclusion

The cellular toxicity profiles of NXY-059, Radicut, and STAZN, three free radical scavenging drugs that have different chemical and pharmacological properties [5, 8, 26, 34, 35, 59–62], were directly compared using the CeeTox assay, a series of standardized assays used to establish the safety of clinical candidates.

NXY-059 was compared to the other two drugs because it has undergone an extensive preclinical and clinical development plan, where NXY-059 was shown to safe in animals and patients, but failed to be efficacious in a double-blind randomized clinical trial when administered either as a monotherapy or in combination with the tPA [13, 18, 63]. Using CeeTox analysis, there were no indications of NXY-059 toxicity to H4IIE cells in culture confirming that safety profile of the drug observed in animals and AIS patients.

Using the hepatoma cells, there were also no indications of toxicity of Radicut, a drug that is formally approved in Japan for the treatment of AIS if administered within 24 h of a stroke [23, 31, 34]. Even though Radicut was not toxic to H4IIE cells in culture, there have been numerous reports of renal toxicity associated with Radicut administration to ischemic stroke patients [64]. It has been reported that approximately 45% of patients with Radicut-induced renal toxicity recover renal function after Radicut treatment is stopped [64]. The renal toxicity may be population specific, since another AIS trial was without toxic effects of Radicut [65].

The last comparison was done using STAZN, a novel nitrone that has not yet achieved clinical trial status, primarily because of the high lipophilicity of the drug and lack of supporting preclinical data package in multiple species. Like NXY-059 and Radicut, STAZN was shown to be without major toxicity effects on H4IIE cells. It is important to note that STAZN would be a good membrane binding compound that may integrate into lipid bilayers. The observation that there was a decrease of ATP at high doses might suggest membrane effects of STAZN and some membrane leakage. However, this hypothesis requires additional investigation.

In conclusion, CeeTox analysis may be a useful de-risking tool. Due to the high cost of the assay, it can only be justified and used if there are possible toxicity concerns associated with a specific chemical structure or class of chemical compound. As demonstrated in this chapter, the assay is not foolproof (see above for Radicut toxicity in patients). The C_{Tox} value was in the range of 260–300 μM for all antioxidants tested. One might question the utility of the assay since three structurally and pharmacologically unique molecules had the same safety profiles. Is this common to all molecules? Clearly, the use of positive controls rotenone and camptothecin provide the necessary control data to show that every component of the assay works as predicted and that toxic molecules will have effects on cellular integrity, metabolism (ATP), and oxidative stress. Ultimately, regardless of the results using the in vitro CeeTox assay, the continued development of any compound will require toxicity screening using a two species GLP toxicity study per FDA requirements [66].

Acknowledgment This chapter was supported by a U01 Translational research grant NS060685 to PAL.

Conflicts of Interest There are no conflicts of interest to disclose.

References

1. Facchinetti F, Dawson VL, Dawson TM. Free radicals as mediators of neuronal injury. Cell Mol Neurobiol. 1998;18(6):667–82.
2. Floyd RA. Antioxidants, oxidative stress, and degenerative neurological disorders. Proc Soc Exp Biol Med. 1999;222(3):236–45.
3. Nakashima M, Niwa M, Iwai T, Uematsu T. Involvement of free radicals in cerebral vascular reperfusion injury evaluated in a transient focal cerebral ischemia model of rat. Free Radic Biol Med. 1999;26(5–6):722–9.
4. Cherubini A, Ruggiero C, Polidori MC, Mecocci P. Potential markers of oxidative stress in stroke. Free Radic Biol Med. 2005;39(7):841–52.
5. Lapchak PA, Araujo DM. Development of the nitrone-based spin trap agent NXY-059 to treat acute ischemic stroke. CNS Drug Rev. 2003;9(3):253–62.
6. Siesjo BK, Katsura K, Zhao Q, Folbergrova J, Pahlmark K, Siesjo P, et al. Mechanisms of secondary brain damage in global and focal ischemia: a speculative synthesis. J Neurotrauma. 1995;12(5):943–56.
7. Siesjo BK, Siesjo P. Mechanisms of secondary brain injury. Eur J Anaesthesiol. 1996;13(3):247–68.
8. Lapchak PA. NXY-059. Centaur. Curr Opin Investig Drugs. 2002;3(12):1758–62.
9. Lapchak PA, Araujo DM. Spin trap agents: a new approach to stroke therapy. Drug News Perspect. 2002;15(4):220–5.
10. Wang J, Dore S. Inflammation after intracerebral hemorrhage. J Cereb Blood Flow Metab. 2007;27(5):894–908.

11. Machatha SG, Yalkowsky SH. Comparison of the octanol/water partition coefficients calculated by ClogP, ACDlogP and KowWin to experimentally determined values. Int J Pharm. 2005;294(1–2):185–92.

12. Ploemen JP, Kelder J, Hafmans T, van de Sandt H, van Burgsteden JA, Saleminki PJ, et al. Use of physicochemical calculation of pKa and CLogP to predict phospholipidosis-inducing potential: a case study with structurally related piperazines. Exp Toxicol Pathol. 2004;55(5):347–55.

13. Shuaib A, Lees KR, Lyden P, Grotta J, Davalos A, Davis SM, et al. NXY-059 for the treatment of acute ischemic stroke. N Engl J Med. 2007;357(6):562–71.

14. Ginsberg MD. Life after cerovive: a personal perspective on ischemic neuroprotection in the post-NXY-059 era. Stroke. 2007;38(6):1967–72.

15. Lapchak PA. Translational stroke research using a rabbit embolic stroke model: a correlative analysis hypothesis for novel therapy development. Transl Stroke Res. 2010;1(2):96–107.

16. Lapchak PA. Emerging therapies: pleiotropic multi-target drugs to treat stroke victims. Transl Stroke Res. 2011;2(2):129–35.

17. Savitz SI. A critical appraisal of the NXY-059 neuroprotection studies for acute stroke: a need for more rigorous testing of neuroprotective agents in animal models of stroke. Exp Neurol. 2007;205(1):20–5.

18. Bath PM, Gray LJ, Bath AJ, Buchan A, Miyata T, Green AR. Effects of NXY-059 in experimental stroke: an individual animal meta-analysis. Br J Pharmacol. 2009;157(7):1157–71.

19. Lapchak PA, Araujo DM, Zivin JA. Comparison of Tenecteplase with Alteplase on clinical rating scores following small clot embolic strokes in rabbits. Exp Neurol. 2004;185(1):154–9.

20. Lapchak PA, Araujo DM, Song D, Wei J, Zivin JA. Neuroprotective effects of the spin trap agent disodium-[(tert- butylimino)methyl]benzene-1,3-disulfonate N-oxide (generic NXY-059) in a rabbit small clot embolic stroke model: combination studies with the thrombolytic tissue plasminogen activator. Stroke. 2002;33(5):1411–6.

21. Edenius C, Strid S, Breitholtz-Emanuelsson A, Dalin L, Jerling M, Fransson B. NXY-059, the first nitrone being developed for stroke, is safe and well tolerated in young and elderly healthy volunteers. Cerebrovasc Dis. 1999;9:Abstract 102.

22. Lees KR, Zivin JA, Ashwood T, Davalos A, Davis SM, Diener HC, et al. NXY-059 for acute ischemic stroke. N Engl J Med. 2006;354(6):588–600.

23. Lapchak PA. A critical assessment of edaravone acute ischemic stroke efficacy trials: is edaravone an effective neuroprotective therapy? Expert Opin Pharmacother. 2010;11(10):1753–63.

24. Lapchak PA, Schubert D, Maher P. De-risking of stilbazulenyl nitrone (STAZN), a lipophilic nitrone to treat stroke using a unique panel of in vitro assays. Transl Stroke Res. 2011;2(2):209–17.

25. Ginsberg MD, Becker DA, Busto R, Belayev A, Zhang Y, Khoutorova L, et al. Stilbazulenyl nitrone, a novel antioxidant, is highly neuroprotective in focal ischemia. Ann Neurol. 2003;54(3):330–42.

26. Becker DA, Ley JJ, Echegoyen L, Alvarado R. Stilbazulenyl nitrone (STAZN): a nitronyl-substituted hydrocarbon with the potency of classical phenolic chain-breaking antioxidants. J Am Chem Soc. 2002;124(17):4678–84.

27. Ley JJ, Belayev L, Saul I, Becker DA, Ginsberg MD. Neuroprotective effect of STAZN, a novel azulenyl nitrone antioxidant, in focal cerebral ischemia in rats: dose-response and therapeutic window. Brain Res. 2007;1180:101–10.

28. Ley JJ, Prado R, Wei JQ, Bishopric NH, Becker DA, Ginsberg MD. Neuroprotective antioxidant STAZN protects against myocardial ischemia/reperfusion injury. Biochem Pharmacol. 2008;75(2):448–56.

29. Ley JJ, Vigdorchik A, Belayev L, Zhao W, Busto R, Khoutorova L, et al. Stilbazulenyl nitrone, a second-generation azulenyl nitrone antioxidant, confers enduring neuroprotection in experimental focal cerebral ischemia in the rat: neurobehavior, histopathology, and pharmacokinetics. J Pharmacol Exp Ther. 2005;313(3):1090–100.

30. Kuroda S, Tsuchidate R, Smith ML, Maples KR, Siesjo BK. Neuroprotective effects of a novel nitrone, NXY-059, after transient focal cerebral ischemia in the rat. J Cereb Blood Flow Metab. 1999;19(7):778–87.

31. Lapchak PA. Translational stroke research using a rabbit embolic stroke model: a correlative analysis hypothesis for novel therapy development. Transl Stroke Res. 2010;1(2):96–107. http://dx.doi.org/10.1007/s12975-010-0018-4.

32. Lapchak PA, Araujo DM, Song D, Wei J, Purdy R, Zivin JA. Effects of the spin trap agent disodium-[tert-butylimino)methyl]benzene-1,3-disulfonate N-oxide (generic NXY-059) on intracerebral hemorrhage in a rabbit large clot embolic stroke model: combination studies with tissue plasminogen activator. Stroke. 2002;33(6):1665–70.

33. Lapchak PA. Development of thrombolytic therapy for stroke: a perspective. Expert Opin Investig Drugs. 2002;11(11):1623–32.

34. Lapchak PA, Zivin JA. The lipophilic multifunctional antioxidant edaravone (Radicut) improves behavior following embolic strokes in rabbits: a combination therapy study with tissue plasminogen activator. Exp Neurol. 2009;215(1):95–100.

35. Becker DA. Azulenyl nitrone spin trapping agents, methods of making and using same; USA Patent 6,291,702. 2008.

36. Belayev L, Becker DA, Alonso OF, Liu Y, Busto R, Ley JJ, et al. Stilbazulenyl nitrone, a novel azulenyl nitrone antioxidant: improved neurological deficit and reduced contusion size after traumatic brain injury in rats. J Neurosurg. 2002;96(6):1077–83.

37. Lapchak PA, KcKim JM. CeeTox™ analysis of CNB-001 a novel curcumin-based neurotrophic/neuroprotective lead compound to treat stroke: comparison with NXY-059 and Radicut. Transl Stroke Res. 2011;2(1):51–9.

38. McKim Jr JM. Building a tiered approach to in vitro predictive toxicity screening: a focus on assays with in vivo relevance. Comb Chem High Throughput Screen. 2010;13(2):188–206.

39. McKim JM, Jr. Ceetox, Inc., assignee. Toxicity screening methods; USA Patent 7615361. 2009.

40. Hopkinson K, Williams EA, Fairburn B, Forster S, Flower DJ, Saxton JM, et al. A MitoTracker green-based flow cytometric assay for natural killer cell activity: variability, the influence of platelets and a comparison of analytical approaches. Exp Hematol. 2007;35(3):350–7.

41. Vickers AE. Use of human organ slices to evaluate the biotransformation and drug-induced side-effects of pharmaceuticals. Cell Biol Toxicol. 1994;10(5–6):407–14.

42. Bernas T, Dobrucki J. Mitochondrial and nonmitochondrial reduction of MTT: interaction of MTT with TMRE, JC-1, and NAO mitochondrial fluorescent probes. Cytometry. 2002;47(4):236–42.

43. Berridge MV, Tan AS. Characterization of the cellular reduction of 3-(4,5-dimethylthiazol-2-yl)-2,5-diphenyltetrazolium bromide (MTT): subcellular localization, substrate dependence, and involvement of mitochondrial electron transport in MTT reduction. Arch Biochem Biophys. 1993;303(2):474–82.

44. Lapchak PA, De Taboada L. Transcranial near infrared laser treatment (NILT) increases cortical adenosine-5'-triphosphate (ATP) content following embolic strokes in rabbits. Brain Res. 2010;1306:100–5.

45. Griffith OW. Determination of glutathione and glutathione disulfide using glutathione reductase and 2-vinylpyridine. Anal Biochem. 1980;106:207–12.

46. Vacchiano CA, Tempel GE. Role of nonenzymatically generated prostanoid, 8-iso-PGF2 alpha, in pulmonary oxygen toxicity. J Appl Physiol. 1994;77(6):2912–7.

47. Muller H, Klinkhammer W, Globisch C, Kassack MU, Pajeva IK, Wiese M. New functional assay of P-glycoprotein activity using Hoechst 33342. Bioorg Med Chem. 2007;15(23):7470–9.

48. Gao GL, Wan HY, Zou XS, Chen WX, Chen YQ, Huang XZ. Relationship between the expression of P-glycoprotein, glutathione S-transferase-pi and thymidylate synthase proteins and adenosine triphosphate tumor chemosensitivity assay in cervical cancer. Zhonghua Fu Chan Ke Za Zhi. 2007;42(3):201–5.

49. Li QY, Wang Y, Yin ZF, Wu MC. Application of the improved MTT assay in predicting the intrinsic drug resistance of liver cancer. Zhonghua Yi Xue Za Zhi. 2007;87(5):333–5.

50. Takeshige K. Superoxide formation and lipid peroxidation by the mitochondrial electron-transfer chain. Rinsho Shinkeigaku. 1994;34(12):1269–71.

51. Ayala A, Venero JL, Cano J, Machado A. Mitochondrial toxins and neurodegenerative diseases. Front Biosci. 2007;12:986–1007.

52. Hartmann JT, Lipp HP. Camptothecin and podophyllotoxin derivatives: inhibitors of topoisomerase I and II—mechanisms of action, pharmacokinetics and toxicity profile. Drug Saf. 2006;29(3):209–30.
53. Erickson-Miller CL, May RD, Tomaszewski J, Osborn B, Murphy MJ, Page JG, et al. Differential toxicity of camptothecin, topotecan and 9-aminocamptothecin to human, canine, and murine myeloid progenitors (CFU-GM) in vitro. Cancer Chemother Pharmacol. 1997;39(5):467–72.
54. Schaeppi U, Fleischman RW, Cooney DA. Toxicity of camptothecin (NSC-100880). Cancer Chemother Rep 3. 1974;5(1):25–36.
55. Friedlander RM, Yuan J. ICE, neuronal apoptosis and neurodegeneration. Cell Death Differ. 1998;5(10):823–31.
56. Tan S, Schubert D, Maher P. Oxytosis: a novel form of programmed cell death. Curr Top Med Chem. 2001;1(6):497–506.
57. D'Amelio M, Cavallucci V, Cecconi F. Neuronal caspase-3 signaling: not only cell death. Cell Death Differ. 2010;17(7):1104–14. doi:10.1038/cdd.2009.180.
58. McKim JM, Jr., inventor. Ceetox, Inc., assignee. Toxicity screening methods; United States Patent Application 20110008823. 2007.
59. Watanabe T, Tahara M, Todo S. The novel antioxidant edaravone: from bench to bedside. Cardiovasc Ther. 2008;26(2):101–14.
60. Higashi Y, Jitsuiki D, Chayama K, Yoshizumi M. Edaravone (3-methyl-1-phenyl-2-pyrazolin-5-one), a novel free radical scavenger, for treatment of cardiovascular diseases. Recent Pat Cardiovasc Drug Discov. 2006;1(1):85–93.
61. Maples KR, Green AR, Floyd RA. Nitrone-related therapeutics: potential of NXY-059 for the treatment of acute ischaemic stroke. CNS Drugs. 2004;18(15):1071–84.
62. Maples KR, Ma F, Zhang YK. Comparison of the radical trapping ability of PBN, S-PPBN and NXY-059. Free Radic Res. 2001;34(4):417–26.
63. Diener HC, Lees KR, Lyden P, Grotta J, Davalos A, Davis SM, et al. NXY-059 for the treatment of acute stroke: pooled analysis of the SAINT I and II trials. Stroke. 2008;39(6):1751–8.
64. Hishida A. Clinical analysis of 207 patients who developed renal disorders during or after treatment with edaravone reported during post-marketing surveillance. Clin Exp Nephrol. 2007;11(4):292–6.
65. Sinha M, Anuradha H, Juyal R, Shukla R, Garg R, Kar A. Edaravone in acute ischemic stroke, an Indian experience. Neurol Asia. 2009;14:7–10.
66. FDA. Guidance for industry M4S: the CTD—safety appendices; 2001. http://www.fda.gov/cder/guidance/index.htm. Accessed Jan 11, 2012.

Part V
Therapy Delivery

Chapter 32
Site-Specific, Sustained-Release Drug Delivery for Subarachnoid Hemorrhage

R. Loch Macdonald

Abstract Drugs approved for use to combat angiographic vasospasm or to improve the sequelae of subarachnoid hemorrhage include nimodipine in North America, Europe, and other countries and fasudil and OKY-046 in Japan. There may be others. Other drugs may be effective, but their efficacy has been difficult to show in clinical trials. The reasons for this are unclear: either they are not effective or they have off-target and systemic side effects when delivered orally or intravenously, the studies are underpowered, the outcomes are insensitive to important clinical benefits, and/or rescue therapies in the placebo groups are effective enough to balance the effect of the drug in the treated patients. This chapter addresses the first issue, which is to deliver drugs that are likely to be effective, yet cannot be delivered enterally or systemically in adequate doses because of dose-limiting side effects, by direct, intracranial methods.

1 Subarachnoid Hemorrhage

Subarachnoid hemorrhage (SAH) is a pathological condition characterized by bleeding into the subarachnoid space. This is the preformed space around the brain that is filled with cerebrospinal fluid (CSF). The causes of SAH include open or closed head injury and a variety of spontaneous causes. The incidence of

R.L. Macdonald, MD, PhD (✉)
Division of Neurosurgery, Labatt Family Centre of Excellence in Brain Injury and Trauma Research, Keenan Research Centre of the Li Ka Shing Knowledge Institute of St. Michael's Hospital, and University of Toronto, Toronto, ON, Canada,
e-mail: macdonaldlo@smh.ca

P.A. Lapchak and J.H. Zhang (eds.), *Translational Stroke Research*, Springer Series in Translational Stroke Research, DOI 10.1007/978-1-4419-9530-8_32,
© Springer Science+Business Media, LLC 2012

aneurysmal SAH is 9/100,000 per year throughout countries in which data are available [20]. The incidence has not declined in the past decades, unlike some other types of stroke. The most common cause of spontaneous SAH is rupture of an intracranial aneurysm, accounting for about 85% of spontaneous SAH cases [103, 104]. Even though aneurysmal SAH accounts for only 6–8% of all strokes, it is associated with a disproportionate amount of death, disability, and cost [46]. Analysis of the National Center for Health Statistics of the United States mortality data from 1979 to 1994 found the median age of death from SAH was 59 years compared to 73 for intracerebral hemorrhage and 81 for ischemic stroke. Although mortality has decreased almost 1% per year over the last 2 decades [69], SAH still accounts for 4% of stroke mortality but 27% of stroke-related years of life lost, compared to 39% for ischemic stroke and 34% for intracerebral hemorrhage. Thus, with the population of the United States estimated at 310,636,575 (www. census.gov-accessed 11/5/10), the 34,000 cases of SAH per year are a major public health problem.

A second major cause of SAH is traumatic brain injury (TBI). TBI is a much more common problem than aneurysmal SAH, affecting 180–250 per 100,000 Americans every year [11]. Traumatic SAH usually only is of importance in the 20% or 50 per 100,000 cases per year of TBI that are classified as moderate or severe. Mortality from TBI was estimated at 50,000 deaths per year in the United States between 1989 and 1998. About 150,000 people are permanently disabled after TBI each year in the United States [70].

The number of cases of traumatic SAH per year depends on the definition since even relatively minor TBI can be associated with some blood in the spinal fluid which would only be detected by lumbar puncture and would not be of any pathophysiologic significance in and of itself. One estimate is that substantial traumatic SAH complicates 33–60% (about 150,000 per year in the United States) of cases of TBI [3, 72]. Traumatic SAH is an important adverse prognostic factor for outcome after TBI. Prognostic factors for outcome among 8,686 patients with TBI who were entered into eight randomized clinical trials and three observational studies were age, motor score on the Glasgow coma score, pupillary responses to light, computed tomographic findings [60], traumatic SAH, hypotension, and hypoxia [68]. Whether SAH contributes pathophysiologically to brain damage after SAH or if it is mainly a marker of the severity of the TBI is controversial. Until an effective treatment was available to selectively treat a mechanism of brain damage due to SAH (which will be unlikely to ever be since most drugs will have multiple actions), the answer to this question will not be known. On the other hand, many studies show the volume and location of blood after traumatic and aneurysmal SAH predict cerebral ischemia and development of infarction [3, 112]. Furthermore, pathophysiologic mechanisms like cortical spreading depolarizations with ischemia (CSI), which are initiated in part by SAH, as well as angiographic vasospasm and microthromboemboli all occur after spontaneous and traumatic SAH and plausibly cause brain damage, especially since patients can appear initially well and then deteriorate.

2   Causes of Poor Outcome After SAH

All cases of SAH in Cincinnati were analyzed for the year 1988 [10]. Of 80 patients, 36 died within 30 days (45%). The time of death was within 2 days in 22 (61% of deaths), and the causes of death were the initial hemorrhage in 23 (29% of cases, 81% of deaths) and rebleeding in 8 (10% of cases, 22% of deaths). Death was attributed to angiographic vasospasm in only 2 patients (3% of cases, 6% of deaths). The results highlight that the initial effects of the SAH cause the most death, usually within 2 days of the ictus. Efforts directed at decreasing this are underway and face the challenges of other acute neuroprotective therapies. Rebleeding can be addressed by timely treatment of the ruptured aneurysm and possibly short courses of antifibrinolytic therapy until the aneurysm can be repaired [41]. Other series have confirmed that death associated with angiographic vasospasm is uncommon in recent series [56]. Angiographic vasospasm and DCI, however, are still major causes of disability, and furthermore, they are delayed in onset and thus potentially more amenable to treatment than the initial effects of the SAH, which could be almost instantaneous [36, 47, 106]. Factors that increased the odds of poor outcome the most in a series of 3,567 patients were cerebral infarction, neurological grade, and age [85]. A path analysis of 413 patients entered into a randomized clinical trial of clazosentan suggested that poor neurological grade, history of hypertension, angiographic vasospasm, neurological worsening, and cerebral infarction each contributed significantly to poor outcome by independent pathways [107].

The pathophysiological basis for morbidity and mortality after SAH includes acute global ischemia due to increased intracranial pressure with or without acute vasoconstriction at the time of the SAH and then developing over days due to brain swelling and other causes. Rebleeding, direct brain damage from intracerebral bleeding, herniations, infarctions from treatment, and DCI and numerous secondary brain insults contribute. Experimentally, the effects of SAH with or without transient global ischemia have been the focus of much investigation and can be considered as early and delayed effects.

3   Early and Delayed Brain Injury

Cahill and Zhang suggested that early brain injury be used to describe the combination of vasospasm plus global ischemia, both as a consequence of SAH. The pathophysiology includes increased intracranial pressure and decreased cerebral perfusion pressure, reduced cerebral blood flow, and decreased brain oxygenation that are due to the SAH and that then cause global ischemia [12, 13]. Global ischemia leads to necrotic cell death and activation of the multiple apoptotic pathways (death receptor, p53, and caspase-dependent). All of these contribute to death of multiple types of brain cells, leading to brain edema and blood-brain barrier disruption. That transient ischemia occurs at the time of SAH in many cases is well known in humans and can be produced in experimental models. Some evidence for a role of apoptosis is derived from experimental studies showing reduced brain edema and preservation of the blood-brain barrier in animals with SAH, treated with inhibitors of caspase 3 (z-VAD-FMK) or p53 (pifithrin) [75, 116]. Interestingly, treatment of rats with SAH with z-VAD-FMK reduced cleaved caspase-3 in the hippocampus and basal cortex and improved behavior scores 24 h after SAH, although there was no obvious effect on neuronal apoptosis observed by TUNEL staining [75]. Whether the changes in the hippocampus and basal cortex could be the substrate for the well-known cognitive deficits that frequently persist after even mild SAH requires further investigation.

Most studies in SAH have focused on angiographic vasospasm and the delayed consequences of SAH. This phenomenon has been clearly linked to the volume, location, and persistence of subarachnoid blood and can be produced experimentally in multiple animal species simply by injecting or placing blood in the subarachnoid space next to the cerebral arteries [82, 100]. The resultant angiographic vasospasm results from constriction of the smooth muscle cells. Secondary degenerative changes occur, including smooth muscle and endothelial cell necrosis, endothelial cell apoptosis, adventitial inflammation and fibrosis, and intimal hyperplasia [33]. Although intimal hyperplasia occurs later after SAH than angiographic vasospasm, there is expression of PCNA (proliferating cell nuclear antigen) in vasospastic basilar arteries of rabbits, and treatment with PD98059, an inhibitor of cell proliferation via ERK1/2, reduced vasospasm [97]. The causes of smooth muscle contraction in vasospasm have been investigated. Overall, contraction must involve pathways that elevate intracellular calcium, alter sensitivity to calcium, or contract through activation of other signal transduction pathways. Calcium sensitivity was not increased in one study of vasospastic arteries [59]. Experiments studying dog basilar arteries after SAH suggest contraction occurs via calcium influx through L-type voltage-gated calcium channels and by a mechanism mediated by transient receptor potential proteins [71, 114].

There are probably multiple extracellular processes that contribute to the vasoconstriction that underlies angiographic vasospasm. Two features are loss of vasodilatory NO that is produced by neuronal NO synthase (NOS) derived from perivascular nerves and endothelial NOS from endothelial cells, as well as increased production of the vasoconstrictor, endothelin-1 (ET-1). The sources of ET-1 include endothelial cells, ischemic brain tissue, and inflammatory cells that are recruited into the subarachnoid space after SAH [32, 79]. Cerebral arteries become more sensitive to ET-1 after SAH [114]. The perivascular blood clot causes inflammation. The clot releases oxyhemoglobin which oxidizes to methemoglobin, forming oxygen-free radicals that produce lipid peroxides and possibly combine with NO to produce peroxynitrite. Further oxidation of heme released from hemoglobin can produce bilirubin oxidation products [15]. These oxidation reactions and products contribute to activation of protein kinase C and Rho kinase in smooth muscle, leading to contraction [78]. Endothelial injury, either due to vasoconstriction causing increased shear stress or by some effect of the perivascular blood, is associated with endothelial cell apoptosis and necrosis leading to loss of NO productoin. There also may be uncoupling of endothelial NOS, so that is produces injurious superoxide anion radical [86].

4 Incidence and Definitions of Angiographic Vasospasm and DCI

Angiographic vasospasm is narrowing of large, conducting arteries of the circle of Willis that are in the subarachnoid space and that is observed on a catheter, computed tomographic, or magnetic resonance angiogram. It occurs 3–14 days after SAH. It may or may not cause neurological deterioration. Neurological worsening during this time is DCI if it is not apparent immediately after the aneurysm-securing procedure (which suggests that the deterioration is a complication of the aneurysm repair) and if there are no other causes by means of clinical assessment or radiological imaging of the brain and appropriate laboratory studies.

The incidence of angiographic vasospasm after SAH depends on how severe it has to be to be considered to be present. Among 3,567 patients entered into randomized clinical trials of tirilazad between 1991 and 1997, 30% developed DCI [58]. In a randomized clinical trial of clazosentan to prevent vasospasm after aneurysmal SAH, 66% of 96 patients in the placebo group developed angiographic vasospasm and 19% developed vasospasm-related cerebral infarction [56]. Dorsch reviewed the literature and found that the incidence of DCI was 29% (6,775 of 23,805 patients) in series reported between 1994 and 2009 [25]. In reports where nimodipine or nicardipine were used routinely, it was 22%. An earlier review found that angiographic vasospasm was present in 67% of patients when angiography was done during the second week after SAH.

Angiographic vasospasm is well documented by catheter angiography or indirectly by transcranial Doppler in patients with TBI [61, 91, 95]. Vasospasm by angiography was reported in 2–41% and by transcranial Doppler in up to 60% of patients with moderate or severe TBI [3]. Vasospasm, as assessed by at least one criterion on transcranial Doppler, was found in 45% of 299 patients with TBI [72]. Vasospasm was more likely in patients with traumatic SAH but could occur in any patient with TBI except those with isolated epi- or subdural hematomas. The time course of vasospasm after TBI shows that it has an earlier onset and shorter duration than after aneurysmal SAH. Since not all patients with vasospasm had SAH, it is likely that other processes contribute to vasospasm after TBI. DCI in patients with TBI is less well characterized, probably in part because it cannot be easily detected in many of these patients who already have impaired consciousness.

5 Causes of DCI

Angiographic vasospasm is highly correlated with development of DCI [16], and severe angiographic vasospasm reduces cerebral blood flow [18]. Nevertheless, in one study, severe vasospasm correlated with the least perfused area in only two-thirds of cases, and half of patients with severe vasospasm do not develop DCI. Thus, the correlation between angiographic vasospasm and DCI is imperfect like

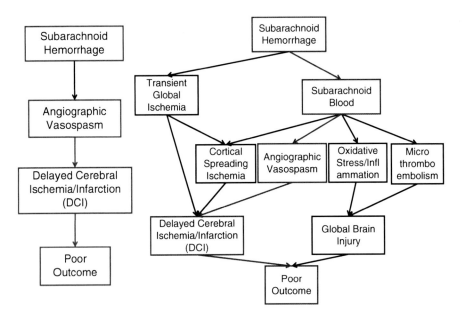

Fig. 32.1 Pathways to poor outcome after SAH. *Left* shows a simple pathway from SAH to angiographic vasospasm to DCI and poor outcome. Current theory is in keeping with right pathway where multiple processes interact to cause DCI and global brain injury, contributing to poor outcome

most correlations in clinical medicine. The other problem with the idea that angiographic vasospasm leads to DCI and poor outcome is that several studies showed that pharmacologic prevention of angiographic vasospasm was not associated with improved patient outcome [30]. This could be because the drugs have toxicity or detrimental off-target effects that counterbalance their benefit on reducing angiographic vasospasm; the studies were inadequately powered; the clinical outcome measure was not sensitive to meaningful improvement in outcome, and/or rescue therapy is used more frequently in the placebo groups and is equally effective as the drug in the treated patients. Data suggest that at least cerebral infarction from DCI is associated with poor outcome [107]. Anyway, the lack of correlation between vasospasm and outcome has led to theories that other processes contribute to DCI in addition to angiographic vasospasm (Fig. 32.1). These include early brain injury (reviewed above), microcirculatory changes, microthromboembolism, and CSI [2, 57, 108, 110]. Each of these processes is less well documented in experimental and clinical studies, and whether reducing them in humans translates to improved outcome is as yet unknown. The only drug convincingly shown to improve outcome after SAH is nimodipine, which reduced cerebral infarction but probably not angiographic vasospasm [23].

Microcirculatory changes are described after SAH. Histological studies of parenchymal microvasculature have yielded conflicting results. One study analyzing parenchymal arteriole changes in the double hemorrhage dog model of SAH found impaired autoregulation with parenchymal artery constriction [73], while another similar study found that intraparenchymal arterioles were dilated [77]. The latter study suggested that this discrepancy may be the product of an age effect or perhaps variation in vessel measurement methodology. Clinical studies to assess this phenomenon also have given inconsistent results. Yundt et al. found that SAH patients with vasospasm had an 18% decrease in cerebral blood volume (CBV), as opposed to patients with carotid occlusion who had an increase of 21% in CBV [115]. This suggested that there was no autoregulatory vasodilation in patients with SAH. On the other hand, other studies found increased CBV after SAH and particularly during vasospasm [38, 42]. Direct observation of cortical microvessels in patients with SAH has been done using orthogonal polarizing spectral imaging. Observation of the cortex in patients with SAH showed that 55% of cortical arterioles were constricted after SAH, with reductions in diameter up to 75% [101]. Furthermore, capillary bed density was significantly reduced. Other studies found an increase in constrictor responses to hyperventilation in 16 patients who were imaged while undergoing surgery for ruptured aneurysms [76]. Thus, there is experimental and clinical support for alterations in the microcirculation after SAH.

Microthromboemboli have been observed in the brains of experimental animals with SAH and in patients who die after SAH [96]. Whether they are thrombi, emboli, or some combination remains a matter of speculation. They have been attributed to activation and amplification of the coagulation cascade which promotes platelet aggregation. Levels of von Willebrand factor, platelet-activating factor, thromboxane B2, and Ð-thromboglobulin are increased in patients with SAH, which will promote platelet aggregation [108]. Thrombin activation, which is suggested by increased fibrinopeptide A, tissue factor, and thrombin-antithrombin complexes that are described after SAH, will stabilize platelet aggregates [96, 108]. Reductions in fibrinolytic activity also are described after SAH and would contribute further to thrombosis after SAH [44, 105]. The microclots may be emboli or thrombi. Clinical studies using TCD report that embolic signals were detected in 16 of 23 (70%) patients with SAH compared to 2 of 11 (18%) control patients [83]. Another study conducted daily TCD in 40 patients with aneurysmal SAH [84]. Embolic signals were detected in 15 patients (38%), and they were more likely in patients who developed DCI. The signals were detected before or on the day of onset of DCI in half of cases. These data suggest that the microclots observed pathologically could be emboli. Also, while microthrombi are associated with infarction, they themselves may not be sufficient to elicit clinically observable ischemic symptoms since many patients have them but do not have obvious DCI, and patients with DCI may not have them detected. Whatever their genesis, intraluminal platelet aggregates can occlude vessels, causing areas of ischemia as well as cause vascular damage and constriction by release of vasoconstricting contents of the platelet granules [89]. Platelets also contain matrix metalloproteinase-9, which degrades collagen IV in the microcirculation, thus potentially mediating BBB breakdown [88, 90].

Clinical trials have targeted platelet aggregation and thrombosis. Meta-analysis of four studies of antiplatelet therapy in patients with SAH showed that there were statistically insignificant effects on poor outcome (RR 0.79, 95% confidence interval [CI] 0.62–1.01), secondary brain ischemia (RR 0.79, 95% CI 0.56–1.22), intracranial hemorrhagic complications (RR 1.36, 95% CI 0.59–3.12), and case fatality (RR 1.01, 95% CI 0.74–1.37) [24]. A clinical trial of acetylsalicylic acid was stopped prematurely after interim analysis showed that the probability of a good outcome was unlikely [102]. It was theorized that the drug dose was inadequate or that aspirin may not be the best antiplatelet drug for SAH. Studies of enoxaparin, an antithrombin III antagonist, are contradictory [93, 113]. Nimodipine and other dihydropyridines have fibrinolytic activity [109]. One theory is they improved outcome in patients with SAH by preventing microthromboemboli.

Cortical spreading depression and CSI are the third main process that is thought to contribute to DCI. Cortical spreading depression is a wave of mass cellular depolarization that is measured as a negative direct current shift and that spreads along cerebral cortex at 2–5 mm/min [52]. The depolarization reduces neuronal activity and produces transient electrocorticographic silence. Spreading depressions occur in normal cerebral cortex in animals and have been observed in humans after ischemic stroke, SAH, and TBI [22]. The physiologic response to cortical spreading depression is a wave of vasodilation that is theorized to supply additional energy to neurons in order to allow them to reverse the transient increase in intracellular calcium (from 60 nM to 25:M) [22]. It has been noted that after stroke, SAH, and TBI, the hyperemic response to cortical spreading depression can be inverted so that there is vasoconstriction or CSI [27]. CSI prevents neurons from recovering, and selective neuronal necrosis and cortical infarction may occur.

After SAH, CSI is hypothesized to be triggered by hemolysis with release of hemoglobin and potassium from the subarachnoid blood clot [28]. Extravascular hemoglobin scavenges and binds NO and, in conjunction with extracellular potassium concentrations over 20 mM, causes vasoconstriction [26]. Subdural grid electrode recordings in patients with aneurysmal SAH and TBI have documented episodes of CSI that are temporally associated with DCI [7, 28]. Experimentally, high local concentrations of nimodipine prevent CSI in experimental models [29].

6 Current Treatment of SAH and DCI

Current management of SAH is to repair the ruptured aneurysm by neurosurgical clipping or endovascular clipping as soon as feasible, usually within 24 h of admission to the hospital. Since there is no treatment for early brain injury and no neuroprotective drugs, the fundamental principle is then to prevent secondary injury to the brain. This is done by maintaining adequate cerebral blood flow with normovolemia, normal hematocrit, adequate blood pressure, oxygenation, and blood glucose and avoidance of factors that reduce cerebral blood flow, constrict arteries, increase cerebral metabolic demand, or reduce supply of oxygen- and glucose-rich blood to the brain. Thus, fever, hypotension, increased intracranial pressure, hypoglycemia, hypomagnesemia, hypovolemia, hypercarbia, seizures, and hypoxia should be prevented. Supportive care such as intubation, ventilation, enteral feeding, venous thromboembolism prophylaxis, anticonvulsants, and CSF drainage for increased intracranial pressure and hydrocephalus may be used. Most patients with aneurysmal SAH are given the dihydropyridine L-type calcium channel antagonist nimodipine, either orally or intravenously. There is no other recommended prophylactic treatment for angiographic vasospasm and DCI.

If a patient deteriorates from DCI, then the treatment is to raise the blood pressure pharmacologically and to infuse vasodilator drugs superselectively into vasospastic arteries or to dilate the arteries with balloons. These maneuvers may be called rescue therapy. They are expensive and time-consuming, and their effectiveness has been questioned, even so far as to suggest that they are ineffective and not recommended [19, 80]. Twenty percent of patients hospitalized with aneurysmal SAH still suffer death or permanent disability from DCI [36].

7 Drugs to Improve Outcome, Potentially by Local Delivery

Numerous drugs have been tested in clinical trials of SAH [111]. So far, few have been directed at treating early brain injury. Vasodilators such as dihydropyridines, endothelin antagonists, and fasudil have been most studied, whereas those directed at other processes thought to contribute to DCI, such as antithrombotics and anti-inflammatory agents, are less well represented. The pathogenesis of SAH starts largely in the subarachnoid space, so it is logical to consider local delivery of drugs here, especially since it is a preformed space. Local delivery might have advantages for drugs that do not cross the blood-brain barrier. This would include some water-soluble drugs as well as proteins and high-molecular-weight compounds. A number of proteins and monoclonal antibodies are approved for human use, for example, monoclonal antibodies that reduce the effects of tumor necrosis factor $\forall$ such as infliximab and etanercept. Some drugs, like nimodipine, however, probably do cross the blood-brain barrier, and thus local delivery is not necessary for this reason. The rationale here is that one can attain high local concentrations next to the cerebral arteries and lower concentrations in the body, thus potentially avoiding hypotension and pulmonary edema. This is important when the drug target is not necessarily modified or much different in the diseased area compared to the rest of the body.

The half-life of most drugs in the subarachnoid space is not known. Unless the disease being treated is of very short duration or the drug half-life was long, multiple administrations into the subarachnoid space would be required. This is technically difficult and could be avoided by using biodegradable polymers that release drug over hours and days [74]. This has been done, and one approved product is carmustine loaded in poly(carboxyphenoxypropane/sebacic acid) anhydride [9].

The properties of a suitable drug carrier would be that it is biodegradable, elicits a local tissue response comparable to other approved medical products, can be loaded to a high degree with the drug, has acceptable initial drug release (burst), and can spread out in the subarachnoid space and then remain there to release drug for probably 10–14 days. Biodegradable carriers include synthetic and natural polymers as well as in situ gelling systems. Synthetic polymers that have been or are being investigated for human use are polyesters (polyglycolide, polylactic acid, poly-caprolactone, and combinations such as poly-[D,L-lactide-co-glycolide][PLG]), polyhydrobutyrate, polyhydroxyvalerate, polyamino biopolymers, tyrosine-derived polycarbonates and polyarylates, polyanhydrides, polyorthoesters, and polyphosphazenes. Naturally occurring polymers include proteins (silk fibroin, elastin, and collagen) and polysaccharides (chitin, hyaluronic acid). The various polymers vary from soft semisolids to hard, brittle materials. The firmer ones, such as PLG, can be formulated into an endless variety of shapes and sizes, including microparticles, or they can be mixed with polyethylene glycol, alginates, or polysaccharides to form a gel. Systems that gel in situ include both nonpolymeric compounds such as sucrose acetate isobutyrate and polymeric systems [53].

Some of the above compounds are approved for systemic use in humans, but placing them in the subarachnoid space raises concerns about potential neurotoxic excipients and host inflammatory response. In addition, the duration of release, which may have to be for 10–14 days, imposes additional restrictions. Most of the polymers are combined with the active drug as a matrix system. Drug release from matrix systems usually follows a square root time profile according to Fick's law. The amount of drug released is affected by the surface area through which the drug delivery is taking place, the diffusion coefficient, the thickness of the polymer, and the time since initiation of drug release. Sustaining drug release over days can be difficult for small molecules, particularly hydrophilic ones. Also, some polymers, such as lower molecular weight hyaluronans, degrade rapidly within hours to a few days.

The effects of only a few polymers have been assessed in the brain. These include PLG [63], alginates [6], poly(carboxyphenoxypropane/sebacic acid) [8], collagens, sucrose acetate isobutyrate [53], and various hydrogels [17]. As mentioned above, poly(carboxyphenoxypropane/sebacic acid) anhydride loaded with carmustine is approved for treatment of human brain tumors [8]. There is a large experience with PLG in humans. These polymers have been used in resorbable sutures since the early 1970s and are now used in many medical devices, in drug delivery products, and as dural substitutes. They elicit minimal inflammatory response and are generally recognized as safe [54]. Tice and colleagues were among the first to use PLG biodegradable polymers implanted in the brain to deliver dopamine for Parkinsonism [63].

8 Prior Experience with Biodegradable PLG Formulations in the Brain

PLG has minimal toxicity in animals. PLG polymers are biocompatible and have established safety over the past 3 decades for various medical applications [55]. Biological degradation of PLG depends largely on water uptake into the polymer. Water is used for hydrolysis of the polymer ester linkages, the rates of which are determined by the physical and chemical characteristics of the PLG, such as the chain length, lactide to glycolide ratio, and polymer chemical end groups [34]. Cellular and enzymatic mechanisms play a smaller role [21, 55]. PLG can be modified to release drug and degrade over times varying from days to months. The ultimate degradation products are the monomers, lactic acid, and glycolic acid and then carbon dioxide and water [43].

PLG pellets and microparticles have been implanted in the subarachnoid space and brain without toxicity [14, 34]. Microparticles are an attractive way to formulate PLG since there are processes to do this already; they can be scaled up and standardized for clinical use. Microparticles between 40 and 100:m in diameter have been shown to hydrolyze in the brain without being removed by macrophages by phagocytosis, and they are small enough to be injectable. Menei et al. studied biodegradation over 4 months of 22:m 50:50 PLG microparticles implanted into the brains of 36 rats [65]. The microparticles were prepared by solvent evaporation using methylene chloride and were sterilized by gamma irradiation. The microparticles were engulfed by activated microglial cells and/or macrophages within a day of implantation. Over days, there were macrophages and occasional foreign body giant cells and reactive astrocytes around the particles. These changes resolved by 2 months. PLG does not seem to be immunogenic [34]. Tice and colleagues studied the response to injection of catecholamine-containing or placebo 65:35 or 50:50 PLG microparticles into the striatum of rats with Parkinsonism [62, 63]. The symptoms of Parkinsonism were reduced, and the same transient inflammatory response described above was observed. Fournier et al. summarized the literature on implantation of PLG in the brains of animals [34]. Ten studies in rats were reviewed and showed that the response to PLG in brain was unrelated to implantation site and carrier shape. It consisted of a moderate, nonspecific inflammatory response due to mechanical trauma. The reaction was maximal 1–2 days after implantation for microglial cells and monocytes/macrophages, at 8–15 days for astrocytes, and decreased markedly after 1 month for both cell types.

PLG microparticles have been formulated with the chemotherapy drug, 5-fluorouracil, and implanted in 56 humans, including 38 who were part of a randomized clinical trial [64, 66].

9 Drugs That Have Been Tested

Drugs that have been studied by local delivery into the subarachnoid space for experimental SAH include papaverine, nicardipine, ibuprofen, NO donors, calcitonin gene-related peptide (CGRP), fasudil, and recombinant tissue plasminogen activator (rt-PA). Interest in papaverine has diminished after reports of toxicity, although this was probably related to the carrier and that it was less effective than nicardipine in a dog model [92, 94]. Many drugs have been instilled into the human subarachnoid space, but sustained-release formulations for SAH seem to have been restricted thus far to nicardipine and rt-PA.

Reduction in the vasodilator NO has been postulated to be important in the pathogenesis of angiographic vasospasm and possibly other aspects of brain injury after SAH. Diethylenetriamine/NO (DETA/NO) was formulated in EVAc and implanted into brains of rats to assess toxicity. Mortality occurred in 50% of animals at a dose of 13 mg/kg, and the 20% lethal dose was 3.4 mg/kg. Histologic examination of the brain in surviving animals showed dose-dependent hemorrhage and ischemic changes up to several millimeters from the implant site. At high doses, there was brain necrosis and dystrophic calcification. A dose of 0.48 mg/kg, however, was not associated with mortality or toxicity and was then administered into the cisterna magna 30 min after SAH in the rabbit single hemorrhage model [37]. The group of investigators led by Tamargo has provided further compelling data on the effect of this approach in a nonhuman primate model of SAH where nontoxic doses of DETA/NO in EVAc reduced angiographic vasospasm [99].

Tamargo and colleagues also studied ibuprofen in ethylene-vinyl acetate copolymer (EVAc) for prevention of vasospasm after SAH in rabbits [35]. Five-mg empty EVAc or EVAc containing 45% ibuprofen was inserted into the cisterna magna at the site of blood injection to create the SAH. If ibuprofen was delivered within 6 h of SAH, vasospasm was reduced, whereas at this dose it had no effect when delivered 12 or more hours after SAH. The effects on inflammation were not assessed, but it was proposed that the beneficial effects were due to ibuprofen-mediated reduction in intercellular adhesion molecule-1 expression on endothelial cells, thus reducing leukocyte recruitment through the artery wall. The rationale for local intracranial delivery was that this pharmacologic effect occurs at high systemic concentrations that are associated with thrombocytopenia, fluid retention, and gastric ulcers. The ibuprofen sustained-release formulation (45% ibuprofen/EVAc polymer, 6 mg/kg ibuprofen) was tested in 14 cynomolgus monkeys that underwent baseline angiography and creation of SAH using a clot placement model [81]. There was no local or systemic toxicity, and the ibuprofen polymers reduced middle cerebral artery narrowing so that the diameter of this artery was 91 ∀ 9% of the baseline diameter 7 days after SAH, compared to 53 ∀ 11% in animals given placebo EVAc polymer. The results support the concept that high doses of drugs can be given intracranially without toxicity that would occur by systemic administration, as well as a role for endothelial cell-leukocyte interactions in the genesis of vasospasm.

The PLG polymer has been used in a sustained-release formulation of CGRP for experimental SAH [1, 45]. Tablets of PLG-CGRP implanted into the cisterna magna of rabbits 2 days after SAH significantly decreased angiographic vasospasm. High CSF concentrations of CGRP were present. It would be interesting to know if CGRP was elevated in plasma and if blood pressure was affected since hypotension occurred in a clinical trial of CGRP for SAH [31]. The same formulation, at a dose of 1.2 mg CGRP, reduced vasospasm in a monkey clot placement model of SAH [45].

Fasudil was encapsulated in liposomes and delivered in doses of 0.42–0.94 mg/kg intrathecally to dogs and rats with SAH [98]. There were no significant changes in mean arterial blood pressure, and intrathecal fasudil did not cause seizures. No pathologic changes occurred in the brains, and vasospasm was reduced.

10   Subarachnoid Dihydropyridines

The most widely studied drugs in terms of subarachnoid delivery are dihydropyridines such as nicardipine. These studies are particularly interesting in view of the evidence that adequate concentrations of nimodipine can prevent or reverse angiographic vasospasm, prevent cortical spreading ischemia in experimental models, and theoretically prevent microthromboemboli by virtue of their fibrinolytic activity [29, 40, 49, 109]. There also is more information on relative plasma and CSF concentrations of these drugs. In general, oral nimodipine (60 mg orally every 4 h) produces a maximal plasma concentration of about 20 ng/ml (50 nM), but there is substantial variability, due partly to variable bioavailability. The CSF concentration at this dose is around 0.2–0.4 ng/ml (0.5–1 nM), extrapolating from Allen et al. [2]. This dose does not have a major effect on angiographic vasospasm. The concentration of nimodipine required to inhibit cerebral artery contractions experimentally is an EC_{50} of 2–5 nM. Blood pressure, however, begins to decrease when plasma concentrations are >30–47 ng/ml (112 nM) [51, 87]. Thus, oral, and probably intravenous, administration of nimodipine lowers the blood pressure when the CSF concentration is still 10 times too low to dilate cerebral arteries.

Intra-arterial infusion of nimodipine or other dihydropyridine calcium channel antagonists have been reported to reverse angiographic vasospasm and improve clinical condition of patients with SAH [5, 40]. Intrathecal and intraventricular injections of nimodipine also may reverse angiographic vasospasm [39]. The limitation of these injections, however, is that the effect is transient and limited by the short half-life of nimodipine (1–2 h in plasma). They do provide some evidence that high, local concentrations of nimodipine could prevent or reverse angiographic vasospasm. Concentrations required to prevent other complications of SAH, such as cortical spreading ischemia and microthromboembolism, are unknown. In rats, intravenous nimodipine, 2 μg/kg/min, transformed cortical spreading ischemia induced by potassium chloride plus hemoglobin or a nitric oxide synthase inhibitor back to cortical spreading hyperemia [27, 29].

A final piece of evidence in support of the concept that high subarachnoid concentrations of nimodipine might be efficacious in SAH can be derived from study of nicardipine, another dihydropyridine calcium channel antagonist that is similar to nimodipine, that was formulated in PLG rods and placed intracranially in 252 patients with aneurysmal SAH, including 16 as part of a randomized, single-blind clinical trial [4, 48, 50]. A randomized trial conducted in Germany allocated 32 patients with SAH who had neurosurgical aneurysm clipping to subarachnoid placement of nicardipine PLG pellets containing 40 mg nicardipine (16 patients) or to surgery but no placement of nicardipine [4]. Angiographic vasospasm was significantly reduced in the group treated with nicardipine (73% control vs. 7% nicardipine). There was a lower incidence of DCI in patients treated with nicardipine (47% control vs. 14% nicardipine), and mortality was reduced from 38% in control to 6% in nicardipine-treated patients. Krischek et al. reviewed the first 100 patients who were treated in Japan with 2–12 nicardipine pellets containing 4 mg each by subarachnoid administration at surgery for clipping of ruptured aneurysms [50].

Severe angiographic vasospasm developed in 11 patients, DCI was observed in 7, and good outcome on the Glasgow outcome score at 3 months was seen in 82. There were no side effects. Finally, a multicenter, cooperative study in Tokyo enrolled 136 patients from 6 hospitals. All patients received subarachnoid nicardipine pellets during surgery for clipping of their aneurysms. DCI, angiographic vasospasm, and cerebral infarctions were seen in 11 of 134 (8%), 32 of 130 patients (25%), and 16 of 129 (12%), respectively. Again, there were no complications attributed to nicardipine.

11   Nimodipine Microparticle Proof of Principle

We have developed nimodipine formulated in PLG microparticles for potential intracisternal and intraventricular use in humans with aneurysmal or traumatic SAH (Fig. 32.2). The formulation releases nimodipine over days in vitro with minimal initial burst. Preliminary study in dogs has been performed. Mongrel dogs were randomly allocated to placebo microparticles, 10 or 30 mg nimodipine microparticles. They underwent baseline cerebral angiography and then cisternal blood and microparticle injection. Cisternal blood injection was repeated 2 days later, and angiograms were done 7 and 14 days after the first blood injection. Blinded analysis of the angiograms showed a dose-dependent reduction in angiographic vasospasm, supporting the hypothesis that site-specific, sustained-release of nimodipine prevents vasospasm (Fig. 32.3). There were no untoward systemic side effects, and histopathology of the brain and subarachnoid space did not show more inflammation than what occurs due to the hemorrhage itself. Plasma concentrations of nimodipine were 1.5 $\forall$ 0.2 (mean $\forall$ standard deviation) and 2.3 $\forall$ 0.8 ng/ml 7 days after administration of 10 and 30 mg nimodipine, respectively, and CSF concentrations were more than 200 ng/ml in the 30-mg group. The plasma concentration that causes hypotension is more than 9 ng/ml (and probably closer to 30–40 ng/ml), and CSF concentrations that are effective must be more than 10 ng/ml [67]. No untoward behavioral effects were observed. The results show that one not once can deliver high doses of nimodipine intracranially and achieve the desired effect while maintaining low systemic concentrations to avoid side effects.

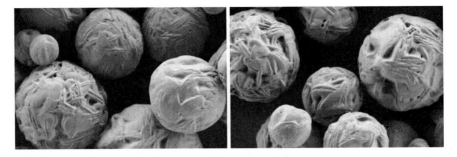

Fig. 32.2 Nimodipine-loaded microparticles composed of PLG

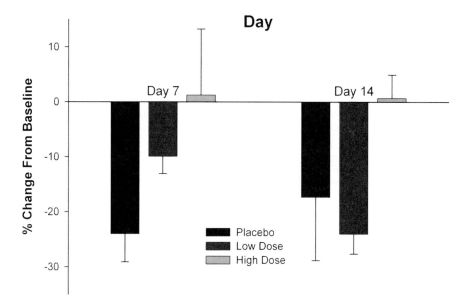

Fig. 32.3 Percent change in diameter of the basilar artery of dogs that had SAH and injection of placebo PLG microparticles or nimodipine-containing PLG microparticles. Demonstrates proof-of-concept that site-specific, sustained-release of a drug intracranially can prevent a disease, in this case, vasospasm ($n = 2$ per group, $p = 0.08$, analysis of variance)

12   Summary

Outcome of patients with aneurysmal SAH has improved, with estimates of decreases in mortality of 0.9% per year over the past 2 decades. Despite this, the mortality still is 20–40%, and many survivors are permanently disabled, particularly by cognitive impairments. A key potentially treatable cause of poor outcome is DCI. DCI may be due to angiographic vasospasm, microcirculatory changes, microthromboemboli, and/or CSI. Clinical trials of a variety of drugs suggest that angiographic vasospasm can be reduced but that clinical outcome is more difficult to improve, at least as measured on the dichotomous Glasgow outcome scale. Why preventing angiographic vasospasm has not translated into improved outcome is unclear, but some reasons include inadequate sample sizes of the trials, insensitivity of the outcome measures, efficacy of rescue therapy in placebo groups counterbalancing drug effect in treated groups, detrimental off-target drug side effects, and contribution of other processes that are not prevented by the drug other than angiographic vasospasm to DCI. There is evidence from intra-arterial and subarachnoid drug delivery studies that some drugs, such as nimodipine and nicardipine, may be more efficacious if they are given in higher doses. They cannot be administered systemically, however, because of side effects such as hypotension and pulmonary edema. We suggest that high doses of drugs like nimodipine, formulated in sustained-release biodegradable PLG microparticles, prevent angiographic vasospasm in dogs and that clinical trials are needed to test this novel hypothesis in humans with aneurysmal and traumatic SAH.

Financial Disclosure/Acknowledgments R.L. Macdonald receives research support from the Physicians Services Incorporated Foundation and the Heart and Stroke Foundation of Ontario. R.L. Macdonald is a consultant for Actelion Pharmaceuticals and Chief Scientific Officer of Edge Therapeutics, Inc.

References

1. Ahmad I, Imaizumi S, Shimizu H, Kaminuma T, Ochiai N, Tajima M, et al. Development of calcitonin gene-related peptide slow-release tablet implanted in CSF space for prevention of cerebral vasospasm after experimental subarachnoid haemorrhage. Acta Neurochir. 1996;138:1230–40.
2. Allen GS, Ahn HS, Preziosi TJ, Battye R, Boone SC, Chou SN, et al. Cerebral arterial spasm–a controlled trial of nimodipine in patients with subarachnoid hemorrhage. N Engl J Med. 1983;308:619–24.
3. Armin SS, Colohan AR, Zhang JH. Traumatic subarachnoid hemorrhage: our current understanding and its evolution over the past half century. Neurol Res. 2006;28:445–52.
4. Barth M, Capelle HH, Weidauer S, Weiss C, Munch E, Thome C, et al. Effect of nicardipine prolonged-release implants on cerebral vasospasm and clinical outcome after severe aneurysmal subarachnoid hemorrhage: a prospective, randomized, double-blind phase IIa study. Stroke. 2007;38:330–6.
5. Biondi A, Ricciardi GK, Puybasset L, Abdennour L, Longo M, Chiras J, et al. Intra-arterial nimodipine for the treatment of symptomatic cerebral vasospasm after aneurysmal subarachnoid hemorrhage: preliminary results. AJNR Am J Neuroradiol. 2004;25:1067–76.
6. Borlongan CV, Skinner SJ, Geaney M, Vasconcellos AV, Elliott RB, Emerich DF. Neuroprotection by encapsulated choroid plexus in a rodent model of Huntington's disease. Neuroreport. 2004;15:2521–5.
7. Bosche B, Graf R, Ernestus RI, Dohmen C, Reithmeier T, Brinker G, et al. Recurrent spreading depolarizations after subarachnoid hemorrhage decreases oxygen availability in human cerebral cortex. Ann Neurol. 2010;67:607–17.
8. Brem H, Ewend MG, Piantadosi S, Greenhoot J, Burger PC, Sisti M. The safety of interstitial chemotherapy with BCNU-loaded polymer followed by radiation therapy in the treatment of newly diagnosed malignant gliomas: phase I trial. J Neurooncol. 1995;26:111–23.
9. Brem H, Piantadosi S, Burger PC, Walker M, Selker R, Vick NA, et al. Placebo-controlled trial of safety and efficacy of intraoperative controlled delivery by biodegradable polymers of chemotherapy for recurrent gliomas. The Polymer-brain Tumor Treatment Group. Lancet. 1995;345:1008–12.
10. Broderick JP, Brott TG, Duldner JE, Tomsick T, Leach A. Initial and recurrent bleeding are the major causes of death following subarachnoid hemorrhage. Stroke. 1994;25:1342–7.
11. Bruns Jr J, Hauser WA. The epidemiology of traumatic brain injury: a review. Epilepsia. 2003;44 Suppl 10:2–10.
12. Cahill J, Calvert JW, Zhang JH. Mechanisms of early brain injury after subarachnoid hemorrhage. J Cereb Blood Flow Metab. 2006;26:1341–53.
13. Cahill J, Zhang JH. Subarachnoid hemorrhage: is it time for a new direction? Stroke. 2009;40:S86–7.
14. Camarata PJ, Suryanarayanan R, Turner DA, Parker RG, Ebner TJ. Sustained release of nerve growth factor from biodegradable polymer microspheres. Neurosurgery. 1992;30:313–9.

15. Clark JF, Sharp FR. Bilirubin oxidation products (BOXes) and their role in cerebral vasospasm after subarachnoid hemorrhage. J Cereb Blood Flow Metab. 2006;26:1223–33.
16. Crowley RW, Medel R, Dumont AS, Ilodigwe D, Kassell NF, Mayer SA, et al. Angiographic vasospasm is strongly correlated with cerebral infarction after subarachnoid hemorrhage. Stroke. 2011;42:919–23.
17. Cui FZ, Tian WM, Hou SP, Xu QY, Lee IS. Hyaluronic acid hydrogel immobilized with RGD peptides for brain tissue engineering. J Mater Sci Mater Med. 2006;17:1393–401.
18. Dankbaar JW, Rijsdijk M, van der Schaaf IC, Velthuis BK, Wermer MJ, Rinkel GJ. Relationship between vasospasm, cerebral perfusion, and delayed cerebral ischemia after aneurysmal subarachnoid hemorrhage. Neuroradiology. 2009;51:813–9.
19. Dankbaar JW, Slooter AJ, Rinkel GJ, van der Schaaf IC. Effect of different components of triple-H therapy on cerebral perfusion in patients with aneurysmal subarachnoid haemorrhage: a systematic review. Crit Care. 2010;14:R23.
20. de Rooij NK, Linn FH, van der Plas JA, Algra A, Rinkel GJ. Incidence of subarachnoid haemorrhage: a systematic review with emphasis on region, age, gender and time trends. J Neurol Neurosurg Psychiatry. 2007;78:1365–72.
21. Devereux DF, O'Connell SM. Biomaterials used in hernia repair, abdominal wall replacement, and the intestinal sling procedure. In: Greco RS, editor. Implantation biology. The host response to biomedical devices. Boca Raton: CRC Press; 1994. p. 229–314.
22. Dietz RM, Weiss JH, Shuttleworth CW. Contributions of Ca^{2+} and Zn^{2+} to spreading depression-like events and neuronal injury. J Neurochem. 2009;109 Suppl 1:145–52.
23. Dorhout Mees SM, Rinkel GJ, Feigin VL, Algra A, van den Bergh WM, Vermeulen M, et al. Calcium antagonists for aneurysmal subarachnoid haemorrhage. Cochrane Database Syst Rev. 2007:CD000277.
24. Dorhout Mees SM, van den Bergh WM, Algra A, Rinkel GJ. Antiplatelet therapy for aneurysmal subarachnoid haemorrhage. Cochrane Database Syst Rev. 2007:CD006184.
25. Dorsch N. A clinical review of cerebral vasospasm and delayed ischaemia following aneurysm rupture. Acta Neurochir Suppl. 2011;110:5–6.
26. Dreier JP. The role of spreading depression, spreading depolarization and spreading ischemia in neurological disease. Nat Med. 2011;17:439–47.
27. Dreier JP, Korner K, Ebert N, Gorner A, Rubin I, Back T, et al. Nitric oxide scavenging by hemoglobin or nitric oxide synthase inhibition by N-nitro-L-arginine induces cortical spreading ischemia when K^+ is increased in the subarachnoid space. J Cereb Blood Flow Metab. 1998;18:978–90.
28. Dreier JP, Major S, Manning A, Woitzik J, Drenckhahn C, Steinbrink J, et al. Cortical spreading ischaemia is a novel process involved in ischaemic damage in patients with aneurysmal subarachnoid haemorrhage. Brain. 2009;132:1866–81.
29. Dreier JP, Windmuller O, Petzold G, Lindauer U, Einhaupl KM, Dirnagl U. Ischemia triggered by red blood cell products in the subarachnoid space is inhibited by nimodipine administration or moderate volume expansion/hemodilution in rats. Neurosurgery. 2002;51:1457–65.
30. Etminan N, Vergouwen MD, Ilodigwe D, Macdonald RL. Effect of pharmaceutical treatment on vasospasm, delayed cerebral ischemia, and clinical outcome in patients with aneurysmal subarachnoid hemorrhage: a systematic review and meta-analysis. J Cereb Blood Flow Metab. 2011;31:1443–51.
31. European CGRP in Subarachnoid Haemorrhage Study Group. Effect of calcitonin-gene-related peptide in patients with delayed postoperative cerebral ischaemia after aneurysmal subarachnoid haemorrhage. Lancet. 1992;339:831–4.
32. Fassbender K, Hodapp B, Rossol S, Bertsch T, Schmeck J, Schutt S, et al. Endothelin-1 in subarachnoid hemorrhage: an acute-phase reactant produced by cerebrospinal fluid leukocytes. Stroke. 2000;31:2971–5.
33. Findlay JM, Weir BK, Kanamaru K, Espinosa F. Arterial wall changes in cerebral vasospasm. Neurosurgery. 1989;25:736–45.
34. Fournier E, Passirani C, Montero-Menei CN, Benoit JP. Biocompatibility of implantable synthetic polymeric drug carriers: focus on brain biocompatibility. Biomaterials. 2003;24:3311–31.

35. Frazier JL, Pradilla G, Wang PP, Tamargo RJ. Inhibition of cerebral vasospasm by intracranial delivery of ibuprofen from a controlled-release polymer in a rabbit model of subarachnoid hemorrhage. J Neurosurg. 2004;101:93–8.

36. Frontera JA, Fernandez A, Schmidt JM, Claassen J, Wartenberg KE, Badjatia N, et al. Defining vasospasm after subarachnoid hemorrhage: what is the most clinically relevant definition? Stroke. 2009;40:1963–8.

37. Gabikian P, Clatterbuck RE, Eberhart CG, Tyler BM, Tierney TS, Tamargo RJ. Prevention of experimental cerebral vasospasm by intracranial delivery of a nitric oxide donor from a controlled-release polymer: toxicity and efficacy studies in rabbits and rats. Stroke. 2002;33: 2681–6.

38. Grubb RLJ, Raichle ME, Eichling JO, Gado MH. Effects of subarachnoid hemorrhage on cerebral blood volume, blood flow, and oxygen utilization in humans. J Neurosurg. 1977;46:446–53.

39. Hanggi D, Beseoglu K, Turowski B, Steiger HJ. Feasibility and safety of intrathecal nimodipine on posthaemorrhagic cerebral vasospasm refractory to medical and endovascular therapy. Clin Neurol Neurosurg. 2008;110:784–90.

40. Hanggi D, Turowski B, Beseoglu K, Yong M, Steiger HJ. Intra-arterial nimodipine for severe cerebral vasospasm after aneurysmal subarachnoid hemorrhage: influence on clinical course and cerebral perfusion. AJNR Am J Neuroradiol. 2008;29:1053–60.

41. Hillman J, Fridriksson S, Nilsson O, Yu Z, Saveland H, Jakobsson KE. Immediate administration of tranexamic acid and reduced incidence of early rebleeding after aneurysmal subarachnoid hemorrhage: a prospective randomized study. J Neurosurg. 2002;97:771–8.

42. Hino A, Mizukawa N, Tenjin H, Imahori Y, Taketomo S, Yano I, et al. Postoperative hemodynamic and metabolic changes in patients with subarachnoid hemorrhage. Stroke. 1989;20:1504–10.

43. Holland SJ, Tighe BJ, Gould PL. Polymers for biodegradable medical devices. 1. The potential of polyesters as controlled macromolecular release systems. J Control Release. 1986;4:155–80.

44. Ikeda K, Asakura H, Futami K, Yamashita J. Coagulative and fibrinolytic activation in cerebrospinal fluid and plasma after subarachnoid hemorrhage. Neurosurgery. 1997;41:344–9.

45. Inoue T, Shimizu H, Kaminuma T, Tajima M, Watabe K, Yoshimoto T. Prevention of cerebral vasospasm by calcitonin gene-related peptide slow-release tablet after subarachnoid hemorrhage in monkeys. Neurosurgery. 1996;39:984–90.

46. Johnston SC, Selvin S, Gress DR. The burden, trends, and demographics of mortality from subarachnoid hemorrhage. Neurology. 1998;50:1413–8.

47. Kassell NF, Torner JC, Haley Jr EC, Jane JA, Adams HP, Kongable GL. The International Cooperative Study on the Timing of Aneurysm Surgery. Part 1: Overall management results. J Neurosurg. 1990;73:18–36.

48. Kasuya H. Clinical trial of nicardipine prolonged-release implants for preventing cerebral vasospasm: multicenter cooperative study in Tokyo. Acta Neurochir Suppl. 2011;110:165–7.

49. Kim JH, Park IS, Park KB, Kang DH, Hwang SH. Intraarterial nimodipine infusion to treat symptomatic cerebral vasospasm after aneurysmal subarachnoid hemorrhage. J Korean Neurosurg Soc. 2009;46:239–44.

50. Krischek B, Kasuya H, Onda H, Hori T. Nicardipine prolonged-release implants for preventing cerebral vasospasm after subarachnoid hemorrhage: effect and outcome in the first 100 patients. Neurol Med Chir (Tokyo). 2007;47:389–94.

51. Laursen J, Jensen F, Mikkelsen E, Jakobsen P. Nimodipine treatment of subarachnoid hemorrhage. Clin Neurol Neurosurg. 1988;90:329–37.

52. Leao AA. Spreading depression of activity in the cerebral cortex. J Neurophysiol. 1944;7:359–90.

53. Lee J, Jallo GI, Penno MB, Gabrielson KL, Young GD, Johnson RM, et al. Intracranial drug-delivery scaffolds: biocompatibility evaluation of sucrose acetate isobutyrate gels. Toxicol Appl Pharmacol. 2006;215:64–70.

54. Lee KH, Lukovits T, Friedman JA. "Triple-H" therapy for cerebral vasospasm following subarachnoid hemorrhage. Neurocrit Care. 2006;4:68–76.

55. Lewis DH. Controlled release of bioactive agents from lactide/glycolide polymers. In: Chasin M, Langer R, editors. Biodegradable polymers as drug delivery systems. New York: Marcel Dekker; 1990. p. 1–41.

56. Macdonald RL, Kassell NF, Mayer S, Ruefenacht D, Schmiedek P, Weidauer S, et al. Clazosentan to overcome neurological ischemia and infarction occurring after subarachnoid hemorrhage (CONSCIOUS-1): randomized, double-blind, placebo-controlled phase 2 dose-finding trial. Stroke. 2008;39:3015–21.

57. Macdonald RL, Pluta RM, Zhang JH. Cerebral vasospasm after subarachnoid hemorrhage: the emerging revolution. Nat Clin Pract Neurol. 2007;3:256–63.

58. Macdonald RL, Rosengart A, Huo D, Karrison T. Factors associated with the development of vasospasm after planned surgical treatment of aneurysmal subarachnoid hemorrhage. J Neurosurg. 2003;99:644–52.

59. Macdonald RL, Zhang ZD, Takahashi M, Nikitina E, Young J, Xie A, et al. Calcium sensitivity of vasospastic basilar artery after experimental subarachnoid hemorrhage. Am J Physiol Heart Circ Physiol. 2006;290:H2329–36.

60. Marshall LF, Marshall SB, Klauber MR. A new classification of head injury based on computed tomography. J Neurosurg. 1991;75(Suppl):S14–20.

61. Martin NA, Doberstein C, Alexander M, Khanna R, Benalcazar H, Alsina G, et al. Posttraumatic cerebral arterial spasm [Review]. J Neurotrauma. 1995;12:897–901.

62. McRae A, Hjorth S, Dahlstrom A, Dillon L, Mason D, Tice T. Dopamine fiber growth induction by implantation of synthetic dopamine-containing microspheres in rats with experimental hemi-parkinsonism. Mol Chem Neuropathol. 1992;16:123–41.

63. McRae A, Ling EA, Hjorth S, Dahlstrom A, Mason D, Tice T. Catecholamine-containing biodegradable microsphere implants as a novel approach in the treatment of CNS neurodegenerative disease. A review of experimental studies in DA-lesioned rats. Mol Neurobiol. 1994;9:191–205.

64. Menei P, Capelle L, Guyotat J, Fuentes S, Assaker R, Bataille B, et al. Local and sustained delivery of 5-fluorouracil from biodegradable microspheres for the radiosensitization of malignant glioma: a randomized phase II trial. Neurosurgery. 2005;56:242–8.

65. Menei P, Daniel V, Montero-Menei C, Brouillard M, Pouplard-Barthelaix A, Benoit JP. Biodegradation and brain tissue reaction to poly(D, L-lactide-co-glycolide) microspheres. Biomaterials. 1993;14:470–8.

66. Menei P, Jadaud E, Faisant N, Boisdron-Celle M, Michalak S, Fournier D, et al. Stereotaxic implantation of 5-fluorouracil-releasing microspheres in malignant glioma. Cancer. 2004;100:405–10.

67. Muck W, Breuel HP, Kuhlmann J. The influence of age on the pharmacokinetics of nimodipine. Int J Clin Pharmacol Ther. 1996;34:293–8.

68. Murray GD, Butcher I, McHugh GS, Lu J, Mushkudiani NA, Maas AI, et al. Multivariable prognostic analysis in traumatic brain injury: results from the IMPACT study. J Neurotrauma. 2007;24:329–37.

69. Nieuwkamp DJ, Setz LE, Algra A, Linn FH, de Rooij NK, Rinkel GJ. Changes in case fatality of aneurysmal subarachnoid haemorrhage over time, according to age, sex, and region: a meta-analysis. Lancet Neurol. 2009;8:635–42.

70. NIH Consensus Development Panel on Rehabilitation of Persons with Traumatic Brain Injury. Consensus conference. Rehabilitation of persons with traumatic brain injury. JAMA. 1999;282:974–83.

71. Nikitina E, Kawashima A, Takahashi M, Zhang ZD, Shang X, Ai J, et al. Alteration in voltage-dependent calcium channels in dog basilar artery after subarachnoid hemorrhage. Laboratory investigation. J Neurosurg. 2010;113:870–80.

72. Oertel M, Boscardin WJ, Obrist WD, Glenn TC, McArthur DL, Gravori T, et al. Posttraumatic vasospasm: the epidemiology, severity, and time course of an underestimated phenomenon: a prospective study performed in 299 patients. J Neurosurg. 2005;103:812–24.

73. Ohkuma H, Suzuki S. Histological dissociation between intra- and extraparenchymal portion of perforating small arteries after experimental subarachnoid hemorrhage in dogs. Acta Neuropathol. 1999;98:374–82.

74. Omeis I, Jayson NA, Murali R, Abrahams JM. Treatment of cerebral vasospasm with biocompatible controlled-release systems for intracranial drug delivery. Neurosurgery. 2008;63:1011–9.

75. Park S, Yamaguchi M, Zhou C, Calvert JW, Tang J, Zhang JH. Neurovascular protection reduces early brain injury after subarachnoid hemorrhage. Stroke. 2004;35:2412–7.

76. Pennings FA, Bouma GJ, Ince C. Direct observation of the human cerebral microcirculation during aneurysm surgery reveals increased arteriolar contractility. Stroke. 2004;35:1284–8.

77. Perkins E, Kimura H, Parent AD, Zhang JH. Evaluation of the microvasculature and cerebral ischemia after experimental subarachnoid hemorrhage in dogs. J Neurosurg. 2002;97: 896–904.

78. Pluta RM, Hansen-Schwartz J, Dreier J, Vajkoczy P, Macdonald RL, Nishizawa S, et al. Cerebral vasospasm following subarachnoid hemorrhage: time for a new world of thought. Neurol Res. 2009;31:151–8.

79. Pluta RM, Oldfield EH, Boock RJ. Reversal and prevention of cerebral vasospasm by intracarotid infusions of nitric oxide donors in a primate model of subarachnoid hemorrhage. J Neurosurg. 1997;87:746–51.

80. Polin RS, Coenen VA, Hansen CA, Shin P, Baskaya MK, Nanda A, et al. Efficacy of transluminal angioplasty for the management of symptomatic cerebral vasospasm following aneurysmal subarachnoid hemorrhage. J Neurosurg. 2000;92:284–90.

81. Pradilla G, Thai QA, Legnani FG, Clatterbuck RE, Gailloud P, Murphy KP, et al. Local delivery of ibuprofen via controlled-release polymers prevents angiographic vasospasm in a monkey model of subarachnoid hemorrhage. Neurosurgery. 2005;57:184–90.

82. Reilly C, Amidei C, Tolentino J, Jahromi BS, Macdonald RL. Clot volume and clearance rate as independent predictors of vasospasm after aneurysmal subarachnoid hemorrhage. J Neurosurg. 2004;101:255–61.

83. Romano JG, Forteza AM, Concha M, Koch S, Heros RC, Morcos JJ, et al. Detection of microemboli by transcranial Doppler ultrasonography in aneurysmal subarachnoid hemorrhage. Neurosurgery. 2002;50:1026–30.

84. Romano JG, Rabinstein AA, Arheart KL, Nathan S, Campo-Bustillo I, Koch S, et al. Microemboli in aneurysmal subarachnoid hemorrhage. J Neuroimaging. 2008;18:396–401.

85. Rosengart AJ, Schultheiss KE, Tolentino J, Macdonald RL. Prognostic factors for outcome in patients with aneurysmal subarachnoid hemorrhage. Stroke. 2007;38:2315–21.

86. Sabri M, Ai J, Knight B, Tariq A, Jeon H, Shang X, et al. Uncoupling of endothelial nitric oxide synthase after experimental subarachnoid hemorrhage. J Cereb Blood Flow Metab. 2011;31:190–9.

87. Schmidt JF, Waldemar G, Paulson OB. The acute effect of nimodipine on cerebral blood flow, its CO_2 reactivity, and cerebral oxygen metabolism in human volunteers. Acta Neurochir (Wien). 1991;111:49–53.

88. Scholler K, Trinkl A, Klopotowski M, Thal SC, Plesnila N, Trabold R, et al. Characterization of microvascular basal lamina damage and blood-brain barrier dysfunction following subarachnoid hemorrhage in rats. Brain Res. 2007;1142:237–46.

89. Sehba FA, Friedrich V. Early micro vascular changes after subarachnoid hemorrhage. Acta Neurochir Suppl. 2011;110:49–55.

90. Sehba FA, Mostafa G, Knopman J, Friedrich Jr V, Bederson JB. Acute alterations in microvascular basal lamina after subarachnoid hemorrhage. J Neurosurg. 2004;101:633–40.

91. Server A, Dullerud R, Haakonsen M, Nakstad PH, Johnsen UL, Magnaes B. Post-traumatic cerebral infarction. Neuroimaging findings, etiology and outcome. Acta Radiol. 2001;42:254–60.

92. Shiokawa K, Kasuya H, Miyajima M, Izawa M, Takakura K. Prophylactic effect of papaverine prolonged-release pellets on cerebral vasospasm in dogs. Neurosurgery. 1998;42: 109–15.

93. Siironen J, Juvela S, Varis J, Porras M, Poussa K, Ilveskero S, et al. No effect of enoxaparin on outcome of aneurysmal subarachnoid hemorrhage: a randomized, double-blind, placebo-controlled clinical trial. J Neurosurg. 2003;99:953–9.

94. Smith WS, Dowd CF, Johnston SC, Ko NU, DeArmond SJ, Dillon WP, et al. Neurotoxicity of intra-arterial papaverine preserved with chlorobutanol used for the treatment of cerebral vasospasm after aneurysmal subarachnoid hemorrhage. Stroke. 2004;35:2518–22.

95. Soustiel JF, Shik V. Posttraumatic basilar artery vasospasm. Surg Neurol. 2004;62:201–6.

96. Stein SC, Browne KD, Chen XH, Smith DH, Graham DI. Thromboembolism and delayed cerebral ischemia after subarachnoid hemorrhage: an autopsy study. Neurosurgery. 2006;59: 781–7.

97. Suzuki H, Hasegawa Y, Kanamaru K, Zhang JH. Mitogen-activated protein kinases in cerebral vasospasm after subarachnoid hemorrhage: a review. Acta Neurochir Suppl. 2011;110: 133–9.

98. Takanashi Y, Ishida T, Meguro T, Kiwada H, Zhang JH, Yamamoto I. Efficacy of intrathecal liposomal fasudil for experimental cerebral vasospasm after subarachnoid hemorrhage. Neurosurgery. 2001;48:894–900.

99. Tierney TS, Pradilla G, Wang PP, Clatterbuck RE, Tamargo RJ. Intracranial delivery of the nitric oxide donor diethylenetriamine/nitric oxide from a controlled-release polymer: toxicity in cynomolgus monkeys. Neurosurgery. 2006;58:952–60.

100. Titova E, Ostrowski RP, Zhang JH, Tang J. Experimental models of subarachnoid hemorrhage for studies of cerebral vasospasm. Neurol Res. 2009;31:568–81.

101. Uhl E, Lehmberg J, Steiger HJ, Messmer K. Intraoperative detection of early microvasospasm in patients with subarachnoid hemorrhage by using orthogonal polarization spectral imaging. Neurosurgery. 2003;52:1307–15.

102. van den Bergh WM, Algra A, Dorhout Mees SM, Van Kooten F, Dirven CM, van Gijn J, et al. Randomized controlled trial of acetylsalicylic acid in aneurysmal subarachnoid hemorrhage: the MASH Study. Stroke. 2006;37:2326–30.

103. van Gijn J, Rinkel GJ. Subarachnoid haemorrhage: diagnosis, causes and management. Brain. 2001;124:249–78.

104. Velthuis BK, Rinkel GJ, Ramos LM, Witkamp TD, van der Sprenkel JW, Vandertop WP, et al. Subarachnoid hemorrhage: aneurysm detection and preoperative evaluation with CT angiography. Radiology. 1998;208:423–30.

105. Vergouwen MD, Bakhtiari K, van Geloven N, Vermeulen M, Roos YB, Meijers JC. Reduced ADAMTS13 activity in delayed cerebral ischemia after aneurysmal subarachnoid hemorrhage. J Cereb Blood Flow Metab. 2009;29:1734–41.

106. Vergouwen MD, de Haan RJ, Vermeulen M, Roos YB. Effect of statin treatment on vasospasm, delayed cerebral ischemia, and functional outcome in patients with aneurysmal subarachnoid hemorrhage: a systematic review and meta-analysis update. Stroke. 2010;41:e47–52.

107. Vergouwen MD, Etminan N, Ilodigwe D, Macdonald RL. Lower incidence of cerebral infarction correlates with improved functional outcome after aneurysmal subarachnoid hemorrhage. J Cereb Blood Flow Metab. 2011;31:1545–53.

108. Vergouwen MD, Vermeulen M, Coert BA, Stroes ES, Roos YB. Microthrombosis after aneurysmal subarachnoid hemorrhage: an additional explanation for delayed cerebral ischemia. J Cereb Blood Flow Metab. 2008;28:1761–70.

109. Vergouwen MD, Vermeulen M, de Haan RJ, Levi M, Roos YB. Dihydropyridine calcium antagonists increase fibrinolytic activity: a systematic review. J Cereb Blood Flow Metab. 2007;27:1293–308.

110. Vergouwen MD, Vermeulen M, van Gijn J, Rinkel GJ, Wijdicks EF, Muizelaar JP, et al. Definition of delayed cerebral ischemia after aneurysmal subarachnoid hemorrhage as an outcome event in clinical trials and observational studies: proposal of a multidisciplinary research group. Stroke. 2010;41:2391–5.

111. Weyer GW, Nolan CP, Macdonald RL. Evidence-based cerebral vasospasm management. Neurosurg Focus. 2006;21:E8.

112. Wong GK, Yeung JH, Graham CA, Zhu XL, Rainer TH, Poon WS. Neurological outcome in patients with traumatic brain injury and its relationship with computed tomography patterns of traumatic subarachnoid hemorrhage. J Neurosurg. 2011;114:1510–5.

113. Wurm G, Tomancok B, Nussbaumer K, Adelwohrer C, Holl K. Reduction of ischemic sequelae following spontaneous subarachnoid hemorrhage: a double-blind, randomized comparison of enoxaparin versus placebo. Clin Neurol Neurosurg. 2004;106:97–103.
114. Xie A, Aihara Y, Bouryi VA, Nikitina E, Jahromi BS, Zhang ZD, et al. Novel mechanism of endothelin-1-induced vasospasm after subarachnoid hemorrhage. J Cereb Blood Flow Metab. 2007;27:1692–701.
115. Yundt KD, Grubb Jr RL, Diringer MN, Powers WJ. Autoregulatory vasodilation of parenchymal vessels is impaired during cerebral vasospasm. J Cereb Blood Flow Metab. 1998;18:419–24.
116. Zhou C, Yamaguchi M, Colohan AR, Zhang JH. Role of p53 and apoptosis in cerebral vasospasm after experimental subarachnoid hemorrhage. J Cereb Blood Flow Metab. 2005;25:572–82.

Chapter 33
Therapeutic Potential of Intranasal Delivery of Drugs and Cells for Stroke and Other Neurological Diseases

Heyu Chen, Caibin Sheng, Weiliang Xia, and Weihai Ying

Abstract Although numerous studies have suggested the pathological mechanisms underlying stroke-induced brain damage, most clinical trials on the drug treatment of ischemic stroke have been unsuccessful. One of the key obstacles for establishing effective therapies for stroke and other neurological diseases is the blockage of entrance of drugs and therapeutic cells into the brain by the blood-brain barriers (BBB). A number of studies have suggested that intranasal drug delivery is a promising approach for effectively delivering drugs into the brain by bypassing the BBB. There may be at least one intracellular transport-mediated route and two extracellular transport-mediated routes for the nose-to-brain delivery. Recent studies have further suggested that intranasal delivery may also deliver therapeutic cells into the brain more effectively and less invasively compared to traditional approaches. However, multiple key questions regarding intranasal drug and cell delivery for treating neurological disorders remain unanswered. Future studies on intranasal delivery in humans as well as the mechanisms underlying the intranasal delivery may suggest novel biological mechanisms and markedly enhance our capacity of treating stroke and other neurological diseases.

1 Introduction

Stroke is one of the leading causes of death and disability around the world. During the last 30 years, numerous studies have significantly advanced our understanding regarding the mechanisms underlying stroke-induced brain injury. However,

H. Chen • C. Sheng • W. Xia • W. Ying, PhD (✉)
School of Biomedical Engineering and Med-X Research Institute, Shanghai Jiao Tong University, 1954 Huashan Road, Shanghai 200030, P.R. China
e-mail: weihaiy@sjtu.edu.cn

P.A. Lapchak and J.H. Zhang (eds.), *Translational Stroke Research*, Springer Series in Translational Stroke Research, DOI 10.1007/978-1-4419-9530-8_33, © Springer Science+Business Media, LLC 2012

nearly all of the clinical trials on the drug therapies for ischemic stroke have failed. These failures have indicated that our understanding regarding the mechanisms underlying stroke-induced brain damage is far from sufficient. Tissue plasminogen activator (tPA), the only FDA-approved drug for treating ischemic stroke, can be used only by a small percentage of the patients due to its short therapeutic window and its toxic side effects [1, 2]. Considering the facts that the aging population is rapidly increasing around the world and the mounting medical cost has been becoming enormous burdens for most families, it is urgent to further expose the mechanisms underlying the pathological mechanisms of stroke and to search for novel therapeutic strategies for the disease.

A key obstacle for developing effective drugs for treating stroke and other neurological diseases is the limitations of drug entrance into the brain by the blood-brain barrier (BBB) [3]. It has been estimated that over 98% of all small-molecule drugs cannot cross the BBB, and approximately 100% of large-molecule drugs cannot cross the BBB. Therefore, it is critically important to search for drug delivery strategies that can efficiently deliver drugs into the CNS.

Delivery of therapeutic cells into the brain is an important strategy for treating neurological diseases, particularly with the rapidly increasing promise of stem cell therapy. For successful clinical applications of this strategy, a pivotal and most challenging task is to develop approaches to effectively and safely deliver cells into the brain. Currently, there are three major approaches for delivering cells into the brain, including cell transplantation, intravenous (i.v.) delivery of cells, and intraarterial (i.a.) delivery of cells. However, all of these approaches have significant limitations. Surgical transplantation of therapeutic cells into the brain can not lead to sufficient survival of transplanted cells, and it can induce inflammation, edema, and reactive gliosis in the brain [4–6]. Compared to surgical transplantation of cells, i.v. or i.a. delivery of cells is less invasive. However, the cells delivered intravenously may be either entrapped in such peripheral organs as the liver or dispersed systemically [7, 8]. Intraarterial administration of cells is a relatively efficient way to deliver cells to targeted brain regions [9]. However, it may lead to occlusion of microvasculatures, thus altering cerebral blood flow.

An increasing number of experimental studies have suggested that intranasal delivery could be used to deliver both drugs and cells into the CNS by bypassing the BBB. Due to the crucial significance of effective delivery of drugs and cells into the brain for treating stroke and other neurological diseases, it is of both theoretical and clinical significance to further determine the efficacy of intranasal delivery as well as the mechanisms underlying the delivery of drugs and cells into the brain by intranasal approach. In this chapter, we provide an overview of the current status of this increasingly interesting and important topic, which may suggest directions for the future research on this topic.

2   Intranasal Drug Delivery into the Brain

Many studies have suggested that intranasal administration may enable substances to enter the brain by pathways involving olfactory epithelium and olfactory bulb since 1970s [10]. In 1995, a study reported that intranasal administration of wheat germ agglutinin-horseradish peroxidase (WGA-HRP) led to significant presence of WGA-HRP in the olfactory bulb of rats, while no detectable amount of WGA-HRP was found in the olfactory bulb after intravenous (i.v.) injection of the same concentration of WGA-HRP [11]. Following this report, a number of studies have supported the notion that intranasal administration is capable of delivering large-sized molecules into the brain, at least in part through direct nose-to-brain routes [10, 12–17]. These molecules include fibroblast growth factor-2, transforming growth factor-β1, erythropoietin, insulin-like growth factor-1 (IGF-1), HIV-1 Tat, insulin, interferon β, and leptin.

Latest studies have further suggested that multiple molecules of various sizes can be effectively delivered into the brain to decrease the brain damage in animal models of stroke. These molecules include capase-9 inhibitor [18], nerve growth factor [19], vascular endothelial growth factor [20, 21], deferoxamine [22], and osteopontin peptide mimetics [23].

Multiple studies using animal models of stroke and other neurological diseases have shown that intranasal delivery of large-sized molecules can produce beneficial effects: Administration of erythropoietin [16] or IGF-I [24] by the intranasal approach can significantly decrease ischemic brain damage, and intranasal delivery of growth factors can also increase neurogenesis in the rat brain [17]. Intranasal NGF administration was also shown to attenuate memory deficits and neurodegeneration in transgenic models of Alzheimer's disease (AD) [25].

Our group has conducted studies regarding the efficacy of intranasal drug delivery for treating brain ischemia, which have suggested the promise of intranasal drug delivery for treating the disorder. We found that intranasal NAD$^+$ administration decreased by approximately 90% the infarct formation when conducted at 2 h after ischemia, which is one of the exceptional protective effects ever reported when drugs are administered at hours after ischemic onset [26]. In contrast, i.v. injection of NAD$^+$ at the same dose did not significantly decrease the ischemic brain damage when the same number of rats was used (data not published). Based on our cell culture studies regarding the protective effects of gallotannin (GT)—an inhibitor of poly(ADP-ribose) glycohydrolase—on oxidative neural cell death [27], we conducted a study that compares the effect of intranasal delivery of GT on ischemic brain injury with that of i.v. injection of GT [28]. We found that intranasal administration of GT produced marked protection against ischemic brain damage in a rat model of transient brain ischemia. In contrast, i.v. injection of GT at the same dose was toxic to the animals, while i.v. injection of GT at lower doses could not produce protective effects.

An important current advance regarding intranasal drug delivery has been made by the studies that apply nanoparticles to further enhance the efficacy of intranasal drug delivery. For example, a latest study compared the efficacy of intranasal delivery of olanzapine and that of intranasal delivery of the drug with concurrent application of poly(lactic-co-glycolic acid) nanoparticles [29]. Their in vivo pharmacokinetic studies showed that, compared to intranasal delivery of olanzapine, significantly higher uptake of the drug was produced by the intranasal delivery of the drug with concurrent application of poly(lactic-co-glycolic acid) nanoparticles. However, there have been few stroke studies that have applied strategies to further enhance the efficacy of intranasal drug delivery.

Our recent study has compared the entrance routes and metabolism pathways of intranasal delivery of fluorescent Cy5.5 in rats with those of i.v. injection of Cy5.5, indicating major differences between these two drug delivery approaches (unpublished observations): First, after the intranasal drug delivery, there was a major increase in Cy5.5 mainly in the stomach; in contrast, after the i.v. injection of Cy5.5, there was Cy5.5 mainly in the liver. Second, only after the intranasal drug delivery, there was the following characteristic change in the Cy5.5 signals: Cy5.5 signals were first increased in the olfactory bulbs, followed by increases in the Cy5.5 signals in the other brain regions. These observations support the proposition that the olfactory bulb is a key point in the routes by which intranasally delivered drugs enter the brain. The observations also suggest that a major portion of intranasally delivered drugs do not enter the blood, thus avoiding the entrapment of the drugs in the liver.

Multiple studies have also suggested that intranasal drug delivery may also produce beneficial effects on human subjects: Intranasal administration of oxytocin can enhance trust in human subjects [30]; insulin administration by the intranasal approach can not only improve the memory and mood of healthy adults [31], but also improve the memory of AD patients without altering blood levels of insulin or glucose [14].

3   Intranasal Delivery of Cells to the Brain

Recent studies have suggested that intranasal delivery could provide a novel and noninvasive approach for cell delivery into the CNS. A study reported that mesenchymal stem cells (MSCs) can be delivered to the brains of rodents by intranasal application, which were detected in the olfactory bulb, thalamus, hippocampus, cortex, and cerebellum. It was further shown that the MSC delivery is enhanced by intranasal application of hyaluronidase prior to the cell administration. The glioma cells can also be delivered into the brains of rats through migration from the nasal mucosa to different regions of the brain such as the olfactory bulb, frontal cortex, and hippocampus. The researchers have suggested two migration pathways: The first is migration into the olfactory bulb and subsequently to the other regions of the brain, and the second is the entry of the cells into the CSF with movement along the surface of the cortex, followed by cell entrance into the brain parenchyma.

Other hypothetical routes for the cell entrance into the brain include the trigeminal route and the perivascular route [32].

It has been reported that intranasal application of bone marrow MSCs significantly improved the outcome and reduced neuronal and white matter loss in a model of hypoxia/ischemia (HI) brain damage. One of the mechanisms underlying the beneficial effects is increases in the levels of multiple growth and differentiation factors such as neuronal growth factor and fibroblast growth factor 2, which can stimulate endogenous repair mechanisms and suppress the expressions of proinflammatory factors [33]. However, it remains unclear if intranasal application of stem cells can produce major beneficial effects on brain ischemia.

Intranasal application of MSCs has also shown therapeutic potential in a PD animal model: The MSCs delivered intranasally to the brain showed long-term survival and proliferation. The population of proliferated cells was located predominantly in the lesioned area (ipsilateral striatum and substantia nigra). Intranasal application of MSCs led to increase in TH and dopamine levels as well as decreases in TUNEL staining signals in the lesioned areas of the host tissues. The cell delivery also significantly improved motor function. The beneficial effects of the intranasal application of MSCs could at least partially result from the anti-inflammatory effects of MSCs since the cell delivery led to decreases in multiple inflammatory cytokines in the lesioned side [34].

4   Potential Mechanisms Underlying the Efficacy of Intranasal Drug Delivery

There are three likely mechanisms underlying the direct nose-to-brain drug delivery [10, 13, 14, 16, 17]. The two likely extracellular transport-based routes could account for the rapid entrance of drugs into the brain, which can occur within minutes of intranasal drug administration [13, 35]. In the first route, substances delivered by intranasal administration could first cross the gaps between the olfactory neurons in olfactory epithelium, which are subsequently transported into olfactory bulb. In the second extracellular transport-based route, intranasally administered substances may be transported along the trigeminal nerve to bypass the BBB [13, 35]. The substances may enter into other brain regions by diffusion after reaching olfactory bulb or trigeminal region. In addition, intranasally administered drugs may also partially enter into the CNS after the drugs enter into the systemic blood circulation from the nose [35]. The intracellular transport-based route is a relatively slow process, taking hours for drugs to reach the olfactory bulb for IN delivery through this route: The olfactory neurons in the olfactory epithelium could uptake the molecules by processes such as endocytosis, which could be transported to the olfactory bulb through axonal transport [10, 13, 14, 16, 17].

5 Conclusion Remarks and Future Perspectives

In summary, cumulating evidence has suggested that intranasal drug administration could enable drugs to directly enter into the CNS through such pathways as olfactory pathways or trigeminal nerve. Recent studies have also suggested that intranasal administration could also deliver therapeutic cells into the CNS. These studies have highlighted great potential of intranasal drug and cell delivery for treating stroke and other neurological diseases.

Intranasal delivery has multiple significant advantages, compared with traditional drug and cell delivery approaches: (1) Both drugs of various sizes and cells could be delivered into the brain by bypassing the BBB, (2) potential side effects of the drugs on the peripheral system could be reduced, (3) the noninvasiveness of intranasal drug delivery can minimize the pain of patients, and (4) smaller amount of drugs is needed to deliver desired concentrations of drugs in the CNS, which could decrease therapeutic cost.

The understanding of the nose-to-brain routes could significantly improve our knowledge regarding the pathways by which substances may enter the CNS. This understanding may also lead to exposures of the pathogenic mechanisms of neurological diseases. For example, the herpes simplex virus type 1 existing in the brain of the carriers of *APOE4* may be a significant risk factor of AD. Therefore, elucidation of the nose-to-brain routes in humans may enhance our understanding on the mechanism of the virus entrance into the CNS.

Although major progress has been made regarding intranasal drug and cell delivery for stroke and other neurological disorders, numerous key questions on this topic remain unanswered. The following studies may be of particular interest and significance:

1. Previous studies have suggested potential pathways to further increase the efficacy of intranasal drug delivery [29]. However, these types of studies are distinctly insufficient. It is warranted to further improve the methods for applying the intranasal drug delivery to treat stroke.
2. To conduct comprehensive studies that compare intranasal drug delivery with other approaches of drug delivery for treating stroke. Multiple aspects of the effects of the drug delivery should be compared, including the efficacy, drug concentrations in the brain, routes, and side effects.
3. To further investigate the mechanisms underlying intranasal drug and cell delivery, particularly those mechanisms in human.
4. It is needed to conduct clinical trials to determine if intranasal drug delivery may be used to treat stroke and neurological diseases. These studies may lead to fundamental breakthrough in therapeutic strategies for the diseases.

Acknowledgments This study was supported by a National Key Basic Research "973 Program" Grant #2010CB834306 (to WY and WX) and a Pujiang Scholar Program Award 09PJ1405900 (to WY).

References

1. Lyden PD. Further randomized controlled trials of tPA within 3 hours are required-not! Stroke. 2001;32:2709–10.
2. Davalos A. Thrombolysis in acute ischemic stroke: successes, failures, and new hopes. Cerebrovasc Dis. 2005;20 Suppl 2:135–9.
3. Pardridge WM. The blood-brain barrier: bottleneck in brain drug development. NeuroRx. 2005;2:3–14.
4. Finsen BR, Sorensen T, Castellano B, Pedersen EB, Zimmer J. Leukocyte infiltration and glial reactions in xenografts of mouse brain tissue undergoing rejection in the adult rat brain. A light and electron microscopical immunocytochemical study. J Neuroimmunol. 1991;32:159–83.
5. McKeon RJ, Schreiber RC, Rudge JS, Silver J. Reduction of neurite outgrowth in a model of glial scarring following CNS injury is correlated with the expression of inhibitory molecules on reactive astrocytes. J Neurosci. 1991;11:3398–411.
6. Perry VH, Andersson PB, Gordon S. Macrophages and inflammation in the central nervous system. Trends Neurosci. 1993;16:268–73.
7. Hauger O, Frost EE, van Heeswijk R, Deminiere C, Xue R, Delmas Y, Combe C, Moonen CT, Grenier N, Bulte JW. MR evaluation of the glomerular homing of magnetically labeled mesenchymal stem cells in a rat model of nephropathy. Radiology. 2006;238:200–10.
8. Kraitchman DL, Tatsumi M, Gilson WD, Ishimori T, Kedziorek D, Walczak P, Segars WP, Chen HH, Fritzges D, Izbudak I, Young RG, Marcelino M, Pittenger MF, Solaiyappan M, Boston RC, Tsui BM, Wahl RL, Bulte JW. Dynamic imaging of allogeneic mesenchymal stem cells trafficking to myocardial infarction. Circulation. 2005;112:1451–61.
9. Walczak P, Zhang J, Gilad AA, Kedziorek DA, Ruiz-Cabello J, Young RG, Pittenger MF, van Zijl PC, Huang J, Bulte JW. Dual-modality monitoring of targeted intraarterial delivery of mesenchymal stem cells after transient ischemia. Stroke. 2008;39:1569–74.
10. Illum L. Transport of drugs from the nasal cavity to the central nervous system. Eur J Pharm Sci. 2000;11:1–18.
11. Thorne RG, Emory CR, Ala TA, Frey II WH. Quantitative analysis of the olfactory pathway for drug delivery to the brain. Brain Res. 1995;692:278–82.
12. Ma YP, Ma MM, Ge S, Guo RB, Zhang HJ, Frey II WH, Xu GL, Liu XF. Intranasally delivered TGF-beta1 enters brain and regulates gene expressions of its receptors in rats. Brain Res Bull. 2007;74:271–7.
13. Thorne RG, Pronk GJ, Padmanabhan V, Frey II WH. Delivery of insulin-like growth factor-I to the rat brain and spinal cord along olfactory and trigeminal pathways following intranasal administration. Neuroscience. 2004;127:481–96.
14. Reger MA, Watson GS, Frey II WH, Baker LD, Cholerton B, Keeling ML, Belongia DA, Fishel MA, Plymate SR, Schellenberg GD, Cherrier MM, Craft S. Effects of intranasal insulin on cognition in memory-impaired older adults: modulation by APOE genotype. Neurobiol Aging. 2006;27:451–8.
15. Pulliam L, Sun B, Rempel H, Martinez PM, Hoekman JD, Rao RJ, Frey II WH, Hanson LR. Intranasal Tat alters gene expression in the mouse brain. JNIP. 2007;2:87–92.
16. Yu YP, Xu QQ, Zhang Q, Zhang WP, Zhang LH, Wei EQ. Intranasal recombinant human erythropoietin protects rats against focal cerebral ischemia. Neurosci Lett. 2005;387:5–10.
17. Jin K, Xie L, Childs J, Sun Y, Mao XO, Logvinova A, Greenberg DA. Cerebral neurogenesis is induced by intranasal administration of growth factors. Ann Neurol. 2003;53:405–9.
18. Akpan N, Serrano-Saiz E, Zacharia BE, Otten ML, Ducruet AF, Snipas SJ, Liu W, Velloza J, Cohen G, Sosunov SA, Frey II WH, Salvesen GS, Connolly Jr ES, Troy CM. Intranasal delivery of caspase-9 inhibitor reduces caspase-6-dependent axon/neuron loss and improves neurological function after stroke. J Neurosci. 2011;31:8894–904.
19. Zhu W, Cheng S, Xu G, Ma M, Zhou Z, Liu D, Liu X. Intranasal nerve growth factor enhances striatal neurogenesis in adult rats with focal cerebral ischemia. Drug Deliv. 2011;18:338–43.

20. Yang JP, Liu HJ, Wang ZL, Cheng SM, Cheng X, Xu GL, Liu XF. The dose-effectiveness of intranasal VEGF in treatment of experimental stroke. Neurosci Lett. 2009;461:212–6.
21. Yang JP, Liu HJ, Liu XF. VEGF promotes angiogenesis and functional recovery in stroke rats. J Investig Surg. 2010;23:149–55.
22. Hanson LR, Roeytenberg A, Martinez PM, Coppes VG, Sweet DC, Rao RJ, Marti DL, Hoekman JD, Matthews RB, Frey WH, Panter II SS. Intranasal deferoxamine provides increased brain exposure and significant protection in rat ischemic stroke. J Pharmacol Exp Ther. 2009;330:679–86.
23. Doyle KP, Yang T, Lessov NS, Ciesielski TM, Stevens SL, Simon RP, King JS, Stenzel-Poore MP. Nasal administration of osteopontin peptide mimetics confers neuroprotection in stroke. J Cereb Blood Flow Metab. 2008;28:1235–48.
24. Liu XF, Fawcett JR, Thorne RG, Frey II WH. Non-invasive intranasal insulin-like growth factor-I reduces infarct volume and improves neurologic function in rats following middle cerebral artery occlusion. Neurosci Lett. 2001;308:91–4.
25. De Rosa R, Garcia AA, Braschi C, Capsoni S, Maffei L, Berardi N, Cattaneo A. Intranasal administration of nerve growth factor (NGF) rescues recognition memory deficits in AD11 anti-NGF transgenic mice. Proc Natl Acad Sci U S A. 2005;102:3811–6.
26. Ying W, Garnier P, Swanson RA. NAD + repletion prevents PARP-1-induced glycolytic blockade and cell death in cultured mouse astrocytes. Biochem Biophys Res Commun. 2003;308:809–13.
27. Ying W, Sevigny MB, Chen Y, Swanson RA. Poly (ADP-ribose) glycohydrolase mediates oxidative and excitotoxic neuronal death. Proc Natl Acad Sci U S A. 2001;98:12227–32.
28. Wei G, Wang D, Lu H, Parmentier S, Wang Q, Panter SS, Frey WH, Ying II W. Intranasal administration of a PARG inhibitor profoundly decreases ischemic brain injury. Front Biosci. 2007;12:4986–96.
29. Seju U, Kumar A, Sawant KK. Development and evaluation of olanzapine-loaded PLGA nanoparticles for nose-to-brain delivery: in vitro and in vivo studies. Acta Biomater. 2011;12:4169–76.
30. Kosfeld M, Heinrichs M, Zak PJ, Fischbacher U, Fehr E. Oxytocin increases trust in humans. Nature. 2005;435:673–6.
31. Benedict C, Hallschmid M, Schultes B, Born J, Kern W. Intranasal insulin to improve memory function in humans. Neuroendocrinology. 2007;86:136–42.
32. Danielyan L, Schafer R, von Ameln-Mayerhofer A, Buadze M, Geisler J, Klopfer T, Burkhardt U, Proksch B, Verleysdonk S, Ayturan M, Buniatian GH, Gleiter CH, Frey II WH. Intranasal delivery of cells to the brain. Eur J Cell Biol. 2009;88:315–24.
33. van Velthoven CT, Kavelaars A, van Bel F, Heijnen CJ. Nasal administration of stem cells: a promising novel route to treat neonatal ischemic brain damage. Pediatr Res. 2010;68:419–22.
34. Danielyan L, Schafer R, von Ameln-Mayerhofer A, Bernhard F, Verleysdonk S, Buadze M, Lourhmati A, Klopfer T, Schaumann F, Schmid B, Koehle C, Proksch B, Weissert R, Reichardt HM, van den Brandt J, Buniatian GH, Schwab M, Gleiter CH, Frey II WH. Therapeutic efficacy of intranasally delivered mesenchymal stem cells in a rat model of Parkinson disease. Rejuvenation Res. 2011;14:3–16.
35. Dhanda DS, Frey II WH, Leopold D, Kompella UB. Nose-to-brain delivery: approaches for drug deposition in the human olfactory epithelium. Drug Del Tech. 2005;5:64–72.

Part VI
Therapy Development

Chapter 34
High-Dose Albumin for Neuroprotection in Acute Ischemic Stroke: From Basic Investigations to Multicenter Clinical Trial

Myron D. Ginsberg

Abstract Albumin (ALB), the plasma's most abundant protein, is a multifunctional molecule with potent antioxidant and intravascular actions. Albumin is in widespread clinical use to provide circulatory support in the settings of shock, burns, and surgery. More recently, we have shown that albumin when administered in high doses acts as a powerful *neuroprotective agent* in acute ischemic stroke and brain injury. In experimental studies of focal cerebral ischemia conducted in physiologically monitored rats, animals treated with ALB (typically 25% solution in doses of 1.25 g/kg and above) showed improved neurological score, substantial reductions of infarct volume, and markedly reduced brain swelling compared to saline placebo-treated rats, with a therapeutic window of 4–5 h. In mechanistic studies, ALB improved local blood flow in the ischemic penumbra; antagonized postischemic microvascular thrombosis; improved perfusion distal to microvascular thrombi; and facilitated delivery of fatty acids to the postischemic brain. In a two-center dose-escalation human pilot clinical trial, ALB was generally well tolerated; the chief adverse event was mild-to-moderate pulmonary edema in 13% of subjects, which could be readily managed. A major phase III multicenter clinical efficacy trial—the ALIAS (Albumin in Acute Stroke) Trial—is now underway in the U.S., Canada, and Israel, employing ALB at the 2 g/kg dose shown in experimental studies to be neuroprotective.

M.D. Ginsberg, MD (✉)
Department of Neurology, University of Miami Miller School of Medicine,
PO Box 016960, Miami, FL 33101, USA
e-mail: mginsberg@med.miami.edu

P.A. Lapchak and J.H. Zhang (eds.), *Translational Stroke Research*, Springer Series
in Translational Stroke Research, DOI 10.1007/978-1-4419-9530-8_34,
© Springer Science+Business Media, LLC 2012

1 Part I: Preclinical and Mechanistic Considerations

1.1 Introduction

In his eloquent Foreword to Peters' comprehensive monograph on albumin, F.W. Putnam writes

> Albumin is the most abundant soluble protein in the body of all vertebrates and is the most prominent protein in plasma. Some of its physiological properties have been recognized since the time of Hippocrates; albumin was named and first studied a century and half ago and was crystallized a century ago. Yet, the recent elucidation of its three-dimensional structure depended on crystallization in the space shuttle and recombinant technology. The physiological functions of albumin were the prime incentive for the intensive wartime program of plasma fractionation beginning in 1942 at Harvard... In peacetime this led to a national program for blood procurement and plasma fractionation and the development of other products for clinical use ... [69].

Thus, parenteral albumin is in widespread clinical use to provide circulatory support in the settings of shock, burns, and surgery [1, 43, 69]. The intent of the present contribution, however, is to focus solely and comprehensively on the recently discovered role of human albumin as a powerful *neuroprotective agent* for acute ischemic stroke and brain injury [31], by describing a scientific journey that began in the experimental animal laboratory and has progressed at present to a definitive multicenter clinical trial for ischemic stroke.

1.2 The Discovery of Albumin-Neuroprotection in Focal Cerebral Ischemia

Our initial observations that high-dose human albumin confers neuroprotection in ischemic stroke arose serendipitously, in the context of an experimental focal-ischemia study conducted at the behest of a corporate sponsor interested in a proprietary, chemically modified form of albumin. Native human albumin and saline were used as controls. This study [11], and our subsequent investigations, incorporated a number of key design features intended to ensure reliability: use of a highly reproducible, minimally invasive model of temporary focal cerebral ischemia produced in rats by modified intraluminal-suture occlusion of the middle cerebral artery (MCA) with initial neurobehavioral testing *during* the occlusion period to verify a neurological deficit [9]; rigorous monitoring and control of physiological variables, including brain temperature [24]; *delayed* administration of albumin (hence, more relevant to the clinical context); suitable survival period (3 days) to engender stable histopathology; blinded assessment of all outcomes; quantification of both infarct volume, brain swelling, and neurological function; and application of a novel image-based method for the regional mapping of group infarct-data, permitting the application of tests of statistical comparison [92, 93].

In the initial study [11], anesthetized, physiologically monitored Sprague–Dawley rats received 2 h of MCA occlusion; either 20% human albumin (ALB, dose, 1% of body weight) or a comparable volume of isotonic saline was administered intravenously (IV) immediately following the 2-h MCA occlusion period. A standardized neurobehavioral battery was applied prior to ischemia and daily thereafter. Brains were perfusion-fixed for quantitative histopathology at 3 days. ALB-treated rats showed improved neurological score compared to saline ($p < 0.01$ at 24 h, trend thereafter) and a significant 34% mean reduction of total (cortical + subcortical) infarct volume (Fig. 34.1) [11]. ALB treatment was also strikingly effective in reducing the extent of brain swelling (by 81%).

We confirmed these findings in a second similar series, in which additional measures were incorporated [22]. Rats again received a 2-h period of MCA suture-occlusion, following which either ALB (25% human albumin, 1% body weight or 2.5 g/kg) or isotonic saline was administered. Diffusion-weighted magnetic resonance imaging (DWI-MRI) at 24 h revealed, in saline-treated animals, the expected ~40% decline in the apparent diffusion coefficient (ADC) within the infarct, but, surprisingly, a preservation of near-normal ADC values within the infarct of ALB-treated animals (Fig. 34.2) [22]. Correspondingly, microscopic histopathology of the infarcted zone showed pan-necrosis in saline-treated rats, but persistence of vascular endothelium and prominent microglial activation within the infarct of ALB animals, suggesting that ALB may have partially preserved the neuropil within zones of residual infarction [22]. (In a confirmatory study, we demonstrated by immunochemistry that the administered albumin could be detected within cortical neurons showing preserved structural features—consistent with ALB-related cellular protection [71]). Total neurological score remained highly impaired throughout the 3-day survival period in saline-treated animals, but improved by one-half at 24 h and beyond in ALB animals [22]. The mean volumes of cortical and subcortical infarction were reduced on average by 84 and 33%, respectively, with ALB therapy. Total infarct volume was reduced by 66% and brain swelling was virtually eliminated by ALB. In a matched series, plasma osmolality was unaffected by ALB (as expected), while colloid oncotic pressure was elevated by 56% at 15 min and by 25% at 24 h in ALB animals. Early hematocrit reductions (hemodilution) of ~40–45% were also observed [22].

Dose–response and therapeutic window: We conducted a third study in the 2-h temporary MCA suture-occlusion rat model of focal ischemia, in order to characterize the dose–response characteristics of ALB therapy and to define the therapeutic window for neuroprotection [15]. In the dose–response series, significant 58 and 67% mean reductions in total infarct volume were seen at ALB doses of 0.63 and 1.25 g/kg, respectively, administered immediately after removal of the occluding MCA suture. The extent of neuroprotection in neocortex was profound, amounting to 66 and 96% at the 0.63 and 1.25 g/kg doses, respectively (Fig. 34.3). Each of these doses led to marked reductions or elimination of brain swelling. To study the therapeutic window, we employed an ALB dose of 1.25 g/kg [15]: Even when ALB therapy was initiated as late as 4 h after onset of MCA occlusion, neurological score was

Fig. 34.1 Frequency distribution of cerebral infarction at nine stereotactic levels in rat brains treated with either saline (**a**) or ALB (**b**) at 2 h after the onset of temporary MCA occlusion (*n* = 8 rats per group). (**c**) Results of Fisher's exact test comparing the data sets of (**a**) and (**b**) on a pixel-by-pixel basis. The *color bars* in (**a**) and (**b**) depict numbers of animals with infarction at each pixel location. In (**c**), the *color bar* shows the level of statistical significance $(1-p)$, which ranges from 0.95 to 1.0 (i.e., $p < 0.05$). A confluent zone of neocortex shows a highly significant reduction of infarct volume in ALB-compared with saline-treated animals (reproduced from Belayev et al. [11] with permission)

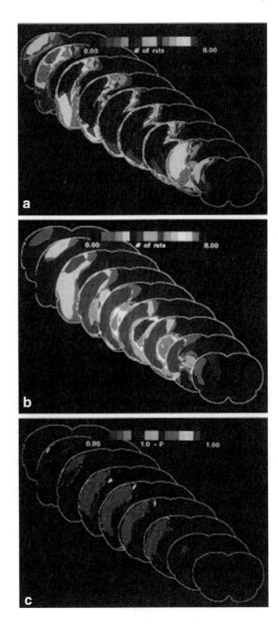

improved and mean infarct volumes were markedly reduced in cortex (68%), subcortical areas (52%), and in toto (61%). Delay of ALB treatment-initiation to 5 h, however, resulted in only minor reductions of total infarct volume and, in general, a failure to obtain significant cortical neuroprotection [15]. From these collective data, we concluded that the therapeutic window for ALB-neuroprotection in temporary focal ischemia began to close between 4 and 5 h after onset of ischemia.

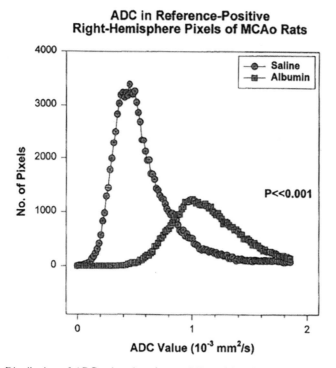

ADC in Reference-Positive Right-Hemisphere Pixels of MCAo Rats

Fig. 34.2 Distribution of ADC values in reference-MR-positive pixels of the ipsilateral hemisphere of rats with prior MCA occlusion. ("Reference-MR-positive" refers to those pixels whose intensity exceeded 2 SD of the contralateral hemisphere values.) Data represent pooled pixel data from $n = 3$ rats per group; pixels with ADC > 1.8×10^{-3} mm²/s were excluded because they were assumed to represent cerebrospinal fluid-containing spaces. The ADC distribution of rats receiving ALB treatment is markedly rightward-shifted compared with the ADC distribution in lesioned pixels of saline-treated rats ($p < 0.001$, Kolmogorov-Smirnov 2-sample test) (reproduced from Belayev et al. [22] with permission)

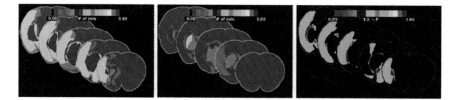

Fig. 34.3 Maps of infarct-frequency at five coronal levels in groups of rats with 2-h MCA occlusion treated with saline (*left*) or ALB (1.25 g/kg; dose–response study) (*middle*). *Right panel* is a map of Fisher's exact test, in which *colored pixels* are regions protected by ALB therapy ($p < 0.05$) (reproduced from Belayev et al. [15] with permission)

Brain swelling: An important and unequivocal finding emerging from these studies is the consistent ability of ALB to ameliorate the brain swelling (edema) associated with acute focal cerebral ischemia and postischemic recirculation [11,15,16,22]. The literature supports a beneficial role of elevated plasma oncotic pressure produced by high-dose albumin, a *colloid*, in the management of brain edema associated with contusion [83] and intracerebral hemorrhage [84]. This salutary action stands in marked contrast to the effect of *crystalloids*, which worsen brain edema and infarction [45,54].

Delayed treatment: We also considered it important to exclude the (albeit unlikely) possibility that ALB treatment of focal ischemia, if initiated in a very *delayed* fashion, might paradoxically worsen outcome. Accordingly, rats with 2-h MCA occlusion were treated with either ALB (2.5 g/kg) or saline at 19 h after the onset of ischemia. Delayed ALB treatment, while ineffective as a neuroprotective therapy, failed to show any adverse influence on behavior or histopathology [14].

Meta-analysis: We conducted a pooled meta-analysis of cortical infarct-volume data derived from several of the above-cited studies of 2-h MCA occlusion [11,15,22], in which neuroprotective ALB doses (1.25–2.5 g/kg) were administered within 2–4 h of the onset of vascular occlusion and histopathology was studied after a 3-day survival [32]. In this analysis, we rank-ordered cortical infarct volumes in individual ALB- and saline-treated rats. Each curve was fitted to a third-order polynomial, and the 95% prediction limits of each curve were computed [28]. As shown in Fig. 34.4, the ALB- and saline-treated populations are entirely distinct—confirming the consistent high-grade cortical neuroprotection conferred by ALB treatment across several series and evident throughout the spectrum of cortical-infarct size [32].

1.3 Albumin-Neuroprotection in Other Experimental Conditions

Permanent focal ischemia: In rats with permanent MCA occlusion by intraluminal suture—an insult that results in a maximal-sized MCA-territory infarction, treatment with ALB (1.25 g/kg) at 2 h after the onset of occlusion reduced the mean areas of infarction at three posterior coronal brain levels by 50–76%. Laser-Doppler perfusion imaging revealed that ALB therapy (but not saline) was associated with significant 1.8-fold increases in relative perfusion [58].

Transient global forebrain ischemia: In a rat model employing temporary (10 min) bilateral common carotid occlusions plus hypotension (50 mmHg) [77], ALB treatment at 5 min after the onset of recirculation significantly improved the neurological score compared to saline treatment throughout the 7-day survival period and also significantly attenuated hippocampal ischemic damage, resulting in a 2.4–5.3-fold increase in the number of surviving CA1 neurons compared to saline-treated animals [20].

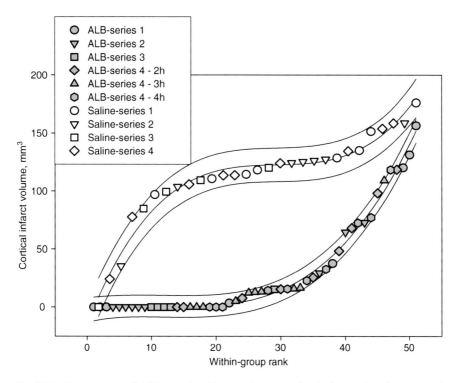

Fig. 34.4 Meta-analysis of ALB-associated histological protection in four series of rats treated with 1.25–2.5 g/kg of ALB, or saline, within the first 4 h after onset of MCA occlusion. Cortical infarct volumes were plotted in ascending rank-order separately for the ALB- and saline-treated groups and were fitted to third-order polynomials; *curves* are shown with their 95% prediction limits. The ALB- and saline-treated groups have virtually nonoverlapping prediction limits except at the lowermost and uppermost extremes (reproduced from Ginsberg et al. [32] with permission)

Traumatic brain injury (TBI): In a model of moderate fluid-percussion brain injury in rats, ALB therapy (2.5 g/kg at 15 min posttrauma) significantly improved the neurological score compared to saline-treated animals at 24 h after TBI and reduced mean contusion area by 51% [10]. As the pathophysiology of TBI involves the acute uncoupling of local cerebral glucose utilization (lCMRglu) and blood flow (lCBF), we conducted autoradiographic studies in rats with TBI followed by either 60 min or 24 h of recovery [36]. When assessed at 60 min post-TBI, saline-treated rats showed bilateral moderate reductions of lCBF but paradoxically normal levels of lCMRglu, resulting in marked bilateral metabolism/blood flow uncoupling (by threefold ipsilaterally and twofold contralaterally). By contrast, ALB treatment diminished the extent of metabolism/flow dissociation by one-half [36].

Intracerebral hematoma (ICH): In studies of acute intrastriatal hematoma produced in rats by the double blood-injection method (15 μL of fresh arterial blood over 3 min, followed 7 min later by 30 μL of blood injected over 5 min), ALB treatment

(1.25 g/kg) at 90 min following ICH resulted in significant neurobehavioral improvement over the subsequent 7-day survival period compared to saline-treated controls [17]. In other ICH studies, 50 μL of autologous nonheparinized blood was injected into the frontal cortex, and rats were treated at 60 thereafter with either ALB (1.25 g/kg) or saline. Neurobehavioral scores measured over the subsequent 48 h were markedly and progressively improved by ALB therapy, beginning within 1 h of treatment. Evans blue spectrofluorometric studies revealed marked tracer extravasation (i.e., blood–brain barrier damage) in the perihematomal cortex of saline-treated rats, but attenuated blood–brain barrier leakage (by 50%) in the ipsilateral forebrain of ALB-treated animals [19].

In a related disorder, acute aneurysmal subarachnoid hemorrhage (SAH), it is of interest that some clinicians routinely administer large amounts of albumin over many days following surgical treatment of the aneurysm, in an effort to prevent delayed cerebral ischemia secondary to vasospasm [56,80]. Suarez and Martin have recently undertaken a pilot-phase dose-finding clinical trial to assess the tolerability and safety of 25% human albumin in SAH patients (ClinicalTrials.gov identifier, NCT00283400) [79], as a prelude to a future, efficacy-oriented placebo-controlled trial for this devastating disorder.

1.4   Mechanisms of ALB-Mediated Neuroprotection

1.4.1   Cerebral Blood Flow

As ALB expands intravascular volume and produces hemodilution [22], it was natural to hypothesize that its therapeutic effect in brain ischemia might be mediated, at least in part, by salutary effects on cerebral perfusion. This was studied in a series of physiologically regulated rats receiving a 2-h period of MCA suture-occlusion, in which local cerebral blood flow (lCBF) was measured autoradiographically [73] after allowing 1 h of postischemic recirculation [48]. Three-dimensional image-alignment and -averaging methods [4,92] were used to align sequential coronal lCBF images into 3D data sets, which were coregistered in replicate animals of the ALB and saline series. These quantitative lCBF data sets were then comapped and compared on a pixel-by-pixel basis with histopathological (infarct-frequency) data sets acquired in two previous series [11,22]. Based upon our previous work in this model [4,21], we defined the *ischemic core* to be those brain regions with lCBF of 0–20% of control values and the *penumbra* to be zones in which lCBF was 20–40% of control. This analysis revealed a significant flow-range × treatment interactive effect, indicating that ALB treatment led to significant upward shifts in the overall lCBF distribution; this was due primarily to an effect of ALB in reducing the size of brain areas lying with the upper ischemic-core and lower penumbral lCBF ranges—the zones in which most infarctive histopathology had been shown to occur (Fig. 34.5) [91]. Inspection of quantitative lCBF difference images revealed that the zone of lCBF significantly improved by ALB lays in a circumferential distribution surrounding

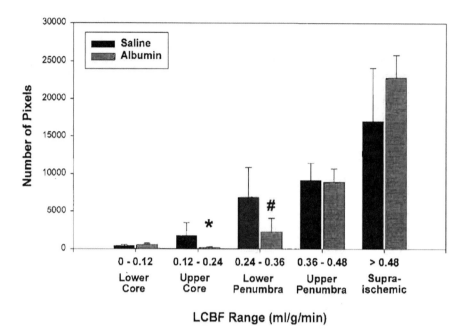

Fig. 34.5 Distribution of lCBF at 1 h of postischemic recirculation following a 2-h period of MCA occlusion, shown for those image pixels that were destined for histopathological infarction in saline-treated rats, but were significantly protected in ALB-treated animals. ALB treatment led to reduced numbers of pixels with lCBF in the upper ischemic-core and lower penumbral blood flow ranges (*$p = 0.033$, #$p = 0.055$ vs. saline group, Kruskal-Wallis tests) (reproduced from Huh et al. [48] with permission)

regions of prior core-ischemia [48]. In summary, this study showed that ALB significantly ameliorated blood flow within zones of (otherwise) critically reduced perfusion. Because of the limited spatial extent of this effect, however, we suspected that mechanisms other than cerebral perfusion also contributed to the neuroprotective effect of ALB in ischemia [48].

1.4.2 Microvascular Hemodynamics

The powerful methods of in vivo laser-scanning confocal microscopy [76] and, more recently, in vivo two-photon microscopy have permitted the real-time visualization and quantitative evaluation of pial and superficial cortical microcirculatory hemodynamics of anesthetized rats in the course of vascular occlusion, recirculation, and ALB therapy. In a collaborative in vivo confocal microscopy study, a closed cranial window was placed over the dorsolateral frontoparietal cortex of physiologically monitored, anesthetized rats receiving 2-h MCA suture-occlusion and recirculation [18]. Fluorescein-isothiocyanate (FITC)-dextran (molecular weight

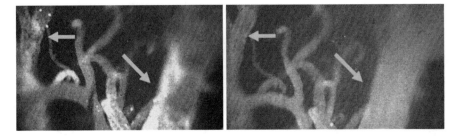

Fig. 34.6 Confocal-microscopic images of the cortical microcirculation during the recirculation period following a 2-h period of MCA occlusion. *Left panel*, at 29 min of recirculation, shows the accumulation of thrombotic material within cortical venules (*arrows*). *Right panel*: at 98 min of recirculation, 68 min after the administration of ALB, 2.5 g/kg, venular perfusion has improved, and the adherent thrombotic material has disappeared (*arrows*) (adapted from Belayev et al. [18] with permission)

70,000) was injected intravenously to visualize microvessels, and FITC-labeled erythrocytes (prepared in vitro) were also injected. Video images of cortical vessels were continually acquired and were digitized off-line to measure vessel diameters and fluorescent red-cell velocities. With the onset of MCA occlusion, arteriolar dilatation and slowing of capillary and venular perfusion were observed. Peri-infarct depolarizations were recorded, confirming the presence of an ischemic penumbra. During the first 15–30 min of recirculation initiated by suture removal following 2 h of MCA occlusion, prominent vascular stasis developed promptly within cortical venules, associated with thrombus-like stagnant foci, as well as corpuscular structures consistent in size with neutrophils adherent to venular endothelium. When ALB (1.25 g/kg) was administered intravenously after 30 min of recirculation, there was prompt improvement of venular and capillary perfusion, partial disappearance of the adherent intravenular thrombotic material, and around threefold mean increases in capillary and venular red-cell flow velocities (Fig. 34.6) [18]. By contrast, saline administration failed to affect these phenomena. These results suggested that ALB was highly effective in reversing stagnation and corpuscular adherence within the postcapillary microcirculation during the early period of postischemic recirculation.

The method of two-photon laser-scanning excitation fluorescence microscopy is also amenable to in vivo studies of the cortical microcirculation and possesses several advantages over confocal microscopy, including increased depth of tissue penetration, higher image contrast, and reduced photobleaching and photodamage to tissue [66,81]. We used this method to obtain confirmatory information as to whether ALB's neuroprotective effect was attributable, in part, to intravascular mechanisms. In this study, a frontoparietal cranial window (with intact dura) was produced in anesthetized, physiologically monitored rats, and the cortical microvasculature was imaged after plasma labeling with FITC-dextran [64]. A single cortical arteriole (20–50 μm diameter) was selected for observation, and its flow velocity was determined by repetitive line scanning along its central longitudinal axis [51].

A thrombus occupying the vessel's full cross-section was induced in this arteriole by pulsed laser irradiation delivered via the two-photon microscope. Animals were then randomly allocated to treatment, 30 min later, with either 25% ALB (2.5 g/kg) or isotonic saline. Microvascular flow velocity immediately distal to the thrombosis was repeatedly measured.

Successful thrombus induction was signaled by focal interruptions of intravascular fluorescence and by fluorescence accumulations in the overlying vessel walls. With the induction of thrombosis, distal arteriolar flow velocity declined to 10–13% of control values in all animals (Fig. 34.7). In saline-treated animals, velocity remained unchanged throughout the 60–90-min observation period. By contrast, ALB treatment caused median flow velocity to rise to 38% of control values within 10 min and to attain 61–67% of control values by 50–60 min (Fig. 34.7) [64]. This result, in the absence of visible changes in thrombus size or morphology, suggested that ALB may have reduced the extent of thrombus adhesion to the underlying endothelium or may have loosened the texture of the thrombus itself, thus permitting distal microvascular flow augmentation.

We used the arteriolar thrombosis/2-photon model in a subsequent study to explore whether an additive or synergistic effect of ALB and thrombolytic therapy might exist [68]. This possibility was initially suggested by a preliminary efficacy analysis of data from our ALIAS (Albumin in Acute Stroke) Pilot Clinical Trial (to be described fully below) [67], in which subjects receiving tissue recombinant activator (tPA) therapy showed a strong suggestion of additional neurological improvement when higher-dose ALB was also administered—a result consistent with the possibility that a component of ALB's protective effect might be mediated within the vascular compartment (e.g., by loosening microvascular thrombi and/or retarding reocclusion after pharmacological thrombolysis) [67]. Thus, we repeated the design of the above-described two-photon study, except that at 30 min after induction of arteriolar thrombosis, all animals received a sublytic dose of the thrombolytic agent, reteplase [61]. In rats receiving reteplase plus ALB at 30 min after thrombosis, median flow velocity increased promptly to 58% of control. In contrast, animals receiving reteplase plus saline showed a lesser median increase (37% of control). Repeated-measures ANOVA over the 1-h posttreatment interval showed a highly significant effect of ALB ($p = 0.01$) (Fig. 34.8) [68]. These results support the conclusion that the coadministration of ALB may confer additional benefit in the context of thrombolytic therapy for ischemic stroke.

1.5 Influence of ALB on Systemic Fatty Acid Responses to Ischemia

As albumin is the chief carrier of free fatty acids (FFAs) in plasma, in collaborative studies we compared the content and composition of plasma FFA in the femoral artery, femoral vein, and jugular vein of rats undergoing 2 h of MCA occlusion and treated with either ALB (1.25 g/kg) or saline at 1 h after the onset of

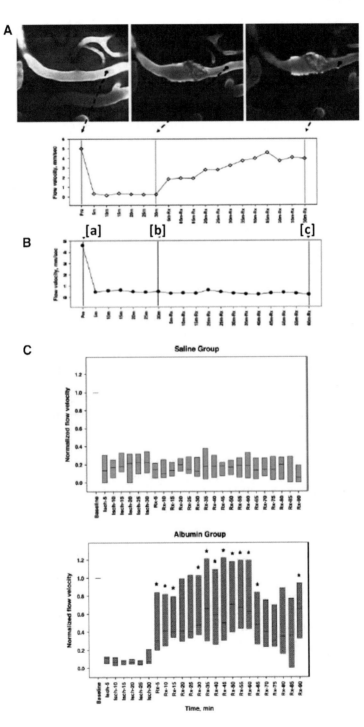

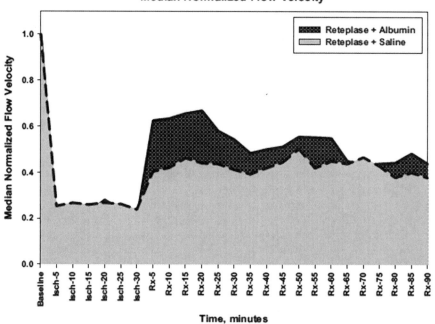

Fig. 34.8 Median arteriolar flow velocity normalized to baseline, in rats prior to and during arteriolar thrombosis followed by treatment with either reteplase plus ALB, or s reteplase plus saline, illustrating the increment in flow velocity conferred by ALB treatment during the first hour post-treatment (reproduced from Park et al. [68] with permission)

Fig. 34.7 Two-photon laser-scanning microscopy of cortical microcirculation. (**A**) *upper panels*: cortical arteriole visualized with FITC-dextran at baseline (*left*), at 30 min after laser-induced thrombosis (*middle*), and at 90 min after ALB treatment (*right*) (20× objective). Thrombosis is evident from the interruption of the plasma column and segmental dilatation. *Lower panel* presents microvascular flow velocities measured by line scanning at 5-min intervals. [**a**] onset of thrombosis; [**b**] ALB administration; [**c**] 60 min after ALB. ALB therapy resulted in a prompt, progressive improvement in microvascular perfusion velocity distal to the arteriolar thrombosis. (**B**) Micro-vascular flow velocities in a representative animal with laser-induced arteriolar thrombosis [**a**] followed 30 min later by saline administration [**b**], which failed to affect microvascular perfusion velocity over the subsequent 60 min [**c**]. (**C**) Arteriolar flow velocity, normalized to each animal's baseline value, measured at baseline, during 30-min thrombosis prior to treatment, and after treatment with either saline (*upper panel*) or ALB (*lower panel*). Bars span the 25th to 75th percentiles; *lines* within each bar denote median values. Thrombus induction led to around 80–95% declines in distal median flow velocity. Saline administration had no effect, while ALB treatment led to a rapid flow-velocity improvement. (The "step" after 60 min is because only five of the initial cohort of eight ALB rats were studied between 65 and 90 min.) *$p < 0.05$ vs. saline, Mann–Whitney rank sum test (reproduced from Nimmagadda et al. [64] with permission)

recirculation [72]. In rats with MCA occlusion and recirculation, as well as in rats with a sham-insult (no ischemia), ALB treatment induced a 1.7-fold increase in total plasma FFA during the 90-min period of observation; this increase consisted mainly of the 16:0, 18:1, and 18:2n-6 FFA species. Of greater interest, in rats with MCA occlusion, ALB stimulated the mobilization of n-3 polyunsaturated fatty acids (PUFA), with early increases in 22:5n-3 (docosapentaenoic acid) and 22:6n-3 (docosahexaenoic acid, DHA) in the femoral artery prior to detectable changes in jugular venous plasma [72]. These findings suggested that ALB mediates the systemic mobilization and supply of FFA to the ischemic brain—a process that would favor the replenishment of PUFA lost from neural membranes during ischemia [90] and/or might help to support the brain's bioenergetic demands.

The observation that MCA occlusion activates the ALB-induced mobilization of n-3 PUFA takes on relevance in view of the fact that DHA, while critical to the normal structure and function of synaptic membranes [6,7], is not synthesized by the brain, but rather must be supplied via the systemic circulation from the liver, where it is actively synthesized [8]. Thus, we hypothesized that, if ALB were complexed with DHA, it might be possible to achieve neuroprotection at lower ALB doses—a desirable outcome inasmuch as high-dose ALB expands intravascular volume and, in susceptible patients, might trigger congestive heart failure (CHF) [34]. To study this, we physically complexed DHA to human albumin to form a stable DHA-albumin complex containing around 2 μmol DHA per mL ALB [16]. Rats with 2-h MCA occlusion were treated at the onset of recirculation with either 0.63 or 1.25 g/kg of ALB, or with 0.63 or 1.25 g/kg of the DHA-albumin complex. In the 0.63 g/kg DHA-albumin group, the improved neurobehavioral score at 72 h significantly exceeded that of the other treatment groups, and the extent of histological protection (86% reduction in cortical infarction) surpassed the degree of protection produced by native ALB at 1.25 g/kg. Lipidomic analysis of DHA-albumin-treated postischemic brains (but not native ALB-treated brains) revealed a large accumulation of the DHA metabolite, 10,17S-docosatriene [16]. This DHA metabolite, termed "neuroprotection-D1," is an overall mediator of ischemic neuroprotection and a potent inhibitor of polymorphonuclear leukocyte infiltration and proinflammatory gene expression [60]. In contrast to lower-dose DHA-albumin, however, the higher-dose (1.25 g/kg) DHA-albumin complex appeared to offer no advantage over native ALB at the same concentration in reducing infarct volume—suggesting that DHA-associated neuroprotection might be dose-dependent. (It is known, for example, that at higher DHA-albumin concentrations, DHA might become a substrate for lipid peroxidation [40] or might lead to uncoupling of mitochondrial respiration [26]).

Belayev et al. have subsequently studied the neuroprotective effects of parenteral DHA administered without exogenous ALB in rats with 2-h MCA occlusion [12]. At low (3.5–7 mg/kg) and medium doses (14–35 mg/kg), DHA markedly reduced infarct volume compared to vehicle-treated rats and improved the neurological score, while no neuroprotection was obtained with the 70 mg/kg DHA dose. In a subsequent study incorporating magnetic resonance imaging, behavior, histopathology, and immunostaining, these authors confirmed the therapeutic effect of DHA

(5 mg/kg) in the 2-h MCA occlusion model; demonstrated a therapeutic window of efficacy extending to 5 h after onset of ischemia; and by lipidomic analysis showed that the synthesis of the neuroprotective metabolite of DHA, neuroprotectin D1, in the ischemic penumbra is potentiated by DHA treatment [13].

1.6   The Multifunctional Properties of the Albumin Molecule: Relevance to Neuroprotection

Albumin is a remarkable and unique plasma protein that has evolved over millions of years and possesses a multitude of attributes, many of which are potentially relevant to its neuroprotective actions in cerebral ischemic and injury. Human serum albumin is composed of 585 amino acids and has a molecular mass of 66.4 kDa [69]. In tertiary structure, it is a highly helical molecule having six subdomains contained within three primary domains [25]; by X-ray crystallography and space-filling models, the molecule is distinctively heart-shaped. It contains 35 half-cystines, which form 17 disulfide bridges that form paired loops [69]; a single thiol moiety is unpaired. Albumin is well known for its ability to bind a great variety of smaller molecules [27]; the reader is referred to the Peters monograph for a schematic showing the many binding-site locations of the albumin molecule [69].

As the most abundant circulating plasma protein, albumin subserves multiple essential homeostatic roles: the maintenance of normal plasma colloid oncotic pressure; the binding and transport of plasma fatty acids [27,85]; the transfer of cholesterol between lipoproteins and cells [94]; and the binding of multiple metabolites and drugs [52,53].

Antioxidant actions: Plasma proteins, chiefly albumin, account for around three fourths of the plasma's total radical-trapping antioxidant activity, for example, fully 10–20 times greater than the activity of vitamin E [87]. As such, albumin is a major antioxidant defense against both endogenous and exogenous oxidizing agents. Several mechanisms account for albumin's potent antioxidant efficacy: its reactive cysteine-34 thiol moiety; its ability to bind redox-active transition metals, in particular copper ions, thereby inhibiting copper ion-dependent lipid peroxidation and hydroxyl-radical formation [29,42]; and its binding of amphipathic species such as fatty acids and heme, which may participate in injurious redox reactions. The N-terminal tetrapeptide of albumin, DAHK (Asp–Ala–His–Lys), constitutes a tight binding site for Cu^{2+} ions [5]; and both human albumin and its N-terminal tetrapeptide have been shown to block oxidant-induced neuronal death in cortical cell culture [41].

Actions on vascular endothelium: Albumin binds to the endothelial glycocalyx and helps to maintain the normal permeability of microvessel walls; by its transcytosis across vascular endothelium, albumin serves as a carrier for small molecules [44]. Fatty acids and other ligands may increase albumin-binding to these receptors and facilitate transendothelial passage [30]. Albumin exerts complex effects on erythro-cyte aggregation, increasing low-shear viscosity but decreasing red-cell sedimenta-tion under conditions of low flow [69].

Albumin is also an inhibitor of platelet aggregation [38,49], increasing the production of the anti-aggregatory prostaglandin PGD2 [38] and binding with high affinity to platelet-activating factor, thereby inhibiting its effects on platelets [2,39].

Plasma albumin also reacts with nitric oxide to form a stable S-nitrosothiol with endothelium-derived relaxing factor-like properties [50]; this mechanism may per-mit albumin to contribute to the regulation of vascular tone. Thrombogenic surfaces coated with S-nitrosylated albumin exhibit reduced platelet adhesion and aggrega-tion [59]. Albumin also inhibits the binding of activated neutrophils to endothelial cells in response to inflammatory stimuli [55] and diminishes neutrophil extravasa-tion to tissue in response to injury [70]. Finally, albumin has been shown to be a specific inhibitor of endothelial-cell apoptosis [95].

Metabolic effects: Albumin exerts major influences on astrocytes, eliciting intercel-lular calcium waves that are inhibited by gap-junction blockers [63] and it acts as an astrocytic mitogen [62]—a property suggesting that albumin might help to stimu-late glial scar formation under pathological conditions. Lactate originating in glial cells is an important energy substrate for the recovery of synaptic function in the settings of neuronal activation and after hypoxia-ischemia [74,75]. Albumin is a major regulator of the enzyme pyruvate dehydrogenase in astrocytes, capable of more than doubling the flux of glucose and lactate [82]. During ischemia, pyruvate dehydrogenase is inhibited, resulting in substrate inhibition and decreased mito-chondrial electron transport [23]. Thus, it is possible that under pathological condi-tions, albumin might help to sustain neuronal metabolism by increasing the export of pyruvate to neurons for metabolism via the Krebs cycle [86].

Hemodilution: It is well known that exogenous albumin may induce hemodilution [22,34]. Thus, it is possible that hemodilution may contribute in part to the neuro-protective effect of administered albumin, although this possibility is not well sup-ported by the experimental or clinical literature. Experimentally, hemodilution with a variety of agents has been reported to result in some degree of ischemic protection in focal ischemia, particularly in models of temporary vascular occlusion [89]. Prior to our own studies, however, high-concentration albumin-hemodilution had been studied in only a single nonrigorous report [57].

Human randomized clinical trials of hemodilution for acute ischemic stroke (including both isovolemic and hypervolemic hemodilution) were analyzed criti-cally in a Cochrane survey; 18 such trials were identified, employing either dextran 40 (12 trials), hydroxyethyl starch (5 trials), or albumin (1 trial); no significant benefits could be documented for any of these agents [3]. The single clinical trial of albumin for acute ischemic stroke, prior to our own studies, was a small con-trolled study in which efficacy was suggested in a subgroup analysis [37]. Taken together, the evidence for the therapeutic efficacy of *non*albumin forms of hemodi-lution in cerebral ischemia is weak and inconsistent. This suggests that the robust neuroprotective efficacy of high-dose albumin cannot be attributed to its hemodi-luting action alone.

2 Part II: Transition to the Clinic

2.1 The ALIAS Pilot Clinical Trial

The preclinical investigations described above, taken together, supported the conclusion that high-dose human albumin administration was strongly neuroprotective in animal models of ischemic stroke and thus provided a compelling rationale to proceed to human clinical trials. The necessary first step in this process was to conduct a pilot-phase trial to demonstrate the *safety* of ALB therapy in human patients with acute ischemic stroke. To this end, the National Institutes of Health funded a pilot clinical trial, termed the ALIAS Pilot Trial, which was initiated in mid-2001 at two clinical sites, the University of Miami and the University of Calgary, Canada [34,67]. The biostatistical and data-management site was at the Medical University of South Carolina. The trial's objective was to employ an open-label, dose-escalation design to ascertain whether subjects with acute ischemic stroke would tolerate 25% human albumin (ALB) when administered at per-kilogram dosages that we had shown to be neuroprotective in preclinical studies, without suffering cardiovascular, neurological, or other adverse events. As the use of intravenous recombinant tissue plasminogen activator (tPA) had come into routine clinical practice for acute ischemic stroke [65], the ALIAS Pilot Trial included both standard-of-care tPA as well as non-tPA subgroups. After informed consent was signed and a CT scan had been obtained to exclude intracranial hemorrhage, subjects (ages 18 or older) were treated within 16 h of stroke onset with ALB and were followed for 3 months to assess the NIH Stroke Scale (NIHSS) score, the modified Rankin Scale (mRS), and the Barthel Index. The trial's major exclusion criteria were designed to exclude patients at potentially heightened risk of cardiovascular adverse events when challenged by ALB-induced volume expansion; these exclusions included a history and/or physical findings of CHF, recent myocardial infarction, arrhythmia if accompanied by hemodynamic instability, and chronic pulmonary disease that interfered with daily activities [34].

In successive groups of subjects, the 25% albumin dose was escalated through six tiers: 0.34, 0.68, 1.03, 1.37, 1.71, and 2.05 g of ALB per kg body weight. The lower three tiers were viewed as likely subtherapeutic, while the highest three tiers were intended to approximate the neuroprotective dose-range established in animal studies [15,22]. ALB was administered by intravenous infusion over 2 h. Following the method of Storer [78], subjects at each dose-tier were adjudicated for adverse events by a safety committee and reviewed by an independent Data and Safety Monitoring Board (DSMB) appointed by the National Institute of Neurological Disorders and Stroke (NINDS) before we were permitted to escalate to the next dose. Midway through the trial, the study protocol was amended to include the administration of the diuretic furosemide as a prophylactic measure unless the investigator judged this treatment not to be necessary.

Eighty-two subjects were enrolled, of whom 42 received standard-of-care intravenous tPA and 40 did not. The mean age was 65 years, and the mean NIHSS score at baseline was 13.6 ± 6.9 (SD). ALB therapy was initiated at 6.5 ± 3.0 h (SD) and at 9.1 ± 3.3 h after stroke onset in the tPA and non-tPA subjects, respectively. ALB infusion produced the expected dose-related increment in plasma albumin level that peaked around 4 h postinfusion and declined toward baseline at 48–72 h; in dose-tier VI, the increment at 4 h averaged 2.0 g/dL. Dose-dependent hemodilution also resulted from ALB infusion, maximal at 4–12 h; in tier VI, the decrease in plasma hematocrit at 4 h averaged 9.8 points, or 23% below baseline [34]. Vital signs were essentially unaffected by ALB administration.

The only adverse event linked to ALB was the occurrence of clinical signs of mild-to-moderate pulmonary edema/CHF arising in the hours or days after ALB administration in 11 of the 82 subjects, or 13.4%. As noted above, prophylactic furosemide was recommended beginning with dose-tier V and was administered in 5 of 24 subjects in tiers V and VI. Only one Serious Adverse Event (SAE) possibly related to ALB occurred—an instance of paroxysmal atrial fibrillation [34].

A major strength of this pilot trial was the ability for subjects to receive and tolerate ALB doses essentially equivalent to those shown to be highly neuroprotective in animals with ischemic stroke. Although this was a noncontrolled trial, we were able to probe the outcome data for preliminary hints of efficacy (a) by comparing our data against historical controls, and (b) by comparing outcomes in the lower three dose-tiers (assumed to be subtherapeutic) to those of the upper three tiers [67] (which lay in the therapeutic range shown in animal studies [15,22]). Overall, comparison of subjects who received ALB in dose-tiers IV, V, and VI with outcome data drawn from the NINDS tPA Stroke Study, Part 2 [65] revealed that the unadjusted probability of good outcome in the present trial (i.e., NIHSS and/or mRS scores of 0–1 at 3 months) was 97% greater than in the historical control population (relative benefit, 1.97; 95% confidence interval (CI), 1.47–2.63), suggesting that ALB therapy at higher doses is highly effective in improving outcome. Similarly, a comparison of the lower- vs. higher-dose ALB groups yielded an overall unadjusted probability of good outcome 77% greater at higher than at lower ALB doses [67]. Subjects of the tPA group who received higher-dose ALB appeared to do particularly well in these comparisons (Fig. 34.9), although it must be remembered that the tPA subjects in this pilot trial received ALB on average 2.6 h earlier than the non-tPA subjects [34] and that, in animal studies of focal ischemia, the therapeutic window of efficacy for ALB began to close after around 4 h [15].

2.2 The ALIAS Multicenter Clinical Trial: Initial Experience

Based on the safety results and suggestion of efficacy emerging from the ALIAS Pilot Trial [34,67], we received planning-grant funding from the National Institutes of Health to design a definitive multicenter efficacy trial of high-dose ALB in acute ischemic stroke, and we were subsequently funded by NIH to conduct the trial itself. The ALIAS Multicenter Trial (ClinicalTrials.gov identifier NCT00235495) was designed as a randomized, double-blind, placebo-controlled trial to assess whether

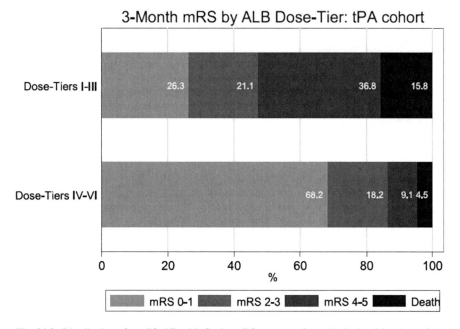

Fig. 34.9 Distribution of modified Rankin Scale (mRS) scores at 3 months in the tPA cohort of the ALIAS Pilot Trial. Lower-dose (tiers I–III) and higher-dose (tiers IV–VI) ALB dose-tiers are compared. A favorable mRS outcome (defined as mRS 0–1) is achieved 2.6 times more often at higher ALB doses (reproduced from Palesch et al. [67] with permission)

ALB therapy (weight-adjusted dose 2 g/kg of 25% albumin; maximal dose 187.5 g) administered within 5 h of the onset of acute ischemic stroke would confer neurological and functional benefit compared to placebo (similar volume of isotonic saline). The major inclusion criteria were acute ischemic stroke, age 18 years and older, baseline NIHSS of 6 or greater, ability to begin study drug within 5 h of stroke onset, and signed and dated informed consent. The major exclusion criteria resembled those of the Pilot Trial and again included a recent history and/or physical findings of CHF, recent myocardial infarction, arrhythmia if accompanied by hemodynamic instability, and chronic pulmonary disease that interfered with daily activities [35]. In the original trial design, it was intended to study two cohorts of subjects, one cohort consisting of subjects who received thrombolytic therapy (either intravenous tPA, intra-arterial tPA, or endovascular mechanical throm-bolysis), and the other cohort consisting of non-thrombolyzed subjects. Within each cohort, subjects were randomized 1:1 to ALB or placebo by means of a centralized, step-forward web-based process [35]. The primary outcome consisted of the NIHSS and mRS scores measured at 90 days postrandomization, with a favorable outcome defined as a score of 0–1 on either or both of those scales. Subjects were followed for 1 year for a variety of secondary outcomes and quality-of-life measures. The trial was powered to detect a 10% absolute effect-size difference in the primary outcome with a power of 80%; the intended sample size was 900 in each cohort, or 1,800 in total. Sixty-two U.S. and Canadian clinical sites were active in subject enrollment [35].

In December, 2007, after 434 subjects had been enrolled and after the first interim analysis of the 3-month outcome had been conducted, the trial's NINDS-appointed DSMB detected a safety concern and recommended that enrollment be suspended; and the DSMB suggested that we consider revising the ALIAS protocol to enable the trial to resume with increased safety. To this end, the ALIAS team spent the following year conducting an extensive unblinded review of the *safety* data, which resulted in major revisions to the ALIAS protocol and analysis plan and the development of a comprehensive site-training program, on the basis of which the DSMB, FDA, and NINDS permitted the trial to resume as an essentially new, separate study termed "the ALIAS Part 2 Trial" [35]. Below, we present the details of the Part 1 safety data analysis.

2.3 ALIAS Part 1 Trial: Safety Analysis

Of the 434 subjects enrolled in Part 1, 327 had received some form of thrombolysis and 97 had not. These cohorts were combined for the safety analysis, which was conducted in the 207 ALB and 217 saline subjects that had received at least 20% of study drug. The mean age of this "safety cohort" was $69.6 + 14.1$ years (maximum, 97); median baseline NIHSS score was 11 [35].

The major safety concern that emerged from our unblinded analysis of the Part 1 safety cohort was an imbalance by treatment in overall deaths between days 5 and 30 postrandomization: these amounted to 27 of 207 subjects and 12 of 217 subjects in the ALB and saline groups, respectively. By contrast, deaths on days 1–4 and following day 30 were virtually identical in the two groups. The overall death rates at 90 days were 43 of 207 (20.8%) for ALB and 29 of 217 (13.4%) for saline. A treatment-blinded adjudication of the primary cause of death revealed that large strokes (with or without medical complications) were the predominant cause of death at days 1–30 in both groups, but no single cause completely explained the differential death rate by treatment [35].

We next explored factors that might have contributed to the differential early deaths by treatment [35]. A multivariate model incorporating treatment assignment, baseline NIHSS score, and age showed a nonsignificant effect of treatment but a significant effect of baseline NIHSS (relative risk 1.10; 95% CI, 1.06–1.14) and of *age* (1.03; 1.01–1.05). Accordingly, we examined the differential death rates by various dichotomized age groups beginning at age 80. We found that 90-day death rates did not differ between the ALB and saline groups in subjects aged 83 or younger, whereas there was a 2.3-fold higher 90-death rate with ALB than saline in subjects *older than 83* [35].

Because ALB is known to expand intravascular volume, we hypothesized that differences in the quantity of intravenous fluids administered to subjects might also

Table 34.1 Analysis of deaths at 90 days in the ALIAS Part 1 safety cohort

| | Entire safety cohort | | Safety cohort with age <84 and 48 h IV fluids ≤4,200 mL and out-of-hospital strokes | |
	ALB	Saline	ALB	Saline
Total subjects	207	217	114	100
No. dead at 90 days	43	29	17	13
Percentage dead at 90 days	20.8%	13.4%	14.8%	12.9%
Relative risk (95% CI)	1.55 (1.01–2.39)		1.15 (0.59–2.24)	

Data from ref. [35]

have contributed to the differential early death rate by treatment. We considered "fluid excess" to be present if a subject received more than 4,200 mL intravenously over the first 48 h. A dichotomized analysis revealed that, in subjects without fluid excess, there was no difference in 90-day deaths by treatment, while in subjects receiving more than 4,200 mL of IV fluids in the first 48 h, death rates were *twofold greater* in the ALB than in the saline group (RR 2.10; CI, 1.10–3.98) [35].

The analyses presented above suggested the need to revise the ALIAS protocol so as to exclude the very elderly and to discourage the administration of excessive IV fluids. (We also decided to exclude in-hospital stroke patients because they tended to be more ill and to undergo more adverse events in general.) We then applied these exclusions to the Part 1 safety cohort (Table 34.1) and demonstrated that the implementation of these measures would be predicted to eliminate the differential early death rate by treatment assignment [35].

Analysis of SAEs in the ALIAS Part 1 Trial revealed the expected difference in the incidence of pulmonary edema in ALB and saline patients (6.8 and 2.8%, respectively), but, as well, an unexpected difference in acute-coronary-syndrome SAEs (8.2 vs. 0.5%). Although we suspected that cardiopulmonary events may have played a role in the differential mortality by treatment, we were unable to demonstrate a specific causal relationship [35].

2.4 Design of the ALIAS Part 2 Multicenter Trial

The ALIAS Part 2 Trial is a standalone study that retains many features of the Part 1 Trial: e.g., 5-h treatment window; 1:1 randomization to ALB and saline; and identical primary outcome measures. The major design changes, as discussed above, were instituted in order to reduce the likelihood of enrolling subjects at higher risk of cardiovascular morbidity and mortality; they consist of the following: (a) An upper age limit of 83 years has been established for eligibility. (b) We now require that the baseline serum troponin level be ≤0.1 μg/L; this is designed to exclude patients with subtle cardiac injury at baseline. (c) Patients with in-hospital strokes are excluded because they tend to have comorbid illnesses and are more prone to

SAEs. (d) We have instituted a guideline that total intravenous fluids within the first 48 h should not exceed 4,200 mL; any fluid overage must be clinically justified, and an adjudication procedure is in place. (e) We mandate the prophylactic administration of a loop diuretic (usually furosemide, 20 mg IV) at 12–24 h after randomization, which may be waived according to the treating physician's judgment if considered unnecessary. (f) A detailed (re-)training module has been developed for clinical staff, which includes a certification examination. (g) The *statistical design* of the Part 2 Trial was modified because it became evident from the Part 1 Trial that we would not be able to enroll sufficient numbers of *non*-thrombolyzed subjects to permit an independent efficacy assessment in that cohort. Thus, in the Part 2 Trial, subjects with and without thrombolytic treatment are combined into a single cohort (rather than being prospectively assigned to two cohorts), and the statistical analysis will search for an interactive effect. This design requires a total sample size of 1,100 subjects in order to provide 80% power and minimize Type I error probability. (h) Finally, after completion of Part 2, we plan to conduct a meta-analysis of the combined data of Parts 1 and 2 [35].

2.5 Exploratory Efficacy Analysis of the ALIAS Part 1 Trial

Following the completion of the Part 1 safety analysis and implementation of the Part 2 Trial, we chose to conduct an *exploratory efficacy analysis* in a subset of Part 1 subjects termed the "target population," which was defined as an intention-to-treat population of Part 1 subjects *who would have fulfilled the eligibility criteria for the Part 2 Trial*: i.e., age 83 or below, baseline serum troponin ≤ 0.1 µg/L or normal baseline creatine kinase CK-MB, and out-of-hospital stroke [46]. Of the 434 subjects studied in Part 1, 311 subjects fulfilled the definitions of the target population (after accounting for missing outcomes); 82% of these had received some form of thrombolysis. In the overall target population, a favorable primary outcome (NIHSS 0–1 and/or mRS 0–1 at 90 days) was achieved in 44.7% of ALB subjects, but only 36.0% of saline subjects, yielding an absolute risk difference of 8.7% (95% CI, −2.2 to 19.5) and a risk ratio (adjusted for baseline NIHSS) of 1.10 (CI, 0.8–1.4). In the *thrombolyzed* target population, a more striking efficacy signal was obtained, with favorable primary outcome in 46.7% of ALB and 36.6% of saline subjects, yielding an absolute risk difference of 10.1% (CI, −2.0 to 20.0) and an adjusted risk ratio of 1.20 (CI, 0.9–1.5). This is encouraging in view of the increasing frequency with which thrombolytic therapy is now being used to treat acute ischemic stroke [88]. Figure 34.10 shows the 90-day mRS scores in the ALB- and saline-treated subjects of the target population: ALB subjects have a 29% relative increase in favorable mRS (i.e., scores of 0 or 1) at 90 days relative to saline subjects [46]. Taken together, these promising preliminary efficacy results strongly support the rationale of the currently ongoing Part 2 Trial.

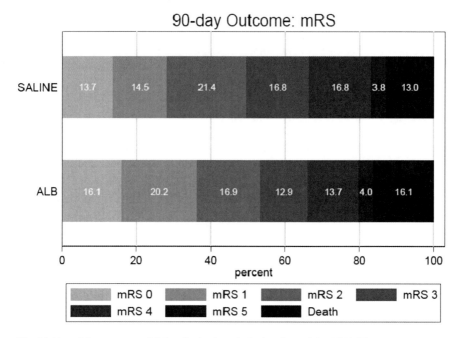

Fig. 34.10 mRS outcomes at 90 days in the thrombolysis cohort of the ALIAS Part 1 target population (i.e., age under 84, normal baseline troponin, out-of-hospital stroke) who received at least
100 mL of study drug. ALB-treated subjects have a 29% relative increase in favorable mRS outcome (i.e., scores of 0 or 1) at 90 days relative to saline subjects (reproduced from Hill et al. [46]
with permission)

2.6   Status Report of the ALIAS Part 2 Trial

As of this writing, the ALIAS Part 2 Trial has enrolled around 530 subjects, or 48%
of the required total of 1,100. Around 70 North American clinical sites are actively
participating, including 15 clinical "hubs" and their associated "spoke" sites belong-
ing to the NIH-funded Neurologic Emergencies Treatment Trials (NETT) Network.
Around ten clinical sites in Israel will soon be joining the trial. As of late 2010, some
baseline descriptors are as follows: mean age, 63 ± 14 years, median age 65. Baseline
NIHSS score, 13.5 ± 6.2. Stroke onset to study-drug initiation is 208 ± 52 min.
Eighty-six percent of subjects have received some form of thrombolysis [33].

2.7   Conclusions

High-dose human albumin has emerged, both from extensive experimental investigations and from preliminary efficacy analyses of early clinical trial data, as a highly
promising neuroprotective agent for use in acute ischemic stroke. The currently

ongoing ALIAS Part 2 Multicenter Clinical Trial, once completed, will provide definitive evidence to whether this therapy is neuroprotective and, as such, should be adopted as a standard-of-care clinical measure in the management of acute ischemic stroke. A recent editorial has congratulated the ALIAS Trialists for their persistent "wisdom and determination" in pursuing this goal [47]. If clinical efficacy is definitively established, this success may, in fact, be a consequence of the many respects in which high-dose albumin satisfies the requirements of a nearly ideal neuroprotectant [31]: (1) Albumin is a *natural biological product* approved for routine clinical use. As such, it does not require additional licensing and is more likely to be acceptable to patients and medical personnel. (2) The experimental evidence for its neuroprotective efficacy is consistent and robust. (3) Albumin possesses many well-documented salutary properties; thus, it likely targets *multiple potential injury-mechanisms*. Although the available evidence does not permit us to identify which actions are most important to albumin's neuroprotective effect, it is quite likely that multiple mechanisms contribute. (4) The experimentally demonstrated therapeutic window for albumin in acute focal ischemia is sufficiently broad to render its clinical use feasible. (5) Albumin's safety profile is clinically acceptable; and, importantly, it can be administered safely to patients *in the same per-kg doses as were shown to be protective experimentally*. (6) Albumin can be administered to patients without the need for complicated laboratory tests.

Acknowledgment Grant support: Dr. Ginsberg is supported by NIH Cooperative Agreement U01 NS040406.

References

1. Alexander MR, Ambre JJ, Liskow BI, Trost DC. Therapeutic use of albumin. JAMA. 1979;241:2527–9.
2. Ammit AJ, O'Neill C. Studies of the nature of the binding by albumin of platelet-activating factor released from cells. J Biol Chem. 1997;272:18772–8.
3. Asplund K. Haemodilution for acute ischaemic stroke. Cochrane Database Syst Rev. 2002;(4):CD000103.
4. Back T, Zhao W, Ginsberg MD. Three-dimensional image analysis of brain glucose metabolism-blood flow uncoupling and its electrophysiological correlates in the acute ischemic penumbra following middle cerebral artery occlusion. J Cereb Blood Flow Metab. 1995;15:566–77.
5. Bar-Or D, Rael LT, Lau EP, Rao NK, Thomas GW, Winkler JV, Yukl RL, Kingston RG, Curtis CG. An analog of the human albumin N-terminus (Asp-Ala-His-Lys) prevents formation of copper-induced reactive oxygen species. Biochem Biophys Res Commun. 2001;284:856–62.
6. Bazan NG. Supply of n-3 polyunsaturated fatty acids and their significance in the central nervous system. In: Wurtman RJ, Wurtman JJ, editors. Nutrition and the brain. New York: Raven; 1990. p. 1–24.
7. Bazan NG, Gordon WC, Rodriguez de Turco EB. The uptake, metabolism and conservation of docosahexaenoic acid (22:6 ω-3) in brain and retina: alterations in liver and retina 22:6 metabolism during inherited progressive retinal degeneration. In: Sinclair A, Gibson R, editors. Essential fatty acids and eicosanoids: invited papers from the third international congress. Champaign, IL: American Oil Chemists' Society; 1992. p. 107–15.

8. Bazan NG, Rodriguez de Turco EB. Alterations in plasma lipoproteins and DHA transport in progressive rod-cone degeneration (PRCD). In: Kato S, Osborne NN, Tamai M, editors. Retinal degeneration and regeneration. Amsterdam: Kugler Publications; 1996. p. 89–97.

9. Belayev L, Alonso OF, Busto R, Zhao W, Ginsberg MD. Middle cerebral artery occlusion in the rat by intraluminal suture. Neurological and pathological evaluation of an improved model. Stroke. 1996;27:1616–22.

10. Belayev L, Alonso OF, Huh PW, Zhao W, Busto R, Ginsberg MD. Posttreatment with high-dose albumin reduces histopathological damage and improves neurological deficit following fluid percussion brain injury in rats. J Neurotrauma. 1999;16:445–53.

11. Belayev L, Busto R, Zhao W, Clemens JA, Ginsberg MD. Effect of delayed albumin hemodilution on infarction volume and brain edema after transient middle cerebral artery occlusion in rats. J Neurosurg. 1997;87:595–601.

12. Belayev L, Khoutorova L, Atkins KD, Bazan NG. Robust docosahexaenoic acid-mediated neuroprotection in a rat model of transient, focal cerebral ischemia. Stroke. 2009;40:3121–6.

13. Belayev L, Khoutorova L, Atkins KD, Eady TN, Hong S, Lu Y, Obenaus A, Bazan NG. Docosahexaenoic acid therapy of experimental ischemic stroke. Transl Stroke Res. 2011;2:33–41.

14. Belayev L, Khoutorova L, Belayev A, Zhang Y, Zhao W, Busto R, Ginsberg MD. Delayed post-ischemic albumin treatment neither improves nor worsens the outcome of transient focal cerebral ischemia in rats. Brain Res. 2004;998:243–6.

15. Belayev L, Liu Y, Zhao W, Busto R, Ginsberg MD. Human albumin therapy of acute ischemic stroke: marked neuroprotective efficacy at moderate doses and with a broad therapeutic window. Stroke. 2001;32:553–60.

16. Belayev L, Marcheselli VL, Khoutorova L, Rodriguez de Turco EB, Busto R, Ginsberg MD, Bazan NG. Docosahexaenoic acid complexed to albumin elicits high-grade ischemic neuroprotection. Stroke. 2005;36:118–23.

17. Belayev L, Obenaus A, Zhao W, Saul I, Busto R, Wu C, Vigdorchik A, Ginsberg MD. Experimental intracerebral hematoma in the rat: characterization by sequential magnetic resonance imaging, behavior, and histopathology. Effect of albumin therapy. Brain Res. 2007;1157:146–55.

18. Belayev L, Pinard E, Nallet H, Seylaz J, Liu Y, Riyamongkol P, Zhao W, Busto R, Ginsberg MD. Albumin therapy of transient focal cerebral ischemia: in vivo analysis of dynamic microvascular responses. Stroke. 2002;33:1077–84.

19. Belayev L, Saul I, Busto R, Danielyan K, Vigdorchik A, Khoutorova L, Ginsberg MD. Albumin treatment reduces neurological deficit and protects blood–brain barrier integrity after acute intracortical hematoma in the rat. Stroke. 2005;36:326–31.

20. Belayev L, Saul I, Huh PW, Finotti N, Zhao W, Busto R, Ginsberg MD. Neuroprotective effect of high-dose albumin therapy against global ischemic brain injury in rats. Brain Res. 1999;845:107–11.

21. Belayev L, Zhao W, Busto R, Ginsberg MD. Transient middle cerebral artery occlusion by intraluminal suture: I. Three-dimensional autoradiographic image-analysis of local cerebral glucose metabolism-blood flow interrelationships during ischemia and early recirculation. J Cereb Blood Flow Metab. 1997;17:1266–80.

22. Belayev L, Zhao W, Pattany PM, Weaver RG, Huh PW, Lin B, Busto R, Ginsberg MD. Diffusion-weighted magnetic resonance imaging confirms marked neuroprotective efficacy of albumin therapy in focal cerebral ischemia. Stroke. 1998;29:2587–99.

23. Bogaert YE, Rosenthal RE, Fiskum G. Postischemic inhibition of cerebral cortex pyruvate dehydrogenase. Free Radic Biol Med. 1994;16:811–20.

24. Busto R, Dietrich WD, Globus MY, Ginsberg MD. The importance of brain temperature in cerebral ischemic injury. Stroke. 1989;20:1113–4.

25. Carter DC, Ho JX. Structure of serum albumin. Adv Protein Chem. 1994;45:153–203.

26. Cha SH, Fukushima A, Sakuma K, Kagawa Y. Chronic docosahexaenoic acid intake enhances expression of the gene for uncoupling protein 3 and affects pleiotropic mRNA levels in skeletal muscle of aged C57BL/6NJcl mice. J Nutr. 2001;131:2636–42.

27. Curry S, Mandelkow H, Brick P, Franks N. Crystal structure of human serum albumin complexed with fatty acid reveals an asymmetric distribution of binding sites. Nat Struct Biol. 1998;5:827–35.

28. Ehrenreich H, Hasselblatt M, Dembowski C, Cepek L, Lewczuk P, Stiefel M, Rustenbeck HH, Breiter N, Jacob S, Knerlich F, Bohn M, Poser W, Ruther E, Kochen M, Gefeller O, Gleiter C, Wessel TC, De Ryck M, Itri L, Prange H, Cerami A, Brines M, Siren AL. Erythropoietin therapy for acute stroke is both safe and beneficial. Mol Med. 2002;8:495–505.

29. Emerson TE. Unique features of albumin: a brief review. Crit Care Med. 1989;17:690–4.

30. Galis Z, Ghitescu L, Simionescu M. Fatty acids binding to albumin increases its uptake and transcytosis by the lung capillary endothelium. Eur J Cell Biol. 1988;47:358–65.

31. Ginsberg MD. Neuroprotection for ischemic stroke: past, present and future. Neuropharmacology. 2008;55:363–89.

32. Ginsberg MD, Belayev L, Bazan NG, Marcheselli VL, Hill MD, Palesch YY, Khoutorova L, Rodriguez de Turco EB, Ryckborst K, Tamariz D, Busto R. Albumin-based neurotherapeutics for acute ischemic stroke: from bench to bedside. In: Krieglstein J, Klumpp S, editors. Pharmacology of cerebral ischemia 2004. Stuttgart: Medpharm Scientific Publishers; 2004. p. 421–33.

33. Ginsberg MD, Hill MD, Palesch YY, Martin RH, Yeatts S, Moy CS, Tamariz D, Ryckborst KJ, et al. ALIAS-part 2 phase III multicenter clinical trial of albumin therapy for neuroprotection in acute ischemic stroke. In: 36th International Stroke Conference, Ongoing Clinical Trial Abstract CT P28; 2011.

34. Ginsberg MD, Hill MD, Palesch YY, Ryckborst KJ, Tamariz T. The ALIAS Pilot Trial: a dose-escalation and safety study of albumin therapy for acute ischemic stroke. I. Physiological responses and safety results. Stroke. 2006;37:2100–6.

35. Ginsberg MD, Palesch YY, Martin RH, Hill MD, Moy CS, Waldman BD, Yeatts SD, Tamariz D, Ryckborst K. The Albumin in Acute Stroke (ALIAS) multicenter clinical trial: safety analysis of part 1 and rationale and design of part 2. Stroke. 2011;42:119–27.

36. Ginsberg MD, Zhao W, Belayev L, Alonso OF, Liu Y, Loor JY, Busto R. Diminution of metabolism/blood flow uncoupling following traumatic brain injury in rats in response to high-dose human albumin treatment. J Neurosurg. 2001;94:499–509.

37. Goslinga H, Eijzenbach V, Heuvelmans JH, van der Laan de Vries E, Melis VM, Schmid-Schonbein H, Bezemer PD. Custom-tailored hemodilution with albumin and crystalloids in acute ischemic stroke. Stroke. 1992;23:181–8.

38. Gresele P, Deckmyn H, Huybrechts E, Vermylen J. Serum albumin enhances the impairment of platelet aggregation with thromboxane synthase inhibition by increasing the formation of prostaglandin D2. Biochem Pharmacol. 1984;33:2083–8.

39. Grigoriadis G, Stewart AG. Albumin inhibits platelet-activating factor (PAF)-induced responses in platelets and macrophages: implications for the biologically active form of PAF. Br J Pharmacol. 1992;107:73–7.

40. Grundt H, Nilsen DW, Mansoor MA, Nordoy A. Increased lipid peroxidation during long-term intervention with high doses of n-3 fatty acids (PUFAs) following an acute myocardial infarction. Eur J Clin Nutr. 2003;57:793–800.

41. Gum ET, Swanson RA, Alano C, Liu J, Hong S, Weinstein PR, Panter SS. Human serum albumin and its N-terminal tetrapeptide (DAHK) block oxidant-induced neuronal death. Stroke. 2004;35:590–5.

42. Halliwell B. Albumin—an important extracellular antioxidant? Biochem Pharmacol. 1988;37:569–71.

43. Haynes GR, Navickis RJ, Wilkes MM. Albumin administration—what is the evidence of clinical benefit? A systematic review of randomized controlled trials. Eur J Anaesthesiol. 2003;20:771–93.

44. He P, Curry FE. Albumin modulation of capillary permeability: role of endothelial cell $[Ca^{2+}]_i$. Am J Physiol. 1993;265:H74–82.

45. Heros RC, Korosue K. Hemodilution for cerebral ischemia. Stroke. 1989;20:423–7.

46. Hill MD, Martin RH, Palesch YY, Tamariz D, Waldman BD, Ryckborst KJ, Moy CS, Barsan WG, Ginsberg MD. The Albumin in Acute Stroke Part 1 Trial: an exploratory efficacy analysis. Stroke. 2011;42:1621–5.

47. Howard GE, Hess DC, Howard VJ. Wisdom and determination in the ongoing pursuit of the ever-elusive neuroprotective agent. Stroke. 2011;42:1505–6.

48. Huh PW, Belayev L, Zhao W, Busto R, Saul I, Ginsberg MD. The effect of high-dose albumin therapy on local cerebral perfusion after transient focal cerebral ischemia in rats. Brain Res. 1998;804:105–13.

49. Jorgensen KA, Stoffersen E. On the inhibitory effect of albumin on platelet aggregation. Thromb Res. 1980;17:13–8.

50. Keaney JFJ, Simon DI, Stamler JS, Jaraki O, Scharfstein J, Vita JA, Loscalzo J. NO forms an adduct with serum albumin that has endothelium-derived relaxing factor-like properties. J Clin Invest. 1993;91:1582–9.

51. Kleinfeld D, Mitra PP, Helmchen F, Denk W. Fluctuations and stimulus-induced changes in blood flow observed in individual capillaries in layers 2 through 4 of rat neocortex. Proc Natl Acad Sci USA. 1998;95:15741–6.

52. Koch-Weser J, Sellers EM. Binding of drugs to serum albumin (first of two parts). N Engl J Med. 1976;294:311–6.

53. Koch-Weser J, Sellers EM. Drug therapy. Binding of drugs to serum albumin (second of two parts). N Engl J Med. 1976;294:526–31.

54. Korosue K, Heros RC, Ogilvy CS, Hyodo A, Tu YK, Graichen R. Comparison of crystalloids and colloids for hemodilution in a model of focal cerebral ischemia. J Neurosurg. 1990;73:576–84.

55. Lang JD Jr, Figueroa M, Chumley P, Aslan M, Hurt J, Tarpey MM, Alvarez B, Radi R, Freeman BA. Albumin and hydroxyethyl starch modulate oxidative inflammatory injury to vascular endothelium. Anesthesiology. 2004;100:51–8.

56. Lennihan L, Mayer SA, Fink ME, Beckford A, Paik MC, Zhang H, Wu YC, Klebanoff LM, Raps EC, Solomon RA. Effect of hypervolemic therapy on cerebral blood flow after subarachnoid hemorrhage: a randomized controlled trial. Stroke. 2000;31:383–91.

57. Little JR, Slugg RM, Latchaw JPJ, Lesser RP. Treatment of acute focal cerebral ischemia with concentrated albumin. Neurosurgery. 1981;9:552–8.

58. Liu Y, Belayev L, Zhao W, Busto R, Belayev A, Ginsberg MD. Neuroprotective effect of treatment with human albumin in permanent focal cerebral ischemia: histopathology and cortical perfusion studies. Eur J Pharmacol. 2001;428:193–201.

59. Maalej N, Albrecht R, Loscalzo J, Folts JD. The potent platelet inhibitory effects of S-nitrosated albumin coating of artificial surfaces. J Am Coll Cardiol. 1999;33:1408–14.

60. Marcheselli VL, Hong S, Lukiw WJ, Tian XH, Gronert K, Musto A, Hardy M, Gimenez JM, Chiang N, Serhan CN, Bazan NG. Novel docosanoids inhibit brain ischemia-reperfusion-mediated leukocyte infiltration and pro-inflammatory gene expression. J Biol Chem. 2003;278:43807–17.

61. Martin U, Kaufmann B, Neugebauer G. Current clinical use of reteplase for thrombolysis. A pharmacokinetic-pharmacodynamic perspective. Clin Pharmacokinet. 1999;36:265–76.

62. Nadal A, Fuentes E, Pastor J, McNaughton PA. Plasma albumin is a potent trigger of calcium signals and DNA synthesis in astrocytes. Proc Natl Acad Sci USA. 1995;92:1426–30.

63. Nadal A, Fuentes E, Pastor J, McNaughton PA. Plasma albumin induces calcium waves in rat cortical astrocytes. Glia. 1997;19:343–51.

64. Nimmagadda A, Park H-P, Prado R, Ginsberg MD. Albumin therapy improves local vascular dynamics in a rat model of primary microvascular thrombosis: a two-photon laser-scanning microscopy study. Stroke. 2008;39:198–204.

65. NINDS rt-PA Stroke Study Group. Tissue plasminogen activator for acute ischemic stroke. N Engl J Med. 1995;333:1581–7.

66. Oheim M, Michael DJ, Geisbauer M, Madsen D, Chow RH. Principles of two-photon excitation fluorescence microscopy and other nonlinear imaging approaches. Adv Drug Deliv Rev. 2006;58:788–808.

67. Palesch YY, Hill MD, Ryckborst KJ, Tamariz D, Ginsberg MD. The ALIAS Pilot Trial: a dose-escalation and safety study of albumin therapy for acute ischemic stroke II. Neurological outcome and efficacy-analysis. Stroke. 2006;37:2107–14.

68. Park HP, Nimmagadda A, DeFazio RA, Busto R, Prado R, Ginsberg MD. Albumin therapy augments the effect of thrombolysis on local vascular dynamics in a rat model of arteriolar thrombosis: a two-photon laser-scanning microscopy study. Stroke. 2008;39:1556–62.
69. Peters Jr T. All about albumin. Biochemistry, genetics, and medical applications. San Diego: Academic press; 1996.
70. Powers KA, Kapus A, Khadaroo RG, He R, Marshall JC, Lindsay TF, Rotstein OD. Twenty-five percent albumin prevents lung injury following shock/resuscitation. Crit Care Med. 2003;31:2355–63.
71. Remmers M, Schmidt-Kastner R, Belayev L, Lin B, Busto R, Ginsberg MD. Protein extravasation and cellular uptake after high-dose human-albumin treatment of transient focal cerebral ischemia in rats. Brain Res. 1999;827:237–42.
72. Rodriguez de Turco EB, Belayev L, Liu Y, Busto R, Parkins N, Bazan NG, Ginsberg MD. Systemic fatty acid responses to transient focal cerebral ischemia: influence of neuroprotectant therapy with human albumin. J Neurochem. 2002;83:515–24.
73. Sakurada O, Kennedy C, Jehle J, Brown JD, Carbin GL, Sokoloff L. Measurement of local cerebral blood flow with iodo [^{14}C] antipyrine. Am J Physiol. 1978;234:H59–66.
74. Schurr A, Miller JJ, Payne RS, Rigor BM. An increase in lactate output by brain tissue serves to meet the energy needs of glutamate-activated neurons. J Neurosci. 1999;19:34–9.
75. Schurr A, Payne RS, Miller JJ, Rigor BM. Brain lactate is an obligatory aerobic energy substrate for functional recovery after hypoxia: further in vitro validation. J Neurochem. 1997;69:423–6.
76. Seylaz J, Charbonne R, Nanri K, Von Euw D, Borredon J, Kacem K, Meric P, Pinard E. Dynamic in vivo measurement of erythrocyte velocity and flow in capillaries and of microvessel diameter in the rat brain by confocal laser microscopy. J Cereb Blood Flow Metab. 1999;19:863–70.
77. Smith ML, Bendek G, Dahlgren N, Rosen I, Wieloch T, Siesjo BK. Models for studying long-term recovery following forebrain ischemia in the rat. 2. A 2-vessel occlusion model. Acta Neurol Scand. 1984;69:385–401.
78. Storer BE. Design and analysis of phase I clinical trials. Biometrics. 1989;45:925–37.
79. Suarez JI, Martin RH. Treatment of subarachnoid hemorrhage with human albumin: ALISAH study. Rationale and design. Neurocrit Care. 2010;13:263–77.
80. Suarez JI, Shannon L, Zaidat OO, Suri MF, Singh G, Lynch G, Selman WR. Effect of human albumin administration on clinical outcome and hospital cost in patients with subarachnoid hemorrhage. J Neurosurg. 2004;100:585–90.
81. Svoboda K, Yasuda R. Principles of two-photon excitation microscopy and its applications to neuroscience. Neuron. 2006;50:823–39.
82. Tabernero A, Medina A, Sanchez-Abarca LI, Lavado E, Medina JM. The effect of albumin on astrocyte energy metabolism is not brought about through the control of cytosolic Ca^{2+} concentrations but by free-fatty acid sequestration. Glia. 1998;25:1–9.
83. Tomita H, Ito U, Tone O, Masaoka H, Tominaga B. High colloid oncotic therapy for contusional brain edema. Acta Neurochir Suppl (Wien). 1994;60:547–9.
84. Tone O, Ito U, Tomita H, Masaoka H, Tominaga B. High colloid oncotic therapy for brain edema with cerebral hemorrhage. Acta Neurochir Suppl (Wien). 1994;60:568–70.
85. Trigatti BL, Gerber GE. A direct role for serum albumin in the cellular uptake of long-chain fatty acids. Biochem J. 1995;308:155–9.
86. Tsacopoulos M, Magistretti PJ. Metabolic coupling between glia and neurons. J Neurosci. 1996;16:877–85.
87. Wayner DD, Burton GW, Ingold KU, Locke S. Quantitative measurement of the total, peroxyl radical-trapping antioxidant capability of human blood plasma by controlled peroxidation. The important contribution made by plasma proteins. FEBS Lett. 1985;187:33–7.
88. Xian Y, Holloway RG, Chan PS, Noyes K, Shah MN, Ting HH, Chappel AR, Peterson ED, Friedman B. Association between stroke center hospitalization for acute ischemic stroke and mortality. JAMA. 2011;305:373–80.

89. Xiong L, Lei C, Wang Q, Li W. Acute normovolaemic haemodilution with a novel hydroxy-ethyl starch (130/0.4) reduces focal cerebral ischaemic injury in rats. Eur J Anaesthesiol. 2008;25:581–8.
90. Zhang JP, Sun GY. Free fatty acids, neutral glycerides, and phosphoglycerides in transient focal cerebral ischemia. J Neurochem. 1995;64:1688–95.
91. Zhao W, Belayev L, Ginsberg MD. Transient middle cerebral artery occlusion by intraluminal suture: II. Neurological deficits, and pixel-based correlation of histopathology with local blood flow and glucose utilization. J Cereb Blood Flow Metab. 1997;17:1281–90.
92. Zhao W, Ginsberg MD, Prado R, Belayev L. Depiction of infarct frequency distribution by computer-assisted image mapping in rat brains with middle cerebral artery occlusion. Comparison of photothrombotic and intraluminal suture models. Stroke. 1996;27:1112–7.
93. Zhao W, Young TY, Ginsberg MD. Registration and three-dimensional reconstruction of auto-radiographic images by the disparity analysis method. IEEE Trans Med Imaging. 1993; 12:782–91.
94. Zhao Y, Marcel YL. Serum albumin is a significant intermediate in cholesterol transfer between cells and lipoproteins. Biochemistry. 1996;35:7174–80.
95. Zoellner H, Hofler M, Beckmann R, Hufnagl P, Vanyek E, Bielek E, Wojta J, Fabry A, Lockie S, Binder BR. Serum albumin is a specific inhibitor of apoptosis in human endothelial cells. J Cell Sci. 1996;109:2571–80.

Chapter 35
The Translation Procedure of Low-Level Laser Therapy in Acute Ischemic Stroke: A Nonpharmaceutics Noninvasive Method

Yair Lampl

Abstract The low-level laser treatment is based on the theory of photon energy absorption by delivery of a selected wavelength of light energy. The transcranial low-level laser device is an appliance which facilitates the insertion of the beam through body cavities. Low-level laser therapy (LLLT) was shown to have a direct effect on adenosine diphosphate (ADP), regulation of the inducible transcription factor 1 and the heatshock protein 70, plus having an antiapoptotic effect. In the ischemic brain, LLLT has an effect on nitric oxide synthase (NOS), brain-derived neurotrophic factor (BDNF), and glial-derived neurotrophic factor (GDNF).

Two very well-designed animal studies were performed in order to examine the efficacy of LLLT treatment on induced stroke in animals—New Zealand RSCEM rabbit model and Wistar and Sprague Dawley rats. The design can be accepted as being the standard recommended model for a nonpharmacological, noninterventional study. However, the time of efficacy was different in both studies—favorable effect after 6 h, but not after 24 h in the rabbit model, and no effect on the early stage and significant better outcome after 24 h in the rats. Translated human clinical studies (NEST1 and NEST2) were designed to confirm the rabbit study hypothesis and showed limited positive findings. The assumption of delayed effect of LLLT was not examined. It would be recommended that an additional study be performed to examine the translational process of the rat study. Problematic issues surrounding this data were discussed.

Y. Lampl, MD (✉)
Neurological Department, Edith Wolfson Medical Center,
Holon Israel and Sackler Medical School, University of Tel-Aviv, Tel-Aviv, Israel
e-mail: y_lampl@hotmail.com

P.A. Lapchak and J.H. Zhang (eds.), *Translational Stroke Research*, Springer Series
in Translational Stroke Research, DOI 10.1007/978-1-4419-9530-8_35,
© Springer Science+Business Media, LLC 2012

1 The Translational Process of Neuroprotection in Acute Stroke

The practical side of the translational process of medicine is based on the congruency between the preclinical animal model and the human disease [1]. The translation of preclinical to clinical model of treatment causes very high interest, but also great skepticism, especially surrounding the issues the neuroprotective effect postischemic stroke. The disappointment from failure of phase 1 and phase 2 of clinical trails after encouraging results from animal model states and failure of efficacy in stage 3 studies after very optimistic data of phase 2 [2–5] leads to frequent discussion and debate concerning these issues [6–8]. A turning point was the very promising results of the SAINT Study (NXY-059) in phase 2 which could not be confirmed in phase 3 [9–12]. Various speculative hypotheses were raised to explain this translation failure. The questions were as follows: (1) is the basic theory of neuroprotection accurate?; (2) are the compounds really active on animal model tissue?; (3) is the effect on animal brain tissue identical to the human brain?; (4) does the compound act differently on different areas of infarct tissue?; and (5) is there a transport of compound from entry site to target location and is this transport according to expected pharmacokinetic (PK) and pharmacodynamic data (PD)?

The neuroprotective theory after stroke is based on the hypothesis of the postischemic stroke cascade effects. This theory describes a dynamic process poststroke, divided into four main time-dependent processes: (1) the hyperacute stage (up to 3–4.5 h) which includes the creation of a dynamic "penumbra" surrounding the necrotic core, changing during the first few hours from potential recovery into nonreversible necrotic tissue; (2) acute stage; (3) the subacute stage; and (4) the regenerative-angiogenetic chronic stage. However, most of the experimental compounds were based on one pathway in each of the variable phases having multiple different complementary mechanisms and pathways. It was, therefore, necessary to find a compound or technique which may have multiple effects or to combine different components of a single effect. The following must be taken into consideration in this multisystem mode: (1) mechanisms of depolarization; (2) oxidative stress; (3) excitotoxicity; (4) transcription factor activity; (5) heat shock protein activity; (6) later inflammatory effect; (7) apoptotic process; (8) astrocyte reactivity; and, last but not the least, (9) the gliosis, angiogenesis, and synaptogenesis.

There is great skepticism concerning the use of rodents as the animal model for poststroke studies in general. The main criticism on rodent studies is based on the significant differences between both species. While the rodent brain contains only among 10% of white matter, the percentage of white matter in humans reaches up to 50%. It cannot be ignored that a main component of nonlacunar brain infarction contains white matter tissue, and only less common, the gray matter. It is also assumed that the impact of ion channels in the propagation of stroke cascade is significantly different in the rodent than in human brain tissue [13, 14].

Different aspects of variability of stroke and transport or compounds were usually neglected in a high percentage of previous human studies. The difference between large vessel lacunar and embolic infarcts, the impact of the integrity of the blood–brain barrier (BBB), and the degree of blood reperfusion must be taken into consideration. In order to overcome all this skepticism and translation failure, five main issues must be addressed: (1) the molecular target validation; (2) the target-component enlargement and interaction; (3) the relationship of the pharmacokinetics (PK) and pharmacodynamic (PD) of the drug; (4) the definition of disease bipolar modification; and (5) the definition of patient selection and stratification [1].

Since the late 1990s, a Stroke Therapy Academic Round Table (called the STAIR consortium) which holds conferences of academic and industry leaders was established to generate universal recommendations for translation of animal studies to human clinic trials and to overcome the translational gap. The official findings are based more on the updated integration hypothesis of action than on concrete stroke guidelines in each step of a study. The recommendations for ensuring good scientific inquiry included sample size calculations, defined inclusion and exclusion criteria, the use of group allocation randomization, allocation concealment, reporting of animals excluded from analysis, blinded assumption of outcome, and the report of conflict of interest. The updated recommendation of the last STAIR VI commentary (2009) included the following: (1) evaluation of dose response; (2) evaluation of therapeutic windows; (3) outcome measurement at least 2–3 weeks or longer after stroke; (4) basic physiological monitory including blood pressure, temperature, blood gases, and blood glucose; and (5) the establishment of at least two species. A separate chapter was given to the issue of linking animal model to clinical stroke [15].

Macleod et al. [16] and later Donnan [10] had published a modified form of the 10 point score based on the STAIR criteria for a standard definition of preclinical studies. They defined the necessity for these criteria in cases where there were the following: (1) too small size of animal (<5), (2) underpowered studies, (3) unblinded studies, (4) nonrandomized studies, and (5) no data about temperature control. Additional criteria are (6) confirmation of results on more than one lab and (7) confirmation of the results in more than one animal model.

The repeated clinical failure of various new medications ranging from NMDA antagonists to anti-inflammatory medication [9–12], as neuroprotectors after acute stroke, focused on the preclinical and clinical trials poststroke guidelines rather than on pharmacogenetic agents and interventional procedures. These nonpharmacogenetic and noninterventional techniques included the use of transcranial low-level laser therapy (LLLT). The practical authority regulations are basically different in the use of nonpharmacogenetic techniques than in the pharmacogenetic compounds.

This section critically discusses the translation criteria of the preclinical to clinical studies of the LLLT, especially in light of absence of a validated disease biomarker, such as is recommended elsewhere [1, 8] and based only on the published data.

2   Pretranslational Steps in LLLT

The criteria of translation of preclinical into clinical phase differ from the true nature of its pharmaceutical components. However, apart from the necessity to follow the recommendations of Good Laboratory Practice (GLP), the preclinical stage of a study must adhere to professional guidelines. The justification for the translation of preclinical into clinical phases is dependent upon the role of the standards of quality assurance.

The various principles that should be taken into consideration for estimation of the value or translation of pharmacogenetic criteria in nonpharmacogenetic and noninterventional device poststroke procedures are as follows:

1. The basic hypothesis, logic, and a clear concept of the LLLT effects on post-stroke neuroprotection.
2. The evidence of physical relevance, the identification of the best wavelength, the depth of insertion, and the ebbing of laser energy by penetration through animal/human bony cranium.
3. The evidence of physical relevance in the ischemic brain tissue.
4. The evidence of a single method with multiple actions, for example, the effect on excitatory neurotransmitter and mediators, the effect on ion channel, the effect on free radicals, the effect on the inflammatory processes, the direct effect on mitochondrial damage, and the activation of the apoptotic process. All the effects are dependent on the windows of time that will be chosen; the angiogenesis and regeneration processes may also be included.
5. All the studies concerning the postischemic stroke effect of LLLT, using the transcranial LLLT device, were to be designed and performed according to the GLP guidelines.
6. The animal studies must fulfill the criteria of confirmation of data in more than one lab, confirmation of data in more than a single animal model species in each study including more than 5 animals, male and female models, publishing of basic physiological monitoring, and outcome of early and late stages.
7. Human strokes—in vitro efficacy in human tissue.
8. Efficacy in novel human models of ischemia.
9. Phase II and phase III studies of acute ischemic stroke, including the aspects of clinical subgroups of emboli, large vessels, and lacunar infarctions and the break-down of BBB and the effect on reperfusion of blood.

3   Acute Stroke and the Cascade of Ischemic Tissue Destruction and Regeneration

"Penumbra," an astronomical term of no light-no shadow areas, was first used by the group of Astrup et al. [17, 18]. They described it for the first time in primates as a phenomenon of dynamic changes around necrotic brain tissue in acute stroke. The dynamic changes were characterized by alterations in the ischemic tissue from reversible hypofused damaged tissue into irreversible necrotic tissue. This phenomenon is composed of three anatomical locations: (1) an irreversible necrotic core which became larger during time; (2) a reversible "penumbra" area which became necrotic in 3–4.5 h; and (3) a hypofused reversible oligemic area. The "penumbra" is characterized by the relative persistence of ATP$^+$, but with inhibition of protein synthesis. There is a progression of events during the period of the next few hours and days after the ischemic injury which develops from enlargement of the cell death area including the implementation of the necessary mediators into neurovascular repair.

The concept of "time is brain" can be defined also as "penumbra" shrinkage and infarct expansion taking several hours and a repair process taking days and weeks. During this devastating period, the rapid energy loss from the neurogenic cell death is associated with an active death pathway process. The regional cerebral blood flow (rCBF) decreases from 35 to 55 mL/100 g/min in the oligemic area to 15–35 mL/100 g/min in the metabolic suppressed, glycolytic, and neurotransmitter disturbed penumbra up to the necrotic core area of the rCBF at lower than 15 mL/100 g/min. Also, the ATP$^+$ reservoir is decreased from 70% in the oligemic area up to 13% in the transitional area from the "penumbra" into the core.

Various processes are also active in parallel during the time of cell death and repair. Among these mechanisms are glutamate accumulation, ionic imbalance, glial and endothelial dysfunction, BBB disruption, activation of neural glial signaling leading to loss of neurotransmitter activity, programmed cell death (apoptosis), and disturbance of the matrix–trophic interaction [19–30].

In the repair phrase, including angiogenesis, neurogenesis, neuroblastic migration, and dendrite remodeling, various systems are activated. The vascular response in acute stroke probably is the result of a two-phase operation—an early one involving the BBB and vascular damage, with a key role played by matrix metalloproteinase (MMP), and a later one involving vascular repair. The new generation of blood vessels in the brain is coupled with other processes featuring neurogenesis, symbiogenesis, dendrite plasticity, and axonal sprouting [31]. The angiogenesis in the brain is regulated by a specific angiogenic signaling system. To the most proangiogenetic factors and regulatory systems belong the vascular endothelial growth factor (VEGF) signaling pathways, including the various members of the VEGF family [32–38] and the system involving the angiopoietin-Tie receptors [39]. The complexity of the brain damage and repair system postischemia, especially in the molecular pathways, raises different questions concerning the necessity to activate an inhibition of different pathways simultaneously and to synchronize each treatment to the exact effective time window. The importance of the synchronization of a pathway to a much defined time window was based on the fact that different systems were found to have a controversial effect in various phases after an ischemic event. It was also shown that N methyl aspartate (NMDA) has a devastating effect on the influx of calcium leading to cell death during the hyperactive phase; whereas, NMDA is necessary at a later stage for the recovery processes, activation of neuroplasticity pathways, inhibition of the apoptosis, and the induction of neurogenesis [40–43]. Matrix metalloproteinase (MMP) 2 and 9, but also 3, 7, and 13, contribute to infarct extension BBB disruption and hemorrhage transformation during the acute stage, but play a role in the angiogenesis in the repair phase [43–46]. The JUN pathway has a role, as well as a trigger, to neuronal damage during the acute phase and a mediator of the caspase and apoptosis in the later phase [47, 48].

The main conclusion of the presented data is the supposition that pharmacological and nonpharmacological therapies for ischemic stroke must take into consideration a direct effect of more than one regulatory pathway happening at a very exact specific time window.

4 Low-Level Laser Technique

4.1 Biophysics of Laser (Light Amplification by Stimulated Emission of Radiation)

This is a mechanism for emitting electromagnetic radiation, visible and nonvisible light, by a stimulated emission technique. The laser-colored light is usually emitted on a narrow low divergence beam. It is characterized by a narrow wavelength electromagnetic light.

The wave is defined as the distance from peak to peak, based on the propagation characteristic in a medium. The photon itself is the particle of light and its energy which is inversely proportional to its wavelength, meaning that it has a higher energy in a shorter wave length. The photon has a quality to excite the position of the atom's electron and raise it to a higher orbital level. The frequency of photons is the number of times a photon oscillates through a full wave in 1 s (equal to the number of wavelengths) per second. The relationship between photon frequency (v) and wavelength (λ) is as follows: $\lambda = c/v$ ($c = 3 \times 10^8$ cm/s = speed of light). The pulse frequency of a device is defined as the number of times the device is turned on and off in 1 s. For example, the typical PRR of the LLLT apparatus (GaSa-904 nm) is in the range of 3,000–5,000 pulses/s = 3–5 µHz.

5 Low-Level Laser Treatment

The light propagation effect is based on the absorption of photon energy by a specific medium and on the fact that the power per unit of area (power density) and the energy of ER photons per unit of area (energy density) of light beam have significant capability of effect especially in the lower level of light. The preferred methods are based on the technique of a selected wavelength of light energy delivered to tissue. This data remains as the basis of the LLLT effect on human tissue (muscle cells, cardiomyocytes), and especially, for nervous system cells, stimulated optimally with an irradiation of 810 nm. The efficacy and safety of the LLLT beam irradiated on biological tissue are dependent upon the total energy, dosage, and pulse therapy.

In Budapest, Hungary, Professor Andre Master investigated the use of low-power laser energy in 1964 and found that low-energy exposure has a stimulating effect on biological systems and high energy had the opposite effect. Experimenting with wounded mice, he discovered rapid healing due to microcoagulation of their blood supply and that it was also a finding with human diabetic sore healing. In 1981, Karu proposed a specific mechanism for the photon therapy [49–52]. It was tried in various medical conditions, especially in short-term pain conditions. The exact dose, wavelength, type of irradiation (continuous or pulsing), and duration have not yet been established and are probably different from one clinical application to another. However, there is a consensus that the medical effect is limited to a specific set of wavelength and that the activation of cells is ineffective in the lower dose range. The most useful average of power level is in the range of 1–500 mW. These are also used in peak power short pulse width devices with a power of 1–100 mW. The most commonly used average beam of wavelength is 600–1,000 mW.

The technique of LLLT has been used previously in dermatology [53–55], rheumatology [56–61], and pain treatment [62–65]. As for neurological concerns, there is a good prospect of the clinical application of transcranial LLLT to be used in an actual functional device. Contrary to other transcranial devices, the transcranial LLLT is not just limited to cranial foramina (foramen magnum and the orbital cavity) or to very thin cranial areas of the temporal bone. The LLLT must be also proven to be effective in full scalp irradiation.

6 The Effectiveness of Power Densities, Type of Frequencies, and Safety of Transcranial LLLT

The low-energy laser has been designated by the American FDA (Food and Drug Administration) to treat soft tissue injury, as based on the safety data defined in a nonsignificant risk device. The power density and the optimal duration on humans for in vivo studies are based on experimental studies in more than one laboratory.

7 Transcranial LLLT Device

Confirmation for the establishment of the transcranial low-level laser, as a medical noninvasive device, was certified according to the US Federal Food, Drug and Cosmetic Act (part 201 [h] definition) and the European Directive according to the Medical Device Directive (MDD 93/42/EEC), including the criteria for the classification of Class 3 (US criteria, 21 CFR part 88) and the classification roles in Annex IX of the European Device Classification—Class III. The device was additionally confirmed according to the US federal regulations code (CFR) and underwent the required clinical investigational phases. It was further examined following the necessary phases of study preparation, conduct, and close investigation using the stage gate process and design control.

The data of delivery of beam in different light spot size (1, 2 cm), the total area power, the power density of scalp and brain, as well as data of light source and efficacy in the wavelength of 630–1,060 (with most optimal light of 780–840 nm, especially 785–835) as well of all technical data of the apparatus for transcranial LLLT are published in "the Patent Application Publication of United States" (Detaboado et al.) Pub No. 2010/0016841A1, patent date January 21, 2010. The studies were devoted to skeletal muscle, satellite cells, according to the myogenic study norm. The optimal effective power density was found to be 180 μm W/cm^3, compared to HeNe laser wavelength of 632.8 nm. This data was published in biological journals. Direct irradiation was found to be only 8 below 50 mW/cm^2. The most effective data were achieved in the same laboratory in 1–2 laser irradiation treatment of 7.5 mW/cm^2 for 2 min (0.9 J/cm^2 energy density). Well-controlled studies have confirmed that there are issues regarding defining the optimal frequency and various biological responses regarding different frequencies modalities. HeNe laser invention study proves a decrease in the extent of muscle regeneration in presence of seven daily postinjury irradiations in comparison to every alternate day treatment.

Depth measurements during a transcranial procedure with photodynamic therapy using an argon-pumped dye laser were researched by Derecki et al. [66] in 62 Fisher rats in order to determine the incidental optimal energy dose. A dura area of 5 mm in diameter of the frontal cortex was activated with a red light of 632 ± 2 nm at 100 mW/cm^2. Tilting the depth of necrosis to a natural log dependence of incidental optimal dosage yielded a slope of 0.83 mm/ln J cm^2.

8   The Safety of Late Effect

The safety concerning the late effect on intact rats was studies in Sprague–Dawley rats [67]. Diode laser (wavelength of 808 nm) with power density of 7.5, 75, and 750 mW/cm^2 was irradiating the two modes of continuous wave (CW) and pulsed wave (CPU vs. Sham). No long-term difference in neurological deficits were found in treated and sham group, up to 70 days posttreatment. Pathological neurological effects were demonstrated only in the CW and power density of 750 mW/cm^2. This power is 100-fold more intense than the optimal dosage. The neurological deficits were assumed to be due to the hyperthermal adverse effects and not to the photon activity itself. In a recent study, histopathological effects were not found after transcranial application of LLLT in rats (10 mW/cm^2) even up to 1-year follow-up [68].

9   Mechanism of Action

The LLLT treatment was shown to affect various mechanisms which may be involved in the neuroprotective process, neurogenesis and angiogenesis. These include (1) the direct effect of ATP (adenosine triphosphate) and anti-mitochondrial inhibition, blockage of cytochrome C release from mitochondria [69, 70], and induction of early cell regulatory proteins (cyclin D, cyclin A, cyclin E demonstrated on pmi 28 mouse satellite cell tissue) (pmi 28) [71], (2) proliferation of cell nuclear antigen (PCNA) (on primary rat satellite cell line) [71], (3) neuroprotective and neurogenesis by regulation of hypoxia-inducible transcription factor 1 (HiFA), endothelial growth factor (VEGF), and nerve growth factor (NGF) [72–74], (4) neuroprotective and angiogenesis by the Heat Shock Protein 70 (HSP 70) [75, 76], (5) effect on the interleukin and tumor necrosis factor [76, 77], (6) activation of signal pathways, activation of MAPK/ERK (nitrogen-activated protein kinase/extracellular signal regulated protein kinase) without activation of p38 and JNK [78], (7) inhibition of apoptosis by hindering NF-kappa and nuclear translocation, increase of Bcl 2 (antiapoptotic) and reduction of BAX (proapoptotic), and reduction of expression of p53 and p21 in the delay phase (24 h) [79, 80], and (8) direct effect on MMP [76].

10 The Regeneration and Proliferation Effects of LLLT in Nonbrain Tissue

LLLT has been used for decades, especially in the areas of tissue healing and inflammatory conditions. It has been shown to have a curative effect on diabetic [81] and postsurgical wounds [82, 83], bones [84, 85], and on hypertrophic scars and keloids [86]. Lately, the LLLT has demonstrated a regenerative effect following acute hepatectomy [87–89] and proliferation of stem cells [90–92]. It has also been proven in the peripheral nervous system in rat sciatic nerve mod after crush injury, neuropathy, and neurotube damage, as a reconstructive procedure in rats [93–95]. Prior to the preclinical animal brain model studies, the regenerative and angiogenic effects were tested on skeletal muscles and postmyocardial infarction tissue.

11 LLLT in Skeletal and Myocardial Muscles

In skeletal muscles on rat gastrocnemius muscles during degeneration, the LLLT (He-Ne technique) was demonstrated to promote the maturation process. In irradiation of 632.8 nm wavelength and 6.0 mW for 2.3 min (immediate after injury and 5 days later), the volume of destructed tissue was reduced significantly [96]. The volume fraction of the young myofibers improved significantly in the irradiated group on days 3–30 [97–99] with the facilitation of myofibroblastic differentiation [100]. In the same animal model and the same laboratory, neoformation of blood vessels was observed on days 9–14 [101]. This finding was confirmed in the same laboratory in a different period of activation and on different experimental models (amphibians and mammalian). In summary, the effects on skeletal muscles were confirmed on different species, in two different laboratories, and on two different target systems (muscles and blood vessels).

12   LLLT on Ischemic Myocardium

The effect of LLLT applied immediately and 2 days postsnake sarafotoxin (cardiotoxic) injection to mice was examined in the heart 8 days after injection of a toxin. The left ventricular dilatation was reduced (up to 62%) and the cytopathological components of cardiomyocytes were decreased significantly [102]. In a randomized controlled study, 50 dogs and 26 rats were irradiated twice (immediately and 3 weeks postmyocardial infarction). At 5–6-weeks follow-up, the infarct size was reduced ($p < 0.0001$) and the mortality was decreased ($p < 0.05$) [103]. In rat cardiomyocytes, the mitochondrial damage was decreased 4 h posttherapy and the ATP content was increased by 7.6-fold [103]. The long-term effect on the rat heart was examined and a reduction of left ventricular was found concomitant with a 3.1-fold increase of newly formed blood vessels [104].

13   Clinical Human Studies

13.1   Nonbrain LLLT Studies

Various studies have been performed in dentistry, dermatology, and pain control. The frequency of double-blind, randomized, and placebo-controlled studies (DRP) was relatively limited before 2001–2009. Since then, an increasing amount of well-designed LLLT has been observed. The efficacy and safety of LLLT have been examined in carpal tunnel syndrome studies. Three main studies with different devices revealed controversial results. Naeser et al. [105] treated 11 patients using two techniques (continuous wave red beam laser 15 mW, 632.8 nm, and infrared pulsed laser 9.0 mW, 904 nm) on acupuncture points. The treatment of 3–30 months at various point locations had efficacy in pain relief and on electrophysiological findings. Ekim et al. [106] and Evcik et al. [107] examined 19 and 81 patients, respectively, with two different devices (860 nm, 6 S/cm^2 or 830 nm, 8.9 S/cm^2)—both at the wrists. In the two studies, no significant benefit was found, although there was some efficacy found concerning pain relief. LLLT was examined in primary Raynaud's phenomenon for 5 days a week for 3 weeks in a DRP crossover study. Among the 48 patients, LLLT reduced frequency and severity of attacks [108]. Rochkind et al. [93] examined LLLT efficacy in peripheral nerve injury. Eighteen patients were randomized for irradiation (780, 250 nm) of 21 days. Irradiation was applied to the injured peripheral nerve (450 S/mm) and the spinal cord (300 S/mm). Significant electrodiagnostic and clinical parameters were observed. In the last 2 years, various LLLT-DRP studies have been performed, such as in acute radiculopathy using a laser with 904 nm, 5,000 Hz frequency, power density 20 mW/cm^2 [109], and in planter fasciitis using a infrared laser with 904 nm, GaAs, in which the authors reported a benefit [110]. On the other hand, in subaromial impingement syndrome using a Ga-As-Al laser with 850 nm [111, 112] and in rheumatoid arthritis using a Ga-As-Al laser with 785 nm, doses of 3 J/cm², and mean power of 70 mW, no efficacy was achieved [113].

14   LLLT in Stroke

14.1   Mechanism of Action Proven Directly on the Ischemic Brain

The irradiation of healthy brain tissue (diode laser, 830 nW, 15 min, 4.8 W/cm³) showed an increase of 19% of the tissue ATP with no effect on the adenosine triphosphate (ADP) [114]. With LLLT irradiation (810 nm, 0.2–68 W/cm³) of olfactory ensheathing cells of the adult rat, an increase of BDNF, glial-derived neurotrophic factor (GDNF), and collagen expression at an irradiation of 68 W/cm³ were shown [115]. By the end of the 1990s, Karageuzyan et al. published results of studies on white rats of the ischemic-reperfusion and acute edema model under oxidative stress. After LLLT treatment, they found a decrease of hyperoxides and malonic dialdehyde and elevation of superoxide dismutase in mitochondrial and microsomal fractions [116]. The effect of LLLT irradiation on nitric oxide synthase (NOS) and transforming growth factor beta 1 (TGF-beta 1) was studied by Leung et al. in an unilateral middle cerebral artery (MCA) occlusion model. They found up-regulation of TGF-beta 1 and suppression of NOS activity. The peak effect was on days 3–4. Recently, two different models of laser irradiation techniques with dissimilar wavelengths and power were examined on cryogenic male Wister rats. An increase of tumor necrosis alpha (TNF alpha) and interleukin 6 (IL6) in blood and a higher IL1 beta in brain compared to the control group was shown [117].

15   LLLT In Vivo Animal Studies

The effects of LLLT on animal models (in vivo studies) were examined in various laboratories. The optimal area for insertion was determined in a study performed by Detabpada et al. [118]. They examined three stroke locations (ipsilateral, contralateral, and bilateral inserta) and found identical favorable effects parallel to each LLLT location. Two main studies by Lapchak et al. examined the effect of transcranial laser therapy in stroke on the New Zealand rabbit emboli model (rabbit small clot emboli stroke model RSCEM) [119, 120]. The other study was a two-step study on rabbits (Sprague–Dawley and Wistar rats) by Oron et al. [121].

Lapchak et al. [119] used a laser power of 7.5 and 25 mW/cm² at 3, 6, 12, and 24 h. The rating score as examined 1–24 h post inducing of stroke. They found an improvement of the behavior performance after early treatment of the laser

Table 35.1 Comparison of both preclinical animal studies

	Rat model	Rabbit model
Animal models	Sprague–Dawley rats pMCAO/craniotomy (Stage 1)	Male New Zealand white rabbits
	Wistar rats pMCAO/filament insertion (Stage 2)	New Zealand rabbit emboli model (rabbit small clot emboli stroke model RSCEM)
Mortality rat	16%	
Neurological clinical score	+	+
Immunohistochemistry	+	−
Timing of laser applications	Stage 1: 4 + 24 h	1, 3, 6, 12, 24 h
	Stage 2: 24 h	
Results	No effect at 4 h	Significant effect within 6 h
	Significant effect at 24 h	No effect at 24 h

(up to 6 h), but not in the delay phases (24 h). The percentage of improvement was in the range of 100–195%.

Oron et al. [121] designed the study in two phases. The first phase was performed in order to determine the appropriate timing of LLLT. Phase two, based on the first phase of the study, was performed in order to prove the therapy's efficacy and safety as confirmed by the use of biochemical biomarkers. The first stage was studied on 43 adult Sprague–Dawley rats, using permanent middle cerebral artery occlusion (pMCAO) with a craniotomy technique. The irradiation (LLLT diode Ge-Al-As; power density 7.5 m/cm^2) was directed contralateral to the stroke area in the early period (4 h) poststroke through to the later ones (24 h). The neurological score included six motor, behavior, and gait parameters. The pretreatment deficits were scored 3 h after inducing stroke. The results showed reduction (32%) of neurological deficits after 24 h, whereas no significant reduction in neurological deficit was observed after 4 h. In the second phase, Wistar rats, after induction of pMCA by filament insertion, were examined using high sensitive neurological clinical methods (Modified Neurological Severity Score [mNSS] from 0 to 18). Immunohistological markers were also investigated. Namely, they were analyzed for bromodeoxyuridine (a marker for proliferating cells), tubulin III 9 (a marker for early neurons), and doublecortin (a marker for migrating neuroblasts in the subventricular zone, SVZ). The irradiation was performed 24 h after the stroke based on the results of the first phase of the study. The results showed a significant clinical improvement with a substantial elevation of newly formed neuronal cells and migrating cells, ipsilateral to the induced beam of the LLLT (Table 35.1).

16   Fulfillment of Preclinical and Clinical Optimal Quality Criteria

All research fulfilled the standards of STAIR I, STAIR II, STAIR III, and STAIR IV [15] criteria with the exception of the time window for confirming efficacy (see below). Therefore, it can be determined that under the limitations of translation from a compound-designed study to a noninterventional compound one, the process algorithms were fulfilled as well.

They included the following:

1. All studies were performed according to the criteria of GLP.
2. Each study involved much more testing animals than the required minimum of ten animal models.
3. Use of the style study criteria by the published studies including (a) confirmation by two different laboratories simultaneously; (b) use of two different stroke phases—early and delay; (c) use of three different types of induced stroke models—(1) pMCAO with the rat craniotomy model; (2) pMCAO with filament insertion rat model; and (3) embolic stroke rabbit model; (d) use of histological biomarkers, as well as clinical and histological markers; (e) the use of more than one different animal model (rabbit and rat).
4. In addition, use of a specific criteria for nonpharmacological noninvasive device translation process in confirming the exact location of irradiation for optimal effect of the device [117].

Although only a relatively few well-designed studies were performed, the design of these fulfilled all the necessary criteria for legitimation of technique translation from animal into human studies.

17   Clinical Trials in Acute Stroke

The treatment of acute ischemic stroke by transcranial LLLT was studied in two complementary research papers called NEST1 (Neurotherma Effectiveness and Safety Trial) and the following NEST2—both prospective, multicenter, double-blind, sham-controlled trials.

In NEST1 [122], 120 patients (79 for treatment and 41 shams) were included. Initiation to treatment was limited to 24 h. Outcomes measured were National Institutes of Health Stroke Scale score (NIHSS for the neurological scores), Modified Rankin Scale (mRS-handicapped score), and Barthel Index (BI-disability score). Primary endpoint was defined as a decrease of NIHSS of 9 points or complete recovery on day 90. The inclusion criteria included NIHSS score of 7–22 (moderate to severe); mild strokes (<7) and very severe ones (>22) were excluded. The blood pressure was included only if lower than 180/80 mmHg. Patients suitable for thrombolysis therapy were also excluded. Medical centers in Peru, Israel, and Sweden were included in the study. The LLLT at a wavelength of 808 nm was activated on 20 predetermined locations on both sides for 2 min. Time to window of treatment ranged from 2 to 24 h. Mean time of treatment was >16 and overall study mean time 18 h. In the treated group, the ones with successful outcome were significantly higher than in the sham group (70 vs. 51%, $p=0.035$, stratified by severity and time to treatment and $p=0.048$, stratified only for severity). In all other parameters, including measurement by change of mean NIHSS and mRS on day 90, the treated group was significantly favorable compared with the sham group. The treated group did not differ from the control group, concerning mortality and serious adverse events.

NEST2 [123] recruited 660 patients (331 treated and 327 sham) from 57 centers in the United States, Germany, Sweden, and Peru. The inclusion and exclusion criteria were similar to NEST1. NIHSS and mRS were used as measurement parameters. The mean time from stroke onset was 14.6 h (range 2.7–32.9 h) and the median time 15.0 h for the LLLT group. In the LLLT group, 36.3% achieved favorable outcome vs. 30.9% in the sham group ($p=0.0948$, OR 1.38). In the other measurements, a tendency toward better outcome was observed, although no statistical significance was achieved. On post hoc analysis, a favorable outcome at day 90 ($p<0.044$) was found on patients with NIHSS< 16. In mortality and serious adverse events, no difference between groups was found. An additional multicenter, randomized, sham, controlled study is ongoing.

18   The Critical Analysis of the Translational Process in LLLT

The two animal studies which precede the human clinical trials were designed on two different animal models—the New Zealand rabbit small clot emboli model (RSCEM) and pMCAO Wistar and Sprague–Dawley rat models. The techniques used in the latter models were craniotomy and a filament insertion method. The validation of the RSCEM model study results was published by Lapchak et al. [124, 125].

Rats are one of the most suitable species for use as animal cerebrovascular models. Contrary to gerbil, cat, and dog, the cranial vascular circulation structure has a high similarity to human. Wistar and Sprague–Dawley rat strains are standardized with no inherent coexisting diseases or uniform cerebral vascular anomalies and are considered the gold standard study model. The techniques and issues surrounding pMCAO were discussed by Tamura et al. [126] and others [127, 128], including the methods of open craniotomy and filament insertions. The comparable scores between animal behavior ones, especially in the mNSS and the human neurological and disabling scores (NIH severity score, NIHSS) and modified Barthel Score (mBS), were validated as well [129–131].

A problem of translation of LLLT animal model studies into LLLT clinical human studies has been raised regarding the question of optimal time of treatment. In his translational stroke research of the RSCEM model, Lapchak [132, 133] also discusses the translational process in the LLLT study. In calculating the therapeutic rat window in the RSCEM and human, a ratio of 2.43–3 was achieved. In the rabbit model, the therapeutic window was 6 h, whereas the peak efficacy in the NEST2 study was 14.6 h. The expected window for opportunity to effect was 14–18 h. The opportunity of treatment effect was expanded to 14–24 h. The results of the rat model studies are more difficult to define. The LLLT end time points were 4 h (with no effect) and 24 h (with significant efficacy). The ARR in this study compared to the NEST2 one was 0.58. This data is inconsistent to the expected ARR, based on the physiological characteristics; a contradiction which can be explained by the presence of another peak of activity for LLLT in the subacute phase of stroke. This window of treatment was not examined in the NEST2 study.

The rationality of this finding can be explained by the existence of different parallel pathways of action that may be involved with transcranial LLLT irradiation. As previously discussed, these pathways are activated at various poststoke cascade time points (NMDA receptor activation, inflammation, apoptosis, heat shock protein mechanisms, and angiogenesis). It can be hypothesized that the NEST1 and NEST2 studies confirmed only the effect on a limited part of pathways in the earlier stroke stages. With activation of other pathways, an additional effect may be the cause of the resulting outcomes in the rat model study at 24 h. Meanwhile, no confirmation of this hypothesis into the human model studies can be determined. The very good equivalence between the human and rat models regarding vessel architecture, but on the other hand, the limited equivalence concerning the brain neuroprotection mechanisms, the great difference in the brain architecture in regard to the white matter/gray matter ratio and the failure of translation of many rat ischemic stroke studies from rats to humans, limited the justification for this assumption, but strengthened the rationalization for designing a new clinical study aimed at a later phase of stroke.

The limitations of the translation process from the preclinical animal studies into the human NEST1 and NEST2 studies are in the determination of the effective window of treatment time. The two animal studies using two different species were designed for two different phases of the poststroke cascade. In the RSCEM New Zealand rabbit model, the time of treatment was 6 h poststroke induction; equivalent to 14.6 h in the human acute stroke stage. Even with limitation of the treated group, the efficacy of treatment was confirmed in both studies at the expected points of time. The rat model studies were designed to evaluate the efficacy at the later acute stage and the early subacute stage. The time point of treatment was 24 h. Calculation of the ratio of human/rat time of treatment in the Sprague–Dawley rat compared to the NEST1/NEST2 studies showed an ARR of ≤1, which was not compatible with the physiological and pathophysiological data. The conclusion of this discussion is that the basic hypothesis of LLLT having a favorable effect in late acute and early subacute (24 h to 3 days) stages in humans has not been determined. For confirmation of this hypothesis, another well-designed study specifying a different time of treatment must be designed and performed.

There are some additional limitations for translation of the LLLT preclinical studies into human ones and are listed as follows:

1. The studies were performed on two different species, and according to the GLP Guidelines, one of these should be primates; so this determination being controversial, primates were not included in the LLLT preclinical design.

2. In all the preclinical studies, healthy and young animal models were examined; for translation of preclinical into clinical studies in the stroke area, an in-between step of examining the data on specific hypertensive, diabetic, and aged animal models is recommended; no such in-between steps were performed.
3. Not the cerebral blood flow, nor the objective changes of the "penumbra" volume were examined using imaging markers such as magnetic resonance imaging (MRI), positron emission tomography (PET), or single positron emission tomography (SPECT); therefore, no evidence toward confirmation of life span extension of the "penumbra" could be determined.

The process of confirmation for the use of transcranial laser therapy from working hypothesis and animal model studies can now be recommended to have the optimal medical developmental steps for implementation of a nonpharmacological, noninterventional device. However, the translation process from animal model studies into human clinical trials has had its handicaps. Two very well-designed animal studies with two different animal species were performed. One of the study results was well translated into a human clinical trial. The other study having different results about the window of action was not confirmed. Perhaps it can be designed as an additional mechanism of treatment in the human trials or there should be a new clinical study based on the data of this preclinical one and on the well-performed translational process.

References

1. Feuerstein GZ, Chavez J. Translational medicine for stroke drug discovery: the pharmaceutical industry perspective. Stroke. 2009;40:S121–5.
2. Kaste M, Fogelholm R, Erila T, Palomaki H, Murros K, Rissanen A, Sama S. A randomized, double-blind, placebo-controlled trial of nimodipine in acute ischemic hemispheric stroke. Stroke. 1994;25(7):1348–53.
3. Albers GW, Goldstein LB, Hall D, Lesko LM. Aptiganel hydrochloride in acute ischemic stroke: a randomized controlled trial. JAMA. 2001;286(21):2673–82.
4. Gandolfo C, Sandercock P, Conti M. Lubeluzole for acute ischaemic stroke. Cochrane Database Syst Rev. 2002;(1):CDOO1924.
5. Furuya K, Takeda H, Azhar S, McCarron RM, Chen Y, Ruetzler CA, Wolcott KM, DeGraba TJ, Rothlein R, Hugli TE, del Zoppo GJ, Hallenbeck JM. Examination of several potential mechanisms for the negative outcome in a clinical stroke trial of enlimomab, a murine anti-human intercellular adhesion molecule-1 antibody: a bedside-to-bench study. Stroke. 2001;32(11):2665–74.
6. Davis SM, Lees KR, Albers GW, Diener HC, Markabi S, Karlsson G, Norris J. Selfotel in acute ischemic stroke: possible neurotoxic effects of an NMDA antagonist. Stroke. 2000;31(2):347–54.
7. Ikonomidou C, Turski L. Why did NMDA receptor antagonists fail clinical trials for stroke and traumatic brain injury? Lancet Neurol. 2002;1(6):383–6.
8. Chavez JC, Hurko O, Barone FC, Feuerstein GZ. Pharmacologic interventions for stroke: looking beyond the thrombolysis time window into the penumbra with biomarkers, not a stopwatch. Stroke. 2009;40(10):e558–63.
9. Savitz SI, Schabitz WR. A critique of SAINT II: wishful thinking, dashed hopes, and the future of neuroprotection for acute stroke. Stroke. 2008;39(4):1389–91.

10. Donnan GA. The 2007 Feinberg lecture: a new road map for neuroprotection. Stroke. 2008;39(1):242.
11. Rother J. Neuroprotection does not work! Stroke. 2008;39(2):523–4.
12. Feuerstein GZ, Zaleska MM, Krams M, Wang X, Day M, Rulkowski JL, Finklestein SP, Pangelos MN, Poole M, Stiles GL, Ruffolo RR, Walsh FL. Missing steps in the STAIR case: a translational medicine perspective on the development of NXY-059 for treatment of acute ischemic stroke. J Cereb Blood Flow Metab. 2008;28(1):217–9.
13. Ransom BR, Stys PK, Waxman SG. The pathophysiology of anoxic injury in central nervous system white matter. Stroke. 1990;21 Suppl 11:1152–7.
14. Ho PW, Reutens DC, Phan TG, Wrightn PM, Markus P, Indra I, Young D, Donnan GA. Is white matter involved in patients entered into typical trials of neuroprotection? Stroke. 2005;36(12):242–4.
15. Saver JL, Albers GW, Dunn B, Johnston KC, Fisher M. Stroke therapy academic industry roundtable (STAIR) recommendations for extended window acute stroke therapy trials. Stroke. 2009;40(7):2594–600.
16. Macleod MR, O'Collins T, Horky LL, Howells DW, Donnan GA. Systemic review and metaanalysis of the efficacy of FK506 in experimental stroke. J Cereb Blood Flow Metab. 2005;25(6):713–21.
17. Astrup J, Symon L, Branston NM, Lassen NA. Cortical evoked potential and extracellular K+ and H+ at critical levels of brain ischemia. Stroke. 1977;8(1):51–7.
18. Astrup J, Siesjo BK, Symon L. Thresholds in cerebral ischemia—the ischemic penumbra. Stroke. 1981;12(6):723–5.
19. Winn HR, Rubio R, Berne RM. Brain adenosine production in the rat during 60 seconds of ischemia. Circ Res. 1979;45(4):486–92.
20. Winn HR, Rubio R, Berne RM. Brain adenosine concentration during hypoxia in rats. Am J Physiol. 1981;241(2):H235–42.
21. Hossmann KA. Viability thresholds and the penumbra of focal ischemia. Ann Neurol. 1994;36(4):557–65.
22. Meyer FB, Anderson RE, Sundt Jr TM, Yaksh TL. Intracellular brain p H, indicator tissue perfusion, electroencephalography, and histology in severe and moderate focal cortical ischemia in the rabbit. J Cereb Blood Flow Metab. 1986;6(1):71–8.
23. Takagi K, Ginsberg MD, Globus MY, Detrich WD, Martinez E, Kraydish S, Busto R. Changes in amino acid neurotransmitters and cerebral blood flow in the ischemic penumbral region following middle cerebral artery occlusion in the rat: correlation with histopathology. J Cereb Blood Flow Metab. 1993;13(4):575–85.
24. Warner DS, Takaoka S, Wu B, Ludwig PS, Pearlstein RD, Binkhous AD, Dexter F. Electroencephalographic burst suppression is not required to elicit maximal neuroprotection from pentobarbital in a rat model of focal cerebral ischemia. Anesthesiology. 1996;84(6):1475–84.
25. Galeffi F, Sinnar S, Schwartz-Bloom RD. Diazepam promotes ATP recovery and prevents cytochrome c release in hippocampal slices after in vitro ischemia. J Neurochem. 2000;75(3):1242–9.
26. Wang J, Chambers G, Cottrell JE, Kass IS. Differential fall in ATP accounts for effects of temperature on hypoxic damage in rat hippocampal slices. J Neurophysiol. 2000;83(6):3462–72.
27. Hata R, Maeda K, Hermann D, Mies G, Hossmann KA. Evolution of brain infarction after focal cerebral ischemia in mice. J Cereb Blood Flow Metab. 2000;20(6):937–46.
28. Selman WR, Lust WD, Pundik S, Zhou Y, Ratcheson RA. Comprised metabolic recovery following spontaneous spreading depression in the penumbra. Brain Res. 2004;999(2):167–74.
29. Bardutzky J, Shen Q, Henninger N, Schwab S, Duong TQ, Fisher M. Characterizing tissue fate after transient cerebral ischemia of varying duration using quantitative diffusion and perfusion imaging. Stroke. 2007;38(4):1336–44.

30. Han JL, Blank T, Schwab S, Kollmar R. Inhibited glutamate release by granulocyte-colony stimulating factor after experimental stroke. Neurosci Lett. 2006;432(3):632–34.

31. Zhang ZG, Chopp M. Neurorestorative therapies for stroke: underlying mechanisms and translation to the clinic. Lancet Neurol. 2009;8(5):491–500.

32. Zhang ZG, Zhang L, Jiang Q, Zhang R, Davies K, Powers C, Bruggen N, Chopp M. VEGF enhances angiogenesis and promotes blood–brain barrier leakage in the ischemic brain. J Clin Invest. 2000;106(7):829–38.

33. Zhang ZG, Zhang L, Tsang W, Soltanian-Zadeh H, Morris D, Zhang R, Goussev A, Powers C, Yeich T, Chopp M. Correlation of VEGF and angiopoietin expression with disruption of blood–brain barrier and angiogenesis after focal cerebral ischemia. J Cereb Blood Flow Metab. 2002;22(4):379–92.

34. Hao Q, Su H, Palmer D, Sun B, Gao P, Yang GY, Young WL. Bone marrow-derived cells contribute to vascular endothelial growth factor-induced angiogenesis in the adult brain by supplying matrix metalloproteinase-9. Stroke. 2011;42(2):453–8.

35. Zhang Z, Chopp M. Vascular endothelial growth factor and angiopoietins in focal cerebral ischemia. Trends Cardiovasc Med. 2002;12(2):62–6.

36. Jin K, Zhu Y, Sun Y, Mao XO, Xie L, Greenberg DA. Vascular endothelial growth factor (VEGF) stimulates neurogenesis in vitro and in vivo. Proc Natl Acad Sci U S A. 2002;99(18):11946–50.

37. Sun Y, Jin K, Xie L, Childs J, Mao XO, Longvinova A, Greenberg DA. VEGF-induced neuroprotection, neurogenesis and angiogenesis after focal cerebral ischemia. J Clin Invest. 2003;111(12):1843–51.

38. Arai K, Jin G, Navaratna D, Lo EH. Brain angiogenesis in developmental and pathological processes: neurovascular injury and angiogenic recovery after stroke. FEBS J. 2009;276(17):4644–52.

39. Zacharek A, Chen J, Cui X, Li A, Li Y, Roberts C, Feng Y, Gao Q, Chopp M. Angioppoietin1/Tie2 and VEGF/FIk1 induced by MSC treatment amplifies angiogenesis and vascular stabilization after stroke. J Cereb Blood Flow Metab. 2007;27(10):1684–91.

40. Young D, Lawlor PA, Leone P, Dragunow MJ, During WJ. Environmental enrichment inhibits spontaneous apoptosis, prevents seizures and is neuroprotective. Nat Med. 1999;5(4):448–53.

41. Ikonomidou C, Stefovska V, Turski L. Neuronal death enhanced by N-methyl-D-aspartate antagonists. Proc Natl Acad Sci U S A. 2000;97(23):12885–90.

42. Arvidsson A, Kokaia Z, Lindvall O. N-methyl-D-aspartate receptor-mediated increase of neurogenesis in adult rat dentate gyrus following stroke. Eur J Neurosci. 2001;14(1):10–8.

43. Kluska MM, Witte OW, Bolz J, Redecker C. Neurogenesis in the adult dentate gyrus after cortical infarcts: effects of infarct location, N-methyl-D-aspartate receptor blockade and anti-inflammatory treatment. Neuroscience. 2005;135(3):723–35.

44. Tang T, Liu XJ, Zhang ZQ, Zhou HJ, Luo JK, Huang JF, Yang QD, Li ZQ. Cerebral angiogenesis after collagenase-induced intracerebral hemorrhage in rats. Brain Res. 2007;1175:134–42.

45. Lee SR, Kim HY, Rogowska J, Zhao BQ, Bhide P, Parent JM, Lo EH. Involvement of matrix metalloproteinase in neuroblast cell migration from the subventricular zone after stroke. J Neurosci. 2006;26(13):3491–5.

46. Zhoa BQ, Wang S, Kim HY, Storrie H, Rosen BR, Mooney DJ, Wang X, Lo EH. Role of matrix metalloproteinases in delayed cortical responses after stroke. Nat Med. 2006;12(4):441–5.

47. Borsello T, Clarke PG, Hirt L, Vercelli A, Repici M, Schorderet DF, Bogousslavsky J, Bonny C. A peptide inhibitor of c-Jun N-terminal kinase protects against excitotoxicity and cerebral ischemia. Nat Med. 2003;9(9):1180–6.

48. Gao Y, Signore AP, Yin W, Cao G, Yin XM, Sun F, Luo Y, Graham SH, Chen J. Neuroprotection against focal ischemic brain injury by inhibition of c-Jun N-terminal kinase and attenuation of the mitochondrial apoptosis-signaling pathway. J Cereb Blood Flow Metab. 2005;25(6):694–712.

49. Karu TI, Kalendo GS, Letokhov VS, Lebko VV. Reaction of proliferating and resting tumor cells to pulsed periodic low-intensity laser UV radiation (article in Russian). Dokl Akad Nauk SSSR. 1982;262(2):1498–501.

50. Karu TI, Fedoseeva GE, Ludakhina EV, Kalendo GS, Lobko VV. Effect of low-intensity periodic-impulse laser UV radiation on the nucleic acid synthesis rate in proliferating and resting cells (article in Russian). Tsitologiia. 1983;25(10):1207–12.

51. Karu TI, Kalendo GS, Letokhov VS. Comparison of the action of powerful pulses of ultra-short UV on the DNA replication and transcription functions in proliferating and resting HeLa cells (article in Russian). Radiobiologiia. 1984;24(1):17–20.

52. Karu TI, Kalendo GS, Lobko VV, Piatibrat LV. Growth kinetics of HeLa tumor cells during subculturing after irradiation with low-intensity red light in the stationary growth phase (article in Russian). Eksp Onkol. 1984;8(4):60–2.

53. Posten W, Wrone DA, Dover JS, Amdt KA, Silapunt S, Alan M. Low-laser therapy for wound healing: mechanism and efficacy. Dermatol Surg. 2005;31(3):334–40.

54. Capon A, Mordon S. Can thermal lasers promote skin wound healing? Am J Clin Dermatol. 2003;4(1):1–12.

55. Whetan HT, Smits Jr RL, Buchman EV, Whelan NT, Turner SG, Margolis DA, Cavenini V, Stinson H, Ignatius R, Martin T, Cwiklinski J, Philppi AF, Graf WR, Hodgson B, Gould L, Kane M, Chen G, Caviness J. Effect of NASE light-emitting diode irradiation on would healing. J Clin Laser Med Surg. 2001;19(6):305–14.

56. Brosseau L, Welch V, Wells G, Tugwell P, de Bie R, Gam A, Haman K, Shea B, Morin M. Low laser therapy for osteoarthritis and rheumatoid arthritis: a metaanalysis. J Rheumatol. 2000;27(8):1961–9.

57. Brosseau L, Robinson V, Wells G, Debie R, Gam A, Haman K, Morin M, Shea B, Tugwell P. Low level laser therapy (Classes I, II and III) for treating rheumatoid arthritis. Cochrane Database Syst Rev. 2005;19(4):CD002049.

58. Bjordal JM, Bogen B, Lopes-Martins RA, Klovning A. Can Cochrane reviews in controversial areas be biased? A sensitivity analysis based on the protocol of a Systematic Cochrane Review of low-level laser therapy on osteoarthritis. Photomed Laser Surg. 2005;23(5):453–8.

59. Brosseau L, Robinson V, Wells G, Debie R, Gam A, Haman K, Morin M, Shea B, Tugwell P. WITHDRAWN: low level laser therapy (Classes III) for treating osteoarthritis. Cochrane Database Syst Rev. 2007;18(1):CD002046.

60. Yousefi-Nooraie R, Schonstein E, Heidari K, Rashidian A, Akbari-Kamrani M, Irani S, Shakiba B, Mortaz Hejri SA, Mortaz Hejri SO, Jonaidi A. Low level laser therapy for nonspecific low-back pain. Cochrane Database Syst Rev. 2007;18(2):CD005107.

61. Bjordal JM, Johnson MI, Lopes-Martins RA, Bogen B, Chow R, Ljunggren AE. Short-term efficacy of physical interventions in osteoarthritic knee pain. A systematic review and meta-analysis of randomised placebo-controlled trials. BMC Musculoskelet Disord. 2007;22(8): 51.

62. Chou R, Huffman LH. Nonpharmacologic therapies for acute and chronic low back pain: a review of the evidence for an American Pain Society/American College of Physicians clinical practice guideline. Ann Intern Med. 2007;147(7):492–504.

63. Hurwitz EL, Carragee EJ, van der Velde G, Carroll LJ, Nordin M, Guzman J, Peloso PM, Holm LW, Cote P, Hogg-Johnson S, Cassidy JD, Haldeman S. Treatment of neck pain: non-invasive interventions: results of the Bone and Joint Decade 2000–2010 Task Force on Neck Pain and Its Associated Disorders. Spine. 2008;33(4 Suppl):S123–52.

64. Guzman J, Haldeman S, Carroll LJ, Carragee EJ, Hurwitz EL, Peloso P, Nordin M, Cassidy JD, Holm LW, Cote P, van der Velde G, Hogg-Johndon S. Clinical practice implications of the Bone and Joint Decade 2000–2010 Task Force on Neck Pain and Its Associated Disorders: from concepts and findings to recommendations. Spine. 2008;33(4 Suppl):S199–213.

65. Chow RT, Johnson MI, Lopes-Martins RA, Bjordal JM. Efficacy of low-level laser therapy in the management of neck pain: a systematic review and meta-analysis of randomized placebo or active-treatment controlled trials. Lancet. 2009;374(9705):1897–908.

66. Derecki MO, Chopp M, Garcia JH, Hetzel FW. Depth measurements and histopathological characterization of photodynamic therapy generated normal brain necrosis as a function of incident optical energy dose. Photochem Photobiol. 1991;54(1):109–12.

67. Ilic S, Leichliter S, Streeter J, Oron A, DeTaboada L, Oron U. Effects of power densities, continuous and pulse frequencies and number of sessions of low-level laser therapy on intact rat brain. Photomed Laser Surg. 2006;24(4):458–66.

68. McCarthy TJ, De Taboada L, Hildebrandt PK, Ziemer EL, Richieri SP, Streeter J. Long-term safety of single and multiple infrared transcranial laser treatments in Sprague-Dawley rats. Photomed Laser Surg. 2010;28(5):663–7.

69. Silveira PC, Streck EL, Pinho RA. Evaluation of mitochondrial respiratory chain activity in wound healing by low-level laser therapy. J Photochem Photobiol B. 2007;86(3):279–82.

70. Oron U, Ilic S, De Taboada L, Streeter J. Ga-As (808 nm) laser irradiation enhances ATP production in human neuronal cells in culture. Photomed Laser Surg. 2007;25(3):180–2.

71. Ben-Dov N, Shefer G, Irintchev A, Wernig A, Oron U, Halevy O. Low-energy laser irradiation affects satellite cell proliferation and differentiation in vitro. Biochim Biophys Acta. 1999;1448(3):372–80.

72. Tuby H, Maltz L, Oron U. Modulation of VEGF and iNOS in the rat heart by low level laser therapy are associated with cardioprotection and enhanced angiogenesis. Lasers Surg Med. 2006;38(7):682–8.

73. Hou JF, Zhang H, Yuan X, Li J, Wei YJ, Hu SS. In vitro effects of low-level laser irradiation for bone marrow mesenchymal stem cells: proliferation, growth factors secretion and myogenic differentiation. Lasers Surg Med. 2008;40(10):726–33.

74. Ferreira MC, Gameiro J, Nagib PR, Brito VN, Vasconcellos Eda C, Verinaud L. Effect of low intensity helium-neon (He-Ne) laser irradiation on experimental paracoccidioidomycotic wound healing dynamics. Photochem Photobiol. 2009;85(1):227–33.

75. Dang Y, Ye X, Weng Y, Tong Z, Ren Q. Effects of the 532-nm and 1,064-nm Q-switched Nd:YAG lasers on collagen turnover cultured human skin fibroblasts: a comparative study. Lasers Med Sci. 2010;25(5):719–26.

76. Novoselova EG, Glushkova OV, Cherenkov DA, Chudnovsky VM, Fesenko EE. Effects of low-power laser radiation on mice immunity. Photodermatol Photoimmunol Photomed. 2006;22(1):33–8.

77. Mesquita-Ferrari RA, Martins MD, Silva Jr JA, da Silva TD, Piovesan RF, Pavesi VC, Bussadroi SK, Fernandos KP. Effects of low-level laser therapy on expression of THF-α and -β in skeletal muscle during the repair process. Lasers Med Sci. 2011;26(3):35–40.

78. Shefer G, Oron U, Irintchev A, Wernig A, Halevy O. Skeletal muscle cell activation by low-energy laser irradiation: a role for the MAPK/ERK pathway. J Cell Physiol. 2001;187(1):73–80.

79. Shefer G, Partridge TA, Heslop L, Gross JG, Oron U, Halevy O. Low-energy laser irradiation promotes the survival and cell cycle entry of skeletal muscle satellite cells. J Cell Sci. 2002;115(Pt 7):1461–9.

80. Aimbire F, Santos FV, Albertini R, Castro-Faria-Neto HC, Mittmann J, Pacheco-Soares C. Low-level laser therapy decreases levels of lung neutrophils anti-apoptotic factors by a NF-kappaB dependent mechanism. Int Immunopharmacol. 2008;8(4):603–5.

81. Jahangiri Noudeh Y, Shabani M, Vatankhah N, Hashemian SJ, Akbari K. A combination of 670 nm and 810 nm diode lasers for wound healing acceleration in diabetic rats. Photomed Laser Surg. 2010;28(5):621–7.

82. Krynicka I, Rutowski R, Staniszewska-Kus J, Fugiel J, Zaleski A. The role of laser biostimulation in early post-surgery rehabilitation and its effect on wound healing. Ortop Traumatol Rehabil. 2010;12(1):67–9.

83. de Oliveira Guirro EC, Montebelo MI, de Lima Bortot B, de Almeida Bortot B, da Costa Betito Torres MA, Polacow ML. Effect of laser (670 nm) on healing of wounds covered with occlusive dressing: a histologic and biomechanical analysis. Photomed Laser Surg. 2010;28(5): 629–34.

84. Medalha CC, Amorim BO, Ferreira JM, Oliveira P, Pereira RM, Tim C, Lirani-Galvao AP, de Silva OL, Renno AC. Comparison of the effects of electrical field stimulation and low-level laser therapy on bone loss in spinal cord-injured rats. Photomed Laser Surg. 2010;28(5): 669–74.

85. Bashardoust Tajali S, Macdermid JC, Houghton P, Grewal R. Effects of low power laser irradiation on bone healing in animals: a meta-analysis. J Orthop Surg Res. 2010;5(1):1.

86. Barolet D, Boucher A. Prophylactic low-level light therapy for the treatment of hypertrophic scars and keloids; a case series. Lasers Surg Med. 2010;42(6):597–601.

87. Liu YH, Chang CC, Ho CC, Pei RJ, Lee KY, Yeh KT, Chan Y, Lai YS. Effects of diode 808 nm GaAlAs low-power laser irradiation on inhibition of the proliferation of human hepatoma cells in vitro and their possible mechanism. Res Commun Mol Pathol Pharmacol. 2004;115–116:185–201.

88. Oliveira AF, Silva TC, Sankarankutty AK, Pacheco EG, Ferreira J, Bagnato VS, Zucoloto S, Silva Ode C. The effect of laser on reminiscent liver tissue after 90% hepatectomy in rats. Acta Cir Bras. 2006;21 Suppl 1:29–32.

89. Oron U, Maltz L, Tuby H, Sorin V, Czerniak A. Enhanced liver regeneration following acute hepatectomy by low-level laser therapy. Photomed Laser Surg. 2010;28(5):675–8.

90. Tuby H, Maltz L, Oron U. Low-level irradiation (LLLI) promotes proliferation of mesenchymal and cardiac stem cells in culture. Lasers Surg Med. 2007;39(4):373–8.

91. Eduardo Fde P, Bueno DF, de Freitas PM, Marques MM, Passos-Bueno MR, Eduardo Cde P, Zatz M. Stem cell proliferation under low intensity laser irradiation: a preliminary study. Lasers Surg Med. 2008;40(6):433–8.

92. Lin F, Josephs SF, Alexandrescu DT, Ramos F, Bogin V, Gammill V, Dasanu CA, De Necochea-Campion R, Patel AN, Carrier E, Koos DR. Lasers, stem cells and COPD. J Transl Med. 2010;8:16.

93. Rochkind S, Drory V, Alon M, Nissan M, Ouaknine GE. Laser phototherapy (780 nm), a new modality in treatment of long-term incomplete peripheral nerve injury: a randomized double-blind placebo-controlled study. Photomed Laser Surg. 2007;25(5):436–42.

94. Rochkind S, Leider-Trejo L, Nissan M, Shamir MH, Kharenko O, Alon M. Efficacy of 780-nm laser phototherapy on peripheral nerve regeneration after neurotube reconstruction procedure (double-blind randomized study). Photomed Laser Surg. 2007;25(3):137–43.

95. Chen YS, Hsu SF, Chiu CW, Lin JG, Chen CT, Yao CH. Effect of low-power pulsed laser on peripheral regeneration in rats. Microsurgery. 2005;25(1):83–9.

96. Bibkova A, Belkin V, Oron U. Enhancement of angiogenesis in regenerating gastronemius muscle of the toad (Bufo viridis) by low-energy laser irradiation. Anat Embryol (Berl). 1994;190(6):597–602.

97. Weiss N, Oron U. Enhancement of muscle regeneration in the rat gastronemius muscle by low energy laser irradiation. Anat Embryol (Bert). 1992;186(5):497–503.

98. Bibkova A, Oron U. Regeneration in denervated toad (Bufo viridis) gastronemius muscle and the promotion of the process by low energy laser irradiation. Anat Rec. 1995;241(1):123–8.

99. Bibkova A, Oron U. Attenuation of the process of muscle regeneration in the toad gastronemius muscle by low energy laser irradiation. Lasers Surg Med. 1994;14(4):355–61.

100. Rubeiro MA, Albuquerque Jr RI, Ramalho LM, Pinheiro AL, Bonjardin LR, Da Cunha SS. Immunohistochemical assessment of myofibroblasts and lymphoid cells during wound healing in rats subjected to laser photobiomodulation at 660 nm. Photomed Laser Surg. 2009;27(1):49–55.

101. Bibikova A, Belkin V, Oron U. Enhancement of angiogenesis in regenerating gastronemius muscle of the toad (Bufo viridis) by low-energy laser irradiation. Anat Embryol (Bert). 1994;190(6):597–602.

102. Yaakov N, Bdolah A, Wollberg Z, Ben-Haim SA, Oron U. Recovery from sarafotoxin-b induced cardiopathological effects in mice following low energy laser irradiation. Basic Res Cardiol. 2000;95(5):385–8.

103. Oron U, Yaakobi T, Oron A, Mordechovitz D, Shofti R, Hayam G, Dror U, Gepstein I, Wolf T, Haudenschild C, Haim SB. Low-energy laser irradiation reduces formation of scar tissue after myocardial infarction in rats and dogs. Circulation. 2001;103(2):296–301.

104. Yaakobi T, Shoshany Y, Levkovitz S, Rubin O, Ben Haim SA, Oron U. Long-term effect of low energy laser irradiation on infarction and reperfusion injury in the rat heart. J Appl Physiol. 2001;90(6):2411–9.

105. Naeser MA, Hahn KA, Lieberman BE, Branco KF. Carpal tunnel syndrome pain treated with low-level laser and microamperes transcutaneous electric nerve stimulation: a controlled study. Arch Phys Med Rehabil. 2002;83(7):978–88.

106. Ekim A, Armagan O, Tascioglu F, Oner C, Colak M. Effect of low level laser therapy in rheumatoid arthritis patients with carpal tunnel syndrome. Swiss Med Wkly. 2007;137(23–24): 347–52.

107. Evcik D, Kavuncu V, Cakir T, Subasi V, Yarman M. Laser therapy in the treatment of carpal tunnel syndrome: a randomized controlled trial. Photomed Laser Surg. 2007;25(1):34–9.

108. Hirschl M, Katzenschlager R, Francesconi C, Kundi M. Low level laser therapy in primary Raynaud's phenomenon—results of a placebo controlled, double blind intervention study. J Rheumatol. 2004;31(12):2406–12.

109. Konstantinovic LM, Kanjuh ZM, Milovanovic AN, Cutovic MR, Djurovic AG, Savic VG, Dragin AS, Milovanovic ND. Acute low back pain with radiculopathy: a double-blind, randomized, placebo-controlled study. Photomed Laser Surg. 2010;28(4):553–60.

110. Kiritsi O, Tsitas K, Malliaropoulos N, Mikroulis G. Ultrasonographic evaluation of plantar fasciitis after low-level laser therapy: results of a double-blind, randomized placebo-controlled study. Lasers Med Sci. 2010;25(2):275–81.

111. Bal A, Eksioglu E, Gurcay E, Gulec B, Karaahmet O, Cakci A. Low-level laser therapy in subacromial impingement syndrome. Photomed Laser Surg. 2009;27(1):31–6.

112. Yeldan I, Cetin E, Ozdincler AR. The effectiveness of low-level laser therapy on shoulder function in subacromial impingement syndrome. Disabil Rehabil. 2009;31(11):935–40.

113. Meireles SM, Jones A, Jennings F, Suda AL, Parizotto NA, Natour J. Assessment of the effectiveness of low-level laser therapy on the hands of patients with rheumatoid arthritis: a randomized, double-blind controlled trial. Clin Rheumatol. 2010;29(5):501–9.

114. Mochizuki-Oda N, Kataoka Y, Cui Y, Yamada H, Heya M, Awazu K. Effects of near-infra-red laser irradiation on adenosine triphosphate and adenosine diphosphate contents of rat brain tissue. Neurosci Lett. 2002;323(3):207–10.

115. Byrnes KR, Wu X, Waynant RW, Ilev IK, Anders JJ. Low power laser irradiation alters gene expression of olfactory ensheathing cells in vitro. Lasers Surg Med. 2005;37(2):161–71.

116. Karageuzyan KG, Sekoyan ES, Karagyan AT, Pogosyan NR, Manucharyan GG, Sekoyan AE, Tunyan AY, Boyajyan VG, Karageuzyan MK. Phospholipid pool, lipid peroxidation, and superoxide dismutase activity under various types of oxidative stress of the brain and effect of low-energy infrared laser irradiation. Biochemistry (Mosc). 1998;63(10):1226–32.

117. Moreira MS, Velasco IT, Ferreira LS, Ariga SK, Barbeiro DF, Meneguzzo DT, Abatepaulo F, Marques MM. Effect of phototherapy with low intensity laser on local and systemic immunomodulation following focal brain damage in rat. J Photochem Photobiol B. 2009;97(3): 145–51.

118. Detabpada L, Ilic S, Leichliter-Martha S, Oron U, Oron A, Streeter J. Transcranial application of low-energy laser irradiation improves neurological deficits in rats following acute stroke. Lasers Surg Med. 2006;38(1):70–3.

119. Lapchak PA, Wei J, Zivin JA. Transcranial infrared laser therapy improves clinical rating scores after embolic strokes in rabbits. Stroke. 2004;35(8):1985–8.

120. Lapchak PA, Salgado KF, Chao CH, Zivin JA. Transcranial near-infrared light therapy improves motor function following embolic strokes in rabbits: an extended therapeutic window study using continuous and pulse frequency delivery modes. Neuroscience. 2007;148(4):907–14.

121. Oron A, Oron U, Chen J, Eilam A, Zhang C, Sadeh M, Lampl Y, Streeter J, De Taboada L, Chopp M. Low-level laser therapy applied transcranially to rats after induction of stroke significantly reduces long-term neurological deficits. Stroke. 2006;37(10):2620–4.

122. Lampl Y, Zivin JA, Fisher M, Lew R, Welin L, Dahlof B, Borenstein P, Anderson B, Perez J, Caparo C, Ilic S, Oron U. Infrared laser therapy for ischemic stroke: a new treatment strategy: results of the Neuro Thera Effectiveness and Safety 1 (NEST-1). Stroke. 2007;38(6): 1843–9.

123. Zivin JA, Albers GW, Bornstein N, Chippendale T, Dahlof B, Devlin T, Fisher M, Hacke W, Holt W, Ilic S, Kasner S, Lew R, Nash M, Perez J, Rymer M, Schellinger P, Schneider D, Schwab S, Veltkamp R, Walker M, Streeter J. Effectiveness and safety of transcranial laser therapy for acute ischemic stroke. Stroke. 2009;40(4):1359–64.

124. Lapchak PA, Araujo DM, Pakola S, Song D, Wei J, Zivin JA. Microplasmin: a novel thrombolytic that improves behavioral outcome after embolic strokes in rabbits. Stroke. 2002;33(9):2279–84.

125. Lapchak PA, Han MK, Salgado KR, Streeter J, Zivin JA. Safety profile of transcranial near-infrared laser therapy administered in combination with thrombolytic therapy to embolized rabbits. Stroke. 2008;39(11):3073–8.

126. Tamura A, Graham DI, McCulloch J, Teasdale GM. Focal cerebral ischaemia in the rat: 1. Description of technique and early neuropathological consequences following middle cerebral artery occlusion. J Cereb Blood Flow Metab. 1981;1(1):53–60.

127. Imai H, McCulloch J, Graham DI, Masayasu H, Macrae IM. New method for the quantitative assessment of axonal damage in focal cerebral ischemia. J Cereb Blood Flow Metab. 2002;22(9):1080–9.

128. Karpiak SE, Tagliavia A, Wakade CG. Animal models for the study of drugs in ischemic stroke. Annu Rev Pharmacol Toxicol. 1989;29:403–14.

129. Chen J, Sanberg PR, Li Y, Wang L, Lu M, Willing AE, Sanchez-Ramos J, Chopp M. Intravenous administration of human umbilical cord blood reduces behavioral deficits after stroke in rats. Stroke. 2001;32(11):2682–8.

130. Modo M, Stroemer RP, Tang E, Veizovic T, Sowniski P, Hodges H. Neurological sequelae and long-term behavioural assessment of rats with transient middle cerebral artery occlusion. J Neurosci Methods. 2000;104(1):99–109.

131. Zhang L, Chen J, Li Y, Zhang ZG, Chopp M. Quantitative measurement of motor and somatosensory impairments after mild (30 min) and severe (2 h) transient middle cerebral artery occlusion in rats. J Neurol Sci. 2000;174(2):141–6.

132. Lapchak PA. Taking a light approach to treating acute ischemic stroke patients: transcranial near-infrared laser therapy translational science. Ann Med. 2010;42(8):576–86.

133. Lapchak PA. Translational stroke research using a rabbit embolic stroke model: a correlative analysis hypothesis for novel therapy development. Transl Stroke Res. 2010;1(2):96–107.

Chapter 36
Use of Microbubbles in Acute Stroke

Marta Rubiera and Carlos A. Molina

Abstract Microbubbles (MB) are gas- or air-filled lipid-shell microspheres in the micron-size range with special acoustic properties that have been used since a long time as diagnostic ultrasound (US) echo-contrasts (Eur J Radiol 27:S157–60, 1998). When an MB passes through a US energy field, it experiences translations and size-oscillations, generating harmonic signals which are able to increase the acoustic impedance mismatch between the blood and surrounding tissue, improving the diagnostic vascular US (Advances in echo imaging using contrast enhancement, 2nd ed. Dordrecht: Kluwer; 1997). However, in the last few years, US therapeutic effects have been demonstrated, and the role of MB as enhancers of therapeutic US has been developed.

1 Sonothrombolysis

Currently, the only approved treatment for acute ischemic stroke is the intravenuous (IV) recombinant tissue plasminogen activator (tPA) administered during the first 4.5 h from symptoms onset [1]. tPA can initiate the intracranial thrombus dissolution leading to an early reperfusion of the brain ischemic tissue and subsequent recovery. However, less than 5% of patients with an ischemic stroke receive IV tPA. Furthermore, only 30–40% of treated patients achieve early recanalization, and only in 18% this recanalization is complete [2–4]. To improve the recanalization rates and clinical outcome of these patients several therapeutic strategies have been developed, and one of them is sonothrombolysis.

M. Rubiera, MD, PhD • C.A. Molina, MD, PhD (✉)
Stroke Unit, Department of Neurosciences, Hospital Universitari Vall d'Hebron,
Barcelona 08035, Spain
e-mail: cmolina@vhebron.net

P.A. Lapchak and J.H. Zhang (eds.), *Translational Stroke Research*, Springer Series
in Translational Stroke Research, DOI 10.1007/978-1-4419-9530-8_36,
© Springer Science+Business Media, LLC 2012

Ultrasound (US) enhanced thrombolysis or sonothrombolysis is the ability of US to increase enzymatic thrombolysis. Although the mechanisms are not fully understood, it is postulated that the thrombolysis enhancement is related to mechanical, nonthermal effects of US [5].

US negative pressure waves in liquid media (as blood) create fluid motion or microstreaming and radiation forces, which can promote tPA circulation and increase the thrombus surface in contact with the enzyme [6]. Furthermore, in vitro studies have shown that US insonation on a clot leads to reversible disaggregation of cross-linked fibrin fibers [7]. US can also increase the exposed plasmin-binding sites and the penetration of fibrinolytic enzymes into the clot [8].

The acoustic cavitation, which is the ability of ultrasound in creating microbubbles of gas from the gasses dissolved in a liquid medium, may also play a role in sonothrombolysis by direct harm to the clot surface or increase in tPA permeation inside the thrombus [9]. However, acoustic cavitation needs high-intensity US energy, which cannot be achieved in human practice because of its harming thermal and mechanical effects.

Several frequencies and intensities of US have been tested in vitro and in animal models [8, 10–13]. The ultrasonic negative pressure is directly related to the US intensity and inversely related to the frequency [13]. Higher US frequencies lead to greater energy attenuation through the skull (up to 90% of energy loss with 1–2 MHz frequencies). On the other hand, an increase in the thermal effects of US is associated with higher intensities [14, 15]. Therefore, the majority of initial in vitro and animal models for sonothrombolysis were developed with low frequency (kHz range) and low intensity (0.2–2 W/cm^2) US [15]. However, a clinical trial of sonothrombolysis in acute stroke with low frequency US had to be stopped because of symptomatic bleeding into the brain [16]. For that reason, the clinical trials at the present time are restricted to low megahertz frequency–low intensity US, similar to that used for diagnostic purposes.

The safety and efficacy of 2 MHz low-intensity (<700 mW/cm^2) diagnostic US settings for sonothrombolysis was evaluated in the CLOTBUST (*Combined Lysis of Thrombus in Brain ischemia using transcranial Ultrasound and Systemic TPA*) trial [4]. This was a randomized international study which tested the combination of 0.9 mg/kg of IV tPA administered within 3 h from symptoms onset and 2 h of continuous monitoring with conventional diagnostic transcranial Doppler (TCD) vs. IV tPA alone. After adjusting for potential confounders, the target group achieved the combined efficacy end-point (complete recanalization on TCD or dramatic clinical recovery defined by total NIHSS score ≤3 points or improvement by ≥10 NIHSS points within 2 h after the TPA bolus) in 49% of cases, as compared with 30% in controls ($p = 0.03$). The SICH rate was 4.8% in both groups, and there was a trend toward a better functional outcome in the US-monitoring group.

This study opened the door to an intense research for improving the effectiveness of sonothrombolysis in acute stroke, and MB potentiated sonothrombolysis was developed.

2   Microbubbles Potentiated Sonothrombolysis

In the last few years, several experimental data have suggested that MB can increase the effect of US in sonothrombolysis [9, 17].

As commented before, acoustic cavitation cannot be achieved by the low-intensity US available in clinical practice. However, extrinsic administrated MB may act as nuclei for the acoustic cavitation, lowering the US intensity threshold for this acoustic phenomenon [9].

There are two types of acoustic cavitation. Static cavitation occurs when low-intensity US waves are applied on MB. Because of the specific acoustic properties of the MB, they experience translations and size-oscillations generating a harmonic emission. This phenomenon also releases energy and agitates the fluid where the MB are dissolved, improving tPA delivery and penetration inside the clot in the case of a vascular occlusion [18]. If the US negative pressure is increased, the bubble collapses (inertial cavitation) leading to intense localized stresses and microjets, which could cause mechanical fragmentation of the thrombus [19]. However, experimental studies have suggested that static cavitation is the main mechanism of MB-potentiated sonothrombolysis on acute stroke, as the US energy needed for the inertial cavitation is not always achieved intracranially because of the US attenuation through the skull [18].

3   Clinical Trials with MB-Enhanced Sonothrombolysis

Multiple animal models have confirmed the boosting effect of MB on sonothrombolysis, either in peripheral or intracranial vessels [20–22].

In human stroke, the first largest study published at date with MB-enhanced sonothrombolysis tested the synergic effect of three bolus of an air-filled galactose-based MB (Levovist®, Shering, Berlin, Germany) associated with 2 h of continuous high-frequency–low-intensity diagnostic TCD monitoring and IV tPA [23]. This was a pilot nonrandomized study, which included patients with an acute MCA occlusion within the 3 h time-window and compared the mentioned protocol with the US + tPA-group and tPA alone-group from CLOTBUST trial. Thirty-eight patients were treated with MB, and investigators demonstrated that 2 h after tPA bolus the tPA-TCD-Levovist group achieved a 55% sustained recanalization rate, compared with 40 and 23% in the tPA-TCD and tPA alone-groups. The rate of intracranial hemorrhage in the target group was 23%, with only 3% of SICH, probably reflecting a higher recanalization rate according to the authors.

The same MB type has been tested with TCCS in a recent study [24]. Patients with a <3 h acute MCA occlusion were randomized to conventional IV tPA alone or associated with 1 h of hand-handle TCCS monitoring and a continuous infusion of Levovist during the same time-period. Complete recanalization in the target group was achieved in 48% of cases, but the study was stopped when only 9 patients have been included because of an unexpected increase on the intracranial hemorrhage rate (78%). However, none of these hemorrhages were symptomatic and it is also unclear whether the hemorrhage increase can be related to the synergistic effect of MS or the TCCS monitoring (which has been suggested previously). This issue was not resolved in the study because of an absence of a non-US group.

Several characteristics may influence the effect of MB on sonothrombolysis. Accumulation and penetration of MB inside the clot can be related to MB-stability, which depends on MB structure, size, and stability in blood-stream [25]. The first generation of MB had bigger sizes and were filled with room-air. Second generation MS are filled with heavier-weighted gases, which confer the MB more stability, and longer life-time. However, a small study comparing first and second generation MB (Levovist® and Sonovue® respectively) showed no difference between both kinds of echocontrast on MB-enhanced sonothrombolysis [26]. Sonovue® was also tested in another small study combined with IV tPA and 1 h TCCS-insonation. There was a significant increase in the recanalization rate and better early clinical outcome, but the long-term clinical information is not available [27].

The safety and feasibility of infusion of new and more stable 1–2-μm C_3F_8 perfultren-lipid MB (MRX 801) was tested in a small randomized pilot clinical trial with patients treated with ultrasound-enhanced thrombolysis [28]. The smaller size of the bubble theoretically allows permeation through the thrombus, and interestingly, in 75% of patients MB permeated to areas with no pretreatment residual flows visible. Fifteen patients were included, and there was not any SICH, with three asymptomatic hemorrhages in the target-group (12 patients).

Based on this study, a multicenter international dose-escalation phase I–II single-blinded trial (TUCSON: Transcranial Ultrasound in Clinical Sonothrombolysis, NCT00504842) was designed [29]. Patients with an acute intracranial occlusion detected by TCD were randomized 2:1 to IV tPA plus continuous infusion of MRX 801 plus TCD monitoring during 90 min or conventional IV tPA. The primary safety outcome was the presence of SICH, and activity outcomes were complete recanalization rate and functional outcome. Thirty-five patients were included, 23 in the target group (Cohort 1 (1.4 mL): 12 patients; Cohort 2 (2.8 mL): 11 patients). There was not any symptomatic hemorrhage in the first Cohort and controls, but 3 in Cohort 2 (27%), so the study was stopped by the sponsor. The recanalization tended to be faster and higher in both treatment groups (67 and 46% of complete recanalization in Cohort 1 and 2, as compare with 33% in controls), and there was a trend toward better functional outcome after 3 months in patients who received the MB. The authors suggest that the increase in SICH in the Cohort 2 may be influenced by imbalances between groups. The Cohort 2 had a higher baseline NIHSS, a longer interval between tPA and MRX-801 infusion, and probably the most important, higher systolic blood pressure with blood pressure control protocol violations. Therefore, a safe dose of 1.4 mL has been identified with a trend toward increased recanalization rate and better functional outcomes, thus additional studies are warranted. Furthermore, the increase in SICH observed in the second dose tier probably will lead to new experiments to understand the mechanisms of interaction between MS, US, and ischemic tissues [30].

In conclusion, a recent meta-analysis have shown that sonothrombolysis is safe and leads to higher rates of complete recanalization when compared to systemic thrombolysis. Furthermore, higher rates of complete recanalization were achieved in patients treated with the combination of tPA, US, and MB (54.3%; 50/92) in comparison to the combination of tPA and US alone (36.6%; 70/191; $P = 0.005$). Further research in sonothrombolysis, specially potentiated with MB, is warranted [31].

4 Intraarterial Application of Sonothrombolysis

The problem of US energy attenuation at the skull bones can be solved by the application of intraarterial US [32]. EKOS NeuroWave™ catheter (EKOS Corporation, Bothell, WA) is a standard microinfusion catheter for intraarterial delivery of thrombolytics with a 2-mm 1.7–2.1 MHz pulsed-wave transducer US (average spatial intensity 400 mW) at its tip.

The EKOS catheter was tested in the Interventional Management of Stroke (IMS) phase II trial [33]. Patients with <3 h acute stroke received 0.6 mg/kg of IV tPA and were transferred to the angiogram suite, where if a proximal occlusion was detected, IA tPA and low intensity US were administered. Small arteries are not accessible to EKOS, and received isolated IA tPA. The safety end-point was SICH, with 9.9% of cases (nonsignificantly superior to 6.6% in the NINDS-rt-PA Stroke Study). The recanalization rate was higher compared to IA tPA in the IMS-I trial, and similar to other embolectomy mechanical devices. The functional outcome measured by the Barthel Index and Global Test Statistic was better than IV-treated patients in the NINDS-rt-PA Stroke Study. These promising results have led to the inclusion of the EKOS catheter in the IMS III trial, which is comparing standard IV tPA therapy with a combination of IV tPA and intraarterial rescue strategies [46].

The combination of endovascular reperfusion techniques, US application, and MB is also tempting. Recently, the results of an experimental study combining EKOS with a MS perfusion have shown a significant increase in thrombus lysis as compared with EKOS alone [34]. In human stroke, a pilot study tested the administration of IV MS (Levovist®) and continuous TCD with IA tPA with a 78% of in-procedure recanalization [35].

5 Other Applications of MB in Acute Stroke

This chapter has been focused to the main use of MB in acute stroke, MB-enhanced sonothrombolysis. However, extensive biomedical and ultrasonic research in the field of MB has been done in the last years. Some of the most promising new advances are related to the development of "intelligent MB": the targeted and loaded MB.

In the targeted MB or immunobubbles, antibodies targeted to a specific tissue have been attached to the surface of the bubble. Therefore, these MB adhere to a specific area of the body, where they can be activated by ultrasound. Several applications of immunobubbles in medicine have been tested lately. In stroke, i.e., immunobubbles have been targeted to thrombus surface using antibodies against the glycoprotein IIb–IIIa of the platelets. This has been used to improve diagnosis of thrombus location, detect microemboli, and also to improve thrombolysis by increasing the number of bubbles attached to the thrombus instead of circulating [22, 36–38].

MB can also be loaded with pharmacoactive drugs. These loaded MB can be specifically activated by an US energy field, and release their contains in a selected part of the body, allowing transference of the drugs to a damaged tissue, i.e., neuroprotectants to the ischemic tissue (allowing administration of neuroprotectants in a systemic lower dose, and reducing potential secondary effects). Of course, the release of the loaded MB can be directed more specifically, if the MB are also targeted to an specific tissue [30, 39].

6 Conclusion

Several applications of MB in acute stroke have been described, but the research is progressively adding very promising possibilities for the use of this noninvasive strategy in the treatment of acute ischemic stroke.

References

1. Hacke W, Kaste M, Bluhmki E, Brozman M, Davalos A, Guidetti D, et al. Thrombolysis with alteplase 3 to 4.5 hours after acute ischemic stroke. N Engl J Med. 2008;359(13):1317–29.
2. Leys D, Ringelstein EB, Kaste M, Hacke W. Facilities available in European hospitals treating stroke patients. Stroke. 2007;38(11):2985–91.
3. Alexandrov AV. Ultrasound identification and lysis of clots. Stroke. 2004;35(11 Suppl 1): 2722–5.
4. Alexandrov AV, Molina CA, Grotta JC, Garami Z, Ford SR, Alvarez-Sabin J, et al. Ultrasound-enhanced systemic thrombolysis for acute ischemic stroke. N Engl J Med. 2004;351(21): 2170–8.
5. Daffertshofer M, Hennerici M. Ultrasound in the treatment of ischaemic stroke. Lancet Neurol. 2003;2(5):283–90.
6. Polak JF. Ultrasound energy and the dissolution of thrombus. N Engl J Med. 2004;351(21): 2154–5.
7. Braaten JV, Goss RA, Francis CW. Ultrasound reversibly disaggregates fibrin fibers. Thromb Haemost. 1997;78(3):1063–8.
8. Siddiqi F, Odrljin TM, Fay PJ, Cox C, Francis CW. Binding of tissue-plasminogen activator to fibrin: effect of ultrasound. Blood. 1998;91(6):2019–25.
9. Tachibana K, Tachibana S. Albumin microbubble echo-contrast material as an enhancer for ultrasound accelerated thrombolysis. Circulation. 1995;92(5):1148–50.

10. Luo H, Steffen W, Cercek B, Arunasalam S, Maurer G, Siegel RJ. Enhancement of thrombolysis by external ultrasound. Am Heart J. 1993;125(6):1564–9.

11. Frenkel V, Oberoi J, Stone MJ, Park M, Deng C, Wood BJ, et al. Pulsed high-intensity focused ultrasound enhances thrombolysis in an in vitro model. Radiology. 2006;239(1):86–93.

12. Lauer CG, Burge R, Tang DB, Bass BG, Gomez ER, Alving BM. Effect of ultrasound on tissue-type plasminogen activator-induced thrombolysis. Circulation. 1992;86(4):1257–12564.

13. Schafer S, Kliner S, Klinghammer L, Kaarmann H, Lucic I, Nixdorff U, et al. Influence of ultrasound operating parameters on ultrasound-induced thrombolysis in vitro. Ultrasound Med Biol. 2005;31(6):841–7.

14. Pfaffenberger S, Devcic-Kuhar B, Kollmann C, Kastl SP, Kaun C, Speidl WS, et al. Can a commercial diagnostic ultrasound device accelerate thrombolysis? An in vitro skull model. Stroke. 2005;36(1):124–8.

15. Suchkova V, Siddiqi FN, Carstensen EL, Dalecki D, Child S, Francis CW. Enhancement of fibrinolysis with 40-kHz ultrasound. Circulation. 1998;98(10):1030–5.

16. Daffertshofer M, Gass A, Ringleb P, Sitzer M, Sliwka U, Els T, et al. Transcranial low-frequency ultrasound-mediated thrombolysis in brain ischemia: increased risk of hemorrhage with combined ultrasound and tissue plasminogen activator: results of a phase II clinical trial. Stroke. 2005;36(7):1441–6.

17. Holland CK, Apfel RE. Thresholds for transient cavitation produced by pulsed ultrasound in a controlled nuclei environment. J Acoust Soc Am. 1990;88(5):2059–69.

18. Prokop AF, Soltani A, Roy RA. Cavitational mechanisms in ultrasound-accelerated fibrinolysis. Ultrasound Med Biol. 2007;33(6):924–33.

19. Dijkmans PA, Juffermans LJ, Musters RJ, van Wamel A, ten Cate FJ, van Gilst W, et al. Microbubbles and ultrasound: from diagnosis to therapy. Eur J Echocardiogr. 2004;5(4):245–56.

20. Nishioka T, Luo H, Fishbein MC, Cercek B, Forrester JS, Kim CJ, et al. Dissolution of thrombotic arterial occlusion by high intensity, low frequency ultrasound and dodecafluoropentane emulsion: an in vitro and in vivo study. J Am Coll Cardiol. 1997;30(2):561–8.

21. Culp WC, Porter TR, McCowan TC, Roberson PK, James CA, Matchett WJ, et al. Microbubble-augmented ultrasound declotting of thrombosed arteriovenous dialysis grafts in dogs. J Vasc Interv Radiol. 2003;14(3):343–7.

22. Culp WC, Porter TR, Lowery J, Xie F, Roberson PK, Marky L. Intracranial clot lysis with intravenous microbubbles and transcranial ultrasound in swine. Stroke. 2004;35(10):2407–11.

23. Molina CA, Ribo M, Rubiera M, Montaner J, Santamarina E, Delgado-Mederos R, et al. Microbubble administration accelerates clot lysis during continuous 2-MHz ultrasound monitoring in stroke patients treated with intravenous tissue plasminogen activator. Stroke. 2006;37(2):425–9.

24. Perren F, Loulidi J, Poglia D, Landis T, Sztajzel R. Microbubble potentiated transcranial duplex ultrasound enhances IV thrombolysis in acute stroke. J Thromb Thrombolysis. 2008;25:219–23.

25. Mizushige K, Kondo I, Ohmori K, Hirao K, Matsuo H. Enhancement of ultrasound-accelerated thrombolysis by echo contrast agents: dependence on microbubble structure. Ultrasound Med Biol. 1999;25(9):1431–7.

26. Rubiera M, Ribo M, Delgado-Mederos R, Santamarina E, Maisterra O, Delgado P, et al. Do bubble characteristics affect recanalization in stroke patients treated with microbubble-enhanced sonothrombolysis? Ultrasound Med Biol. 2008;34(10):1573–7.

27. Larrue V, Viguier A, Arnaud C. Trancranial ultrasound combined with intravenous microbubbles and tissue plasminogen activator for acute ischemic stroke: a randomized controlled study. Stroke. 2007;38:472 (abstract).

28. Alexandrov AV, Mikulik R, Ribo M, Sharma VK, Lao AY, Tsivgoulis G, et al. A pilot randomized clinical safety study of sonothrombolysis augmentation with ultrasound-activated perflutren-lipid microspheres for acute ischemic stroke. Stroke. 2008;39(5):1464–9.

29. Molina CA, Barreto AD, Tsivgoulis G, Sierzenski P, Malkoff MD, Rubiera M, Gonzales N, Mikulik R, Pate G, Ostrem J, Singleton W, Manvelian G, Unger EC, Grotta JC, Schellinger PD, Alexandrov AV. Transcranial ultrasound in clinical sonothrombolysis (TUCSON) trial. Ann Neurol. 2009;66:28–38.

30. Hu YZ, Zhu JA, Jiang YG, Hu B. Ultrasound microbubble contrast agents: application to therapy for peripheral vascular disease. Adv Ther. 2009;26(4):425–34.

31. Tsivgoulis G, Eggers J, Ribo M, Perren F, Saqqur M, Rubiera M, et al. Safety and efficacy of ultrasound-enhanced thrombolysis: a comprehensive review and meta-analysis of randomized and nonrandomized studies. Stroke. 2010;41(2):280–7.

32. Atar S, Luo H, Nagai T, Siegel RJ. Ultrasonic thrombolysis: catheter-delivered and transcutaneous applications. Eur J Ultrasound. 1999;9(1):39–54.

33. IMS II Trial Investigators. The Interventional Management of Stroke (IMS) II study. Stroke. 2007;38(7):2127–35.

34. Soltani A, Singhal R, Obtera M, Roy RA, Clark WM, Hansmann DR. Potentiating intra-arterial sonothrombolysis for acute ischemic stroke by the addition of the ultrasound contrast agents (Optison & SonoVue((R))). J Thromb Thrombolysis. 2011;31:71–84.

35. Ribo M, Molina CA, Alvarez B, Rubiera M, Alvarez-Sabin J, Matas M. Intra-arterial administration of microbubbles and continuous 2-MHz ultrasound insonation to enhance intra-arterial thrombolysis. J Neuroimaging. 2010;20(3):224–7.

36. Martin MJ, Chung EM, Goodall AH, Della Martina A, Ramnarine KV, Fan L, et al. Enhanced detection of thromboemboli with the use of targeted microbubbles. Stroke. 2007;38(10): 2726–32.

37. Alonso A, Della Martina A, Stroick M, Fatar M, Griebe M, Pochon S, et al. Molecular imaging of human thrombus with novel abciximab immunobubbles and ultrasound. Stroke. 2007;38(5):1508–14.

38. Tachibana K. Ultrasound therapy for stroke and regenerative medicine. Int Congr Ser. 2004;1274:153–8.

39. Ferrara K, Pollard R, Borden M. Ultrasound microbubble contrast agents: fundamentals and application to gene and drug delivery. Annu Rev Biomed Eng. 2007;9:415–47.

40. Calliada F, Campani R, Bottinelli O, Bozzini A, Sommaruga MG. Ultrasound contrast agents: basic principles. Eur J Radiol. 1998;27 Suppl 2:S157–60.

41. Nanda NC, Schlief R, Goldberg BB. Advances in echo imaging using contrast enhacement. 2nd ed. Dordrecht: Kluwer; 1997.

Chapter 37
Transcranial High-Intensity Focused Ultrasound for Sonothrombolysis in Stroke

Thilo Hölscher, Golnaz Ahadi, Daniel Lotz, Cheryl Schendel, David Fisher, and Arne Voie

1 Introduction

1.1 *History of Focused Ultrasound in the Brain*

The first experimental animal studies using focused ultrasound for targeted energy delivery in the brain were performed in the 1940s [1]. Whereas early results were not necessarily encouraging, the belief was that therapeutic applications of focused ultrasound could be possible if the skull could be removed. The Fry brothers were able to demonstrate first results showing that it is possible to generate brain lesions [2, 3] and to open the blood–brain barrier [4] in targeted areas of the brain. Main goal to use focused ultrasound was, and still is in most applications, to create temperature elevations to damage brain tissue in a dedicated area of interest [5]. Besides the Fry brothers, Lele and his group were the pioneers during the early years of focused ultrasound. In 1957, the first clinical trial was initiated by the Fry brothers at the University of Illinois. In this first trial, focused ultrasound was used to treat Parkinson's patients [6]. To overcome the problem of the skull bone trepanation windows were created. The work by the Fry brothers provided substantial information on the ability to treat within the brain and to establish quantitative thresholds to

T. Hölscher, MD (✉)
Department of Radiology, University of California, 200 West Arbor Drive,
San Diego, CA 92103, USA

Department of Neurosciences, University of California, 200 West Arbor Drive,
San Diego, CA 92103, USA
e-mail: thoelscher@ucsd.edu

G. Ahadi • D. Lotz • C. Schendel • D. Fisher • A. Voie
Department of Radiology, University of California, 200 West Arbor Drive,
San Diego, CA 92103, USA

P.A. Lapchak and J.H. Zhang (eds.), *Translational Stroke Research*, Springer Series
in Translational Stroke Research, DOI 10.1007/978-1-4419-9530-8_37,
© Springer Science+Business Media, LLC 2012

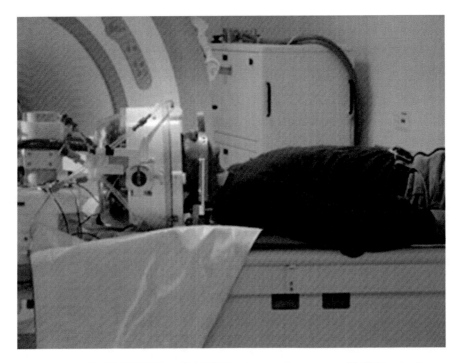

Fig. 37.1 ExAblate™ 4000 MRI-guided HIFU headsystem (by courtesy of InSightec, Inc.)

induce permanent changes in brain tissue [7]. Whereas more clinical applications and technological development were investigated throughout the following years, craniectomies were still common procedures since the problem of the intact skull bone itself could not be overcome at this time. In the mid-1970s, the Fry brothers focused their research interest on the transskull insonation problem and were able to show in 1980 that focusing through the intact skull is possible [8], however, with the occurrence of beam distortion. Strategies to correct for beam distortion were developed by Hynynen and Jolesz in the late 1990s who became the leading scientists in the field of transskull phase aberration correction [9]. In 1999, the first MRI compatible focused ultrasound transducers were developed [10]. MRI became the preferred monitoring tool because it provided the possibility to not only monitor a therapeutic effect but to monitor temperature elevation as well as focus navigation based on MRI derived thermometry [11]. The term *MRI-guided focused ultrasound*, short *MRgFUS*, is commonly being used since then. In 2002, the first clinical brain MRgFUS system was introduced (ExAblate™ 3000, InSightec, Inc., Tirat Carmel, Israel) [12]. The current system (ExAblate™ 4000) which has been used for the transcranial sonothrombolysis work at the UCSD Brain Ultrasound Research Laboratory is very similar to the one mentioned above. Figure 37.1 shows the actual set up of the ExAblate™ 4000 headsystem prior to the patient's transition into the MRI bore.

1.2   HIFU Thrombolysis

The use of high-intensity focused ultrasound (HIFU) for thrombolysis has been described earlier for nontranscranial applications and mostly in combination with tissue plasminogen activator (tPA).

In 2006, Frenkel et al. [13] tested the impact of HIFU on thrombolysis in a dedicated in vitro thrombolysis model, using artificial human blood clots. Four experimental groups were evaluated: (1) control (nontreated) clots, (2) clots treated with pulsed HIFU only, (3) clots treated with tPA only, and (4) clots treated with pulsed HIFU in combination with tPA. They could show that HIFU alone had no effect on clot lysis compared with the control clots. The clots treated with HIFU in combination with tPA, however, showed a statistically significant greater weight loss compared to tPA alone. Further, they could demonstrate that increasing pulse repetition frequency as well as total acoustic power yielded increasing thrombolysis rates. The author's conclusion was that the rate of tPA-mediated thrombolysis can be enhanced by using HIFU in vitro.

Stone et al. [14] tested (1) pulsed-HIFU exposures combined with tPA boluses in comparison to treatment with (2) tPA alone, (3) HIFU alone, and (4) control in an in vivo clot model, using the rabbit marginal ear vein. They could demonstrate that the combination pulsed-HIFU and tPA lead to complete vessel recanalization, whereas only partial recanalization was seen with tPA treatment alone. Similar to the control group, the effect of pulsed-HIFU alone showed no significant difference. Post-treatment, neither endothelial nor extravascular tissue damage was assessed histologically. Stone et al. concluded that tPA-mediated thrombolysis can significantly be enhanced when combined with noninvasive pulsed-HIFU exposures in vivo.

Different from Frenkel and Stone, Rosenschein et al. [15] could demonstrate that efficient thrombolysis can be achieved with HIFU alone and in absence of tPA. They used a therapeutic transducer which was built using an acoustic lens and which had an integrated ultrasound imaging transducer. In vitro clots were inserted into bovine arterial segments and sonicated under real-time ultrasound imaging guidance in a water tank. Using pulsed-wave ultrasound, the group could show that the total sonication time correlated with thrombolysis efficiency and was dependent on preferred ultrasound operating parameter combinations. With the combined ultrasound imaging system Rosenschein et al. could visualize cavitation clouds during HIFU insonation Suggesting that HIFU-induced cavitation mechanisms might play a key role in sonothrombolysis. Regarding clot fragmentation, it could be shown that 93% of the clot material was in the subcapillary size range. In vitro safety studies, however, showed arterial damage when an intensity (I_{SPTA}) of 45 W/cm^2 was used for periods of $\geq$300 s.

Our aim was to test whether HIFU sonothrombolysis efficacy can be achieved transcranially in absence of tPA using a HIFU head system which has been developed for the applications in humans.

1.3 HIFU Sonothrombolysis: Significance

Stroke is the second common cause of death worldwide. 800,000 US citizens suffer from stroke every year. tPA is the only FDA-approved drug for stroke treatment. Less than 3% of all stroke victims are currently receiving tPA therapy and less than 40% of stroke patients treated with tPA truly benefit on the long term. First data showing that the use of transcranial ultrasound in combination with tPA administration improves early recanalization is very promising [16–25]. However, the patient population which might benefit is still less than 3% of all stroke victims. The focus of our stroke research is the investigation of optional, HIFU-based thrombolytic strategies in absence of tPA, without or with the use of ultrasound microbubbles. Using a diagnostic, nonfocused ultrasound device, we could demonstrate earlier that ultrasound-induced thrombolysis can be augmented in presence of microbubbles and without tPA [26]. Further, we could demonstrate recently that noninvasive transcranial-HIFU enables clot lysis within seconds and without the use of further drugs (*publication pending*). However, potential HIFU applications in the brain immediately raise safety concerns, especially with regard to potential cavitation and thermal effects. The latter is of special concern since the vast majority of current HIFU applications in the human body are based on thermo-ablative effects [27–30]. Controlled cavitation, however, could be used to enhance the therapeutic effects of HIFU without damaging healthy brain tissue, due to either mechanical disruption or HIFU induced tissue heating or heating clot surrounding tissue [31]. The expected outcome of the current HIFU + microbubbles project is to demonstrate that (a) microbubbles can be used as "thrombolytic amplifiers," (b) the effect can be increased by using thrombus-specific targeting microbubbles, (c) cavitation can be controlled and be used as a mechanism to accelerate clot lysis, and, by doing so, (d) the acoustic energy exposure can be reduced significantly. The novelty of this concept is the noninvasive and non-tPA related approach to potentially treat stroke patients using transcranial-HIFU and to induce thrombolysis within seconds. The major clinical significance would be that millions of stroke victims worldwide could potentially be treated who are now not eligible for tPA treatment.

1.4 HIFU Sonothrombolysis: Innovation

The idea to use HIFU and clot targeting microbubbles for sonothrombolysis is a very recent development and very much enhanced by the ability to use magnetic resonance (MR) imaging to aim the acoustic energy at very precise regions in the brain. The truly new idea put forth with HIFU + microbubbles is that cavitation, once feared because of its potential harmful effects, now can be detected, characterized and controlled, in order to minimize the time to restore blood flow to ischemic regions of the brain following stroke [19, 32]. We believe this is the final missing piece to bring a new therapy paradigm to reality, one that consists of a technological partnership of HIFU, MR guidance, thrombus-targeting microbubbles, and now controlled cavitation. The idea is to thoroughly examine the interplay between HIFU presentation (i.e., pulse width, duty cycle, intensity, and sonication time), microbubbles (targeting and nontargeting), and cavitation (stable, inertial, and combined). The aim is to establish which acoustic parameter combinations cause these cavitation varieties, and which of these lead to the fastest and most complete reflow results in a rabbit carotid artery model. We will conduct safety studies to determine whether the therapy results in completely dissolved clots or pieces that may cause further distal occlusion of smaller vessels. The safety studies additionally will expose the rabbit brain to HIFU and microbubbles to determine if any harmful bio-effects such as thermal damage or tissue disruption result.

While work in this field of research has been intense, comprehensive data remain scarce. To date, no study has been published that examines sonothrombolysis by HIFU and microbubbles, and its relationship to cavitation type using advanced cavitation detection techniques. This project will yield a wealth of data from both in vitro and in vivo experiments and will greatly add to and advance the body of knowledge on the subject. We will learn which type of cavitation (stable, inertial, or a combination) best promotes sonothrombolysis, how to detect it by appropriate signal acquisition and analysis, and how to control this cavitation by adjusting sonication parameters. In turn, these data and techniques will guide the development of noninvasive extracranial instrumentation for the purpose of cavitation control during the application of therapy for acute ischemic stroke.

2 Materials and Methods

2.1 Technical Principle

An ExAblate™ 4000 HIFU headsystem (InSightec, Inc., Tirat Carmel, Israel), equipped with a transcranial transducer, has been installed in our research lab recently. The system has been developed for brain applications. Key component of this system is a hemispheric-phased array transducer with 1,000 single piezo elements which can be operated independently. Different from conventional ultrasound systems the ExAblate™ 4000 does not produce a single beam which has to pass the skull at a certain location (i.e., temporal bone window) and travel through brain tissue until it hits the target from one direction. The geometry of the HIFU system has been chosen to insonate through the entire skull with 1,000 individual beams. Each individual beam has a significantly lower intensity compared to a conventional ultrasound beam. However, the sum of all beams leads to a high intensity, three-dimensional focus beam in the center of the transducer. A sharp focus (radius: 2.0 mm in x/y- and 3.0 mm in z-orientation) can be generated, located in the center of the transducer and steered electronically in a radius of 3.0 cm in any direction without losing its shape (Fig. 37.2). Due to the three dimensionality of the focus beam, the target (i.e., thrombus) is insonated from all directions.

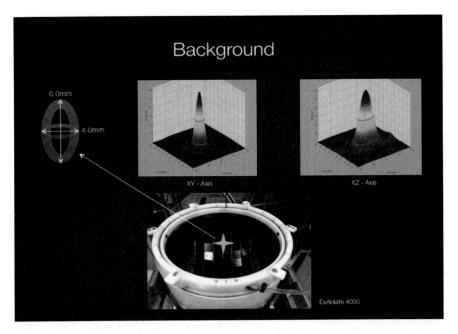

Fig. 37.2 *Bottom*: *top* view inside the hemispheric transducer. *Upper left*: represents the dimensions of the focus beam. *Upper center/right*: representative parametric sound field images in *xy*- and *xz*-axis

2.2 Experimental Considerations: In Vitro

To mimic transcranial insonation, a human cadaveric skull is mounted, upside down, to the bottom of an acrylic plate which covers the water filled hemispheric transducer (Fig. 37.3). The plate has a hole in its center (16 cm).

2.2.1 Thrombus Preparation

Venous whole blood is drawn from healthy, unmedicated human donors and transferred into sodium citrate tubes. 0.5 mL sodium citrate blood is mixed with 40 μL CaCl$_2$ (210 mmol/L) and transferred into a boro-silicated glass tube, which has a silk thread inside. The thrombi have an average weight of 0.2519 g ± 7%. The average length of each thrombus is about 2.5 cm (Fig. 37.4). Thrombi are incubated for 3.0 h in a preheated (37°C) waterbath. After incubation blood clots are then transferred into a polyethylene (PE) test tube (Advanced Polymers, Inc., Salem, NH).

2.2.2 Flow System

The PE tubing has an inner diameter of 4.3 mm, a wall thickness of 25.4 μm and preferred acoustic properties for US testing. The PE tube with the clot inside is

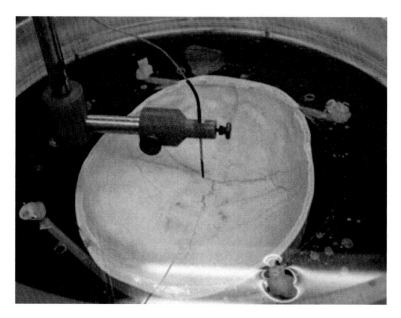

Fig. 37.3 *Top view*: showing the mounting of the cadaveric skull specimen (*upside down*) inside the degassed water filled hemispheric HIFU transducer. The pointing device (*needle tip*) is directed towards the natural focus location

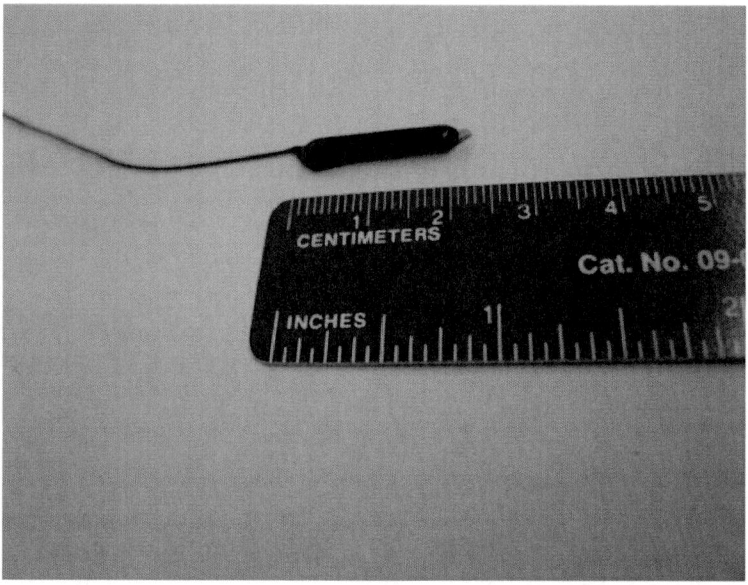

Fig. 37.4 Human blood clot organized around a silk thread

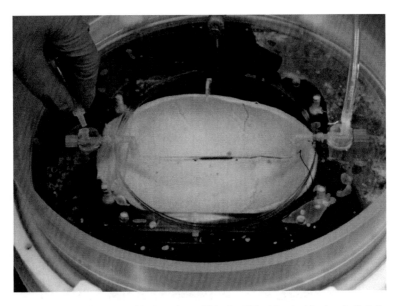

Fig. 37.5 Experimental set up with thrombus within the PE tubing inside the skull, being connected to the flow system

connected to the flow system (Fig. 37.5). The loose end of the silk thread is fixed upstream inside the flow system to keep the thrombus in place. Manually, the PE tube is positioned in such a way that the center of thrombus is aligned with the natural HIFU focus location inside the hemispheric transducer. The PE tube is connected to a unidirectional noncirculating flow system, using a peristaltic pump. Phosphate buffered saline (PBS) is used as a fluid medium, using a flow rate of 10 mL/min.

2.2.3 Data Analysis

To assess thrombolytic efficacy we have chosen to weigh the clots before and after each experiment. The thrombi are weighed pre/post experiment and the weight loss documented in percent. After insonation, the PE tubing is flushed for another 2 min and the solution collected in a beaker. Before entering the beaker the solution passes a serial filtration, with three differently sized pore mesh filters (Schweizer Seidengarne, Zürich, Switzerland) of 180, 60 and 11 µm, to capture clot fragments. The amount of clot fragmentation per filter size is calculated by subtracting the pre-wet filter weights from the post-wet filter weights with the difference documented in percent.

2.3   Experimental Considerations: In Vivo

2.3.1   Rabbit Carotid Artery Model: Efficacy

A rabbit carotid artery model has been chosen because of two main reasons: (a) the carotid artery model provides a bifurcation (common—internal/external carotid artery) and arterial dimensions that are comparable to either a smaller-sized proximal segment of a human middle cerebral artery (MCA) or a normal-sized MCA branch. (b) The model includes the brain as the end organ, including its microcirculatory system, which can be studied for potential clot fragmentation and consecutive secondary downstream vessel occlusion related to this. Since the rabbit carotid artery is positioned inside the cavity of the cadaveric human skull, transcranial-HIFU insonation is performed in each case.

We have considered a transcranial rabbit model to test for efficacy. All attempts to image and localize the MCA in a controlled fashion have failed. Further, due to the anatomy of the rabbit MCA the thrombi causing an occlusion would be a fraction in size compared to a M1 or M2 occlusion in humans. However, we have learned that thrombolysis efficiency is related to clot size as well.

For the surgical preparation, as well as for the positioning of the animal inside the hemispheric transducer, a customized restrainer has been developed specifically for the purpose of transcranial-HIFU sonothrombolysis experiments (Fig. 37.6). Using a precision jag, the animal can be positioned in such a way that the thrombosed carotid artery will be exposed to the focus of the ExAblate™. To date, feasibility of this model has been shown and a small number of animals have been studied. The knowledge gained during these studies will be discussed further below.

2.3.2   Rabbit Brain Model: Safety

In the past we learned that a transcranial sonothrombolysis animal model which might allow to test for both, *efficacy* and *safety*, is not feasible. Therefore, we decided to develop an animal model to test for *efficacy* only and an additional model to test for *safety* only. The latter is a rabbit model as well, using the same strain. In essence, the rabbits will be anesthetized and their heads will be shaved. The rabbits actually do not undergo a true surgery, i.e., comparable to the one described for the efficacy model. Once the heads are shaved and the animals are placed in supine position into the customized mounting device, aiming the positioning of the right hemisphere of the brain in the focal area of the transducer. To maintain a transcranial insonation, the animal's head will be placed inside the cavity of the human skull calvaria, similar to how the carotid artery will be placed inside the skull. We intend not to create a trepanation defect into the rabbit skull itself, because we know from other investigators (Kullervo Hynynen, verbal report) that the rabbit skull,

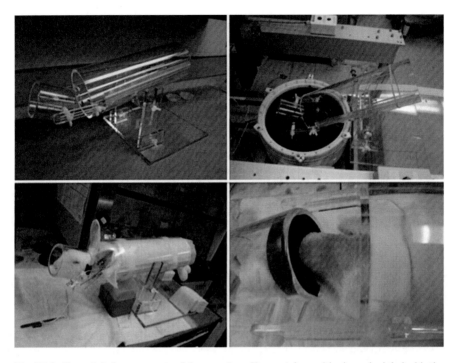

Fig. 37.6 *Upper left*: first prototype of the restrainer. *Upper right*: positioning principle inside the hemispheric transducer. *Lower left*: rabbit inside the device. *Lower right*: exposure of the rabbit neck prior to surgery

due to its marginal thickness, can be neglected for HIFU experiments with regard to signal absorption. After each insonation the animals are allowed to regain consciousness and kept for the following 10 days for behavioral assessment. After this period, the animals are euthanized and the brains harvested for further histochemistry analysis.

The basic idea is to insonate, instead of the carotid artery, the right hemisphere of the rabbit brain using the same parameters which have been used for the *efficacy* testing. By doing so, we aim to test whether the same ultrasound, which has been used for the in vivo *efficacy* testing causes potential brain tissue damage, such as blood–brain barrier disrupture, intracranial hemorrhages, edema, inflammation, etc. In this model we do not intend to lyse blood clots, but to investigate solely whether the ultrasound intensity, duration etc. which is used for the carotid artery model, is safe with regard to brain tissue since intracranial clot lysis in stroke is the ultimate goal of this project. To do so, behavioral monitoring as well as brain histology will be performed. Safety studies using this model have not been performed.

3 Results

3.1 HIFU Sonothrombolysis Efficacy: In Vitro

To date, about $N=2{,}000$ in vitro HIFU sonothrombolysis experiments have been performed. We have learned that HIFU sonothrombolysis can be achieved (Fig. 37.7) and that efficacy is dependent on acoustic output power, operating parameter combinations, skull bone and clot characteristics as well as flow mechanics. In the following, we would like to present the in vitro data more detailed.

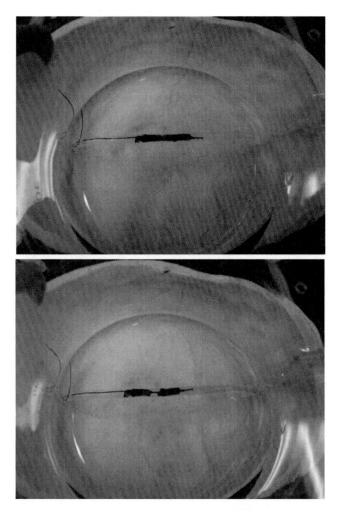

Fig. 37.7 Clot in transcranial in vitro HIFU system pre insonation (*top*) and partially lysed clot post insonation (*bottom*). Clot lysis occurred exactly at the natural location of the HIFU focus with a width of 4–5 mm (focus dimension in *xy* orientation: 4 mm)

Table 37.1 Percent clot weight loss: clot lysis in relation to intensity

Acoustic output power (W)	Number	Mean weight loss in (mg)	STDV	Mean weight loss in %	STDV	p-Value
0	60	0.0000	0.0096	1.746	4.0075	>0.05
50	62	0.0100	0.0108	4.547	4.6205	>0.05
100	66	0.0100	0.0085	5.625	3.2149	>0.05
125	63	0.0200	0.0154	8.971	6.2981	<0.001
150	65	0.0300	0.0243	12.900	9.7526	<0.001
200	61	0.0700	0.0457	28.220	18.7010	<0.001
235	61	0.1000	0.0543	41.410	20.2991	<0.001
270	62	0.1500	0.0394	61.040	14.5640	<0.001
400	61	0.1800	0.0305	74.830	10.1207	<0.001

STDV standard deviation

Table 37.2 Focal point acoustic measurements: for each output power the spatial peak temporal average intensity (I_{SPTA}), rarefractional pressure (P_{NEG}) and mechanical index (MI) were measured at the focus, using a needle hydrophone

AC power (W)	I_{SPTA} (W/cm²)	P_{NEG} (MPa)	P_{POS} (MPa)	Energy (kJ)	MI
0	0.00	0.00	0.00	0.00	0.00
50	33.08	1.562	1.416	0.158	2.972
100	66	2.221	2.001	0.315	4.226
125	84.15	2.514	2.270	0.394	4.783
150	102.05	2.734	2.514	0.473	5.201
200	134.8	3.173	2.929	0.630	6.037
235	160.6	3.515	3.125	0.74	6.687
270	145.25	3.808	3.320	0.851	7.245
400	215.185	4.635	4.041	1.26	8.818

3.1.1 Dependency on Acoustic Output Power

To study the effect of acoustic output power on sonothrombolysis efficacy we used a standard operating parameter setting (duty cycle 50%, pulse width 200 ms, duration 30 s) with increasing acoustic output powers. For all this experiments the same calvarium was used. Depending on the acoustic intensity at the focus, the clot weight loss differed between 1.7% (0 W) and 74.83% (400 W) (Table 37.1). Whereas no statistical significant weight loss could be seen for the low output powers between 0 and 100 W, significant ($p < 0.001$) clot lysis could be shown for all output power levels >125 W.

For each output power value the acoustic parameters spatial peak temporal average intensity (I_{SPTA}), rarefractional pressure (P_{NEG}) and mechanical index (MI) were measured at the focus, using a needle hydrophone. The values are displayed in Table 37.2.

3.1.2 Dependency on HIFU Operating Parameters

To test the impact of various duty cycles and pulse widths on thrombolysis the decision was made to retain an insonation duration of 30 s and an acoustic output power

PW >		.1 ms	1 ms	10 ms	100 ms
DC					
5%		GP 1	GP 2	GP 3	GP 4
	N	25	25	25	25
	% Wt Loss	6.0	15.3	22.7	26.0
	Std Dev	7.5	5.3	13.5	23.1
	Std Err	1.5	1.1	2.7	4.6
	Loss/kJoule	16.9	43.5	64.5	73.9
10%		GP 5	GP 6	GP 7	GP 8
	N	25	25	25	25
	% Wt Loss	15.6	21.0	27.0	49.1
	Std Dev	9.1	12.9	15.9	27.1
	Std Err	1.8	2.6	3.2	5.4
	Loss/kJoule	22.1	29.7	38.3	69.7
20%		GP 9	GP 10	GP 11	GP 12
	N	25	25	25	25
	% Wt Loss	27.7	29.4	33	59.1
	Std Dev	20.0	16.9	20	20.6
	Std Err	4.0	3.4	4.1	4.1
	Loss/kJoule	19.7	20.8	23.4	41.9
50%		GP 13	GP 14	GP 15	GP 16
	N	25	25	25	25
	% Wt Loss	48.8	51.9	48.6	64.7
	Std Dev	24.9	21.7	24.9	16.0
	Std Err	5.0	4.3	5.0	3.2
	Loss/kJoule	13.8	14.7	13.8	18.4

Fig. 37.8 The combination of four different pulse widths (PW) and four different duty cycles (DC) leads to a total of 16 study groups, which can be identified as GP1 to GP16. For each group the total number (*N*) of experiments performed, the percent weight loss (% Wt Loss), the standard deviation (Std Dev), the standard error (Std Err), and the percent weight loss per kilo Joule (loss/kJ) are provided

of 235 W for all studies. The latter parameters were chosen due to own historical data showing that transcranial thrombolysis can be achieved sufficiently using these parameters. Four different duty cycles were tested in combination with four different pulse widths (Figs. 37.8 and 37.9). It appeared that sonothrombolytic efficacy increased with longer duty cycles, longer pulse widths, or, preferably, the combination of both. Increasing the duty cycle had a higher impact on thrombolysis efficacy than increasing the pulse width. With regard to percent weight loss per kilo Joule energy, however, the combinations of short duty cycles and long pulse widths seemed to have the highest potential to increase sonothrombolysis rate.

For each ultrasound presentation the acoustic parameters I_{SPTA}, P_{NEG} and MI were measured at the focus, using a needle hydrophone. The values are displayed in Table 37.3.

3.1.3 HIFU Sonothrombolysis and Clot Fragmentation

To test for clot fragmentation using HIFU we performed series of experiments for a total of nine acoustic power settings, whereas other operating parameters, such as duty cycle (50%), pulse width (200 ms), and insonation duration (30 s) remained

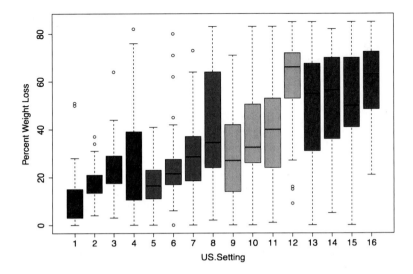

Fig. 37.9 Percent weight loss: Boxplots by US settings

Table 37.3 Acoustic measurements at focus for all 16 experimental groups

Experimental group	Pulse width (ms)	Pulse period (ms)	No. of sonications	I_{SPTA} (W/cm²)	Duty cycle (%)
1	0.1	2	15,000	13.7	5
2	1	20	1,500	13.7	5
3	10	200	150	13.7	5
4	100	2,000	15	13.7	5
5	0.1	1	30,000	27.4	10
6	1	10	3,000	27.4	10
7	10	100	300	27.4	10
8	100	1,000	30	27.4	10
9	0.1	0.5	60,000	54.7	20
10	1	5	6,000	54.7	20
11	10	50	600	54.7	20
12	100	500	60	54.7	20
13	0.1	0.2	150,000	136.8	50
14	1	2	15,000	136.8	50
15	10	20	1,500	136.8	50
16	100	200	150	136.8	50

Transmit frequency = 220 kHz for all groups, transmit duration = 30 s for all groups, peak positive pressure = 3.1 MPa for all groups, peak negative pressure = 2.8 MPa for all groups

the same. For all clot fragmentation experiments the same skull specimen was used. Three different pore-sized mesh filters (180, 60, 11 μm) were used. Compared to the control group (intensity: 0 W), mesh filter weights did not increase with increasing acoustic power or with increasing clot weight loss (Fig. 37.10). For all three mesh filter sizes no statistical significant differences in pre/post weight could be seen, independent from the acoustic output power.

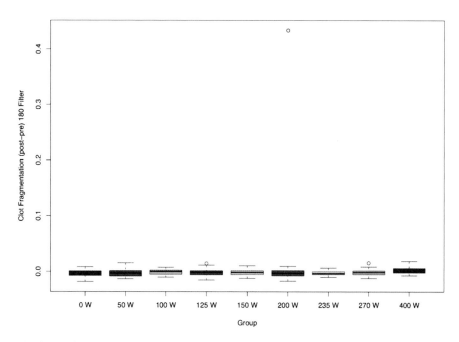

Fig. 37.10 Clot fragmentation boxplot for 180 micron mesh filters for each group

3.1.4 Dependency on Skull Bone Thickness/Density

To investigate the impact of skull bone density and skull bone thickness of sono-thrombolysis efficacy we performed experiments using three different human calvaria. For comparison, the HIFU operating parameter settings remained the same for the three samples:

Acoustic output power (AP)	270 W
Duty cycle (DC)	50%
Pulse width (PW)	200 ms
HIFU duration (ID)	30 s

For Skull #1, the average weight loss was $42.1 \pm 16.2\%$, compared to $35.6 \pm 14.2\%$ for Skull #2 and $10.3 \pm 4.4\%$ for Skull #3. For all three series individually, the average weight loss was statistically significant ($p < 0.001$). Clots treated using Skull #1 and #2 lost significantly more weight compared to Skull #3 clots (both $p < 0.001$). A smaller percent weight loss was seen in the Skull #2 compared to the Skull #1. This comparison was not found to be statistically significant, adjusting for multiple comparisons ($p = 0.051$).

Analyzing the CT data, we found that Skull #1 and Skull #2 were very similar in both bone thickness (mean 5.51 vs. 5.49 mm) and radio density, expressed in this case by the amount of Hounsfield units (HU) (mean 868 vs. 850 HU). CT analysis of Skull #3, revealed significantly greater bone thickness (mean 6.22 mm) and higher radio density (mean 1,208 HU) compared to the others (Fig. 37.11).

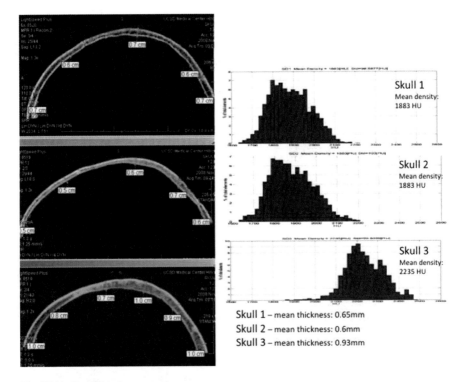

Fig. 37.11 Bone density comparisons

Table 37.4 Acoustic measurements at focus[a]

	Intensity (W/cm²)	Peak negative pressure (MPa)	Peak positive pressure (MPa)
No skull	888	8.01	7.39
Skull #1	111	2.81	2.75
Skull #2	160	3.33	3.22
Skull #3	79	2.39	2.29

[a]Measured at AP: 270 W, DC: 50%, PW: 200 ms

Similar to the differences in bone thickness and radio density, the acoustic measurements of the three samples showed higher values for focus intensity, peak negative and peak positive pressure for Skull #1 and #2 compared to Skull #3. The acoustic data is shown in Table 37.4.

3.1.5 Impact of Microbubbles

To study whether microbubbles may increase thrombolysis efficacy we performed experiments using a low (50 W) and a high (200 W) acoustic output power setting. For both values all other operating parameters, such as duty cycle (50%), pulse width (200 ms), and insonation duration (30 s) remained unchanged. The same skull specimen was used for all studies. For the low power setting we could demonstrate that the use of nontargeted microbubbles (DEFINITY™, Lantheus Medical Imaging, Billerica, MA) lead to statistically significant greater weight loss. Whereas, no weight loss ($0.05 \pm 2.2\%$) was seen in the control group (no microbubbles), a $10.88 \pm 3.49\%$ weight loss could be achieved when microbubbles were added. A similar, more accentuated observation was made in the high power comparison. Here, a weight loss of $18.47 \pm 17.68\%$ in the non microbubble could be achieved which increased to $60.57 \pm 14.05\%$ in the microbubble group.

3.1.6 Dependency on Flow Mechanics

The experimental set up using a pulsatile flow pump producing a low velocity flow (10 mL/min) might occur in vivo, such as in case of partial recanalization or subtotal occlusion. The set up, however, does not mimic the event of flow stasis during acute thrombotic vessel occlusion. To produce flow stasis we simply turned off the flow pump during HIFU insonation. For this experiment we used the same skull and tested for sonothrombolysis efficacy in two groups:

- Group 1: AP 270 W, DC 50%, PW 200 ms, ID 30 s, *flow 10 mL/min*
- Group 2: AP 270 W, DC 50%, PW 200 ms, ID 30 s, *no flow*

Whereas a weight loss of 42.12% could be achieved in Group 1, the weight loss dropped to 18.14% in Group 2 during no flow conditions.

3.2 HIFU Sonothrombolysis Efficacy: In Vivo

3.2.1 First Experiences

To date, $N=46$ animals have been performed. In $N=1$ case fully recanalization, in $N=3$ partial recanalization, and in $N=1$ case clot reduction without recanalization could be achieved. In all other experiments the carotid artery remained occluded with/without visual increase in clot size. Enthused by the promising in vitro results, the relatively poor success rate after translation into the animal model was not necessarily expected. It is not conclusive what kind of parameter combination promoted recanalization best. We have seen recanalization using lower as well as higher output powers, pulsed wave as well as continuous wave mode and used three different clot preparation recipes. In our efforts to understand why the in vivo success rate to achieve recanalization has been limited we decided to investigate whether the clot preparation itself might be of concern. Initially, we used rabbit whole blood to generate the clots. To better understand the possible differences between human (whole blood, citrate blood) and rabbit clots we decided to perform scanning electron microscopy (SEM) of clots.

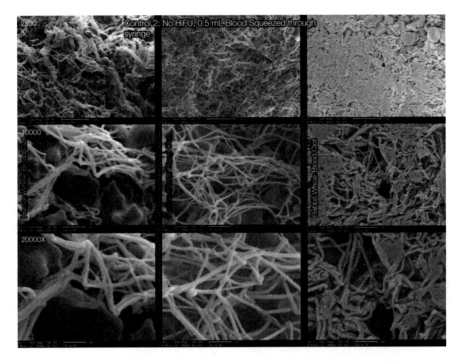

Fig. 37.12 From *top* to *bottom* the magnification increases (*top*: 2,500×, *center*: 10,000×, *bottom*: 20,000×). *Left column*: SEM's of human whole blood clots. *Center column*: represent human citrate blood clots. *Right column*: display rabbit whole blood clots

3.2.2 Scanning Electron Microscopy

Prior to HIFU insonation it appears that the fibrin fiber density between human whole blood clots and human citrate blood clots does not seem to vary much. Showing at the most a slightly higher density in the citrate clot preparations. It appears to be obvious, however, that the density in rabbit clots is much higher than in human clots. Noteworthy as well, the individual fibers are much more windy, rather than straight as seen in human clots, and they seem to be organized in a more "chaotic" pattern compared to the fiber organization in human clots (Fig. 37.12). In post HIFU insonation, the samples seem to be covered by "dust," which is most likely cellular matrix, due to HIFU-induced cell lysis. This observation was most striking in human clots. Most imminent is the difference in fibrin fiber density which is much higher in rabbit clots (Fig. 37.13). Based on this observation we draw the conclusion that rabbit thrombi are more difficult to be lysed because of their higher fibrin density and the complex organization of fibrin fibers compared to human clots. This could explain, in parts, the limited in vivo success rate.

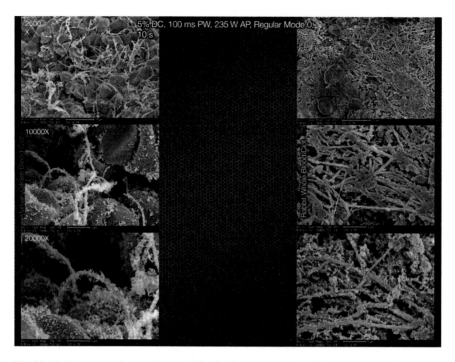

Fig. 37.13 From *top* to *bottom* the magnification increases (*top*: 2,500×, *center*: 10,000×, *bottom*: 20,000×). *Left column*: represent human citrate blood clots. *Right column*: display rabbit whole blood clots. These clots were exposed to an acoustic power output of 235 W for a duration of 10 s to Group 4 US parameters, 5% DC, 100 ms PW

3.2.3 Platelet Aggregation

In several cases we got the impression that the thrombus rather grew than getting lysed. This occurred primarily in those animals in which longer duty cycles were used and we associated this observation with potential thermal effects, causing clot formation. Therefore switching to shorter duty cycles seemed to be reasonable, in fact, the first successful recanalization was done using a 5% duty cycle. However, this result could not be reproduced and the observation of increasing thrombus size was still apparent in many cases, whether or not we used thrombin to generate the thrombotic occlusion. We searched the literature thoroughly with regard to potential effects of ultrasound on coagulation. Based on selected publications, which are presented later and our own understanding, now, we believe that sufficient evidence exists in support of the hypothesis that HIFU induces thrombus generation, using higher acoustic energies. HIFU, the way it has been used until today, during this project promotes platelet aggregation in vivo, due to creation of mechanical shear stresses. Shear stress is caused by inertial cavitation and leads to the expression of platelet factors such as β-thromboglobulin (β-TG) which are responsible for the aggregation of platelets. Williams et al. [33] studied the effects of therapeutic intensities of ultrasound on human platelets in whole blood and monitored the release of the platelet-specific protein β-TG. More β-TG was released as the intensity of the ultrasound was increased, as well as when the driving frequency was decreased from 3.0 to 0.75 MHz. Some β-TG was released at low spatially-averaged intensities as low as 0.6 W/cm^2 at 0.75 MHz, a value significantly lower than that observed for the onset of aggregation of platelet rich plasma (obtained from the same volunteer) in the same exposure system. Liberation of β-TG by ultrasound was diminished but not abolished in the presence of inhibitors which rendered the platelets functionally inert. Our data might suggest that β-TG is liberated in two ways: firstly, as a result of platelet disruption by cavitation, and subsequently, by potent aggregating agents, liberated in parallel with β-TG, inducing the physiological release reaction in adjacent platelets. The low therapeutic intensities and short exposure times (30 s or less) necessary to liberate β-TG from normal human platelets in vitro, suggest that patients with abnormally sensitive platelets and/or "hypercoagulable state" could be at risk if subjected to high therapeutic intensities of ultrasound. Similar results were shown by Kornowski et al. [34] who studied the effect of ultrasound on thrombolysis in a rabbit thrombosis model. Although time of initial reflow was shortened by ultrasound, it was associated with less reperfusion and more reocclusion. A possible explanation for these results is ultrasound-induced platelet activation counterbalancing its thrombolysis-accelerating effect. Poliachik et al. [35] showed that HIFU is capable of producing "primary acoustic hemostasis" in the form of ultrasound-induced platelet activation, aggregation, and adhesion to a collagen-coated surface. They could show that increased cavitation activity lowers the intensity threshold to produce platelet aggregation and decreased cavitation activity in an overpressure system raises the intensity threshold for platelet aggregation.

3.2.4 Clot Elasticity

Each time we confirm thrombotic vessel occlusion using high frequency duplex ultrasound we observe the striking compression/decompression of the clot itself due to the "water hammer effect," caused by the pulsed cardiac blood pressure wave. We focused on this phenomenon more detailed and collected pre/post HIFU video clips. These video clips display quite obviously the mechanics which are taking place within the clot itself and its surrounding tissue. In principle, we could observe two different behaviors:

Scenario 1: The clot is attached to the vessel wall in its entire circumference. During systole, the clot does not detach from the vessel wall but is compressed significantly along its longitudinal axis. During diastole, the clot extends to its original length (Fig. 37.14).

Scenario 2: The clot detaches partially from the vessel wall during systole and attaches back during diastole. We observed in this case that the clot dimensions do not change during the cardiac cycle and the clot remains in its full extension.

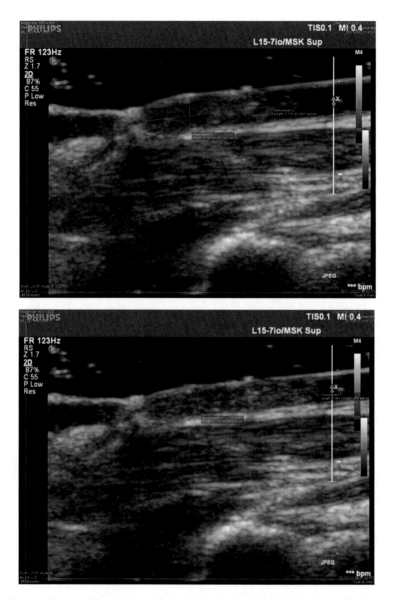

Fig. 37.14 Rabbit carotid artery occluding, hyperechogenic (*bright*) thrombus. The images are screenshots of a real-time movie clip showing clot compression, during a heart cycle, along its longitudinal axis from 1.0 cm in the diastole to 0.7 cm in the systole (*upper image*). Along its vertical axis the diameter enlarges from 0.12 cm (diastole) to 0.15 cm (systole) (*lower image*)

The great amount of thrombus elasticity is confirmed by other [36] and underscores our impression that even using high acoustic energies certain, thrombi, predominantly fibrin rich, are very difficult to break mechanically.

4   Comment

To use transcranial-HIFU for sonothrombolysis in stroke is an exciting new research field. The potential of ultrasound-induced thrombolysis within seconds, noninvasively and without the use of a thrombolytic agent, such as tPA, could open a new avenue of how stroke patients might be treated in the future. Especially if the use of tPA could be spared, the approach of transcranial-HIFU sonothrombolysis could provide a true therapeutic option for millions, the mass majority, of stroke patients who are today not eligible for tPA or neurointerventional therapies. The first in vitro experiences using this new technology are very encouraging and raise the expectations that this approach might broaden the spectrum of efficient stroke therapies, which are so sparse today. The translation into the animal model, however, dampened to some extent the enthusiasm, despite the fact that the amount of data acquired until today is minor and the animal model chosen for this purpose is still in a developmental stage. Besides the technical challenges we encountered during the translation from in vitro into in vivo, the latter reveals the high complexity of thrombolysis research in a live organism.

With regard to the future potential of transcranial-HIFU sonothrombolysis the question might be asked 'whether these first observations are rather discouraging or encouraging. First and foremost, we have learned that platelet aggregation most likely plays a critical part and might counterbalance the potential effects ultrasound has on thrombolysis. However, we have learned as well that high acoustic energies are the cause of platelet activation. As a consequence, it seems to be reasonable to suggest that if the acoustic energies applied to the thrombus can be lowered to such an extent that platelet activation does not occur the efficiency of HIFU induced sonothrombolysis could be increased significantly. Hence, in the near future the expressive term "high-intensity focused ultrasound" might be exchanged by "high-focused ultrasound," suggesting that high acoustic energies are not needed to achieve thrombolysis in vivo. Further, to prevent platelet aggregation the add-on administration of a platelet aggregation blocker during insonation might become the therapy of choice. To date, the use of platelet aggregation blockers in stroke therapy is restricted only to the postacute phase for secondary prophylaxis. In this context, the use of targeting/nontargeting microbubbles is of special interest, as well. Due to their special acoustic properties they might be used as "thrombolytic amplifiers" by promoting the mechanical impact on the clot itself without the need for high acoustic energies. Microbubbles' potential use as drug carriers (i.e., tPA, platelet blockers) is of great scientific interest, way beyond the application of stroke therapy.

Future challenges are the need for a better understanding of basic mechanisms of ultrasound induced clot lysis, to increase the knowledge about clot characteristics, the potential therapeutic use of microbubbles, and to better understand the impact of the human skull bone on transcranial sonothrombolysis.

Acknowledgments This work was supported by InSightec, Inc., Tirat Carmel, Israel.

References

1. Lynn JG, Zwemer RL, Chick AJ, Miller AE. A new method for the generation and use of focused ultrasound in experimental biology. J Gen Physiol. 1942;26(2):179–93.
2. Fry WJ, Barnard JW, Fry EJ, Krumins RF, Brennan JF. Ultrasonic lesions in the mammalian central nervous system. Science. 1955;122(3168):517–8.
3. Fry WJ, Barnard JW, Fry FJ, Brennan JF. Ultrasonically produced localized selective lesions in the central nervous system. Am J Phys Med. 1955;34(3):413–23.
4. Bakay L, Ballantine Jr HT, Hueter TF, Sosa D. Ultrasonically produced changes in the blood–brain barrier. AMA Arch Neurol Psychiatry. 1956;76(5):457–67.
5. Lele PP. A simple method for production of trackless focal lesions with focused ultrasound: physical factors. J Physiol. 1962;160:494–512.
6. Fry WJ, Fry FJ. Fundamental neurological research and human neurosurgery using intense ultrasound. IRE Trans Med Electron. 1960;ME-7:166–81.
7. Fry FJ, Kossoff G, Eggleton RC, Dunn F. Threshold ultrasonic dosages for structural changes in the mammalian brain. J Acoust Soc Am. 1970;48(6:Suppl 2):1413–7.
8. Fry FJ, Goss SA. Further studies of the transskull transmission of an intense focused ultrasonic beam: lesion production at 500 kHz. Ultrasound Med Biol. 1980;6(1):33–8.
9. Clement GT, Hynynen K. A non-invasive method for focusing ultrasound through the human skull. Phys Med Biol. 2002;47(8):1219–36.
10. Clement GT, White PJ, King RL, McDannold N, Hynynen K. A magnetic resonance imaging-compatible, large-scale array for trans-skull ultrasound surgery and therapy. J Ultrasound Med. 2005;24(8):1117–25.
11. Hynynen K, Vykhodtseva NI, Chung AH, Sorrentino V, Colucci V, Jolesz FA. Thermal effects of focused ultrasound on the brain: determination with MR imaging. Radiology. 1997;204(1):247–53.
12. Hynynen K, McDannold N, Clement G, et al. Pre-clinical testing of a phased array ultrasound system for MRI-guided noninvasive surgery of the brain–a primate study. Eur J Radiol. 2006;59(2):149–56.
13. Frenkel V, Oberoi J, Stone MJ, et al. Pulsed high-intensity focused ultrasound enhances thrombolysis in an in vitro model. Radiology. 2006;239(1):86–93.
14. Stone MJ, Frenkel V, Dromi S, et al. Pulsed-high intensity focused ultrasound enhanced tPA mediated thrombolysis in a novel in vivo clot model, a pilot study. Thromb Res. 2007;121(2):193–202.
15. Rosenschein U, Furman V, Kerner E, Fabian I, Bernheim J, Eshel Y. Ultrasound imaging-guided noninvasive ultrasound thrombolysis: preclinical results. Circulation. 2000;102(2):238–45.
16. Braaten JV, Goss RA, Francis CW. Ultrasound reversibly disaggregates fibrin fibers. Thromb Haemost. 1997;78(3):1063–8.
17. Devcic-Kuhar B, Pfaffenberger S, Gherardini L, et al. Ultrasound affects distribution of plasminogen and tissue-type plasminogen activator in whole blood clots in vitro. Thromb Haemost. 2004;92(5):980–5.
18. Datta S, Ammi AY, Coussios CC, Holland CK. Monitoring and simulating stable cavitation during ultrasound-enhanced thrombolysis. J Acoust Soc Am. 2007;122:3052.
19. Datta S, Coussios CC, McAdory LE, et al. Correlation of cavitation with ultrasound enhancement of thrombolysis. Ultrasound Med Biol. 2006;32(8):1257–67.
20. Prokop AF, Soltani A, Roy RA. Cavitational mechanisms in ultrasound-accelerated fibrinolysis. Ultrasound Med Biol. 2007;33(6):924–33.
21. Eggers J, Koch B, Meyer K, Konig I, Seidel G. Effect of ultrasound on thrombolysis of middle cerebral artery occlusion. Ann Neurol. 2003;53(6):797–800.
22. Eggers J, Konig IR, Koch B, Handler G, Seidel G. Sonothrombolysis with transcranial color-coded sonography and recombinant tissue-type plasminogen activator in acute middle cerebral artery main stem occlusion: results from a randomized study. Stroke. 2008;39(5):1470–5.

23. Alexandrov AV. Ultrasound identification and lysis of clots. Stroke. 2004;35(11 Suppl 1): 2722–5.
24. Alexandrov AV, Mikulik R, Ribo M, et al. A pilot randomized clinical safety study of sono-thrombolysis augmentation with ultrasound-activated perflutren-lipid microspheres for acute ischemic stroke. Stroke. 2008;39(5):1464–9.
25. Alexandrov AV, Molina CA, Grotta JC, et al. Ultrasound-enhanced systemic thrombolysis for acute ischemic stroke. N Engl J Med. 2004;351(21):2170–8.
26. Holscher T, Raman R, Ernstrom K, et al. In vitro sonothrombolysis with duplex ultrasound: first results using a simplified model. Cerebrovasc Dis. 2009;28(4):365–70.
27. Martin E, Jeanmonod D, Morel A, Zadicario E, Werner B. High-intensity focused ultrasound for noninvasive functional neurosurgery. Ann Neurol. 2009;66(6):858–61.
28. Damianou C, Ioannides K, Hadjisavvas V, Mylonas N, Couppis A, Iosif D. In vitro and in vivo brain ablation created by high-intensity focused ultrasound and monitored by MRI. IEEE Trans Ultrason Ferroelectr Freq Control. 2009;56(6):1189–98.
29. ter Haar GR. High intensity focused ultrasound for the treatment of tumors. Echocardiography. 2001;18(4):317–22.
30. Jolesz FA, McDannold N. Current status and future potential of MRI-guided focused ultra-sound surgery. J Magn Reson Imaging. 2008;27(2):391–9.
31. Maxwell AD, Cain CA, Duryea AP, Yuan L, Gurm HS, Xu Z. Noninvasive thrombolysis using pulsed ultrasound cavitation therapy—histotripsy. Ultrasound Med Biol. 2009;35(12):1982–94.
32. Datta S, Coussios CC, Ammi AY, Mast TD, de Courten-Myers GM, Holland CK. Ultrasound-enhanced thrombolysis using Definity as a cavitation nucleation agent. Ultrasound Med Biol. 2008;34(9):1421–33.
33. Williams AR, Chater BV, Allen KA, Sherwood MR, Sanderson JH. Release of beta-thrombo-globulin from human platelets by therapeutic intensities of ultrasound. Br J Haematol. 1978; 40(1):133–42.
34. Kornowski R, Meltzer RS, Chernine A, Vered Z, Battler A. Does external ultrasound acceler-ate thrombolysis? Results from a rabbit model. Circulation. 1994;89(1):339–44.
35. Poliachik SL, Chandler WL, Mourad PD, et al. Effect of high-intensity focused ultrasound on whole blood with and without microbubble contrast agent. Ultrasound Med Biol. 1999;25(6): 991–8.
36. Marder VJ, Chute DJ, Starkman S, et al. Analysis of thrombi retrieved from cerebral arteries of patients with acute ischemic stroke. Stroke. 2006;37(8):2086–93.

Chapter 38
Cellular Therapy for Ischemic Stroke

Todd Deveau, Shan Ping Yu, and Ling Wei

Abstract Ischemic stroke is a leading cause of death and disability in the United States. Currently, tPA remains the first (and only) FDA-approved therapy for the treatment of ischemic stroke. Although tPA was a breakthrough, it has a very limited window of administration and a large list of exclusion criteria, and only a small percentage of patients can receive tPA treatment. Additionally, no therapy currently exists to promote CNS repair and recovery following ischemic stroke. The state of clinical therapy and research to date highlights the need for novel approaches for the treatment of ischemic stroke. Cellular therapy, specifically the use of stem cells for cellular augmentation and/or replacement, is a promising novel approach for treatment. Cellular therapy holds promise by providing multiple mechanisms of action combined with a prolonged therapeutic time window of administration to promote CNS repair and recovery. From the sources of stem cells to their mechanisms of action, this chapter outlines the current state of knowledge regarding cellular therapy for stroke.

1 Introduction

Ischemic stroke is a leading cause of death and disability in the United States. Currently, therapies for ischemic stroke are vessel and blood based [1]. No FDA-approved treatments exist to provide neuroprotection or promote CNS recovery.

T. Deveau • S.P. Yu
Department of Anesthesiology, Emory University, Atlanta, GA 30322, USA

L. Wei (✉)
Department of Anesthesiology, Emory University, Atlanta, GA 30322, USA
e-mail: lwei7@emory.edu

P.A. Lapchak and J.H. Zhang (eds.), *Translational Stroke Research*, Springer Series
in Translational Stroke Research, DOI 10.1007/978-1-4419-9530-8_38,
© Springer Science+Business Media, LLC 2012

Intense investigation into neuroprotective agents for ischemia has failed to produce a single FDA-approved compound for neuroprotection following stroke. Trials involving protein therapy for recovery have fared better, with the 1995 NINDS tPA trial producing the only FDA-approved drug (alteplase) for stroke therapy. Although the use of tPA clinically is a great step forward, its benefits extend to few patients in practice due to limitations of its use. The failure of clinical neuroprotection trials to date and the limited efficacy of thrombolytics highlight the need for not only novel treatments but also novel approaches to stroke treatment and recovery. Cellular therapy, specifically the use of stem cells for circuit replacement and/or augmentation, represents a promising new avenue for stroke therapy. This chapter discusses the need for novel and innovative treatment strategies and the current state of knowledge regarding the use stem cells for the treatment of ischemic stroke.

2 Current Treatment Options: The Need for New Treatments

Every year in the United States, approximately 795,000 people suffer strokes. In 2007, it was estimated that the direct and indirect costs of stroke were $40.9 billion and stroke was responsible for 1 out of every 18 deaths in the US [2]. Ischemic stroke, defined as the reduction of blood flow to an area of the brain due to vascular occlusion by the artery that supplies the affected area [3], accounts for the vast majority of stroke cases [2]. Cerebrovascular occlusion initiates a complex cellular cascade of events, involving multiple mechanisms and pathways, leading to irreversible tissue injury and cell death [4, 5]. Irreversible gray and white matter tissue injury in this context is also known as cerebral infarction. The impact of ischemic stroke is overwhelming, yet to date clinicians have few tools at their disposal for successful neurological therapy.

Current treatments for stroke focus on prevention, clinical management of blood flow, and rehabilitation. Timely clinical vessel recanalization and long-term rehabilitation remain the best treatment strategies for stroke patients. Recent meta-analysis of recanalization on ischemic stroke outcome suggests that timely recanalization is the most robust predictor of increased clinical prognosis following ischemia [6, 7]. To date, no proven clinical treatments exist that can provide acute neuroprotection for ischemic stroke.

Following cerebral ischemia, thrombolytic therapy (to be discussed in more detail in the following section) involving intravenous administration of tPA remains the best, and only, FDA-approved therapy for ischemic stroke to date. Thrombolytic therapy with tPA aims to recanalize vessels through clot dissolution. tPA has many limitations and only a small percentage of patients receive the drug in the clinic [8]. A large amount of exclusion criteria, a short temporal window for therapeutic administration (3–4.5 h after the onset of stroke attack), and risk of hemorrhage development remain major hurdles for the widespread use of thrombolytic therapy.

Advances in imaging have allowed for the increased use of endovascular therapy for mechanical clot clearance. Though not without risks, endovascular therapy offers many advantages for clot disruption over thrombolytic therapy. The use of the FDA-approved devices the Merci Receiver and the Penumbra System allows for the mechanical clearance of clots. Endovascular clearance of clots often increases the rate of recanalization over thrombolytic therapy without the large risk of hemorrhage development [8, 9]. Although greatly beneficial for some patients, the use of endovascular therapy for stroke remains limited. The clinical infrastructure needed to successfully image the affected area and identify patients who would benefit largely impedes the widespread adoption of endovascular therapy for ischemic stroke. Additionally, incomplete removal of clots could lead to clot disruption, clot spreading, and occlusion at other sites within the brain. More research on endovascular therapy is needed [10], and, as of 2007, the AHA/ASA Stroke council only recommends the use of endovascular therapy in clinical trial settings [11]. A recent scientific statement on mechanical clot removal by the AHA Stroke Council states that it may be beneficial but echoes the sentiments that more research is needed [12].

Rehabilitation remains the only option available to stroke patients for therapy associated with long-term recovery. Most patients survive ischemic stroke and ultimately immense burdens fall on the patients and their families due to long-term consequences of disability [13]. Despite the advances made in the clinical management of stroke, without effective medical treatments "most post-stroke care will continue to rely on rehabilitation interventions" [13, 14]. There is strong evidence that suggests task-oriented training assists innate recovery following ischemia; however, more research is needed on the process that underlies learning following recovery [13]. Further details of rehabilitation lie beyond the scope of this chapter. For more information, see Miller et al. [15] and Langhorne et al. [13].

3   The Search for New Translational Therapies for Stroke

3.1   Neuroprotection: 2 Decades of Failure

Since the early 1990s, much focus on preclinical research and clinical trials has centered on finding therapeutic agents that provide neuroprotection for ischemic tissues. Neuroprotective strategies aim to "antagonize, interrupt, or slow the sequence of injurious biochemical and molecular events that, if left unchecked, would eventuate in irreversible ischemic damage" [16]. Acute neuroprotective intervention could feasibly reduce the size of the infarction while delayed administration of neuroprotective agents would be beneficial for salvaging injured tissue within the ischemic penumbra. Any successful neuro-protective strategy would undoubtedly improve prognosis and increase the potential for and speed of long-term recovery.

A wide variety of targets have been examined for their putative neuroprotective properties. Examples of targets range from neurotransmitter receptors (AMPA, NMDA, GABA), ion channels (e.g., Na^+, K^+, Ca^{2+} channels), reactive oxygen species (ROS), cyclooxygenases, phospholipids, and beyond (specific details are beyond the scope of this chapter, see [17] for more detailed information). Preclinical data for many agents has been promising. Experimental evidence in animal models has demonstrated that it is possible to achieve large reductions in ischemic brain injury through early intervention via neuroprotective mechanisms [16] and to improve functional outcome. Despite the preclinical evidence, no neuroprotective drug exists which has "demonstrated unequivocal efficacy in clinical trials" and "fulfilled regulatory requirements for approval" [5].

It is unclear whether the failure of neuroprotective agents lies in the variability in quality and reliability of preclinical data [16], shortcomings in clinical trial design, or simply failure of the preclinical agent to translate to human therapy. Regardless of the source of failure, future putative neuroprotective agents will require rigorous preclinical testing, carefully designed clinical trials, and possibly even multiple mechanisms of action if neuroprotective strategies are to be successful.

Examples of issues surrounding the quality of preclinical data for many putative neuroprotective agents to date include incomplete pharmacokinetics data, lack of an adequate dose–response curve, poorly defined therapeutic window, and failure to test compounds in a variety of models with a variety of confounding factors [5]. Doubt has even been cast on the validity of animal models for ischemic stroke. Evidence, however, suggests that animals in animal models undergo similar physiological changes to humans throughout ischemia and reperfusion, suggesting that animal models should have predictive validity for evaluating neuroprotective agents [18]. In an effort to improve the quality of preclinical data, the STAIR guidelines have been proposed. First outlined in 1999 (STAIR Roundtable [19]) and recently updated [20], the STAIR guidelines provide researchers with a checklist of data that should be generated before a putative neuroprotective agent goes to clinical trials. The effect of the STAIR trial guidelines on the translational success of preclinical neuroprotective agents is yet to be determined [5, 16].

Confounding factors of the trials themselves bear some of the burden of the state of neuroprotection trials to date. Inherent problems with study design have impeded the ability of clinical trials to reach appropriate conclusions in some instances. The proper timing of drug delivery for efficacy, sample sizes, heterogeneous patient populations, and lack of penumbral imaging are examples of the issues that have surrounded clinical trials for neuroprotection to date [5]. Future clinical trials will require careful design and rigorous design adherence in order to adequately deter-mine any potential benefit of putative neuroprotective agents.

Another potential underlying factor in the failure of neuroprotection trials to date is that most neuroprotective agents investigated act on one pathway or have one mechanism of action. Although neuroprotection trials have failed in bringing neuroprotective agents into widespread use, their impact overall is not negative. Numerous trials to date have elucidated the roles of many different pathways in the context of ischemia. For example, we now know ischemic cell death involves glutamate-mediated excitotoxicity through anoxic depolarization, disruption of ion and neurotransmitter gradients, aberrant intracellular Ca^{2+} signaling, oxidative stress and ROS, intrinsic and extrinsic neuroinflammation, protease induction, DNA damage through caspase activation, death of different cell types, and many other factors [4]. Through animal models and the failure of clinical trials, we have learned stroke is a heterogeneous injury that affects heterogeneous populations. Successful therapies for stroke will therefore need to involve multiple mechanisms to account for disease heterogeneity, and a need exists to investigate therapies with multiple mechanisms of action.

3.2 Thrombolytic Research: Thrombolytics and tPA

The use of thrombolytic agents to treat coronary occlusion in the cardiovascular field has been widespread for decades. Because the vast majority of strokes are ischemic in nature (~85%), thrombolytic agents used in the cardiovascular field inevitably found a place in the treatment of ischemic stroke as well [21]. Initial studies focused on the use of urokinase and streptokinase, but studies utilizing intravenous tPA finally provided a breakthrough for acute ischemic stroke therapy. Newer studies on synthetic thrombolytics and xenogenic thrombolytics are currently underway. These newer thrombolytic agents aim to improve the speed of recanalization and to extend the time window for treatment. Unfortunately thrombolytic therapy also carries a large risk of hemorrhage development, a hurdle that must always be taken into account when researching new thrombolytic therapies. Although thrombolytic therapy for stroke has provided huge benefits for some patients, its application is limited highlighting again a need for novel approaches.

Early clinical trials for thrombolytic therapy and ischemic stroke focused on the use of streptokinase [22]. Trials involving streptokinase demonstrated its efficacy in acute myocardial infarction in the 1970s and 1980s [21]. The use of streptokinase moved toward the treatment of cerebral ischemia in the 1980s and 1990s. Case reports demonstrated the efficacy of streptokinase in recanalization; however, the use was cautioned due to risks of hemorrhage. The MAST-I (MAST I Group [23]), MAST-E (MAST E Group [24]), and ASK [25] trials in the 1990s examined the use of streptokinase but were terminated early due to unacceptable increases in patient mortality [22, 26].

To date, the NINDS tPA trials in the mid-1990s have been the only Phase III clinical trials to produce an FDA-approved therapy for stroke. These trials found that patients receiving intravenous administration of tPA within 3 h of ischemic onset were 30% more likely to exhibit minimal or no disability compared to placebo 90 days after drug administration (NINDS Study Group [9, 27]). Recent research on tPA has focused on extending the therapeutic window of tPA administration.

The European ECASS-III trial formally demonstrated that tPA has benefit in some patients if the time window is extended from 3 to 4.5 h [28]. The results of the ECASS-III trial and the Australian EPITHET trial [29] resulted in the AHA/ASA Stroke Council changing their stance on tPA use and extending the recommended time window to 4.5 h in some patients [30].

Further research has examined the use of new thrombolytics. Newer agents aim to increase thrombolytic efficacy by providing longer half-lives and/or greater specificity for clot components. Tenecteplase and desmoteplase are two examples of so-called "third generation" thrombolytics currently being tested in early clinical trials [3]. Tenecteplase is a modified form of tPA designed to increase half-life and fibrin specificity and desmoteplase is a venom from vampire bat saliva [21]. Data from preclinical work and early trials are currently mixed with further studies on these compounds needed before they can be recommended for therapy for ischemic stroke [3, 21, 31].

While the use of thrombolytics has been groundbreaking, thrombolytic therapy is not without limitations. Largely as a result of the short window of time in which treatment can be administered, it is estimated that only as many as 3–4% of eligible patients throughout the US and Europe receive tPA [8]. Current research examining intra-arterial (IA) administration, the effects of combinatorial thrombolytic administration and pharmacotherapy, and thrombolytic administration plus endovascular treatments are ongoing. Whether or not IA thrombolysis or combinatorial thrombolytic treatment will increase recanalization rates, extend time windows, and decrease risks of hemorrhage is yet to be seen.

3.3 The Need for Novel Approaches: Cell-Based Therapy

Current treatment options for stroke highlight the need for novel treatments and innovative strategic interventions to improve patient prognosis. The failures of research to date (other than tPA) to generate FDA-approved treatments that promote CNS neuroprotection or recovery demonstrate the urgent need for effective treatments for stroke. A potential new approach to stroke treatment lies in stem-cell based cellular therapy. Interest in the use of stem cells for the treatment of CNS disorders and CNS injury has been growing since the 1980s, when the first cells were transplanted into Parkinson's patients. Since the initial Parkinson's trials, stem cell transplantation has been used in other CNS disorders, such as Huntington's disease, spinal cord injury, and stroke patients. It is now widely recognized that stem cell therapy provides a unique approach to brain repair and functional recovery in the post-ischemic brain [1]. Stem cell-based therapy involves multiple mechanisms of action, can respond to changes in the surrounding cellular microenvironment, and substantially increases the time window within which treatment can be administered. Cellular therapy targets aspects of nervous system repair by modulating endogenous neurogenesis, inflammation, trophic factor expression, or providing exogenous cells for cellular replacement and/or augmentation [1].

The remainder of the chapter is devoted to cellular therapy for stroke. From the source of cells, putative mechanisms of action, and the timing and delivery method of cells, we discuss the state of knowledge regarding cellular therapy for stroke. Reviews such as this aim to provide investigators with the tools to direct their research and advance the field of cellular therapy, not only for stroke, but for other CNS disorders as well.

4 Post-ischemic Cell-Based Therapy: The Next Frontier

The past 2 decades have seen an explosion of research on the use of stem cells as a therapy for ischemic stroke. Embryonic stem cells (ESCs) are often the first cell type that comes to mind; however, it is clear that other sources of non-ESCs exist. Non-embryonic cells can be isolated from bone marrow, umbilical cords, and even the adult CNS. Embryonic and non-embryonic cells isolated from any of these sources all carry therapeutic potential for CNS disorders, especially ischemic stroke. Additionally, it is now clear that neurogenesis in the subgranular zone (SGZ) and subventricular zone (SVZ) is regulated by ischemia, and increasing the proliferation and survival of adult neural progenitor cells in these regions may provide therapeutic benefit. Cells from different sources do not necessarily have the same properties, and an understanding of the properties of different stem cells from different sources is necessary when considering which stem cell type to use for stroke therapy.

4.1 Cellular Therapy: Sources of Cells

Cellular therapy can be divided into exogenous and endogenous. Exogenous cellular therapy involves the transplantation of xenogenic, allogenic, or autologous cells through a variety of means. Endogenous therapy involves the stimulation or amplification of innate neurogenesis in the ischemic brain through extrinsic and/or intrinsic factors. We now consider different sources of cells for cellular therapy, different mechanisms of action, and other considerations regarding the use of cellular therapy for ischemia.

4.2 Exogenous Cells

The nervous system is a complex organ made up of neurons and glial cells, which surround and support neurons [32, 33]. Neurons send signals that provide the driving force (in one way or another) for numerous functions including, but not limited to, thought processes and movement. Glial cells, including astrocytes, microglia, and oligodendrocytes are necessary for the normal function of the nervous system. For example, oligodendrocytes act to enhance the conduction velocity of neural signals. Neural signals, such as descending motor commands into the spinal cord, sometimes must traverse great distances. Support from oligodendrocytes is necessary for efficient neural signal transduction across time and space. The loss of any of these cell types may have grave implications for brain function. Most of the advances in stem cell research to date have been directed at treating degenerative diseases through cellular replacement. Following ischemic stroke, multiple cell types are lost in the infarcted region. Cellular therapy for ischemic stroke holds promise in the potential for cell replacement across different cell types, although research shows stem cells can also exert therapeutic benefit by providing support for injured neurons and glia in the peri-infarct region. Whether aiming for cellular replacement or cellular augmentation, cellular therapy holds great promise for the treatment of ischemic stroke. Currently, stem cell research plays an important role in pushing the boundaries for therapy of neurological disorders [34–36].

4.3   Neural Stem Cells (NSCs)

Beginning with work done published by Joseph Altman in the 1960s [37–39], it is now well established that multipotent neural stem cells (NSCs) exist in several regions of the brain that have the capacity to self-renew. The two primary niches of neurogenesis in the adult mammalian brain are the SVZ lining the lateral ventricle and the SGZ in the dentate gyrus of hippocampus [32]. Neurogenesis may occur in other brain regions, such as the subcallosal zone (SCZ), which lies between the corpus collasum and hippocampus [40]. New neurons and glia are generated from these areas and are thought to contribute to neuronal plasticity, possibly even neuronal repair following CNS injury [41]. NSCs from the adult mammalian brain have the potential to differentiate into astrocytes, neurons, and oligodendrocytes [42]. Neurogenic niches in the adult mammalian CNS are described in more detail in the following section.

The behavior of NSCs is controlled by many intrinsic and extrinsic factors [43–45]. An example of intrinsic regulation of NSC differentiation and proliferation is the binding of transcription factors, such as the neuron-restrictive silencing factor/RE-1 silencing transcription (REST) factor system and retinoic acid [44, 46]. Transcription factors exert changes in the transcriptome profiles of NSCs, changes that are tightly associated with NSC "state" [44]. DNA modifications themselves are also known to affect NSC behavior. For example, chromatin remodeling and associated changes in methylation state of CpGs affect gene expression patterns in NSCs in vitro [44, 46]. In addition to intrinsic factors, extrinsic factors can also regulate behavior of NSCs [44, 45, 47]. For example, exposure to experimental hypoxia can modulate NSC differentiation and proliferation [48]. Growth factors, such as EGF and FGF-2, are another example of extrinsic factors that also enhance the proliferation and differentiation of NSCs [44, 46].

It is known that neurogenesis in the adult mammalian brain is enhanced following CNS injury, and that new neuronal cells are generated and migrate to the sites of injuries [1, 42, 49–51]. The contribution of endogenous adult neurogenesis to nonpathological and pathological CNS function is currently unclear. Correlations have been seen with hippocampal neurogenesis and learning [52]. Potential mechanisms behind putative increases in cognition as a result of adult hippocampal neurogenesis include computational theories [53], reduced interference between memories [54], and as a mechanism to encode temporal aspects of memory [55]. Despite correlative evidence, a clear causal relationship between adult neurogenesis and learning and memory remains elusive [32, 33, 43, 52, 56].

NSCs can be isolated from embryonic, fetal, and adult mammalian brains and cultured in vitro as neurospheres [44]. In vitro, neurospheres from NSCs provide a tool with which to study neurogenesis. NSC neurosphere cultures can also be expanded in vitro and used for transplantation studies. NSCs in vitro have the capacity to self-renew and differentiate into neurons, astrocytes, and oligodendrocytes [57]. NSC transplantation has been studied in several animal models of CNS pathologies, such as Parkinson's disease (PD), multiple sclerosis (MS), cerebral ischemia, and intracerebral hemorrhage (ICH) [49, 57]. In animal models, adult NSC transplantation can aid in neural repair [49, 57]. NSCs show a lot of promise and can easily be differentiated into neurons and glia experimentally. However, NSC sources lie deep in the brain, favoring the use of other cell types that are more accessible over NSCs.

4.4 Mesenchymal Stem Cells (MSCs)

Friedenstein et al. first described mesenchymal stem cells (MSCs) derived from bone marrow in the 1960s [58–60]. MSCs, also known as "multipotent mesenchymal stromal cells" [61], have since been found in other tissues and other sources including periosteum, muscle connective tissue, umbilical cord blood, placenta and amniotic fluid, adipose tissue, and fetal tissue [59, 62]. As defined in 2005, the minimal criteria for a cell to be considered as a multipotent mesenchymal stromal cell are as follows: (1) cell must be adherent to plastic under standard culture conditions; (2) exhibit positive expression of the cell surface markers CD105, CD73, and CD90 and be negative for hematopoietic markers CD34, CD45, CD11a, CD19, HLA-DR; and (3) differentiate into adipocytes, osteocytes, and chondrocytes in vitro under specific conditions [61, 63]. MSCs represent a small fraction of nucleated cells in the parent tissue. It is estimated that in bone marrow, MSCs represent around 1 in 10,000 nucleated cells in a newborn, declining to around 1 in 1,000,000 by age 80 [59, 64]. Despite the relatively low abundance of MSCs in mammalian tissue, MSCs can be isolated and expanded in vitro.

In vitro, MSCs in culture rapidly self-renew and can be passaged for many passages, although they are not immortal [65]. MSC cultures are considered a heterogenic mix of committed and uncommitted progenitors. Cells in culture generally have the potential to form multiple mesenchymal cell types, with some progenitors capable of multilineage differentiation that can form endodermal and ectodermal cells [65]. Evidence shows that MSCs can be differentiated into functional neurons; however, this is far from conclusive due to cell fusion and spontaneous expression of neuronal markers by undifferentiated cells [66, 67].

Despite the controversy surrounding differentiation of MSCs into functional neurons [66], MSCs may not need to differentiate into functional neurons to exhibit therapeutic benefits. Intravenous MSC-NPC transplantation has been tested extensively in animal models of cerebral ischemia and shown to home to the ischemic region through chemokine–chemokine receptor interactions [68], providing benefit for repair and functional recovery. Although the precise mechanisms behind IV-MSC transplantation are not yet fully understood, many studies show benefits of MSC transplantation through angiogenesis [69] and potential paracrine mechanisms such as: increases in trophic factor expression [70] and immune modulation [71]. Additionally, clinical trials have also been performed with IV transplanted MSCs following ischemia. Trials to date demonstrate not only the feasibility of IV transplantation of autologous MSCs, but also that the approach is clinically safe and may provide some benefit for functional recovery [72]. Last, although most transplantation studies to date have focused on intravenous MSC transplantation, other transplantation routes with MSCs have shown benefit in animal models. For example, intracerebral human MSC transplantation has also provided benefit in gerbils and nonhuman primates following ischemia [67, 71]

Most work on MSCs to date has been done in bone marrow-derived MSCs. More concrete studies are needed on their ability to generate functional cells of the nervous system. Additionally, due to the fact that abundance decreases with age along with their proliferation/differentiation capacity, and the fact that bone marrow cells must be obtained through a painful and invasive procedure, more studies on the use of alternate sources of MSCs for ischemic stroke therapy [62].

4.5 Embryonic Stem Cells (ESCs)

Due to the ethical, moral, and political implications surrounding their harvest and use, ESCs are most often what comes to mind when stem cells are discussed. ESCs are derived from the inner cell mass (ICM) of the blastocyst (an early-stage embryo) from a variety of sources, including mouse, nonhuman primates, and humans [73–75]. Thomson et al. first established human ES (hES) cell lines in 1998. ICMs of several human embryos produced for clinical purposes through in vitro fertilization (IVF) were isolated and expanded/maintained in culture indefinitely, providing researchers with H1, H7, H9, H13, and H14 hES cell lines still in use today [73]. hES cells are distinct from other embryonal pluripotent cells, such as teratocarcinoma-derived pluripotent embryonal carcinoma (EC) cells [73]. hES cells are pluripotent and are able to differentiate into all cellular derivatives of the three primary germ layers: ectoderm, endoderm, and mesoderm [74]. hES cells are powerful cells with much potential for CNS therapy. Serious moral and ethical issues are raised with the isolation of hES cells as collection of the ICM destroys the blastocyst and the fertilized embryo [76–78]. Additionally, hES cells are surrounded by political controversy in the US, as exemplified by the recent ruling of Chief Justice Lambert of the District of Columbia Federal District Court in August of 2010 who put an injunction on the use of federal funding for hES research [79].

Grafted ES cells have been seen to differentiate into neurons and functionally integrate into the recipient brain in several models of CNS pathologies, including Parkinson's disease and stroke [80–83]. In vivo studies have shown benefit of ES cell transplantation specifically for stroke as well. Transplantation of neural precursors derived from ES cells improved functional recovery in models of focal and global ischemia in rats [84–89]. Although ES cells have shown benefit in animal models of ischemia, current transplantation approaches may be suboptimal due to the inherently low survival of cells as a result of the toxic post-ischemic microenvironment in the peri-infarct cortex. Pre-transplantation cellular modifications have been demonstrated to improve therapeutic potential of transplanted ES cells, ostensibly through increased graft survival. For example, overexpression of the antiapoptotic gene BCL-2 and exposure to sublethal hypoxic insults before transplantation have been shown to improve the benefits of ES cell transplantation for ischemia [88, 89]. Further studies should examine the optimization of ES cell transplantation for ischemic stroke.

Research to date on the use of ES cells for CNS recovery following ischemia is promising. The use of ES cells of any origin has advantages over other cell types relating to their capacity for multilineage differentiation and self-renewal ability. These properties do not come without a price and the use of ES cells carries a high risk of teratoma formation (for clarification of teratoma nomenclature, please see [90]) "hindering the fulfillment of the clinical potential" [91]. Several studies have shown that ES cell grafts develop teratomas when transplanted into severely combined immunodeficient (SCID) mice, regardless of the species cells were derived from [73, 91–94]. The post-ischemic environment itself has even been observed to influence teratoma formation of transplanted hES-derived neural progenitor grafts following MCAO in rats [94]. ES cell grafts also carry the possibility of immune rejection by the recipient. Possible strategies to minimize immune rejection for ES cell grafts include genetic manipulations of cells to reduce potential for rejection and banking of cells with defined major histocompatibility complex (MHC) backgrounds [73]. Although ES cells represent powerful tools for cellular therapy and despite the ethical and moral issues that abound, more studies are needed to ensure the use of ES cells can be a safe therapy for ischemia.

4.6 Induced Pluripotent Stem Cells (iPS Cells)

Researchers at Kyoto University in Japan first described induced pluripotent stem cells (iPSCs) in 2006. In a seminal paper titled "Induction of Pluripotent Stem Cells from Mouse Embryonic and Adult Fibroblast Cultures by Defined Factors," Takahashi and Yamanaka demonstrated that adult fibroblasts could be "reprogrammed" to an ES cell-like state by the introduction of four factors: Oct3/4, Sox2, c-myc, and Klf-4 [95]. Their groundbreaking discovery challenged conventional thought about the plasticity of somatic cells [91]. iPS cells generated in their study exhibited ES cell morphology, ES growth properties, expressed ES cell markers, and showed similar patterns of gene expression and DNA methylation [95]. Furthermore, when transplanted into nude mice, pluripotent capacity was demonstrated because they generated tumors consisting of cells derived from all three germ layers [95]. Although initial work was done in murine cells, Takahashi also demonstrated the reprogramming of adult human somatic cells with the same four factors [96]. Work with human cells is important in demonstrating the principle that hES-like cells can be obtained without destruction of embryos. Because iPSCs do not rely on embryonic sources, iPSCs do not face the ethical, moral, and legal issues faced by ES cells. Since their discovery, there has been an explosion of research aimed at improving reprogramming techniques and implementing cells into research on animal models of CNS pathologies [97–102].

Although iPSC research is still in its infancy, iPS cells have immense potential as tools for modeling human disease, drug discovery, and cell transplantation therapies [103, 104]. In addition to exhibiting ES cell-like properties, iPSCs are of particular interest for transplantation therapies to researchers and clinicians because of the potential to generate autologous, patient-specific cells. The use of autologous cells could feasibly side step the hurdle of immune rejection that currently faces the use of clinical ES cells because donor MHCs should be identical to recipient. As a proof of principle, researchers have already demonstrated that autologous iPS cells can be generated from patients with a variety of CNS pathologies, such as amyotrophic lateral sclerosis (ALS) and schizophrenia [104, 105].

The use of iPSCs for ischemic stroke therapy has great potential. Studies on the transplantation of iPSCs for ischemic stroke therapy in animal models have already begun, but have exhibited mixed results thus far. To be discussed more in the following section, iPSCs not only exhibit the putative therapeutic properties of ES cells, but also share their pitfalls, such as tumor formation [91]. Kawai et al. demonstrated the risks associated with the intracerebral transplantation of undifferentiated iPSCs in an MCAO rodent model of stroke. Animals that received the transplants developed teratomas and fared worse in behavioral tests than sham animals [106]. Other studies have been more promising. Also in 2010, Chen et al. saw functional improvements after subdural iPSC transplantation with fibrin glue (to increase iPS graft adhesion) in rats following MCAO occlusion. Suzuki et al. saw improvements after iPSC transplantation in a rodent model of hindlimb ischemia. While we are enthusiastic about the study of Chen et al., the decision to pretreat animals in the study with iPS cell grafts before the onset of ischemia raises questions about the clinical relevance of their approach [107, 108]. More recently, a 2011 study by Jiang et al. used intracerebral transplantation of iPSCs reprogrammed from human fibroblasts in an MCAO model of stroke. iPSCs were injected ipsi- and contralateral to the infarct in female rats following MCAO. Groups receiving iPSC transplants showed a smaller infarct volume and improved sensorimotor function compared with animals receiving non-reprogrammed primary fibroblasts [109].

As stated, the potential to transplate iPSC therapy into clinical applications is hindered by the primary factor that also hinders ES cells: tumorigenicity [91]. Tumorigenicity is an issue inherent in initial reprogramming methods because the four reprogramming factors used to reprogram cells have all been associated with tumors or are oncogenes themselves [91]. The traditional reliance on retroviruses for reprogramming runs the additional risk of reactivation at a later time point following transplantation. Research into alternative reprogramming methods is ongoing, which may provide efficient methods to generate iPSCs while diminishing their tumorigenic capacity at the same time [97–102]. Initial research on comparisons of iPSCs and ESCs suggested high levels of genetic similarities; however, it is becoming clear that substantial differences in DNA methylation, copy number variation, and genetic imprinting exist between iPSCs and ESCs [91, 110–114]. Some work has shown that early passage iPSCs retain an epigenetic profile similar to the donor tissue, suggesting iPSCs have the capacity to retain epigenetic "memory" [114, 115]. This may actually be beneficial for establishing in vitro model systems for CNS pathologies with strong epigenetic components. Regardless, more work is needed on the relationship between iPSC genetics and epigenetics and the phenotypic/functional behavior of iPSCs before they can be translated to clinical studies involving transplantation [113].

4.7   Endogenous NPCs, Neurogenesis, and NPC Migration

Dating back to Cajal in the early twentieth century, the prevailing dogma was that no new cell growth occurs in the adult CNS. However, early twentieth century views on cell division and growth in the adult brain were limited by the technology of the time. Since Cajal's time, tools like [H^3]-thymidine and BrdU have come along making the study of adult neurogenesis possible [116]. In 1962, Joseph Altman published the first study challenging canonical thinking regarding neurogenesis in the adult mammalian brain [37, 117]. Using radiolabeled thymidine to mark dividing cells, he noticed neuroblasts (and even neurons) incorporated the radiolabeled thymidine following electrolytic lesions to the lateral geniculate body. Early studies on postnatal neurogenesis were met with skepticism. In 1977, Kaplan and Hinds provided the first concrete evidence of neurogenesis with electron microscopy in 90-day-old rats. They showed cells in the dentate gyrus and olfactory bulb that incorporated radiolabeled thymidine were indeed neurons [117, 118]. Acceptance of neurogenesis in adult vertebrates was further solidified by work done in the songbird field in the early 1980s by Nottebohm and Goldman [117, 119, 120]. Since then, adult neurogenesis has been demonstrated in many species, including nonhuman primates [121] and humans [122].

We now know unequivocally that new proliferating cells are born every day in several restricted niches in the adult mammalian CNS: the SGZ of the dentate gyrus of the hippocampus, the SVZ, and the recently described SCZ. It is known that neurogenesis in these areas are regulated by a variety of CNS pathologies in what appears to be an attempt at self-repair. While the innate neurogenic response to ischemia seems incapable of being functionally efficacious, bolstering this response may be a potential therapy for ischemic stroke.

4.8   Neurogenesis in the Sub-granular Zone (SGZ)

The SGZ of the hippocampus was the first region in the adult mammalian brain where dividing neurons were concretely identified [38, 118]. Neurogenic cells in the SGZ have a limited capacity to self-renew [117], and it is debated whether or not cells in the SGZ truly meet the full criteria of "stem cells" [52]. The exact identity of progenitor parent population in the SGZ is unclear; however, progenitor cells express GFAP and are postulated to be radial glial cells or the remnants thereof [117, 123]. Nevertheless, it is estimated that approximately 9,000 new cells are generated in the hippocampus of the adult rat everyday [124] and lower numbers in the mouse that appear to vary by strain [125, 126]. The numbers in the human hippocampus are not known, but it is known that hippocampal neurogenesis does occur in humans [122]. SGZ-derived proliferating progenitors ultimately migrate to the granule cell layer (GCL) where the majority becomes excitatory granule neurons. A smaller population, observed at about 14% of the total proliferating population in adult transgenic rats, becomes GABAergic basket cells [127]. Further studies are needed to determine the mechanisms by which newly proliferating cells become granule neurons or GABAergic neurons, but evidence suggests that progenitors form a somewhat heterogeneous neuronal population in the GCL [41].

Much work has been done to characterize newborn cells in the GCL. In rodents, new cells have been shown to receive synaptic input, express neuronal markers, and extend processes into the CA3 region of the hippocampus [123]. Proliferating SGZ-derived neural progenitors mature, migrate, and differentiate in a stepwise fashion. Immature proliferating granule cells first express the markers DCX and PSA-NCAM and show expression of GABA, AMPA, and NMDA receptors [41]. As they mature, DCX and PSA-NCAM marker expression fades, giving way to NeuN and Calretinin expression with a growth in excitatory GABAergic synaptic input [41]. NeuN and Calretinin positive immature granule cells exhibit small, minimally repetitive action potentials [41]. Adult SGZ-derived granule neurons reach maturity around 28 days after birth in the rodent. Mature new granule cells: express the markers NeuN and Calbindin; express GABA, AMPA, and NMDA receptors; receive synaptic input from GABA receptors, AMPA receptors, and NMDA receptors; and display large, repetitive, action potentials [41]. Although in rodent models granule cells are mature at around 28 days, a recent study by Kohler et al. suggests that in primates maturation time is longer, exceeding 6 months [128].

A large percentage of new-born granule cells die in the GCL within 28 days [123, 129]. Neurogenesis in the SGZ persists throughout life; however, there is an age-dependent decrease throughout the life of animals [130]. The exact function of new neurons in the GCL from SGZ-derived progenitors is unclear. Strong evidence suggests that SGZ neurogenesis plays an important role in learning and memory; however, a definitive causal relationship has remained elusive [52]. Animal models in which proliferating cells in the adult SGZ can be selectively ablated will be needed in order to establish definitively the role of SGZ neurogenesis.

4.9 Neurogenesis in the Subventricular Zone (SVZ)

Cortical neurons in the embryonic rodent arise from NSCs in the ventricular zone (VZ) [50]. Following embryonic cortical neurogenesis, an ependymal layer replaces the ventricular zone, but the SVZ persists throughout life [50, 131]. A subpopulation of radial glia from the VZ transform into adult astrocytes in the SVZ that form the basis of NSCs in the adult SVZ [50, 132, 133]. Each day, more than 30,000 neuroblasts are generated in the adult SVZ [41, 50, 134]. SVZ-derived neuroblasts form chains that are ensheathed by astrocytes and migrate relatively great distances through the rostral migratory stream (RMS) toward the olfactory bulb under non-pathological conditions [39, 41, 50, 52, 134]. In the olfactory bulb, these cells differentiate into granule and periglomular interneurons [41].

Doetsch et al. performed ultrastructural studies on the cellular composition of the SVZ in 1997 [135]. Type A, B, C, D, and E cells are all present in the SVZ as confirmed by electron microscopy and serial reconstruction of ultra-thin sections. Type A cells represent migrating neuroblasts, type B cells represent astrocytic parent cells, type C cells putative neural precursors, type D tanycytes, and type E large ependymal cells (for a detailed description of morphological characteristics of different cell types based on electron micrographs, please see [135]). Type A, B, and C cells all incorporated [H^3]-thymidine in the SVZ of adult mice, suggesting that all three cell types are dividing cells.

SVZ-derived granule and periglomular neurons differentiate and mature in a stepwise fashion similar to SGZ-derived granule neurons. SVZ-derived granule cells are mature around 14 days after birth in the rodent [41, 136]. Immature granule cells express the markers DCX and PSA-NCAM which subsequently give way to TUJ1, TUC-4, and finally NeuN expression throughout maturation. Granule cells express GABA, AMPA, and NMDA receptors and receive synaptic input from GABA, AMPA, and NMDA receptors. Sodium currents are observed only after synaptic contacts are established. SVZ-derived periglomular cells take longer to mature, around 28 days in the adult rodent. Immature periglomular cells first express the markers DCX and PSA-NCAM which give way to TUJ1 and TUC-4 expression, followed by NeuN/GAD/Calbindin/Calretinin expression, and finally TH expression sequentially through maturation. They express GABA, AMPA, and NMDA receptors. Synaptic input is first provided by GABA neurotransmission about mid-way through maturation, followed by AMPA- and NMDA-mediated inputs. Periglomular cells exhibit sodium currents before synaptic contacts are made. [41].

It is suggested that as many as 50% of SVZ-derived granule cells that reach the olfactory bulb die and that sensory input is critical for those that survive after 15 days [136].

SVZ neurogenesis has been best studied and characterized in rodents [52]. Evidence exists for the generation of new neurons and the RMS in nonhuman primates [137, 138] and in the adult human brain [137, 139]. Characterization of SVZ-NPCs and the RMS in humans has remained controversial [52, 140], although a recent study by Wang et al. seems to support initial observations by Curtis et al. that dividing cells are present in the adult human SVZ and migrate toward the olfactory bulb in the RMS [137]. The functional role of SVZ neurogenesis in rodents is yet to be definitely defined but "correlative evidence" suggests "a role in olfactory memory and discrimination" [52].

It is known, in animal models, that following ischemia, SVZ-derived NPCs migrate laterally away from the RMS toward infarcted tissue by way of chemotaxic gradients such as the SDF-1α and CXCR4 axis (to be discussed in more detail in the following section). Migration of SVZ-derived NPCs toward injured tissue following ischemia suggests an attempt at self-repair by the adult brain; however, a large majority of migrating cells die [141]. It is yet to be seen if NPCs from the SVZ can make a meaningful contribution to neuronal recovery following ischemia. A much greater number of neuroblasts are generated than survive [141], suggesting that migration of SVZ-derived NPCs has great potential for stroke therapy.

4.10 Neurogenesis in the Subcallosal Zone (SCZ)

Most discussions on neurogenesis in the adult mammalian brain have focused mainly on the SGZ and the SVZ. Evidence for neurogenesis has been described in other areas of the brain such as the neocortex [116] and the substantia nigra [142], but whether or not neurogenesis truly occurs in these regions remains controversial. One recently characterized region where neurogenesis appears to occur is the SCZ. Anatomically, the SCZ lies between the hippocampus and the corpus callosum (CC) in the adult rodent. It extends medial and caudal from the SVZ and is comprised of a "lamina of proliferating cells" that is "sandwiched" between axon bundles of the hippocampus and CC [40].

Cells in the SCZ have similar properties to those of the SVZ, and are comprised of type E ependymal cells, type A migrating cells, type B astrocytes, and type C dividing cells [40, 135]. In culture, they can form neurospheres, and SCZ-derived NSCs can differentiate into neurons, astrocytes, and oligodendrocytes. In vitro experiments suggest that cells from the SCZ form "multipotent, self-renewing" stem cells. SCZ-derived cells differ from SVZ-derived cells in that they form less primary neurospheres in culture and form isolated PSA-NCAM positive clusters, unlike the PSA-NCAM positive elongated chains that cells from the SVZ form [40].

The exact function of SCZ-derived NPCs and their potential to contribute to self-repair is unknown. *In vivo* microtransplantation experiments using GFP-labeled SCZ cells from adult B-actin:GFP mice transplanted into the SCZ of adult CD1 mice suggest that SCZ-derived NPCs become Rip-1 positive oligodendrocytes and GFAP positive astrocytes that migrate into the corpus collasum (CC) [40]. A recent study by Kim et al. suggests that neurons are formed by SCZ-NPCs, but these newborn neurons undergo massive Bax-mediated programmed cell death [143]. Interestingly, a 2002 study by Nakatomi et al. showed that proliferation was increased the pPV (area within the recently described SCZ) following ischemia in rodents and that pPV-derived cells contributed to neuronal replacement in the CA1 region of the hippocampus, a process that can be amplified with ICV growth factor infusion [144]. While it seems that the SCZ has the potential to contribute to self-repair after ischemia, more studies are needed to determine the role of SCZ-derived NPCs in ischemia and the capacity of SCZ-NPCs to contribute to neuronal self-repair.

4.11 Strategies and Challenges for Endogenous Cellular Therapy Applications

Discoveries that neurogenesis occurs in the adult mammalian brain have been groundbreaking. Although the exact functional contribution of adult neurogenesis under nonpathological conditions is unclear, it is clear in several animal models of disease that neurogenesis is upregulated in the adult brain under pathological conditions (especially ischemic stroke). Furthermore, studies have demonstrated that this pathological upregulation can be modulated by intrinsic and extrinsic factors such as growth factors, pharmacologic agents, and environmental enrichment. A challenge that faces the use of endogenous cellular therapy for ischemia is to overcome the low rate of surviving newborn cells. To be discussed in more detail in the following section, therapeutic strategies that can increase innate pathological neurogenic upregulation and improve the survival of newborn cells in injured tissue may prove beneficial for the treatment of ischemic stroke.

5 Cellular Therapy: Mechanisms of Action

The previous section described exogenous and endogenous cellular therapy and the types of cells that can be utilized for each. Despite the varying sources of cells, all types of cellular therapy share common goals. Cellular therapy aims to promote CNS recovery through cellular replacement, cellular augmentation, or any combination of the two. In order to determine the most appropriate context for cellular therapy in ischemia, a basic understanding of the causes of ischemic cellular injury is needed. The rest of the chapter will be devoted to work done in animal models and our current understanding of how cellular therapy may be beneficial following ischemic stroke.

5.1   Causes of Ischemic Injury

No discussion on therapy for stroke would be complete without some mention of causes of cellular injury and loss following vessel occlusion and reperfusion. Excellent reviews have summarized the mechanism of ischemia-induced cell death and brain injury; for example, see [4, 145] for more information. Here, we provide a brief review on some recent aspects of inflammatory activities and its role in brain stroke-induced brain injury.

5.2   Neuroinflammation

The body of literature elucidating the importance of inflammation in the nervous system (neuroinflammation) under a variety of CNS pathologies has seen a lot of growth over the past decade. Neuroinflammation in stroke is a very complex phenomenon and not yet fully understood in relation to ischemia. Following ischemia, production of pro- and anti-inflammatory cytokines and chemokines is upregulated in endothelial cells, neurons, and glia in the brain [146]. Additionally, ischemia causes the upregulation of extracellular matrix remodeling proteins (such as the matrix metalloproteinases or MMPs [147, 148]), and the breakdown of the blood–brain barrier. "Leaky" regions of the blood–brain barrier allow for the infiltration of peripheral immune cells such as lymphocytes, t-lymphocytes, macrophages, and neutrophils [149]. These peripheral immune cells also contribute to cytokine production following ischemia.

Much work on inflammation in ischemia has focused on the pro-inflammatory cytokines tumor necrosis-alpha (TNF-α), interleukin-six (IL-6), and interleukin one-beta (IL-1β) in addition to the anti-inflammatory cytokine interleukin-10 (IL-10). In human patients, increased IL-6 levels in serum, plasma, and CSF following cerebral ischemia have been correlated with increased neurological worsening [150–152]. Increased plasma and serum IL-10 levels have been associated with less neurological worsening in humans [150, 151]. Experimental studies on TNF-α and ischemia have been conflicting, with some reporting improvement with administration of TNF-α binding proteins [153] and others reporting worsening with TNF-α receptor knockout [154, 155]. Experiments involving IL-1β knockout mice show that KO mice exhibit far smaller infarcts compared to wild-type controls following MCAO [156], suggesting a injurious role of IL-1β following ischemia [146].

Modulation of neuroinflammation may prove to be beneficial for ischemic treatment and CNS recovery. Care must be taken with conclusions from immune research, however, as many inflammatory mediators have dual actions. TNF-α, for example, acts predominantly through the p55 (TNFR1) receptor. p55-mediated TNF-α signaling can initiate both pro-apoptotic and pro-survival intracellular cascades [157, 158]. Administration of TNF-α binding proteins can reduce ischemic injury in rats [153]; however, ischemic injury is exacerbated in TNF receptor knockout mice [154, 155]. Microglial-derived TNF-α has also been shown to be neuroprotective [159]. MMPs are another example of the dual actions of immune mediators. In the acute phase of ischemia, upregulated MMPs (such as MMP-2 and MMP-9) are detrimental to the CNS; however, in the chronic phase extracellular remodeling by MMPs are necessary for axonal growth, angiogenesis, and other repair mechanisms [160]. Future studies on neuroinflammation and ischemia therefore should take into account the multiplicity of inflammatory mechanisms, the temporal and spatial regulation of neuroinflammatory factors, and the consequences of the interplay of inflammation within the neurovascular unit [4].

6   Underlying Mechanisms for Cellular Therapy

Stem cells hold great promise for novel stroke therapeutics because of the extended window of time for efficacy, multiple mechanisms of action, and the ability to adapt to their environment. The original goal of cellular therapy for stroke was mostly cell replacement. Researchers aimed to utilize stem cells to replace lost tissue and regenerate lost neuronal circuits. As the field matures, much of the interests have shifted toward the use of stem cells for trophic support and stimulation of endogenous regenerative mechanisms. Some studies documented functional improvement in rats following ischemia despite the fact that the murine NSC grafts were undetectable at 6 months, suggesting that recovery was mediated by some indirect mechanisms [161]. It is now clear that exogenous and endogenous cell delivery can improve prognosis in animal models as a result of trophic support, angiogenic stimulation, increases in endogenous neurogenesis/angiogenesis, and inflammatory modulation. Meanwhile, as the novel strategies are proposed to improve cell survival and optimize directed cell differentiation, cell replacement therapy still holds huge potential for future treatment of stroke.

6.1   Trophic Support

Trophic factors such as vascular endothelial growth factor (VEGF), nerve growth factor (NGF), and brain-derived neurotrophic factor (BDNF) are important for the growth of vascular and neuronal tissue in the CNS. VEGF is important for vascular growth and angiogenesis [162]. Under nonpathological conditions, neurotrophins such as NGF and BDNF "promote survival, differentiation, and neurite extension in many types of mammalian central nervous system neurons" [163]. Following ischemia, VEGF and BDNF are upregulated among others [164, 165]. Exogenous administration of growth factors themselves has been demonstrated to be beneficial in animal models of ischemia [163, 164, 166]. Furthermore, studies involving cellular therapy (MSCs especially) have demonstrated that transplanted cells exert beneficial effects, at least in part, through increased secretion of trophic factors [70]. Increases in trophic factor production through cellular therapy may prove beneficial for cerebral ischemia [167]. For example, increases in trophic factor are seen in several studies following MSC transplantation post-ischemia. Increases in VEGF expression are observed in peri-infarct tissue 14 days after ischemia (pMCAO) in rats treated with IV or intracarotid BM-MSCs compared with non-treated [168]. A study by Wakabayashi et al. demonstrated that animals receiving IV hMSC transplants following transient MCAO had smaller infarcts, improved recovery, and increased expression of VEGF, EGF, and bEGF compared with controls [70].

Evidence to date suggests that increasing trophic factor expression through cellular transplantation may exert therapeutic efficacy in animal models of ischemia. Trophic factor actions can be detrimental to prognosis, depending on the timing. VEGF, for example, can increase BBB permeability and the risk of hemorrhagic transformation [164]. Therefore, future studies are needed on specific windows of time across cell types that trophic factor induction is beneficial.

6.2   Promotion of Angiogenesis

Following stroke in rodent models, there is an innate increase in angiogenesis in the ischemic border. Surface collateral growth and parenchymal reperfusion increases are observed, indicative of post-stroke angiogenesis [162]. BrdU/endothelial co-labeling also increases, peaking at 7 days in the rat [162]. Increases in angiogenesis are also observed in human patients following ischemia [169]. Boosting the innate angiogenic response following ischemia would ostensibly be beneficial for ischemic stroke therapy by increasing blood flow, reducing infarct volume, and supporting neural vascular networks for quicker neurological recovery [170]. Additionally, increases in angiogenesis in non-infarcted tissue may be beneficial after ischemia as there is a well-known correlation with post-ischemic dementia and reduced blood flow in non-infarcted areas of the brain [170, 171].

Several studies have shown the effect of stem cell, especially MSC, transplantation and increased angiogenesis in several animal models of disease. A 2007 study showed that MSC transplantation in an excisional wound splinting model in mice increased capillary density after transplantation [69]. Transplantation of allogenic fetal membrane-derived and bone marrow-derived MSCs in a rat model of hindlimb ischemia increased capillary density 3 weeks after injection [172]. Hu et al. demonstrated that cardiac transplantation of murine BM-MSCs in a rodent model of myocardial infarction increased vessel density/area fraction in the heart 6 weeks after transplantation [173]. Suzuki et al. showed increases in hindlimb blood flow in a model of hindlimb ischemia after transplantation of iPS cell-derived Flk-1 positive cells [108]. Increases in angiogenesis and capillary density have also been observed in rat MCAO models following intravenous allogenic BM-MSC transplantation [174].

Literature suggests a relationship between angiogenesis and improved prognosis in animal models and in human patients, highlighting the putative importance of angiogenesis in post-ischemic recovery. Evidence to date suggests that stem cell transplantation can increase angiogenesis. Future studies are needed to fully elucidate the mechanisms behind stem cell-mediated increases in angiogenesis and any possible relationship between stem cell-mediated angiogenesis and functional recovery.

6.3 Modulation of Endogenous Neurogenesis

As mentioned previously, the adult CNS is capable of generating new neurons and glia from progenitor cells in several regions. Of particular interest in cerebral ischemia are neural progenitors from the sub-ventricular zone (SVZ). Under nonpathological conditions, neural progenitors from the SVZ migrate long distances along the RMS where they reach the olfactory bulb. In the olfactory bulb, SVZ-derived NPCs differentiate into inhibitory granule and periglomular interneurons [41].

It is now known that endogenous neurogenesis in the adult SVZ is regulated by stroke itself (for a review see [50, 117]). Ischemia not only increases innate neurogenesis in the SVZ but also diverts migrating NPCs away from the RMS toward the infarct. Unfortunately, the amount of new cells that survive is small, and only a minute amount of dead neurons are replaced [175]. One study by Arvidsson et al. demonstrated in a rat MCAO model of ischemia that only about 60% of migrating neural progenitors survived in the striatum 6 weeks after ischemia, replacing an estimated 0.2% of the dead neurons [141]. Ischemia has also been known to regulate neurogenesis in the SGZ and SCZ. Upregulation of neurogenesis in the SGZ has been observed following cerebral ischemia in different animal models of ischemia [71, 183, 207, 208]. Nakatomi et al. demonstrated the contribution of cells from the pPV (part of the SCZ) in repopulating hippocampal cells following ischemia [144].

Several cytokine and chemotaxic gradients are known to regulate NPC upregulation and infarct migration following ischemia. The CXCR4/SDF-1α (CXCL12) axis is one chemotactic gradient involved in migration of neural progenitors to the infarct. Following ischemia, stromal cell-derived factor 1-α is upregulated, attracting neural precursor cells expressing the complement receptor CXCR4 [51, 176]. Increasing the strength and/or duration of chemotactic gradients may be one method to improve the efficacy of endogenous NPCs following ischemia.

A number of studies have demonstrated that endogenous neurogenesis and migration can be upregulated following ischemia by a number of mechanisms. For example, use of cytokines [177], growth factors [178], small molecules [179], environmental enrichment [180], and peripheral stimulation [181] has all demonstrated increases in endogenous neurogenesis following ischemia. Transplantation of exogenous cells also increases innate neurogenesis [182].

The exact contribution of SGZ neurogenesis to ischemic stroke recovery is unclear. However, Li et al. recently demonstrated that reducing SVZ and SGZ neurogenesis following ICV Ara-C infusion (a cell proliferation inhibitor) increased the size of the infarct and neurological deficits following ischemia were worsened [183]. It is reasonable to suggest that post-ischemia neurogenesis would be beneficial for prognosis. Future studies should examine methods to increase the innate neurogenic response, increase the migration of NPCs, and increase the survival of NPCs.

6.4 Immune Modulation

The effects of pro- and anti-inflammatory cytokines were discussed in the previous sections. Therapies for stroke that can reduce pro-inflammatory cytokines and/or increase anti-inflammatory cytokines at certain times following ischemia could be beneficial for prognosis. Several studies have demonstrated the effect of exogenous cell transplantation and modulation of neuroinflammation following ischemia in different cell types. iPS transplantation with fibrin glue can reduce pro-inflammatory cytokines (IL-1β, TNF-α, IL-6, IL-2) and increase anti-inflammatory cytokine production (IL-10, iNOS, IL-4) 1 week after transplantation in an MCAO model in 8-week-old rats [107]. Nonhuman primate ischemia models have also seen immune influence by transplanted stem cells. Li et al. demonstrated in *Macaca fascicularis* males that intracerebral human BM-MSC transplantation following ischemia decreases GFAP reactivity and increases IL-10 positive cells and mRNA levels [71].

7 Cell-Based Stroke Therapy: Timing, Delivery, Survival, and Other Considerations

There are many considerations that must be made if cell-based therapy is to translate into clinical success. Proper dosages (cell numbers/densities/percentages of different types of cells or progenitors), source of cells, timing of delivery, and route of delivery are some key issues that need to be considered among other things [1]. Clinical trials to date have demonstrated as a proof of principle that stem cell transplantation for stroke is clinically safe [72, 184]. Cell-based therapy for stroke is still in its infancy and more work is needed to optimize the approach. Learning from the pitfalls of translation from neuroprotection studies, the American Stroke Association has made early attempts to address issues and considerations that need to be taken into account in order for cell-based therapy to successfully translate to the clinic. More details on considerations can be found in the Stem Cell Therapies as an Emerging Therapy for Stroke (STEPS) guideline publications [185, 186]. We now present examples of other considerations that should be taken in account for stem cell-based therapy for stroke.

7.1 Timing of Intervention

As with any therapy for ischemic stroke, the timing of delivery must be considered if the desired therapeutic efficacy is to be obtained. There are several phases of stroke recovery and therefore several windows of time within which cell-based therapy can be administered. Cell-based therapy holds promise for longer windows of time than traditional neuroprotective and thrombolytic approaches. However, it is imperative for a cell-based therapy to identify the therapeutic target[s] in order to determine the best time window for transplantation of cells. For example, therapies aimed at neuroprotection should usually be delivered acutely (within a few hours after the onset of stroke), while delayed administration is justified for a therapy that aims to promote post-injury regenerative processes. Additionally, the timing of post-ischemia scar formation should be considered for delayed therapies (especially transplantation therapies). Another significant concern is the microenvironment in the ischemic brain acutely, subacutely, and chronically after cerebral ischemia because the activated injurious and responsive signals and genes are differentially regulated at different post-stroke stages.

7.2 Intracranial Delivery

Intracranial delivery is the most direct route for cellular delivery, capable of delivering the most cells to the infarct [187]. Clinical studies to date have demonstrated the safety of intracranial delivery methods for stroke [184]; however, it is unclear if patients receive any functional benefit. As stated, the ischemic brain is a hostile environment as a result of pro-inflammatory cytokines, oxidative stress, and glutamate-mediated excitotoxicity and transplanted cell survival is low. In acute/subacute treatments, intracerebral transplantation routes may be best suited after the injurious factors such as ROS production subsided (usually several days after stroke) in the post-ischemia brain. Furthermore, intracerebral grafts may require a physical scaffold to adhere to due to loss of tissue following ischemia [187].

7.3 Intravenous Delivery

Intravenous delivery of stem cells has been studied at great length, especially with MSCs. As discussed, it is known that exogenous stem cells can exert therapeutic benefit without replacing cells. Current body of evidence actually supports the notion that intravenous administration of MSCs does not act through cell replacement, rather through modulation of neuroinflammation, growth factor production, endogenous neurogenesis, and angiogenesis [187]. To date, clinical trials of IV MSCs have shown to be safe and even provide some therapeutic benefit [72]. IV or IA administration may provide the most benefit during time periods when cytokines and chemokines are still upregulated in order to home cells toward the infarct [188]. Caution must be taken with IV or IA delivery. Hematically delivered cells have the potential to obstruct pulmonary and arterial vasculature and, since administration is systemic, there is potential of buildup of cells in other organs that may cause detrimental side effects [187, 188]. Increasing the homing of transplanted cells to the infarct brain region and increasing the survival of these cells once they arrive at the ischemic site remain to be the most challenging issues in intravenous delivery of cells and perhaps in other delivery methods.

7.4 Intranasal Delivery

Intranasal administration (INA) of cells has recently surfaced as a novel method for cellular delivery to the CNS that sidesteps the BBB. Intranasal delivery of peptides and other therapeutic agents has previously shown therapeutic benefit for several diseases in animal models and in humans [189]. Danielyan et al. first demonstrated that cells could be administered into the CNS through INA [190]. Since then, intranasal delivery of MSCs has shown to be beneficial in rodent models of hypoxia/ ischemia strokes [191, 192] and in rodent models of Parkinsons [193]. The exact route taken by intranasally administered agents is not well understood; however, it is thought that the RMS plays an important role [194]. INA seems to be a promising novel route for cellular delivery; however, more studies examining the precise mechanism of intranasal delivery and new strategies of improved migration and homing are needed.

7.5 Cell Survival and Tumorigenesis

The ischemic brain is a hostile environment and as a result, survival of exogenous and endogenous cells at the ischemic border is often low. Future studies examining methods to increase cell survival in the ischemic brain would be of great benefit to the cellular therapy field. For example, genetic modifications have shown great potential to improve cell survival after transplantation and functional recovery in experimental stroke models [89]. In this event, permanent modification of cell death/ survival genes adds another layer of complexity and should be evaluated carefully in future investigations. Alternatively, utilization of endogenous protective mechanisms using preconditioned cells for enhanced tolerance and regenerative potential of cells has drawn increasing attention and provided encouraging results. For example, hypoxic preconditioning has been sought to improve cell survival without increasing long-term concern of tumorigenesis after transplantation (see next section for details).

A long existing concern of stem cell transplantation is that the transplantation of multipotent cells carries an inherent risk of tumorigenesis, and longitudinal studies are needed across all cell types and routes of administration to ensure cellular therapy does not induce tumorigenesis following administration. Proper dosage studies may be needed in order to determine the best number of cells to administer that provides the greatest benefit while minimizing the risk of tumorigenesis.

8 Cellular Therapy: Modifications and Looking Toward the Future

Studies to date utilizing cellular therapy for ischemia have laid the foundation as a proof of principle and demonstrated beneficial effects of exogenous cell transplantation and enhancement of innate neurogenesis. As stated, the cellular microenvironment of the peri-infarct is a hostile environment for cells, and as a result, cell survival is low and perhaps not reaching its full potential. Early studies show promise in improving the effects of cellular therapy *in vitro* and in animal models by optimizing cells for transplant. We now present some early evidence that suggests optimization of cellular therapy and combinatorial therapy may be beneficial for cellular therapy. Because ischemic stroke is a heterogeneous insult, treatment strategies that provide combinations of pharmacology, gene therapy, cellular therapy, and rehabilitation may be needed for future translational success.

8.1 Genetic Modifications: Overexpression

Studies involving genetically modified stem cells and ischemia go back to at least the mid-2000s. Several groups have shown that stem cells overexpressing genes of interest improve outcome measures over non-modified cells in several animal models of disease. Genetic modification of stem cells may be beneficial in improving survival and homing to sites of ischemic injury. Additionally, genetically modified cells have the potential to serve as drug delivery vectors, eliminating the need for additional drug administration.

We tested the hypothesis that genetic modification of transplanted cells may enhance recovery after stroke. Mouse ES cells were modified to overexpress the antiapoptotic gene BCL-2 and transplanted into the ischemic hemisphere of rats 7 days after transient MCAO. Cell survival and neuronal differentiation of modified mES cells were increased over normal mES cells, and animals that received modified mES cells showed greater long-term functional recovery 21 and 35 days after transplantation compared to animals that received normal mES cells [89]. Ikeda et al. transplanted rat BM-MSCs overexpressing human FGF-2 into the striatum of rats 24 h after MCAO. They observed decreased infarct volume and improved neurological scores at days 7, 14, and 21 in the FGF-2-MSC group over non-overexpressing MSC group [195]. More recently, Wu et al. transplanted NSCs from embryonic rat hippocampi that were engineered to overexpress hypoxia-inducible factor-1 alpha (HIF-1α) into adult rats after transient MCAO. They observed that animals that received the modified cells showed more functional improvement 7, 14, 21, and 28 days after MCAO than any other of their experimental groups [196].

Other animal models have been examined as well. There are many parallels between cellular damage in myocardial ischemia following myocardial infarction and cerebral ischemia following cerebrovascular occlusion. Cellular therapy for myocardial infarction faces difficulties in cellular survival similar to ischemic stroke. Using MSCs overexpressing the antiapoptotic gene BCL-2, Li et al. demonstrated the benefit of cells overexpressing BCL-2. They saw that transplantation of BCL-2 MSCs into the myocardium decreased the number of TUNEL positive cells, increased VEGF secretion, increased vessel density, and decreased scar size compared to vector MSCs in a rodent model of myocardial infarction [197]. Benefit has also been seen in models of spinal cord injury. Transplantation of hNSCs overexpressing antiapoptotic BCL-X_L in female rats 7 days after spinal cord injury increased graft survival, increased white matter volume, decreased cavity volume, and improved functional recovery compared to transplanted normal hNSCs [198].

Research in animal models to date suggests that genetically modifying cells to overexpress proteins may be beneficial for not only graft survival but also functional recovery. Only a handful of candidate proteins have been tested, leaving countless others to be investigated. Future studies should also examine the effect of protein overexpression in different cell types. The effects of overexpression of a certain gene of interest may not carry over across various cell types. Last, care must be taken with cells overexpressing proteins such as growth factors, which are normally strictly temporally and spatially regulated endogenously. Long-term longitudinal studies are needed for transplantation studies using cells overexpressing such genes of interest to demonstrate the safety of this approach.

8.2   Hypoxic Preconditioning of Transplanted Cells

Ischemia induces a hostile environment in brain tissues. Not only do endogenous cells die as a result of necrosis and apoptosis, but also microenvironment in the ischemic penumbra is also hostile to migrating endogenous NPCs and exogenous cellular grafts. As a result, the survival of migrating and transplanted cells is often very low. By utilizing the innate oxygen-sensing system of the cell, preconditioning cells with a sublethal hypoxic insult can be protective and even increase therapeutic efficacy after transplant.

We recently demonstrated that hypoxic preconditioning of murine BM-MSCs provided benefits over normoxic BM-MSCs in a rodent model of myocardial infarction and cerebral ischemia [88, 173, 199, 200]. HP-BM-MSCs had a greater rate of survival, and HP-BM-MSC transplanted animals showed smaller infarcts and greater cardiac vascular density after transplantation than normoxic BM-MSC transplanted animals [173]. In vitro studies using human embryonic cells have shown that, over control, hypoxic preconditioning increases neural differentiation, action potential amplitude and number of APs, cell survival following ischemic insult, and delayed upregulation of erythropoietin, VEGF, and the pro-survival gene BCL-2 [200]. The benefits of hypoxic preconditioning for stem cells have been demonstrated in other animal models of ischemia, and have been shown to benefit ischemic tissue regrowth in rodent hindlimb ischemia models [201]. Hypoxic preconditioning has demonstrable benefits for other stem cell types in animal models of ischemia as well, such as ES cells [88].

Hypoxic preconditioning represents a promising method to improve exogenous graft survival. While hypoxic preconditioning undoubtedly involves activation of the cellular HIF system, further studies should investigate the mechanistic interplay between hypoxic-preconditioned grafts, host tissue, and cellular survival.

8.3 Combinatorial Therapies: Cellular Therapy and Environmental Enrichment

As previously stated, successful clinical therapies for stroke will most likely encompass multiple mechanisms of action to provide neuroprotection and promote CNS recovery. Additionally, stroke is a debilitating injury that will most likely rely, not only on acute clinical treatment, but also on long-term rehabilitation to promote recovery and significantly improve the quality of life for patients. Successful stroke therapy will not only involve multiple mechanisms of action on the cellular level but also combine clinical care with another modality such as rehabilitation or sensory stimulation to promote systems-level recovery.

Previous studies have examined the effect of environmental enrichment and rehabilitation in animal models of ischemia. Environmental enrichment can enhance the intrinsic self-repair of the CNS following ischemia. Environmental enrichment is seen to increase dendritic arborization [202], endogenous neurogenesis [203], and growth factor production [85, 204]. We showed that an enriched environment created by peripheral (whisker) stimulation markedly increased endogenous angiogenesis and functional recovery after ischemic stroke to the barrel cortex [205, 206]. Hicks et al. have demonstrated the beneficial effects of environmental enrichment and adult murine NPC transplantation following ischemia in rats [180] and the trend of environmental enrichment on increasing hES cell survival following transplantation after ischemia in rats [85].

Clinically, many patients undergo some sort of rehabilitation therapy following stroke. It seems prudent, therefore, to examine the effect of environmental enrichment and rehabilitation combined with cellular therapy. Initial studies on environmental enrichment and cellular therapy have shown potential. Future studies should examine the effect of environmental enrichment and different task/context-specific rehabilitation paradigms combined with cellular therapy on CNS recovery.

9 Conclusion

Cell-based therapy for stroke represents an exciting new avenue for stroke treatment. Early preclinical and clinical data is promising for a variety of cell types from a variety of sources through a variety of administration routes suggesting that "continued development of stem cell therapies may ultimately lead to viable treatment options for ischemic brain injury" [52]. By learning from evidence to date, there is hope that cell-based stroke therapy may one day be translated into a successful therapy for ischemic stroke.

References

1. Sahota P, Savitz SI. Investigational therapies for ischemic stroke: neuroprotection and neurorecovery. Neurotherapeutics. 2011;8(3):434–51.
2. Roger VL, Go AS, Lloyd-Jones DM, Adams RJ, Berry JD, Brown TM, Carnethon MR, Dai S, De Simone G, Ford ES, Fox CS, Fullerton HJ, Gillespie C, Greenlund KJ, Hailpern SM, Heit JA, Ho PM, Howard VJ, Kissela BM, Kittner SJ, Lackland DT, Lichtman JH, Lisabeth LD, Makuc DM, Marcus GM, Marelli A, Matchar DB, Mcdermott MM, Meigs JB, Moy CS, Mozaffarian D, Mussolino ME, Nichol G, Paynter NP, Rosamond WD, Sorlie PD, Stafford RS, Turan TN, Turner MB, Wong ND, Wylie-Rosett J, Roger VL, Turner MB. Heart disease and stroke statistics—2011 update: a report from the American Heart Association. Circulation. 2011;123:e18–209.
3. Wechsler LR. Intravenous thrombolytic therapy for acute ischemic stroke. N Engl J Med. 2011;364:2138–46.
4. Lo EH, Dalkara T, Moskowitz MA. Mechanisms, challenges and opportunities in stroke. Nat Rev Neurosci. 2003;4:399–415.
5. Fisher M. New approaches to neuroprotective drug development. Stroke. 2011;42:S24–7.
6. Rha JH, Saver JL. The impact of recanalization on ischemic stroke outcome: a meta-analysis. Stroke. 2007;38:967–73.
7. Stankowski JN, Gupta R. Therapeutic targets for neuroprotection in acute ischemic stroke: lost in translation? Antioxid Redox Signal. 2011;14:1841–51.
8. Grunwald IQ, Wakhloo AK, Walter S, Molyneux AJ, Byrne JV, Nagel S, Kuhn AL, Papadakis M, Fassbender K, Balami JS, Roffi M, Sievert H, Buchan A. Endovascular stroke treatment today. AJNR Am J Neuroradiol. 2011;32(2):238–43.
9. Gandhi CD, Christiano LD, Prestigiacomo CJ. Endovascular management of acute ischemic stroke. Neurosurg Focus. 2009;26:E2.
10. Baker WL, Colby JA, Tongbram V, Talati R, Silverman IE, White CM, Kluger J, Coleman CI. Neurothrombectomy devices for the treatment of acute ischemic stroke: state of the evidence. Ann Intern Med. 2011;154:243–52.
11. Adams Jr HP, del Zoppo G, Alberts MJ, Bhatt DL, Brass L, Furlan A, Grubb RL, Higashida RT, Jauch EC, Kidwell C, Lyden PD, Morgenstern LB, Qureshi AI, Rosenwasser RH, Scott PA, Wijdicks EF. Guidelines for the early management of adults with ischemic stroke: a guideline from the American Heart Association/American Stroke Association Stroke Council, Clinical Cardiology Council, Cardiovascular Radiology and Intervention Council, and the Atherosclerotic Peripheral Vascular Disease and Quality of Care Outcomes in Research Interdisciplinary Working Groups: the American Academy of Neurology affirms the value of this guideline as an educational tool for neurologists. Stroke. 2007;38:1655–711.
12. Meyers PM, Schumacher HC, Higashida RT, Barnwell SL, Creager MA, Gupta R, Mcdougall CG, Pandey DK, Sacks D, Wechsler LR. Indications for the performance of intracranial

endovascular neurointerventional procedures: a scientific statement from the American Heart Association Council on Cardiovascular Radiology and Intervention, Stroke Council, Council on Cardiovascular Surgery and Anesthesia, Interdisciplinary Council on Peripheral Vascular Disease, and Interdisciplinary Council on Quality of Care and Outcomes Research. Circulation. 2009;119:2235–49.

13. Langhorne P, Bernhardt J, Kwakkel G. Stroke rehabilitation. Lancet. 2011;377:1693–702.

14. Kwakkel G, Kollen B, Twisk J. Impact of time on improvement of outcome after stroke. Stroke. 2006;37:2348–53.

15. Miller EL, Murray L, Richards L, Zorowitz RD, Bakas T, Clark P, Billinger SA. Comprehensive overview of nursing and interdisciplinary rehabilitation care of the stroke patient: a scientific statement from the American Heart Association. Stroke. 2010;41:2402–48.

16. Ginsberg MD. Current status of neuroprotection for cerebral ischemia: synoptic overview. Stroke. 2009;40:S111–4.

17. Ginsberg MD. Neuroprotection for ischemic stroke: past, present and future. Neuropharmacology. 2008;55:363–89.

18. Richard Green A, Odergren T, Ashwood T. Animal models of stroke: do they have value for discovering neuroprotective agents? Trends Pharmacol Sci. 2003;24:402–8.

19. Stroke Therapy Academic Industry Roundtable (STAIR). Recommendations for standards regarding preclinical neuroprotective and restorative drug development. Stroke. 1999;30:2752–8.

20. Fisher M, Feuerstein G, Howells DW, Hurn PD, Kent TA, Savitz SI, Lo EH. Update of the stroke therapy academic industry roundtable preclinical recommendations. Stroke. 2009;40:2244–50.

21. Barreto AD. Intravenous thrombolytics for ischemic stroke. Neurotherapeutics. 2011;8(3): 388–99.

22. Blakeley JO, Llinas RH. Thrombolytic therapy for acute ischemic stroke. J Neurol Sci. 2007;261:55–62.

23. Multicentre Acute Stroke Trial-Italy (MAST-I) Group. Randomised controlled trial of streptokinase, aspirin, and combination of both in treatment of acute ischaemic stroke. Lancet. 1995;346:1509–14.

24. The Multicenter Acute Stroke Trial—Europe Study Group. Thrombolytic therapy with streptokinase in acute ischemic stroke. N Engl J Med. 1996;335:145–50.

25. Donnan GA, Davis SM, Chambers BR, Gates PC, Hankey GJ, McNeil JJ, Rosen D, Stewart-Wynne EG, Tuck RR. Streptokinase for acute ischemic stroke with relationship to time of administration: Australian Streptokinase (ASK) Trial Study Group. JAMA. 1996;276: 961–6.

26. Hommel M, Boissel JP, Cornu C, Boutitie F, Lees KR, Besson G, Leys D, Amarenco P, Bogaert M. Termination of trial of streptokinase in severe acute ischaemic stroke. MAST Study Group. Lancet. 1995;345:57.

27. The National Institute of Neurological Disorders and Stroke rt-PA Stroke Study Group. Tissue plasminogen activator for acute ischemic stroke. N Engl J Med. 1995;333:1581–7.

28. Hacke W, Kaste M, Bluhmki E, Brozman M, Dávalos A, Guidetti D, Larrue V, Lees KR, Medeghri Z, Machnig T, Schneider D, von Kummer R, Wahlgren N, Toni D, Investigators E. Thrombolysis with alteplase 3 to 4.5 hours after acute ischemic stroke. N Engl J Med. 2008;359:1317–29.

29. Davis SM, Donnan GA, Parsons MW, Levi C, Butcher KS, Peeters A, Barber PA, Bladin C, De Silva DA, Byrnes G, Chalk JB, Fink JN, Kimber TE, Schultz D, Hand PJ, Frayne J, Hankey G, Muir K, Gerraty R, Tress BM, Desmond PM, EPITHET Investigators. Effects of alteplase beyond 3 h after stroke in the Echoplanar Imaging Thrombolytic Evaluation Trial (EPITHET): a placebo-controlled randomised trial. Lancet Neurol. 2008;7:299–309.

30. Del Zoppo GJ, Saver JL, Jauch EC, Adams Jr HP. Expansion of the time window for treatment of acute ischemic stroke with intravenous tissue plasminogen activator: a science advisory from the American Heart Association/American Stroke Association. Stroke. 2009; 40:2945–8.

31. Wu T-C, Grotta JC. Stroke treatment and prevention: five new things. Neurology. 2010; 75:S16–21.

32. Ming GL, Song H. Adult neurogenesis in the mammalian brain: significant answers and significant questions. Neuron. 2011;70:687–702.

33. Ming GL, Song H. Adult neurogenesis in the mammalian central nervous system. Annu Rev Neurosci. 2005;28:223–50.

34. Lindvall O, Kokaia Z. Stem cells for the treatment of neurological disorders. Nature. 2006;441:1094–6.

35. Orlacchio A, Bernardi G, Martino S. Stem cells and neurological diseases. Discov Med. 2010;9:546–53.

36. Shimada IS, Spees JL. Stem and progenitor cells for neurological repair: minor issues, major hurdles, and exciting opportunities for paracrine-based therapeutics. J Cell Biochem. 2011;112:374–80.

37. Altman J. Are new neurons formed in the brains of adult mammals? Science. 1962;135: 1127–8.

38. Altman J, Das GD. Autoradiographic and histological evidence of postnatal hippocampal neurogenesis in rats. J Comp Neurol. 1965;124:319–35.

39. Altman J. Autoradiographic and histological studies of postnatal neurogenesis. IV. Cell proliferation and migration in the anterior forebrain, with special reference to persisting neurogenesis in the olfactory bulb. J Comp Neurol. 1969;137:433–57.

40. Seri B, Herrera DG, Gritti A, Ferron S, Collado L, Vescovi A, Garcia-Verdugo JM, Alvarez-Buylla A. Composition and organization of the SCZ: a large germinal layer containing neural stem cells in the adult mammalian brain. Cereb Cortex. 2006;16 Suppl 1:i103–11.

41. Lledo P-M, Alonso M, Grubb MS. Adult neurogenesis and functional plasticity in neuronal circuits. Nat Rev Neurosci. 2006;7:179–93.

42. Taupin P. Therapeutic potential of adult neural stem cells. Recent Pat CNS Drug Discov. 2006;1:299–303.

43. Lie DC, Song H, Colamarino SA, Ming GL, Gage FH. Neurogenesis in the adult brain: new strategies for central nervous system diseases. Annu Rev Pharmacol Toxicol. 2004; 44:399–421.

44. Massirer KB, Carromeu C, Griesi-Oliveira K, Muotri AR. Maintenance and differentiation of neural stem cells. Wiley Interdiscip Rev Syst Biol Med. 2011;3:107–14.

45. Hsu YC, Lee DC, Chiu IM. Neural stem cells, neural progenitors, and neurotrophic factors. Cell Transplant. 2007;16:133–50.

46. Kazanis I, Lathia JD, Moss L, ffrench-Constant C. The neural stem cell microenvironment (August 31, 2008). Cambridge: Harvard Stem Cell Institute; 2008.

47. Qu Q, Shi Y. Neural stem cells in the developing and adult brains. J Cell Physiol. 2009;221:5–9.

48. De Filippis L, Delia D. Hypoxia in the regulation of neural stem cells. Cell Mol Life Sci. 2011;68:2831–44.

49. Pluchino S, Zanotti L, Deleidi M, Martino G. Neural stem cells and their use as therapeutic tool in neurological disorders. Brain Res Brain Res Rev. 2005;48:211–9.

50. Zhang RL, Zhang ZG, Chopp M. Ischemic stroke and neurogenesis in the subventricular zone. Neuropharmacology. 2008;55:345–52.

51. Imitola J, Raddassi K, Park KI, Mueller F-J, Nieto M, Teng YD, Frenkel D, Li J, Sidman RL, Walsh CA, Snyder EY, Khoury SJ. Directed migration of neural stem cells to sites of CNS injury by the stromal cell-derived factor 1alpha/CXC chemokine receptor 4 pathway. Proc Natl Acad Sci USA. 2004;101:18117–22.

52. Burns TC, Verfaillie CM, Low WC. Stem cells for ischemic brain injury: a critical review. J Comp Neurol. 2009;515:125–44.

53. Becker S. A computational principle for hippocampal learning and neurogenesis. Hippocampus. 2005;15:722–38.

54. Wiskott L, Rasch MJ, Kempermann G. A functional hypothesis for adult hippocampal neurogenesis: avoidance of catastrophic interference in the dentate gyrus. Hippocampus. 2006;16:329–43.

55. Aimone JB, Wiles J, Gage FH. Potential role for adult neurogenesis in the encoding of time in new memories. Nat Neurosci. 2006;9:723–7.
56. Jordan JD, Ming GL, Song H. Adult neurogenesis as a potential therapy for neurodegenerative diseases. Discov Med. 2006;6:144–7.
57. Bithell A, Williams BP. Neural stem cells and cell replacement therapy: making the right cells. Clin Sci (Lond). 2005;108:13–22.
58. Friedenstein AJ, Petrakova KV, Kurolesova AI, Frolova GP. Heterotopic transplants of bone marrow. Analysis of precursor cells for osteogenic and hematopoietic tissues. Transplantation. 1968;6:230–47.
59. Bernardo ME, Locatelli F, Fibbe WE. Mesenchymal stromal cells. Ann N Y Acad Sci. 2009;1176:101–17.
60. Friedenstein AJ, Piatetzky II S, Petrakova KV. Osteogenesis in transplants of bone marrow cells. J Embryol Exp Morphol. 1966;16:381–90.
61. Horwitz EM, Le Blanc K, Dominici M, Mueller I, Slaper-Cortenbach I, Marini FC, Deans RJ, Krause DS, Keating A. Clarification of the nomenclature for MSC: The International Society for Cellular Therapy position statement. Cytotherapy. 2005;7:393–5.
62. Malgieri A, Kantzari E, Patrizi MP, Gambardella S. Bone marrow and umbilical cord blood human mesenchymal stem cells: state of the art. Int J Clin Exp Med. 2010;3:248–69.
63. Salem HK, Thiemermann C. Mesenchymal stromal cells: current understanding and clinical status. Stem Cells. 2010;28:585–96.
64. Caplan AI. The mesengenic process. Clin Plast Surg. 1994;21:429–35.
65. Kemp KC, Hows J, Donaldson C. Bone marrow-derived mesenchymal stem cells. Leuk Lymphoma. 2005;46:1531–44.
66. Thomas MG, Stone L, Evill L, Ong S, Ziman M, Hool L. Bone marrow stromal cells as replacement cells for Parkinson's disease: generation of an anatomical but not functional neuronal phenotype. Transl Res. 2011;157:56–63.
67. Xu H, Miki K, Ishibashi S, Inoue J, Sun L, Endo S, Sekiya I, Muneta T, Inazawa J, Dezawa M, Mizusawa H. Transplantation of neuronal cells induced from human mesenchymal stem cells improves neurological functions after stroke without cell fusion. J Neurosci Res. 2010;88:3598–609.
68. Rosenkranz K, Kumbruch S, Lebermann K, Marschner K, Jensen A, Dermietzel R, Meier C. The chemokine SDF-1/CXCL12 contributes to the "homing" of umbilical cord blood cells to a hypoxic-ischemic lesion in the rat brain. J Neurosci Res. 2010;88(6):1223–33.
69. Wu Y, Chen L, Scott PG, Tredget EE. Mesenchymal stem cells enhance wound healing through differentiation and angiogenesis. Stem Cells. 2007;25:2648–59.
70. Wakabayashi K, Nagai A, Sheikh AM, Shiota Y, Narantuya D, Watanabe T, Masuda J, Kobayashi S, Kim SU, Yamaguchi S. Transplantation of human mesenchymal stem cells promotes functional improvement and increased expression of neurotrophic factors in a rat focal cerebral ischemia model. J Neurosci Res. 2010;88:1017–25.
71. Li J, Zhu H, Liu Y, Li Q, Lu S, Feng M, Xu Y, Huang L, Ma C, An Y, Zhao RC, Wang R, Qin C. Human mesenchymal stem cell transplantation protects against cerebral ischemic injury and upregulates interleukin-10 expression in *Macaca fascicularis*. Brain Res. 2010; 1334:65–72.
72. Lee JS, Hong JM, Moon GJ, Lee PH, Ahn YH, Bang OY, STARTING Collaborators. A long-term follow-up study of intravenous autologous mesenchymal stem cell transplantation in patients with ischemic stroke. Stem Cells. 2010;28:1099–106.
73. Thomson JA, Itskovitz-Eldor J, Shapiro SS, Waknitz MA, Swiergiel JJ, Marshall VS, Jones JM. Embryonic stem cell lines derived from human blastocysts. Science. 1998;282:1145–7.
74. Odorico JS, Kaufman DS, Thomson JA. Multilineage differentiation from human embryonic stem cell lines. Stem Cells. 2001;19:193–204.
75. Thomson JA, Kalishman J, Golos TG, Durning M, Harris CP, Becker RA, Hearn JP. Isolation of a primate embryonic stem cell line. Proc Natl Acad Sci USA. 1995;92:7844–8.
76. Chu G. Embryonic stem-cell research and the moral status of embryos. Intern Med J. 2003;33:530–1.

77. Devolder K. Human embryonic stem cell research: why the discarded-created-distinction cannot be based on the potentiality argument. Bioethics. 2005;19:167–86.
78. Jain KK. Ethical and regulatory aspects of embryonic stem cell research. Expert Opin Biol Ther. 2005;5:153–62.
79. Harris G. U.S. Judge rules against Obama's stem cell policy [Epub Newspaper]. 2010. http://www.nytimes.com/2010/08/24/health/policy/24stem.html. Accessed 1 July 2011.
80. Buhnemann C, Scholz A, Bernreuther C, Malik CY, Braun H, Schachner M, Reymann KG, Dihne M. Neuronal differentiation of transplanted embryonic stem cell-derived precursors in stroke lesions of adult rats. Brain. 2006;129:3238–48.
81. Zhang P, Li J, Liu Y, Chen X, Kang Q. Transplanted human embryonic neural stem cells survive, migrate, differentiate and increase endogenous nestin expression in adult rat cortical peri-infarction zone. Neuropathology. 2009;29:410–21.
82. Bjorklund LM, Sanchez-Pernaute R, Chung S, Andersson T, Chen IY, McNaught KS, Brownell AL, Jenkins BG, Wahlestedt C, Kim KS, Isacson O. Embryonic stem cells develop into functional dopaminergic neurons after transplantation in a Parkinson rat model. Proc Natl Acad Sci USA. 2002;99:2344–9.
83. Guillaume DJ, Johnson MA, Li XJ, Zhang SC. Human embryonic stem cell-derived neural precursors develop into neurons and integrate into the host brain. J Neurosci Res. 2006;84:1165–76.
84. Kang HC, Kim DS, Kim JY, Kim HS, Lim BY, Kim HD, Lee JS, Eun BL, Kim DW. Behavioral improvement after transplantation of neural precursors derived from embryonic stem cells into the globally ischemic brain of adolescent rats. Brain Dev. 2010;32:658–68.
85. Hicks AU, Lappalainen RS, Narkilahti S, Suuronen R, Corbett D, Sivenius J, Hovatta O, Jolkkonen J. Transplantation of human embryonic stem cell-derived neural precursor cells and enriched environment after cortical stroke in rats: cell survival and functional recovery. Eur J Neurosci. 2009;29:562–74.
86. Daadi MM, Maag AL, Steinberg GK. Adherent self-renewable human embryonic stem cell-derived neural stem cell line: functional engraftment in experimental stroke model. PLoS One. 2008;3:e1644.
87. Takahashi K, Yasuhara T, Shingo T, Muraoka K, Kameda M, Takeuchi A, Yano A, Kurozumi K, Agari T, Miyoshi Y, Kinugasa K, Date I. Embryonic neural stem cells transplanted in middle cerebral artery occlusion model of rats demonstrated potent therapeutic effects, compared to adult neural stem cells. Brain Res. 2008;1234:172–82.
88. Theus MH, Wei L, Cui L, Francis K, Hu X, Keogh C, Yu SP. In vitro hypoxic preconditioning of embryonic stem cells as a strategy of promoting cell survival and functional benefits after transplantation into the ischemic rat brain. Exp Neurol. 2008;210:656–70.
89. Wei L, Cui L, Snider BJ, Rivkin M, Yu SS, Lee C-S, Adams LD, Gottlieb DI, Johnson EM, Yu SP, Choi DW. Transplantation of embryonic stem cells overexpressing Bcl-2 promotes functional recovery after transient cerebral ischemia. Neurobiol Dis. 2005;19:183–93.
90. Damjanov I, Andrews PW. The terminology of teratocarcinomas and teratomas. Nat Biotechnol. 2007;25:1212; discussion 1212.
91. Ben-David U, Benvenisty N. The tumorigenicity of human embryonic and induced pluripotent stem cells. Nat Rev Cancer. 2011;11:268–77.
92. Shih CC, Forman SJ, Chu P, Slovak M. Human embryonic stem cells are prone to generate primitive, undifferentiated tumors in engrafted human fetal tissues in severe combined immunodeficient mice. Stem Cells Dev. 2007;16:893–902.
93. Asano T, Sasaki K, Kitano Y, Terao K, Hanazono Y. In vivo tumor formation from primate embryonic stem cells. Methods Mol Biol. 2006;329:459–67.
94. Seminatore C, Polentes J, Ellman D, Kozubenko N, Itier V, Tine S, Tritschler L, Brenot M, Guidou E, Blondeau J, Lhuillier M, Bugi A, Aubry L, Jendelova P, Sykova E, Perrier AL, Finsen B, Onteniente B. The postischemic environment differentially impacts teratoma or tumor formation after transplantation of human embryonic stem cell-derived neural progenitors. Stroke. 2010;41:153–9.

95. Takahashi K, Yamanaka S. Induction of pluripotent stem cells from mouse embryonic and adult fibroblast cultures by defined factors. Cell. 2006;126:663–76.

96. Takahashi K, Tanabe K, Ohnuki M, Narita M, Ichisaka T, Tomoda K, Yamanaka S. Induction of pluripotent stem cells from adult human fibroblasts by defined factors. Cell. 2007; 131:861–72.

97. Hockemeyer D, Soldner F, Cook EG, Gao Q, Mitalipova M, Jaenisch R. A drug-inducible system for direct reprogramming of human somatic cells to pluripotency. Cell Stem Cell. 2008;3:346–53.

98. Maherali N, Ahfeldt T, Rigamonti A, Utikal J, Cowan C, Hochedlinger K. A high-efficiency system for the generation and study of human induced pluripotent stem cells. Cell Stem Cell. 2008;3:340–5.

99. Feng B, Ng J-H, Heng J-CD, Ng H-H. Molecules that promote or enhance reprogramming of somatic cells to induced pluripotent stem cells. Cell Stem Cell. 2009;4:301–12.

100. Lin T, Ambasudhan R, Yuan X, Li W, Hilcove S, Abujarour R, Lin X, Hahm HS, Hao E, Hayek A, Ding S. A chemical platform for improved induction of human iPSCs. Nat Methods. 2009;6:805–8.

101. Okita K, Nakagawa M, Hyenjong H, Ichisaka T, Yamanaka S. Generation of mouse induced pluripotent stem cells without viral vectors. Science. 2008;322:949–53.

102. Patel M, Yang S. Advances in reprogramming somatic cells to induced pluripotent stem cells. Stem Cell Rev. 2010;6:367–80.

103. Okita K. iPS cells for transplantation. Curr Opin Organ Transplant. 2011;16(1):96–100.

104. Brennand KJ, Simone A, Jou J, Gelboin-Burkhart C, Tran N, Sangar S, Li Y, Mu Y, Chen G, Yu D, McCarthy S, Sebat J, Gage FH. Modelling schizophrenia using human induced pluripotent stem cells. Nature. 2011;473:221–5.

105. Dimos JT, Rodolfa KT, Niakan KK, Weisenthal LM, Mitsumoto H, Chung W, Croft GF, Saphier G, Leibel R, Goland R, Wichterle H, Henderson CE, Eggan K. Induced pluripotent stem cells generated from patients with ALS can be differentiated into motor neurons. Science. 2008;321:1218–21.

106. Kawai H, Yamashita T, Ohta Y, Deguchi K, Nagotani S, Zhang X, Ikeda Y, Matsuura T, Abe K. Tridermal tumorigenesis of induced pluripotent stem cells transplanted in ischemic brain. J Cereb Blood Flow Metab. 2010;30:1487–93.

107. Chen S-J, Chang C-M, Tsai S-K, Chang Y-L, Chou S-J, Huang S-S, Tai L-K, Chen Y-C, Ku H-H, Li H-Y, Chiou S-H. Functional improvement of focal cerebral ischemia injury by subdural transplantation of induced pluripotent stem cells with fibrin glue. Stem Cells Dev. 2010;19:1757–67.

108. Suzuki H, Shibata R, Kito T, Ishii M, Li P, Yoshikai T, Nishio N, Ito S, Numaguchi Y, Yamashita JK, Murohara T, Isobe K. Therapeutic angiogenesis by transplantation of induced pluripotent stem cell-derived Flk-1 positive cells. BMC Cell Biol. 2010;11:72.

109. Jiang M, Lv L, Ji H, Yang X, Zhu W, Cai L, Gu X, Chai C, Huang S, Sun J, Dong Q. Induction of pluripotent stem cells transplantation therapy for ischemic stroke. Mol Cell Biochem. 2011;354:67–75.

110. Meissner A. Epigenetic modifications in pluripotent and differentiated cells. Nat Biotechnol. 2010;28:1079–88.

111. Gore A, Li Z, Fung HL, Young JE, Agarwal S, Antosiewicz-Bourget J, Canto I, Giorgetti A, Israel MA, Kiskinis E, Lee JH, Loh YH, Manos PD, Montserrat N, Panopoulos AD, Ruiz S, Wilbert ML, Yu J, Kirkness EF, Izpisua Belmonte JC, Rossi DJ, Thomson JA, Eggan K, Daley GQ, Goldstein LS, Zhang K. Somatic coding mutations in human induced pluripotent stem cells. Nature. 2011;471:63–7.

112. Lister R, Pelizzola M, Kida YS, Hawkins RD, Nery JR, Hon G, Antosiewicz-Bourget J, O'Malley R, Castanon R, Klugman S, Downes M, Yu R, Stewart R, Ren B, Thomson JA, Evans RM, Ecker JR. Hotspots of aberrant epigenomic reprogramming in human induced pluripotent stem cells. Nature. 2011;471:68–73.

113. Barrero MJ, Izpisua Belmonte JC. iPS cells forgive but do not forget. Nat Cell Biol. 2011;13:523–5.

114. Kim K, Doi A, Wen B, Ng K, Zhao R, Cahan P, Kim J, Aryee MJ, Ji H, Ehrlich LI, Yabuuchi A, Takeuchi A, Cunniff KC, Hongguang H, McKinney-Freeman S, Naveiras O, Yoon TJ, Irizarry RA, Jung N, Seita J, Hanna J, Murakami P, Jaenisch R, Weissleder R, Orkin SH, Weissman IL, Feinberg AP, Daley GQ. Epigenetic memory in induced pluripotent stem cells. Nature. 2010;467:285–90.

115. Sullivan GJ, Bai Y, Fletcher J, Wilmut I. Induced pluripotent stem cells: epigenetic memories and practical implications. Mol Hum Reprod. 2010;16:880–5.

116. Rakic P. Neurogenesis in adult primate neocortex: an evaluation of the evidence. Nat Rev Neurosci. 2002;3:65–71.

117. Wiltrout C, Lang B, Yan Y, Dempsey RJ, Vemuganti R. Repairing brain after stroke: a review on post-ischemic neurogenesis. Neurochem Int. 2007;50:1028–41.

118. Kaplan MS, Hinds JW. Neurogenesis in the adult rat: electron microscopic analysis of light radioautographs. Science. 1977;197:1092–4.

119. Goldman SA, Nottebohm F. Neuronal production, migration, and differentiation in a vocal control nucleus of the adult female canary brain. Proc Natl Acad Sci USA. 1983; 80:2390–4.

120. Nottebohm F. A brain for all seasons: cyclical anatomical changes in song control nuclei of the canary brain. Science. 1981;214:1368–70.

121. Gould E, Reeves AJ, Graziano MS, Gross CG. Neurogenesis in the neocortex of adult primates. Science. 1999;286:548–52.

122. Eriksson PS, Perfilieva E, Björk-Eriksson T, Alborn AM, Nordborg C, Peterson DA, Gage FH. Neurogenesis in the adult human hippocampus. Nat Med. 1998;4:1313–7.

123. Gould E, Vail N, Wagers M, Gross CG. Adult-generated hippocampal and neocortical neurons in macaques have a transient existence. Proc Natl Acad Sci USA. 2001;98:10910–7.

124. Cameron HA, McKay RD. Adult neurogenesis produces a large pool of new granule cells in the dentate gyrus. J Comp Neurol. 2001;435:406–17.

125. Hayes NL, Nowakowski RS. Dynamics of cell proliferation in the adult dentate gyrus of two inbred strains of mice. Brain Res Dev Brain Res. 2002;134:77–85.

126. Christie BR, Cameron HA. Neurogenesis in the adult hippocampus. Hippocampus. 2006; 16:199–207.

127. Liu S, Wang J, Zhu D, Fu Y, Lukowiak K, Lu YM. Generation of functional inhibitory neurons in the adult rat hippocampus. J Neurosci. 2003;23:732–6.

128. Kohler SJ, Williams NI, Stanton GB, Cameron JL, Greenough WT. Maturation time of new granule cells in the dentate gyrus of adult macaque monkeys exceeds six months. Proc Natl Acad Sci USA. 2011;108(25):10326–31.

129. Dayer AG, Ford AA, Cleaver KM, Yassaee M, Cameron HA. Short-term and long-term survival of new neurons in the rat dentate gyrus. J Comp Neurol. 2003;460:563–72.

130. Kuhn HG, Dickinson-Anson H, Gage FH. Neurogenesis in the dentate gyrus of the adult rat: age-related decrease of neuronal progenitor proliferation. J Neurosci. 1996;16:2027–33.

131. Morshead CM, Craig CG, van der Kooy D. In vivo clonal analyses reveal the properties of endogenous neural stem cell proliferation in the adult mammalian forebrain. Development. 1998;125:2251–61.

132. Tramontin AD, García-Verdugo JM, Lim DA, Alvarez-Buylla A. Postnatal development of radial glia and the ventricular zone (VZ): a continuum of the neural stem cell compartment. Cereb Cortex. 2003;13:580–7.

133. Bonfanti L, Peretto P. Radial glial origin of the adult neural stem cells in the subventricular zone. Prog Neurobiol. 2007;83:24–36.

134. Alvarez-Buylla A, García-Verdugo JM, Tramontin AD. A unified hypothesis on the lineage of neural stem cells. Nat Rev Neurosci. 2001;2:287–93.

135. Doetsch F, García-Verdugo JM, Alvarez-Buylla A. Cellular composition and three-dimensional organization of the subventricular germinal zone in the adult mammalian brain. J Neurosci. 1997;17:5046–61.

136. Petreanu L, Alvarez-Buylla A. Maturation and death of adult-born olfactory bulb granule neurons: role of olfaction. J Neurosci. 2002;22:6106–13.

137. Wang C, Liu F, Liu Y-Y, Zhao C-H, You Y, Wang L, Zhang J, Wei B, Ma T, Zhang Q, Zhang Y, Chen R, Song H, Yang Z. Identification and characterization of neuroblasts in the subventricular zone and rostral migratory stream of the adult human brain. Cell Res. 2011; 21(11):1534–50.

138. Pencea V, Bingaman KD, Freedman LJ, Luskin MB. Neurogenesis in the subventricular zone and rostral migratory stream of the neonatal and adult primate forebrain. Exp Neurol. 2001;172:1–16.

139. Curtis MA, Kam M, Nannmark U, Anderson MF, Axell MZ, Wikkelso C, Holtås S, van Roon-Mom WMC, Björk-Eriksson T, Nordborg C, Frisén J, Dragunow M, Faull RLM, Eriksson PS. Human neuroblasts migrate to the olfactory bulb via a lateral ventricular extension. Science. 2007;315:1243–9.

140. Sanai N, Berger MS, Garcia-Verdugo JM, Alvarez-Buylla A. Comment on "Human neuroblasts migrate to the olfactory bulb via a lateral ventricular extension." Science. 2007;318:393; author reply 393.

141. Arvidsson A, Collin T, Kirik D, Kokaia Z, Lindvall O. Neuronal replacement from endogenous precursors in the adult brain after stroke. Nat Med. 2002;8:963–70.

142. Zhao M, Momma S, Delfani K, Carlen M, Cassidy RM, Johansson CB, Brismar H, Shupliakov O, Frisen J, Janson AM. Evidence for neurogenesis in the adult mammalian substantia nigra. Proc Natl Acad Sci USA. 2003;100:7925–30.

143. Kim WR, Chun SK, Kim TW, Kim H, Ono K, Takebayashi H, Ikenaka K, Oppenheim RW, Sun W. Evidence for the spontaneous production but massive programmed cell death of new neurons in the subcallosal zone of the postnatal mouse brain. Eur J Neurosci. 2011;33:599–611.

144. Nakatomi H, Kuriu T, Okabe S, Yamamoto S-I, Hatano O, Kawahara N, Tamura A, Kirino T, Nakafuku M. Regeneration of hippocampal pyramidal neurons after ischemic brain injury by recruitment of endogenous neural progenitors. Cell. 2002;110:429–41.

145. Lipton P. Ischemic cell death in brain neurons. Physiol Rev. 1999;79:1431–568.

146. Lakhan SE, Kirchgessner A, Hofer M. Inflammatory mechanisms in ischemic stroke: therapeutic approaches. J Transl Med. 2009;7:97.

147. Clark AW, Krekoski CA, Bou SS, Chapman KR, Edwards DR. Increased gelatinase A (MMP-2) and gelatinase B (MMP-9) activities in human brain after focal ischemia. Neurosci Lett. 1997;238:53–6.

148. Romanic AM, White RF, Arleth AJ, Ohlstein EH, Barone FC. Matrix metalloproteinase expression increases after cerebral focal ischemia in rats: inhibition of matrix metalloproteinase-9 reduces infarct size. Stroke. 1998;29:1020–30.

149. Gelderblom M, Leypoldt F, Steinbach K, Behrens D, Choe C-U, Siler DA, Arumugam TV, Orthey E, Gerloff C, Tolosa E, Magnus T. Temporal and spatial dynamics of cerebral immune cell accumulation in stroke. Stroke. 2009;40:1849–57.

150. Basic Kes V, Simundic AM, Nikolac N, Topic E, Demarin V. Pro-inflammatory and anti-inflammatory cytokines in acute ischemic stroke and their relation to early neurological deficit and stroke outcome. Clin Biochem. 2008;41:1330–4.

151. Vila N, Castillo J, Davalos A, Chamorro A. Proinflammatory cytokines and early neurological worsening in ischemic stroke. Stroke. 2000;31:2325–9.

152. Vila N, Castillo J, Davalos A, Esteve A, Planas AM, Chamorro A. Levels of anti-inflammatory cytokines and neurological worsening in acute ischemic stroke. Stroke. 2003;34:671–5.

153. Nawashiro H, Martin D, Hallenbeck JM. Neuroprotective effects of TNF binding protein in focal cerebral ischemia. Brain Res. 1997;778:265–71.

154. Bruce AJ, Boling W, Kindy MS, Peschon J, Kraemer PJ, Carpenter MK, Holtsberg FW, Mattson MP. Altered neuronal and microglial responses to excitotoxic and ischemic brain injury in mice lacking TNF receptors. Nat Med. 1996;2:788–94.

155. Gary DS, Bruce-Keller AJ, Kindy MS, Mattson MP. Ischemic and excitotoxic brain injury is enhanced in mice lacking the p55 tumor necrosis factor receptor. J Cereb Blood Flow Metab. 1998;18:1283–7.

156. Boutin H, LeFeuvre RA, Horai R, Asano M, Iwakura Y, Rothwell NJ. Role of IL-1alpha and IL-1beta in ischemic brain damage. J Neurosci. 2001;21:5528–34.

157. Chen G, Goeddel DV. TNF-R1 signaling: a beautiful pathway. Science. 2002;296:1634–5.
158. Schütze S, Tchikov V, Schneider-Brachert W. Regulation of TNFR1 and CD95 signalling by receptor compartmentalization. Nat Rev Mol Cell Biol. 2008;9:655–62.
159. Lambertsen KL, Clausen BH, Babcock AA, Gregersen R, Fenger C, Nielsen HH, Haugaard LS, Wirenfeldt M, Nielsen M, Dagnaes-Hansen F, Bluethmann H, Faergeman NJ, Meldgaard M, Deierborg T, Finsen B. Microglia protect neurons against ischemia by synthesis of tumor necrosis factor. J Neurosci. 2009;29:1319–30.
160. Yong VW, Power C, Forsyth P, Edwards DR. Metalloproteinases in biology and pathology of the nervous system. Nat Rev Neurosci. 2001;2:502–11.
161. Ramos-Cabrer P, Justicia C, Wiedermann D, Hoehn M. Stem cell mediation of functional recovery after stroke in the rat. PLoS One. 2010;5:e12779.
162. Wei L, Erinjeri JP, Rovainen CM, Woolsey TA. Collateral growth and angiogenesis around cortical stroke. Stroke. 2001;32:2179–84.
163. Schäbitz WR, Schwab S, Spranger M, Hacke W. Intraventricular brain-derived neurotrophic factor reduces infarct size after focal cerebral ischemia in rats. J Cereb Blood Flow Metab. 1997;17:500–6.
164. Kaya D, Gürsoy-Ozdemir Y, Yemisci M, Tuncer N, Aktan S, Dalkara T. VEGF protects brain against focal ischemia without increasing blood–brain permeability when administered intracerebroventricularly. J Cereb Blood Flow Metab. 2005;25:1111–8.
165. Zhang Z, Chopp M. Vascular endothelial growth factor and angiopoietins in focal cerebral ischemia. Trends Cardiovasc Med. 2002;12:62–6.
166. Schäbitz WR, Sommer C, Zoder W, Kiessling M, Schwaninger M, Schwab S. Intravenous brain-derived neurotrophic factor reduces infarct size and counterregulates Bax and Bcl-2 expression after temporary focal cerebral ischemia. Stroke. 2000;31:2212–7.
167. Caplan AI, Dennis JE. Mesenchymal stem cells as trophic mediators. J Cell Biochem. 2006;98:1076–84.
168. Gutiérrez-Fernández M, Rodríguez-Frutos B, Álvarez-Grech J, Vallejo-Cremades MT, Expósito-Alcaide M, Merino J, Roda JM, Díez-Tejedor E. Functional recovery after hematic administration of allogenic mesenchymal stem cells in acute ischemic stroke in rats. Neuroscience. 2011;175:394–405.
169. Krupinski J, Kaluza J, Kumar P, Kumar S, Wang JM. Role of angiogenesis in patients with cerebral ischemic stroke. Stroke. 1994;25:1794–8.
170. Navaratna D, Guo S, Arai K, Lo EH. Mechanisms and targets for angiogenic therapy after stroke. Cell Adh Migr. 2009;3:216–23.
171. Schmidt R, Schmidt H, Fazekas F. Vascular risk factors in dementia. J Neurol. 2000; 247:81–7.
172. Ishikane S, Ohnishi S, Yamahara K, Sada M, Harada K, Mishima K, Iwasaki K, Fujiwara M, Kitamura S, Nagaya N, Ikeda T. Allogeneic injection of fetal membrane-derived mesenchymal stem cells induces therapeutic angiogenesis in a rat model of hind limb ischemia. Stem Cells. 2008;26:2625–33.
173. Hu X, Yu SP, Fraser JL, Lu Z, Ogle ME, Wang J-A, Wei L. Transplantation of hypoxia-preconditioned mesenchymal stem cells improves infarcted heart function via enhanced survival of implanted cells and angiogenesis. J Thorac Cardiovasc Surg. 2008;135:799–808.
174. Komatsu K, Honmou O, Suzuki J, Houkin K, Hamada H, Kocsis JD. Therapeutic time window of mesenchymal stem cells derived from bone marrow after cerebral ischemia. Brain Res. 2010;1334:84–92.
175. Haas S, Weidner N, Winkler J. Adult stem cell therapy in stroke. Curr Opin Neurol. 2005;18:59–64.
176. Robin AM, Zhang ZG, Wang L, Zhang RL, Katakowski M, Zhang L, Wang Y, Zhang C, Chopp M. Stromal cell-derived factor 1alpha mediates neural progenitor cell motility after focal cerebral ischemia. J Cereb Blood Flow Metab. 2006;26:125–34.
177. Gonzalez-Perez O, Quiñones-Hinojosa A, Garcia-Verdugo JM. Immunological control of adult neural stem cells. J Stem Cells. 2010;5:23–31.

178. Schäbitz W-R, Steigleder T, Cooper-Kuhn CM, Schwab S, Sommer C, Schneider A, Kuhn HG. Intravenous brain-derived neurotrophic factor enhances poststroke sensorimotor recovery and stimulates neurogenesis. Stroke. 2007;38:2165–72.

179. Tanaka Y, Tanaka R, Liu M, Hattori N, Urabe T. Cilostazol attenuates ischemic brain injury and enhances neurogenesis in the subventricular zone of adult mice after transient focal cerebral ischemia. Neuroscience. 2010;171:1367–76.

180. Hicks AU, Hewlett K, Windle V, Chernenko G, Ploughman M, Jolkkonen J, Weiss S, Corbett D. Enriched environment enhances transplanted subventricular zone stem cell migration and functional recovery after stroke. Neuroscience. 2007;146:31–40.

181. Li W-L, Yu SP, Ogle ME, Ding XS, Wei L. Enhanced neurogenesis and cell migration following focal ischemia and peripheral stimulation in mice. Dev Neurobiol. 2008;68:1474–86.

182. Jin K, Xie L, Mao X, Greenberg MB, Moore A, Peng B, Greenberg RB, Greenberg DA. Effect of human neural precursor cell transplantation on endogenous neurogenesis after focal cerebral ischemia in the rat. Brain Res. 2011;1374:56–62.

183. Li B, Piao CS, Liu XY, Guo WP, Xue YQ, Duan WM, Gonzalez-Toledo ME, Zhao LR. Brain self-protection: the role of endogenous neural progenitor cells in adult brain after cerebral cortical ischemia. Brain Res. 2010;1327:91–102.

184. Kondziolka D, Steinberg GK, Wechsler L, Meltzer CC, Elder E, Gebel J, Decesare S, Jovin T, Zafonte R, Lebowitz J, Flickinger JC, Tong D, Marks MP, Jamieson C, Luu D, Bell-Stephens T, Teraoka J. Neurotransplantation for patients with subcortical motor stroke: a phase 2 randomized trial. J Neurosurg. 2005;103:38–45.

185. Savitz SI, Chopp M, Deans R, Carmichael ST, Phinney D, Wechsler L. Stem Cell Therapy as an Emerging Paradigm for Stroke (STEPS) II. Stroke. 2011;42:825–9.

186. The STEPS Participants. Stem Cell Therapies as an Emerging Paradigm in Stroke (STEPS): bridging basic and clinical science for cellular and neurogenic factor therapy in treating stroke. Stroke. 2009;40:510–5.

187. Hess DC, Borlongan CV. Cell-based therapy in ischemic stroke. Expert Rev Neurother. 2008;8:1193–201.

188. Bliss TM, Andres RH, Steinberg GK. Optimizing the success of cell transplantation therapy for stroke. Neurobiol Dis. 2009;37:275–83.

189. Henkin RI. Intranasal delivery to the brain. Nat Biotechnol. 2011;29:480.

190. Danielyan L, Schäfer R, von Ameln-Mayerhofer A, Buadze M, Geisler J, Klopfer T, Burkhardt U, Proksch B, Verleysdonk S, Ayturan M, Buniatian GH, Gleiter CH, Frey WH. Intranasal delivery of cells to the brain. Eur J Cell Biol. 2009;88:315–24.

191. van Velthoven CTJ, Kavelaars A, van Bel F, Heijnen CJ. Nasal administration of stem cells: a promising novel route to treat neonatal ischemic brain damage. Pediatr Res. 2010;68:419–22.

192. Wei N, Liu X, Chau M, Mohamad O, Wei L, Yu SP. Intranasal delivery of bone marrow stem cells into the cerebral ischemic lesion of mice. Soc Neurosci Abstr. 2010;57:10/T15.

193. Danielyan L, Schäfer R, Von Ameln-Mayerhofer A, Bernhard F, Verleysdonk S, Buadze M, Lourhmati A, Klopfer T, Schaumann F, Schmid B, Koehle C, Proksch B, Weissert R, Reichardt HM, Van Den Brandt J, Buniatian GH, Schwab M, Gleiter CH, Frey WH. Therapeutic efficacy of intranasally delivered mesenchymal stem cells in a rat model of Parkinson disease. Rejuvenation Res. 2011;14:3–16.

194. Scranton RA, Fletcher L, Sprague S, Jimenez DF, Digicaylioglu M. The rostral migratory stream plays a key role in intranasal delivery of drugs into the CNS. PLoS One. 2011;6:e18711.

195. Ikeda N, Nonoguchi N, Zhao MZ, Watanabe T, Kajimoto Y, Furutama D, Kimura F, Dezawa M, Coffin RS, Otsuki Y, Kuroiwa T, Miyatake S-I. Bone marrow stromal cells that enhanced fibroblast growth factor-2 secretion by herpes simplex virus vector improve neurological outcome after transient focal cerebral ischemia in rats. Stroke. 2005;36:2725–30.

196. Wu W, Chen X, Hu C, Li J, Yu Z, Cai W. Transplantation of neural stem cells expressing hypoxia-inducible factor-1α (HIF-1α) improves behavioral recovery in a rat stroke model. J Clin Neurosci. 2010;17:92–5.

197. Li W, Ma N, Ong L-L, Nesselmann C, Klopsch C, Ladilov Y, Furlani D, Piechaczek C, Moebius JM, Lützow K, Lendlein A, Stamm C, Li R-K, Steinhoff G. Bcl-2 engineered MSCs inhibited apoptosis and improved heart function. Stem Cells. 2007;25:2118–27.
198. Lee SI, Kim BG, Hwang DH, Kim HM, Kim SU. Overexpression of Bcl-XL in human neural stem cells promotes graft survival and functional recovery following transplantation in spinal cord injury. J Neurosci Res. 2009;87:3186–97.
199. Hu X, Wei L, Taylor TM, Wei J, Zhou X, Wang JA, Yu SP. Hypoxic preconditioning enhances bone marrow mesenchymal stem cell migration via Kv2.1 channel and FAK activation. Am J Physiol Cell Physiol. 2011;301:C362–72.
200. Francis KR, Wei L. Human embryonic stem cell neural differentiation and enhanced cell survival promoted by hypoxic preconditioning. Cell Death Dis. 2010;1:e22.
201. Leroux L, Descamps B, Tojais NF, Séguy B, Oses P, Moreau C, Daret D, Ivanovic Z, Boiron J-M, Lamazière J-MD, Dufourcq P, Couffinhal T, Duplàa C. Hypoxia preconditioned mesenchymal stem cells improve vascular and skeletal muscle fiber regeneration after ischemia through a Wnt4-dependent pathway. Mol Ther. 2010;18:1545–52.
202. Johansson BB, Belichenko PV. Neuronal plasticity and dendritic spines: effect of environmental enrichment on intact and postischemic rat brain. J Cereb Blood Flow Metab. 2002;22:89–96.
203. Komitova M, Mattsson B, Johansson BB, Eriksson PS. Enriched environment increases neural stem/progenitor cell proliferation and neurogenesis in the subventricular zone of stroke-lesioned adult rats. Stroke. 2005;36:1278–82.
204. Ickes BR, Pham TM, Sanders LA, Albeck DS, Mohammed AH, Granholm AC. Long-term environmental enrichment leads to regional increases in neurotrophin levels in rat brain. Exp Neurol. 2000;164:45–52.
205. Li WL, Yu SP, Ogle ME, Ding XS, Wei L. Enhanced neurogenesis and cell migration following focal ischemia and peripheral stimulation in mice. Dev Neurobiol. 2008;68:1474–86.
206. Whitaker VR, Cui L, Miller S, Yu SP, Wei L. Whisker stimulation enhances angiogenesis in the barrel cortex following focal ischemia in mice. J Cereb Blood Flow Metab. 2007;27:57–68.
207. Liu J, Solway K, Messing RO, Sharp FR. Increased neurogenesis in the dentate gyrus after transient global ischemia in gerbils. J Neurosci Res. 1998;18(19):7768–78.
208. Jin K, Minami M, Lan JQ, Mao XO, Batteur S, Simon RP, Greenberg DA. Neurogenesis in dentate subgranular zone and rostral subventricular zone after focal cerebral ischemia in the rat. Proc Natl Acad Sci USA. 2001;98(8):4710–5.

Part VII
Clinical Trial Design

Chapter 39
Repair-Based Therapies After Stroke

Steven C. Cramer

Abstract Stroke is a leading cause of human disability. Most patients show some degree of spontaneous recovery, but this is generally incomplete. Studies on the neurobiology of this recovery are leading to therapeutic interventions that aim to improve patient outcomes. A number of classes of such restorative therapies exist and are reviewed below. Most have a time window measured in days or weeks and so have the potential to help a large fraction of patients. These therapies are best given with certain principles of brain repair in mind. Implications for clinical trial design are also considered. Restorative therapies have the potential to substantially improve outcomes in a large fraction of patients with stroke.

1 Brain Repair

Brain repair can be defined as a process—spontaneous or therapeutically induced—that restores some aspect of brain structure or function after an insult. Repair contrasts with other therapeutic strategies in cerebrovascular disease such as prevention or approaches that aim to limit the injury such as neuroprotection or reperfusion. Instead, repair is focused on regrowth, repair, restoration, rewiring, and rehabilitation.

The burden of disability after stroke is large. Stroke is the leading neurologic cause of lost disability-adjusted life years in middle- and high-income countries around the world [1]. An estimated 6,400,000 American adults have had a symptomatic stroke, with a prevalence that increases with age. Note too that an estimated

S.C. Cramer, MD (✉)
Departments of Neurology and Anatomy & Neurobiology, University of California, Irvine, CA, USA

UC Irvine Medical Center, 101 The City Drive South, Building 53 Room 203, Orange, CA 92868-4280, USA
e-mail: scramer@uci.edu

P.A. Lapchak and J.H. Zhang (eds.), *Translational Stroke Research*, Springer Series in Translational Stroke Research, DOI 10.1007/978-1-4419-9530-8_39, © Springer Science+Business Media, LLC 2012

13 million people in the USA have had a silent stroke. Each year, 795,000 people experience a stroke, of which 610,000 are first-ever symptomatic stroke. The mean survival after stroke is 7 years, with approximately 85% of patients living past the first year of stroke [2].

Current therapies for a new stroke reduce disability in only a subset of patients. The only drug approved to treat acute stroke is tissue plasminogen activator (tPA) [3, 4]. A limited fraction of patients receive this medicine [5], in large part due to the narrow time window for safe drug administration. Despite recent data supporting administration of intravenous tPA up to 4.5 h after ischemic stroke onset, it continues to be true that only a minority of acute stroke patients receive this drug. Moreover, of those so treated, half or more have significant long-term disability [3, 4]. Because most repair-based approaches have a time window measured in days rather than hours, any repair-based approach that achieves regulatory approval will likely have the potential to help a large proportion of patients affected by stroke.

2 Spontaneous Stroke Recovery

Preclinical studies have characterized the neurobiology of spontaneous stroke recovery. After an experimental infarct, brain regions become excitable, in some cases showing GABA receptor downregulation and increased NMDA receptor binding. Expression changes for a number of genes, for example, resulting in increased levels of several growth factors. Angiogenesis is accompanied by structural changes in axons, dendrites, and synapses. These changes are often preferentially seen in the area surrounding an infarct and in areas with network connections to injured zones. In many cases, parallels exist with normal development and with learning. Preclinical studies provide mechanistic insights and also suggest therapeutic targets for improving recovery.

In human subjects, direct cellular and molecular measures are difficult to obtain. Noninvasive neuroimaging studies have provided insights in human subjects and in general have been concordant with findings in animals. A number of methods have been used to study brain function, structure, physiology, and metabolism. Focal injury reduces local tissue function and also has distant effects on activity in a number of brain areas connected within a distributed network, indeed recent studies emphasize the importance of network interactions after stroke [6–8]. After a unilateral supratratentorial stroke, changes can been seen within brain areas within the affected hemisphere, both deep and cortical. Diaschisis refers to a depression in activity seen in areas that are not injured but which share a connection with injured zones [9]. Over time, resolution of diaschisis can be followed by overactivation of network nodes, a pattern reminiscent of the brain during increased effort in healthy subjects. Changes in cortical map representations are also seen in relation to behavioral recovery, with numerous patterns having been described [10–15].

Changes in the function of brain networks can also be seen within the contralesional hemisphere. In general, the contribution of the contralesional hemisphere to spontaneous behavioral recovery after stroke seems largest in subjects with the greatest injury and deficits [16–18]. Attention has therefore been paid to laterality of activity as a measure of altered brain function. Changes in laterality are related to stroke-induced changes in interhemispheric interactions [19], a finding that suggests that improved function of ipsilesional brain regions might in part be facilitated by normalizing interhemispheric interactions [20].

3 Therapies to Promote Improved Recovery

A number of categories of therapy are being examined in relation to promoting brain repair [21]. Many have reached the point of human study, though in general most are in early-phase trials. The current emphasis is on the study of single agents, a key step to understanding therapeutic efficacy. Over time, it is likely that combinations of therapy will be examined as well. Some therapies are started in the early days after stroke onset, aiming to amplify innate repair mechanisms. Other therapies are initiated months after stroke onset, at a time when most spontaneous behavioral recovery has occurred, and so have as the goal induction of new neural plasticity.

A large number of growth factors have been examined in preclinical studies. Many have shown a significant effect on behavioral outcome when initiated 24 h poststroke or longer. Growth factors are appealing as a restorative agent because of their established role in normal development. A preponderance of human stroke studies to date has examined the hematopoietic growth factors, such as granulocyte colony-stimulating factor [22] and erythropoietin [23]. Growth factors are generally large proteins, for which CNS ingress is limited. Strategies to overcome this and allow the growth factors to cross the blood–brain barrier include conjugation to a molecular Trojan horse [24] and transfection of an exogenous stem cell with a gene encoding for a growth factor such as fibroblast growth factor-2 [25], GDNF [26], BDNF [27], VEGF [28], PIGF [29], or HGF [30].

Other large molecules are also under study. One set of therapies focuses on neutralizing factors that inhibit growth in the adult CNS. Thus, a key reason neurons do not regenerate after CNS injury is lack of a permissive growth environment. Three major inhibitors of such growth have been described, myelin-associated glycoprotein (MAG), oligo-myelin glycoprotein, and Nogo-A. Blockade of these inhibitors, such as with a monoclonal antibody, promotes axonal growth [31, 32] and is being examined in human studies as a means to improve outcomes.

Numerous small molecules have been examined to improve outcome after stroke. In many cases, these represent drugs that have already been approved for other indications and are now being examined for effects on stroke recovery. Some of these drugs, such as amphetamine, levodopa, ropinirole, and escitalopram, target specific neurotransmitter systems, while others, such as inosine, sildenafil, and niacin, do not. There have been some noteworthy positive studies in this category, such as the FLAME study, which found that fluoxetine started within 10 days of stroke had a favorable effect on long-term motor outcome in 118 patients [33]; the study by Jorge et al., which found that escitalopram started within 3 months of stroke onset improved cognitive outcomes at 1 year in 129 patients [34]; and the study by Scheidtmann et al., which found that levodopa started within 6 months of stroke onset improved motor function in 53 patients [35].

Cell-based therapies are receiving increased attention, with numerous types under consideration. Some therapies target endogenous neural stem cells, while others administer exogenous cells including transformed tumor cells, adult stem cells such as marrow stromal cells, stem cells with modified genes or a bioscaffold, umbilical cord cells, placental cells, fetal stem cells, and embryonic stem cells. Exogenous cells can be autologous, allogeneic, and xenografts. Promising results in human subjects with stroke have been reported with the administration of intravenous marrow stromal cells [36, 37]. Cellular therapies hold great promise but can introduce complexities not always encountered with other restorative therapies, such stability of therapy during manufacturing and delivery, the long-term fate of the administered cells, ethical concerns raised in procurement of some cells, and the potential effects of any concomitant immunosuppression that might be required.

A number of intensive activity-based therapy regimens have been studied. A key example is constraint-induced therapy, an approach that trains the affected limb while restraining the nonaffected limb, in order to overcome learned disuse of the affected limb. In the EXCITE trial, constraint-induced therapy was associated with significant gains in motor outcome in 222 patients enrolled 3–9 months after stroke onset [38]. This approach is also being studied in nonmotor domains, such as aphasia. The VECTORS trial examined constraint-induced therapy early after stroke and found that in 52 patients enrolled within 1 month of stroke onset, higher intensity of therapy was associated with poorer behavioral outcome at day 90 [39]—too much too soon can be harmful. The LEAPS trial compared two therapies focused on gait, examining 408 patients within 2 months of stroke, and found that treadmill training with body-weight support did not differ from progressive exercise at home managed by a physical therapist in effects on walking ability 1 year after stroke [40]. Important questions remain for activity-based regimens focused on intensive therapy, such as the optimal timing and amount of training, how to design the content of training in order to maximize generalizability of therapy effects across real-life challenges, adjusting content of training in relation to time after stroke, and the potential to maximize treatment effects by combining with pharmacological therapies.

The effect of therapy delivered by robotic devices has been examined. Numerous robotic devices are under study [41–44]. Robotic devices offer potential advantages, such as consistent and long-lasting output, programmability, utility for virtual reality applications, high precision, great potential for telerehabilitation and so reaching underserved regions, and the potential for an improved therapist to patient ratio. However, challenges remain. Lo et al., in 127 patients who were in the chronic phase of stroke, found that robot-assisted therapy did not significantly improve motor function after 12 weeks, as compared with usual care or intensive therapy, though in secondary analyses, robot-assisted therapy improved outcomes over 36 weeks as compared with usual care but not with intensive therapy [45]. Factors that might represent avenues for improving the impact of robotic therapy include more fully defining the relationship between robotic therapy and traditional physiotherapy and matching the right patients with the right robotic devices and protocols.

The brain is an electrical organ, and brain stimulation is also being studied to improve outcomes after stroke. Many forms of stimulation have been examined, including repetitive transcranial magnetic stimulation, theta burst stimulation, epidural cortical stimulation, transcranial direct current stimulation, and stimulation via a laser-based device. There is precedence for a focus on brain stimulation, as the gold standard therapy for major depression remains a form of brain stimulation, electroconvulsive therapy. Some results have been favorable [46, 47] while others have not [48, 49]. Some therapies based on brain stimulation can have bidirectional effects and so, for example, can be set to reduce activity in brain areas that might be interfering with recovery or to increase activity in dormant brain areas where increased activity is desired. Greater effects might one day be seen as brain stimulation protocols take advantage of the potential of this approach to stimulate different areas in a distributed network with high temporal resolution.

Humans are cognitive creatures and so the potential exists to enhance outcomes with approaches focused on cognitive skills. Athletes who envision successful performances in their minds can improve their scores. Building on the neurobiology of mirror neuron systems, therapies have been devised that incorporate motor imagery and motor observation [50]. Other cognitive strategies include those that incorporate music [51] and those focused on overcoming neglect and hemi-inattention [52].

4 Principles of Brain Repair After Stroke

The effectiveness of restorative therapies can be maximized with attention to certain principles [53]. First, brain repair is time sensitive. Treating stroke recovery is a four-dimensional issue. Some biological targets are only relevant during a specific time period after stroke. Indeed, some therapies can have different effects on stroke depending on timing. For example, long-term effects of a GABA agonist or NMDA receptor blocker can be favorable if administered in the early hours after stroke [54, 55] but deleterious if initiated days later [56–58], and the reverse may be true for matrix metalloproteinases [59]. Thus, an important consideration is that the cellular and biochemical underpinnings of recovery, many of which are potential therapeutic targets, evolve over time [14, 60, 61].

Second, brain repair is experience dependent. Since the classic study by Feeney et al. [62], which showed that a stimulant promoted improved motor outcome after experimental stroke in rodents only when the drug was paired with training, increasing evidence suggests that a restorative therapy needs reinforcement and shaping through experience to produce best behavioral results. A number of studies support this conclusion [63–67]. As pharmacological and cellular therapies gain traction, increased attention will be needed to the accompanying experiences whose effects influence therapeutic efficacy.

Fig. 39.1 Extent of injury to a specific motor tract predicts gains in arm motor function from a course of robotic therapy in subjects with chronic stroke [69]. Two examples of stroke injury to the corticospinal tract descending from primary motor cortex are provided. The subject in (**a**) had only 37.5% of this tract injured by stroke and had a gain of 11 points on the Fugl-Meyer scale, while the subject in (**b**) had 93.4% of this tract injured by stroke and had a gain of only 1 point. In this study, tract-specific injury was stronger than infarct volume or baseline clinical status at predicting gains. Such findings might be incorporated into clinical trials, for example, as an entry criterion, for identifying subjects with sufficient biological substrate to improve from therapy

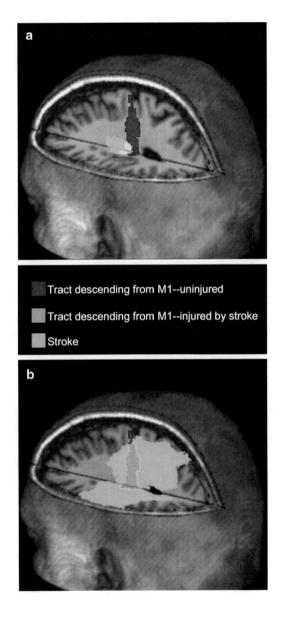

Third, patient stratification is likely important to studies of poststroke brain repair. Numerous variables have been found to be potential predictors of stroke outcome, including location and size of injury [68, 69], genotype [70], measures of brain function [71], and affective disorders [72, 73]. Such measures may be of pivotal value in defining the population most likely to benefit from a given therapy (see Fig. 39.1), as stroke is a very heterogeneous condition. This point is further considered below in the discussion of biomarkers.

Fourth, modality-specific measures might be useful to measure treatment effects when the target is brain repair [74]. The reasons for this approach revolve around the fact that restorative therapies achieve their effect by improving the function of specific neural systems. Improvement is seen in neural systems where surviving substrate is amenable to repair: a behavior whose underlying brain regions are destroyed is less likely to improve than a behavior whose underlying regions are accessible to a restorative therapy. For example, a domain whose neural underpinnings are partially spared, such as arm motor function or language, might show substantial gains in response to a restorative therapy, with only modest effects on global measures of poststroke outcome. Such gains might be considered worthwhile by many patients and so important to measure in clinical trials. Modality-specific endpoints are aligned with this strategy, akin to their utility for understanding spontaneous stroke recovery, where the rate and degree of gains often vary widely across behavioral modalities [75–77]. Improvement in global clinical status is of course a goal of paramount importance, but a treatment that provides gains by promoting neuroplasticity might demonstrate maximum effect in brain networks that have subtotal injury, and endpoints are needed that measure these specific treatment effects. A number of modality-specific endpoints have been successfully incorporated into clinical trials and indeed have been the basis for regulatory approval [78].

Fifth, the nature of brain organization *prior* to stroke influences *poststroke* brain plasticity. For example, in healthy subjects, some behaviors such as language or hand movement tend to be highly lateralized (i.e., generation of the behavior involves mainly one hemisphere), and other behaviors such as bulbar and facial movement tend to be less lateralized (i.e., generation of the behavior involves both hemispheres). These differences remain apparent after stroke. Cramer and Crafton [79] found that face movement is more bilaterally organized than is shoulder or arm movement in healthy subjects (i.e., before any stroke) and that this remained true after stroke. Such a difference could have functional implications. Hamdy et al. found that the cortical representation for swallowing is normally bilaterally organized [80]. Not surprisingly, therefore, dysphagic patients who recovered after stroke showed an increase in their cortical pharyngeal map size within the contralesional hemisphere, whereas patients who remained dysphagic did not show this change [81]. For reorganization of brain maps after stroke, the pattern of brain reorganization can influence behavioral status, and this pattern is at times constrained by features of normal brain organization.

5 Implications for Clinical Trials of Brain Repair After Stroke

The rapidly expanding knowledge base about brain repair and its application to humans, as above, inform the design of restorative trials, a topic that has been reviewed in detail [82].

In the design of restorative stroke clinical trials, some points can be gleaned from lessons learned in the setting of acute stroke trials [83–87]. Topics include methodological quality of preclinical programmatic development [87, 88], choice of entry criteria [89], and choice of outcome measures [90]. The value of modality-specific endpoints for studies of restorative agents has been discussed above. Dose–response relationships are important to define and might change with increasing time after stroke.

In contrast with the neuroprotective and reperfusion therapies examined in many acute stroke trials, repair-based therapies are influenced by experience, training, and environment. Acute stroke therapies such as tPA generally exert their effect rapidly, around the time of therapy administration, while for a repair-based trial, many of the biological events that improve final behavioral outcome occur days to weeks after therapy initiation. A subject's experiences, concomitant training, and environment can each interact with a repair-based therapy in defining the final therapy effects. Issues such as caregiver status [91, 92], affective state [72, 93, 94], and socioeconomic factors [95] might also significantly influence restorative therapy effects. Both the quantity and the quality of such influences can affect brain repair and behavioral outcome [96–103]. Genetic factors might also be an important consideration to studies aiming to therapeutically induce brain plasticity from stroke [70, 104–108]. Also, certain medications can *adversely* influence brain plasticity, repair, and recovery from stroke [62, 109–113], and so are covariates of interest to a restorative stroke trial.

Influences that arise between the time that a repair-based therapy is initiated and the time of final clinical trial outcome assessment thus require substantial consideration in a repair-based trial because they can interact with repair-based therapy and modify its effects. There are multiple ways that study design might address these issues. In some cases, these influences can be precisely controlled, such as via entry criteria or by carefully controlling the details of any study-administered concomitant therapy such as physiotherapy or speech therapy. When external influences cannot be controlled, study design can insure that they are at least measured. Such an approach provided useful insights in one recent repair-based stroke clinical trial, where the amount of "outside" physiotherapy (i.e., physiotherapy occurring in parallel with trial participation, but prescribed by private physicians, outside of trial jurisdiction) was found to differ significantly between active and placebo treatment arms [114]. Such measures can then be treated as planned covariates of interest in statistical analyses.

Repair-based therapies have the potential to examine within-subject treatment effects, as a baseline behavioral measure can be obtained prior to initiating therapy. Even when a restorative therapy is introduced early after stroke, the time window is generally at least 24 h. Measuring within-subject behavioral change offers potential statistical advantages over cross-sectional outcome assessments, and so are a consideration in the design of restorative stroke trials.

Biomarkers have the potential to strengthen clinical trials of brain repair after stroke. A biomarker can be defined as a measure that provides insight into a tissue state or disease state, and in a clinical trial context would provide information beyond that available from bedside exam. Biomarkers have the potential to identify patients most likely to respond to a treatment (Fig. 39.1) and so might reduce variance and increase study power [115–117]. A biomarker might also provide insight into a treatment's mechanism of action [118–120], which can provide useful insights at the stage of protocol development. There are important caveats in the selection of a biomarker; for example, the utility of a biomarker is highest when its relationships with the disease process and with the therapy are well understood [121, 122].

A number of specific measures are potentially available to serve as biomarkers in the context of a restorative trial, including blood tests [123, 124]. Imaging-based methods can provide anatomical measures of injury [69, 125], tissue status such as cortical thickness [126], white matter tract integrity [68, 127], regional brain function [71, 128, 129], network interactions [8], or chemical state [130, 131]. Physiological assessments may be useful [132] and indeed complementary [68, 133]. Measures of injury to a predefined functional brain region, such as the extent of insult to the hand region of primary motor cortex [134], white matter cholinergic projections [135], or left temporal language areas [136], might provide useful insights into the likelihood that a particular therapy will be able to promote repair in a specific target region.

6 Conclusions

Preclinical and human studies are providing increased insight into the neurobiology of spontaneous recovery after stroke. This information is opening the door to a number of potential restorative therapies, many of which are in human clinical trials. When applied according to selected neurobiological principles, these therapies have the potential to improve outcome for many patients after stroke.

References

1. Johnston SC, Hauser SL. Neurological disease on the global agenda. Ann Neurol. 2008;64(1):A11–2.
2. Lloyd-Jones D, Adams RJ, Brown TM, Carnethon M, Dai S, De Simone G, et al. Heart disease and stroke statistics—2010 update: a report from the American Heart Association. Circulation. 2010;121(7):e46–215.
3. Hacke W, Kaste M, Bluhmki E, Brozman M, Davalos A, Guidetti D, et al. Thrombolysis with alteplase 3 to 4.5 hours after acute ischemic stroke. N Engl J Med. 2008;359(13):1317–29.
4. The National Institute of Neurological Disorders and Stroke rt-PA Stroke Study Group. Tissue plasminogen activator for acute ischemic stroke. N Engl J Med. 1995;333(24): 1581–7.
5. Reed S, Cramer S, Blough D, Meyer K, Jarvik J. Treatment with tissue plasminogen activator and inpatient mortality rates for patients with ischemic stroke treated in community hospitals. Stroke. 2001;32(8):1832–40.
6. Sharma N, Baron JC, Rowe JB. Motor imagery after stroke: relating outcome to motor network connectivity. Ann Neurol. 2009;66(5):604–16.

7. Grefkes C, Nowak DA, Eickhoff SB, Dafotakis M, Kust J, Karbe H, et al. Cortical connectivity after subcortical stroke assessed with functional magnetic resonance imaging. Ann Neurol. 2008;63(2):236–46.

8. Carter AR, Astafiev SV, Lang CE, Connor LT, Rengachary J, Strube MJ, et al. Resting interhemispheric functional magnetic resonance imaging connectivity predicts performance after stroke. Ann Neurol. 2010;67(3):365–75.

9. von Monakow C. Diaschisis, 1914. In: Pribram K, editor. Brain and behavior 1 mood, states and mind. Baltimore: Penguin; 1969. p. 26–34.

10. Nudo R. Recovery after damage to motor cortical areas. Curr Opin Neurobiol. 1999;9(6): 740–7.

11. Cramer S, Chopp M. Recovery recapitulates ontogeny. Trends Neurosci. 2000;23(6): 265–71.

12. Ward NS, Cohen LG. Mechanisms underlying recovery of motor function after stroke. Arch Neurol. 2004;61(12):1844–8.

13. Yozbatiran N, Cramer SC. Imaging motor recovery after stroke. NeuroRx. 2006;3(4):482–8.

14. Li S, Carmichael ST. Growth-associated gene and protein expression in the region of axonal sprouting in the aged brain after stroke. Neurobiol Dis. 2006;23(2):362–73.

15. Wieloch T, Nikolich K. Mechanisms of neural plasticity following brain injury. Curr Opin Neurobiol. 2006;16(3):258–64.

16. Netz J, Lammers T, Homberg V. Reorganization of motor output in the non-affected hemisphere after stroke. Brain. 1997;120:1579–86.

17. Turton A, Wroe S, Trepte N, Fraser C, Lemon R. Contralateral and ipsilateral EMG responses to transcranial magnetic stimulation during recovery of arm and hand function after stroke. Electroencephalogr Clin Neurophysiol. 1996;101:316–28.

18. Heiss WD, Thiel A. A proposed regional hierarchy in recovery of post-stroke aphasia. Brain Lang. 2006;98(1):118–23.

19. Murase N, Duque J, Mazzocchio R, Cohen L. Influence of interhemispheric interactions on motor function in chronic stroke. Ann Neurol. 2004;55(3):400–9.

20. Floel A, Cohen LG. Recovery of function in humans: cortical stimulation and pharmacological treatments after stroke. Neurobiol Dis. 2010;37(2):243–51.

21. Cramer SC. Repairing the human brain after stroke II. Restorative therapies. Ann Neurol. 2008;63(5):549–60.

22. Schabitz WR, Laage R, Vogt G, Koch W, Kollmar R, Schwab S, et al. AXIS: a trial of intravenous granulocyte colony-stimulating factor in acute ischemic stroke. Stroke. 2010;41(11):2545–51.

23. Cramer SC, Fitzpatrick C, Warren M, Hill MD, Brown D, Whitaker L, et al. The beta-hCG + erythropoietin in acute stroke (BETAS) study: a 3-center, single-dose, open-label, noncontrolled, phase IIa safety trial. Stroke. 2010;41(5):927–31.

24. Zhang Y, Pardridge WM. Blood–brain barrier targeting of BDNF improves motor function in rats with middle cerebral artery occlusion. Brain Res. 2006;1111(1):227–9.

25. Kim BO, Tian H, Prasongsukarn K, Wu J, Angoulvant D, Wnendt S, et al. Cell transplantation improves ventricular function after a myocardial infarction: a preclinical study of human unrestricted somatic stem cells in a porcine model. Circulation. 2005;112(9 Suppl):I96–104.

26. Horita Y, Honmou O, Harada K, Houkin K, Hamada H, Kocsis JD. Intravenous administration of glial cell line-derived neurotrophic factor gene-modified human mesenchymal stem cells protects against injury in a cerebral ischemia model in the adult rat. J Neurosci Res. 2006;84(7):1495–504.

27. Zhao L-X, Zhang J, Cao F, Meng L, Wang D-M, Li Y-H, et al. Modification of the brain-derived neurotrophic factor gene: a portal to transform mesenchymal stem cells into advantageous engineering cells for neuroregeneration and neuroprotection. Exp Neurol. 2004;190(2): 396–406.

28. Iwase T, Nagaya N, Fujii T, Itoh T, Murakami S, Matsumoto T, et al. Comparison of angiogenic potency between mesenchymal stem cells and mononuclear cells in a rat model of hindlimb ischemia. Cardiovasc Res. 2005;66(3):543–51.

29. Liu H, Honmou O, Harada K, Nakamura K, Houkin K, Hamada H, et al. Neuroprotection by PlGF gene-modified human mesenchymal stem cells after cerebral ischaemia. Brain. 2006;129(Pt 10):2734–45.

30. Zhao MZ, Nonoguchi N, Ikeda N, Watanabe T, Furutama D, Miyazawa D, et al. Novel therapeutic strategy for stroke in rats by bone marrow stromal cells and ex vivo HGF gene transfer with HSV-1 vector. J Cereb Blood Flow Metab. 2006;26(9):1176–88.

31. Domeniconi M, Filbin MT. Overcoming inhibitors in myelin to promote axonal regeneration. J Neurol Sci. 2005;233(1–2):43–7.

32. Buchli AD, Schwab ME. Inhibition of Nogo: a key strategy to increase regeneration, plasticity and functional recovery of the lesioned central nervous system. Ann Med. 2005;37(8):556–67.

33. Chollet F, Tardy J, Albucher JF, Thalamas C, Berard E, Lamy C, et al. Fluoxetine for motor recovery after acute ischaemic stroke (FLAME): a randomised placebo-controlled trial. Lancet Neurol. 2011;10(2):123–30.

34. Jorge RE, Acion L, Moser D, Adams Jr HP, Robinson RG. Escitalopram and enhancement of cognitive recovery following stroke. Arch Gen Psychiatry. 2010;67(2):187–96.

35. Scheidtmann K, Fries W, Muller F, Koenig E. Effect of levodopa in combination with physiotherapy on functional motor recovery after stroke: a prospective, randomised, double-blind study. Lancet. 2001;358:787–90.

36. Bang OY, Lee JS, Lee PH, Lee G. Autologous mesenchymal stem cell transplantation in stroke patients. Ann Neurol. 2005;57(6):874–82.

37. Honmou O, Houkin K, Matsunaga T, Niitsu Y, Ishiai S, Onodera R, et al. Intravenous administration of auto serum-expanded autologous mesenchymal stem cells in stroke. Brain. 2011;134(6):1790–807.

38. Wolf SL, Winstein CJ, Miller JP, Taub E, Uswatte G, Morris D, et al. Effect of constraint-induced movement therapy on upper extremity function 3 to 9 months after stroke: the EXCITE randomized clinical trial. JAMA. 2006;296(17):2095–104.

39. Dromerick A, Lang C, Powers W, Wagner J, Sahrmann S, Videen T, et al. Very early constraint-induced movement therapy (VECTORS): Phase II trial results. Stroke. 2007;38:465.

40. Duncan PW, Sullivan KJ, Behrman AL, Azen SP, Wu SS, Nadeau SE, et al. Body-weight-supported treadmill rehabilitation after stroke. N Engl J Med. 2011;364(21):2026–36.

41. Brewer BR, McDowell SK, Worthen-Chaudhari LC. Poststroke upper extremity rehabilitation: a review of robotic systems and clinical results. Top Stroke Rehabil. 2007;14(6):22–44.

42. Volpe BT, Huerta PT, Zipse JL, Rykman A, Edwards D, Dipietro L, et al. Robotic devices as therapeutic and diagnostic tools for stroke recovery. Arch Neurol. 2009;66(9):1086–90.

43. Reinkensmeyer D, Emken J, Cramer S. Robotics, motor learning, and neurologic recovery. Annu Rev Biomed Eng. 2004;6:497–525.

44. Balasubramanian S, Klein J, Burdet E. Robot-assisted rehabilitation of hand function. Curr Opin Neurol. 2010;23(6):661–70.

45. Lo AC, Guarino PD, Richards LG, Haselkorn JK, Wittenberg GF, Federman DG, et al. Robot-assisted therapy for long-term upper-limb impairment after stroke. N Engl J Med. 2010;362(19):1772–83.

46. Lindenberg R, Renga V, Zhu LL, Nair D, Schlaug G. Bihemispheric brain stimulation facilitates motor recovery in chronic stroke patients. Neurology. 2011;75(24):2176–84.

47. Ackerley SJ, Stinear CM, Barber PA, Byblow WD. Combining theta burst stimulation with training after subcortical stroke. Stroke. 2010;41(7):1568–72.

48. Levy R, Benson R, Winstein C, for the Everest Study Investigators, editors. Cortical stimulation for upper-extremity hemiparesis from ischemic stroke: Everest study primary endpoint results. International stroke conference, New Orleans, LA; 2008.

49. Pomeroy VM, Cloud G, Tallis RC, Donaldson C, Nayak V, Miller S. Transcranial magnetic stimulation and muscle contraction to enhance stroke recovery: a randomized proof-of-principle and feasibility investigation. Neurorehabil Neural Repair. 2007;21(6):509–17.

50. Small SL, Buccino G, Solodkin A. The mirror neuron system and treatment of stroke. Dev Psychobiol.; 2010.
51. Sarkamo T, Tervaniemi M, Laitinen S, Forsblom A, Soinila S, Mikkonen M, et al. Music listening enhances cognitive recovery and mood after middle cerebral artery stroke. Brain. 2008;131(Pt 3):866–76.
52. Ramachandran VS, Altschuler EL. The use of visual feedback, in particular mirror visual feedback, in restoring brain function. Brain. 2009;132(Pt 7):1693–710.
53. Cramer SC, Sur M, Dobkin BH, O'Brien C, Sanger TD, Trojanowski JQ, et al. Harnessing neuroplasticity for clinical applications. Brain. 2011;134(Pt 6):1591–609.
54. Green AR, Hainsworth AH, Jackson DM. GABA potentiation: a logical pharmacological approach for the treatment of acute ischaemic stroke. Neuropharmacology. 2000;39(9): 1483–94.
55. Ovbiagele B, Kidwell CS, Starkman S, Saver JL. Neuroprotective agents for the treatment of acute ischemic stroke. Curr Neurol Neurosci Rep. 2003;3(1):9–20.
56. Kozlowski D, Jones T, Schallert T. Pruning of dendrites and restoration of function after brain damage: role of the NMDA receptor. Restor Neurol Neurosci. 1994;7:119–26.
57. Wahlgren N, Martinsson L. New concepts for drug therapy after stroke. Can we enhance recovery? Cerebrovasc Dis. 1998;8 Suppl 5:33–8.
58. Barth T, Hoane M, Barbay S, Saponjic R. Effects of glutamate antagonists on the recovery and maintenance of behavioral function after brain injury. In: Goldstein L, editor. Restorative neurology: advances in pharmacotherapy for recovery after stroke. Armonk: Futura; 1998.
59. Zhao BQ, Tejima E, Lo EH. Neurovascular proteases in brain injury, hemorrhage and remodeling after stroke. Stroke. 2007;38(2 Suppl):748–52.
60. Stroemer R, Kent T, Hulsebosch C. Enhanced neocortical neural sprouting, synaptogenesis, and behavioral recovery with D-amphetamine therapy after neocortical infarction in rats. Stroke. 1998;29(11):2381–95.
61. Jones T, Schallert T. Overgrowth and pruning of dendrites in adult rats recovering from neocortical damage. Brain Res. 1992;581:156–60.
62. Feeney D, Gonzalez A, Law W. Amphetamine, Halperidol, and experience interact to affect the rate of recovery after motor cortex injury. Science. 1982;217:855–7.
63. Garcia-Alias G, Barkhuysen S, Buckle M, Fawcett JW. Chondroitinase ABC treatment opens a window of opportunity for task-specific rehabilitation. Nat Neurosci. 2009;12(9): 1145–51.
64. Fang PC, Barbay S, Plautz EJ, Hoover E, Strittmatter SM, Nudo RJ. Combination of NEP 1–40 treatment and motor training enhances behavioral recovery after a focal cortical infarct in rats. Stroke. 2010;41(3):544–9.
65. Starkey ML, Schwab ME. Anti-Nogo-A and training: can one plus one equal three? Experimental Neurol; 2011.
66. Hovda D, Feeney D. Amphetamine with experience promotes recovery of locomotor function after unilateral frontal cortex injury in the cat. Brain Res. 1984;298:358–61.
67. Adkins DL, Hsu JE, Jones TA. Motor cortical stimulation promotes synaptic plasticity and behavioral improvements following sensorimotor cortex lesions. Exp Neurol. 2008;212(1): 14–28.
68. Stinear CM, Barber PA, Smale PR, Coxon JP, Fleming MK, Byblow WD. Functional potential in chronic stroke patients depends on corticospinal tract integrity. Brain. 2007;130 (Pt 1):170–80.
69. Riley JD, Le V, Der-Yeghiaian L, See J, Newton JM, Ward NS, et al. Anatomy of stroke injury predicts gains from therapy. Stroke. 2011;42(2):421–6.
70. Siironen J, Juvela S, Kanarek K, Vilkki J, Hernesniemi J, Lappalainen J. The Met allele of the BDNF Val66Met polymorphism predicts poor outcome among survivors of aneurysmal subarachnoid hemorrhage. Stroke. 2007;38(10):2858–60.
71. Cramer SC, Parrish TB, Levy RM, Stebbins GT, Ruland SD, Lowry DW, et al. Predicting functional gains in a stroke trial. Stroke. 2007;38(7):2108–14.

72. Lai SM, Duncan PW, Keighley J, Johnson D. Depressive symptoms and independence in BADL and IADL. J Rehabil Res Dev. 2002;39(5):589–96.
73. Gillen R, Tennen H, McKee TE, Gernert-Dott P, Affleck G. Depressive symptoms and history of depression predict rehabilitation efficiency in stroke patients. Arch Phys Med Rehabil. 2001;82(12):1645–9.
74. Cramer SC, Koroshetz WJ, Finklestein SP. The case for modality-specific outcome measures in clinical trials of stroke recovery-promoting agents. Stroke. 2007;38(4):1393–5.
75. Hier D, Mondlock J, Caplan L. Recovery of behavioral abnormalities after right hemisphere stroke. Neurology. 1983;33:345–50.
76. Marshall R, Perera G, Lazar R, Krakauer J, Constantine R, DeLaPaz R. Evolution of cortical activation during recovery from corticospinal tract infarction. Stroke. 2000;31(3):656–61.
77. Markgraf C, Green E, Hurwitz B, Morikawa E, Dietrich W, McCabe P, et al. Sensorimotor and cognitive consequences of middle cerebral artery occlusion in rats. Brain Res. 1992;575(2):238–46.
78. Goodman AD, Brown TR, Edwards KR, Krupp LB, Schapiro RT, Cohen R, et al. A phase 3 trial of extended release oral dalfampridine in multiple sclerosis. Ann Neurol. 2010;68(4): 494–502.
79. Cramer SC, Crafton KR. Somatotopy and movement representation sites following cortical stroke. Exp Brain Res. 2006;168(1–2):25–32.
80. Hamdy S, Aziz Q, Rothwell JC, Singh KD, Barlow J, Hughes DG, et al. The cortical topography of human swallowing musculature in health and disease. Nat Med. 1996;2(11):1217–24.
81. Hamdy S, Aziz Q, Rothwell J, Power M, Singh K, Nicholson D, et al. Recovery of swallowing after dysphagic stroke relates to functional reorganization in the intact motor cortex. Gastroenterology. 1998;115(5):1104–12.
82. Cramer SC. Issues in clinical trial methodology for brain repair after stroke. In: Cramer SC, Nudo RJ, editors. Brain repair after stroke. Cambridge: Cambridge University Press; 2010. p. 173–82.
83. Grotta J, Bratina P. Subjective experiences of 24 patients dramatically recovering from stroke. Stroke. 1995;26(7):1285–8.
84. Fisher M, Ratan R. New perspectives on developing acute stroke therapy. Ann Neurol. 2003;53(1):10–20.
85. Gladstone D, Black S, Hakim A. Toward wisdom from failure: lessons from neuroprotective stroke trials and new therapeutic directions. Stroke. 2002;33(8):2123–36.
86. Fisher M, Feuerstein G, Howells DW, Hurn PD, Kent TA, Savitz SI, et al. Update of the Stroke Therapy Academic Industry Roundtable Preclinical recommendations. Stroke. 2009;40(6):2244–50.
87. Philip M, Benatar M, Fisher M, Savitz SI. Methodological quality of animal studies of neuroprotective agents currently in phase II/III acute ischemic stroke trials. Stroke. 2009;40(2):577–81.
88. Savitz SI, Fisher M. Future of neuroprotection for acute stroke: in the aftermath of the SAINT trials. Ann Neurol. 2007;61(5):396–402.
89. Uchino K, Billheimer D, Cramer S. Entry criteria and baseline characteristics predict outcome in acute stroke trials. Stroke. 2001;32(4):909–16.
90. Duncan P, Jorgensen H, Wade D. Outcome measures in acute stroke trials: a systematic review and some recommendations to improve practice. Stroke. 2000;31(6):1429–38.
91. Smith J, Forster A, Young J. Cochrane review: information provision for stroke patients and their caregivers. Clin Rehabil. 2009;23(3):195–206.
92. Glass TA, Matchar DB, Belyea M, Feussner JR. Impact of social support on outcome in first stroke. Stroke. 1993;24(1):64–70.
93. Jonsson AC, Lindgren I, Hallstrom B, Norrving B, Lindgren A. Determinants of quality of life in stroke survivors and their informal caregivers. Stroke. 2005;36(4):803–8.
94. Mukherjee D, Levin RL, Heller W. The cognitive, emotional, and social sequelae of stroke: psychological and ethical concerns in post-stroke adaptation. Top Stroke Rehabil. 2006;13(4):26–35.

95. McFadden E, Luben R, Wareham N, Bingham S, Khaw KT. Social class, risk factors, and stroke incidence in men and women: a prospective study in the European prospective investigation into cancer in Norfolk cohort. Stroke. 2009;40(4):1070–7.
96. Kwakkel G. Impact of intensity of practice after stroke: issues for consideration. Disabil Rehabil. 2006;28(13–14):823–30.
97. Kwakkel G, Wagenaar R, Twisk J, Lankhorst G, Koetsier J. Intensity of leg and arm training after primary middle-cerebral-artery stroke: a randomised trial. Lancet. 1999;354(9174): 191–6.
98. Dobkin B. The clinical science of neurologic rehabilitation. New York: Oxford University Press; 2003.
99. Van Peppen RP, Kwakkel G, Wood-Dauphinee S, Hendriks HJ, Van der Wees PJ, Dekker J. The impact of physical therapy on functional outcomes after stroke: what's the evidence? Clin Rehabil. 2004;18(8):833–62.
100. Cicerone KD, Dahlberg C, Malec JF, Langenbahn DM, Felicetti T, Kneipp S, et al. Evidence-based cognitive rehabilitation: updated review of the literature from 1998 through 2002. Arch Phys Med Rehabil. 2005;86(8):1681–92.
101. Bhogal S, Teasell R, Speechley M. Intensity of aphasia therapy, impact on recovery. Stroke. 2003;34(4):987–93.
102. Jones T, Chu C, Grande L, Gregory A. Motor skills training enhances lesion-induced structural plasticity in the motor cortex of adult rats. J Neurosci. 1999;19(22):10153–63.
103. Johansson B. Brain plasticity and stroke rehabilitation. The Willis lecture. Stroke. 2000;31(1):223–30.
104. Kleim JA, Chan S, Pringle E, Schallert K, Procaccio V, Jimenez R, et al. BDNF val66met polymorphism is associated with modified experience-dependent plasticity in human motor cortex. Nat Neurosci. 2006;9(6):735–7.
105. Alberts MJ, Graffagnino C, McClenny C, DeLong D, Strittmatter W, Saunders AM, et al. ApoE genotype and survival from intracerebral haemorrhage. Lancet. 1995;346(8974):575.
106. Niskakangas T, Ohman J, Niemela M, Ilveskoski E, Kunnas TA, Karhunen PJ. Association of apolipoprotein E polymorphism with outcome after aneurysmal subarachnoid hemorrhage: a preliminary study. Stroke. 2001;32(5):1181–4.
107. Cramer S, Warren M, Enney L, Sanaee N, Hancock S, Procaccio V, editors. BDNF polymorphism and clinical outcome in the GAIN trials. International stroke conference, San Diego, CA; 2009.
108. Pearson-Fuhrhop KM, Cramer SC. Genetic influences on neural plasticity. PM R. 2010;2 (12 Suppl 2):S227–40.
109. Butefisch C, Davis B, Wise S, Sawaki L, Kopylev L, Classen J, et al. Mechanisms of use-dependent plasticity in the human motor cortex. Proc Natl Acad Sci U S A. 2000;97(7): 3661–5.
110. Goldstein L. Sygen in Acute Stroke Study Investigators. Common drugs may influence motor recovery after stroke. Neurology. 1995;45:865–71.
111. Troisi E, Paolucci S, Silvestrini M, Matteis M, Vernieri F, Grasso MG, et al. Prognostic factors in stroke rehabilitation: the possible role of pharmacological treatment. Acta Neurol Scand. 2002;105(2):100–6.
112. Conroy B, Zorowitz R, Horn SD, Ryser DK, Teraoka J, Smout RJ. An exploration of central nervous system medication use and outcomes in stroke rehabilitation. Arch Phys Med Rehabil. 2005;86(12 Suppl 2):S73–81.
113. Lazar R, Fitzsimmons B, Marshall R, Berman M, Bustillo M, Young W, et al. Reemergence of stroke deficits with midazolam challenge. Stroke. 2002;33(1):283–5.
114. Cramer S, Dobkin B, Noser E, Rodriguez R, Enney L. A randomized, placebo-controlled, double-blind study of ropinirole in chronic stroke. Stroke. 2009;40(9):3034–8.
115. Toth G, Albers GW. Use of MRI to estimate the therapeutic window in acute stroke: is perfusion-weighted imaging/diffusion-weighted imaging mismatch an EPITHET for salvageable ischemic brain tissue? Stroke. 2009;40(1):333–5.

116. Donnan GA, Baron JC, Ma H, Davis SM. Penumbral selection of patients for trials of acute stroke therapy. Lancet Neurol. 2009;8(3):261–9.

117. Feuerstein GZ, Zaleska MM, Krams M, Wang X, Day M, Rutkowski JL, et al. Missing steps in the STAIR case: a Translational Medicine perspective on the development of NXY-059 for treatment of acute ischemic stroke. J Cereb Blood Flow Metab. 2008;28(1):217–9.

118. Carey J, Kimberley T, Lewis S, Auerbach E, Dorsey L, Rundquist P, et al. Analysis of fMRI and finger tracking training in subjects with chronic stroke. Brain. 2002;125(Pt 4):773–88.

119. Johansen-Berg H, Dawes H, Guy C, Smith S, Wade D, Matthews P. Correlation between motor improvements and altered fMRI activity after rehabilitative therapy. Brain. 2002;125 (Pt 12):2731–42.

120. Koski L, Mernar T, Dobkin B. Immediate and long-term changes in corticomotor output in response to rehabilitation: correlation with functional improvements in chronic stroke. Neurorehabil Neural Repair. 2004;18(4):230–49.

121. Fleming T, DeMets D. Surrogate end points in clinical trials: are we being misled? Ann Intern Med. 1996;125(7):605–13.

122. Bucher H, Guyatt G, Cook D, Holbrook A, McAlister F. Users' guides to the medical literature: XIX. Applying clinical trial results. A. How to use an article measuring the effect of an intervention on surrogate end points. Evidence-Based Medicine Working Group. JAMA. 1999;282(8):771–8.

123. Geiger S, Holdenrieder S, Stieber P, Hamann GF, Bruening R, Ma J, et al. Nucleosomes as a new prognostic marker in early cerebral stroke. J Neurol. 2007;254(5):617–23.

124. Yip HK, Chang LT, Chang WN, Lu CH, Liou CW, Lan MY, et al. Level and value of circulating endothelial progenitor cells in patients after acute ischemic stroke. Stroke. 2008;39(1): 69–74.

125. Brott T, Marler J, Olinger C, Adams H, Tomsick T, Barsan W, et al. Measurements of acute cerebral infarction: lesion size by computed tomography. Stroke. 1989;20(7):871–5.

126. Schaechter JD, Moore CI, Connell BD, Rosen BR, Dijkhuizen RM. Structural and functional plasticity in the somatosensory cortex of chronic stroke patients. Brain. 2006;129(Pt 10): 2722–33.

127. Ding G, Jiang Q, Li L, Zhang L, Zhang ZG, Ledbetter KA, et al. Magnetic resonance imaging investigation of axonal remodeling and angiogenesis after embolic stroke in sildenafil-treated rats. J Cereb Blood Flow Metab. 2008;28(8):1440–8.

128. Hodics T, Cohen LG, Cramer SC. Functional imaging of intervention effects in stroke motor rehabilitation. Arch Phys Med Rehabil. 2006;87(12 Suppl):36–42.

129. Richards LG, Stewart KC, Woodbury ML, Senesac C, Cauraugh JH. Movement-dependent stroke recovery: a systematic review and meta-analysis of TMS and fMRI evidence. Neuropsychologia. 2008;46(1):3–11.

130. Parsons M, Li T, Barber P, Yang Q, Darby D, Desmond P, et al. Combined (1)H MR spectroscopy and diffusion-weighted MRI improves the prediction of stroke outcome. Neurology. 2000;55(4):498–505.

131. Pendlebury S, Blamire A, Lee M, Styles P, Matthews P. Axonal injury in the internal capsule correlates with motor impairment after stroke. Stroke. 1999;30(5):956–62.

132. Talelli P, Greenwood RJ, Rothwell JC. Arm function after stroke: neurophysiological correlates and recovery mechanisms assessed by transcranial magnetic stimulation. Clin Neurophysiol. 2006;117(8):1641–59.

133. Nouri S, Cramer SC. Anatomy and physiology predict response to motor cortex stimulation after stroke. Neurology. 2011;77(11):1076–83.

134. Crafton K, Mark A, Cramer S. Improved understanding of cortical injury by incorporating measures of functional anatomy. Brain. 2003;126(Pt 7):1650–9.

135. Bocti C, Swartz RH, Gao FQ, Sahlas DJ, Behl P, Black SE. A new visual rating scale to assess strategic white matter hyperintensities within cholinergic pathways in dementia. Stroke. 2005;36(10):2126–31.

136. Hillis AE, Gold L, Kannan V, Cloutman L, Kleinman JT, Newhart M, et al. Site of the ischemic penumbra as a predictor of potential for recovery of functions. Neurology. 2008;71(3):184–9.

Chapter 40
A Critical Review of Stroke Trial Analytical Methodology: Outcome Measures, Study Design, and Correction for Imbalances*

Pitchaiah Mandava, Chase S. Krumpelman,
Santosh B. Murthy, and Thomas A. Kent

Abstract Despite considerable advances in understanding the pathophysiology of stroke, there has been a lack of success in identifying new therapies to improve outcome. Our work suggests that the execution of stroke trials is not the primary issue. Here, we consider the *analysis* of clinical trials as a potential source of error. We review several components of stroke trial analysis. We conclude that many of these trials have been plagued by inappropriate use of complex statistical analytical methods that have not considered the underlying assumptions required for their valid application. Unfortunately, many of these methods have been encouraged by publishing, regulatory, granting, and pharmaceutical entities, yet continue to generate flawed results, usually discovered when early results are not confirmed in subsequent large trials. Because these errors may be just as likely to occur when early studies appear negative and so potentially reflect a missed opportunity to identify an effective therapy, we urge a reassessment of these analytical principles and provide some alternative approaches.

1 Introduction

Despite tremendous advancements in the understanding of ischemic stroke, from both a basic and clinical perspective, there is yet no approved pharmacological therapy in the USA to improve outcome other than intravenous recombinant tissue plasminogen activator (rt-PA). The study to support this approval was published in

*Funding for the Stroke Outcomes Laboratory provided by the Institute of Clinical and Translational Research Pilot Grant Program, Baylor College of Medicine.

P. Mandava, MD, PhD, MSEE (✉) • C.S. Krumpelman, PhD, MSEE
• S.B. Murthy, MD, MPH • T.A. Kent, MD
Department of Neurology and The Stroke Outcomes Laboratory (SOuL), Baylor College of Medicine and The Michael E. DeBakey VA Medical Center Comprehensive Stroke Program, 2002 Holcombe Blvd (127), Houston, TX 77030, USA
e-mail: pmandava@bcm.edu

P.A. Lapchak and J.H. Zhang (eds.), *Translational Stroke Research*, Springer Series in Translational Stroke Research, DOI 10.1007/978-1-4419-9530-8_40,
© Springer Science+Business Media, LLC 2012

833

1995, over 15 years ago. There are many potential explanations for the inability to identify new therapies, including whether preclinical conditions were adequately incorporated into the clinical design and the clinical relevance of models [1]. But one very interesting aspect to this issue is why so many therapies were thought to have the potential to be effective at the point in their development when they reached the clinical trial stage, yet subsequently failed in Phase 3 [2]. We believe that a better understanding of why so many failed trials were taken to advanced clinical stages would be instructive in both avoiding these expensive propositions in the future, but also because failure to identify negative therapies may mean there is also an inability to identify effective therapies, suggesting that the process of progression through the phases of drug development is flawed. While many potential reasons have been explored for these Phase 3 failures, we review here our contention that it is the manner in which clinical trials are designed and analyzed, rather than their execution, that is the main problem.

Stroke is a heterogeneous disorder. This heterogeneity involves nearly every aspect of the illness, including underlying pathophysiology, race, gender, the baseline medical and psychological condition of the stroke patient, size, severity and location of the stroke, timing, and treatments provided. From this perspective, it actually seems quite remarkable that even one treatment, intravenous rt-PA, was found to be effective and able to generate positive results beyond the statistical noise that such heterogeneity will inevitably generate. Yet, even the NINDS trial generated considerable controversy, primarily related to the way that baseline imbalances that favored the treatment arm was handled [3]. Despite these problems, we concluded that intravenous rt-PA is an effective treatment, but our analysis, using methods discussed below, found that the benefit was modestly overstated by approximately 20% because of baseline imbalances [4].

The potential for erroneous results from an experiment is referred to as either a Type 1 (false positive) or Type 2 (false negative) error. In the enthusiasm for moving potential blockbuster therapies into clinical trials, it seems to us that over the last 2 decades, there has been more of a tendency to accept a Type 1 error than a Type 2 error. Hence, early Phase 1 or 2 studies, which may have had uneven results or required statistical corrections to adjust for imbalances because of the heterogeneity of baseline conditions, have been relatively readily accepted, particularly if there is preclinical data to support its potential benefit. We have seen this pattern at virtually all levels of investigation, from journal reviewers and editors, to granting agencies, and pharmaceutical industry. For example, there is a well-described phenomenon called "publication bias" [5]; this reflects an apparent bias toward favorable reviews of small trials if they are positive than if they are negative. The argument has been made that if a small trial is positive, then the treatment more likely than not has a large positive effect, but if it is negative, that the size of the sample was too small to overcome statistical noise, and editors are less liable to accept a negative small study, but more likely to accept a positive small study. However, we are unaware of any data to support the contention that small positive trials in stroke have a greater likelihood to be positive when definitively studied than if the initial trial was negative. It would be difficult to test this question since early negative trials rarely go on to be definitively tested. But the consistent failure of follow-up trials to confirm early results argues the initial trial result is not adequately predictive of future success. By no means is this an issue exclusive to stroke; indeed for many neurological conditions, initial positive results are not subsequently confirmed. For example, lack of confirmation of early benefit has been an issue in ALS and Alzheimer's disease [6, 7]. In the latter case, pharmaceutical companies have acknowledged their difficulty and put together their control populations in the hope that novel ideas will be generated to adjust for natural history of the disease [8].

The complexity of stroke and the magnitude of the effort of the world stroke community to find new therapies have made the consistent failure of clinical efforts particularly discouraging. We believe that it is premature to discount the potential of translating preclinical breakthroughs in the understanding of the pathophysiology of stroke to clinically effective therapies because we do not believe that many of the therapies have been given a proper assessment. The lack of suitable appreciation for the complexity that stroke heterogeneity imposes on conventional clinical trial analysis provides an opportunity to reassess the potential utility of many therapeutic approaches and to avoid the same mistakes going forward.

We discuss various features of stroke clinical trials that offer opportunities for more rigorous analysis or have not been widely appreciated. We first describe the nature of heterogeneity in recruited stroke patients, using the NINDS rt-PA trial as an example. With this background, we consider the choice of outcome measures, how to apply these (dichotomization or continuous measures), trial design and the consideration of "adaptive" design to stroke, and finally the methods to accommodate that heterogeneity.

2   Imbalances in Stroke Trials: NINDS rt-PA as a Prototype

Baseline imbalances in NIH stroke scale (NIHSS), age, and glucose persist in decades old [9, 10] as well as more recent stroke trials [11–13]. Baseline NIHSS strongly predicts outcome [14, 15]. In the NINDS rt-PA trial, the baseline imbalances in NIHSS were particularly striking and favored the rt-PA group because a higher proportion of treated patients had milder strokes [16], see Table 40.1. It would therefore not be surprising that a higher percentage of treated patients had a favorable outcome (assuming that rt-PA did not worsen their outcome). Imbalances in NIHSS, age, and glucose in some other acute stroke trials are shown in Table 40.2.

Imbalances in factors that affect outcomes have lead to contentious debates regarding completed stroke trials and their validity [16–18]. Trialists usually address the question of imbalances by applying correction factors for the baseline variables. The actual magnitudes of these correction factors are rarely reported and more importantly the traditional methods of corrections make a number of assumptions [19, 20]: (1) The baseline variables are represented by (univariate and multivariate) normal distributions. (2) Linear relationships exist between different baseline variables. (3) Linearity extends across the whole range of values of variables. (4) Functions

Table 40.1 Imbalances across NIHSS quintiles between treatment and placebo arms in the NINDS rt-PA trial

NIHSS	0–5	6–10	11–15	16–20	>20	p^*
Placebo	16	83	66	70	77	<0.001
t-PA	42	68	66	73	63	

Table 40.2 Imbalances in NIHSS, age, and glucose in several acute stroke trials [9–13]

	Treatment arm	Control arm	p
NINDS: median NIHSS	14	15	0.09
NINDS: age (mean ± SD)	68.0 ± 11.3	65.9 ± 11.9	0.03
ECASS-III: median NIHSS	9	10	0.03
EPITHET: median NIHSS	14	10	NA
CLEAR: median NIHSS	14	10	0.01
PROACT-II: glucose (mean ± SD)	132 ± 45	156 ± 73	0.007

representing outcomes are monotonically increasing or decreasing. (5) There exists a linear relationship between baseline variables and outcome.

However, rarely, if ever, are these assumptions verified. To visually verify if the univariate and bivariate functions follow normal distributions we used kernel density estimation methods to display the distributions of the variables in multidimensions [21]. Using the NINDS database, functions were created for NIHSS alone (Fig. 40.1a, b) and NIHSS and Glucose (Fig. 40.2a, b), for both the treatment and placebo arms.

The kernel density function for NIHSS in treatment arm has two peaks at 7.83 and 15.46 (Fig. 40.1a), and the function for placebo arm has peaks at 10.13 and 14.87 (Fig. 40.1b; since the curves shown are a smoothed continuous function peaks are expressed to two decimal points). It is clear from these visualizations that the distributions among these variables are very complex, not normally distributed and hence will not be amenable to traditional statistical analysis. Moreover, the distributions are not identical between control and treatment arms, making even more suspect attempts to correct for imbalances among the groups.

2.1 Baseline Variables and Outcomes

Baseline variables such as NIHSS and glucose are not linearly associated with outcome. From TOAST trial individual subject data [14], good outcome in terms of Barthel Index 60–100 and Glasgow Outcome Scale (GOS) score of 1 and 2 can be modeled by a fourth degree polynomial function (Fig. 40.3). It is apparent that there would not be a simple linear correction between different trials or among cohorts within a trial depending on which portion of this function the different arms lay, either the flat or steep portion of the curve.

Ntaois et al. have recently shown that modeling baseline glucose values fits best with a J-shaped association with outcome [23]. Low and high glucose were associated

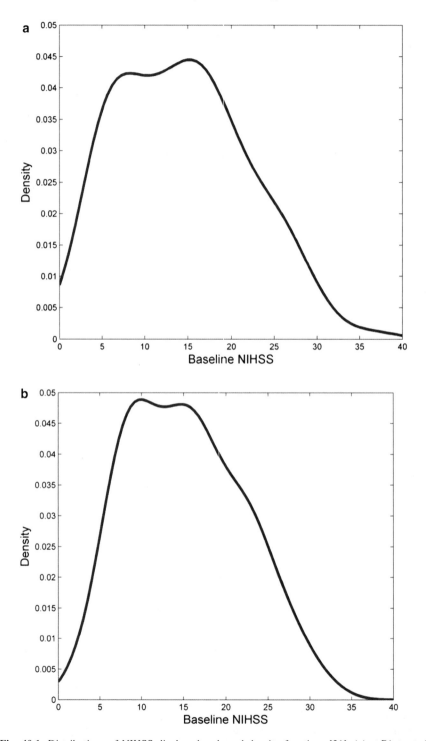

Fig. 40.1 Distributions of NIHSS displayed as kernel density functions [21]. (**a**) rt-PA treated subjects. (**b**) Placebo subjects

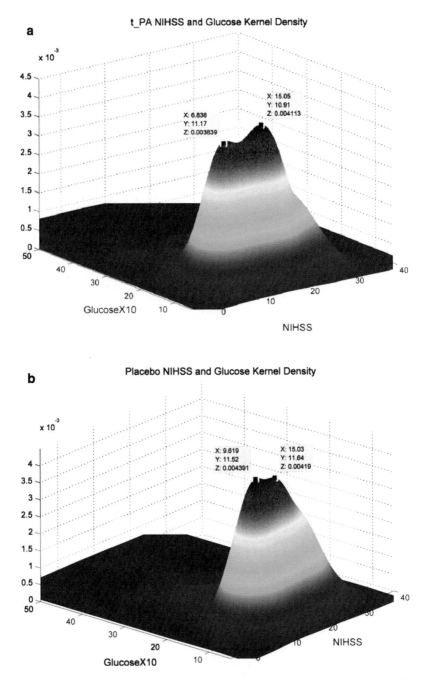

Fig. 40.2 (a) Kernel density function of NIHSS and Glucose, rt-PA treated subjects. (b) Kernel density function of NIHSS and glucose, placebo treated subjects. Note that neither function is normally distributed and that the distributions differ between control and treatment groups, indicating that the typical analytical methodology will not be valid that assume normal distributions

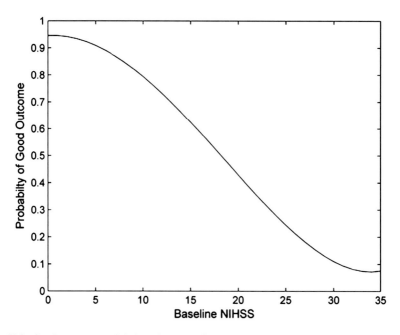

Fig. 40.3 Good outcome modeled as a function of NIHSS from TOAST data (taken from Mandava and Kent [22]). Note that this function is nonlinear

with high modified Rankin Score and moderate glucose values were associated with good outcomes (Fig. 1 of Ntaois et al. [23]). We need to point out that the J-shaped curve is a result of the fractional polynomials chosen (2,–2) and other nonlinear associations may result from a different set of fractional polynomials. Also, the subjects in Ntaois et al. study had glucose values substantially lower than NINDS placebo group (115.2 ± 32 vs. 151 ± 77.9; $p = 0.0001$) and may not be representative of the range of glucose values typically found in acute stroke trials. Nevertheless, this is one of the few attempts to apply nonlinear modeling to a factor related to outcome. As such it illustrates the complexity of such modeling, when focus on a single factor shows nonlinear characteristics, the complexity of multiplies when other factors are simultaneously considered. Typically, multivariate regression or similar methods are employed [24], while the assumptions on which such methods depend are typically not verified. In the context of regression analysis, a particularly critical assumption is the relationship among variables is linear throughout the range of those variables, which is relatively unlikely to occur in as complex a condition as stroke [25].

The implication of this heterogeneity is that traditional statistical methods, such as multivariate or logistic regression, to relate factors such as age, sex, race, etc. to outcome, are suspect and not surprisingly, often provide contradictory results. Indeed, taking just one factor, gender, a recent review [26] confirmed that no conclusion can yet be made on the influence of gender on outcome with rt-PA.

Given this background on heterogeneity, we review specific issues in clinical stroke trials.

3   Outcome Measures

A number of outcome assessment scales exist in the stroke literature [27]. These range from a five point GOS to a complex >50 item Fugl-Meyer Scale. Most formal assessments in randomized stroke trials are done 90 days after a stroke to allow for recuperation and stay in rehabilitation facility. There are rare reports of assessments performed at 6 months and a year. Because of the relative ease in obtaining them, short-term outcomes (e.g., mortality, discharge home, or rehabilitation facility) have been used in many case reports and single-arm series [28]. There is a modest correlation between these end points and 90 day outcomes.

Rankin working with 206 patients in a Scottish hospital introduced a six point scale as a measure of disability after a stroke [29, 30]. The original Rankin Scale (RS) ranges from 0 to 5. A zero on the scale indicates no disability, 1 no disability despite symptoms, 2 a slight disability, 3 a moderate disability, 4 a moderate to severe disability, and a 5 indicates a bedridden state requiring constant nursing care [31]. During the course of time the scale was modified to range from 0 to 6 with 6 indicating death [32]. The modified Rankin Score is one of two measures of functional outcome (the other being Barthel Index) that is widely adopted in acute stroke trials [33]. Despite this wide adoption the scale has some weaknesses. The rating of 0, 1 and 5 and 6 are not usually in question, but the scale in mid-range is known to be error prone with substantial interrater variability [34–37]. The interrater variability in the mRS range of 2–4 persists in spite of repeated training with video training [38] The clustering of disagreement in mid-grades of 2, 3, and 4 [34, 38] call into question the use of the complete range of mRS and use of state transitions from mid-range mRS to lower mRS as measures of efficacy [39].

The Barthel Index (BI) is an assessment instrument of activities of daily living (ADL) after a stroke [40, 41]. BI is a ten-item scale that deals with a patient's ability to walk, bathe, toilet, dress, and control bladder and bowel. Scores on individual items as in NINDS trial or total scores ranging from 0 to 100 for each subject are tabulated. A subject with a BI of 100 is considered completely independent and subjects who died are given a BI of 0. Depending on the median NIHSS of a trial, U-distributions signifying both floor and ceiling effects with clustering of subjects at a BI of 0 and 100 have been reported [42, 43].

The Functional Independence Measure (FIM™) is an often used measure of ADLs in the setting of rehabilitative medicine [44]. FIM is an 18 item scale with total scores ranging from 18 to 126. The highest score of 126 corresponds to a patient who is completely independent.

The Fugl-Meyer (FM) scale is a measure of patient function in five domains: motor, sensory, balance, joint-pain, and range of motion. This highly detailed evaluation in the motor (50 items) and sensory (6 items) domains has been suggested as an alternate scale for use in stroke trials and used in a recently completed stroke trial [45, 46]. However, the detailed nature and length of time needed to evaluate a patient has limited its utility in large scale clinical trials.

The Stroke Impact Scale (SIS) and the shorter version of SIS the SF-36 have been proposed to assess quality of life issues poststroke and incorporated into the ongoing FAST-MAG trial [47, 48]. GOS a five point scale similar to mRS has been used in the past in a few acute stroke trials and traumatic brain injury trial [9, 49–52]. GOS as initially proposed for use in traumatic brain injury, ranges from 1 ("dead") to 5 ("good recovery") [53]. Some stroke trials such as TOAST have reversed the order of the scale to range from 1 ("good recovery") to 5 ("dead") [14], but in general the GOS is not widely used in stroke.

The Scandinavian Stroke Scale (SSS) is another measure of baseline stroke severity and the change from baseline to 90 days has been used to establish efficacy of treatment [54].

NIHSS introduced as a measure of stroke severity and screening tool for entry into stroke trials has been used as an outcome measure in several trials [9]. The NIHSS ranging in possible values of 0–38 has essentially replaced the other scales such as SSS, Orgogozo scale and others as a selection metric. The scale is fairly robust in reproducibility and takes less than 150 s to assess. But its relationship to infarcted tissue and clinical outcomes depends on factors that underscore the heterogeneity in stroke. Right and left middle cerebral artery strokes with same amount of tissue damage will produce a different NIHSS due to increased significance given to language [55]. And posterior circulation strokes with seemingly low NIHSS may progress into coma and certain death [56]. Despite these inherent weaknesses some authors have suggested using 90 day NIHSS as an outcome measure, and other have included a "correction" factor for these biases [57]. Our concern with these and other correction factors is that most assume linear relationships, which may or may not be the case in the particular circumstance. Some authors have employed the NIHSS as a before and after measure. While this approach appears reasonable, it is not clear that it is linear throughout its range so that changes are equivalent either in magnitude (either a certain point drop) or percentage throughout the range.

3.1   What Is a Good Outcome?

For the prespecified primary analysis acute stroke trials used a variety of definitions of a good outcome and a variety of statistical tests. Simple dichotomizations, trichotomization, combined measures of dichotomization across different scales, responder analysis, and use of complete range of ordinal scales are some of the methods used to establish efficacy.

3.1.1   Dichotomization

Dichotomization defines a threshold at which subjects either achieves a specified end point or not. The advantage of this approach is simplicity and the ability to define a level of function that can be readily understood. There are, however, many disadvantages to this approach beyond the inherent noise in mid ranges of the scales. Studies have selected different outcome values, which makes comparison among them difficult. Investigators also tend to select the threshold for positive outcome based on expected severity of subjects to be recruited. The target population may not necessarily be achieved and the outcome threshold may subsequently be either too stringent or lenient. Note that larger trials include a number of secondary end points, although the more variables analyzed, the greater the likelihood of random false positives.

Some examples of thresholds chosen follow: Subjects achieving mRS of 0 or 1 were considered to have a good outcome in ECASS II [58]. Proportions of subjects achieving mRS of 0–1 in the treatment and control arms were compared to establish treatment effectiveness. Subjects achieving mRS of 0–2 were considered to have a good outcome in trials such as PROACT II that had a high baseline NIHSS of 17 [10]. Hemicraniectomy studies DESTINY and DECIMAL with highest baseline NIHSS of any studies (21 and 24), compared proportions achieving an mRS of 0–4 to establish effectiveness of treatment [59, 60]. This approach appears justified based on the severity of the patients requiring hemicraniectomy, but the expected range of outcomes needs to be clear when explaining the study to potential patients and families.

3.1.2 Combined Measure of Efficacy

The global outcome measure uses logistic regression and the link-logit function as the underlying basis and adjustments are made for stratifying factors [9]. These methods relate outcome measures to each other based on their correlations and have the advantage of convergence of multiple measures. However, it is not clear that each of these measures is independent of each other, which may then potentially exacerbate errors in measurements rather than provide a more valid outcome measure. The method has been applied to several trials. For example, the NINDS rt-PA trial was analyzed with a global outcome measure which takes into account subjects achieving a combined end point of mRS 0–1, BI 95–100, GOS of 1, and NIHSS of 0–1 [61]. TOAST with median NIHSS of 7 used a similar combined measure of GOS of 1 and BI of 90–100 as an indication of excellent outcome [14]. Sometimes, these combined measures are developed post hoc because neither one individually was positive.

3.1.3 Responder Analysis (Stratified Dichotomization)

In ABESTT subjects were divided into three strata depending on baseline NIHSS [62]. To be classified as having achieved a good outcome, subjects with a baseline NIHSS of 4–7 needed an mRS of 0. The dichotomization cutoff for subjects with NIHSS of 8–14 was an mRS of 1. For subjects with NIHSS >14 the dichotomization cutoff was an mRS of 2. The advantage of responder analysis is that outcome can be stratified to what is considered reasonable for that individual's severity of illness. However, stratification essentially generates subgroups; the likelihood of imbalances increases with increased parsing of the population, particularly when the overall study numbers are relatively small, as is the case in many stroke trials. As an example of potential difficulties with responder analysis particularly in a situation in which baseline imbalances complicate the results, the NINDS trial was reanalyzed using responder analysis and produced what appears to be a relatively implausible result that showed that treatment of patients with rt-PA in the 90–180 min window yields better outcomes (OR 1:80 CI:1.11–2.91) compared to patients treated in the 0–90 min window (OR 1.56 CI:0.96–2.52) [63]. We are in general not in favor of subgrouping or stratification unless the balance between treatment and placebo arms can be confirmed. We discuss the outcomes from AbESTT in more detail below.

3.1.4 Considering the Whole Range of mRS 0–6 Scale and the "Shift"

The European Medical Agency (Food and Drug Administration [FDA] equivalent) has suggested that when a range of scores is used to measure outcome that the whole range be considered in establishing efficacy [64]. A number of randomized studies (SAINT I, SAINT II) have incorporated this outcome, without however, confirming the statistical bases of comparing the whole range of the mRS scale as discussed in detail below [65, 66]. Various authors have termed this method "Shift analysis" [67–69]. Note that this form of "Shift Analysis" is different than the original proposed "Shift analysis" as proposed by Lai and Duncan of shift from baseline to 90 days of a subject's mRS value or the shift in NIHSS from baseline to 90 days as suggested by Bruno et al. [39, 70].

Authors who promote the "Shift analysis" [71–73] invoke Shannon [74], Altman and Royston [75], and Federov et al. [76] in support of their position against dichotomization. Royston and Altman were specifically referring to continuous variables such as blood pressure, age or weight to question the relevance of dichotomization [75]. Federov et al. suggest that increased sample sizes are needed for trials that use dichotomization, hence supporting the use of "shift" for studies with relatively small numbers of subjects [76]. Federov et al. refer to continuous variables or discrete variables in their modeling and not ordinal or categorical variables such as mRS [76].

Howard [77] has cautioned against the use of full ordinal scale of the mRS and applying Cochran-Maentel-Hanschel (CMH) test as done in SAINT I and SAINT II. He points out that the CMH test assumes uniformity of treatment effect over the complete range of mRS so that is there is an implicit assumption that treatment effect is the same over the mRS 0–1 as it is over the mRS of 2–3, etc. Palesch has reiterated similar concerns regarding the use of full ordinal scale and the underlying assumptions [78]. These concerns apply even after application of complex transformations such as RIDIT that some investigators have claimed improves its validity [68, 78, 79].

Shannon, another author cited in support of the "Shift Analysis," in his "Theory of Communication" laid the ground work for the nascent field of information theory [74]. His interest was in quantifying the limits of information transfer in terms of bits per second and the amount of information transferred [74]. Shannon's Figure 1 had a noise source in the path from a transmitter to a listener. Noise in the model referred to by Shannon applies uniformly and is not "band-limited." However, as shown by van Swieten [34] and others [35, 36] errors of misclassification (or noise) are primarily in the mid-range of mRS and not spread evenly across the complete range of mRS. Our simulation of 16 acute stroke trials shows that percentage of error is nearly 26% and predominantly or wholly in the mRS 2–4 range [80]. Thus if the signal/outcome is mRS 0 and 1 or mRS 5 and 6 then it has higher fidelity with less noise or error. Given this nonuniform distribution of "noise" across the range of mRS, dichotomization is preferred over considering the full range. Perhaps, if sufficiently high numbers of subjects, possibly in thousands, are recruited with no systematic bias in noise, it is conceivable that considering the full range of mRS would be a reasonable choice since errors would be similarly randomly distributed between both groups.

We are therefore not certain that "shift" as more recently applied as a stroke outcome is a valid concept when used with ordinal outcome measures such as the mRS. Moreover, it does nothing to adjust for imbalances, and may indeed accentuate these imbalances. If, as occurred in the NINDS rt-PA trial, there is a larger proportion of mild stroke enrolled in the treatment arm, it is not surprising that the "shift" analysis between active treatment and placebo would favor the treatment as a larger proportion will achieve lower mRS values after treatment. We consider it suspect for any assessment of the potential relative advantages of the "shift" analysis that does not consider that influence of imbalances [68].

4 Adaptive Design of Clinical trials

The FDA, realizing that clinical trials have become prohibitive in terms of cost, length of time taken to complete a trial, paucity of patients willing to be enrolled in blinded trials, and potential lack of "power" in some studies, has suggested that adaptive trial design be considered [81]. The concept of adaptive design of clinical trials is quite simple and appealing. Baseline data such as NIHSS and age, outcome data, or drug kinetic data (in case of a dose ranging study) and drug toxicity data of subjects recruited into incomplete phase 2 trial is utilized to guide the future recruitment of subjects into phase 3 trial or to trim the number of arms in a dose finding trial [82, 83]. Alternately, interim unblinded data in an ongoing phase 3 trial can be used to guide recruitment of later recruited subjects. It is hoped that adaptive designs will have several benefits: (1) Creating balanced groups of subjects in the treatment and control arms. (2) There may be potential decrease in the number of subjects needed in a trial. (3) Ability to modify exclusion and inclusion criteria.

ALIAS part 2 is being conducted on principles of adaptive design. The initial plan calls for recruiting 1,100 subjects [83]. It is not clear if indeed adaptive design will speed recruitment if a sizable number of subjects are not eligible, although any method to reduce heterogeneity is welcome. However, it is not established how many factors adaptive design can consider, and ultimately balance for all the factors associated with outcome will be difficult and some post hoc correction method is likely going to be needed.

On the other hand, we are very concerned about using unblinded outcome data to guide future enrollment, since imbalances among recruited subjects may be the more important factor in outcome rather than the therapy itself and adaptive enrollment based on early outcomes may artifactually reduce the median stroke severity of the treatment arm.

The adaptive design suggested above has similarities to Bayesian analysis. In Bayesian analysis a model (or a prior) is continually updated to predict the future outcomes (or generate a *posterior*) and assign a probability value to an outcome [84]. In a Bayesian scenario, the prior defines the distribution over outcomes when there is no knowledge of some affecting event. Once the knowledge of the affecting event is incorporated, the resulting distribution is called the posterior.

Adaptive dose finding designs have been modified to incorporate Bayesian principles. Such a technique was used in an acute stroke trial ASTIN [54] which started with 15 dose regimens and ended with four regimens. This trial was evaluated for treatment efficacy using Bayesian techniques intending to show a benefit in terms of differences in mean treatment effect in change in SSS from randomization to 90 days. Despite the identification of promising regimens, the ASTIN trial failed to show any benefit of treatment. Notably, the Bayesian technique in ASTIN trial used an underlying linear regression model with covariate adjustment for SSS, age, and time to treatment that may have led to later difficulties given that these factors are unlikely to be linearly related.

5   Pooled Placebo Functions as a Basis for Comparison Among Trials (pPREDICTS©)

We [85] and others [86] have shown that functional outcomes and mortality of acute stroke trials are dependent on baseline NIHSS and age in a nonlinear fashion. Good functional outcome in terms of proportion of patients achieving an mRS of 0–2 and mortality can be predicted in terms of NIHSS and age (Fig. 40.4a, b).

The models in Fig. 40.4 were obtained by Freeman-Tukey transformation of outcomes of 28 acute stroke trials representing ~7,000 patients. Freeman-Tukey transformation linearizes a nonlinear association, stabilizes the variance and additionally applies corrections to the outcomes based on number of subjects in a trial [87]. This transformation and the associated arc-sine-square root transformation are well suited when trials have wide ranges in number of subjects, if the ratio of subjects in the smallest to largest trials is greater than 5.

With this family of functions, outcomes are always asymptotic to 0 and 100% at extremes of the dependent variables of NIHSS and age. In Fig. 40.4a, b the surfaces in the middle are model functions and the bounding surface on either side are the ±95% prediction interval surfaces for mRS 0–2 and for mortality. The outcome variance explained for mRS by NIHSS and age was 90% and for mortality 83%, indicating that these factors are major, but not exclusive, determinants of outcome. Note that balancing for only these factors may therefore not account for other important factors, as discussed above in the case of adaptive design.

While we believe that randomization and blinding are *necessary* for a definitive assessment of a new therapy, we do not believe that randomization alone is *sufficient* to achieve a reliable result because the arms may be imbalanced. One potential use

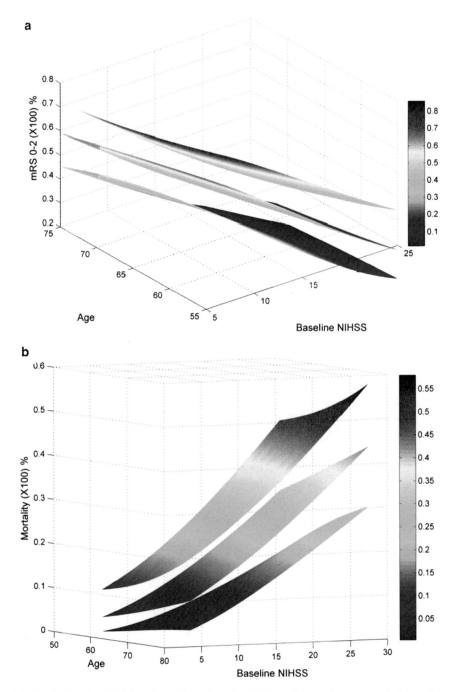

Fig. 40.4 pPREDICTS© functions relating baseline NIHSS and age to the percent of subjects that achieve an mRS 0–2 (*top*) and mortality (*bottom*). The middle function is the fitted function and is bounded by 95% prediction intervals

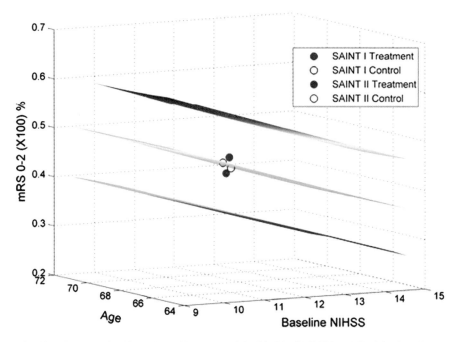

Fig. 40.5 Percent of subjects that achieved an mRS of 0–2 in SAINT I and II trials plotted onto pPREDICTS© model. Note the reproducibility between the two trials in terms of where both lie on the function with respect to baseline variables (NIHSS and age) and the overlap in outcomes (mRS 0–2, z axis)

for these pPREDICTS© functions is that for any treatment trial one could plot the Freeman-Tukey transformed outcome at its own baseline NIHSS and age and compare outcomes to this pooled placebo function that may be more representative of the stroke population at large. In addition, one can plot the placebo arm and determine how representative it is of the more general stroke population. Using this concept, we showed that four treatment arms (NINDS, NEST 1, CLOTBUST, and MELT; Refs. [9, 88–90]) had outcomes better than +95% prediction interval and thus showed potential benefit. The graphical/mathematical tool pPREDICTS is a tool designed to compare outcomes of treatment arms against the model outcome and to check for imbalances between treatment and control arms.

5.1   Analysis of SAINT I and SAINT II

"Shift Analysis" was used as a means of determining efficacy of treatment in SAINT I and SAINT II [65, 66]. Shift analysis uses Cochran Maentel Haenszel (CMH) test with Van Elteren modification [68]. The CMH test was proposed to account for imbalances in larger groups [91]. It was felt that breaking up into smaller groups and stratification into groups would reduce the imbalances. The intent in suggesting this test is to generate nearly perfectly matched samples [91].

As applied in SAINT I and SAINT II, the whole range of mRS is entered into analysis after adjustments to baseline NIHSS, side of infarction, and use of rt-PA. Using this analysis SAINT I was considered positive by "shift analysis" but not by comparing the proportions of patients who achieved mRS of 0–2 or NIHSS of 0–1 etc. The correction factors or adjustments are not available to the public and we are not aware of how the adjustments were made. Also, the baseline factors such as NIHSS, age, glucose, side of stroke, type of stroke, and others for the different strata and the two arms are not reported. In the absence of data on these factors, balances that are fundamental to CMH cannot be verified. Koziol and Feng [92, 93] expressed these reservations soon after publication of the SAINT I results. And applying Mann–Whitney test they showed that there is no benefit of treatment with NXY-059 [92]. They also correctly predicted even before the results of SAINT II were available that it would be negative.

If we were to plot SAINT I and SAINT II treatment and control arm outcomes onto the pPREDICTS model several things are immediately apparent (Fig. 40.5). First, all four arms are clustered around the natural history surface in the middle well within ±95% surfaces and essentially what would be expected from best medicine therapy. This finding indicates to us that the execution of both studies was outstanding, as reflected in the overlapping outcome in the placebo arms of SAINT I and SAINT II. Second, there is a very small separation between SAINT I treatment and control arm outcomes. Because of smaller size, subgrouping often leads to imbalances not evident in the larger groups (e.g., see Table 4a lacunar stroke group in Ref. [4]). We suggest that stratifying into subgroups favored the SAINT I treatment arm resulting in a spurious positive result.

In spite of the failure of SAINT II, consideration of using the shift analysis continues. Several authors have applied the shift analysis to thrombolytic trails to show the benefit of treatment with rt-PA and thus use this as validation to promote future use of shift analysis [73]. As discussed above, the impetus for the move to abandon dichotomization in favor of shift analysis comes from papers by Altman and Royston [75] and Federov et al. [76].

Altman and Royston were referring to continuous variables and thus their suggestions do not apply to any of the outcome variables in stroke trials. Federov et al. are referring to continuous variables and discrete variables that can be modeled by Poisson distributions. Distributions of mRS (a categorical variable) from 16 trials (Fig. 40.6a) do not bear resemblance to any of a family of Poisson distributions (Fig. 40.6b), thus violating one of the basis for the original consideration of a shift analysis. Additionally, we need to be cautious when entering the intermediate values of mRS into any analysis given that there is high degree of inter-rater variability in mRS 2–4.

"Shift analysis" as used in SAINT I and SAINT II [65, 66] is not the only shift method mentioned in literature. Predating SAINT I and SAINT II, Lai and Duncan [39] suggested assessing mRS prestroke and 90 day poststroke. They suggested that any "shift" or change in mRS be used as a measure of efficacy of treatment. While this may be appealing in its simplicity, note that the assessment of midrange of mRS 2–4 is unreliable [34] and transitions between mRS 5–3 do not have the same meaning or relevance as a transition from 2 to 0.

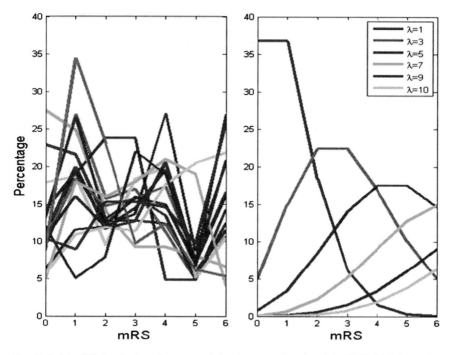

Fig. 40.6 (**a**) mRS distributions from several placebo arms of stroke trials. (**b**) Valid Poisson distributions with different λ values. Note that the actual stroke trial data does not follow a Poisson distribution, as assumed by the underlying theories supporting continuous vs. dichotomous outcomes

5.2   Analysis of ABESTT

The initial randomized AbESTT trial comparing cardiac doses of the GP IIb/IIIa antagonist, abciximab to placebo with a 6 h treatment window suggested benefit in a subgroup of patients with moderate stroke although with minimal signal of efficacy for the group as a whole [62]. Note the overlapping treatment arms from the different AbESTT trials when plotted on pPREDICTS© outcome model (Fig. 40.7), indicating that the three trials were remarkably reproducible when accounting for these baseline factors and functional outcome. While the overall study groups shows good balance on baseline NIHSS, the baseline characteristics of this initially positive subgroup was not provided, so it is conceivable that baseline imbalances in the subgroup may have been responsible for the erroneous signal of efficacy that was not subsequently confirmed. As we have indicated above, subgrouping by baseline factors increases the likelihood of imbalances in other factors as the number of subjects dwindles, and use of stratified outcomes has the unintended consequence of further subgrouping and increases the chances of imbalances of factors other than baseline NIHSS, such as gender, stroke subtype, etc. The net effect is to make early subgroup analysis suspect. Indeed, the follow-up randomized trial was halted early

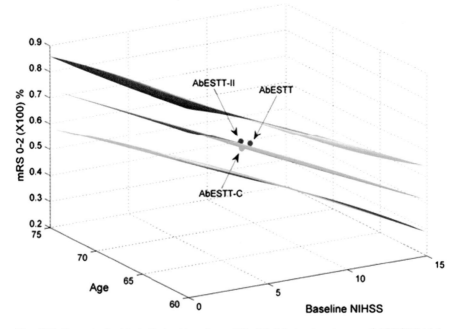

Fig. 40.7 Percent of subjects that achieved an mRS of 0–2 in treatment arms of ABESTT trials plotted onto pPREDICTS model. Note how similar each outcome is both in terms of where the different arms lie with respector baseline factors (baseline NIHSS and age) as well as overlap on the percent that achieve the functional outcome measure (z axis mRS 0–2%)

because of lack of benefit and increased hemorrhage [94]. Note also that investigation into the use of this class of agent is continuing based on different doses [95] and routes of administration (e.g., intra-arterial).

6   A Different Approach: Post hoc Matching of Subjects. Comparison of Methods

Imbalances between treatment and control arms in terms of baseline factors such as NIHSS [11, 12], age [9], and glucose [10] are common. These imbalances make comparisons of outcomes difficult requiring complex correction factors. Comparisons of different trials such as ECASS III [11] and ATLANTIS [96] are difficult if not impossible to do without ignoring critical assumptions. pPREDICTS offers a solution but it is limited to using widely used outcome measures, which may not be the primary outcome measure for the studies of interest.

To correct for imbalances in NINDS, we suggested that a subject in rt-PA arm be matched with a subject in placebo arm by adopting a nearest-neighbor method in the

3D Euclidean space of NIHSS, age, and glucose [4, 97, 98]. We call this method pPAIRS©. The weighted Euclidean distance is calculated using the formula:

$$Dist = sqrt(((wt_N * (NIHSS1 - NIHSS2))^2$$
$$+ (wt_a * (Age1 - Age2))^2$$
$$+ (wt_g * (Glucose1 - Glucose2))^2) \tag{40.1}$$

Weights of 9.9 for NIHSS, 2.24 for age, and 1 for glucose were obtained by taking ratio of means of glucose, age, and NIHSS. Multiplication by weight mitigates the effect of glucose and age which range over larger values (glucose: 70–500, age: 18–90) compared to NIHSS that ranges between 0 and 38. To eliminate extreme values of distances, a threshold described by Tukey that depends on 75th and 25th percentile was chosen [99]:

$$Threshold = 1.5 * Dist\_IQR \tag{40.2}$$

A competing method of matching is based on calculating "propensity scores" and matching two subjects with the smallest differences in propensity scores [100]. In order to calculate propensity score for each subject logistic regression is performed on NIHSS, age, and glucose as independent variables and assigning a 1 to a dependent variable if a subject is assigned to treatment arm or a zero if assigned to the placebo arm. The coefficients obtained from logistic regression are used to calculate for each subject the probability value ranging from 0 to 1 that a subject is assigned to treatment arm. The probability value is the propensity score. After determining propensity scores for all subjects in the treatment and placebo arm, a search is made to find pairs with the smallest difference between their propensity scores. Again, outliers with extreme differences in propensity scores are eliminated using (40.2).

6.1   A Common Measure to Determine the Best Matching Method

We compared the two methods using the NINDS rt-PA trial data. City-block distance, also called the Manhattan distance or Manhattan length [101], between a matched subject from the rt-PA arm and the subject's pair from the placebo arm is an indication of how close the methods identified a matching subject based on the baseline factors of choice. In this case, we examined the sum of absolute differences of NIHSS, Age, and Glucose and expressed the distance as

$$C\_dist = abs(NIHSS1 - NIHSS2) + abs(Age1 - Age2)$$
$$+ abs(Glucose1 - Glucose2) \tag{40.3}$$

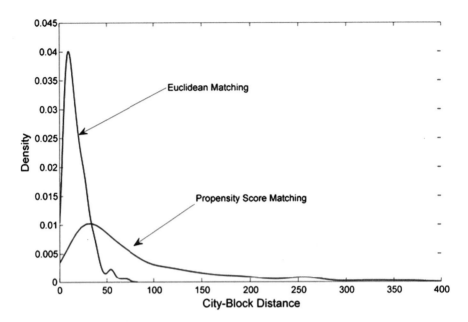

Fig. 40.8 Density distributions of city-block distances with Euclidean matching (*blue*) and with propensity score matching (*red*). Far smaller distances are evident with Euclidean method, suggesting that closer "matches" were found than that with the more widely used propensity score matching

By this method city-block distance of zero between two matched points would suggest a perfect match. In NINDS trial age was expressed to the fourth decimal place and glucose ranged over large values thus we did not expect a number of perfect matches. Distributions of city-block distances for both matching methods are shown in Fig. 40.8.

Table 40.3 shows the baseline variables prematch and postmatch by both methods. As shown in the last row the city-block distances are lower for the Euclidean nearest-neighbor method. We suspect that the propensity score matching does not yield good matches because the propensity scores span a narrow range of 0.36–0.6 and for proper utilization of the method, the scores should span the entire range from 0 to 1.

Using two independent variables of NIHSS and age and glucose did not increase the range of propensity scores. In addition to the narrow range of propensity scores we mentioned in our earlier paper [4] three additional conditions have to be met for valid application of Propensity Score methods: (1) Large samples in both the treatment and control arms. (2) Distributions have to be Gaussian. (3) The variances and means have to be nearly the same in both groups. We suspect that these conditions are not satisfied in acute stroke trials.

Table 40.3 Comparison of baseline variables prematch and post-Euclidean and propensity score match

	Prematch			Post-Euclidean match			Post-prop-score match		
	rt-PA (n=312)	Placebo (n=312)	p	rt-PA (n=283)	Placebo (n=283)	p	rt-PA (n=259)	Placebo (n=259)	p
Median NIHSS	14	15	0.09	14	14	0.62	14	15	0.31
NIHSS mean+SD	14.4±7.5	15.2±6.8	0.14	14.4±7.3	14.7±6.7	0.68	14.6±7.7	15.1±6.8	0.44
Age mean+SD	68.0±11.3	65.9±11.9	0.03	67.9±11.4	66.5±11.8	0.15	66.8±11.6	67.1±11.8	0.77
Glucose mean+SD	149±70.7	151±77.9	0.78	144±67.4	145±61.4	0.90	150±74.4	150±77.1	0.99
City block distance				17.7±12.45			80.61±76.41		<0.001

7 Measuring the Distribution of Baseline Factors Among Subjects: A Novel Adaptation of Kolmogorov–Smirnov

Customarily when multiple baseline variables are reported for two arms in a trial, the variables are compared one at a time. For example, median NIHSS between two arms of a study are often compared with Wilcoxon Rank-Sum test, mean NIHSS, age, and glucose are compared with Students t-test. We are proposing that multidimensional tests be employed to consider NIHSS and age or NIHSS, age, and glucose of the two populations because the distributions among several key variables may differ than when each is looked at individually [102].

Kolmogorov–Smirnov test is designed to determine the similarity of distributions among two groups, but was originally developed for a single factor. Peacock made a brilliant extension of a one-dimensional Kolmogorov–Smirnov test to two dimensions [103]. His intended application, in the field of astronomy, was to compare two populations and check if the two are distinct populations. This method has been expanded to fields other than astronomy. Fasano and Fraceschini simplified the computation of the two-dimensional measure and extended the test to three dimensions [104]. The result of the test is expressed as a distance between two populations and the significance (p value) of the distance is calculated. We modified the Fasano-Franceschino algorithm and show an example (Table 40.4) of calculating the distances between two populations derived from NINDS trial in normoglycemic subjects with large vessel stroke in the control and treatment arms. The left half of the table shows the comparisons of median NIHSS, mean NIHSS and age prematch and the right half shows postmatch values. The last row shows the two-dimensional Kolmogorov–Smirnov distance (KSD) and the p value before the match and postmatch.

By individual tests of baseline variables $p > 0.05$, as usually performed, there is no difference between the populations prematch or postmatch. However, the KSD is

Table 40.4 Comparison of Kolmogorov–Smirnov distance (KSD) prematch and postmatch

	Prematch (large vessel, glucose <150)			Postmatch (large vessel, glucose <150)		
	rt-PA ($n = 76$)	Placebo ($n = 99$)	p	rt-PA ($n = 76$)	Placebo ($n = 76$)	p
Median NIHSS	14.0	16.0	0.276	14.0	14.0	0.861
Mean NIHSS STD	14.7 ± 6.70	15.4 ± 6.04	0.472	14.7 ± 6.70	14.6 ± 6.20	0.950
Mean age STD	67.4 ± 11.9	65.0 ± 11.4	0.171	67.4 ± 11.9	65.5 ± 10.4	0.282
KSD	0.283		0.008	0.217		0.175

The usual method to compare populations by indivdual factors is not significant, while KSD is, indicating that the populations are different when two factors are considered simultaneously. It can be seen that post-Euclidean matching (pPAIRS©), the KS distributions are no longer significantly different ($p = 0.175$)

higher and the *p* value significant prematch thus indicating that the two populations are indeed different when two factors are considered simultaneously. Post-Euclidean match (pPAIRSC©) KSD is lower and the *p* value >0.05, indicating the populations are similar and matching has been successful. This is an example of use of 2D KSD estimation. The method can be expanded to 3D KSD estimation.

This example indicates that the comparison of individual factors when assessing for balance among populations is unable to capture the complexity of these distributions in multiple dimensions. This finding raises the possibility that apparently similar groups or trials may in fact differ in important ways and comparison of outcomes will therefore be potentially misleading.

8   Conclusion

We have provided a critical review of the analysis of clinical trial data in stroke. We believe that the methodology used to analyze results has not kept pace with advances in preclinical discoveries, nor in the standardization of clinical trial execution. The Phase 3 SAINT trial and the AbESTT trials show how reproducible clinical trial outcomes can be with virtually identical results among separate trials. While clinical trial analysis has become similarly standardized, we do not believe it is correctly conceived and adequately addresses the heterogeneity of the clinical presentation of stroke. The central issues going forward as seen by us are the following:

- Will regulatory agencies recognize the inadequacies of their preferred methods to analyze clinical trial results? We certainly hope that they will be open to new ideas both going forward as well as going backward to see whether there is sufficient evidence of the potential for Type 2 errors to warrant new studies for apparently negative studies. At the same time, we would encourage caution in adopting too many permutations of the simple randomized control trial and appropriately selected dichotomous outcome measures without further experience.
- Given the number of false leads, will granting agencies be more discriminating and question the validity of the assumptions behind statistical correction methodologies when making funding decisions? There should be broad agreement that the assumptions be addressed explicitly. Recognizing the likelihood that these assumptions will not usually be met in the small to mid-range trials, one solution to this issue is to fund only large trials in which correction methods would be less likely to be needed. But focusing on large trials means that proper analysis of the early data becomes even more critical to give reasonable probability of the success of the larger and more expensive trial. Many negative examples even in the secondary prevention realm give additional urgency to increased attention to the manner in which early trial results are assessed, particularly with regard to baseline balances and subgroups. At the minimum, we urge that the full set of baseline characteristics of subgroups be available to the public, rather than only for the group as a whole as is the more common practice.

- Will pharmaceutical agencies regain an enthusiasm for new therapy trials in acute ischemic stroke? We believe that this will happen only if the academic community is willing to acknowledge weaknesses in their prior approach and propose new methods so as not to repeat the previous mistakes.
- Is there any cause for optimism that that new therapies can be developed? We believe that there are several promising therapeutic approaches [85] as well as others not yet formally analyzed by these methods such as edaravone [105]; although evidence so far suggests that benefit from some of these therapies may ultimately be restricted to certain subtypes of stroke or other inclusion/exclusion factors. The following question then arises: will there be enough interest in pursuing an agent with a more limited range of potential benefit? We certainly hope so. Any beneficial effect on outcome for even a restricted type of stroke patient would provide a meaningful improvement in the life of many individuals.

References

1. Fisher M, Feuerstein G, Howells DW, Hurn PD, Kent TA, Savitz SI, et al. Update of the stroke therapy academic industry roundtable preclinical recommendations. Stroke. 2009;40(6): 2244–50.
2. Kidwell CS, Liebeskind DS, Starkman S, Saver JL. Trends in acute ischemic stroke trials through the 20th century. Stroke. 2001;32(6):1349–59.
3. Mann J. NINDS reanalysis committee's reanalysis of the NINDS trial. Stroke. 2005;36(2): 230–1.
4. Mandava P, Kalkonde YV, Rochat RH, Kent TA. A matching algorithm to address imbalances in study populations: application to the National Institute of Neurological Diseases and Stroke Recombinant Tissue Plasminogen Activator acute stroke trial. Stroke. 2010;41(4):765–70.
5. Egger M, Davey Smith G, Schneider M, Minder C. Bias in meta-analysis detected by a simple, graphical test. BMJ. 1997;315(7109):629–34.
6. A controlled trial of recombinant methionyl human BDNF in ALS: the BDNF study group (phase III). Neurology. 1999;52(7):1427–33.
7. Doody RS. Evolving early (pre-dementia) Alzheimer's disease trials: suit the outcomes to the population and study design. J Nutr Health Aging. 2010;14(4):299–302.
8. C-path online depository. http://www.c-path.org/CAMD.cfm. Accessed 3 Aug 2011.
9. Tissue plasminogen activator for acute ischemic stroke. The national institute of neurological disorders and stroke rt-PA stroke study group. N Engl J Med. 1995;333(24):1581–7.
10. Furlan A, Higashida R, Wechsler L, Gent M, Rowley H, Kase C, et al. Intra-arterial prourokinase for acute ischemic stroke The PROACT II study: a randomized controlled trial. Prolyse in acute cerebral thromboembolism. JAMA. 1999;282(21):2003–11.
11. Hacke W, Kaste M, Bluhmki E, Brozman M, Davalos A, Guidetti D, et al. Thrombolysis with alteplase 3 to 4.5 hours after acute ischemic stroke. N Engl J Med. 2008;359(13):1317–29.
12. Davis SM, Donnan GA, Parsons MW, Levi C, Butcher KS, Peeters A, et al. Effects of alteplase beyond 3 h after stroke in the Echoplanar Imaging Thrombolytic Evaluation Trial (EPITHET): a placebo-controlled randomised trial. Lancet Neurol. 2008;7(4):299–309.
13. Pancioli AM, Broderick J, Brott T, Tomsick T, Khoury J, Bean J, et al. The combined approach to lysis utilizing eptifibatide and rt-PA in acute ischemic stroke: the CLEAR stroke trial. Stroke. 2008;39(12):3268–76.
14. Adams Jr HP, Davis PH, Leira EC, Chang KC, Bendixen BH, Clarke WR, et al. Baseline NIH stroke scale score strongly predicts outcome after stroke: a report of the trial of Org 10172 in acute stroke treatment (TOAST). Neurology. 1999;53(1):126–31.

15. Weimar C, Konig IR, Kraywinkel K, Ziegler A, Diener HC. Age and national institutes of health stroke scale score within 6 hours after onset are accurate predictors of outcome after cerebral ischemia: development and external validation of prognostic models. Stroke. 2004;35(1):158–62.

16. Ingall TJ, O'Fallon WM, Asplund K, Goldfrank LR, Hertzberg VS, Louis TA, et al. Findings from the reanalysis of the NINDS tissue plasminogen activator for acute ischemic stroke treatment trial. Stroke. 2004;35(10):2418–24.

17. Grotta JC. The NINDS stroke study group response. J Stroke Cerebrovasc Dis. 2002;11(3–4):121–4.

18. Clark WM, Madden KP. Keep the three hour TPA window: the lost study of Atlantis. J Stroke Cerebrovasc Dis. 2009;18(1):78–9.

19. Crager MR. Analysis of covariance in parallel-group clinical trials with pretreatment baselines. Biometrics. 1987;43(4):895–901.

20. Koch GG, Tangen CM, Jung JW, Amara IA. Issues for covariance analysis of dichotomous and ordered categorical data from randomized clinical trials and non-parametric strategies for addressing them. Stat Med. 1998;17(15–16):1863–92.

21. Martinez WL, Martinez AR. Computational statistics with Matlab. Boca Raton: Chapman and Hall/CRC; 2002. p. 285–8.

22. Mandava P, Kent TA. Intra-arterial therapies for acute ischemic stroke. Neurology. 2007;68:2132–9.

23. Ntaios G, Egli M, Faouzi M, Michel P. J-shaped association between serum glucose and functional outcome in acute ischemic stroke. Stroke. 2010;41(10):2366–70.

24. Lyden P, Shuaib A, Ng K, Levin K, et al. Clomethiazole acute stroke study in ischemic stroke. Final results. Stroke. 2002;33:122–9.

25. Garvey JE, Marschall EA, Wright RA. From star chars to stoenflies: detecting relationships in continuous bivariate data. Ecology. 1998;79:442–7.

26. Bushnell CD. Stroke and the female brain. Nat Clin Pract Neurol. 2008;4(1):22–33.

27. Kasner SE. Clinical interpretation and use of stroke scales. Lancet Neurol. 2006;5:603–12.

28. Shaltoni HM, Albright KC, Gonzales NR, Weir RU, et al. Is intra-arterial thrombolysis safe after full-dose intravenous recombinant tissue plasminogen activator for acute ischemic stroke? Stroke. 2007;38:80–4.

29. Rankin J. Cerebral vascular accidents in patients over the age of 60 II. Prognosis. Scott Med J. 1957;2(5):200–15.

30. Quinn TJ, Dawson J, Walters M. Dr John Rankin; his life, legacy and the 50th anniversary of the Rankin Stroke Scale. Scott Med J. 2008;53(1):44–7.

31. Duncan PW, Sullivan KJ, Behrman AL, Azen SP, Wu SS, Nadeau SE, et al. Protocol for the locomotor experience applied post-stroke (LEAPS) trial: a randomized controlled trial. BMC Neurol. 2007;7:39.

32. Farrell B, Godwin J, Richards S, Warlow C. The United Kingdom transient ischaemic attack (UK-TIA) aspirin trial: final results. J Neurol Neurosurg Psychiatry. 1991;54(12):1044–54.

33. Quinn TJ, Dawson J, Walters MR, Lees KR. Functional outcome measures in contemporary stroke trials. Int J Stroke. 2009;4(3):200–5.

34. van Swieten JC, Koudstaal PJ, Visser MC, Schouten HJ, van Gijn J. Interobserver agreement for the assessment of handicap in stroke patients. Stroke. 1988;19(5):604–7.

35. Wilson JT, Hareendran A, Grant M, Baird T, Schulz UG, Muir KW, et al. Improving the assessment of outcomes in stroke: use of a structured interview to assign grades on the modified Rankin scale. Stroke. 2002;33(9):2243–6.

36. Wilson JT, Hareendran A, Hendry A, Potter J, Bone I, Muir KW. Reliability of the modified Rankin scale across multiple raters: benefits of a structured interview. Stroke. 2005;36(4):777–81.

37. Quinn TJ, Dawson J, Walters MR, Lees KR. Exploring the reliability of the modified Rankin scale. Stroke. 2009;40(3):762–6.

38. Quinn TJ, Dawson J, Walters MR, Lees KR. Variability in modified Rankin scoring across a large cohort of international observers. Stroke. 2008;39(11):2975–9.

39. Lai SM, Duncan PW. Stroke recovery profile and the modified Rankin assessment. Neuroepidemiology. 2001;20(1):26–30.

40. Mahoney FI, Barthel DW. Functional evaluation: the Barthel Index. Md State Med J. 1965;14:61–5.

41. Kwon S, Hartzema AG, Duncan PW, Min-Lai S. Disability measures in stroke: relationship among the Barthel Index, the functional independence measure, and the modified Rankin scale. Stroke. 2004;35(4):918–23.

42. Young FB, Lees KR, Weir CJ. Strengthening acute stroke trials through optimal use of disability end points. Stroke. 2003;34(11):2676–80.

43. Balu S. Differences in psychometric properties, cut-off scores, and outcomes between the Barthel Index and modified Rankin scale in pharmacotherapy-based stroke trials: systematic literature review. Curr Med Res Opin. 2009;25(6):1329–41.

44. Wright J, Bushnik T, O'Hare P. The center for outcome measurement in brain injury (COMBI): an internet resource you should know about. J Head Trauma Rehabil. 2000;15(1):734–8.

45. Sullivan KJ, Tilson JK, Cen SY, Rose DK, Hershberg J, Correa A, et al. Fugl-Meyer assessment of sensorimotor function after stroke: standardized training procedure for clinical practice and clinical trials. Stroke. 2011;42(2):427–32.

46. Chollet F, Tardy J, Albucher JF, Thalamas C, Berard E, Lamy C, et al. Fluoxetine for motor recovery after acute ischaemic stroke (FLAME): a randomised placebo-controlled trial. Lancet Neurol. 2011;10(2):123–30.

47. Lai SM, Perera S, Duncan PW, Bode R. Physical and social functioning after stroke: comparison of the Stroke Impact Scale and Short Form-36. Stroke. 2003;34(2):488–93.

48. The field administration of stroke therapy: magnesium phase 3 clinical trial. http://www.fastmag.info/. Accessed 19 Jan, 2012.

49. The Publications Committee for the Trial of ORG 10172 in Acute Stroke Treatment (TOAST) Investigators. Low molecular weight heparinoid, ORG 10172 (danaparoid), and outcome after acute ischemic stroke: a randomized controlled trial. JAMA. 1998;279(16):1265–72.

50. Teasdale GM, Pettigrew LE, Wilson JT, Murray G, Jennett B. Analyzing outcome of treatment of severe head injury: a review and update on advancing the use of the Glasgow outcome scale. J Neurotrauma. 1998;15(8):587–97.

51. Cooper DJ, Rosenfeld JV, Murray L, Arabi YM, Davies AR, D'Urso P, et al. Decompressive craniectomy in diffuse traumatic brain injury. N Engl J Med. 2011;364(16):1493–502.

52. Roozenbeek B, Lingsma HF, Perel P, Edwards P, Roberts I, Murray GD, et al. The added value of ordinal analysis in clinical trials: an example in traumatic brain injury. Crit Care. 2011;15(3):R127.

53. Jennett B, Bond M. Assessment of outcome after severe brain damage. Lancet. 1975;7905: 480–4.

54. Krams M, Lees KR, Hacke W, Grieve AP, Orgogozo JM, Ford GA. Acute stroke therapy by inhibition of neutrophils (ASTIN): an adaptive dose–response study of UK-279,276 in acute ischemic stroke. Stroke. 2003;34(11):2543–8.

55. Woo D, Broderick JP, Kothari RU, Lu M, Brott T, Lyden PD, NINDS t-PA Stroke Study Group, et al. Does the National Institutes of Health Stroke Scale favor left hemisphere strokes? Stroke. 1999;30(11):2355–9.

56. Sato S, Toyoda K, Uehara T, Toratani N, Yokota C, Moriwaki H, et al. Baseline NIH stroke scale score predicting outcome in anterior and posterior circulation strokes. Neurology. 2008;70(24 Pt 2):2371–7.

57. Lyden P, Claesson L, Havstad S, Ashwood T, Lu M. Factor analysis of the National Institutes of Health Stroke Scale in patients with large strokes. Arch Neurol. 2004;61(11):1677–80.

58. Hacke W, Kaste M, Fieschi C, von Kummer R, Davalos A, Meier D, et al. Randomised double-blind placebo-controlled trial of thrombolytic therapy with intravenous alteplase in acute ischaemic stroke (ECASS II). Second European-Australasian Acute Stroke Study Investigators. Lancet. 1998;352(9136):1245–51.

59. Juttler E, Schwab S, Schmiedek P, Unterberg A, Hennerici M, Woitzik J, et al. Decompressive surgery for the treatment of malignant infarction of the middle cerebral artery (DESTINY): a randomized, controlled trial. Stroke. 2007;38(9):2518–25.

60. Vahedi K, Vicaut E, Mateo J, Kurtz A, Orabi M, Guichard JP, et al. Sequential-design, multi-center, randomized, controlled trial of early decompressive craniectomy in malignant middle cerebral artery infarction (DECIMAL trial). Stroke. 2007;38(9):2506–17.

61. Tilley BC, Marler J, Geller NL, Lu M, Legler J, Brott T, et al. Use of a global test for multiple outcomes in stroke trials with application to the National Institute of Neurological Disorders and Stroke t-PA Stroke Trial. Stroke. 1996;27(11):2136–42.

62. Abciximab Emergent Stroke Treatment Trial (AbESTT) Investigators. Emergency administration of abciximab for treatment of patients with acute ischemic stroke: results of a randomized phase 2 trial. Stroke. 2005;36(4):880–90.

63. Saver JL, Yafeh B. Confirmation of tPA treatment effect by baseline severity-adjusted end point reanalysis of the NINDS-tPA stroke trials. Stroke. 2007;38(2):414–6.

64. Points to consider on clinical investigation of medicinal products for the treatment of acute stroke. The European agency for the evaluation of medicinal products. 2001. http://www.ema.europa.eu/docs/en_GB/document_library/Scientific_guideline/2009/09/WC500003342.pdf. Accessed 19 Jan, 2012.

65. Lees KR, Zivin JA, Ashwood T, Davalos A, Davis SM, Diener HC, et al. NXY-059 for acute ischemic stroke. N Engl J Med. 2006;354(6):588–600.

66. Shuaib A, Lees KR, Lyden P, Grotta J, Davalos A, Davis SM, et al. NXY-059 for the treatment of acute ischemic stroke. N Engl J Med. 2007;357(6):562–71.

67. Saver JL. Novel end point analytic techniques and interpreting shifts across the entire range of outcome scales in acute stroke trials. Stroke. 2007;38(11):3055–62.

68. Savitz SI, Lew R, Bluhmki E, Hacke W, Fisher M. Shift analysis versus dichotomization of the modified Rankin scale outcome scores in the NINDS and ECASS-II trials. Stroke. 2007;38(12):3205–12.

69. Mishra NK, Lyden P, Grotta JC, Lees KR. Thrombolysis is associated with consistent functional improvement across baseline stroke severity: a comparison of outcomes in patients from the Virtual International Stroke Trials Archive (VISTA). Stroke. 2010;41(11):2612–7.

70. Bruno A, Saha C, Williams LS. Using change in the National Institutes of Health Stroke Scale to measure treatment effect in acute stroke trials. Stroke. 2006;37(3):920–1.

71. Bath PM, Gray LJ, Collier T, Pocock S, Carpenter J. Can we improve the statistical analysis of stroke trials? Statistical reanalysis of functional outcomes in stroke trials. Stroke. 2007;38(6):1911–5.

72. Saver JL, Gornbein J. Treatment effects for which shift or binary analyses are advantageous in acute stroke trials. Neurology. 2009;72(15):1310–5.

73. Saver JL. Optimal end points for acute stroke therapy trials: best ways to measure treatment effects of drugs and devices. Stroke. 2011;42(8):2356–62.

74. Shannon CE. A mathematical theory of communication. Bell Sys Tech J. 1948;27:379–423. http://www.alcatel-lucent.com/bstj/vol27-1948/articles/bstj27-4-623.pdf. Accessed 19 Jan, 2012.

75. Altman DG, Royston P. The cost of dichotomising continuous variables. BMJ. 2006;332(7549):1080.

76. Federov V, Mannino F, Zhang R. Consequences of dichotomization. Pharm Stat. 2009;8:50–61.

77. Howard G. Nonconventional clinical trial designs: approaches to provide more precise estimates of treatment effects with a smaller sample size, but a cost. Stroke. 2007;38:804–8.

78. Hall CE, Mirski M, Palesch YY, Diringer MN, Qureshi AI, Robertson CS, et al. First neurocritical care research conference investigators. Clinical trial design in the neurocritical care unit. Neurocrit Care. 2012;16(1):6–19.

79. Bross IDJ. How to use RIDIT analysis. Biometrics. 1958;14:18–38.

80. Krumpelman CS, Mandava P, Kent TA. Error rate estimates for the modified Rankin Score shift analysis using information theory modeling. International Stroke Conference 2012. Stroke. 43:P290.

81. The Food and Drug Administration. Guidance for industry. Adaptive design clinical trials for drugs and biologics. 2010. http://www.fda.gov/downloads/Drugs/GuidanceCompliance RegulatoryInformation/Guidances/ucm201790.pdf. Accessed 19 Jan, 2012.

82. Elkind MS, Sacco RL, MacArthur RB, Fink DJ, Peerschke E, Andrews H, et al. The neuro-protection with Statin Therapy for Acute Recovery Trial (NeuSTART): an adaptive design phase I dose-escalation study of high-dose lovastatin in acute ischemic stroke. Int J Stroke. 2008;3(3):210–8.

83. Ginsberg MD, Palesch YY, Martin RH, Hill MD, Moy CS, Waldman BD, et al. The albumin in acute stroke (ALIAS) multicenter clinical trial: safety analysis of part 1 and rationale and design of part 2. Stroke. 2011;42(1):119–27.

84. Howard G, Coffey CS, Cutter GR. Is Bayesian analysis ready for use in phase III randomized clinical trials? Beware the sound of the sirens. Stroke. 2005;36(7):1622–3.

85. Mandava P, Kent TA. A method to determine stroke trial success using multidimensional pooled control functions. Stroke. 2009;40(5):1803–10.

86. Uchino K, Billheimer D, Cramer SC. Entry criteria and baseline characteristics predict out-come in acute stroke trials. Stroke. 2001;32(4):909–16.

87. Zar JH. Biostatistical analysis. 3rd ed. Upper Saddle: Prentice Hall; 1996. p. 282–3.

88. Lampl Y, Zivin JA, Fisher M, Lew R, Welin L, Dahlof B, et al. Infrared laser therapy for ischemic stroke: a new treatment strategy: results of the NeuroThera Effectiveness and Safety Trial-1 (NEST-1). Stroke. 2007;38(6):1843–9.

89. Alexandrov AV, Molina CA, Grotta JC, Garami Z, Ford SR, Alvarez-Sabin J, et al. Ultrasound-enhanced systemic thrombolysis for acute ischemic stroke. N Engl J Med. 2004;351: 2170–8.

90. Ogawa A, Mori E, Minematsu K, Taki W, Takahashi A, Nemoto S, et al. Randomized trial of intra-arterial infusion of urokinase within 6 hours of middle cerebral artery stroke: the middle cerebral artery embolism local fibrinolytic intervention trial (MELT) Japan. Stroke. 2007;38(10):2633–9.

91. Lehman EL, D'Abrera HJM. Blocked comparisons for two treatments. Chapter 3 in Nonparametrics. Statistical methods based on ranks. San Francisco: Holden-Day Inc; 1975. p. 120–45.

92. Koziol JA, Feng AC. On the analysis and interpretation of outcome measures in stroke clini-cal trials: lessons from the SAINT I study of NXY-059 for acute ischemic stroke. Stroke. 2006;37(10):2644–7.

93. Koziol JA, Feng AC. On the analysis and interpretation of outcome measures in stroke clini-cal trials: lessons from the SAINT I study of NXY-059 for acute ischemic stroke. Response to letter by Saver. Stroke. 2007;38:258.

94. Adams Jr HP, Effron MB, Torner J, Dávalos A, Frayne J, Teal P, AbESTT-II Investigators, et al. Emergency administration of abciximab for treatment of patients with acute ischemic stroke: results of an international phase III trial: Abciximab in Emergency Treatment of Stroke Trial (AbESTT-II). Stroke. 2008;39(1):87–99.

95. Mandava P, Dalmeida W, Anderson JA, Thiagarajan P, Fabian RH, Weir RU, et al. A Pilot trial of low-dose intravenous abciximab and unfractionated heparin for acute ischemic stroke: translating GP IIb/IIIa receptor inhibition to clinical practice. Transl Stroke Res. 2010;1:170–7.

96. Clark WM, Wissman S, Albers GW, Jhamandas JH, Madden KP, Hamilton S. Recombinant tissue-type plasminogen activator (alteplase) for ischemic stroke 3 to 5 hours after symptom onset: the ATLANTIS study: a randomized controlled trial: alteplase thrombolysis for acute noninterventional therapy in ischemic stroke. JAMA. 1999;282(21):2019–26.

97. Bergstralh EJ, Kosanke JL. Computerized matching of cases to controls. Technical report 56. http://www.mayoresearch.mayo.edu/mayo/research/biostat/upload/56.pdf. Accessed 19 Jan, 2012.

98. Mandava P, Sarma AK, Martini SR, Kent TA. Evaluation of subject matching methods to adjust for imbalances in stroke trials. (Submitted).

99. NIST/SEMATECH e-Handbook of statistical methods. http://www.itl.nist.gov/div898/hand-book/. Accessed 19 Jan, 2012.

100. Egorova N, Giacovelli J, Greco G, Gelijns A, Kent CK, McKinsey JF. National outcomes for the treatment of ruptured abdominal aortic aneurysm: comparison of open versus endovascular repairs. J Vasc Surg. 2008;48(5):1092–100, 100 e1–2.
101. Black PE. Manhattan distance, in dictionary of algorithms and data structures (online). In: Black PE, editors. U.S. National Institute of Standards and Technology. http://www.nist.gov/dads/HTML/manhattanDistance.html. Accessed 31 May 2006.
102. Mandava P, Brooks M, Krumpelman C, Kent TA. A new more sensitive method to assess balance among stroke trial populations. International Stroke Conference 2012. Stroke. 43:P295.
103. Peacock JA. Two-dimensional goodness-of-fit testing in astronomy. Roy Astron Soc. 1983;202:615–27.
104. Fasano G, Franceschini A. A multidimensional version of the Kolmogorov-Smirnov test. Roy Astron Soc. 1987;225:155–70.
105. Lapchak PA. A critical assessment of edaravone acute ischemic stroke efficacy trials: is edaravone an effective neuroprotective therapy? Expert Opin Pharmacother. 2010;11(10):1753–63.

Chapter 41
Metabolic Imaging in Translational Stroke Research

Krishna A. Dani and Keith W. Muir

Abstract Recent technological developments have improved understanding of not only the structural changes following cerebral arterial occlusion, but have also highlighted the dynamic evolution of stroke pathophysiology. Although altered cerebral tissue metabolism is a central feature of such changes following stroke, direct imaging correlates of metabolic activity are currently not used for therapeutic decision making. However, metabolic imaging may potentially improve targeting of therapies to those patients with a relevant tissue substrate. In this chapter we discuss imaging techniques which may provide metabolic information in acute stroke. Although positron emission tomography is the gold standard technique for metabolic imaging in acute stroke in a research environment, magnetic resonance techniques such as spectroscopy, ^{17}O imaging, and deoxyhaemoglobin weighted imaging may have have potential clinical utility. However, despite measuring relative concentrations of metabolites directly, the application of magnetic resonance spectroscopy has been limited by issues surrounding quantification and signal to noise ratio. ^{17}O and deoxyhemoglobin weighted imaging are promising, but require to be validated in acute stroke. In this chapter we discuss potential future directions of these metabolic imaging in translational stroke research.

1 Introduction

To date, the major aims of imaging in acute ischaemic stroke have been to exclude intracranial haemorrhage and major established infarction prior to treatment decisions such as thrombolytic therapy administration, and to some extent to exclude

K.A. Dani, MRCP (✉) • K.W. Muir, MD, FRCP
Institute of Neuroscience and Psychology, University of Glasgow,
Glasgow, G12 8QB, Scotland
e-mail: krishna.dani@glasgow.ac.uk

P.A. Lapchak and J.H. Zhang (eds.), *Translational Stroke Research*, Springer Series
in Translational Stroke Research, DOI 10.1007/978-1-4419-9530-8_41,
© Springer Science+Business Media, LLC 2012

those stroke mimics that have positive imaging features. Although the benefit of reperfusion therapies is generally accepted to derive from the restoration of cerebral blood flow (CBF) to regions of "at risk" tissue, termed the ischaemic penumbra, current guidelines are constrained by the imaging that was employed in the pivotal randomised, controlled clinical trials that constitute the evidence base for such treatment, and do not require positive identification of such tissue. Instead, a "time since onset" threshold is conventionally employed in the selection of patients for this therapy; administration of therapy to subjects within 4.5 h of onset, and for whom intra-cerebral haemorrhage (ICH) was excluded by CT was proven to be effective in the ECASS III trial [1]. The application of a "time window" is biologically plausible, since the proportion of subjects with salvageable tissue steadily declines with time. However, the sensitivity and specificity of this approach could be improved; tissue with imaging features consistent with the ischaemic penumbra may exist for up to 48 h [2], or may be entirely absent by 3 h. Indeed, the lack of requirement for positive identification of the target tissue simply reflects the technology available in the early 1990s when the seminal stroke thrombolytic trial (NINDS [3]) was conducted; although positron emission tomography (PET) could demonstrate the metabolic penumbra, logistical issues meant that CT was the only feasible imaging modality for evaluation of subjects in the hyperacute phase of stroke at that time. However, since the publication of the NINDS trial, there have been significant advances in the imaging field. First, the rate of progress in technological advances has been rapid, particularly with respect to structural and functional magnetic resonance imaging techniques. Second, the understanding of appropriate definitions for irreversibly infarcted tissue (core) and ischaemic penumbra with MRI and multimodal CT, and the potential impact on treatment selection, has improved [4, 5]. Multi-modal magnetic resonance imaging including diffusion weighted imaging (DWI) and perfusion weighted imaging (PWI) and the application of the perfusion–diffusion mismatch concept [6] as a surrogate marker for the penumbra have provided much of these data, but despite encouraging signals from cohort and observational studies, there are no definitive data to support the superiority of penumbral imaging in patient selection. Only one clinical trial has explicitly set out to evaluate the hypothesis that diffusion–perfusion mismatch was associated with treatment response to intra-venous rtPA (EPITHET), and this was unable to confirm the hypothesis due to a combination of small patient numbers and sub-optimal analytical approaches.

One possible explanation for the failure to confirm the mismatch hypothesis is that despite providing valuable data, including haemodynamics, these techniques do not provide direct metabolic information. Given that the penumbra has traditionally been defined using metabolic indices such as the cerebral metabolic rate for oxygen ($CMRO_2$) and oxygen extraction fraction (OEF), metabolic data may further improve the definition of the MRI penumbra and improve patient selection for therapies. With respect to patient selection, new techniques may be judged by new goals for imaging. First, can we define a favourable substrate for reperfusion therapy? Second, can we identify profiles which identify increased the risk of complications of reperfusion therapies. Third, is implementation feasible in acutely unwell patients? Finally, from a research perspective, can advanced imaging techniques be of use as biomarkers and surrogate end points in clinical trials?

In this chapter, we review novel metabolic imaging techniques and discuss their potential utility in the hyperacute ischaemic stroke setting. The chapter begins by reviewing the data from gold standard penumbral imaging technique (PET) before considering novel MRI techniques. Although potentially very important, we do not consider techniques which image the consequences of disturbed tissue metabolism, such as sodium imaging or DWI, nor consider techniques which may elicit other vital aspects of pathophysiology, such as perfusion weighted imaging, permeability imaging, pH imaging, or carotid plaque imaging, since these techniques do not directly measure tissue metabolism, and therefore, discussion of these techniques is beyond the remit of this chapter.

2 Positron Emission Tomography

Data from PET is derived from the simultaneous detection of pairs of photons which are emitted after the collision and subsequent annihilation of a positron emitted from a decaying radioligand with an electron. This allows spatial localisation of the source of the annihilation, which may be as little as 3 mm from the nucleus, the latter distance being the factor which determines spatial resolution. Data can then be modelled to provide quantitative measures of physiological parameters by assuming the kinetics of a freely diffusible tracer. The physiological data modelled depend on the radiotracer used. For example, ^{18}F-deoxyglucose may be used to study glucose metabolism, and multi-tracer ^{15}O may be used to study oxidative metabolism with quantitative measures of oxygen consumption (cerebral metabolic rate of oxygen, $CMRO_2$) and OEF, CBF and cerebral blood volume (CBV) being provided. ^{15}O labelled water may be used to provide measures of CBF, ^{15}O labelled carbon monoxide/dioxide for CBV, and O-15-O for OEF. These data can then be used to compute $CMRO_2$ Multi-tracer ^{15}O labelled PET remains the gold standard imaging technique for the ischaemic penumbra. Although logistic limitations such as arterial puncture, relatively long imaging time (about 1 h) and scarce PET scanner and cyclotron availability have precluded its clinical use in acute stroke imaging, the insights provided from PET have been invaluable.

Human PET studies have complemented animal work in the determination that the ischaemic penumbra is present in man. Using PET, Marchal and colleagues demonstrated that a combination of CBF and $CMRO_2$ imaging identified three distinct clinical patterns. Simultaneously decreased CBF and decreased $CMRO_2$ signifying irreversible tissue damage, was associated with uniformly poor clinical outcome. Conversely, preserved CBF and $CMRO_2$ were associated with uniformly good clinical outcome. The "penumbral" pattern, of reduced CBF but preserved $CMRO_2$, was associated with variable clinical outcome [7]. Moreover, PET studies have provided data on the duration of existence of the penumbra, with some compromised but viable tissue being present for up to 48 h post ictus [2]. More importantly, PET data confirmed the hypothesis that penumbral tissue could be genuinely salvaged with clinical benefit, by showing that the volume of salvaged penumbral tissue was correlated to clinical improvement [8]. Therefore, PET studies not only confirmed that penumbra could be detected in man, but suggested that reperfusion therapies may offer a clinical benefit in human acute stroke. Currently accepted PET criteria for penumbra include a preserved $CMRO_2$, OEF >0.7 and CBF in the penumbra range of approximately 7–22 mL/100 g/min [9].

As a result of the practical difficulties associated with acquisition and analysis of ^{15}O multitracer PET, other ligands have been investigated as penumbral markers. ^{18}F fluoromisonidazole (F-MISO) is a compound which is selectively trapped by hypoxic tissues which still retain some metabolic activity and is therefore an attractive candidate for use as a penumbral marker. Using this compound as a PET tracer, it has been shown that there is intracellular trapping within 48 h of stroke onset in human subjects, but not beyond, consistent with the known temporal profile of the penumbra [10]. This technique has also allowed tracking of the evolution of the penumbra, whereby tracer binding was found in the centre of the stroke region in hyperacute patients imaged at less than 6 h post stroke, compared to later time points when such trapping was found predominantly in the periphery of the lesions, allowing conclusions that the evolution of penumbra in man is analogous to that in experimental models [11]. Animal models have confirmed these findings and have allowed for a greater understanding of the technique including aspects concerning the timing of binding of the tracer [12, 13]. However, recent animal data suggest that binding of tracer occurs not only in penumbral tissue but also in regions that retain some metabolic activity but which are destined to die ("early infarct core"), suggesting that further validation of this technique is required before it can be considered a truly "penumbral" imaging technique [13].

In addition, just as F-MISO binds to hypoxic tissue and may potentially identify the penumbra, the central benzodiazepine receptor ligand ^{11}C-flumazenil may distinguish cortical core tissue from other tissue compartments, since it binds only to viable cortical tissue. Whilst preliminary studies in humans with acute ischaemic stroke suggested that ^{13}C-flumazenil may predict morphological outcome of the stroke [14], such studies have been limited by small numbers of subjects and hetero-geneous patterns and evolution of tissue compartments. A recent rodent study has questioned the sensitivity in identifying early infarction. No decrease in ^{11}C-flumazenil binding was detected up to 24 h in a transient occlusion-reperfusion model of rodent stroke, despite cortical and striatal infarction [15]. Therefore, for both radiolabelled fluoromisonidazole and flumazenil, animal studies have highlighted potential limitations in interpretation in human stroke.

Does PET have a direct role in clinical trials? PET is impractical in the acute setting and therefore could only be considered in stable populations. Even with this is mind, the limited availability, even in well-resourced centres, and the exposure to low doses of radiation, impose both practical and ethical restrictions on the widespread implementation of this modality in the clinical trial setting. The relevance to clinical trials, therefore, is likely to be indirect, for example as a validation for other tools which may ultimately be used in the determination of end points.

3   Magnetic Resonance Imaging

3.1   Magnetic Resonance Spectroscopy

Magnetic resonance spectroscopy (MRS) is a modality which measures the relative concentration of metabolites directly, and thus is potentially well placed to provide metabolic data in stroke research studies. As with other MRI techniques, clinical MRS has traditionally measured the signal from protons because of the abundance of water, but others such as carbon-13, have been investigated. MRS exploits the principle of chemical shift, whereby the nucleus of a given metabolite with unpaired electrons is shielded from the main magnetic field (B_0) by the "electron shield" which is generated as a consequence of the negative charge of the electrons in a magnetic field. This electron shield lowers the "effective field" experienced by the nucleus which therefore lowers the resonance frequencies resulting in the phenomenon of chemical shift. Given that the nuclei of different metabolites are surrounded by different chemical environments, they therefore experience different magnitudes of shielding from electrons, and can therefore be identified by their characteristic chemical shift. This chemical shift is expressed in parts per million (PPM) on an abscissa which traditionally runs from right to left (Fig. 41.1). Confusingly, the "PPM scale" does not refer to a concentration, but rather is an expression of the chemical shift independent of B_0. Most data from stroke studies have been derived from a single voxel placed over a region on interest, for example an infarct (SVS, single voxel spectroscopy), but this approach is insensitive to the spatial heterogeneity of stroke. With improvements in technology, more recent studies have reported individual spectra from multiple voxels (MVS, multi-voxel spectroscopy) selected from a grid. Therefore, MRS is not, strictly speaking, an imaging technique. However, for pictorial reasons, an "image" of metabolite concentrations may be produced from MVS. Whilst this approach gives a convenient summary map, detailed inspection of individual spectra which contribute to the image is still mandatory.

With the continued escalation of possible field strengths, an increasing number of metabolites may now be imaged, especially with the use of spectral editing techniques. However, even the standard metabolites measured by standard clinical scanners and post-processed by most commercial software packages are of potential relevance to stroke and its research needs.

3.2   What Happens to Metabolites in Stroke?

The "major" metabolites which may be derived from standard MRS packages are N-acetyl-L-aspartate (NAA) (marker of neuronal integrity), lactate (anaerobic glycolysis), choline (cell membrane marker) and creatine (marker of energy stores). The temporal profile of the major metabolite changes after stroke has been well established by both animal and human studies. Indeed, this has been the major focus of many

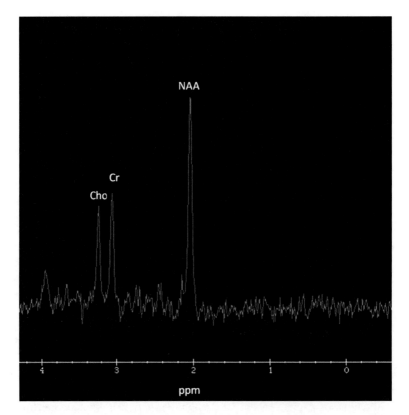

Fig. 41.1 Spectra from a voxel placed on "normal" tissue from a patient at the Institute of Neurological Sciences Glasgow. (Kindly provided by John McLean, Clinical Physics) The *x*-axis runs from *right* to *left*. Peaks above the baseline indicate measurable metabolite concentrations. *Cho* choline; *Cr* Creatine; *NAA* N-acetyl-L-aspartate; *ppm* parts per million

of the MRS studies in stroke. Both human and animal studies have confirmed that NAA and creatine levels diminish after stroke [16–18], lactate levels rise after stroke [16, 19] and the changes in choline levels are more variable [17, 18, 20] (Fig. 41.2). Most human studies have been performed at time points beyond the acute phase, and therefore, these findings have not yet provided much diagnostic utility, since established infarction can be more readily delineated on non contrast CT and standard MR sequences. However, reports that NAA levels decrease steadily over the 2 weeks following stroke onset [17, 21] have given rise to recent suggestions that there may be a prolonged window for intervention for therapies other than those aimed at reperfusion. At the other end of the spectrum, animal studies have allowed insights to changes in metabolite levels in the hyperacute phase. In such studies, lactate levels have been reported to rise within minutes [16], and although NAA levels may also decrease in minutes [16], such decreases in the hyperacute phase are minor. These observations have given rise to the hypothesis that the penumbra may be defined by a "normal" NAA level and elevated lactate level [22], but to date this has not been

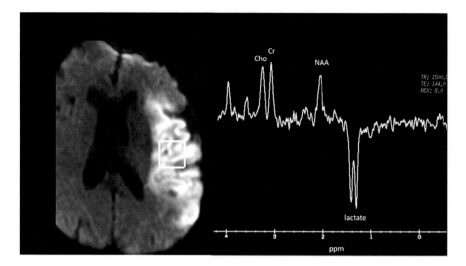

Fig. 41.2 Spectra from a voxel placed over a region of tissue infarction from a patient from the Institute of Neurological Sciences Glasgow. (Kindly provided by Mr. John McLean, Clinical Physics) The brain scan shows a hyperintense lesion on diffusion weighted imaging (DWI) indicating cytotoxic oedema. A voxel is placed over this region. The resulting spectrum is shown on the *left*. Note the reduction in NAA and the appearance of lactate, with its characteristic doublet peak extending below the baseline, acquired with an intermediate echo time (TE). The *x*-axis runs from *right* to *left*. Peaks above the baseline indicate measurable metabolite concentrations. Cho choline; *Cr* Creatine; *NAA* N-acetyl-L-aspartate; *ppm* parts per million

confirmed. Given the limitations of MRS imaging, it is unlikely that this profile alone could ever be used to identify penumbra and select patients for clinical trials or treatment. Used in combination with other advanced imaging techniques such as PWI–DWI, it could potentially refine the MRI definition of penumbra and aid patient selection for therapy and trials. Indeed, it has been shown that a rise in lactate can precede the appearance of the DWI lesion in hypoperfused regions in human stroke [23].

3.3 MRS in Stroke Research: Potential Pitfalls

Despite the years of research, and many promising signals of utility, MRS has so far failed to find its niche in the evaluation of stroke, in both clinical and research settings. Although absolute quantification is theoretically possible, for example by using water as a reference signal, these approaches require a number of assumptions and are not routinely offered by standard commercial MRS packages. Therefore, most studies to date have reported metabolite values as a ratio to another metabolite within the same voxel. Although this approach would be feasible in a disease state where only one metabolite is expected to change, it produces unclear results when used after stroke, where all major metabolites may change concentration. Using ratios of a metabolite compared to the same metabolite in the contralateral hemisphere is an alternative option, but requires a completely homogeneous magnetic field, and identical tissue composition for accurate results. Other problems in stroke are related to motion artefacts. Subjects with stroke are often clinically unstable, have communication impaired, and are liable to move within the magnet. Not only may this degrade the signal quality from MRS, but it can also introduce signal contamination from scalp lipid. This can interfere with the interpretation of the spectra, particularly the lactate peak. Indeed, MRS data may frequently reflect a composite measure of signal from tissue parenchyma, lipid and cerebrospinal fluid (CSF).

3.4   MRS in Clinical Trials

If MRS is to find its niche, it should complement other advanced MRI techniques such as DWI, PWI and fluid attenuated inversion recovery (FLAIR) sequences. MRS is not as accurate as DWI in the delineation of cytotoxic oedema and is unlikely, by itself, to be able to accurately delineate the ischaemic penumbra. However, used with PWI and DWI, it offers potential to more accurately determine DWI lesions which are truly core (possibly using NAA), and which PWI–DWI mismatch regions are truly penumbra (possibly using lactate). This may help to refine clinical trial design when the evaluation of the efficacy of a therapy, such as reperfusion therapy, depends on the relative and absolute volumes of core and penumbra. In this way, MRS still offers potential for use in subject selection.

In addition to subject selection, MRS still offers potential for use as a biomarker in clinical trials where the intervention is postulated to influence a specific metabolic marker. The association of hyperglycaemia with poor outcome after stroke has been widely and repeatedly demonstrated. Associations with acute infarct evolution are supported by greater infarct growth being correlated with sustained elevation of glucose [24], higher probability of haemorrhagic transformation after intravenous thrombolysis in the presence of early hyperglycaemia, and reduced probability of penumbral rescue that correlates with blood glucose concentration [25] A leading mechanistic hypothesis for the detrimental effect of hyperglycaemia on infarct evolution is that excess substrate supply to tissue with insufficient oxygen generates lactic acid as a consequence of anaerobic glycolysis, and that lactic acidosis is neurotoxic. Some support for this was derived from a study that confirmed a relationship between blood glucose concentration and lactate levels on single voxel MRS, where the voxel was placed in the infarct core (DWI lesion) [25]. Following this, spectroscopy was employed as a biomarker in the Spectroscopic Evaluation of Lesion Evolution in Stroke: Trial of Insulin for Acute Lactic Acidosis (SELESTIAL) study [26], in which lactate levels were measured in hyperglycaemic patients randomised to an insulin infusion regime or control. Although brain lactate levels were significantly lower in subjects who received glucose–potassium–insulin infusions compared to control subjects, this was not reflected in attenuation of infarct growth.

This study is not just an example of the use of MRS in clinical trials, but is also an example of how MRS can aid our understanding of clinically relevant concepts of pathophysiology in acute stroke. With the expanding number of metabolites which may be measured, other metabolic effects may become amenable to direct imaging. For example, one of the limitations of the neuroprotectant trials is that an effect on the target substrate could never be demonstrated [27]. With the advent of the ability to detect glutathione [28, 29], and detect changes in concentration after stroke, MRS offers an attractive tool for use as an endpoint in potential therapies which aim to have an effect on redox status. These, and similar approaches, could potentially be employed in early phase clinical trials.

3.5 Imaging of Oxidative Metabolism Using O^{17}

Although conventional proton MRS is entirely non-invasive, the inhalation of $^{17}O_2$ gas is also being investigated. ^{17}O is a stable isotope of oxygen and is present in low levels in the blood and may be used to investigate oxidative metabolism. ^{17}O studies have been classified as either "direct" where spectroscopy detects the signal from ^{17}O in $H_2{}^{17}O$, or indirect methods which detect T2 changes of water through the coupling of protons (^1H) and ^{17}O. The use of ^{17}O in the determination of oxidative metabolism is analogous to that of ^{15}O in PET, but has additional theoretical advantages. After ^{17}O is inhaled, it remains MR invisible whilst part of the oxyhaemoglobin complex in the blood, and also when it dissociates from haemoglobin and diffuses across the blood–brain barrier to enter the mitochondria. It only becomes MR visible when in the form of $H_2{}^{17}O$, when water is produced as part of the terminal event of oxidative phosphorylation. This gives an accurate measurement of water production consequent upon oxidative metabolism, which is not confounded by detection of the ^{17}O isotope at other stages of the metabolic pathway, unlike in PET. Quantification is aided by the fact that signal intensities are proportional to the concentrations of the ^{17}O, and the natural abundance of $H_2{}^{17}O$ is known. Thereafter, determination of $CMRO_2$ may be calculated with the aid of mathematical modelling. Modelling has previously required invasive determination of CBF and the arterial input function for $H_2{}^{17}O$ to calculate $CMRO_2$. However, more recently, studies have suggested that non-invasive modelling with a number of assumptions may be reliable and robust [30].

A number of small animal studies have demonstrated the feasibility of measuring $CMRO_2$ and producing a map capable of demonstrating spatial heterogeneity in healthy animals [31, 32, 33]. Furthermore, using an ^{17}O injection technique, changes in $H_2{}^{17}O$ were demonstrated in rodents in a region of focal cerebral ischaemia, with a $CMRO_2$ map showing good spatial correlation with other MRI parameters [31]. In addition to the small animal data, large animal data have recently been reported, which took into account the recirculation delay seen in large but not small animals [34]. Using the "indirect" method, imaging of $CMRO_2$ in the swine was performed using clinical scanners and changes in metabolism precipitated by a mitochondrial uncoupling agent were detected.

With respect to human scanning, there have been limited studies to date. Two published studies have demonstrated the feasibility of this approach and have suggested methods to account for the spatial heterogeneity in oxygen consumption and brain tissue mass [35, 36]. These studies produced maps of $CMRO_2$ in humans and are therefore encouraging. In addition, it has been shown that [17]O studies do not cause adverse effects in human volunteers [37]. However, these techniques still require further validation and investigation in the context of cerebral ischaemia, which is lacking currently.

Despite the encouraging signals from these data, there are a number of limitations. First, [17]O gas is expensive. However, as use of [17]O increases this issue has the potential to diminish in importance, particularly as the isotope is relatively stable. In addition, cost-efficient administration systems are being developed [38]. Second, low signal to noise ratio is an issue, particularly with the "direct" methods, necessitating high magnetic fields and longer acquisition times. For example, a recent study in humans was performed at 9.4T [36]. However, even though such ultrahigh field strengths have shown to have considerable advantage compared to even 4.7T [39], a human study has been successfully been carried out at 1.5T [35].

In summary, the use of [17]O imaging offers an MR alternative to PET for the direct measurement of oxidative metabolism, and indeed may have several advantages over PET. However, despite 2 decades of research in animals, human studies are only just emerging. The techniques require robust validation in humans in healthy and disease states before they can be considered as a viable addition to the stroke research imager's armamentarium.

3.6   Deoxyhaemoglobin Weighted Imaging

In an effort to detect changes in oxidative metabolism there has been recent interest in techniques which are sensitive to changes in the concentration of deoxyhaemo-globin. MRI is ideally suited to such a task, since T2*-weighted MRI techniques are sensitive to paramagnetic compounds, which include deoxyhaemoglobin. The effect of deoxyhaemoglobin on the T2*-weighted signal was described by Ogawa and colleagues, who termed it the "Blood Oxygenation Level Dependent" effect [40]. It was first described in rodents, when it was noticed that cerebral veins appeared as dark lines when animals were breathing 20% oxygen, but these lines disappeared when the oxygen concentration was increased. An increase in CBF could also precipitate such an effect. It became clear that the effect was dependent on the balance between paramagnetic deoxyhaemoglobin and diamagnetic oxyhae-moglobin. Deoxyhaemoglobin generates magnetisation when placed in a magnetic field, such as an MRI scanner. This generates a susceptibility difference between vessels and parenchyma, which manifests as a reduction in T2*-weighted signal intensity in the vessel. It has been shown that T2*-weighted signal intensity is related to the deoxyhaemoglobin concentration. The relationship is quadratic, although at oxygen extraction fractions greater than 0.2 it has been concluded that a linear fit is reasonable [41–43]. These results are important, since they suggest that an MRI technique may be able to probe local oxygen dynamics. Given that the deoxyhaemoglobin concentration will vary according to OEF, and that OEF is intrinsically related to the interplay between CBF and $CMRO_2$, this technique has been considered to hold promise in providing insights into cerebral oxidative metabolism. It should be noted that the BOLD effect is also exploited by functional MRI techniques which will not be discussed here.

There have been a number of approaches to exploitation of the BOLD signal in the stroke field. First, T2*-weighted images may be interpreted qualitatively, in a manner analogous to other sequences such as DWI. Second, T2*-weighted data may be modelled to provide quantification. Finally, dynamic changes in T2*-weighted signal intensity may be provoked by stimuli, and the magnitude of change can be recorded.

All three possibilities have employed a translational approach, with animal experiments informing human studies and vice versa. Qualitative analysis of animal models of anoxia and middle cerebral artery occlusion (MCAO) have shown a temporary decline in T2*-weighted parenchymal signal followed by reversal and temporary overshoot when reperfusion has been established [44, 45, 46]. Animal studies have also highlighted potential complexities in the interpretation of the signal changes, such as regional heterogeneity of response [47] and differences between grey and white matter [44, 48]. Similar qualitative differences have been observed in human studies. These have included hypointensity of the ipsilateral hemisphere on T2*-weighted PWI sequences [49, 50]. Parenchymal hypointensity and vessel hypointensity is frequently seen on gradient echo sequences in hyperacute stroke, which are conventionally employed for the detection of ICH. However, it is unlikely that simple qualitative evaluation of T2*-weighted images will be sufficient to influence clinical decision-making. Geisler and colleagues [51] employed T2'-weighted sequences (T2* corrected for spin–spin [T2] effects) and showed that although there were signal differences between operationally defined penumbra, core and normal tissues, wide confidence intervals precluded the determination of discriminatory thresholds.

A separate research group have investigated the possibility of deriving measures of MRI derived cerebral metabolic rate of oxygen ($CMRO_2$). They exploited the "static dephasing regime" of Yablonskiy and Hacke [52] and employed a sequence with combined spin and gradient echo properties to model the T2'-weighted signal as dependent on oxygen extraction fraction, magnetic susceptibility of blood, venous CBV, haematocrit and static magnetic field. They modelled animal data to derive a value for OEF of around 40%, consistent with the literature values [53]. They also modelled data from human volunteers [54]. After deriving values for OEF, they used CBF values from other sequences, and could therefore derive a value for MRI derived $CMRO_2$—which they termed MR cerebral oxygen metabolism index (COMI). Since then, animal experiments have confirmed that the lowest values of MR COMI are in the MRI defined infarct core, and that the MR COMI values decline with time [55]. Human data have also been encouraging. In a study published in 2003 [56], which was of seven patients imaged at an average of 7.5 h post ictus, MR COMI values were also lowest in the MRI-defined core. Interestingly, in a region of matched PWI–DWI deficit, there was a gradient of values for MRI COMI, suggesting potential additional utility of this technique compared to conventional stroke MRI sequences. It should be noted that this research group focussed on the extra-vascular signal loss of T2*, and other groups have focussed on modelling the intravascular loss of T2-weighted signal, with similar results [57, 58]. This technique has recently suggested that OEF may be elevated in a proportion of subjects with carotid artery steno-occlusive disease, sometimes in the absence of elevated CBV [59]. Further work with these techniques is ongoing and will be required to demonstrate that values of OEF and $CMRO_2$ are robust and are accurate under all pathophysiological conditions.

An alternative approach which the authors have pursued is to use transient hyperoxia to yield a dynamic evaluation of oxygen metabolism. The technique assumes that in infarct core, there would be little or no oxygen extraction and therefore no significant change in deoxyhaemoglobin concentration in the vascular pool and therefore no signal change during increased oxygen administration. In healthy tissues, with an OEF of ~0.3–0.4, it was hypothesised there would be a T2*-weighted signal increase with supplemental oxygen, since the deoxyhaemoglobin produced after oxygen extraction by metabolising tissue would combine with oxygen forming oxyhaemoglobin, thereby eliminating deoxyhaemoglobin's paramagnetic effect, leading to an increase in T2*-weighted signal intensity. Finally, it was hypothesised that in penumbral tissue, where OEF is high, and CBV is elevated, we would see an exaggerated T2*-weighted signal increase. These hypotheses have been investigated in both animal and human studies.

Animal studies confirmed our hypotheses, with absent, intermediate and exaggerated T2*-weighted signal increases after a period of hyperoxia, being seen in core, healthy and penumbral tissue, respectively, in a model of permanent MCAO [60]. Further work in a rodent model has shown a rapid transition from the "penumbral" signature for oxygen challenge data to a response consistent with normal tissue, following reperfusion [61]. Furthermore, 2-deoxyglucose autoradiography has confirmed the presence of glucose metabolism within the normal range in tissues which have the "exaggerated" increase in T2*-weighted signal intensity after hyperoxia [62] (Fig. 41.3).

A single human study has confirmed the potential of this method as an imaging technique in acute stroke patients and produced results that are generally consistent with rodent MCAO models [63]. T2*-weighted increases after hyperoxia were diminished in the DWI lesion (Fig. 41.4) and intermediate in healthy tissue (see Fig. 41.4). Although the number of subjects with an operationally defined penumbra was small, there were exaggerated increases in T2*-weighted signal in the regions considered penumbral on the basis of diffusion–perfusion mismatch or into which DWI lesion expanded at hyperacute time points. In addition, a diminished T2*-weighted signal response after hyperoxia was seen in the contralateral cerebellar hemisphere, consistent with crossed cerebellar diaschisis [64]. Although encouraging, there are, at present, some remaining issues. For example, what is the optimal concentration of oxygen to use for probing metabolism, and does this vary according to the tissue compartment of interest? In addition, whether quantification of signal change has a direct physiological correlate remains to be determined.

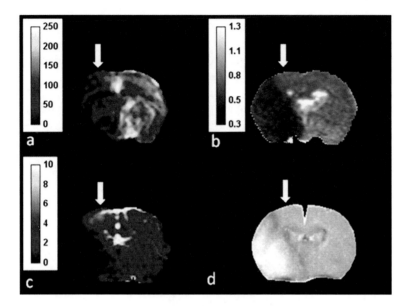

Fig. 41.3 Preserved glucose metabolism in a region of exaggerated T2*-weighted signal increase after hyperoxia in a rodent model of middle cerebral artery occlusion. Images provided by Craig Robertson, University of Glasgow. (**a**) Map of cerebral blood flow (scale in mL/100 g/min), (**b**) Map of the apparent diffusion coefficient (scale in $\times 10^{-3}$ mm^2/s), (**c**) Percentage change map for the T2*-weighted signal after hyperoxia (scale in units of % signal change), (**d**) [14C]2-deoxyglucose (2-DG) autoradiography map (hyperintense regions show regions of low glucose metabolism). *White arrows* indicate the region of interest, which is hypoperfused, with normal ADC, exaggerated T2*-weighted signal increase after hyperoxia and preserved glucose metabolism

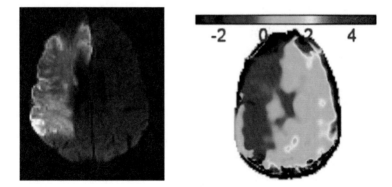

Fig. 41.4 Diminution of the T2*-weighted signal increase in a region of likely infarct core. The *left* most image is a diffusion weighted image (DWI) acquired at the Institute of Neurological Sciences at 18.5 h post ictus. The *right* most image is a percentage change map corresponding to the T2*-weighted signal data. The scale at the *top* of the image is in units of % change

4 Summary

In summary, metabolic imaging methods may complement current brain imaging techniques in acute stroke, with the potential to refine the definition of the ischaemic penumbra, or act as biomarkers of therapeutic effect. We have reviewed promising MR techniques against the background of the gold standard PET methodology. MR spectroscopy studies have provided correlates with other imaging modalities, but it is yet to be proven that they provide additional utility to current MR sequences. Nonetheless, newer approaches such as ^{17}O imaging are encouraging. Development of the deoxyhaemoglobin weighted imaging techniques is in its infancy, and it is unclear which approach for these techniques is most appropriate for use in stroke patients. These MR techniques may be complementary to other emerging techniques, not discussed in this chapter, such as optical imaging. In addition, other MRI techniques, although not purely metabolic imaging, but are indirectly related, such as pH imaging and sodium imaging, may also provide complementary roles. One way or another, there are promising signals that metabolic imaging has a role to play in the clinical and research evaluation of stroke. Much of these best data has been derived from complementary animal and clinical studies, with a truly bench to bedside to bench approach.

References

1. Hacke W, Kaste M, Bluhmki E, Brozman M, Davalos A, Guidetti D, et al. Thrombolysis with alteplase 3 to 4.5 hours after acute ischemic stroke. N Engl J Med. 2008;359:1317–29.
2. Heiss WD, Huber M, Fink GR, Herholz K, Pietrzyk U, Wagner R, et al. Progressive derangement of periinfarct viable tissue in ischemic stroke. J Cereb Blood Flow Metab. 1992;12:193.
3. The National Institute of Neurological Disorders and Stroke rt-PA Stroke Study Group. Tissue plasminogen activator for acute ischemic stroke. N Engl J Med. 1995;333:1581.
4. Ma H, Zavala JA, Teoh H, Churilov L, Gunawan M, Ly J, et al. Penumbral mismatch is underestimated using standard volumetric methods and this is exacerbated with time. J Neurol Neurosurg Psychiatry. 2009;80:991–6.
5. Kakuda W, Lansberg MG, Thijs VN, Kemp SM, Bammer R, Wechsler LR, et al. Optimal definition for PWI/DWI mismatch in acute ischemic stroke patients. J Cereb Blood Flow Metab. 2008;28:887–91.
6. Schlaug G, Benfield A, Baird AE, Siewert B, Lovblad KO, Parker RA, et al. The ischemic penumbra: operationally defined by diffusion and perfusion MRI. Neurology. 1999;53:1528.
7. Marchal G, Serrati C, Rioux P, Petit-Taboue MC, Viader F, de la Sayette V, et al. Pet imaging of cerebral perfusion and oxygen consumption in acute ischaemic stroke: relation to outcome. Lancet. 1993;341:925.
8. Furlan M, Marchal G, Viader F, Derlon JM, Baron JC. Spontaneous neurological recovery after stroke and the fate of the ischemic penumbra. Ann Neurol. 1996;40:216.
9. Muir KW, Buchan A, von Kummer R, Rother J, Baron J-C. Imaging of acute stroke. Lancet Neurol. 2006;5:755–68.
10. Read SJ, Hirano T, Abbott DF, Sachinidis JI, Tochon-Danguy HJ, Chan JG, et al. Identifying hypoxic tissue after acute ischemic stroke using PET and 18F-fluoromisonidazole. Neurology. 1998;51:1617.

11. Markus R, Reutens DC, Kazui S, Read S, Wright P, Chambers BR, et al. Topography and temporal evolution of hypoxic viable tissue identified by 18F-fluoromisonidazole positron emission tomography in humans after ischemic stroke. Stroke. 2003;34:2646–52.
12. Spratt NJ, Donnan GA, Howells DW. Characterisation of the timing of binding of the hypoxia tracer FMISO after stroke. Brain Res. 2009;1288:135–42.
13. Spratt NJ, Donnan GA, McLeod DD, Howells DW. 'Salvaged' stroke ischaemic penumbra shows significant injury: studies with the hypoxia tracer FMISO. J Cereb Blood Flow Metab. 2011;31:934–43.
14. Heiss WD, Kracht L, Grond M, Rudolf J, Bauer B, Wienhard K, et al. Early [C-11]flumazenil/ H_2O positron emission tomography predicts irreversible ischemic cortical damage in stroke patients receiving acute thrombolytic therapy. Stroke. 2000;31:366–9.
15. Rojas S, Martin A, Pareto D, Herance JR, Abad S, Ruiz A, et al. Positron emission tomography with C-11-flumazenil in the rat shows preservation of binding sites during the acute phase after 2 h-transient focal ischemia. Neuroscience. 2011;182:208–16.
16. Higuchi T, Fernandez EJ, Maudsley AA, Shimizu H, Weiner MW, Weinstein PR. Mapping of lactate and N-acetyl-L-aspartate predicts infarction during acute focal ischemia: in vivo 1 h magnetic resonance spectroscopy in rats. Neurosurgery. 1996;38:121–9. discussion 129–130.
17. Saunders DE, Howe FA, van den Boogaart A, McLean MA, Griffiths JR, Brown MM. Continuing ischemic damage after acute middle cerebral artery infarction in humans demonstrated by short-echo proton spectroscopy. Stroke. 1995;26:1007.
18. Duijn JH, Matson GB, Maudsley AA, Hugg JW, Weiner MW. Human brain infarction—proton MR spectroscopy. Radiology. 1992;183:711–8.
19. Nicoli F, Lefur Y, Denis B, Ranjeva JP, Confort-Gouny S, Cozzone PJ. Metabolic counterpart of decreased apparent diffusion coefficient during hyperacute ischemic stroke: a brain proton magnetic resonance spectroscopic imaging study. Stroke. 2003;34:e82.
20. Gideon P, Henriksen O, Sperling B, Christiansen P, Olsen TS, Jorgensen HS, et al. Early time course of N-acetylaspartate, creatine and phosphocreatine, and compounds containing choline in the brain after acute stroke. A proton magnetic resonance spectroscopy study. Stroke. 1992;23:1566.
21. Maniega SM, Cvoro V, Chappell FM, Armitage PA, Marshall I, Bastin ME, et al. Changes in NAA and lactate following ischemic stroke a serial MR spectroscopic imaging study. Neurology. 2008;71:1993–9.
22. Gillard JH, Barker PB, van Zijl PC, Bryan RN, Oppenheimer SM. Proton MR spectroscopy in acute middle cerebral artery stroke. AJNR Am J Neuroradiol. 1996;17:873.
23. Dani KA, An L, Shen J, Warach S. Magnetic resonance spectroscopy may be helpful in acute stroke. Cerebrovasc Dis. 2010;29:1015–9770.
24. Baird TA, Parsons MW, Phanh T, Butcher KS, Desmond PM, Tress BM, et al. Persistent post-stroke hyperglycemia is independently associated with infarct expansion and worse clinical outcome. Stroke. 2003;34:2208.
25. Parsons MW. Acute hyperglycemia adversely affects stroke outcome: a magnetic resonance imaging and spectroscopy study. Ann Neurol. 2002;52:20–8.
26. McCormick M, Hadley D, McLean JR, Macfarlane JA, Condon B, Muir KW. Randomized, controlled trial of insulin for acute poststroke hyperglycemia. Ann Neurol. 2010;67:570–8.
27. Muir KW. Heterogeneity of stroke pathophysiology and neuroprotective clinical trial design. Stroke. 2002;33:1545.
28. An L, Dani KA, Shen J, Warach S, Investigator NNHS. Pilot results of in vivo brain glutathione measurements in stroke patients using magnetic resonance spectroscopy. Stroke.41:E386-E386
29. An L, Zhang Y, Thomasson DM, Latour LL, Baker EH, Shen J, et al. Measurement of glutathione in normal volunteers and stroke patients at 3T using J-difference spectroscopy with minimized subtraction errors. J Magn Reson Imaging. 2009;30:263–70.
30. Zhang NY, Zhu XH, Lei H, Ugurbil K, Chen W. Simplified methods for calculating cerebral metabolic rate of oxygen based on o-17 magnetic resonance spectroscopic imaging measurement during a short o-17(0) inhalation. J Cereb Blood Flow Metab. 2004;24:840–8.
31. de Crespigny AJ, D'Arceuil HE, Engelhorn T, Moseley ME. MRI of focal cerebral ischemia using O-17-labeled water. Magn Reson Med. 2000;43:876–83.

32. Fiat D, Kang SH. Determination of the rate of cerebral oxygen-consumption and regional cerebral blood-flow by noninvasive O-17 in vivo NMR-spectroscopy and magnetic-resonance-imaging. 2. Determination of CMRO2 for the rat by O-17 NMR, and CMRO2, RCBF and the partition-coefficient for the cat by O-17 MRI. Neurol Res. 1993;15:7–22.
33. Zhu XH, Zhang Y, Tian RX, Lei H, Zhang NY, Zhang XL, et al. Development of O-17 NMR approach for fast imaging of cerebral metabolic rate of oxygen in rat brain at high field. Proc Natl Acad Sci USA. 2002;99:13194–9.
34. Mellon EA, Beesam RS, Elliott MA, Reddy R. Mapping of cerebral oxidative metabolism with MRI. Proc Natl Acad Sci U S A. 2010;107:11787–92.
35. Fiat D, Hankiewicz J, Liu SY, Trbovic S, Brint S. O-17 magnetic resonance imaging of the human brain. Neurol Res. 2004;26:803–8.
36. Atkinson IC, Thulborn KR. Feasibility of mapping the tissue mass corrected bioscale of cerebral metabolic rate of oxygen consumption using 17-oxygen and 23-sodium MR imaging in a human brain at 9.4 T. Neuroimage. 2010;51:723–33.
37. Atkinson IC, Sonstegaard R, Pliskin NH, Thulborn KR. Vital signs and cognitive function are not affected by 23-sodium and 17-oxygen magnetic resonance imaging of the human brain at 9.4 T. J Magn Reson Imaging. 2010;32:82–7.
38. Hoffman S, Begovatz P, Nagel A, Umathum R, Schommer K, Bachert P, et al. A measurement setup for direct 17O MRI at 7T. Magn Reson Med. 2011;000:000.
39. Zhu XH, Merkle H, Kwag JH, Ugurbil K, Chen W. O-17 relaxation time and NMR sensitivity of cerebral water and their field dependence. Magn Reson Med. 2001;45:543–9.
40. Ogawa S, Lee TM, Kay AR, Tank DW. Brain magnetic resonance imaging with contrast dependent on blood oxygenation. Proc Natl Acad Sci U S A. 1990;87:9868.
41. Thulborn KR, Waterton JC, Matthews PM, Radda GK. Oxygenation dependence of the transverse relaxation-time of water protons in whole-blood at high-field. Biochim Biophys Acta. 1982;714:265–70.
42. Wright GA, Hu BS, Macovski A. Estimating oxygen-saturation of blood in vivo with MR imaging at 1.5 T. J Magn Reson Imaging. 1991;1:275–83.
43. Li DB, Wang Y, Waight DJ. Blood oxygen saturation assessment in vivo using T-2* estimation. Magn Reson Med. 1998;39:685–90.
44. Turner R, Lebihan D, Moonen CTW, Despres D, Frank J. Echo-planar time course MRI of cat brain oxygenation changes. Magn Reson Med. 1991;22:159–66.
45. Decrespigny AJ, Wendland MF, Derugin N, Kozniewska E, Moseley ME. Real-time observation of transient focal ischemia and hyperemia in cat brain. Magn Reson Med. 1992;27:391–7.
46. Roussel SA, Vanbruggen N, King MD, Gadian DG. Identification of collaterally perfused areas following focal cerebral-ischemia in the rat by comparison of gradient-echo and diffusion-weighted MRI. J Cereb Blood Flow Metab. 1995;15:578–86.
47. Dunn JF, Wadghiri YZ, Meyerand ME. Regional heterogeneity in the brain's response to hypoxia measured using bold MR imaging. Magn Reson Med. 1999;41:850–4.
48. Jones RA, Muller TB, Haraldseth O, Baptista AM, Oksendal AN. Cerebrovascular changes in rats during ischemia and reperfusion: A comparison of bold and first pass bolus tracking techniques. Magn Reson Med. 1996;35:489–96.
49. Tamura H, Hatazawa J, Toyoshima H, Shimosegawa E, Okudera T, Tamura H, et al. Detection of deoxygenation-related signal change in acute ischemic stroke patients by T2*-weighted magnetic resonance imaging. Stroke. 2002;33:967–71.
50. Wardlaw JM, von Heijne A. Increased oxygen extraction demonstrated on gradient echo (T2*) imaging in a patient with acute ischaemic stroke. Cerebrovasc Dis. 2006;22:456–8.
51. Geisler BS, Brandhoff F, Fiehler J, Saager C, Speck O, Rother J, et al. Blood-oxygen-level-dependent MRI allows metabolic description of tissue at risk in acute stroke patients. Stroke. 2006;37:1778–84.
52. Yablonskiy DA, Haacke EM. Theory of NMR signal behavior in magnetically inhomogeneous tissues: the static dephasing regime. Magn Reson Med. 1994;32:749–63.
53. An H, Lin W. Quantitative measurements of cerebral blood oxygen saturation using magnetic resonance imaging. J Cereb Blood Flow Metab. 2000;20:1225–36.

54. An H, Lin W, Celik A, Lee YZ. Quantitative measurements of cerebral metabolic rate of oxygen utilization using MRI: a volunteer study. NMR Biomed. 2001;14:441–7.
55. An H, Liu Q, Chen Y, Lin W, An H, Liu Q, et al. Evaluation of MR-derived cerebral oxygen metabolic index in experimental hyperoxic hypercapnia, hypoxia, and ischemia. Stroke. 2009;40:2165–72.
56. Lee JM, Vo KD, An H, Celik A, Lee Y, Hsu CY, et al. Magnetic resonance cerebral metabolic rate of oxygen utilization in hyperacute stroke patients. Ann Neurol. 2003;53:227–32.
57. Kavec M, Grohn OH, Kettunen MI, Silvennoinen MJ, Penttonen M, Kauppinen RA. Use of spin echo T(2) bold in assessment of cerebral misery perfusion at 1.5T. MAGMA. 2001;12:32–9.
58. Kettunen MI, Grohn OH, Silvennoinen MJ, Penttonen M, Kauppinen RA, Kettunen MI, et al. Quantitative assessment of the balance between oxygen delivery and consumption in the rat brain after transient ischemia with T2-BOLD magnetic resonance imaging. J Cereb Blood Flow Metab. 2002;22:262–70.
59. Kavec M, Usenius JP, Tuunanen PI, Rissanen A, Kauppinen RA, Kavec M, et al. Assessment of cerebral hemodynamics and oxygen extraction using dynamic susceptibility contrast and spin echo blood oxygenation level-dependent magnetic resonance imaging: applications to carotid stenosis patients. Neuroimage. 2004;22:258–67.
60. Santosh C, Brennan D, McCabe C, Macrae IM, Holmes WM, Graham DI, et al. Potential use of oxygen as a metabolic biosensor in combination with T2*-weighted MRI to define the ischemic penumbra. J Cereb Blood Flow Metab. 2008;28:1742–53.
61. Robertson C, McCabe C, Gallagher L, Lopez-Gonzalez M, Condon B, Muir K, et al. Stroke penumbra defined by an MRI-based oxygen challenge technique: 2. Validation based on the consequences of reperfusion. J Cereb Blood Flow Metab. 2011;31(8):1788–98.
62. Robertson C, McCabe C, Gallagher L, Lopez-Gonzalez M, Condon B, Muir K, et al. Stroke penumbra defined by an MRI-based oxygen challenge technique: 1. Validation using[14c]2-deoxyglucose autoradiography. J Cereb Blood Flow Metab. 2011;31(8):1778–87.
63. Dani KA, Santosh C, Brennan D, McCabe C, Holmes WM, Condon B, et al. T2*-weighted magnetic resonance imaging with hyperoxia in acute ischemic stroke. Ann Neurol. 2010;68:37–47.
64. Dani K, Santosh C, Brennan D, McCabe C, Holmes W, Condon B, et al. Oxygen challenge MRI may detect crossed cerebellar diaschsis. Stroke. 2011;42:E199.

Chapter 42
Computational Analysis: A Bridge to Translational Stroke Treatment

Nirmalya Ghosh, Yu Sun, Christine Turenius, Bir Bhanu, Andre Obenaus, and Stephen Ashwal

Abstract Objective rapid quantification of injury using computational methods can improve the assessment of the degree of stroke injury, aid in the selection of patients for early or specific treatments, and monitor the evolution of injury and recovery. In this chapter, we use neonatal ischemia as a case-study of the application of several computational methods that in fact are generic and applicable across the age and disease spectrum. We provide a summary of current computational approaches used for injury detection, including Gaussian mixture models (GMM), Markov random fields (MRFs), normalized graph cut, and K-means clustering. We also describe more recent automated approaches to segment the region(s) of ischemic injury including hierarchical region splitting, support vector machine, a brain symmetry/asymmetry integrated model, and a watershed method that are robust at different developmental stages. We conclude with our assessment of probable future research directions in the field of computational noninvasive stroke analysis such as automated detection of the ischemic core and penumbra, monitoring

N. Ghosh (✉) • C. Turenius, PhD • S. Ashwal, MD
Department of Pediatrics, Loma Linda University, Loma Linda, CA, USA
e-mail: NGhosh@llu.edu

Y. Sun, PhD • B. Bhanu, PhD
Center for Research in Intelligent Systems, University of California, Riverside, CA, USA

A. Obenaus, PhD
Department of Pediatrics, Loma Linda University, Loma Linda, CA, USA

Department of Radiology, Loma Linda University, Loma Linda, CA, USA

Department of Radiation Medicine, Loma Linda University, Loma Linda, CA, USA

Department of Biophysics and Bioengineering, Loma Linda University,
Loma Linda, CA, USA

P.A. Lapchak and J.H. Zhang (eds.), *Translational Stroke Research*, Springer Series
in Translational Stroke Research, DOI 10.1007/978-1-4419-9530-8_42,
© Springer Science+Business Media, LLC 2012

of implanted neuronal stem cells in the ischemic brain, injury localization specific to different brain anatomical regions, and quantification of stroke evolution, recovery and spatiotemporal interactions between injury volume/severity and treatment. Computational analysis is expected to open a new horizon in current clinical and translational stroke research by exploratory data mining that is not detectable using the standard "methods" of visual assessment of imaging data.

1 Introduction

Focal and global cerebrovascular insults remain a common and devastating disorder in adults and children. The last 2 decades have seen a rapid progression of scientific and clinical advances due in large part to the implementation and application of neuroimaging, including magnetic resonance imaging (MRI), computed tomography (CT), and positron emission tomography (PET). Numerous reviews (including chapters in this book) highlight the importance of neuroimaging for diagnosis and assessment of therapeutic effectiveness [1, 2]. Visual assessment is used primarily to guide clinical decisions for patient care, but what is lacking in clinical and experimental studies is the ability to rapidly extract quantitative data. We summarize the current state of computational analytical approaches particularly, as they apply to MRI to evaluate focal and global cerebrovascular injuries. Although substantial information can be acquired from visual assessment of MR images, it is readily apparent that application of computational analytical methods can have substantial clinical usefulness (Table 42.1). Since these computational advances provide rapid and quantitative data, they may assist clinicians in treatment decisions including which patients may be candidates for specific treatments and to assess where in the brain particular treatments could be targeted (e.g., location of stem cell implantation or drug injection). Our research efforts have focused on newborns with global or focal ischemic injuries in experimental animal model systems, as well as in term neonates. However, the computational approaches we describe have broad applicability across a spectrum of ages and diseases.

Application of computational assessments to translational stroke research can be separated into the following sequential steps: (a) detection of injury and extraction

Table 42.1 Benefits of computational methods in cerebrovascular disease

Minimize or eliminate observer bias for lesion detection
Objective quantification of lesion volume compared to manual methods
Ability to rapidly acquire computational results compared to manual methods
Ability to quantitatively and serially determine changes in lesion volume and extent
Differentiate ischemic core from penumbra
Quantify regional anatomical injury severity (by use of templates and parcellation)
Quantify evolution of injury over space and time
Quantify injury–treatment interaction and recovery over space and time
Utilizing these information in candidate and treatment selection for therapeutics

of lesion volume, (b) estimation of a lesion's core and penumbra, (c) quantification of a lesion's region-specific information based on anatomy, (d) target drug treatment or other therapeutic advances (e.g., implantation of neural stem cells: NSC), (e) detection of implanted MR-labeled stem cells, (f) serial imaging and monitoring of the interaction(s) between the stroke lesion and treatment effects in space and time, and (g) analyzing treatment effectiveness on tissues associated with a lesion. Computational methods based on image processing, computer vision, machine learning, and pattern recognition techniques have the ability to improve data accuracy and yield clinically relevant information. The contents of this chapter are focused on current automated methods that are limited primarily to lesion detection but also describe several recent approaches to lesion quantification including estimation of core–penumbral tissues. We have also applied some of these computational approaches to monitoring of implanted stem cells in experimental models. Finally, we explore several different approaches to future research using brain atlases for anatomy-specific spatiotemporal monitoring of stroke and stem cell interactions.

2   Current Computational Approaches for Lesion Detection

Visual and manual lesion assessments which are commonly used [1, 3–6] suffer from the misperception of contrast due to the parallax effects of neighboring pixels/ voxels—pixels with the same intensity values may look different due to its neighboring pixels [7, 8]. Issues such as user fatigue related errors in time-consuming manual assessment methods and intra- and interobserver variability potentially compromise data comparisons. Simplified semiquantitative scoring systems to grade injury severity have been useful in grading injury severity in clinical patients (e.g., [9–12]) and in experimental models (e.g., rat pup severity score, RPSS [13]). Although useful, these simplified scores are grossly subjective and only qualitatively measure the extent of injury. Unfortunately, they do not provide objective and accurate quantification of injury volumes. Some investigators have developed automated lesion segmentation methods that use manual thresholds [14], cross-correlations based on ad hoc templates [15], Otsu's algorithm [16] on similarity maps of intensity and proximity [17], or use manually derived percentage of maximum values from apparent diffusion coefficient (ADC) maps in neonates with stroke [6, 18]. In most cases, these approaches are ad hoc and are not robust to MR signal and noise level variations across different datasets. What has not been developed are computational methods, based on reliable mathematical models that objectively analyze medical images in a reproducible, quantifiable, and accurate manner at near real time speed.

Some computational methods have been developed for other diseases such as multiple sclerosis (MS) [19, 20], focal cortical dysplasias (FCD) [21, 22], and white matter lesions (WML) [23]. Since these methods use MRI signal contrast to detect abnormalities, they can be modified or extended to evaluate lesion characteristics seen with ischemia or stroke [23, 24]. Reviewed below are some of the different available approaches, focused on automated methods such as GMMs [25], MRFs [26], normalized graph cut [27], and K-means clustering [28] for the detection of brain abnormalities. Although many of these methods have considerable similarities and are sometimes indistinguishable, we describe them separately.

2.1 Gaussian Mixed Models

MR signal intensity of similar brain tissues are expected to be the same. Image noise, including the noise within MR images is primarily Gaussian in nature. This and the central limit theorem [29] (which states that, as the number of data points increase toward infinity, the distribution of data points gradually becomes close to Gaussian) imply that different tissues, such as gray matter (GM), white matter (WM), cerebrospinal fluid (CSF), or lesioned tissues, are expected to aggregate under different Gaussian curves with their means distinctly separated for easy differentiation. Each pixel/voxel has a particular probability to be classified as WM, GM, CSF, or lesion, and the highest probability defines the tissue type of the pixel/voxel. These variations within tissues are modeled by a GMM where every pixel/voxel is represented as a weighted sum of Gaussian distributions parameterized by (a) weights (or prior probability), (b) means, and (c) standard deviations of the individual Gaussian (i.e., tissue) components. With appropriate selection of the MRI modality and tissue types, a training set of data can be used to learn these parameters by use of an Expectation-Maximization (EM) algorithm [7]. EM iteratively modifies the parameters namely weights, means and standard deviations of the Gaussian distributions to fit the training data for which the tissue class of each pixel/voxel is known. In other words, the EM algorithm maximizes the expectation that the data came from the estimated GMM model. A key research focus has been on automated MRI segmentation of brain anatomical structures (GM, WM, and CSF) using GMM and its derivatives (e.g., constrained GMM; CGMM). Greenspan and colleagues have used a CGMM where global intensity with local spatial characteristics has been used with EM based parameter learning [30]. Connected component based top-down splitting of the MR regions is done to get tissue classes (WM, GM, and CSF) having different mean MR intensities. As GMM does not utilize an atlas-based registration it can be applied to other disease states or to data where atlases are not yet available. An example of this relates to the developing neonatal brain where reliable age-matched atlases have not yet been published. These Gaussian distributions can be closely spaced, i.e., the mean MR signal intensity of different tissues are close, so that on occasion, one needs to consider fuzzy connectedness where class membership values are like weights in GMM but with "fuzzyness," in which the probabilities of a voxel being in tissue A and tissue B are assigned [31]. However, histogram fitting to a normalized histogram, as adapted by Fan and colleagues, cannot be used when large lesions affect the shape of the histogram [32].

In some cases, the lesion has been modeled as a "reject" class in normal tissue models (GM, WM, CSF) where the voxels, not satisfying the GMMs derived from the EM algorithm, are detected as lesion [33]. Agam and colleagues have used a mixture-parametric probabilistic model where optimized parameters are found by EM-based incomplete log-likelihood maximization for detecting chronic stroke lesions [24]. Probabilistic priors are computed from registered control data sets and deviation from these priors in multimodal fused data sets (T1, T2, and diffusion tensor imaging (DTI)) is used to detect the lesion. A key disadvantage of GMM based methods is that they break down in analysis of imaging data with a magnetic bias field (an unwanted baseline magnetic field in the MRI scanner) as this shifts the Gaussian means of the tissues linearly or nonlinearly leading to incorrect tissue classification. Bias field correction methods can reduce such problems [27]. Also GMM suffers badly when there are data with large voxel sizes, such as magnetic resonance spectroscopy (MRS), where partial volume effects may affect classification of voxels in tissue borders. Superresolution methods could be used to improve performance [34].

2.2 Markov Random Fields

In Markov models, the data points (e.g., MRI voxels or regions) and their different possible classes (e.g., WM, GM, CSF, lesion) are represented as nodes of a graph where links between the nodes carry a probabilistic association for each data point to different classes. MRFs are generally multidimensional where the connected lines of the graph are not directed. Tissue types classified as normal/abnormal or WM–GM–CSF (or some other user-based classification) are designated as nodes. Voxel nodes are defined as the observed variables while tissue nodes are the estimated or hidden classes. Association of voxel nodes to different classes are controlled by joint probability distributions (i.e., probability of a particular voxel being classified as a particular tissue) and represented by lines connecting these nodes. Atlas prior probabilities and MRFs have been used to discard partial volume effects and reduce outliers in patients with multiple sclerosis lesions [33]. Kabir and colleagues have used multimodal MR data (T2, fluid attenuated inversion recovery (FLAIR), diffusion sequences) and longitudinal data (6 h, 5 and 30 days) that were coregistered to form a multimodal MRF [35]. Using this novel approach, they were able to demonstrate that in adults with stroke, the lesion distribution follows the vascular territory of an occluded blood vessel. They also generated normalized vascular territory maps that could be utilized for stroke classification and improved clinical interventions. The significance of their work is that they demonstrated that stroke territories shift over time. Hidden Markov models (HMM), a variant of multidimensional MRF, have been used to find WM–GM–CSF tissue types [26]. In another recent study, 3D Hilbert–Peano mapping was used to convert 3D T1 and FLAIR data into 1D data for computational ease [19]. A 1D MRF (Hidden Markov Chain; HMC) was then used in the preprocessing step to account for neighborhood information to improve intratissue signal homogeneity and intertissue signal contrast. A likelihood estimator was then used to prune and detect regions not falling in any of the tissue class (outliers) to identify the lesion as an MS lesion. Finally a probabilistic tissue atlas was used to reject false detection.

As the number of MR regions and tissue classes increase, the complexities of the joint probability distributions in MRF become intractable. Often prior knowledge that a particular type of lesion is only localized to a certain brain region simplifies the connectivity of the graph and the associated joint distributions. Similarly, an MR region highly associated with a tissue class (e.g., CSF) precludes its possible association with another tissue class (e.g., GM), as generally they are not adjacent. This type of prior knowledge leads to probabilistic conditional independence with a known condition, as one MR region cannot be associated with different tissue or anatomy classes. This brings about cause and effect relations between graphical nodes and removes some of the connected links in the MRF (based on conditional independence).The remaining links are manifested as arrows pointing from a "cause" node to an "effect" node of the graph. Bayesian networks (BNs) are such a probabilistic graphical model with directed acyclic graphs (DAG) [36]. BNs also have been used to find abnormal tissues [37]. In these cases, network parameters (e.g., which region(s) are connected to which tissue type) are to be learned from training data sets with manually derived tissue classification available for each MR voxel/region [36].

2.3   Normalized Graph Cuts

In MR images, voxels with similar signal intensity form regions. The similarity or contrast of signal (i.e., coherence or difference) defines how strongly two voxels are alike (or different). Based on such intervoxel signal comparisons, an MR image can be represented as a graph with each pixel/voxel as a node of the graph and connecting (undirected) lines are weighted to represent how strongly the two voxels are similar (i.e., from the same tissue class). The strengths of these links are computed from the similarity in features of corresponding voxels, such as MR signal intensity, texture (in a close neighborhood), and location proximities. This graph can be represented as a multidimensional matrix and its Eigen vectors can be computed by singular value decomposition (SVD). Based on graph mathematical theory, the second weakest Eigen vector, sometimes called "Fiddler vector," defines the "weakest link" between two strongly connected "subgraphs" [38], for example a separating plane between WM and GM tissue types in T1/T2 weighted MRI. If we apply normalized graph cuts (N-Cut) for solving MRFs, the basic graph cut algorithm ensures near global minimum of the energy function of the MRF. As N-Cut algorithms often suffer from oversegmentation due to artifacts and noise, age-matched control atlases or probabilistic tissue models can be useful [27]. Such prior tissue models and N-Cut methods have been used to separate WM, GM, and CSF regions where iterative algorithms alternate between N-Cut based segmentation and atlas based inhomogeneity corrections. In other words, small outlier regions that are detected inside large correctly classified tissue (due to signal inhomogeneity) are modified to have a better match with the atlas.

2.4 K-Means Clustering

GMM and MRF typically require training datasets with manually detected lesions (ground truth) to assist in supervised learning. When such ground truths are not available, unsupervised learning can be used by evaluating MR image clusters based on the cohesiveness of voxel features such as MR signal intensity, location, texture, etc. Cluster cohesiveness can be measured as the ratio between the intracluster standard deviation and the intercluster mean distances, that is, how distant the clusters are from each other [29]. Freifeld and colleagues have used K-means clustering to detect multiple sclerosis lesions from T2-weighted images [25] where they have used a multimodal approach, starting with CGMM-based initial grouping of tissue types (WM, GM, CSF) that iteratively split the tissue regions [30]. Then they have used expert rule based and multimodal fusion of MR images (T1, T2, and proton density MRI) to robustly identify small lesions. Since multimodal fusion often blurs the boundaries between the regions they utilized active contours to sharpen the boundaries.

In some instances, after CGMM based initial tissue segmentation, deformable models along with K-means clustering have been used to improve MRI tissue classification within the brain so as to generate a tissue type atlas [28]. In these heavily model driven approaches, lesions of a known shape can be readily identified if the lesion encompasses an entire brain region (e.g., putamen) but this approach does not work when the lesions cross anatomical boundaries, as is the case in the majority of stroke patients. When data clusters are too close to one another, partial or fuzzy association to neighboring clusters (i.e., classes) can be used, just like the previously described fuzzy concept in the GMM approach. Spatially constrained fuzzy kernel clustering methods have been used to estimate MR bias fields, which need to be corrected for improved brain atlas generation or GMM-based lesion detection [39]. This also accelerates the usually slow kernel-clustering method in high dimensional feature space [39]. Before applying these types of bias correction methods, it should be kept in mind that in some situations, small lesions can be classified as an artifact and hence removed as an outlier. Fuzzy clustering has also been used to detect gray matter atrophy in MS lesions using a fusion of probability maps (from brain morphology), signal intensity (from MRI), and brain anatomy (from atlas) [40]. A similar method, K-nearest neighbor method, has been used for WM–GM–CSF classification after tissue probability models or atlases were registered to the data [23]. The MRI was intensity-rescaled after truncating high-low outliers. WM lesions were detected from the histogram of the WM only.

2.5 Other Methods

Lesion detection can be improved by preprocessing the MR images using spatial regularization or congruity by MRI textures [41, 42], probabilistic models [24], morphological operations [43], and physical model estimation [44]. In some instances, multiple MRI modalities can be used sequentially to achieve better insights into the clinical relevance of the imaging abnormalities. Dugas-Phocion and colleagues used EM based temporary grouping of tissue types (WM, GM, CSF), thresholded the voxel-wise MR signal differences to find supratentorial MS lesions in T2 FLAIR, and then proposed T1/T2 lesion differences as the possible internal structure of the lesion [20]. They registered all the modalities for voxel wise comparisons. For multimodal fusion, a key requirement is coregistration to a standard space which often severely suffers when different modalities have different resolutions. For example, functional MRI (fMRI) data or MRS derived metabolite ratios in different anatomical brain regions are often acquired at a coarse resolution (i.e., large voxel size) compared to T1 or T2 anatomical data [45]. The common method of down sampling high resolution T1/T2 data and working on the coarse functional space is relatively simple but associated with loss of small variations due to the large voxel size. Attempts have been made to improve the resolution of fMRI to overlay onto high resolution anatomical spaces using ad hoc methods [45]. Application of superresolution technologies that have been used in the computer vision field may help to overcome partial volume effects when merging low to high resolution MR data [34].

3 Recently Emerging Approaches for Automated Lesion Detection

Several disadvantages of current approaches to lesion detection have hampered their widespread translational use. First, computationally intensive mathematical models are generally used that hinder rapid estimation of the size and nature of the lesion which is time sensitive when used clinically [19, 21, 25, 30, 39]. Second, many current methods use anatomical brain atlases and a priori probabilistic tissue models [23] to facilitate lesion detection and reject outliers [19, 33]. However, neither atlases nor probabilistic models are always statistically valid for the data being evaluated (e.g., stroke data may have distortions and structural uniqueness that cannot be mapped well to an atlas) and may lose information during spatial registration of the injured brain to the atlas brain. Since ischemic and stroke lesions often cross anatomical boundaries, the use of modeling or atlases or priors is likely not justified. In addition, none of the current research has the capability of analyzing the internal structure of an ischemic/stroke lesion which ultimately may be relevant for clinical treatment decision making (see [20] for an exception). Finally, compared to other neurological disorders, there has been little research dedicated to automated detection and quantification of ischemic injury [35].

3.1 Hierarchical Region Splitting

Hierarchical region splitting (HRS) is an automated region segmentation method that splits MR images recursively to generate a binary tree-like structure (Fig. 42.1).

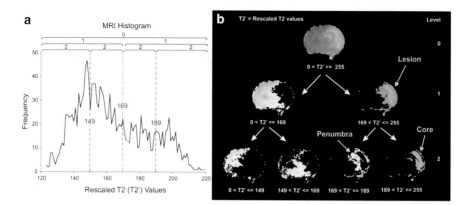

Fig. 42.1 HRS methodology. (**a**) T2 histogram: T2WI is rescaled (T2′) to intensity range (0–255). The HRS method fits a bimodal distribution and detects a valley at $T2' = 169$ as the threshold to split the histogram. This splitting is repeated recursively (next level thresholds are $T2' = 149$ and 189). (**b**) HRS tree: Segmenting the T2WI into regions with T2′ values in these particular ranges in (**a**) form the HRS tree. Subimages in (**b**) are rescaled for enhanced visualization. HRS automatically detects the right image in Level 1 as the ischemic lesion. HRS further segments the lesion into the core and the penumbra regions in Level 2. Note that only part of the complete HRS tree is shown

The subimages are based on the uniformity and contrast in MR signal (intensity or quantitative values) that are representative of the underlying tissue characteristics (e.g., normal vs. injured brain). Recently, we have reported that HRS can detect and extract neonatal hypoxic ischemic lesion volumes, with excellent volumetric correlations ($r^2 = 0.95$; $P = 8.6 \times 10^{-7}$) and overlap (sensitivity: 0.82, specificity: 0.86, similarity: 1.47) to manually extracted results [32]. Although the methods and results of HRS were obtained from MR images, HRS is a generic method for any medical image (e.g., MRI, PET, CT) where contrast in the medical image is used to detect abnormalities.

The key steps for HRS detection of ischemic/stroke lesions (Fig. 42.1) are: (1) *MR image rescaling*: MR images are rescaled to an image intensity range [0–255] and the pixel wise conversion factors are stored to map the automatically derived results back to their original MR values or intensities. (2) *Derive image histogram*: A histogram of the MRI is computed. (3) *Compute adaptive segmentation threshold*: The image histogram is then modeled as a bimodal distribution with two distinct and distant peaks, which segregates the MR images into two different tissue regions (Fig. 42.1a). The valley between the histogram is then computed and acts as an adaptive threshold that is used to split the image into two subimage regions that have uniform image intensity or values [16]. Each histogram peak is representative of a region with a minimum intraregional image variance and a maximum interregional image variance. (4) *Recursive bimodal segmentation*: The adaptive image segmentation step is now used recursively to split each of the resultant subregions (subimages) to generate a tree-like hierarchical data structure (called HRS

Table 42.2 T2 and ADC values for tissue-types over time in rat pups

	ADC (×10⁻⁵ mm²/s)			T2 (ms)		
Age (days)[a]	NABM	1–2 days lesion	Fe-labeled NSC[b]	NABM	3–7 days lesion	Fe-labeled NSC
10–12	60–150	<40	–	30–50	>80	<20
13–17	60–120	>150	–	50–80	>100	<40
>17	60–100	>120	–	50–100	>120	<50

The MRI range of values were used as prior knowledge in HRS based stroke and stem cell detection.

[a]HII was induced at postnatal day 10, followed by serial imaging (see Obenaus et al. [73]).

[b]Iron-labeled NSC ADC values were not extracted as T2 is a more sensitive modality.

tree; Fig. 42.1b). Each segmented region at any level of the HRS tree is the MRI data within two threshold values. (5) *Criteria for stopping segmentation*: Recursive image segmentation is continued until each of the resultant subregions/subimages have uniform image intensity based on these criteria: (a) individual connected regions are small and unlikely to be partitioned further into separate subregions (i.e., different tissues); (b) the image intensity or value for each subregion has a low standard deviation value (i.e., if from MRI, uniform MR physical properties); and (c) the image histogram of the segmented region has a low kurtosis value (i.e., a sharp histogram peak that cannot be split further).

When the medical image contains bone (e.g., skull) and CSF regions, HRS can be employed using a different set of parametric values to separate these unwanted regions leaving only brain tissue for recursive splitting. Use of automated HRS avoids manual skull stripping which is time-consuming. To detect small lesions, HRS further splits the skull-stripped brain using a different set of parameters. For this, as the HRS criteria for stopping segmentation, smaller limits for area, standard deviation and kurtosis are used that provide a deep-rooted HRS tree with small subregions. Finally, HRS can use previously established normative data (from published studies or experience) to provide a generic range of image values (a) for normal appearing brain matter (NABM) and injured tissues (e.g., stroke), (b) for the particular neuroimaging modality being used (e.g., MRI T2, diffusion weighted, etc.), and (c) for the appropriate imaging time point (e.g., relevant to the time point post ischemia/stroke onset as neuroimaging values change). Because T2 relaxation and diffusion coefficient (ADC) values change with brain development or with injury, or both, they need to be adjusted for accurate lesion detection.

Based on our application of HRS in a neonatal model of hypoxic ischemic injury we have determined an approximate range of MR image values that are useful for temporal discrimination of normal appearing brain matter and lesion (Table 42.2). Means of the subregions from the HRS tree are compared with these a priori known tissue values. The subregions with their mean values within the MR property range for the injury in Table 42.2 are merged to obtain the final HRS detected lesion [32]. This is repeated for each imaging modality and at each imaging time point considered in the study. This HRS approach results in excellent HRS detected ischemic lesions (Figs. 42.1b and 42.2a) at two time points (4 and 14 days post hypoxia ischemia). In other neurological diseases, the HRS threshold values are likely to vary; however, typically the image value ranges between NABM and abnormal regions are widely separated and can be used to automatically distinguish abnormal from normal tissue. Compared to current lesion detection approaches (see above) that depend heavily on tissue models, HRS is a generic method that adapts to any type of medical image, injury and across the age spectrum.

3.2 Support Vector Machine

At the present time, HRS has utilized only the MR signal or MR value as the primary image feature to classify pixels/voxels. However, other image features from multiple (registered) modalities such as signal intensities, textures [41, 42], proximity [17], shape indices, probabilistic indices from anatomical atlases, and prior knowledge [23] can be used to classify any voxel. The feature sets from each pixel (or region) from each MR image creates a singular data point within a multidimensional feature space.

Our ultimate goal is to define a set of image features that robustly separate ischemic/stroke volume from normal brain tissues. In mathematical terms, within the feature space, we wish to find a surface, one side of which has data points from the normal brain while the other side having data points from the lesion. The data points in both classes that are closest to separating this hyperplane are the most error prone points and define the "functional margin" [46]. Instead of defining ad hoc the location of a hyperplane, it is best to identify this separator adaptively from the data itself. One approach is to use a "training data" sample that has the features of interest, as well as a manually determined classification (e.g., stroke, NABM). Once a hyperplane or classifier is trained using a supervised learning method (e.g., EM algorithm), a new data point without any manual detection (i.e., "testing data") can be classified based on the output of the discriminating function (learned during the training phase) by determining on which side of the classifier surface the new data point falls.

Support vector machine (SVM) is one such supervised machine-learning algorithm that searches for nonlinear classifier surfaces within the feature space that separates different classes [47]. For example, after registering T2 and ADC maps of an injured brain, T2 relaxation values and diffusion coefficients are used as the voxel feature in SVM (Fig. 42.2b), and normal brain tissue, lesion core, and lesion penumbra are separated by nonlinear curves. If the feature space does not reliably separate tissue classes, different combinations of the features can be used to generate more complex features that would assist in differentiating the tissue features. This process is done by projecting current feature-set to a higher dimensional feature space using kernel functions that are nonlinear functions of the simple features such as the T2 and ADC values. The underlying assumption is that a better nonlinear hyperplane (separating tissue classes) could be learned in this new feature space. SVM attempts to maximize the "functional margin" of the hyperplane or a separation that maximizes accuracy and distance between the separating plane and

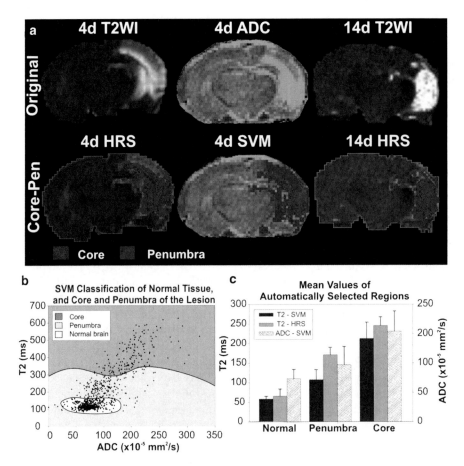

Fig. 42.2 Lesion core and penumbra detection by HRS and SVM: (**a**) HRS detected core (*red*) and penumbra (*blue*) at 4 and 14 days post ischemia using T2 maps. SVM detected core and penumbra from T2 and ADC maps (after spatial coregistration) at 4 days post injury. The core–penumbra relative percentages and locations change considerably from 4 to 14 days, illustrating a temporal evolution of the ischemic lesion. (**b**) SVM feature-space with T2 and ADC values of the voxels in 4d (post ischemia) data in (**a**) and nonlinear classification surfaces learned by segmentation of normal tissue (*white*), core (*pink*) and penumbra (*yellow*) of the lesion. (**c**) In addition to core–penumbra overlap, T2 and ADC means for normal tissues, core, and penumbra that were detected by HRS and SVM reveal excellent concordance

most data points [46], leading to improved tissue identification. Recently, SVM has been used to detect FCD from texture features derived from MRI [21]. Initially, gray matter was found by registering to an atlas and gray matter textures were derived from statistical gray level cooccurrence and "run length" features (i.e., for how many pixels/voxels the same classification continues in different image directions). In the texture space, SVM is then trained and used to separate relatively small gray matter lesions.

Higher "functional margins" would be associated with lower generalization errors in the test set. Incremental learning can then be used where classification for the test data is validated by an expert (e.g., a physician, researcher) and this feedback may be incorporated to improve the SVM classifier. At different imaging time points, different sets of features may be more effective in classifying the same tissue (normal tissue, ischemic/stroke lesion or presence of implanted stem cells). Different training data sets may be needed to train different classifiers for different time points to better extract the evolving lesion. Following coregistration of ADC and T2 maps in a neonatal model of ischemia, an SVM based classifier effectively separated the lesion from NABM as well as discriminating putative ischemic core and penumbra regions (Fig. 42.2a, b). The advantage of SVM compared to HRS lies in its ability to adaptively learn the classifier by a training set of data and to utilize multiple features (e.g., intensity, texture, shape, etc.), but at the cost of being more computationally intensive (than HRS) for training, that can limit throughput. Comparative results in Fig. 42.2a, c show that HRS and SVM detected lesion regions overlap almost identically. Thus, identification of similar regions using two completely different computational methods to detect the lesion strengthens the validity of both results. However, HRS will be more effective for single feature based classifications, whereas SVM will be more useful when multiple features are used to classify the data.

3.3 Brain Symmetry-Based Approaches

Normal brains are highly symmetric and this has been used for automated brain segmentation [48]. However, in stroke and other diseases (as well as in translational models), injury areas are often asymmetric (Fig. 42.3a), i.e., the region contralateral to a stroke is often normal, albeit with significantly different MRI signals. Asymmetry itself can be used as a prominent feature for identification of brain injuries. Examples of asymmetry detection in brain injury and disease include tumors from T2-FLAIR images [49], MS plaques from T1 and T2 weighted images [48], and FCD using textures in T1 weighted image [22].

Recent studies have reported a fully automated symmetry integrated detection method for stroke [50, 51]. Like most current automated (and manual) methods, a brain region is detected as abnormal when regional properties such as symmetry deviate from those for NABM. The regions of the brain that are not symmetric are hypothesized as the regions involved in the injury and are then further tested for the presence/absence of injury. Key steps of this approach are summarized in Fig. 42.3. From the input MR image (Fig. 42.3a), the axis of symmetry (AoS) is first estimated (Fig. 42.3b) and a "symmetry affinity matrix" (Fig. 42.3c) is computed based on intensity similarity between mirror-symmetric locations of the two sides of the AoS. A symmetry integrated region growing method is then applied to segment symmetric and asymmetric regions (Fig. 42.3d) followed by kurtosis and skewness measurements of symmetry affinity to extract the asymmetric region(s) (Fig. 42.3e). For increased robustness to variations in MRI and lesions, the symmetry affinity

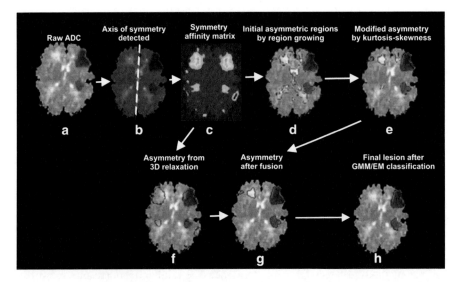

Fig. 42.3 Symmetry-based lesion detection. Results are shown of a symmetry-integrated lesion detection method in a term newborn with stroke using ADC. From (**a**) the original ADC, (**b**) the axis of symmetry (AoS) is detected (*dotted line*). (**c**) A symmetry affinity matrix is computed where brighter (*yellow*) regions are more asymmetric across the AoS. (**d**) Symmetry-integrated region-growing extracts these initial asymmetric regions (**e**) that are then modified using kurtosis–skewness measures of the regions. (**f**) Separately, asymmetric clusters are also detected from 3D gradient relaxation algorithm. (**g**) Robust asymmetric regions are computed from the fusion of the detected regions in (**e, f**). (**h**) Finally GMM/EM classifies the stroke regions from the asymmetric regions in (**g**)

matrix (Fig. 42.3c) is also separately processed using a 3D gradient relaxation algorithm for clustering and identification of asymmetric groups (Fig. 42.3f). The results from Fig. 42.3e, f, are then fused to obtain refined asymmetric regions (Fig. 42.3g). Finally a GMM/EM based supervised classifier is trained (from "training data") and applied to classify asymmetric regions into injured and healthy tissues using image signal intensity and 3D asymmetry volumes to quantify the stroke lesion (Fig. 42.3h). Using this approach, we found that automated symmetry identified lesion volumes varied only by an average of 7.53%, compared to manually identified lesion volumes [50, 51].

The key advantages of the symmetry method are that it is not dependent on the patient or stroke age or a priori knowledge of NABM and lesion ranges (see Table 42.2), and it does not require registration with an atlas or prior models. However, disadvantages include that it is more computationally intensive than HRS, can fail to identify an AoS when a severe injury causes bilateral hemispheric alterations, and can fail to detect injury regions if the stroke is diffused or global in nature when mirror-symmetric regions from both hemispheres are injured [32]. In these cases, other types of "prior knowledge" may be used and currently are being investigated.

3.4 Watershed-Based Segmentation Approaches

Watershed approaches to injury detection attempt to model a medical image as a topographic map and find the regional (local) maxima in the altitude as segmentation boundaries. For instance, the signal values in a brain MR image can be thought of as the altitudes on a topographic map. If a drop of water falls on this relief, it can take different paths to finally reach a local minimum, called a "catchment basin." For MRI, the lowest signal regions represent these local minima and regional maxima form the crests or watersheds. Watersheds are the limits of the adjacent catchment basins and represent the image segmentation boundaries. The path of the downward water flow in the watershed model is similar to the minimization of signal energy (from MRI intensity or its contrast), that can be modeled by established methods such as simulated annealing, gradient descent, genetic algorithm, etc. [29]. The watershed method has a rich history of success in image segmentation [52–56]. In general, MRI contrast based edges are estimated during image preprocessing and based on the MR image gradients, the contrast edges are merged to define tissue segmentation. Improved watershed methods have been used successfully to segment knee cartilage and normal brain tissues (GM, WM, CSF) with a tissue probability map (prior knowledge) registration [57].

Recently we have used a modified watershed method [58] to detect experimental and human hypoxic ischemic lesions. Initially, a multiparameter calculation was done to compute pixel features such as image contrast edges (by Sobel edge detector [7]), gray-scale intensity values, and mean gradient values of the pixel with its 8-connected neighborhoods (i.e., the local contrast) [7]. These pixel features are considered as the altitudes of the water basin. The water starts at the local maxima and flows along certain paths that have been defined by a genetic algorithm (GA) based energy minimizer to finally reach the local minima [59]. To reduce oversegmentation effects, which are a critical drawback of the watershed method, a similarity based region-merging was then performed. Finally, stroke/ischemic lesions were extracted by utilizing the prior knowledge of mean and standard deviations of stroke regions from MRI, based on similar experimental or clinical neuroimaging data. This process readily detects the ischemic injury in a rodent brain (Fig. 42.4a) and some false positive regions that illustrate the oversegmentation that occurs using the watershed approach.

The GA [59] that was utilized as an energy minimizer in the watershed model [58], allows pixel features (edge, signal intensity, and local contrast) to form an energy topography of the brain MRI. GA starts with a set of randomly selected pixels and iteratively searches for pixels with the lowest variation around it (i.e., pixels with the lowest contrast energy). For any particular iteration, each pixel X is compared with its neighboring pixels and is then replaced by a neighboring pixel Y that is most dissimilar to the original pixel. This process continues until little variation is found in the neighborhood of the pixels contained in the searching set and results in reaching the local energy minima (catchment basins). Small outliers (shallow basins) may contain a few pixels within the current searching set

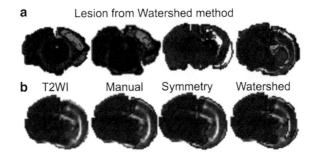

Fig. 42.4 Detection of ischemic injury using the watershed method. (**a**) Ischemic injury in a neo-natal rat brain can be detected (*red outline*) with some outlier regions being identified as injured tissues due to oversegmentation inherent in watershed method. (**b**) Comparative results for the symmetry-based and watershed methods along with the manually detected ground-truth ("gold standard") show that the symmetry-integrated approach performs better than the watershed method

but that do not fulfill a trend or criteria (called "fitness" in GA terminology). These outlier pixels are replaced by the same number of new pixels. Replacement is done using random or rule-based techniques that mimic biological evolutionary processes, such as "crossover" (rule-based replacement), "mutation" (random replacement), etc. In this way, the GA based energy minimizer gradually finds the catchment basins from which the watersheds (crests) are computed that segment the medical image into normal and abnormal tissue regions.

Disadvantages of the watershed method include oversegmentation, sensitivity to signal-to-noise ratio (SNR), or poor detection of thin or low SNR brain structures. A comparative example (Fig. 42.4b) illustrates that the modified watershed method detects outliers while the symmetry-integrated method performs much better. In addition, prior information using brain atlases or tissue probability masks can improve segmentation results [57]. However, maturational age-dependence and registration related complexities and uncertainties, specifically in injured brains, may limit the ability of watershed methods in medical image segmentation and lesion detection. Research continues to progress towards solving such limitations.

4   Future Directions in Computational Stroke Assessment

The above overview describes well-established mathematical approaches for the detection of ischemic lesions. Further adaptation could identify signatures within globally injured tissues. While little has been reported, the approaches described below could further delineate tissue level changes.

4.1 Detection of the Ischemic Core and Penumbra

Assessment of salvageable/nonsalvageable tissues and their evolution after a stroke is paramount for candidate selection for therapeutic interventions as well as for monitoring recovery and improving outcome prediction. Tissues within an ischemic lesion are not homogenous. Improved MRI-based estimation of the irreversibly injured core and the potentially salvageable penumbra may lead to better approaches for stroke treatment. Postmortem histology [60] is still considered to be the most reliable method for identifying the core and penumbra but is not clinically applicable. Instead, we have pursued an MRI-based noninvasive estimation of the core/penumbra. We have recently reported in a model of adult rodent ischemia that the expression of the astrocytic marker, glial fibrillary acidic protein (GFAP), was reduced in the penumbra. The penumbra region of interest (ROI) in GFAP pictures was initially localized manually but then confirmed and quantified semiautomatically using manually derived intensity-based threshold, morphological cleaning, then automated volume computation [61]. While we initially used immunohistochemical approaches to identify the penumbra, we sought to extend these studies for use to in vivo neuroimaging.

Current approaches to differentiate the core from penumbra rely on mismatch between diffusion weighted imaging (DWI) and perfusion weighted imaging (PWI) in relevant brain regions and this measure has been used clinically for more than a decade [62, 63]. However, a recent study has detailed several critical drawbacks of this method [64]. When using DWI and PWI, visual (subjective) estimation of the penumbra volume is performed [62, 63], or ROIs detected from one modality are manually traced on the other to estimate the mismatch [65], or manual registration between these two imaging modalities is performed to measure the mismatch [66]. These approaches are neither statistically reliable nor objective. What is required is an automated multimodality coregistration method between DWI and PWI [67]. Previously, one study has proposed T1/T2 lesion differences as possible measure of internal structure of supratentorial lesion [20], but this approach remains limited as it requires multimodality registration and has not been further validated.

We are investigating how variations between MR signal intensity/values within the ischemic/stroke lesion from a single MRI modality (e.g., T2WI or ADC) could discriminate core from penumbra. Because the human eye often cannot distinguish these subtle MR signal variations within the stroke, automated methods such as HRS or SVM (Fig. 42.2a) may be more sensitive, time-saving, and accurate. We have been able to verify our initial automated core/penumbra detection using immunohistochemical validation (Fig. 42.5). In the HRS tree (Fig. 42.1b), we first detect the ischemic lesion and then continue recursive lesion splitting into more uniform subregions. This subdivision allows us to detect the core (higher T2 values) as well as the penumbra (lower T2 values; Fig. 42.1b). The core/penumbra separation is automatically determined from the data based on the actual T2 means of the lesion subregions and prior expert knowledge (Table 42.2). Results from our animal data demonstrate that core/penumbral regions extracted using the HRS method matched

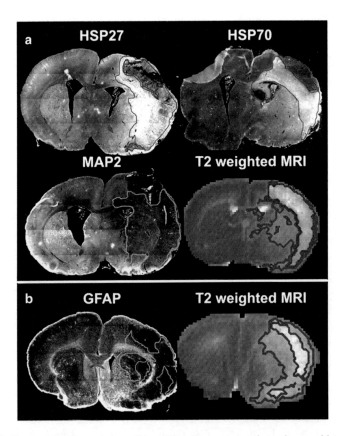

Fig. 42.5 Computational core–penumbra delineation compared to immunohistochemistry. (**a**) HSP27 immunoreactivity (–ir) (*bright region*) and activated HSP70-ir (*bright region*) illustrate the tissue encompassed by the penumbra (see text). Regions of hypointensity (*dark regions*) MAP2 staining reveals the lesion core. The core (*red*) and penumbra (*blue*) detection by HRS on T2WI matches well with corresponding immunohistochemical detections. (**b**) The core (*darkest region*) and penumbra (*less dark area* around the core) manually detected in GFAP stained section also match well with the HRS detected core (*red*) and penumbra (*blue*) in a corresponding T2WI

remarkably well with immunohistochemical staining (Fig. 42.5). We used four molecular markers that have been previously characterized in stroke [60]; (1) heat shock protein 27 (HSP27) is a stress-induced protein that can be found in both astrocytes and neurons within the penumbral tissues, (2) HSP70 (also known as HSP72) is restricted mostly to penumbra tissue, primarily in neurons, (3) glial fibrillary acidic protein (GFAP) as a marker for reactive astrocytes, that are often found in the penumbra, and (4) microtubule associated protein 2 (MAP2) which stains primarily for infarct core tissues and is a global marker for cytoskeletal degradation, independent of cell type.

After semiautomated image registration of immunohistochemical images and MRI of injured brains, good overlap was obtained for the core (sensitivity: 0.60, specificity: 0.95, similarity: 0.83) and the penumbra (sensitivity: 0.59, specificity: 0.89,

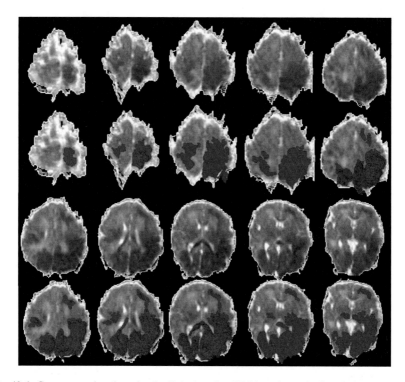

Fig. 42.6 Core–penumbra detection in clinical stroke. HRS based core (*red*) and penumbra (*blue*) detection in a neonatal brain 3 days after stroke. Core regions might be disconnected across consecutive MR slices and a particular MR slice may have only penumbra (*no core*) visible

similarity: 0.65) regions of the ischemic injury [68]. Our findings suggest that HRS has the ability to detect the core/penumbra regions from a single MR modality (as compared to two modalities when DWI/PWI mismatch is used) and also avoids the error prone registration step with DWI/PWI. We performed HRS detection of core/penumbral regions from T2 and ADC maps in several different animal models of ischemia [5, 13]. In addition, we have been able to detect core/penumbra in term neonates suffering from ischemic perinatal stroke (Fig. 42.6).

4.2   Detection of Implanted Stem Cells in Ischemic Brain

Given the current and future interest regarding stem cell therapy, computational approaches to rapidly, objectively and automatically identify these cells in the brain or other organs would have unprecedented utility. We have applied computational approaches to identify neural stem cells (NSCs) implanted following rodent neonatal hypoxia ischemia. NSCs inhibit scar formation, diminish inflammation, promote angiogenesis, provide neurotrophic and neuroprotective support, and stimulate host regenerative processes [2, 69]. Behavioral and anatomical improvements (lesion volume reduction) after NSC implantation in ischemic brain have been reported [70] and immunohistochemical data have shown that NSCs integrate into injured tissues and develop functionally active connections [70].

To noninvasively monitor implanted NSCs, several labeling techniques have been utilized including, MRI T1-shortening agents (i.e., gadolinium-diethylene tri-amine pentaacetic acid (Gd-DTPA) labeling) [71], positron emitting isotopes in PET [72], and MRI T2 shortening contrast agents (i.e., Feridex, iron oxide labeling) [73]. We have utilized iron oxide labeling of NSCs to noninvasively visualize these cells as they migrate and replicate in ischemic tissue in translational HII models, even as long as 58 weeks after injury [2, 73]. However, the majority of current NSC detection methods are performed by visual [74] or ad hoc thresholds or templates [15]. NSC locations are currently manually evaluated by observation of MR visible signal voids (iron labeled NSC) at some distance from the implantation site [2]. Reported studies on replication (i.e., proliferation) of NSCs, including our own calculated proliferation index (CPI), also lack region- and cluster-specific quantification [2, 73]. NSC proliferation or replication is defined as an increase in the number of NSCs (quantified by volume and cell density), within a particular NSC cluster over time. However, cellular density cannot be adequately determined using visual methods, hampering true quantitative measurements of replication. Recent studies have attempted to quantify the amount of iron, hence the number of NSCs, using susceptibility weighted images (SWI) and maps [75]. While location is important, the number, density, and replication of these NSCs are equally critical. A recent review on NSC therapy noted that the lack of computational NSC monitoring tools has hindered finding spatiotemporal characteristics of implanted cells [76]. Objective (computational) cellular tracking using feature based matching of NSC clusters (correspondence) at different time points will be required. In addition to NSC volumes, a feature vector developed for individual NSC clusters should contain characteristics that include MR signal intensity or values, 2D/3D shapes, cellular density (measure of signal void), and differences in the gradients of cell density inside (subregions) and outside (gradients across boundary) for each NSC cluster. NSC characteristics (e.g., detectability, density, migration, and replication) will be dependent on their location relative to the HII lesion and the time after implantation. Computational methods could noninvasively quantify and derive the speed/direction of NSC migration, the correlation between migration and cell density variations in NSC clusters, and when, where, and how many NSCs replicate. Such "online" data extraction and analytical approaches are necessary to assure that NSCs have been appropriately implanted, that they remain viable, reach their injury targets, and that host safety after implantation is not compromised.

We have used a deep-rooted HRS tree paradigm (Fig. 42.1b) that can segment very small brain regions based on low kurtosis and low standard deviation values from T2 images and T2 maps. Using previous knowledge (Table 42.2), we reliably detected iron-labeled implanted NSCs from T2WI (Fig. 42.7). Our HRS automated results of NSC detection at and near the implantation site at different imaging time points (Fig. 42.7) are promising and are currently being validated using immunohistochemical staining [77, 78]. We have extended this method to detect iron-labeled NSCs from SWI which we expect will have increased detectability

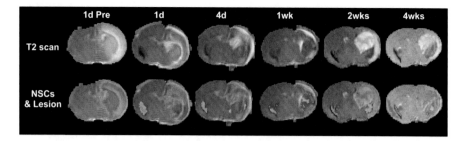

Fig. 42.7 Automatic derivation of NSC and ischemic lesion volumes from MRI. In the Rice-Vanucci model of neonatal ischemia, T2WI were used to delineate normal appearing brain matter, NSCs, and ischemic lesions. The *top row* shows raw T2WI data from serial neuroimaging. The *bottom row* superimposes the HRS extracted lesion (*red*) and NSC (*yellow*) locations onto the T2 maps

due to the exquisite sensitivity of SWI to enhance detection of iron particles within NSCs [75].

4.3 Anatomy-Specific Lesion and Stem Cell Quantification

Stroke within certain anatomical regions (e.g., posterior limb of the internal capsule) can cause significant functional damage, independent of the size of the lesion. In addition, stem cells might have preferential paths in the brain (e.g., through white matter tracts) to reach sites close to the ischemic injury. Better assessment of stroke severity combined with enhanced monitoring of therapeutic interventions could be further strengthened by automated regional and anatomical segmentation. Some treatments may preferentially target cortical or subcortical regions. Specific brain disorders that are known to affect certain anatomical regions (e.g., Parkinson disease) could use an automated atlas or model-based method to minimize false positives. Few studies have been undertaken in this arena.

Previous studies have used brain anatomical templates to first segment brain tissue and then attempt detection of ischemic tissue within these segmented regions [23]. In contrast, we propose a different sequence, whereby the lesions (or implanted cells) are detected first and are then followed by localization based on an anatomical atlas. We believe this would restrict anatomical registration-related errors, as we are not detecting the injury in this step but only quantifying the anatomy-specific proportions of the injured tissues.

Numerous adult brain atlases exist but there are only few available pediatric atlases. A small sample of adult anatomical templates includes: (a) Laboratory of Neuroimaging at the University of California, Los Angeles, CA, USA (http://www.loni.ucla.edu/Atlases/), (b) Biodiversity Bank at Michigan State University, Hickory Corners, MI, USA (https://www.msu.edu/~brains/brains/human/index.html), (c) BrainWeb at MacGill University, Montreal, QC, Canada (http://mouldy.bic.mni.mcgill.ca/brainweb/), and (d) Imperial College, London, UK (http://biomedic.doc.ic.ac.uk/brain-development/index.php?n=Main.Adult). Some pediatric atlases

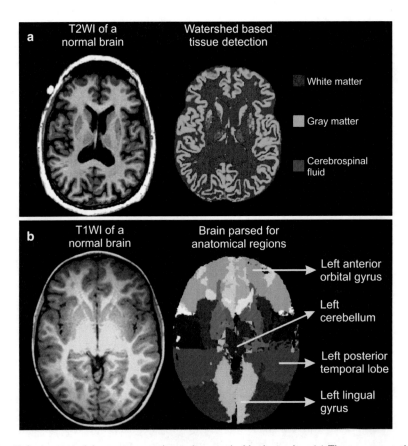

Fig. 42.8 Automated tissue segmentation and anatomical brain parsing. (**a**) Tissue segmentation: The watershed tissue segmentation method of a normal brain MRI into different tissues: white matter (*red*), gray matter (*green*), cerebrospinal fluid (*blue*). Skull-stripping was performed as well. (**b**) Anatomical brain parsing: Anatomical region mapping in an 8-year-old normal pediatric brain using a 2-year-old pediatric brain template from Imperial College, London, UK available online (http://biomedic.doc.ic.ac.uk/brain-development/index.php?n=Main.Neonatal2)

include, (a) Imperial College, London, UK (http://biomedic.doc.ic.ac.uk/brain-development/index.php?n=Main.Neonatal2), (b) Seattle Children's Hospital, Seattle, WA, USA (http://www.seattlechildrens.org/healthcare-professionals/education/radiology/pediatricbrainatlas/), and (c) University of North Carolina, Chapel Hill, NC, USA (https://www.med.unc.edu/bric/ideagroup/free-softwares/unc-infant-0-1-2-atlases).

It is known that one anatomical region may contain multiple tissues (e.g., GM and WM), and injury susceptibility may vary within a single anatomical region also. Thus, anatomical brain parsing results can be further complimented by tissue detections (GM, WM, CSF). We have used our modified watershed method (Sect. 3.4) to extract tissue based segmentation of human brain (Fig. 42.8a). Our automated registration of a normal 8-year old pediatric brain to a pediatric brain atlas containing 83 brain regions (Imperial College atlas, comprised of 33 two-year old normal children [79]) is illustrated in Fig. 42.8b. Utilizing the Pipeline environment developed by the Laboratory of Neuroimaging at UCLA (http://pipeline.loni.ucla.edu/; [80]) we have found that even without age-matching, anatomical brain regions could be mapped accurately. Beside the already available brain atlases, case-specific atlases can be developed for particular studies to improve computational performance. Such atlases can be generated from age-matched control brains by: (a) coregistering to a standard 3D space, (b) averaging out the individual subtleties, and (c) manually demarking specific brain regions. Such atlases can be developed separately for a particular MR modality (e.g., T2WI, ADC, Diffusion Tensor Imaging, etc.) or images from different modalities can be registered (multimodal registration) to the atlas space (e.g., generally T1/T2 MRI) to acquire anatomical information. Finally, the injured brain can be registered (intra- or intermodality registration) to find region-specific statistical details of the injury. Severe internal or external brain distortions in the diseased brain may create problems with atlas registration, but this can be reduced if only the normal brain areas of the injured brain (after removing the lesion areas) are utilized to compute the image registration matrix [7]. This registration matrix can be used for the entire brain, including injured tissues, for registering it to an atlas followed by extraction of anatomy-specific measures of the injury (or labeled stem cells if implanted).

By using data from core/penumbra regions, GM/WM/CSF tissue regions, and atlas-based brain anatomical localization, one could further extract and quantify region-specific information for the above described data sets, including mean MR values (e.g., T2, ADC, etc.), standard deviations, textures, shapes, kurtosis, skewness, etc. Such detailed complimentary information would be extremely useful in understanding the spatiotemporal evolution of stroke, reparative activities of implanted stem cells and their interactions, and additional therapeutics. These types of assessments cannot be reasonably conducted manually due to the large volume of data and the subjectivity associated with current manual methods.

4.4 Monitoring Spatiotemporal Interactions Between Stroke Tissues and Stem Cells

A final frontier in stroke lesion analysis is automated computation of the process of injury recovery with or without treatment. Beyond simplistic assessment of stem cell volumes, automated algorithms should be able assess stem cell related therapeutic activities such as migration, replication, localization as well as the interaction between the lesion tissues and NSC activities for better understanding of this complex dynamic process. One should be able to determine whether an anatomical region is recovering faster than another, whether the lesion is shifting in the anatomical space, how the MR statistics (e.g., mean, standard deviation) of the lesion changes across anatomy and over time, etc. Despite research using automated

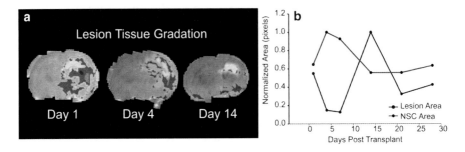

Fig. 42.9 Lesion evolution and interaction. (**a**) Spatiotemporal changes in lesion composition: T2 images from the same RVM rat pup imaged at days 1, 4, and 14 after HII. Each panel shows how the MR values change within the lesion, including the area and shape of the lesion and its subregions over time. Mean T2 relaxation times of the subregions are significantly different: *red* (199±2 ms), *green* (208±3 ms), *blue* (217±3 ms), and *yellow* (231±6 ms). These changes likely reflect different degrees of necrosis/apoptosis and may also reflect differential tissue salvageability. (**b**) Interrelations between temporal evolution of HII lesion and NSC areas. 2D areas of HII lesion and NSC at the same brain level vary over time. The trends are inversely correlated, implying that initially decreasing lesion area gives more viable tissues for NSCs to replicate and as lesion volume increases later, either NSCs die or differentiate into other cells

methods to identify brain lesions, very limited research has considered the importance of the spatiotemporal evolution of the lesion [35]. Kabir and colleagues have qualitatively shown that adult stroke shifts from one anatomical region to other regions without further quantitative evaluation. Estimating or predicting these dynamic changes will expedite advanced monitoring of injury progression and of optimizing treatment strategies.

As ischemic/stroke tissues provide chemoattractive/chemorepulsive cellular and molecular cues [2, 73] combined with differentiation [70] signals to stem cells for injury repair, there is likely a strong interrelationship between these two processes that mutually drive one another. "Toxicity" at the lesional or perilesional regions from dying cells in a compromised vascular microenvironment might be expected to negatively affect NSC viability [2] and we believe that this, as well as the characteristics of homing signals from the lesion, will influence stem cell viability, the extent of cellular migration, replication (i.e., proliferation), and differentiation along with tissue recovery.

Such spatiotemporal interactions are visually and qualitatively reported [81] and the necessity of objective quantification is highlighted in a recent stem cell research review [82]. Our recent work shows that HRS-derived subregions in experimental ischemic lesions (Fig. 42.9a) reveal considerable cellular dynamics in space and time. The HRS method [32] automatically detects ischemic lesions and implanted iron-labeled stem cells (at the implantation site and close to the lesion) over the time course of several weeks (Fig. 42.7). Examining the interaction between injured tissues and stem cell interactions would suggest that stem cell replication is enhanced when lesion volume is reduced (Fig. 42.9b) in the subacute phase (4–5 days after stroke). As the ischemic lesion volume starts to increase toward its final size, stem cells start to die, presumably due to a more toxic environment. Discovering such patterns using regression analysis or data mining strategies using computational pattern recognition [29] likely will be a key direction for future translational stroke research.

5 Conclusions

Our own studies and those of others suggest that significant advances in computational analytical methods of MRI, CT, and PET data will result in clinically useful applications that have the potential to greatly improve the treatment of children and adults who suffer from focal and global ischemic injuries. The computational approaches we have described provide a glimpse of some of the future directions of automated assessments from imaging data. These approaches with some adaptation can be used for other acute, chronic, and acquired brain injuries.

Acknowledgments Part of this work has been funded by NIH NINDS 1R01NS059770-01A2, National Medical Test Bed (NMTB), LLU Pediatric Research Fund, and an anonymous donation to the Loma Linda University School of Medicine. Part of the research by Drs. Bhanu and Sun has been funded by NSF grants 0641076, 0727129, and 0903667. We are grateful to Dr. Samuel Barnes (LLU) for SVM results, Beatriz Tone and Dr. Hui Rou Tian (LLU) for surgical procedures, Kamal Ambadipudi and Sonny Kim (LLU) for technical assistance with MRI acquisition, Dr. Jerome Badaut (LLU) for use of histochemical equipment, Dr. Evan Y. Snyder (Sanford-Burnham Medical Research Institute, La Jolla, CA, USA) for iron-labeled stem cells, Dr. Ivo Dinov and Dr. Alen Zamanyan (Laboratory of Neuroimaging, UCLA, Los Angeles, USA) for assistance on using LONI Pipeline and brain parsing.

References

1. Ashwal S, Tone B, Tian HR, Chong S, Obenaus A. Serial magnetic resonance imaging in a rat pup filament stroke model. Exp Neurol. 2006;202:294–301.
2. Ashwal S, Obenaus A, Snyder EY. Neuroimaging as a basis for rational stem cell therapy. Pediatr Neurol. 2009;40:227–36.
3. Saunders DE, Clifton AG, Brown MM. Measurement of infarct size using MRI predicts prognosis in middle cerebral artery infarction. Stroke. 1995;26:2272–6.
4. Schiemanck SK, Post MWM, Kwakkel G, Witkamp TD, Kappelle LJ, Prevo AJH. Ischemic lesion volume correlates with long-term functional outcome and quality of life of middle cerebral artery stroke survivors. Restor Neurol Neurosci. 2005;23:257–63.
5. Vannucci RC, Vannucci SJ. Perinatal hypoxic-ischemic brain damage: evolution of an animal model. Dev Neurosci. 2005;27:81–6.
6. Coats JS, Freeberg A, Pajela EG, Obenaus A, Ashwal S. Meta-analysis of apparent diffusion coefficients in the newborn brain. Pediatr Neurol. 2009;41:263–74.
7. Shapiro LG. Stockman GC. Computer Vision: Prentice Hall; 2001.
8. Niimi T, Imai K, Maeda H, Ikeda M. Information loss in visual assessments of medical images. Eur J Radiol. 2007;61:362–6.
9. Barkovich AJ, Westmark K, Partridge C, Sola A, Ferriero DM. Perinatal asphyxia: MR findings in the first 10 days. AJNR Am J Neuroradiol. 1995;16:427–38.

10. Barkovich AJ, Hajnal BL, Vigneron D, Sola A, Partridge JC, Allen F, Ferriero DM. Prediction of neuromotor outcome in perinatal asphyxia: evaluation of MR scoring systems. AJNR Am J Neuroradiol. 1998;19:143–9.

11. Haataja L, Mercuri E, Guzzetta A, Rutherford M, Counsell S, Flavia Frisone M, Cioni G, Cowan F, Dubowitz L. Neurologic examination in infants with hypoxic-ischemic encephalopathy at age 9 to 14 months: use of optimality scores and correlation with magnetic resonance imaging findings. J Pediatr. 2001;138:332–7.

12. Rutherford MA, Pennock JM, Counsell SJ, Mercuri E, Cowan FM, Dubowitz LM, Edwards AD. Abnormal magnetic resonance signal in the internal capsule predicts poor neurodevelopmental outcome in infants with hypoxic-ischemic encephalopathy. Pediatrics. 1998; 102:323–8.

13. Recker R, Adami A, Tone B, Tian HR, Lalas S, Hartman RE, Obenaus A, Ashwal S. Rodent neonatal bilateral carotid artery occlusion with hypoxia mimics human hypoxic-ischemic injury. J Cereb Blood Flow Metab. 2009;29:1305–16.

14. Jiang Q, Zhang ZG, Ding GL, Zhang L, Ewing JR, Wang L, Zhang R, Li L, Lu M, Meng H, Arbab AS, Hu J, Li QJ, Pourabdollah Nejad DS, Athiraman H, Chopp M. Investigation of neural progenitor cell induced angiogenesis after embolic stroke in rat using MRI. Neuroimage. 2005;28:698–707.

15. Mills PH, Wu Y-JL, Ho C, Ahrens ET. Sensitive and automated detection of iron-oxide-labeled cells using phase image cross-correlation analysis. Magn Reson Imaging. 2008;26:618–28.

16. Otsu N. A threshold selection method from gray-level histograms. IEEE Trans Syst Man Cybern. 1979;9:62–6.

17. Flexman JA, Cross DJ, Kim Y, Minoshima S. Morphological and parametric estimation of fetal neural stem cell migratory capacity in the rat brain. Conference Proceedings: Annual International Conference of the IEEE Engineering in Medicine and Biology Society IEEE Engineering in Medicine and Biology Society Conference. 2007; p. 4464–7.

18. Ashwal S, Caots JS, Bianchi A, Bhanu B, Obenaus A. Semi-automated segmentation of ADC maps reliably defines ishchemic perinatal stroke injury. Ecquevilly, France: Sixth Hershey conference on developmental brain injury; 2008.

19. Bricq S, Collet C, Armspach JP. Markovian segmentation of 3D brain MRI to detect multiple sclerosis lesions. Proc of 15th IEEE international conference on image processing (ICIP). San Diego, CA; 2008. p. 733–6.

20. Dugas-Phocion G, Gonzalez MA, Lebrun C, Chanalet S, Bensa C, Malandain G, Ayache N. Hierarchical segmentation of multiple sclerosis lesions in multi-sequence MRI. Proc of IEEE International Symposium on Biomedical Imaging 2007 (ISBI 2007). Arlington, VA, USA; 2007. p. 157–60.

21. Loyek C, Woermann FG, Nattkemper TW. Detection of focal cortical dysplasia in MRI using textural features. Workshop on algorithm, systems and Anwendungen. Berlin: Springer; 2008. p. 432–6.

22. Bergo FPG, Falcao AX, Yasuda CL, Cendes F. FCD segmentation using texture asymmetry of MR-T1 images of the brain. Proc 5th IEEE Intl Symp Biomed Img: from Nano to Macro (ISBI). Paris, France; 2008. p. 424–7.

23. de Boer R, Der Lijn F, Vrooman H, Vernooij M, Ikram M, Breteler M, Niessen W. Automatic segmentation of brain tissue and whitematter lesions in MRI. Proc of IEEE International Symposium on Biomedical Imaging 2007 (ISBI 2007). Arlington, VA; 2007. p. 652–5.

24. Agam G, Weiss D, Soman M, Arfanakis K. Probabilistic brain lesion segmentation in DT-MRI. Proc of IEEE Intl Conf on Image Processing (ICIP). Atlanta Marriott Marquis, Atlanta, GA; 2006. p. 89–92.

25. Freifeld O, Greenspan H, Goldberger J. Lesion detection in noisy MR brain images using constrained GMM and active contours. Proc of IEEE international symposium on biomedical imaging 2007 (ISBI 2007). Arlington, VA; 2007. p. 596–9.

26. Ibrahim M, John N, Kabuka M, Younis A. Hidden Markov models-based 3D MRI brain segmentation. Image Vis Comput. 2006;24:1065–79.

27. Song Z, Tustison N, Avants B, Gee J. Adaptive graph cuts with tissue priors for brain MRI segmentation. Proc of IEEE international symposium on biomedical imaging (ISBI). Arlington, VA; 2006. p. 762–5.

28. He Q, Karsch K, Duan Y. A novel algorithm for automatic brain structure segmentation from MRI. Advances in visual computing. Berlin: Springer; 2008. p. 552–61.

29. Duda RO, Hart PE, Stork DG. Pattern classification. Hoboken: Wiley-Interscience; 2000.

30. Greenspan H, Ruf A, Goldberger J. Constrained Gaussian mixture model framework for automatic segmentation of MR brain images. IEEE Trans Med Imaging. 2006;25:1233–45.

31. Fan L-W, Lin S, Pang Y, Lei M, Zhang F, Rhodes PG, Cai Z. Hypoxia-ischemia induced neurological dysfunction and brain injury in the neonatal rat. Behav Brain Res. 2005;165:80–90.

32. Ghosh N, Recker R, Shah A, Bhanu B, Ashwal S, Obenaus A. Automated ischemic lesion detection in a neonatal model of hypoxic ischemic injury. J Magn Reson Imaging. 2011;33:772–81.

33. Rouainia M, Medjram MS, Doghmance N. Brain MRI segmentation and lesions detection by EM algorithm. Proc of World Academy of Science, Engineering and Technology; 2006. p. 301–4.

34. Yu J, Bhanu B. Super-resolution restoration of facial images in video. Proc of IEEE Intl Conf on pattern recognition (ICPR). Hong Kong, China; 2006. p. 342–5.

35. Kabir Y, Dojat M, Scherrer B, Forbes F, Garbay C. Multimodal MRI segmentation of ischemic stroke lesions. Conference proceedings: annual international conference of the IEEE Engineering in Medicine and Biology Society IEEE Engineering in Medicine and Biology Society Conference; 2007. p. 1595–8.

36. Korb K, Nicholson AE. Bayesian artificial intelligence. Boca Raton, FL: Chapman & Hall; 2003.

37. Chen R, Herskovits EH. A Bayesian network classifier with inverse tree structure for voxelwise magnetic resonance image analysis. Proceeding of the eleventh ACM SIGKDD international conference. Chicago, IL; 2005. p. 4.

38. Shi J, Malik J. Normalized cuts and image segmentation. IEEE Trans Pattern Anal Mach Intell. 2000;22:888–905.

39. Liao L, Lin T, Li B. MRI brain image segmentation and bias field correction based on fast spatially constrained kernel clustering approach. Pattern Recognition Letters. 2008;29: 1580–8.

40. Nakamura K, Fisher E. Segmentation of brain magnetic resonance images for measurement of gray matter atrophy in multiple sclerosis patients. Neuroimage. 2009;44:769–76.

41. Castellano G, Bonilha L, Li LM, Cendes F. Texture analysis of medical images. Clin Radiol. 2004;59:1061–9.

42. Antel SB, Collins DL, Bernasconi N, Andermann F, Shinghal R, Kearney RE, Arnold DL, Bernasconi A. Automated detection of focal cortical dysplasia lesions using computational models of their MRI characteristics and texture analysis. Neuroimage. 2003;19:1748–59.

43. Dokladal P, Bloch I, Couprie M, Ruijters D, Urtasun R, Garnero L. Topologically controlled segmentation of 3D magnetic resonance images of the head by using morphological operators. Pattern Recognition. 2003;36:2463–78.

44. Prastawa M, Gerig G. Brain lesion segmentation through physical model estimation. Advances in visual computing. Berlin: Springer; 2008 p. 562–71.

45. Kang X, Yund EW, Herron TJ, Woods DL. Improving the resolution of functional brain imaging: analyzing functional data in anatomical space. Magn Reson Imaging. 2007;25:1070–8.

46. Vapnik VN. Statistical learning theory. New York: Wiley-Blackwell; 1998.

47. Lao Z, Shen D, Liu D, Jawad AF, Melhem ER, Launer LJ, Bryan RN, Davatzikos C. Computer-assisted segmentation of white matter lesions in 3D MR images using support vector machine. Acad Radiol. 2008;15:300–13.

48. Saha S, Bandyopadhyay S. MRI brain image segmentation by fuzzy symmetry based genetic clustering technique. Proc of IEEE Cong on evolutionary computation, Singapore; 2007. p. 4417–24.

49. Ray N, Greiner R, Murtha A. Using symmetry to detect abnormalities in brain MRI. Proc Comp Soc Ind Comm. 2008;31:7–10.
50. Sun Y, Bhanu B. Symmetry integrated region-based image segmentation. Proc IEEE Conf on computer vision and pattern recognition (CVPR). Miami, FL; 2009. p. 826–31.
51. Sun Y, Bhanu B, Bhanu S. Automatic symmetry-integrated brain injury detection in MRI sequences. Proc IEEE CVPR workshop on mathematical methods in biomedical image analysis. Miami, FL; 2009. p. 79–86.
52. Beucher S. The watershed transformation applied to image segmentation. Microscopy and Microanalysis: Pfefferkorn Conf on Signal and Image Processing in; 1991.
53. Beucher S, Meyer F. The morphological approach to segmentation: the watershed transform. In: Dougherty ER, editor. Mathematical morphology in image processing. New York, NY: Marcel Dekker; 1993. p. 433–81.
54. Vincent L, Soille P. Watersheds in digital spaces: an efficient algorithm based on immersion simulations. IEEE Trans Pattern Anal Mach Intell. 1991;13:583–98.
55. Nguyen HT, Ji Q. Improved watershed segmentation using water diffusion and local shape priors. Proc IEEE Conf computer vision and pattern recognition. New York, NY; 2006. p. 985–92.
56. Cousty J, Bertrand G, Najman L, Couprie M. Watershed cuts: thinnings, shortest path forests, and topological watersheds. IEEE Trans Pattern Anal Mach Intell. 2010;32:925–39.
57. Grau V, Mewes AU, Alcañiz M, Kikinis R, Warfield SK. Improved watershed transform for medical image segmentation using prior information. IEEE Trans Med Imaging. 2004;23:447–58.
58. Sun Y, Ghosh N, Obenaus A, Ashwal S, Bhanu B. Automated symmetry-integrated brain ROI detection in MRI sequences: a comparison. *IEEE Trans Med Imaging* (TMI) (submitted).
59. Bhanu B, Lee S. Genetic learning for adaptive image segmentation. Boston, MA: Kluwer; 1994.
60. Popp A, Jaenisch N, Witte OW, Frahm C. Identification of ischemic regions in a rat model of stroke. PLoS One. 2009;4:e4764.
61. Titova E, Ostrowski RP, Adami A, Badaut J, Lalas S, Ghosh N, Vlkolinsky R, Zhang JH, Obenaus A. Brain irradiation improves focal cerebral ischemia recovery in aged rats. J Neurol Sci. 2011;306:143–53.
62. Wechsler LR. Imaging evaluation of acute ischemic stroke. Stroke. 2011;42:S12–5.
63. Olivot J-M, Albers GW. Diffusion-perfusion MRI for triaging transient ischemic attack and acute cerebrovascular syndromes. Curr Opin Neurol. 2011;24:44–9.
64. Wardlaw JM. Neuroimaging in acute ischaemic stroke: insights into unanswered questions of pathophysiology. J Intern Med. 2010;267:172–90.
65. Straka M, Albers GW, Bammer R. Real-time diffusion-perfusion mismatch analysis in acute stroke. J Magn Reson Imaging. 2010;32:1024–37.
66. Schlaug G, Benfield A, Baird AE, Siewert B, LÃvblad KO, Parker RA, Edelman RR, Warach S. The ischemic penumbra: operationally defined by diffusion and perfusion MRI. Neurology. 1999;53:1528–37.
67. Ma H, Zavala JA, Teoh H, Churilov L, Gunawan M, Ly J, Wright P, Phan T, Arakawa S, Davis SM, Donnan GA. Penumbral mismatch is underestimated using standard volumetric methods and this is exacerbated with time. J Neurol Neurosurg Psychiatry. 2009;80:991–6.
68. Ghosh N, Turenius CI, Tone B, Snyder EY, Obenaus A, Ashwal S. Automated core-penumbra quantification in neonatal ischemic brain injury. Stroke (submitted).
69. Singec I, Jandial R, Crain A, Nikkhah G, Snyder EY. The leading edge of stem cell therapeutics. Annu Rev Med. 2007;58:313–28.
70. Park KI, Himes BT, Stieg PE, Tessler A, Fischer I, Snyder EY. Neural stem cells may be uniquely suited for combined gene therapy and cell replacement: evidence from engraftment of neurotrophin-3-expressing stem cells in hypoxic-ischemic brain injury. Exp Neurol. 2006;199:179–90.
71. Adler ED, Bystrup A, Briley-Saebo KC, Mani V, Young W, Giovanonne S, Altman P, Kattman SJ, Frank JA, Weinmann HJ, Keller GM, Fayad ZA. In vivo detection of embryonic stem

cell-derived cardiovascular progenitor cells using Cy3-labeled Gadofluorine M in murine myocardium. JACC Cardiovasc Imaging. 2009;2:1114–22.

72. Qiao H, Zhang H, Zheng Y, Ponde DE, Shen D, Gao F, Bakken AB, Schmitz A, Kung HF, Ferrari VA, Zhou R. Embryonic stem cell grafting in normal and infarcted myocardium: serial assessment with MR imaging and PET dual detection. Radiology. 2009;250:821–9.

73. Obenaus A, Dilmac N, Tone B, Tian HR, Hartman R, Digicaylioglu M, Snyder EY, Ashwal S. Long-term magnetic resonance imaging of stem cells in neonatal ischemic injury. Ann Neurol. 2011;69:282–91.

74. Guzman R, Bliss T, De Los Angeles A, Moseley M, Palmer T, Steinberg G. Neural progenitor cells transplanted into the uninjured brain undergo targeted migration after stroke onset. J Neurosci Res. 2008;86:873–82.

75. Kressler B, de Rochefort L, Liu T, Spincemaille P, Jiang Q, Wang Y. Nonlinear regularization for per voxel estimation of magnetic susceptibility distributions from MRI field maps. IEEE Trans Med Imaging. 2009;29:273–81.

76. Kraitchman DL, Gilson WD, Lorenz CH. Stem cell therapy: MRI guidance and monitoring. J Magn Reson Imaging. 2008;27:299–310.

77. Ghosh N, Turenius CI, Tone B, Obenaus A, Ashwal S. MRI-based automated monitoring of activities of implanted stem cells in neonatal ischemic injury. Ann Neurol (submitted).

78. Turenius CI, Ghosh N, Dulcich M, Denham CM, Tone B, Hartman R, Snyder EY, Obenaus A, Ashwal S. Iron toxicity and gender based study of implanted hNSC in neonatal ischemic injury. Exp Neurol (submitted).

79. Gousias IS, Rueckert D, Heckemann RA, Dyet LE, Boardman JP, Edwards AD, Hammers A. Automatic segmentation of brain MRIs of 2-year-olds into 83 regions of interest. Neuroimage. 2008;40:672–84.

80. Dinov ID, Van Horn JD, Lozev KM, Magsipoc R, Petrosyan P, Liu Z, Mackenzie-Graham A, Eggert P, Parker DS, Toga AW. Efficient, distributed and interactive neuroimaging data analysis using the LONI pipeline. Front Neuroinform. 2009;3:22.

81. Faiz M, Acarin L, Villapol S, Schulz S, Castellano B, Gonzalez B. Substantial migration of SVZ cells to the cortex results in the generation of new neurons in the excitotoxically damaged immature rat brain. Mol Cell Neurosci. 2008;38:170–82.

82. Kim D, Hong KS, Song J. The present status of cell tracking methods in animal models using magnetic resonance imaging technology. Mol Cells. 2007;23:132–7.

Chapter 43
Innovations in Stroke Clinical Trial Design

Karen L. Furie and Michael K. Parides

Abstract The pace of development of new drugs and devices for the treatment of stroke is dependent upon the rate that promising compounds can be tested in humans. While this final stage of drug development is the product of years of preclinical work, it is a critical step not only because these trials are extremely expensive and take many years to complete but also because poorly designed clinical trials can derail a promising therapy or promote an ineffective one.

Historically, the stroke field has suffered from a poor track record in drug development. The only approved therapy for acute stroke is intravenous tPA (N Engl J Med 333:1581–87). Stroke preventive therapies include antithrombotic agents and statins (Eur Heart J. 2008;29:1082–3; BMJ. 2002;324:71–86; N Engl J Med. 2006;355:549–59). There are no proven therapies for stroke recovery. There are numerous obstacles inherent to developing therapies for stroke, such as the need to deliver drug to the affected blood vessels and injured brain tissue, the narrow window for intervention, and the heterogeneity of the patient population.

1 Clinical Trial Phases

Before any interventions are assessed in humans, preclinical experiments must have established that the proposed interventions are sufficiently safe to warrant investigation in humans. A variety of nonclinical information is necessary to support informed

K.L. Furie, MD, MPH (✉)
J. Philip Kistler MGH Stroke Research Center,
175 Cambridge Street, Suite 300, Boston, MA 02114, USA
e-mail: kfurie@partners.org

M.K. Parides, PhD
Director, Mount Sinai Center for Biostatistics, Director of Biostatistics of the International
Center for Health Outcomes and Innovation Research (InCHOIR), Professor, Health Evidence
and Policy, Mount Sinai School of Medicine, New York

P.A. Lapchak and J.H. Zhang (eds.), *Translational Stroke Research*, Springer Series
in Translational Stroke Research, DOI 10.1007/978-1-4419-9530-8_43,
© Springer Science+Business Media, LLC 2012

Table 43.1 Trial classification by study objective

Study type	Study objective	Examples
Human pharmacology	• Assess tolerance • Define pharmacokinetics (PK) and pharmacodynamics (PD) • Explore drug metabolism and interactions • Determine drug activity	• Dose-tolerance studies • PK and PD studies (single or multiple dose) • Drug interaction studies
Therapeutic exploratory	• Explore use for targeted indication • Dose estimation for subsequent trials • Determine feasibility of pivotal trials • Develop and refine end points and methods for future trials	• Dose–response trials • Early trials of short duration in well-defined populations using surrogate or clinical end points
Therapeutic confirmatory	• Demonstrate (confirm) efficacy • Establish safety • Establish dose–response	• Randomized parallel groups trials • Clinical safety trials • Trials of clinical end points • Mortality, morbidity • Large simple trials
Therapeutic use	• Refine understanding of benefit–risk relationship in general or special populations and/or environments • Identify less common adverse reactions • Refine dosing recommendation	• Comparative effectiveness studies • Studies of mortality/morbidity outcomes • Studies of additional end points • Large simple trials • Pharmacoeconomic studies

clinical trial design; including toxicology, pharmacology, and pharmacokinetics. Barring any reasonable safety concerns, clinical trials may begin at this juncture. No matter what the phase of a clinical trial, the primary objective must be clearly and explicitly stated, and the trial must be designed to ensure that the objective is achieved.

The rationale for a phased, or staged, process is that the design of later trials should be informed by the results of previous trials. This nomenclature (Phase I–Phase IV) is not immutable and alternative classification schemes have been proposed. In addition, the drug development paradigm does not correspond neatly to the development of devices, or to investigation of surgical or procedural interventions. An alternative classification scheme in terms of trial objectives is arguably more informative. The classification scheme proposed by the *International Conference on Harmonization of Technical Requirements for Registration of Pharmaceutical for Human Use* is presented in Table 43.1 [5].

2 Adaptive Designs

Rather than being locked into a rigid plan from the start of human trials, adaptive designs are prospectively designed to allow for modification of one or more study aspects based on interim analyses [6]. The primary benefits of adaptation are early identification of treatment failures so resources may be redirected, and the flexibility to revise a design based on assumptions, or other elements, discovered to be errone-ous, thereby salvaging a trial that might otherwise fail due to design flaws. However, these putative benefits come at a cost. Repeated examination of accumulated data, and decisions based on them, may introduce subtle, unknown biases into the trial due to changes in the characteristics of patients recruited, administration of the intervention, end point assessment, changes in investigator enthusiasm, and disrup-tion of equipoise. While these potential biases are not unique to adaptive approaches the risks are greater. Adaptive designs generally require meticulous planning to ensure that interim results are examined without introducing bias and that adaptations are performed so as not to increase the probability of falsely declaring treatment benefit when none exists.

3 Examples of Adaptive Designs

An adaptive design is defined as a multistage study design that uses accumulating data to decide on how to modify aspects of the study without undermining the *validity* and *integrity* of the trial. Maintaining trial validity refers to correct statistical inference (such as computation of correctly adjusted *p* values and confidence inter-vals, and unbiased estimation of treatment effects), assuring consistency between different stages of the study, and minimizing operational bias. Maintaining trial integrity means that the results are clearly interpretable, convincing, and acceptable to the broader scientific community.

Adaptively designed trials are generally conducted in multiple stages with potential changes made to successive stages based on examination of the accumulated data. All data are used in the final analysis, i.e., all stages are combined. Using terminology adopted by the PhRMA working group, an adaptive design may have one or more of the following rules applied at an interim look [7]:

- *Allocation rule*: Describes how patients will be allocated to the arms of the trial. Randomization allocations may be fixed (typically 1:1) throughout a trial, or it may be adaptive with the randomization ratio changing from stage to stage, or even from randomization to randomization, based on accruing data. This includes also the decision to drop or add treatment arms.
- *Sampling rule*: Describes the sample size for the subsequent stage. The sample size typically depends on accrual to this point, or on estimates of nuisance param-eters, e.g., variance, or even on estimates of treatment effect. In dose-escalation studies, this is the cohort size per stage.

- *Stopping rule*: Should the trial proceed or be halted? Typically, this decision is based on an examination of a trials current and future potential based on risk–benefit summary. Trials may be halted for efficacy, for harm, for futility, or for safety.
- *Decision rule*: All other decisions; including changing the primary end point, changing the primary analysis, and restrictions to the patient population being enrolled.

At any stage, data may be analyzed and subsequent stages redesigned taking into account all available data. Stages may be fixed as part of the design or may be flexibly determined themselves.

While adaptive designs for clinical trials have attracted a great deal of recent attention, the underlying idea is not new, and the basic ideas have been adopted and practiced for some time. Note that group sequential designs, for which the only design revision is stopping a trial early for sufficiently strong or weak evidence of a treatment effect, are an adaptive design. The concept of adaptive randomization, which increases the probability of assignment of superior performing treatments, is not new [8]. *Sample size reestimation,* the recalculation of sample size based on interim information about the values of nuisance parameters, is not an uncommon practice. Although not a new concept, recent adaptive approaches have considered bolder implementation including changing the target treatment difference for which the study is powered, changing the primary end point, varying the form of the primary analysis, and restricting the patient population to a subset of that originally targeted. Also new is the increasing use of Bayesian approaches, which naturally mesh with the idea of adapting (or updating) based on observed results.

3.1 Types of Adaptive Designs

The adaptations we describe are prospective in the sense that the adaptation has been anticipated, planned for, and will proceed according to a preformulated plan. This distinguishes them from the so-called concurrent or retrospective adaptations which are generally not anticipated but determined necessary during trial conduct. Concurrent changes are not unusual and include refinements to elements such as eligibility criteria, treatment administration, duration of follow-up, and sample size. Note that group sequential approaches are prospectively adaptive according to this distinction.

A variety of adaptive designs have been classified, based on the use of one or more of the rules described above. These include, but are not limited to, (1) group sequential designs, (2) adaptive randomization designs, (3) designs with sample-size reestimation, (4) adaptive dose finding (e.g., dose escalation), (5) designs dropping treatment arms (drop the loser designs), (6) seamless designs (generally seamless Phase II/III), (7) biomarker-adaptive or patient enrichment designs.

4 Experience with Adaptive Designs in Stroke Trials

ASTIN (Acute Stroke Therapy by Inhibition of Neutrophils) was an adaptive Phase 2 dose–response finding, proof-of-concept study conducted between November 2000 and November 2001 to establish whether an experimental treatment, UK-279,276, improved recovery in acute ischemic stroke [9]. This trial marked the first implementation of an adaptive dose response design in a large multicenter multinational stroke study. A Bayesian sequential design with real-time data capture and continuous reassessment of the dose response enabled double-blind, randomized, adaptive allocation to 1 of 15 doses (dose range, 10–120 mg) or placebo, and early termination for efficacy or futility. Dose allocation was based on a computer algorithm which determined the "optimal" dose. Each time the posterior estimate of the dose–response curve was calculated, the estimate of the effect over placebo at the ED_{95} was also evaluated to enable application of the stopping rule. In all, 966 patients were randomized in ASTIN. As soon as 90-day outcome data from 500 evaluable patients were available, futility was declared. The steering committee concurred and the sponsor terminated the trial.

More recently, the Neuroprotection with Statin Therapy for Acute Recovery Trial (NeuSTART) used a continuous reassessment method (CRM) to select the maximal tolerated dose (MTD) of lovastatin in acute ischemic stroke [10]. NeuSTART was a Phase 1B dose-escalation and dose-finding study testing escalating doses of short-term high-dose lovastatin (1, 3, 6, 8, or 10 mg/kg/day for 3 days). The operating characteristics of the CRM were established using 5,000 simulated trials with explicit assumptions regarding expected rates of DLT at each dose level. A sample size of 33 subjects yielded an 82% probability to detect any unexpected toxicity occurring at a rate as slow as 5%. Because the focus was on safety, the conservative model had no more than an 18% probability of choosing a dose with 25% or higher likelihood of toxicity. The trial successfully enrolled 33 subjects and the maximum tolerated dose, to be used in the Phase II trial, was estimated to be 8 mg/kg/day, a lower dose than originally anticipated by the investigators [11].

5 Summary

The appeal of adaptive designs is understandable and motivates the current high level of interest in this topic. Adaptive designs are not always better. One can adapt too quickly and may not answer the question "faster" (i.e., may require a larger sample size), or at all—the question itself may become unclear. It is imperative to understand the operating characteristics of the adaptive design by conducting simulations to assess performance under realistic scenarios. Issues relating to monitoring of accruing data, restriction of knowledge of interim results, and the processes of data review, decision-making, and implementation are likely to be critical in determining the extent and shaping the nature of adaptive design utilization in clinical trials.

References

1. The National Institute of Neurological Disorders and Stroke rt-PA Stroke Study Group. Tissue plasminogen activator for acute ischemic stroke. N Engl J Med. 1995;333:1581–87.
2. Baumgartner RW. Network meta-analysis of antiplatelet treatments for secondary stroke prevention. Eur Heart J. 2008;29:1082–3.
3. Antithrombotic Trialists' Collaboration. Collaborative meta-analysis of randomised trials of antiplatelet therapy for prevention of death, myocardial infarction, and stroke in high risk patients. BMJ. 2002;324:71–86.
4. Amarenco P, et al. High-dose atorvastatin after stroke or transient ischemic attack. N Engl J Med. 2006;355:549–59.
5. Food and Drug Administration. Guidelines for clinical trials. Guideline, I.H.T. FDA: Washington, DC; 1998.
6. FDA Draft Guidance for Industry. Draft guidance for industry on adaptive design clinical trials for drugs and biologics. 2010. http://www.fda.gov/downloads/drugs. Acessed February 2010.
7. Gallo P, et al. Adaptive designs in clinical drug development—an executive summary of the PhRMA Working Group. J Biopharm Stat. 2006;16:275–283; discussion 285–91, 293–8, 311–2.
8. Rosenberger WF, Lachin JM. The use of response-adaptive designs in clinical trials. Control Clin Trials. 1993;14:471–84.
9. Krams M, et al. Acute Stroke Therapy by Inhibition of Neutrophils (ASTIN): an adaptive dose–response study of UK-279,276 in acute ischemic stroke. Stroke. 2003;34:2543–8.
10. Elkind MS, et al. The Neuroprotection with Statin Therapy for Acute Recovery Trial (NeuSTART): an adaptive design phase I dose-escalation study of high-dose lovastatin in acute ischemic stroke. Int J Stroke. 2008;3:210–8.
11. Elkind MS, et al. High-dose lovastatin for acute ischemic stroke: results of the phase I dose escalation neuroprotection with statin therapy for acute recovery trial (NeuSTART). Cerebrovasc Dis. 2009;28:266–75.

Index

Printed by Publishers' Graphics LLC
SO20120731